Advances in Neurobiology

Volume 3

Series Editor
Abel Lajtha

For further volumes, go to
http://www.springer.com/series/8787

Ralph A. Nixon · Aidong Yuan
Editors

Cytoskeleton of the Nervous System

Springer

Editors
Ralph A. Nixon
Nathan S. Kline Institute for
Psychiatric Research
Orangeburg, NY 10962, USA

Aidong Yuan
Nathan S. Kline Institute for
Psychiatric Research
Orangeburg, NY 10962, USA

ISSN 2190-5215
e-ISSN 2190-5223
ISBN 978-1-4419-6786-2
e-ISBN 978-1-4419-6787-9
DOI 10.1007/978-1-4419-6787-9
Springer New York Dordrecht Heidelberg London

Printed on acid-free paper

Springer is part of Springer Science+Business Media (www.springer.com)

Preface

Paul Wintrebert, a French embryologist, was the first to introduce the concept and the term of cytoskeleton (cytosquelette in French) in 1931. Since then, much has been learned about the organization and structure of the cytoskeleton in many types of cells. In no cell is the cytoskeleton more complex and well developed than in the nervous system, where the remarkable polarity of neurons and the extensive process formation in several types of glial cells are especially reliant on this superstructure. *The Cytoskeleton of the Nervous System* is intended to help bring together key advances from the enormous body of information relevant specifically to the cytoskeleton in the nervous system. The cytoskeleton of neurons is comprised of three main components - microtubules, neurofilaments, and actin filaments - which are important for axonal transport along long distances, establishing the uniquely polar cellular shapes, and providing a scaffold for regulating the topography and function of axonal and synaptic proteins and organelles. Alterations of cytoskeleton function in neurons are linked to the development of major neurodegenerative disorders such as Alzheimer's disease, Parkinson's disease, amyotrophic lateral sclerosis, and a host of others. Glial fibrillary acidic protein replaces neurofilaments and is the main intermediate filament component of the cytoskeleton in astrocytes, and mutations in this gene product cause Alexander disease. New technologies such as live-cell imaging and genetic engineering to create an array of in vivo models have powered the exponential development of our knowledge of the cytoskeleton in the nervous system. The topics included in the areas covered were selected to capture this technical progress and conceptual development.

This monograph begins with a general description of the cytoskeleton in axonal development and pathology and then moves to more detailed descriptions of particular components, including microtubules and associated proteins, neurofilaments and interacting proteins, actin and its binding proteins, and glial fibrillary acidic protein. The later chapters focus on the functional significance of the neuronal cytoskeleton in axonal transport and its regulation in health and disease states. Because each chapter is intended to read as a self-contained review of a specific topic, some unavoidable overlaps between them may exist. We hope that, by bringing relevant data together, "The Cytoskeleton of the Nervous System" will encourage further development of unifying principles and stimulate new conceptual and technical approaches toward a better understanding of cytoskeleton functions in health and disease.

Orangeburg, NY Ralph A. Nixon
Orangeburg, NY Aidong Yuan

Contents

Contributors

Niranjana D. Amin Laboratory of Neurochemistry, National Institute of Neurological Disorders and Stroke, National Institutes of Health, Bethesda, MD 20892-4130, USA, aminn@ninds.nih.gov

Jesús Avila Centro de Biología Molecular Severo Ochoa (CSIC-UAM), Universidad Autónoma de Madrid, 28049 Madrid, Spain, javila@cbm.uam.es

Devin M. Barry Department of Biological Sciences, Bond Life Sciences Centre, University of Missouri-Columbia, Columbia, MO 65211, USA, dmb3k5@mail.mizzou.edu

Alicia Brusco Instituto de Biología Celular y Neurociencias "Prof. Eduardo De Robertis", Facultad de Medicina, Universidad de Buenos Aires, Paraguay 2155 3er piso, (C1121ABG) Buenos Aires, Argentina, hbrusco@fmed.uba.ar

Luc Buée Université Lille Nord de France, F-59000 Lille, France; Inserm, U837, 59045 Lille, France; UDSL, Faculté de Médecine, Institut de Médecine prédictive et de Recherche thérapeutique, Centre de Recherches Jean-Pierre Aubert, 59045 Lille, France; CHU-Lille, 59000 Lille, France, luc.buee@inserm.fr

Brian Callaghan Department of Neurology, University of Michigan, Ann Arbor, MI 48109-2200, USA, bcallagh@umich.edu

Hsinlin T. Cheng Department of Neurology, University of Michigan, Ann Arbor, MI 48109-2200, USA, chengt@umich.edu

Donna J. Cross Department of Radiology, University of Washington, Seattle, WA, USA, dcross@u.washington.edu

Nikki Margarita Curthoys Department of Pharmacology, School of Medical Sciences, University of New South Wales, Sydney, NSW 2052, Australia, n.curthoys@unsw.edu.au

Jacqueline R. Dauch Department of Neurology, University of Michigan, Ann Arbor, MI 48109-2200, USA, jdauch@umich.edu

Douglas L. Eng Department of Pathology, Stanford University School of Medicine, Stanford, CA, USA, engdl@yahoo.com

Lawrence F. Eng Department of Pathology, Stanford University School of Medicine, Stanford, CA, USA, lfeng@stanford.edu

Sergio G. Evrard Instituto de Biología Celular y Neurociencias "Prof. Eduardo De Robertis", Facultad de Medicina, Universidad de Buenos Aires and Hospital Neuropsiquiátrico "Braulio A. Moyano", Paraguay 2155 3er piso, (C1121ABG) Buenos Aires, Argentina, sgevrard@yahoo.com.ar

Joël Eyer Laboratoire Neurobiologie et Transgenèse, UPRES-EA3143, INSERM, Bâtiment Monteclair, CHU, 49033 Angers, France, joel.eyer@univ-angers.fr

Thomas Fath Department of Anatomy, School of Medical Sciences, University of New South Wales, Sydney, NSW 2052, Australia, t.fath@unsw.edu.au

Eva L. Feldman Department of Neurology, University of Michigan, Ann Arbor, MI 48109-2200, USA, efeldman@umich.edu

Nobuyuki Fukushima Department of Life Science, Kinki University, Higashiosaka, Japan, nfukushima@life.kindai.ac.jp

Michael L. Garcia Department of Biological Sciences, Bond Life Sciences Center, University of Missouri-Columbia, Columbia, MO 65211, USA, garciaml@missouri.edu

Takahiro Gotow Laboratory of Cell Biology, College of Nutrition, Koshien University, Takarazuka, Hyogo 665-0006, Japan, gotow@koshien.ac.jp

Elizabeth Gray Department of Neurology, Institute of Clinical Neurosciences, Clinical Sciences North Bristol, University of Bristol, Bristol BS8 1TH, UK, elizabeth.gray@bristol.ac.uk

Andrew J. Grierson Department of Neuroscience, University of Sheffield, Sheffield S10 2RX, UK, a.j.grierson@sheffield.ac.uk

Peter William Gunning Department of Pharmacology, School of Medical Sciences, University of New South Wales, Sydney, NSW 2052, Australia, p.gunning@unsw.edu.au

Robin L. Haynes Department of Pathology, Harvard Medical School, Children's Hospital Boston, Boston, MA 02115, USA, robin.haynes@childrens.harvard.edu

Lawrence J. Hayward Department of Neurology, University of Massachusetts Medical School, Worcester, MA 01655, USA, lawrence.hayward@umassmed.edu

Félix Hernández Centro de Biología Molecular Severo Ochoa (CSIC-UAM), Universidad Autónoma de Madrid, 28049 Madrid, Spain, fhernandez@cbm.uam.es

Shin-ichi Hisanaga Department of Biological Sciences, Graduate School of Science, Tokyo Metropolitan University, Hachioji, Tokyo 192-0397, Japan, hisanaga-shinichi@tmu.ac.jp

Jean-Pierre Julien Department of Anatomy and Physiology, Centre de Recherche du Centre Hospitalier, Universitaire de Québec, Laval University, Sainte-Foy, QC, Canada G1V 4G2, jean-pierre.julien@crchul.ulaval.ca

Hiroyuki Kamiguchi Laboratory for Neuronal Growth Mechanisms, RIKEN Brain Science Institute, Saitama 351-0198, Japan, kamiguchi@brain.riken.jp

Jyotshnabala Kanungo Laboratory of Neurochemistry, National Institute of Neurological Disorders and Stroke, National Institutes of Health, Bethesda, MD 20892-4130, USA, kanungo@fda.hhs.gov

Edward J. Kasarskis Department of Neurology, University of Kentucky and Department of Veterans Affairs Medical Centers, Lexington, KY 40536, USA, ejkas@uky.edu

Hannah C. Kinney Department of Pathology, Harvard Medical School, Children's Hospital Boston, Boston, MA 02115, USA, hannah.kinney@childrens.harvard.edu

Susumu Kotani Department of Biological Sciences, Faculty of Science, Kanagawa University, Hiratsuka, Kanagawa 259-1293, Japan, kotani-bio@kanagawa-u.ac.jp

Rashmi Kothary Ottawa Hospital Research Institute, Ottawa, ON, Canada K1H 8L6; Department of Cellular and Molecular Medicine, University of Ottawa, Ottawa, ON, Canada; Department of Medicine, University of Ottawa, Ottawa, ON, Canada, rkothary@ohri.ca

Kazuyuki Matsushima Department of Biological Sciences, Faculty of Science, Kanagawa University, Hiratsuka, Kanagawa 259-1293, Japan, pt125432ul@kanagawa-u.ac.jp

Jesse R. McLean Center for Neuroregeneration Research, McLean Hospital/Harvard Medical School, Belmont, MA 02478, USA, jmclean@mclean.harvard.edu

Christopher C.J. Miller Department of Neuroscience, King's College London, The Institute of Psychiatry, MRC Centre for Neurodegeneration Research, London, SE5 8AF, UK, chris.miller@kcl.ac.uk

K.E. Miller Department of Anatomy and Cell Biology, Oklahoma State University CHS, Tulsa, OK 74107, USA, kenneth.miller@okstate.edu

Satoshi Minoshima Department of Radiology, University of Washington, Seattle, WA, USA; Department of Bioengineering, University of Washington, Seattle, WA, USA, minoshim@u.washington.edu

Bibhutibhushan Mishra Laboratory of Developmental Neurobiology, Georgetown University School of Medicine, Medical/Dental Building, Washington, DC 20007, USA, bm72@georgetown.edu

Hiroyuki Nakagawa Division of Biology, Faculty of Science, Fukuoka University, Fukuoka 814-0180, Japan, hnakagawa@fukuoka-u.ac.jp

Ralph A. Nixon Center for Dementia Research, Nathan S. Kline Institute for Psychiatric Research, Orangeburg, NY 10962, USA; Departments of Psychiatry and Cell Biology, New York University Langone Medical Center and School of Medicine, New York, NY 10016, USA, nixon@nki.rfmh.org

Harish C. Pant Laboratory of Neurochemistry, National Institute of Neurological Disorders and Stroke, National Institutes of Health, Bethesda, MD 20892-4130, USA, panth@ninds.nih.gov

David A.D. Parry Institute of Fundamental Sciences and Riddet Institute, Massey University, Palmerston North 4442, New Zealand, d.parry@massey.ac.nz

Rodolphe Perrot Department of Psychiatry and Neuroscience, Laval University, Research Centre of CHUQ, 2705, boulevard Laurier, T-2-50, Québec (Québec), Canada, rodolphe.perrot@crchul.ulaval.ca

Janice Robertson Centre for Research in Neurodegenerative Diseases, University of Toronto, Toronto, ON, Canada M5S 3H2, jan.robertson@utoronto.ca

Parvathi Rudrabhatla Laboratory of Neurochemistry, National Institutes of Neurological Disorders and Stroke, National Institutes of Health, Bethesda, MD 20892, USA, parvathir@ninds.nih.gov

Takahiro Sasaki Department of Demyelinating Disease and Aging, National Institute of Neuroscience, Ogawa-Higashi, Kodaira, Tokyo, Japan, tsasaki@ncnp.go.jp

R. Schechter Department of Anatomy and Cell Biology, Oklahoma State University CHS, Tulsa, OK 74107, USA, ruben.schechter@oustate.edu

Nicolas Sergeant Univ Lille Nord de France, F-59000, Lille, FranceInserm, U837, 59045 Lille, France; UDSL, Faculté de Médecine, Institut de Médecine prédictive et de Recherche thérapeutique, Centre de Recherches Jean-Pierre Aubert, 59045 Lille, France, nicolas.sergeant@inserm.fr

Anna-Lena Ström Department of Neurochemistry, University of Stockholm, Stockholm, Sweden, anna-lena.strom@neurochem.su.se

Michael J. Strong Molecular Brain Research Group, Department of Clinical Neurological Sciences, Robarts Research Institute, The University of Western Ontario, London, ON N6A 5K8, Canada, mstrong@uwo.ca

Ben G. Szaro Department of Biological Sciences, Neuroscience Research Center, University at Albany, Albany, NY 12222, USA, bgs86@albany.edu

Akihiko Takashima Laboratory for Alzheimer's Disease, Brain Science Institute, Saitama, Japan, kenneth@brain.riken.jp

Takuro Tojima Laboratory for Neuronal Growth Mechanisms, RIKEN Brain Science Institute, Saitama 351-0198, Japan, tojima@brain.riken.jp

Kiyotaka Tokuraku Department of Chemical Science and Engineering, Miyakonojo National College of Technology, Miyakonojo, Miyazaki 885-8567, Japan, tokuraku@cc.miyakonojo-nct.ac.jp

Atsuko Uchida Center for Molecular Neurobiology, The Ohio State University, 1060 Carmack Rd. Columbus, OH, uchida.11@osu.edu

Alastair Wilkins Department of Neurology, Institute of Clinical Neurosciences, Clinical Sciences North Bristol, University of Bristol, Bristol BS8 1TH, UK, alastair.wilkins@bristol.ac.uk

Kevin G. Young National Research Council of Canada, Ottawa, ON K1A 0R6, Canada, kevin.young@nrc.gc.ca

Aidong Yuan Center for Dementia Research, Nathan S. Kline Institute for Psychiatric Research, Orangeburg, NY 10962, USA; Department of Psychiatry, New York University Langone Medical Center and School of Medicine, New York, NY 10016, USA, yuan@nki.rfmh.org

Ya-li Zheng Laboratory of Neurochemistry, National Institute of Neurological Disorders and Stroke, National Institutes of Health, Bethesda, MD 20892-4130, USA, yalinew@yahoo.com

Haining Zhu Department of Molecular and Cellular Biochemistry, University of Kentucky, Lexington, KY 40536, USA, haining@uky.edu

Chapter 1
Central Axonal Development and Pathology in Early Life

Robin L. Haynes and Hannah C. Kinney

Abstract Critical to brain development is the proper growth and maturation of the axons underlying neuronal circuitry and communication throughout the nervous system. During development, the axon undergoes a remarkable process of outgrowth from the cell body, elongation, pathfinding to its target, differentiation of its synaptic junction, and establishment of connections with dendritic and/or somatic synapses. In the human brain, the most dramatic period of axonal development is midgestation to the end of the second year of life (i.e., the period when the brain itself attains approximately 80% of its adult weight). During this period, the developing axon is especially vulnerable to different types of injury as it undergoes rapid elongation and differentiation. In this chapter, we review critical aspects of axonal development and pathology in early human life with a focus on the central nervous system. These aspects include the initial growth of the axon, the pathfinding mechanisms by which the axon locates the proper target, elimination of axonal connections to ensure the correct axon-to-target balance, and maturation of the axons to ensure proper transmission of signal, as well as the susceptibility of the developing axon to injury and selected disorders of axons in the developing human brain.

Keywords Growth cone · Axonal guidance · Myelination · Extracellular matrix · Antegrade axonal injury · Retrograde axonal injury

Contents

R.L. Haynes (✉)
Department of Pathology, Harvard Medical School, Children's Hospital Boston, Boston, MA 02115, USA
e-mail: robin.haynes@childrens.harvard.edu

R.A. Nixon, A. Yuan (eds.), *Cytoskeleton of the Nervous System*,
Advances in Neurobiology 3, DOI 10.1007/978-1-4419-6787-9_1,

1.1 Introduction

The central axon is a smooth, thin process of variable length that extends from the polarized neuronal cell body and propagates action potentials, thereby assisting in synaptic communication between neurons. Axonal development involves complex processes of growth, pathfinding, target selection, and maturation, each of which requires specific proteins or "cues" to ensure proper connectivity and function. In this chapter, we review critical aspects of each of these processes as well as axonal pathology in early human life. To establish a basic understanding of axons, we first begin with a brief review of the structure and function of the mature axon and how this structure serves as the basis for signal propagation. In this chapter, we focus upon the axons of the central nervous system (CNS), given the broad scope of the biological and pathologic aspects of central and peripheral axons in combination.

1.2 General Structure and Function of Axons

Though the diameter of the axon is relatively uniform in caliber, it is composed of morphologically and functionally distinct regions. These regions include the axon hillock, the initial segment, and the nodal segment.

1.2.1 Axonal Components

At the site where the axon emerges from the soma of the neuron, there is a slightly expanded region known as the axon hillock and identified by light microscopy by the absence of Nissl substance (i.e., rough endoplasmic reticulum at the ultrastructural level) (Graham and Lantos, 2002); it is the interface between the soma and the initial segment of the axon. The structural differences in the hillock and the initial segment are subtle and detected only at the ultrastructural level. At the intersection of the hillock and the initial segment, a fine granular material containing electron-dense granulations spaced at regular intervals appears under the plasma membrane (Palay et al., 1968); this material extends from the initial segment to the origin of the myelin sheath. A second differentiating feature is the appearance of rosettes or clusters of polyribosomes that appear in the hillock but disappear along the length of the initial segment (Graham and Lantos, 2002; Palay et al., 1968). Whereas morphologically the axon hillock and initial segment are distinct, electrophysiologically they are similar. At the hillock/initial segment, the membrane potential reaches a triggering threshold, and an action potential propagates through the remainder of the axon. The importance of the site is based on evidence of a high density of sodium (Na^+) channels (Farinas and DeFelipe, 1991; Palay et al.,1968). Nevertheless, Na^+ channels in the initial segment may contribute little to the initiation of action potentials, whereas Na^+ channels located in the axon beyond the initial segment are likely to be critical for determining threshold (Colbert and Johnston, 1996; Colbert and Pan, 2002; Stuart et al., 1997). Beyond the initial segment, the axon is ensheathed by phospholipid-rich myelin, which serves to insulate the axon and increase the velocity at which the nerve impulse travels down the axon. The myelin sheath is interrupted at regular intervals by the nodes of Ranvier, which are rich in Na^+ channels and therefore generate an intense depolarizing inward Na^+ current in response to depolarization down the myelinated component of the axon (Rasband and Trimmer, 2001). This structural organization acts to periodically intensify the amplitude of the action potential to ensure the complete transfer of signal from origin to destination without loss of "strength."

1.2.2 Cytoskeletal Structure of Axons

The neuronal cytoskeleton is composed of three different types of filaments: actin microfilaments, microtubules, and intermediate filaments (IFs). Each of these filaments serves a distinct purpose in organization and function in adult axons as well as in various stages of axonal growth and development (see later). The most abundant of these filaments is the IF. They are stable polymers and consist of three different subunits: neurofilament (NF) light (NFL), NF medium (NFM), and NF heavy (NFH) with molecular mass of 68, 160, and 200 kDa, respectively (Graham and Lantos, 2002). Together these NFs form a cytoskeletal core with a backbone running parallel along the axon that is composed predominately of the amino terminus of NFL. The shell covering the backbone is composed of the carboxy-terminal side arms of NFM and NFH (Kumar et al., 2002; Lee and Cleveland, 1996). Phosphorylation of these side arms is associated with a decrease in NF transport rate, as well as an increase in axonal caliber during axonal development (see later) (Sihag et al., 2007). The second cytoskeletal component of axons is the microfilaments, of which the predominant type is actin. Actin filaments compose the cytoskeletal meshwork of filaments immediately underlying the axolemma. Whereas neurofilaments are stable with very low turnover, the actin component of the cytoskeleton is dynamic and is adjusted in response to physiologic cues, a property that underlies its prominent role in growth cone function during axonal growth and pathfinding (see later). The third component of the axon cytoskeleton includes microtubules and the microtubule-associated proteins (MAPs). Within the axon, they are long, tube-like structures with a diameter of 25 nm and are composed of 13 protofilaments consisting of dimers of α- and β-tubulin arranged side by side to form the microtubule walls. Associated with the microtubule structures are the MAP proteins, which serve as side arms or cross-links that interconnect microtubule structures to each other, as well as to other cytoskeletal components. The most prominent MAPs located within the axon are MAP1A, MAP1B, MAP3, MAP5, and Tau (Halpain and Dehmelt, 2006; Huber et al., 1985; Paglini et al., 2000; Riederer et al., 1986). Of note, MAP2 is restricted to neuronal cell bodies and dendrites and is not expressed in the axon (Dehmelt and Halpain, 2005).

1.2.3 Axonal Transport

Due to the variable and oftentimes long distance between the axonal terminal (site of the synapse) and cell body (site of production of the secretion molecules necessary at the synapse), a mechanism for transport down the axon is in place that involves a system of both anterograde transport from the cell body to the terminal, as well as retrograde transport from the terminal back to the cell body. Once produced in the cell body, membrane-bound organelles, including mitochondria and secretory vesicles, are transported down the length of the axon using microtubules and the motor molecule kinesin, which is thought to "walk" molecules along the microtubule (Amaratunga et al., 1993; Brady et al., 1990; Hirokawa et al., 1991; Hirokawa and Takemura, 2004; Muresan, 2000). Other cytoskeletal proteins are subject to anterograde transport to the axonal terminal but by a mechanism known as slow axonal transport. To achieve microtubules movement, transport along the actin cytoskeleton involves the protein dynein as a motor (Dillman et al., 1996a, b; Hirokawa and Takemura, 2004; Pfister, 1999). In addition to anterograde transport, axons participate in the retrograde transport of membrane-bound organelles involved in the endocytotic pathway (e.g., in neurotransmitter packaging and release), which are delivered back to the cell body for reuse. In addition, ligand–receptor complexes interacting at the axonal terminal are endocytosed at the site of action and carried by retrograde transport to the neuronal cell body (Howe and Mobley, 2005; Ibanez, 2007; Zweifel et al., 2005). This retrograde signaling is thought to be particularly important in mediating the action of the neurotrophin nerve growth factor (NGF)–TrkA receptor complex (Howe and Mobley, 2005; Ibanez, 2007; Zweifel et al., 2005). Retrograde trafficking of membrane-bound organelles as well as ligand–receptor complexes is thought to involve dynein as a motor (Ha et al., 2008; Motil et al., 2006; Schnapp and Reese, 1989).

1.3 Axons in Development

1.3.1 Axonal Outgrowth

1.3.1.1 Growth Cone and Cytoskeleton

One of the most notable features of axons during development is their ability to navigate long distances and accurately target a precise destination; the key structure underlying their outgrowth is the motile growth cone. The growth cone is formed at the tip of the growing axon and is responsible for "sensing" environmental cues, determining the direction of growth, and guiding the growing axon in the proper direction. The mechanism underlying its formation and movement is the dynamic nature of cytoskeletal elements, which are continuously changing during outgrowth, with retraction of the filopodia and lamellipodia (i.e., narrow cylindrical extensions and veil-like extensions, respectively) protruding from the periphery of the growth cone. During axonal outgrowth, the growth cone participates in three distinct stages termed (1) protrusion, (2) engorgement, and (3) consolidation (Dent and Gertler, 2003; Goldberg and Burmeister, 1986). Protrusion involves the extension of filopodia and lamellipodia associated with the rapid rearrangement and polymerization of actin (Letourneau, 1983; Okabe and Hirokawa, 1991). If the filopodia "senses" a cue to turn, the actin selectively polymerizes on the side toward the cue, with collapse of the filopodia and lamellipodia on the opposite side (Bentley and O'Connor, 1994; Dent and Gertler, 2003; Lin et al., 1994). Once the filopodia have been extended, engorgement occurs when microtubules along with membranous vesicles and organelles from the central core of the axon advance into the growth cone (Dailey and Bridgman, 1991; Gordon-Weeks, 1991; Lin et al., 1996). These microtubules are transported into the extended filopodia by dynein-driven forces either against the actin network or against longer microtubules (Grabham et al., 2007; Myers et al., 2006). This invasion of microtubules into the growth cone and into individual direction-oriented filopodia is thought to turn the growth cone toward a directional cue (Myers et al., 2006; Zhou and Cohan, 2004). The final stage of axonal growth involves the depolymerization of actin around the neck of the growth cone to generate a new segment of axon (Goldberg and Burmeister, 1986).

A notable phenomenon associated with the protrusion phase of growth cone extension involves the retrograde flow of actin away from the tip of the growth cone toward its central core (Lin et al., 1996; Mitchison and Kirschner, 1988; Okabe and Hirokawa, 1991). This retrograde F-actin flow is inversely proportional to the advance of the growth cone (Lin and Forscher, 1995) and involves a "clutch" mechanism that links this retrograde flow to forward motion of individual filopodia through interaction with the substrate or matrix in which it travels (Fig. 1.1) (Jay, 2000; Mitchison and Kirschner, 1988). In stationary filopodium, individual actin subunits are added to an actin filament on the membrane-associated side or tip of the filopodium. Upon addition of actin subunits on one end, the entire filament is translocated via the action of myosin motors (Lin et al., 1996; Medeiros et al., 2006) toward the center of the growth cone where actin subunits and fragments are removed from the opposite end and recycled (Fig. 1.1a). For filopodia to move forward, attachments to the substrate or matrix must be made (Fig. 1.1b). These attachments consist of ligand–receptor interactions involving members of the integrin family of extracellular matrix proteins and the cadherin family of cell adhesion molecules, as well as other molecules. When the filopodium makes contact with the substrate, a clutch is engaged, thereby preventing retroactive flow, fixing the actin filament in place, and allowing the forward movement of materials such

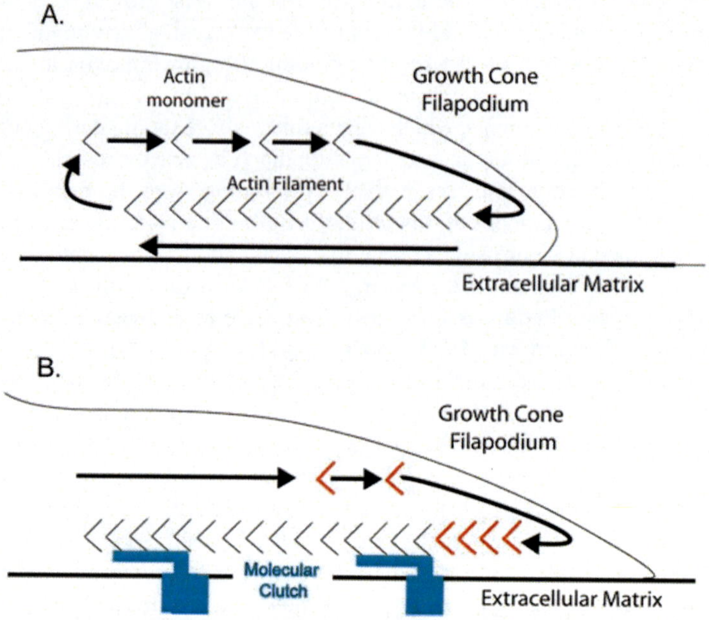

Fig. 1.1 Retrograde actin flow and the clutch hypothesis. **a** Stationary filopodium undergoes a cycle of actin polymerization at the tip of the filopodium and retrograde flow of the actin filament toward the center of the growth cone where the filament is broken down into individual actin subunits. **b** Contact between the filopodium and the matrix mediated by ligand–receptor interactions engages a "molecular clutch" that acts to prevent the retroactive flow of actin. Simultaneous addition of actin at the tip (shown in *red*) results in the elongation of the actin filament and subsequent growth cone advancement

as membrane and actin monomers to supply the growing tip (Fig. 1.1b) (Jay, 2000; Mitchison and Kirschner, 1988).

1.3.1.2 Axonal Guidance

The temporal and spatial movement of the axon to its appropriate target is critical to the proper development of neuronal circuits. This movement is controlled by a number of different environmental cues that direct the axons via interactions with the growth cone receptor. Environmental cues are categorized as either permissive or inhibitory and can be divided into two separate categories: nondiffusible and diffusible signals (Table 1.1).

Nondiffusible Signals for Axonal Guidance

Nondiffusible signals include a number of molecules categorized into three families of ligands – cell adhesion molecules (CAMs), extracellular matrix (ECM) molecules, and ephrins – each of which acts through receptor-mediated interactions on the cell surface of the growth cone.

Table 1.1 Axonal Guidance Molecules in the Central Nervous System

	NONDIFFUSIBLE Interactions with	Primary effect
CELL ADHESION MOLECULES (CAM)		
Cadherins	Cadherins (homophilic)	Permissive
L1-CAM	L1-CAM (homophilic) CAMS (heterophilic)	Permissive
NCAM	NCAM (homophilic) L1-CAM and others (heterophilic)	Permissive
EXTRACELLULAR MATRIX MOLECULES (ECM)		
Hyaluronan	Glycoprotein receptor CD44	Permissive
Chondroitin sulfate proteoglycans	Hyaluronan, Tenascins, NCAM	Inhibitory
Glycoproteins		
Tenascins	Integrin receptors, CAMS, other ECMs	Both
Laminin	Integrin receptors	Permissive
Heparin-binding growth-associated molecule	heparin sulfate proteoglycan	Permissive
EPHRINS		
Ephrin	Eph Receptor	Both
	DIFFUSIBLE	
Netrins	DCC, Neogenin, Unc5a-Unc5d	Both
Slit	Robo	Inhibitory
Semaphorins	neurophilin, and plexin	Inhibitory
Neurotrophins		
NGF	TrkA	Permissive
BDNF	TrkB	Permissive

Legend: This list represents a subset of molecules that signal axonal guidance. The primary effect of each of these molecules is listed; however, there are circumstances for many of these molecules where the opposite effect on axonal guidance is seen (see text).

Cell Adhesion Molecules and Axonal Guidance

CAMs are transmembrane receptors with three characteristic structural components: a transmembrane domain, an intracellular domain that interacts directly with the elements of the cytoskeleton, and an extracellular domain that interacts with CAM proteins of the same kind or different kinds by homophilic or heterophilic binding, respectively. Upon either homophilic or heterophilic binding, CAMs function in growth cone mobility by activating intracellular signaling cascades, rather than by directly modulating the adhesion of the growth cone to cellular substrates (Doherty et al., 2000). Below, we focus on three different classes of CAMs – cadherins, L1-CAM, and NCAM – each of which is implicated in the control of axonal growth and guidance.

Cadherins. During axonal development and growth, cellular contact results from the specificity of cadherin binding between the cadherin extracellular domain on axonal growth cones and those on the substrate on which the growth cone travels (Ranscht, 2000). The mechanisms underlying the function of cadherins as a CAM and the role of cadherins in axonal elongation are not completely understood. One proposed mechanism involves fibroblast growth factor receptor (FGFR) signaling within the growth cone (Fig. 1.2a). In this mechanism, N-cadherin on the growth cone of an axon interacts with an N-cadherin molecule on a substrate for axonal growth (Williams et al., 1994). This interaction results in a cascade that involves the activation of FGFR, a downstream production of arachidonic acid, and a subsequent influx of calcium (Ca^{2+}). Upon Ca^{2+} influx, multiple downstream effectors are activated that contribute to the cytoskeletal changes involved

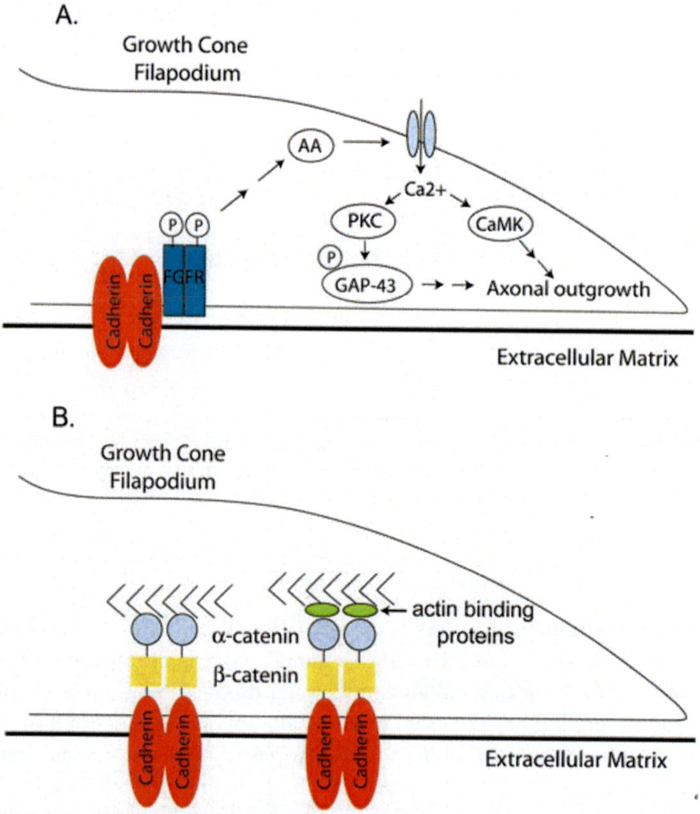

Fig. 1.2 Mechanisms of cadherin function in growth cone advancement. **a** Cadherins function through intracellular signaling involving the interactions of cadherins (as well as other CAMs) with FGFR. Upon interaction, FGFR activation and phosphorylation ultimately leads to arachidonic acid (AA) release and Ca^{2+} influx. Increased cytoplasmic Ca^{2+} activates CAMKII as well as GAP-43, both of which are important for axonal outgrowth. **b** Cadherins couple to cytoskeletal actin filaments either directly though interactions with cytoplasmic catenins or indirectly though actin binding proteins

in outgrowth. One such effector is CaM kinaseII, whose activation has been shown to be important for axonal growth in vivo (Bolsover, 2005; Doherty et al., 2000). A second downstream effector of Ca^{2+} is GAP-43, an actin-capping protein that becomes phosphorylated by protein kinase C upon Ca^{2+} influx, leading to a cessation of actin-capping activity (He et al., 1997) and subsequent initiation of actin polymerization and axonal outgrowth (Fig. 1.2a) (Bolsover, 2005; Doherty et al., 2000). Of note, while FGFR signaling is detailed here with cadherins, similar FGFR signaling occurs upon interaction with other CAMs including L1-CAM and NCAM (Williams et al., 1994).

A second mechanism by which cadherins function in growth cone movement involves the direct coupling of cadherins to the actin cytoplasm through the interaction of cadherins with catenins (Fig. 1.2b). Here, homophilic cadherin binding leads to the interaction of the cytoplasmic tail of cadherin with catenin molecules that interact with actin filaments either directly or indirectly through actin binding proteins such as ZO-1 and α-actinin (Fig. 1.2b) (Gates and Peifer, 2005; Goodwin and Yap, 2004; Itoh et al., 1997; Knudsen et al., 1995). This model has recently been expanded upon to suggest that cadherin receptors are linked to actin flow via catenin interactions and are involved in a molecular clutch within the growth cone to drive growth cone advance (see earlier) (Fig. 1.1) (Bard et al., 2008).

L1-CAM. L1 is a cell adhesion molecule concentrated in the axon that plays a role in growth cone mobility and axonal extension. The structure of L1 consists of six immunoglobin (Ig) domains, a single transmembrane domain, and a highly conserved cytoplasmic domain. L1 can participate in both homophilic binding with other L1 molecules and heterophilic binding with other CAMs to mediate its role in axonal growth and extension (Burden-Gulley et al., 1997). The mechanism of L1-mediated axonal outgrowth involves indirect interactions with the cytoskeletal protein actin, whose retrograde flow generates traction force, which in turn drives process outgrowth forward (see earlier) (Fig. 1.1). One of the best characterized mediators of the L1–actin interaction is the cytosolic protein ankyrin, which binds to the L1 cytoplasmic tail (Burden-Gulley et al., 1997; Davis and Bennett, 1994; Gil et al., 2003; Nishimura et al., 2003). Ankyrin plays a critical role in L1-mediated axonal growth based on the following observations: (1) L1-CAM and ankyrins are colocalized in premyelinated axon tracts during development (Gil et al., 2003); and (2) $ankyrin_b$ (isoform of ankyrin) deficient mice exhibit axon tract hypoplasia (Gil et al., 2003). Although it has been established that $ankyrin_b$ couples actin retrograde flow to L1 during neurite initiation (Nishimura et al., 2003), the role of the L1–ankyrin interaction in neurite outgrowth and axonal extension is now thought to be inhibitory, based on ankyrin binding inhibition studies that show stimulation of retrograde actin movement (Gil et al., 2003; Nishimura et al., 2003). Other proteins may mediate the L1–actin traction force involved in axonal outgrowth, including the adaptor protein ezrin (Sakurai et al., 2008) and the brain-specific protein shootin (Shimada et al., 2008). Experimental data show inhibition of retrograde actin transport upon inhibition or reduction of actin and/or L1 binding to these proteins (Sakurai et al., 2008; Shimada et al., 2008).

L1-induced axonal outgrowth is not only dependent on L1 adhesion and subsequent interaction with actin but also on several signaling cascades. Similar to cadherin, L1 activates FGFR, resulting in a downstream cascade, an increase in Ca^{2+}, and ultimately cytoskeletal changes associated with axonal outgrowth (see earlier) (Burden-Gulley et al., 1997; Doherty et al., 1996). In addition to signaling through FGFR, L1 is endocytosed, resulting in a signaling cascade that leads to the activation of extracellular signal-regulated kinases (ERK1/2) and subsequent neurite outgrowth (Schmid et al., 2000). In this cascade, L1 endocytosis results in the activation of PI3 kinase, Rac-1, mitogen-activated protein kinase kinase (MEK), and mitogen activated protein kinase (MAPK) or ERK1/2 (a member of the MAPK family) (Bearer, 2001; Kiryushko et al., 2004; Schmid et al., 2000). The regulation of axonal growth mediated by ERK1/2 likely involves both local, growth cone–specific mechanisms, as well as distant events, such as regulation of transcription (Perron and Bixby, 1999).

NCAM. The neural cell adhesion molecule (NCAM) protein, highly expressed in the mammalian nervous system, is similar to L1-CAM in that it is a member of the Ig superfamily of adhesion molecules. While there exists only one NCAM gene, multiple forms of NCAM are generated through alternative splicing, including NCAM-180, NCAM-140, and NCAM-120. These variants differ not only in molecular weight but also in the presence of a transmembrane domain and differences in the cytoplasmic domain (Kiryushko et al., 2004; Rinaldi et al., 2001). As with other CAMs, NCAM mediates cell adhesion primarily through homophilic binding to other NCAM molecules, although heterophilic interactions with other molecules, including L1-CAM, have been reported (Kadmon et al., 1990; Kiryushko et al., 2004; Milev et al., 1996). Similar to L1-CAM, NCAM mediates neurite outgrowth through activation of FGFR and subsequent Ca^{2+} release (see earlier) (Fig. 1.2) (Doherty et al., 2000; Kiryushko et al., 2004). At least one of the NCAMs, NCAM-140, has also been shown to participate in neurite outgrowth mediated by the activation of the MAPK (ERK1/2) cascade discussed earlier. Though the NCAM-140 and L1-CAM signaling pathways converge to activate MEK and ERK, NCAM-140 activation involves recruitment of $p125^{fak}$ to a membrane-associated complex consisting of NCAM-140 and $p59^{fyn}$, rather than endocytosis and subsequent activation of PI3 kinase as seen in L1-CAM (Schmid et al., 1999).

In addition to alternative splicing, NCAM is also subject to posttranslational modification. The most notable modification is the attachment of carbohydrate polysialic acid (PSA) exclusively to NCAM to generate PSA–NCAM, a molecule highly expressed in the nervous system. Unlike NCAM, which is widely expressed in the entire CNS throughout development and adulthood, PSA–NCAM is developmentally regulated and is widely expressed in the embryonic and early postnatal brain, a period of time correlating with axon outgrowth, branching, and contact formation (Bonfanti, 2006; Ronn et al., 1998). Though PSA–NCAM expression is thought to negatively regulate the process of myelination (see later), it also plays an important role in the development of the corticospinal tract and specifically in the formation of collateral branches (Daston et al., 1996), as well as in the fasciculation of mossy fiber axons of the hippocampus (Cremer et al., 1997).

Extracellular Matrix Molecules and Axonal Guidance

The second category of nondiffusible growth signals includes components of the ECM. During brain development, the ECM is primarily formed from the linear polysaccharide hyaluronan, which promotes aggregates of chondroitin sulfate proteoglycans. These proteoglycans include members of the lectican family aggrecan, versican, neurocan, and brevican. Lectican proteins then bind, via their C-terminal ends, to glycolipids and glycoproteins, which in turn bind to and cross-link other lecticans. One of the most prominent glycoproteins in the nervous system is tenascin R. Other components of the CNS extracellular matrix include the laminin family of glycoproteins, heparin-binding growth-associated molecule (HB-GAM), and heparin sulfates.

Hyaluronan. Although hyaluronan is present in high quantities in the developing brain, little is known as to the contribution of hyaluronan in axonal outgrowth. Recent studies examining hyaluronan and its glycoprotein receptor CD44 in the developing optic chiasm suggest a role for this interaction in the routing process of axons crossing at the midline. In these studies, disruption of the hyaluronan/CD44 ligand–receptor interaction results in a failure of midline crossing of newly generated optic axons (Chan et al., 2007a; Lin et al., 2007).

Chondroitin sulfate proteoglycans. The second group of ECM proteins thought to play a role in axonal guidance is the chondroitin sulfate proteoglycans (CSPGs). In the developing CNS, these molecules display primarily inhibitory effects but have also been shown to display growth-promoting effects depending on the neuronal cell type and the way in which they are presented to the neuron (Friedlander et al., 1994; Garwood et al., 1999; Iijima et al., 1991; Snow et al., 1990; Wu et al., 2004; Yamada et al., 1997; Yamaguchi, 2000; Zacharias and Rauch, 2006). The primary inhibitory role of CSPGs is that of a barrier molecule that blocks axonal growth (Carulli et al., 2005; Yamaguchi, 2000). This inhibitory effect is primarily attributed to the chondroitin sulfate moiety (Friedlander et al., 1994; Masuda et al., 2004; Yamaguchi, 2000), although the core proteins of these molecules also have an inhibitory effect on neurite outgrowth (Ughrin et al., 2003; Yamaguchi, 2000). Although the mechanism of the inhibitory effect of CSPGs is not clearly understood, it is most commonly explained by the CSPGs' ability to interfere with binding of the neurite to the matrix on which it is extending, resulting in decreased adhesion to the matrix and subsequent immobility (Friedlander et al., 1994; Viapiano and Matthews, 2006). There is evidence, however, suggesting that CSPGs inhibit neurite outgrowth directly through cellular signaling involving activation of the epidermal growth factor receptor and signaling through the GTP binding proteins of the Rho-ROCK pathway (Koprivica et al., 2005; Monnier et al., 2003; Viapiano and Matthews, 2006).

Tenascin. The tenascins are a family of glycoproteins that show very discreet patterns of expression particularly in the developing embryo. Specific to the CNS are tenascin C and tenascin R, with tenascin C being the best characterized. The tenascins mediate their action during development by modulating or cross-linking homophilic and heterophilic binding between families of CAMs. These interactions are further modulated by additional interactions with the lectican family of

CSPGs. Through the various interactions, tenascins mediate both restrictive and growth-promoting effects on neurite and axonal extension (Jones and Jones, 2000). Of particular interest is the finding that tenascin C protein expression in the developing rat brain is restricted in early development to the marginal and subplate zones, extending with development into the cortical plate, a pattern in close correlation with afferent innervation (Gotz et al., 1997). In vitro studies perturbing tenascin C function suggest a growth-promoting effect on both thalamic and cortical axons as they grow and innervate the cortex (Gotz et al., 1997).

Laminins. The laminins are a family of extracellular matrix glycoproteins expressed throughout developing neural tissue (Powell and Kleinman, 1997). Laminins interact with cells to mediate a response primarily through the integrin receptor family. In the CNS, laminin expression has been observed in a spatial and temporal pattern correlating with growth of axonal tracts (Zhou, 1990). Expression also occurs in the developing visual pathway, cerebellum, cerebral cortex, and spinal cord (Cohen et al., 1987; Hunter et al., 1992; Liesi and Silver, 1988; McLoon et al., 1988; Vuolteenaho et al., 1994). In vitro studies showing neurite outgrowth-promoting activity support a role of laminins in axonal growth (Lein et al., 1992; Lein and Higgins, 1989; Powell and Kleinman, 1997). Recent in vivo studies suggests that the laminin isoform, laminin $\alpha 1$, influences neuronal polarity and directional guidance (Wolman et al., 2008).

Heparin-binding growth-associated molecule. HB-GAM plays a role in CNS axonal outgrowth. The molecule HB-GAM is localized to certain axonal tracts, including the commissural pathways and the corpus callosum, as well as the thalamocortical pathways in embryonic and postnatal tissues (Kinnunen et al., 1998, 1999; Rauvala et al., 1994). HB-GAM promotes neurite outgrowth in a variety of neuronal cell types via interactions with heparin sulfate proteoglycan, another ECM protein (Kinnunen et al., 1999, 1996; Raulo et al., 2005; Rauvala et al., 1994).

Ephrins and Axonal Guidance

Ephrins and their Eph receptors are membrane-bound proteins that together constitute an important nondiffusible family of guidance molecules. The Eph receptors serve as tyrosine kinases and interact with ligand through an extracellular N-terminal domain. Ephrin ligands, categorized as A or B, bind Eph receptors A and B, respectively, with a limited amount of cross-binding between the two groups (Martinez and Soriano, 2005). The axonal pathfinding cues established by the interactions between Eph receptors and Eph ligands during development are primarily repulsive, resulting in growth cone collapse (Drescher et al., 1995; Wang and Anderson, 1997). The opposite effect with receptor–ligand interaction resulting in enhancement of axonal growth has also been shown in certain cell types (Hansen et al., 2004; Knoll et al., 2001), suggesting cell-specificity of axonal response.

Upon interaction of Eph receptors and Eph ligands, high-affinity heterodimers are formed, and downstream signaling is initiated. Interestingly, this signaling is seen in multiple forms including forward and reverse signaling (Cowan and Henkemeyer, 2002; Davy and Soriano, 2005; Martinez and Soriano, 2005)

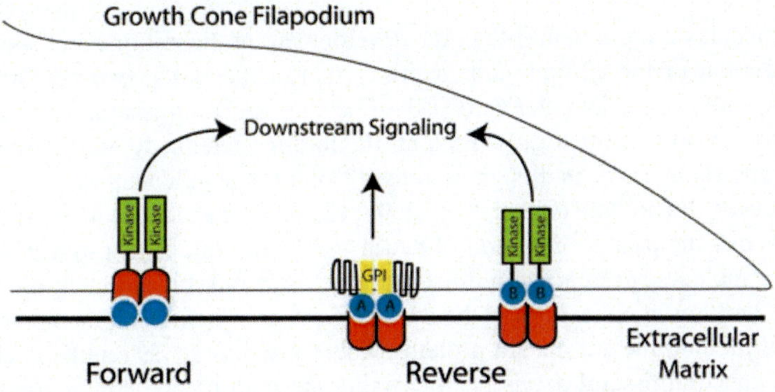

Fig. 1.3 Forward and reverse signaling of the Eph–ephrin interaction Upon the Eph–ephrin interaction, downstream signaling results in the response of the growth cone, either repulsive or attractive, to its environment. This signaling can be either in a forward direction mediated through Eph receptors on the growth cone or in a reverse direction mediated through ephrin ligands on the growth cone. See text for details

(Fig. 1.3). In forward signaling, the binding of ephrin ligand on the matrix in which the axon travels by Eph receptors on the axon triggers receptor activation and downstream signaling through activation of an intrinsic catalytic tyrosine kinase and subsequent modulation of actin cytoskeletal organization and axonal growth cone mobility (Cowan and Henkemeyer, 2002; Davy and Soriano, 2005; Martinez and Soriano, 2005) (Fig. 1.3). In reverse signaling, the binding of ephrin ligands, which have no intrinsic catalytic activity, on the axon by Eph receptors on the matrix results in downstream signaling in one of two ways depending on the class of ephrin. In the case of ephrin B, signaling results from the association of the cytoplasmic domain of ephrin B with intracellular kinases. In the case of ephrin A, which lacks a cytoplasmic domain, signaling results from the association of transmembrane proteins capable of inducing signaling pathways with the glycosylphosphatidylinositol (GPI)-linkage attaching ephrin A to the cell membrane (Bruckner et al., 1997; Kalo et al., 2001; Martinez and Soriano, 2005). Similar to forward signaling, the activation of ephrin results in cytoskeletal modulation (Fig. 1.3).

Diffusible Signals for Axonal Guidance

Several diffusible molecules guide axonal growth by generating either attractive or repulsive forces to guide the axon to the appropriate target. Prominent among the diffusible factors are the netrins, Slits, and semaphorins.

Netrins. Netrins are guidance cues that can trigger either attraction or repulsion of the growing axon depending on the context in which the cue is encountered. Netrins have multiple known receptors on the surface of the growth cone that specify the axonal response. In vertebrates, netrin receptors include DCC (Deleted in colorectal cancer) and Neogenin, which mediate primarily attractive signals, and Unc5a–Unc5d, which mediate repulsion signals (Baker et al., 2006; Round and Stein, 2007).

Netrins play a prominent role in the development of the midline and specifically the formation of the commissural pathway that connects the two cerebral hemispheres. A netrin gradient along the path of developing commissural axons supports a role for netrin in axonal growth along the midline (Kennedy et al., 2006). This netrin expression has been thought to provide long-range guidance cues to direct the axons toward the midline (Baker et al., 2006; Kennedy et al., 2006). Recent evidence suggests that the primary function of netrins in the midline is that of a short-range guidance molecule promoting midline crossing (Brankatschk and Dickson, 2006).

Slit. Working in opposition to the attractive effect of netrins on the commissural axons of the midline are the slit proteins. These proteins act as repulsion cues that prevent noncommissural axons from crossing the midline and commissural axons from crossing back over the midline once they have emerged on the opposite side (Dickson and Gilestro, 2006). The Slit family is a group of proteins found in both vertebrates and invertebrates that are expressed in the midline and act as the ligand to the cell receptor Robo (Dickson and Gilestro, 2006). The posttranscriptional regulation of Robo on the growth cone of commissural and noncommissural axons mediates the role of Slit on commissural development (Kidd et al., 1998). On noncommissural axons, Robo expression is high, thus mediating a strong repulsive effect of Slit in the midline and preventing the axon from crossing the midline (Kidd et al., 1998). On commissural axons, Robo expression is low, allowing the axon to cross the midline. Upon crossing the midline, however, Robo expression is upregulated in the growth cone of commissural axons, thus preventing the axon from crossing back over to the ipsilateral side (Kidd et al., 1998). The contribution of Slit to the development of the midline is underscored by genetic mutants of Slit in *Drosophilia*, which show collapse of axonal tracts onto the midline, and in mice, which show aberrant crossing or stalling of axons at the midline (Kidd et al., 1999; Long et al., 2004; Rothberg et al., 1990).

Semaphorins. The third category of diffusible signals includes the semaphorins, which are a large family of secreted and transmembrane proteins that act primarily as repulsive guidance cues during axonal pathfinding, although examples of semaphorins exerting a chemoattractant role have also been reported (Bagnard et al., 1998; de Castro et al., 1999; Dickson, 2002; Falk et al., 2005; Mann et al., 2007; Mann and Rougon, 2007; Polleux et al., 2000). In the CNS, the semaphorins are involved in the development of many neuronal networks including the hippocampus, olfactory system, and cerebral cortex (Chedotal et al., 1998; Polleux et al., 1998; Schwarting et al., 2000; Taniguchi et al., 2003). The most prominent semaphorin in vertebrate development is that of the Sema 3 class of semaphorins, which induces growth cone collapse in vitro (Luo et al., 1993; Messersmith et al., 1995; Puschel et al., 1995) and provides short-range inhibitory cues for axonal growth in vivo (Cheng et al., 2001; Dickson, 2002; Raper, 2000). The specific members of the Sema 3 family of semaphorins selectively bind members of the cell surface glycoprotein neurophilin, although the actions of Sema 3 are mediated by Sema 3 binding to a family of signaling receptors identified as the Plexin family (Mann et al., 2007; Zhou et al., 2008). The spatiotemporal coexpression of the various Sema, neurophilin, and plexin members likely results in differential effects of Sema on the axonal growth in various brain regions during different stages of development (Chilton and Guthrie, 2003; Cohen et al., 2005).

Neurotrophins. Neurotrophins, a class of proteins known to regulate neuronal growth, survival, and neurotransmission, have also been shown to signal axonal growth as well as provide cues for axonal guidance. This family of proteins includes nerve growth factor (NGF), neurotrophin-3 (NT-3), neurotrophin-4/5 (NT-4/5), and brain-derived neurotrophic factor (BDNF). The actions of these molecules are mediated by two types of membrane receptors: Trk tyrosine kinase receptors (Trk) and neurotrophic receptor p75 (p75NTR), with very specific interactions between the particular neurotrophin and members of these receptor families. Although there are many examples of neurotrophins as axonal guidance cues, the most prominent is that of the chemoattractant effect of NGF (Letourneau, 1978; Menesini Chen et al., 1978). This effect is mediated through the interaction of NGF with TrkA receptors on the cell surface, which results in receptor dimerization, activation of TrkA's intrinsic tyrosine kinase activity, and the subsequent autophosphorylation of TrkA (Ming et al., 1999). The downstream signaling resulting in the attractant effect of NGF involves the activation of phosphatidylinositol-3 kinase (PI3-K) and phospholipase C (PLC-γ), the latter of which, through further downstream signaling involving other kinases [RhoA kinase (ROCK), Akt, and MEK], effects the actin cytoskeleton of growth cones to promote growth cone motility (Lykissas et al., 2007; Tornieri et al., 2006).

A second neurotrophin BDNF is also known to be a chemoattractant with this activity mediated through the interaction with TrkB (Lykissas et al., 2007). Studies examining this attractant property of BDNF, however, have revealed insight into the complexity of the mechanisms behind neurotrophin-induced growth cone activity. For example, varying the cellular levels of two second messengers, cyclic AMP (cAMP) and protein kinase A, can affect the response of growth cones to BDNF to yield a repulsive rather than an attractive effect (Song et al., 1997; Wang and Zheng, 1998). Similar to NGF, BDNF affects the dynamics of the actin cytoskeleton to promote growth cone motility (Avwenagha et al., 2003).

1.3.2 Target Selection

Once the axon has traveled, using the above molecules as guidance cues, to its approximate destination, it must then invade the region of its final address or target, find its proper topographic location, and locate the correct individual cell in which it will form a synaptic connection. To ensure proper connectivity, this invasion must be accurate with the factors imparting this accuracy being numerous and highly specific for the type of pre- and postsynaptic connection. While the mechanisms that govern this specificity are poorly understood for many of the connections of the brain, there are certain systems in which this mechanism has been elucidated. One of the best characterized examples of target selection is that of the migration of the ganglion axon to the optic tectum. Here axons from ganglion cells of the retina travel along the optic nerve toward the optic chiasm where they either cross to the contralateral side or remain on the ipsilateral side to continue traveling to the optic tectum (superior colliculus in mammals) of the midbrain. This path is mediated by many of the factors discussed above including cadherins, NCAM, netrin, and ECM molecule condroitin sulfate proteoglycan. Once the axons enter the highly layered structure of the tectum, specific target selection begins as they migrate from the tectal surface, descend into the tectum navigating to their appropriate layer after which they turn and navigate through that layer to find their appropriate synaptic partner. The mechanism by which these events occurs involves gradients of ephrin A ligand and EphA receptor in the tectum and along ganglion axons, respectively (Ernst et al., 1998; Mann et al., 2004), with these gradients dependent on the axis of development (anteroposterior axis versus dorsoventral axis) and the origin of the growing axon. Given that the interaction of ephrin A and EphA is inhibitory to axonal growth, these gradients likely provide a stop signal for the growing axon. More recently, a gradient of ephrin B and EphB has been discovered that provides permissive cues for the growing axon thus creating, along with other yet unknown signals, a chemical map on which the axon responds to locate its specific target (Mann et al., 2004, 2002).

Development of thalamocortical projections is also thought to involve a similar ephrin concentration gradient controlling the trajectory of thalamic axons into the developing telencephalon and ultimately into specific cortical lamina (Maruyama et al., 2008; Uziel et al., 2006). Ephrins work in combination with other guidance molecules, including semaphorins and slits, to guide axons initiating in the thalamus, exiting the thalamus, penetrating the internal capsule, and innervating the cerebral cortex (Maruyama et al., 2008). Ephrin receptors exist in a high rostromedial to low caudolateral expression gradient with a complementary gradient of ephrin ligand in the ventral telencephalon (Dufour et al., 2003). Ephrin expression also influences layer-specific wiring of the thalamocortical connections, with ephrin A5 expressed in the upper layers of the cortex working in cooperation with other repulsive molecules to suppress thalamic fibers from growing past their intended target of layers 4 and 6 (Maruyama et al., 2008; Uziel et al., 2006).

1.3.3 Axonal Pruning

The ultimate goal of axonal target selection is connection specificity, which is met using the rather indirect approach of initial axonal excess followed by axonal refinement or pruning. Much of what is known about axonal pruning comes from studies of the peripheral neural muscular junction (NMJ), as well as studies of the *Drosophila* mushroom body (MB), a brain structure implicated in learning and memory. There are examples, however, in the vertebrate CNS that show conservation of the mechanisms used in the NMJ and MB. These mechanisms exist in at least two known variations, terminal arbor pruning and stereotyped axon pruning. In terminal arbor pruning, a single target is innervated by multiple terminal arbors. As the target develops, however, activity-dependent competition occurs, and weaker arbors are eliminated leaving the strongest input to maintain its synaptic connectivity. In the case of the NMJ, terminal arbor pruning is associated with distinct changes in the morphology of the axon to be eliminated. These changes include thinning of the temporary axon prior to elimination and the appearance of retraction bulbs, rounded structures filled with vesicles, cytoskeletal components, and damaged mitochondria (Low and Cheng, 2005, 2006; Riley, 1981). In the CNS, a similar pattern of axonal elimination occurs in the development of the cerebellum, where several climbing fibers extend terminal arbors to a single Purkinje cell, and with development the arbors in excess are eliminated leaving a singe connection (Hashimoto and Kano, 2003; Mason and Gregory, 1984).

The second type of axonal elimination, stereotyped axonal pruning, involves axonal elimination on a much larger scale relative to terminal arbor pruning in terms of the length of axon to be eliminated. Here an axon extends to a target region and branches to a second, distinct, target region. During the process of development, the inappropriate connection is eliminated leaving the proper axon–target connection (Low and Cheng, 2005, 2006). The most notable example of this in the CNS involves axons of layer V cortical neurons from the motor and visual cortex that extend initially to overlapping areas of the spinal cord and superior colliculus. With development, the axons innervating the inappropriate target (i.e., those originating in the motor cortex and extending to the superior colliculus and those originating in the visual cortex and extending to the spinal cord) are eliminated (O′Leary and Stanfield, 1989; Stanfield and O′Leary, 1985; Stanfield et al., 1982). Although the mechanisms underlying this elimination are unclear, there is evidence for the role of transcription factor *Otx1*, which is expressed in the cytoplasm of layer V neurons prior to axonal pruning but then translocates to the nucleus upon axon elimination (Weimann et al., 1999). These data, in addition to data showing a lack of axonal pruning in *Otx* loss-of-function mutants, suggests that the genes involved in axonal refinement are transcriptional regulated (Weimann et al., 1999). Most recently, semaphorin signaling through plexin receptors has been implicated in the refinement and pruning of axons originating in the visual but not motor cortex, suggesting that the mechanisms of refinement are specific to axonal origin (Low et al., 2008).

1.3.4 Myelination

Once an axon makes synaptic contact with its target cell, it is necessary to ensure a fast and efficient action potential from the presynaptic to the postsynaptic cell. In the CNS, this is done through the myelination of axons by glial cells, specifically oligodendrocytes (OLs), which establish contact with the axon, synthesize myelin membrane components, and deliver these components to the axon contact site where they act to wrap or ensheath the axon. Here we divide the process of myelination into two distinct phases: (1) axonal and OL maturation and (2) active axonal ensheathment.

1.3.4.1 Axonal and Oligodendrocyte Maturation

Prior to myelination, the maturation of the OL as well as the axon is necessary. Axonal maturation involves an increase in axonal caliber, which is associated with the phosphorylation of the carboxy-terminus side arms of neurofilaments, particularly NFM and NFH (Haynes et al., 2005; Martin et al., 1999). OL maturation from an OL progenitor into a myelinating OL involves a sequence of developmental stages, each characterized by a progressively complex morphology, as well as by the expression of stage-specific markers. This progression includes the following OL-developmental stages in increasing order of maturation: (1) A2B5-expressing premyelinating OL; (2) O4-expressing precursor OLs; (3) O1-expressing immature OLs; and (4) myelin basic protein (MBP)-expressing mature OL (Back et al., 2002, 2007). It is the MBP-expressing OL that marks a fully differentiated OL capable of myelination with this differentiation being associated with a redistribution of MBP from the OL cell body to the OL processes (Butt and Berry, 2000; Butt et al., 1997).

1.3.4.2 Active Axonal Ensheathment

Oligodendrocyte Process Growth and Axonal Contact

In the CNS, one mature OL with a large network of processes is capable of making contact with and myelinating multiple axons. Proper positioning and contact of the OL process on the axon is critical for ensuring complete myelination and is highly complex requiring a numerous number of signals, many of which are unknown, generated by both the axon and the OL. The initial contact between the axon and the OL process requires OL process extension, which, similar to the lamellipodia of a growth cone, involves the dynamic remodeling of cytoskeletal elements regulated by signaling molecules at the leading edges of the OL process. In OL process protrusion toward the axon, polymerization of actin monomers into actin filaments is mediated by the actin nucleation factor ARP2/3 complex. Two proteins, neural-Wiskott Aldrich syndrome protein (N-WASP) and WASP family verprolin homologous (WAVE), have recently been identified as important for myelination, specifically for formation and protrusion of OL processes, via that direct activation of the ARP2/3 complex and subsequent actin motility (Bacon et al., 2007; Kim et al., 2006). Inhibition of N-WASP decreases the initial ensheathment of axons in optic nerve samples, and mice lacking WAVE1 show hypomyelination in the corpus callosum and optic nerve (Bacon et al., 2007; Kim et al., 2006). In the human, OL contact with the axon involves the extension of immature, O4+/O1+/MBP–, "pioneer" processes that extend longitudinally along the length of an axon (Back et al., 2002) (Fig. 1.4). The distal ends of these processes then divide to form finger-like projections that make contact along the longitudinal extent of the axon and are hypothesized to anchor the OL before the initiation of myelination and the insertion of MBP into mature myelin sheets (Back et al., 2002).

Cell Adhesion Molecules in Axonal Ensheathment

Upon axon–OL contact, ensheathment begins. Important to the onset and maintenance of myelination are CAMs, which play very distinct roles specific to the stage of myelination.

Myelin-associated glycoprotein. One of the most prominent adhesion molecules is myelin-associated glycoprotein (MAG), which is expressed by actively

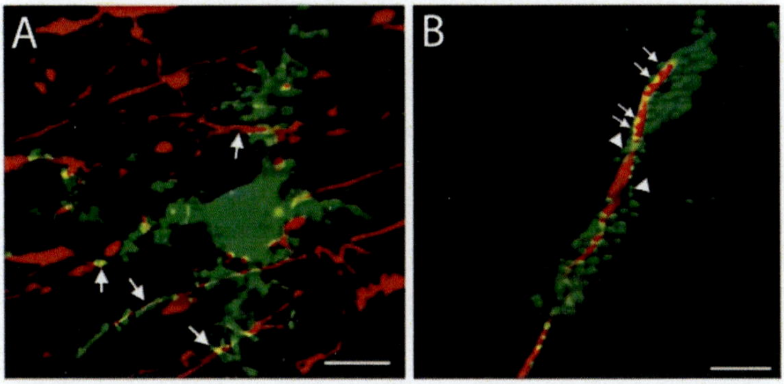

Fig. 1.4 Myelination in human cerebral white matter. **a** The relationship of OLs to axon as myelination is initiated is visualized with double-label immunocytochemistry (ICC) using the O1 marker of immature OLs (*green*) and pan-axonal marker SMI 312 (*red*) in the cerebral white matter of a case at 18 postconceptional weeks. OL processes are longitudinally oriented in close association with individual axons (*arrows*). **b** Morphologic features of axonal contact by an immature OL are demonstrated by the three-dimensional reconstruction of a region of the optic radiation in a case at 30 postconceptional weeks. An O4-labeled immature OL (*green*) spirals around an axon labeled with SMI 312 (*red*) making multiple points of contact (*arrows*, shown in *yellow*) (Back et al., 2002)

myelinating OLs and plays a role throughout the process of myelination, including OL–axon contact, initiation of myelination, and stabilization of axon–myelin contacts (Bartsch, 2003). Studies of MAG-deficient mice show an increased number of unmyelinated axons at adult ages (Bartsch et al., 1997; Li et al., 1998), significant delays in the formation of myelin sheaths (Montag et al., 1994), and alterations in myelin structure, specifically the length of the periaxonal cytoplasmic collar, the cytoplasm-filled inner loop of the myelin sheath (Li et al., 1994). In addition to myelin deficits, MAG-deficient mice also show degeneration of distal OL processes with pathologic changes similar to the "dying-back" oligodendrogliopathy in mouse models of demyelination, thus suggesting that MAG also acts to maintain OL integrity (Lassmann et al., 1997).

Polysialic acid–neural cell adhesion molecule. A second cell adhesion molecule, polysialic acid–neural cell adhesion molecule (PSA–NCAM), is localized on the surface of axons during development and negatively regulates the onset of myelination. PSA–NCAM represents the cell adhesion molecule NCAM that has undergone the posttranslational attachment of a polysialic polymer. During development in the rodent, as well as the human, the expression of PSA–NCAM is detected on unmyelinated cortical axonal tracts (Jakovcevski et al., 2007; Oumesmar et al., 1995). Interestingly, the loss of PSA–NCAM expression on these tracts is coincident with the appearance of MBP expression, suggesting that PSA–NCAM inhibits initiation of myelination and that the downregulation of PSA–NCAM is necessary for its onset. Although the mechanism of this inhibition is unknown, it has been suggested to involve prevention by PSA–NCAM of the initial attachment of OL to the axon (Coman et al., 2005).

The Role of Electrical Stimulation in Axonal Ensheathment

In addition to molecular influences upon both the axon and the OL, electrical activity generated by the neuron is necessary for the proliferation of OLs and initiation of myelination (Barres and Raff, 1993; Demerens et al., 1996). Prior to the onset of myelination, the activity-induced release of adenosine from the axon induces the differentiation of immature OLs through purinergic receptors on the OLs ultimately resulting in an increase in myelination (Stevens et al., 2002). Most recently, neuronal activity has been shown to stimulate myelination in mature OLs through a mechanism involving activity-dependent release of ATP. Interestingly, the ATP released from the axons' firing action potentials acts indirectly on the mature OL through the stimulation of astrocytes to release the cytokine leukemia inhibitory factor (LIF) (Ishibashi et al., 2006; Spiegel and Peles, 2006). This study provides insight on not only the role of cytokines but also on the role of astrocytes and the interaction of both with OLs in the process of myelination.

In addition to adenosine and ATP, axonal activity also induces the release of glutamate, which in turn acts via glutamate receptors on the OL cell body (AMPA and kainate) and OL processes (NMDA) to inhibit OL proliferation (Gallo et al., 1996; Karadottir and Attwell, 2007; Salter and Fern, 2005; Yuan et al., 1998). This inhibition may serve to control the number of OLs to ensure a proper ratio of OLs to axons in the process of myelination (Karadottir and Attwell, 2007). Selective expression of NMDA receptors on OL processes suggests their involvement in the process of myelination and particularly in mediating the axon–OL interaction (Benarroch, 2009). In addition to glutamate receptors, developing OLs also express the glutamate transporter EAAT2 during the period prior to the initiation of myelination (Desilva et al., 2007). It is possible that the OL glutamate transporter, which clears extracellular glutamate, helps to maintain the proper balance of extracellular glutamate and OL glutamate receptor activation during the process of myelination.

Polarization of Oligodendrocytes in Myelination

Once axonal contact has been made and myelination initiated, OLs deliver myelin components, proteins and lipids, to the myelin membrane. Whereas the plasma membranes of the OL cell body and the myelin sheath are continuous, the composition of each is significantly different, thus necessitating the polarization of OLs and transport of membrane components from the cell body to the myelin. Many myelin proteins, such as 2′3′-cyclic nucleotide 3′-phosphodiesterase (CNP) and myelin basic protein (MBP), are transported from the OL cell body to the cytoplasm of the myelin sheath and are transported either as protein in association with cytoskeletal components or mRNA incorporated into granules and transported along microtubules. Other proteins such as proteolipid protein (PLP) and MAG, as well as lipids such as galactosylceramide (GalCer) and sulfatide, are transported to and incorporated into the membrane component of myelin. The mechanisms of trafficking of proteins and lipids to the myelin membrane likely includes synthesis at the endoplasmic reticulum followed by either direct endosomal transport to the myelin membrane or indirect endosomal transport to the myelin membrane after initial transport to the plasma membrane (Maier et al., 2008; Simons and Trotter, 2007).

1.4 Disorders of Central Axons in Early Human Life

1.4.1 Tissue Assessment of Axonal Pathology

Axonal pathology in early life is characterized by (1) absent, altered, or delayed development, including impaired decussation of fiber tracts; or (2) secondary damage to the fully or partially mature axon with resultant degeneration (Fig. 1.5). Prior to the complete maturation of axons, neural injury affects the developmental programs of axonal elongation, growth cone formation, diameter increase, guidance, and wrapping by myelin sheaths, resulting in fiber tract aplasia, hypoplasia, abnormal decussation, and/or hypo- or delayed myelination. For the assessment of developmental axonal pathology, normative quantitative standards are needed to establish total/partial agenesis or aberrant crossing of an axonal tract and secondary myelin disturbances. Yet, the normative sequences of axonal development in different fiber tracts at a *quantitative* level have yet to be established in the human brain. They are best extrapolated, however, from the well-studied sequences of myelination because axonal sequences are thought to follow upon those of myelination, given that axons are known to differentiate first and in turn initiate myelin sheath wrapping (Brody et al., 1987; Kinney et al., 1988) (Table 1.2). In this regard, the general rules of myelination are likely obeyed by axonal maturation; that is, caudal to rostral, proximal to distal, motor before sensory fibers, primary before association fibers, projection before association fibers, central sulcus to poles, occipital pole before temporal pole before frontal pole, and posterior telencephalic before anterior telencephalic sites (Kinney et al., 1988) (Table 1.3). In an illustrative study of central axonal development, the temporal and spatial profile of the expression of different neurofilaments (Fig. 1.6) and GAP-43 (marker of axonal elongation) (Fig. 1.7) was determined from midgestation through early infancy in the parietal white matter using immunocytochemistry and Western blot analysis (Haynes et al., 2005). This study demonstrated that axonal elongation, as assessed by GAP-43 expression, is not complete in the telencephalic white matter until after infancy, indicating that axonal formation occurs in the human brain throughout the last half of gestation and well after birth (Fig. 1.7). Normative information about axonal number, width, and myelin sheath thickness at the ultrastructural level is of major interest but is extraordinarily difficult to obtain due to the need for special (electron microscopic) fixatives and very short (rarely attainable) postmortem intervals in autopsy cases and is not currently available for central axons. Currently, the means to establish aplasia or hypoplasia of a fiber tract at autopsy is by *qualitative* comparison with age-matched controls; consequently, subtle deficiencies in fiber tract size or crossing in the forebrain (e.g., corpus callosum, anterior commissure) and brain stem (e.g., corticospinal tract) are likely underdiagnosed at autopsy due to the lack of rigorous quantitative methods and special cellular markers. As specific products of genes

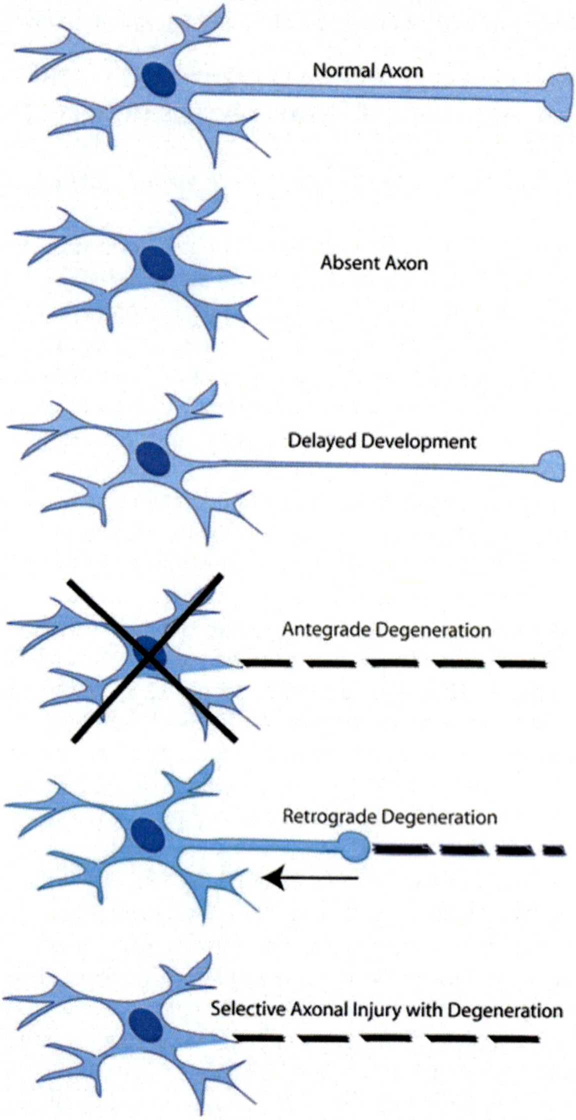

Fig. 1.5 Forms of axonal pathology in CNS development. Axonal pathology in early life involves developmental abnormalities or delays, seen here as an absent axon or a delay in the developmental increase in axonal caliber. It also involves injury to the axon either through injury to the neuronal cell body and subsequent antegrade degeneration of the axon, injury to the distal axon resulting in retrograde degeneration without obligatory loss of the neuronal cell body, or primary injury directly to the axon

are increasingly identified that are critical for axonal elongation and guidance, for example, GAP-43 (Kinney et al., 1993), KIF2IA (Chan et al., 2007b; Yamada et al., 2003), and AHI1 (Ferland et al., 2004; Jiang et al., 2002) (see later), their analysis with immunocytochemical, Western blot, and other methods in human axonal disorders will provide important new insights into pediatric disorders of axonal formation.

The morphologic assessment of degeneration of fully or partially mature axons relies upon conventional and silver stains. Stereotypical features include spheroids,

Table 1.2 Sequence of myelination in human brain development *before* birth

Neural system	<7 months	7.5–17 months	20–24 months	>24 months
Sensory: visual	Optic tract Optic chiasm			
Sensory: auditory			Brachium Inferior colliculus	
Pyramidal	Posterior limb Midbrain CST Pons CST	Pyramid	Cervical CST Thoracic CST Lumbar CST	
Central white matter				
Commissures				
Limbic		Stria medullaris thalami		

White matter sites that begin to myelinate *before* birth. The time period in postnatal months represents the median age at which tract reaches mature myelin (grade 3 of 4). Abbreviation: CST, corticospinal tract (Kinney et al., 1988).

Table 1.3 Sequence of myelination in human brain development *after* birth

Neural system	<7 months	7.5–17 months	20–24 months	>24 months
Sensory: Visual	Proximal OR Distal OR	SAF calcarine cortex		Stripe of Gennari
Sensory: Auditory	Proximal AR	Heschl's gyrus		
Pyramidal				
Central white matter		Posterior frontal Posterior parietal Occipital pole	Temporal lobe at LGN Temporal pole Frontal pole SAF all sites	
Commissures	Body corpus callosum SPLENIUM	Rostrum	Anterior commissure	
Limbic		Cingulum	MMT, alveus, fimbria	Medial fornix Lateral fornix

White matter sites that begin to myelinate *after* birth. The time period in postnatal months represents the median age at which tract reaches mature myelin (grade 3 of 4). Abbreviation: CST, corticospinal tract; OR, optic radiation; AR, auditory radiation; MMT, mammillothalamic tract; SAF, subcortical association fibers; LGN, lateral geniculate nucleus (Kinney et al., 1988).

beading, and fragmentation (Fig. 1.8). The classic silver stains for axons, for example, demonstrate swelling due to the interruption of their fast transport system and the proximal accumulation of organelles and fluid (Medana and Esiri, 2003). Upon transaction of an axon, the distal part (separated from the cell nucleus and body) degenerates, so-called Wallerian degeneration. This axonal disintegration is followed in turn by degeneration of the myelin sheath, which is readily visualized

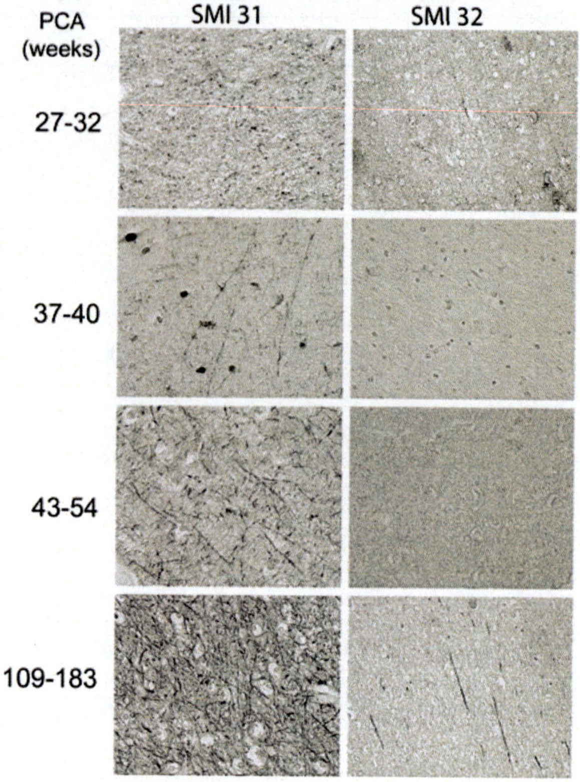

Fig. 1.6 Patterns of NF immunostaining in human cerebral white matter. Immunostaining to SMI 31 and 32 indicate the developmental expression from 27 postconceptional weeks to 183 postconceptional weeks (approximately 3 years) of phosphorylated and nonphosphorylated NFH, respectively (Haynes et al., 2005)

with myelin stains specific for degenerating (not normal) myelin (e.g., the Marchi stain). An attempt at axonal repair in the developing human brain is suggested by the demonstration of the expression of GAP-43, a marker of axonal outgrowth and regeneration, in spheroids in the focally necrotic foci of periventricular leukomalacia (PVL) (Haynes et al., 2008) (see later).

1.4.2 Mechanisms of Axonal Degeneration

Axonal degeneration may occur in early life secondary to the death of the parent neuron (i.e., antegrade degeneration) (Fig. 1.5), as illustrated in this chapter by the infantile ascending hereditary spastic paralysis, an entity of motor neurons that results in central axonal degeneration in the corticospinal and corticobulbar tracts (see later). Yet, this chapter is focused upon selective injury to axons that typically occurs without the death of the parent neuron or somatic/dendritic alterations. Central axonal disorders in early life, for example, can result from distal or retrograde degeneration without obligatory loss of the neuronal soma (Fig. 1.5). This "dying-back" phenomenon is thought to be due to a deficiency of unspecified trophic factors upon the dying axon that are generated by the neuron to which it projects; if the projecting axon is traumatically or otherwise severed from this "next-in-line" neuron, the trophic influences are lost, and retrograde degeneration results. In

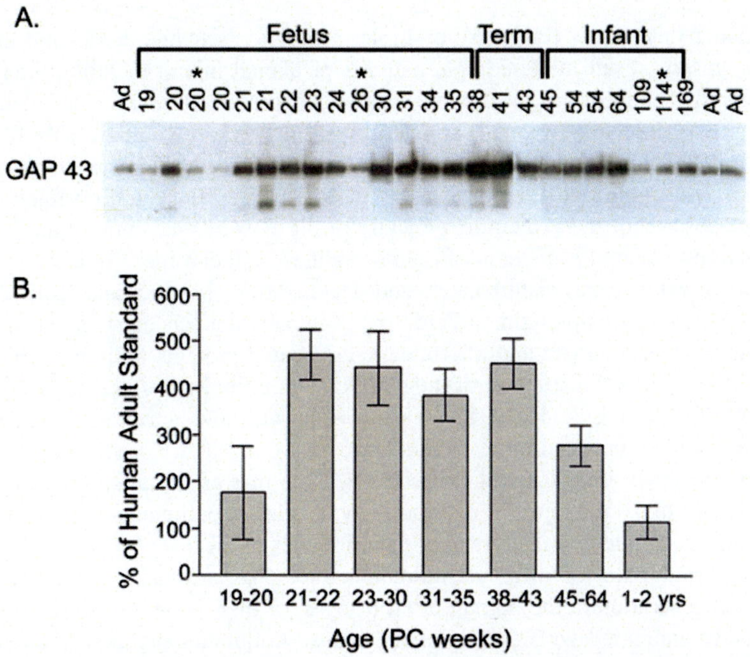

Fig. 1.7 GAP-43 expression in developing human cerebral white matter. **a** GAP-43 expression levels were determined in the cerebral white matter by Western blot analysis. **b** GAP-43 levels were grouped according to ages reflecting epochs of human brain development and plotted as a percentage of adult human standard. GAP-43 levels remain high, relative to adult levels, beyond birth and into infancy (Haynes et al., 2005)

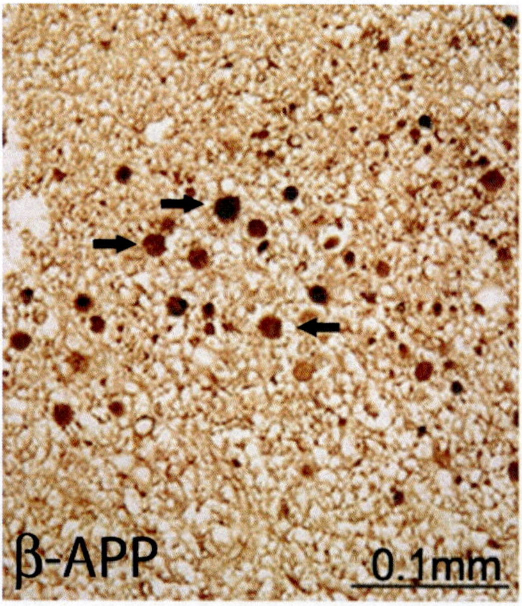

Fig. 1.8 Axonal spheroids in PVL. Axonal spheroids (indicated by *arrows*), as detected by β-APP immunostaining, are seen in the focal necrotic lesion of PVL (Haynes et al., 2008)

selective axonal injury (Fig. 1.5), multiple metabolic, ischemic, toxic, and inflammatory insults result in stereotypic patterns of axonal injury, including increased permeability, abnormal intracellular cascades, and disturbed transport (Medana and Esiri, 2003). Increased axonal permeability results from pathologic elevations in intra-axonal calcium and sodium resulting, for example, from energy deficits and acidosis (LoPachin and Lehning, 1994; Wolf et al., 2001). Such changes in the ion homeostasis of the axolemma typically result in excess calcium influx, leading to calcium-mediated mitochondrial abnormalities, cytochrome c release, caspase activation, cytoskeletal disturbances, and degeneration (Medana and Esiri, 2003). Caspase-3 cleaves calpastatin, an inhibitor of calpain that acts on such cytoskeletal proteins as neurofilament and microtubule-associated proteins (Schumacher et al., 1999); the structural disruption of the cytoskeleton ultimately affects the neuron as a whole (Fitzpatrick et al., 1998). Reactive astrocytes and activated microglia are a source of cytokines and other factors toxic to axons that are produced in inflammatory processes (Medana and Esiri, 2003). Note that understanding of the basic mechanisms of axonal injury is based mainly on adult experimental paradigms, and mechanisms unique to immature axons are not well delineated.

The introduction of different immunomarkers has advanced the analysis of axonal degeneration in the human brain because of their increased sensitivity and specificity compared with that of silver stains. Antibodies are now available to certain neurofilaments indicative of axonal injury (Trapp et al., 1998), β-amyloid precursor protein (β-APP) (Haynes et al., 2008; Medana and Esiri, 2003), and fractin (Haynes et al., 2008). A constitutive component of neurons, β-APP accumulates in axons to a level detectable by immunocytochemistry due to the disruption of fast axonal transport systems, typically upon axonal transaction (Medana and Esiri, 2003). In transected axons, β-APP accumulates in the proximal axonal ends (i.e., axonal end bulbs) before conventional morphologic evidence of axonal damage occurs (Medana and Esiri, 2003), underscoring the increased sensitivity of the β-APP immunocytochemistry to conventional methods. In the human brain, β-APP immunopositive axons are identified 1–3 h after head trauma and remain positive for 1 month, perhaps as long as 3 months (Medana and Esiri, 2003). Fractin, an antibody to the 32-kDa product of caspase-cleaved actin, is a marker of apoptosis-related events. In human developmental abnormalities, it is expressed in degenerating axons in periventricular leukomalacia (Haynes et al., 2008) and apoptotic cell bodies in pontosubicular necrosis due to perinatal hypoxic–ischemic injury (Haynes et al., 2008).

1.4.3 Neuroimaging of Central Axons

The introduction of diffusion tensor imaging (DTI) combined with tractography is leading to major advances in our understanding of pediatric disorders of central axonal pathology, including in perinatal brain damage, autism, agenesis of the corpus callosum, and temporal lobe epilepsy (Conturo et al., 2008; Govindan et al., 2008; Thomas et al., 2005; Vangberg et al., 2006; Wahl et al., 2009). DTI is based on the pattern of the diffusion of water molecules, that is, whether it is free in all directions (isotropic diffusion) or restricted to certain directions (anisotropic diffusion); anisotropic water diffusion is a characteristic property of white matter and is highly sensitive to maturational and pathologic conditions (Govindan et al., 2008). In combination with tractography, DTI has become a powerful tool to subdivide cerebral white matter into tracts so that their diffusion properties can be analyzed and their development and pathology can be assessed in vivo (Govindan et al., 2008). Definitive correlations, however, between pathologic findings concerning human axonal injury and DTI data cannot be made without DTI data obtained at or close to the time of autopsy and correlated with postmortem brain findings, information that requires major coordination of clinical and pathology services at the time of death. Nevertheless, DTI and tractography – with its capability for three-dimensional and regional visualization of axonal damage – will clearly direct future morphologic analyses of axons in pediatric neurologic disorders, thereby leading to important new insights. It should be emphasized that it is possible to determine the anatomy of fiber pathways in the *postmortem* human brain with the use of 1,1′-dioctadecyl-3,3.3′,3′-tetramethylindocarbocyanine perchlorate (DiI) track-tracing methods, as demonstrated in the optic radiation of the visual system (Hevner, 2000), the entorhinal–hippocampal pathway (Hevner and Kinney, 1996), and brainstem interconnections related to cardiorespiratory control (Fig. 1.9) (Zec et al., 1997b; Zec and Kinney, 2001, 2003). The major drawback, however, is that this technique depends upon the diffusion of the dye through unmyelinated fibers, as myelin is known to inhibit this process, and thus only fetal (unmyelinated) specimens can be used that may not be truly representative of the mature tract of interest.

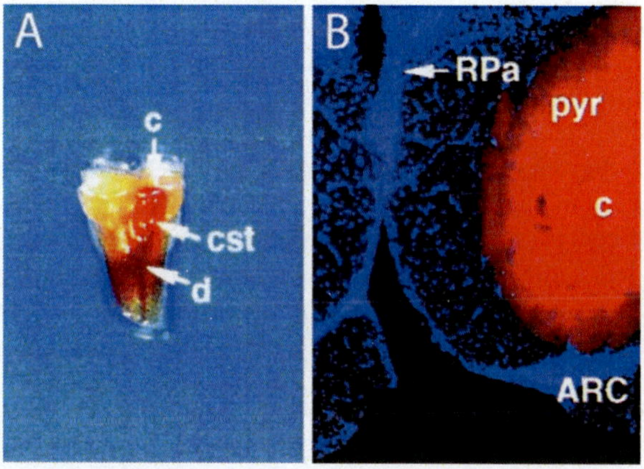

Fig. 1.9 DiI track-tracing in the human postmortem brain. **a** The results of placement of DiI crystal (c) in the pyramid at the cut surface are shown. The ventral view of the human fetal medulla at midgestation shows labeling of the corticospinal tract (cst) with the crossing of the DiI-labeled corticospinal fibers to the contralateral side at the decussation (d). **b** The DiI crystal (c) placed in the pyramid produced labeling of the corticospinal tract (cst) in the pyramid (pyr) adjacent to the crystal. There was no labeling in the arcuate nucleus (ARC) or caudal raphé (RPa, raphé pallidus) (DiI/DAPI double exposure, ×19) (Zec et al., 1997a)

1.4.4 Axonal Disorders in Early Human Life

Axonal injury occurs in multiple pediatric disorders of the central and peripheral nervous systems leading to a spectrum of clinical disability. A comprehensive review of all pediatric disorders with axonal injury is beyond the scope of this chapter. Rather, we focus here upon *representative* central disorders in which axonal injury is *primary, selective,* or *predominant* and which illustrate the spectrum of etiologies and mechanisms at work in axonal pathology in early life.

1.4.4.1 Developmental Disorders

Agenesis of the corpus callosum (ACC). The corpus callosum is the major interhemispheric fiber bundle in the human brain and is composed of approximately 200 million axons (i.e., 2–3% of all cortical fibers) (Paul et al., 2007; Schell-Apacik et al., 2008). Agenesis of this structure is among the most frequent human brain malformations with an incidence of 0.5–20/10,000 live births (Schell-Apacik et al., 2008); in children with developmental disorders, the incidence is as high as 230/10,000 (Jeret et al., 1987). The clinical spectrum of ACC varies from asymptomatic to profound neuropsychological disabilities (Glass et al., 2008; Paul et al., 2007). It is a heterogeneous condition due to multiple genetic and environmental disorders, including acardia syndrome, Meckel–Gruber syndrome, CRASH syndrome, Walker–Warburg syndrome, oro-facial-digital syndrome, chromosomal rearrangements, X-linked lissencephaly, familial white matter hypoplasia, and fetal alcohol syndrome (Curatolo et al., 1993; Glass et al., 2008; Paul et al., 2007; Schell-Apacik et al., 2008; Tang et al., 2009). In as many as 68–75% of cases of ACC, the cause remains unknown (Paul et al., 2007; Schell-Apacik et al., 2008).

A complex disorder, ACC results from disruption of one or more of the multiple steps in the formation of the corpus callosum, which involves formation of the telencephalic hemispheres, birth and specification of commissural neurons, midline glial patterning, axonal growth, and axonal guidance across the midline to the final target in the contralateral hemisphere (Paul et al., 2007). Midline crossing is a major event as fibers passing through the corpus callosum link homologous cortical regions in the left and right hemispheres according to an anteroposterior organization. The human corpus callosum begins to form around 6 gestational weeks when axons destined to cross the midline grow medially within the hemispheres (Schell-Apacik et al., 2008). At 11–12 gestational weeks, the first fibers cross the midline, and by 18–20 weeks, the corpus callosum assumes its final configuration (Schell-Apacik et al., 2008). Several glial populations play a role in the guidance of the crossing of the callosal fibers (e.g., midline zipper glia) (Paul et al., 2007; Schell-Apacik et al., 2008), as well as a "sling" of midline migratory neurons (Paul et al., 2007). In instances in which axons form but do not cross the midline, large aberrant (noncrossing) bundles occur along the midline, so-called Probst bundles. Certain genes in humans and/or animal models have been linked to ACC, including schizophrenia 1 (*Disc1*) (Clapcote and Roder, 2006; Paul et al., 2007), L1 cell adhesion molecule (L1-CAM) (Demyanenko et al., 1999), meckelin (Smith et al.,

2006), and aristaless-related homeobox gene (*ARX*) (Sherr, 2003). Likewise implicated in ACC are molecules involved in attracting and repelling axons that cross the midline; for example, NFIA (Paul et al., 2007) and Rac1 (member of the Rho family of GTPases) (Kassai et al., 2008). In addition, ACC is associated with major forebrain malformations with an onset prior to the formation of the callosal anlage (e.g., holoprosencephaly). In addition to genetic influences, environmental factors are important in the pathogenesis of ACC. Prenatal alcohol exposure, for example, adversely affects gliogenesis and neuronal–glial interactions (i.e., processes crucial for callosal development) (Paul et al., 2007). The interplay of genetic and environmental factors in ACC is well-illustrated by the role of L1-CAM in its pathogenesis. Mutations in the L1-CAM gene are associated with ACC, corticospinal tract hypoplasia, hydrocephalus, and lower-limb spasticity in humans (Burden-Gulley et al., 1997; Rosenthal et al., 1992; Wong et al., 1995). Prenatal alcohol exposure, which is also associated with ACC, interacts with L1-CAM–mediated functions (e.g., it disrupts L1-mediated cell–cell adhesion in various cell types) (Charness et al., 1994; Ramanathan et al., 1996; Wilkemeyer and Charness, 1998) and decreases neurite outgrowth in cerebellar granule cell cultures (Bearer et al., 1999). The identification of novel proteins involved in callosal formation provides important means for the study of human ACC, for it is now possible to use relevant markers (e.g., antibodies, mRNA probes) in tissue sections [e.g., markers to products of the gene LIM domain only 4 (*Lmow4*)] to assess specific callosal projection neurons in the cerebral cortex (Arlotta et al., 2005).

Agenesis of the corticospinal tracts. This rare condition occurs in relative isolation or in association with diverse disorders, including X-linked congenital aqueductal stenosis, holoprosencephaly, anencephaly, porencephaly, hydranencephaly, Meckel syndrome, Moebius syndrome, agyria, microcephaly with arthrogryposis and renal hypoplasia, schizencephaly, Walker–Warburg syndrome, and hypoxic–ischemic lesions of the cerebral cortex (Bubis and Landau, 1964; Chow et al., 1985; Coad et al., 1997; Hori et al., 1986; Nardelli et al., 1982). Agenesis of the corticospinal tract is best assessed in transverse sections of the medulla where the pyramids are distinctly defined. It is characterized by bilateral absence in an otherwise intact medulla, including the presence of the arcuate nucleus and external arcuate fibers overlying the ventral medullary surface (Bubis and Landau, 1964; Coad et al., 1997). The cerebral peduncles in midbrain sections are small, and the longitudinal fibers in pontine sections are reduced. The corticospinal tracts mediate voluntary motor activity, and their absence in early life results in reduced movement and, in some instances, the fetal akinesia deformation sequence (Coad et al., 1997). The role of the corticospinal tract in motor dysfunction is underscored by the report of two sibling infants in whom relatively isolated agenesis of the corticospinal tract was associated with delayed motor development, head lag, and spasticity (Roessmann et al., 1990).

Little is known about the underlying molecular cues in the development of the human corticospinal tract; moreover, specific mutations have not been identified in genes responsible for its axonal elongation and/or guidance. The netrin 1 receptors *Unc5h3* and *Dcc* are of particular interest as they are necessary at multiple "choice" points in the guidance of corticospinal tract axons, and *netrin 1* mutant mice demonstrate corticospinal tract abnormalities (Finger et al., 2002). The time-table of the growth of the human corticospinal tract is well delineated: its axons reach the lower medulla by 10–12 gestational weeks, where they apparently stop until around 14 weeks, then decussate and descend slowly (Humphrey, 1966; Sidman and Rakic, 1982); by 17 gestational weeks, the axons reach the midthoracic level, and by 29 weeks, they reach their final targets in the lower spinal cord (Humphrey, 1966; Sidman and Rakic, 1982). The expression of GAP-43, a marker of axonal elongation, is extremely intense in the corticospinal tracts in the human pons and medulla at 19–20 gestational weeks; it decreases over the last half of gestation and at term but persists into infancy (beyond the time the lower spinal cord targets are reached) at low levels in the medullary pyramids (Fig. 1.10) (Kinney et al., 1993). This persistence in expression suggests a possible role for GAP-43 in axons beyond the period of targeting, perhaps in retaining plastic capacities for increase in diameter, elongation, and/or sprouting (Kinney et al., 1993). This idea seems particularly relevant because the human corticospinal tract must continue to elongate over long distances, even after its spinal targets are reached, to accommodate the enormous changes in the somatic length of the growing infant and child (Kinney et al., 1993). The possibility that the human corticospinal tract originates solely within the rolandic region of the cerebral cortex, including area 4, is supported by the report of agenesis of the corticospinal tracts in association with schizencephalic clefts isolated to the rolandic cortex in a 6-year-old child (Bubis and Landau, 1964).

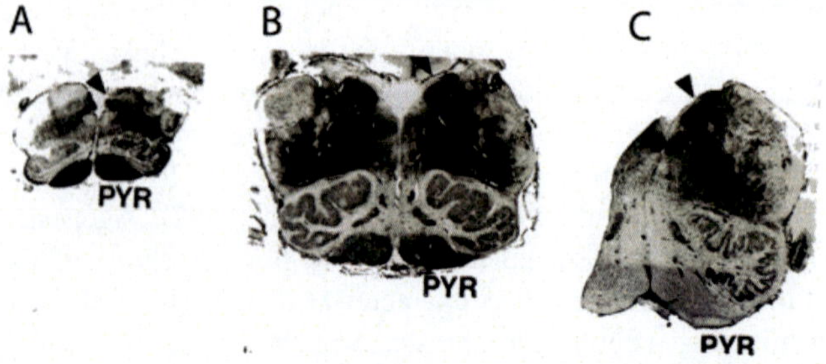

Fig. 1.10 GAP-43 expression in corticospinal tract. The relative distribution of GAP-43 immunostaining in the corticospinal tract in the human medulla across development. Bars = 10 mm. **a** Fetus at 21 gestational weeks. **b** Infant at 5 postnatal months. **c** Adult at 52 years. Immunostaining is extremely intense in the corticospinal tract at the level of the pyramid (PYR) at 21 gestational weeks and mildly intense at 5 postnatal months; by adulthood, it has completely disappeared (Kinney et al., 1993)

Agenesis of the corticospinal tract is due to either developmental defects or to destructive lesions rostral to the medulla before its outgrowth and elongation (i.e., in the embryonic period or early part of the second trimester). In anencephaly and holoprosencephaly (i.e., disorders of cortical formation in the embryonic period), for example, agenesis of the corticospinal tract is a constant feature (Chow et al., 1985). Destructive lesions to the precentral gyrus due to prenatal hypoxia–ischemia are also found (Bubis and Landau, 1964). Yet, this destructive damage likely occurs early in gestation (i.e., the first trimester or early second trimester) in order to result in anomalous development of the corticospinal tract and not secondary degeneration of axonal fibers, because the histologic hallmarks of acquired damage (i.e., gliosis, macrophagocytic infiltration, and/or Wallerian degeneration) are not found. Congenital hemiplegia due to corticospinal tract abnormalities occurs in early life in association with multiple developmental and acquired lesions, including multicystic encephalomalacia, venous infarction, polymicrogyria, and periventricular leukomalacia (Bleyenheuft et al., 2007; Glenn et al., 2007), and occurs in one-third of patients with the clinical diagnosis of cerebral palsy (Glenn et al., 2007). The severity of motor dysfunction in such patients correlates with the degree of asymmetry in diffusion measures in the pyramidal tracts, which are defined by diffuse tensor MRI tractography (Glenn et al., 2007).

Abnormal central axonal development associated with congenital encephalopathy, cardiomyopathy, and myopathy. This familial entity was first reported in three infants, two of whom were siblings, who died at 5, 10, and 16 months after a complicated course of progressive psychomotor retardation, microcephaly, cataracts, long tract signs, hypotonia, failure to thrive, cardiomyopathy, and necrotizing myopathy that began around the time of birth (Lyon et al., 1990). All three cases had neuropathologic examination at autopsy that was remarkable for an extremely thin corpus callosum without Probst bundles, reduced white matter volume in the cerebral hemispheres, diffuse axonal swellings throughout gray and white matter structures, including in the corpus callosum, and absence of the corticospinal tract at the level of the medullary pyramid (Lyon et al., 1990). A causative gene mutation has not been described to date. The pathogenesis of this disorder is thought to result from a primary disorder of axonal development resulting in a thin (hypoplastic) corpus callosum and agenesis of the corticospinal tract as there is no morphologic evidence of the secondary features of axonal degeneration (i.e., gliosis and myelin breakdown). The possibility of abnormal axonal elimination is a consideration as axonal swellings are associated with this regressive developmental process in experimental models (Lyon et al., 1990).

1.4.4.2 Hypoxia–Ischemia

Periventricular leukomalacia (PVL). Periventricular leukomalacia is the major neuropathologic substrate of cerebral palsy and cognitive deficits in survivors of prematurity (Kinney and Volpe, 2009; Volpe, 2009). It is a problem of major public health significance because 10% of premature infants less than 1,500 g born in the United States each year develop cerebral palsy, and 25–50% develop cognitive deficits (Volpe, 2009). Periventricular leukomalacia is a disorder of the immature telencephalic white matter characterized by (1) focal necrosis in the periventricular white matter; (2) diffuse reactive gliosis and microglial activation in the surrounding white matter; and (3) relative (albeit not complete) sparing of the overlying cerebral cortex (Kinney and Volpe, 2009). The major cause of PVL is cerebral ischemia–reperfusion oftentimes complicated in some cases by maternofetal infection (Kinney and Volpe, 2009). A complex interplay of vascular factors also predisposes to human periventricular white matter injury, including the presence of vascular end-zones and a propensity for the sick premature infant to exhibit a pressure-passive circulation reflecting disturbance of cerebral autoregulation (Kinney and Volpe, 2009). The greatest period of risk for PVL is the second half of gestation (24–32 weeks), although it also occurs in full-term neonates, particularly those with congenital cardiac or pulmonary disease. The peak window of vulnerability coincides with the period of dominance of premyelinating OLs (Back et al., 2001) and active axonal elongation (Haynes et al., 2005) in the cerebral white matter, and thus developing OLs and axons are postulated to underlie the selective susceptibility of the immature white matter to ischemic injury (Kinney and Volpe, 2009). The white matter damage is further characterized in the diffuse component by deficits in myelin formation (without loss of OL cell bodies) that is likely due to combined free radical, excitotoxic, and cytokine toxicity to developing OLs but without a loss of OL cell bodies (Billiards et al., 2008; Kinney and Volpe, 2009).

Axonal injury with spheroids has long been recognized in the periventricular necrotic foci of PVL and is part of the destruction of all cellular elements, not only axons, in the periventricular "core" infarcts in the arterial end zones (Kinney and Volpe, 2009) (Fig. 1.8) Intra-axonal accumulation of β-APP, indicative of disturbances in axoplasmic transport, are present within and at the edge of the necrotic foci (Fig. 1.8). Recently, *diffuse and widespread* axonal injury was demonstrated in the white matter surrounding and distant from the focally necrotic foci (i.e., in the penumbra of ischemic injury) (Fig. 1.11). Long-standing axonal damage in the diffuse white matter may contribute to the reduced white matter volume and callosal thinning found in severe and end-stage PVL in combination with myelin deficits (Haynes et al., 2008). Indeed, the myelin deficits may be secondary to the axonal damage that results in impaired axonal–OL interactions in the initiation and

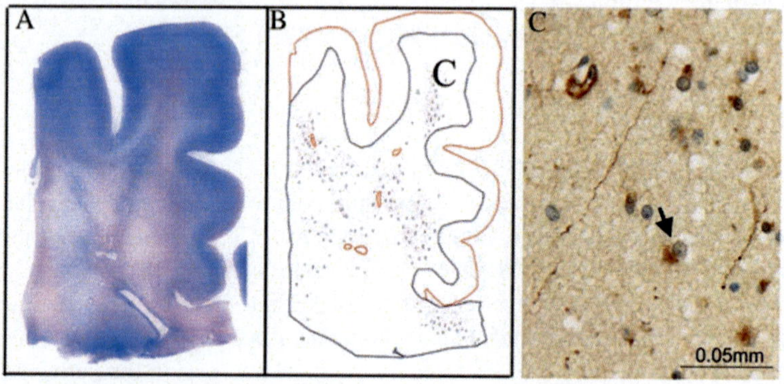

Fig. 1.11 Diffuse axonal injury in PVL. **a** In a 39-postconceptional-week case with organizing PVL, a hematoxylin and eosin stain indicates the presence of multiple cysts within the deep white matter and widespread rarefaction of the diffuse gliotic component. **b** A Neurolucida image of this same section stained for fractin indicates focal microcysts (*red circles*) and fractin-positive axons (*blue lines*) (C) that can be detected throughout the white matter, distant from the focal necrosis, and into the intragyral white matter and corpus callosum (Haynes et al., 2008)

maintenance of myelination (Haynes et al., 2008). The finding of extensive axonal pathology suggests the possibility that PVL is not only a "primary" disorder of the premyelinating OL but also of developing axons. This primary axonal injury in the diffuse component of PVL is likely due to hypoxia–ischemia based on the recognized hypoxic–ischemic insults to developing axons in animal models (McCarran and Goldberg, 2007; Underhill and Goldberg, 2007); it may also reflect secondary degeneration of thalamocortical afferents based on the recent report of thalamic damage in association with PVL (Ligam et al., 2008). Injury to axons throughout the diffuse and focal components of PVL may also lead to architectonic changes in the overlying cortex, as demonstrated in Golgi preparations (Marin-Padilla, 1997). Afferent fibers appear to fail to reach their cortical targets due to the white matter lesions; in addition, cortical pyramidal neurons from which long-projecting axons originate undergo transformations due to severance of these axons via a dying-back mechanism, and the pyramidal neurons change from long-projecting cells to local circuit interneurons (Marin-Padilla, 1997). Indeed, the neurologic sequelae following perinatal PVL may be, at least in part, a direct consequence of postinjury gray matter transformations that involve axons (Marin-Padilla, 1997). DTI of very low birth weight infants, as early as term equivalent age (Thomas et al., 2005; Vangberg et al., 2006), and then later in childhood (Prayer et al., 2001) demonstrate diminished relative anisotropy, especially in infants with cerebral white matter injury. This impairment in anisotropic diffusion could be caused by either a disturbance of axonal number, size, packing, or of axonal membranes or intracellular constituents or, alternatively, a disturbance in axonal ensheathment by OLs.

Of major interest is the demonstration by diffusion tensor MRI of alterations in the microstructure of the genu of the corpus callosum in young adult females who were born very preterm (<33 gestational weeks) compared with that of those born at term (Kontis et al., 2009). Moreover, these callosal alterations were associated with reduced performance IQ in the affected women (Kontis et al., 2009). The genu contains axonal fibers that connect the right and left prefrontal lobes, and pathologic changes in their organization may adversely affect prefrontal function (Kontis et al., 2009). These observations highlight the potential long-term consequences of diffuse axonal injury in PVL in the perinatal period (Haynes et al., 2008).

1.4.4.3 Trauma

Shaken baby syndrome. Head injury due to homicide typically involves children less than 3 years of age and is the most common cause of traumatic death in this population (Reichard et al., 2003); its main neuropathologic feature is diffuse axonal injury (Geddes et al., 2001; Reichard et al., 2003; Shannon et al., 1998). In the early 1970s, Caffey reported that violent shaking of infants caused severe neurologic damage (i.e., the shaken baby syndrome) (Caffey, 1972, 1974). This syndrome typically presents as a seemingly inexplicable, catastrophic decline in the infant's neurologic status leading to coma and oftentimes cardiopulmonary arrest and death. It is associated with acute subdural hematomas, subarachnoid hemorrhages, cerebral edema with or without transtentorial herniation, and retinal hemorrhages with no or minimal evidence of blunt head trauma (Shannon et al., 1998). The typical history is that the infant experienced minor trauma, fell in the bath tub or from the bed, or was found unresponsive or dead (Reichard et al., 2003). Shaking of the infant is thought to result in flexion/extension injury of the weakly supported infant neck and acceleration/deceleration of the infant brain with tearing of vessels and intracranial hemorrhage (Shannon et al., 1998). In shaken baby syndrome, as in adult entities of traumatic head injury, immunoreactive β-APP spheroids are found in the cerebral hemispheric white matter, including the corpus callosum and internal capsule, and in the brain stem, including the superior cerebellar peduncle and corticospinal tracts. The pathogenesis of traumatic axonal injury appears to involve (1) vascular compromise secondary to apnea and/or cardiopulmonary arrest resulting in hypoxic brain injury and increased intracranial pressure (Geddes et al., 2001; Reichard et al., 2003; Shannon et al., 1998); and/or (2) acceleration-induced shearing injury of axons (Geddes et al., 2001; Shannon et al., 1998). A primary role for hypoxic-induced axonal injury is emphasized because axonal spheroids/bulbs are not specific to traumatic head injury but also occur in children and adults dying with atraumatic, hypoxic–ischemic encephalopathy in which axonal injury occurs in a similar frequency and distribution to that in trauma (Geddes et al., 2001; Kaur et al., 1999; Shannon et al., 1998). Noteworthy in the shaken baby syndrome compared with hypoxic–ischemic encephalopathy is the common occurrence of cervical cord and root injury in the former disorder (Geddes et al., 2001; Shannon et al., 1998). Indeed, in shaken baby syndrome, β-APP immunoreactivity is present at the glial head of the cervical spinal nerve roots, suggesting that this lesion results from stretch-induced injury secondary to cervical hyperextension/flexion; damage to this region could account in turn for apnea and cardiopulmonary arrest leading to brain hypoxia, edema, and diffuse axonal damage in the forebrain and brain stem (Geddes et al., 2001; Shannon et al., 1998).

1.4.4.4 Infection

Infantile botulism. This life-threatening disorder of flaccid paralysis is due to intestinal infection with *Clostridium botulinum* in infants, resulting in the production of botulinum neurotoxin. It presents in infants with weakness, hypotonia, poor feeding, and constipation typically between 2 weeks and 6 months of age (Volpe, 2008b). It is further characterized by a neurologic syndrome of ptosis, facial diplegia, weak suck, impaired swallowing and gag, pupillary dysfunction, peripheral neuropathy, and hypotonia, with the paralytic process progressing in a descending direction (Volpe, 2008b). The most severe cases require mechanical ventilation and tube feeding. The typical duration of infantile botulism is 1–2 months; it is rarely fatal. Diagnosis depends primarily upon the isolation of *Clostridium botulinum* from the stool in conjunction with the clinical picture (Brook, 2007). Infantile botulism illustrates the principle that toxic interactions at the presynaptic terminal of the axon can produce disease without a morphologic "footprint." The botulinum toxin prevents exocytosis of presynaptic vesicles that contain the neurotransmitter acetylcholine both in central and peripheral axons (Brooks, 1956; Brunger et al., 2008), thereby inhibiting cholinergic release and transmission and leading to paralysis of cranial and somatic motor nerves. The toxin specifically blocks neurotransmitter release by proteolytic cleavage of SNAREs (soluble *N*-ethylmaleimide–sensitive factor attachment protein receptor) that play a key role in calcium-triggered neurotransmitter release in the synaptic vesicle fusion machinery (Brunger et al., 2008).

Acquired immunodeficiency syndrome (AIDS). Neurologic disorders are well-recognized complications of AIDS in infants and children, resulting from direct HIV infection and secondary infections due to immune compromise (Dickson et al., 1989; Kinney and Armstrong, 2002). In the majority of pediatric AIDS cases, HIV infection is vertically transmitted from mother to infant during gestation. Neurologic manifestations of direct HIV infection in infants and young children include primary HIV encephalopathy, with microcephaly and subsequent cognitive impairment in the congenital cases (Dickson et al., 1989; Kinney and Armstrong, 2002). In a neuropathologic study of pediatric AIDS, corticospinal tract degeneration was noted in 15 of 20 examined spinal cords at autopsy; in 6 cases the tract degenerations were consistent with delayed myelination with little or no axonal injury and in the remaining cases with axonal and Wallerian degeneration likely secondary to encephalitic lesions in the forebrain (Dickson et al., 1989). The characteristic long-tract signs in pediatric AIDS (i.e., hyper-reflexia, clonus, Babinski sign, paraparesis, and quadriparesis) are considered the neurologic consequences of this corticospinal injury. The precise role of HIV in producing delayed myelination or primary long-tract injury of the corticospinal tract is unknown.

1.4.4.5 Genetic Disorders

Infantile ascending hereditary spastic paralysis (IAHSP). Presenting at 1–2 years of life with lower-limb spasticity, this autosomal recessive disorder is characterized by severe degeneration of the pyramidal tracts; it progresses slowly to affect the upper limbs and bulbar muscles in the second decade resulting in wheelchair confinement in association with intact cognition (Lesca et al., 2003). The clinical, neuroimaging, and neurophysiologic findings are consistent with a relatively selective involvement of the corticospinal and corticobulbar pathways, and ISHSP is therefore classified as a motor neuron disease (i.e., selective dysfunction and/or loss of upper and/or lower motor neurons). It is reviewed here as representative of motor neuron diseases in very early life that selectively involve central axons and to emphasize the principle that injury/death to the cell body of the central neuron can result in antegrade axonal loss and tract degeneration. The neuropathology of ISHSP, however, has yet to be well characterized. Causative mutations in the gene *ALS2* have now been reported in a juvenile autosomal recessive form of motor neuron diseases, including IAHSP, amyotrophic lateral sclerosis (ALS) 2, and juvenile primary lateral sclerosis (Chinnery et al., 2004; Eymard-Pierre et al., 2002; Hadano et al., 2007; Panzeri et al., 2006); to date, 12 independent *ALS2* mutations (small deletion, non-sense mutation, or mis-sense mutation) have been reported (Hadano et al., 2007). A loss of functions in the *ALS2*-coded protein thus accounts for the motor dysfunction and/or degeneration of the *ALS2*-linked motor neuron diseases (Devon et al., 2005; Hadano et al., 2007; Panzeri et al., 2006). The *ALS2* gene encodes the protein ALS2 or alsin, which activates Rab5 small GTPases and is involved in endosome/membrane trafficking and fusion at the synapse, the promotion of neurite outgrowth in cell culture, and neuroprotection against cytotoxicity generated by oxidative stress and excitotoxicity (Devon et al., 2005; Hadano et al., 2007). The protein is expressed in neurons but not glia; it is present in motor neurons and the cerebellum (Devon et al., 2005; Hadano et al., 2007). The precise mechanism(s) and basis of motor neuron selectivity underlying ALS2 mutations in ISHSP are under active investigation.

Infantile neuroaxonal dystrophy (IND). This autosomal recessive disorder, also known as connatal Seitelberger disease, is a familial, progressive neurodegenerative disease that originates in utero, presents at birth or within the first 2 years of postnatal life, and involves neurodevelopmental regression, hypotonia, tetraplegia, visual impairments, dementia, and death in the first decade (Chow and Padfield, 2008). The disease gene has been mapped to a 1.17-Mb locus on chromosome 22q13.1, and an underlying mutation has been identified in PLA2G6; this gene encodes phospholipase A2 group VI, suggesting a role for phospholipase in the pathogenesis of the axonal defects (Khateeb et al., 2006). In IND, axonal swellings and spheroids are found through the central and peripheral nervous system; central involvement includes the cerebral and cerebellar cortex, deep cerebellar nuclei, brain-stem tegmentum, and posterior columns of the spinal cord. Associated findings in occasional cases are microcephaly and arthrogryposis (Chow and Padfield, 2008).

Congenital cranial disinnervation disorders. This rare group of genetic disorders is defined by congenital oculomotility syndromes that result from mutations in genes that are critical to the development and connectivity of cranial motoneurons (Engle, 2007). Human homeobox A1 (HOXA1) syndromes are part of the congenital cranial disinnervation disorders due to disruption of early motor neuron development (Engle, 2007). Germane to our focus are two disorders affecting axons predominately: (1) horizontal gaze palsy with progressive scoliosis in which there is aberrant axonal targeting onto the motor neurons of cranial nerve VI (abducens); and (2) congenital fibrosis of the extraocular muscles type I (CFEOM1) in which there is aberrant axonal targeting onto the extraocular muscles (Engle, 2007). In CFEOM1, there is bilateral ptosis and ophthalmoplegia with a fixed downward gaze. An autosomal recessive disorder, the mutated gene is now known to be *KIF21A,* a member of the kinesin family of molecular motors (Chan et al., 2007b; Yamada et al., 2003). Kinesins are responsible for anterograde transport in axons, moving cargo along microtubules from the neuronal cell body to the growing or mature synapse (Engle, 2007). There are at least 45 human kinesins that transport different cargoes, including mitochondria, vesicles, and protein complexes; the cargo of KIF21A is currently unknown (Engle, 2007). The *KIF21A* mutation nevertheless may disrupt the protein's ability to bind to specific cargo critical to the development of oculomotor axons (Engle, 2007).

Joubert syndrome. This syndrome is a rare, autosomal recessive, neurodevelopmental disorder that is characterized by ataxia, hyperpnea/apnea, abnormal eye movements, hypotonia, and cognitive deficits; variable features include retinal dystrophy and renal cystic disease. The pathologic features of Joubert syndrome include aplasia or hypoplasia of the cerebellar vermis, olivary anomalies, and cranial nerve nuclei abnormalities of cranial nerves V and X (Yachnis and Rorke, 1999). Of major interest here is the occurrence of abnormalities in axonal tracts that cross the posterior midline of the brain (e.g., superior cerebellar peduncles and corticospinal tracts) (Yachnis and Rorke, 1999). Not all midline crossing tracts are abnormal, however, in Joubert syndrome; the corpus callosum and optic chiasm, for example, are intact. Differences in the clinical and neuropathologic characteristics among patients suggest that this disorder is indeed a syndrome composed of heterogeneous disorders (Doering et al., 2008). Mutations in the *Aberson-helper integration site-1* gene (*AHI1*), mapped to chromosomal region 6q23.3, cause at least some cases of Joubert syndrome; this gene encodes a cytoplasmic protein that is likely a signaling or scaffolding protein involved in protein–protein interactions (Ferland et al., 2004; Jiang et al., 2002). In mice, the Ahi1 protein is expressed in the cytoplasm, dendrites, and axons of neurons of the forebrain and brain stem (Doering et al., 2008); its expression appears around embryonic day 10 and persists into adulthood. In the human brain, Ahi1 protein expression, as determined by Western blot analysis, is present in the mature cerebellar vermis, cerebral peduncles, and medulla (Doering et al., 2008). Its mRNA is expressed in the neurons that give rise to the crossing axons of the corticospinal tract and superior cerebellar peduncle (Ferland et al., 2004; Utsch et al., 2006). Jouberin, the protein encoded by *AHI1,* is involved in multiple functions, including signal transduction and cytoskeleton assembly (Jiang et al., 2002). It is reasoned that the *AHI1* gene is important in midline crossing of axons such that mutations within it are related to the pathogenesis of the decussation defects in Joubert syndrome (Meng et al., 2009). Of note, up to 40% of children with Joubert syndrome are diagnosed with autism spectrum disorder, and several linkage studies in autism have implicated the region on 6q where *AHI1* resides; recently, common variants in AHI1 have also been associated with autism (Alvarez Retuerto et al., 2008).

Down syndrome. This autosomal disorder is the leading cause of inherited mental retardation and is due to trisomy 21. The neuropathology is complex, including an underweight brain, hypoplasia of the superior temporal lobe, cytoarchitectonic anomalies, and dendritic abnormalities in the cerebral cortex (Volpe, 2008a). The report of abnormalities in proteins related to axonal targeting in Down syndrome is of interest here (Weitzdoerfer et al., 2001). Based on a proteomic approach, five members of the semaphorin/collapsin family have been found to be significantly decreased in the cerebral cortex of aborted fetuses (around 19 weeks) with Down syndrome compared with that of controls (Weitzdoerfer et al., 2001). These proteins are involved in the guidance of axons in fetal life (Weitzdoerfer et al., 2001). This proteomic study suggests that possibility that they are involved in the abnormal wiring underlying cognitive impairment in Down syndrome in conjunction with neuronal, dendritic, and myelin impairments. Further research is needed into central axonal pathology in Down syndrome.

1.4.4.6 Neurodegenerative Disorders of Unknown Etiology

Central nervous system degeneration is defined as progressive deterioration in a neural system(s) that is oftentimes familial and the cause of which is unknown. It typically affects a neural system (e.g., motor, sensory, and/or autonomic pathways) in a bilateral and symmetric distribution. Pathologically it is characterized by neuronal loss, axonal loss, and/or gliosis, oftentimes without a distinct histologic hallmark. Neurodegenerative disorders are due to genetic, toxic, metabolic, or other causes: Once the cause is determined, the entity is labeled in regard to the specific gene defect or other causative factor and is no longer categorized as a "neurodegenerative" disorder. Degeneration implies the deterioration of a neural system that is fully mature, and thus the possibility of its occurrence in early life (i.e., before complete maturation) seems counterintuitive. Yet, certain disorders of the developing nervous system meet the criteria of a neurodegenerative disease (Folkerth et al., 1993; Salman et al., 2009), particularly after 20–22 gestational weeks when adult-like gliosis is first recognized and the embryonic ground-plan of neural systems is complete. In addition, a combination of developmental and degenerative features may be present in the same case as found in pontocerebellar hypoplasia (Barth, 1993), suggesting that a particular insult in early life may simultaneously affect neural systems that are under development and that are completely developed, thereby leading to a mixture of developmental and degenerative features. Moreover, degenerative processes may be superimposed on developmental programs as the insult extends beyond the developmental period for a particular system. It should be emphasized that the developmental sequences of axonal maturation are incompletely unknown, as are the cellular and molecular factors that constitute axonal "maturity."

Arthrogryposis multiplex congenita (AMC) with posterior column degeneration and peripheral neuropathy is an example of a complex neurodegenerative disorder associated with a malformative process (i.e., AMC) in early life. In a report of this entity, a 9-week-old, full-term infant was born with multiple fixed joints (AMC), hypotonia, areflexia, failure to sustain respiration at birth, and ventilatory dependence from birth (Folkerth et al., 1993); a sural (sensory) nerve biopsy was devoid of myelinated axons. At autopsy, there was severe, bilateral, and symmetric degeneration of the posterior columns with gliosis in association with posterior horn gliosis, posterior root atrophy, and degeneration of peripheral axons (Fig. 1.12). This case underscores the occurrence of a bilateral and symmetric neurodegeneration of a neural (sensory) system that begins in utero, presents at birth, and leads to death in infancy. The neuropathologic features suggest a dying-back neuropathy of the kinesthetic pathway in utero leading to AMC as the structurally intact posterior columns demonstrated histopathologic features consistent with secondary degeneration (i.e., gliosis and Wallerian degeneration) (Fig. 1.12) (Folkerth et al., 1993). In this case, AMC is likely a consequence of arrested joint modeling during its critical period of 14–16 gestational weeks. Joint modeling is dependent upon fetal movement, and AMC is well recognized to result from depressed or absent movement due to multiple intrauterine motor disorders (Folkerth et al., 1993). The case of AMC with

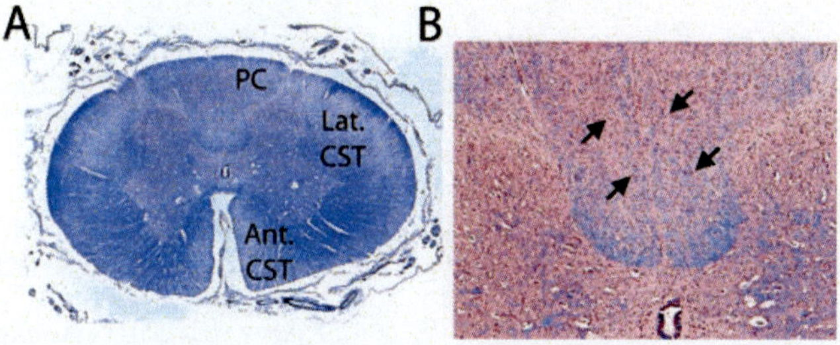

Fig. 1.12 Posterior column degeneration. There is bilateral, symmetric degeneration of the posterior columns (PC) with severe myelin pallor and gliosis at all spinal levels, including the thoracic level. **a** At the age of the patient (9 weeks), the posterior columns are normally completely myelinated and stain intensely blue with the Luxol-fast-blue stain for myelin. The myelin pallor of the anterior (Ant CST) and lateral corticospinal (Lat CST) tracts reflects age-appropriate immaturity of myelination. At high power, there is intense gliosis in the posterior column in association with myelin loss. **b** A mild decrease in axonal number in the posterior columns was demonstrated by a Bodian stain (not shown) (Luxol-fast-blue/hematoxylin and eosin, ×1)

posterior column degeneration, however, suggests that intact kinesthetic, as well as motor, innervation is essential for normal joint modeling (Folkerth et al., 1993) and thus exemplifies the interface between malformations and degenerative processes in central axons in early life.

1.5 Summary

In this chapter, we reviewed the development and pathology of central axons in early life. Though much is now known about the processes involved in axonal growth, maturation, and myelination, there remain many unanswered questions, particularly regarding the mechanisms and signaling pathways underlying axonal function and pathology. To gain further insight, the integrated study of human disorders, the molecular biology of axons, and experimental paradigms of axonal function and dysfunction continue to be essential

Acknowledgments The authors appreciate the help of Kate Ackerson in the preparation of the manuscript. We are grateful for the careful reading of the manuscript by Dr. Joseph J. Volpe.

This work was supported by grants from the National Institute of Neurological Diseases and Stroke (PO1-NS38475), the National Institute of Child Health and Development (Children's Hospital Developmental Disabilities Research Center) (P30-HD18655), the William Randolph Hearst Award, and the National Institute of Alcohol Abuse and Alcoholism (K01AA015373).

References

Alvarez Retuerto AI, Cantor RM, Gleeson JG, Ustaszewska A, Schackwitz WS, Pennacchio LA, Geschwind DH (2008) Association of common variants in the Joubert syndrome gene (AHI1) with autism. Hum Mol Genet 17:3887–3896

Amaratunga A, Morin PJ, Kosik KS, Fine RE (1993) Inhibition of kinesin synthesis and rapid anterograde axonal transport in vivo by an antisense oligonucleotide. J Biol Chem 268:17427–17430

Arlotta P, Molyneaux BJ, Chen J, Inoue J, Kominami R, Macklis JD (2005) Neuronal subtype-specific genes that control corticospinal motor neuron development in vivo. Neuron 45:207–221

Avwenagha O, Campbell G, Bird MM (2003) Distribution of GAP-43, beta-III tubulin and F-actin in developing and regenerating axons and their growth cones in vitro, following neurotrophin treatment. J Neurocytol 32:1077–1089

Back SA, Luo NL, Borenstein NS, Levine JM, Volpe JJ, Kinney HC (2001) Late oligodendrocyte progenitors coincide with the developmental window of vulnerability for human perinatal white matter injury. J Neurosci 21:1302–1312

Back SA, Luo NL, Borenstein NS, Volpe JJ, Kinney HC (2002) Arrested oligodendrocyte lineage progression during human cerebral white matter development: dissociation between the timing of progenitor differentiation and myelinogenesis. J Neuropathol Exp Neurol 61:197–211

Back SA, Riddle A, McClure MM (2007) Maturation-dependent vulnerability of perinatal white matter in premature birth. Stroke 38:724–730

Bacon C, Lakics V, Machesky L, Rumsby M (2007) N-WASP regulates extension of filopodia and processes by oligodendrocyte progenitors, oligodendrocytes, and Schwann cells-implications for axon ensheathment at myelination. Glia 55:844–858

Bagnard D, Lohrum M, Uziel D, Puschel AW, Bolz J (1998) Semaphorins act as attractive and repulsive guidance signals during the development of cortical projections. Development 125:5043–5053

Baker KA, Moore SW, Jarjour AA, Kennedy TE (2006) When a diffusible axon guidance cue stops diffusing: roles for netrins in adhesion and morphogenesis. Curr Opin Neurobiol 16:529–534

Bard L, Boscher C, Lambert M, Mege RM, Choquet D, Thoumine O (2008) A molecular clutch between the actin flow and N-cadherin adhesions drives growth cone migration. J Neurosci 28:5879–5890

Barres BA, Raff MC (1993) Proliferation of oligodendrocyte precursor cells depends on electrical activity in axons. Nature 361:258–260

Barth PG (1993) Pontocerebellar hypoplasias. An overview of a group of inherited neurodegenerative disorders with fetal onset. Brain Dev 15:411–422

Bartsch U (2003) Neural CAMS and their role in the development and organization of myelin sheaths. Front Biosci 8:d477–d490

Bartsch S, Montag D, Schachner M, Bartsch U (1997) Increased number of unmyelinated axons in optic nerves of adult mice deficient in the myelin-associated glycoprotein (MAG). Brain Res 762:231–234

Bearer CF (2001) L1 cell adhesion molecule signal cascades: targets for ethanol developmental neurotoxicity. Neurotoxicology 22:625–633

Bearer CF, Swick AR, O'Riordan MA, Cheng G (1999) Ethanol inhibits L1-mediated neurite outgrowth in postnatal rat cerebellar granule cells. J Biol Chem 274:13264–13270

Benarroch EE (2009) Oligodendrocytes: susceptibility to injury and involvement in neurologic disease. Neurology 72:1779–1785

Bentley D, O'Connor TP (1994) Cytoskeletal events in growth cone steering. Curr Opin Neurobiol 4:43–48

Billiards SS, Haynes RL, Folkerth RD, Borenstein NS, Trachtenberg FL, Rowitch DH, Ligon KL, Volpe JJ, Kinney HC (2008) Myelin abnormalities without oligodendrocyte loss in periventricular leukomalacia. Brain Pathol 18:153–163

Bleyenheuft Y, Grandin CB, Cosnard G, Olivier E, Thonnard JL (2007) Corticospinal dysgenesis and upper-limb deficits in congenital hemiplegia: a diffusion tensor imaging study. Pediatrics 120:e1502–e1511

Bolsover SR (2005) Calcium signalling in growth cone migration. Cell Calcium 37:395–402

Bonfanti L (2006) PSA-NCAM in mammalian structural plasticity and neurogenesis. Prog Neurobiol 80:129–164

Brady ST, Pfister KK, Bloom GS (1990) A monoclonal antibody against kinesin inhibits both anterograde and retrograde fast axonal transport in squid axoplasm. Proc Natl Acad Sci USA 87:1061–1065

Brankatschk M, Dickson BJ (2006) Netrins guide Drosophila commissural axons at short range. Nat Neurosci 9:188–194

Brody BA, Kinney HC, Kloman AS, Gilles FH (1987) Sequence of central nervous system myelination in human infancy. I. An autopsy study of myelination. J Neuropathol Exp Neurol 46:283–301

Brook I (2007) Infant botulism. J Perinatol 27:175–180

Brooks VB (1956) An intracellular study of the action of repetitive nerve volleys and of botulinum toxin on miniature end-plate potentials. J Physiol 134:264–277

Bruckner K, Pasquale EB, Klein R (1997) Tyrosine phosphorylation of transmembrane ligands for Eph receptors. Science 275:1640–1643

Brunger AT, Jin R, Breidenbach MA (2008) Highly specific interactions between botulinum neurotoxins and synaptic vesicle proteins. Cell Mol Life Sci 65:2296–2306

Bubis JJ, Landau WM (1964) Agenesis of the pyramidal tracts associated with schizencephalic clefts in rolandic cortex. Neurology 14:821–824

Burden-Gulley SM, Pendergast M, Lemmon V (1997) The role of cell adhesion molecule L1 in axonal extension, growth cone motility, and signal transduction. Cell Tissue Res 290:415–422

Butt AM, Berry M (2000) Oligodendrocytes and the control of myelination in vivo: new insights from the rat anterior medullary velum. J Neurosci Res 59:477–488

Butt AM, Ibrahim M, Berry M (1997) The relationship between developing oligodendrocyte units and maturing axons during myelinogenesis in the anterior medullary velum of neonatal rats. J Neurocytol 26:327–338

Caffey J (1972) On the theory and practice of shaking infants. Its potential residual effects of permanent brain damage and mental retardation. Am J Dis Child 124:161–169

Caffey J (1974) The whiplash shaken infant syndrome: manual shaking by the extremities with whiplash-induced intracranial and intraocular bleedings, linked with residual permanent brain damage and mental retardation. Pediatrics 54:396–403

Carulli D, Laabs T, Geller HM, Fawcett JW (2005) Chondroitin sulfate proteoglycans in neural development and regeneration. Curr Opin Neurobiol 15:116–120

Chan WM, Andrews C, Dragan L, Fredrick D, Armstrong L, Lyons C, Geraghty MT, Hunter DG, Yazdani A, Traboulsi EI, Pott JW, Gutowski NJ, Ellard S, Young E, Hanisch F, Koc F, Schnall B, Engle EC (2007b) Three novel mutations in KIF21A highlight the importance of the third coiled-coil stalk domain in the etiology of CFEOM1. BMC Genet 8:26

Chan CK, Wang J, Lin L, Hao Y, Chan SO (2007a) Enzymatic removal of hyaluronan affects routing of axons in the mouse optic chiasm. Neuroreport 18:1533–1538

Charness ME, Safran RM, Perides G (1994) Ethanol inhibits neural cell-cell adhesion. J Biol Chem 269:9304–9309

Chedotal A, Del Rio JA, Ruiz M, He Z, Borrell V, de Castro F, Ezan F, Goodman CS, Tessier-Lavigne M, Sotelo C, Soriano E (1998) Semaphorins III and IV repel hippocampal axons via two distinct receptors. Development 125:4313–4323

Cheng HJ, Bagri A, Yaron A, Stein E, Pleasure SJ, Tessier-Lavigne M (2001) Plexin-A3 mediates semaphorin signaling and regulates the development of hippocampal axonal projections. Neuron 32:249–263

Chilton JK, Guthrie S (2003) Cranial expression of class 3 secreted semaphorins and their neuropilin receptors. Dev Dyn 228:726–733

Chinnery PF, Keers SM, Holden MJ, Ramesh V, Dalton A (2004) Infantile hereditary spastic paraparesis due to codominant mutations in the spastin gene. Neurology 63:710–712

Chow CW, Halliday JL, Anderson RM, Danks DM, Fortune DW (1985) Congenital absence of pyramids and its significance in genetic diseases. Acta Neuropathol 65:313–317

Chow G, Padfield CJ (2008) A case of infantile neuroaxonal dystrophy: connatal Seitelberger disease. J Child Neurol 23:418–420

Clapcote SJ, Roder JC (2006) Deletion polymorphism of Disc1 is common to all 129 mouse substrains: implications for gene-targeting studies of brain function. Genetics 173:2407–2410
Coad JE, Angel C, Pierpont ME, Gorlin RJ, Anderson ML (1997) Microcephaly with agenesis of corticospinal tracts and arthrogryposis, hypospadias, single umbilical artery, hypertelorism, and renal and adrenal hypoplasia–previously undescribed syndrome. Am J Med Genet 71:458–462
Cohen J, Burne JF, McKinlay C, Winter J (1987) The role of laminin and the laminin/fibronectin receptor complex in the outgrowth of retinal ganglion cell axons. Dev Biol 122:407–418
Cohen S, Funkelstein L, Livet J, Rougon G, Henderson CE, Castellani V, Mann F (2005) A semaphorin code defines subpopulations of spinal motor neurons during mouse development. Eur J Neurosci 21:1767–1776
Colbert CM, Johnston D (1996) Axonal action-potential initiation and Na+ channel densities in the soma and axon initial segment of subicular pyramidal neurons. J Neurosci 16:6676–6686
Colbert CM, Pan E (2002) Ion channel properties underlying axonal action potential initiation in pyramidal neurons. Nat Neurosci 5:533–538
Coman I, Barbin G, Charles P, Zalc B, Lubetzki C (2005) Axonal signals in central nervous system myelination, demyelination and remyelination. J Neurol Sci 233:67–71
Conturo TE, Williams DL, Smith CD, Gultepe E, Akbudak E, Minshew NJ (2008) Neuronal fiber pathway abnormalities in autism: an initial MRI diffusion tensor tracking study of hippocampo-fusiform and amygdalo-fusiform pathways. J Int Neuropsychol Soc 14:933–946
Cowan CA, Henkemeyer M (2002) Ephrins in reverse, park and drive. Trends Cell Biol 12:339–346
Cremer H, Chazal G, Goridis C, Represa A (1997) NCAM is essential for axonal growth and fasciculation in the hippocampus. Mol Cell Neurosci 8:323–335
Curatolo P, Cilio MR, Del Giudice E, Romano A, Gaggero R, Pessagno A (1993) Familial white matter hypoplasia, agenesis of the corpus callosum, mental retardation and growth deficiency: a new distinctive syndrome. Neuropediatrics 24:77–82
Dailey ME, Bridgman PC (1991) Structure and organization of membrane organelles along distal microtubule segments in growth cones. J Neurosci Res 30:242–258
Daston MM, Bastmeyer M, Rutishauser U, O'Leary DD (1996) Spatially restricted increase in polysialic acid enhances corticospinal axon branching related to target recognition and innervation. J Neurosci 16:5488–5497
Davis JQ, Bennett V (1994) Ankyrin binding activity shared by the neurofascin/L1/NrCAM family of nervous system cell adhesion molecules. J Biol Chem 269:27163–27166
Davy A, Soriano P (2005) Ephrin signaling in vivo: look both ways. Dev Dyn 232:1–10
de Castro F, Hu L, Drabkin H, Sotelo C, Chedotal A (1999) Chemoattraction and chemorepulsion of olfactory bulb axons by different secreted semaphorins. J Neurosci 19:4428–4436
Dehmelt L, Halpain S (2005) The MAP2/Tau family of microtubule-associated proteins. Genome Biol 6:204
Demerens C, Stankoff B, Logak M, Anglade P, Allinquant B, Couraud F, Zalc B, Lubetzki C (1996) Induction of myelination in the central nervous system by electrical activity. Proc Natl Acad Sci USA 93:9887–9892
Demyanenko GP, Tsai AY, Maness PF (1999) Abnormalities in neuronal process extension, hippocampal development, and the ventricular system of L1 knockout mice. J Neurosci 19:4907–4920
Dent EW, Gertler FB (2003) Cytoskeletal dynamics and transport in growth cone motility and axon guidance. Neuron 40:209–227
Desilva TM, Kinney HC, Borenstein NS, Trachtenberg FL, Irwin N, Volpe JJ, Rosenberg PA (2007) The glutamate transporter EAAT2 is transiently expressed in developing human cerebral white matter. J Comp Neurol 501:879–890
Devon RS, Schwab C, Topp JD, Orban PC, Yang YZ, Pape TD, Helm JR, Davidson TL, Rogers DA, Gros-Louis F, Rouleau G, Horazdovsky BF, Leavitt BR, Hayden MR (2005) Cross-species characterization of the ALS2 gene and analysis of its pattern of expression in development and adulthood. Neurobiol Dis 18:243–257
Dickson BJ (2002) Molecular mechanisms of axon guidance. Science 298:1959–1964

Dickson DW, Belman AL, Park YD, Wiley C, Horoupian DS, Llena J, Kure K, Lyman WD, Morecki R, Mitsudo S et al (1989) Central nervous system pathology in pediatric AIDS: an autopsy study. APMIS Suppl 8:40–57

Dickson BJ, Gilestro GF (2006) Regulation of commissural axon pathfinding by slit and its Robo receptors. Annu Rev Cell Dev Biol 22:651–675

Dillman JF 3rd, Dabney LP, Karki S, Paschal BM, Holzbaur EL, Pfister KK (1996a) Functional analysis of dynactin and cytoplasmic dynein in slow axonal transport. J Neurosci 16:6742–6752

Dillman JF 3rd, Dabney LP, Pfister KK (1996b) Cytoplasmic dynein is associated with slow axonal transport. Proc Natl Acad Sci USA 93:141–144

Doering JE, Kane K, Hsiao YC, Yao C, Shi B, Slowik AD, Dhagat B, Scott DD, Ault JG, Page-McCaw PS, Ferland RJ (2008) Species differences in the expression of Ahi1, a protein implicated in the neurodevelopmental disorder Joubert syndrome, with preferential accumulation to stigmoid bodies. J Comp Neurol 511:238–256

Doherty P, Smith P, Walsh FS (1996) Shared cell adhesion molecule (CAM) homology domains point to CAMs signalling via FGF receptors. Perspect Dev Neurobiol 4:157–168

Doherty P, Williams G, Williams EJ (2000) CAMs and axonal growth: a critical evaluation of the role of calcium and the MAPK cascade. Mol Cell Neurosci 16:283–295

Drescher U, Kremoser C, Handwerker C, Loschinger J, Noda M, Bonhoeffer F (1995) In vitro guidance of retinal ganglion cell axons by RAGS, a 25 kDa tectal protein related to ligands for Eph receptor tyrosine kinases. Cell 82:359–370

Dufour A, Seibt J, Passante L, Depaepe V, Ciossek T, Frisen J, Kullander K, Flanagan JG, Polleux F, Vanderhaeghen P (2003) Area specificity and topography of thalamocortical projections are controlled by ephrin/Eph genes. Neuron 39:453–465

Engle EC (2007) Oculomotility disorders arising from disruptions in brainstem motor neuron development. Arch Neurol 64:633–637

Ernst AF, Jurney WM, McLoon SC (1998) Mechanisms involved in development of retinotectal connections: roles of Eph receptor tyrosine kinases, NMDA receptors and nitric oxide. Prog Brain Res 118:115–131

Eymard-Pierre E, Lesca G, Dollet S, Santorelli FM, di Capua M, Bertini E, Boespflug-Tanguy O (2002) Infantile-onset ascending hereditary spastic paralysis is associated with mutations in the alsin gene. Am J Hum Genet 71:518–527

Falk J, Bechara A, Fiore R, Nawabi H, Zhou H, Hoyo-Becerra C, Bozon M, Rougon G, Grumet M, Puschel AW, Sanes JR, Castellani V (2005) Dual functional activity of semaphorin 3B is required for positioning the anterior commissure. Neuron 48:63–75

Farinas I, DeFelipe J (1991) Patterns of synaptic input on corticocortical and corticothalamic cells in the cat visual cortex. II. The axon initial segment. J Comp Neurol 304:70–77

Ferland RJ, Eyaid W, Collura RV, Tully LD, Hill RS, Al-Nouri D, Al-Rumayyan A, Topcu M, Gascon G, Bodell A, Shugart YY, Ruvolo M, Walsh CA (2004) Abnormal cerebellar development and axonal decussation due to mutations in AHI1 in Joubert syndrome. Nat Genet 36:1008–1013

Finger JH, Bronson RT, Harris B, Johnson K, Przyborski SA, Ackerman SL (2002) The netrin 1 receptors Unc5h3 and Dcc are necessary at multiple choice points for the guidance of corticospinal tract axons. J Neurosci 22:10346–10356

Fitzpatrick MO, Dewar D, Teasdale GM, Graham DI (1998) The neuronal cytoskeleton in acute brain injury. Br J Neurosurg 12:313–317

Folkerth RD, Guttentag SH, Kupsky WJ, Kinney HC (1993) Arthrogryposis multiplex congenita with posterior column degeneration and peripheral neuropathy: a case report. Clin Neuropathol 12:25–33

Friedlander DR, Milev P, Karthikeyan L, Margolis RK, Margolis RU, Grumet M (1994) The neuronal chondroitin sulfate proteoglycan neurocan binds to the neural cell adhesion molecules Ng-CAM/L1/NILE and N-CAM, and inhibits neuronal adhesion and neurite outgrowth. J Cell Biol 125:669–680

Gallo V, Zhou JM, McBain CJ, Wright P, Knutson PL, Armstrong RC (1996) Oligodendrocyte progenitor cell proliferation and lineage progression are regulated by glutamate receptor-mediated K+ channel block. J Neurosci 16:2659–2670

Garwood J, Schnadelbach O, Clement A, Schutte K, Bach A, Faissner A (1999) DSD-1-proteoglycan is the mouse homolog of phosphacan and displays opposing effects on neurite outgrowth dependent on neuronal lineage. J Neurosci 19:3888–3899

Gates J, Peifer M (2005) Can 1000 reviews be wrong? Actin, alpha-Catenin, and adherens junctions. Cell 123:769–772

Geddes JF, Vowles GH, Hackshaw AK, Nickols CD, Scott IS, Whitwell HL (2001) Neuropathology of inflicted head injury in children. II. Microscopic brain injury in infants. Brain 124:1299–1306

Gil OD, Sakurai T, Bradley AE, Fink MY, Cassella MR, Kuo JA, Felsenfeld DP (2003) Ankyrin binding mediates L1CAM interactions with static components of the cytoskeleton and inhibits retrograde movement of L1CAM on the cell surface. J Cell Biol 162:719–730

Glass HC, Shaw GM, Ma C, Sherr EH (2008) Agenesis of the corpus callosum in California 1983–2003: a population-based study. Am J Med Genet A 146A:2495–2500

Glenn OA, Ludeman NA, Berman JI, Wu YW, Lu Y, Bartha AI, Vigneron DB, Chung SW, Ferriero DM, Barkovich AJ, Henry RG (2007) Diffusion tensor MR imaging tractography of the pyramidal tracts correlates with clinical motor function in children with congenital hemiparesis. AJNR Am J Neuroradiol 28:1796–1802

Goldberg DJ, Burmeister DW (1986) Stages in axon formation: observations of growth of Aplysia axons in culture using video-enhanced contrast-differential interference contrast microscopy. J Cell Biol 103:1921–1931

Goodwin M, Yap AS (2004) Classical cadherin adhesion molecules: coordinating cell adhesion, signaling and the cytoskeleton. J Mol Histol 35:839–844

Gordon-Weeks PR (1991) Evidence for microtubule capture by filopodial actin filaments in growth cones. Neuroreport 2:573–576

Gotz M, Bolz J, Joester A, Faissner A (1997) Tenascin-C synthesis and influence on axonal growth during rat cortical development. Eur J Neurosci 9:496–506

Govindan RM, Makki MI, Sundaram SK, Juhasz C, Chugani HT (2008) Diffusion tensor analysis of temporal and extra-temporal lobe tracts in temporal lobe epilepsy. Epilepsy Res 80:30–41

Grabham PW, Seale GE, Bennecib M, Goldberg DJ, Vallee RB (2007) Cytoplasmic dynein and LIS1 are required for microtubule advance during growth cone remodeling and fast axonal outgrowth. J Neurosci 27:5823–5834

Graham D, Lantos P (2002) Greenfield's neuropathology. Arnold, London

Ha J, Lo KW, Myers KR, Carr TM, Humsi MK, Rasoul BA, Segal RA, Pfister KK (2008) A neuron-specific cytoplasmic dynein isoform preferentially transports TrkB signaling endosomes. J Cell Biol 181:1027–1039

Hadano S, Kunita R, Otomo A, Suzuki-Utsunomiya K, Ikeda JE (2007) Molecular and cellular function of ALS2/alsin: implication of membrane dynamics in neuronal development and degeneration. Neurochem Int 51:74–84

Halpain S, Dehmelt L (2006) The MAP1 family of microtubule-associated proteins. Genome Biol 7:224

Hansen MJ, Dallal GE, Flanagan JG (2004) Retinal axon response to ephrin-as shows a graded, concentration-dependent transition from growth promotion to inhibition. Neuron 42:717–730

Hashimoto K, Kano M (2003) Functional differentiation of multiple climbing fiber inputs during synapse elimination in the developing cerebellum. Neuron 38:785–796

Haynes RL, Billiards SS, Borenstein NS, Volpe JJ, Kinney HC (2008) Diffuse axonal injury in periventricular leukomalacia as determined by apoptotic marker fractin. Pediatr Res 63: 656–661

Haynes RL, Borenstein NS, Desilva TM, Folkerth RD, Liu LG, Volpe JJ, Kinney HC (2005) Axonal development in the cerebral white matter of the human fetus and infant. J Comp Neurol 484:156–167

He Q, Dent EW, Meiri KF (1997) Modulation of actin filament behavior by GAP-43 (neuromodulin) is dependent on the phosphorylation status of serine 41, the protein kinase C site. J Neurosci 17:3515–3524

Hevner RF (2000) Development of connections in the human visual system during fetal midgestation: a DiI-tracing study. J Neuropathol Exp Neurol 59:385–392

Hevner RF, Kinney HC (1996) Reciprocal entohinal-hippocampal connections established by human fetal midgestation. J Comp Neurol 372:384–394

Hirokawa N, Sato-Yoshitake R, Kobayashi N, Pfister KK, Bloom GS, Brady ST (1991) Kinesin associates with anterogradely transported membranous organelles in vivo. J Cell Biol 114: 295–302

Hirokawa N, Takemura R (2004) Molecular motors in neuronal development, intracellular transport and diseases. Curr Opin Neurobiol 14:564–573

Hori A, Bardosi A, Goebel HH, Roessmann U (1986) Muscular alteration in agyria with pyramidal tract anomaly. Brain Dev 8:624–630

Howe CL, Mobley WC (2005) Long-distance retrograde neurotrophic signaling. Curr Opin Neurobiol 15:40–48

Huber G, Alaimo-Beuret D, Matus A (1985) MAP3: characterization of a novel microtubule-associated protein. J Cell Biol 100:496–507

Humphrey T (1966) The development of the pyramidal tracts in human fetuses, correlated with cortical differentiation. In: Tower D, Schade J (Eds) Structure and function of the cerebral cortex. Elsevier, Amsterdam, pp 93–103

Hunter DD, Llinas R, Ard M, Merlie JP, Sanes JR (1992) Expression of s-laminin and laminin in the developing rat central nervous system. J Comp Neurol 323:238–251

Ibanez CF (2007) Message in a bottle: long-range retrograde signaling in the nervous system. Trends Cell Biol 17:519–528

Iijima N, Oohira A, Mori T, Kitabatake K, Kohsaka S (1991) Core protein of chondroitin sulfate proteoglycan promotes neurite outgrowth from cultured neocortical neurons. J Neurochem 56:706–708

Ishibashi T, Dakin KA, Stevens B, Lee PR, Kozlov SV, Stewart CL, Fields RD (2006) Astrocytes promote myelination in response to electrical impulses. Neuron 49: 823–832

Itoh M, Nagafuchi A, Moroi S, Tsukita S (1997) Involvement of ZO-1 in cadherin-based cell adhesion through its direct binding to alpha catenin and actin filaments. J Cell Biol 138: 181–192

Jakovcevski I, Mo Z, Zecevic N (2007) Down-regulation of the axonal polysialic acid-neural cell adhesion molecule expression coincides with the onset of myelination in the human fetal forebrain. Neuroscience 149:328–337

Jay DG (2000) The clutch hypothesis revisited: ascribing the roles of actin-associated proteins in filopodial protrusion in the nerve growth cone. J Neurobiol 44:114–125

Jeret JS, Serur D, Wisniewski KE, Lubin RA (1987) Clinicopathological findings associated with agenesis of the corpus callosum. Brain Dev 9:255–264

Jiang X, Hanna Z, Kaouass M, Girard L, Jolicoeur P (2002) Ahi-1, a novel gene encoding a modular protein with WD40-repeat and SH3 domains, is targeted by the Ahi-1 and Mis-2 provirus integrations. J Virol 76:9046–9059

Jones FS, Jones PL (2000) The tenascin family of ECM glycoproteins: structure, function, and regulation during embryonic development and tissue remodeling. Dev Dyn 218: 235–259

Kadmon G, Kowitz A, Altevogt P, Schachner M (1990) The neural cell adhesion molecule N-CAM enhances L1-dependent cell-cell interactions. J Cell Biol 110:193–208

Kalo MS, Yu HH, Pasquale EB (2001) In vivo tyrosine phosphorylation sites of activated ephrin-B1 and ephB2 from neural tissue. J Biol Chem 276:38940–38948

Karadottir R, Attwell D (2007) Neurotransmitter receptors in the life and death of oligodendrocytes. Neuroscience 145:1426–1438

Kassai H, Terashima T, Fukaya M, Nakao K, Sakahara M, Watanabe M, Aiba A (2008) Rac1 in cortical projection neurons is selectively required for midline crossing of commissural axonal formation. Eur J Neurosci 28:257–267

Kaur B, Rutty GN, Timperley WR (1999) The possible role of hypoxia in the formation of axonal bulbs. J Clin Pathol 52:203–209

Kennedy TE, Wang H, Marshall W, Tessier-Lavigne M (2006) Axon guidance by diffusible chemoattractants: a gradient of netrin protein in the developing spinal cord. J Neurosci 26:8866–8874

Khateeb S, Flusser H, Ofir R, Shelef I, Narkis G, Vardi G, Shorer Z, Levy R, Galil A, Elbedour K, Birk OS (2006) PLA2G6 mutation underlies infantile neuroaxonal dystrophy. Am J Hum Genet 79:942–948

Kidd T, Bland KS, Goodman CS (1999) Slit is the midline repellent for the robo receptor in Drosophila. Cell 96:785–794

Kidd T, Brose K, Mitchell KJ, Fetter RD, Tessier-Lavigne M, Goodman CS, Tear G (1998) Roundabout controls axon crossing of the CNS midline and defines a novel subfamily of evolutionarily conserved guidance receptors. Cell 92:205–215

Kim HJ, DiBernardo AB, Sloane JA, Rasband MN, Solomon D, Kosaras B, Kwak SP, Vartanian TK (2006) WAVE1 is required for oligodendrocyte morphogenesis and normal CNS myelination. J Neurosci 26:5849–5859

Kinney H, Armstrong D (2002) Perinatal neuropathology. In: Graham D, Lantos P (Eds) Greenfield's neuropathology. Arnold, London, pp 519–606

Kinney HC, Brody BA, Kloman AS, Gilles FH (1988) Sequence of central nervous system myelination in human infancy. II. Patterns of myelination in autopsied infants. J Neuropathol Exp Neurol 47:217–234

Kinney HC, Rava LA, Benowitz LI (1993) Anatomic distribution of the growth-associated protein GAP-43 in the developing human brainstem. J Neuropathol Exp Neurol 52:39–54

Kinney HC, Volpe JJ (2009) Perinatal panencephalopathy in the premature infant: is it due to hypoxia-ischemia? In: Haddad GG, Ping YS (Eds) Brain hypoxia and ischemia. Vol. Contemporary clinical neuroscience. Humana Press, New York, NY

Kinnunen A, Kinnunen T, Kaksonen M, Nolo R, Panula P, Rauvala H (1998) N-syndecan and HB-GAM (heparin-binding growth-associated molecule) associate with early axonal tracts in the rat brain. Eur J Neurosci 10:635–648

Kinnunen A, Niemi M, Kinnunen T, Kaksonen M, Nolo R, Rauvala H (1999) Heparan sulphate and HB-GAM (heparin-binding growth-associated molecule) in the development of the thalamocortical pathway of rat brain. Eur J Neurosci 11:491–502

Kinnunen T, Raulo E, Nolo R, Maccarana M, Lindahl U, Rauvala H (1996) Neurite outgrowth in brain neurons induced by heparin-binding growth-associated molecule (HB-GAM) depends on the specific interaction of HB-GAM with heparan sulfate at the cell surface. J Biol Chem 271:2243–2248

Kiryushko D, Berezin V, Bock E (2004) Regulators of neurite outgrowth: role of cell adhesion molecules. Ann N Y Acad Sci 1014:140–154

Knoll B, Zarbalis K, Wurst W, Drescher U (2001) A role for the EphA family in the topographic targeting of vomeronasal axons. Development 128:895–906

Knudsen KA, Soler AP, Johnson KR, Wheelock MJ (1995) Interaction of alpha-actinin with the cadherin/catenin cell-cell adhesion complex via alpha-catenin. J Cell Biol 130: 67–77

Kontis D, Catani M, Cuddy M, Walshe M, Nosarti C, Jones D, Wyatt J, Rifkin L, Murray R, Allin M (2009) Diffusion tensor MRI of the corpus callosum and cognitive function in adults born preterm. Neuroreport 20:424–428

Koprivica V, Cho KS, Park JB, Yiu G, Atwal J, Gore B, Kim JA, Lin E, Tessier-Lavigne M, Chen DF, He Z (2005) EGFR activation mediates inhibition of axon regeneration by myelin and chondroitin sulfate proteoglycans. Science 310:106–110

Kumar S, Yin X, Trapp BD, Hoh JH, Paulaitis ME (2002) Relating interactions between neurofilaments to the structure of axonal neurofilament distributions through polymer brush models. Biophys J 82:2360–2372

Lassmann H, Bartsch U, Montag D, Schachner M (1997) Dying-back oligodendrogliopathy: a late sequel of myelin-associated glycoprotein deficiency. Glia 19:104–110

Lee MK, Cleveland DW (1996) Neuronal intermediate filaments. Annu Rev Neurosci 19:187–217

Lein PJ, Banker GA, Higgins D (1992) Laminin selectively enhances axonal growth and accelerates the development of polarity by hippocampal neurons in culture. Brain Res Dev Brain Res 69:191–197

Lein PJ, Higgins D (1989) Laminin and a basement membrane extract have different effects on axonal and dendritic outgrowth from embryonic rat sympathetic neurons in vitro. Dev Biol 136:330–345

Lesca G, Eymard-Pierre E, Santorelli FM, Cusmai R, Di Capua M, Valente EM, Attia-Sobol J, Plauchu H, Leuzzi V, Ponzone A, Boespflug-Tanguy O, Bertini E (2003) Infantile ascending hereditary spastic paralysis (IAHSP): clinical features in 11 families. Neurology 60: 674–682

Letourneau PC (1978) Chemotactic response of nerve fiber elongation to nerve growth factor. Dev Biol 66:183–196

Letourneau PC (1983) Differences in the organization of actin in the growth cones compared with the neurites of cultured neurons from chick embryos. J Cell Biol 97:963–973

Li C, Trapp B, Ludwin S, Peterson A, Roder J (1998) Myelin associated glycoprotein modulates glia-axon contact in vivo. J Neurosci Res 51:210–217

Li C, Tropak MB, Gerlai R, Clapoff S, Abramow-Newerly W, Trapp B, Peterson A, Roder J (1994) Myelination in the absence of myelin-associated glycoprotein. Nature 369:747–750

Liesi P, Silver J (1988) Is astrocyte laminin involved in axon guidance in the mammalian CNS? Dev Biol 130:774–785

Ligam P, Haynes R, Folkerth R, Liu L, Yang M, Volpe J, Kinney H 2008. Thalamic damage in pervientricular leukomalacia: novel pathologic observations relevant to cognitive deficits in survivors of prematurity. Pediatr Res. 2009 May;65(5):524–9

Lin CH, Espreafico EM, Mooseker MS, Forscher P (1996) Myosin drives retrograde F-actin flow in neuronal growth cones. Neuron 16:769–782

Lin CH, Forscher P (1995) Growth cone advance is inversely proportional to retrograde F-actin flow. Neuron 14:763–771

Lin CH, Thompson CA, Forscher P (1994) Cytoskeletal reorganization underlying growth cone motility. Curr Opin Neurobiol 4:640–647

Lin L, Wang J, Chan CK, Chan SO (2007) Effects of exogenous hyaluronan on midline crossing and axon divergence in the optic chiasm of mouse embryos. Eur J Neurosci 26:1–11

LoPachin RM Jr., Lehning EJ (1994) Acrylamide-induced distal axon degeneration: a proposed mechanism of action. Neurotoxicology 15:247–259

Long H, Sabatier C, Ma L, Plump A, Yuan W, Ornitz DM, Tamada A, Murakami F, Goodman CS, Tessier-Lavigne M (2004) Conserved roles for Slit and Robo proteins in midline commissural axon guidance. Neuron 42:213–223

Low LK, Cheng HJ (2005) A little nip and tuck: axon refinement during development and axonal injury. Curr Opin Neurobiol 15:549–556

Low LK, Cheng HJ (2006) Axon pruning: an essential step underlying the developmental plasticity of neuronal connections. Philos Trans R Soc Lond B Biol Sci 361:1531–1544

Low LK, Liu XB, Faulkner RL, Coble J, Cheng HJ (2008) Plexin signaling selectively regulates the stereotyped pruning of corticospinal axons from visual cortex. Proc Natl Acad Sci USA 105:8136–8141

Luo Y, Raible D, Raper JA (1993) Collapsin: a protein in brain that induces the collapse and paralysis of neuronal growth cones. Cell 75:217–227

Lykissas MG, Batistatou AK, Charalabopoulos KA, Beris AE (2007) The role of neurotrophins in axonal growth, guidance, and regeneration. Curr Neurovasc Res 4:143–151

Lyon G, Arita F, Le Galloudec E, Vallee L, Misson JP, Ferriere G (1990) A disorder of axonal development, necrotizing myopathy, cardiomyopathy, and cataracts: a new familial disease. Ann Neurol 27:193–199
Maier O, Hoekstra D, Baron W (2008) Polarity development in oligodendrocytes: sorting and trafficking of myelin components. J Mol Neurosci 35:35–53
Mann F, Chauvet S, Rougon G (2007) Semaphorins in development and adult brain: Implication for neurological diseases. Prog Neurobiol 82:57–79
Mann F, Harris WA, Holt CE (2004) New views on retinal axon development: a navigation guide. Int J Dev Biol 48:957–964
Mann F, Ray S, Harris W, Holt C (2002) Topographic mapping in dorsoventral axis of the Xenopus retinotectal system depends on signaling through ephrin-B ligands. Neuron 35:461–473
Mann F, Rougon G (2007) Mechanisms of axon guidance: membrane dynamics and axonal transport in semaphorin signalling. J Neurochem 102:316–323
Marin-Padilla M (1997) Developmental neuropathology and impact of perinatal brain damage. II: white matter lesions of the neocortex. J Neuropathol Exp Neurol 56:219–235
Martin R, Door R, Ziegler A, Warchol W, Hahn J, Breitig D (1999) Neurofilament phosphorylation and axon diameter in the squid giant fibre system. Neuroscience 88:327–336
Martinez A, Soriano E (2005) Functions of ephrin/Eph interactions in the development of the nervous system: emphasis on the hippocampal system. Brain Res Brain Res Rev 49: 211–226
Maruyama T, Matsuura M, Suzuki K, Yamamoto N (2008) Cooperative activity of multiple upper layer proteins for thalamocortical axon growth. Dev Neurobiol 68:317–331
Mason CA, Gregory E (1984) Postnatal maturation of cerebellar mossy and climbing fibers: transient expression of dual features on single axons. J Neurosci 4:1715–1735
Masuda T, Fukamauchi F, Takeda Y, Fujisawa H, Watanabe K, Okado N, Shiga T (2004) Developmental regulation of notochord-derived repulsion for dorsal root ganglion axons. Mol Cell Neurosci 25:217–227
McCarran WJ, Goldberg MP (2007) White matter axon vulnerability to AMPA/kainate receptor-mediated ischemic injury is developmentally regulated. J Neurosci 27:4220–4229
McLoon SC, McLoon LK, Palm SL, Furcht LT (1988) Transient expression of laminin in the optic nerve of the developing rat. J Neurosci 8:1981–1990
Medana IM, Esiri MM (2003) Axonal damage: a key predictor of outcome in human CNS diseases. Brain 126:515–530
Medeiros NA, Burnette DT, Forscher P (2006) Myosin II functions in actin-bundle turnover in neuronal growth cones. Nat Cell Biol 8:215–226
Menesini Chen MG, Chen JS, Levi-Montalcini R (1978) Sympathetic nerve fibers ingrowth in the central nervous system of neonatal rodent upon intracerebral NGF injections. Arch Ital Biol 116:53–84
Meng XF, Luo Y, Xiao W, Li M, Shi J (2009) Cloning and characterization of the promoter of the human AHI1 gene. Biochem Genet 47:427–438
Messersmith EK, Leonardo ED, Shatz CJ, Tessier-Lavigne M, Goodman CS, Kolodkin AL (1995) Semaphorin III can function as a selective chemorepellent to pattern sensory projections in the spinal cord. Neuron 14:949–959
Milev P, Maurel P, Haring M, Margolis RK, Margolis RU (1996) TAG-1/axonin-1 is a high-affinity ligand of neurocan, phosphacan/protein-tyrosine phosphatase-zeta/beta, and N-CAM. J Biol Chem 271:15716–15723
Ming G, Song H, Berninger B, Inagaki N, Tessier-Lavigne M, Poo M (1999) Phospholipase C-gamma and phosphoinositide 3-kinase mediate cytoplasmic signaling in nerve growth cone guidance. Neuron 23:139–148
Mitchison T, Kirschner M (1988) Cytoskeletal dynamics and nerve growth. Neuron 1:761–772
Monnier PP, Sierra A, Schwab JM, Henke-Fahle S, Mueller BK (2003) The Rho/ROCK pathway mediates neurite growth-inhibitory activity associated with the chondroitin sulfate proteoglycans of the CNS glial scar. Mol Cell Neurosci 22:319–330

Montag D, Giese KP, Bartsch U, Martini R, Lang Y, Bluthmann H, Karthigasan J, Kirschner DA, Wintergerst ES, Nave KA et al (1994) Mice deficient for the myelin-associated glycoprotein show subtle abnormalities in myelin. Neuron 13:229–246

Motil J, Chan WK, Dubey M, Chaudhury P, Pimenta A, Chylinski TM, Ortiz DT, Shea TB (2006) Dynein mediates retrograde neurofilament transport within axons and anterograde delivery of NFs from perikarya into axons: regulation by multiple phosphorylation events. Cell Motil Cytoskeleton 63:266–286

Muresan V (2000) One axon, many kinesins: What's the logic? J Neurocytol 29:799–818

Myers KA, He Y, Hasaka TP, Baas PW (2006) Microtubule transport in the axon: re-thinking a potential role for the actin cytoskeleton. Neuroscientist 12:107–118

Nardelli E, Vio M, Ghersini L, Rizzuto N (1982) Mobius-like syndrome due to multiple cerebral abnormalities including hypoplasia of the descending tracts. A case report. J Neurol 227:11–19

Nishimura K, Yoshihara F, Tojima T, Ooashi N, Yoon W, Mikoshiba K, Bennett V, Kamiguchi H (2003) L1-dependent neuritogenesis involves ankyrinB that mediates L1-CAM coupling with retrograde actin flow. J Cell Biol 163:1077–1088

Okabe S, Hirokawa N (1991) Actin dynamics in growth cones. J Neurosci 11:1918–1929

O'Leary DD, Stanfield BB (1989) Selective elimination of axons extended by developing cortical neurons is dependent on regional locale: experiments utilizing fetal cortical transplants. J Neurosci 9:2230–2246

Oumesmar BN, Vignais L, Duhamel-Clerin E, Avellana-Adalid V, Rougon G, Baron-Van Evercooren A (1995) Expression of the highly polysialylated neural cell adhesion molecule during postnatal myelination and following chemically induced demyelination of the adult mouse spinal cord. Eur J Neurosci 7:480–491

Paglini G, Peris L, Mascotti F, Quiroga S, Caceres A (2000) Tau protein function in axonal formation. Neurochem Res 25:37–42

Palay SL, Sotelo C, Peters A, Orkand PM (1968) The axon hillock and the initial segment. J Cell Biol 38:193–201

Panzeri C, De Palma C, Martinuzzi A, Daga A, De Polo G, Bresolin N, Miller CC, Tudor EL, Clementi E, Bassi MT (2006) The first ALS2 missense mutation associated with JPLS reveals new aspects of alsin biological function. Brain 129:1710–1719

Paul LK, Brown WS, Adolphs R, Tyszka JM, Richards LJ, Mukherjee P, Sherr EH (2007) Agenesis of the corpus callosum: genetic, developmental and functional aspects of connectivity. Nat Rev Neurosci 8:287–299

Perron JC, Bixby JL (1999) Distinct neurite outgrowth signaling pathways converge on ERK activation. Mol Cell Neurosci 13:362–378

Pfister KK (1999) Cytoplasmic dynein and microtubule transport in the axon: the action connection. Mol Neurobiol 20:81–91

Polleux F, Giger RJ, Ginty DD, Kolodkin AL, Ghosh A (1998) Patterning of cortical efferent projections by semaphorin-neuropilin interactions. Science 282:1904–1906

Polleux F, Morrow T, Ghosh A (2000) Semaphorin 3A is a chemoattractant for cortical apical dendrites. Nature 404:567–573

Powell SK, Kleinman HK (1997) Neuronal laminins and their cellular receptors. Int J Biochem Cell Biol 29:401–414

Prayer D, Barkovich AJ, Kirschner DA, Prayer LM, Roberts TP, Kucharczyk J, Moseley ME (2001) Visualization of nonstructural changes in early white matter development on diffusion-weighted MR images: evidence supporting premyelination anisotropy. AJNR Am J Neuroradiol 22:1572–1576

Puschel AW, Adams RH, Betz H (1995) Murine semaphorin D/collapsin is a member of a diverse gene family and creates domains inhibitory for axonal extension. Neuron 14:941–948

Ramanathan R, Wilkemeyer MF, Mittal B, Perides G, Charness ME (1996) Alcohol inhibits cell-cell adhesion mediated by human L1. J Cell Biol 133:381–390

Ranscht B (2000) Cadherins: molecular codes for axon guidance and synapse formation. Int J Dev Neurosci 18:643–651

Raper JA (2000) Semaphorins and their receptors in vertebrates and invertebrates. Curr Opin Neurobiol 10:88–94

Rasband MN, Trimmer JS (2001) Developmental clustering of ion channels at and near the node of Ranvier. Dev Biol 236:5–16

Raulo E, Tumova S, Pavlov I, Pekkanen M, Hienola A, Klankki E, Kalkkinen N, Taira T, Kilpelainen I, Rauvala H (2005) The two thrombospondin type I repeat domains of the heparin-binding growth-associated molecule bind to heparin/heparan sulfate and regulate neurite extension and plasticity in hippocampal neurons. J Biol Chem 280:41576–41583

Rauvala H, Vanhala A, Castren E, Nolo R, Raulo E, Merenmies J, Panula P (1994) Expression of HB-GAM (heparin-binding growth-associated molecules) in the pathways of developing axonal processes in vivo and neurite outgrowth in vitro induced by HB-GAM. Brain Res Dev Brain Res 79:157–176

Reichard RR, White CL 3rd, Hladik CL, Dolinak D (2003) Beta-amyloid precursor protein staining of nonaccidental central nervous system injury in pediatric autopsies. J Neurotrauma 20: 347–355

Riederer B, Cohen R, Matus A (1986) MAP5: a novel brain microtubule-associated protein under strong developmental regulation. J Neurocytol 15:763–775

Riley DA (1981) Ultrastructural evidence for axon retraction during the spontaneous elimination of polyneuronal innervation of the rat soleus muscle. J Neurocytol 10:425–440

Rinaldi M, Barrera G, Spinsanti P, Pizzimenti S, Ciafre SA, Parella P, Farace MG, Signori E, Dianzani MU, Fazio VM (2001) Growth inhibition and differentiation induction in murine erythroleukemia cells by 4-hydroxynonenal. Free Radic Res 34:629–637

Roessmann U, Horwitz SJ, Kennell JH (1990) Congenital absence of the corticospinal fibers: pathologic and clinical observations. Neurology 40:538–541

Ronn LC, Hartz BP, Bock E (1998) The neural cell adhesion molecule (NCAM) in development and plasticity of the nervous system. Exp Gerontol 33:853–864

Rosenthal A, Jouet M, Kenwrick S (1992) Aberrant splicing of neural cell adhesion molecule L1 mRNA in a family with X-linked hydrocephalus. Nat Genet 2:107–112

Rothberg JM, Jacobs JR, Goodman CS, Artavanis-Tsakonas S (1990) slit: an extracellular protein necessary for development of midline glia and commissural axon pathways contains both EGF and LRR domains. Genes Dev 4:2169–2187

Round J, Stein E (2007) Netrin signaling leading to directed growth cone steering. Curr Opin Neurobiol 17:15–21

Sakurai T, Gil OD, Whittard JD, Gazdoiu M, Joseph T, Wu J, Waksman A, Benson DL, Salton SR, Felsenfeld DP (2008) Interactions between the L1 cell adhesion molecule and ezrin support traction-force generation and can be regulated by tyrosine phosphorylation. J Neurosci Res 86:2602–2614

Salman MS, Marles SL, Booth FA, Del Bigio MR (2009) Early-onset neurodegenerative disease of the cerebellum and motor axons. Pediatr Neurol 40:365–370

Salter MG, Fern R (2005) NMDA receptors are expressed in developing oligodendrocyte processes and mediate injury. Nature 438:1167–1171

Schell-Apacik CC, Wagner K, Bihler M, Ertl-Wagner B, Heinrich U, Klopocki E, Kalscheuer VM, Muenke M, von Voss H (2008) Agenesis and dysgenesis of the corpus callosum: clinical, genetic and neuroimaging findings in a series of 41 patients. Am J Med Genet A 146A: 2501–2511

Schmid RS, Graff RD, Schaller MD, Chen S, Schachner M, Hemperly JJ, Maness PF (1999) NCAM stimulates the Ras-MAPK pathway and CREB phosphorylation in neuronal cells. J Neurobiol 38:542–558

Schmid RS, Pruitt WM, Maness PF (2000) A MAP kinase-signaling pathway mediates neurite outgrowth on L1 and requires Src-dependent endocytosis. J Neurosci 20: 4177–4188

Schnapp BJ, Reese TS (1989) Dynein is the motor for retrograde axonal transport of organelles. Proc Natl Acad Sci USA 86:1548–1552

Schumacher PA, Eubanks JH, Fehlings MG (1999) Increased calpain I-mediated proteolysis, and preferential loss of dephosphorylated NF200, following traumatic spinal cord injury. Neuroscience 91:733–744

Schwarting GA, Kostek C, Ahmad N, Dibble C, Pays L, Puschel AW (2000) Semaphorin 3A is required for guidance of olfactory axons in mice. J Neurosci 20:7691–7697

Shannon P, Smith CR, Deck J, Ang LC, Ho M, Becker L (1998) Axonal injury and the neuropathology of shaken baby syndrome. Acta Neuropathol 95:625–631

Sherr EH (2003) The ARX story (epilepsy, mental retardation, autism, and cerebral malformations): one gene leads to many phenotypes. Curr Opin Pediatr 15:567–571

Shimada T, Toriyama M, Uemura K, Kamiguchi H, Sugiura T, Watanabe N, Inagaki N (2008) Shootin1 interacts with actin retrograde flow and L1-CAM to promote axon outgrowth. J Cell Biol 181:817–829

Sidman R, Rakic P (1982) Development of the human central nervous system. In: Haymaker W, Adams R (Eds) Histology and histopathology of the nervous system, vol 1. Charles C. Thomas, Springfield, IL, pp 83–92

Sihag RK, Inagaki M, Yamaguchi T, Shea TB, Pant HC (2007) Role of phosphorylation on the structural dynamics and function of types III and IV intermediate filaments. Exp Cell Res 313:2098–2109

Simons M, Trotter J (2007) Wrapping it up: the cell biology of myelination. Curr Opin Neurobiol 17:533–540

Smith UM, Consugar M, Tee LJ, McKee BM, Maina EN, Whelan S, Morgan NV, Goranson E, Gissen P, Lilliquist S et al (2006) The transmembrane protein meckelin (MKS3) is mutated in Meckel-Gruber syndrome and the wpk rat. Nat Genet 38:191–196

Snow DM, Lemmon V, Carrino DA, Caplan AI, Silver J (1990) Sulfated proteoglycans in astroglial barriers inhibit neurite outgrowth in vitro. Exp Neurol 109:111–130

Song HJ, Ming GL, Poo MM (1997) cAMP-induced switching in turning direction of nerve growth cones. Nature 388:275–279

Spiegel I, Peles E (2006) A new player in CNS myelination. Neuron 49:777–778

Stanfield BB, O'Leary DD (1985) Fetal occipital cortical neurones transplanted to the rostral cortex can extend and maintain a pyramidal tract axon. Nature 313:135–137

Stanfield BB, O'Leary DD, Fricks C (1982) Selective collateral elimination in early postnatal development restricts cortical distribution of rat pyramidal tract neurones. Nature 298:371–373

Stevens B, Porta S, Haak LL, Gallo V, Fields RD (2002) Adenosine: a neuron-glial transmitter promoting myelination in the CNS in response to action potentials. Neuron 36:855–868

Stuart G, Schiller J, Sakmann B (1997) Action potential initiation and propagation in rat neocortical pyramidal neurons. J Physiol 505(Pt 3):617–632

Tang PH, Bartha AI, Norton ME, Barkovich AJ, Sherr EH, Glenn OA (2009) Agenesis of the corpus callosum: an MR imaging analysis of associated abnormalities in the fetus. AJNR Am J Neuroradiol 30:257–263

Taniguchi M, Nagao H, Takahashi YK, Yamaguchi M, Mitsui S, Yagi T, Mori K, Shimizu T (2003) Distorted odor maps in the olfactory bulb of semaphorin 3A-deficient mice. J Neurosci 23:1390–1397

Thomas B, Eyssen M, Peeters R, Molenaers G, Van Hecke P, De Cock P, Sunaert S (2005) Quantitative diffusion tensor imaging in cerebral palsy due to periventricular white matter injury. Brain 128:2562–2577

Tornieri K, Welshhans K, Geddis MS, Rehder V (2006) Control of neurite outgrowth and growth cone motility by phosphatidylinositol-3-kinase. Cell Motil Cytoskeleton 63:173–192

Trapp BD, Peterson J, Ransohoff RM, Rudick R, Mork S, Bo L (1998) Axonal transection in the lesions of multiple sclerosis. N Engl J Med 338:278–285

Ughrin YM, Chen ZJ, Levine JM (2003) Multiple regions of the NG2 proteoglycan inhibit neurite growth and induce growth cone collapse. J Neurosci 23:175–186

Underhill SM, Goldberg MP (2007) Hypoxic injury of isolated axons is independent of ionotropic glutamate receptors. Neurobiol Dis 25:284–290

Utsch B, Sayer JA, Attanasio M, Pereira RR, Eccles M, Hennies HC, Otto EA, Hildebrandt F (2006) Identification of the first AHI1 gene mutations in nephronophthisis-associated Joubert syndrome. Pediatr Nephrol 21:32–35

Uziel D, Garcez P, Lent R, Peuckert C, Niehage R, Weth F, Bolz J (2006) Connecting thalamus and cortex: the role of ephrins. Anat Rec A Discov Mol Cell Evol Biol 288:135–142

Vangberg TR, Skranes J, Dale AM, Martinussen M, Brubakk AM, Haraldseth O (2006) Changes in white matter diffusion anisotropy in adolescents born prematurely. Neuroimage 32:1538–1548

Viapiano MS, Matthews RT (2006) From barriers to bridges: chondroitin sulfate proteoglycans in neuropathology. Trends Mol Med 12:488–496

Volpe JJ (2008a) Down syndrome. In: Fletcher J (Eds) Neurology of the newborn. Saunders, Philadelphia, PA, p 92

Volpe JJ (2008b) Infantile botulism. In: Fletcher J (Eds) Neurology of the newborn. Saunders, Philadelphia, PA, pp 791–793

Volpe JJ (2009) Brain injury in premature infants: a complex amalgam of destructive and developmental disturbances. Lancet Neurol 8:110–124

Vuolteenaho R, Nissinen M, Sainio K, Byers M, Eddy R, Hirvonen H, Shows TB, Sariola H, Engvall E, Tryggvason K (1994) Human laminin M chain (merosin): complete primary structure, chromosomal assignment, and expression of the M and A chain in human fetal tissues. J Cell Biol 124:381–394

Wahl M, Strominger Z, Jeremy RJ, Barkovich AJ, Wakahiro M, Sherr EH, Mukherjee P (2009) Variability of homotopic and heterotopic callosal connectivity in partial agenesis of the corpus callosum: a 3T diffusion tensor imaging and Q-ball tractography study. AJNR Am J Neuroradiol 30:282–289

Wang HU, Anderson DJ (1997) Eph family transmembrane ligands can mediate repulsive guidance of trunk neural crest migration and motor axon outgrowth. Neuron 18:383–396

Wang Q, Zheng JQ (1998) cAMP-mediated regulation of neurotrophin-induced collapse of nerve growth cones. J Neurosci 18:4973–4984

Weimann JM, Zhang YA, Levin ME, Devine WP, Brulet P, McConnell SK (1999) Cortical neurons require Otx1 for the refinement of exuberant axonal projections to subcortical targets. Neuron 24:819–831

Weitzdoerfer R, Fountoulakis M, Lubec G (2001) Aberrant expression of dihydropyrimidinase related proteins-2,-3 and -4 in fetal Down syndrome brain. J Neural Transm Suppl 61:95–107

Wilkemeyer MF, Charness ME (1998) Characterization of ethanol-sensitive and insensitive fibroblast cell lines expressing human L1. J Neurochem 71:2382–2391

Williams EJ, Furness J, Walsh FS, Doherty P (1994) Activation of the FGF receptor underlies neurite outgrowth stimulated by L1, N-CAM, and N-cadherin. Neuron 13:583–594

Wolf JA, Stys PK, Lusardi T, Meaney D, Smith DH (2001) Traumatic axonal injury induces calcium influx modulated by tetrodotoxin-sensitive sodium channels. J Neurosci 21:1923–1930

Wolman MA, Sittaramane VK, Essner JJ, Yost HJ, Chandrasekhar A, Halloran MC (2008) Transient axonal glycoprotein-1 (TAG-1) and laminin-alpha1 regulate dynamic growth cone behaviors and initial axon direction in vivo. Neural Develop 3:6–20

Wong EV, Kenwrick S, Willems P, Lemmon V (1995) Mutations in the cell adhesion molecule L1 cause mental retardation. Trends Neurosci 18:168–172

Wu Y, Sheng W, Chen L, Dong H, Lee V, Lu F, Wong CS, Lu WY, Yang BB (2004) Versican V1 isoform induces neuronal differentiation and promotes neurite outgrowth. Mol Biol Cell 15:2093–2104

Yachnis AT, Rorke LB (1999) Neuropathology of Joubert syndrome. J Child Neurol 14:655–659; discussion 669–72

Yamada K, Andrews C, Chan WM, McKeown CA, Magli A, de Berardinis T, Loewenstein A, Lazar M, O'Keefe M, Letson R et al (2003) Heterozygous mutations of the kinesin KIF21A in congenital fibrosis of the extraocular muscles type 1 (CFEOM1). Nat Genet 35:318–321

Yamada H, Fredette B, Shitara K, Hagihara K, Miura R, Ranscht B, Stallcup WB, Yamaguchi Y (1997) The brain chondroitin sulfate proteoglycan brevican associates with astrocytes

ensheathing cerebellar glomeruli and inhibits neurite outgrowth from granule neurons. J Neurosci 17:7784–7795
Yamaguchi Y (2000) Lecticans: organizers of the brain extracellular matrix. Cell Mol Life Sci 57:276–289
Yuan X, Eisen AM, McBain CJ, Gallo V (1998) A role for glutamate and its receptors in the regulation of oligodendrocyte development in cerebellar tissue slices. Development 125: 2901–2914
Zacharias U, Rauch U (2006) Competition and cooperation between tenascin-R, lecticans and contactin 1 regulate neurite growth and morphology. J Cell Sci 119:3456–3466
Zec N, Filiano JJ, Kinney HC (1997a) Anatomic relationships of the human arcuate nucleus of the medulla: a DiI-labeling study. J Neuropathol Exp Neurol 56:509–522
Zec N, Filiano JJ, Kinney HC (1997b) Anatomic relationships of the human arcuate nucleus of the medulla: a Dil labeling study. J Neuropathol Exp Neurol 56:509–522
Zec N, Kinney HC (2001) Anatomic relationships of the human nucleus paragigantocellularis lateralis: a DiI labeling study. Auton Neurosci 89:110–124
Zec N, Kinney HC (2003) Anatomic relationships of the human nucleus of the solitary tract in the medulla oblongata: a DiI labeling study. Auton Neurosci 105:131–144
Zhou FC (1990) Four patterns of laminin-immunoreactive structure in developing rat brain. Brain Res Dev Brain Res 55:191–201
Zhou FQ, Cohan CS (2004) How actin filaments and microtubules steer growth cones to their targets. J Neurobiol 58:84–91
Zhou Y, Gunput RA, Pasterkamp RJ (2008) Semaphorin signaling: progress made and promises ahead. Trends Biochem Sci 33:161–170
Zweifel LS, Kuruvilla R, Ginty DD (2005) Functions and mechanisms of retrograde neurotrophin signalling. Nat Rev Neurosci 6:615–625

Chapter 2
Microtubules in the Nervous System

Nobuyuki Fukushima

Abstract Neurons undergo various morphologic changes during development, including neuritogenesis, neurite outgrowth, neurite branching, and neurite retraction. Many studies have examined how microtubules (MTs) are reorganized or transported within the neurites of developing neurons and have revealed that MT dynamics are regulated by MT-interacting proteins and motor proteins, in concert with actin microfilaments. Here, I will describe recent progress in research on the behavior of MTs in the nervous system.

Keywords Microtubule · Tubulin · Actin · Neurite · Growth cone · Polarity · Transport · Microtubule-associated protein

Contents

2.1 Introduction

Microtubules (MTs) are one component of the cytoskeleton and are composed of heterodimers of α- and β-tubulin. These α/β heterodimers are polymerized in a

N. Fukushima (✉)
Department of Life Science, Kinki University, Higashiosaka, Japan
e-mail: nfukushima@life.kindai.ac.jp

R.A. Nixon, A. Yuan (eds.), *Cytoskeleton of the Nervous System*,
Advances in Neurobiology 3, DOI 10.1007/978-1-4419-6787-9_2,

head-to-tail fashion, resulting in a polarity to the MT, with one end designated the plus end and the other the minus end. The plus ends of the MTs exhibit cycles of growth and shrinkage, a behavior that is termed *dynamic instability*. In contrast, the minus ends are relatively unstable and shrink without stabilization. MT functions are highly regulated by the intrinsic GTPase activity of tubulins as well as MT-interacting proteins, including MT-associated proteins (MAPs), MT-severing proteins, MT plus-end tracking proteins (+TIPs), and motor proteins (Dent and Gertler, 2003; Rodriguez et al., 2003; Akhmanova and Steinmetz, 2008; Jaworski et al., 2008). The interactions between MTs and MT-interacting proteins are influenced by diverse posttranslational modifications (PTMs), which include acetylation, tyrosination, detyrosination, and polyglutamylation, and play a role in neuronal development (Westermann and Weber, 2003; Hammond et al., 2008). Although I will discuss some topics related to tubulin PTMs in this chapter, the current progress in PTM research will be reviewed in detail elsewhere.

For many years, MTs and actin microfilaments (MFs) (Fukushima et al., 2009) were viewed as functionally separate. In neurons, MTs serve as a scaffold for organelle transport and structural components of the neurite that play an important role in neurite elongation. In contrast, MFs regulate growth cone morphology and motility and play a role in proper axon guidance. Many studies over the past 20 years have provided clear evidence that MTs structurally and functionally interact with MFs in developing neurons. For example, MTs themselves are transported in an axon or a growth cone in an actin-dependent manner. Furthermore, a growing number of actin–MT cross-linking proteins, which include some of the known MAPs and +TIPs, have been identified. In this chapter, I will address how MTs are involved in the development of neurites and also review recent reports showing how interactions between actin and MTs regulate the shape and motility of neurites.

2.2 Tubulin in the Brain

In mice, there are seven α-tubulin genes and eight β-tubulin genes. Each α-tubulin isoform shows more than 90% amino acid identity to other α-tubulins, and β-tubulins, with the exception of β1-tubulin, show more than 90% amino acid identity to other β-tubulins, but only approximately 78% to β1-tubulin. The expression of the tubulin isoforms in brain is developmentally regulated. The α1-, α2-, α4-, α6-, and α8-tubulin and the β2-, β3-, β4-, and β5-tubulin isoforms are expressed in the brain (Lewis et al., 1985; Villasante et al., 1986; Stanchi et al., 2000). Levels of α1- and α2-tubulin expression remain constant during postnatal brain development. In contrast, expression of α4-, β3-, and β4-tubulin increases during postnatal brain development, whereas that of β2- and β5-tubulin decreases (Lewis et al., 1985; Villasante et al., 1986). The developmental expression profile of α8-tubulin remains to be determined. Because any combination of α- and β-tubulin isoforms can copolymerize, the combination of α/β-tubulin heterodimers is thought to vary during brain development. However, the actual isoforms constituting individual MTs in developing neurons have not been directly demonstrated.

2.3 Microtubules in Neuronal Development

Neurons display various morphologic changes during development in culture (da Silva and Dotti, 2002). After neuronal commitment, spherical neurons make membrane sprouts, which are transformed into neurites (neuritogenesis) and are extended as the neurons differentiate (neurite outgrowth) (Fig. 2.1). Extending neurites may generate branches (neurite branching), corresponding with axon collaterals or dendritic arbors, or exhibit transient retraction (neurite retraction) (Fig. 2.1). Each event overlaps functionally and structurally with other events, and there are currently no clear boundaries that have been used to classify each step. However, for the purposes

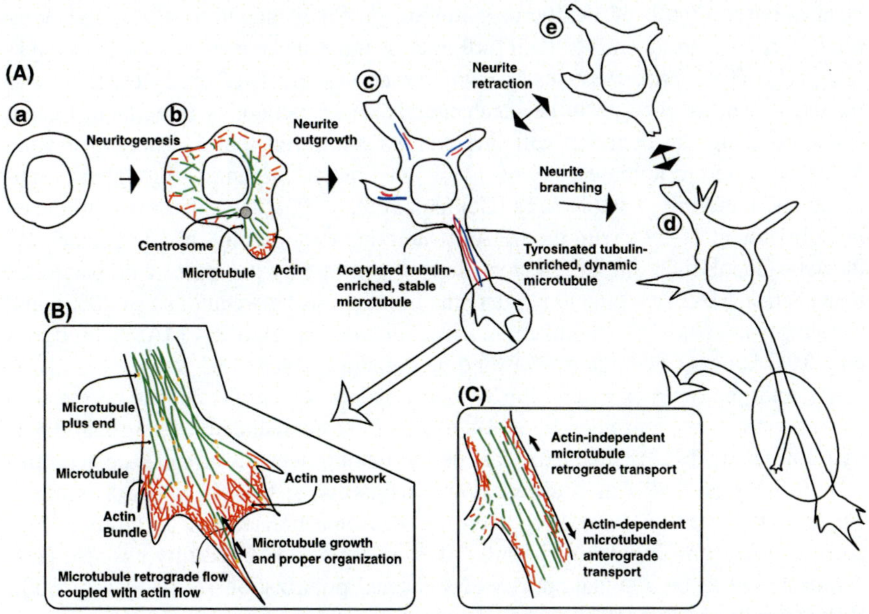

Fig. 2.1 Microtubule organization and dynamics in the developing neuron in culture. **a** a and b; Shortly after plating, the spherical neurons produce protrusive membrane structures containing actin microfilaments (*red*) at the leading edges and microtubules (*green*) behind actin meshwork (neuritogenesis). c; These protrusions develop into several immature neurites and are guided by the advancement of motile growth cones as the neurons differentiate (neurite outgrowth). d and e; Extending neurites generate branches (neurite branching) and display transient retraction intrinsically or in response to extracellular repulsive signals (neurite retraction). Centrosome (*green circle* in b) localization or the presence of higher levels of stable microtubules (magenta in c) than dynamic microtubules (*blue* in c) is involved in the establishment of neuronal polarity. **b** The growth cone contains bundled microtubules in the center, an actin meshwork (*red thin lines*) in the lamellipodia, and actin bundles in filopodia. Dynamic microtubules extending from the C domain, which are oriented with the plus-ends (*yellow*) pointing to the periphery, grow along filopodial actin bundles (*red thick lines*) and move rearward coupled with actin retrograde flow. **c** Microtubules are bidirectionally transported within the axon. The anterograde microtubule transport is actin dependent, whereas the retrograde transport is actin independent. Short microtubules accumulate at the branching sites

of the current discussion, I will describe MT organization separately for these four events: neuritogenesis, neurite outgrowth, neurite branching, and neurite retraction.

2.3.1 Neuritogenesis

The roles of the cytoskeleton in neuritogenesis have been investigated in cultured neurons. Neurons initially form membrane sprouts consisting of MF-rich lamellipodia (Fig. 2.1) (Dotti et al., 1988; da Silva and Dotti, 2002). The actin cytoskeleton and its upstream modulators, including Rho, Rho-associated, coiled-coil containing protein kinase (ROCK), and profilin, have been well documented to be negative regulators of membrane sprouting (da Silva and Dotti, 2002). The membrane sprouts subsequently extend further to become short neurites containing MTs (Fig. 2.1). These processes clearly involve rearrangements of the cytoskeletal elements, which are thought to be mechanically and functionally integrated. Recent evidence using the neuronal cell line Neuro-2a showed that MAP2c, a juvenile MAP expressed in immature brains, is involved in neuritogenesis through accumulation and bundling of stable MTs (Dehmelt et al., 2003). This effect is reproduced by cotreatment of cells with the MT-polymerizing (stabilizing) drug paclitaxel and the actin-depolymerizing drug cytochalasin D, but not by treatment with paclitaxel alone. Thus, MAP2c seems to regulate the interactions between MTs and MFs, and not simply to cause MT stabilization. Indeed, it was reported that MAP2c binds not only MTs but also MFs via its MT-binding domain (Roger et al., 2004).

As multiple neurites extend during neuritogenesis, neurons choose one neurite as an axon and others as dendrites, leading to the establishment of neuronal polarity (Dotti et al., 1988). Such axon fate determination depends on intrinsic activities of both MTs and MFs in neurites. Local instability of the MF network within a single growth cone is known to determine neuronal polarization in cultured hippocampal neurons (Bradke and Dotti, 1999). Recently, MT stability has also been demonstrated to be a signal specifying neuronal polarization (Witte et al., 2008). Before the establishment of neuronal polarity or during neuritogenesis, the future axon contains more stable (acetylated) MTs versus dynamic (tyrosinated) MTs than do other neurites (Fig. 2.1). After establishment of polarity, the single axon still maintains higher levels of stable MTs. When neurons are treated with a glycogen synthase kinase-3β (GSK-3β) inhibitor that is a known inducer of multiple axons, many axon-like neurites contain stable MTs. In contrast, synapses of the amphid defective (SAD) kinase–deficient neurons that exhibit a loss of neuronal polarity have decreased levels of stable MTs in their neurites. Conversely, a low dose of paclitaxel induces the formation of multiple axons. In contrast, a low dose of the MT destabilizer nocodazole reduces dendrite formation without affecting axon formation, presumably due to the presence of more nocodazole-resistant, stable MTs in the axon and less in the dendrites. These observations suggest that changes in MT dynamics in neurites play a crucial role in the initial polarization of neuronal cells. Thus, the behavior of MTs and MFs must be coordinated to determine neuronal polarity. Although there is an increasing number of signaling molecules that regulate the dynamics of MTs and MFs to affect neuronal polarity (Arimura and Kaibuchi, 2007), the molecular mechanisms are not fully understood.

Another study has demonstrated that the location of the centrosome, rather than cytoskeletal dynamics, may play an important role in establishment of neuronal polarity in the very early phase of neuritogenesis (de Anda et al., 2005). Shortly after plating hippocampal neurons on a substrate in culture, centrosomes, together with the Golgi apparatus and clusters of endosomes, accumulate beneath the first neurite that later develops into an axon (Fig. 2.1). Furthermore, suppression of centrosome function inhibits polarization and, conversely, neurons having multiple centrosomes develop multiple axons. Thus, localization of a functional centrosome is a key determinant in polarization, and MTs may be delivered only to the first neurite from the centrosome. However, the location of the centrosome is regulated by phosphoinositide 3-kinase (PI3K), Cdc42, and dynein (Palazzo et al., 2001; Etienne-Manneville and Hall, 2003; Arimura and Kaibuchi, 2007). Whether the location of the centrosome is cause or consequence is still under debate.

2.3.2 Neurite Outgrowth

2.3.2.1 Axon Outgrowth

Because extending axons are guided by the advancement of motile growth cones, it is important to understand the morphologic changes and their regulation through cytoskeletal rearrangements. In a growth cone, there are three distinct domains: the peripheral (P) domain composed of MF-rich filopodia and lamellipodia; the transitional (T) domain, an interface zone between the P domain and the central (C) domain; and the C domain composed of a thicker, MT-rich region containing organelles and vesicles and connected to the axon shaft (Fig. 2.1). Early studies have demonstrated that MFs are the primary elements that alter the growth cone shape and are essential for proper axon guidance, whereas MTs are essential to stabilize the axon structure and play an important role in axon elongation (Yamada et al., 1970, 1971).

Growing evidence suggests that MF–MT interactions in the growth cone play a role in the shape changes of the growth cone and in axon outgrowth (Dent et al., 1999; Rodriguez et al., 2003; Kalil and Dent, 2005). In the P domain, dynamic MTs are present and sometimes penetrate into and retract from the filopodia (Fig. 2.1). This restricted MT organization in the P domain is likely to be regulated by actin assemblies. Forscher and Smith showed that, in an *Aplysia* bag cell neuron, MTs rapidly advance into the P domain when actin networks are completely abolished by treatment with cytochalasin B, suggesting an inhibitory role for MF structures in MT advance (Forscher and Smith, 1988). Further studies demonstrated that MTs advance along actin bundles within the filopodia, and retrograde actin flow is coupled with MT retrograde movements observed in the P domain (Schaefer et al., 2002). Closer analyses have revealed that selective loss of filopodia, but not the actin network in the lamellipodia, by treatment with low doses of cytochalasin B induces increased lateral MT movements and a more randomized MT distribution in the P domain (Burnette et al., 2007). These results indicate that actin bundles in the filopodia are essential for proper MT organization, but not MT advancement, in the P domain. Interestingly, when the actin bundles are locally disrupted in the P domain, growth cones respond with repulsive turning (Zhou et al., 2002). This turning is accompanied by reorientation of dynamic MTs within the P domain from areas where actin bundles are lacking into the regions where actin bundles remain.

MTs exploring the P domain exhibit properties of dynamic instability (Schaefer et al., 2002). Although the precise mechanisms of MT invasion into the P domain are not well understood, MF–MT interactions are presumably involved in MT growth toward the leading edge in the direction of the growth cone extension and axon outgrowth. Dynamic MTs invading the P domain are oriented with their plus-ends pointing toward the leading edge of the growth cone and preferentially interact with plus-end tracking proteins (+TIPs) (Akhmanova and Steinmetz, 2008; Jaworski et al., 2008) (Fig. 2.1). Indeed, growth cones have been shown to contain several +TIPs, including CLIP-associated protein 1 (CLASP1), CLASP2, adenomatous polyposis coli (APC), dynein, and Miller-Dieker lissencephaly-1 (LIS1) (Lee et al., 2004; Zhou et al., 2004; Grabham et al., 2007). CLASPs stabilize MTs and interact with other +TIPs, such as end-binding protein-1 (EB1) and cytoplasmic linker protein-170 (CLIP-170), and MFs (Mimori-Kiyosue et al., 2005). APC also has MT-stabilizing properties and the ability to interact with actin or signaling molecules that modulate actin polymerization (Zhou et al., 2004). Direct evidence implicating +TIPs in MT advance in the P domain comes from a study showing that dynein and LIS1 inhibition reduced penetration of MTs into the P domain during laminin-induced axon outgrowth (Grabham et al., 2007). This suggests that dynein and LIS1 cooperatively play a prominent role in MT advance within the growth cones during axon outgrowth. Whether these +TIPs bind MF remains unknown. Rather, dynein interacts with the actin cytoskeleton, presumably via dynactin, and LIS1 regulates MF organization by affecting signaling molecules, such as IQ motif-containing GTPase-activating protein (IQGAP) and Cdc42 (Ahmad et al., 2000; Kholmanskikh et al., 2006). Thus, +TIPs may regulate direct and indirect MF–MT interactions crucial for growth cone extension or turning.

In addition to growth cone advancement, generation of a new segment of the axon is essential for axon outgrowth and is achieved by bundling of MTs at the growth cone neck (Dent and Gertler, 2003). Very recently, myosin II–mediated lateral movements of actin arcs, contractile structures present in the T domain, have been shown to promote MT bundling and to transport the MTs into the C domain, resulting in the formation of a new segment of the axon shaft contiguous with the growth cone neck (Burnette et al., 2008). Although these experiments employed pharmacologic manipulations of *Aplysia* bag cell neurons, similar observations have been obtained using local application of the Ig superfamily cell surface molecule *aplysia* cell adhesion molecule (apCAM), a physiologic guidance cue, to the growth cone of *Aplysia* neurons (Schaefer et al., 2008). Stimulation of growth cones with apCAM induced activation of Rho/Rho kinase/myosin II signaling pathways and resulted in MT movements in the growth cone neck and subsequent axonal outgrowth. This response is in contrast with the well-known observations that the Rho pathway is required for repulsive cue–induced neurite retraction (see later). Thus, this pathway must be locally and precisely controlled within the growth cones.

Recent studies have also revealed a new aspect of MT behavior in the shaft of the growing axon. A live-cell imaging study using rat sympathetic neurons showed that short MTs (a few micrometers in length) are transported down the axon and that the transport is bidirectional, rapid, infrequent, and highly asynchronous (Wang and Brown, 2002). Baas and colleagues demonstrated that in chick sensory neurons, the anterograde transport of short MTs is mediated by cytoplasmic dynein in concert with actin, whereas the retrograde transport is presumably dependent on the kinesin-5 family, but not actin (Fig. 2.1) (Hasaka et al., 2004; He et al., 2005; Baas et al., 2006). Interestingly, the motor activities of dynein and kinesin are in part regulated by tau, a MAP present in axons (Dixit et al., 2008). Retarded axonal extension in tau-deficient hippocampal neurons may be due to reduced MT transport by lack of tau-mediated regulation of motor protein activities (Dawson et al., 2001).

2.3.2.2 Dendrite Outgrowth

In contrast with our understanding of the mechanisms regulating development of the axon, our knowledge of the roles of MTs in dendrite development is limited (Georges et al., 2008). As discussed above, axon/dendrite fate determination is partially mediated through selective MT stabilization in the future axon and localization of the centrosome beneath the cell surface where the future axon will extend (de Anda et al., 2005; Witte et al., 2008). However, whether dendrite fate is determined by a default mechanism, as a result of axon fate determination, or an active mechanism, which selectively decreases MT stability in the future dendrites, remains unclear.

Unlike axonal MTs, which exhibit a unidirectional polarity with the plus-ends directed away from the cell body, dendrite MTs are oriented bidirectionally, although more distal dendrites contain unipolar MTs oriented similarly to those in axons (Baas et al., 1988). Thus, the cytoskeletal mechanisms underlying dendrite outgrowth may differ from those in axons. Because dendrite outgrowth and development is associated with dendrite branching, I will discuss this issue in the next section.

2.3.3 Neurite Branching

The formation of neurite branches is essential for the establishment of neuronal networks and requires remodeling of the cytoskeleton. Branch formation can occur by different modes: splitting, delayed, and interstitial (Acebes and Ferrus, 2000; Dent and Gertler, 2003). The splitting mode involves branch formation via concomitant growth cone bifurcation and is likely to be related to dendritic branch formation. The delayed mode is mechanistically linked to behaviors of the growth cone. In this mode, the primary growth cones of axons enlarge during prolonged pausing behaviors to leave remnants behind on the axon shaft after the axon resumes forward extension, and these remnants subsequently give rise to branches. The interstitial mode consists of development of sprouts from a stable axon shaft.

Dent et al. examined MT reorganization during axonal branch formation, which resembled the delayed mode of neurite branching, in hamster cortical neurons (Dent and Kalil, 2001). Time-lapse imaging of neurons co-injected with fluorescently labeled tubulin and phalloidin revealed that dynamic MTs colocalize with MFs in transition regions of growth cones and at axon branch points. When neurons were treated with low doses of paclitaxel or nocodazole to inhibit MT dynamic instability, branch formation was prevented, whereas axonal elongation was not substantially affected. Similarly, treatment with cytochalasin D and latrunculin A, which attenuate MF dynamics, inhibited neurite branching. These data suggest that the coordinated dynamics of MTs and MFs regulate the delayed mode of branch formation.

In contrast, when chick sensory neurons are treated with low doses of paclitaxel or nocodazole, the growth of the axonal shaft is markedly suppressed, whereas interstitial branch formation is only minimally affected (Gallo and Letourneau, 1999). In addition, vinblastine treatment to inhibit MT repolymerization after depletion of MTs by a high dose of nocodazole results in the reappearance of branches. Therefore, branch formation is unlikely to be dependent on MT dynamics. Rather, transport of MT polymers appears to be involved in branch formation. Thus, the mechanism of MT reorganization in the interstitial mode of branch formation may be different from that of the delayed mode. However, the possibility that the mode of MT reorganization in axon branch formation varies depending on the neuronal cell type cannot be excluded.

Detailed analyses using electron micrographs and time-lapse imaging revealed that large numbers of short MTs are present, and that long MTs are absent, at the sites of axonal branches (Yu et al., 1994), suggesting that short MTs are transported into axon branches (Fig. 2.1). The questions to be resolved were the source of short MTs and the mechanism of their production. Baas and colleagues examined the behavior of spastin, an MT-severing protein concentrated at the sites of branch formation (Yu et al., 2008). They showed that exogenous expression of spastin results not only in production of short MT polymers but also in increases of the number of axonal branches. Generation of short MTs gives rise to an abundance of new MT ends, which could interact with various MT-interacting proteins, +TIPs, and motor proteins (Akhmanova and Steinmetz, 2008; Jaworski et al., 2008). These proteins may mediate anterograde transport of short MTs into branches, in concert with actin, which has been observed in the axon or growth cone (Hasaka et al., 2004; Grabham et al., 2007).

How guidance cues influence the cytoskeleton in axonal branch formation is still unknown. However, one report shows that some guidance cues change the organization and dynamics of the cytoskeleton at the growth cone and the axon shaft, resulting in promotion or suppression of branch formation without stimulation of axon elongation (Dent et al., 2004). FGF and netrin-1, which stimulate branch formation, increase actin polymerization and formation of MT loops and splaying in growth cones. In contrast, semaphorin 3A, which inhibits branch formation, stimulated actin depolymerization and attenuated MT dynamics in growth cones. However, the signaling mechanisms by which these guidance cues regulate cytoskeletal dynamics and produce axon branching remain unclear.

Dendrites of most types of neurons are highly elaborated, and their outgrowth during development involves branch formation. A recent study has shown that stathmin, an MT destabilizer, is involved in dendrite branch formation (Ohkawa et al., 2007). Overexpression of stathmin in Purkinje cells leads to suppression of dendritic arborization. In contrast, inhibition of stathmin by neural activity and/or phosphorylation by calcium/calmodulin-dependent protein kinase increases dendritic arborization. However, knockdown of stathmin reduces dendrite growth. Thus, proper regulation of stathmin expression and activity is likely to be important in the development of dendritic arbors. Another study from the same laboratory has demonstrated that acetylated tubulin, present in stable MTs, plays a crucial role in dendrite branch formation (Ohkawa et al., 2008). These researchers identified an enzyme responsible for tubulin acetylation, an *N*-acetyltransferase complex consisting of ARD1 and NAD1, and found that inhibition of the complex limited dendrite branch formation in cortical neurons. Both studies implicate the importance of MT stability in dendrite branch formation. However, these results are in contrast with findings that MT stability determines the fate of the axon (Witte et al., 2008). Further investigation will be needed to understand how MTs are organized during dendrite growth.

2.3.4 Neurite Retraction

Transient neurite retraction is a fundamental process in neuronal network formation. This morphologic change occurs not only intrinsically but also in response to extracellular repulsive signals. Many studies have been conducted to further our understanding of the regulation of cytoskeletal rearrangements during repulsive signal–induced neurite retraction. Baas and colleagues have proposed an interesting model in which two motor proteins, dynein and myosin, are counterbalanced in their activities, and alterations in this balance determine whether the axon elongates or retracts (Baas and Ahmad, 2001; Baas et al., 2006). According to this model, dynein-dependent anterograde MT transport is predominant in growing neurites, and when neurites encounter repulsive signals, transient activation of actomyosin systems and the relative inhibition of dynein may result in enhancement of retrograde MT transport and neurite retraction. Indeed, nitric oxide (NO), a physiologic candidate for repulsive signals, induces neurite retraction that is reminiscent of that induced by dynein inhibition (Ahmad et al., 2000; He et al., 2002). Treatment of sensory neurons with an NO donor induces a rapid retraction of axons, characterized by an enlarged distal region, a thin trailing remnant, and unique sinusoidal MT structures. The NO-induced retraction is also accompanied by a lack of detectable changes in MT amounts and MT retrograde transport, and these effects are not blocked by the MT stabilizer paclitaxel. These responses are similar to those observed in dynein-inhibited neurons (Ahmad et al., 2000). However, whether NO inhibits dynein or stimulates myosin remained to be determined.

A similar neurite retraction response is observed in cortical neurons treated with lysophosphatidic acid (LPA), a major lysophospholipid that exerts diverse cellular responses in many cell types (Fukushima et al., 2002; Fukushima and Morita, 2006). LPA causes neurite retraction through the Rho-actomyosin pathway in various neuronal cells (Fukushima, 2004). This effect is observed within 1~60 min after LPA exposure, depending on the cell types. Exposure of young cortical neurons that are initiating neurite extension to LPA results in neurite retraction accompanied by the rearrangement of MTs in neurites and the accumulation of MTs in cell bodies, without significant changes in the total amount of MTs in the cytoskeletal fraction of cultured neurons (Fukushima and Morita, 2006). These effects of LPA on MTs are blocked by pretreatment with inhibitors of the actomyosin and Rho pathways (cytochalasin D, blebbistatin and Y27632), but not by paclitaxel, although paclitaxel inhibits neurite retraction and MT depolymerization induced by nocodazole (Fig. 2.2). These observations suggest that LPA-induced MT rearrangement is not due to depolymerization of MTs and is dependent on actin polymerization during neurite remodeling.

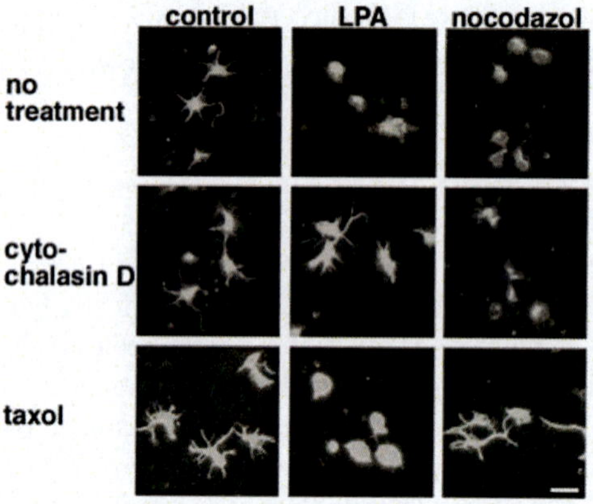

Fig. 2.2 Lysophosphatidic acid (LPA)-induced microtubule rearrangements in cultured cortical neurons. Cortical neurons were pretreated with 2 μM cytochalasin D or 10 μM paclitaxel for 5 min and exposed to vehicle, 1 μM LPA, or 10 μM nocodazole for 15 min. Then, cells were stained for α-tubulin. LPA-induced neurite retraction and concomitant microtubule rearrangements depend on actin polymerization. Bar = 20 μm

Interestingly, two recent reports using MAP1B-deficient neurons have shown that MAP1B, which can cross-link MTs and MFs, is required for MT transport during NO- and LPA-induced neurite retraction (Bouquet et al., 2007; Stroissnigg et al., 2007). Exposure of neurons to an NO donor induces S-nitrosylation of MAP1B, leading to enhanced interaction with MTs. This interaction interferes with dynein motor function, resulting in neurite retraction. Whether LPA induces S-nitrosylation is unknown. Rather, phosphorylation of MAP1B may be an important signal in LPA-induced neurite retraction, because LPA can activate many protein kinases including glycogen synthase kinase 3β, which phosphorylates MAP1B, and phosphorylated MAP1B then alters the interaction between LIS1 and dynein (Jimenez-Mateos et al., 2005; Trivedi et al., 2005; Shano et al., 2008). Further investigation is needed to unveil the underlying mechanisms for LPA-induced neurite retraction.

2.4 Microtubules in Neuronal Injuries and Diseases

When injured, neurons respond with neurite regeneration or degeneration, depending on the property or intensity of the injury or the type of neuron. Generally, after injury, axons in the central nervous system (CNS) do not regenerate well, whereas those in the peripheral nervous system easily regenerate. A recent report has demonstrated that the state of MTs in the neurite tip after injury determines the ability of the neurite to regenerate (Erturk et al., 2007). When CNS axons are injured, the tips form swelled structures designated *retraction bulbs*, which contain disorganized MTs. MT stabilization inhibits the formation of retraction bulbs in CNS axons, resulting in axonal growth, whereas MT destabilization causes growth cones to transform into retraction bulb–like structures, resulting in restriction of axonal growth. The molecular mechanisms by which the MT state is controlled during nerve injury and regeneration remain unknown.

Another study has demonstrated that axotomy of *Aplysia* neurons induces the formation of swelling about 100 μm proximal to the cut end (Erez et al., 2007). With time, the cut axonal end retracts and produces a flat lamellipodium, and a swelling develops at the center of the newly formed growth cone. During this process, MTs are reconstructed and play a role in vesicle transport into the swelling.

Many investigations have revealed that MTs and MAPs are involved in the onset and/or progression of neuronal diseases. For example, hyperphosphorylated tau is a well-known component of neurofibrillary tangles detected in Alzheimer's disease (Higuchi et al., 2002; Stoothoff and Johnson, 2005). Excellent reviews have discussed many other examples that have differential aspects of MTs in neuronal diseases, such as MT-based transport, MAP-related functions, and MT-targeted drugs (Benitez-King et al., 2004; Gerdes and Katsanis, 2005; Michaelis et al., 2005; Chevalier-Larsen and Holzbaur, 2006; Feng, 2006; Stokin and Goldstein, 2006; El-Kadi et al., 2007; Kerjan and Gleeson, 2007; Riederer, 2007; Sonnenberg and Liem, 2007). Here, I will discuss tubulin acetylation and mutations related to neuronal diseases.

Several lines of evidence have indicated that tubulin acetylation may be involved in neurodegenerative diseases, such as Huntington's disease (HD) and Parkinson's disease (PD) (Dompierre et al., 2007; Outeiro et al., 2007; Suzuki and Koike, 2007). HD is a neurodegenerative disorder characterized by cognitive and motor deficits. In HD, the MT-dependent transport of vesicles containing brain-derived neurotrophic factor (BDNF) is reduced, resulting in a reduction of trophic support and subsequent neuronal cell death. Pharmacologic inhibition of histone deacetylase

6 (HDAC6), a tubulin deacetylase, increases the level of acetylated MTs and concomitantly enhances BDNF transport in neuronal cell lines (Dompierre et al., 2007). Because this increased MT acetylation led to enhanced recruitment of kinesin-1 and dynein, these molecular motors are likely to be involved in BDNF transport. Interestingly, acetylated MT levels are reduced in HD patients, and the HDAC inhibitor trichostatin A (TSA) rescues the transport defect in neuronal cells carrying a polyglutamine expansion model of HD. These observations imply that HDAC inhibition and regulation of MT acetylation may be a therapeutic target in HD.

PD is characterized by motor deficits and loss of dopaminergic neurons, which may be induced by accumulation of α-synuclein in the midbrain. Inhibition of sirtuin 2/a mammalian homolog of yeast silent information regulator 2 (SIRT2), another tubulin deacetylase, results in protection of neurons from α-synuclein–induced cell death (Outeiro et al., 2007). SIRT2 inhibition also causes resistance to axonal degeneration in cerebellar granule neurons of mutant mice displaying slow Wallerian degeneration (Suzuki and Koike, 2007). However, in both cases, the mechanisms by which SIRT2 inhibition protects neurons from cell death remain unknown.

A recent mutagenesis study using mice has revealed that mutation in α1-tubulin causes developmental brain abnormalities and that α1-tubulin (*TUBA1A*) is a gene responsible for lissencephaly in human (Keays et al., 2007). Investigators screened mice treated with *N*-ethyl-*N*-nitrosourea, a strong mutagen, and identified a mutant mouse line that showed disorganized neuronal architecture of the hippocampus and cortex, impaired neuronal migration, and abnormal behavior, which resembled features observed in the lissencephaly mouse models (e.g., *Lis1*, *Dcx* mutants) (Paylor et al., 1999; Corbo et al., 2002). A mutation in the GTP-binding domain in α1-tubulin was identified. Interestingly, several mutations in the human tubulin homolog were also found in patients with lissencephaly (Keays et al., 2007; Bahi-Buisson et al., 2008; Fallet-Bianco et al., 2008; Morris-Rosendahl et al., 2008), characterized by abnormal migration of neural progenitor cells in the cerebral cortex, although the location of these mutations differs from that in the mutant mice. The mutations in human may affect interactions of tubulin with MAPs.

2.5 Concluding Remarks

During the past two to three decades, growing evidence has changed our understanding of the roles and functions of MTs. MTs are more dynamic cytoskeletal components than previously understood and produce a variety of changes in neuronal morphology; they are not simple, stable, static structures to merely support neuronal shapes. As I described in this chapter, MTs function in concert with MFs and are transported in an MF-dependent manner. Similar interactions with intermediate filaments (IFs) also play a crucial role in MT functions. This observation is supported by the finding of MT–IF cross-linkers, such as the plakin family proteins (Leung et al., 2002; Sonnenberg and Liem, 2007). Moreover, the identification and analyses of various types of molecules mediating the interactions of MTs with

MFs or other signaling molecules, such as +TIPs, suggest that MTs are the final acceptors of extracellular signals (Gundersen and Cook, 1999; Hollenbeck, 2001; Galjart, 2005; Akhmanova and Steinmetz, 2008). A good example has recently been published from two independent laboratories, which showed that MTs are also present in MF-rich dendritic spines that are correlated with neural development and activity-dependent plasticity, and the localization of MTs is mediated via the +TIP EB3 (Gu et al., 2008; Hu et al., 2008). A more systematic and integrated analysis will help us to understand MT functions and dynamics in the nervous system in development and disease.

References

Acebes A, Ferrus A (2000) Cellular and molecular features of axon collaterals and dendrites. Trends Neurosci 23:557–565

Ahmad FJ, Hughey J, Wittmann T, Hyman A, Greaser M, Baas PW (2000) Motor proteins regulate force interactions between microtubules and microfilaments in the axon. Nat Cell Biol 2: 276–280

Akhmanova A, Steinmetz MO (2008) Tracking the ends: a dynamic protein network controls the fate of microtubule tips. Nat Rev Mol Cell Biol 9:309–322

Arimura N, Kaibuchi K (2007) Neuronal polarity: from extracellular signals to intracellular mechanisms. Nat Rev Neurosci 8:194–205

Baas PW, Ahmad FJ (2001) Force generation by cytoskeletal motor proteins as a regulator of axonal elongation and retraction. Trends Cell Biol 11:244–249

Baas PW, Deitch JS, Black MM, Banker GA (1988) Polarity orientation of microtubules in hippocampal neurons: uniformity in the axon and nonuniformity in the dendrite. Proc Natl Acad Sci USA 85:8335–8339

Baas PW, Vidya Nadar C, Myers KA (2006) Axonal transport of microtubules: the long and short of it. Traffic 7:490–498

Bahi-Buisson N, Poirier K, Boddaert N, Saillour Y, Castelnau L, Philip N, Buyse G, Villard L, Joriot S, Marret S, Bourgeois M, Van Esch H, Lagae L, Amiel J, Hertz-Pannier L, Roubertie A, Rivier F, Pinard JM, Beldjord C, Chelly J (2008) Refinement of cortical dysgeneses spectrum associated with TUBA1A mutations. J Med Genet 45:647–653

Benitez-King G, Ramirez-Rodriguez G, Ortiz L, Meza I (2004) The neuronal cytoskeleton as a potential therapeutical target in neurodegenerative diseases and schizophrenia. Curr Drug Targets CNS Neurol Disord 3:515–533

Bouquet C, Ravaille-Veron M, Propst F, Nothias F (2007) MAP1B coordinates microtubule and actin filament remodeling in adult mouse Schwann cell tips and DRG neuron growth cones. Mol Cell Neurosci 36:235–247

Bradke F, Dotti CG (1999) The role of local actin instability in axon formation. Science 283: 1931–1934

Burnette DT, Ji L, Schaefer AW, Medeiros NA, Danuser G, Forscher P (2008) Myosin II activity facilitates microtubule bundling in the neuronal growth cone neck. Dev Cell 15:163–169

Burnette DT, Schaefer AW, Ji L, Danuser G, Forscher P (2007) Filopodial actin bundles are not necessary for microtubule advance into the peripheral domain of *Aplysia* neuronal growth cones. Nat Cell Biol 9:1360–1369

Chevalier-Larsen E, Holzbaur EL (2006) Axonal transport and neurodegenerative disease. Biochim Biophys Acta 1762:1094–1108

Corbo JC, Deuel TA, Long JM, LaPorte P, Tsai E, Wynshaw-Boris A, Walsh CA (2002) Doublecortin is required in mice for lamination of the hippocampus but not the neocortex. J Neurosci 22:7548–7557

da Silva JS, Dotti CG (2002) Breaking the neuronal sphere: regulation of the actin cytoskeleton in neuritogenesis. Nat Rev Neurosci 3:694–704
Dawson HN, Ferreira A, Eyster MV, Ghoshal N, Binder LI, Vitek MP (2001) Inhibition of neuronal maturation in primary hippocampal neurons from tau deficient mice. J Cell Sci 114:1179–1187
de Anda FC, Pollarolo G, Da Silva JS, Camoletto PG, Feiguin F, Dotti CG (2005) Centrosome localization determines neuronal polarity. Nature 436:704–708
Dehmelt L, Smart FM, Ozer RS, Halpain S (2003) The role of microtubule-associated protein 2c in the reorganization of microtubules and lamellipodia during neurite initiation. J Neurosci 23:9479–9490
Dent EW, Barnes AM, Tang F, Kalil K (2004) Netrin-1 and semaphorin 3A promote or inhibit cortical axon branching, respectively, by reorganization of the cytoskeleton. J Neurosci 24:3002–3012
Dent EW, Callaway JL, Szebenyi G, Baas PW, Kalil K (1999) Reorganization and movement of microtubules in axonal growth cones and developing interstitial branches. J Neurosci 19: 8894–8908
Dent EW, Gertler FB (2003) Cytoskeletal dynamics and transport in growth cone motility and axon guidance. Neuron 40:209–227
Dent EW, Kalil K (2001) Axon branching requires interactions between dynamic microtubules and actin filaments. J Neurosci 21:9757–9769
Dixit R, Ross JL, Goldman YE, Holzbaur EL (2008) Differential regulation of dynein and kinesin motor proteins by tau. Science 319:1086–1089
Dompierre JP, Godin JD, Charrin BC, Cordelieres FP, King SJ, Humbert S, Saudou F (2007) Histone deacetylase 6 inhibition compensates for the transport deficit in Huntington's disease by increasing tubulin acetylation. J Neurosci 27:3571–3583
Dotti CG, Sullivan CA, Banker GA (1988) The establishment of polarity by hippocampal neurons in culture. J Neurosci 8:1454–1468
El-Kadi AM, Soura V, Hafezparast M (2007) Defective axonal transport in motor neuron disease. J Neurosci Res 85:2557–2566
Erez H, Malkinson G, Prager-Khoutorsky M, De Zeeuw CI, Hoogenraad CC, Spira ME (2007) Formation of microtubule-based traps controls the sorting and concentration of vesicles to restricted sites of regenerating neurons after axotomy. J Cell Biol 176:497–507
Erturk A, Hellal F, Enes J, Bradke F (2007) Disorganized microtubules underlie the formation of retraction bulbs and the failure of axonal regeneration. J Neurosci 27:9169–9180
Etienne-Manneville S, Hall A (2003) Cell polarity: Par6, aPKC and cytoskeletal crosstalk. Curr Opin Cell Biol 15:67–72
Fallet-Bianco C, Loeuillet L, Poirier K, Loget P, Chapon F, Pasquier L, Saillour Y, Beldjord C, Chelly J, Francis F (2008) Neuropathological phenotype of a distinct form of lissencephaly associated with mutations in TUBA1A. Brain 131:2304–2320
Feng J (2006) Microtubule: a common target for parkin and Parkinson's disease toxins. Neuroscientist 12:469–476
Forscher P, Smith SJ (1988) Actions of cytochalasins on the organization of actin filaments and microtubules in a neuronal growth cone. J Cell Biol 107:1505–1516
Fukushima N (2004) LPA in neural cell development. J Cell Biochem 92:993–1003
Fukushima N, Furuta D, Hidaka Y, Moriyama R, Tsujiuchi T (2009) Posttranslational modifications of tubulin in the nervous system. J Neurochem 109:683–693
Fukushima N, Morita Y (2006) Actomyosin-dependent microtubule rearrangement in lysophosphatidic acid-induced neurite remodeling of young cortical neurons. Brain Res 1094:65–75
Fukushima N, Weiner JA, Kaushal D, Contos JJA, Rehen SK, Kingsbury MA, Kim K-Y, Chun J (2002) Lysophosphatidic acid influences the morphology and motility of young, postmitotic cortical neurons. Mol Cell Neurosci 20:271–282
Galjart N (2005) CLIPs and CLASPs and cellular dynamics. Nat Rev Mol Cell Biol 6:487–498
Gallo G, Letourneau PC (1999) Different contributions of microtubule dynamics and transport to the growth of axons and collateral sprouts. J Neurosci 19:3860–3873

Georges PC, Hadzimichalis NM, Sweet ES, Firestein BL (2008) The yin-yang of dendrite morphology: unity of actin and microtubules. Mol Neurobiol 38:270–284

Gerdes JM, Katsanis N (2005) Small molecule intervention in microtubule-associated human disease. Hum Mol Genet 14(Spec No 2):R291–R300

Grabham PW, Seale GE, Bennecib M, Goldberg DJ, Vallee RB (2007) Cytoplasmic dynein and LIS1 are required for microtubule advance during growth cone remodeling and fast axonal outgrowth. J Neurosci 27:5823–5834

Gu J, Firestein BL, Zheng JQ (2008) Microtubules in dendritic spine development. J Neurosci 28:12120–12124

Gundersen GG, Cook TA (1999) Microtubules and signal transduction. Curr Opin Cell Biol 11:81–94

Hammond JW, Cai D, Verhey KJ (2008) Tubulin modifications and their cellular functions. Curr Opin Cell Biol 20:71–76

Hasaka TP, Myers KA, Baas PW (2004) Role of actin filaments in the axonal transport of microtubules. J Neurosci 24:11291–11301

He Y, Francis F, Myers KA, Yu W, Black MM, Baas PW (2005) Role of cytoplasmic dynein in the axonal transport of microtubules and neurofilaments. J Cell Biol 168:697–703

He Y, Yu W, Baas PW (2002) Microtubule reconfiguration during axonal retraction induced by nitric oxide. J Neurosci 22:5982–5991

Higuchi M, Lee VM, Trojanowski JQ (2002) Tau and axonopathy in neurodegenerative disorders. Neuromolecular Med 2:131–150

Hollenbeck P (2001) Cytoskeleton: microtubules get the signal. Curr Biol 11:R820–R823

Hu X, Viesselmann C, Nam S, Merriam E, Dent EW (2008) Activity-dependent dynamic microtubule invasion of dendritic spines. J Neurosci 28:13094–13105

Jaworski J, Hoogenraad CC, Akhmanova A (2008) Microtubule plus-end tracking proteins in differentiated mammalian cells. Int J Biochem Cell Biol 40:619–637

Jimenez-Mateos EM, Wandosell F, Reiner O, Avila J, Gonzalez-Billault C (2005) Binding of microtubule-associated protein 1B to LIS1 affects the interaction between dynein and LIS1. Biochem J 389:333–341

Kalil K, Dent EW (2005) Touch and go: guidance cues signal to the growth cone cytoskeleton. Curr Opin Neurobiol 15:521–526

Keays DA, Tian G, Poirier K, Huang GJ, Siebold C, Cleak J, Oliver PL, Fray M, Harvey RJ, Molnar Z, Pinon MC, Dear N, Valdar W, Brown SD, Davies KE, Rawlins JN, Cowan NJ, Nolan P, Chelly J, Flint J (2007) Mutations in alpha-tubulin cause abnormal neuronal migration in mice and lissencephaly in humans. Cell 128:45–57

Kerjan G, Gleeson JG (2007) Genetic mechanisms underlying abnormal neuronal migration in classical lissencephaly. Trends Genet 23:623–630

Kholmanskikh SS, Koeller HB, Wynshaw-Boris A, Gomez T, Letourneau PC, Ross ME (2006) Calcium-dependent interaction of Lis1 with IQGAP1 and Cdc42 promotes neuronal motility. Nat Neurosci 9:50–57

Lee H, Engel U, Rusch J, Scherrer S, Sheard K, Van Vactor D (2004) The microtubule plus end tracking protein Orbit/MAST/CLASP acts downstream of the tyrosine kinase Abl in mediating axon guidance. Neuron 42:913–926

Leung CL, Green KJ, Liem RK (2002) Plakins: a family of versatile cytolinker proteins. Trends Cell Biol 12:37–45

Lewis SA, Lee MG, Cowan NJ (1985) Five mouse tubulin isotypes and their regulated expression during development. J Cell Biol 101:852–861

Michaelis ML, Seyb KI, Ansar S (2005) Cytoskeletal integrity as a drug target. Curr Alzheimer Res 2:227–229

Mimori-Kiyosue Y, Grigoriev I, Lansbergen G, Sasaki H, Matsui C, Severin F, Galjart N, Grosveld F, Vorobjev I, Tsukita S, Akhmanova A (2005) CLASP1 and CLASP2 bind to EB1 and regulate microtubule plus-end dynamics at the cell cortex. J Cell Biol 168: 141–153

Morris-Rosendahl DJ, Najm J, Lachmeijer AM, Sztriha L, Martins M, Kuechler A, Haug V, Zeschnigk C, Martin P, Santos M, Vasconcelos C, Omran H, Kraus U, Van der Knaap MS, Schuierer G, Kutsche K, Uyanik G (2008) Refining the phenotype of *α-1a Tubulin* (*TUBA1A*) mutation in patients with classical lissencephaly. Clin Genet 74: 425–433

Ohkawa N, Fujitani K, Tokunaga E, Furuya S, Inokuchi K (2007) The microtubule destabilizer stathmin mediates the development of dendritic arbors in neuronal cells. J Cell Sci 120: 1447–1456

Ohkawa N, Sugisaki S, Tokunaga E, Fujitani K, Hayasaka T, Setou M, Inokuchi K (2008) N-acetyltransferase ARD1-NAT1 regulates neuronal dendritic development. Genes Cells 13:1171–1183

Outeiro TF, Kontopoulos E, Altmann SM, Kufareva I, Strathearn KE, Amore AM, Volk CB, Maxwell MM, Rochet JC, McLean PJ, Young AB, Abagyan R, Feany MB, Hyman BT, Kazantsev AG (2007) Sirtuin 2 inhibitors rescue alpha-synuclein-mediated toxicity in models of Parkinson's disease. Science 317:516–519

Palazzo AF, Joseph HL, Chen YJ, Dujardin DL, Alberts AS, Pfister KK, Vallee RB, Gundersen GG (2001) Cdc42, dynein, and dynactin regulate MTOC reorientation independent of Rho-regulated microtubule stabilization. Curr Biol 11:1536–1541

Paylor R, Hirotsune S, Gambello MJ, Yuva-Paylor L, Crawley JN, Wynshaw-Boris A (1999) Impaired learning and motor behavior in heterozygous Pafah1b1 (Lis1) mutant mice. Learn Mem 6:521–537

Riederer BM (2007) Microtubule-associated protein 1B, a growth-associated and phosphorylated scaffold protein. Brain Res Bull 71:541–558

Rodriguez OC, Schaefer AW, Mandato CA, Forscher P, Bement WM, Waterman-Storer CM (2003) Conserved microtubule-actin interactions in cell movement and morphogenesis. Nat Cell Biol 5:599–609

Roger B, Al-Bassam J, Dehmelt L, Milligan RA, Halpain S (2004) MAP2c, but not tau, binds and bundles F-actin via its microtubule binding domain. Curr Biol 14:363–371

Schaefer AW, Kabir N, Forscher P (2002) Filopodia and actin arcs guide the assembly and transport of two populations of microtubules with unique dynamic parameters in neuronal growth cones. J Cell Biol 158:139–152

Schaefer AW, Schoonderwoert VT, Ji L, Mederios N, Danuser G, Forscher P (2008) Coordination of actin filament and microtubule dynamics during neurite outgrowth. Dev Cell 15:146–162

Shano S, Hatanaka K, Ninose S, Moriyama R, Tsujiuchi T, Fukushima N (2008) A lysophosphatidic acid receptor lacking the PDZ-binding domain is constitutively active and stimulates cell proliferation. Biochim Biophys Acta 1783:748–759

Sonnenberg A, Liem RK (2007) Plakins in development and disease. Exp Cell Res 313:2189–2203

Stanchi F, Corso V, Scannapieco P, Ievolella C, Negrisolo E, Tiso N, Lanfranchi G, Valle G (2000) TUBA8: a new tissue-specific isoform of α-tubulin that is highly conserved in human and mouse. Biochem Biophys Res Commun 270:1111–1118

Stokin GB, Goldstein LS (2006) Axonal transport and Alzheimer's disease. Annu Rev Biochem 75:607–627

Stoothoff WH, Johnson GV (2005) Tau phosphorylation: physiological and pathological consequences. Biochim Biophys Acta 1739:280–297

Stroissnigg H, Trancikova A, Descovich L, Fuhrmann J, Kutschera W, Kostan J, Meixner A, Nothias F, Propst F (2007) S-nitrosylation of microtubule-associated protein 1B mediates nitric-oxide-induced axon retraction. Nat Cell Biol 9:1035–1045

Suzuki K, Koike T (2007) Mammalian Sir2-related protein (SIRT) 2-mediated modulation of resistance to axonal degeneration in slow Wallerian degeneration mice: a crucial role of tubulin deacetylation. Neuroscience 147:599–612

Trivedi N, Marsh P, Goold RG, Wood-Kaczmar A, Gordon-Weeks PR (2005) Glycogen synthase kinase-3β phosphorylation of MAP1B at Ser1260 and Thr1265 is spatially restricted to growing axons. J Cell Sci 118:993–1005

Villasante A, Wang D, Dobner P, Dolph P, Lewis SA, Cowan NJ (1986) Six mouse α-tubulin mRNAs encode five distinct isotypes: testis-specific expression of two sister genes. Mol Cell Biol 6:2409–2419

Wang L, Brown A (2002) Rapid movement of microtubules in axons. Curr Biol 12:1496–1501

Westermann S, Weber K (2003) Post-translational modifications regulate microtubule function. Nat Rev Mol Cell Biol 4:938–947

Witte H, Neukirchen D, Bradke F (2008) Microtubule stabilization specifies initial neuronal polarization. J Cell Biol 180:619–632

Yamada KM, Spooner BS, Wessells NK (1970) Axon growth: roles of microfilaments and microtubules. Proc Natl Acad Sci USA 66:1206–1212

Yamada KM, Spooner BS, Wessells NK (1971) Ultrastructure and function of growth cones and axons of cultured nerve cells. J Cell Biol 49:614–635

Yu W, Ahmad FJ, Baas PW (1994) Microtubule fragmentation and partitioning in the axon during collateral branch formation. J Neurosci 14:5872–5884

Yu W, Qiang L, Solowska JM, Karabay A, Korulu S, Baas PW (2008) The microtubule-severing proteins spastin and katanin participate differently in the formation of axonal branches. Mol Biol Cell 19:1485–1498

Zhou FQ, Waterman-Storer CM, Cohan CS (2002) Focal loss of actin bundles causes microtubule redistribution and growth cone turning. J Cell Biol 157:839–849

Zhou FQ, Zhou J, Dedhar S, Wu YH, Snider WD (2004) NGF-induced axon growth is mediated by localized inactivation of GSK-3β and functions of the microtubule plus end binding protein APC. Neuron 42:897–912

Chapter 3
Tau Phosphorylation

Jesús Avila and Félix Hernández

Abstract Tau protein is a brain microtubule–associated protein having 79 putative phosphorylatable sites that could be modified by serine/threonine protein kinases. This phosphorylation can be divided in two types, depending whether the modified residue is phosphorylated by proline-directed or by non-proline-directed protein kinases. In neurodegenerative processes (tauopathies), tau is mainly (but not only) modified by a proline-directed protein kinase, glycogen synthase kinase 3 (GSK3). In this chapter, we will review that phosphorylation at serine and threonine residues, the modification that can take place at tyrosine residues, and the dephosphorylation of the modified residues by tau phosphatases. In addition, we will comment on the toxicity of phosphotau in disorders such as Alzheimer's disease and the development of possible therapies to prevent tau phosphorylation.

Keywords Alzheimer's disease · Tau phosphorylation · GSK3

Contents

J. Avila (✉)
Centro de Biología Molecular Severo Ochoa (CSIC-UAM), Universidad Autónoma de Madrid, 28049 Madrid, Spain
e-mail: javila@cbm.uam.es

R.A. Nixon, A. Yuan (eds.), *Cytoskeleton of the Nervous System*, Advances in Neurobiology 3, DOI 10.1007/978-1-4419-6787-9_3,

3.1 Introduction

Tau protein was discovered more than 30 years ago as a brain microtubule–associated protein (Fellous et al., 1977; Weingarten et al., 1975) at the same time that other microtubule-associated proteins (MAPs) were found in brain cell extracts (Shelanski et al., 1973).

By performing experiments in vitro or with cell cultures, a general function for tau and the other MAPs as microtubule stabilizers of assembled microtubules or as promoters of microtubule polymerization in vitro was indicated (Cleveland et al., 1977). Because tau protein was abundant in brain, a cDNA for tau was first isolated from a mouse brain expression library (Lee et al., 1988), and the sequence of the translated protein was obtained (Lee et al., 1988). Since then, the sequence of tau protein from different organisms, including human (Goedert et al., 1989), has been reported. The human tau gene is present on chromosome 17 (Neve et al., 1986) where it occupies more than 100 kb and contains at least 16 exons. This unique tau gene is transcribed into nuclear RNA that, by alternative splicing, yields several mRNA species. The translation of these different mRNAs results in the expression of different isoforms with different functional characteristics. Also, the expressions of these different isoforms are developmentally regulated (Kosik et al., 1989). At early developmental brain stages, mainly tau isoforms lacking exon 10 are present, whereas exon 10 is alternatively spliced in adult brain (Kosik et al., 1989). This is of interest because exon 10 encodes for one of the four repeated regions involved in the binding of tau to microtubules (see later). Thus, alternative splicing of exon 10 produces tau isoforms with either three (tau 3R) or four (tau 4R) microtubule binding regions. In the central nervous system (CNS), alternative splicing of exons 2, 3, and 10 results in the appearance of six tau isoforms. The expressed tau isoforms can be posttranslationally modified, mainly by phosphorylation, yielding additional tau isoforms.

Tau protein is mainly expressed in the frontal and temporal lobes in the brain, being present in a higher proportion in neurons (Santa-Maria et al., 2005; Trojanowski et al., 1989), where it can be associated with the plasma membrane, although it has, preferentially, a cytoplasmic localization (Arrasate et al., 2000; Brandt et al., 1995). In differentiated neurons, the majority of tau isoforms are located at the axons, although highly phosphorylated tau isoforms are present in the somatodendritic compartment (Dotti et al., 1987).

Tau protein (441 residues, the longest CNS isoform) can be divided in four regions. The first region, the amino-terminal 1–100 residues, is rich in acidic residues. This part of the molecule can be involved in the interaction with plasma membrane (Brandt et al., 1995), an interaction that can be modulated by phosphorylation (Arrasate et al., 2000), and with mitochondria (Atlante et al., 2008). Also, it has been reported that a peptide containing residues 26–44 of tau protein impairs mitochondrial oxidative phosphorylation acting at the level of the adenine nucleotide translocation (Atlante et al., 2008). In addition, tau protein, through its N-terminal regions, can interact with dynactin complex (Magnani et al., 2007). A second region, rich in proline, accounts for residues 101–239, and this region may be involved in the interaction of tau protein with proteins containing Src homology 3 (SH3) domains. Phosphorylation regulates tau interaction with SH3 domains of phosphatidylinositol 3-kinase, phospholipase cγ1, and Src family kinases (Reynolds et al., 2008).

However, the most important tau binding protein is tubulin, the main component of microtubules. This protein binds to tau through the third region containing residues 240–370. This is a basic region rich in lysine (and arginine) residues that contains four similar, but not identical, repeated sequences directly involved in the binding to tubulin (Avila et al., 2004; Lee et al., 1988). One of these sequences is codified by exon 10, as previously indicated. Because exon 10 can be alternatively spliced, there are tau 3R and tau 4R isoforms, the binding of tau 4R to microtubules being of higher affinity than that of tau 3R (Avila et al., 2004). Finally, there is a fourth region comprising the C-terminal residues from amino acid 371 to the C-terminal residue (amino acid 441). This is an acidic region that is rich in serines. Through this region, tau may interact with muscarinic receptors M1 and M3 (Gomez-Ramos et al., 2008).

3.2 Tau Phosphorylation

Tau is a protein rich in serine and threonine, having 79 putative phosphorylation sites (in the longest CNS isoform) that could be modified by serine or threonine protein kinases. In addition, phosphorylation at its tyrosine residues has also been reported (Williamson et al., 2002). Focusing on serine–threonine protein kinases, tau phosphorylation has been divided in two types: (1) the modified residue is phosphorylated by proline-directed protein kinases (PDPKs) such as tau kinase I (GSK3β) (Ishiguro et al., 1993), tau kinase II (cdk5) (Ishiguro et al., 1994), or other kinases belonging to the family of MAP kinases; (2) the modified residue is phosphorylated by a non-proline-directed protein kinase (NPDPK) such as protein kinase A (PKA), microtubule affinity regulating kinases (MARK), partition defective 1 (PAR1), protein kinase C (PKC), or other kinases (Avila et al., 2004). The phosphorylation carried out by these kinases mainly takes place in regions flanking the tubulin-binding region domain, although phosphorylation at the residues present in that tubulin-binding region could also take place mainly by NPDPK. In many cases, phosphorylation regulates the binding of tau to microtubules, and therefore, phosphorylation could be a way in which tau function can be regulated. Also, tau phosphorylation can modulate its subcellular localization in neurons (Dotti et al., 1987), the binding of tau to membranes (Arrasate et al., 2000), or the possible transport of tau to the cell nucleus (Greenwood and Johnson, 1995).

3.2.1 Tau Phosphorylation by GSK3

Tau phosphorylation changes during development. Thus, fetal tau is more highly phosphorylated than adult tau, and the degree of phosphorylation of tau isoforms decreases with age (Alonso et al., 2008; Mawal-Dewan et al., 1994). However, in neuronal disorders such as Alzheimer's disease or other tauopathies, tau isoforms are in a hyperphosphorylated form (Avila et al., 2004). In any case, tau has been defined as a phosphoprotein that can be modified, as shown above, by several kinases. Here, we will focus on its phosphorylation by GSK3 because this phosphorylation has different characteristics. This kinase could modify tau in two different ways: one way requires that tau protein must be previously phosphorylated (primed) by other kinase; the other way is that tau can be modified by unprimed phosphorylation. GSK3 can not only modify tau in monomeric form, but also it has been shown that assembled tau filaments can be phosphorylated by GSK3 to form large tangle-like structures (Rankin et al., 2008).

On one hand, tau phosphorylation could be affected by the presence or absence of specific proteins. The absence of the signaling adapter p62 leads to accumulation of hyperphosphorylated tau (Ramesh Babu et al., 2008). Also, the p53 family member p73 regulates the levels of tau phosphorylation (Wetzel et al., 2008). The presence of heat shock protein Hsp90 (Tortosa et al., 2009) may also affect tau modification. Also, there are compounds such as nicotinamide (Green et al., 2008) that reduce the phosphorylation at those sites modified by GSK3. In addition, there are several small molecules that can specifically inhibit GSK3 (see later).

On the other hand, it has been discussed that the development of tau pathology can correlate with that of Alzheimer's disease (Braak and Braak, 1991) and can be followed by analyzing tau phosphorylation (Delacourte et al., 1999). Tau pathology initiates at the entorhinal cortex; from there, it propagates to the limbic regions and afterwards to the cortex, in a process in which extracellular tau (coming from dead neuron) could play a role (Gomez-Ramos et al., 2008).

3.2.2 Tau Phosphorylation at Tyrosine Residues

Tau phosphorylation can also take place at tyrosine residues (Williamson et al., 2002), with tyrosine 394 identified as the major site of tyrosine phosphorylation in tau (Derkinderen et al., 2005) when tyrosine kinase Abl was used, whereas tyrosine 18 appears to be the major site for Fyn (another tyrosine kinase) phosphorylation (Lee et al., 2004). However, it is not clearly known if the tyrosine residue amino acid 310 is modified or not by phosphorylation, although it has been suggested that if phosphorylation occurs at amino acid 310, this may have an important influence in tau fibrillation (Inoue et al., 2008).

3.3 Tau Phosphatases

The phosphorylation level of protein like tau is the consequence of equilibrium between its phosphorylation (by kinase) and its dephosphorylation (by phosphatases). The majority of brain cellular phosphatase activity on phosphoserine and phosphothreonine residues is due to the action of four main phosphatase classes: PP1, PP2A, PP2B (or calcineurin) and PP2C (Cohen, 1989). PP2A is the most implicated in phosphotau dephosphorylation (Gong et al., 1994b; Wang et al., 1995), whereas PP2C appears to dephosphorylate tau only when it is phosphorylated by

PKA in vitro (Gong et al., 1994). PP2A is composed of three subunits, with the B subunit being involved in the binding to the substrate (Ramesh Babu et al., 2008). It has been suggested that PP2A binds to tau through its tubulin-binding region (Goedert et al., 2000). The B subunits consist of four families (Janssens and Goris, 2001), with the Bα subunit being the one that facilitates tau dephosphorylation (Drewes et al., 1993). PP2A activity could be regulated by at least two inhibitors, I_1^{PP2A} and I_2^{PP2A} (Chen et al., 2008). Phosphorylation of tau by proline-directed protein kinase takes place on serine or threonine that precedes a proline, and it may alter the rate of prolyl-isomerization. There exists a chaperone, Pin-1, that could restore this isomerization (Lu et al., 1999) facilitating the action of PP2A to dephosphorylate tau protein.

3.4 Toxicity of Phosphotau

There is converging evidence, through use of cell culture and animal (invertebrate and vertebrate) models, that an aberrant tau hyperphosphorylation can cause neurodegeneration in the absence of tau aggregate formation (for review, see Brandt et al., 2005). Examples can be found in a mouse model overexpressing human tau. This tau overexpression induces tau phosphorylation, neural dysfunction, and memory deficits without tau aggregation (Kimura et al., 2007). In another model, tau phosphorylation and synapse loss precede formation of aggregates (Kimura T, et al., 2010). Also, in a mouse overexpressing GSK3β (Engel et al., 2006b), phosphotau appears to be toxic in the absence of tau aggregation; a result that correlates well with a previous observation (Santacruz et al., 2005). Also, there are other models, mainly in *Drosophila melanogaster*, indicating phosphotau toxicity (Steinhilb et al., 2007).

A possible mechanism to explain tau toxicity is based on the fact that phosphorylated tau can recruit unmodified tau (and other MAPs) from microtubules, which may result in disruption of microtubules (Alonso et al., 1996). Another possible mechanism could be related to an impairment of axonal transport, although this possibility is now under discussion (Yuan et al., 2008).

3.4.1 Tau Phosphorylation and Alzheimer's Disease

The best known tauopathy is Alzheimer's disease (AD), in which neurofibrillary tangles are made up of paired helical filaments formed by hyperphosphorylated tau (Grundke-Iqbal et al., 1986). As previously indicated, there are several tau kinases, but among them a major role has been assigned to GSK3. For a transgenic *Drosophila* model, it has been described that tau phosphorylation by GSK3 facilitates tau aggregation (Ishihara et al., 2001). Similar results were found for mouse models (Engel et al., 2006; Perez et al., 2003) and by use of cell culture and in vitro models (Perez et al., 2000; Perez et al., 2002).

It is known that there are two types of AD: familial AD (FAD) and sporadic AD (SAD). In FAD, mutations in three genes (app, ps-1, and ps-2) result in the appearance of the disease (Price and Sisodia, 1998). The mutations may yield an increase in the amount of β-amyloid peptide (Aβ), a peptide that could increase GSK3 activity because it acts as an antagonist of insulin/IGF-1 receptor (Townsend et al., 2007). Also, an increase in intracellular calcium that can occur in AD can increase the activity of calpain protease and the activity not only of GSK3 (Goni-Oliver et al., 2007) but also of other tau kinases such as cdk5 (Patrick et al., 1999; Yoo and Lubec, 2001). Interestingly, it has been recently indicated that an intron, non-codifying GSK3β polymorphism shows association with AD (Schaffer et al., 2008).

Also, inflammatory cytokines such as interleukin 18 produced by activated microglia, an event occurring in AD, may increase tau phosphorylation by GSK3 or cdk5 (Ojala et al., 2008). Both kinases could also be activated in mice treated with 3,4-methylenedioxymethamphetamine ("Ecstasy") (Busceti et al., 2008).

On the other hand, it has been described that tau in polymer form can be phosphorylated by GSK3 to form large, tangle-like structures (Rankin et al., 2008).

3.5 Therapies to Prevent Tau Phosphorylation

Because tau phosphorylation could be toxic for neurons, some strategies have been developed to attempt to inhibit tau phosphorylation in a direct or in an indirect way (Schneider and Mandelkow, 2008). A direct way to inhibit tau kinases such as GSK3 could involve the use of iRNA (Yu et al., 2003) or the use of small molecules such as (for GSK3) lithium (Noble et al., 2005); indirubins (Leclerc et al., 2001); maleimides (Smith et al., 2001); paullones (Leost et al., 2000); thiazoles (Bhat et al., 2003), or thiadiazolidinones (Martinez et al., 2002).

Also, a reciprocal relationship between phosphorylation and O-GlcNAC modification of tau has been indicated (Yuzwa et al., 2008).

Indirect ways to inhibit tau phosphorylation have also been indicated. For example, it has been reported that use of folate decreases tau phosphorylation through its regulation of PP2A (Sontag et al., 2008) and that use of nicotinamide reduces tau phosphorylation through a mechanism involving sirtuin inhibition (Green et al., 2008). Also, inhibition of calpain (Trinchese et al., 2008), which could activate GSK3 (Goni-Oliver et al., 2007), may prevent tau phosphorylation. Inhibitors of PP2A inhibitors (Tsujio et al., 2005) could also be used for decreasing tau phosphorylation.

References

Alonso AC, Grundke-Iqbal I, Iqbal K (1996) Alzheimer's disease hyperphosphorylated tau sequesters normal tau into tangles of filaments and disassembles microtubules. Nat Med 2:783–787

Alonso AC, Li B, Grundke-Iqbal I, Iqbal K (2008) Mechanism of tau-induced neurodegeneration in Alzheimer disease and related tauopathies. Curr Alzheimer Res 5:375–384

Arrasate M, Perez M, Avila J (2000) Tau dephosphorylation at tau-1 site correlates with its association to cell membrane. Neurochem Res 25:43–50

Atlante A, Amadoro G, Bobba A, de Bari L, Corsetti V, Pappalardo G, Marra E, Calissano P, Passarella S (2008) A peptide containing residues 26–44 of tau protein impairs mitochondrial oxidative phosphorylation acting at the level of the adenine nucleotide translocator. Biochim Biophys Acta 1777:1289–1300

Avila J, Lucas JJ, Perez M, Hernandez F (2004) Role of tau protein in both physiological and pathological conditions. Physiol Rev 84:361–384

Bhat R, Xue Y, Berg S, Hellberg S, Ormo M, Nilsson Y, Radesater AC, Jerning E, Markgren PO, Borgegard T, Nylof M, Gimenez-Cassina A, Hernandez F, Lucas JJ, Diaz-Nido J, Avila J (2003) Structural insights and biological effects of glycogen synthase kinase 3-specific inhibitor AR-A014418. J Biol Chem 278:45937–45945

Braak H, Braak E (1991) Neuropathological stageing of Alzheimer-related changes. Acta Neuropathol 82:239–259

Brandt R, Leger J, Lee G (1995) Interaction of tau with the neural plasma membrane mediated by tau's amino-terminal projection domain. J Cell Biol 131:1327–1340

Brandt R, Hundelt M, Shahani N (2005) Tau alteration and neuronal degeneration in tauopathies: mechanisms and models. Biochim Biophys Acta 1739:331–354

Busceti CL, Biagioni F, Riozzi B, Battaglia G, Storto M, Cinque C, Molinaro G, Gradini R, Caricasole A, Canudas AM, Bruno V, Nicoletti F, Fornai F (2008) Enhanced tau phosphorylation in the hippocampus of mice treated with 3,4-methylenedioxymethamphetamine ("Ecstasy"). J Neurosci 28:3234–3245

Chen S, Li B, Grundke-Iqbal I, Iqbal K (2008) I1PP2A affects tau phosphorylation via association with the catalytic subunit of protein phosphatase 2A. J Biol Chem 283:10513–10521

Cleveland DW, Hwo SY, Kirschner MW (1977) Purification of tau, a microtubule-associated protein that induces assembly of microtubules from purified tubulin. J Mol Biol 116:207–225

Cohen P (1989) The structure and regulation of protein phosphatases. Annu Rev Biochem 58: 453–508

Delacourte A, David JP, Sergeant N, Buee L, Wattez A, Vermersch P, Ghozali F, Fallet-Bianco C, Pasquier F, Lebert F, Petit H, Di Menza C (1999) The biochemical pathway of neurofibrillary degeneration in aging and Alzheimer's disease. Neurology 52:1158–1165

Derkinderen P, Scales TM, Hanger DP, Leung KY, Byers HL, Ward MA, Lenz C, Price C, Bird IN, Perera T, Kellie S, Williamson R, Noble W, Van Etten RA, Leroy K, Brion JP, Reynolds CH, Anderton BH (2005) Tyrosine 394 is phosphorylated in Alzheimer's paired helical filament tau and in fetal tau with c-Abl as the candidate tyrosine kinase. J Neurosci 25: 6584–6593

Dotti CG, Banker GA, Binder LI (1987) The expression and distribution of the microtubule-associated proteins tau and microtubule-associated protein 2 in hippocampal neurons in the rat in situ and in cell culture. Neuroscience 23:121–130

Drewes G, Mandelkow EM, Baumann K, Goris J, Merlevede W, Mandelkow E (1993) Dephosphorylation of tau protein and Alzheimer paired helical filaments by calcineurin and phosphatase-2A. FEBS Lett 336:425–432

Engel T, Hernandez F, Avila J, Lucas JJ (2006a) Full reversal of Alzheimer's disease-like phenotype in a mouse model with conditional overexpression of glycogen synthase kinase-3. J Neurosci 26:5083–5090

Engel T, Lucas JJ, Gomez-Ramos P, Moran MA, Avila J, Hernandez F (2006b) Cooexpression of FTDP-17 tau and GSK-3beta in transgenic mice induce tau polymerization and neurodegeneration. Neurobiol Aging 27:1258–1268

Fellous A, Francon J, Lennon AM, Nunez J (1977) Microtubule assembly in vitro. Purification of assembly-promoting factors. Eur J Biochem 78:167–174

Goedert M, Satumtira S, Jakes R, Smith MJ, Kamibayashi C, White CL 3rd, Sontag E (2000) Reduced binding of protein phosphatase 2A to tau protein with frontotemporal dementia and parkinsonism linked to chromosome 17 mutations. J Neurochem 75:2155–2162

Goedert M, Spillantini MG, Potier MC, Ulrich J, Crowther RA (1989) Cloning and sequencing of the cDNA encoding an isoform of microtubule-associated protein tau containing four tandem repeats: differential expression of tau protein mRNAs in human brain. EMBO J 8:393–399

Gomez-Ramos A, Diaz-Hernandez M, Rubio A, Miras-Portugal MT, Avila J (2008) Extracellular tau promotes intracellular calcium increase through M1 and M3 muscarinic receptors in neuronal cells. Mol Cell Neurosci 37:673–681

Gong CX, Grundke-Iqbal I, Damuni Z, Iqbal K (1994a) Dephosphorylation of microtubule-associated protein tau by protein phosphatase-1 and -2C and its implication in Alzheimer disease. FEBS Lett 341:94–98

Gong CX, Grundke-Iqbal I, Iqbal K (1994b) Dephosphorylation of Alzheimer's disease abnormally phosphorylated tau by protein phosphatase-2A. Neuroscience 61:765–772

Goni-Oliver P, Lucas JJ, Avila J, Hernandez F (2007) N-terminal cleavage of GSK-3 by calpain: a new form of GSK-3 regulation. J Biol Chem 282:22406–22413

Green KN, Steffan JS, Martinez-Coria H, Sun X, Schreiber SS, Thompson LM, LaFerla FM (2008) Nicotinamide restores cognition in Alzheimer's disease transgenic mice via a mechanism involving sirtuin inhibition and selective reduction of Thr231-phosphotau. J Neurosci 28:11500–11510

Greenwood JA, Johnson GV (1995) Localization and in situ phosphorylation state of nuclear tau. Exp Cell Res 220:332–337

Grundke-Iqbal I, Iqbal K, Tung YC, Quinlan M, Wisniewski HM, Binder LI (1986) Abnormal phosphorylation of the microtubule-associated protein tau (tau) in Alzheimer cytoskeletal pathology. Proc Natl Acad Sci USA 83:4913–4917

Inoue M, Hirata A, Tainaka K, Morii T, Konno T (2008) Charge-pairing mechanism of phosphorylation effect upon amyloid fibrillation of human tau core peptide. Biochemistry 47:11847–11857

Ishiguro K, Kobayashi S, Omori A, Takamatsu M, Yonekura S, Anzai K, Imahori K, Uchida T (1994) Identification of the 23 kDa subunit of tau protein kinase II as a putative activator of cdk5 in bovine brain. FEBS Lett 342:203–208

Ishiguro K, Shiratsuchi A, Sato S, Omori A, Arioka M, Kobayashi S, Uchida T, Imahori K (1993) Glycogen synthase kinase 3 beta is identical to tau protein kinase I generating several epitopes of paired helical filaments. FEBS Lett 325:167–172

Ishihara T, Zhang B, Higuchi M, Yoshiyama Y, Trojanowski JQ, Lee VM (2001) Age-dependent induction of congophilic neurofibrillary tau inclusions in tau transgenic mice. Am J Pathol 158:555–562

Janssens V, Goris J (2001) Protein phosphatase 2A: a highly regulated family of serine/threonine phosphatases implicated in cell growth and signalling. Biochem J 353:417–439

Kimura T, Yamashita S, Fukuda T, Park JM, Murayama M, Mizoroki T, Yoshiike Y, Sahara N, Takashima A (2007) Hyperphosphorylated tau in parahippocampal cortex impairs place learning in aged mice expressing wild-type human tau. EMBO J 26:5143–5152

Kimura T, Fukuda T, Sahara N, Yamashita S, Murayama M, Mizoroki T, Yoshiike Y, Lee B, Sotiropoulos I, Maeda S, Takashima A (2010) Aggregation of detergent-insoluble tau is involved in neuronal loss but not in synaptic loss. J Biol Chem 285:38692–38699

Kosik KS, Orecchio LD, Bakalis S, Neve RL (1989) Developmentally regulated expression of specific tau sequences. Neuron 2:1389–1397

Leclerc S, Garnier M, Hoessel R, Marko D, Bibb JA, Snyder GL, Greengard P, Biernat J, Wu YZ, Mandelkow EM, Eisenbrand G, Meijer L (2001) Indirubins inhibit glycogen synthase kinase-3 beta and CDK5/p25, two protein kinases involved in abnormal tau phosphorylation in Alzheimer's disease. A property common to most cyclin-dependent kinase inhibitors? J Biol Chem 276:251–260

Lee G, Cowan N, Kirschner M (1988) The primary structure and heterogeneity of tau protein from mouse brain. Science 239:285–288

Lee G, Thangavel R, Sharma VM, Litersky JM, Bhaskar K, Fang SM, Do LH, Andreadis A, Van Hoesen G, Ksiezak-Reding H (2004) Phosphorylation of tau by fyn: implications for Alzheimer's disease. J Neurosci 24:2304–2312

Leost M, Schultz C, Link A, Wu YZ, Biernat J, Mandelkow EM, Bibb JA, Snyder GL, Greengard P, Zaharevitz DW, Gussio R, Senderowicz AM, Sausville EA, Kunick C, Meijer L (2000) Paullones are potent inhibitors of glycogen synthase kinase-3beta and cyclin-dependent kinase 5/p25. Eur J Biochem 267:5983–5994

Lu PJ, Wulf G, Zhou XZ, Davies P, Lu KP (1999) The prolyl isomerase Pin1 restores the function of Alzheimer-associated phosphorylated tau protein. Nature 399:784–788

Magnani E, Fan J, Gasparini L, Golding M, Williams M, Schiavo G, Goedert M, Amos LA, Spillantini MG (2007) Interaction of tau protein with the dynactin complex. EMBO J 26:4546–4554

Martinez A, Alonso M, Castro A, Perez C, Moreno FJ (2002) First non-ATP competitive glycogen synthase kinase 3 beta (GSK-3beta) inhibitors: thiadiazolidinones (TDZD) as potential drugs for the treatment of Alzheimer's disease. J Med Chem 45:1292–1299

Mawal-Dewan M, Henley J, Van de Voorde A, Trojanowski JQ, Lee VM (1994) The phosphorylation state of tau in the developing rat brain is regulated by phosphoprotein phosphatases. J Biol Chem 269:30981–30987

Neve RL, Harris P, Kosik KS, Kurnit DM, Donlon TA (1986) Identification of cDNA clones for the human microtubule-associated protein tau and chromosomal localization of the genes for tau and microtubule-associated protein 2. Brain Res 387:271–280

Noble W, Planel E, Zehr C, Olm V, Meyerson J, Suleman F, Gaynor K, Wang L, LaFrancois J, Feinstein B, Burns M, Krishnamurthy P, Wen Y, Bhat R, Lewis J, Dickson D, Duff K (2005) Inhibition of glycogen synthase kinase-3 by lithium correlates with reduced tauopathy and degeneration in vivo. Proc Natl Acad Sci USA 102:6990–6995

Ojala JO, Sutinen EM, Salminen A, Pirttila T (2008) Interleukin-18 increases expression of kinases involved in tau phosphorylation in SH-SY5Y neuroblastoma cells. J Neuroimmunol 205:86–93

Patrick GN, Zukerberg L, Nikolic M, de la Monte S, Dikkes P, Tsai LH (1999) Conversion of p35 to p25 deregulates Cdk5 activity and promotes neurodegeneration. Nature 402: 615–622

Perez M, Cuadros R, Smith MA, Perry G, Avila J (2000) Phosphorylated, but not native, tau protein assembles following reaction with the lipid peroxidation product, 4-hydroxy-2-nonenal. FEBS Lett 486:270–274

Perez M, Hernandez F, Gomez-Ramos A, Smith M, Perry G, Avila J (2002) Formation of aberrant phosphotau fibrillar polymers in neural cultured cells. Eur J Biochem 269:1484–1489

Perez M, Hernandez F, Lim F, Diaz-Nido J, Avila J (2003) Chronic lithium treatment decreases mutant tau protein aggregation in a transgenic mouse model. J Alzheimers Dis 5:301–308

Price DL, Sisodia SS (1998) Mutant genes in familial Alzheimer's disease and transgenic models. Annu Rev Neurosci 21:479–505

Ramesh Babu J, Lamar Seibenhener M, Peng J, Strom AL, Kemppainen R, Cox N, Zhu H, Wooten MC, Diaz-Meco MT, Moscat J, Wooten MW (2008) Genetic inactivation of p62 leads to accumulation of hyperphosphorylated tau and neurodegeneration. J Neurochem 106:107–120

Rankin CA, Sun Q, Gamblin TC (2008) Pre-assembled tau filaments phosphorylated by GSK-3b form large tangle-like structures. Neurobiol Dis 31:368–377

Reynolds CH, Garwood CJ, Wray S, Price C, Kellie S, Perera T, Zvelebil M, Yang A, Sheppard PW, Varndell IM, Hanger DP, Anderton BH (2008) Phosphorylation regulates tau interactions with Src homology 3 domains of phosphatidylinositol 3-kinase, phospholipase Cgamma1, Grb2, and Src family kinases. J Biol Chem 283:18177–18186

Santa-Maria I, Hernandez F, Smith MA, Perry G, Avila J, Moreno FJ (2005) Neurotoxic dopamine quinone facilitates the assembly of tau into fibrillar polymers. Mol Cell Biochem 278:203–212

Santacruz K, Lewis J, Spires T, Paulson J, Kotilinek L, Ingelsson M, Guimaraes A, DeTure M, Ramsden M, McGowan E, Forster C, Yue M, Orne J, Janus C, Mariash A, Kuskowski M, Hyman B, Hutton M, Ashe KH (2005) Tau suppression in a neurodegenerative mouse model improves memory function. Science 309:476–481

Schaffer BA, Bertram L, Miller BL, Mullin K, Weintraub S, Johnson N, Bigio EH, Mesulam M, Wiedau-Pazos M, Jackson GR, Cummings JL, Cantor RM, Levey AI, Tanzi RE, Geschwind

DH (2008) Association of GSK3B with Alzheimer disease and frontotemporal dementia. Arch Neurol 65:1368–1374
Schneider A, Mandelkow E (2008) Tau-based treatment strategies in neurodegenerative diseases. Neurotherapeutics 5:443–457
Shelanski ML, Gaskin F, Cantor CR (1973) Microtubule assembly in the absence of added nucleotides. Proc Natl Acad Sci USA 70:765–768
Smith DG, Buffet M, Fenwick AE, Haigh D, Ife RJ, Saunders M, Slingsby BP, Stacey R, Ward RW (2001) 3-Anilino-4-arylmaleimides: potent and selective inhibitors of glycogen synthase kinase-3 (GSK-3). Bioorg Med Chem Lett 11:635–639
Sontag JM, Nunbhakdi-Craig V, Montgomery L, Arning E, Bottiglieri T, Sontag E (2008) Folate deficiency induces in vitro and mouse brain region-specific downregulation of leucine carboxyl methyltransferase-1 and protein phosphatase 2A B(alpha) subunit expression that correlate with enhanced tau phosphorylation. J Neurosci 28:11477–11487
Steinhilb ML, Dias-Santagata D, Fulga TA, Felch DL, Feany MB (2007) Tau phosphorylation sites work in concert to promote neurotoxicity in vivo. Mol Biol Cell 18:5060–5068
Tortosa E, Santa-Maria I, Moreno F, Lim F, Perez M, Avila J (2009) Binding of Hsp90 to tau promotes a conformational change and aggregation of tau protein. J Alzheimers Dis 17:319–25
Townsend M, Mehta T, Selkoe DJ (2007) Soluble Abeta inhibits specific signal transduction cascades common to the insulin receptor pathway. J Biol Chem 282:33305–33312
Trinchese F, Fa M, Liu S, Zhang H, Hidalgo A, Schmidt SD, Yamaguchi H, Yoshii N, Mathews PM, Nixon RA, Arancio O (2008) Inhibition of calpains improves memory and synaptic transmission in a mouse model of Alzheimer disease. J Clin Invest 118:2796–2807
Trojanowski JQ, Schuck T, Schmidt ML, Lee VM (1989) Distribution of tau proteins in the normal human central and peripheral nervous system. J Histochem Cytochem 37:209–215
Tsujio I, Zaidi T, Xu J, Kotula L, Grundke-Iqbal I, Iqbal K (2005) Inhibitors of protein phosphatase-2A from human brain structures, immunocytological localization and activities towards dephosphorylation of the Alzheimer type hyperphosphorylated tau. FEBS Lett 579:363–372
Wang JZ, Gong CX, Zaidi T, Grundke-Iqbal I, Iqbal K (1995) Dephosphorylation of Alzheimer paired helical filaments by protein phosphatase-2A and -2B. J Biol Chem 270:4854–4860
Weingarten MD, Lockwood AH, Hwo SY, Kirschner MW (1975) A protein factor essential for microtubule assembly. Proc Natl Acad Sci USA 72:1858–1862
Wetzel MK, Naska S, Laliberte CL, Rymar VV, Fujitani M, Biernaskie JA, Cole CJ, Lerch JP, Spring S, Wang SH, Frankland PW, Henkelman RM, Josselyn SA, Sadikot AF, Miller FD, Kaplan DR (2008) p73 regulates neurodegeneration and phospho-tau accumulation during aging and Alzheimer's disease. Neuron 59:708–721
Williamson R, Scales T, Clark BR, Gibb G, Reynolds CH, Kellie S, Bird IN, Varndell IM, Sheppard PW, Everall I, Anderton BH (2002) Rapid tyrosine phosphorylation of neuronal proteins including tau and focal adhesion kinase in response to amyloid-beta peptide exposure: involvement of Src family protein kinases. J Neurosci 22:10–20
Yoo BC, Lubec G (2001) p25 protein in neurodegeneration. Nature 411:763–764;discussion 764–765
Yu JY, Taylor J, DeRuiter SL, Vojtek AB, Turner DL (2003) Simultaneous inhibition of GSK3alpha and GSK3beta using hairpin siRNA expression vectors. Mol Ther 7:228–236
Yuan A, Kumar A, Peterhoff C, Duff K, Nixon RA (2008) Axonal transport rates in vivo are unaffected by tau deletion or overexpression in mice. J Neurosci 28:1682–1687
Yuzwa SA, Macauley MS, Heinonen JE, Shan X, Dennis RJ, He Y, Whitworth GE, Stubbs KA, McEachern EJ, Davies GJ, Vocadlo DJ (2008) A potent mechanism-inspired O-GlcNAcase inhibitor that blocks phosphorylation of tau in vivo. Nat Chem Biol 4:483–490

Chapter 4
Tau Pathology

Nicolas Sergeant and Luc Buée

Abstract Tau pathology refers to molecular mechanisms leading to the intracellular aggregation of abnormally modified tau protein isoforms and to the propagation of this degenerating process along neuronal circuitry. Tau proteins belong to the family of microtubule-associated proteins. They are essential for the assembly of tubulin dimers into microtubules and their parallel-ordered organization. Tau is strongly expressed in neurons, localized in the axon, and is essential for neuronal architecture, plasticity, and network. From the gene to the protein, several post-transcriptional and posttranslational modifications regulate tau. At each of these levels, mechanisms can be deregulated and associated with the development of tau pathology such as in Alzheimer's disease and in several neurologic disorders – the so-called tauopathies. In animal models, overexpression of mutated tau proteins is often used to induce a tau pathology leading to motor and/or cognitive dysfunction. Tau pathology is certainly a good therapeutic target, but untangling tau remains a major therapeutic challenge.

Keywords Alzheimer's disease · Neurofibrillary degeneration · Microtubule-associated protein tau · Tauopathies · Tau pathology

Contents

L. Buée (✉)
Université Lille Nord de France, F-59000, Lille, France; Inserm, U837, 59045 Lille, France; UDSL, Faculté de Médecine, Institut de Médecine prédictive et de Recherche thérapeutique, Centre de Recherches Jean-Pierre Aubert, 59045 Lille, France; CHU-Lille, 59000 Lille, France
e-mail: luc.buee@inserm.fr

R.A. Nixon, A. Yuan (eds.), *Cytoskeleton of the Nervous System*,
Advances in Neurobiology 3, DOI 10.1007/978-1-4419-6787-9_4,

4.1 General Introduction

In Alzheimer's disease (AD), microtubule-associated protein tau is the major component of paired helical filaments inside degenerating neurons referred to as neurofibrillary tangles (NFTs). However, aggregation of tau inside neurons or glial cells is observed in more than 20 neurologic disorders, which together form the so-called tauopathies. The *Tau/MAPT* gene locus is associated with the development of frontotemporal dementia, and microdeletion in the locus region is associated with mental retardation. Besides its most well-known function as being a microtubule-polymerizing and -stabilizing factor, tau is implicated in the architecture and organization of the parallel order of microtubules and is also important for the regulation of axonal transport through the regulation of motor proteins and vesicular transport. When aggregated, one of the major questions that remain to be answered is whether there is a loss or gain of toxic functions. Recent development of transgenic animal models may be useful to address this question. In this chapter, several of those important aspects of tau biology will be addressed as well as the use of tau proteins as potential diagnostic markers.

4.2 Microtubule-Associated Tau

4.2.1 Tau/MAPT *Gene*

The human *Tau/MAPT* gene is unique and contains 16 exons located over more than 150 kb on the long arm of chromosome 17 at band position 17q21 (Neve et al., 1986; Andreadis et al., 1995) (Fig. 4.1b). The restriction analysis and sequencing of the gene shows that it contains two CpG islands, one associated with the promoter region, the other within exon 9 (Andreadis et al., 1995). The CpG island in the putative MAPT promoter region resembles a previously described neuron-specific promoter. Two regions homologous to the mouse Alu-like sequence are present. The sequence of the promoter region also reveals a TATA-less sequence that is likely to be related to the presence of multiple initiation sites, typical of housekeeping genes. Three SP1-binding sites that are important in directing transcription initiation in other TATA-less promoter are also found in the proximity of the first transcription initiation site (Andreadis et al., 1996). The SP1-binding sites are suggested to control neuronal-specific expression of tau (Heicklen-Klein and Ginzburg, 2000). Although the mechanism underlying the neuronal expression of tau is not known, the expression of Cre recombinase under the control of the murine tau promoter results in an almost complete restricted expression of the recombinase in neurons (Muramatsu et al., 2008). In human, tau expression is also found in muscle and scarcely in other tissues or organs (Castle et al., 2008). Several polymorphisms are identified and found to be in complete linkage disequilibrium with one another defining an extended haplotype that covers the entire *MAPT* gene (Schraen-Maschke et al., 2004) and even spans to a region covering ~1.8 Mb (Pittman et al., 2004) (Fig. 4.1a). The H2 haplotype is much more rare than the H1 haplotype (73–79%) in healthy individuals and results from H1 by the inversion of a ~970-kb segment (Stefansson et al., 2005). The H2 haplotype is associated with an increased risk of de novo microdeletion and developmental delay and learning disability (Shaw-Smith et al., 2006; Zody et al., 2008). Specific H1 sub-haplotypes are associated with the risk of developing AD, parkinsonian syndromes such as corticobasal degeneration, and progressive supranuclear palsy and likely influence the profile of tau isoform expression (Rademakers et al., 2005; Myers et al., 2007; Caffrey et al., 2008).

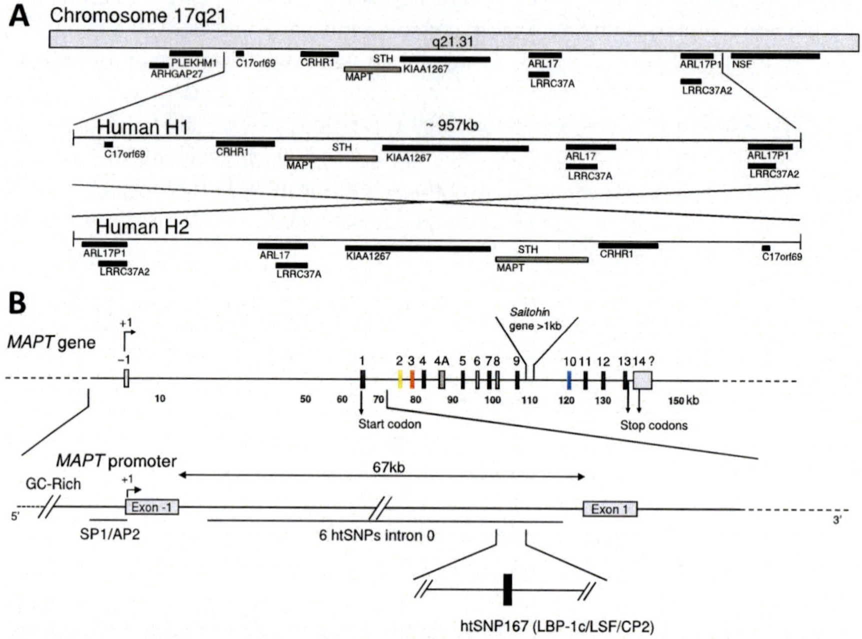

Fig. 4.1 Chromosome localization and microtubule-associated MAPT/tau gene. **a** Chromosome localization and haplotypes of tau. *Tau* gene is located on chromosome 17 at position q21.31. The locus of the human *Tau* gene occurs as two haplotypes H1 and the inverted version of H1, H2. NO recombination is observed between H1 and H2 over a region of approximately 1.5 Mb. Tau gene locus and the surrounding genes are represented. H1 and its ~970-kb inverted H2 haplotype are represented. **b** Tau gene. Tau gene spans more than 130 kb and is composed of 16 exons. Transcript initiation starts at exon −1, and the start codon is located in exon 1. They are two alternate stop codons located following exon 13 or inside exon 14. The 3′ ending of exon 14 is not completely characterized in humans. The initiation of transcription is indicated by +1. The promoter region of tau encompasses the sequence upstream exon −1, the exon −1 and the intron 0. Intron 0 contains many haplotype tagging single-nucleotide polymorphisms; among those, htSNP167 modulates tau promoter function depending on the allele. The promoter is GC-rich and contains SP1 and Activator protein-2 (AP2) binding-motifs proximal to exon −1. The SP1-binding sites are suggested to control neuronal specific expression of tau (Heicklen-Klein and Ginzburg, 2000). The A allele of the haplotype tagging SNP167 abolishes putative binding sites for LBP-1c/LSF/CP2. The G allele of htSNP167 promotes a higher expression of luciferase in mouse and human neuroblastoma cells (Rademakers et al., 2005). It is noteworthy that polymorphism in LBP-1c/LSF/CP2 gene is potentially a genetic determinant for AD (Lambert et al., 2000; Bertram et al., 2005)

4.2.2 Posttranscriptional Modifications

The tau primary transcript contains 16 exons (Fig. 4.2a). In the central nervous system (CNS), two transcripts of 2 and 6 kb arise from the use of two alternative polyadenylation sites: the 2 kb mRNA targets tau to the nucleus, and the 6 kb encodes the major form in axons (Andreadis, 2005; Gallo et al., 2007). In the human brain, tau exons 4A and 8 are skipped. Exon 4A is found in bovine, human, and

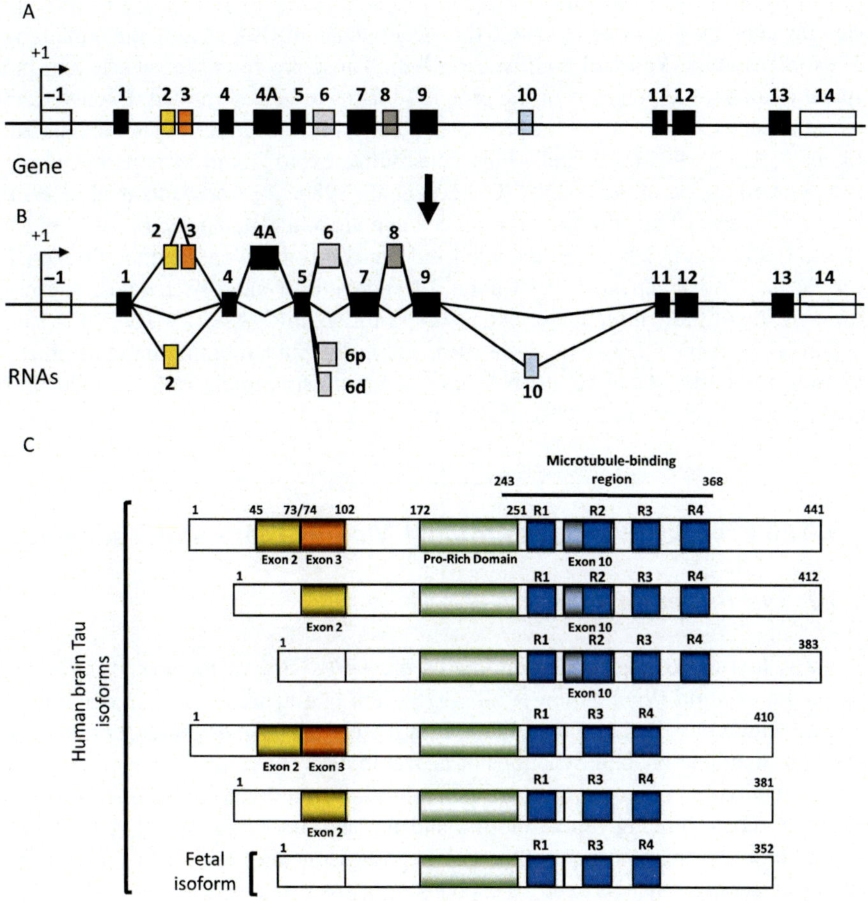

Fig. 4.2 Alternative splicing and tau proteins in the central nervous system. **a** Alternatively spliced cassettes. Several mRNAs are generated by alternative splicing of exons 2, 3, 4A, and 10. The 6p and 6d are cryptic splicing sites rarely used. The selection of those sites is supposed to generate carboxy-truncated tau proteins (Luo et al., 2004) (for review, see Andreadis, 2005). Exon 4A is excluded in the brain and included in the spinal cord and peripheral nervous system. Exon 8 is found included only in muscle and found also in other species such as bovine tau. **b** Tau protein isoforms in the central nervous system. In the human brain, six major tau isoforms are generated from the alterative splicing of exons 2, 3, and 10. Exon 3 is always included with exon 2. Exon 10 encodes an additional microtubule-binding motif numbered R1–R4 that together define the microtubule-binding region. Half of tau proteins contain three microtubule-binding motifs (*light gray*) and the other half (*dark gray*) has four microtubule-binding motifs. Upstream the microtubule-binding is the proline-rich domain containing several phosphorylation sites that regulate tau binding to microtubules

rodent peripheral tissues with a high degree of homology. Exon 8 is alternative and found in striated muscle spinal cord and pituitary (Castle et al., 2008). Notably, tau expression is 10-fold higher in brain and spinal cord than in skeletal muscle. Cryptic splicing sites are described in exon 6 that generate tau mRNA lacking the remaining 3′ exon cassettes (Wei and Andreadis, 1998). Those are found in muscle and the spinal cord, but the presence of the protein remains to be determined. Exon−1 and the adjacent intron are part of the promoter. Exon−1 is transcribed but not translated. Exons 1, 4, 5, 7, 9, 11, 12, and 13 are constitutive exons. Exon 14 is part of the 3′ untranslated region of tau mRNA (Goedert et al., 1989a, b; Andreadis et al., 1992). Exons 2, 3, and 10 are alternatively spliced and are adult brain specific (Andreadis et al., 1995). Exon 3 is never included independently of exon 2 although exon 2 can be included independently of exon 3 (Andreadis et al., 1992). Thus, alternative splicing of these three major exons allows for six mRNAs (2–3–10–; 2+3–10–; 2+3+10–; 2–3–10+; 2+3–10+; 2+3+10+), and in the human brain, the tau primary transcript is composed of six mRNAs translated into six protein isoforms (Goedert et al., 1989a, b; Himmler, 1989) (Fig. 4.2b).

4.3 Tau Proteins: Posttranslational Modifications and Functions

4.3.1 Tau Functions

Tau was first discovered in the middle of the 1970s as a factor promoting tubulin polymerization (Weingarten et al., 1975). Tau is a neuronal protein essentially located within the axonal compartment. Its structure makes it essential for the organization, stabilization, and dynamics of microtubules. The primary sequence of tau can be subdivided in an amino-terminal region followed by a proline-rich domain, the microtubule-binding repeat motifs, and the carboxy-terminal tail (Fig. 4.2b). Regarding the primary structure, the polypeptide sequences encoded by exons 2/3 add to tau acidity, whereas exon 10 encodes a positively charged sequence that adds to the basic character of tau. More generally, the amino-terminal region has a pI of 3.8 followed by the proline domain, which has a pI of 11.4, and the carboxy-terminal region is also positively charged with a pI of 10.8. Tau is rather a dipole with two domains with opposite charge, which can be modulated by posttranslational modifications. Structural analysis of human tau using several biophysical methods showed that in solution, tau behaves as an unfolded protein (Schweers et al., 1994). However, functions of tau are distributed both in the amino- and the repeat-domains.

The amino-terminal region together with the proline-rich domain is referred to as the *projection domain*. This unstructured and negatively charged region detaches from the surface of microtubules (Hirokawa et al., 1988) and can interact with the plasma membrane or cytoskeletal proteins (Brandt et al., 1995). Tau may also contribute to the spacing in between microtubule lattice and to the parallel-ordered organization of microtubules in axons (Chen et al., 1992). The spacing may be dependent upon the presence of additional amino-terminal sequences such as exons 2, 3, or 4A. The latter is included only in the spinal cord in peripheral nerve tissue (Georgieff et al., 1993). More recently, using a biophysical assay, Rosenberg and colleagues suggest that the amino-terminal regions of two tau molecules, each one individually binding to a microtubule, form an electrostatic “zipper” (Rosenberg et al., 2008). The amino-terminal region of tau also interacts with a growing panel of proteins including motor proteins such as kinesin-1 (Utton et al., 2005) and dynactin/dynein complex (Magnani et al., 2007), SH3 containing tyrosine kinases such as the phospholipase C-gamma 1, or the p85α subunit of PI3K (Reynolds et al., 2008). Recently, it has been established that tau regulates the motility of dynein and kinesin motor proteins by an isoform-dependent mechanism (Dixit et al., 2008). Indeed, the shortest tau isoform lacking exons 2, 3, and 10 impedes the motility of both kinesin and dynein, whereas the longest tau isoforms with all exons have less effect on motor protein motility (Dixit et al., 2008). Thus, the axonal transport of vesicles may be finely tuned by the ratio of tau isoforms expressed. In neurodegenerative disorders, a modified pattern of tau isoform expression/ratio, due to tau aggregation for instance, may profoundly affect the axonal transport and could possibly lead to neurodegeneration (Crosby, 2003). Interactors of tau proteins may also regulate tau function. Tau has been shown to interact with thioredoxin-like protein RdCVFL in the retinal neurons (Fridlich et al., 2009). One isoform of RdCVFL having the thioredoxin-like function likely regulates the phosphorylation and degradation of tau.

The carboxy-terminal region, which is the basic region of tau protein, is characterized by the presence of three or four repeat motifs, depending on the inclusion or not of the exon 10 encoding sequence. The repeat motifs corresponding with the microtubule-binding domains. Apart from binding to microtubules, the repeat-domains of tau interact with the histone deacetylase 6 (HDAC6), and HDAC6 is suggested to regulate tau phosphorylation. The inhibition of the proteasome activity by MG132 enhances the interaction of HDAC6 with tau. Interestingly, in this condition HDAC6–tau complexes are observed in perinuclear aggresomes (Ding et al., 2008). The proteasomal degradation of tau is also sensitive to chaperone and co-chaperones Hsc70 and Bcl2-associated athanogene-1 (BAG-1), respectively, which are interactors with tau proteins (Elliott et al., 2007; 2009). The repeat-domains also interact more strongly with apolipoprotein E3 than the E4 isovariant (Huang et al., 1995). Apolipoprotein E4 genotype is a major risk factor in AD (for review, see Lambert and Amouyel, 2007). The interaction of tau and ApoE is reduced by phosphorylation of tau at serine 262 residue. The functionality of the tau–ApoE interaction remains unknown. However, a triple transgenic mouse model of AD crossed with a knock-in ApoE4 mouse showed a strong influence of ApoE4 upon the topographical distribution of amyloid deposits, but surprisingly, the tau pathology was strongly reduced (Oddo et al., 2008). Together, tau interacting partners may profoundly regulate its function but also modulate the tau pathology. The physiologic and pathologic function of tau is also regulated by posttranslational modifications such as phosphorylation.

4.3.2 Tau Phosphorylation

Tau microtubule-associated proteins are phosphoproteins (Butler and Shelanski, 1986). There are 85 potential phosphorylation sites on the longest tau isoform (see Fig. 4.3). Phosphorylation sites were characterized using phospho-dependent tau

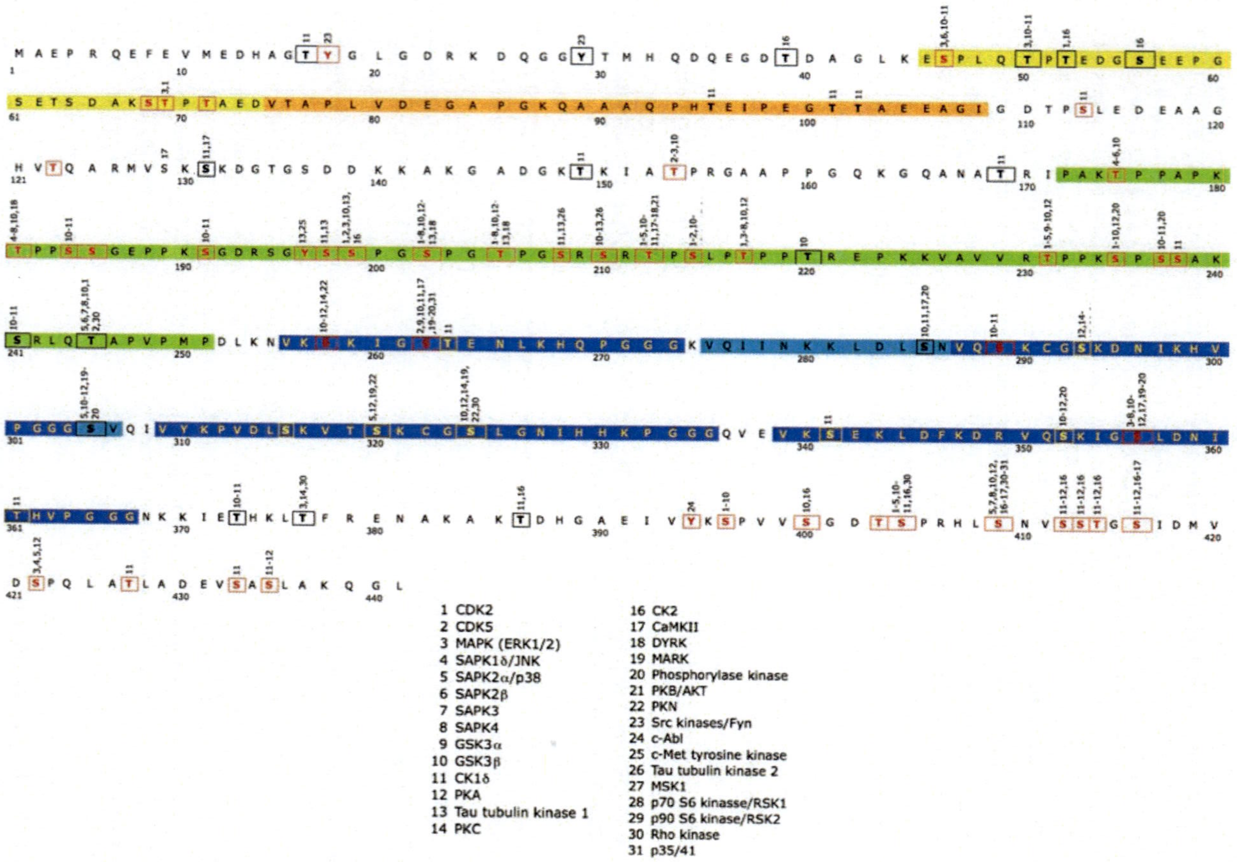

Fig. 4.3 Phosphorylation sites on tau and PHF-tau. The longest tau isoform sequence is represented with the encoded sequences of exons 2, 3, and 10 in *yellow*, *orange*, and *light blue*, respectively. In *dark blue* are the repeat domains of tau. The phosphorylated sites characterized are in *bold*, and those in *red* were found on aggregated tau obtained from AD brain tissue. The kinases known to phosphorylate tau in vitro or in vivo are numbered, and the serines, threonines, or tyrosines that are phosphorylated by those kinases are indicated by the number position above the amino acid

antibodies, phospho-peptide mapping, mass spectrometry, or NMR. According to the latest extensive analysis of tau phosphorylation (Hanger et al., 2007) and that of a previously published review (Buee et al., 2000), 71 among the 85 putative phosphorylation sites can be phosphorylated in physiologic or pathologic conditions (for review, see Hanger et al., 2009). Most of the phosphorylation sites surround the microtubule-binding domains in the proline-rich region and carboxy-terminal region of tau. A total of more than 20 protein kinases can phosphorylate tau proteins. This include four groups of protein kinases: (1) the proline-directed protein kinases (PDPK), which phosphorylate tau on serines or threonines that are followed by a proline residue. This group includes cyclin-dependent kinase 2 and 5 (Baumann et al., 1993; Vincent et al., 1997; Hamdane et al., 2003a), mitogen-activated protein kinase (MAPK) (Holzer et al., 2001), and several stress-activated protein kinases (SAPKs) (Goedert et al., 1997; Buee-Scherrer and Goedert, 2002). (2) The non-PDPK group includes tau-tubulin kinases 1 and 2 (Tomizawa et al., 2001; Sato et al., 2006; Kitano-Takahashi et al., 2007), casein kinases 1, 2, and 1δ (Greenwood et al., 1994; Hanger et al., 2007), Dual specificity tyrosine-phosphorylation-regulated kinase 1A (DYRK1A) (Drewes et al., 1997; Woods et al., 2001; Drewes, 2004; Liu et al., 2008), the phosphorylase kinase (Paudel, 1997), Rho kinase (Amano et al., 2003), and protein kinase A (PKA), protein kinase B (PKB/AKT), protein kinase C (PKC), protein kinase N (PKN) (for review, see Sergeant et al., 2008). (3) The third group includes protein kinases that phosphorylate tau on serine or threonine residues followed or not by a proline. Glycogen protein kinases (GSK3α and GSK3β) belong to this group and have a recognition motif (SXXXS or SXXXD/E) (Hanger et al., 1992). This group also includes mitogen and stress-activated protein kinase MSK1, which belongs to the AGC kinases group AGC kinases group (for protein kinases A, G, and C). AGC kinases preferentially phosphorylate serine and threonine residues that lie in RXRXXS/T motifs (Virdee et al., 2007). The S6 kinases p70 and p90 (RSK1 and RSK2) also phosphorylate tau as well as serum- and glucocorticoid-induced kinase-1 (SGK1) (Virdee et al., 2007). (4) The fourth group corresponds with tyrosine protein kinases such as Src kinases, C-abl, and c-met (Kinoshita et al., 2008). (The phosphorylation sites for each kinase are represented on the longest tau isoform in Fig. 4.3; see also Diane Hanger's Web site, http://cnr.iop.kcl.ac.uk/hangerlab/tautable.)

These different groups share or have specific structural functions. The PDPK and GSK families are involved in modulation of tau binding to microtubules (Thr231 and Ser396) and also in the pathologic process of aggregation (for review, see Buee et al., 2000). The second group includes a number of kinases involved in signal transduction where tau may act as a linker or modulator. In contrast, microtubule-affinity regulating kinase (MARK) kinases are strictly involved in the regulation of tau binding to microtubules by phosphorylating specific motifs (KXGS) within the microtubule-binding domains (R1–R4 encoded by exons 9–12) (Mandelkow et al., 2004). Finally, regarding the fourth group, studies have determined that human tau Tyr18 and Tyr29 are phosphorylated by the src family tyrosine kinase fyn (Williamson et al., 2002; Lee et al., 2004). The same proline-rich region of tau proteins is likely involved in the interaction with phospholipase C-γ (PLC-γ) isozymes (Hwang et al., 1996; Jenkins and Johnson, 1998). Hwang and colleagues have demonstrated in vitro that tau proteins complex specifically with the SH3 domain of PLC-γ and enhance PLC-γ activity in the presence of unsaturated fatty acids such as arachidonic acid. These results suggest that in cells that express tau proteins, receptors coupled to cytosolic phospholipase A2 may activate PLC-γ indirectly, in the absence of the usual tyrosine phosphorylation, through the hydrolysis of phosphatidylcholine to generate arachidonic acid (Hwang et al., 1996; Jenkins and Johnson, 1998). Altogether, these data indicate that tau proteins may also play a role in the signal transduction pathway involving PLC-γ (for review, see Rhee, 2001; Rhee, 2001). Interestingly, tau is phosphorylated by c-met (Kinoshita et al., 2008), the tyrosine kinase receptor of hepatocyte growth and scatter factor (HGF/SF) (Sergeant et al., 2000). This tyrosine receptor is also proteolyzed by the same proteases as the amyloid protein precursor (APP) (Foveau et al., 2009). Yet, it remains unknown whether HGF/SF could also modulate tau function.

Tau-tubulin kinase 1 (TTBK1) phosphorylates tau on tyrosine 197 and therefore belongs to the fourth group of tau protein kinases (Sato et al., 2006). TTBK1 shows high similarity with tau-tubulin kinase 2 (TTBK2) that is supposed to phosphorylate tau at serine 208 and 210 as TTBK1 (Tomizawa et al., 2001; Kitano-Takahashi et al., 2007). Notably, a very recent study from Holden and collaborators has discovered a mutation of *TTBK2* gene associated with spinocerebellar ataxia of type 11. Neuropathologic examination of the brain demonstrated the accumulation of tau in degenerating neurons of the medullary tegmentum of brain stem (Houlden et al., 2007). TTBK1 and TTBK2 belong to the casein kinase 1 (CK1) phylogenetic tree, and TTBK1 shows more than 30% similarity with CK1δ. Although the phosphorylation sites were characterized using mass spectrometry, TTBK1 and CK1δ do not phosphorylate tau at identical residues. Tau is phosphorylated by CK1δ on 37 residues, not including a tyrosine residue. Only serines 198, 208, and 210 are phosphorylatable by TTBK1 and CK1δ.

Finally, kinases may also regulate the posttranscriptional maturation of tau RNAs. Thus, several kinases, including GSK3β, DYRK1A, and CLK2 or p25/CDK5 indirectly modulate the splicing of tau exon 10 through the phosphorylation of splicing factors (Hartmann et al., 2001; Hernandez et al., 2004; Shi et al., 2008). Phosphorylation or dephosphorylation of tau may also contribute to the cell localization of tau. Phosphorylation of tau by GSK3β regulates its axonal transport by reducing its interaction with kinesin (Cuchillo-Ibanez et al., 2008). In sharp contrast, dephosphorylated tau is located to the cell nucleus (Loomis et al., 1990) and is suggested to contribute to nucleolar organization (Sjoberg et al., 2006) and may also contribute to chromosome stability (Rossi et al., 2008).

4.3.3 Tau Prolyl Isomerization

The peptidyl-prolyl cis/trans isomerase Pin1 isomerizes the peptide bond of a phosphorylated-serine or phosphorylated-threonine followed by a proline (pSer-/pThr-Pro). Pin1 potentiates the cis-/trans-isomerization of its substrates, consequently changing the accessibility of the phospho site for further dephosphorylation, ultimately leading to a modified activity of the substrate or degradation. Through isomerization of pSer-/pThr-Pro, Pin1 therefore regulates a number of proteins including mitotic and proapoptotic proteins. Pin1 consists of two functional domains. The amino-terminal region is a group IV WW domain that specifically binds to pSer-/pThr-Pro motifs. The carboxyl-terminal region is the catalytic domain that isomerizes the peptidyl-prolyl bond between pSer-/pThr and proline (for reviews, see Landrieu et al., 2006; Lippens et al., 2007). During neuronal differentiation, Pin1 may be a key actor because its expression is strongly increased at both mRNA (Zhang et al., 2004) and protein levels (Hamdane et al., 2006). Pin1 levels are also high in mature neurons (Galas et al., 2006; Hamdane et al., 2006).

Pin1 is of particular interest in AD because it was first described as a new regulator of phosphorylation and conformation of tau proteins (Lu et al., 1999). In vitro, Pin1 would bind to two pThr-Pro motifs (Thr212 and Thr231) (for reviews, see Landrieu et al., 2006; Lippens et al., 2007). Pin1 may isomerize peptide bonds and facilitate dephosphorylation of these pThr-Pro motifs in trans-conformation by protein phosphatase 2A (Zhou et al., 2000). Thus, in vitro, the dephosphorylation of prephosphorylated tau by brain extracts obtained from Pin1 knocked-out mice is attenuated when compared with that of control brain extracts (Yotsumoto et al., 2009). Notably, when using mutated tau instead of wild-type tau, the lack of Pin1 had no effect (see later). In situ, in SY5Y human neuroblastoma cells and mouse primary neuronal cortical cultures, Pin1 facilitates a differential dephosphorylation of tau at pThr231 (Galas et al., 2006; Hamdane et al., 2006). Furthermore, some hypotheses state that Pin1 could regulate some cell cycle markers that are altered in AD (Hamdane et al., 2003b; Dourlen et al., 2007). Thus, all of these data suggest a potential neuroprotective function of Pin1 against AD but likely not in frontotemporal dementia linked to tau mutations. However, Pin1 is oxidized and downregulated in the hippocampus of AD (Sultana et al., 2006) and patients with mild cognitive impairment (Butterfield et al., 2006). Furthermore, soluble Pin1 is depleted in AD brains (Lu et al., 1999), and a subcellular mislocalization of Pin1 is observed in AD and frontotemporal dementias (Lu et al., 1999; Thorpe et al., 2001; Holzer et al., 2002; Thorpe et al., 2004). Thorpe and collaborators hypothesized that the mislocalization of Pin1 could be due to an excessive amount of hyperphosphorylated tau (Thorpe et al., 2001). An abnormal regulation of Pin1 in AD might participate in the disease process.

4.3.4 Tau O-glycosylation

Phosphorylation is regulated by kinases and phosphatases, but growing evidence shows that phosphorylation of serine or threonine residues interplay with O-glycosylation. O-glycosylation of cytosolic proteins is a dynamic and abundant posttranslational modification that is characterized by the addition of an O-linked *N*-acetylglucosamine (O-GlcNAc) residue on serine or threonine in the vicinity of Pro residues (Haltiwanger et al., 1997) by an O-GlcNAc transferase (Kreppel et al., 1997). Although the functional significance of O-GlcNAcylation is not yet fully understood, it is implicated in transcriptional regulation, protein degradation, cell activation, cell cycle regulation, and the proper assembly of multimeric protein complexes (for review, see Hart, 1997). It occurs in neurofilaments (Dong et al., 1997) and in microtubule-associated proteins including microtubule-associated protein 2 (MAP2) and tau proteins (Lefebvre et al., 2003b). The number of O-GlcNAcylated sites on tau proteins is lower than the number of phosphorylation sites. Site-specific or stoichiometric changes in O-GlcNAcylation may modulate tau function. In fact, phosphorylation and O-GlcNAcylation may have opposite effects. For instance, O-GlcNAcylation of tau proteins and other microtubule-associated proteins suggest a role for O-GlcNAc in mediating their interactions with tubulin. Alternatively, O-GlcNAcylation could be implicated in the nuclear shuttling of hypophosphorylated tau proteins. Thus, O-GlcNAcylation may also play a role in the subcellular localization and degradation of tau proteins (for reviews, see Lefebvre et al., 2003a). However, as for phosphorylation, it would be essential to determine precisely the serine or threonine residues on tau that can be O-GlcNAcylated by the O-GlcNAc transferase. Phosphorylation and O-GlcNAcylation are both labile upon ionization likely explaining the challenge to identify the sites that are targets for both phosphorylation and N-GlcNAcylation. Nuclear magnetic resonance (NMR) could be an alternate approach to investigate further the O-glycosylation of tau and help to decipher the structure–function relationship of N-GlcNAcylation of tau.

4.3.5 Other Posttranslational Modifications

Apart from phosphorylation and glycosylation, tau proteins have also been shown to exhibit ubiquitination (Morishima-Kawashima et al., 1993), nitration (Mailliot et al., 2002), glycation, and modification by advanced glycation end-products (AGEs) (Munch et al., 2003). Those posttranslational modifications were essentially evidenced on aggregated tau (for review, see Avila et al., 2004).

4.4 The Microtubule Assembly Domain

Tau proteins bind microtubules through repetitive regions in their C-terminal part (Fig. 4.2b). These repetitive regions are the repeat domains (R1–R4) encoded by exons 9–12 (Lee et al., 1989). The three (3R) or four (4R) copies are made of a highly conserved 18-amino-acid repeat (Lee et al., 1988; Goedert et al., 1989a; Himmler, 1989; Lee et al., 1989) separated from each other by less conserved 13- or 14-amino-acid inter-repeat domains. Tau proteins are known to act as promoters of tubulin polymerization in vitro (Weingarten et al., 1975; Cleveland et al., 1977a, b; Brandt and Lee, 1993). They have been shown to increase the rate of microtubule polymerization and to inhibit the rate of depolymerization (Drechsel et al., 1992). Tau also protects microtubules from being severed by severing proteins such katanin (Qiang et al., 2006).

The binding of tau to microtubules implicates the repeat domains. The 18-amino-acid repeats bind to microtubules through a flexible array of distributed weak sites (Lee et al., 1989; Butner and Kirschner, 1991). Tau isoforms with 4R (R1–R4) are more efficient at promoting microtubule assembly than the fetal isoform with 3R

(R1, R3, R4) (Goedert and Jakes, 1990; Butner and Kirschner, 1991; Gustke et al., 1994). Interestingly, the most potent polypeptide to induce dimers of tubulin polymerization is the inter-region between repeats 1 and 2 (R1–R2 inter-region) and more specifically peptide 274KVQIINKK281 within this sequence. This R1–R2 inter-region is unique to 4R tau (because it occurs between exons 9 and 10), adult specific, and responsible for differences in the binding affinities between 3R and 4R tau (Goode and Feinstein, 1994). Using a cysteine mutant of the repeat domain of tau enabling nanogold-labeling of the microtubule-binding domain, tau protein is shown to decorate the inner surface of the microtubule close to the Taxol (Paclitaxel) binding site on -tubulin (Kar et al., 2003a, b).

Recent evidence supports a role for the microtubule-binding domain in the modulation of the phosphorylation state of tau proteins. A direct and competitive binding has been demonstrated between residues 224–236 (according to the numbering of the longest isoform) and microtubules on one hand and residues 224–236 and protein phosphatase 2A (PP2A) on the other hand (Sontag et al., 1999). As a consequence, microtubules could inhibit PP2A activity by competing for binding to tau at the microtubule-binding domains.

4.5 Tau Pathology

4.5.1 Tau in Neurofibrillary Degeneration

Neurofibrillary degeneration, as first described by Aloïs Alzheimer, consists of the intraneuronal accumulation of proteinaceous fibrils forming flame-shaped neurons. Fibrils were later demonstrated to correspond with paired helical filaments (PHF) indicating the well-structured organization of the molecular constituent (Kidd, 1964). This material is insoluble, and this biochemical property was used to isolate the material and to generate antibodies. The major antigen of PHF was shown to correspond with tau protein, namely PHF-tau (Brion et al., 1985). Similarly, antibodies raised against isolated PHF stained tau proteins (Delacourte and Defossez, 1986). Proteomic isolation and characterization of the most aggregated components of PHF showed that the core of PHF was mostly composed of the microtubule-binding motif lacking the exon 10 encoding sequences (Kondo et al., 1988; Wischik et al., 1988; Jakes et al., 1991).

The most striking difference between postmortem tau and PHF-tau proteins is the molecular weight, as tau protein derived from postmortem control individuals is resolved as 6 main bands (45–67 kDa), whereas PHF-tau comprises four main bands between 60 and 74 kDa. Two-dimensional gel electrophoresis followed by Western blotting is a useful method to demonstrate that PHF-tau is more acidic than normal tau (Ksiezak-Reding et al., 1988; Bretteville et al., 2009). Increased tau acidity is due to hyper- and abnormal phosphorylation in degenerating neurons. A number of studies, using phosphatase, antibodies against PHF-tau (thereafter shown to be phospho-dependent tau antibodies) and mass spectrometry analyses, together demonstrated that PHF-tau is hyperphosphorylated on "native" phospho-sites and abnormally phosphorylated on pathologically related sites (phosphorylation sites on PHF-tau corresponds with the red "P" in Fig. 4.3). The former observation was demonstrated using phospho-dependent antibodies and two-dimensional gel electrophoresis on human brain tissue obtained from tumor resections and, therefore, without postmortem/surgery delay (Matsuo et al., 1994; Sergeant et al., 1995). Abnormal phospho-sites were demonstrated using antibodies that recognize PHF-tau and not normal or "native" tau, such as AT100 or AP422, which recognize the phosphorylated threonines 212–214 and serine 217 or the serine 422, respectively. Nonetheless, to generate the pathologic phospho-sites at AT100 epitope, a sequential phosphorylation by two kinases (Yoshida and Goedert, 2006) is necessary suggesting that several kinases may be deregulated in AD. It is also interesting to note that the AT100 epitope includes a Pin1 binding site (see earlier).

4.6 Tau Pathology Spreading from "Normal" Aging to Alzheimer's Disease

The prospective and multidisciplinary study of tau pathology in multiple human brain regions either by neuropathologic or biochemical approaches has shown that with aging, the entorhinal cortex and hippocampal formation are prone to develop a tau pathology. Phenotypically, this tau pathology is not different from the one observed in AD. It is characterized by a quadruplet of abnormal tau proteins and is systematically present in variable amounts in the entorhinal and hippocampal regions of nondemented patients aged over 75 years (Vermersch et al., 1992; Lace et al., 2009). Tau pathology affects the cholinergic system in AD, but the nucleus basalis of Meynert, which contains most cholinergic axonal projections to the cortex, is the only basal forebrain structure displaying an age-related tau pathology (Mesulam et al., 2004).

In AD disease, tau pathology is found in brain areas affected with aging, and several additional brains areas are also affected. However, the tau pathology follows cortico-cortical connections with a stereotyped, sequential, and hierarchical pathway affecting large pyramidal neurons. Braak and Braak proposed a six-stage spatiotemporal progression of tau pathology in AD (Braak and Braak, 1991). A similar scheme of spatiotemporal spreading into 10 stages is also described according to the brain regions sequentially affected: transentorhinal cortex (S1), entorhinal (S2), hippocampus (S3), anterior temporal cortex (S4), inferior temporal cortex (S5), mid temporal cortex (S6), polymodal association areas (prefrontal, parietal inferior, temporal superior) (S7), unimodal areas (S8), primary motor (S9a) or sensory (S9b, S9c) areas, and all neocortical areas (S10). Up to stage 6, the disease could be asymptomatic. In all cases of our study, stage 7 individuals with two polymodal association areas affected by tau pathology were cognitively impaired (Delacourte et al., 1999).

Yet, the cortical factors that trigger the cortical spreading of tau pathology in AD remain ill-defined (Delacourte et al., 1999; Duyckaerts et al., 1999; Delacourte et al., 2002). However, recent data suggest that neurofibrillary degeneration cortical spreading could follow a transmissible prion-like process. In fact, purified aggregates of PHF-tau were purified from transgenic mice. Intracranial injection of this preparation was done in a mouse model that overexpresses human tau protein but does not display tau pathology. After the injection, development of a tau pathology was observed. This pathology progressed from the injection site to neighboring brain structures (Clavaguera et al., 2009).

4.7 Tau Pathology Classification

In several neurodegenerative disorders, referred to as tauopathies, abnormally and hyperphosphorylated tau proteins aggregate. Comparative biochemistry of the tau aggregates shows that they differ in both tau isoform phosphorylation and content, which enables a molecular classification of tauopathies. In previous reviews, we proposed a classification composed of five classes of tauopathies, defined depending on the type of tau aggregates that constitute the "bar code" for neurodegenerative disorders (Delacourte et al., 2007; Sergeant et al., 2008). This classification contains four classes because the class 0 corresponding with frontal lobe dementia of non-Alzheimer, non-Pick type is genetically linked to mutations in the progranulin gene (Baker et al., 2006; Cruts et al., 2006). Brain morphologic changes comprise a neuronal cell loss, spongiosis, and gliosis mainly in the superficial cortical layers of the frontal and temporal cortex. No tau aggregates are observed although a loss of tau protein expression is observed in this disorder (Delacourte and Buee, 1997; Zhukareva et al., 2001, 2003). In contrast, ubiquitin-positive nuclear inclusion composed of TDP-43/TAR DNA-binding Protein (TARDBP) is a characteristic neuropathologic lesion of this disease, which is considered a novel proteinopathy (for review, see Forman et al., 2007). The relationship between progranulin loss of function and intranuclear inclusion of TDP-43 and loss of tau protein expression remains unknown.

4.7.1 Class I: All Brain Tau Isoforms Are Aggregated

Class I is characterized by a pathologic tau quartet at 60, 64, and 69 kDa and a minor pathologic tau at 72/74 kDa. This pathologic tau quartet corresponds with the aggregation of the six tau isoforms (Goedert et al., 1992; Hanger et al., 1992; Sergeant et al., 1997a, b). The pathologic tau 60 is composed of the shortest tau isoform (2–3–10–). The pathologic tau 64 and 69 are each composed of two tau isoforms. Tau isoforms with either the exon 2 or exon 10 alone compose the pathologic tau 64. The pathologic tau 69 is made of tau isoforms either with exon 2+10 or exon 2+3. The longest tau isoform containing exons 2, 3, and 10 (2+3+10) constitute the 72/74 kDa pathologic component, as determined by two-dimensional gel electrophoresis coupled to Western blotting using exon-specific tau antibodies (Sergeant et al., 1997a). This typical tau profile was first characterized in AD but now includes nine additional neurologic disorders listed on Fig. 4.3 (for review, see Buee et al., 2000).

4.7.2 Class II: Tau Isoforms Containing the Exon 10 Encoding Sequence Aggregate

The class II profile is characterized essentially by the aggregation of tau with four microtubule-binding domains (Fig. 4.3). This pathologic tau profile is observed in corticobasal degeneration (CBD), argyrophilic grain dementia (AGD), progressive supranuclear palsy (PSP), and frontotemporal dementia linked to chromosome 17 due to tau gene mutations (Sergeant et al., 1999; Tolnay et al., 2002). PSP is a late-onset atypical parkinsonian disorder described by Steele, Richardson, and Olszewski in 1964 (Richardson et al., 1963; Steele et al., 1964). CBD was first described in 1967 and referred to as corticodentatonigral degeneration with neuronal achromasia (Rebeiz et al., 1967, 1968). It is a rare, sporadic, and slowly progressive late-onset neurodegenerative disorder characterized clinically by cognitive disturbances and extrapyramidal motor dysfunction. AGD was described in 1987, when Braak and co-workers reported a series of eight patients with a non-Alzheimer, late onset dementia (Braak and Braak, 1987). Clinically, AGD is characterized by behavioral disturbances such as personality change and emotional imbalance, as well as memory and cognitive impairment. The Consortium for Frontotemporal Lobar Degeneration is giving an update on clinical and neuropathologic criteria for the pathology diagnosis in neurologic disorders grouped as frontotemporal lobar degeneration (Cairns et al., 2007).

A series of 10 polymorphisms within the tau gene, including 8 single nucleotide polymorphism (SNP) and 1 deletion, have been found to be inherited in complete linkage disequilibrium with each other and with the dinucleotide polymorphism 116507$(TG)_n$ defining two extended haplotypes (H1 and H2) that cover at least the entire tau gene (Baker et al., 1999; for review, see Schraen-Maschke et al., 2004). The H1/H1 haplotype is more frequent in PSP/CBD patients than in controls or other tauopathies (Baker et al., 1999; Hutton, 2000; Rademakers et al., 2004). H1/H1 may be more frequent in AGD patients than in controls, though recent studies have failed to establish a statistically significant difference (Ishizawa et al., 2002; Miserez et al., 2003).

4.7.3 Class III: Tau Isoforms Lacking the Exon 10 Encoding Sequence Aggregate

Class III of tauopathies includes Pick's disease and autosomal dominant inherited frontotemporal dementia with parkinsonism linked to chromosome 17 (FTDP-17) (Fig. 4.3). Pick's disease is a rare form of neurodegenerative disorder characterized by a progressive dementing process. Early in the clinical course, patients show signs of frontal disinhibition (Brion et al., 1991). Neuropathologically, Pick's disease is characterized by the presence of typical spheroid inclusions in the soma of neurons called Pick bodies (Buee Scherrer et al., 1996; Delacourte et al., 1996). Pick bodies are labeled by tau antibodies, with a higher density in the hippocampus than in the neocortex. The pathologic tau profile of Pick's disease contrasts with that of class II tauopathies, with the pathologic tau isoforms consisting essentially of the 3R tau isoforms. In addition, aggregated tau proteins in Pick's disease are not detected by the monoclonal antibody 12E8 raised against the phosphorylated residue Ser262/Ser356, whereas in other neurodegenerative disorders, this phosphorylation site is detected (Delacourte et al., 1996; Probst et al., 1996). The lack of phosphorylation at Ser262 and 356 sites is likely to be related to either a kinase inhibition in neurons that degenerate in Pick's disease or an absence of these kinases within degenerating neurons (Mailliot et al., 1998).

4.7.4 Class IV: Tau Isoform Lacking Exons 2, 3, and 10 Principally Aggregate

Class IV is represented by myotonic dystrophy of type I and type II (Fig. 4.3). Myotonic dystrophy (DM) is the most prevalent form of adult-onset muscular dystrophy. Many systems including the central nervous system (cognitive and neuropsychiatric impairments), the heart (cardiac conduction defects), the genital tract (testicular atrophy), eyes (cataracts), ears (deafness), the gastrointestinal tract (smooth muscle), and the endocrine system (insulin resistance and type 2 diabetes) are affected, thus leading to a wide and variable complex panel of symptoms (Harper, 1997; Reardon and Harper, 1992). DMs are inherited autosomal dominant disorders caused by dynamically instable CTG or CCTG expansions in the 3′UTR of *DPMK* (Brook et al., 1992) or the first intron of *ZNF9* in DM1 and DM2, respectively (Ranum et al., 1998). Cognitive impairment, including memory, facial-emotion recognition, visuospatial recall, verbal scale, with cortical atrophy of the frontal and the temporal lobe and white matter lesions are often described in both DM1 and DM2 (Meola, 2000; Meola and Sansone, 2007; Sansone et al., 2007). Neuropathologic lesions, such as neurofibrillary tangles, have been observed in adult DM1 individuals over 50 years of age (Yoshimura, 1990; Yoshimura et al., 1990; Vermersch et al., 1996). The pathologic tau profile of DM1 is characterized by a strong pathologic tau band at 60 kDa and to a lesser extent a pathologic tau component at 64 and 69 kDa. This typical pathologic tau profile is reflected by a reduced number of tau isoform expression in the brain of individuals with DM1, at both the protein and mRNA levels (Sergeant et al., 2001). In addition, tau protein expression is also altered in transgenic mice with human DM1 locus (Seznec et al., 2001; Gomes-Pereira et al., 2007). The analysis of multiple brain regions of one genetically confirmed DM2 patient aged 71 years showed some neurofibrillary degenerating processes. Using specific immunologic probes against amino acid sequences corresponding to exons 2 and 3, the neurofibrillary lesions were shown to be devoid of tau isoforms with N-terminal inserts (Maurage et al., 2005). An altered splicing of tau characterized by a reduced expression of tau isoforms containing the N-terminal inserts characterizes both DM1 and DM2. Overall, it demonstrates that the central nervous system is affected and that DMs are real tauopathies. The direct relationship between the altered splicing of tau and neurofibrillary degeneration in DM remains to be established. Indeed, such an altered splicing of tau is commonly observed in FTDP-17 and considered reminiscent of neurofibrillary degeneration and tauopathies.

4.8 Inherited Frontotemporal Dementia Linked to Chromosome 17 and Tau Gene Mutations

Historically, frontotemporal dementia (FTD) was often classified as a form of Pick's disease, even when Pick cells or Pick bodies were not found (Constantinidis et al., 1974). However, this denomination may involve different subgroups of pathologies, and the Lund and Manchester groups published in 1994 a consensus on Clinical and Neuropathological Criteria for Frontotemporal Dementia. This publication clarified the position of Pick's disease within FTD, and several of the reported cases of familial Pick's disease were probably cases of familial FTD. Indeed, it is difficult to ascertain families that have the classic pathologic features of Pick's disease in the literature (Brown, 1992) because they often have unusual clinical features.

In 1994, Wilhelmsen and colleagues described an autosomal dominantly inherited disease related to familial FTD, characterized by adult-onset behavioral disturbances, frontal lobe dementia, parkinsonism, and amyotrophy (Wilhelmsen et al., 1994). They demonstrated a genetic linkage between this pathology, designated disinhibition–dementia–parkinsonism–amyotrophy complex (DDPAC), and chromosome 17q21-22 (Wilhelmsen et al., 1994). Since then, several families sharing strong clinical and pathologic features and for which there is a linkage with chromosome 17q22-22 have been described (Wijker et al., 1996; Bird et al., 1997; Heutink et al., 1997; Murrell et al., 1997). They have been included in a group of pathologies referred to as frontotemporal dementia with parkinsonism linked to chromosome 17 (FTDP-17) (Foster et al., 1997).

Although clinical heterogeneity exists between and within the families with FTDP-17, usual symptoms include behavioral changes, loss of frontal executive functions, language deficit, and hyperorality. Parkinsonism and amyotrophy are described in some families but are not consistent features. Neuropathologically, brains of FTD patients exhibit an atrophy of frontal and temporal lobes, severe neuronal cell loss, gray and white matter gliosis, and a superficial laminar spongiosis. One of the main characteristics is the filamentous pathology affecting the neuronal cells, or both neuronal and glial cells in some cases. The absence of amyloid aggregates is usually established (Foster et al., 1997; Spillantini et al., 1998c).

Fig. 4.4 (continued) intracytoplasmic tau aggregation in astrocytes (*arrow, left panel*). Tau E2 and tau E10 show glial tau pathology as well, astrocytic (*upper right, arrow*) or oligodendrial (*arrowhead, lower right*). The "globoid" neurofibrillary tangles are seen here in the substantia nigra (*upper right, asterisk*) and are stained by both tau E2 and tau E10. **c** Tau isoforms with three microtubule-binding domains are found in Pick's disease and few FTDP-17, corresponding with class III diseases. Pick bodies are stained by anti-pTau and tau E2 (*arrows* showing spherical 10-μm bodies) but not by tau E10 (in the *lower right panel*, a Pick body is readily seen but unstained, *asterisk*). **d** At last, class IV is characterized by the aggregation of tau isoforms lacking sequences encoded by exons 2 and 3. To date, the only known class IV diseases are DM1 and DM2. On the *left panel*, aggregates of hyperphosphorylated tau are labeled with AD2. On the *right panel*, only unphosphorylated, nonaggregated normal tau are stained by anti–tau E2, which in turn labels no aggregate

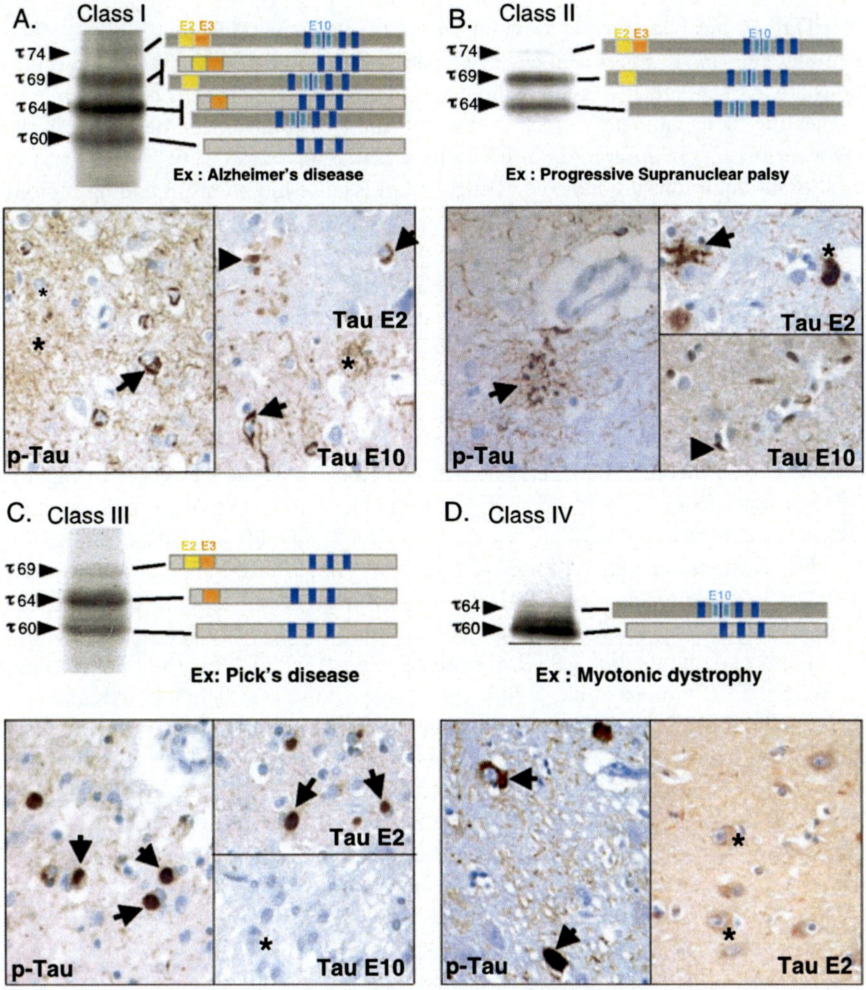

Fig. 4.4 The "bar code" of neurodegenerative diseases. Aggregated tau proteins from the brain tissue of patients suffering from different neurodegenerative disorders were resolved by one-dimensional gels. Four main patterns of tau bands were observed, and the isoform content was determined using tau exon-specific antibodies or two-dimensional gel electrophoresis. A classification is proposed according to tau isoforms composing four biochemical patterns. **a** Class I, which encompasses the largest number of degenerative diseases with tau aggregation, is characterized by the aggregation of six tau isoforms. This class is characterized by the occurrence on the brain samples of the hallmarks of AD, that is, (1) intraneuronal somatic neurofibrillary tangles (NFTs) as shown on the *left panel* by an antibody against pSer396–404 of tau (AD2, *arrows*); (2) neuropil threads (NTs), corresponding with neuritic processes filled with aggregated tau (*asterisks*); and (3) neuritic plaques (NPs). Antibodies raised against the sequences coded either by exon 2 (tau E2, *upper right panel*) or by exon 10 (tau E10, *lower right*) stain similarly NFTs (*arrows*), NTs (*asterisks*), and NPs (*arrowhead*) in AD and in other class I tauopathies. **b** Only tau isoforms containing four microtubule-binding domains aggregate in Class II disorders. They are represented here by a PSP case. AD2 anti-pTau labels the so-called astrocytic tufts, corresponding with

FTDP-17 has been related to mutations on the tau gene (Dumanchin et al., 1998; Hutton et al., 1998; Poorkaj et al., 1998; Spillantini and Goedert, 2001). Tau mutations always segregate with the pathology and are not found in the control subjects, suggesting their pathogenic role. To date, 17 mutations have been described in the tau gene among the different families with cases diagnosed as FTDP-17 (Fig. 4.4). A total of 44 mutations are indexed. Thirty-one missense mutations in coding regions R5H (Hayashi et al., 2002), R5L (Poorkaj et al., 2002), K257T (Pickering-Brown et al., 2000; Rizzini et al., 2000), I260V (Grover et al., 2003), L266V (Kobayashi et al., 2003), G272V (Hutton et al., 1998; Spillantini et al., 1998c), N279K (Delisle et al., 1999; Hasegawa et al., 1999; Yasuda et al., 1999; Wszolek et al., 2000), N296H (Iseki et al., 2001) and deltaN 296 (Yoshida et al., 2002), G303V (Ros et al., 2005), P301L (Heutink et al., 1997; Dumanchin et al., 1998; Poorkaj et al., 2001), P301S (Bugiani et al., 1999; Sperfeld et al., 1999), P301T (Llado et al., 2007), S305N (Hasegawa et al., 1999; Iijima et al., 1999), S305I (Kovacs et al., 2008), S305S (Stanford et al., 2000), L315R (Rosso et al., 2003; van Herpen et al., 2003), K317M (Zarranz et al., 2005), S320F (Rosso et al., 2002), G335V (Neumann et al., 2005), G335S (Spina et al., 2007; Ghetti et al., 2008), Q336R (de Silva et al., 2006), V337M (Hasegawa et al., 1998), E342V (Lippa et al., 2000), S352L (Nicholl et al., 2003), K369I (Neumann et al., 2001), G389R (Murrell et al., 1999), R406W (Hutton et al., 1998), N424K (A. Brice, personal communication), T427M (Giaccone et al., 2005), three silent mutation L284L (D'Souza et al., 1999), N296N (Spillantini et al., 2000), S305S (Stanford et al., 2000), two single amino acid deletions ΔK280 and ΔN296 (D'Souza et al., 1999; Rizzu et al., 1999), and nine intronic mutations in the splicing region following exon 10 at position +3 (Spillantini et al., 1998c), + 11, +12, +13, +14 (Hutton et al., 1998), +16 (Goedert et al., 1999), +19, +29, and +33 have been reported (Stanford et al., 2003) (Fig. 4.5). More recently, a tau gene deletion encompassing exons 6–9 has been described in a French family and associated with the development of FTD (Rovelet-Lecrux et al., 2009).

4.9 Mutations Affecting Tau Splicing

Depending on their functional effects, tau protein mutations may be divided into two groups: mutations affecting the alternative splicing of exon 10, and leading to changes in the proportion of 4R- and 3R-tau isoforms; and mutations modifying tau interactions with microtubules. In patients with FTDP-17 mutations affecting splicing include intronic mutations in the splicing region following exon 10 (+3, +13, +14, +16) and some missense mutations. Intronic mutations disturb a stem loop structure in the 5′ splice site of exon 10 that stabilizes this region of the pre-mRNA (Hutton et al., 1998; Spillantini et al., 1998c; Grover et al., 1999; Varani et al., 1999; Stanford et al., 2003). Sequence analysis of this splicing region in different animals indicates that the lack of the stem loop structure is associated with an increase in tau mRNAs containing exon 10 (Grover et al., 1999). Indeed, without this stem loop, access of U1snRNP to this site may be facilitated, increasing the formation of exon

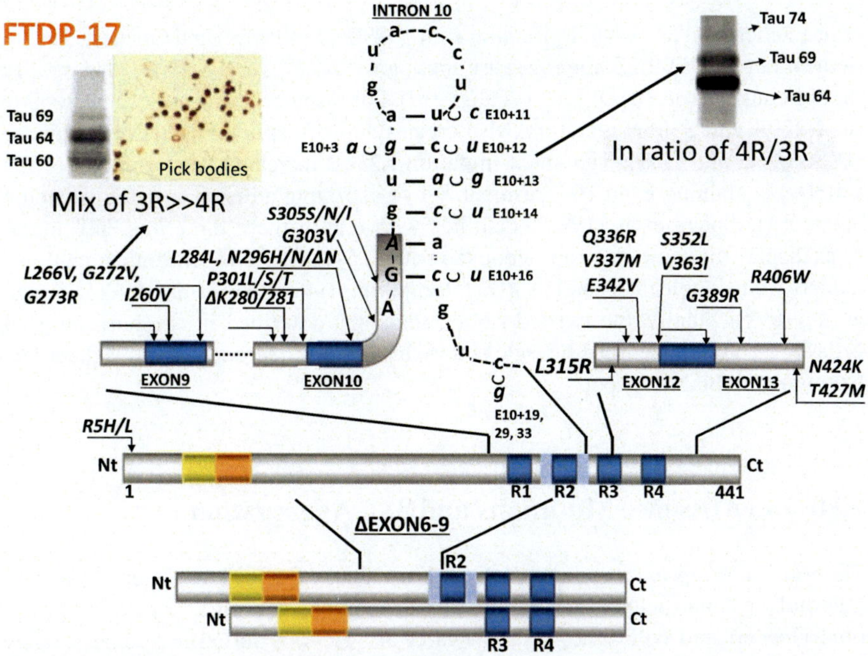

Fig. 4.5 Schematic representation of the tau gene mutations and deletion involved in FTDP-17. There are currently 44 *Tau* gene mutations characterized. *Tau* gene mutations located either in exon 10 and outside the second repeat domain R2 (N279K, ΔK280/281, P301L/P/S/T, L284L, N296H/N/ΔN, S305N/S/I) (slashed frame) or in intron 10 (E10+3, +11, +12, +13, +14, +16, +19) (*dotted line*) affect the alternative splicing of exon 10. Intronic mutations modify the regulatory elements of the splice site or disturb the stem loop structure in the 5′ end of the site, respectively. The final effect is a modification of the 4R-tau/3R-tau proportion leading to PSP or CBD Western-blot profile. Patients bearing the mutation G272V develop a hereditary form of Pick's disease with Pick bodies and increased inclusion of tau protein lacking the exon 10 encoding sequence (Bronner et al., 2005). Mutations located in tau region common to all six isoforms, outside exon 10 (R5H/L, K257T, I260V, L266V, G272V, G273R, L315R, K317M, S320F, G335S, Q336R, V337M, E342V, S352L, V363I, K369I, G389R, R406W, N424K, T427M) (Giaccone et al., 2005), affect the function of the six tau proteins in microtubules. The missense mutations located in the sequence of exon 10 corresponding with the R2 domain (P301L, P301S) (*black box*) affect only the 4R-tau binding. The mutation ΔK280 modulates the ratio 4R-tau/3R-tau and also affects the interaction with the microtubules. Mutation R406W modifies the phosphorylation of tau in residues surrounding the mutation. The *Tau* gene deletion Δexon 6–9 produces tau proteins with a lower or lack of affinity for microtubules depending on the presence or not of exon 10 (Rovelet-Lecrux et al., 2009). Mutations located in exons are indicated as peptidic sequence (framed part), and intronic mutations are indicated as nucleotidic sequence (*dotted part*). Updates of *Tau* gene mutations and polymorphisms are referenced in Christine Van Broeckhoven's laboratory and Marc Cruts' Web site (http://www.molgen.ua.ac.be/ADMutations)

10+ tau mRNAs and thus the 4R-tau isoform (D'Souza et al., 1999; Grover et al., 1999; Varani et al., 1999). Interestingly, in these families, abnormally phosphorylated 4R-tau isoforms aggregate into filaments and display a tau electrophoretic profile similar to the major tau doublet at 64 and 69 kDa found in PSP and CBD

(Buee Scherrer et al., 1996; Spillantini et al., 1998a; Spillantini and Goedert, 1998b; Sergeant et al., 1999). Some missense mutations (N279K and S305N) also modify the splicing of exon 10 (D'Souza et al., 1999). For instance, the change in nucleotide for N279K and S305N mutations also creates an exon-splicing enhancer sequence (D'Souza et al., 1999). The silent mutation L284L increases the formation of tau mRNAs containing exon 10, presumably by destroying an exon splicing silencing element (D'Souza et al., 1999). Families with one of these three missense mutations display the same electrophoretic tau pattern as those having intronic mutations, namely a tau doublet at 64 and 69 kDa (Hong et al., 1998; Reed et al., 1998; D'Souza et al., 1999). Finally, the twisted ribbon filaments described in neurons and glial cells are a common neuropathologic feature in all of the neurodegenerative disorders belonging to this first group.

4.10 Tau Missense Mutations and Tau Aggregation

The second group of tau mutations found in FTDP-17 affects tau interaction with microtubules and includes several missense mutations as well as a deletion mutation. Goedert and colleagues (Hasegawa et al., 1998) reported the consequences of G272V, P301L, P301S, V337M, G389R, and R406W mutations in an in vitro system of microtubule assembly. They showed that mutated tau isoforms bind microtubules to a lesser extent than wild-type isoforms. They suggest that the mutated isoforms may induce microtubule disassembly (Hasegawa et al., 1998; Goedert et al., 1999). These data were confirmed by a number of studies (Hong et al., 1998; D'Souza et al., 1999; Yen et al., 1999b). When missense mutations are located in tau regions common to all isoforms, outside exon 10 (G272V, V337M, G389R, R406W), the six tau isoforms do not bind properly to microtubules. These proteins aggregate into paired helical filaments and straight filaments similar to those described in AD and are present in neuronal cells. Their biochemical characterization shows a tau electrophoretic profile similar to the AD tau-triplet. Conversely, when missense mutations are located in exon 10 (P301L, P301S), 4R-tau isoforms are affected, and these do not bind to microtubules, but instead aggregate into twisted ribbon filaments. This type of filamentous inclusion is described in both neurons and glial cells. The biochemical characterization shows a tau electrophoretic profile similar to the major tau doublet encountered in PSP and CBD.

The ΔK280 mutation is particular, because it modifies the ratio 4R-tau/3R-tau ratio and could also affect the interaction between tau and microtubules. This mutation may decrease the formation of tau mRNAs containing exon 10 and thus enhance the formation of 3R-tau isoforms. Moreover, this deletion mutation is also responsible for a considerably reduced ability of tau to promote microtubule assembly and is stronger than the effect of the P301L mutation. Finally, this mutation has a strong proaggregating effect in both cell and animal models (Barghorn et al., 2000; Mocanu et al., 2008). No data are currently available on the biochemistry of tau aggregates in this family (D'Souza et al., 1999; Rizzu et al., 1999).

Mutations of the tau gene and their involvement in FTDP-17 emphasize the fact that abnormal tau proteins may play a central role in the etiopathogenesis of neurodegenerative disorders, without any implication of the amyloid cascade. The functional effects of the mutations suggest that a reduced ability of tau to interact with microtubules may be upstream of hyperphosphorylation and aggregation. These mutations may also lead to an increase in free cytoplasmic tau (especially the 4R-tau isoforms), therefore facilitating their aggregation into filaments (Yen et al., 1999b). Finally, some mutations may have a direct effect on tau fibrillogenesis (Chiti et al., 2003).

4.10.1 Tau Missense Mutations and In Vivo Microtubule Stabilization

Regarding the functional effects of tau mutants, it was shown that mutations G272V, ΔK280, P301L, P301S, and V337M cause a decreased ability of tau to promote microtubule assembly in vitro (Hasegawa et al., 1998). However, the magnitude of the observed effects and the reported rank order of potency of individual mutations have been variable. Overexpression of mutant tau in transfected cells has given inconsistent results as far as effects on microtubule binding and stability are concerned (Hasegawa et al., 1999; Rizzu et al., 1999; Yen et al., 1999a). Such studies are confounded by the problem that high expression of mutant tau may override any effects that are present at more physiologic levels.

For in vivo investigation, recombinant tau 3R, 4R, or mutated tau are injected in *Xenopus* oocytes, and maturation of oocytes is used as an indicator of microtubule function (Delobel et al., 2002). Normal oocyte maturation can be visualized by the appearance of a white spot, which indicates that oocyte meiosis driven by microtubules is normal. Wild-type 4R tau inhibits maturation of oocytes in a concentration-dependent manner, whereas 3R tau has no effect. These data suggest that there is a direct interaction of 4R tau and microtubules that interferes with oocyte maturation. Whatever the concentration of the 4R mutant tau (G272V, P301L, P301S, and V337M) injected in the oocyte, it fails to affect oocyte maturation although few changes in the organization of meiotic spindles are observed. In contrast with wild-type 4R tau, these mutations are likely to reduce the interaction of tau with microtubules. Two additional mutations (R406W and S305N) have been analyzed and found to perturb oocyte maturation. Differential phosphorylation cannot explain the data obtained for the first group of tau mutants because wild-type 3R and 4R tau are found to be phosphorylated at the same extent as the G272V, P301L, P301S, and V337M 4R tau mutants. Therefore, these mutations strongly reduce the ability of tau to interact with microtubules (Delobel et al., 2002).

Regarding the R406W mutation, there are some controversial effects: state of phosphorylation and reduced microtubule binding. At low concentrations, R406W tau mutant strongly interferes with oocyte maturation even when no effect is observed with wild-type 4R tau. Conversely, at high concentrations similar to those of injected wild-type 4R tau that block maturation, oocyte maturation is never completely abolished. Altogether, these data suggest that at low concentrations, the R406W mutant tau is less phosphorylated and thus shows a better microtubule binding, whereas at high concentrations, the phosphorylation state is not sufficient to thwart the reduction of microtubule binding of this tau mutant (Delobel et al., 2002).

The most pronounced inhibitory effect on oocyte maturation is observed with the S305N tau, even following microinjections of small amounts of tau protein. This effect is independent of phosphorylation. It was shown that S305N tau has a slightly increased ability to promote microtubule assembly in vitro compared with wild-type protein (Hasegawa et al., 1999).

Altogether, our data demonstrate that in vivo tau missense mutations either strongly reduce interactions with microtubules or increase these interactions. The observed phenotype is dependent on the combined effect of tau phosphorylation and concentration. It also demonstrates that the *Xenopus* oocyte is an interesting heterologous model system for following perturbations in microtubule function.

4.11 Tau Catabolic Products as Biomarkers of Tau Pathology

The presence of tau in cerebrospinal fluid (CSF) was first described in 1993 (Vandermeeren et al., 1993) and subsequently independently confirmed by several groups using different tau antibodies (Mori et al., 1995; Motter et al., 1995; Zemlan et al., 2003). A first attempt to characterize tau in CSF via Western blot was done by Johnson et al. (1997) and further confirmed and extended by both Sjogren and Borroni (Sjogren et al., 2001; Borroni et al., 2007). Tau in CSF consists essentially of two polypeptides of 55 kDa and 33 kDa. The 55-kDa band contains a large domain of tau extending from the N-terminal region (amino acid 159) to the carboxy-terminal region (amino acid 432). In contrast, the 33-kDa band shares a similar amino-terminal epitope with that of the 55-kDa band but is truncated at the C-terminus of tau. Together, those studies suggest that tau in CSF essentially consists of partially proteolyzed fragments of tau, mainly at the carboxy-terminus of tau. Those tau fragments were shown to be present in the brain tissue. Zemlan and collaborators using C-terminal tau antibodies (Zemlan et al., 2003) also evidenced a 14-kDa fragment of tau. Using our C-terminal tau antibody, a C-terminal fragment of tau is also observed in the cerebral cortex. Therefore, it cannot be excluded that several other proteolytic fragments of tau are present in the CSF, and for a complete characterization of these, a large panel of tau antibodies should be used. The use of phospho-dependent antibodies showed that, in CSF, tau polypeptides are phosphorylated at many sites, such as serine 181, 199 (Hampel et al., 2004), 202 (Maccioni et al., 2006), 231 (Hampel et al., 2004; 2005), 396, and 404 (Hu et al., 2002). Together these data show that microtubule-associated tau is recovered in the CSF and that several phospho-epitopes are present (for review, see Blennow, 2004).

4.12 Modeling Tau Pathology

4.12.1 Aged Mammal Individuals

As seen above, tau pathology is encountered in many neurodegenerative disorders in humans but also in aging. In some non-human mammals, tau pathology is also encountered with aging. For instance, hyperphosphorylated tau proteins (AT8 immunoreactivity) were identified in baboon, spectacled bear, bison, guanaco, Campbell's guenon, rabbit, reindeer, and rhesus monkey. These neuropathologic lesions were striking in old bisons with a clear neuritic pathology. Moreover, in old rhesus monkeys, NFTs were identified in the hippocampal formation and entorhinal cortex, which are known as starting points for tangle spreading in the cortex of AD patients (Hartig et al., 2000). Finally, similar data are also observed in aged baboons with an age-related NFT progression close to that encountered in humans (Schultz et al., 2000). Altogether, the data suggest that tau pathology may be encountered not only in great apes but also in other aged mammals. Nevertheless, the use of such models for the understanding of etiopathogenic factors involved in tau pathology will be difficult, but they remain valuable models for validating present and future tau pathology therapeutics. Moreover, they can be valuable to evaluate tau pathology brain imaging tracers.

4.13 Overexpression Models in Non-mammal Individuals

4.13.1 *Caenorhabditis elegans*

This nematode has a short life span. Because a large number of genetic mutants can be made and are available, it is also an easy tool for studying genetic modifiers (Kraemer et al., 2006). Pan-neuronal overexpression of tau [wild-type and FTDP-17 or pseudo-hyperphosphorylated mutants (Ser/Thr sites mutated in Glu residues)] was induced to model tauopathy. It leads to a progressive uncoordinated locomotion (Unc), which is characteristic of nervous system defects in worms. Hyperphosphorylated tau proteins accumulate leading to axonal vacuolization and subsequent neurodegeneration. FTDP-17 and pseudo-hyperphosphorylated tau mutants lead to more severe axonopathy and Unc phenotype (presynaptic defect) compared with that in lines overexpressing wild-type tau (Kraemer et al., 2003; Brandt et al., 2009).

In *C. elegans*, the most recent data suggest that tau pathology may lead to specific interference with intracellular mechanisms of axonal outgrowth and pathfinding.

4.13.2 *Drosophila melanogaster*

This fruit fly has been widely used for studying genetic interactors of different neurodegenerative disorders. Regarding tauopathies, different groups developed fly models by overexpressing wild-type and mutant forms of human tau. Transgenic flies showed key features of these human disorders: adult onset, progressive neurodegeneration, early death, enhanced toxicity of mutant tau, accumulation of abnormal tau, and relative anatomic selectivity. However, neurodegeneration often occurred without the neurofibrillary tangle formation that is seen in human disease and some rodent tauopathy models. These fly models allow for a genetic analysis of the cellular pathways underlying tau neurotoxicity including wingless pathway, kinases/phosphatases equilibrium, actin-binding proteins, microtubule dynamics, and so forth (Wittmann et al., 2001; Jackson et al., 2002; Shulman and Feany, 2003).

Among the kinases, GSK3β, mammalian target of rapamycin (mTOR), and protease-activated receptor-1 (Par-1) (MARK) have been involved in tau pathology in fly models, whereas cdk5 has no effect (Mudher et al., 2004; Nishimura et al., 2004). For instance, it is possible in a fly model to reverse the pathologic phenotype (axonal transport defects and behavior) by inhibiting GSK3β (Mudher et al., 2004). Altogether, these data should allow for the identification of a tau phosphorylation pattern involved in neurotoxicity.

Another interest in fly models is to better understand tau biology through its proteolysis or its binding partners. For instance, the role of proteasome has been investigated, and actin-binding proteins were identified in genetic screens in fly models of tauopathies (Blard et al., 2006, 2007).

Altogether, although they lack real tau aggregation, fly models may allow for a better understanding of tau biology in neurodegeneration.

4.13.3 *Danio rerio*

Due to their optical transparency and small size, zebrafish larvae are well suited for in vivo imaging and developmental gene modifier identification. Only two studies were reported on zebrafish tau models. Tomasiewicz et al. reported that FTDP-17 mutant form of human tau expressed in zebrafish neurons produced a cytoskeletal disruption that closely resembled the NFT in human disease (Tomasiewicz et al., 2002). Paquet et al. generated fluorescently labeled tau transgenic zebrafish, which rapidly recapitulated key pathologic features of tauopathies, including phosphorylation and conformational changes of human tau protein, tangle formation, neuronal and behavioral disturbances, and cell death (Paquet et al., 2009). This model is particularly useful for drug screening.

4.13.4 *Petromyzon marinus*

A unique cellular model of tau filament/NFT formation for studying the cytopathologic changes that accompany chronic overexpression in human was also developed. This cellular model of tau-induced NFT consists of giant neurons [anterior bulbar cells (ABCs)] in the hindbrain of the ammocoete sea lamprey, *Petromyzon marinus*, which have been induced to overexpress human tau. With this system, overexpression of the shortest human tau isoform can cause the incorporation of human tau into filaments, tau hyperphosphorylation, and gross degenerative changes. Moreover, this sequence of degenerative changes is spatiotemporally correlated with the appearance of several AD-related phosphoepitopes. These results are reminiscent of cellular degeneration sequences proposed for human NFT (Hall et al., 2001).

4.14 Tau Transgenic Mice

4.14.1 Tau Knockout Mice

Tau expression was suppressed in different mouse models by MAPT deletion or invalidation (Harada et al., 1994; Dawson et al., 2001; Tucker et al., 2001). All of them appear physically normal and are able to reproduce. They do not display any change in central or peripheral nervous systems (Dawson et al., 2001; Tucker et al., 2001). In one model, slight changes were seen in axonal diameter (Harada et al., 1994), and motor and cognitive deficits were also reported (Ikegami et al., 2000). Moreover, delay in axonal maturation was also reported in primary neuronal cell cultures from tau-deficient mice (Dawson et al., 2001). Tau deficiency is likely to be compensated by other microtubule-associated proteins such as MAP1A.

4.14.2 Wild-Type Tau Transgenic Mice

When 3R-tau isoforms are overexpressed in transgenic mice, hyperphosphorylated tau mainly accumulates in spinal cord neurons. The main pathology observations were axonal degeneration, diminished number of microtubules, and reduced axonal transport (Brion et al., 1999; Ishihara et al., 1999). Similar data were observed in transgenic mice expressing the longest human brain tau isoform under the control of the human Thy-1 promoter. Hyperphosphorylated human tau protein was present in nerve cell bodies, axons, and dendrites (Gotz et al., 1995; Spittaels et al., 1999; Probst et al., 2000).

Interestingly, transgenic mice bearing the human MAPT mini-gene were also developed (Duff et al., 2000). This model overexpresses human tau proteins by 2- to 3-fold higher than murine tau but it does not exhibit any tau pathology. However, surprisingly, when this model was crossbred with the murine MAPT gene deletion background model (Tucker et al., 2001), the offsprings displayed tau pathology in a time-dependent manner suggesting that murine tau proteins may act as inhibitors of tau aggregation (Andorfer et al., 2005). Independently, crossbreeding between other tau-deficient mice (Dawson et al., 2001) and transgenic mice with MAPT mini-gene was also used for understanding tau splicing in early developmental stages (from 3R to 4R isoforms) (McMillan et al., 2008). There is no data yet available on aged animals in these strains.

4.14.3 Mutated Tau Transgenic Mice

With the discovery of FTDP-17 mutations on MAPT, numerous transgenic models using these mutations were developed. They all allow for the development of tau pathology characterized by tau aggregation and neurofibrillary degeneration. Tau transgene is under various promoters [2′,3′-cyclic nucleotide 3′-phosphodiesterase, calcium calmodulin protein kinase II (CaMKII), platelet-derived growth factor (PDGF), prion, or Thy1.2) under an inducible system or not.

They display various phenotypes with the most prominent one being motor deficits. In Table 4.1, we summarize the general phenotypes of these models. In the next section, we describe four representative mutated tau transgenic mice: rTg4510, $Tau_{RD}/\Delta K280$, K3, and Thy-Tau22 (Santacruz et al., 2005; Schindowski et al., 2006; Ittner et al., 2008; Mocanu et al., 2008), which allow for better understanding of tau pathology.

In rTg4510 Tau transgenic mice, P301L mutant tau protein is expressed in an inducible way. After P301L Tau expression, these mice develop NFTs, neuron loss, and behavioral impairment in a time-dependent manner (Santacruz et al., 2005). However, it is possible to switch off the expression of the transgene at a given time. Suppression of P301L tau expression in rTg4510 tau transgenic mice, which normally express the mutant protein at a high level, reverses behavioral impairments in these mice, although NFT formation continues. It suggested that NFT formation could be dissociated from neuronal dysfunction. In fact, soluble tau rather than NFTs may be neurotoxic (Santacruz et al., 2005). Thus, NFTs are unlikely to be the major toxic tau species, at least in the early stages of tauopathy. Two forms of tau multimers (140 and 170 kDa), whose molecular weight suggests an oligomeric aggregate, were recently described in rTg4510 tau transgenic mice. They accumulate early in the pathogenic cascade, and their levels correlated consistently with memory loss at various ages in the rTg4510 mouse model (Berger et al., 2007). These tau oligomers were also found in the brains of JNPL3 mice and patients presenting with FTDP-17 (Berger et al., 2007).

$Tau_{RD}/\Delta K280$ is also an inducible mouse model of tauopathy that is based on the expression of only the 4R tau microtubule-binding domains carrying the ΔK280 mutation (deletion of both amino- and carboxy-terminal parts of tau protein). The ΔK280 mutation accelerates tau aggregation and is referred to as a proaggregation mutant. Low expression of ΔK280 tau repeats leads to tau aggregation followed by astrogliosis and loss of synapses and neurons (Mocanu et al., 2008). Thus, tau aggregates are required for neuronal death. It is interesting to note that expression of ΔK280 tau repeats in $Tau_{RD}/\Delta K280$ is very low (about 70% of endogenous murine tau). When the transgene is switched off, ΔK280 tau repeats disappear and only aggregated murine tau proteins are found. Thus, this construct acts as a nucleation factor for tau aggregation (Mocanu et al., 2008).

The K3 transgenic mouse strain expresses human tau carrying the K369I mutation under the Thy1 promoter (Ittner et al., 2008). The K369I tau mutation was found in a family of patients presenting with Pick's disease without parkinsonism

Table 4.1 Main Tau transgenic mouse models

Strain name	Promoter	Mutation tau construct	Phenotype	References
JNPL3	PrP	P301L 0N4R	NFT, Mot	Lewis et al. (2000)
VLW	Thy1	G272V P301L R406W	NFT	Lim et al. (2001)
pR5	Thy1.2	P301L 2N4R	NFT, Cogn	Götz et al., (2001)
Tg214	PDGF	V337M 2N4R (myc/FlAG-tagged)	NFT, Cogn	Tanemura et al. (2001)
P301S Tau	Thy1.2	P301S 0N4R	NFT, Mot	Allen et al. (2002)
R406W Tg	CaMKII	R406W 2N4R (myc/FlAG-tagged)	NFT, Cogn	Tatebayashi et al. (2002)
rTg4510	(Inducible) CaMKII-driven rTA+tetOp (binary system)	P301L 0N4R	NFT, Cogn, Mot	Santacruz et al. (2005)
Tau-P301L	Thy1.2	P301L 2N4R	NFT, Mot	Terwel et al. (2005)
(line 12) Pl Tg	2′,3′-cyclic nucleotide 3′-phosphodiesterase	P301L 1N4R	GFT, Mot	Higuchi et al. (2005)
Tg-TauP301L	PrP	P301L 2N4R	NFT, GFT, Cogn	Murakami et al. (2006)
Thy-Tau22	Thy1.2	G272V/P301S 1N4R	NFT, Cogn	Schindowski et al. (2006)
Thy-Tau30	Thy1.2	G272V/P301S 1N4R	NFT, Mot	Leroy et al. (2007)
T-279	MAPT	N279K 2N4R	NFT, GFT, Mot	Dawson et al. (2007)
PS19	PrP	P301S 1N4R	NFT, Mot	Yoshiyama et al. (2007)
TauV337M	Thy1	V337M 2N3R	NFT, Cogn	Lambourne et al. (2007)
TauRD/ΔK280 proaggregation	CaMKII-driven rTA+tetOp (binary system)	ΔK280, 4R construct	NFT, Mot	Mocanu et al. (2008)
TauRD/ΔK280/2P antiaggregation	CaMKII-driven rTA+tetOp (binary system)	ΔK280/I277P/I308P, 4R construct	4R construct expression	Mocanu et al. (2008)
K3	Thy1.2	K369I 1N4R	NFT, Mot, Park	Ittner et al. (2008)

Abbreviations: 4R, four microtubule-binding domains; CaMKII, calcium calmodulin protein kinase II; Cogn, cognitive impairment; GFT, gliofibrillary tangles; MAPT, MAPT/tau gene; Mot, motor deficits; NFT, neurofibrillary or neuronal-like tangles; Park, parkinsonism; PDGF, platelet-derived growth factor; PrP, prion protein.

and amyotrophy (Neumann et al., 2001). K3 mice show early-onset memory impairment, Pick body–like inclusions, amyotrophy, and parkinsonism in the absence of overt neurodegeneration. Amyotrophy is likely to be related to tau expression in the sciatic nerve as described in Leroy et al. (2007). Moreover, tau transgene is mainly expressed in the substantia nigra, and such expression leads to an early-onset parkinsonism phenotype. Interestingly, motor performance of young, but not old K3 mice improves upon L-DOPA treatment (Ittner et al., 2008). This model emphasizes that tau mutations are responsible for some phenotypes (Pick body–like inclusions and memory impairment) but not others, which are likely to be related to the insertion site of the transgene in the murine genome.

Thy-Tau22 mouse transgenic line exhibits progressive neuron-specific AD-like tau pathology devoid of any motor deficits. In this model, a progressive development of NFT is observed in the hippocampus and amygdala, which parallels behavioral impairments as well as electrophysiologic alterations (Schindowski et al., 2006; Van der Jeugd et al., submitted). These latter changes are observed despite any striking loss of neuronal/synaptic markers until 12 months of age in the hippocampus (Van der Jeugd et al., submitted). In addition, in the hippocampus, hyper- and abnormally phosphorylated tau species accumulate within the somatodendritic area (Schindowski et al., 2006, 2008), supporting a possible influence on hippocampal-dependent plasticity always confirmed by behavioral and electrophysiologic evaluations. Indeed, a careful evaluation of learning and memory at 12 months of age indicates alterations in various spatial and nonspatial tests. These alterations were sustained by long-term depression and depotentiation defects within the hippocampus (Van der Jeugd et al., submitted).

Interestingly, at that time point, Thy-Tau22 mice exhibit septo-hippocampal tau pathology accompanied by altered retrograde transport from hippocampus to medial septum and the loss of cholinergic neurons within this latter area (Belarbi et al., 2009).

In summary, NFD with accumulation of abnormal tau species in the Thy-Tau22 transgenic model leads to synaptic plasticity defects underlying learning and memory alterations without striking neuronal loss and is thus a unique model to study molecular dysfunctions resulting from NFD.

Altogether, this section indicates that a large number of animal models are available. They allow for both understanding tau pathology and developing innovative therapeutic strategies. All models have strengths and weaknesses, but they are still valuable tools for studying tau biology and pathology.

4.15 Conclusions

In this review, we indicate that tau is clearly first a microtubule-associated protein and then a key actor in dementia. Most of the research has focused on the role of tau proteins in microtubule dynamics and its dysfunctions. However, the role of tau in neurodegeneration may be trickier than a simple defect in axonal transport due to microtubule depolymerization. However, different therapeutic strategies have emerged from this hypothesis. For instance, overexpression of wild-type tau in transgenic mice leads from 3 to 12 months of age to hyperphosphorylation of tau proteins accompanied by reduced microtubule numbers, impaired fast axonal transport, and motor deficits (Ishihara et al., 1999). By using the microtubule-stabilizing drug paclitaxel, such mice exhibit improvement in motor behavior, but cognitive functions were not analyzed (Zhang et al., 2005). Others have chosen to target GSK3 to modulate microtubule dynamics and rescue axonopathy (Spittaels et al., 2000; Noble et al., 2005; Engel et al., 2008). Nevertheless, until now, these approaches did not show any promising follow-ups in clinical trials (Hampel et al., 2009; Checler and Buée, 2009).

From the data obtained on myotonic dystrophies and amino-terminal part of the protein, tau is more than a microtubule-associated protein and is likely involved in signal transduction and cytoskeleton dynamics (Brandt et al., 1995; Hwang et al., 1996; Lee et al., 1998; Sergeant et al., 2001; Blard et al., 2007; Kraemer and Schellenberg, 2007; Dhaenens et al., 2008; Reynolds et al., 2008; Guthrie et al., 2009). This function has not been analyzed in the scope of the tau pathology but it may be more promising.

Finally, the way of propagation of neurofibrillary degeneration in tauopathies is a real question. Is it related to specific neuronal subpopulations, genetics, or other mechanisms? Recent data have suggested that tau aggregates may act as a prion-like agent. In their experiment, the authors performed intracranial injections of PHF from P301S transgenic brain extracts in hTau40 transgenic mice. They observed tau aggregation at the injection site and then cortical spreading of neurofibrillary degeneration in a time-dependent manner (Clavaguera et al., 2009). This shows that, similar to prion disease, exogenously induced tauopathy depends on both the host and the source of the agent, suggesting the existence of polymorphic tau-PHF strains reminiscent of prion strains. It is known that tau may be secreted and then captured by neighboring neurons (Frost and Diamond, 2009a; Frost et al., 2009b). Such mechanisms have interesting outcomes for therapeutic and especially tau immunotherapy (Asuni et al., 2007; Kayed and Jackson, 2009; Sigurdsson, 2009).

More specifically, in AD, tau pathology was thought to be downstream of amyloid pathology. However, higher 4R tau levels increase risk of developing neurodegeneration (Myers et al., 2005; Tobin et al., 2008). Conversely, when human amyloid precursor protein (hAPP)-expressing mice were crossed onto tau knockout backgrounds, this prevented behavioral deficits, without altering the high Aβ levels (Roberson et al., 2007). Reducing tau levels may be a relevant solution to postpone tau pathology and the toxicity associated with the amyloid pathology (Rapoport et al., 2002; Roberson et al., 2007).

Which tau species are toxic? Are toxic tau species phosphorylated? It was shown that any stress will lead to tau phosphorylation (amyloid, anesthesia, temperature, infection, etc.) (Hof et al., 1992; Buee-Scherrer et al., 1997; Bussiere et al., 1999; Planel et al., 2007, 2008, 2009), but a nucleation factor is needed for tau aggregation (Mocanu et al., 2008). As many aggregating proteins, tau has the property to self-aggregate. In vivo, stabilizing factors may exist, and the loss of function of these factors may also trigger tau aggregation. Several factors are potentiating tau aggregation such as heparan sulfate (Goedert et al., 1996), fatty acids (Wilson et al., 1997), or polyanions (Friedhoff et al., 1998). Thus, anti-aggregating agents may also be of interest.

Thus, in conclusion, tau is more than a microtubule-associated protein: it is a sensor, a reservoir of phosphorylation, and a killer. It has likely other functions, which may be of interest in understanding its biology and pathophysiology. It is clear that more studies are needed to understand fully all of its facets.

Acknowledgments This work was supported by Inserm, CNRS, IMPRT, University Lille 2, Lille County Hospital (CHR-Lille), Région Nord/Pas-de-Calais, FEDER, ADERMA, PRIM, DN2M, and grants from Association Française contre les Myopathies (AFM2006-1579 and AFM2007-1043), ANR-05-BLANC-0320-01, ANR-08-MNPS-002 AMYTOXTAU, France Alzheimer and Fédération pour la Recherche sur le Cerveau and from the European Community: APOPIS (contract LSHM-CT-2003-503330), MEMOSAD (contract 200611), and cNEUPRO (contract LSHM-CT-2007-037950). We also thank patients and their families and all members of the laboratory.

Disclosure Statement The authors have reported no conflict of interest.

References

Allen B, Ingram E, Takao M, Smith MJ, Jakes R, Virdee K, Yoshida H, Holzer M, Craxton M, Emson PC, Atzori C, Migheli A, Crowther RA, Ghetti B, Spillantini MG, Goedert M (2002) Abundant tau filaments and nonapoptotic neurodegeneration in transgenic mice expressing human P301S tau protein. J Neurosci 22:9340–9351

Amano M, Kaneko T, Maeda A, Nakayama M, Ito M, Yamauchi T, Goto H, Fukata Y, Oshiro N, Shinohara A, Iwamatsu A, Kaibuchi K (2003) Identification of Tau and MAP2 as novel substrates of Rho-kinase and myosin phosphatase. J Neurochem 87:780–790

Andorfer C, Acker CM, Kress Y, Hof PR, Duff K, Davies P (2005) Cell-cycle reentry and cell death in transgenic mice expressing nonmutant human tau isoforms. J Neurosci 25: 5446–5454

Andreadis A (2005) Tau gene alternative splicing: expression patterns, regulation and modulation of function in normal brain and neurodegenerative diseases. Biochim Biophys Acta 1739: 91–103

Andreadis A, Broderick JA, Kosik KS (1995) Relative exon affinities and suboptimal splice site signals lead to non-equivalence of two cassette exons. Nucleic Acids Res 23:3585–3593

Andreadis A, Brown WM, Kosik KS (1992) Structure and novel exons of the human tau gene. Biochemistry 31:10626–10633

Andreadis A, Wagner BK, Broderick JA, Kosik KS (1996) A tau promoter region without neuronal specificity. J Neurochem 66:2257–2263

Asuni AA, Boutajangout A, Quartermain D, Sigurdsson EM (2007) Immunotherapy targeting pathological tau conformers in a tangle mouse model reduces brain pathology with associated functional improvements. J Neurosci 27:9115–9129

Avila J, Lucas JJ, Perez M, Hernandez F (2004) Role of tau protein in both physiological and pathological conditions. Physiol Rev 84:361–384

Baker M, Litvan I, Houlden H, Adamson J, Dickson D, Perez-Tur J, Hardy J, Lynch T, Bigio E, Hutton M (1999) Association of an extended haplotype in the tau gene with progressive supranuclear palsy. Hum Mol Genet 8:711–715

Baker M, Mackenzie IR, Pickering-Brown SM, Gass J, Rademakers R, Lindholm C, Snowden J, Adamson J, Sadovnick AD et al (2006) Mutations in progranulin cause tau-negative frontotemporal dementia linked to chromosome 17. Nature 442:916–919

Barghorn S, Zheng-Fischhöfer Q, Ackmann M, Biernat J, von Bergen M, Mandelkow EM, Mandelkow E (2000) Structure, microtubule interactions, and paired helical filament aggregation by tau mutants of frontotemporal dementias. Biochemistry 39:11714–21.

Baumann K, Mandelkow EM, Biernat J, Piwnica-Worms H, Mandelkow E (1993) Abnormal Alzheimer-like phosphorylation of tau-protein by cyclin-dependent kinases cdk2 and cdk5. FEBS Lett 336:417–424

Belarbi K, Schindowski K, Burnouf S, Caillierez R, Grosjean ME, Demeyer D, Hamdane M, Sergeant N, Blum D, Buee L (2009) Early Tau pathology involving the septo-hippocampal pathway in a Tau transgenic model: relevance to Alzheimer's disease. Curr Alzheimer Res 6:152–157

Berger Z, Roder H, Hanna A, Carlson A, Rangachari V, Yue M, Wszolek Z, Ashe K, Knight J, Dickson D, Andorfer C, Rosenberry TL, Lewis J, Hutton M, Janus C (2007) Accumulation of pathological tau species and memory loss in a conditional model of tauopathy. J Neurosci 27:3650–3662

Bertram L, Parkinson M, McQueen MB, Mullin K, Hsiao M, Menon R, Moscarillo TJ, Blacker D, Tanzi RE (2005) Further evidence for LBP-1c/CP2/LSF association in Alzheimer's disease families. J Med Genet 42:857–62.

Bird TD, Wijsman EM, Nochlin D, Leehey M, Sumi SM, Payami H, Poorkaj P, Nemens E, Rafkind M, Schellenberg GD (1997) Chromosome 17 and hereditary dementia: linkage studies in three non-Alzheimer families and kindreds with late-onset FAD. Neurology 48:949–954

Blard O, Feuillette S, Bou J, Chaumette B, Frebourg T, Campion D, Lecourtois M (2007) Cytoskeleton proteins are modulators of mutant tau-induced neurodegeneration in Drosophila. Hum Mol Genet 16:555–566

Blard O, Frebourg T, Campion D, Lecourtois M (2006) Inhibition of proteasome and Shaggy/Glycogen synthase kinase-3beta kinase prevents clearance of phosphorylated tau in Drosophila. J Neurosci Res 84:1107–1115

Blennow K (2004) Cerebrospinal fluid protein biomarkers for Alzheimer's disease. NeuroRx 1:213–225

Borroni B, Premi E, Di Luca M, Padovani A (2007) Combined biomarkers for early Alzheimer disease diagnosis. Curr Med Chem 14:1171–1178

Braak H, Braak E (1987) Argyrophilic grains: characteristic pathology of cerebral cortex in cases of adult onset dementia without Alzheimer changes. Neurosci Lett 76:124–127

Braak H, Braak E (1991) Neuropathological stageing of Alzheimer-related changes. Acta Neuropathol 82:239–259

Brandt R, Gergou A, Wacker I, Fath T, Hutter H (2009) A Caenorhabditis elegans model of tau hyperphosphorylation: induction of developmental defects by transgenic overexpression of Alzheimer's disease-like modified tau. Neurobiol Aging 30:22–33

Brandt R, Lee G (1993) The balance between tau protein's microtubule growth and nucleation activities: implications for the formation of axonal microtubules. J Neurochem 61:997–1005

Brandt R, Leger J, Lee G (1995) Interaction of tau with the neural plasma membrane mediated by tau's amino-terminal projection domain. J Cell Biol 131:1327–1340

Bretteville A, Ando K, Ghestem A, Loyens A, Begard S, Beauvillain JC, Sergeant N, Hamdane M, Buee L (2009) Two-dimensional electrophoresis of tau mutants reveals specific phosphorylation pattern likely linked to early tau conformational changes. PLoS One 4:e4843

Brion JP, Couck AM, Passareiro E, Flament-Durand J (1985) Neurofibrillary tangles of Alzheimer's disease: an immunohistochemical study. J Submicrosc Cytol 17:89–96

Brion S, Plas J, Jeanneau A (1991) Pick's disease. Anatomo-clinical point of view. Rev Neurol (Paris) 147:693–704

Brion JP, Tremp G, Octave JN (1999) Transgenic expression of the shortest human tau affects its compartmentalization and its phosphorylation as in the pretangle stage of Alzheimer's disease. Am J Pathol 154:255–270

Bronner IF, ter Meulen BC, Azmani A, Severijnen LA, Willemsen R, Kamphorst W, Ravid R, Heutink P, van Swieten JC (2005) Hereditary Pick's disease with the G272V tau mutation shows predominant three-repeat tau pathology. Brain 128:2645–2653

Brook JD, McCurrach ME, Harley HG, Buckler AJ, Church D, Aburatani H, Hunter K, Stanton VP, Thirion JP, Hudson T et al (1992) Molecular basis of myotonic dystrophy: expansion of a trinucleotide (CTG) repeat at the 3' end of a transcript encoding a protein kinase family member. Cell 69:385

Brown J (1992) Pick's disease. Baillieres Clin Neurol 1:535–557

Buee L, Bussiere T, Buee-Scherrer V, Delacourte A, Hof PR (2000) Tau protein isoforms, phosphorylation and role in neurodegenerative disorders. Brain Res Brain Res Rev 33:95–130

Buee-Scherrer V, Buee L, Leveugle B, Perl DP, Vermersch P, Hof PR, Delacourte A (1997) Pathological tau proteins in postencephalitic parkinsonism: comparison with Alzheimer's disease and other neurodegenerative disorders. Ann Neurol 42:356–359

Buee-Scherrer V, Goedert M (2002) Phosphorylation of microtubule-associated protein tau by stress-activated protein kinases in intact cells. FEBS Lett 515:151–154

Buee Scherrer V, Hof PR, Buee L, Leveugle B, Vermersch P, Perl DP, Olanow CW, Delacourte A (1996) Hyperphosphorylated tau proteins differentiate corticobasal degeneration and Pick's disease. Acta Neuropathol 91:351–359

Bugiani O, Murrell JR, Giaccone G, Hasegawa M, Ghigo G, Tabaton M, Morbin M, Primavera A, Carella F, Solaro C, Grisoli M, Savoiardo M, Spillantini MG, Tagliavini F, Goedert M, Ghetti B (1999) Frontotemporal dementia and corticobasal degeneration in a family with a P301S mutation in tau. J Neuropathol Exp Neurol 58:667–677

Bussiere T, Hof PR, Mailliot C, Brown CD, Caillet-Boudin ML, Perl DP, Buee L, Delacourte A (1999) Phosphorylated serine422 on tau proteins is a pathological epitope found in several diseases with neurofibrillary degeneration. Acta Neuropathol 97:221–230

Butler M, Shelanski ML (1986) Microheterogeneity of microtubule-associated tau proteins is due to differences in phosphorylation. J Neurochem 47:1517–1522

Butner KA, Kirschner MW (1991) Tau protein binds to microtubules through a flexible array of distributed weak sites. J Cell Biol 115:717–730

Butterfield DA, Abdul HM, Opii W, Newman SF, Joshi G, Ansari MA, Sultana R (2006) Pin1 in Alzheimer's disease. J Neurochem 98:1697–1706

Caffrey TM, Joachim C, Wade-Martins R (2008) Haplotype-specific expression of the N-terminal exons 2 and 3 at the human MAPT locus. Neurobiol Aging 29:1923–1929

Cairns NJ, Bigio EH, Mackenzie IR, Neumann M, Lee VM, Hatanpaa KJ, White CL 3rd, Schneider JA, Grinberg LT, Halliday G, Duyckaerts C, Lowe JS, Holm IE, Tolnay M, Okamoto K, Yokoo H, Murayama S, Woulfe J, Munoz DG, Dickson DW, Ince PG, Trojanowski JQ, Mann DM (2007) Neuropathologic diagnostic and nosologic criteria for frontotemporal lobar degeneration: consensus of the Consortium for Frontotemporal Lobar Degeneration. Acta Neuropathol 114:5–22

Castle JC, Zhang C, Shah JK, Kulkarni AV, Kalsotra A, Cooper TA, Johnson JM (2008) Expression of 24, 426 human alternative splicing events and predicted cis regulation in 48 tissues and cell lines. Nat Genet 40:1416–1425

Checler F, Buee L (2009) Fundamental data on the pathologies amyloid and Tau in Alzheimer's disease: which therapeutic perspectives? Ann Pharm Fr 67(2):136–153

Chen J, Kanai Y, Cowan NJ, Hirokawa N (1992) Projection domains of MAP2 and tau determine spacings between microtubules in dendrites and axons. Nature 360:674–677

Chiti F, Stefani M, Taddei N, Ramponi G, Dobson CM (2003) Rationalization of the effects of mutations on peptide and protein aggregation rates. Nature 424:805–808

Clavaguera F, Bolmont T, Crowther RA, Abramowski D, Frank S, Probst A, Fraser G, Stalder AK, Beibel M, Staufenbiel M, Jucker M, Goedert M, Tolnay M (2009) Transmission and spreading of tauopathy in transgenic mouse brain. Nat Cell Biol 11:909–913

Cleveland DW, Hwo SY, Kirschner MW (1977a) Physical and chemical properties of purified tau factor and the role of tau in microtubule assembly. J Mol Biol 116:227–247

Cleveland DW, Hwo SY, Kirschner MW (1977b) Purification of tau, a microtubule-associated protein that induces assembly of microtubules from purified tubulin. J Mol Biol 116: 207–225

Constantinidis J, Richard J, Tissot R (1974) Pick's disease. Histological and clinical correlations. Eur Neurol 11:208–217

Crosby AH (2003) Disruption of cellular transport: a common cause of neurodegeneration? Lancet Neurol 2:311–316

Cruts M, Gijselinck I, van der Zee J, Engelborghs S, Wils H, Pirici D, Rademakers R, Vandenberghe R, Dermaut B, Martin JJ, van Duijn C, Peeters K, Sciot R, Santens P, De Pooter T, Mattheijssens M, Van den Broeck M, Cuijt I, Vennekens K, De Deyn PP, Kumar-Singh S, Van Broeckhoven C (2006) Null mutations in progranulin cause ubiquitin-positive frontotemporal dementia linked to chromosome 17q21. Nature 442:920–924

Cuchillo-Ibanez I, Seereeram A, Byers HL, Leung KY, Ward MA, Anderton BH, Hanger DP (2008) Phosphorylation of tau regulates its axonal transport by controlling its binding to kinesin. FASEB J 22:3186–3195

Dawson HN, Cantillana V, Chen L, Vitek MP (2007) The tau N279K exon 10 splicing mutation recapitulates frontotemporal dementia and parkinsonism linked to chromosome 17 tauopathy in a mouse model. J Neurosci 27:9155–9168

Dawson HN, Ferreira A, Eyster MV, Ghoshal N, Binder LI, Vitek MP (2001) Inhibition of neuronal maturation in primary hippocampal neurons from tau deficient mice. J Cell Sci 114:1179–1187

Delacourte A, Buee L (1997) Normal and pathological Tau proteins as factors for microtubule assembly. Int Rev Cytol 171:167–224

Delacourte A, David JP, Sergeant N, Buee L, Wattez A, Vermersch P, Ghozali F, Fallet-Bianco C, Pasquier F, Lebert F, Petit H, Di Menza C (1999) The biochemical pathway of neurofibrillary degeneration in aging and Alzheimer's disease. Neurology 52:1158–1165

Delacourte A, Defossez A (1986) Biochemical characterization of an immune serum which specifically marks neurons in neurofibrillary degeneration in Alzheimer's disease. C R Acad Sci III 303:439–444

Delacourte A, Robitaille Y, Sergeant N, Buee L, Hof PR, Wattez A, Laroche-Cholette A, Mathieu J, Chagnon P, Gauvreau D (1996) Specific pathological Tau protein variants characterize Pick's disease. J Neuropathol Exp Neurol 55:159–168

Delacourte A, Sergeant N, Buee L (2007) Neurodegenerative diseases: tau proteins in neurodegenerative diseases other than Alzheimer's disease. In: Fojo, T (Ed) The role of microtubules in cell biology, neurobiology, and oncology. Cancer Drug Discovery and Development, vol XVIII, Chapter 20. Humana Press, Totowa, NJ

Delacourte A, Sergeant N, Champain D, Wattez A, Maurage CA, Lebert F, Pasquier F, David JP (2002) Nonoverlapping but synergetic tau and APP pathologies in sporadic Alzheimer's disease. Neurology 59:398–407

Delisle MB, Murrell JR, Richardson R, Trofatter JA, Rascol O, Soulages X, Mohr M, Calvas P, Ghetti B (1999) A mutation at codon 279 (N279K) in exon 10 of the Tau gene causes a tauopathy with dementia and supranuclear palsy. Acta Neuropathol 98:62–77

Delobel P, Flament S, Hamdane M, Delacourte A, Vilain JP, Buee L (2002) Modelling Alzheimer-specific abnormal Tau phosphorylation independently of GSK3beta and PKA kinase activities. FEBS Lett 516:151–155

de Silva R, Lashley T, Strand C, Shiarli AM, Shi J, Tian J, Bailey KL, Davies P, Bigio EH, Arima K et al (2006) An immunohistochemical study of cases of sporadic and inherited frontotemporal lobar degeneration using 3R- and 4R-specific tau monoclonal antibodies. Acta Neuropathol 111:329–340

Dhaenens CM, Schraen-Maschke S, Tran H, Vingtdeux V, Ghanem D, Leroy O, Delplanque J, Vanbrussel E, Delacourte A, Vermersch P, Maurage CA, Gruffat H, Sergeant A, Mahadevan MS, Ishiura S, Buee L, Cooper TA, Caillet-Boudin ML, Charlet-Berguerand N, Sablonniere B, Sergeant N (2008) Overexpression of MBNL1 fetal isoforms and modified splicing of Tau in the DM1 brain: two individual consequences of CUG trinucleotide repeats. Exp Neurol 210: 467–478

Ding H, Dolan PJ, Johnson GV (2008) Histone deacetylase 6 interacts with the microtubule-associated protein tau. J Neurochem 106:2119–2130

Dixit R, Ross JL, Goldman YE, Holzbaur EL (2008) Differential regulation of dynein and kinesin motor proteins by tau. Science 319:1086–1089

Dong M, Owens IS, Sheen YY (1997) Cloning and expression of human liver UDP-glucuronosyltransferase cDNA, UDPGTh2. Arch Pharm Res 20:459–464

Dourlen P, Ando K, Hamdane M, Begard S, Buee L, Galas MC (2007) The peptidyl prolyl cis/trans isomerase Pin1 downregulates the inhibitor of apoptosis protein survivin. Biochim Biophys Acta 1773:1428–1437

Drechsel DN, Hyman AA, Cobb MH, Kirschner MW (1992) Modulation of the dynamic instability of tubulin assembly by the microtubule-associated protein tau. Mol Biol Cell 3: 1141–1154

Drewes G (2004) MARKing tau for tangles and toxicity. Trends Biochem Sci 29:548–555

Drewes G, Ebneth A, Preuss U, Mandelkow EM, Mandelkow E (1997) MARK, a novel family of protein kinases that phosphorylate microtubule-associated proteins and trigger microtubule disruption. Cell 89:297–308

D'Souza I, Poorkaj P, Hong M, Nochlin D, Lee VM, Bird TD, Schellenberg GD (1999) Missense and silent tau gene mutations cause frontotemporal dementia with parkinsonism-chromosome 17 type, by affecting multiple alternative RNA splicing regulatory elements. Proc Natl Acad Sci USA 96:5598–5603

Duff K, Knight H, Refolo LM, Sanders S, Yu X, Picciano M, Malester B, Hutton M, Adamson J, Goedert M, Burki K, Davies P (2000) Characterization of pathology in transgenic mice over-expressing human genomic and cDNA tau transgenes. Neurobiol Dis 7:87–98

Dumanchin C, Camuzat A, Campion D, Verpillat P, Hannequin D, Dubois B, Saugier-Veber P, Martin C, Penet C, Charbonnier F, Agid Y, Frebourg T, Brice A (1998) Segregation of a missense mutation in the microtubule-associated protein tau gene with familial frontotemporal dementia and parkinsonism. Hum Mol Genet 7:1825–1829

Duyckaerts C, Colle MA, Delatour B, Hauw JJ (1999) Alzheimer's disease: lesions and their progression. Rev Neurol (Paris) 155(Suppl 4):S17–S27

Elliott E, Laufer O, Ginzburg I (2009) BAG-1 M is upregulated in hippocampus of Alzheimer's disease patients and associates with tau and APP proteins. J Neurochem 109(4): 1168–1178

Elliott E, Tsvetkov P, Ginzburg I (2007) BAG-1 associates with Hsc70.Tau complex and regulates the proteasomal degradation of Tau protein. J Biol Chem 282:37276–37284

Engel T, Goni-Oliver P, Gomez-Ramos P, Moran MA, Lucas JJ, Avila J, Hernandez F (2008) Hippocampal neuronal subpopulations are differentially affected in double transgenic mice overexpressing frontotemporal dementia and parkinsonism linked to chromosome 17 tau and glycogen synthase kinase-3beta. Neuroscience 157:772–780

Forman MS, Trojanowski JQ, Lee VM (2007) TDP-43: a novel neurodegenerative proteinopathy. Curr Opin Neurobiol 17(5):548–555

Foster NL, Wilhelmsen K, Sima AA, Jones MZ, D'Amato CJ, Gilman S (1997) Frontotemporal dementia and parkinsonism linked to chromosome 17: a consensus conference. Conference Participants. Ann Neurol 41:706–715

Foveau B, Ancot F, Leroy C, Petrelli A, Reiss K, Vingtdeux V, Giordano S, Fafeur V, Tulasne D (2009) Down-regulation of the met receptor tyrosine kinase by presenilin-dependent regulated intramembrane proteolysis. Mol Biol Cell 20:2495–2507

Fridlich R, Delalande F, Jaillard C, Lu J, Poidevin L, Cronin T, Perrocheau L, Millet-Puel G, Niepon ML, Poch O, Holmgren A, Van Dorsselaer A, Sahel JA, Leveillard T (2009) The thioredoxin-like protein RdCVFL interacts with Tau and inhibits its phosphorylation in the retina. Mol Cell Proteomics 8(6):1206–1218

Friedhoff P, Schneider A, Mandelkow EM, Mandelkow E (1998) Rapid assembly of Alzheimer-like paired helical filaments from microtubule-associated protein tau monitored by fluorescence in solution. Biochemistry 37:10223–10230

Frost B, Diamond MI (2009a) The expanding realm of prion phenomena in neurodegenerative disease. Prion 3:74–77

Frost B, Jacks RL, Diamond MI (2009b) Propagation of tau misfolding from the outside to the inside of a cell. J Biol Chem 284:12845–12852

Galas MC, Dourlen P, Begard S, Ando K, Blum D, Hamdane M, Buee L (2006) The peptidylprolyl cis/trans-isomerase Pin1 modulates stress-induced dephosphorylation of Tau in neurons. Implication in a pathological mechanism related to Alzheimer disease. J Biol Chem 281:19296–19304

Gallo JM, Noble W, Martin TR (2007) RNA and protein-dependent mechanisms in tauopathies: consequences for therapeutic strategies. Cell Mol Life Sci 64:1701–1714

Georgieff IS, Liem RK, Couchie D, Mavilia C, Nunez J, Shelanski ML (1993) Expression of high molecular weight tau in the central and peripheral nervous systems. J Cell Sci 105(Pt 3):729–737

Ghetti B, Spina S, Murrell JR, Huey ED, Pietrini P, Sweeney B, Wassermann EM, Keohane C, Farlow MR, Grafman J (2008) In vivo and postmortem clinicoanatomical correlations in frontotemporal dementia and parkinsonism linked to chromosome 17. Neurodegener Dis 5:215–217

Giaccone G, Rossi G, Farina L, Marcon G, Di Fede G, Catania M, Morbin M, Sacco L, Bugiani O, Tagliavini F (2005) Familial frontotemporal dementia associated with the novel MAPT mutation T427M. J Neurol 252:1543–1545

Goedert M, Hasegawa M, Jakes R, Lawler S, Cuenda A, Cohen P (1997) Phosphorylation of microtubule-associated protein tau by stress-activated protein kinases. FEBS Lett 409: 57–62

Goedert M, Jakes R (1990) Expression of separate isoforms of human tau protein: correlation with the tau pattern in brain and effects on tubulin polymerization. EMBO J 9:4225–4230

Goedert M, Jakes R, Spillantini MG, Hasegawa M, Smith MJ, Crowther RA (1996) Assembly of microtubule-associated protein tau into Alzheimer-like filaments induced by sulphated glycosaminoglycans. Nature 383:550–553

Goedert M, Spillantini MG, Cairns NJ, Crowther RA (1992) Tau proteins of Alzheimer paired helical filaments: abnormal phosphorylation of all six brain isoforms. Neuron 8:159–168

Goedert M, Spillantini MG, Crowther RA, Chen SG, Parchi P, Tabaton M, Lanska DJ, Markesbery WR, Wilhelmsen KC, Dickson DW, Petersen RB, Gambetti P (1999) Tau gene mutation in familial progressive subcortical gliosis. Nat Med 5:454–457

Goedert M, Spillantini MG, Jakes R, Rutherford D, Crowther RA (1989a) Multiple isoforms of human microtubule-associated protein tau: sequences and localization in neurofibrillary tangles of Alzheimer's disease. Neuron 3:519–526

Goedert M, Spillantini MG, Potier MC, Ulrich J, Crowther RA (1989b) Cloning and sequencing of the cDNA encoding an isoform of microtubule-associated protein tau containing four tandem repeats: differential expression of tau protein mRNAs in human brain. EMBO J 8: 393–399

Gomes-Pereira M, Foiry L, Nicole A, Huguet A, Junien C, Munnich A, Gourdon G (2007) CTG trinucleotide repeat "big jumps": large expansions, small mice. PLoS Genet 3:e52

Goode BL, Feinstein SC (1994) Identification of a novel microtubule binding and assembly domain in the developmentally regulated inter-repeat region of tau. J Cell Biol 124:769–782

Gotz J, Chen F, Barmettler R, Nitsch RM (2001) Tau filament formation in transgenic mice expressing P301L tau. J Biol Chem 276:529–534

Gotz J, Probst A, Spillantini MG, Schafer T, Jakes R, Burki K, Goedert M (1995) Somatodendritic localization and hyperphosphorylation of tau protein in transgenic mice expressing the longest human brain tau isoform. EMBO J 14:1304–1313

Greenwood JA, Scott CW, Spreen RC, Caputo CB, Johnson GV (1994) Casein kinase II preferentially phosphorylates human tau isoforms containing an amino-terminal insert. Identification of threonine 39 as the primary phosphate acceptor. J Biol Chem 269:4373–4380

Grover A, England E, Baker M, Sahara N, Adamson J, Granger B, Houlden H, Passant U, Yen SH, DeTure M, Hutton M (2003) A novel tau mutation in exon 9 (1260 V) causes a four-repeat tauopathy. Exp Neurol 184:131–140

Grover A, Houlden H, Baker M, Adamson J, Lewis J, Prihar G, Pickering-Brown S, Duff K, Hutton M (1999) 5' splice site mutations in tau associated with the inherited dementia FTDP-17 affect a stem-loop structure that regulates alternative splicing of exon 10. J Biol Chem 274: 15134–15143

Gustke N, Trinczek B, Biernat J, Mandelkow EM, Mandelkow E (1994) Domains of tau protein and interactions with microtubules. Biochemistry 33:9511–9522

Guthrie C, Schellenberg GD, Kraemer BC (2009) SUT-2 potentiates tau-induced neurotoxicity in C. elegans. Hum Mol Genet. 18(10):1825–1838

Hall GF, Lee VM, Lee G, Yao J (2001) Staging of neurofibrillary degeneration caused by human tau overexpression in a unique cellular model of human tauopathy. Am J Pathol 158:235–246

Haltiwanger RS, Busby S, Grove K, Li S, Mason D, Medina L, Moloney D, Philipsberg G, Scartozzi R (1997) O-glycosylation of nuclear and cytoplasmic proteins: regulation analogous to phosphorylation? Biochem Biophys Res Commun 231:237–242

Hamdane M, Delobel P, Sambo AV, Smet C, Begard S, Violleau A, Landrieu I, Delacourte A, Lippens G, Flament S, Buee L (2003a) Neurofibrillary degeneration of the Alzheimer-type: an alternate pathway to neuronal apoptosis? Biochem Pharmacol 66:1619–1625

Hamdane M, Dourlen P, Bretteville A, Sambo AV, Ferreira S, Ando K, Kerdraon O, Begard S, Geay L, Lippens G, Sergeant N, Delacourte A, Maurage CA, Galas MC, Buee L (2006) Pin1 allows for differential Tau dephosphorylation in neuronal cells. Mol Cell Neurosci 32: 155–160

Hamdane M, Sambo AV, Delobel P, Begard S, Violleau A, Delacourte A, Bertrand P, Benavides J, Buee L (2003b) Mitotic-like tau phosphorylation by p25-Cdk5 kinase complex. J Biol Chem 278:34026–34034

Hampel H, Buerger K, Zinkowski R, Teipel SJ, Goernitz A, Andreasen N, Sjoegren M, DeBernardis J, Kerkman D, Ishiguro K, Ohno H, Vanmechelen E, Vanderstichele H, McCulloch C, Moller HJ, Davies P, Blennow K (2004) Measurement of phosphorylated tau epitopes in the differential diagnosis of Alzheimer disease: a comparative cerebrospinal fluid study. Arch Gen Psychiatry 61:95–102

Hampel H, Burger K, Pruessner JC, Zinkowski R, DeBernardis J, Kerkman D, Leinsinger G, Evans AC, Davies P, Moller HJ, Teipel SJ (2005) Correlation of cerebrospinal fluid levels of tau protein phosphorylated at threonine 231 with rates of hippocampal atrophy in Alzheimer disease. Arch Neurol 62:770–773

Hampel H, Ewers M, Burger K, Annas P, Mortberg A, Bogstedt A, Frolich L, Schroder J, Schonknecht P, Riepe MW, Kraft I, Gasser T, Leyhe T, Moller HJ, Kurz A, Basun H (2009) Lithium trial in Alzheimer's disease: a randomized, single-blind, placebo-controlled, multicenter 10-week study. J Clin Psychiatry 70:922–931

Hanger DP, Anderton BH, Noble W (2009) Tau phosphorylation: the therapeutic challenge for neurodegenerative disease. Trends Mol Med 15:112–119

Hanger DP, Byers HL, Wray S, Leung KY, Saxton MJ, Seereeram A, Reynolds CH, Ward MA, Anderton BH (2007) Novel phosphorylation sites in tau from Alzheimer brain support a role for casein kinase 1 in disease pathogenesis. J Biol Chem 282: 23645–23654

Hanger DP, Hughes K, Woodgett JR, Brion JP, Anderton BH (1992) Glycogen synthase kinase-3 induces Alzheimer's disease-like phosphorylation of tau: generation of paired helical filament epitopes and neuronal localisation of the kinase. Neurosci Lett 147:58–62

Harada A, Oguchi K, Okabe S, Kuno J, Terada S, Ohshima T, Sato-Yoshitake R, Takei Y, Noda T, Hirokawa N (1994) Altered microtubule organization in small-calibre axons of mice lacking tau protein. Nature 369:488–491

Harper PS (1997) Trinucleotide repeat disorders. J Inherit Metab Dis 20(2):122–124

Hart GW (1997) Dynamic O-linked glycosylation of nuclear and cytoskeletal proteins. Annu Rev Biochem 66:315–335

Hartig W, Klein C, Brauer K, Schuppel KF, Arendt T, Bruckner G, Bigl V (2000) Abnormally phosphorylated protein tau in the cortex of aged individuals of various mammalian orders. Acta Neuropathol 100:305–312

Hartmann AM, Rujescu D, Giannakouros T, Nikolakaki E, Goedert M, Mandelkow EM, Gao QS, Andreadis A, Stamm S (2001) Regulation of alternative splicing of human tau exon 10 by phosphorylation of splicing factors. Mol Cell Neurosci 18:80–90

Hasegawa M, Smith MJ, Goedert M (1998) Tau proteins with FTDP-17 mutations have a reduced ability to promote microtubule assembly. FEBS Lett 437:207–210

Hasegawa M, Smith MJ, Iijima M, Tabira T, Goedert M (1999) FTDP-17 mutations N279K and S305N in tau produce increased splicing of exon 10. FEBS Lett 443:93–96

Hayashi S, Toyoshima Y, Hasegawa M, Umeda Y, Wakabayashi K, Tokiguchi S, Iwatsubo T, Takahashi H (2002) Late-onset frontotemporal dementia with a novel exon 1 (Arg5His) tau gene mutation. Ann Neurol 51:525–530

Heicklen-Klein A, Ginzburg I (2000) Tau promoter confers neuronal specificity and binds Sp1 and AP-2. J Neurochem 75:1408–1418

Hernandez F, Perez M, Lucas JJ, Mata AM, Bhat R, Avila J (2004) Glycogen synthase kinase-3 plays a crucial role in tau exon 10 splicing and intranuclear distribution of SC35. Implications for Alzheimer's disease. J Biol Chem 279:3801–3806

Heutink P, Stevens M, Rizzu P, Bakker E, Kros JM, Tibben A, Niermeijer MF, van Duijn CM, Oostra BA, van Swieten JC (1997) Hereditary frontotemporal dementia is linked to chromosome 17q21-q22: a genetic and clinicopathological study of three Dutch families. Ann Neurol 41:150–159

Higuchi M, Zhang B, Forman MS, Yoshiyama Y, Trojanowski JQ, Lee VM (2005) Axonal degeneration induced by targeted expression of mutant human tau in oligodendrocytes of transgenic mice that model glial tauopathies. J Neurosci 25:9434–9443

Himmler A (1989) Structure of the bovine tau gene: alternatively spliced transcripts generate a protein family. Mol Cell Biol 9:1389–1396

Hirokawa N, Shiomura Y, Okabe S (1988) Tau proteins: the molecular structure and mode of binding on microtubules. J Cell Biol 107:1449–1459

Hof PR, Charpiot A, Delacourte A, Buee L, Purohit D, Perl DP, Bouras C (1992) Distribution of neurofibrillary tangles and senile plaques in the cerebral cortex in postencephalitic parkinsonism. Neurosci Lett 139:10–14

Holzer M, Gartner U, Klinz FJ, Narz F, Heumann R, Arendt T (2001) Activation of mitogen-activated protein kinase cascade and phosphorylation of cytoskeletal proteins after neurone-specific activation of p21ras. I. Mitogen-activated protein kinase cascade. Neuroscience 105:1031–1040

Holzer M, Gartner U, Stobe A, Hartig W, Gruschka H, Bruckner MK, Arendt T (2002) Inverse association of Pin1 and tau accumulation in Alzheimer's disease hippocampus. Acta Neuropathol 104:471–481

Hong M, Zhukareva V, Vogelsberg-Ragaglia V, Wszolek Z, Reed L, Miller BI, Geschwind DH, Bird TD, McKeel D, Goate A, Morris JC, Wilhelmsen KC, Schellenberg GD, Trojanowski JQ, Lee VM (1998) Mutation-specific functional impairments in distinct tau isoforms of hereditary FTDP-17. Science 282:1914–1917

Houlden H, Johnson J, Gardner-Thorpe C, Lashley T, Hernandez D, Worth P, Singleton AB, Hilton DA, Holton J, Revesz T, Davis MB, Giunti P, Wood NW (2007) Mutations in TTBK2, encoding a kinase implicated in tau phosphorylation, segregate with spinocerebellar ataxia type 11. Nat Genet 39:1434–1436

Hu YY, He SS, Wang X, Duan QH, Grundke-Iqbal I, Iqbal K, Wang J (2002) Levels of nonphosphorylated and phosphorylated tau in cerebrospinal fluid of Alzheimer's disease patients: an ultrasensitive bienzyme-substrate-recycle enzyme-linked immunosorbent assay. Am J Pathol 160:1269–1278

Huang DY, Weisgraber KH, Goedert M, Saunders AM, Roses AD, Strittmatter WJ (1995) ApoE3 binding to tau tandem repeat I is abolished by tau serine262 phosphorylation. Neurosci Lett 192:209–212

Hutton M (2000) Molecular genetics of chromosome 17 tauopathies. Ann N Y Acad Sci 920: 63–73

Hutton M, Lendon CL, Rizzu P, Baker M, Froelich S, Houlden H, Pickering-Brown S, Chakraverty S, Isaacs A, Grover A et al (1998) Association of missense and 5'-splice-site mutations in tau with the inherited dementia FTDP-17. Nature 393:702–705

Hwang SC, Jhon DY, Bae YS, Kim JH, Rhee SG (1996) Activation of phospholipase C-gamma by the concerted action of tau proteins and arachidonic acid. J Biol Chem 271:18342–18349

Iijima M, Tabira T, Poorkaj P, Schellenberg GD, Trojanowski JQ, Lee VM, Schmidt ML, Takahashi K, Nabika T, Matsumoto T, Yamashita Y, Yoshioka S, Ishino H (1999) A distinct familial presenile dementia with a novel missense mutation in the tau gene. Neuroreport 10: 497–501

Ikegami S, Harada A, Hirokawa N (2000) Muscle weakness, hyperactivity, and impairment in fear conditioning in tau-deficient mice. Neurosci Lett 279:129–132

Iseki E, Matsumura T, Marui W, Hino H, Odawara T, Sugiyama N, Suzuki K, Sawada H, Arai T, Kosaka K (2001) Familial frontotemporal dementia and parkinsonism with a novel N296H mutation in exon 10 of the tau gene and a widespread tau accumulation in the glial cells. Acta Neuropathol 102:285–292

Ishihara T, Hong M, Zhang B, Nakagawa Y, Lee MK, Trojanowski JQ, Lee VM (1999) Age-dependent emergence and progression of a tauopathy in transgenic mice overexpressing the shortest human tau isoform. Neuron 24:751–762

Ishizawa T, Ko LW, Cookson N, Davias P, Espinoza M, Dickson DW (2002) Selective neurofibrillary degeneration of the hippocampal CA2 sector is associated with four-repeat tauopathies. J Neuropathol Exp Neurol 61:1040–1047

Ittner LM, Fath T, Ke YD, Bi M, van Eersel J, Li KM, Gunning P, Gotz J (2008) Parkinsonism and impaired axonal transport in a mouse model of frontotemporal dementia. Proc Natl Acad Sci USA 105:15997–16002

Jackson GR, Wiedau-Pazos M, Sang TK, Wagle N, Brown CA, Massachi S, Geschwind DH (2002) Human wild-type tau interacts with wingless pathway components and produces neurofibrillary pathology in Drosophila. Neuron 34:509–519

Jakes R, Novak M, Davison M, Wischik CM (1991) Identification of 3- and 4-repeat tau isoforms within the PHF in Alzheimer's disease. EMBO J 10:2725–2729

Jenkins SM, Johnson GV (1998) Tau complexes with phospholipase C-gamma in situ. Neuroreport 9:67–71

Johnson GV, Seubert P, Cox TM, Motter R, Brown JP, Galasko D (1997) The tau protein in human cerebrospinal fluid in Alzheimer's disease consists of proteolytically derived fragments. J Neurochem 68:430–433

Kar S, Fan J, Smith MJ, Goedert M, Amos LA (2003a) Repeat motifs of tau bind to the insides of microtubules in the absence of taxol. EMBO J 22:70–77

Kar S, Florence GJ, Paterson I, Amos LA (2003b) Discodermolide interferes with the binding of tau protein to microtubules. FEBS Lett 539:34–36

Kayed R, Jackson GR (2009) Prefilament tau species as potential targets for immunotherapy for Alzheimer disease and related disorders. Curr Opin Immunol 21:359–363

Kidd M (1964) Alzheimer's Disease–an Electron Microscopical Study. Brain 87:307–320

Kinoshita E, Kinoshita-Kikuta E, Matsubara M, Yamada S, Nakamura H, Shiro Y, Aoki Y, Okita K, Koike T (2008) Separation of phosphoprotein isotypes having the same number of phosphate groups using phosphate-affinity SDS-PAGE. Proteomics 8:2994–3003

Kitano-Takahashi M, Morita H, Kondo S, Tomizawa K, Kato R, Tanio M, Shirota Y, Takahashi H, Sugio S, Kohno T (2007) Expression, purification and crystallization of a human tau-tubulin kinase 2 that phosphorylates tau protein. Acta Crystallogr Sect F Struct Biol Cryst Commun 63:602–604

Kobayashi T, Ota S, Tanaka K, Ito Y, Hasegawa M, Umeda Y, Motoi Y, Takanashi M, Yasuhara M, Anno M, Mizuno Y, Mori H (2003) A novel L266V mutation of the tau gene causes frontotemporal dementia with a unique tau pathology. Ann Neurol 53:133–137

Kondo J, Honda T, Mori H, Hamada Y, Miura R, Ogawara M, Ihara Y (1988) The carboxyl third of tau is tightly bound to paired helical filaments. Neuron 1:827–834

Kovacs GG, Pittman A, Revesz T, Luk C, Lees A, Kiss E, Tariska P, Laszlo L, Molnar K, Molnar MJ, Tolnay M, de Silva R (2008) MAPT S305I mutation: implications for argyrophilic grain disease. Acta Neuropathol 116:103–118

Kraemer BC, Burgess JK, Chen JH, Thomas JH, Schellenberg GD (2006) Molecular pathways that influence human tau-induced pathology in Caenorhabditis elegans. Hum Mol Genet 15: 1483–1496

Kraemer BC, Schellenberg GD (2007) SUT-1 enables tau-induced neurotoxicity in C. elegans. Hum Mol Genet 16:1959–1971

Kraemer BC, Zhang B, Leverenz JB, Thomas JH, Trojanowski JQ, Schellenberg GD (2003) Neurodegeneration and defective neurotransmission in a Caenorhabditis elegans model of tauopathy. Proc Natl Acad Sci USA 100:9980–9985

Kreppel LK, Blomberg MA, Hart GW (1997) Dynamic glycosylation of nuclear and cytosolic proteins. Cloning and characterization of a unique O-GlcNAc transferase with multiple tetratricopeptide repeats. J Biol Chem 272:9308–9315

Ksiezak-Reding H, Binder LI, Yen SH (1988) Immunochemical and biochemical characterization of tau proteins in normal and Alzheimer's disease brains with Alz 50 and Tau-1. J Biol Chem 263:7948–7953

Lace G, Savva GM, Forster G, de Silva R, Brayne C, Matthews FE, Barclay JJ, Dakin L, Ince PG, Wharton SB (2009) Hippocampal tau pathology is related to neuroanatomical connections: an ageing population-based study. Brain 132(Pt 5):1324–1334

Lambert JC, Amouyel P (2007) Genetic heterogeneity of Alzheimer's disease: complexity and advances. Psychoneuroendocrinology 32(Suppl 1):S62–S70

Lambert JC, Goumidi L, Vrièze FW, Frigard B, Harris JM, Cummings A, Coates J, Pasquier F, Cottel D, Gaillac M, St Clair D, Mann DM, Hardy J, Lendon CL, Amouyel P, Chartier-Harlin MC (2000) The transcriptional factor LBP-1c/CP2/LSF gene on chromosome 12 is a genetic determinant of Alzheimer's disease. Hum Mol Genet 9: 2275–80.

Lambourne SL, Humby T, Isles AR, Emson PC, Spillantini MG, Wilkinson LS (2007) Impairments in impulse control in mice transgenic for the human FTDP-17 tauV337M mutation are exacerbated by age. Hum Mol Genet 16:1708–1719

Landrieu I, Smet C, Wieruszeski JM, Sambo AV, Wintjens R, Buee L, Lippens G (2006) Exploring the molecular function of PIN1 by nuclear magnetic resonance. Curr Protein Pept Sci 7: 179–194

Lee G, Cowan N, Kirschner M (1988) The primary structure and heterogeneity of tau protein from mouse brain. Science 239:285–288

Lee G, Neve RL, Kosik KS (1989) The microtubule binding domain of tau protein. Neuron 2: 1615–1624

Lee G, Newman ST, Gard DL, Band H, Panchamoorthy G (1998) Tau interacts with src-family non-receptor tyrosine kinases. J Cell Sci 111(Pt 21):3167–3177

Lee G, Thangavel R, Sharma VM, Litersky JM, Bhaskar K, Fang SM, Do LH, Andreadis A, Van Hoesen G, Ksiezak-Reding H (2004) Phosphorylation of tau by fyn: implications for Alzheimer's disease. J Neurosci 24:2304–2312

Lefebvre T, Caillet-Boudin ML, Buee L, Delacourte A, Michalski JC (2003a) O-GlcNAc glycosylation and neurological disorders. Adv Exp Med Biol 535:189–202

Lefebvre T, Ferreira S, Dupont-Wallois L, Bussiere T, Dupire MJ, Delacourte A, Michalski JC, Caillet-Boudin ML (2003b) Evidence of a balance between phosphorylation and O-GlcNAc glycosylation of Tau proteins–a role in nuclear localization. Biochim Biophys Acta 1619: 167–176

Leroy K, Bretteville A, Schindowski K, Gilissen E, Authelet M, De Decker R, Yilmaz Z, Buee L, Brion JP (2007) Early axonopathy preceding neurofibrillary tangles in mutant tau transgenic mice. Am J Pathol 171:976–992

Lewis J, McGowan E, Rockwood J, Melrose H, Nacharaju P, Van Slegtenhorst M, Gwinn-Hardy K, Paul Murphy M, Baker M, Yu X, Duff K, Hardy J, Corral A, Lin WL, Yen SH, Dickson DW, Davies P, Hutton M (2000) Neurofibrillary tangles, amyotrophy and progressive motor disturbance in mice expressing mutant (P301L) tau protein. Nat Genet 25: 402–405

Lim F, Hernandez F, Lucas JJ, Gomez-Ramos P, Moran MA, Avila J (2001) FTDP-17 mutations in tau transgenic mice provoke lysosomal abnormalities and Tau filaments in forebrain. Mol Cell Neurosci 18:702–714

Lippa CF, Zhukareva V, Kawarai T, Uryu K, Shafiq M, Nee LE, Grafman J, Liang Y, St George-Hyslop PH, Trojanowski JQ, Lee VM (2000) Frontotemporal dementia with novel tau pathology and a Glu342Val tau mutation. Ann Neurol 48:850–858

Lippens G, Landrieu I, Smet C (2007) Molecular mechanisms of the phospho-dependent prolyl cis/trans isomerase Pin1. FEBS J 274:5211–5222

Liu F, Liang Z, Wegiel J, Hwang YW, Iqbal K, Grundke-Iqbal I, Ramakrishna N, Gong CX (2008) Overexpression of Dyrk1A contributes to neurofibrillary degeneration in Down syndrome. FASEB J 22:3224–3233

Llado A, Ezquerra M, Sanchez-Valle R, Rami L, Tolosa E, Molinuevo JL (2007) A novel MAPT mutation (P301T) associated with familial frontotemporal dementia. Eur J Neurol 14:e9–e10

Loomis PA, Howard TH, Castleberry RP, Binder LI (1990) Identification of nuclear tau isoforms in human neuroblastoma cells. Proc Natl Acad Sci USA 87:8422–8426

Lu PJ, Wulf G, Zhou XZ, Davies P, Lu KP (1999) The prolyl isomerase Pin1 restores the function of Alzheimer-associated phosphorylated tau protein. Nature 399:784–788

Luo MH, Leski ML, Andreadis A (2004) Tau isoforms which contain the domain encoded by exon 6 and their role in neurite elongation. J Cell Biochem 91:880–895

Maccioni RB, Lavados M, Guillon M, Mujica C, Bosch R, Farias G, Fuentes P (2006) Anomalously phosphorylated tau and Abeta fragments in the CSF correlates with cognitive impairment in MCI subjects. Neurobiol Aging 27:237–244

Magnani E, Fan J, Gasparini L, Golding M, Williams M, Schiavo G, Goedert M, Amos LA, Spillantini MG (2007) Interaction of tau protein with the dynactin complex. EMBO J 26:4546–4554

Mailliot C, Sergeant N, Bussiere T, Caillet-Boudin ML, Delacourte A, Buee L (1998) Phosphorylation of specific sets of tau isoforms reflects different neurofibrillary degeneration processes. FEBS Lett 433:201–204

Mailliot C, Trojanowski JQ, Lee VM (2002) Impaired tau protein function following nitration-induced oxidative stress in vitro and in vivo. Neurobiol Aging 23(1):S415

Mandelkow EM, Thies E, Trinczek B, Biernat J, Mandelkow E (2004) MARK/PAR1 kinase is a regulator of microtubule-dependent transport in axons. J Cell Biol 167:99–110

Matsuo ES, Shin RW, Billingsley ML, Van de Voorde A, O'Connor M, Trojanowski JQ, Lee VM (1994) Biopsy-derived adult human brain tau is phosphorylated at many of the same sites as Alzheimer's disease paired helical filament tau. Neuron 13:989–1002

Maurage CA, Udd B, Ruchoux MM, Vermersch P, Kalimo H, Krahe R, Delacourte A, Sergeant N (2005) Similar brain tau pathology in DM2/PROMM and DM1/Steinert disease. Neurology 65:1636–1638

McMillan P, Korvatska E, Poorkaj P, Evstafjeva Z, Robinson L, Greenup L, Leverenz J, Schellenberg GD, D'Souza I (2008) Tau isoform regulation is region- and cell-specific in mouse brain. J Comp Neurol 511:788–803

Meola G (2000) Clinical and genetic heterogeneity in myotonic dystrophies. Muscle Nerve 23:1789–1799

Meola G, Sansone V (2007) Cerebral involvement in myotonic dystrophies. Muscle Nerve 36: 294–306

Mesulam M, Shaw P, Mash D, Weintraub S (2004) Cholinergic nucleus basalis tauopathy emerges early in the aging-MCI-AD continuum. Ann Neurol 55:815–828

Miserez AR, Clavaguera F, Monsch AU, Probst A, Tolnay M (2003) Argyrophilic grain disease: molecular genetic difference to other four-repeat tauopathies. Acta Neuropathol 106:363–366

Mocanu MM, Nissen A, Eckermann K, Khlistunova I, Biernat J, Drexler D, Petrova O, Schonig K, Bujard H, Mandelkow E, Zhou L, Rune G, Mandelkow EM (2008) The potential for beta-structure in the repeat domain of tau protein determines aggregation, synaptic decay, neuronal loss, and coassembly with endogenous Tau in inducible mouse models of tauopathy. J Neurosci 28:737–748

Mori H, Hosoda K, Matsubara E, Nakamoto T, Furiya Y, Endoh R, Usami M, Shoji M, Maruyama S, Hirai S (1995) Tau in cerebrospinal fluids: establishment of the sandwich ELISA with antibody specific to the repeat sequence in tau. Neurosci Lett 186:181–183

Morishima-Kawashima M, Hasegawa M, Takio K, Suzuki M, Titani K, Ihara Y (1993) Ubiquitin is conjugated with amino-terminally processed tau in paired helical filaments. Neuron 10: 1151–1160

Motter R, Vigo-Pelfrey C, Kholodenko D, Barbour R, Johnson-Wood K, Galasko D, Chang L, Miller B, Clark C, Green R et al (1995) Reduction of beta-amyloid peptide42 in the cerebrospinal fluid of patients with Alzheimer's disease. Ann Neurol 38:643–648

Mudher A, Shepherd D, Newman TA, Mildren P, Jukes JP, Squire A, Mears A, Drummond JA, Berg S, MacKay D, Asuni AA, Bhat R, Lovestone S (2004) GSK-3beta inhibition reverses axonal transport defects and behavioural phenotypes in Drosophila. Mol Psychiatry 9:522–530

Munch G, Kuhla B, Luth HJ, Arendt T, Robinson SR (2003) Anti-AGEing defences against Alzheimer's disease. Biochem Soc Trans 31:1397–1399

Murakami T, Paitel E, Kawarabayashi T, Ikeda M, Chishti MA, Janus C, Matsubara E, Sasaki A, Kawarai T, Phinney AL et al (2006) Cortical neuronal and glial pathology in TgTauP301L transgenic mice: neuronal degeneration, memory disturbance, and phenotypic variation. Am J Pathol 169:1365–1375

Muramatsu K, Hashimoto Y, Uemura T, Kunii M, Harada R, Sato T, Morikawa A, Harada A (2008) Neuron-specific recombination by Cre recombinase inserted into the murine tau locus. Biochem Biophys Res Commun 370:419–423

Murrell JR, Koller D, Foroud T, Goedert M, Spillantini MG, Edenberg HJ, Farlow MR, Ghetti B (1997) Familial multiple-system tauopathy with presenile dementia is localized to chromosome 17. Am J Hum Genet 61:1131–1138

Murrell JR, Spillantini MG, Zolo P, Guazzelli M, Smith MJ, Hasegawa M, Redi F, Crowther RA, Pietrini P, Ghetti B, Goedert M (1999) Tau gene mutation G389R causes a tauopathy with abundant pick body-like inclusions and axonal deposits. J Neuropathol Exp Neurol 58: 1207–1226

Myers AJ, Kaleem M, Marlowe L, Pittman AM, Lees AJ, Fung HC, Duckworth J, Leung D, Gibson A, Morris CM, de Silva R, Hardy J (2005) The H1c haplotype at the MAPT locus is associated with Alzheimer's disease. Hum Mol Genet 14:2399–2404

Myers AJ, Pittman AM, Zhao AS, Rohrer K, Kaleem M, Marlowe L, Lees A, Leung D, McKeith IG, Perry RH, Morris CM, Trojanowski JQ, Clark C, Karlawish J, Arnold S, Forman MS, Van Deerlin V, de Silva R, Hardy J (2007) The MAPT H1c risk haplotype is associated with increased expression of tau and especially of 4 repeat containing transcripts. Neurobiol Dis 25:561–570

Neumann M, Diekmann S, Bertsch U, Vanmassenhove B, Bogerts B, Kretzschmar HA (2005) Novel G335V mutation in the tau gene associated with early onset familial frontotemporal dementia. Neurogenetics 6:91–95

Neumann M, Schulz-Schaeffer W, Crowther RA, Smith MJ, Spillantini MG, Goedert M, Kretzschmar HA (2001) Pick's disease associated with the novel Tau gene mutation K369I. Ann Neurol 50:503–513

Neve RL, Harris P, Kosik KS, Kurnit DM, Donlon TA (1986) Identification of cDNA clones for the human microtubule-associated protein tau and chromosomal localization of the genes for tau and microtubule-associated protein 2. Brain Res 387:271–280

Nicholl DJ, Greenstone MA, Clarke CE, Rizzu P, Crooks D, Crowe A, Trojanowski JQ, Lee VM, Heutink P (2003) An English kindred with a novel recessive tauopathy and respiratory failure. Ann Neurol 54:682–686

Nishimura I, Yang Y, Lu B (2004) PAR-1 kinase plays an initiator role in a temporally ordered phosphorylation process that confers tau toxicity in Drosophila. Cell 116:671–682

Noble W, Planel E, Zehr C, Olm V, Meyerson J, Suleman F, Gaynor K, Wang L, LaFrancois J, Feinstein B, Burns M, Krishnamurthy P, Wen Y, Bhat R, Lewis J, Dickson D, Duff K (2005) Inhibition of glycogen synthase kinase-3 by lithium correlates with reduced tauopathy and degeneration in vivo. Proc Natl Acad Sci USA 102:6990–6995

Oddo S, Caccamo A, Cheng D, Laferla FM (2008) Genetically altering abeta distribution from the brain to the vasculature ameliorates tau pathology. Brain Pathol 19(3):421–430

Paquet D, Bhat R, Sydow A, Mandelkow EM, Berg S, Hellberg S, Falting J, Distel M, Koster RW, Schmid B, Haass C (2009) A zebrafish model of tauopathy allows in vivo imaging of neuronal cell death and drug evaluation. J Clin Invest 119(5):1382–1395

Paudel HK (1997) The regulatory Ser262 of microtubule-associated protein tau is phosphorylated by phosphorylase kinase. J Biol Chem 272:1777–1785

Pickering-Brown S, Baker M, Yen SH, Liu WK, Hasegawa M, Cairns N, Lantos PL, Rossor M, Iwatsubo T, Davies Y, Allsop D, Furlong R, Owen F, Hardy J, Mann D, Hutton M (2000) Pick's disease is associated with mutations in the tau gene. Ann Neurol 48:859–867

Pittman AM, Myers AJ, Duckworth J, Bryden L, Hanson M, Abou-Sleiman P, Wood NW, Hardy J, Lees A, de Silva R (2004) The structure of the tau haplotype in controls and in progressive supranuclear palsy. Hum Mol Genet 13:1267–1274

Planel E, Bretteville A, Liu L, Virag L, Du AL, Yu WH, Dickson DW, Whittington RA, Duff KE (2009) Acceleration and persistence of neurofibrillary pathology in a mouse model of tauopathy following anesthesia. FASEB J 23:2595

Planel E, Krishnamurthy P, Miyasaka T, Liu L, Herman M, Kumar A, Bretteville A, Figueroa HY, Yu WH, Whittington RA, Davies P, Takashima A, Nixon RA, Duff KE (2008) Anesthesia-induced hyperphosphorylation detaches 3-repeat tau from microtubules without affecting their stability in vivo. J Neurosci 28:12798–12807

Planel E, Richter KE, Nolan CE, Finley JE, Liu L, Wen Y, Krishnamurthy P, Herman M, Wang L, Schachter JB, Nelson RB, Lau LF, Duff KE (2007) Anesthesia leads to tau hyperphosphorylation through inhibition of phosphatase activity by hypothermia. J Neurosci 27:3090–3097

Poorkaj P, Bird TD, Wijsman E, Nemens E, Garruto RM, Anderson L, Andreadis A, Wiederholt WC, Raskind M, Schellenberg GD (1998) Tau is a candidate gene for chromosome 17 frontotemporal dementia. Ann Neurol 43:815–825

Poorkaj P, Grossman M, Steinbart E, Payami H, Sadovnick A, Nochlin D, Tabira T, Trojanowski JQ, Borson S, Galasko D, Reich S, Quinn B, Schellenberg G, Bird TD (2001) Frequency of tau gene mutations in familial and sporadic cases of non-Alzheimer dementia. Arch Neurol 58:383–387

Poorkaj P, Muma NA, Zhukareva V, Cochran EJ, Shannon KM, Hurtig H, Koller WC, Bird TD, Trojanowski JQ, Lee VM, Schellenberg GD (2002) An R5L tau mutation in a subject with a progressive supranuclear palsy phenotype. Ann Neurol 52:511–516

Probst A, Gotz J, Wiederhold KH, Tolnay M, Mistl C, Jaton AL, Hong M, Ishihara T, Lee VM, Trojanowski JQ, Jakes R, Crowther RA, Spillantini MG, Burki K, Goedert M (2000) Axonopathy and amyotrophy in mice transgenic for human four-repeat tau protein. Acta Neuropathol 99:469–481

Probst A, Tolnay M, Langui D, Goedert M, Spillantini MG (1996) Pick's disease: hyperphosphorylated tau protein segregates to the somatoaxonal compartment. Acta Neuropathol 92:588–596

Qiang L, Yu W, Andreadis A, Luo M, Baas PW (2006) Tau protects microtubules in the axon from severing by katanin. J Neurosci 26:3120–3129

Rademakers R, Cruts M, van Broeckhoven C (2004) The role of tau (MAPT) in frontotemporal dementia and related tauopathies. Hum Mutat 24:277–295

Rademakers R, Melquist S, Cruts M, Theuns J, Del-Favero J, Poorkaj P, Baker M, Sleegers K, Crook R, De Pooter T et al (2005) High-density SNP haplotyping suggests altered regulation of tau gene expression in progressive supranuclear palsy. Hum Mol Genet 14: 3281–3292

Ranum LP, Rasmussen PF, Benzow KA, Koob MD, Day JW (1998) Genetic mapping of a second myotonic dystrophy locus. Nat Genet 19:196–198

Rapoport M, Dawson HN, Binder LI, Vitek MP, Ferreira A (2002) Tau is essential to beta -amyloid-induced neurotoxicity. Proc Natl Acad Sci USA 99:6364–6369

Reardon W, Harper PS (1992) Advances in myotonic dystrophy: a clinical and genetic perspective. Curr Opin Neurol Neurosurg 5:605–609

Rebeiz JJ, Kolodny EH, Richardson EP Jr. (1967) Corticodentatonigral degeneration with neuronal achromasia: a progressive disorder of late adult life. Trans Am Neurol Assoc 92:23–26

Rebeiz JJ, Kolodny EH, Richardson EP Jr. (1968) Corticodentatonigral degeneration with neuronal achromasia. Arch Neurol 18:20–33

Reed LA, Schmidt ML, Wszolek ZK, Balin BJ, Soontornniyomkij V, Lee VM, Trojanowski JQ, Schelper RL (1998) The neuropathology of a chromosome 17-linked autosomal dominant parkinsonism and dementia "pallido-ponto-nigral degeneration". J Neuropathol Exp Neurol 57:588–601

Reynolds CH, Garwood CJ, Wray S, Price C, Kellie S, Perera T, Zvelebil M, Yang A, Sheppard PW, Varndell IM, Hanger DP, Anderton BH (2008) Phosphorylation regulates tau interactions with Src homology 3 domains of phosphatidylinositol 3-kinase, phospholipase Cgamma1, Grb2, and Src family kinases. J Biol Chem 283:18177–18186

Rhee SG (2001) Regulation of phosphoinositide-specific phospholipase C. Annu Rev Biochem 70:281–312

Richardson JC, Steele J, Olszewski J (1963) Supranuclear ophthalmoplegia, pseudobulbar palsy, nuchal dystonia and dementia. A clinical report on eight cases of "heterogenous system degeneration". Trans Am Neurol Assoc 88:25–29

Rizzini C, Goedert M, Hodges JR, Smith MJ, Jakes R, Hills R, Xuereb JH, Crowther RA, Spillantini MG (2000) Tau gene mutation K257T causes a tauopathy similar to Pick's disease. J Neuropathol Exp Neurol 59:990–1001

Rizzu P, Van Swieten JC, Joosse M, Hasegawa M, Stevens M, Tibben A, Niermeijer MF, Hillebrand M, Ravid R, Oostra BA, Goedert M, van Duijn CM, Heutink P (1999) High prevalence of mutations in the microtubule-associated protein tau in a population study of frontotemporal dementia in the Netherlands. Am J Hum Genet 64:414–421

Roberson ED, Scearce-Levie K, Palop JJ, Yan F, Cheng IH, Wu T, Gerstein H, Yu GQ, Mucke L (2007) Reducing endogenous tau ameliorates amyloid beta-induced deficits in an Alzheimer's disease mouse model. Science 316:750–754

Ros R, Thobois S, Streichenberger N, Kopp N, Sanchez MP, Perez M, Hoenicka J, Avila J, Honnorat J, de Yebenes JG (2005) A new mutation of the tau gene, G303V, in early-onset familial progressive supranuclear palsy. Arch Neurol 62:1444–1450

Rosenberg KJ, Ross JL, Feinstein HE, Feinstein SC, Israelachvili J (2008) Complementary dimerization of microtubule-associated tau protein: implications for microtubule bundling and tau-mediated pathogenesis. Proc Natl Acad Sci USA 105:7445–7450

Rossi G, Dalpra L, Crosti F, Lissoni S, Sciacca FL, Catania M, Di Fede G, Mangieri M, Giaccone G, Croci D, Tagliavini F (2008) A new function of microtubule-associated protein tau: involvement in chromosome stability. Cell Cycle 7:1788–1794

Rosso SM, Donker Kaat L, Baks T, Joosse M, de Koning I, Pijnenburg Y, de Jong D, Dooijes D, Kamphorst W, Ravid R, Niermeijer MF, Verheij F, Kremer HP, Scheltens P, van Duijn CM, Heutink P, van Swieten JC (2003) Frontotemporal dementia in The Netherlands: patient characteristics and prevalence estimates from a population-based study. Brain 126:2016–2022

Rosso SM, van Herpen E, Deelen W, Kamphorst W, Severijnen LA, Willemsen R, Ravid R, Niermeijer MF, Dooijes D, Smith MJ, Goedert M, Heutink P, van Swieten JC (2002) A novel tau mutation, S320F, causes a tauopathy with inclusions similar to those in Pick's disease. Ann Neurol 51:373–376

Rovelet-Lecrux A, Lecourtois M, Thomas-Anterion C, Le Ber I, Brice A, Frebourg T, Hannequin D, Campion D (2009) Partial deletion of the MAPT gene: a novel mechanism of FTDP-17. Hum Mutat 30:E591–E602

Sansone V, Gandossini S, Cotelli M, Calabria M, Zanetti O, Meola G (2007) Cognitive impairment in adult myotonic dystrophies: a longitudinal study. Neurol Sci 28:9–15

Santacruz K, Lewis J, Spires T, Paulson J, Kotilinek L, Ingelsson M, Guimaraes A, DeTure M, Ramsden M, McGowan E, Forster C, Yue M, Orne J, Janus C, Mariash A, Kuskowski M, Hyman B, Hutton M, Ashe KH (2005) Tau suppression in a neurodegenerative mouse model improves memory function. Science 309:476–481

Sato S, Cerny RL, Buescher JL, Ikezu T (2006) Tau-tubulin kinase 1 (TTBK1), a neuron-specific tau kinase candidate, is involved in tau phosphorylation and aggregation. J Neurochem 98:1573–1584

Schindowski K, Belarbi K, Buee L (2008) Neurotrophic factors in Alzheimer's disease: role of axonal transport. Genes Brain Behav 7(Suppl 1):43–56

Schindowski K, Bretteville A, Leroy K, Begard S, Brion JP, Hamdane M, Buee L (2006) Alzheimer's disease-like tau neuropathology leads to memory deficits and loss of functional synapses in a novel mutated tau transgenic mouse without any motor deficits. Am J Pathol 169:599–616

Schraen-Maschke S, Dhaenens CM, Delacourte A, Sablonniere B (2004) Microtubule-associated protein tau gene: a risk factor in human neurodegenerative diseases. Neurobiol Dis 15: 449–460

Schultz C, Hubbard GB, Rub U, Braak E, Braak H (2000) Age-related progression of tau pathology in brains of baboons. Neurobiol Aging 21:905–912

Schweers O, Schonbrunn-Hanebeck E, Marx A, Mandelkow E (1994) Structural studies of tau protein and Alzheimer paired helical filaments show no evidence for beta-structure. J Biol Chem 269:24290–24297

Sergeant N, Bretteville A, Hamdane M, Caillet-Boudin ML, Grognet P, Bombois S, Blum D, Delacourte A, Pasquier F, Vanmechelen E, Schraen-Maschke S, Buee L (2008) Biochemistry of Tau in Alzheimer's disease and related neurological disorders. Expert Rev Proteomics 5:207–224

Sergeant N, Bussiere T, Vermersch P, Lejeune JP, Delacourte A (1995) Isoelectric point differentiates PHF-tau from biopsy-derived human brain tau proteins. Neuroreport 6:2217–2220

Sergeant N, David JP, Goedert M, Jakes R, Vermersch P, Buee L, Lefranc D, Wattez A, Delacourte A (1997a) Two-dimensional characterization of paired helical filament-tau from Alzheimer's disease: demonstration of an additional 74-kDa component and age-related biochemical modifications. J Neurochem 69:834–844

Sergeant N, David JP, Lefranc D, Vermersch P, Wattez A, Delacourte A (1997b) Different distribution of phosphorylated tau protein isoforms in Alzheimer's and Pick's diseases. FEBS Lett 412:578–582

Sergeant N, Lyon M, Rudland PS, Fernig DG, Delehedde M (2000) Stimulation of DNA synthesis and cell proliferation of human mammary myoepithelial-like cells by hepatocyte growth factor/scatter factor depends on heparan sulfate proteoglycans and sustained phosphorylation of mitogen-activated protein kinases p42/44. J Biol Chem 275:17094–17099

Sergeant N, Sablonniere B, Schraen-Maschke S, Ghestem A, Maurage CA, Wattez A, Vermersch P, Delacourte A (2001) Dysregulation of human brain microtubule-associated tau mRNA maturation in myotonic dystrophy type 1. Hum Mol Genet 10:2143–2155

Sergeant N, Wattez A, Delacourte A (1999) Neurofibrillary degeneration in progressive supranuclear palsy and corticobasal degeneration: tau pathologies with exclusively "exon 10" isoforms. J Neurochem 72:1243–1249

Seznec H, Agbulut O, Sergeant N, Savouret C, Ghestem A, Tabti N, Willer JC, Ourth L, Duros C, Brisson E, Fouquet C, Butler-Browne G, Delacourte A, Junien C, Gourdon G (2001) Mice transgenic for the human myotonic dystrophy region with expanded CTG repeats display muscular and brain abnormalities. Hum Mol Genet 10:2717–2726

Shaw-Smith C, Pittman AM, Willatt L, Martin H, Rickman L, Gribble S, Curley R, Cumming S, Dunn C, Kalaitzopoulos D, Porter K, Prigmore E, Krepischi-Santos AC, Varela MC, Koiffmann CP, Lees AJ, Rosenberg C, Firth HV, de Silva R, Carter NP (2006) Microdeletion encompassing

MAPT at chromosome 17q21.3 is associated with developmental delay and learning disability. Nat Genet 38:1032–1037

Shi HR, Zhu LQ, Wang SH, Liu XA, Tian Q, Zhang Q, Wang Q, Wang JZ (2008) 17beta-estradiol attenuates glycogen synthase kinase-3beta activation and tau hyperphosphorylation in Akt-independent manner. J Neural Transm 115:879–888

Shulman JM, Feany MB (2003) Genetic modifiers of tauopathy in Drosophila. Genetics 165: 1233–1242

Sigurdsson EM (2009) Tau-Focused Immunotherapy for Alzheimer's Disease and Related Tauopathies. Curr Alzheimer Res 6:446–50

Sjoberg MK, Shestakova E, Mansuroglu Z, Maccioni RB, Bonnefoy E (2006) Tau protein binds to pericentromeric DNA: a putative role for nuclear tau in nucleolar organization. J Cell Sci 119:2025–2034

Sjogren M, Davidsson P, Gottfries J, Vanderstichele H, Edman A, Vanmechelen E, Wallin A, Blennow K (2001) The cerebrospinal fluid levels of tau, growth-associated protein-43 and soluble amyloid precursor protein correlate in Alzheimer's disease, reflecting a common pathophysiological process. Dement Geriatr Cogn Disord 12:257–264

Sontag E, Nunbhakdi-Craig V, Lee G, Brandt R, Kamibayashi C, Kuret J, White CL 3rd, Mumby MC, Bloom GS (1999) Molecular interactions among protein phosphatase 2A, tau, and microtubules. Implications for the regulation of tau phosphorylation and the development of tauopathies. J Biol Chem 274:25490–25498

Sperfeld AD, Collatz MB, Baier H, Palmbach M, Storch A, Schwarz J, Tatsch K, Reske S, Joosse M, Heutink P, Ludolph AC (1999) FTDP-17: an early-onset phenotype with parkinsonism and epileptic seizures caused by a novel mutation. Ann Neurol 46:708–715

Spillantini MG, Bird TD, Ghetti B (1998a) Frontotemporal dementia and Parkinsonism linked to chromosome 17: a new group of tauopathies. Brain Pathol 8:387–402

Spillantini MG, Goedert M (1998b) Tau protein pathology in neurodegenerative diseases. Trends Neurosci 21:428–433

Spillantini MG, Goedert M (2001) Tau gene mutations and tau pathology in frontotemporal dementia and parkinsonism linked to chromosome 17. Adv Exp Med Biol 487:21–37

Spillantini MG, Murrell JR, Goedert M, Farlow MR, Klug A, Ghetti B (1998c) Mutation in the tau gene in familial multiple system tauopathy with presenile dementia. Proc Natl Acad Sci USA 95:7737–7741

Spillantini MG, Yoshida H, Rizzini C, Lantos PL, Khan N, Rossor MN, Goedert M, Brown J (2000) A novel tau mutation (N296N) in familial dementia with swollen achromatic neurons and corticobasal inclusion bodies. Ann Neurol 48:939–943

Spina S, Murrell JR, Yoshida H, Ghetti B, Bermingham N, Sweeney B, Dlouhy SR, Crowther RA, Goedert M, Keohane C (2007) The novel Tau mutation G335S: clinical, neuropathological and molecular characterization. Acta Neuropathol 113:461–470

Spittaels K, Van den Haute C, Van Dorpe J, Bruynseels K, Vandezande K, Laenen I, Geerts H, Mercken M, Sciot R, Van Lommel A, Loos R, Van Leuven F (1999) Prominent axonopathy in the brain and spinal cord of transgenic mice overexpressing four-repeat human tau protein. Am J Pathol 155:2153–2165

Spittaels K, Van den Haute C, Van Dorpe J, Geerts H, Mercken M, Bruynseels K, Lasrado R, Vandezande K, Laenen I, Boon T, Van Lint J, Vandenheede J, Moechars D, Loos R, Van Leuven F (2000) Glycogen synthase kinase-3beta phosphorylates protein tau and rescues the axonopathy in the central nervous system of human four-repeat tau transgenic mice. J Biol Chem 275:41340–41349

Stanford PM, Halliday GM, Brooks WS, Kwok JB, Storey CE, Creasey H, Morris JG, Fulham MJ, Schofield PR (2000) Progressive supranuclear palsy pathology caused by a novel silent mutation in exon 10 of the tau gene: expansion of the disease phenotype caused by tau gene mutations. Brain 123(Pt 5):880–893

Stanford PM, Shepherd CE, Halliday GM, Brooks WS, Schofield PW, Brodaty H, Martins RN, Kwok JB, Schofield PR (2003) Mutations in the tau gene that cause an increase in three repeat tau and frontotemporal dementia. Brain 126:814–826

Steele JC, Richardson JC, Olszewski J (1964) Progressive supranuclear palsy: a heterogeneous degeneration involving the brain stem, basal ganglia and cerebellum with vertical gaze and pseudobulbar palsy, nuchal dystonia and dementia. Arch Neurol 10:333–359

Stefansson H, Helgason A, Thorleifsson G, Steinthorsdottir V, Masson G, Barnard J, Baker A, Jonasdottir A, Ingason A, Gudnadottir VG et al (2005) A common inversion under selection in Europeans. Nat Genet 37:129–137

Sultana R, Boyd-Kimball D, Poon HF, Cai J, Pierce WM, Klein JB, Markesbery WR, Zhou XZ, Lu KP, Butterfield DA (2006) Oxidative modification and down-regulation of Pin1 in Alzheimer's disease hippocampus: a redox proteomics analysis. Neurobiol Aging 27:918–925

Tanemura K, Akagi T, Murayama M, Kikuchi N, Murayama O, Hashikawa T, Yoshiike Y, Park JM, Matsuda K, Nakao S, Sun X, Sato S, Yamaguchi H, Takashima A (2001) Formation of filamentous tau aggregations in transgenic mice expressing V337M human tau. Neurobiol Dis 8:1036–1045

Tatebayashi Y, Miyasaka T, Chui DH, Akagi T, Mishima K, Iwasaki K, Fujiwara M, Tanemura K, Murayama M, Ishiguro K, Planel E, Sato S, Hashikawa T, Takashima A (2002) Tau filament formation and associative memory deficit in aged mice expressing mutant (R406W) human tau. Proc Natl Acad Sci USA 99:13896–13901

Terwel D, Lasrado R, Snauwaert J, Vandeweert E, Van Haesendonck C, Borghgraef P, Van Leuven F (2005) Changed conformation of mutant Tau-P301L underlies the moribund tauopathy, absent in progressive, nonlethal axonopathy of Tau-4R/2 N transgenic mice. J Biol Chem 280:3963–3973

Thorpe JR, Morley SJ, Rulten SL (2001) Utilizing the peptidyl-prolyl cis-trans isomerase pin1 as a probe of its phosphorylated target proteins. Examples of binding to nuclear proteins in a human kidney cell line and to tau in Alzheimer's diseased brain. J Histochem Cytochem 49:97–108

Thorpe JR, Mosaheb S, Hashemzadeh-Bonehi L, Cairns NJ, Kay JE, Morley SJ, Rulten SL (2004) Shortfalls in the peptidyl-prolyl cis-trans isomerase protein Pin1 in neurons are associated with frontotemporal dementias. Neurobiol Dis 17:237–249

Tobin JE, Latourelle JC, Lew MF, Klein C, Suchowersky O, Shill HA, Golbe LI, Mark MH, Growdon JH, Wooten GF et al (2008) Haplotypes and gene expression implicate the MAPT region for Parkinson disease: the GenePD Study. Neurology 71:28–34

Tolnay M, Sergeant N, Ghestem A, Chalbot S, De Vos RA, Jansen Steur EN, Probst A, Delacourte A (2002) Argyrophilic grain disease and Alzheimer's disease are distinguished by their different distribution of tau protein isoforms. Acta Neuropathol 104:425–434

Tomasiewicz HG, Flaherty DB, Soria JP, Wood JG (2002) Transgenic zebrafish model of neurodegeneration. J Neurosci Res 70:734–745

Tomizawa K, Omori A, Ohtake A, Sato K, Takahashi M (2001) Tau-tubulin kinase phosphorylates tau at Ser-208 and Ser-210, sites found in paired helical filament-tau. FEBS Lett 492: 221–227

Tucker KL, Meyer M, Barde YA (2001) Neurotrophins are required for nerve growth during development. Nat Neurosci 4:29–37

Utton MA, Noble WJ, Hill JE, Anderton BH, Hanger DP (2005) Molecular motors implicated in the axonal transport of tau and alpha-synuclein. J Cell Sci 118:4645–4654

Van der Jeugd A, Ahmed T, Burnouf S, Belarbi S, Hamdame M, Grosjean ME, Humez S, Balschun D, Blum D, Buée L, D'hooge R. Hippocampal tauopathy in tau transgenic mice coincides with impaired hippocampus-dependent learning and memory, and attenuated late-phase long-term depression of synaptic transmission. Neurobiol Learn Mem

Vandermeeren M, Mercken M, Vanmechelen E, Six J, van de Voorde A, Martin JJ, Cras P (1993) Detection of tau proteins in normal and Alzheimer's disease cerebrospinal fluid with a sensitive sandwich enzyme-linked immunosorbent assay. J Neurochem 61:1828–1834

van Herpen E, Rosso SM, Serverijnen LA, Yoshida H, Breedveld G, van de Graaf R, Kamphorst W, Ravid R, Willemsen R, Dooijes D, Majoor-Krakauer D, Kros JM, Crowther RA, Goedert M, Heutink P, van Swieten JC (2003) Variable phenotypic expression and extensive tau pathology in two families with the novel tau mutation L315R. Ann Neurol 54:573–581

Varani L, Hasegawa M, Spillantini MG, Smith MJ, Murrell JR, Ghetti B, Klug A, Goedert M, Varani G (1999) Structure of tau exon 10 splicing regulatory element RNA and destabilization by mutations of frontotemporal dementia and parkinsonism linked to chromosome 17. Proc Natl Acad Sci USA 96:8229–8234

Vermersch P, Frigard B, David JP, Fallet-Bianco C, Delacourte A (1992) Presence of abnormally phosphorylated Tau proteins in the entorhinal cortex of aged non-demented subjects. Neurosci Lett 144:143–146

Vermersch P, Sergeant N, Ruchoux MM, Hofmann-Radvanyi H, Wattez A, Petit H, Dwailly P, Delacourte A (1996) Specific tau variants in the brains of patients with myotonic dystrophy. Neurology 47:711–717

Vincent I, Jicha G, Rosado M, Dickson DW (1997) Aberrant expression of mitotic cdc2/cyclin B1 kinase in degenerating neurons of Alzheimer's disease brain. J Neurosci 17:3588–3598

Virdee K, Yoshida H, Peak-Chew S, Goedert M (2007) Phosphorylation of human microtubule-associated protein tau by protein kinases of the AGC subfamily. FEBS Lett 581:2657–2662

Wei ML, Andreadis A (1998) Splicing of a regulated exon reveals additional complexity in the axonal microtubule-associated protein tau. J Neurochem 70:1346–1356

Weingarten MD, Lockwood AH, Hwo SY, Kirschner MW (1975) A protein factor essential for microtubule assembly. Proc Natl Acad Sci USA 72:1858–1862

Wijker M, Wszolek ZK, Wolters EC, Rooimans MA, Pals G, Pfeiffer RF, Lynch T, Rodnitzky RL, Wilhelmsen KC, Arwert F (1996) Localization of the gene for rapidly progressive autosomal dominant parkinsonism and dementia with pallido-ponto-nigral degeneration to chromosome 17q21. Hum Mol Genet 5:151–154

Wilhelmsen KC, Lynch T, Pavlou E, Higgins M, Nygaard TG (1994) Localization of disinhibition-dementia-parkinsonism-amyotrophy complex to 17q21-22. Am J Hum Genet 55:1159–1165

Williamson R, Scales T, Clark BR, Gibb G, Reynolds CH, Kellie S, Bird IN, Varndell IM, Sheppard PW, Everall I, Anderton BH (2002) Rapid tyrosine phosphorylation of neuronal proteins including tau and focal adhesion kinase in response to amyloid-beta peptide exposure: involvement of Src family protein kinases. J Neurosci 22:10–20

Wilson DM, Binder LI (1997) Free fatty acids stimulate the polymerization of tau and amyloid beta peptides. In vitro evidence for a common effector of pathogenesis in Alzheimer's disease. Am J Pathol 150:2181–2195

Wischik CM, Novak M, Edwards PC, Klug A, Tichelaar W, Crowther RA (1988) Structural characterization of the core of the paired helical filament of Alzheimer disease. Proc Natl Acad Sci USA 85:4884–4888

Wittmann CW, Wszolek MF, Shulman JM, Salvaterra PM, Lewis J, Hutton M, Feany MB (2001) Tauopathy in Drosophila: neurodegeneration without neurofibrillary tangles. Science 293: 711–714

Woods YL, Cohen P, Becker W, Jakes R, Goedert M, Wang X, Proud CG (2001) The kinase DYRK phosphorylates protein-synthesis initiation factor eIF2Bepsilon at Ser539 and the microtubule-associated protein tau at Thr212: potential role for DYRK as a glycogen synthase kinase 3-priming kinase. Biochem J 355:609–615

Wszolek ZK, Tsuboi Y, Uitti RJ, Reed L (2000) Two brothers with frontotemporal dementia and parkinsonism with an N279K mutation of the tau gene. Neurology 55:1939

Yasuda M, Kawamata T, Komure O, Kuno S, D'Souza I, Poorkaj P, Kawai J, Tanimukai S, Yamamoto Y, Hasegawa H, Sasahara M, Hazama F, Schellenberg GD, Tanaka C (1999) A mutation in the microtubule-associated protein tau in pallido-nigro-luysian degeneration. Neurology 53:864–868

Yen S, Easson C, Nacharaju P, Hutton M, Yen SH (1999a) FTDP-17 tau mutations decrease the susceptibility of tau to calpain I digestion. FEBS Lett 461:91–95

Yen SH, Hutton M, DeTure M, Ko LW, Nacharaju P (1999b) Fibrillogenesis of tau: insights from tau missense mutations in FTDP-17. Brain Pathol 9:695–705

Yoshida H, Crowther RA, Goedert M (2002) Functional effects of tau gene mutations deltaN296 and N296H. J Neurochem 80:548–551

Yoshida H, Goedert M (2006) Sequential phosphorylation of tau protein by cAMP-dependent protein kinase and SAPK4/p38delta or JNK2 in the presence of heparin generates the AT100 epitope. J Neurochem 99:154–164

Yoshimura N (1990) Alzheimer's neurofibrillary changes in the olfactory bulb in myotonic dystrophy. Clin Neuropathol 9:240–243

Yoshimura N, Otake M, Igarashi K, Matsunaga M, Takebe K, Kudo H (1990) Topography of Alzheimer's neurofibrillary change distribution in myotonic dystrophy. Clin Neuropathol 9:234–239

Yoshiyama Y, Higuchi M, Zhang B, Huang SM, Iwata N, Saido TC, Maeda J, Suhara T, Trojanowski JQ, Lee VM (2007) Synapse loss and microglial activation precede tangles in a P301S tauopathy mouse model. Neuron 53:337–351

Yotsumoto K, Saito T, Asada A, Oikawa T, Kimura T, Uchida C, Ishiguro K, Uchida T, Hasegawa M, Hisanaga SI (2009) Effect of PIN1 or microtubule binding on dephosphorylation of FTDP-17 mutant TAU. J Biol Chem 284(25):16840–16847

Zarranz JJ, Ferrer I, Lezcano E, Forcadas MI, Eizaguirre B, Atares B, Puig B, Gomez-Esteban JC, Fernandez-Maiztegui C, Rouco I, Perez-Concha T, Fernandez M, Rodriguez O, Rodriguez-Martinez AB, de Pancorbo MM, Pastor P, Perez-Tur J (2005) A novel mutation (K317M) in the MAPT gene causes FTDP and motor neuron disease. Neurology 64: 1578–1585

Zemlan FP, Mulchahey JJ, Gudelsky GA (2003) Quantification and localization of kainic acid-induced neurotoxicity employing a new biomarker of cell death: cleaved microtubule-associated protein-tau (C-tau). Neuroscience 121:399–409

Zhang B, Maiti A, Shively S, Lakhani F, McDonald-Jones G, Bruce J, Lee EB, Xie SX, Joyce S, Li C, Toleikis PM, Lee VM, Trojanowski JQ (2005) Microtubule-binding drugs offset tau sequestration by stabilizing microtubules and reversing fast axonal transport deficits in a tauopathy model. Proc Natl Acad Sci USA 102:227–231

Zhang J, Moseley A, Jegga AG, Gupta A, Witte DP, Sartor M, Medvedovic M, Williams SS, Ley-Ebert C, Coolen LM, Egnaczyk G, Genter MB, Lehman M, Lingrel J, Maggio J, Parysek L, Walsh R, Xu M, Aronow BJ (2004) Neural system-enriched gene expression: relationship to biological pathways and neurological diseases. Physiol Genomics 18(2):167–183

Zhou XZ, Kops O, Werner A, Lu PJ, Shen M, Stoller G, Kullertz G, Stark M, Fischer G, Lu KP (2000) Pin1-dependent prolyl isomerization regulates dephosphorylation of Cdc25C and tau proteins. Mol Cell 6:873–883

Zhukareva V, Sundarraj S, Mann D, Sjogren M, Blenow K, Clark CM, McKeel DW, Goate A, Lippa CF, Vonsattel JP, Growdon JH, Trojanowski JQ, Lee VM (2003) Selective reduction of soluble tau proteins in sporadic and familial frontotemporal dementias: an international follow-up study. Acta Neuropathol 105:469–476

Zhukareva V, Vogelsberg-Ragaglia V, Van Deerlin VM, Bruce J, Shuck T, Grossman M, Clark CM, Arnold SE, Masliah E, Galasko D, Trojanowski JQ, Lee VM (2001) Loss of brain tau defines novel sporadic and familial tauopathies with frontotemporal dementia. Ann Neurol 49:165–175

Zody MC, Jiang Z, Fung HC, Antonacci F, Hillier LW, Cardone MF, Graves TA, Kidd JM, Cheng Z, Abouelleil A, Chen L, Wallis J, Glasscock J, Wilson RK, Reily AD, Duckworth J, Ventura M, Hardy J, Warren WC, Eichler EE (2008) Evolutionary toggling of the MAPT 17q21.31 inversion region. Nat Genet 40:1076–1083

Chapter 5
Tauopathy and Brain Aging

Akihiko Takashima

Abstract Rigorous studies investigating tauopathies have revealed that the processes involved in neurofibrillary tangle (NFT) formation, not NFTs themselves, play a key role in neuronal dysfunction and neuronal loss in tauopathies. In neurodegeneration, tau becomes hyperphosphorylated and dislodged from microtubules. In the cytoplasm, the naïve structure of hyperphosphorylated tau transforms into a pathologic structure, which may have a greater affinity for engaging in tau–tau interactions than that of microtubule-bound tau. These tau–tau interactions promote the formation of tau oligomers. When tau oligomers form a β-sheet structure, a granular form of tau appears in the form of sarcosyl-insoluble tau. Granular tau molecules bind together and form tau fibrils. In this sense, changes in tau – such as hyperphosphorylation and the formation of different aggregate species – may affect neuronal function, and NFTs can be thought of as a tombstone of neuronal dysfunction. Changes in tau may be involved in brain aging and may consequently lead to the formation of NFTs in the entorhinal cortex. This chapter introduces the mechanisms underlying neurodegeneration and brain aging that occur through tau dysfunction.

Keywords Tau · Hyperphosphorylation · Neurofibrillary tangle (NFT) · Neurodegenerative disease · Brain aging · GSK-3β

Contents

A. Takashima (✉)
Laboratory for Alzheimer's Disease, Brain Science Institute, Saitama, Japan
e-mail: kenneth@brain.riken.jp

R.A. Nixon, A. Yuan (eds.), *Cytoskeleton of the Nervous System*,
Advances in Neurobiology 3, DOI 10.1007/978-1-4419-6787-9_5,

5.1 Introduction

5.1.1 Tauopathies

Tauopathies are a class of neurodegenerative diseases stemming from the pathologic aggregation of tau protein, which in the human brain is called neurofibrillary tangles (NFTs). NFTs are observed in many age-related neurodegenerative diseases, such as Alzheimer's disease, Pick's disease, progressive supranuclear palsy (PSP), and corticobasal degeneration. Because all NFT-related diseases are associated with neurologic disorders characterized by neuronal dysfunction, NFTs are considered to be a common pathologic marker of neurologic disorders. However, some NFTs in the entorhinal cortex are also present even in clinically nondemented aged persons. When the distribution of NFTs also includes frontotemporal regions and other parts of the neocortex, brain function is impaired, which in terms of clinical diagnosis is manifested as dementia. Thus, distribution of NFTs is closely related to brain dysfunction.

Alois Alzheimer first described NFTs in the brain of one of his demented patients, a disorder now known as Alzheimer's disease (AD). The major component of NFTs is hyperphosphorylated tau. In its normal state, tau is linked to microtubules as a microtubule-associated protein (MAP). In neurodegeneration, tau is hyperphosphorylated, has a reduced affinity for microtubules, and forms aggregates of hyperphosphorylated tau, which is an insoluble form referred to as paired helical filaments (PHFs).

In tauopathies, neuronal death is observed in the same brain regions as NFTs, and both neuronal death and NFTs correlate with the duration and severity of illness in AD, although the amount of neuronal death is many times more than the number of NFTs (Gomez-Isla et al., 1997). Recently, a tau mutation was identified as the causal factor in frontotemporal dementia parkinsonism-17 (FTDP-17), a dementing disease characterized by NFT formation and neuronal loss (Spillantini and Goedert, 1998,

Goedert and Spillantini, 2000; Hutton, 2000, 2001; Ingram and Spillantini, 2002; van Swieten et al., 2004). The elucidation of mutated tau in FTDP-17 conclusively demonstrated that tau dysfunction or abnormality alone could induce neurodegeneration characterized by NFTs and neuronal death, leading to a clinical dementia similar to that found in AD.

Tau dysfunction may also arise through modification of the tau protein in addition to genetic mutation. Phosphorylation is a major modification of the tau protein found in NFTs, and hyperphosphorylation of tau is thought to be the critical step in NFT formation. Although tau abnormalities do not appear to be a single causative factor in AD, they do appear to be a critical intermediate step in at least one of the pathways involved in the ultimate pathology and neuronal dysfunction. Thus, investigations of tau and NFTs are thought to be useful in elucidating intracellular events that occur during neurodegeneration and in increasing our understanding of mechanisms and pathways of neurodegeneration.

5.2 Function of Tau

Tau is a member of the MAP family, and the major role of tau is thought to be to bind and stabilize microtubules. Microtubules have a dynamic structure, which flexibly change in response to cellular signals such as cell division, migration, and differentiation (Caceres and Kosik, 1990). Tau is expressed in several different tissues, but the brain is the major site of tau expression. Here tau is mainly localized in axons (Kosik and Finch, 1987; Kosik and Caceres 1991). In human brain, tau protein has six isoforms (generated through alternative splicing), varying in length from 352 to 441 amino acids (Goedert et al., 1989; Goedert and Jakes, 1990). The N-terminal region has three alternatively spliced forms with exon 2 and/or exon 3 and is required for interacting with the cellular membrane through phospholipase C (Shimohama et al., 1995; Hwang et al., 1996) or the src family kinase, fyn, a non-receptor-type tyrosine kinase (Lee et al., 1998; Lee et al., 2004; Lee, 2005). Moreover, the N-terminal region affects axon diameter by determining the distance between microtubules (Chen et al., 1992).

In the C-terminal region of tau, there are two alternatively spliced forms, with or without exon 10. Exon 10 encodes part of the repeat sequence in the microtubule-binding domain consisting of 31–32 amino acid core sequences with a PGGG motif, called R1, R2, R3, and R4. R1 consists of residues Q244–K274, R2 consists of residues V275–S305, R3 consists of residues V306–Q336, and R4 consists of residues V337–N368. Tau encoded by exon 10 is called 4-repeat tau, whereas tau encoded without exon 10 is called 3-repeat tau. Because exon 10 encodes the R2 amino acid sequence, 3-repeat tau is formed by R1, R3, and R4, whereas 4-repeat tau is formed by R1, R2, R3, and R4. The core region for microtubule binding is R1–R2 in 4-repeat tau and R1–R3 in 3-repeat tau. Because the amino acid sequence 274–281 (R1–R2) shows the highest affinity for microtubules, 3-repeat tau is less able to associate with tubulin and promote the formation of microtubules compared with 4-repeat tau. The P301L and G272V FTDP-17 mutations localize in R1 and

R2 so that P301L and G272V mutant tau are less capable of promoting microtubule formation.

Tau accelerates tubulin polymerization and stabilizes microtubules by binding to them. Whereas tau gene–deficient mice do not show significant morphologic changes – because another MAP, MAP1b, compensates for normal tau function in microtubule formation and stabilization – behavioral analysis of these mice indicates that they have muscle weakness, are hyperactive in a novel environment, and are impaired in fear-conditioning memory. This suggests that tau may have other functions besides promoting microtubule formation and stabilization.

5.3 Phosphorylation of Tau

5.3.1 Tau Phosphorylation Sites

The longest tau isoform (tau 40), which is encoded by exons 2, 3, and 10, contains 45 presumed serine (Ser) phosphorylation sites, 35 presumed threonine (Thr) phosphorylation sites, and 5 presumed tyrosine (Tyr) phosphorylation sites. Mass spectral analysis and analyses employing phosphorylation site-specific antibodies to tau have shown that PHF-tau contains at least 30 phosphorylated Ser and Thr residues (Morishima and Ihara, 1994). Hyperphosphorylated tau is normally present in the brain during late embryonic to early postnatal periods (Brion et al., 1993), when the selection of synapses and destruction of axons is greatest and the resulting naturally occurring neuronal death occurs. During brain development, tau is highly phosphorylated, supporting the idea that flexibility of the microtubule system underlies neural plasticity. Although the PHF-1 epitope (phospho-Ser396, Ser404) is found in distal regions of the axon, the phosphorylated tau-1 epitope (Ser195, Ser198, Ser199, and Ser202) is found in proximal regions of the axon (Otvos et al., 1994; Mandell and Banker, 1996).

In mature brain, tau-1 immunoreactivity is found in axons, and, on average, two phosphates are incorporated into one tau molecule. However, an average of seven phosphates modifies one tau molecule in the developing brain, whereas an average of eight phosphates are incorporated into one tau molecule in the AD brain (Ksiezak-Reding et al., 1992; Kenessey and Yen, 1993; Kopke et al., 1993). Fetal tau, by contrast, is highly phosphorylated, but tau accumulations, such as those observed in AD brain, have not been observed in the developing brain. Furthermore, whereas PHF-tau completely loses the ability to bind to microtubules, highly phosphorylated fetal tau retains this ability, although the extent of microtubule binding is significantly reduced compared with that of tau in the normal mature brain (Bramblett et al., 1993; Yoshida and Ihara, 1993). In AD brain, phosphorylation of tau at some specific sites is observed, including at the Ser262, Ser422, and the AT100 epitope (phospho-Thr212/Ser214) (Morishima and Ihara, 1994). The phosphorylation of these sites might contribute to the formation of tau fibrils and the loss of microtubule-binding ability. A combination of kinases might be required to account for all of the phosphorylation sites of tau that have been observed in PHF-tau, because single kinases phosphorylate specific sites of tau. For example, Ser262 of tau is phosphorylated by partition-defective or microtubule-affinity regulating kinases (PAR-1/MARK), Ser422 is phosphorylated by c-Jun N-terminal kinase (JNK), and the AT100 epitope (phospho-Ser212/Ser214) is formed by being phosphorylated by a combination of JNK and glycogen synthase kinase 3β (GSK-3β) (Sato et al., 2002).

5.3.2 PAR-1/MARK

PAR-1/MARK can phosphorylate tau at Ser262 and Ser356, sites that are within the microtubule-binding domain of tau. Indeed, the phosphorylation of these tau sites reduces the ability of tau to bind microtubules (Schneider et al., 1999). However, phosphorylated tau at Ser262 and Ser356 also reduces tau self-aggregation.

Recently, it was shown that inhibition of PAR-1 in a *Drosophila* model that overexpressed human tau with the FTDP-17 mutation R406W attenuated tau-induced toxicity (Schneider et al., 1999; Nishimura et al., 2004). This attenuation of tau-induced toxicity was accompanied by less phosphorylation at Ser262 and Ser356. The overexpression of human mutant tau containing substitutions of Ser262 and Ser356 to alanine entirely abolished tau-induced neurotoxicity (Chatterjee et al., 2009). Surprisingly, this mutant tau overexpression also inhibited phosphorylation of several other tau sites known to be phosphorylated by different kinases, suggesting that tau phosphorylation by PAR-1/MARK may act as a trigger for subsequent multistep phosphorylation by other kinases. Although the mechanism would need to be clarified, because PAR-1/MARK has recently been reported to be phosphorylated by GSK-3β, which inhibits its activity (Timm et al., 2008), it is conceivable that the activation of PAR-1/MARK dislodges tau from microtubules and primes tau for further phosphorylation, first by JNK, then by GSK-3β. Activation of GSK-3β inhibits PAR-1/MARK and dephosphorylates Ser262 and Ser356 sites of hyperphosphorylated tau, leading to tau aggregation and neurodegeneration.

5.3.3 JNK

Tau phosphorylation at Ser422 is specifically observed in AD brain; however, neither GSK-3 nor PAR-1/MARK phosphorylates Ser422. Although extracellular signal-regulated kinase (ERK) is known to be capable of phosphorylating tau at Ser422 in vitro, the activation of ERK tends to inhibit tau phosphorylation through the inhibition of GSK-3, and ERK itself does not phosphorylate tau in vivo (Lovestone et al., 1994; Latimer et al., 1995). This led to the search of other kinases responsible for phosphorylating Ser422.

JNK was first cloned as a stress-activating kinase that responded to cellular stress, such as exposure to ultraviolet light or tumor necrosis factor (TNF) (Yan et al., 1994). JNK has three isoforms and belongs to the MAP kinase family. In vitro, it was confirmed that JNK could phosphorylate tau at Ser422 (Reynolds et al., 1997). When JNK and the JNK activator, delta MAP kinase kinase kinase (MEKK), are overexpressed with tau in COS7 cells, active JNK robustly phosphorylated tau at Ser202, Thr205, and Ser422; moderately at Thr212, Ser214, Thr181, Thr231, Ser235, and Ser396; and not at all at Ser199, Ser262, and Ser404. When active JNK and GSK-3 are coexpressed with tau in COS7 cells, all potential phosphorylation sites of tau become highly phosphorylated (Sato et al., 2002). Moreover, the resulting phospho-tau has AT100 immunoreactivity, which is specifically observed in PHF-tau. In addition, highly phosphorylated tau was recovered in the sodium dodecyl sulfate SDS-insoluble fraction, suggestive of aggregation, and observation by electron microscopy revealed that tau aggregates formed in the cytoplasm (Sato et al., 2002). These results suggest that hyperphosphorylation of tau is sufficient to trigger the transition from soluble tau to insoluble tau fibril formation.

GSK-3β has been shown to phosphorylate and activate MEKK1, consequently leading to activation of JNK (Kim et al., 2003). Thus, when neurons begin to receive fewer extracellular signals, such as neurotrophic factors and neurotransmitters, GSK-3β may be activated through inactivation of Ser/Thr protein kinase (AKT) signaling in response to weak extracellular survival stimuli, and activation of GSK-3β may simultaneously activate JNK. Activating both kinases may then induce high levels of tau phosphorylation resulting in identical epitopes to those of PHF-tau. Interestingly, active JNK is observed at tau granules in early stages of NFT formation (Lagalwar et al., 2006).

5.3.4 GSK-3β

GSK-3β was initially called tau protein kinase 1 (TPK1). It was isolated from the microtubule fraction of bovine brain as a tau kinase and was found to induce hyperphosphorylation of tau at the same epitopes as PHF-tau (Ishiguro et al., 1992). Previously, Woodgett and colleagues cloned a kinase for glycogen synthase, called GSK-3, which was a rate-limiting enzyme for glycogen biosynthesis (Woodgett, 1990). Cloning of TPK1 revealed that TPK1 was identical to GSK-3β (Ishiguro et al., 1993).

Expression levels of GSK-3β are higher in developing brains than in mature brains (Takahashi et al., 1994; Takahashi et al., 2000). Although GSK-3β is found predominately in the perikarya and proximal parts of neuronal dendrites in adults, it can also be found in embryonic axons (Takahashi et al., 1994; Leroy and Brion, 1999). Importantly, GSK-3β is activated by amyloid beta (Aβ) in hippocampal primary cell cultures, leading to hyperphosphorylation of tau and subsequent neuronal death (Takashima et al., 1993). This was confirmed in mutant tau-overexpressing mice treated with a GSK-3 inhibitor (Perez et al., 2003; Noble et al., 2005).

Unlike many other kinases, GSK-3β is constitutively active in resting cells, and this activity is regulated by Ser9 phosphorylation at the phosphorylation site of protein kinase A (PKA), PKB (also called AKT), PKC, p90Rsk, and p70 S6 kinase (Grimes and Jope, 2001). Cell survival signals are known to activate PKA, PKB, and PKC, thereby inactivating GSK-3β. Thus, GSK-3β can be activated by cell death signals, such as the experimental reduction of growth factors.

The phosphoserine-binding protein 14-3-3 forms a complex that connects the inactivated form of GSK-3β, phospho-Ser9 GSK-3β, to tau, which stimulates GSK-3β phosphorylation of tau (Yuan et al., 2004). The active form of GSK-3β can phosphorylate many substrates and participates in various cellular events, including metabolism, signaling, and transcription (Grimes and Jope, 2001). Although GSK-3 is categorized as a proline-directed kinase, most GSK-3 targets are phosphorylated by another kinase before they can be phosphorylated by GSK-3β. GSK-3β binds to the resulting phosphate at position P_0+4 through the GSK-3β "priming pocket" and transfers a phosphate to position P_0 (Frame et al., 2001; Harwood, 2001). This phosphate transfer requires Arg96 to form the priming pocket and is an important feature of GSK-3β (Frame et al., 2001; Hagen et al., 2002). For example, after Ser404 is phosphorylated, GSK-3β recognizes this phosphorylation and sequentially phosphorylates Ser400 and Ser396 of tau. Thus, GSK-3β is a potent tau kinase in terms of forming highly phosphorylated, PHF-tau.

5.3.5 Effects of Tau Phosphorylation

Phosphorylation of tau at Ser262 and Ser356 by PAR-1/MARK dislodges tau from microtubules. In *Drosophila* neurons, when Ser262 and Ser356 of tau are substituted with Ala, tau is not highly phosphorylated, thus reducing cell death (Nishimura et al., 2004; Chatterjee et al., 2009). Therefore, kinases may be prevented from accessing tau on microtubules but can phosphorylate tau in the cytoplasm. In this way, hyperphosphorylated tau may gain a toxic function.

Conformation of tau may be affected by phosphorylation state. Jeganathan and colleagues reported that tau in solution adopts a "paperclip" conformation (Jeganathan et al., 2008), and pseudophosphorylation of the AT8 site or PHF-1 sites opened up the paperclip conformation. Pseudophosphorylation of both AT8 and PHF-1 sites causes tau to form a compact paperclip conformation, and additional phosphorylation induces a pathologic conformational change in tau (Jeganathan et al., 2008). This tau conformational change may lead to the formation of NFTs.

5.4 Tau Aggregation and Neurodegeneration

5.4.1 Tau Fibril Formation Process

One FTDP-17 mouse model displays age-related NFTs, neuronal death, and behavioral deficits. FTDP-17 mice overexpress P301L mutant tau under the regulation of a tetracycline-inducible promoter. Although inhibiting mutant tau overexpression in these mice blocks neuronal death and improves memory, NFTs continue to form (Santacruz et al., 2005). NFT-bearing neurons are associated with caspase activation but do not show acute apoptosis (Spires-Jones et al., 2008). These findings suggest that NFTs are not themselves toxic, but instead that the mechanism of NFT formation is shared by the process underlying neuronal death and neuronal dysfunction.

Tau fibril formation has been studied extensively in vitro. Anionic surfactants accelerate fibril formation of tau protein in vitro (Barghorn and Mandelkow, 2002). Fibril formation can be monitored with thioflavin (ThT) fluorescence; ThT recognizes aggregations having a β-sheet conformation. We applied atomic force microscopy (AFM) to tau in solution to track structural changes in tau. To understand the relationship between these different tau aggregates, we investigated how tau assembly in vitro changes over time by measuring ThT fluorescence and by using AFM (Maeda et al., 2007).

Before ThT fluorescence intensity increases, tau forms oligomers (dimers to octamers) as incubation time increases; however, these oligomers are not visible with AFM (Sahara et al., 2007). These tau oligomers were originally analyzed under nonreducing SDS-gel conditions. However, under reducing conditions, tau oligomers (except dimers) disappear, suggesting that tau oligomers form through disulfide bonds and other SDS-resistant tau–tau associations. At increased ThT fluorescence intensities, two forms of tau aggregate are observed under AFM. One is a granular form of tau aggregate, and the other is a fibrillar form of tau aggregate. Laser light scattering analysis of granular tau aggregates purified by sucrose gradient centrifugation indicated that granular tau aggregates consist of approximately 40 tau molecules. Because increased concentrations of granular tau aggregate induce tau fibril formation, granular tau is thought to be an intermediate form of tau fibril (Maeda et al., 2007). As shown in Fig. 5.1, monomeric tau first binds together through disulfide bonds and SDS-resistant interactions to form tau oligomers, which

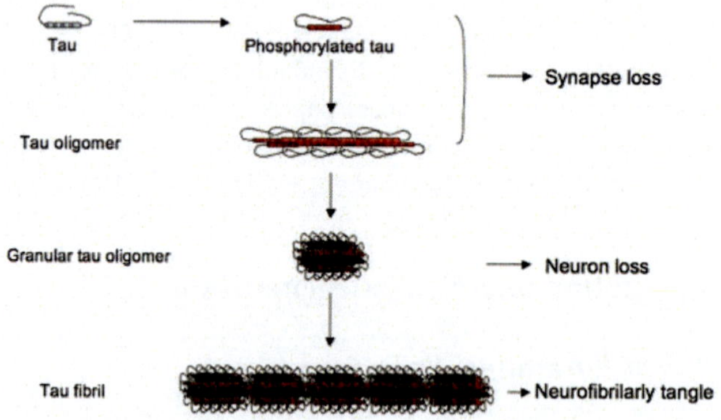

Fig. 5.1 Tau fibril formation process and its corresponding role in neurodegeneration. Hyperphosphorylated tau dislodges from microtubules and then binds together to form tau oligomers (i.e., oligomers that are detected on SDS-PAGE but are not detectable by ThT staining or AFM). Next, granular tau oligomers form before NFTs are formed. In vivo studies showed that hyperphosphorylated tau induces neuronal dysfunction, but NFTs do not. Therefore, hyperphosphorylated tau oligomers induce synapse loss, whereas granular tau oligomers may represent a potent toxic species of tau that is responsible for neuron loss

are not visible with AFM. Forty tau molecules bind together and take on a β-sheet structure; these aggregates appear granular in shape with AFM. When granular tau levels increase, granular tau aggregates stick together and form tau fibrils. Therefore, tau forms two different types of aggregate – (1) tau oligomers, which are not detectable by AFM, and (2) granular tau oligomers – before forming tau fibrils.

Increased levels of granular tau oligomer are observed in frontal cortex during Braak stage I, whereas NFTs do not appear in frontal cortex until Braak stage V, suggesting that increased granular tau oligomer formation precedes NFT formation (Maeda et al., 2006). Even nondemented elderly individuals display reduced frontal cortical function, and frontal lobe volume decreases with aging (Buckner, 2004). Therefore, granular tau oligomers might be involved in neuronal loss.

5.4.2 Mouse Model of Tauopathy

Understanding the role of tau on neurodegeneration can be achieved by analyzing tau Tg Transgenic Mouse. The occurrence of tau mutations in FTDP-17 strongly supports the hypothesis that tau abnormalities induce the NFT formation and neuronal loss that ultimately lead to dementia (Goedert and Spillantini, 2000; Hutton, 2000; Spillantini et al., 2000). Therefore, studying and understanding the mechanisms underlying NFT formation may lead to new strategies for preventing tauopathy-associated dementias (Gotz et al., 2000; Lewis et al., 2000; Gotz et al., 2001; Lewis et al., 2001; Tanemura et al., 2001; Tanemura et al., 2002; Tatebayashi et al., 2002; Gotz et al., 2004; Egashira et al., 2005; Lee et al., 2005; Taniguchi et al., 2005; Yoshiyama et al., 2005; Murakami et al., 2006). Indeed, Tg mice that harbor FTDP-17 mutant tau display NFTs, neuronal death, and behavioral deficits.

Although most mouse models of FTDP-17 mutant tau exhibit NFTs, neuronal loss, and behavioral deficits, one model does not: 12-month-old mice expressing mutant N279K tau exhibit behavioral deficits but do not display NFTs and neuronal loss (Taniguchi et al., 2005). These findings suggest that behavioral deficits might occur before NFTs form and neuronal death occurs.

Kimura and colleagues (Kimura et al., 2007) generated a Tg mouse line that expresses wild-type human tau. These mice exhibit impaired place learning and memory in the absence of NFT formation or neuronal loss at old age. To determine which brain region's activity is linked to impairment of place learning in this mouse model, we analyzed the relationship between age-dependent tau-induced changes in behavior and the corresponding activity in all brain regions using Mn-enhanced MRI (MEM) methods. MEM showed that the activity of the parahippocampal area strongly correlates with the decline of memory as assessed by the Morris water maze. Moreover, hyperphosphorylated tau and synapse loss are found in the same regions where hyperphosphorylated tau and synapse loss normally occur in old age. Taken together, the accumulation of hyperphosphorylated tau in aged mice, leading to inhibition of neural activity due to synapse loss in parahippocampal areas including the entorhinal cortex, may underlie place learning impairment in Tg mice that express wild-type human tau. Thus, the accumulation of hyperphosphorylated tau that occurs before NFT formation in the entorhinal cortex may contribute to neuronal dysfunction at old age.

5.5 Brain Aging

5.5.1 Tau and Brain Aging

Although amyloid deposition, which is a hallmark of AD, is observed in older individuals, it is relatively uncommon in the brains of individuals who died without showing signs of dementia. On the other hand, NFTs, a tombstone of neuronal dysfunction, are rarely seen in nondemented elderly people. As individuals grow older, they experience memory loss and cognitive slowing that can interfere with their daily routine. Thus, a person can experience memory loss and cognitive slowing even though they do not have dementia. Two types of brain dysfunction are commonly observed in the elderly: One is diminished memory recall and disruption of executive functions, which rely on prefrontal cortex, and the other is decline of long-term memory, which relies on the entorhinal cortex and hippocampus. NFTs are commonly observed in most elderly people over the age of 75 years. When NFTs occur in the entorhinal cortex, increased granular tau levels are also observed in prefrontal cortex, suggesting that the tau aggregation process may underlie manifestations associated with age-related brain dysfunction.

5.5.2 Development of NFTs and Brain Dysfunction

Braak and colleagues defined the progression of disease in terms of six stages based on the distribution of NFTs in the brain (Braak and Braak, 1996). In Braak stage I, NFTs are formed in the transentorhinal cortex and the CA1 region of the hippocampus. The number of NFTs increase in Braak stage II, and Braak stages I and II together are called the transentorhinal stage. Braak stages I and II are classified as normal aging. In Braak stages III and IV, called the limbic stage, many ghost tangles appear in the entorhinal cortex, and NFTs are found throughout the entire limbic system, in hippocampal regions CA1–4, and the amygdala. In the limbic stage, patients show various AD-specific symptoms, such as memory impairment, reduced spatial cognition, and reduced desire, as a result of neural dysfunction in the limbic system. In Braak stages V and VI, called the isocortical stage, NFTs in the cerebral cortex impair cortical neural function, resulting in dementia. This increasing extension of NFTs correlates with increasing impairment of brain function.

Figure 5.2 summarizes the relationship between NFTs and Aβ deposition during aging. During the transentorhinal stage, Aβ deposition varies, sometimes with no Aβ deposition. The prevalence of the transentorhinal stage increases with age, independent of Aβ deposition (which peaks at approximately 75 years of age, and then decreases). However, the prevalence of NFTs in Braak stages III–VI increases after Aβ deposition. Thus, NFT formation in normal aging is not dependent on Aβ

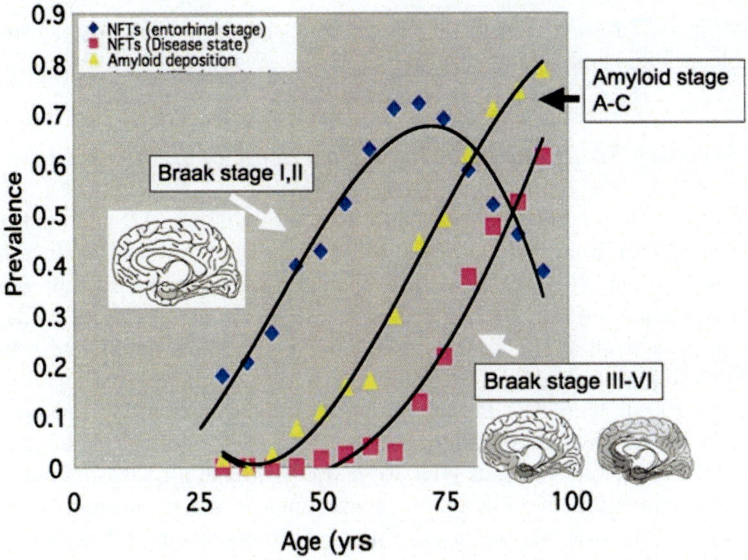

Fig. 5.2 Relationship between NFT formation and Aβ deposition

deposition. After NFTs form in the entorhinal cortex, the spread of NFTs through the limbic system and neocortex seen in AD may require Aβ deposition. Thus, the chronology of the pathologic changes seen in AD seems to start with the appearance of NFTs in the entorhinal cortex independent of Aβ deposition, which may be caused by aging-related factors. This is followed by Aβ deposition, which triggers the spread of NFTs into the limbic system and neocortex (Fig. 5.3). This chronology

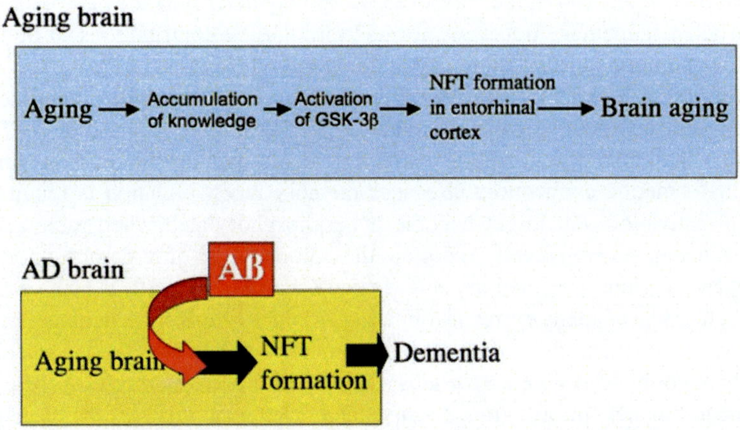

Fig. 5.3 Development of AD requires brain aging. Aβ accelerates the aging process, leading to dementia

suggests that NFT formation in the entorhinal cortex may reflect brain aging and that Aβ accelerates this brain aging, resulting in AD.

5.5.3 Memory Maintenance May Be a Factor of Brain Aging

GSK-3β is a tau kinase that induces the formation of NFTs in AD and tauopathies (Bhat et al., 2004). In an animal model of AD, increased Aβ levels induce memory impairment due to learning deficits that are accompanied by inhibition of long term potentiation (LTP) (Chapman et al., 1999). In a cell culture model, Aβ was shown to activate GSK-3β, leading to tau hyperphosphorylation and neuronal death (Takashima et al., 1993; Takashima et al., 1996; Takashima et al., 1998; Takashima, 2006). Moreover, the inhibition of GSK-3β rescued Aβ-induced neuronal death and tau hyperphosphorylation. In an FTDP-17 animal model, lithium, which inhibits GSK-3β, prevents NFT formation (Perez et al., 2003; Noble et al., 2005). Overexpression of GSK-3β in mouse induces tau hyperphosphorylation and decline in cognitive performance, which is accompanied by inhibition of LTP (Hernandez et al., 2002; Hooper et al., 2007). Cognitive performance decline is due to impairment of learning. Therefore, Aβ constitutively activates GSK-3β, leading to impairment of learning through inhibition of LTP. Even in normal aging, NFTs occur in the entorhinal cortex without Aβ deposition. This may represent a very early pathologic change that occurs in sporadic AD, suggesting that GSK-3β may be activated also in the aged brain through aging factors.

Analyses of GSK-3β heterozygote mice have helped us to understand the role of GSK-3β activation in the brain, showing that GSK-3β is required for memory reconsolidation (Kimura et al., 2008). Recently, GSK-3β activation has been reported to be required for NMDA-dependent long term depression (LTD) induction (Peineau et al., 2007). Because activation of GSK-3β can induce LTD but inhibit LTP (Hooper et al., 2007), the state of GSK-3β may determine the nature of synaptic plasticity. If memory traces are formed through the contributions of both LTP and LTD, memory consolidation may preferentially depend on LTP for the ultimate formation of stable memory traces but on LTD to maintain prior potentiated circuits through competitive synaptic maintenance (Diamond et al., 2005). GSK-3β activation may also play a role in protecting stable memory traces from forming additional synaptic connections. Because memory reconsolidation is required for GSK-3β activation, tau might become hyperphosphorylated when memories are reconsolidated. As we recently reported, the accumulation of hyperphosphorylated tau impairs learning and memory in old age (Kimura et al., 2007). Therefore, frequent activation of memory reconsolidation in older brains may damage learning and memory processes through the hyperphosphorylation of tau. Thus, when elderly people are confronted with a new idea, they recall related memories to help them understand this new information, because memories accumulate as the elderly age. The frequent need to recall and reconsolidate memories relies on the activation of GSK-3β, which leads to tau phosphorylation and aggregation and ultimately to brain aging.

5.6 Conclusions

Rigorous studies of tauopathies have demonstrated that processes underlying NFT formation, not NFTs themselves, play a role in neuronal dysfunction and neuronal loss in tauopathies. In this sense, some changes in tau may affect neuronal function without resulting in the formation of pathologic tau aggregates. During the aging process, tau may be involved in brain aging, during which changes in tau consequently lead to the formation of NFTs in the entorhinal cortex. In addition, function of the prefrontal cortex and the entorhinal cortex and hippocampus become partially impaired. Granular tau appears in the prefrontal cortex when NFTs occur in the entorhinal cortex, suggesting that tau changes may play a role in brain dysfunction with aging. However, elderly people with NFTs in the entorhinal cortex are clinically normal. This suggests the existence of a compensatory mechanism whose function is to maintain normal brain function. This hypothesis is consistent with recent findings showing that although older adults have reduced functional connectivity between the hippocampus and parietotemporal network, they have increased connectivity between rhinal and frontal cortical network (Grady, 2008), indicating that older adults compensate for hippocampal deficits by relying more on the rhinal cortex, possibly through top-down frontal modulation. This type of compensation plays an important role in maintaining normal brain function in the elderly. More studies are needed to examine the mechanisms underlying brain compensation. Understanding this mechanism may pave the way toward developing strategies to resist brain aging.

Currently, most studies on neurodegenerative diseases, including AD, tauopathies, Parkinson's disease, and Huntington's disease, focus on the molecular mechanisms underlying neurodegeneration. However, this approach seems to overlook mechanisms that underlie disease progression ultimately leading to dementia. Explaining the progression of disease is definitely an important target for the study of neurodegenerative diseases, as doing so may provide a new therapeutic way to treat these diseases from a different approach.

References

Barghorn S, Mandelkow E (2002) Toward a unified scheme for the aggregation of tau into Alzheimer paired helical filaments. Biochemistry 41:14885–14896

Bhat RV, Budd Haeberlein SL, Avila J (2004) Glycogen synthase kinase 3: a drug target for CNS therapies. J Neurochem 89:1313–1317

Braak H, Braak E (1996) Evolution of the neuropathology of Alzheimer's disease. Acta Neurol Scand Suppl 165:3–12

Bramblett GT, Goedert M, Jakes R, Merrick SE, Trojanowski JQ, Lee. VM (1993) Abnormal tau phosphorylation at Ser396 in Alzheimer's disease recapitulates development and contributes to reduced microtubule binding. Neuron 10:1089–1099

Brion JP, Smith C, Couck AM, Gallo JM, Anderton. BH (1993) Developmental changes in tau phosphorylation: fetal tau is transiently phosphorylated in a manner similar to paired helical filament-tau characteristic of Alzheimer's disease. J Neurochem 61:2071–2080

Buckner RL (2004) Memory and executive function in aging and AD: multiple factors that cause decline and reserve factors that compensate. Neuron 44:195–208

Caceres A, Kosik. KS (1990) Inhibition of neurite polarity by tau antisense oligonucleotides in primary cerebellar neurons. Nature 343:461–463

Chapman PF, White GL, Jones MW, Cooper-Blacketer D, Marshall VJ, Irizarry M, Younkin L, Good MA, Bliss TV, Hyman BT, Younkin SG, Hsiao. KK (1999) Impaired synaptic plasticity and learning in aged amyloid precursor protein transgenic mice. Nat Neurosci 2:271–276

Chatterjee S, Sang TK, Lawless GM, Jackson. GR (2009) Dissociation of tau toxicity and phosphorylation: role of GSK-3beta, MARK and Cdk5 in a Drosophila model. Hum Mol Genet 18:164–177

Chen J, Kanai Y, Cowan NJ, Hirokawa. N (1992) Projection domains of MAP2 and tau determine spacings between microtubules in dendrites and axons. Nature 360:674–677

Diamond DM, Park CR, Campbell AM, Woodson. JC (2005) Competitive interactions between endogenous LTD and LTP in the hippocampus underlie the storage of emotional memories and stress-induced amnesia. Hippocampus 15:1006–1025

Egashira N, Iwasaki K, Takashima A, Watanabe T, Kawabe H, Matsuda T, Mishima K, Chidori S, Nishimura R, Fujiwara. M (2005) Altered depression-related behavior and neurochemical changes in serotonergic neurons in mutant R406W human tau transgenic mice. Brain Res 1059:7–12

Frame S, Cohen P, Biondi. RM (2001) A common phosphate binding site explains the unique substrate specificity of GSK3 and its inactivation by phosphorylation. Mol Cell 7:1321–1327

Goedert M, Jakes. R (1990) Expression of separate isoforms of human tau protein: correlation with the tau pattern in brain and effects on tubulin polymerization. EMBO J 9:4225–4230

Goedert M, Spillantini. MG (2000) Tau mutations in frontotemporal dementia FTDP-17 and their relevance for Alzheimer's disease. Biochim Biophys Acta 1502:110–121

Goedert M, Spillantini MG, Jakes R, Rutherford D, Crowther. RA (1989) Multiple isoforms of human microtubule-associated protein tau: sequences and localization in neurofibrillary tangles of Alzheimer's disease. Neuron 3:519–526

Gomez-Isla T, Hollister R, West H, Mui S, Growdon JH, Petersen RC, Parisi JE, Hyman. BT (1997) Neuronal loss correlates with but exceeds neurofibrillary tangles in Alzheimer's disease. Ann Neurol 41:17–24

Gotz J, Barmettler R, Ferrari A, Goedert M, Probst A, Nitsch. RM (2000) In vivo analysis of wild-type and FTDP-17 tau transgenic mice. Ann N Y Acad Sci 920:126–133

Gotz J, Chen F, van Dorpe J, Nitsch. RM (2001) Formation of neurofibrillary tangles in P301l tau transgenic mice induced by Abeta 42 fibrils. Science 293:1491–1495

Gotz J, Streffer JR, David D, Schild A, Hoerndli F, Pennanen L, Kurosinski P, Chen. F (2004) Transgenic animal models of Alzheimer's disease and related disorders: histopathology, behavior and therapy. Mol Psychiatry 9:664–683

Grady CL (2008) Cognitive neuroscience of aging. Ann N Y Acad Sci 1124:127–144

Grimes CA, Jope. RS (2001) The multifaceted roles of glycogen synthase kinase 3beta in cellular signaling. Prog Neurobiol 65:391–426

Hagen T, Di Daniel E, Culbert AA, Reith. AD (2002) Expression and characterization of GSK-3 mutants and their effect on beta-catenin phosphorylation in intact cells. J Biol Chem 277:23330–23335

Harwood AJ (2001) Regulation of GSK-3: a cellular multiprocessor. Cell 105:821–824

Hernandez F, Borrell J, Guaza C, Avila J, Lucas. JJ (2002) Spatial learning deficit in transgenic mice that conditionally over-express GSK-3beta in the brain but do not form tau filaments. J Neurochem 83:1529–1533

Hooper C, Markevich V, Plattner F, Killick R, Schofield E, Engel T, Hernandez F, Anderton B, Rosenblum K, Bliss T, Cooke SF, Avila J, Lucas JJ, Giese KP, Stephenson J, Lovestone. S (2007) Glycogen synthase kinase-3 inhibition is integral to long-term potentiation. Eur J Neurosci 25:81–86

Hutton M (2000) Molecular genetics of chromosome 17 tauopathies. Ann N Y Acad Sci 920:63–73

Hutton M (2001) Missense and splice site mutations in tau associated with FTDP-17: multiple pathogenic mechanisms. Neurology 56:S21–S25

Hwang SC, Jhon DY, Bae YS, Kim JH, Rhee. SG (1996) Activation of phospholipase C-gamma by the concerted action of tau proteins and arachidonic acid. J Biol Chem 271:18342–18349
Ingram EM, Spillantini. MG (2002) Tau gene mutations: dissecting the pathogenesis of FTDP-17. Trends Mol Med 8:555–562
Ishiguro K, Shiratsuchi A, Sato S, Omori A, Arioka M, Kobayashi S, Uchida T, Imahori. K (1993) Glycogen synthase kinase 3 beta is identical to tau protein kinase I generating several epitopes of paired helical filaments. FEBS Lett 325:167–172
Ishiguro K, Takamatsu M, Tomizawa K, Omori A, Takahashi M, Arioka M, Uchida T, Imahori. K (1992) Tau protein kinase I converts normal tau protein into A68-like component of paired helical filaments. J Biol Chem 267:10897–10901
Jeganathan S, Hascher A, Chinnathambi S, Biernat J, Mandelkow EM, Mandelkow. E (2008) Proline-directed pseudo-phosphorylation at AT8 and PHF1 epitopes induces a compaction of the paperclip folding of Tau and generates a pathological (MC-1) conformation. J Biol Chem 283:32066–32076
Kenessey A, Yen. SH (1993) The extent of phosphorylation of fetal tau is comparable to that of PHF-tau from Alzheimer paired helical filaments. Brain Res 629:40–46
Kim JW, Lee JE, Kim MJ, Cho EG, Cho SG, Choi. EJ (2003) Glycogen synthase kinase 3 beta is a natural activator of mitogen-activated protein kinase/extracellular signal-regulated kinase kinase kinase 1 (MEKK1). J Biol Chem 278:13995–14001
Kimura T, Yamashita S, Fukuda T, Park JM, Murayama M, Mizoroki T, Yoshiike Y, Sahara N, Takashima. A (2007) Hyperphosphorylated tau in parahippocampal cortex impairs place learning in aged mice expressing wild-type human tau. EMBO J 26:5143–5152
Kimura T, Yamashita S, Nakao S, Park JM, Murayama M, Mizoroki T, Yoshiike Y, Sahara N, Takashima. A (2008) GSK-3beta is required for memory reconsolidation in adult brain. PLoS One 3:e3540
Kopke E, Tung YC, Shaikh S, Alonso AC, Iqbal K, Grundke-Iqbal. I (1993) Microtubule-associated protein tau. Abnormal phosphorylation of a non-paired helical filament pool in Alzheimer disease. J Biol Chem 268:24374–24384
Kosik KS, Caceres. A (1991) Tau protein and the establishment of an axonal morphology. J Cell Sci Suppl 15:69–74
Kosik KS, Finch. EA (1987) MAP2 and tau segregate into dendritic and axonal domains after the elaboration of morphologically distinct neurites: an immunocytochemical study of cultured rat cerebrum. J Neurosci 7:3142–3153
Ksiezak-Reding H, Liu WK, Yen. SH (1992) Phosphate analysis and dephosphorylation of modified tau associated with paired helical filaments. Brain Res 597:209–219
Lagalwar S, Guillozet-Bongaarts AL, Berry RW, Binder. LI (2006) Formation of phospho-SAPK/JNK granules in the hippocampus is an early event in Alzheimer disease. J Neuropathol Exp Neurol 65:455–464
Latimer DA, Gallo JM, Lovestone S, Miller CC, Reynolds CH, Marquardt B, Stabel S, Woodgett JR, Anderton. BH (1995) Stimulation of MAP kinase by v-raf transformation of fibroblasts fails to induce hyperphosphorylation of transfected tau. FEBS Lett 365:42–46
Lee G (2005) Tau and src family tyrosine kinases. Biochim Biophys Acta 1739:323–330
Lee VM, Kenyon TK, Trojanowski. JQ (2005) Transgenic animal models of tauopathies. Biochim Biophys Acta 1739:251–259
Lee G, Newman ST, Gard DL, Band H, Panchamoorthy. G (1998) Tau interacts with src-family non-receptor tyrosine kinases. J Cell Sci 111(Pt 21):3167–3177
Lee G, Thangavel R, Sharma VM, Litersky JM, Bhaskar K, Fang SM, Do LH, Andreadis A, Van Hoesen G, Ksiezak-Reding. H (2004) Phosphorylation of tau by fyn: implications for Alzheimer's disease. J Neurosci 24:2304–2312
Leroy K, Brion. JP (1999) Developmental expression and localization of glycogen synthase kinase-3beta in rat brain. J Chem Neuroanat 16:279–293
Lewis J, Dickson DW, Lin WL, Chisholm L, Corral A, Jones G, Yen SH, Sahara N, Skipper L, Yager D, Eckman C, Hardy J, Hutton M, McGowan. E (2001) Enhanced

neurofibrillary degeneration in transgenic mice expressing mutant tau and APP. Science 293: 1487–1491

Lewis J, McGowan E, Rockwood J, Melrose H, Nacharaju P, Van Slegtenhorst M, Gwinn-Hardy K, Paul Murphy M, Baker M, Yu X, Duff K, Hardy J, Corral A, Lin WL, Yen SH, Dickson DW, Davies P, Hutton. M (2000) Neurofibrillary tangles, amyotrophy and progressive motor disturbance in mice expressing mutant (P301L) tau protein. Nat Genet 25:402–405

Lovestone S, Reynolds CH, Latimer D, Davis DR, Anderton BH, Gallo JM, Hanger D, Mulot S, Marquardt B, Stabel S et al (1994) Alzheimer's disease-like phosphorylation of the microtubule-associated protein tau by glycogen synthase kinase-3 in transfected mammalian cells. Curr Biol 4:1077–1086

Maeda S, Sahara N, Saito Y, Murayama S, Ikai A, Takashima. A (2006) Increased levels of granular tau oligomers: an early sign of brain aging and Alzheimer's disease. Neurosci Res 54:197–201

Maeda S, Sahara N, Saito Y, Murayama M, Yoshiike Y, Kim H, Miyasaka T, Murayama S, Ikai A, Takashima. A (2007) Granular tau oligomers as intermediates of tau filaments. Biochemistry 46:3856–3861

Mandell JW, Banker. GA (1996) Microtubule-associated proteins, phosphorylation gradients, and the establishment of neuronal polarity. Perspect Dev Neurobiol 4:125–135

Morishima M, Ihara. Y (1994) Posttranslational modifications of tau in paired helical filaments. Dementia 5:282–288

Murakami T, Paitel E, Kawarabayashi T, Ikeda M, Chishti MA, Janus C, Matsubara E, Sasaki A, Kawarai T, Phinney AL et al (2006) Cortical neuronal and glial pathology in TgTauP301L transgenic mice: neuronal degeneration, memory disturbance, and phenotypic variation. Am J Pathol 169:1365–1375

Nishimura I, Yang Y, Lu. B (2004) PAR-1 kinase plays an initiator role in a temporally ordered phosphorylation process that confers tau toxicity in Drosophila. Cell 116:671–682

Noble W, Planel E, Zehr C, Olm V, Meyerson J, Suleman F, Gaynor K, Wang L, LaFrancois J, Feinstein B, Burns M, Krishnamurthy P, Wen Y, Bhat R, Lewis J, Dickson D, Duff. K (2005) Inhibition of glycogen synthase kinase-3 by lithium correlates with reduced tauopathy and degeneration in vivo. Proc Natl Acad Sci USA 102:6990–6995

Otvos L Jr, Feiner L, Lang E, Szendrei GI, Goedert M, Lee. VM (1994) Monoclonal antibody PHF-1 recognizes tau protein phosphorylated at serine residues 396 and 404. J Neurosci Res 39:669–673

Peineau S, Taghibiglou C, Bradley C, Wong TP, Liu L, Lu J, Lo E, Wu D, Saule E, Bouschet T, Matthews P, Isaac JTR, Bortolotto ZA, Wang YT, Collingridge. GL (2007) LTP Inhibits LTD in the hippocampus via regulation of gSK3[beta]. Neuron 53:703–717

Perez M, Hernandez F, Lim F, Diaz-Nido J, Avila. J (2003) Chronic lithium treatment decreases mutant tau protein aggregation in a transgenic mouse model. J Alzheimers Dis 5:301–308

Reynolds CH, Utton MA, Gibb GM, Yates A, Anderton. BH (1997) Stress-activated protein kinase/c-jun N-terminal kinase phosphorylates tau protein. J Neurochem 68:1736–1744

Sahara N, Maeda S, Murayama M, Suzuki T, Dohmae N, Yen SH, Takashima. A (2007) Assembly of two distinct dimers and higher-order oligomers from full-length tau. Eur J Neurosci 25: 3020–3029

Santacruz K, Lewis J, Spires T, Paulson J, Kotilinek L, Ingelsson M, Guimaraes A, DeTure M, Ramsden M, McGowan E, Forster C, Yue M, Orne J, Janus C, Mariash A, Kuskowski M, Hyman B, Hutton M, Ashe. KH (2005) Tau suppression in a neurodegenerative mouse model improves memory function. Science 309:476–481

Sato S, Tatebayashi Y, Akagi T, Chui DH, Murayama M, Miyasaka T, Planel E, Tanemura K, Sun X, Hashikawa T, Yoshioka K, Ishiguro K, Takashima. A (2002) Aberrant tau phosphorylation by glycogen synthase kinase-3beta and JNK3 induces oligomeric tau fibrils in COS-7 cells. J Biol Chem 277:42060–42065

Schneider A, Biernat J, von Bergen M, Mandelkow E, Mandelkow. EM (1999) Phosphorylation that detaches tau protein from microtubules (Ser262, Ser214) also protects it against aggregation into Alzheimer paired helical filaments. Biochemistry 38:3549–3558

Shimohama S, Perry G, Richey P, Praprotnik D, Takenawa T, Fukami K, Whitehouse PJ, Kimura. J (1995) Characterization of the association of phospholipase C-delta with Alzheimer neurofibrillary tangles. Brain Res 669:217–224

Spillantini MG, Goedert. M (1998) Tau protein pathology in neurodegenerative diseases. Trends Neurosci 21:428–433

Spillantini MG, Van Swieten JC, Goedert. M (2000) Tau gene mutations in frontotemporal dementia and parkinsonism linked to chromosome 17 (FTDP-17). Neurogenetics 2:193–205

Spires-Jones TL, de Calignon A, Matsui T, Zehr C, Pitstick R, Wu HY, Osetek JD, Jones PB, Bacskai BJ, Feany MB, Carlson GA, Ashe KH, Lewis J, Hyman. BT (2008) In vivo imaging reveals dissociation between caspase activation and acute neuronal death in tangle-bearing neurons. J Neurosci 28:862–867

Takahashi M, Tomizawa K, Ishiguro. K (2000) Distribution of tau protein kinase I/glycogen synthase kinase-3beta, phosphatases 2A and 2B, and phosphorylated tau in the developing rat brain. Brain Res 857:193–206

Takahashi M, Tomizawa K, Kato R, Sato K, Uchida T, Fujita SC, Imahori. K (1994) Localization and developmental changes of tau protein kinase I/glycogen synthase kinase-3 beta in rat brain. J Neurochem 63:245–255

Takashima A (2006) GSK-3 is essential in the pathogenesis of Alzheimer's disease. J Alzheimers Dis 9:309–317

Takashima A, Honda T, Yasutake K, Michel G, Murayama O, Murayama M, Ishiguro K, Yamaguchi. H (1998) Activation of tau protein kinase I/glycogen synthase kinase-3beta by amyloid beta peptide (25–35) enhances phosphorylation of tau in hippocampal neurons. Neurosci Res 31:317–323

Takashima A, Noguchi K, Michel G, Mercken M, Hoshi M, Ishiguro K, Imahori. K (1996) Exposure of rat hippocampal neurons to amyloid beta peptide (25–35) induces the inactivation of phosphatidyl inositol-3 kinase and the activation of tau protein kinase I/glycogen synthase kinase-3 beta. Neurosci Lett 203:33–36

Takashima A, Noguchi K, Sato K, Hoshino T, Imahori. K (1993) Tau protein kinase I is essential for amyloid beta-protein-induced neurotoxicity. Proc Natl Acad Sci USA 90:7789–7793

Tanemura K, Akagi T, Murayama M, Kikuchi N, Murayama O, Hashikawa T, Yoshiike Y, Park JM, Matsuda K, Nakao S, Sun X, Sato S, Yamaguchi H, Takashima. A (2001) Formation of filamentous tau aggregations in transgenic mice expressing V337M human tau. Neurobiol Dis 8:1036–1045

Tanemura K, Murayama M, Akagi T, Hashikawa T, Tominaga T, Ichikawa M, Yamaguchi H, Takashima. A (2002) Neurodegeneration with tau accumulation in a transgenic mouse expressing V337M human tau. J Neurosci 22:133–141

Taniguchi T, Doe N, Matsuyama S, Kitamura Y, Mori H, Saito N, Tanaka. C (2005) Transgenic mice expressing mutant (N279K) human tau show mutation dependent cognitive deficits without neurofibrillary tangle formation. FEBS Lett 579:5704–5712

Tatebayashi Y, Miyasaka T, Chui DH, Akagi T, Mishima K, Iwasaki K, Fujiwara M, Tanemura K, Murayama M, Ishiguro K, Planel E, Sato S, Hashikawa T, Takashima. A (2002) Tau filament formation and associative memory deficit in aged mice expressing mutant (R406W) human tau. Proc Natl Acad Sci USA 99:13896–13901

Timm T, Balusamy K, Li X, Biernat J, Mandelkow E, Mandelkow. EM (2008) Glycogen synthase kinase (GSK) 3beta directly phosphorylates Serine 212 in the regulatory loop and inhibits microtubule affinity-regulating kinase (MARK) 2. J Biol Chem 283:18873–18882

van Swieten JC, Rosso SM, van Herpen E, Kamphorst W, Ravid R, Heutink. P (2004) Phenotypic variation in frontotemporal dementia and parkinsonism linked to chromosome 17. Dement Geriatr Cogn Disord 17:261–264

Woodgett JR (1990) Molecular cloning and expression of glycogen synthase kinase-3/factor A. EMBO J 9:2431–2438

Yan M, Dai T, Deak JC, Kyriakis JM, Zon LI, Woodgett JR, Templeton. DJ (1994) Activation of stress-activated protein kinase by MEKK1 phosphorylation of its activator SEK1. Nature 372:798–800

Yoshida H, Ihara. Y (1993) Tau in paired helical filaments is functionally distinct from fetal tau: assembly incompetence of paired helical filament-tau. J Neurochem 61:1183–1186

Yoshiyama Y, Uryu K, Higuchi M, Longhi L, Hoover R, Fujimoto S, McIntosh T, Lee VM, Trojanowski. JQ (2005) Enhanced neurofibrillary tangle formation, cerebral atrophy, and cognitive deficits induced by repetitive mild brain injury in a transgenic tauopathy mouse model. J Neurotrauma 22:1134–1141

Yuan Z, Agarwal-Mawal A, Paudel HK (2004) 14-3-3 binds to and mediates phosphorylation of microtubule-associated tau protein by Ser9-phosphorylated glycogen synthase kinase 3beta in the brain. J Biol Chem 279:26105–26114

Chapter 6
Microtubule-Associated Protein 4

Structural and Functional Features

Kiyotaka Tokuraku, Kazuyuki Matsushima, Hiroyuki Nakagawa, and Susumu Kotani

Abstract In contrast with neural microtubule-associated proteins (MAPs) such as MAP1, MAP2, and tau, MAP4 is identified as a ubiquitous MAP. MAP4 stimulates tubulin polymerization and stabilizes polymerized-microtubules as do neural MAPs. Because MAP2, MAP4, and tau are structurally similar, the three MAPs are considered to constitute a superfamily. The architecture of the superfamily proteins consists of an amino-terminal projection domain and a carboxyl-terminal microtubule-binding domain, and the microtubule-binding domain is further divided into three subdomains: the *Pro-rich region*, the *repeat region*, and the *tail region*. Recent studies have revealed the functions of these domains/subdomains in the MAP4 molecule: the projection domain keeps individual microtubules separated by suppressing the bundle-forming ability of the microtubule-binding domain, the Pro-rich region promotes the nucleation of microtubules, the repeat region promotes their elongation, and the tail region may contribute to the proper folding of the molecule. Isoforms of MAP4 are produced from the single MAP4 gene by alternative RNA splicing, thereby five isoforms, with a deletion in the Pro-rich region or the repeat region, are expressed. The expression of these isoforms depends on the tissue type and the developmental stage, suggesting that the function of MAP4 is elaborately regulated by alternative splicing. In this chapter, we will address the structural and functional futures of MAP4 focusing on the functions of domains/subdomains and add some insights into the roles of the protein in neurons.

Keywords MAP1 · MAP2 · MAP3 · MAP4 · MAP5 · Tau · Microtubule

Contents

K. Tokuraku (✉)
Department of Chemical Science and Engineering, Miyakonojo National College of Technology, Miyakonojo, Miyazaki 885-8567, Japan
e-mail: tokuraku@cc.miyakonojo-nct.ac.jp

R.A. Nixon, A. Yuan (eds.), *Cytoskeleton of the Nervous System*, Advances in Neurobiology 3, DOI 10.1007/978-1-4419-6787-9_6,

6.1 Introduction

From the 1970s up to the 1980s, several microtubule-associated proteins (MAPs) – MAP1, MAP2, MAP3, MAP4, MAP5, and tau – were found in mammalian cells (Murphy and Borisy, 1975; Weingarten et al., 1975; Sloboda et al., 1976; Huber et al., 1985; Riederer et al., 1986; Kotani et al., 1988; Tokuraku et al., 2002; Matsushima et al., 2005; Matsushima and Kotani, 2007). Subsequent studies revealed that MAP3 is identical to MAP4 (Kobayashi et al., 2000) and MAP5 to MAP1B (Garner et al., 1990). Because these MAPs stimulate microtubule assembly and stabilize assembled microtubules, they are believed to play important roles in the regulation of microtubule dynamics in eukaryotic cells (for review, see Tokuraku et al., 2002).

In contrast with brain-specific MAPs such as MAP1, MAP2, and tau, MAP4 is identified as a major ubiquitous MAP. Bulinski and Borisy isolated and characterized a 210-kDa heat-stable MAP from HeLa cells (Bulinski and Borisy, 1979, 1980a, 1980b, 1980c), and Olmsted and Lyon identified a 215-kDa MAP in neuroblastoma cells, and named it MAP4 (Olmsted and Lyon, 1981; Parysek et al., 1984). We rather chose mammalian organs than cultured cells as starting materials, and successfully isolated a 190-kDa MAP from bovine adrenal gland (Kotani et al., 1986; Murofushi et al., 1986). MAPs similar to the 190-kDa MAP were also found in human fibroblast cells, HeLa cells, and rat liver (Kotani et al., 1987; Murofushi et al., 1987; Kotani et al., 1988). We compared the characteristics of these MAPs, including the HeLa 210-kDa MAP and the neuroblastoma 215-kDa MAP, and concluded that they should be considered as orthologs (Kotani et al., 1988). Subsequent primary structure analyses (Aizawa et al., 1990; Chapin and Bulinski, 1991; West et al., 1991) supported this idea, and now these MAPs are considered to be members of the MAP4 family. MAP4 family member proteins are ubiquitously expressed in various cells and organs, suggesting that this MAP is important for microtubule-dependent fundamental cellular functions.

MAP3 was originally found in rat brain as a 180-kDa MAP (Huber et al., 1985). Anti-MAP3 antibody staining showed that MAP3 is present in neurofilament-rich axons and glial processes in the adult brain (Bernhardt et al., 1985; Huber et al., 1985). MAP3 was induced in nerve growth factor (NGF)-treated PC12 cells, and the timing of this induction coincided with the bundling of microtubules and neurite outgrowth (Brugg and Matus, 1988), suggesting that MAP3 plays an important role in regulating the growth of the neuronal process. Because MAP3 was mainly identified as an antigen of a monoclonal antibody, its biochemical nature had long been uncertain, and its similarity to MAP4 was pointed out at an early stage (Kotani et al., 1988). MAP3 was eventually identified as MAP4 (Kobayashi et al., 2000), as described above. Consequently, the reported characteristics of MAP3 are referred to as those of MAP4 in this chapter.

Early reports revealed that MAP2 consisted of two domains: a microtubule-binding domain and a projection domain (Vallee and Borisy, 1977; Vallee, 1980) (Fig. 6.1a). Subsequently, MAP4 (Aizawa et al., 1987) and tau (Aizawa et al., 1988) were also revealed to have the two domains (Fig. 6.1a). Studies of the primary structures (Drubin et al., 1984; Goedert et al., 1988; Lee et al., 1988; Lewis et al., 1988; Himmler et al., 1989; Aizawa et al., 1990; Kindler et al., 1990; Chapin and Bulinski, 1991; West et al., 1991; Andreadis et al., 1992; Kalcheva et al., 1995) showed that these MAPs have a homologous sequence of 18-amino-acid residues (AP sequence) in the microtubule-binding domain responsible for microtubule-binding and assembly-promoting activities (Aizawa et al., 1989; Ennulat et al., 1989; Joly et al., 1989). The AP sequences imperfectly repeat from three to five times in the microtubule-binding domain of these MAPs, constituting the repeat region (Fig. 6.1b). In any of these MAPs, the repeat region is flanked by a Pro-rich region, a region rich in proline and basic residues, and a tail region, a short C-terminal region with hydrophobic and acidic residues. Because the domain organization of

Fig. 6.1 Schematic structures of the MAP2/MAP4/tau superfamily proteins. **a** Side view of a MAP-bound microtubule to show how each domain operates in binding to microtubules. PD and MBD show a projection domain and a microtubule-binding domain, respectively. **b** Domain organizations of the three proteins. Each protein has a projection domain and a microtubule-binding domain. The projection domain of MAP4 contains a KDM-repeat region. The microtubule-binding domain is further divided into three subdomains, the Pro-rich region (P), the repeat region (R), and the tail region (T)

the three MAPs is similar, researchers have considered that these MAPs constitute a MAP2/MAP4/tau superfamily (Tokuraku et al., 2002).

Over the past few decades, the domain functions in these MAP superfamily proteins have been revealed. In this chapter, we will mainly address the structural and functional features of MAP4 protein focusing on the domain functions and will add discussion about the functions of the protein in neurons.

6.2 Overall Structure

MAP4, MAP2, and tau proteins are encoded by their respective single genes (Tokuraku et al., 2002). For the molecular phylogenetics of MAP4, the MAP4 genes on the Web sites PubMed Home (http://www.ncbi.nlm.nih.gov/sites/entrez), BLAST (http://blast.ncbi.nlm.nih.gov/Blast.cgi), and NCBI Conserved Domain Database (http://blast.ncbi.nlm.nih.gov/Blast.cgi) were surveyed, and MAP4-like genes from 17 species were successfully found. Figure 6.2 shows the evolutionary tree of the MAP4 genes. The tree shows that the molecular evolution of MAP4 is almost consistent with the traditional theory of eukaryotic evolution. It also reveals

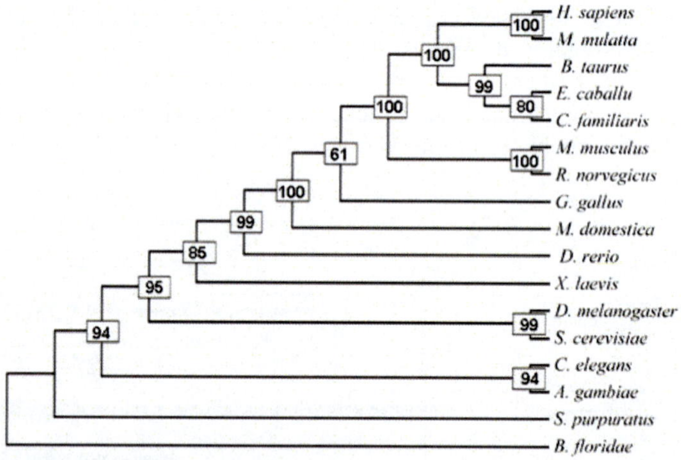

Fig. 6.2 An evolutionary tree of MAP4 molecules. The protein sequence data were analyzed by DINASIS version 2.07 (Hitachi Software Engineering, Tokyo, Japan) using the Clustal W algorithm. The boxed numbers are bootstrap values. Seventeen species of MAP4 were analyzed: *Homo sapiens* (NP_002366), *Macaca mulatta* (XP_001113017), *Bos taurus* (BAC07258 and NP_776530), *Equus caballus* (XP_001495345), *Canis lupus familiaris* (XP_541897), *Mus musculus* (AAH55364), *Rattus norvegicus* (EDL77097), *Gallus gallus* (XP_418480), *Monodelphis domestica* (XP_001375442), *Danio rerio* (XP_001920671), *Xenopus laevis* (BAA36221), *Drosophila melanogaster* (NP_733465), *Saccharomyces cerevisiae* (NP_012493), *Caenorhabditis elegans* (AAB97090), *Anopheles gambiae* (XP_321540), *Strongylocentrotus purpuratus* (XP_788973), and *Branchiostoma floridae* (XP_002238608)

that the MAP4 structures of mammalian species, except for *Monodelphis domestica* (*Homo sapiens–Rattus norvegicus*), are significantly different from those of other species (bootstrap value, 61%), suggesting that MAP4 serves some specific functions in mammalian cells.

The MAP4 gene in a genome invariably consists of plural exons. Figure 6.3 shows the human MAP4 gene with 18 exons. Isoforms are derived from single MAP4 gene by alternative RNA splicing (Chapin et al., 1995), thereby at least five MAP4 variants are expressed (Matsushima and Kotani, 2007) (Fig. 6.4). The structural features of the MAP4 variants are a shortening of the Pro-rich region or the repeat region in the microtubule-binding domain. The alternative splicing in general is now expected to account for the proteomic and functional diversity of higher eukaryotic organisms. The expression of MAP4 isoforms depended on the tissue type and the developmental stage (Chapin et al., 1995; Matsushima et al., 2005), suggesting that the functions of MAP4 are regulated by alternative splicing.

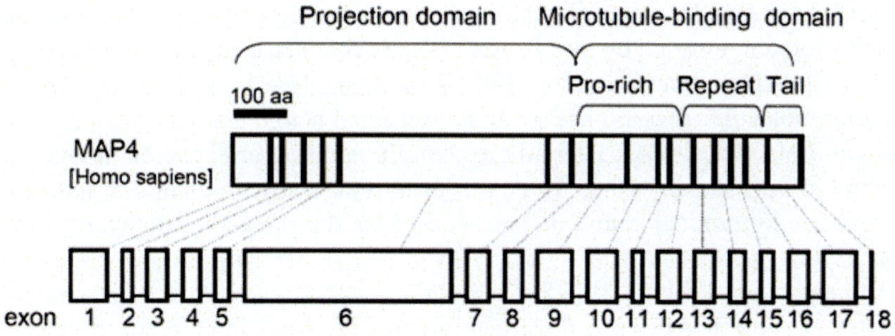

Fig. 6.3 The genomic structure of *Homo sapiens* MAP4 (GeneBank Accession No. NT005825), consisting of 18 exons (CCDS database; http://www.ncbi.nlm.nih.gov/CCDS)

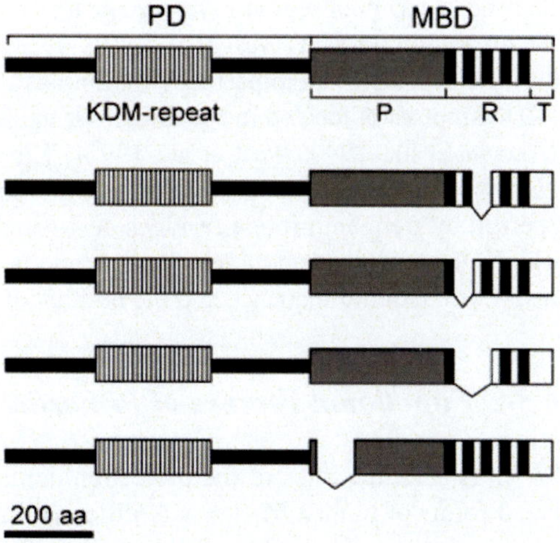

Fig. 6.4 Schematic structures of the MAP4 isoforms

As mentioned in Section 6.1, the overall structure of MAP4 consists of an amino-terminal projection domain and a carboxyl-terminal microtubule-binding domain, and the microtubule-binding domain is further divided into three subdomains: the Pro-rich region, the repeat region, and the tail region (Fig. 6.1b). The detailed structures and functions of the domains will be elaborated in the following sections.

6.3 Projection Domain

6.3.1 Structural Feature of Projection Domain

Electron microscopic observation revealed that microtubules reconstituted from purified tubulin and MAP4 possessed lateral projections extending from their surface (Murofushi et al., 1986) (Fig. 6.1a). Because microtubules had no lateral projections when the proteolytic microtubule-binding fragment was used instead of full-length MAP4 (Aizawa et al., 1987), the lateral projection seemed to correspond with the domain removed by the cleavage. The primary structure of MAP4 (Aizawa et al., 1990; Chapin and Bulinski, 1991; West et al., 1991) revealed that MAP4 is a bipolar molecule consisting of an amino-terminal acidic domain and a carboxy-terminal basic domain. A bacterially expressed amino-terminal acidic domain did not bind to microtubule, whereas a carboxy-terminal basic domain did, indicating that the amino-terminal acidic domain constructs the projection domain (Aizawa et al., 1991).

A comparison of MAP2, MAP4, and tau sequences revealed that the projection domains of these MAPs have no structural characteristics in common (Lee et al., 1988; Lewis et al., 1988; Aizawa et al., 1990). Consequently, it is generally assumed that each projection domain plays a role specific to each MAP. The interspecific homology of extreme N-terminal 80-amino-acid residues in MAP4 projection domains was higher (>90%) than that of the other regions in the domain, suggesting that the N-terminal region is important for specifying some unknown functions of the projection domain (West et al., 1991). A repetitive region containing 18–26 consecutive repeats of an imperfect 14-amino-acid KD(M/V)X(L/P)(P/L)XETEVALA (KDM-repeat) is located in the middle of the MAP4 projection domain (Fig. 6.1b) (Aizawa et al., 1990; West et al., 1991). The KDM-repeat is predicted to form a highly flexible filamentous structure composed of a series of short α-helixes interspersed by frequent proline residues between the KDM-repeat (West et al., 1991). The KDM-repeats from mouse, human, and bovine sources vary considerably with respect to both the sequence and the number of repeat units.

6.3.2 Functional Feature of Projection Domain

To investigate the roles of the projection domain, Iida et al. (2002) prepared truncated forms of human MAP4 with different lengths of the projection domains and examined their effects on bundle formation of microtubules. The projection domain suppressed the bundle-forming ability intrinsic to the microtubule-binding domain of MAP4 by keeping individual microtubules separated. The suppressive activity was correlated with the length of the projection domain but independent of the amino acid sequence (Iida et al., 2002). Truncation of the projection domain also led to the attenuation of microtubule dynamic instability (Permana et al., 2005). Although the projection domains of MAPs have long been expected to possess their own functions other than the regulation of microtubule dynamics, neither in vitro nor in vivo experiments have yet clarified the interacting counterpart of the MAP4 projection domain. When a green fluorescent protein (GFP)-fused projection domain was introduced into living baby hamster kidney cells, the transfected proteins displayed diffuse distributions and showed no interactions with either microtubules or other organelles (Olson et al., 1995). The MAP4 projection domain may regulate intrinsic properties of microtubules rather than interact with other components.

6.4 Pro-Rich Region of Microtubule-Binding Domain

6.4.1 Structural Feature of Pro-Rich Region

The Pro-rich region of bovine MAP4 (residues 641–880) contains a high proline content (19%) (Aizawa et al., 1990) as do the Pro-rich regions of mouse MAP2 (residues: 1,420–1,673) or bovine tau (residues 110–262) (13% and 19%, respectively) (Lewis et al., 1988; Himmler et al., 1989). The amino acid composition of the three Pro-rich regions showed an additional similarity, that is, the abundance of the three limited amino acid species, Lys, Thr, and Ser, yet no significant sequence homology was observed among them. Although the MAP2/MAP4/tau superfamily proteins regulate their activity by producing multiple splicing variants as mentioned in Section 6.2, the Pro-rich region variants were not found until our recent identification of a MAP4 isoform lacking 72 consecutive amino acid residues, encoded by exon 9 (Matsushima et al., 2005). The amino acid sequence of the missing region was highly conserved (about 85% identity/similarity) among the corresponding region of bovine, human, mouse, and rat MAP4, suggesting the functional significance of this region. The newly identified isoform is especially interesting from a neurobiological viewpoint, as it is exclusively expressed in the brain and related tissues. This issue will be discussed later in Section 6.7.

6.4.2 Functional Feature of Pro-Rich Region

To study the function of the Pro-rich region per se, the microtubule-binding and microtubule assembly–promoting activities of a bacterially expressed protein fragment exclusively containing the Pro-rich region of MAP4 was observed (Katsuki et al., 1999). Although the fragment bound to preformed microtubules, it induced the assembly of tubulin into amorphous aggregates, suggesting that the Pro-rich region alone is incompetent at forming normal microtubules. Because the presence of the Pro-rich region greatly enhanced the microtubule-nucleating activity of the truncated MAP4 fragments, as revealed by molecular dissection studies, we proposed a hypothesis in which the Pro-rich region of MAP4 promotes microtubule nucleation (Tokuraku et al., 1999). The Pro-rich regions of MAP2 and tau are presumed to function in a similar manner despite the low sequence homology.

Three Pro-rich region–related regulatory mechanisms, phosphorylation, binding of septins, and alternative splicing, have been reported. Ser-696 and Ser-787 in the Pro-rich region of MAP4 were phosphorylated by cyclin B-cdc2 kinase, and the phosphorylation rendered more dynamic microtubules by decreasing the frequency of the rescue process in microtubule dynamic instability (Ookata et al., 1995, 1997). Of the two phosphorylation sites, Ser-787 was critical (Kitazawa et al., 2000). Septins are a family of GTP-binding proteins conserved among eukaryotes, except plants (Hartwell, 1971; Longtine et al., 1996; Nguyen et al., 2000). The SEPT2:6:7 heterotrimer bound directly to the Pro-rich region of MAP4, blocked its binding to microtubules, and consequently regulated microtubule dynamics (Kremer et al., 2005). The Pro-rich region variant lacking exon 9 showed lower microtubule-stabilizing activity than the full-length MAP4 variant (Matsushima et al., 2005). Fluorescence recovery after photobleaching analyses revealed that the turnover of the MAP4 variant on cytoplasmic microtubules is more rapid than that of the full-length variant (Hasan et al., 2006).

6.5 Repeat Region of Microtubule-Binding Domain

6.5.1 Structural Feature of Repeat Region

The repeat region contains three to five imperfect repeats, each consisting of 18-amino-acid residues responsible for microtubule-binding activity (AP sequence). The AP sequences of MAP4, MAP2, and tau are more than 70% identical, and each AP sequence is flanked by a 13- to 20-amino-acid-long inter-repeat sequence (Fig. 6.5). Phylogenic implications of the repeat region are discussed elsewhere (Tokuraku et al., 2002). We reported four MAP4 variants, which differed from each other in both the number and the arrangement of the AP sequences in bovine cells (Tokuraku et al., 2003) (Fig. 6.4). These variants are generated by skipping exon 14 and/or exon 13 (Fig. 6.3) by alternative splicing. The expression of these variants depended on the tissue type and the developmental stage (Chapin et al., 1995), suggesting that the repeat region variants play a role in tissue type–specific and/or developmental stage–specific processes.

The number of AP sequences on genes varies among species (Fig. 6.6). From 1 to 10 AP sequences are found in the repeat region of MAP4 (Fig. 6.6), suggesting that neither the set number of AP sequences nor the tandem repeat are essential for MAP4 function. The splicing-dependent regulatory mechanism may be acquired in the later stages of evolution.

	Repeat sequences	Inter-repeat sequence
MAP4	(935) VRSKVGSTENIKHQPGGG	RAKVEKKTEAAATTRKPESNA
	(974) VTKTAGPIASAQKQPAG-	KVQIVSKKVSYSH
	(1004) IQSKCGSKDNIKHVPGGG	NVQIQNKKVDISK
	(1035) VSSKCGSKANIKHKPGGG	DVKIESQKLNFKEK
	(1067) AQAKVGSLDNVGHLPAGG	
MAP2	(1673) VKSKIGSTDNIKYQPKGG	QVQIVTKKIDLSH
	(1704) VTSKCGSLKNI-DRPGGG	RVKIESVKLDFKEK
	(1735) VQAKVGSLDNAHHVPGGG	
tau	(198) VKSKIGSTENLKHQPGGG	KVQIINKKLDLSN
	(229) VQSKCGSKDNIKHVPGGG	SVQIVYKPVDLSK
	(260) VTSKCGSLGNIHHKPGGG	QVEVKSEKLDFKDR
	(292) VQSKIGSLDNITHVPGGG	
consensus	V-SK-GS--NIKH-PGGG	

Fig. 6.5 Alignment of the repeat region of human MAP4, MAP2, and tau. The consensus of the AP sequences is shown at the bottom cell

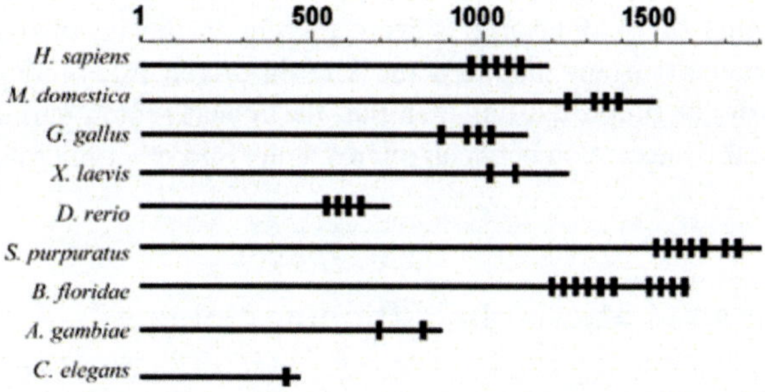

Fig. 6.6 Alignment of MAP4 species with respect to the number and the positions of AP sequences. Nine species are represented: *Homo sapiens*, *Monodelphis domestica*, *Gallus gallus*, *Xenopus laevis*, *Danio rerio*, *Strongylocentrotus purpuratus*, *Branchiostoma floridae*, *Anopheles gambiae*, and *Caenorhabditis elegans*. Closed boxes indicate AP sequences. The *ruler* at the *top* shows the amino acid residue numbers

6.5.2 Functional Feature of Repeat Region

Because synthetic polypeptides, containing one AP sequence, bound to microtubules and stimulated tubulin polymerization, and the maximum binding of the synthetic polypeptide reached 1.2 mol of the polypeptide/mol of tubulin dimer (Aizawa et al., 1989), a theory was once widely accepted that multiple AP sequences in the repeat region bind to multiple tubulin molecules to make nuclei for microtubule assembly (Aizawa et al., 1990). However, this traditional hypothesis was not supported by further studies. The subsequent stoichiometric studies agreed that only one or two AP sequences are active in binding to tubulin molecules (Tokuraku et al., 1999; 2002). We also showed by molecular dissection studies that the nucleating activity does not reside in the repeated region but in the Pro-rich region (Tokuraku et al., 1999).

Why are the AP sequences tandemly arranged in one MAP molecule? To answer this question, we examined the functional differences of MAP4 variants, which differed from each other in both the number and arrangement of the AP sequences (Tokuraku et al., 2003). The microtubule-binding domain fragments of the MAP4 variants showed a similar degree of microtubule assembly promoting activity and microtubule binding affinity, suggesting that variation in the repeat region is not important for the control of microtubule dynamics, the main function of MAP4, which agrees with the phylogenic consideration (Section 6.5.1). On the other hand, the microtubule bundle–forming activity differed among the variants (Tokuraku et al., 2003). Consequently, we proposed a hypothesis that the role of the variations in the repeat region is to alter the microtubule surface properties. As a result, the interaction between microtubules and other microtubule-binding proteins, such as microtubule-dependent motor proteins, may be regulated. Thus, we examined whether the MAP4 variants affect the movement of kinesin by an in vitro gliding assay (Tokuraku et al., 2007). When the microtubule-binding domain fragments of MAP4 variants were present in the in vitro gliding assay, only the five-repeat variant inhibited the kinesin-driven microtubule movement. The results strongly supported our idea that the repeat region variation is generated to regulate the interaction between microtubules and other microtubule-binding factors.

6.6 Tail Region of Microtubule-Binding Domain

6.6.1 Structural Feature of Tail Region

The MAP2/MAP4/tau superfamily has a slightly acidic and hydrophobic tail region (about 70-amino-acid length) on the extreme carboxyl-terminal part of the microtubule-binding domain (Fig. 6.1). MAP2 and tau share a homologous tail region (Lewis et al., 1988), whereas MAP4 does not (Aizawa et al., 1990). Basic residues accounted for 4.5% in the tail region of MAP4 but 14.5% in those of MAP2 and tau, indicating that the MAP4 tail region is more acidic than those of MAP2 and tau (Katsuki et al., 1997). The acidity of the MAP4 tail region implies that the region interacts with basic Pro-rich/repeat regions.

6.6.2 Functional Feature of Tail Region

Because the bacterial expression of a fragment exclusively containing the tail region was not successful, we fused fragments with thioredoxin (Katsuki et al., 1999). The thioredoxin-fused tail region fragment did not bind to microtubules, indicating that the tail region lacks microtubule-binding activity. In an earlier report, we argued that the tail region is indispensable for the nucleation step, based on the result that removal of the tail region greatly reduced the nucleating activity of one of the MAP4 microtubule-binding domain fragments (Katsuki et al., 1997). The tail region affected nucleating activity despite its lack of microtubule-binding activity, suggesting that the tail region contributes to the proper folding of the molecule by an intramolecular interaction with other subdomains by an electrostatic interaction between the acidic tail region and the basic Pro-rich/repeat regions.

6.7 Function of MAP4 in Neurons

Although MAP4 has long been known as a ubiquitous MAP and its presence in the brain was also demonstrated at an early stage, a histochemical study reported that MAP4 was expressed only in glia and was absent from neurons in the brain (Parysek et al., 1985). However MAP3, which was subsequently proved to be identical to MAP4 (Kobayashi et al., 2000), was present in both glia and neurons (Bernhardt et al., 1985; Huber et al., 1985). The discrepancy of MAP4 localization may be ascribed to the specificity of the antibodies, as various MAP4 variants are expressed, depending on the tissue type and the developmental stage (Chapin et al., 1995; Matsushima et al., 2005). Either or both of the studies may have used antibodies specific to a certain subtype of MAP4. We also performed immunohistochemistry with anti-MAP4 antibody and have confirmed that MAP4 is expressed not only in primary cultured neurons but also in many types of neurons in the brain (Tokuraku et al., 2010).

What are the functions of MAP4 in neurons? In this respect, early studies of Matus and his co-workers (Bernhardt et al., 1985; Huber et al., 1985; Brugg and Matus, 1988) can provide some insights (MAP4 was designated as MAP3 in their studies). MAP4 transitorily appeared in granule cells and their parallel fiber axons during the developmental stage of the cerebellum. The expression of MAP4 also coincided with the development of neurites in cultured PC12 cells in response to NGF treatment, with a concomitant proliferation of microtubule bundles. In the adult brain, MAP4-bound microtubules were usually co-localized with neurofilaments. These studies suggest the roles of MAP4 not only in developing neurites but also in mature neurofilament-rich axons.

As described in Section 6.4, we have found a new MAP4 variant specific to neural cells. The expression of this isoform was also induced by NGF treatment in PC12 cells (Matsushima et al., 2005). Since this manner of induction is similar to that reported by Brugg and Matus (1988), the antigen detected in their work may be the neuron-specific variant of MAP4. The diffused distribution of NGF-induced MAP4 in the cytoplasm (Brugg and Matus, 1988) may be related to the rapid turnover of the neuron-specific variant (Hasan et al., 2006). The rapid MAP turnover results in an increase in the number of dynamic microtubules, which may contribute to neural tissue differentiation and neurite dynamics.

In neurons, the long-distance transportation by the microtubule motors plays an important role. As described in Section 6.5.2, the five-repeat variant of MAP4 inhibited the movement of kinesin motor in vitro (Tokuraku et al., 2007).

Meanwhile, overexpression of the five-repeat MAP4 variant reportedly resulted in the inhibition of organelle and receptor traffic (Bulinski et al., 1997; Nguyen et al., 1997; Cheng et al., 2002, 2005), suggesting that the five-repeat variant also inhibits movement of motor proteins in vivo. Consequently, the presence of the five-repeat variant seems to be unfavorable to long-distance transportation; this may explain the lack of five-repeat versions in the neuron-specific MAPs, MAP2 and tau (Fig. 6.5). During brain differentiation, the expression ratio of the five-repeat isoform decreased with an increase in that of the four-repeat isoform (Chapin et al., 1995). The replacement of MAP4 isoforms may facilitate the long-distance and smooth transportation in mature neurons.

In the past few decades, a large number of studies have revealed the structure and function of MAP4. Although MAP4 has been studied in the light of its ubiquitous nature, various MAP4 isoforms have been revealed to be expressed in the brain in a site- and stage-specific manner. To understand the roles of MAP4 in neurons, we should not only characterize the isoforms in vitro but also develop techniques to trace the behavior of each isoform.

References

Aizawa H, Emori Y, Mori A et al (1991) Functional analyses of the domain structure of microtubule-associated protein-4 (MAP-U). J Biol Chem 266:9841–9846

Aizawa H, Emori Y, Murofushi H et al (1990) Molecular cloning of a ubiquitously distributed microtubule-associated protein with Mr 190,000. J Biol Chem 265:13849–13855

Aizawa H, Kawasaki H, Murofushi H et al (1988) Microtubule-binding domain of tau proteins. J Biol Chem 263:7703–7707

Aizawa H, Kawasaki H, Murofushi H et al (1989) A common amino acid sequence in 190-kDa microtubule-associated protein and tau for the promotion of microtubule assembly. J Biol Chem 264:5885–5890

Aizawa H, Murofushi H, Kotani S et al (1987) Limited chymotryptic digestion of bovine adrenal 190,000-Mr microtubule-associated protein and preparation of a 27,000-Mr fragment which stimulates microtubule assembly. J Biol Chem 262:3782–3787

Andreadis A, Brown WM, Kosik KS (1992) Structure and novel exons of the human tau gene. Biochemistry 31:10626–10633

Bernhardt R, Huber G, Matus A (1985) Differences in the developmental patterns of three microtubule-associated proteins in the rat cerebellum. J Neurosci 5:977–991

Brugg B, Matus A (1988) PC12 cells express juvenile microtubule-associated proteins during nerve growth factor-induced neurite outgrowth. J Cell Biol 107:643–650

Bulinski JC, Borisy GG (1979) Self-assembly of microtubules in extracts of cultured HeLa cells and the identification of HeLa microtubule-associated proteins. Proc Natl Acad Sci USA 76:293–297

Bulinski JC, Borisy GG (1980a) Immunofluorescence localization of HeLa cell microtubule-associated proteins on microtubules in vitro and in vivo. J Cell Biol 87:792–801

Bulinski JC, Borisy GG (1980b) Microtubule-associated proteins from cultured HeLa cells. Analysis of molecular properties and effects on microtubule polymerization. J Biol Chem 255:11570–11576

Bulinski JC, Borisy GG (1980c) Widespread distribution of a 210,000 mol wt microtubule-associated protein in cells and tissues of primates. J Cell Biol 87:802–808

Bulinski JC, McGraw TE, Gruber D et al (1997) Overexpression of MAP4 inhibits organelle motility and trafficking in vivo. J Cell Sci 110(Pt 24):3055–3064

Chapin SJ, Bulinski JC (1991) Non-neuronal 210 × 10(3) Mr microtubule-associated protein (MAP4) contains a domain homologous to the microtubule-binding domains of neuronal MAP2 and tau. J Cell Sci 98(Pt 1):27–36

Chapin SJ, Lue CM, Yu MT et al (1995) Differential expression of alternatively spliced forms of MAP4: a repertoire of structurally different microtubule-binding domains. Biochemistry 34:2289–2301

Cheng G, Iijima Y, Ishibashi Y et al (2002) Inhibition of G protein-coupled receptor trafficking in neuroblastoma cells by MAP 4 decoration of microtubules. Am J Physiol Heart Circ Physiol 283:H2379–H2388

Cheng G, Qiao F, Gallien TN et al (2005) Inhibition of beta-adrenergic receptor trafficking in adult cardiocytes by MAP4 decoration of microtubules. Am J Physiol Heart Circ Physiol 288:H1193–H1202

Drubin DG, Caput D, Kirschner MW (1984) Studies on the expression of the microtubule-associated protein, tau, during mouse brain development, with newly isolated complementary DNA probes. J Cell Biol 98:1090–1097

Ennulat DJ, Liem RK, Hashim GA et al (1989) Two separate 18-amino acid domains of tau promote the polymerization of tubulin. J Biol Chem 264:5327–5330

Garner CC, Garner A, Huber G et al (1990) Molecular cloning of microtubule-associated protein 1 (MAP1A) and microtubule-associated protein 5 (MAP1B): identification of distinct genes and their differential expression in developing brain. J Neurochem 55:146–154

Goedert M, Wischik CM, Crowther RA et al (1988) Cloning and sequencing of the cDNA encoding a core protein of the paired helical filament of Alzheimer disease: identification as the microtubule-associated protein tau. Proc Natl Acad Sci USA 85:4051–4055

Hartwell LH (1971) Genetic control of the cell division cycle in yeast. IV. Genes controlling bud emergence and cytokinesis. Exp Cell Res 69:265–276

Hasan MR, Jin M, Matsushima K et al (2006) Differences in the regulation of microtubule stabilily by the pro-rich region variants of microtubule-associated protein 4. FEBS Lett 580:3505–3510

Himmler A, Drechsel D, Kirschner MW et al (1989) Tau consists of a set of proteins with repeated C-terminal microtubule-binding domains and variable N-terminal domains. Mol Cell Biol 9:1381–1388

Huber G, Alaimo-Beuret D, Matus A (1985) MAP3: characterization of a novel microtubule-associated protein. J Cell Biol 100:496–507

Iida J, Itoh TJ, Hotani H et al (2002) The projection domain of MAP4 suppresses the microtubule-bundling activity of the microtubule-binding domain. J Mol Biol 320:97–106

Joly JC, Flynn G, Purich DL (1989) The microtubule-binding fragment of microtubule-associated protein-2: location of the protease-accessible site and identification of an assembly-promoting peptide. J Cell Biol 109:2289–2294

Kalcheva N, Albala J, O'Guin K et al (1995) Genomic structure of human microtubule-associated protein 2 (MAP-2) and characterization of additional MAP-2 isoforms. Proc Natl Acad Sci USA 92:10894–10898

Katsuki M, Tokuraku K, Murofushi H et al (1999) Functional analysis of microtubule-binding domain of bovine MAP4. Cell Struct Funct 24:337–344

Katsuki M, Tokuraku K, Nakagawa H et al (1997) The 'assembly-promoting sequence region' of microtubule-associated protein 4 failed to promote microtubule assembly. FEBS Lett 418:35–38

Kindler S, Schulz B, Goedert M et al (1990) Molecular structure of microtubule-associated protein 2b and 2c from rat brain. J Biol Chem 265:19679–19684

Kitazawa H, Iida J, Uchida A et al (2000) Ser787 in the proline-rich region of human MAP4 is a critical phosphorylation site that reduces its activity to promote tubulin polymerization. Cell Struct Funct 25:33–39

Kobayashi N, Heid HW, Sakai T et al (2000) Molecular characterization reveals identity of microtubule-associated proteins MAP3 and MAP4. Biochem Biophys Res Commun 268: 306–309

Kotani S, Murofushi H, Maekawa S et al (1986) Characterization of microtubule-associated proteins isolated from bovine adrenal gland. Eur J Biochem 156:23–29

Kotani S, Murofushi H, Maekawa S et al (1987) Identification of the 190 kD microtubule-associated protein in cultured fibroblasts and its association with interphase and mitotic microtubules. Cell Struct Funct 12:1–9

Kotani S, Murofushi H, Maekawa S et al (1988) Isolation of rat liver microtubule-associated proteins. Evidence for a family of microtubule-associated proteins with molecular mass of around 200,000 which distribute widely among mammalian cells. J Biol Chem 263:5385–5389

Kremer BE, Haystead T, Macara IG (2005) Mammalian septins regulate microtubule stability through interaction with the microtubule-binding protein MAP4. Mol Biol Cell 16:4648–4659

Lee G, Cowan N, Kirschner M (1988) The primary structure and heterogeneity of tau protein from mouse brain. Science 239:285–288

Lewis SA, Wang DH, Cowan NJ (1988) Microtubule-associated protein MAP2 shares a microtubule binding motif with tau protein. Science 242:936–939

Longtine MS, DeMarini DJ, Valencik ML et al (1996) The septins: roles in cytokinesis and other processes. Curr Opin Cell Biol 8:106–119

Matsushima K, Aosaki M, Tokuraku K et al (2005) Identification of a neural cell specific variant of microtubule-associated protein 4. Cell Struct Funct 29:111–124

Matsushima K, Kotani S (2007) Neuronal microtubule-associated proteins. Insights into their structures and functions. NeuroQuantology 5:9–31

Murofushi H, Kotani S, Aizawa H et al (1986) Purification and characterization of a 190-kD microtubule-associated protein from bovine adrenal cortex. J Cell Biol 103:1911–1919

Murofushi H, Kotani S, Aizawa H et al (1987) Comparison of a major heat-stable microtubule-associated protein in HeLa cells and 190-kDa microtubule-associated protein in bovine adrenal cortex. J Biochem 102:1101–1112

Murphy DB, Borisy GG (1975) Association of high-molecular-weight proteins with microtubules and their role in microtubule assembly in vitro. Proc Natl Acad Sci USA 72:2696–2700

Nguyen HL, Chari S, Gruber D et al (1997) Overexpression of full- or partial-length MAP4 stabilizes microtubules and alters cell growth. J Cell Sci 110(Pt 2):281–294

Nguyen TQ, Sawa H, Okano H et al (2000) The C. elegans septin genes, unc-59 and unc-61, are required for normal postembryonic cytokineses and morphogenesis but have no essential function in embryogenesis. J Cell Sci 113(Pt 21):3825–3837

Olmsted JB, Lyon HD (1981) A microtubule-associated protein specific to differentiated neuroblastoma cells. J Biol Chem 256:3507–3511

Olson KR, McIntosh JR, Olmsted JB (1995) Analysis of MAP 4 function in living cells using green fluorescent protein (GFP) chimeras. J Cell Biol 130:639–650

Ookata K, Hisanaga S, Bulinski JC et al (1995) Cyclin B interaction with microtubule-associated protein 4 (MAP4) targets p34cdc2 kinase to microtubules and is a potential regulator of M-phase microtubule dynamics. J Cell Biol 128:849–862

Ookata K, Hisanaga S, Sugita M et al (1997) MAP4 is the in vivo substrate for CDC2 kinase in HeLa cells: identification of an M-phase specific and a cell cycle-independent phosphorylation site in MAP4. Biochemistry 36:15873–15883

Parysek LM, Asnes CF, Olmsted JB (1984) MAP 4: occurrence in mouse tissues. J Cell Biol 99:1309–1315

Parysek LM, del Cerro M, Olmsted JB (1985) Microtubule-associated protein 4 antibody: a new marker for astroglia and oligodendroglia. Neuroscience 15:869–875

Permana S, Hisanaga S, Nagatomo Y et al (2005) Truncation of the projection domain of MAP4 (microtubule-associated protein 4) leads to attenuation of microtubule dynamic instability. Cell Struct Funct 29:147–157

Riederer B, Cohen R, Matus A (1986) MAP5: a novel brain microtubule-associated protein under strong developmental regulation. J Neurocytol 15:763–775

Sloboda RD, Dentler WL, Rosenbaum JL (1976) Microtubule-associated proteins and the stimulation of tubulin assembly in vitro. Biochemistry 15:4497–4505

Tokuraku K, Katsuki M, Kotani S (2002) Structural and functional analyses of microtubule-associated protein 4. Recent Res Devel Biochem 3:315–333
Tokuraku K, Katsuki M, Nakagawa H et al (1999) A new model for microtubule-associated protein (MAP)-induced microtubule assembly. The Pro-rich region of MAP4 promotes nucleation of microtubule assembly in vitro. Eur J Biochem 259:158–166
Tokuraku K, Matsushima K, Matui T et al (2003) The number of repeat sequences in microtubule-associated protein 4 affects the microtubule surface properties. J Biol Chem 278:29609–29618
Tokuraku K, Noguchi TQ, Nishie M et al (2007) An isoform of microtubule-associated protein 4 inhibits kinesin-driven microtubule gliding. J Biochem 141:585–591
Tokuraku K, Okuyama S, Matsushima K et al (2010) Distinct neuronal localization of microtubule-associated protein 4 in the mammalian brain. Neurosci Lett 484:143–147
Vallee R (1980) Structure and phosphorylation of microtubule-associated protein 2 (MAP 2). Proc Natl Acad Sci USA 77:3206–3210
Vallee RB, Borisy GG (1977) Removal of the projections from cytoplasmic microtubules in vitro by digestion with trypsin. J Biol Chem 252:377–382
Weingarten MD, Lockwood AH, Hwo SY et al (1975) A protein factor essential for microtubule assembly. Proc Natl Acad Sci USA 72:1858–1862
West RR, Tenbarge KM, Olmsted JB (1991) A model for microtubule-associated protein 4 structure. Domains defined by comparisons of human, mouse, and bovine sequences. J Biol Chem 266:21886–21896

Chapter 7
Structure of Neural Intermediate Filaments

David A.D. Parry

Abstract The basic structure of neural intermediate filament (IF) molecules is known, as are the modes of interaction that direct these molecules into highly specific and fully functional IF. It has been established that the IF molecule is a dimer and that its two constituent chains lie parallel to each other and in axial register. The molecule has a tripartite structure with a central, α-helical, coiled-coil–rich region (the rod domain) separating the head (the region N-terminal to the rod) from the tail (the region C-terminal to the rod). A regularity in the linear dispositions of the charged residues in the longest coiled-coil segments aids assembly through the formation of numerous intermolecular ionic interactions. By analogy with the well-defined surface lattice structure of trichocyte keratin (deduced from the extensive x-ray diffraction data available), the surface lattice structure of neural IF can also be deduced, primarily from the pattern of cross-links induced between molecules. Using scanning transmission electron microscopy (STEM) and cryo-tomographic data, the neural IF are likely to have a four-protofibril structure containing 32 chains in section. Assembly occurs through a rapid lateral aggregation of about eight tetramers to form a unit-length-filament (ULF), with each of these tetramers consisting of a half-staggered, antiparallel pair of molecules. Elongation occurs through the axial aggregation of ULF to form immature IF about 16 nm in diameter and many micrometers in length. Radial compaction then takes place resulting in close packing of the molecular filaments and the formation of IF with diameter in the range 8–12 nm.

Keywords Glial fibrillary acidic protein · Vimentin · Peripherin · α-Internexin · Neurofilaments · Nestin · Synemin

D.A.D. Parry (✉)
Institute of Fundamental Sciences and Riddet Institute, Massey University, Palmerston North 4442, New Zealand
e-mail: d.parry@massey.ac.nz

R.A. Nixon, A. Yuan (eds.), *Cytoskeleton of the Nervous System*,
Advances in Neurobiology 3, DOI 10.1007/978-1-4419-6787-9_7,

Abbreviations

IF	Intermediate filament
STEM	Scanning transmission electron microscopy
DST	Disulfosuccinimidyl tartrate
NF-L, NF-M, NF-H	Neurofilament-light, -medium and -heavy chain respectively

Contents

7.1 Introduction

Intermediate filaments (IF) are a ubiquitous component of a wide diversity of eukaryotic cells. They are circular in cross section with diameters lying in the range 8–12 nm and have lengths that are generally measured in micrometers. As such, the IF represent a family of closely related, nonpolar cytoskeletal filaments intermediate in size between the other two filaments found in the cell – the microtubules (25-nm diameter) and the actin-containing microfilaments (7-nm diameter). Both of these filaments are polar structures. Unlike the microtubules and the microfilaments, however, the IF have low solubilities and are frequently phosphorylated (Omary et al., 2006). Examples of cytoplasmic IF include the keratins, which are found in trichocytes and epithelial cells, and desmin found in myogenic cells. The lamin IF are, in contrast, located in the cell nucleus.

In this contribution, discussion will be concentrated on the structure of neural IF, and reference will only be made to other IF where structural information is available for them and not, as yet, for those of neural origin. The family of neural IF include the glial fibrillary acidic proteins (found in glial cells such as astrocytes, the majority cell type in the central nervous system), peripherin (in peripheral neurons), vimentin (in cells of mesenchymal origin), synemin (in astrocytes), the neurofilament-light (NF-L), neurofilament-medium (NF-M), and neurofilament-heavy (NF-H) chains (in both central and peripheral neurons), α-internexin (in central neurons), and nestin (in neuroepithelial cells). The structures of these chains, their alignment, and the interactions they make with one another allow molecular formation to proceed. In turn, the assembly of these molecules into fully functioning IF can be described in terms of several, well-defined modes of molecular interaction.

The thrust of this contribution is primarily structural, and only limited reference to functional attributes will be made, as these aspects are covered by others elsewhere in this book. Nonetheless, it is worth noting that useful background reading is available in earlier reviews of IF structure and function. These include those written by Lazarides (1982); Parry and Steinert (1995, 1999); Herrmann et al. (2003); Herrmann and Aebi (2000, 2004); Parry (2005); Parry et al. (2007). In addition, several books and special editions of journals have been produced on various aspects of the structure and function of IF. These include one in *Methods in Cell Biology* (M.B. Omary and P.A. Coulombe, eds., 2004, volume 78) and, most recently, that published in *Experimental Cell Research* (M.B. Omary, H. Herrmann, and U. Lendahl, eds., 2007, volume 313).

7.2 Neural IF Structure

7.2.1 Chain Structure

Although IF chains differ significantly from one another in terms of molecular mass, it has been shown above that all of them have highly conserved sequence characteristics over a region that is about 310–315 residues long (Geisler et al., 1982; Crewther et al., 1983). This is approximately "centrally located" in the chains and is predicted to be α-helix-rich, a point confirmed from biochemical data and from crystallographic studies on IF fragments from non-neural IF molecules (see, e.g., Herrmann et al., 2000; Smith et al., 2002; Strelkov et al., 2002, 2004) (Fig. 7.1). Importantly, the sequences contain a heptad repeat of the form $(a\text{-}b\text{-}c\text{-}d\text{-}e\text{-}f\text{-}g)_n$, where positions *a* and *d* are generally occupied by apolar residues (about 70–75% frequency). Using the heptad repeats as a criterion, the central domain sequence can be subdivided into four segments 1A, 1B, 2A, and 2B (35, 101, 19, and 121 residues long, respectively) and three (generally) nonhelical linkers L1, L12, and L2 (typically 8–10, 16–22, and 8 residues long, respectively) (Table 7.1).

At the center of segment 2B, a "stutter" occurs in the sequence. This is equivalent to a three-residue deletion from an otherwise continuous heptad repeat (Dowling et al., 1983). Interestingly, a short region comprising segment 2A and linker L2 also displays a hendecad substructure (11 residue quasi-repeat) (Parry, 2006). This is a variant of the heptad repeat and has the form $(a\text{-}b\text{-}c\text{-}d\text{-}e\text{-}f\text{-}g\text{-}h\text{-}i\text{-}j\text{-}k)_n$, where apolar residues lie in key positions (*a*, *d*, *e*, and *h*) akin to those in the heptad repeat.

Further characteristics of all IF sequences are the highly conserved sequences found at the N-terminal end of segment 1A and the C-terminal end of segment 2B. These are known respectively as the helix initiation and the helix termination motifs. Their consensus sequences are, respectively, L-N-D-R-F-A-S-Y-I-D/E-K-V-R-F-L-E and E-Y-Q-X-L-L-D/N-V-K-X-R/A-L-D/E-X-E-I-A-T-Y-R-K/R-L-L-E-G-E-E/D-X-R-L/N/I. Segments 1B and 2B both display a regular linear disposition of acidic and basic residues as first described by Parry et al. (1977) for two fragments

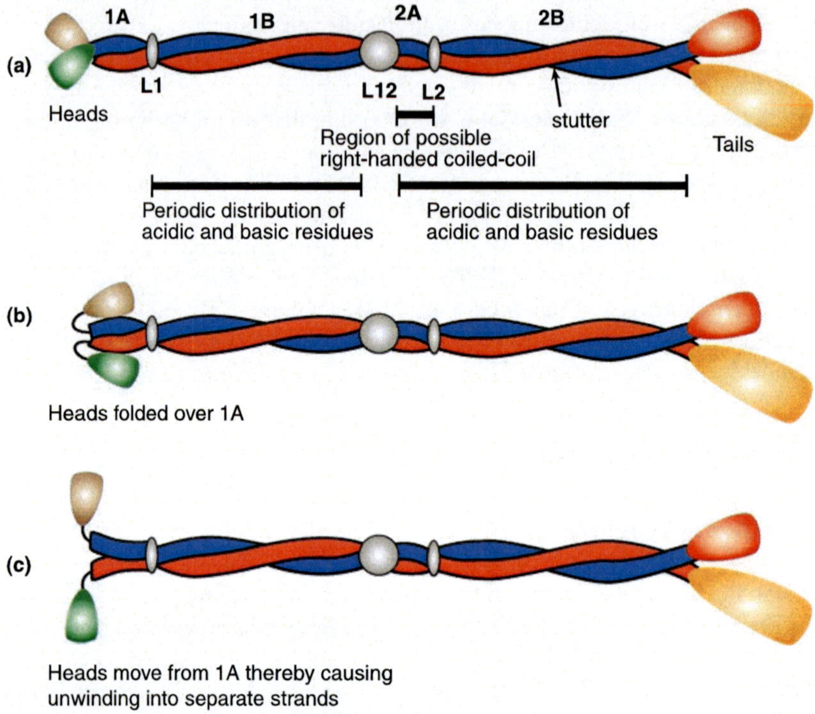

Fig. 7.1 **a** Schematic representation of the two-stranded structure of the NF-L/NF-H heterodimer. The large tail domain corresponds with that of NF-H. The rod domain contains four coiled-coil segments (1A, 1B, 2A, and 2B), and these are separated from one another by three linkers (L1, L12, and L2). A stutter occurs in the sequence at the center of segment 2B. This corresponds with a three-residue deletion from an otherwise continuous heptad repeat. **b** The head domains are shown lying over segment 1A, thereby stabilizing its two-stranded coiled-coil conformation. **c** When the head domains move away from segment 1A, it has been suggested that the latter will become less stable and that the coiled coil will unwind into two separate strands. This would allow the heads a much greater degree of spatial freedom and, hence, interaction potential

from wool keratin. For human keratin IF, as a whole these have periods of 9.51 and 9.87 residues in segments 1B and 2B, respectively (Smith and Parry, 2007). Comparable values are found in the far smaller number of neural IF chains that have been studied to date. This periodicity specifies various modes of molecular aggregation (see Section 7.5) that occur through the formation of ionic interactions between antiparallel molecules that are appropriately aligned axially.

The domains that are N- and C-terminal to the central region are known as the head and tail domains, respectively, and, in contrast with the central region, have a wide range of sizes, sequence characteristics, structures, and functions that are dependent on cell type. The (limited) structural information on these regions is described in Sections 7.4.1 and 7.4.2, respectively.

Table 7.1 Segment boundaries in human neuronal IF chains

Chain	Type	N	1A	L1	1B	L12	2A	L2	2B	C
Glial fibrillary acidic protein	III	1–69	70–104	105–112	113–213	214–229	230–248	249–256	257–377	378–432
Vimentin	III	1–103	104–138	139–146	147–247	248–263	264–282	283–290	291–411	412–466
Peripherin	III	1–97	98–132	133–140	141–241	242–259	260–278	279–286	287–407	408–470
Synemin α	III	1–11	12–46	47–56	57–157	158–178	179–193	194–201	202–322	323–1,565
α-Internexin	IV	1–94	95–129	130–139	140–240	241–259	260–278	279–286	287–407	408–499
Neurofilament-L	IV	1–90	91–125	126–135	136–236	237–252	253–271	272–279	280–400	401–543
Neurofilament-M	IV	1–101	102–136	137–146	147–247	248–264	265–283	284–291	292–412	413–916
Neurofilament-H	IV	1–97	98–132	133–142	143–243	244–265	266–284	285–292	293–413	414–1,020
Nestin	IV	1–8	9–43	44–52	53–153	154–173	174–192	–	193–313	314–1,621

Numbers in Table 7.1 refer to the residue numbers in the amino acid sequences that define the boundaries of the segments that characteristise the intermediate filament chain

7.2.2 Chain Typing

On the basis of sequence homology within the central domain, IF sequences may be classified into five different but nonetheless closely related types. In addition to the keratins, which are defined as type I (net acidic, pI values 4.5–5.5 and molecular masses 40–64 kDa) or type II chains (neutral-basic, pI values 6.5–7.5 and molecular weights 52–68 kDa), desmin, vimentin, glial fibrillary acidic protein, peripherin, syncoilin, and synemin are all classified as type III (molecular masses 52–55 kDa except for human synemin α and β, which are much larger at 173 and 140 kDa, respectively). Human type IV chains include NF-L, NF-M, NF-H, α-internexin, and nestin (molecular masses of about 62, 102, 112, 55, and 177 kDa, respectively), and type V chains are those from the nuclear lamina (lamins A/C, B1, and B2; molecular masses 62–78 kDa). Syncoilin and nestin differ from other IF chains in that they lack linkers L1 (Kemp et al., 2009) and L2, respectively. The lengths of the coiled-coil segments in IF chains are conserved in all cases. Consensus sequences have been derived for types I–V chains (Smith et al., 2002).

7.3 Rod Domain Structure

The heptad substructure, first predicted in the early 1950s and subsequently recognized in the sequences of many fibrous and globular proteins, has 3.5 residues per turn on average (Crick, 1953). As this is less than the average number of residues per turn in a standard α-helix (typically 3.60–3.63), this implies a coiled-coil structure in which right-handed α-helical chains will wind around one another in a left-handed manner (with pitch lengths now known to be typically in the range 14–17 nm for a two-stranded structure). This type of assembly internalizes the apolar residues and stabilizes the conformation adopted through close packing of these residues. For segment 2A and linker L2, however, there are 11 residues in three turns of α-helix (3.67 residues per turn on average) (Parry, 2006). As 3.67 is slightly greater than the average number of residues per turn in a standard α-helix (3.60–3.63), the right-handed α-helical chains will wind around one another in a right-handed manner with a very long period pitch length (possibly infinite). Thus, whereas segments 1A, 1B, and 2B will form left-handed coiled-coils (a point confirmed from x-ray crystallographic studies of IF coiled-coil fragments from non-neural sources), segment 2A and L2 may form either a short piece of right-handed coiled-coil with a very long pitch length (say 100 nm) or, alternatively, a region in which the component α-helices lie parallel to one another (Parry, 2006).

Cross-linking, biochemical data, and theoretical modeling have confirmed that there are two chains in all IF molecules, that these lie in axial register, and that the chains are parallel to each other [see, e.g., Steinert et al. (1993a, b, c) for epidermal keratins, Quinlan and Franke (1983) for heteropolymers of vimentin and glial fibrillary acidic protein, and Carden and Eagles (1983) and Parry et al. (1985) for neurofilaments]. The molecules may be in the form of homodimers or obligate heterodimers, depending on the situation (see Parry, 2005 for a summary). Homodimers have been formed from vimentin (III/III), glial fibrillary acidic protein (III/III), peripherin (III/III), synemin (III/III), NF-L chains (IV/IV), α-internexin (IV/IV), and nestin (VI/VI), and heterodimers have been observed between various chain combinations that include the following: peripherin (III)/NF-L (IV), glial fibrillary acidic protein (III)/NF-L (IV), nestin (IV)/vimentin (III), nestin (IV)/α-internexin (IV), NF-L (IV)/NF-M (IV), and NF-L (IV)/NF-H (IV). Heteropolymers, some of which are almost certainly in the form of heterodimers, have also been shown to exist for desmin (III) with synemin (III), and for vimentin (III) with synemin (III). There are no examples of neural IF chains that form only homodimers or, put in a different way, all neural IF chains are capable of participating in heterodimer formation. Two chains – NF-M and NF-H – are present only in heterodimers.

At the center of segment 2B, a stutter (Brown et al., 1996) is observed in the sequences of all IF protein chains (Dowling et al., 1983). This corresponds with a deletion of three residues from an otherwise continuous heptad repeat. Structurally, this results in the chains unwinding a little over a short distance axially, thereby causing the local pitch length of the coiled coil to increase (see crystal structure of vimentin: Strelkov and Burkhard, 2002). This effectively allows some relative reorientation to occur between the two parts of coiled-coil segment 2B that lie on either side of the stutter.

Parry et al. (2002) and Smith et al. (2002) have suggested that the two-stranded coiled coil comprising segment 1A depends for its structural integrity on the head domain folding back over it to stabilize it. It follows that if the head domain moves away from segment 1A, the latter would become destabilized, and the two α-helical strands would separate (Fig. 7.1). An advantage of this would be that the heads would have a much greater opportunity of interacting with other IF molecules or associated molecules. Two observations provided support for this hypothesis. First, it was shown that a vimentin dimer was formed only in the presence of the head domain (Strelkov et al., 2001), and second, the crystal structure of segment 1A alone was not a two-stranded coiled coil but displayed instead individual coiled α-helical strands (Strelkov et al., 2002).

The IF molecule as a whole thus has a polar, tripartite structure with a two-stranded coiled-coil rod domain about 45.5 nm long, that is, 310–315 residues multiplied by 0.1485 nm, the average unit rise per residue in a coiled-coil conformation. Segment 1A, however, may be stable as a two-stranded coiled-coil rope only when the head domains fold back over this region. In their absence, it has been suggested that segment 1A is less stable and that the coiled coil will separate into two distinct α-helical strands connected to the remaining rod domain through the flexible linker L1.

7.4 Terminal Domain Structures

No tertiary (and only very little secondary) structural information is available for either the head or the tail domains of any neural IF chain. Fortunately, the amino acid sequences of the neural IF chains are well known, and this does provide a source of data from which conformational inferences can be made. Sequence repeats have been observed in the heads and tails of some neural proteins and, allied to biochemical data, it has sometimes been possible to ascribe them functional and/or structural attributes, albeit imperfectly. The head domains and tail domains thus remain major areas of challenge and continue to impose serious limitations to the depth of understanding of IF structure and function that is desired by those working in the field. The limited information currently at hand is listed below.

7.4.1 Head Domain

The number of residues in the head domains of human neural IF chains show relatively little variation (69–103 residues). There are, however, two exceptions with human nestin and synemin containing just 8 and 11 residues, respectively (Table 7.1). The sequences of the head domains as a whole vary considerably, but in types III and IV IF chains, a conserved nonapeptide of the form S-S-Y-R-R-T-F-G-G has been observed. Furthermore, it has been shown that its maintenance is important if assembly is to proceed normally (Herrmann et al., 1992).

Phosphorylation is known to be an important regulator of IF dynamics and, as such, is involved in modulating the organization of IF networks, the distribution of IF proteins in the cell, and their functions. This has been characterized in some detail for both type III and IV IF (Ivaska et al., 2007; Sihag et al., 2007). Vimentin, for example, contains multiple phosphorylation sites – generally serine residues – in both its head and its tail domains though the sites are much more prevalent in the former than in the latter (for a summary of the sites, see Ivaska et al., 2007). Those in the head domain are specific for various cellular states including mitosis, differentiation, and stress. The phosphorylation events vary with the cell cycle and are directly involved in both the assembly of IF molecules into IF and, conversely, in the disassembly of the IF back into IF molecules. In this regard, peripherin has many similarities to vimentin (Huc et al., 1989; Giasson and Mushynski, 1998). Head phosphorylation is also known to affect the phosphorylation of sites in the tail domains.

Postsynthetic modifications of a different sort occur to both NF-L and NF-M chains in the form of O-glycosylation (Dong et al., 1993). This affects the state of assembly of the IF both in vivo and in vitro.

No crystal structures are known for any head domain at the present time.

7.4.2 Tail Domain

The tail domains, in contrast with the heads, differ widely in size. They may contain as few as 55 residues – as in glial fibrillary acidic protein – but as many as 607 residues in the human neurofilament heavy chain, 1,243 residues in human synemin α, and an even greater number in nestin (1,308 and 1,683 residues in human and hamster nestin, respectively; Table 7.1).

The tail domains of the type IV IF chains are net basic and are thought to adopt a folded structure containing many β-turns. They can be subdivided into several clearly defined subdomains on the basis of residue type and/or sequence repeat. These include those that are glutamic acid–rich (E-segments in NF-L, NF-M,

```
GFGPIPFSLPEGLPKIPSVSTHIKVKSEEKIKVVEKSEKETVIV
EEQTEETQVTEEVTEEEEKEAKEEEGKEEEGGEEEEAEGGEEE          (E-segment)

TKSPPAEEAASPEKE
AKSPVKEE
AKSPA  EAKSPEKEE
AKSPA  EVKSPEK
AKSPAKEE
AKSPP  EAKSPEKEE
AKSPA  EVKSPEK
AKSPAKEE
AKSPA  EAKSPEK
AKSPVKEE
AKSPAE
AKSPVKEE
AKSPA  EVKSPEK                        (KSP-containing segment)
AKSPTKEEAKSPEK
        AKSPEK
      EEAKSPEK
AKSPVKAEAKSPEK
AKSPVKAEAKSPEK
AKSPVKEEAKSPEK
AKSPVKEEAKSPEK
AKSPVKEEAKTPEK
AKSPVKEEAKSPEK
        AKSPEK
AKTL   DVKSPE
AKTPAKEE
ARSPA   DKFPEK
AKSPVKEEVKSPEK
AKSPLKEDAKAPEK
         EIPKKEE
VKSPVKEE

EKPQEVKVKEPPKKAEEEKAPATPKTEEKKDSKKEEAPKKEAPKPKVEEK    (KEP-segment)
KEPAVEKPKESKVEAKKEEAEDKKKVPTPEKEAPAKVEVKEDAKPKEKTE
VAKKEPDDAKAKEPSKPAEKKEAAPEKKDTKEEKAKKPEEKPKTEAKAKE
DDKTLSKEPSKPKAEKAEKSSSTDQKDSKPPEKATEDKAAKGK
```

Fig. 7.2 Amino sequence of the tail domain of the human NF-H chain. This shows a glutamic acid–rich region (the E-segment), the lysine-serine-proline–containing repeat region (the KSP-repeats), and the lysine, glutamic acid, and proline-rich region (KEP-segment). Regions of the KSP-repeat segment that display the consensus repeats A-K-S-P-A/V-K and A/V-K-S-P-E-K are colored yellow and red, respectively. They are separated from each other by glutamic acid residues (0, 1, or 2) and are colored green. The repeats are degenerate

NF-H, and α-internexin), those that are rich in lysine and glutamic acid residues (KE segments in NF-M and α-internexin), and those that contain frequent lysine-serine-proline repeats (KSP segments in NF-M and NF-H) (Fig. 7.2). In addition, a lysine–glutamic acid–proline-rich region (KEP segment) exists at the C-terminal ends in NF-M, NF-H, and α-internexin (Fig. 7.3). It has been proposed that the E-segments may be involved in IF assembly (Birkenberger and Ip, 1990).

In human NF-H, the K-S-P sequences are present in degenerate repeats of the form (A-K-S-P-A/V-K) $(E)_{0,1,2}$ (A/V-K-S-P-E-K) $(E)_{0,1,2}$. A strong pseudo-halving is clearly evident, and whereas both "halves" often occur consecutively in the sequence, it is not uncommon for one or the other to be absent (Fig. 7.2). The KSP repeats vary in number between one species and another, but, in all cases, they represent major sites of phosphorylation (for a recent summary, see Omary et al.,

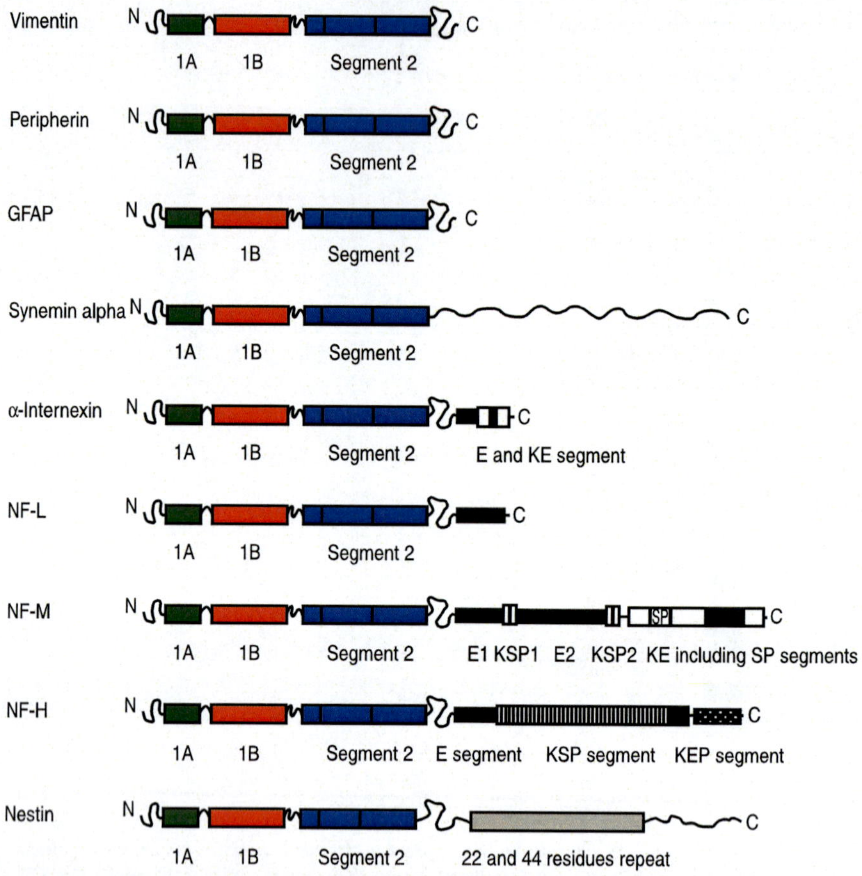

Fig. 7.3 Domain structure of nine neural IF molecules. Four of these are type III molecules [peripherin, vimentin, glial fibrillary acidic protein (GFAP), and synemin α], and five are type IV (α-internexin, NF-L, NF-M, NF-H, and nestin). The E and KEP segments are rich in glutamic acid residues and in lysine, glutamic acid, and proline residues, respectively. The KSP sequences are embedded in homologous pairs of hexapeptide repeats. The 22- and 44-residue repeats in nestin vary substantially in extent depending on the species involved

2006). They are thus of particular importance functionally and are involved in such features as transport and associations with other components in the cytoskeleton (for a summary of earlier work, see Parry and Steinert, 1995).

The tail domains of type IV nestin contain quasi-repeats 11 residues long based on repeats of 22 residues in human (consensus sequence E-E-D/N-L/Q-E-S/T-L-R/K-X-L-E-K-E-N-Q-E-P/S-L-R/K-S-L-E) and 44 residues in hamster (E-E-D-Q-L/R-V-E-R/T-L-I-E-K-E-G-Q-E-S-L-S-S-P-E-E-E-D-Q-E-T-D-R-P-L-E-K-E-N-G-E-P-L-K-P-V-E) and rat (E-E-D/E-Q-E-A/I-X-R-P-L-E-K-E-N-Q-E-S-L-G-Y/S-L-E-A-E-D-Q-M-L-E-R-L-L-E-K-E-S-Q-X-S-L-R/K-S-P-E) (Parry and Steinert, 1999; Steinert et al., 1999b). The structural implications of these repeats are, as yet, unknown.

Disappointingly, the crystal structures of tail domains from IF proteins have remained elusive, though the tail domain of human lamin A has been solved at atomic resolution (Krimm et al., 2002; Dhe-Paganon et al., 2002). In this case, the conformation adopted is closely akin to an Ig-fold.

7.5 Modes of Molecular Assembly

The presence of a regularity in the linear distribution of the acidic and basic residues in α-helical, coiled-coil segments 1B and 2 (2A+L2+2B) provided a strong indication that intermolecular ionic interactions would likely drive the assembly of IF molecules into fully functional IF. On this assumption, it was proposed by Crewther et al. (1983) that three modes of alignment between antiparallel molecules were likely to occur. In the first of these, it was suggested that the 1B segments would be largely overlapped (this was subsequently termed the A_{11} mode), in the second it was believed that the 2B segments would be largely overlapped (the A_{22} mode), and in the third it was thought that the molecules as a whole would be largely overlapped (the A_{12} mode). In each case, the alignments were proposed on the assumption that maximization of intermolecular interactions between two 1B or two 2B segments (or a combination of both) was probable. A fourth mode of molecular alignment (A_{CN}) involving a short head-to-tail overlap between similarly oriented (parallel) molecules was subsequently defined (Steinert et al., 1993a) and arose directly, in effect, from the consecutive application of the A_{11} and A_{22} modes. Indicative evidence to support the existence of the A_{11} mode came from work on epidermal keratin (Parry et al., 1985; Steinert, 1991a, b) and wool (Woods and Inglis, 1984; Sparrow et al., 1989), but definitive data were only obtained some years later using disulfosuccinimidyl tartrate (DST) cross-linking between spatially adjacent lysine residues in epidermal keratins K1/K10 (Steinert et al., 1993a) and K5/K14 (Steinert et al., 1993b). In all cases, their dispositions were consistent with the A_{11}, A_{22}, A_{12}, and A_{CN} modes of alignment. Several intramolecular cross-links were also found, thereby providing additional evidence that the two chains in the molecule were indeed parallel to each other and in axial register. Using the cross-link data, least squares analyses was performed to determine the axially projected lengths of the linker regions (L1, L12, and L2) and the precise values of A_{11}, A_{22}, A_{12}, and A_{CN}.

In addition to keratin data that involve only type I and type II chains, equivalent parameters derived from cross-linking have been obtained for type III homodimers (vimentin: Steinert et al., 1993c; Parry et al., 2001), type IV homodimers (α-internexin: Steinert et al., 1999a), and type III/type IV copolymers (vimentin/α-internexin: Steinert et al., 1999) (Fig. 7.4). The A_{11} mode of interaction was also supported from the work of Mücke et al. (2004) on vimentin. It follows, therefore, that type III, IV, and VI IF chains (in a variety of well-specified combinations) assemble in a common manner and that this is different from that adopted in both the "reduced" and "oxidized" states of trichocyte keratin.

There are no x-ray diffraction data pertaining to the surface lattice structure of any vertebrate neural IF. There are, however, some data from neurofilaments in

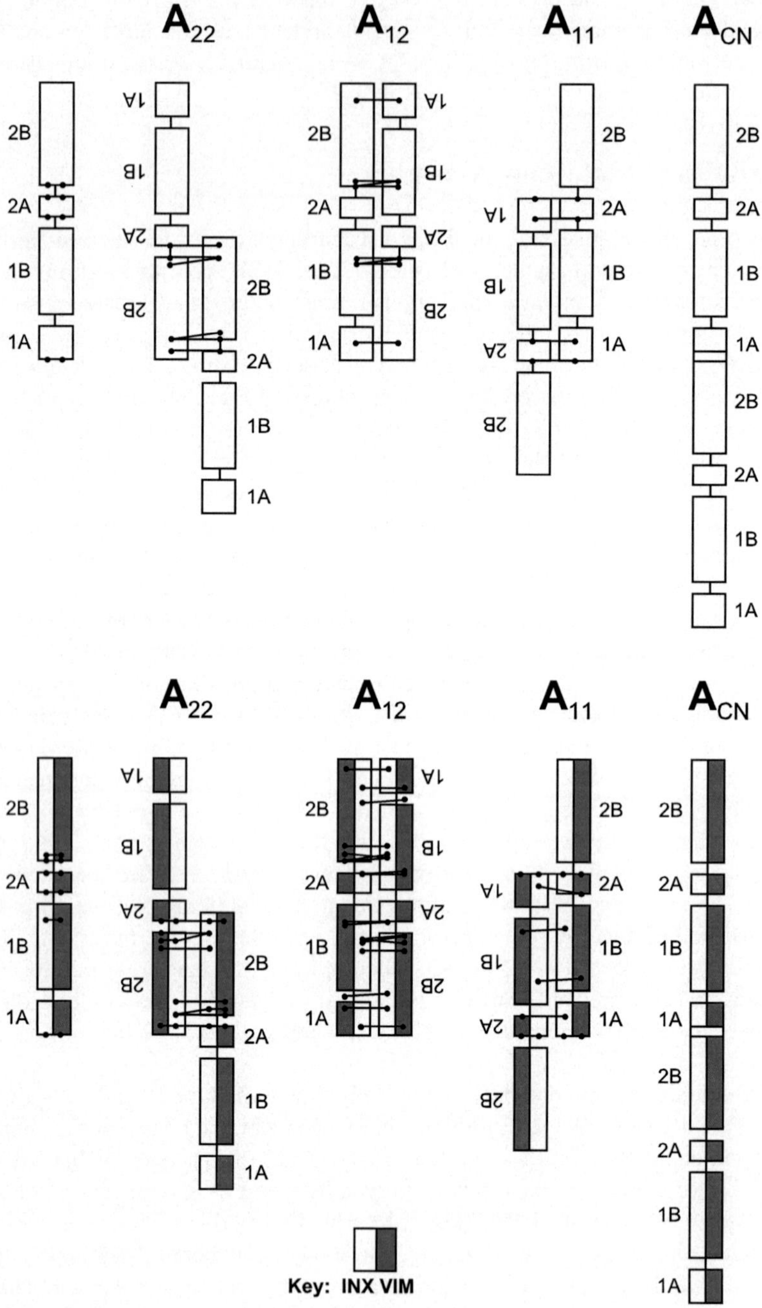

Fig. 7.4 (continued)

the invertebrate *Myxicola* (Wais-Steider et al., 1983) that shows an axial period of 25.2 nm. This can be compared with the repeat of 47 nm observed in trichocyte keratin. (In this case, there are extensive x-ray data available, and a surface lattice structure has been deduced; Fraser and MacRae, 1985, 1988; Watts et al., 2002; Fraser et al., 2003; Fraser and Parry, 2005, 2007). The invertebrate neurofilament chains do differ significantly from the vertebrate ones, however, in having a lamin-like chain structure with a 42-residue insert in segment 1B. The lack of detailed x-ray diffraction data from neural IF means that it is only from the cross-linking data that a surface lattice structure can be deduced (Fig. 7.5). From cross-link data for all IF, it has been possible to show that the assembly of IF molecules into filaments leads to four closely related structures. These arise from the small but very characteristic sequence differences that are associated with chain type. Refined values of the parameters corresponding with the three most closely related structures are listed in Table 7.2 (Parry et al., 2007). The fourth relates to lamin IF, which, as noted earlier, has a 42-residue insert in segment 1B. By definition, it must therefore have a structure different from the others. It follows that the differing structures for IF provide a ready mechanism that permits assembly and turnover between allowed molecules while preventing that between those molecules that are not functionally appropriate.

Evidence supports the concept that all IF, with the exception of lamin IF, are formed by a common mechanism. This involves three steps. First, there is a rapid lateral aggregation of about eight tetramers in the A_{12} mode of interaction to form a ULF. Each of these tetramers consists of a half-staggered, antiparallel pair of molecules assembled (in all probability) in the A_{11} rather than the A_{22} mode (Fig. 7.6). The latter mode cannot be totally excluded, however. The ULF is thus about 55 nm in length. In the well-characterized case of epidermal keratin, the assembly at the two- to four-molecule level is facilitated through interactions involving the H1 zones in the head domain of type II chains (Steinert et al., 1993a; Steinert and Parry, 1993; Hatzfeld and Burba, 1994). The equivalent sequences are present in the trichocyte keratin chains thereby suggesting that they have the same role in assembly as that shown experimentally for the epidermal keratins. Second, an elongation step occurs that involves the axial aggregation of ULF through (in all probability) the A_{22} mode to form immature IF about 16 nm in diameter and many micrometers in length. Aggregation through the A_{11} mode remains a possibility.

Fig. 7.4 (continued) Molecular alignments deduced by the DST cross-linking method for (*top*) α-internexin IF and (*lower*) α-internexin–vimentin copolymer IF. In the latter case, the vimentin chains are colored green. In each of these two cases, some of the observed cross-links are intramolecular (*left*) thereby showing that the two chains must be parallel and in axial register. A_{22}, A_{12}, and A_{11} show antiparallel arrays in which, respectively, the molecules are half-staggered with their 2B segments largely overlapped, with their entire lengths largely overlapped, and with their 1B segments largely overlapped. A_{CN} shows that there is a parallel, short, head-to-tail overlap of the ends of segment 1A and 2B. This arises in effect from the consecutive applications of A_{11} and A_{22}. Reproduced in a modified form from Steinert et al. (1999a)

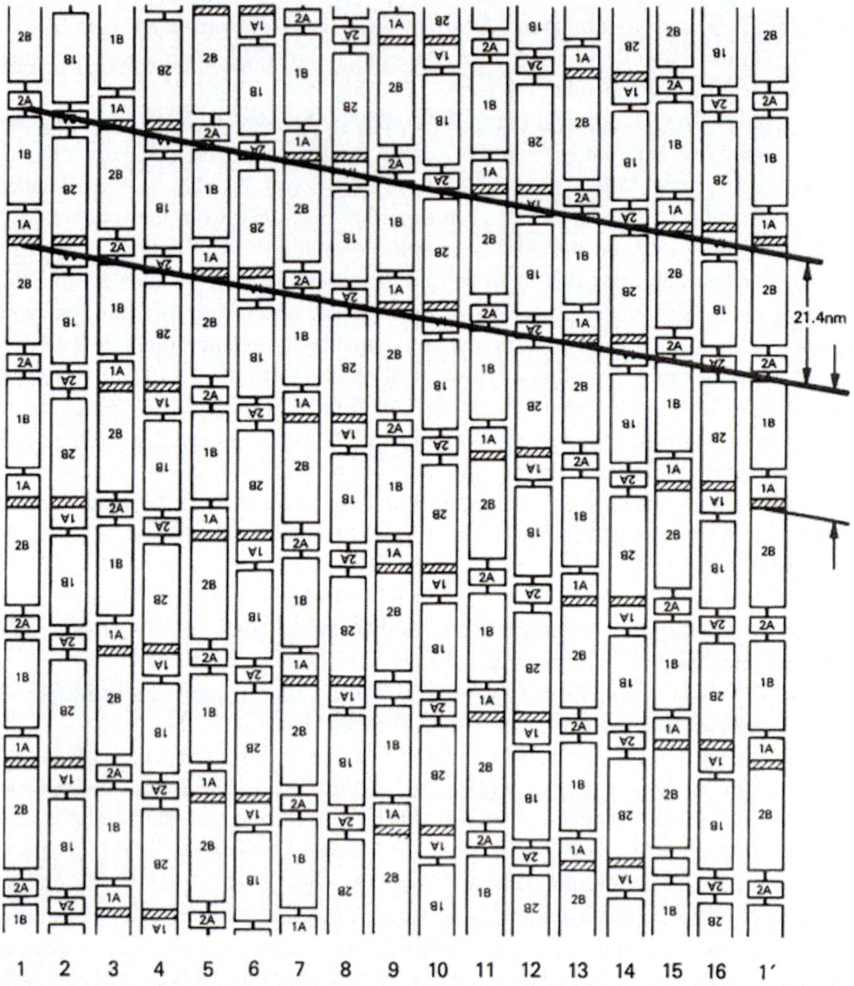

Fig. 7.5 The two-dimensional surface lattice model for vimentin is based on the cross-linking data derived by Steinert et al. (1993c). The STEM data indicate a best estimate of 16 molecules in cross section, hence the numbering along the *bottom* of the figure. The model comprises alternating rows of antiparallel in-register (A_{12}) and antiparallel staggered (A_{11} and A_{22}) molecules. There is a small head-to-tail overlap (shown *hatched*) between similarly directed molecules. The intermediate filament has an axial repeat of about 45 nm with equivalent points occurring at axial intervals of about half this value (heavy *parallel diagonal lines*), interrupted at the dislocation. The second *diagonal line* is obtained by connecting equivalent points on the molecules, and the dislocation between 16 and 1′ is denoted between the *arrows*. In a 16-molecule IF, the discontinuity will almost match when the lattice is folded in three dimensions. If there are more or less than 16 molecules in cross section, the magnitudes of the discontinuities will, necessarily, differ also. Reproduced from Steinert et al. (1993c) with permission

Dimers within this structure are believed to undergo some reorganization at this stage. Third, radial compaction takes place that results in close packing of the molecular filaments and the formation of IF with diameter in the range 8–12 nm (Eichner

Table 7.2 Structural parameters of IF derived from cross-linking studies

Fixed	Vimentin	α-Internexin	Type III, type IV, and type III/IV copolymers	Reduced α-keratin		Oxidized α-keratin		
				L1, L12, L2	L2, z_a/z_b	L1, L12, L2		z_a, z_b
z_a	23.89	–	25.06	78.15	74.48	50.00	47.14	50.03
z_b	139.53	137.00	136.43	112.18	111.78	133.30	130.71	133.33
Repeat	302.95	–	297.93	302.51	298.05	316.61	308.48	316.63
Overlap	4.31	10.00	6.99	5.67	2.42	−8.43	−11.72	−13.67
A_{12}	5.50	7.00	4.19	3.30	−0.34	6.98	1.55	2.56
A_{11}	−134.03	−130.00	−132.24	−108.88	−112.12	−126.32	−129.16	−130.77
A_{22}	168.92	–	165.68	193.63	185.93	190.28	179.33	185.86
L1	11.47	14.50	10.99	14.84	8.96	14.84	7.21	9.56
L12	11.99	–	8.99	12.57	10.03	12.57	9.43	11.85
L2	7.79	9.00	8.94	4.77	5.52	4.77	4.11	5.52
Δz(1BU,1BD)	−20.72	–	−20.30	−1.38	−0.54	−18.82	−17.85	−17.98
Δz(2BU,2BD)	−17.34	−19.00	−18.24	6.45	6.46	3.10	3.57	3.91
Δz(1BU, 2D)	−40.97	−42.50	−41.80	−46.54	−44.31	−42.86	−40.74	−42.02

Values are expressed in multiples of the mean axial rise per residue in a coiled coil conformation (0.1485 nm). The terms, z_a and z_b, refer to the axial projection of the vectors a and b that define the surface lattice cell. Δz(1BU,1BD), Δz(2BU,2BD), Δz(1BU,2D) are the axial staggers between antiparallel 1B segments, between antiparallel 2 segments (2A+L2+2B) and between antiparallel 1B and 2 segments respectively.

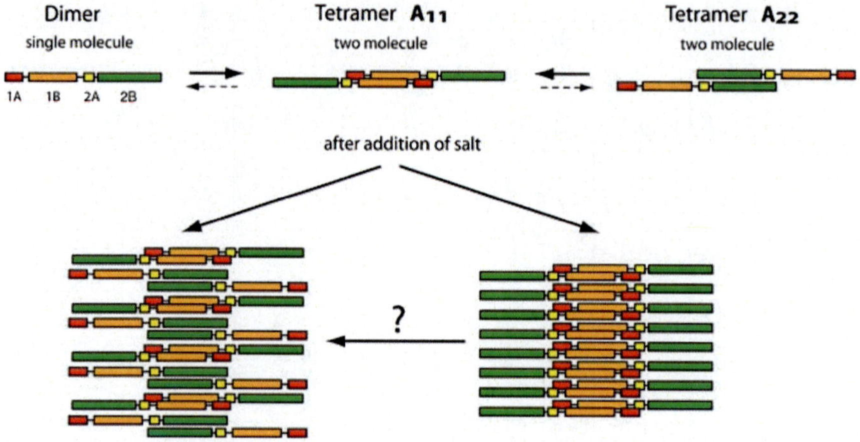

Fig. 7.6 Schematic figure illustrating the proposed mechanism of assembly of IF molecules into functional IF. Initially, pairs of antiparallel molecules align with their 1B segments largely overlapped (A_{11} tetramer) and/or (possibly) their 2B segments largely overlapped (A_{22}). After the addition of salt, eight tetramers rapidly aggregate laterally in an antiparallel manner and are stabilized by A_{12} interactions. These are termed unit-length-filaments (ULFs). These then assemble axially through antiparallel A_{22} and/or possibly A_{11} interactions to form wide, loosely packed filaments of considerable length. Finally, these undergo compaction to form IF of normal diameter. Reproduced from Mücke et al. (2004) with permission

et al., 1985; Herrmann et al., 1999, 2002; Wang et al., 2000; Parry et al., 2007). Equivalent data for most of the neural IF is lacking at present, but it is widely believed that the fundamentals listed here are applicable to them also.

It is appropriate at this point to describe the terminology commonly used to describe specific postassembly aggregates. A molecular strand is defined as a filament containing similarly directed molecules linked by a short head-to-tail overlap or gap in many cases (A_{CN}). It is thus one molecule or two chains wide (excluding the gap/overlap regions). Pairs of antiparallel molecular strands are termed protofilaments (two molecules or four chains in width) and are defined by A_{11} and A_{22} interactions. Finally, pairs of protofilaments are defined as a protofibril (four molecules or eight chains in width) and are linked by A_{12} interactions. It should be emphasized that these subfibrillar elements are not functional intermediates in assembly but are a convenient way to define postassembly IF substructure.

STEM provides a method that readily allows the mass-per-unit-length of filaments to be determined experimentally. For native vimentin IF, the data indicate that the majority of filaments have a mass-per-unit-length corresponding with about 32 chains in cross section (Steven et al., 1982), though a small number contain only about 22 chains in section. Reconstituted vimentin IF showed a similar mass-per-unit-length distribution except that larger polymorphs were also seen with about 40 and 48 chains in section (Herrmann et al., 1996; Herrmann and Aebi, 2000). Heins and Aebi (1994) have undertaken similar STEM studies on NF-L chains (Fig. 7.7).

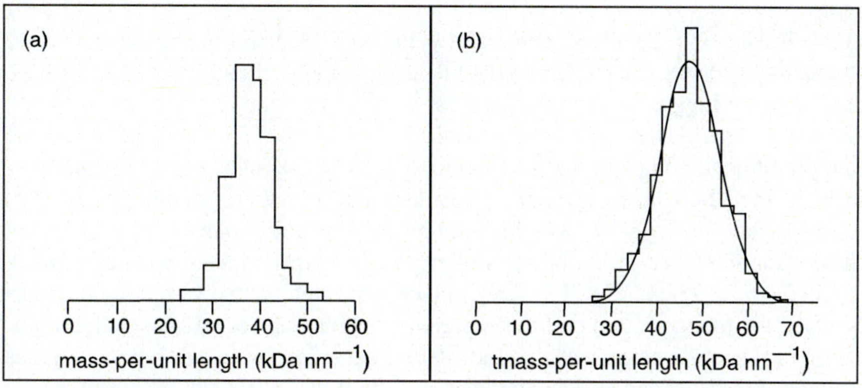

Fig. 7.7 Mass-per-unit-length distributions determined by STEM for (**a**) native human vimentin (Herrmann et al., 1996) and (**b**) native mouse NF-L (Heins et al., 1993). Both distributions are unimodal. Their peak values are 37 kDa/nm (30.5 chains in cross section) and 47 kDa/nm (33.8 chains in section), respectively. However, it is not uncommon for in vitro IF assemblies to display a multimodal mass distribution. The observed peaks generally differ from one another by the equivalent of eight chains in section. In these cases, the numbers of chains in section are about 16, 24, 32, 40, and 48. These provide evidence that an eight-chain protofibril may exist as an identifiable substructure postassembly. Redrawn from the original figures

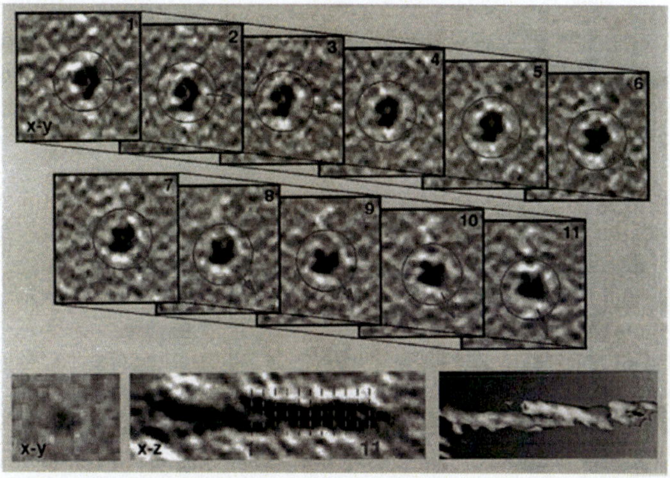

Fig. 7.8 Cryo-electron tomography has allowed the supertwist of protofibrils in vimentin IF to be visualized and characterized. **a** Sequential 1.8-nm-thick slices along a vimentin IF reveal a right-handed supertwist of the protofibrils with a period of about 96 nm. The filament was followed over a distance of 18 nm axially. Over this length, the filament showed a 70-degree rotation to the *right*. Variability between filaments suggests that the supertwist is less than perfectly regular and that the axial interactions along an individual protofibril are considerably stronger than the lateral ones. **b** Side view of the same filament with the slices used in (**a**) illustrated by *dashed lines*. **c** A three-dimensional model. Reproduced from Goldie et al. (2007) with permission

In this instance, the data were consistent with about 33 chains in section in vivo, 41 chains in vitro, and, for NF-L lacking the head domains, about 24 chains. Overall, these data are suggestive of preferred mass modes corresponding with about 24, 32, 40, and 48 chains, and hence the existence of an eight-chain subfibrillar element (the protofibril). Although the interpretation of these results is not totally straightforward, the 32-chain structure in section does appear to be the dominant one. This is consistent with cryo-tomographic studies on vimentin (Goldie et al., 2007) that show a four-stranded protofibrillar structure with a right-handed twist and a pitch length of about 96 nm (Fig. 7.8). It was noted, nonetheless, that variability did exist between the filaments and that the supertwist length was not invariant. The implication of this observation was that the lateral interactions within the IF were not as strong and well-defined as the axial ones (Goldie et al., 2007).

7.6 Summary

Through a combined theoretical and experimental approach, a consistent model for the structure of the neural IF molecule has been deduced, as has the assembly of such molecules into fully functional IF. In some cases, however, aspects of the model could only be deduced through the use of complementary data obtained from non-neural IF. The IF molecule has been characterized as a dimer with the two chains parallel to each other and in axial register. A central, α-helical, coiled-coil–rich region (the rod domain) separates the head (the region N-terminal to the rod) from the tail (the region C-terminal to the rod). The molecule thus has a tripartite structure with the rod domains having especial (but not sole) importance in specifying the assembly of the molecules into intact IF. In contrast, most of the functional attributes of the IF molecules lie with the heads and tails. These occur predominately on the surface of the IF and are in positions that render them particularly accessible to interactions with other IF-associated molecules. Regular linear dispositions of charged residues in the main coiled-coil segments provides a mechanism to aid assembly through the formation of intermolecular ionic interactions. Although there are no x-ray diffraction data to specify the surface lattice structure of any vertebrate neural IF, the latter can be deduced from the pattern of cross-links induced between molecules. Allied to STEM and cryo-tomographic data, it seems probable that neural IF have a four-protofibril structure containing 32 chains in section. The assembly occurs in a three-part process. First, a rapid lateral aggregation of about eight tetramers occurs through the A_{12} mode of interaction to form a ULF. Each of these tetramers consists of a half-staggered, antiparallel pair of molecules assembled, in all probability, in the A_{11} mode. Second, an elongation step occurs that involves the axial aggregation of ULF to form immature IF about 16 nm in diameter and many micrometers in length. In this case, the A_{22} mode appears to be the dominant one. Dimers within this structure are believed to undergo some reorganization at this stage. Third, radial compaction takes place that results in close packing of the molecular filaments and the formation of IF with diameter in the range 8–12 nm.

References

Birkenberger L, Ip W (1990) Properties of the desmin tail domain: studies using synthetic peptides and antipeptide antibodies. J Cell Biol 111:2063–2075

Brown JH, Cohen C, Parry DAD (1996) Heptad breaks in α-helical coiled coils: stutters and stammers. Proteins Struct Funct Genet 26:134–145

Carden MJ, Eagles PAM (1983) Neurofilaments from ox spinal nerves: isolation, disassembly, reassembly and crosslinking properties. Biochem J 215:227–237

Crewther WG, Dowling LM, Steinert PM, Parry DAD (1983) Structure of intermediate filaments. Int J Biol Macromol 5:267–274

Crick FHC (1953) The packing of α-helices: simple coiled-coils. Acta Cryst 6:689–697

Dhe-Paganon S, Werner ED, Chi YI, Shoelson SE (2002) Structure of the globular tail of nuclear lamin. J Biol Chem 277:17381–17384

Dong DL-Y, Xu Z-S, Chevrier MR, Cotter RJ, Cleveland DW, Hart GW (1993) Glycosylation of mammalian neurofilaments: localization of multiple O-linked N-acetylglucosamine moieties on neurofilament peptides L and M. J Biol Chem 268:16679–16687

Dowling LM, Parry DAD, Sparrow LG (1983) Structural homology between hard α-keratin and the intermediate filament proteins desmin and vimentin. Biosci Rep 3:73–78

Eichner R, Rew P, Engel A, Aebi U (1985) Human epidermal keratin filaments: studies on their structure and assembly. Ann N Y Acad Sci 455:381–402

Fraser RDB, MacRae TP (1985) Intermediate filament structure. Biosci Rep 5:573–579

Fraser RDB, MacRae TP (1988) Surface lattice in α-keratin filaments. Int J Biol Macromol 10: 178–184

Fraser RDB, Parry DAD (2005) The three-dimensional structure of trichocyte (hard α-) keratin intermediate filaments: features of the molecular packing deduced from the sites of induced crosslinks. J Struct Biol 151:171–181

Fraser RDB, Parry DAD (2007) Structural changes in the trichocyte intermediate filaments accompanying the transition from the reduced to the oxidized form. J Struct Biol 159:36–45

Fraser RDB, Steinert PM, Parry DAD (2003) Structural changes in trichocyte keratin intermediate filaments during keratinization. J Struct Biol 142:266–271

Geisler N, Kaufmann E, Weber K (1982) Protein-chemical characterization of three structurally distinct domains along the protofilament unit of 10 nm filaments. Cell 30:277–286

Giasson BI, Mushynski WE (1998) Intermediate filament disassembly in cultured dorsal root ganglion neurons is associated with amino-terminal head domain phosphorylation of specific subunits. J Neurochem 70:1869–1875

Goldie KN, Wedig T, Mitra AK, Aebi U, Herrmann H, Hoenger A (2007) Dissecting the 3-D structure of vimentin intermediate filaments by cryo-electron tomography. J Struct Biol 158:378–385

Hatzfeld M, Burba M (1994) Function of type I and type II keratin head domains: their role in dimer, tetramer and filament formation. J Cell Sci 107:1959–1972

Heins S, Aebi U (1994) Making heads and tails of intermediate filament assembly, dynamics and networks. Curr Opin Cell Biol 6:25–33

Heins S, Wong P-C, Müller S, Goldie K, Cleveland DW, Aebi U (1993) The rod domain of NF-L determines neurofilament architecture, whereas the end domains specify filament assembly and network formation. J Cell Biol 123:1517–1533

Herrmann H, Aebi U (2000) Intermediate filaments and their associates: multi-talented structural elements specifying cytoarchitecture and cytodynamics. Curr Opin Cell Biol 12:79–90

Herrmann H, Aebi U (2004) Intermediate filaments: molecular structure, assembly mechanism, and integration into functionally distinct intracellular scaffolds. Annu Rev Biochem 73: 749–789

Herrmann H, Häner M, Brettel M, Ku NO, Aebi U (1999) Characterization of distinct early assembly units of different intermediate filament proteins. J Mol Biol 286: 1403–1420

Herrmann H, Häner M, Brettel M, Müller SA, Goldie KN, Fedtke B, Lustig A, Franke WW, Aebi U (1996) Structure and assembly properties of the intermediate filament protein vimentin: the role of its head, rod and tail domains. J Mol Biol 264:933–953

Herrmann H, Hesse M, Reichenzeller M, Aebi U, Magin TM (2003) Functional complexity of intermediate filament cytoskeletons: from structures to assembly to gene ablation. Int Rev Cytol 223:83–175

Herrmann H, Hofmann I, Franke WW (1992) Identification of a nonapeptide motif in the vimentin head domain involved in intermediate filament assembly. J Mol Biol 223:637–650

Herrmann H, Strelkov SV, Feja B, Rogers KR, Brettel M, Lustig A, Häner M, Parry DAD, Steinert PM, Burkhard P, Aebi U (2000) The intermediate filament protein consensus motif of segment 2B: its atomic structure and contribution to assembly. J Mol Biol 298:817–832

Herrmann H, Wedig T, Porter RM, Lane EB, Aebi U (2002) Characterization of early assembly intermediates of recombinant human keratins. J Struct Biol 137:82–96

Huc C, Escurat M, Djabali K, Derer M, Landon F, Gros F, Portier MM (1989) Phosphorylation of peripherin, an intermediate filament protein, in mouse neuroblastoma NIE 115 cell line and in sympathetic neurons. Biochem Biophys Res Commun 160:772–779

Ivaska J, Pallari H-M, Nevo J, Eriksson JE (2007) Novel functions of vimentin in cell adhesion, migration, and signaling. Exp Cell Res 313:2050–2062

Kemp M, Edwards B, Burgess M, Clarke WE, Nicholson G, Parry DAD, Davies KE (2009) Syncoilin isoform organization and differential expression in murine striated muscle. J Struct Biol 165:196–203

Krimm I, Ostlund C, Gilquin B, Couprie J, Hossenlopp P, Mornon JP, Bonne G, Courvalin JC, Worman HJ, Zinn-Justin S (2002) The Ig-like structure of the C-terminal domain of lamin A/C, mutated in muscular dystrophies, cardiomyopathy, and partial lipodystrophy. Structure 10:811–823

Lazarides E (1982) Intermediate filaments: a chemically heterogeneous, developmentally regulated class of proteins. Annu Rev Biochem 51:219–250

Mücke N, Wedig T, Bürer A, Marekov LN, Steinert PM, Langowski J, Aebi U, Herrmann H (2004) Molecular and biophysical characterization of assembly-starter units of human vimentin. J Mol Biol 340:97–114

Omary MB, Ku N-O, Tao G-F, Toivola DM, Liao J (2006) "Heads and tails" of intermediate filament phosphorylation: multiple sites and functional insights. Trends Biochem Sci 31:383–394

Parry DAD (2005) Microdissection of the sequence and structure of intermediate filament chains. Adv Pro Chem 70:113–142

Parry DAD (2006) Hendecad repeat in segment 2A and linker L2 of intermediate filament chains implies the possibility of a right-handed coiled-coil structure. J Struct Biol 155: 370–374

Parry DAD, Crewther WG, Fraser RDB, MacRae TP (1977) Structure of α-keratin: structural implication of the amino acid sequences of the type I and type II chains segments. J Mol Biol 113:449–454

Parry DAD, Marekov LN, Steinert PM (2001) Subfilamentous protofibril structures in fibrous proteins: cross-linking evidence for protofibrils in intermediate filaments. J Biol Chem 276:39253–39258

Parry DAD, Marekov LN, Steinert PM, Smith TA (2002) A role for the 1A and L1 rod domain segments in head domain organization and function of intermediate filaments: structural analysis of trichocyte keratin. J Struct Biol 137:97–108

Parry DAD, Steinert PM (1995) Intermediate filament structure. Springer, Heidelberg

Parry DAD, Steinert PM (1999) Intermediate filaments: molecular architecture, assembly, dynamics and polymorphism. Quart Rev Biophys 32:99–187

Parry DAD, Steven AC, Steinert PM (1985) The coiled-coil molecules of intermediate filaments consist of two parallel chains in exact axial register. Biochem Biophys Res Commun 127: 1012–1018

Parry DAD, Strelkov SV, Burkhard P, Aebi U, Herrmann H (2007) Towards a molecular description of intermediate filament structure and assembly. Exp Cell Res 313:2204–2216

Quinlan RA, Franke WW (1983) Molecular interactions in intermediate-sized filaments revealed by chemical crosslinking: heteropolymers of vimentin and glial filament protein in cultured human glioma cells. Eur J Biochem 132:477–484

Sihag RK, Inagaki M, Yamaguchi T, Shea TB, Pant HC (2007) Role of phosphorylation on the structural dynamics and function of types III and IV intermediate filaments. Exp Cell Res 313:2098–2109

Smith TA, Parry DAD (2007) Sequence analyses of type I and type II chains in human hair and epithelial keratin intermediate filaments: promiscuous obligate heterodimers, type II template for molecule formation and a rationale for heterodimer formation. J Struct Biol 158: 344–357

Smith TA, Strelkov SV, Burkhard P, Aebi U, Parry DAD (2002) Sequence comparisons of intermediate filament chains: evidence of a unique functional/structural role for the coiled-coil segment 1A and linker L1. J Struct Biol 137:128–145

Sparrow LG, Dowling LM, Loke VY, Strike PM (1989) Amino acid sequence in wool keratin IF proteins. In: Rogers GE, Reis PJ, Ward KA, Marshall RC (eds) The biology of wool and hair. Chapman and Hall, London, pp 145–155

Steinert PM (1991a) Organization of coiled-coil molecules in native keratin 1/keratin 10 intermediate filaments: evidence for alternating rows of antiparallel in-register and antiparallel molecules. J Struct Biol 107:157–174

Steinert PM (1991b) Analysis of the mechanism of assembly of mouse keratin 1/keratin 10 intermediate filaments in vitro suggests that intermediate filaments are built from multiple oligomeric units rather than a unique tetrameric building block. J Struct Biol 107:175–188

Steinert PM, Marekov LN, Fraser RDB, Parry DAD (1993a) Keratin intermediate filament structure: crosslinking studies yield quantitative information on molecular dimensions and mechanism of assembly. J Mol Biol 230:436–452

Steinert PM, Marekov LN, Parry DAD (1993b) Conservation of the structure of keratin intermediate filaments: molecular mechanism by which different keratin molecules integrate into pre-existing keratin intermediate filaments during differentiation. Biochemistry 32: 10046–10056

Steinert PM, Marekov LN, Parry DAD (1993c) Diversity of intermediate filament structure. Evidence that the alignment of coiled-coil molecules in vimentin is different from that in keratin intermediate filaments. J Biol Chem 268:24916–24925

Steinert PM, Marekov LN, Parry DAD (1999a) Molecular parameters of type IV α-internexin and type IV – type III α-internexin-vimentin copolymer intermediate filaments. J Biol Chem 274:1657–1666

Steinert PM, Chou YH, Prahlad V, Parry DAD, Marekov LN, Wu KC, Jang SI, Goldman RD (1999b) A high molecular weight intermediate filament-associated protein in BHK-21 cells is nestin, a type VI intermediate filament protein: limited co-assembly *in vitro* to form heteropolymers with type III vimentin and type IV α-internexin. J Biol Chem 274:9881–9890

Steinert PM, Parry DAD (1993) The conserved H1 domain of the type II keratin 1 chains plays an essential role in the alignment of nearest neighbor molecules in mouse and human keratin 1/keratin 10 intermediate filaments at the two- to four-molecule level of structure. J Biol Chem 268:2878–2887

Steven AC, Wall JS, Hainfeld JF, Steinert PM (1982) Structure of fibroblastic intermediate filaments: analysis by scanning transmission electron microscopy. Proc Natl Acad Sci USA 79:3101–3105

Strelkov SV, Burkhard P (2002) Analysis of α-helical coiled coils with the program TWISTER reveals a structural mechanism for stutter compensation. J Struct Biol 137:54–64

Strelkov SV, Herrmann H, Geisler N, Lustig A, Ivaninskii S, Zimbelmann R, Burkhard P, Aebi U (2001) Divide-and-conquer crystallographic approach towards an atomic structure of intermediate filaments. J Mol Biol 306:773–781

Strelkov SV, Herrmann H, Geisler N, Wedig T, Zimbelmann R, Aebi U, Burkhard P (2002) Conserved segments 1A and 2B of the intermediate filament dimer: their atomic structures and role in assembly. EMBO J 21:1255–1266

Strelkov SV, Schumacher J, Burkhard P, Aebi U, Herrmann H (2004) Crystal structure of the human lamin A coil 2B dimer: implications for the head-to-tail association of nuclear lamins. J Mol Biol 343:1067–1080

Wais-Steider C, Eagles PAM, Gilbert DS, Hopkins JM (1983) Structural similarities and differences amongst neurofilaments. J Mol Biol 165:393–400

Wang H, Parry DAD, Jones LN, Idler WW, Marekov LN, Steinert PM (2000) In vitro assembly and structure of trichocyte keratin intermediate filaments: a novel role for stabilization by disulfide bonding. J Cell Biol 151:1459–1468

Watts NR, Jones LN, Cheng N, Wall JS, Parry DAD, Steven AC (2002) Cryo-electron microscopy of trichocyte (hard α-keratin) intermediate filaments reveals a low-density core. J Struct Biol 137:109–118

Woods EF, Inglis A (1984) Organization of the coiled-coils in the wool microfibril. Int J Biol Macromol 6:277–283

Chapter 8
Alpha-Internexin: The Fourth Subunit of Neurofilaments in the Mature CNS

Aidong Yuan and Ralph A. Nixon

Abstract Because α-internexin is expressed in embryonic brain and can form homopolymers in vitro, it was originally considered to be an independent filament system critical mainly during brain development and distinct from that assembled in neurons from neurofilament (NF) triplet proteins. Analysis of mice engineered to eliminate individual or different combinations of NF triplet or α-internexin has proved that α-internexin and NF proteins are constituents of the same NF and functionally interdependent. α-Internexin satisfies all criteria previously used to establish the NF triplet proteins as subunits of a single intermediate filament and can now be considered to be a fourth subunit of NF in the adult CNS. Its role as a subunit of CNS NF explains the close relationship of α-internexin with the pathology of CNS diseases associated with NF accumulation.

Keywords Alpha-internexin · Neurofilament · Intermediate filament · Cytoskeleton · Cytoskeletal protein · Axonal transport · Neurofilament inclusion disease · Tropical spastic paraparesis

Contents

A. Yuan (✉)
Nathan Kline Institute, Center for Dementia Research, Orangeburg, NY 10962, USA;
Department of Psychiatry, New York University School of Medicine, New York, NY 10016, USA
e-mail: yuan@nki.rfmh.org

R.A. Nixon, A. Yuan (eds.), *Cytoskeleton of the Nervous System*,
Advances in Neurobiology 3, DOI 10.1007/978-1-4419-6787-9_8,

8.1 Discovery of α-Internexin

α-Internexin, together with β-internexin (Napolitano et al., 1985), was originally discovered as a 66-kDa intermediate filament-binding protein, rather than as an intermediate filament protein, by Liem and colleagues (Pachter and Liem, 1985). Internexins were so named because of their close association with isolated intermediate filaments and by their inability to assemble into 10-nm filaments under conditions that promote the assembly of glial filaments and neurofilaments from their individual purified subunits (Liem and Hutchison, 1982; Pachter and Liem, 1985). Analyses of the cDNAs and predicted amino acid sequences of internexins later proved that β-internexin is a microtubule-associated protein identical to the 70-kDa heat-shock cognate protein and the clathrin-uncoating ATPase (Green and Liem, 1989) and that α-internexin is a neuronal intermediate filament protein (Fliegner et al., 1990; Kaplan et al., 1990) probably identical to a 66-kDa neuronal intermediate filament protein reported earlier (Chiu et al., 1989).

The subsequent cloning of the α-internexin gene and characterization of its exon–intron organization and sequence establish α-internexin as a member of the type IV intermediate filament (IF) gene family (Ching and Liem, 1991), which includes the genes encoding the neurofilament (NF) triplet proteins, NF-L, NF-M, and NF-H, and also genes encoding nestin, syncoilin, and synemin (Yuan, 2007). Its specificity for neurons and abundance in axons led early investigators to suspect that α-internexin, like peripherin, may be a subunit of NFs (Portier et al., 1983; Chiu et al., 1989; Kaplan et al., 1990) although some early evidence seemed to suggest that α-internexin might function independently of the triplet proteins. α-Internexin was shown to form homopolymers (Ching and Liem, 1993), unlike the NF triplets, which are obligate heteropolymers requiring NF-L and NF-M or NF-H (Lee et al., 1993; Lee and Cleveland, 1996). α-Internexin gene expression was also found to be highest before birth and to decline somewhat before the expression of the triplet protein sharply rises postnatally (Kaplan et al., 1990; Fliegner et al., 1994). In some exceptional cases, such as mature cerebellar granules cells, α-internexin was reported to exist in the absence of detectable NF triplet proteins (Chien et al., 1996). These studies contributed to the view that α-internexin forms a filament system distinct from NFs composed of triplet proteins. More recent studies involving mice lacking individual NF subunits or combinations of subunits, however, have now established that α-internexin is present in stoichiometric amounts with the NF triplet proteins in mature CNS NFs and is critical for transport of the NF triplet proteins, providing strong evidence that α-internexin is, in fact, an authentic subunit of NFs in the mature CNS.

8.2 α-Internexin Structure

Like other intermediate filament proteins, α-internexin has a homologous central rod domain of 321 amino acid residues flanked by an amino-terminal head region of 87 residues and a carboxyl tail of 91 residues (Fig. 8.1). The rod domain of

A

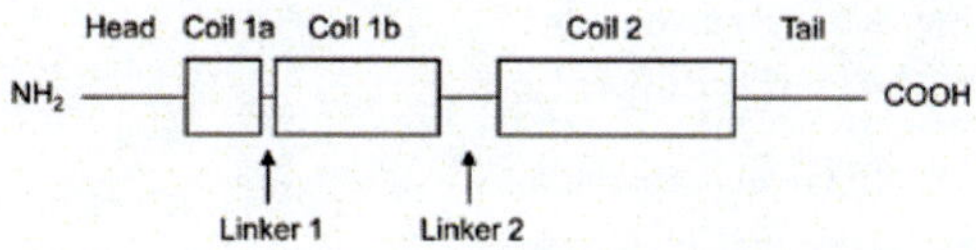

B

Domain	From	To	Length
Head	1	87	87
Rod	88	408	321
Coil 1a	88	129	42
Linker 1	130	142	12
Coil 1b	143	238	96
Linker 2	239	262	24
Coil 2	263	408	146
Tail	409	499	91

Fig. 8.1 Schematic representation of α-internexin domains and their amino acid length. Like other intermediate filament proteins, α-internexin consists of three domains, an amino-terminal globular head, a conserved α-helical rod, and a carboxy-terminal tail (**A**). The rod domain can be subdivided into coil 1a, linker 1, coil 1b, linker 2, and coil 2. The length of each domain is shown (**B**)

α-internexin is organized into coil 1a, 1b, and 2 segments similar to those in NF-L, peripherin, and vimentin (Fliegner et al., 1990). The tail domain of α-internexin is enriched in glutamate and lysines, and both head and tail domains contain sequences highly homologous to NF-M (Fliegner et al., 1990). This unique hybrid nature of α-internexin may contribute to its role both in neuronal development and in the mature CNS (Fliegner et al., 1994; Yuan et al., 2006). The structure of α-internexin is highly conserved among mice (501aa), rats (505aa), and humans (499aa) (Chien and Liem, 1994; Chan and Chiu, 1996). Although α-internexin is able to self-assemble into homopolymers in the absence of a preexisting filament network in SW13vim^{-} cells (Ching and Liem, 1993), its transport and content in axons in the mature CNS is largely dependent upon the presence of other NF subunits, especially NF-M (Yuan et al., 2006). Moreover, α-internexin is also critical for NF-M transport and content in axons in the absence of NF-L and NF-H (Yuan et al., 2006). These studies suggest the NF-L, NF-M, and NF-H form obligate heteropolymers with α-internexin in the mature CNS. Obligatory 10-nm NF formation is believed to begin with the formation of heterodimers from two parallel, unstaggered subunit chains via their α-helical rod domains (Fuchs and Weber, 1994). Two heterodimers then align to form an antiparallel heterotetramer (protofilament) of four subunit chains. Two heterotetramers are then packed together to form hetero-octamers of protofibrils and ultimately 10-nm NF (Figs. 8.2 and 8.3) (Wen and Wisniewski, 1984; Cohlberg et al., 1995).

8.3 α-Internexin Expression

When neuronal precursor cells become postmitotic, the expression of vimentin and nestin is downregulated (Cochard and Paulin, 1984; Lendahl et al., 1990),

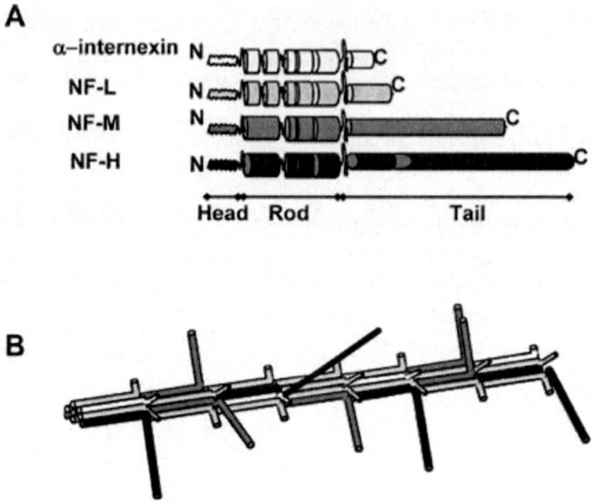

Fig. 8.2 Schematic representation of four NF subunits and their assembly. (**A**) NF proteins consist of head and tail domains separated by a coiled-coil rod domain. (**B**) A schematic showing that NFs are heteropolymers composed of NF-L, NF-M, NF-H, and α-internexin with the protrusions of NF-H and NF-M to form cross-bridges

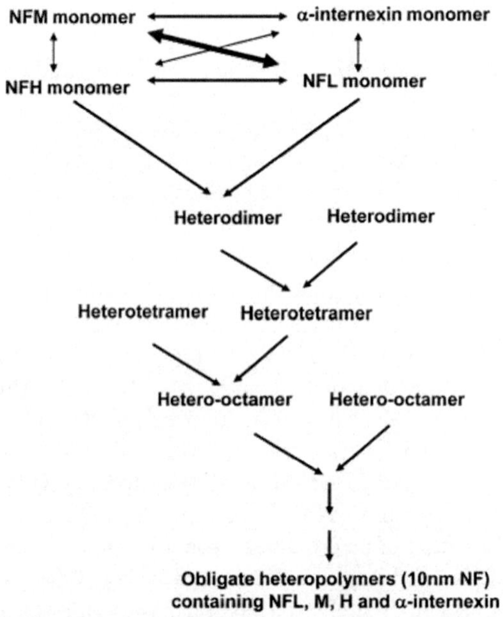

Fig. 8.3 Current model of mature NF fiber construction in the CNS. NF subunit monomer pairs with a different monomer to form a heterodimer in which the conserved central rod domains are aligned in parallel and wound together into coiled-coil. Two heterodimers then line up side by side to form an antiparallel heterotetramer of four polypeptide chains. Two heterotetramers are packed together to form a hetero-octamer, which eventually associates in a helical array to form the final 10-nm ropelike obligate NF composed of NF-L, NF-M, NF-H, and α-internexin. Note that NF-M shows the strongest partnership with NF-L among the four subunits (indicated by the thickness of the *double arrow*)

whereas α-internexin expression rises. Although it is possible that α-internexin forms homopolymeric filaments at an early stage in development (Ching and Liem, 1993; Fliegner et al., 1994), it is important to note that, even during the peak of late embryonic expression of α-internexin, NF-M, the critical partner for α-internexin transport, is also expressed albeit at low levels (Faussone-Pellegrini et al., 1999; Undamatla and Szaro, 2001). The expression of NF-M has been reported to precede that of NF-L and NF-H both in rat cerebral cortex and in cultured cortical neurons (Giasson and Mushynski, 1997). Similar earlier expression of NF-M than NF-L and NF-H has also been observed in the developing nervous system of *Xenopus laevis* (Szaro et al., 1989), chicken (Bennett and DiLullo, 1985), mouse (Faussone-Pellegrini et al., 1999), rat (Schlaepfer and Bruce, 1990), and human (Tohyama et al., 1991). Because NF-M protein by itself does not assemble into homopolymers (Ching and Liem, 1993; Lee et al., 1993; Nakagawa et al., 1995), detection of NF-M in the absence of other triplet proteins NF-L and NF-H during embryonic development implies that it forms heteropolymers with α-internexin. In fact, α-internexin has been shown to be critical for NF-M transport in the absence of NF-L and NF-H (Yuan et al., 2003). α-Internexin is highly expressed in the developing CNS and is only modestly reduced in expression postnatally (Fliegner et al., 1990; 1994). Despite somewhat lowered expression of α-internexin mRNA (Fliegner et al., 1994; Usui et al., 1996), the abundance of α-internexin protein in the mature CNS remains in the same range as that of the NF triplet proteins, which are among the most abundant proteins in brain (Chiu et al., 1989; Chan et al., 1997; Yuan et al., 2006). Although α-internexin was previously believed to be the only neuronal intermediate filament protein expressed in the parallel fibers of developing cerebellar granule neurons (Chien et al., 1996), genetic deletion of NF M and H subunits in cerebellar granule cells blocks the efficient axonal transport and function of α-internexin (Yuan et al., 2006). These results are consistent with previous findings showing that limited amounts of NF mRNA or protein may be present in cerebellar granule cells (Liesi et al., 1986; Gilad et al., 1989; Vitadello and Denis-Donini, 1990; Dahlstrand et al., 1995; Riederer et al., 1996). Although α-internexin is initially expressed in peripheral axons, its expression is markedly downregulated during embryonic development and is expressed at very low levels in most adult peripheral nerve tissues. The developmental downregulation of α-internexin in the PNS coincides with the appearance of another intermediate filament protein, peripherin, which could interact with NFs in vivo and possibly serve a role in the PNS analogous to α-internexin (Portier et al., 1983).

8.4 α-Internexin As a Fourth Subunit of CNS Neurofilament

NFs were originally believed to be composed of only three subunits, NF-L together with NF-M and NF-H (Lee et al., 1993; Lee and Cleveland, 1996; Al-Chalabi and Miller, 2003; Lariviere and Julien, 2004). However, recent studies provide strong evidence that α-internexin functions in the adult CNS principally as a fourth subunit of NFs based on every criterion previously used to establish the NF "triplet proteins" as subunits of the NFs as well as on the basis of novel genetic criteria (Yuan et al., 2006). α-Internexin is as abundant as the NF triplet proteins in the adult CNS and exists in isolated NFs in a fixed stoichiometry with the triplet. α-Internexin is transported along axons at the identical rate to the triplet proteins, and all four subunits exhibit the same turnover rates in optic axons. α-Internexin co-localizes ultrastructurally with NF-M on single NFs in vivo and co-assembles with all three NF proteins into a single network of filaments in quadruple-transfected SW13vim^{-} cells (Yuan et al., 2009). Moreover, in the absence of NF-H and NF-L, NF-M forms filaments with α-internexin that disappear if α-internexin is deleted (Yuan et al., 2003; Yuan and Nixon, manuscript in preparation). Genetically deleting NF-M alone or together with NF-H in mice dramatically reduces α-internexin transport and content in axons throughout the CNS. Similarly, α-internexin is required for axonal transport of NF-M in the absence of NF-H and NF-L in optic axons, and deleting α-internexin potentiates the effects of NF-M deletion in reducing the export of NF-H and NF-L into axons (Yuan et al., 2003). Overexpressing an NF-H- LacZ fusion protein in mice induces α-internexin and NF triplet to aggregate in neuronal perikarya and greatly reduces their transport and content selectively in axons. These findings establish that α-internexin and the NF proteins are functionally interdependent in the mature CNS. More recently, α-internexin has been shown to be an integral and abundant component of NFs in cultured sympathetic neurons (Yan et al., 2007).

8.5 α-Internexin Function

As a fourth subunit of NFs, α-internexin influences the structure and function of CNS NFs. Some of the serine/threonine residues in the head domain of α-internexin are phosphorylated by second messenger–dependent kinases such as protein kinase A and C, which may modulate filament assembly (Tanaka et al., 1993). α-Internexin is a crucial factor in the transport of these assemblies. In the absence of NF-H and NF-L, α-internexin mediates NF-M axonal transport and content in CNS axons (Yuan et al., 2003). In addition to partnering with the NF-M subunit, some α-internexin also partners with NF-H and NF-L subunits (Yuan et al., 2006). A second function of α-internexin is likely to be in facilitating the integration of NFs with the membrane-associated cytoskeleton (Jing-Ping et al., 2005) and with receptor components (Ehlers et al., 1998; Kim et al., 2002; Ratnam and Teichberg, 2005). The higher proportions of α-internexin in dendritic spines (Benson et al., 1996) and at postsynaptic densities (Suzuki et al., 1997) are consistent with this role. A third function of α-internexin may be to confer greater plasticity to the NF network (Nixon and Shea, 1992; Giasson and Mushynski, 1997), perhaps by modifying the cross-linking capabilities of NF populations (Giasson and Mushynski, 1997). The relatively high proportions of α-internexin in the NFs of small-caliber axons (Kaplan et al., 1990; Fliegner et al., 1994; Yuan et al., 2006), which are highly dynamic in vivo (Stettler et al., 2006), are compatible with this proposed role. α-Internexin may also facilitate axonal outgrowth. Gelfilin and zefiltin, homologs of α-internexin in zebrafish and *Xenopus laevis*, respectively, are highly upregulated during optic axon growth and regeneration (Glasgow et al., 1994; Zhao and Szaro, 1997a, b). Similar upregulation of α-internexin is also observed in injured rat motoneurons (McGraw et al., 2002). Consistent with this role, overexpressing α-internexin in PC12 cell enhances neurite outgrowth (Chien et al., 2005), whereas knocking down the expression of α-internexin by antisense oligonucleotides decreases axonal outgrowth in differentiated NB2a/d1neuroblastoma cells (Shea and Beermann, 1999). Although the deletion of α-internexin gene in mouse had no visible effect on development of the nervous system (Levavasseur et al., 1999), it may well be that the function of α-internexin is compensated by other proteins or not fully revealed yet.

8.6 α-Internexin Disease Associations

As a fourth subunit of NFs, dysfunction of α-internexin in disease states impairs NF function. Abnormal neuronal aggregates of α-internexin along with NF triplet proteins NF-L, NF-M, and NF-H have been identified as the pathologic hallmarks of NF inclusion disease, a human neurodegenerative disease associated with frontal temporal dementia and motor disability (Cairns et al., 2004b; Budrys et al., 2007). In tropical spastic paraparesis, a degenerative neurologic disorder, α-internexin interacts selectively with the human T-cell leukemia virus transcriptional transactivator Tax leading to cytoskeletal disorganization and marked NF accumulation (Wu et al., 1993; Liberski et al., 1994; Reddy et al., 1998; Liberski et al., 1999). α-Internexin abnormalities have also been observed in other human neurodegenerative disorders such as Alzheimer's disease (Dickson et al., 2005), dementia with Lewy bodies, and amyotrophic lateral sclerosis (motor neuron disease) (Cairns et al., 2004a). α-Internexin together with NF triplet proteins are also among the first cytoskeletal proteins lost during the course of Wallerian degeneration (Soifer et al., 1981), and their axonal transport is similarly impaired in giant axonal neuropathy (Monaco et al., 1985). Overexpression of α-internexin in mice causes NF accumulations in perikarya and axons of Purkinje cells, leading to the progressive neurodegeneration and ultimate death of the neurons and motor coordination deficits (Ching et al., 1999). Overexpression of α-internexin in PC12 cells upregulates the expression of NF-L and NF-M, leading to abnormal accumulations of NFs, progressive neurodegeneration, and cell death (Chien et al., 2005). The evidence that α-internexin is a fourth subunit of NF in CNS provides a basis for its close relationship to NF pathobiology in these disease states.

8.7 Conclusion

NFs were once believed to be composed of only the triplet subunits NF-L, NF-M, and NF-H. Recent studies now provide strong evidence, however, that α-internexin is a functional subunit of NF fibers in the mature CNS based on all criteria previously used to establish the NF "triplet proteins" as subunits of the same filament. α-Internexin and the NF proteins are functionally interdependent in mature CNS. The concept of α-internexin as a CNS NF subunit provides a new framework within which to interpret the interrelationship among α-internexin and the NF proteins previously observed during axon outgrowth, regeneration, and degeneration. Given that α-internexin is the pathologic hallmark of NF inclusion disease, a form of frontal temporal dementia with motor disability, it will be critical to define further whether α-internexin dysfunction is the key pathogenic factor in this disorder or acts by disrupting one or more functions of NFs. Understanding this role may shed light on emerging functions of NF in neuronal processes that modulate learning, memory, and movement.

Acknowledgments We would like to thank our colleagues Drs. Jean-Pierre Julien, Mala Rao, Ronald K. Liem, Takahiro Sasaki, Asok Kumar, Yuanxin Chen, Veeranna, Panaiyur Mohan, Joel Eyer, and Alan C. Peterson for their collaborative efforts in the NF studies. We also thank Nicole Piorkowski for manuscript preparation. This work was supported by Grant AG05604 from the National Institutes on Aging.

References

Al-Chalabi A, Miller CC (2003) Neurofilaments and neurological disease. Bioessays 25:346–355

Bennett GS, DiLullo C (1985) Transient expression of a neurofilament protein by replicating neuroepithelial cells of the embryonic chick brain. Dev Biol 107:107–127

Benson DL, Mandell JW, Shaw G, Banker G (1996) Compartmentation of alpha-internexin and neurofilament triplet proteins in cultured hippocampal neurons. J Neurocytol 25:181–196

Budrys V, Skullerud K, Petroska D, Lengveniene J, Kaubrys G (2007) Dementia and art: neuronal intermediate filament inclusion disease and dissolution of artistic creativity. Eur Neurol 57: 137–144

Cairns NJ, Uryu K, Bigio EH, Mackenzie IR, Gearing M, Duyckaerts C, Yokoo H, Nakazato Y, Jaros E, Perry RH, Arnold SE, Lee VM, Trojanowski JQ (2004a) alpha-Internexin aggregates are abundant in neuronal intermediate filament inclusion disease (NIFID) but rare in other neurodegenerative diseases. Acta Neuropathol (Berl) 108:213–223

Cairns NJ, Zhukareva V, Uryu K, Zhang B, Bigio E, Mackenzie IR, Gearing M, Duyckaerts C, Yokoo H, Nakazato Y, Jaros E, Perry RH, Lee VM, Trojanowski JQ (2004b) alpha-Internexin is present in the pathological inclusions of neuronal intermediate filament inclusion disease. Am J Pathol 164:2153–2161

Chan SO, Chiu FC (1996) The 66-kDa neurofilament protein (NF-66): sequence analysis and evolution. Neurochem Res 21:449–455

Chan SO, Peng D, Chiu FC (1997) Heterogeneous expression of neurofilament proteins in forebrain and cerebellum during development: clinical implications for spinocerebellar ataxia. Brain Res 775:107–118

Chien CL, Liem RK (1994) Characterization of the mouse gene encoding the neuronal intermediate filament protein alpha-internexin. Gene 149:289–292

Chien CL, Liu TC, Ho CL, Lu KS (2005) Overexpression of neuronal intermediate filament protein alpha-internexin in PC12 cells. J Neurosci Res 80:693–706

Chien CL, Mason CA, Liem RK (1996) alpha-Internexin is the only neuronal intermediate filament expressed in developing cerebellar granule neurons. J Neurobiol 29:304–318

Ching GY, Chien CL, Flores R, Liem RK (1999) Overexpression of alpha-internexin causes abnormal neurofilamentous accumulations and motor coordination deficits in transgenic mice. J Neurosci 19:2974–2986

Ching GY, Liem RK (1991) Structure of the gene for the neuronal intermediate filament protein alpha-internexin and functional analysis of its promoter. J Biol Chem 266:19459–19468
Ching GY, Liem RK (1993) Assembly of type IV neuronal intermediate filaments in nonneuronal cells in the absence of preexisting cytoplasmic intermediate filaments. J Cell Biol 122: 1323–1335
Chiu FC, Barnes EA, Das K, Haley J, Socolow P, Macaluso FP, Fant J (1989) Characterization of a novel 66 kd subunit of mammalian neurofilaments. Neuron 2:1435–1445
Cochard P, Paulin D (1984) Initial expression of neurofilaments and vimentin in the central and peripheral nervous system of the mouse embryo in vivo. J Neurosci 4:2080–2094
Cohlberg JA, Hajarian H, Tran T, Alipourjeddi P, Noveen A (1995) Neurofilament protein heterotetramers as assembly intermediates. J Biol Chem 270:9334–9339
Dahlstrand J, Lardelli M, Lendahl U (1995) Nestin mRNA expression correlates with the central nervous system progenitor cell state in many, but not all, regions of developing central nervous system. Brain Res Dev Brain Res 84:109–129
Dickson TC, Chuckowree JA, Chuah MI, West AK, Vickers JC (2005) alpha-Internexin immunoreactivity reflects variable neuronal vulnerability in Alzheimer's disease and supports the role of the beta-amyloid plaques in inducing neuronal injury. Neurobiol Dis 18:286–295
Ehlers MD, Fung ET, O'Brien RJ, Huganir RL (1998) Splice variant-specific interaction of the NMDA receptor subunit NR1 with neuronal intermediate filaments. J Neurosci 18: 720–730
Faussone-Pellegrini MS, Matini P, DeFelici M (1999) The cytoskeleton of the myenteric neurons during murine embryonic life. Anat Embryol (Berl) 199:459–469
Fliegner KH, Ching GY, Liem RK (1990) The predicted amino acid sequence of alpha-internexin is that of a novel neuronal intermediate filament protein. EMBO J 9:749–755
Fliegner KH, Kaplan MP, Wood TL, Pintar JE, Liem RK (1994) Expression of the gene for the neuronal intermediate filament protein alpha-internexin coincides with the onset of neuronal differentiation in the developing rat nervous system. J Comp Neurol 342:161–173
Fuchs E, Weber K (1994) Intermediate filaments: structure, dynamics, function, and disease. Annu Rev Biochem 63:345–382
Giasson BI, Mushynski WE (1997) Developmentally regulated stabilization of neuronal intermediate filaments in rat cerebral cortex. Neurosci Lett 229:77–80
Gilad GM, Gilad VH, Dahl D (1989) Expression of neurofilament immunoreactivity in developing rat cerebellum in vitro and in vivo. Neurosci Lett 96:7–12
Glasgow E, Druger RK, Fuchs C, Lane WS, Schechter N (1994) Molecular cloning of gefiltin (ON1): serial expression of two new neurofilament mRNAs during optic nerve regeneration. EMBO J 13:297–305
Green LA, Liem RK (1989) beta-Internexin is a microtubule-associated protein identical to the 70-kDa heat-shock cognate protein and the clathrin uncoating ATPase. J Biol Chem 264: 15210–15215
Jing-Ping Z, Tian QB, Sakagami H, Kondo H, Endo S, Suzuki T (2005) p55 protein is a member of PSD scaffold proteins in the rat brain and interacts with various PSD proteins. Brain Res Mol Brain Res 135:204–216
Kaplan MP, Chin SS, Fliegner KH, Liem RK (1990) alpha-Internexin, a novel neuronal intermediate filament protein, precedes the low molecular weight neurofilament protein (NF-L) in the developing rat brain. J Neurosci 10:2735–2748
Kim OJ, Ariano MA, Lazzarini RA, Levine MS, Sibley DR (2002) Neurofilament-M interacts with the D1 dopamine receptor to regulate cell surface expression and desensitization. J Neurosci 22:5920–5930
Lariviere RC, Julien JP (2004) Functions of intermediate filaments in neuronal development and disease. J Neurobiol 58:131–148
Lee MK, Cleveland DW (1996) Neuronal intermediate filaments. Annu Rev Neurosci 19:187–217
Lee MK, Xu Z, Wong PC, Cleveland DW (1993) Neurofilaments are obligate heteropolymers in vivo. J Cell Biol 122:1337–1350

Lendahl U, Zimmerman LB, McKay RD (1990) CNS stem cells express a new class of intermediate filament protein. Cell 60:585–595

Levavasseur F, Zhu Q, Julien JP (1999) No requirement of alpha-internexin for nervous system development and for radial growth of axons. Brain Res Mol Brain Res 69:104–112

Liberski PP, Buczynski J, Yanagihara R, Mora C, Gibbs CJ, Gajdusek C, Cartier L, Verdugo A, Araya F, Castillo L (1999) Ultrastructural pathology of a Chilean case of tropical spastic paraparesis/human T-cell lymphotropic type I-associated myelopathy (TSP/HAM). Ultrastruct Pathol 23:157–162

Liberski PP, Rodgers-Johnson P, Yanagihara R, Lee JW, Kramer BS, Piccardo P, Mora CA, Gibbs CJ Jr, Gajdusek DC (1994) Ultrastructural pathology of human T-cell lymphotropic virus type I encephalomyelopathy in a white patient with adult T-cell leukemia/lymphoma. Ultrastruct Pathol 18:511–518

Liem RK, Hutchison SB (1982) Purification of individual components of the neurofilament triplet: filament assembly from the 70 000-dalton subunit. Biochemistry 21:3221–3226

Liesi P, Julien JP, Vilja P, Grosveld F, Rechardt L (1986) Specific detection of neuronal cell bodies: in situ hybridization with a biotin-labeled neurofilament cDNA probe. J Histochem Cytochem 34:923–926

McGraw TS, Mickle JP, Shaw G, Streit WJ (2002) Axonally transported peripheral signals regulate alpha-internexin expression in regenerating motoneurons. J Neurosci 22:4955–4963

Monaco S, Autilio-Gambetti L, Zabel D, Gambetti P (1985) Giant axonal neuropathy: acceleration of neurofilament transport in optic axons. Proc Natl Acad Sci USA 82:920–924

Nakagawa T, Chen J, Zhang Z, Kanai Y, Hirokawa N (1995) Two distinct functions of the carboxyl-terminal tail domain of NF-M upon neurofilament assembly: cross-bridge formation and longitudinal elongation of filaments. J Cell Biol 129:411–429

Napolitano EW, Pachter JS, Chin SS, Liem RK (1985) beta-Internexin, a ubiquitous intermediate filament-associated protein. J Cell Biol 101:1323–1331

Nixon RA, Shea TB (1992) Dynamics of neuronal intermediate filaments: a developmental perspective. Cell Motil Cytoskeleton 22:81–91

Pachter JS, Liem RK (1985) alpha-Internexin, a 66-kD intermediate filament-binding protein from mammalian central nervous tissues. J Cell Biol 101:1316–1322

Portier MM, de Nechaud B, Gros F (1983) Peripherin, a new member of the intermediate filament protein family. Dev Neurosci 6:335–344

Ratnam J, Teichberg VI (2005) Neurofilament-light increases the cell surface expression of the N-methyl-D-aspartate receptor and prevents its ubiquitination. J Neurochem 92: 878–885

Reddy TR, Li X, Jones Y, Ellisman MH, Ching GY, Liem RK, Wong-Staal F (1998) Specific interaction of HTLV tax protein and a human type IV neuronal intermediate filament protein. Proc Natl Acad Sci USA 95:702–707

Riederer BM, Porchet R, Marugg RA (1996) Differential expression and modification of neurofilament triplet proteins during cat cerebellar development. J Comp Neurol 364:704–717

Schlaepfer WW, Bruce J (1990) Simultaneous up-regulation of neurofilament proteins during the postnatal development of the rat nervous system. J Neurosci Res 25:39–49

Shea TB, Beermann ML (1999) Neuronal intermediate filament protein alpha-internexin facilitates axonal neurite elongation in neuroblastoma cells. Cell Motil Cytoskeleton 43: 322–333

Soifer D, Iqbal K, Czosnek H, De Martini J, Sturman JA, Wisniewski HM (1981) The loss of neuron-specific proteins during the course of Wallerian degeneration of optic and sciatic nerve. J Neurosci 1:461–470

Stettler DD, Yamahachi H, Li W, Denk W, Gilbert CD (2006) Axons and synaptic boutons are highly dynamic in adult visual cortex. Neuron 49:877–887

Suzuki T, Mitake S, Okumura-Noji K, Shimizu H, Tada T, Fujii T (1997) Excitable membranes and synaptic transmission: postsynaptic mechanisms. Localization of alpha-internexin in the postsynaptic density of the rat brain. Brain Res 765:74–80

Szaro BG, Lee VM, Gainer H (1989) Spatial and temporal expression of phosphorylated and non-phosphorylated forms of neurofilament proteins in the developing nervous system of Xenopus laevis. Brain Res Dev Brain Res 48:87–103

Tanaka J, Ogawara M, Ando S, Shibata M, Yatani R, Kusagawa M, Inagaki M (1993) Phosphorylation of a 62 kd porcine alpha-internexin, a newly identified intermediate filament protein. Biochem Biophys Res Commun 196:115–123

Tohyama T, Lee VM, Rorke LB, Trojanowski JQ (1991) Molecular milestones that signal axonal maturation and the commitment of human spinal cord precursor cells to the neuronal or glial phenotype in development. J Comp Neurol 310:285–299

Undamatla J, Szaro BG (2001) Differential expression and localization of neuronal intermediate filament proteins within newly developing neurites in dissociated cultures of Xenopus laevis embryonic spinal cord. Cell Motil Cytoskeleton 49:16–32

Usui H, Ichikawa T, Miyazaki Y, Nagai S, Kumanishi T (1996) Isolation of cDNA clones of the rat mRNAs expressed preferentially in the prenatal stages of brain development. Brain Res Dev Brain Res 97:185–193

Vitadello M, Denis-Donini S (1990) Expression of neurofilament proteins in granule cells of the cerebellum. Brain Res 509:47–54

Wen GY, Wisniewski HM (1984) Substructures of neurofilaments. Acta Neuropathol 64:339–343

Wu E, Dickson DW, Jacobson S, Raine CS (1993) Neuroaxonal dystrophy in HTLV-1-associated myelopathy/tropical spastic paraparesis: neuropathologic and neuroimmunologic correlations. Acta Neuropathol (Berl) 86:224–235

Yan Y, Jensen K, Brown A (2007) The polypeptide composition of moving and stationary neurofilaments in cultured sympathetic neurons. Cell Motil Cytoskeleton 64:299–309

Yuan A (2007) Neurofilament protein partnership, export, transport, phosphorylation and neurodegeneration. In: Arlen RK (ed) New research on neurofilament proteins. Nova Science Publishers, Inc, New York, NY, pp 53–79

Yuan A, Rao MV, Kumar A, Julien JP, Nixon RA (2003) Neurofilament transport in vivo minimally requires hetero-oligomer formation. J Neurosci 23:9452–9458

Yuan A, Rao MV, Sasaki T, Chen Y, Kumar A, Veeranna, Liem RK, Eyer J, Peterson AC, Julien JP, Nixon RA (2006) alpha-Internexin is structurally and functionally associated with the neurofilament triplet proteins in the mature CNS. J Neurosci 26:10006–10019

Yuan A, Sasaki T, Rao MV, Kumar A, Kanumuri V, Dunlop DS, Liem RK, Nixon RA (2009) Neurofilaments form a highly stable stationary cytoskeleton after reaching a critical level in axons. J Neurosci 29:11316–11329

Zhao Y, Szaro BG (1997a) Xefiltin, a new low molecular weight neuronal intermediate filament protein of Xenopus laevis, shares sequence features with goldfish gefiltin and mammalian alpha-internexin and differs in expression from XNIF and NF-L. J Comp Neurol 377:351–364

Zhao Y, Szaro BG (1997b) Xefiltin, a Xenopus laevis neuronal intermediate filament protein, is expressed in actively growing optic axons during development and regeneration. J Neurobiol 33:811–824

Chapter 9
Peripherin Pathology

Jesse R. McLean and Janice Robertson

Abstract Peripherin, a 58-kDa, type III intermediate filament (IF) protein expressed in the peripheral nervous system (PNS) and in defined subsets of neurons in the central nervous system (CNS), is thought to play an integral role in the development of the highly plastic neuronal cytoskeleton and to provide mechanical support to terminally differentiated neurons. Peripherin expression is a dynamically regulated process that involves mediation by a number of *cis*- and *trans*-acting factors throughout the gene region. Peripherin expression is characterized by a stoichiometric ratio of functionally distinct isoforms that arise as a result of alternative splicing and translation. Pathologically, peripherin has been implicated in several human diseases but is best known for its descriptions in the literature of the devastating motor neuron disease, amyotrophic lateral sclerosis (ALS). Here, peripherin is a component of the major pathologic hallmarks of both familial and sporadic ALS – intracellular aggregates – and is found upregulated and abnormally spliced in areas of disease. Moreover, peripherin mutations have been associated with sporadic ALS cases, and transgenic overexpression of peripherin in mice recapitulates some of the clinical and pathologic aspects of ALS. Peripherin pathology represents a point-of-convergence for both the familial and sporadic forms of ALS; however, despite our increasing knowledge about peripherin biology, little is known about the nature of its involvement in disease pathogenesis. This chapter is aimed at providing the first intensive review of peripherin biology and its involvement in disease.

Keywords Peripherin · Intermediate filament · Cytoskeleton · Amyotrophic lateral sclerosis · ALS · Motor neuron · Aggregate · Alternative splicing · Translation · Epilepsy · Cancer · Diabetes

---

J. Robertson (✉)
Centre for Research in Neurodegenerative Diseases, University of Toronto, Toronto, ON, Canada M5S 3H2
e-mail: jan.robertson@utoronto.ca

R.A. Nixon, A. Yuan (eds.), *Cytoskeleton of the Nervous System*, Advances in Neurobiology 3, DOI 10.1007/978-1-4419-6787-9_9,

Contents

9.1 Structure and Development

Peripherin, a 58-kDa, Triton-X–insoluble protein, was first identified by its antigenic and filamentous similarity to the larger class of intermediate filament (IF) proteins (Portier et al., 1983). Along with vimentin, desmin, and glial fibrillary acidic protein (GFAP), peripherin is classified as a type III IF protein, which share >70% nucleotide sequence identity with each other (Fuchs and Weber, 1994). Like all IFs, peripherin shares the same tripartite protein structure, composed of a central conserved α-helical rod domain flanked by non-α-helical N-terminal "head" and C-terminal "tail" domains, which show variation among different IF members (Steinert and Roop, 1988). The α-helical rod domain is further subdivided into four coil domains (1a, 1b, 2a, and 2b) that are joined together by three non-α-helical linker sequences. Hydrophobic residues, as part of a sequential heptad repeat pattern, characterize the α-helical domain and provide a template for dimerization between two IF polypeptides. These dimers associate to form staggered and antiparallel tetramers, which are the building blocks for further lateral and longitudinal growth, until a complete ~10-nm-wide filament is achieved. The N terminus is involved in the regulation of polymer elongation and assembly, and the function of the C terminus is largely unknown but may play a role in establishing interactions with other cellular components. Peripherin is capable of self-assembly into homopolymers, however, it can be co-expressed with other IFs, particularly the type IV neurofilaments, to form mature filament networks (Parysek et al., 1991; Cui et al., 1995; Athlan and Mushynski, 1997; Beaulieu et al., 1999b).

Peripherin assembly and organization into cytoplasmic networks is associated with the acquisition of a terminally differentiated neuronal phenotype. During development, peripherin, α-internexin, and vimentin are expressed early to form an initial IF protein scaffold that is temporally and spatially regulated (Chien et al., 1998; Merigo et al., 2005). Differentiation into a mature cytoskeletal network is a tightly integrated process that requires peripherin to form homopolymers or to heteropolymerize with neurofilament subunit proteins (Athlan and Mushynski, 1997; Athlan et al., 1997). The sequential and transitional appearance of different IF proteins during development is related to cytoskeletal plasticity during neurite outgrowth and is considered a mechanism to enhance the stability of existing filament networks (Giasson and Mushynski, 1997). Unlike the ubiquitous neurofilaments, peripherin expression during development is limited to motor neurons, autonomic preganglionic and ganglionic neurons of the retina and optic nerves, sensory neurons of the dorsal and cranial root ganglia (DRG and CRG, respectively), sympathetic neurons from the neural crest, and some neuronal placodes, including olfactory axons and acoustic ganglia (Escurat et al., 1990). In the mature nervous system, peripherin is predominately expressed in small-caliber fibers of the PNS but is also present at lower levels in defined neuronal populations of the CNS, particularly of efferent spinal motor neurons and their projections (Table 9.1).

9.2 Regulation and Function

The human peripherin gene (*PRPH*) contains nine exons spanning an ~3.9-kb region within q12–q13 of chromosome 12 (Moncla et al., 1992; Foley et al., 1994). Gene sequencing of the human PRPH and other mammalian species has revealed several highly conserved coding, intragenic, and transcriptional motifs (Thompson and Ziff, 1989; Karpov et al., 1992; Foley et al., 1994), suggesting important evolutionarily defined developmental and tissue-specific regulatory elements. The major contributor to neuron type–specific peripherin expression is the presence of an intact intron 1 sequence (Belecky-Adams et al., 1993; Uveges et al., 2002), and the activation of the peripherin gene, itself, appears to be the result of dynamic *cis*- and *trans*-acting interactions between different regions mapped to within a 5.8-kb region upstream of the 5′-flanking sequence, with the strongest determinants of promoter activity situated just upstream of the transcriptional start site (Desmarais et al., 1992; Ferrari et al., 1995). The identification of a negative regulatory element (NRE) and two positive regulatory elements within the 5′-flanking region has fueled speculation that complete activation of peripherin is based on the dissociation of inhibitory elements at the NRE, and thus, derepression of the gene (Chang and Thompson, 1996). The normal repression of peripherin in undifferentiated and nonneuronal cells may be regulated by several DNA-binding proteins, such as NF1-L, a member of the CCAAT-binding transcription factor/nuclear factor 1 (CTF/NF-1) transcription factor family (Adams et al., 1995), and Sp1 (Ferrari et al., 1995; Chang and Thompson, 1996). As of yet, the identity of the repressor(s) is unknown; however, blockage of axoplasmic transport in intact nerves with vincristine suggests a distal, retrogradely transported inhibitory signal (Terao et al., 2000). These findings highlight the importance of environmental cues and neuron-target interactions in peripherin regulation and, indeed, a few in vitro studies have identified nerve growth factor (NGF) (Portier et al., 1983; Leonard et al., 1987; Parysek and Goldman, 1987; Aletta et al., 1988; Leonard et al., 1988; Thompson et al., 1992), fibroblast growth factor (FGF) (Choi et al., 2001), leukemia inhibitory factor (LIF) (Lecomte et al., 1998), and interleukin-6 (IL-6) (Sterneck et al., 1996) as transcriptional inducers of peripherin expression. These factors, acting alone or synergistically, have been shown to elicit peripherin-associated neurite outgrowth and differentiation by triggering the tyrosine kinase (Trk) receptor signaling cascade through the JAK protein family, which in turn stimulates the activity of STAT transcription factors (Djabali et al., 1993; Sterneck et al., 1996; Lecomte et al., 1998.

Table 9.1 Peripherin expression in culture and tissue

Tissue/culture	Cell line (common use)	Species	Features	References
Tumorogenesis				
Cutaneous lesions		h	Differential marker of Schwann cell-derived NFib, SCH, PEN, and NMN	Prieto et al. (1997)
Extraskeletal myxoid chondrosarcoma		h	Not expressed in epitheloid sarcoma	Cummings et al. (2000)
Insulinoma	RIN5F	r	Islet of Langerhans pancreatic beta cell line	Escurat et al. (1991)
Neuroblastoma (NB)	NUB-7 NB-1 IMR32	h	Variable expression; induced by dbcAMP and RA; prominent perinuclear distribution	Pedersen et al. (1993)
	SK-N-BE(2)C		PKCε activation induces peripherin aggregation and apoptosis	Sunesson et al. (2008)
	NIE 115	m	Lamin β as a binding partner	Djabali et al. (1991) Landon et al. (2000)
	N1A-103	m	Adriamycin induced expression	Larcher et al. (1992)
	Neuro-2A	m		Djabali et al. (1999)
Neuroendocrine (NE)				
Merkel cell		h	Perinuclear aggregation in large tumor cells	Baudoin et al. (1993) Alvarez-Gago et al. (1996)
Pheochromocytoma	PC12	r	Expression increased by IL-6 and Trk/NGF	Sterneck et al. (1996)
Neuroepithelioma (NE)	NUB-20 SK-N-MC	h	Variable expression; less expression than NB	Pedersen et al. (1993)
Central Nervous System				
Brainstem				
Cranial nerve roots		r	Fine and large-caliber nerve fibers and cell body expression	Parysek and Goldman (1988)
Vestibular ganglion cells		c, r, m	Support notion that central axons through vestibular nucleus are afferents	Lysakowksi et al. (1999)

Table 9.1 (continued)

Tissue/culture	Cell line (common use)	Species	Features	References
Vestibular tract and nucleus		gb, r	Supports notion that apex of the cristae ampullaris are efferents	Leonard and Kevetter (2006)
Cerebellum		r	Granular layer and corticospinal tract	Parysek and Goldman (1988)
Cerebrum				
Temporal/parietal ctx.		mq	Astrocytic expression during encephalitis	Mathew et al. (2001)
Hippocampal neurites		m	Increased expression after KA-induced seizures	Kriz et al. (2005)
Thalamic neurons		m	Increased expression after lesion injury and cerebral ischemia	Beaulieu et al. (2002)
Cochlea				
Spiral ganglion neurons (SGNs) and projections		r	Expressed in type II SGNs or neomycin-induced type 1 and III SGNs	Wang et al. (2006); Lallemend et al. (2007)
Olfactory axons		r	Minimal labeling	Escurat et al. (1990)
Spinal cord		r	Strong expression in dorsal columns; weak labeling in ventral colums	Parysek and Goldman (1988)
Pituitary glandr		r	Expression mostly in posterior lobe fibers	Bäck et al. (1995)
Retina and optic nerves		r	Neural tube origin	Escurat et al. (1990)
Peripheral Nervous System				
Dorsal root ganglia (DRG)		r	Variable expression; mixed populations	Parysek and Goldman (1988)
Enteric nervous system				Goldstein et al. (1991)
Myenteric plexus to submucosa		h, r, m	Increased expression from esophagus to distal colon; temporal from E12.5	Faussone-Pellegrini et al. (1999) Rauch et al. (2006) Matini et al. (1997)
Sciatic nerve (SN)		r	Intense small caliber fibre labeling; variable labeling for large caliber fibres	Parysek and Goldman (1988)

Although there are currently ~70 individual proteins constituting the IF superfamily (http://www.interfil.org), with an abundance of literature on their structural, biochemical, histochemical, regulatory, and interactive properties (Herrmann et al., 2007; Kim and Coulombe, 2007; Goldman et al., 2008), surprisingly, very little is known about their exact biological role. Unlike other cytoskeletal filaments, such as microtubules and actin that are polymers of globular subunits, IFs as fibrillar complexes appear suited for maintaining the structural integrity of cells in response to mechanical stress. In addition, the widespread cellular distribution and associative properties of IFs make them ideally suited for signal transduction at the receptor and subcellular levels. Elucidating the role of peripherin in vitro has come from studies limiting peripherin expression through the use of antisense oligonucleotides and small interfering RNAs (siRNAs) in rat pheochromocytoma cells (PC12) stimulated with NGF (Troy et al., 1992; Helfand et al., 2003). These studies found that whereas peripherin is not required for neurite formation, its absence was associated with an inability to initiate, extend, and maintain neuritic outgrowth. The generation of peripherin knockout ($Per^{-/-}$) mice by Julien and colleagues demonstrated that a complete loss of peripherin results in a substantial decline (~34%) of unmyelinated sensory fibers in the dorsal horn (Lariviere et al., 2002). Although these mice developed normally, with no overt behavioral phenotype, a significant increase in α-internexin was observed in the ventral horn, suggesting that homeostatic mechanisms are in place to minimize the effects of peripherin loss in response to abnormal congenital alterations.

9.3 Traumatic Neuronal Injury

Considering the importance of peripherin in providing cytoskeletal stability during development and in the adult nervous system, and that peripherin expression can be induced by neurotrophins and cytokines, several groups have assessed the effects of nerve injury on peripherin expression and assembly. Examination of DRG after unilateral sciatic nerve axotomy revealed that peripherin is substantially upregulated in both small and large DRG neurons (Oblinger et al., 1989b; Troy et al., 1990). Moreover, using ^{35}S-methionine to label proteins in newly regenerating axons, Parysek and colleagues also demonstrated that, in primed conditions, where neurons are preconditioned with a prior axotomy, increased amounts of peripherin can be observed in transit en route to regenerating axonal sprouts (Oblinger et al., 1989b). These studies are in stark contrast with the neurofilaments, where there is a significant downregulation in synthesis and transport after peripheral nerve injury (Hoffman et al., 1987; Wong and Oblinger, 1987; Goldstein et al., 1988; Oblinger et al., 1989a).

Despite being transcriptionally silent in the majority of neuronal populations in the brain, peripherin expression can be induced at the site of traumatic injury, including stab-like lesions and cerebral focal ischemia (Beaulieu et al., 2002). This response was shown to be enhanced in adjacent regions when IL-6 and LIF were administered at the site of injury. The relationship between peripherin and neuronal injury was subsequently extended to other pathologies by systemic injections of the excitatory amino acid, kainate (Kriz et al., 2005), which invokes an epileptogenic-like response through limbic seizures (Ben-Ari and Cossart, 2000). Increased peripherin expression can be observed after kainic acid (KA)-induced seizures in the cortex, hippocampus, and thalamus. Interestingly, in transgenic peripherin overexpressing mice (described later; Beaulieu et al., 1999a), KA induces an enhanced peripherin response, which is associated with alterations in synaptic plasticity, as measured by short- and long-term potentiation in the cornu ammonis (CA) fields of the hippocampus (Kriz et al., 2005). The significance of these findings is certainly not lost in humans, as increased IF expression is observed to occur in focal cortical dysplasia – the most common cause of intractable epilepsy among children. Here, peripherin increases occur predominately within dysplastic "ballooned" neurons, suggesting that abnormal IF expression may contribute to the cytoarchitectural disorganization in the cortices of these patients (Taylor et al., 2001).

9.4 Posttranscriptional and Posttranslational Modifications

After transcription, the ability of peripherin to respond accordingly to the requirements of the neuron may, in part, be regulated through posttranscriptional and posttranslational modifications. Like other IFs, peripherin is constitutively phosphorylated, and, depending on the cell line or tissue of interest, different degrees of peripherin phosphorylation in both the amino- and carboxyl-terminals are observed (Aletta et al., 1989; Huc et al., 1989; Angelastro et al., 1998; Konishi et al., 2007). Enhanced peripherin phosphorylation can be demonstrated in PC12 cells when exposed to NGF, activators of protein kinases A and C, and even depolarizing levels of K^+ (Aletta et al., 1989). Peripherin was recently identified to be a novel IF substrate for phosphatidylinositol 3-kinase (P13K)-Akt–mediated phosphorylation at Ser^{66} in PC12 and HEK 293T cells, with phosphorylation being particularly pronounced in the cell body of damaged hypoglossal motor neurons undergoing regeneration (Konishi et al., 2007). Akt (also known as protein kinase B) is thought to mediate various cellular processes, such as cell survival, proliferation, and differentiation (Franke, 2008). Considering that the phosphorylation of IF head domains has been associated with disassembly (Omary et al., 2006), Akt may be a valuable mechanism for sensitive regulation of newly synthesized peripherin. In addition to phosphorylation, peripherin residues Tyr^{17} and Tyr^{376} were recently shown to undergo nitration after NGF-induced differentiation in PC12 cells (Tedeschi et al., 2007). These authors also found that nitration of peripherin is correlated with increased microtubule stability even in the presence of depolymerizing agents such as nocodazole.

Peripherin is unique among neuronal IFs in that several alternative protein isoforms have been characterized. The constitutive and most typically described peripherin isoform expressed from the peripherin gene is Per-58, which is encoded by all nine exons of the gene (Thompson and Ziff, 1989; Parysek et al., 1991;

Karpov et al., 1992). The first nondescript identification of peripherin isoforms was made by Alletta and colleagues when several peripherin electrophoretic bands were localized on two-dimensional IEF × SDS-PAGE gels by a polyclonal antiserum (Aletta et al., 1989). While the identity of these prospective isoforms was unknown, they noticed that isoform-specific expression was changed in response to NGF stimulation, suggesting that these isoforms play a role in peripherin function. In addition to Per-58, two alternative isoforms have been identified in human (Per-45 and Per-28) and three in mouse (Per-61, Per-56, and Per-45) (Fig. 9.1). The

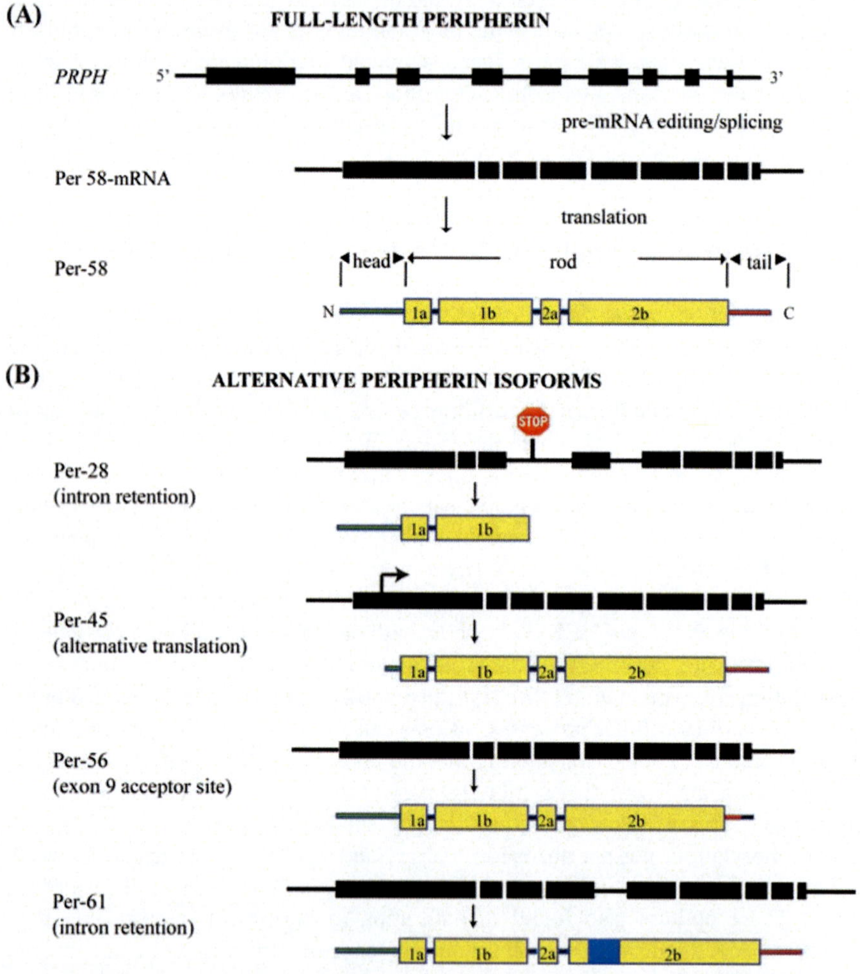

Fig. 9.1 Peripherin isoforms. **a** The constitutively expressed, full-length peripherin isoform, Per-58, is generated as a result of complete intron splicing to produce the typical intermediate filament protein structure with a molecular weight of ~58 kDa. **b** Alternative peripherin isoforms: Per-28 arises from full retention of introns 3 and 4 with premature stop codon to generate an ~28-kDa protein product; Per-45 results from internal translation of an in-frame start codon to generate an N-terminally truncated protein of ~45 kDa; Per-56 has a cryptic acceptor site in exon 9 leading to replacement of the COOH-terminal 21-aa of Per 58 with a unique 8-aa sequence; Per-61 arises from full retention of intron 4, which leads to an in-frame amino acid insertion into coil 2b

events generating these isoforms have been determined; briefly, Per-61 is generated by an in-frame retention of intron 4 that introduces a unique 32-amino-acid insertion; Per-56 is generated by a cryptic acceptor site at the start of exon 9, resulting in an open reading frame shift that replaces the COOH-terminal 21 amino acids of Per 58 with a unique 8-amino-acid sequence (Landon et al., 1989; 2000); Per-45 is an N-terminally truncated protein that is generated in both human and mouse as a result of in-frame alternative translation from internal initiation codons (McLean et al., 2008); and finally, Per-28 is a C-terminal truncated protein that arises from a premature stop codon created via the in-frame retention of intron 3 (Xiao et al., 2008). There has also been a higher molecular weight disulfide peripherin dimer of ~130 kDa identified in rat sciatic nerve and DRG that is dramatically increased after sciatic crush (Chadan et al., 1994).

Although the functional relevance of these isoforms is unknown, distinct morphologies and assembly properties can be observed when expressed in SW13vim$^-$ cells (Fig. 9.2) (Robertson et al., 2003; McLean et al., 2008; Xiao et al., 2008), a human adrenal carcinoma cell line that lacks endogenous cytoplasmic IF proteins (Sarria et al., 1990). Though these findings are similar to the morphologic changes that are observed in peripherin assembly studies using truncated or mutated cDNA transcripts to elucidate the functional roles of the head, tail, and rod domains (Cui et al., 1995; Ho et al., 1995; 1998), peripherin isoforms represent a relevant physiologic process capable of modulating filament structure, and, as such,

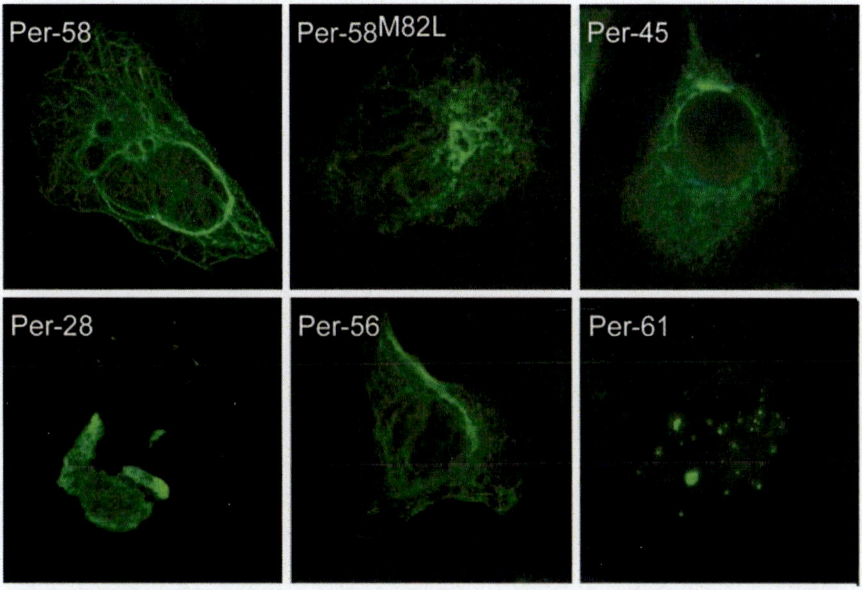

Fig. 9.2 Peripherin isoform expression. Distinct peripherin morphologic phenotypes can be observed when the cDNA encoding each isoform is transfected into SW13vim$^-$ cells. The full-length cDNA encoding Per-58 produces a filament network, but also produces Per-45. Per-58 expression alone (Per-58^{M86L}) is filamentous, but with shortened fibrils and a collapsed filament network; Per-45 is assembly incompetent; Per-28 forms large inclusions; Per-56 forms normal filament networks; and Per-61 forms small punctuate aggregates

function. Several new avenues of peripherin research have arisen as a result of the characterization of alternative isoforms. For example, understanding the roles of these isoforms may lie partly in the recognition that peripherin is capable of forming intra-isoform associations. We have recently established that the coexpression of these isoforms, notably Per-58 and Per-45, constitute an intra-isoform ratio that is associated with the structural integrity of peripherin filament networks; changes to the normal expression pattern of these isoforms are associated with malformed filaments and/or intracellular inclusions (McLean et al., 2008; Xiao et al., 2008). Importantly, these ratios are seen in vivo exhibiting regional variation, with, for example, Per-45 being the predominant species expressed in the brain, whereas Per-58 is the major species expressed in the spinal cord (McLean et al., 2008).

A surprising point to emerge from studies on Per-45 was the finding that the loss of the N-terminal head domain does not prevent intra-isoform polymerization. Prior to these findings, it was thought that the head domain was an absolute requirement for filament assembly while the C-terminal tail domain imparted specific functional properties to different IF proteins (Kreplak et al., 2004). Though Per-45, itself, is unable to form homopolymers, its incorporation within peripherin filaments is part of an emerging trend of type III IFs whereby assembly-compromised subunits or splice variants are incorporated within filaments to modulate IF structure and/or modify intracellular signaling pathways (Steinert et al., 1999; Schweitzer et al., 2001; Sahlgren et al., 2006; Perng et al., 2008). For the peripherin isoforms Per-56, Per-61, and Per-28, the generation of novel protein sequences as a result of the splicing process may represent important protein binding regions and/or provide a gain-of-function toxicity. Indeed, Per-61 and Per-28 have been identified to contribute to the neurodegenerative mechanism in human mutant $Cu^{(2+)}/Zn^{(2+)}$ superoxide dismutase (mtSOD1) transgenic mice and in amyotrophic lateral sclerosis (ALS), respectively (Robertson et al., 2003; Xiao et al., 2008).

9.5 Amyotrophic Lateral Sclerosis

ALS, also known as motor neuron disease (MND) or Lou Gehrig's disease, is an age-dependent fatal neurodegenerative disease that manifests as a result of progressive and selective death of upper and lower motor neurons of the CNS. The majority of individuals are afflicted during their fifth and sixth decades of life and manifest symptoms of muscle weakness, spasticity, and atrophy, ultimately culminating in death by denervation of respiratory muscles within <5 years of symptom onset (Strong, 2003; Goodall and Morrison, 2006). The majority of ALS cases (90–95%) are considered sporadic with no apparent genetic linkage (Broom et al., 2004). A familial inheritance pattern is observed in the remaining patients, with ~10–15% of these cases caused by mutations in the genes encoding the antioxidant protein, copper/zinc superoxide dismutase-1 (SOD1) (Rosen et al., 1993), and the nuclear factors TAR DNA binding protein (TDP-43) (Kabashi et al., 2008; Sreedharan et al., 2008) and fused in sarcoma/translated in liposarcoma gene (FUS/TLS) (Kwiatkowski et al., 2009; Vance et al., 2009). As yet, there is no treatment that significantly prolongs life as both symptomatic and disease-targeting therapies remain largely ineffective (Aggarwal and Cudkowicz, 2008). The clinical diagnosis of ALS is considered difficult due to the phenotypic heterogeneity of the disease, including the manifestation of extra motor symptoms (van der Graaff et al., 2009) and possible clinical resemblance or overlap with other neurodegenerative diseases (Majoor-Krakauer et al., 1994; Wokke, 2000; Annesi et al., 2005; Vance et al., 2006). The El Escorial diagnostic criteria for ALS was established by the World Federation of Neurology (Brooks, 1994; Brooks et al., 2000) to increase diagnostic consistency and to classify patients into various levels of certainty, ranging from possible to definite ALS. To date, there is no single diagnostic marker or test available for differential diagnosis (Turner et al., 2009).

As several neurodegenerative disorders may mimic the clinical symptoms of ALS, including disorders like myasthenia gravis, Kennedy's disease, multifocal motor neuropathy, and certain myelopathies (Traynor et al., 2000; Wokke, 2000), only a post-mortem pathology review can provide a definitive diagnosis. Three neuropathologic hallmarks characterize ALS and assist with diagnosis: (1) the finding of selective degeneration of populations of motor neurons, particularly in the descending supraspinal motor pathways and their neurons of origin, and of brain-stem and spinal motor neurons (Strong and Rosenfeld, 2003); (2) neuroinflammation, predominately characterized by reactive gliosis, as well as infiltrating macrophages and lymphocytes (Lampson et al., 1990; Troost et al., 1990; Appel et al., 1993; Nagy et al., 1994; Schiffer et al., 1996; Schwab et al., 1996; Henkel et al., 2004); and (3) the presence of abnormal intracellular inclusions within the perikarya and axons of motor neurons (Averback, 1981; Delisle and Carpenter, 1984; Munoz et al., 1988; Murayama et al., 1990; 1992; Wong et al., 2000; He and Hays, 2004). In the latter, peripherin is found to be one of the major protein constituents in ubiquitinated inclusions (Corbo and Hays, 1992; Migheli et al., 1993; Wong et al., 2000; He and Hays, 2004), co-localizing with TDP-43 in compact and Lewy-body–like inclusions of sporadic and familial forms of ALS (Sanelli et al., 2007), as well as with neurofilaments in axonal spheroids and hyaline conglomerate inclusions, the last of which appear specific to ALS cases caused by mutations in SOD1 (Fig. 9.3) (Xiao et al., 2006). In addition, there is an overall general upregulation of peripherin mRNA and protein levels in spinal cords of ALS patients when normalized to neuron-specific markers (Robertson et al., 2003; Strong et al., 2004; McLean et al., 2008; Xiao et al., 2008), although interindividual variations in expression are observed (Corbo and Hays, 1992; Wong et al., 2000; Xiao et al., 2008, McLean et al., 2010).

Although the mechanism of inclusion formation in ALS remains unknown, cytoskeletal abnormalities are prominent features of most neurofilament and peripherin transgenic mice (Robertson et al., 2002). It is believed that perturbations in the normal IF stoichiometry through targeted deletion or overexpression leads to unstable filament associations, ultimately causing a collapse in the filament network and the formation of cytoplasmic inclusions. Whereas some of these mice demonstrate the co-occurrence of a number of specific pathologies reminiscent of ALS (Cote et al., 1993; Lee et al., 1994; Zhu et al., 1997; Kriz et al., 2000; McLean et al., 2005), only peripherin-overexpressing mice, under the control of the wild-type (Per) or the

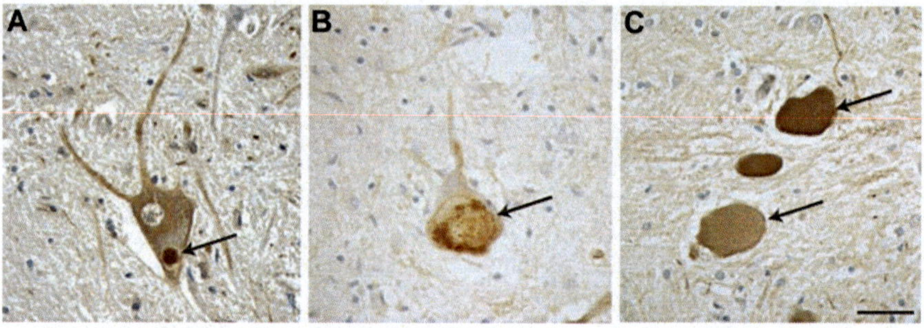

Fig. 9.3 Peripherin pathology in ALS. Immunolabeling of lumbar spinal cord tissue from ALS cases with peripherin antibody. **a** Motor neuron containing a compact/round inclusion. **b** Motor neuron from a mutant A4V SOD1 ALS case containing a hyaline conglomerate inclusion. These types of inclusion appear specific for mutant SOD1 cases. **c** Axonal spheroids. Note size can be as great as the motor neuron perikaryon. Scale bar = 35 μm

human Thy-1 (TPer) gene promoters, have been shown to develop an age-dependent motor neuron disease (Beaulieu et al., 1999a). Inclusions bodies, appearing as disorganized 10-nm filaments entangled with various membranes and mitochondria, were found in the presymptomatic stages of these mice, with disease onset occurring around ~2 years of age and coinciding with an ~35% loss in motor neurons of the lumbar ventral roots. Though few studies have used these mice to study motor neuron degeneration because of the long latency of disease onset, crossing these mice with the low-molecular-weight neurofilament (NF-L) knockout ($NFL^{-/-}$) mice ($Per;L^{-/-}$), mimicking more closely the neuronal IF conditions found in ALS (Bergeron et al., 1994), augments the formation of inclusions and disease onset to ~6–8 months, with ventral root loss observed as early as 4 months (Beaulieu et al., 1999a). The inclusion bodies in Per and in TPer mice were also abnormally immunoreactive for the medium- and high-molecular-weight neurofilament subunits (NF-M and NF-H, respectively), with the majority of NF-H being hyperphosphorylated. To evaluate further the mechanisms underlying peripherin-mediated motor neuron degeneration, $Per;L^{-/-}$ mice were crossed with mice overexpressing human NF-H, a mouse model with massive perikaryal accumulations of human NF-H with no motor neuron loss ($hH;TPer;L^{-/-}$) (Beaulieu and Julien, 2003). Although peripherin expression was not changed in these mice, per se, there was a dramatic redistribution of peripherin accumulation from the axon to the perikarya, and a reorganization of peripherin ultrastructure, from 10-nm filaments in $TPer;L^{-/-}$ mice to nonfilamentous protein aggregates in $hH;TPer;L^{-/-}$ mice. The benefits associated with these changes may reflect a reduction in the burden associated with dysfunctional axonal transport seen in the TPer and $TPer;L^{-/-}$ mice (Millecamps et al., 2006); however, this has yet to be confirmed.

Although the nature of intracellular inclusions in transgenic neuronal IF mice and in ALS remains a hotly debated topic, it is clear that the in vivo associations of neuronal IFs are crucial for motor neuron viability (Julien and Beaulieu, 2000; Robertson et al., 2002; Bruijn et al., 2004; Xiao et al., 2006). Peripherin abnormalities are also found in mtSOD1 transgenic mice, which develop early-onset motor neuron degeneration (Shibata, 2001). In addition to the immunodetection of peripherin inclusions in mtSOD1^{G37R} (Beaulieu et al., 1999a; Lariviere et al., 2003), there is also abnormal expression of the Per-61 isoform in motor neurons of these mice (Robertson et al., 2003). The expression of Per-61 is far from innocuous, having been shown to disrupt neuronal IF associations, induce neurofilament aggregation, and cause cell death when Per-61 is expressed via intranuclear microinjection of plasmid cDNA into motor neurons in culture (Robertson et al., 2003). Some of our more recent work has demonstrated that Per-61 is expressed after the presymptomatic stages of the disease in mtSOD-1^{G93A} mice (McLean et al., 2010), suggesting that the progression of motor neuron degeneration as a result of the toxic effects of mutant SOD1 may trigger abnormal alternative splicing. The proof that abnormal splicing of peripherin occurs in ALS came when we identified peripherin mRNA transcripts that retain introns 3 and 4, leading to the generation of a C-terminally truncated protein of 28 kDa, termed Per-28 (Xiao et al., 2008). Per-28 is aggregate-prone in SW13vim^{-} cells, is associated with inclusion body formation, and is upregulated in ALS cases. Considering that peripherin expression is characterized by the generation of multiple isoforms, we were able to observe distinct peripherin isoform expression patterns in various neuronal tissues, during traumatic neuronal injury, in mouse models of motor neuron disease, and in ALS (McLean et al., 2008; Xiao et al., 2008, McLean et al., 2010). The consequences associated with abnormal changes in the isoform ratio, such as the generation of Per-61 and Per-28, are likely to have unwanted cellular effects (McLean et al., 2008; Xiao et al., 2008). The recent findings of peripherin mutations in sporadic ALS (sALS) have reinforced the importance of peripherin as a prospective causative or propagative factor of ALS. Direct evidence for peripherin genetic changes in ALS has come from three studies that have identified at least three pathogenic mutations associated with sporadic ALS cases (Gros-Louis et al., 2004; Leung et al., 2004; Corrado et al., 2010). One variant consists of a nucleotide deletion (PRPH^{228delC}) resulting in a frameshift within exon 1 and a stop codon predicted to generate a truncated peripherin species of 85 aa encompassing the head domain; a second variant is a homozygous nonconservative D141Y mutation within the first linker sequence of the rod domain; and the third is a missense R133P mutation. Although the mechanism of their toxicity is unknown, the former two mutations are associated with aggregate formation, suggesting that their toxicity arises from isoform-mediated disruption of the cytoskeletal network.

Our current understanding of the cause and propagation of ALS is now intimately linked to the non-cell autonomous theory of disease pathogenesis, which states that the selective vulnerability of motor neurons in ALS is derived from damage incurred to both neurons and their non-neuronal neighboring cells (Boillee et al., 2006a; Rothstein, 2009). From experiments using restricted promoters in transgenic mice to selectively express mtSOD1 in various cell types, it has been shown that astrocytes, Schwann cells, and microglia are determinants of disease progression, while toxicity arising within motor neurons initiates disease onset (Gong et al., 2000; Lino et al., 2002; Clement et al., 2003; Boillee et al., 2006b; Lobsiger et al.,

2009). Recent evidence suggests that the activation of microglia and astrocytes in ALS leads to a disturbance in the normal interrelationships, or "cross-talk," between glia and motor neurons (Van Den Bosch and Robberecht, 2008). This has particular relevance to our discussion about peripherin in ALS as certain neurotrophins (NGF and FGF) and proinflammatory cytokines (LIF and IL-6) are capable of eliciting a peripherin response (Portier et al., 1983; Leonard et al., 1987; Parysek and Goldman, 1987; Aletta et al., 1988; Leonard et al., 1988; Thompson et al., 1992; Choi et al., 2001). As these factors are robustly secreted by glial cells in ALS (Giess et al., 2000; Moreau et al., 2005; Pehar et al., 2005), it is reasonable to speculate that neuroinflammation may play a role in modulating peripherin expression. In fact, Robertson et al. (2001) identified that inclusion-bearing DRG derived from TPer mice were uniquely susceptible to apoptosis upon exposure to the proinflammatory environment of dissociated spinal cord cultures, which are rich in activated microglia and astrocytes. The proapoptotic effect could be blocked by the administration of tumor necrosis factor-α (TNF-α)–neutralizing antibody, showing that this selective effect was downstream of receptor-mediated signaling pathways. Although the exact mechanisms of these interactions remain elusive, they suggest that peripherin maintains specific properties that can lead to or enhance neuronal degeneration. Our recent work, combined with previous studies on neuronal intermediate filament (nIF) transgenic mice, suggests that the environmental conditions found among degenerating motor neurons in ALS patients, including oxidative stress and neuroinflammation, may have a direct impact on peripherin isoform expression and formation of peripherin-immunoreactive aggregates (McLean et al., unpublished results). Although very few studies have attempted to identify relationships between neuroinflammation and IF expression, it is possible that peripherin abnormalities may be part of a vicious cycle of disease propagation involving both neuronal and non-neuronal cells (Fig. 9.4). Whereas the cause and propagation of ALS are

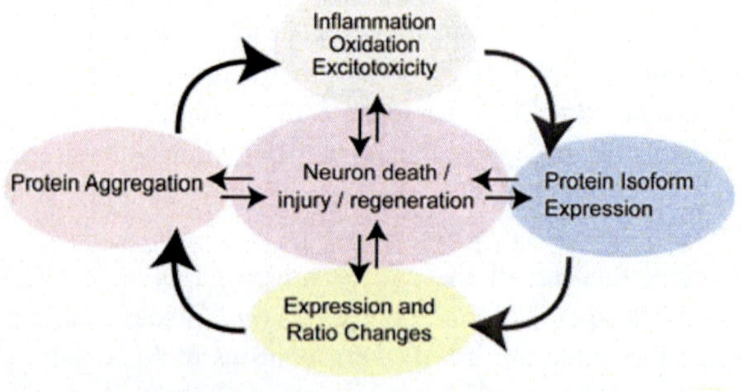

Fig. 9.4 Model of peripherin involvement in neuronal degeneration or regeneration. The effects of peripherin on neuronal fate may be dependent on both autonomous and non-cell autonomous factors. For example, in ALS, inflammation, oxidation, and excitotoxicity may alter peripherin isoform expression. These changes may involve the generation of abnormal splice variants, such as Per-28, or modify the normal isoform ratio. These ratio changes have been associated with aggregate formation, which, in turn, may interfere with neuronal metabolism and elicit a further inflammatory response or cause oxidative stress

believed to be separate events and likely to involve multiple causative factors, it may be argued that there is no other protein in the ALS literature that is more representative of a pathologic point-of-convergence for both the autonomous and non-autonomous causes of motor neuron degeneration.

9.6 Cancer and Diabetes

The utility of peripherin as a marker for disease can be found in its neuron-specific expression patterns that originate from different lineages (neural tube, neural crest, placodes) and with different functions (motor, sensory, and sympathetic). This has been best demonstrated in its use as an enteric marker of gastrointestinal abnormalities (Eaker and Sallustio, 1994; Szabolcs et al., 1996; Eaker, 1997; Petchasuwan and Pintong, 2000; Ganns et al., 2006) and for delineating the origin and differentiation status of various cancer cell lines (Ho and Liem, 1996; Prasad et al., 1999; Mattiazzi et al., 2002), including, among others, neuroendocrine carcinomas (Baudoin et al., 1993; Alvarez-Gago et al., 1996), neuroblastomas (Pedersen et al., 1993; Foley et al., 1994; Willoughby et al., 2008), and melanocytic lesions (Moll et al., 1996; Prieto et al., 1997b; a; Kanitakis et al., 1998). Although the use of peripherin as a marker for tumor identification is of particular value, caution should be taken in assigning neuronal origin to tumors based solely on IF expression patterns, considering that, as we have seen, peripherin expression can be dramatically influenced by exogenous factors. Common genetic or chromosomal alterations associated with cancer growth and differentiation may also play a role in peripherin expression and further confound its nosological and diagnostic use.

Recently, peripherin has received considerable attention for its role in diabetogenesis. In type 1 diabetes, insulin-producing β-cells are destroyed by the immune system by an, as of yet, unknown mechanism. The expression of both peripherin and NF-L in the rat insulinoma Rin5F cell line was the first substantial link between peripherin and the endocrine system suggesting an ectodermal origin for the β-cells of the islets of Langerhans (Escurat et al., 1991). A more direct role for peripherin came from the discovery that several anti-I-A^{g7} autoantibodies circulating in the Rin5F cells and islets in sera from nonobese diabetic (NOD) mice were cross-reactive (Boitard et al., 1992). The antigenic repertoire of peripherin was further established when islet-infiltrating B lymphocytes from different mouse NOD strains, which develop different degrees of insulitis and incidence of diabetes, were found to secrete antibodies against peripherin (Puertas et al., 2007). These findings are relevant to the early development of peripheral neuropathy in the diabetogenic process, considering that autoreactive B lymphocytes have been shown to be primed against peripherin early in the course of the disease (Carrillo et al., 2008). The relationship here further highlights the importance of peripherin and its interactions with other intracellular proteins, including nuclear lamin B (Djabali et al., 1991) and bullous pemphigoid antigen 1 (BPAG1-n) (Leung et al., 1999), which may act as attachment sites and/or mediators of peripherin functional specificity.

9.7 Conclusions

As we have seen, the increasing knowledge about peripherin biology, from gene regulation to protein expression patterns and interactions, has provided an exceptional opportunity from which to understand peripherin involvement in a number of human diseases, both neurologic and systemic. The recent identification and characterization of normal and abnormal peripherin isoforms tells us that peripherin expression is a complex process capable of responding to a number of cellular events in a highly specific way. Several questions, however, regarding the nature of peripherin expression, remain: How and why does peripherin become part of intracellular inclusions? Is the regenerative or degenerative capacity of peripherin encoded in the specific set of isoforms expressed in response to stress or disease? To what extent do exogenous factors regulate peripherin isoform-specific expression? Does TDP-43 or FUS/TLS modify peripherin isoform expression? What signal transduction pathways is peripherin a component of? These questions, and many left unsaid, are at the heart of an ongoing debate regarding peripherin involvement in ALS and other serious human diseases, and, in time, our understanding of the protein may yield useful biomarkers and/or biotherapeutic applications.

References

Adams AD, Choate DM, Thompson MA (1995) NF1-L is the DNA-binding component of the protein complex at the peripherin negative regulatory element. J Biol Chem 270: 6975–6983

Aggarwal S, Cudkowicz M (2008) ALS drug development: reflections from the past and a way forward. Neurotherapeutics 5:516–527

Aletta JM, Angeletti R, Liem RK, Purcell C, Shelanski ML, Greene LA (1988) Relationship between the nerve growth factor-regulated clone 73 gene product and the 58-kilodalton neuronal intermediate filament protein (peripherin). J Neurochem 51:1317–1320

Aletta JM, Shelanski ML, Greene LA (1989) Phosphorylation of the peripherin 58-kDa neuronal intermediate filament protein. Regulation by nerve growth factor and other agents. J Biol Chem 264:4619–4627

Alvarez-Gago T, Bullon MM, Rivera F, Velasco A, Mayo A (1996) Intermediate filament aggregates in mitoses of primary cutaneous neuroendocrine (Merkel cell) carcinoma. Histopathology 28:349–355

Angelastro JM, Ho CL, Frappier T, Liem RK, Greene LA (1998) Peripherin is tyrosine-phosphorylated at its carboxyl-terminal tyrosine. J Neurochem 70:540–549

Annesi G, Savettieri G, Pugliese P et al (2005) DJ-1 mutations and parkinsonism-dementia-amyotrophic lateral sclerosis complex. Ann Neurol 58:803–807

Appel SH, Smith RG, Engelhardt JI, Stefani E (1993) Evidence for autoimmunity in amyotrophic lateral sclerosis. J Neurol Sci 118:169–174

Athlan ES, Mushynski WE (1997) Heterodimeric associations between neuronal intermediate filament proteins. J Biol Chem 272:31073–31078

Athlan ES, Sacher MG, Mushynski WE (1997) Associations between intermediate filament proteins expressed in cultured dorsal root ganglion neurons. J Neurosci Res 47:300–310

Averback P (1981) Unusual particles in motor neuron disease. Arch Pathol Lab Med 105:490–493

Bäck N, Tyynelä M, Portier MM, Virtanen I, Soinila S (1995) Distribution of neurofilament proteins and peripherin in the rat pituitary gland. Neurosci Res 22:267–275

Baudoin C, Meneguzzi G, Portier MM, Demarchez M, Bernerd F, Pisani A, Ortonne JP (1993) Peripherin, a neuronal intermediate protein, is stably expressed by neuroendocrine carcinomas

of the skin, their xenograft on nude mice, and the corresponding primary cultures. Cancer Res 53:1175–1181
Beaulieu JM, Julien JP (2003) Peripherin-mediated death of motor neurons rescued by overexpression of neurofilament NF-H proteins. J Neurochem 85:248–256
Beaulieu JM, Kriz J, Julien JP (2002) Induction of peripherin expression in subsets of brain neurons after lesion injury or cerebral ischemia. Brain Res 946:153–161
Beaulieu JM, Nguyen MD, Julien JP (1999a) Late onset of motor neurons in mice overexpressing wild-type peripherin. J Cell Biol 147:531–544
Beaulieu JM, Robertson J, Julien JP (1999b) Interactions between peripherin and neurofilaments in cultured cells: disruption of peripherin assembly by the NF-M and NF-H subunits. Biochem Cell Biol 77:41–45
Belecky-Adams T, Wight DC, Kopchick JJ, Parysek LM (1993) Intragenic sequences are required for cell type-specific and injury-induced expression of the rat peripherin gene. J Neurosci 13:5056–5065
Ben-Ari Y, Cossart R (2000) Kainate, a double agent that generates seizures: two decades of progress. Trends Neurosci 23:580–587
Bergeron C, Beric-Maskarel K, Muntasser S, Weyer L, Somerville MJ, Percy ME (1994) Neurofilament light and polyadenylated mRNA levels are decreased in amyotrophic lateral sclerosis motor neurons. J Neuropathol Exp Neurol 53:221–230
Boillee S, Vande Velde C, Cleveland DW (2006a) ALS: a disease of motor neurons and their nonneuronal neighbors. Neuron 52:39–59
Boillee S, Yamanaka K, Lobsiger CS, Copeland NG, Jenkins NA, Kassiotis G, Kollias G, Cleveland DW (2006b) Onset and progression in inherited ALS determined by motor neurons and microglia. Science 312:1389–1392
Boitard C, Villa MC, Becourt C, Gia HP, Huc C, Sempe P, Portier MM, Bach JF (1992) Peripherin: an islet antigen that is cross-reactive with nonobese diabetic mouse class II gene products. Proc Natl Acad Sci USA 89:172–176
Brooks BR (1994) El Escorial World Federation of Neurology criteria for the diagnosis of amyotrophic lateral sclerosis. Subcommittee on Motor Neuron Diseases/Amyotrophic Lateral Sclerosis of the World Federation of Neurology Research Group on Neuromuscular Diseases and the El Escorial "Clinical limits of amyotrophic lateral sclerosis" workshop contributors. J Neurol Sci 124(Suppl):96–107
Brooks BR, Miller RG, Swash M, Munsat TL (2000) El Escorial revisited: revised criteria for the diagnosis of amyotrophic lateral sclerosis. Amyotroph Lateral Scler Other Motor Neuron Disord 1:293–299
Broom WJ, Parton MJ, Vance CA et al (2004) No association of the SOD1 locus and disease susceptibility or phenotype in sporadic ALS. Neurology 63:2419–2422
Bruijn LI, Miller TM, Cleveland DW (2004) Unraveling the mechanisms involved in motor neuron degeneration in ALS. Annu Rev Neurosci 27:723–749
Carrillo J, Puertas MC, Planas R et al (2008) Anti-peripherin B lymphocytes are positively selected during diabetogenesis. Mol Immunol 45:3152–3162
Chadan S, Moya KL, Portier MM, Filliatreau G (1994) Identification of a peripherin dimer: changes during axonal development and regeneration of the rat sciatic nerve. J Neurochem 62:1894–1905
Chang L, Thompson MA (1996) Activity of the distal positive element of the peripherin gene is dependent on proteins binding to an Ets-like recognition site and a novel inverted repeat site. J Biol Chem 271:6467–6475
Chien CL, Lee TH, Lu KS (1998) Distribution of neuronal intermediate filament proteins in the developing mouse olfactory system. J Neurosci Res 54:353–363
Choi DY, Toledo-Aral JJ, Lin HY et al (2001) Fibroblast growth factor receptor 3 induces gene expression primarily through Ras-independent signal transduction pathways. J Biol Chem 276:5116–5122
Clement AM, Nguyen MD, Roberts EA et al (2003) Wild-type nonneuronal cells extend survival of SOD1 mutant motor neurons in ALS mice. Science 302:113–117

Corbo M, Hays AP (1992) Peripherin and neurofilament protein coexist in spinal spheroids of motor neuron disease. J Neuropathol Exp Neurol 51:531–537

Corrado L, Carlomagno Y, Falasco L, Mellone S, Godi M, Cova E, Cereda C, Testa L, Mazzini L, D'Alfonso S (2010) A novel peripherin gene (PRPH) mutation identified in one sporadic amyotrophic lateral sclerosis patient. Neurobiol Aging. 2010 Apr 1. [Epub ahead of print]

Cote F, Collard JF, Julien JP (1993) Progressive neuronopathy in transgenic mice expressing the human neurofilament heavy gene: a mouse model of amyotrophic lateral sclerosis. Cell 73: 35–46

Cui C, Stambrook PJ, Parysek LM (1995) Peripherin assembles into homopolymers in SW13 cells. J Cell Sci 108:3279–3284

Cummings TJ, Shea CR, Reed JA, Burchette JL, Prieto VG (2000) Expression of the intermediate filament peripherin in extraskeletal myxoid chondrosarcoma. J Cutan Pathol 27:141–146

Delisle MB, Carpenter S (1984) Neurofibrillary axonal swellings and amyotrophic lateral sclerosis. J Neurol Sci 63:241–250

Desmarais D, Filion M, Lapointe L, Royal A (1992) Cell-specific transcription of the peripherin gene in neuronal cell lines involves a cis-acting element surrounding the TATA box. EMBO J 11:2971–2980

Djabali K, Piron G, de Néchaud B, Portier MM (1999) alphaB-crystallin interacts with cytoplasmic intermediate filament bundles during mitosis. Exp Cell Res 253:649–662

Djabali K, Portier MM, Gros F, Blobel G, Georgatos SD (1991) Network antibodies identify nuclear lamin B as a physiological attachment site for peripherin intermediate filaments. Cell 64:109–121

Djabali K, Zissopoulou A, de Hoop MJ, Georgatos SD, Dotti CG (1993) Peripherin expression in hippocampal neurons induced by muscle soluble factor(s). J Cell Biol 123:1197–1206

Eaker EY (1997) Neurofilament and intermediate filament immunoreactivity in human intestinal myenteric neurons. Dig Dis Sci 42:1926–1932

Eaker EY, Sallustio JE (1994) The distribution of novel intermediate filament proteins defines subpopulations of myenteric neurons in rat intestine. Gastroenterology 107:666–674

Escurat M, Djabali K, Gumpel M, Gros F, Portier MM (1990) Differential expression of two neuronal intermediate-filament proteins, peripherin and the low-molecular-mass neurofilament protein (NF-L), during the development of the rat. J Neurosci 10:764–784

Escurat M, Djabali K, Huc C, Landon F, Becourt C, Boitard C, Gros F, Portier MM (1991) Origin of the beta cells of the islets of Langerhans is further questioned by the expression of neuronal intermediate filament proteins, peripherin and NF-L, in the rat insulinoma RIN5F cell line. Dev Neurosci 13:424–432

Faussone-Pellegrini MS, Matini P, DeFelici M (1999) The cytoskeleton of the myenteric neurons during murine embryonic life. Anat Embryol (Berl) 199:459–469

Ferrari N, Desmarais D, Royal A (1995) Transcriptional activation of the neuronal peripherin-encoding gene depends on a G + C-rich element that binds Sp1 in vitro and in vivo. Gene 159:159–165

Foley J, Ley CA, Parysek LM (1994) The structure of the human peripherin gene (PRPH) and identification of potential regulatory elements. Genomics 22:456–461

Franke TF (2008) PI3K/Akt: getting it right matters. Oncogene 27:6473–6488

Fuchs E, Weber K (1994) Intermediate filaments: structure, dynamics, function, and disease. Annu Rev Biochem 63:345–382

Ganns D, Schrodl F, Neuhuber W, Brehmer A (2006) Investigation of general and cytoskeletal markers to estimate numbers and proportions of neurons in the human intestine. Histol Histopathol 21:41–51

Giasson BI, Mushynski WE (1997) Developmentally regulated stabilization of neuronal intermediate filaments in rat cerebral cortex. Neurosci Lett 229:77–80

Giess R, Beck M, Goetz R, Nitsch RM, Toyka KV, Sendtner M (2000) Potential role of LIF as a modifier gene in the pathogenesis of amyotrophic ateral sclerosis. Neurology 54:1003–1005

Goldman RD, Grin B, Mendez MG, Kuczmarski ER (2008) Intermediate filaments: versatile building blocks of cell structure. Curr Opin Cell Biol 20:28–34

Goldstein ME, House SB, Gainer H (1991) NF-L and peripherin immunoreactivities define distinct classes of rat sensory ganglion cells. J Neurosci Res 30:92–104
Goldstein ME, Weiss SR, Lazzarini RA, Shneidman PS, Lees JF, Schlaepfer WW (1988) mRNA levels of all three neurofilament proteins decline following nerve transection. Brain Res 427:287–291
Gong YH, Parsadanian AS, Andreeva A, Snider WD, Elliott JL (2000) Restricted expression of G86R Cu/Zn superoxide dismutase in astrocytes results in astrocytosis but does not cause motoneuron degeneration. J Neurosci 20:660–665
Goodall EF, Morrison KE (2006) Amyotrophic lateral sclerosis (motor neuron disease): proposed mechanisms and pathways to treatment. Expert Rev Mol Med 8:1–22
Gros-Louis F, Lariviere R, Gowing G, Laurent S, Camu W, Bouchard JP, Meininger V, Rouleau GA, Julien JP (2004) A frameshift deletion in peripherin gene associated with amyotrophic lateral sclerosis. J Biol Chem 279:45951–45956
He CZ, Hays AP (2004) Expression of peripherin in ubiquinated inclusions of amyotrophic lateral sclerosis. J Neurol Sci 217:47–54
Helfand BT, Mendez MG, Pugh J, Delsert C, Goldman RD (2003) A role for intermediate filaments in determining and maintaining the shape of nerve cells. Mol Biol Cell 14:5069–5081
Henkel JS, Engelhardt JI, Siklos L et al (2004) Presence of dendritic cells, MCP-1, and activated microglia/macrophages in amyotrophic lateral sclerosis spinal cord tissue. Ann Neurol 55: 221–235
Herrmann H, Bar H, Kreplak L, Strelkov SV, Aebi U (2007) Intermediate filaments: from cell architecture to nanomechanics. Nat Rev Mol Cell Biol 8:562–573
Ho CL, Chien CL, Brown K, Liem RK (1998) Ultrastructural studies of carboxyl-terminal truncation mutants of the neuronal intermediate filament protein peripherin. J Neurochem 70:1916–1924
Ho CL, Chin SS, Carnevale K, Liem RK (1995) Translation initiation and assembly of peripherin in cultured cells. Eur J Cell Biol 68:103–112
Ho CL, Liem RK (1996) Intermediate filaments in the nervous system: implications in cancer. Cancer Metastasis Rev 15:483–497
Hoffman PN, Cleveland DW, Griffin JW, Landes PW, Cowan NJ, Price DL (1987) Neurofilament gene expression: a major determinant of axonal caliber. Proc Natl Acad Sci USA 84: 3472–3476
Huc C, Escurat M, Djabali K, Derer M, Landon F, Gros F, Portier MM (1989) Phosphorylation of peripherin, an intermediate filament protein, in mouse neuroblastoma NIE 115 cell line and in sympathetic neurons. Biochem Biophys Res Commun 160:772–779
Julien JP, Beaulieu JM (2000) Cytoskeletal abnormalities in amyotrophic lateral sclerosis: beneficial or detrimental effects? J Neurol Sci 180:7–14
Kabashi E, Valdmanis PN, Dion P et al (2008) TARDBP mutations in individuals with sporadic and familial amyotrophic lateral sclerosis. Nat Genet 40:572–574
Kanitakis J, Bourchany D, Faure M, Claudy A (1998) Expression of the intermediate filament peripherin in skin tumors. Eur J Dermatol 8:339–342
Karpov V, Landon F, Djabali K, Gros F, Portier MM (1992) Structure of the mouse gene encoding peripherin: a neuronal intermediate filament protein. Biol Cell 76:43–48
Kim S, Coulombe PA (2007) Intermediate filament scaffolds fulfill mechanical, organizational, and signaling functions in the cytoplasm. Genes Dev 21:1581–1597
Konishi H, Namikawa K, Shikata K, Kobatake Y, Tachibana T, Kiyama H (2007) Identification of peripherin as a Akt substrate in neurons. J Biol Chem 282:23491–23499
Kreplak L, Aebi U, Herrmann H (2004) Molecular mechanisms underlying the assembly of intermediate filaments. Exp Cell Res 301:77–83
Kriz J, Beaulieu JM, Julien JP, Krnjevic K (2005) Up-regulation of peripherin is associated with alterations in synaptic plasticity in CA1 and CA3 regions of hippocampus. Neurobiol Dis 18:409–420

Kriz J, Meier J, Julien JP, Padjen AL (2000) Altered ionic conductances in axons of transgenic mouse expressing the human neurofilament heavy gene: a mouse model of amyotrophic lateral sclerosis. Exp Neurol 163:414–421

Kwiatkowski TJ Jr, Bosco DA, Leclerc AL et al (2009) Mutations in the FUS/TLS gene on chromosome 16 cause familial amyotrophic lateral sclerosis. Science 323:1205–1208

Lallemend F, Vandenbosch R, Hadjab S, Bodson M, Breuskin I, Moonen G, Lefebvre PP, Malgrange B (2007) New insights into peripherin expression in cochlear neurons. Neuroscience 150:212–222

Lampson LA, Kushner PD, Sobel RA (1990) Major histocompatibility complex antigen expression in the affected tissues in amyotrophic lateral sclerosis. Ann Neurol 28:365–372

Landon F, Lemonnier M, Benarous R, Huc C, Fiszman M, Gros F, Portier MM (1989) Multiple mRNAs encode peripherin, a neuronal intermediate filament protein. EMBO J 8: 1719–1726

Landon F, Wolff A, de Nechaud B (2000) Mouse peripherin isoforms. Biol Cell 92:397–407

Larcher JC, Cordeau-Lossouarn L, Romey G, Gros F, Croizat B, Vayssiere JL (1992) Adriamycin promotes neurite outgrowth in the "neurite-minus" N1A-103 mouse neuroblastoma cell line. Exp Cell Res 203:72–79

Lariviere RC, Beaulieu JM, Nguyen MD, Julien JP (2003) Peripherin is not a contributing factor to motor neuron disease in a mouse model of amyotrophic lateral sclerosis caused by mutant superoxide dismutase. Neurobiol Dis 13:158–166

Lariviere RC, Nguyen MD, Ribeiro-da-Silva A, Julien JP (2002) Reduced number of unmyelinated sensory axons in peripherin null mice. J Neurochem 81:525–532

Lecomte MJ, Basseville M, Landon F, Karpov V, Fauquet M (1998) Transcriptional activation of the mouse peripherin gene by leukemia inhibitory factor: involvement of STAT proteins. J Neurochem 70:971–982

Lee MK, Marszalek JR, Cleveland DW (1994) A mutant neurofilament subunit causes massive, selective motor neuron death: implications for the pathogenesis of human motor neuron disease. Neuron 13:975–988

Leonard DG, Gorham JD, Cole P, Greene LA, Ziff EB (1988) A nerve growth factor-regulated messenger RNA encodes a new intermediate filament protein. J Cell Biol 106:181–193

Leonard RB, Kevetter GA (2006) Vestibular efferents contain peripherin. Neurosci Lett 408: 104–107

Leonard DG, Ziff EB, Greene LA (1987) Identification and characterization of mRNAs regulated by nerve growth factor in PC12 cells. Mol Cell Biol 7:3156–3167

Leung CL, He CZ, Kaufmann P, Chin SS, Naini A, Liem RK, Mitsumoto H, Hays AP (2004) A pathogenic peripherin gene mutation in a patient with amyotrophic lateral sclerosis. Brain Pathol 14:290–296

Leung CL, Sun D, Liem RK (1999) The intermediate filament protein peripherin is the specific interaction partner of mouse BPAG1-n (dystonin) in neurons. J Cell Biol 144:435–446

Lino MM, Schneider C, Caroni P (2002) Accumulation of SOD1 mutants in postnatal motoneurons does not cause motoneuron pathology or motoneuron disease. J Neurosci 22: 4825–4832

Lobsiger CS, Boillee S, McAlonis-Downes M, Khan AM, Feltri ML, Yamanaka K, Cleveland DW (2009) Schwann cells expressing dismutase active mutant SOD1 unexpectedly slow disease progression in ALS mice. Proc Natl Acad Sci USA 106:4465–4470

Lysakowski A, Alonto A, Jacobson L (1999) Peripherin immunoreactivity labels small diameter vestibular 'bouton' afferents in rodents. Hear Res 133:149–154

Majoor-Krakauer D, Ottman R, Johnson WG, Rowland LP (1994) Familial aggregation of amyotrophic lateral sclerosis, dementia, and Parkinson's disease: evidence of shared genetic susceptibility. Neurology 44:1872–1877

Mathew JS, Westmoreland SV, Alvarez X, Simon MA, Pauley DR, MacKey JJ, Lackner AA (2001) Expression of peripherin in the brain of macaques (Macaca mulatta and Macaca fascicularis) occurs in astrocytes rather than neurones and is associated with encephalitis. Neuropathol Appl Neurobiol 27:434–443

Matini P, Mayer B, Faussone-Pellegrini MS (1997) Neurochemical differentiation of rat enteric neurons during pre- and postnatal life. Cell Tissue Res 288:11–23

Mattiazzi M, D'Aurelio M, Gajewski CD, Martushova K, Kiaei M, Beal MF, Manfredi G (2002) Mutated human SOD1 causes dysfunction of oxidative phosphorylation in mitochondria of transgenic mice. J Biol Chem 277:29626–29633

McLean JR, Sanelli TR, Leystra-Lantz C, He BP, Strong MJ (2005) Temporal profiles of neuronal degeneration, glial proliferation, and cell death in hNFL(+/+) and NFL(–/–) mice. Glia 52: 59–69

McLean J, Xiao S, Miyazaki K, Robertson J (2008) A novel peripherin isoform generated by alternative translation is required for normal filament network formation. J Neurochem 104:1663–1673

McLean J, Liu HN, Miletic D, Weng YC, Rogaeva E, Zinman L, Kriz J, Robertson J (2010) Distinct biochemical signatures characterize peripherin isoform expression in both traumatic neuronal injury and motor neuron disease. J Neurochem 114:1177–92

Merigo F, Mucignat-Caretta C, Zancanaro C (2005) Timing of neuronal intermediate filament proteins expression in the mouse vomeronasal organ during pre- and postnatal development. An immunohistochemical study. Chem Senses 30:707–717

Migheli A, Pezzulo T, Attanasio A, Schiffer D (1993) Peripherin immunoreactive structures in amyotrophic lateral sclerosis. Lab Invest 68:185–191

Millecamps S, Robertson J, Lariviere R, Mallet J, Julien JP (2006) Defective axonal transport of neurofilament proteins in neurons overexpressing peripherin. J Neurochem 98:926–938

Moll I, Paus R, Moll R (1996) Merkel cells in mouse skin: intermediate filament pattern, localization, and hair cycle-dependent density. J Invest Dermatol 106:281–286

Moncla A, Landon F, Mattei MG, Portier MM (1992) Chromosomal localisation of the mouse and human peripherin genes. Genet Res 59:125–129

Moreau C, Devos D, Brunaud-Danel V, Defebvre L, Perez T, Destee A, Tonnel AB, Lassalle P, Just N (2005) Elevated IL-6 and TNF-alpha levels in patients with ALS: inflammation or hypoxia? Neurology 65:1958–1960

Munoz DG, Greene C, Perl DP, Selkoe DJ (1988) Accumulation of phosphorylated neurofilaments in anterior horn motoneurons of amyotrophic lateral sclerosis patients. J Neuropathol Exp Neurol 47:9–18

Murayama S, Bouldin TW, Suzuki K (1992) Immunocytochemical and ultrastructural studies of upper motor neurons in amyotrophic lateral sclerosis. Acta Neuropathol 83:518–524

Murayama S, Mori H, Ihara Y, Bouldin TW, Suzuki K, Tomonaga M (1990) Immunocytochemical and ultrastructural studies of lower motor neurons in amyotrophic lateral sclerosis. Ann Neurol 27:137–148

Nagy D, Kato T, Kushner PD (1994) Reactive astrocytes are widespread in the cortical gray matter of amyotrophic lateral sclerosis. J Neurosci Res 38:336–347

Oblinger MM, Szumlas RA, Wong J, Liuzzi FJ (1989a) Changes in cytoskeletal gene expression affect the composition of regenerating axonal sprouts elaborated by dorsal root ganglion neurons in vivo. J Neurosci 9:2645–2653

Oblinger MM, Wong J, Parysek LM (1989b) Axotomy-induced changes in the expression of a type III neuronal intermediate filament gene. J Neurosci 9:3766–3775

Omary MB, Ku NO, Tao GZ, Toivola DM, Liao J (2006) "Heads and tails" of intermediate filament phosphorylation: multiple sites and functional insights. Trends Biochem Sci 31:383–394

Parysek LM, Goldman RD (1987) Characterization of intermediate filaments in PC12 cells. J Neurosci 7:781–791

Parysek LM, Goldman RD (1988) Distribution of a novel 57 kDa intermediate filament (IF) protein in the nervous system. Neuron 1:395–401

Parysek LM, McReynolds MA, Goldman RD, Ley CA (1991) Some neural intermediate filaments contain both peripherin and the neurofilament proteins. J Neurosci Res 30:80–91

Pedersen WA, Becker LE, Yeger H (1993) Expression and distribution of peripherin protein in human neuroblastoma cell lines. Int J Cancer 53:463–470

Pehar M, Vargas MR, Cassina P, Barbeito AG, Beckman JS, Barbeito L (2005) Complexity of astrocyte-motor neuron interactions in amyotrophic lateral sclerosis. Neurodegener Dis 2: 139–146

Perng MD, Wen SF, Gibbon T, Middeldorp J, Sluijs J, Hol EM, Quinlan RA (2008) Glial fibrillary acidic protein filaments can tolerate the incorporation of assembly-compromised GFAP-delta, but with consequences for filament organization and alphaB-crystallin association. Mol Biol Cell 19:4521–4533

Petchasuwan C, Pintong J (2000) Immunohistochemistry for intestinal ganglion cells and nerve fibers: aid in the diagnosis of Hirschsprung's disease. J Med Assoc Thai 83:1402–1409

Portier MM, de Nechaud B, Gros F (1983) Peripherin, a new member of the intermediate filament protein family. Dev Neurosci 6:335–344

Prasad S, Soldatenkov VA, Srinivasarao G, Dritschilo A (1999) Intermediate filament proteins during carcinogenesis and apoptosis. Int J Oncol 14:563–570

Prieto VG, McNutt NS, Lugo J, Reed JA (1997a) Differential expression of the intermediate filament peripherin in cutaneous neural lesions and neurotized melanocytic nevi. Am J Surg Pathol 21:1450–1454

Prieto VG, McNutt NS, Lugo J, Reed JA (1997b) The intermediate filament peripherin is expressed in cutaneous melanocytic lesions. J Cutan Pathol 24:145–150

Puertas MC, Carrillo J, Pastor X et al (2007) Peripherin is a relevant neuroendocrine autoantigen recognized by islet-infiltrating B lymphocytes. J Immunol 178:6533–6539

Rauch U, Klotz M, Maas-Omlor S, Wink E, Hänsgen A, Hagl C, Holland-Cunz S, Schäfer KH (2006) Expression of intermediate filament proteins and neuronal markers in the human fetal gut. J Histochem Cytochem 54:39–46

Robertson J, Beaulieu JM, Doroudchi MM, Durham HD, Julien JP, Mushynski WE (2001) Apoptotic death of neurons exhibiting peripherin aggregates is mediated by the proinflammatory cytokine tumor necrosis factor-alpha. J Cell Biol 155:217–26

Robertson J, Doroudchi MM, Nguyen MD, Durham HD, Strong MJ, Shaw G, Julien JP, Mushynski WE (2003) A neurotoxic peripherin splice variant in a mouse model of ALS. J Cell Biol 160:939–949

Robertson J, Kriz J, Nguyen MD, Julien JP (2002) Pathways to motor neuron degeneration in transgenic mouse models. Biochimie 84:1151–1160

Rosen DR, Siddique T, Patterson D et al (1993) Mutations in Cu/Zn superoxide dismutase gene are associated with familial amyotrophic lateral sclerosis. Nature 362:59–62

Rothstein JD (2009) Current hypotheses for the underlying biology of amyotrophic lateral sclerosis. Ann Neurol 65(Suppl 1):S3–S9

Sahlgren CM, Pallari HM, He T, Chou YH, Goldman RD, Eriksson JE (2006) A nestin scaffold links Cdk5/p35 signaling to oxidant-induced cell death. EMBO J 25:4808–4819

Sanelli T, Xiao S, Horne P, Bilbao J, Zinman L, Robertson J (2007) Evidence that TDP-43 is not the major ubiquitinated target within the pathological inclusions of amyotrophic lateral sclerosis. J Neuropathol Exp Neurol 66:1147–1153

Sarria AJ, Nordeen SK, Evans RM (1990) Regulated expression of vimentin cDNA in cells in the presence and absence of a preexisting vimentin filament network. J Cell Biol 111: 553–565

Schiffer D, Cordera S, Cavalla P, Migheli A (1996) Reactive astrogliosis of the spinal cord in amyotrophic lateral sclerosis. J Neurol Sci 139(Suppl):27–33

Schwab C, Steele JC, McGeer PL (1996) Neurofibrillary tangles of Guam parkinson-dementia are associated with reactive microglia and complement proteins. Brain Res 707:196–205

Schweitzer SC, Klymkowsky MW, Bellin RM, Robson RM, Capetanaki Y, Evans RM (2001) Paranemin and the organization of desmin filament networks. J Cell Sci 114:1079–1089

Shibata N (2001) Transgenic mouse model for familial amyotrophic lateral sclerosis with superoxide dismutase-1 mutation. Neuropathology 21:82–92

Sreedharan J, Blair IP, Tripathi VB et al (2008) TDP-43 mutations in familial and sporadic amyotrophic lateral sclerosis. Science 319:1668–1672

Steinert PM, Chou YH, Prahlad V, Parry DA, Marekov LN, Wu KC, Jang SI, Goldman RD (1999) A high molecular weight intermediate filament-associated protein in BHK-21 cells is nestin, a type VI intermediate filament protein. Limited co-assembly in vitro to form heteropolymers with type III vimentin and type IV alpha-internexin. J Biol Chem 274:9881–9890

Steinert PM, Roop DR (1988) Molecular and cellular biology of intermediate filaments. Annu Rev Biochem 57:593–625

Sterneck E, Kaplan DR, Johnson PF (1996) Interleukin-6 induces expression of peripherin and cooperates with Trk receptor signaling to promote neuronal differentiation in PC12 cells. J Neurochem 67:1365–1374

Strong MJ (2003) The basic aspects of therapeutics in amyotrophic lateral sclerosis. Pharmacol Ther 98:379–414

Strong MJ, Leystra-Lantz C, Ge WW (2004) Intermediate filament steady-state mRNA levels in amyotrophic lateral sclerosis. Biochem Biophys Res Commun 316:317–322

Strong M, Rosenfeld J (2003) Amyotrophic lateral sclerosis: a review of current concepts. Amyotroph Lateral Scler Other Motor Neuron Disord 4:136–143

Sunesson L, Hellman U, Larsson C (2008) Protein kinase Cepsilon binds peripherin and induces its aggregation, which is accompanied by apoptosis of neuroblastoma cells. J Biol Chem 283:16653–16664

Szabolcs MJ, Visser J, Shelanski ML, O'Toole K, Schullinger JN (1996) Peripherin: a novel marker for the immunohistochemical study of malformations of the enteric nervous system. Pediatr Pathol Lab Med 16:51–70

Taylor JP, Sater R, French J, Baltuch G, Crino PB (2001) Transcription of intermediate filament genes is enhanced in focal cortical dysplasia. Acta Neuropathol 102:141–148

Tedeschi G, Cappelletti G, Nonnis S, Taverna F, Negri A, Ronchi C, Ronchi S (2007) Tyrosine nitration is a novel post-translational modification occurring on the neural intermediate filament protein peripherin. Neurochem Res 32:433–441

Terao E, Janssens S, van den Bosch de Aguilar P, Portier M, Klosen P (2000) In vivo expression of the intermediate filament peripherin in rat motoneurons: modulation by inhibitory and stimulatory signals. Neuroscience 101:679–688

Thompson MA, Lee E, Lawe D, Gizang-Ginsberg E, Ziff EB (1992) Nerve growth factor-induced derepression of peripherin gene expression is associated with alterations in proteins binding to a negative regulatory element. Mol Cell Biol 12:2501–2513

Thompson MA, Ziff EB (1989) Structure of the gene encoding peripherin, an NGF-regulated neuronal-specific type III intermediate filament protein. Neuron 2:1043–1053

Traynor BJ, Codd MB, Corr B, Forde C, Frost E, Hardiman O (2000) Amyotrophic lateral sclerosis mimic syndromes: a population-based study. Arch Neurol 57:109–113

Troost D, Van den Oord JJ, Vianney de Jong JM (1990) Immunohistochemical characterization of the inflammatory infiltrate in amyotrophic lateral sclerosis. Neuropathol Appl Neurobiol 16:401–410

Troy CM, Greene LA, Shelanski ML (1992) Neurite outgrowth in peripherin-depleted PC12 cells. J Cell Biol 117:1085–1092

Troy CM, Muma NA, Greene LA, Price DL, Shelanski ML (1990) Regulation of peripherin and neurofilament expression in regenerating rat motor neurons. Brain Res 529:232–238

Turner MR, Kiernan MC, Leigh PN, Talbot K (2009) Biomarkers in amyotrophic lateral sclerosis. Lancet Neurol 8:94–109

Uveges TE, Shan Y, Kramer BE, Wight DC, Parysek LM (2002) Intron 1 is required for cell type-specific, but not injury-responsive, peripherin gene expression. J Neurosci 22: 7959–7967

Van Den Bosch L, Robberecht W (2008) Crosstalk between astrocytes and motor neurons: what is the message? Exp Neurol 211:1–6

van der Graaff MM, de Jong JM, Baas F, de Visser M (2009) Upper motor neuron and extra-motor neuron involvement in amyotrophic lateral sclerosis: a clinical and brain imaging review. Neuromuscul Disord 19:53–58

Vance C, Al-Chalabi A, Ruddy D et al (2006) Familial amyotrophic lateral sclerosis with frontotemporal dementia is linked to a locus on chromosome 9p13.2-21.3. Brain 129:868–876
Vance C, Rogelj B, Hortobagyi T et al (2009) Mutations in FUS, an RNA processing protein, cause familial amyotrophic lateral sclerosis type 6. Science 323:1208–1211
Wang Y, Liu H, Shen Y, Wang Z, Li H (2006) Peripherin as a marker for degeneration of spiral ganglion neurons after aminoglycoside ototoxicity. Acta Otolaryngol 126:1128–1133
Willoughby V, Sonawala A, Werlang-Perurena A, Donner LR (2008) A comparative immunohistochemical analysis of small round cell tumors of childhood: utility of peripherin and alpha-internexin as markers for neuroblastomas. Appl Immunohistochem Mol Morphol 16:344–348
Wokke JH (2000) Confounding effects of mimicking disorders in the early diagnosis of amyotrophic lateral sclerosis. Amyotroph Lateral Scler Other Motor Neuron Disord 1(Suppl 1): S61–S63
Wong NK, He BP, Strong MJ (2000) Characterization of neuronal intermediate filament protein expression in cervical spinal motor neurons in sporadic amyotrophic lateral sclerosis (ALS). J Neuropathol Exp Neurol 59:972–982
Wong J, Oblinger MM (1987) Changes in neurofilament gene expression occur after axotomy of dorsal root ganglion neurons: an in situ hybridization study. Metab Brain Dis 2:291–303
Xiao S, McLean J, Robertson J (2006) Neuronal intermediate filaments and ALS: a new look at an old question. Biochim Biophys Acta 1762:1001–1012
Xiao S, Tjostheim S, Sanelli T, McLean JR, Horne P, Fan Y, Ravits J, Strong MJ, Robertson J (2008) An aggregate-inducing peripherin isoform generated through intron retention is upregulated in amyotrophic lateral sclerosis and associated with disease pathology. J Neurosci 28:1833–1840
Zhu Q, Couillard-Despres S, Julien JP (1997) Delayed maturation of regenerating myelinated axons in mice lacking neurofilaments. Exp Neurol 148:299–316

Chapter 10
Neurofilament Cross-Bridge – A Structure Associated Specifically with the Neurofilament Among the Intermediate Filament Family

Takahiro Gotow

Abstract The neurofilament (NF) is a member of the intermediate filament family, which is composed of different proteins but all having the same width, approximately 10 nm diameter. Among intermediate filaments, NFs have unique chemical and structural properties. They are composed of three distinctly different proteins called *triplet proteins* (NF-L, NF-M, and NF-H) and are constituted morphologically of two distinct domains in vivo: core filaments and cross-bridges laterally interconnecting the core filaments. Core filaments are ~10 nm thick, and cross-bridges are much thinner, 3–5 nm thick. Cross-bridges are constructed from carboxy-terminal tail domains of NF-M and NF-H. When NFs are isolated in vitro, they are observed to be constructed of core filaments and long projections extending vertically from the core filaments. The projections, 3–5 nm in width, correspond morphologically with the tail domains of NF-M and NF-H and are much longer than cross-bridges, but with similar widths. The projections have ramified meshwork-like profiles, whereas the cross-bridges are smooth and straight; however, the projections are the structural scaffolding of cross-bridges, although the mechanism whereby projections are converted to cross-bridges is unknown. Although the tail domain of NF-H is longer and can be phosphorylated more extensively than that of NF-M, curiously NF-M appears to be more essential to form cross-bridges that are related to orienting core filaments parallel and increasing axonal calibers. However, cross-bridges are still constructed even in the presence of the NF-H tail alone without the NF-M tail, and more importantly, the cross-bridges are almost normal when having a phosphorylation-incompetent NF-M tail and an intact NF-H tail. I still have a question whether the NF-M tail or the NF-H tail is essential for cross-bridge formation but emphasize in this chapter that both tails are involved in the bridge formation and are able to compensate for each other when either protein is absent. In this reciprocity between NF-M and NF-H, phosphorylated tail domains of both proteins would be necessary. That is, normal cross-bridges are not as frequent as in

T. Gotow (✉)
Laboratory of Cell Biology, College of Nutrition, Koshien University, Takarazuka, Hyogo 665-0006, Japan
e-mail: gotow@koshien.ac.jp

R.A. Nixon, A. Yuan (eds.), *Cytoskeleton of the Nervous System*,
Advances in Neurobiology 3, DOI 10.1007/978-1-4419-6787-9_10,

either NF-M or NF-H tail-less mice when compared with wild-type mice expressing both tails, suggesting strongly that both tail domains are necessary for typical cross-bridges. When cross-bridges are not formed in the absence of both NF-M and NF-H tails, core filaments without cross-bridges are irregular in alignment, resulting in the impairment of axonal transport of membrane-bound organelles, even where microtubules are normal in appearance. Although it remains unresolved how projections are converted to cross-bridges, it seems certain that cross-bridges are an essential structure in axons, especially in long projection axons where NFs are extremely numerous. Cross-bridges are critical to enhance resistance of NFs to mechanical stress in elongated, nonrigid axoplasm and also to create a constant space between core filaments for the axonal transport of various organelles and molecules regulated by microtubules and their associated proteins.

Keywords Neurofilament · Cross-bridge · Core filament · Projection · Neurofilament triplet proteins · Carboxyl-terminal tail domain · Phosphorylation · Deletion of tail domains · Axoplasm

Contents

10.1 Introduction

The cytoskeletal system is well developed in the mature neuron – which is divided into three distinct cytoplasmic domains (i.e., cell body, dendrite, and axon) – and maintains the complicated cellular shape (especially that of the extremely elongated axon, which is the most essential) and unique structure of this cell. In the neuron, among the major cytoskeletons, such as actin filament, microtubule, and intermediate filament, the intermediate filaments are most conspicuous in the most essential compartment of the neuron (i.e., the axon) in comparison with other neuronal compartments (cf. Matus, 1988; Nixon and Sihag, 1991; Cleveland, 1996; Julien, 1997; Shea and Chan, 2008). The intermediate filament appearing in the adult mammalian central nervous system is called the *neurofilament* (NF). NFs are therefore most abundant in the axon, in which they are densely and regularly oriented along the axon, and occupy almost all cytoplasm, especially in large-caliber myelinated axons. In these orderly bundled NFs, features called *cross-bridges* are conspicuous. That is, NFs are constituted of an elongated filamentous backbone, called *core filaments*. Cross-bridges interconnect adjacent core filaments to create a constant space between the filaments.

Visualization of cross-bridges of NFs is dependent on the technical methods used in electron microscopy because they are very fine in structure as well as delicate in nature, and their dimension is nearly at the resolution level of the electron microscope. They are difficult to see with the most common electron microscopic technique (i.e., the thin-section method) because they are too fine in profile to be distinguished from the background texture when embedded in resin. Quick-freeze deep etching using fresh or mildly chemically fixed tissues, isolated or reassembled NF proteins, is the only method at present by which the cross-bridges are convincingly detectable (Hirokawa, 1982; Tsukita et al., 1982; Gotow et al., 1992). In this electron microscopic technique, the morphologic organization of NFs is more distinctly recognized than with any other technical approaches. Although cross-bridges are very fine, they are essential for structural organization and function of NFs and appear only in NFs and not in other intermediate filaments (Hirokawa, 1982; Tsukita et al., 1982; Gotow and Hashimoto, 1988). To reveal the structural and biochemical mystery of the cross-bridges, the structure of whole NFs at the macromolecular level will be described first.

10.2 Macromolecular Structure of Neurofilament

Neurons, or nerve cells, are, as described earlier, morphologically intricate due to the presence of two different types of processes extending from the cell bodies. The cellular processes are composed of dendrites and axons, and axons are mostly more elongated than the dendrites, sometimes reaching approximately 1 m in length in the human. Due to the intricate structure brought by numerous cellular processes in the neuron, cytoskeletons are well developed in this cell compared with those of other types of cells in mammals. As described earlier, among the major cytoskeletons – actin filaments (~6 nm in diameter in thin-section electron microscopy), microtubules (~25 nm), and intermediate filaments (~10 nm, thus also called 10-nm filaments) – the intermediate filaments, that is, NFs, are the most conspicuous cytoskeleton in the neuron. The three kinds of cytoskeletons are a little thicker when they are seen with replica techniques for electron microscopy, such as the quick-freeze deep-etch method, because of the coating of biological structures by platinum. NFs are especially abundant in long, large-caliber myelinated axons, occupying most parts of the axoplasm (Figs. 10.1 and 10.2).

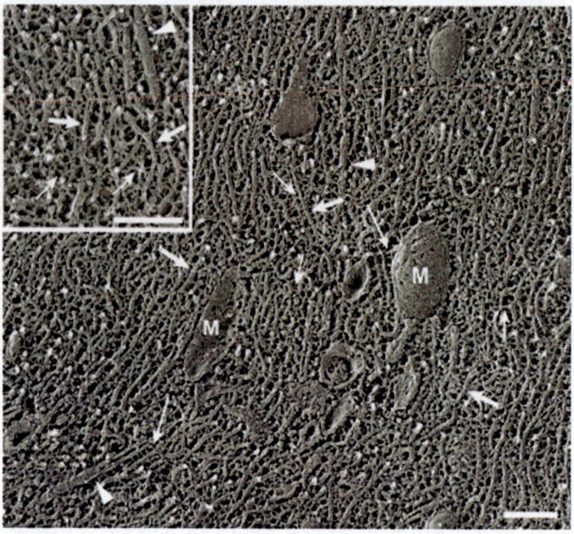

Fig. 10.1 Electron micrographs of quick-frozen and deep-etched axoplasm of myelinated axon in rat spinal cord that is directly frozen by impacting against a copper block cooled to 4 K by liquid helium from a living animal. Axoplasm is occupied with regularly oriented core filaments (*thick arrows*) of NFs. Microtubules (*arrowheads*) are also visible among NF core filaments. Because the axoplasm is not either extracted or incubated with buffer solution, the space between NF core filaments is filled with soluble granular proteins. Due to this condition, cross-bridges (*thin arrows*) between core filaments are sometimes difficult to see in this specimen. Membrane-bound organelles (M), such as mitochondria and vesicles, are located among NF core filaments. They are also seen between NF core filaments and microtubules or membrane-bound organelles (*long thin arrows*). Inset shows higher magnification from another axon. Here, ramified cross-bridges (*thin arrows*) are detectable between widely spaced core filaments. Scale bars, 0.2 μm

NFs are different from other intermediate filaments, unique in both molecular and morphologic organization, although their core filaments are 12–13 nm in diameter, the same as those of other intermediate filaments when seen by quick-freeze deep-etch electron microscopy (~10 nm in width in the thin section) (Hirokawa, 1982; Hirokawa et al., 1984; Gotow and Hashimoto, 1988; Gotow et al., 1994, 1995, 1999; Gotow, 1995, 2000). First, morphologically NFs are composed of two distinct domains, which is clearly recognized when viewed by the deep-etch replica method: a central 12- to 13-nm filament backbone (core filaments) and peripheral 4- to 5-nm thin filaments protruding laterally from the former (Figs. 10.1 and 10.2). Thin filaments protruding from the core filaments construct cross-bridges interconnecting the core filaments when NFs are close enough to orient parallel to each other (Figs. 10.1 and 10.2) (Hirokawa, 1982; Hirokawa et al., 1984; Gotow and Tanaka, 1994; Gotow, 1995, 2000; Gotow et al., 1999), and also to other kinds of cytoskeletons, such as microtubules (Fig. 10.1) (Hirokawa et al., 1988). Among intermediate filaments, the most unique morphologic feature in the NF is the cross-bridge, which is never detectable in other intermediate filaments (Figs. 10.1, 10.2, and 10.3) (Gotow and Hashimoto, 1988; Gotow, 1995), and must be related to a

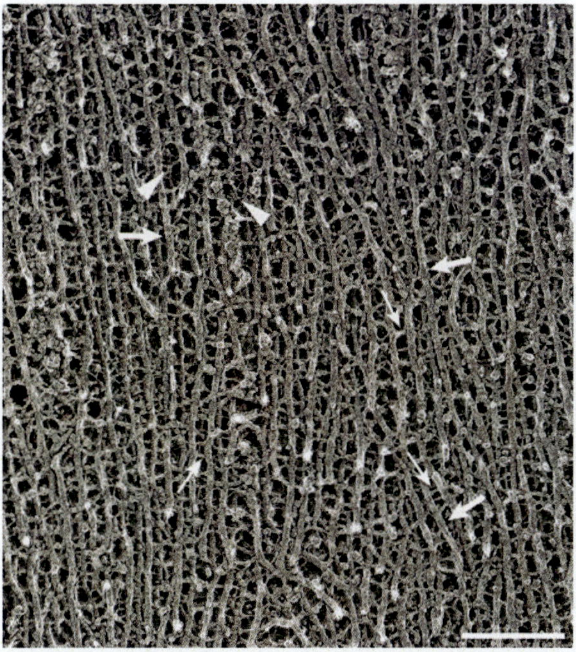

Fig. 10.2 Axoplasm of myelinated axon in rat spinal cord that is chemically fixed with 2% buffered paraformaldehyde before quick freezing. Because soluble proteins between core filaments (*thick arrows*) are mostly removed, cross-bridges (*thin arrows*) are clearly visible here. Ramified profiles of cross-bridges are still detectable. Scale bar, 0.2 μm. Reproduced from Gotow (2000), with permission of the publisher

specialized function of this intermediate filament, especially in longer and thicker axons.

Second, biochemically, although other intermediate filaments are usually constituted of proteins with similar molecular weights, around 50–70 kDa (Shaw, 1986, 1991), NFs are constituted of three proteins with distinctly different molecular weights, called *NF triplet proteins*; that is, according to their lighter-to-heavier molecular weights, light (called NF-L), medium (NF-M), and heavy (NF-H) subunit proteins. The molecular weights of these three proteins are, from NF-L to NF-H, approximately 68–70, 140–160, and 200 kDa as determined by SDS-PAGE (Leterrier and Eyer, 1987; Nixon and Sihag, 1991; Gotow et al., 1992, 1994), but they are, from complete amino acid sequence data, actually around 60–68, 90–100, and 110–115 kDa, respectively (cf. Shaw, 1991; Cleveland, 1996; Julien, 1997). The increase in molecular weight of NF subunit proteins, especially of NF-M and NF-H measured by SDS-PAGE, may be due to the presence of their extremely extended carboxyl-tail domains having abundant KSP (Lys-Ser-Pro) and its related repeat or charged motifs. In SDS-PAGE, according to the data from the dephosphorylation of NF proteins (Eyer and Leterrier, 1988; Hisanaga and Hirokawa, 1989; Gotow

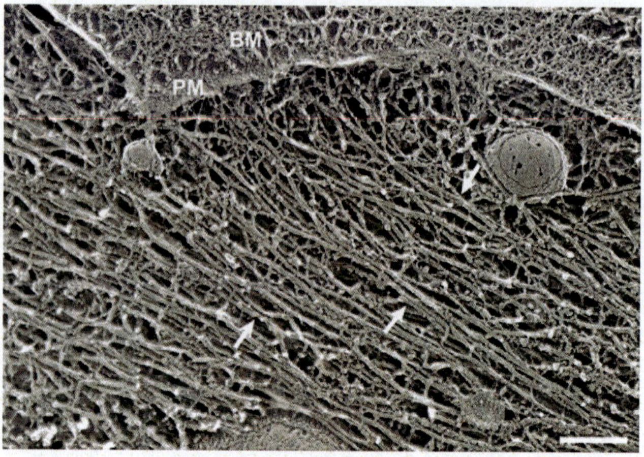

Fig. 10.3 Glial filaments in astrocyte, covered by the basement membrane (BM) and forming the glial limitans at the pial surface, quick frozen after incubation with artificial cerebrospinal fluid (CSF). Although some soluble proteins are still visible between the filaments, images corresponding with cross-bridges are hardly visible, and filaments are smooth in contour and attached frequently to each other (*arrows*). Cross-bridge–like profiles are visible around or close to the plasma membrane (PM) covered by basement membrane, but they may be plasma membrane–associated cytoskeletons, such as actin filaments. Scale bar, 0.2 μm

et al., 1994), the highly phosphorylated condition of the KSP motif of NF-H contributes to a significant increase of the weight, but that of NF-M is hardly related to the increase of the molecular weight, due to much smaller number of KSP repeat and charged motifs in NF-M than in NF-H. The major factor for the increase in molecular weight of both NF-M and NF-H must be due to the extended molecular conformation of their tail domains.

10.3 Carboxyl-Terminal Tail Domains of NF-M and NF-H

Carboxyl-tail domains of NF-M and NF-H can be detectable morphologically as long, thin strands extending vertically from core filaments when isolated or reassembled NFs are viewed by low-angle rotary-shadow electron microscopy (Fig. 10.4a) (Hisanaga and Hirokawa, 1988, 1989; Gotow et al., 1992, 1994). With this electron microscopic technique, core filaments are about 20 nm in width, twice as much as those measured by samples obtained from thin-section or negative-stain electron microscopy, due to the decoration of glycerol (which is mixed with NF proteins during preparation of the sample) around the core filaments (Hisanaga and Hirokawa et al., 1988; Gotow et al., 1992, 1994). Different from other intermediate filaments, these peripheral thin strands, called *projections*, are the structure specific to NFs, and they are not decorated by glycerol, possibly because their too-thin, 3- to 5-nm width cannot be adhered to by glycerol (Fig. 10.4). Curiously, profiles corresponding exactly with projections are not detectable in axons in vivo. To detect

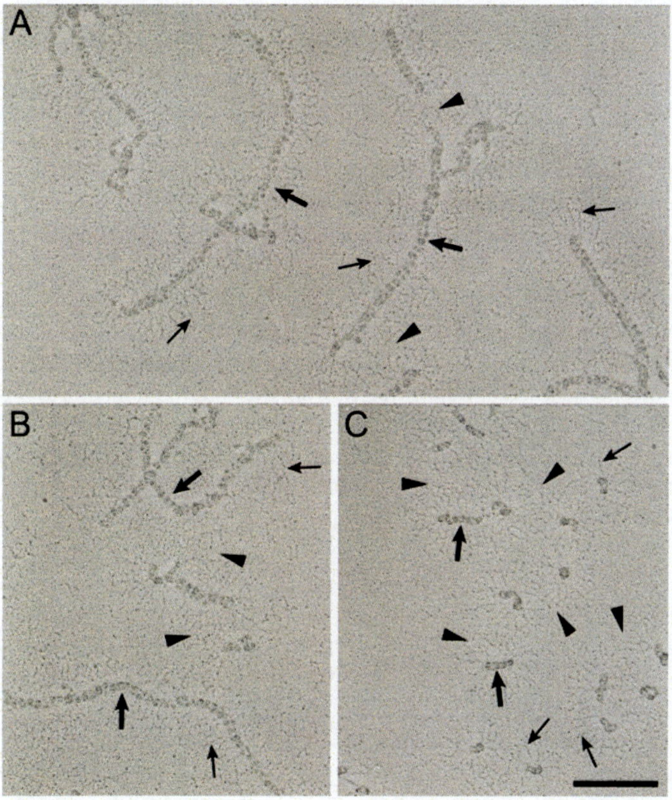

Fig. 10.4 Isolated (**a**) and reassembled bovine NFs (**b**, **c**) revealed by low-angle rotary-shadow electron microscopy. NF proteins with 50% glycerol sprayed onto mica and shadowed with platinum at low angle (6 degrees). NFs reassembled in vitro by triplet proteins (**b**) and filaments by NF-H alone (**c**). All filaments are composed of core filaments (*thick arrows*), thicker (~20 nm) than those revealed by quick-freeze deep-etching due to the decoration of glycerol, and projections (*thin arrows*). Projections of NF-H alone appear to be similar in thickness and length to those of isolated or triplet proteins, although core filaments of NF-H are much shorter and thinner. Where projections from, adjacently located core filaments are close enough to touch each other, they are ramified or irregularly aligned in profile (*arrowheads*). Scale bar, 0.2 μm. Image (**a**) is reproduced from Gotow (2000), with permission of the publisher

projections, it is necessary to isolate NFs from axons or the nervous tissue. Filaments reconstructed from NF-L and NF-M or those from NF-L and NF-H also show similar projections, but both are slightly shorter than those of NFs isolated directly from the nervous tissue or reconstructed in vitro from the triplet proteins (Hisanaga and Hirokawa, 1988). Widths of projections of filaments reassembled by NF-L and NF-M or NF-H are somewhat different between investigators; slightly thinner (~3 nm) (Hisanaga and Hirokawa, 1988) than or similar (4–5 nm) (Gotow et al., 1992) to those of isolated NFs.

Because NF-L lacks a long carboxyl-tail domain, filaments formed by NF-L alone are not expected to be provided with projections (Hisanaga and Hirokawa, 1988; Gotow et al., 1992), though NF-L alone cannot form filamentous profiles in vivo (Lee et al., 1993). Although the carboxyl-tail domain of NF-L is much shorter than that of NF-M or NF-H, it is still significantly longer (provided with a glutamate-rich domain) than those of other kinds of intermediate filament proteins (cf. Shaw, 1986, 1991; Nixon and Sihag, 1991; Julien, 1997). It is thus also expected that NF-L filaments may be provided with very short projections. However, we cannot see any profiles corresponding with projections by use of any technical methods. Such possible short projections may be inside the glycerol decoration around the core filaments by rotary-shadow electron microscopy (Hisanaga and Hirokawa, 1988; Gotow et al., 1992) or may be difficult to extend straight from the core filament due to the absence of KSP phosphorylatable domains, even in the presence of charged motif by glutamates, in the quick-freeze deep-etch electron microscopy (Gotow et al., 1992; Gou et al., 1998). This subject of the NF-L tail and projections will again be discussed in a later section of the chapter (see Section 10.5).

10.4 Morphologic Difference Between Projection and Cross-Bridge

Projections formed by long tail domains of NF-M and NF-H are the structural scaffolding of the cross-bridges interconnecting the core filaments, although the mechanism by which they form the cross-bridges, which are frequent in native NFs in axons (Figs. 10.1 and 10.2), is still controversial (Gotow et al., 1992; Leterrier et al., 1996; Gou et al., 1998; Rao et al., 2002, 2003). Are cross-bridges between core filaments formed by single projections combined by two opposed projections from adjacent core filaments? Also, are both projections of NF-M and NF-H necessary for cross-bridges or are single projections from either NF-M or NF-H enough for the bridges? In rotary-shadow electron microscopy, where isolated NFs or filaments reconstructed with NF-L and NF-M and/or NF-H are diluted with 50% glycerol and sprayed on the mica, projections are easily seen but cross-bridges are not, even where plural core filaments are close, located within 50 nm in distance (enough to form them), because structural images of projections or cross-bridges in the space between core filaments cannot be disclosed distinctly due to too low an angle of platinum shadowing (6 degrees). However, the projection structure in such space, where core filaments are separated more than 100 nm from each other, can be seen as meshwork profiles with irregularly ramified strands (Fig. 10.4b, c).

Such images formed by projections are quite different from cross-bridges in reassembled NF-H and NFs, isolated NFs, and NFs in the axon visualized by quick-freeze deep-etch electron microscopy, where the cross-bridges are simple (i.e., smooth and straight in both in vitro and in vivo conditions) (Figs. 10.1, 10.2, and

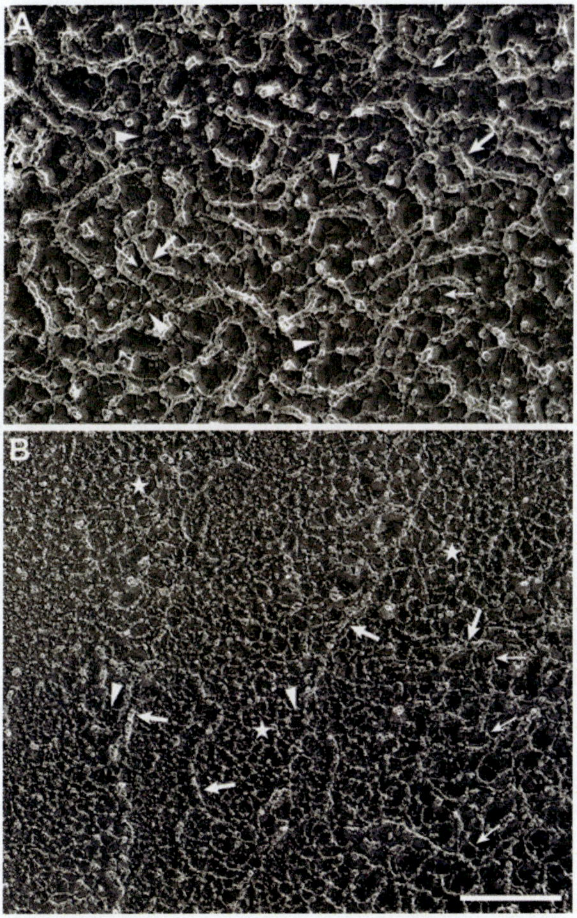

Fig. 10.5 Quick-freeze deep-etch images of filaments reassembled by NF-H alone. As described in Fig. 10.4, core filaments (*thick arrows*) are shorter and thinner, but cross-bridges (*thin arrows*) are similar to those in isolated NFs or reassembled by NF triplet proteins (Gotow et al., 1992) and also to in vivo NFs in the axon (Figs. 10.1 and 10.2). **a** Cross-bridges form ramified profiles (*arrowheads*) where core filaments are separated far from each other. These profiles are similar to the ramified structure between projections revealed by rotary shadowing. **b** When NF-H filaments are almost attached to the mica, but slightly separated from it, they show the same profile as projections. Here, these projections are ramified, except portions close to the core filaments are smooth and straight (*arrowheads*), to make a meshwork structure (*asterisks*) between core filaments. These ramified projections are considered to convert into typical cross-bridges (*thin arrows*) when core filaments are located close to each other. Scale bar, 0.2 μm. Reproduced from Gotow et al. (1992), with permission of the publisher

10.5a). Closer analysis, however, shows the appearance of extended ramified cross-bridges in isolated, reassembled (Fig. 10.5a), or native NFs in the axon (Figs. 10.1 and 10.2), where core filaments are separated a little wider from each other. These images are, although fewer in appearance in deep etching, similar somewhat to those revealed by rotary shadowing. This suggests that these ramified cross-bridges are flexible in structure and alterable by the change in distance between core filaments. This idea is also related to the opinion that ramified-distal portions are important for binding of the projections to form cross-bridges. That is, it might be concluded that single cross-bridges are constructed by plural projections, but importantly, we still do not understand how ramified portions of projections are incorporated into a part of cross-bridges.

Curiously, when core filaments together with projections are directly attached to the mica, different from the case of low-angle rotary shadowing, images corresponding with cross-bridges disappear, although their very proximal parts, which are similar to images of projections, protrude from the surface of core filaments (Gotow et al., 1992). This may be due to the extremely hydrophilic nature of the mica, to which NF-M and NF-H tail domains with higher ionic charges (KEP (Lys-Glu-Pro) and phosphorylated KSP domains) (Gou et al., 1998) may strongly bind. Such possible strong electrostatic binding may make the conformational change of NF-M and NF-H tail domains impossible to distinguish from the mica surface in the absence of glycerol. This physicochemical property of both NF-M and NF-H tails is important for the tails to interact with each other and may provide a clue to the mechanism by which longer projections are converted to shorter cross-bridges (Gou et al., 1998). Such manner of electrostatic binding could be carried out when two opposed projections are involved in the formation of single cross-bridges, brought about by conformational changes of NF-M and NF-H tails, between the core filaments.

When core filaments and projections are slightly separated from the mica surface, however, detailed analysis with quick-freeze deep etching using mica flakes shows complicated networks of projections similar to those revealed by low-angle rotary shadowing. With this deep-etch analysis, a meshwork pattern of projections and their relationship to core filaments is more clearly visible, probably due to higher shadowing angle (25 degrees) than that for rotary shadowing (6 degrees) and also to the absence of glycerol decoration (Figs. 10.4b, c and 10.5b) (Gotow et al., 1992). If core filaments are located much closer to each other within 40–50 nm, ramified projections may be straight and smooth in profile (i.e., typical or normal cross-bridges must be established between core filaments) (Fig. 10.5b). Ramified projections extending from respective core filaments closely located to each other may interact through distal portions of tail domains of NF-M and NF-H, or these projections extending from single core filaments may interact directly with portions of core filaments themselves located adjacently (Nakagawa et al., 1995; Chen et al., 2000), resulting in formation of the cross-bridge structure. Such images of projections disclosed by the deep etching and rotary shadowing favor the concept that opposed or plural long projections from respective core filaments interact to form a meshwork pattern resulting in the formation of smooth cross-bridges between core filaments. Length of the projections shown by rotary shadowing varies, 60–90 nm, according to the composition of NF subunit proteins in isolated or reconstructed NFs (Fig. 10.4) (Hisanaga and Hirokawa et al., 1988; Gotow et al., 1992), but that of cross-bridges revealed by deep etching varies, 20–50 nm, in native NFs in axons and also in the filaments formed by in vitro reassembly (Figs. 10.1, 10.2, and 10.5) (Hirokawa, 1982; Gotow et al., 1992, 1994, 1999) or by transfection of recombinant baculovirus vectors encoding NF-L and NF-M or NF-H into the Sf9 cell devoid of endogenous intermediate filament (Nakagawa et al., 1995; Chen et al., 2000).

10.5 Is the NF-M or the NF-H Tail Domain Essential for Formation of Cross-Bridges?

10.5.1 How Are Projections Structurally Converted into Cross-Bridges?

With regard to the idea of the formation of a single cross-bridge by two different proteins, a cross-bridge may be formed by combination of two opposed projections, which are formed by NF-M and NF-H tails, respectively, derived from adjacent core filaments with local charge interactions brought by KSP and KEP or KE (Lys-Glu) domains (Shaw, 1991; Gou et al., 1998). If this is the case, according to the difference in length between the projection and cross-bridge, it is unaccountable why a single cross-bridge formed by antiparallel-aligned projections becomes much shorter (20–50 nm) than completely overlapped domains of the projections (60–90 nm). Although we cannot explain such difference in length between projections (visible only when NFs are isolated or reassembled in vitro) and cross-bridges (detectable most clearly by the deep-etching method with either in vitro or in vivo conditions), overlong projections would be necessary for interactions between opposed projections formed from NF-M and/or NF-H tail domains. In in vitro reassembly and transfection experiments, the cross-bridges formed from NF-L plus NF-M are slightly (Nakagawa et al., 1995; Chen et al., 2000) or significantly (Gotow et al., 1992) shorter than those formed from NF-L plus NF-H, but both bridges are much shorter than single projections formed by the respective combination of NF triplet proteins. Because the projections formed from NF-L plus NF-M are slightly shorter than those from NF-L plus NF-H (Hisanaga and Hirokawa, 1988), the length of projections is somewhat proportionally related to that of cross-bridges. Indeed, the genetic deletion of some parts of NF-H or NF-M carboxyl tail domains shows clearly the reduction in length of cross-bridges (Nakagawa et al., 1995; Chen et al., 2000). Curiously, in the cell transfection experiment, the essential domain to form cross-bridges appears not to be identical in NF-M and NF-H tails; the proximal to middle portion of NF-M (Nakagawa et al., 1995) and the distal portion of NF-H (Chen et al., 2000) might be essential for the cross-bridge formation. In the filaments constructed with NF-L plus various deletion mutants of the NF-H tail, cross-bridges are still formed in the presence of only a distal portion corresponding with the KEP domain of NF-H (Chen et al., 2000). However, in the filaments constructed of NF-L plus deleted portions of the

NF-M tail, cross-bridges can be formed in the presence of only a short proximal portion of the KSP domain or the remainder without the former (Nakagawa et al., 1995). Thus, some restricted portions of projections may be enough to be the structural scaffolding for cross-bridges, but this does not explain why long meshwork projections are converted to short, straight cross-bridges. It is again speculated that the unusually hydrophilic property of projection domains is related to this molecular conformation between reciprocal change of projections and cross-bridges.

Nevertheless, we hypothesize that the tail domain of projections associates with the rod domain of core filaments located adjacently, although it seems impossible that hydrophilic distal projection domains can intrude into a hydrophobic α-helical central rod domain (Shaw, 1991) of core filaments. Even in this case, the discrepancy in the difference in length between projections and cross-bridges still appears to be unresolved, because we cannot explain how 60- to 90-nm-length projections will be a half-length of cross-bridges and also even if possible how the distal half portion of projections interacts with the coiled-coil rod domain of core filaments, associating on the surface or intruding into the inside of the latter. However, this seems to be supported by the involvement of NF-M and NF-H tails in the elongation of core filaments when combined with NF-L (Nakagawa et al., 1995; Chen et al., 2000). The possibility that distal portions of projections are directly associated with other core filaments will be discussed later in this chapter (see Section 10.5).

10.5.2 NF-M Appears To Be More Essential for Forming the Cross-Bridge than NF-H

Surprisingly, however, in vivo evidence demonstrates that complete deletion of the tail domain of NF-H in the mutant mouse, in which a gene expressing tail-less NF-H is inserted resulting in deletion of endogenous NF-H (Rao et al., 2002), does not alter the NF organization, especially with appearance of cross-bridges (Fig. 10.6). In NF-H tail–deleted mice, cross-bridges are formed from the NF-M tail domain alone. Cross-bridges constructed of the NF-M tail in the absence of NF-H tail are similar in both length and width to those in wild-type mice. Closer analysis indicates that cross-bridges in NF-H tail-less mice are not as frequent as those in wild-type animals, and core filaments are closely apposed with each other in many places along the filaments (Fig. 10.6b). Nevertheless, inter-core-filament spacing observed in NF-H tail-less mice is similar in interval to that in wild-type mice, making core filaments align parallel with each other. That is, NF-M alone has an ability to form cross-bridges by its tail in in vivo condition, suggesting that the extensively phosphorylated tail domain of NF-H is not essential for forming normal bridges. In this condition, NF-M enhances significantly the phosphorylation degree in the tail domain without change in expression ratio of the NF triplet proteins (Rao et al., 2002). Such exceptionally highly phosphorylated condition of NF-M tail domain may contribute to construction of normal cross-bridges in the axon.

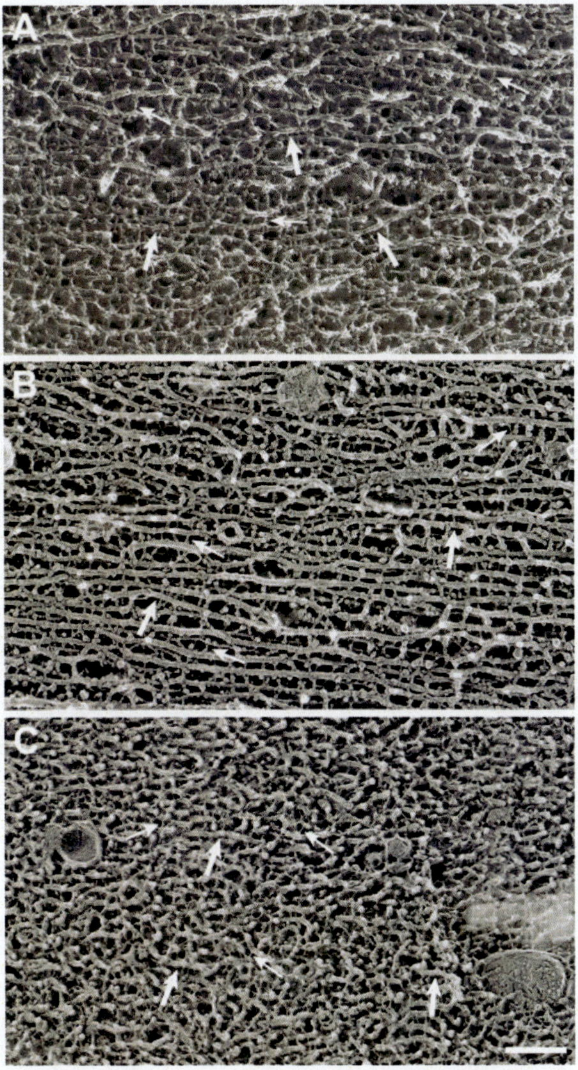

Fig. 10.6 Deep-etch images of NFs in myelinated axons of the sciatic nerve from wild-type (**a**), NF-H tail-less (**b**), and NF-M tail-less (**c**) mice that were quick frozen after incubation with artificial CSF. The following micrographs (Figs. 10.7, 10.8, and 10.9) were obtained in the same technical way. Compared with the NFs in the wild-type mouse, cross-bridges are less conspicuous in both NF-H and NF-M tail-less mice and a little more frequent in NF-M tail-less than in NF-H tail-less mice. Interspace between core filaments appears to be similar in NF-H tail-less but narrower in NF-M tail-less mice compared with that in wild-type mouse, indicating that cross-bridges formed by the NF-M tail are longer than those formed by the NF-H tail. This suggests that the NF-M tail plays a more essential role in formation of cross-bridges than the NF-H tail. Scale bar, 0.2 μm. These micrographs are related to Garcia et al. (2003) and Rao et al. (2003)

When NF-M tail domain alone is deleted in a similar gene-replacement method as in the case for the NF-H tail-less, however, NF-M-tail-less mutant mice display significantly altered NF organization in axons (Fig. 10.6c) (Garcia et al., 2003; Rao et al., 2003). Cross-bridges are still visible, somewhat more frequent, but significantly shorter here than in the case of NF-H tail-less, resulting in reduced spacing between core filaments in the former. That is, cross-bridges formed from the NF-H tail alone are shorter than those from the NF-M tail alone. This is unexpected, because, as mentioned earlier, the projections of NF-H are longer than those of NF-M (Hisanaga and Hirokawa, 1988), and cross-bridges formed by the NF-H tail are also longer than those formed by the NF-M tail in reassembly (Gotow et al., 1992) or transfection (Nakagawa et al., 1995; Chen et al., 2000) in vitro analyses. The in vivo situation for cross-bridge length is quite the reverse of the results from in vitro experiments.

The length of the cross-bridge might not be exactly related to that of the projection in the in vivo condition. The projection domain essential to form cross-bridges appears to be only a limited part that is not identical in both NF-H tail and NF-M tail by the transfection experiment using deletion mutants of each tail domain, that is, as pointed out earlier, portions essential for formation of cross-bridges are proximal for NF-M tail (Nakagawa et al., 1995) and distal for NF-H tail (Chen et al., 2000). Distal portions of projections are frequently ramified, and this unique complicated structure might be important to form cross-bridges through interaction with the same frame of opposed projections extending from adjacent core filaments. This story corresponds well with the case for NF-H tail but not for the NF-M tail, whose distal portion would not be essential for cross-bridge formation. Instead, in the case of NF-M, its proximal or middle portion might be important for cross-bridge formation. In filamentous profiles reconstructed in vitro by NF-M or NF-M plus NF-L, ramified distal structures formed from NF-M tail are not conspicuous compared with those formed from NF-H or NF-H plus NF-L (Hisanaga and Hirokawa, 1988; Gotow et al., 1992). The difference in such structural organization and also in amino acid sequences between NF-M and NF-H tails may be related to the difference in essential portion for the cross-bridge formation in both proteins. This may reflect the difference in ability to form cross-bridges between NF-M and NF-H tails in the axon.

It may be difficult for NF-H tail alone to form longer cross-bridges in the absence of whole tail of NF-M in vivo. This may be associated with the reduction of axon caliber by the compaction of core filaments in axoplasm without change in number of NFs (Garcia et al., 2003; Rao et al., 2003). This also suggests that NF-M tail is more essential to form normal cross-bridges than NF-H tail, related to the difference in caliber of axons, which corresponds well with results from mice deficient in whole NF-M or NF-H (Fig. 10.7) (Rao et al., 1998, 2002; Elder et al., 1998, 1999), although the detailed organization of cross-bridges in NF-M–deficient mice is not analyzed by the deep-etching method. NF-M, more specifically NF-M tail, instead of NF-H or NF-H tail, is therefore essential for cross-bridge construction. Compared with cross-bridges formed by NF-H, NF-M cross-bridges are closer in structure to normal bridges constructed with both NF-M and NF-H tails and also contribute more to increase of axon calibers to the normal level, especially in myelinated axons. In the neuron, cross-bridges are constructed with both NF-M and NF-H and are more developed than those formed by each protein individually. However, results from genetic manipulation experiments (Rao et al., 1998; Elder et al., 1998, 1999) suggest that NF-M is indispensable for formation of cross-bridges and NF-H is not always necessary for them, although the phosphorylation level of NF-M or NF-H tail domain is significantly increased in full-length NF-H–deficient or NF-M–deficient mice (Rao et al., 1998; Elder et al., 1998, 1999), as in NF-H or NF-M tail-less mice, respectively (Rao et al., 2002, 2003). The phosphorylated KSP motif

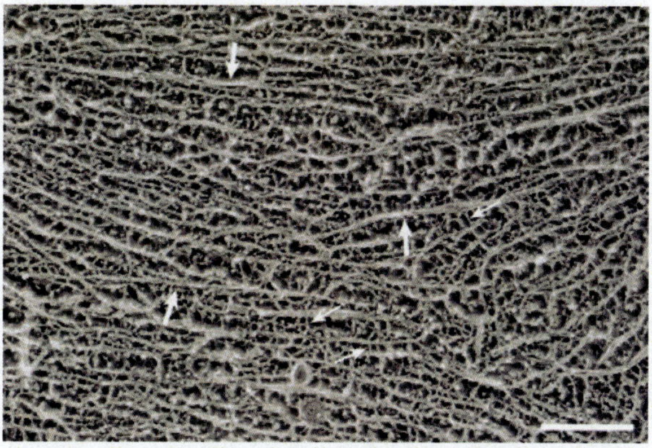

Fig. 10.7 Myelinated axon in the sciatic nerve from the NF-H–deficient mouse. Cross-bridges here are similar in organization to those of wild-type mouse, resulting in similar organization of core filaments in both mice. The cross-bridges are formed only by NF-M that is more highly phosphorylated than in the wild-type condition (Rao et al., 1998; Zhu et al., 1998), similar to the case of NF-H tail-less (Rao et al., 2002), but the total amount of NF-M itself is more than in the latter case (Rao et al., 1998, 2002), suggesting that NF-M compensates greatly for NF-H in biochemical function when whole NF-H is deleted instead of deletion of its tail alone. This suggests a difference in NF organization between the NF-H deficiency and NF-H tail-less mice. Scale bar, 0.5 μm. This micrograph is related to Rao et al. (2002)

and its related sequence including the serine residue in NF-M tail may be required for normal cross-bridge formation and also for maintaining normal radial axonal growth (Garcia et al., 2003; Rao et al., 2003).

10.5.3 Which Is Really Essential for the Cross-Bridge Formation, NF-M Tail or NF-H Tail?

The constitutive phosphorylation of NF-M is considered to be essential for cross-bridge formation; however, this hypothesis is distinctly excluded as follows. When all serines of KSP repeats in NF-M tail domain are genetically replaced with phosphorylation-incompetent alanines, mice provided with such phospho-incompetent NF-M tail do not show any significant alteration in cross-bridge organization and axonal caliber (Fig. 10.8) (Garcia et al., 2009). This means that phosphorylation of NF-M tail is not essential for cross-bridge formation. However, because these mutant mice devoid of serines in the KSP repeats of NF-M tail express intact NF-H provided with its tail, this NF-H tail, whose phosphorylation degree is slightly increased (Garcia et al., 2009), may contribute to formation of normal cross-bridges together with the phospho-incompetent NF-M tail, although significant increase in phosphorylation level of NF-H cannot form normal cross-bridges, different from the case of mice with phospho-incompetent NF-M, when the NF-M tail is completely deleted (Garcia et al., 2003; Rao et al., 2003). The tail domain itself of NF-M may be necessary to make up normal cross-bridges, either in a phosphorylation-competent or -incompetent manner, and NF-H may support the property of NF-M to form cross-bridges by the augmentation of its phosphorylation level. It will be of interest to observe how the cross-bridges are altered when NF-H tail domain is deleted from the mice expressing phospho-incompetent NF-M.

The NF-H tail is significantly longer than the NF-M tail, but the former forms shorter cross-bridges than the latter in the axon when the latter is absent, resulting in the reduction of axonal caliber. The NF-H tail alone, however, still forms cross-bridges, and, more importantly, allows the core filaments to align regularly, suggesting the axonal transport of molecules or cellular organelles including NFs are unaltered even in NF-M tail-less mice (Rao et al., 2003). That is, both NF-M and NF-H may compensate with each other, and when each of them is deleted, the other subunit would maintain, in compensatory manner, the axonal structure and function, although NF-M appears to be more basic in property. However, the nature of NF-H is still mysterious, because its tail is provided with much more numerous phospho-competent KSP repeats than that of the NF-M tail and is phosphorylated later in the developing stage, nearly corresponding with the termination of myelination (Willard and Simon, 1983; Cleveland, 1996; Sánchez et al., 2000; Garcia et al., 2003). Such NF-H with unusually heavily phosphorylated tail must be important for the ordered organization of core filaments, extensively related to maintaining specific structure and function for the axon, especially in long projectional axons, such as those in the pyramidal tract and sciatic nerve of large mammals including the human, which could not be clarified significantly in smaller mammals, such as rodents.

10.6 Importance of Cross-Bridges in Axon of the Neuron

When NF-M and NF-H are both genetically deleted from mice, NFs or intermediate filaments are not constructed at all in neurons, although NF-L protein is still expressed (Jacomy et al., 1999). The absence of NFs in the neuron is also produced when NF-L gene alone is deleted (Zhu et al., 1997). That is, although NF-L is more important for the NF backbone, the construction of NFs requires NF-L with either NF-M or NF-H in vivo (Lee et al., 1993). It is well known that NF-L is the most essential protein to form core filaments in vitro (Hisanaga and Hirokawa, 1988; Gotow et al., 1992), but it demands either of the other two NF subunit proteins to form core filaments in vivo (Jacomy et al., 1999), suggesting that formation of NFs in neurons is always associated with the presence of cross-bridges formed by either NF-M or NF-H tail. That is, in the neuron, NFs or core filaments cannot be formed without cross-bridges or projections, meaning that any intermediate filaments made from NF proteins in vivo should be provided with cross-bridges. Cross-bridges must therefore be indispensable for NF architecture and function, and, as mentioned above, also more important for large-caliber, long myelinated axons.

To address the significance of cross-bridges in the axon, we made the condition of absence of cross-bridges in vivo from both NF-M and NF-H tail-less mice

(Garcia et al., 2003). These double tail-less mice, in which both NF-M and NF-H tails are absent, as expected, reveal core filaments alone without cross-bridges in axons, whose calibers are somewhat more reduced than those in NF-M tail-less mice (Garcia et al., 2003). Surprisingly, these mutant mice appear to live normally without any behavioral problem (Garcia et al., 2003), like genetically engineered mice in which NFs are completely deleted (Zhu et al., 1997; Jacomy et al., 1999). The importance of NFs and also cross-bridges might be related to much longer axons in mammals larger in size and also with longer life spans than rodents. When cross-bridges are absent, core filaments, which are the same in diameter to normal NF core filaments, arrange irregularly and are frequently attached to each other (Fig. 10.9a), like other kinds of intermediate filaments, such as glial filaments shown here (Fig. 10.3). These core filaments are completely smooth in contour, which also proves distinctly that the NF-L tail does not contribute to the formation of either projection or cross-bridges even between closely apposed filaments. Membrane-bound organelles, such as mitochondria and vacuoles/vesicles, are frequently visible among core filaments, even though microtubules appear to be increased in distribution density here (Garcia et al., 2003). This suggests that the axonal transport capacity, which may be related to not only microtubules but also NF cross-bridges, is significantly impaired, due possibly to the lack of regular inter-core-filament space that should be formed by cross-bridges. More importantly, abnormal axoplasmic organization in the absence of both NF-M and NF-H tails supports strongly the idea that a well-phosphorylated condition of these tails, contributing intimately to the formation of cross-bridges, is necessary for orderly oriented NF bundling related to normally organized axoplasm (Gotow et al., 1992, 1999; Gou et al., 1998; Gotow, 2000; Shea and Chan, 2008).

In these mutant mice, cross-bridges between microtubules or microtubules and NFs are still detectable (Fig. 10.9b). These cross-bridges must be formed by microtubule-associated proteins (MAPs) including kinesin superfamily proteins essential for the axonal transport of many kinds of organelles (cf. Hirokawa and Noda, 2008; Shea and Chan, 2008). Nevertheless, as indicated earlier, many organelles are retained in the axoplasm in double tail-less mice (Fig. 10.8), suggesting again that both NF-M and NF-H tails contribute greatly to the axonal transport for membrane-bound organelles, although the possibility that the ordered organization of NFs modulated by NF-M and NF-H tails is required for functions of MAP-related motor proteins to transport organelles and molecules is also conceivable.

Cross-bridges appearing between NF core filaments and microtubules in double tail-less axons must be formed by single projections of MAPs, such as tau (Hirokawa et al., 1988; Chen et al., 1992). This means that single side ends of MAP projections extending from microtubules combine with or intrude into the surface or inside of NF core filaments, respectively. Although projections of MAPs are proportionally related in length to their cross-bridges between microtubules (Chen et al., 1992), in a similar manner, NF-M and NF-H tails also form cross-bridges by their single projections extending from each of two core filaments, as pointed out earlier. It therefore seems that the interaction of two opposed projections is not

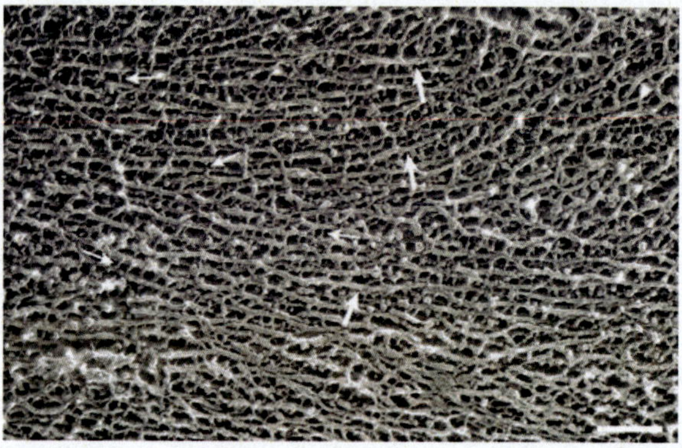

Fig. 10.8 Myelinated axon in the sciatic nerve from the mouse in which serines of KSP repeats in the NF-M tail are genetically replaced by alanines, meaning that this animal has a phospho-incompetent NF-M tail but normal NF-H tail. NFs here are almost similar to those of the wild-type mouse: cross-bridges are well developed and normal in appearance (long, smooth, straight, and frequent), and core filaments are aligned regularly. This suggests that phosphorylation of the NF-M tail domain is not required to form a normal cross-bridge. Scale bar, 0.2 μm. This micrograph is related to Garcia et al. (2009)

necessary for formation of cross-bridges between two adjacently located core filaments. If this is the case, half the number of projections remains as projections with free endings instead of forming cross-bridges, or the frequency of cross-bridges along core filaments is twice as much as that of projections. In the former case, we do not see such images: Projections located between core filaments always appear as cross-bridges connecting the latter in vitro–assembled filaments (Gotow et al., 1992, 1994) and in vivo axonal NFs (Gotow and Tanaka, 1994; Gotow et al., 1995, 1999; Rao et al., 2002, 2003; Garcia et al., 2003, 2009) with the presence of NF-M and/or NF-H. In the latter case, we see only two-dimensional images of core filaments with projections, in vitro, in the rotary-shading method, where all projections protruding from core filaments are visible, but we see three-dimensional organization, in vivo, in the deep-etching method, where cross-bridges are broken, when freeze fractured before coating of platinum, if they are located closer to the fractured face, or invisible if they are located behind core filaments. This technical problem might result in reducing the frequency of cross-bridges to approximately half that of projections. That is, four projections are considered to protrude from the same sites of core filaments when NFs are isolated and viewed with rotary shadowing (Hisanaga and Hirokawa, 1988, 1990), but eight cross-bridges could be formed around the same sites of core filaments that are close to each other.

NFs themselves are more resistant to mechanical stress than other intermediate filaments due to the presence of cross-bridges, which might be regulated by phosphorylation degree (Eyer and Leterrier, 1988; Lewis and Nixon, 1988; Nixon and Sihag, 1991; de Waegh et al., 1992; Gotow and Tanaka, 1994; Leterrier et al., 1996; Julien, 1997; Sánchez et al., 2000; Leterrier, 2001). This mechanical rigidity of NFs must contribute greatly to the physical scaffolding of long projection

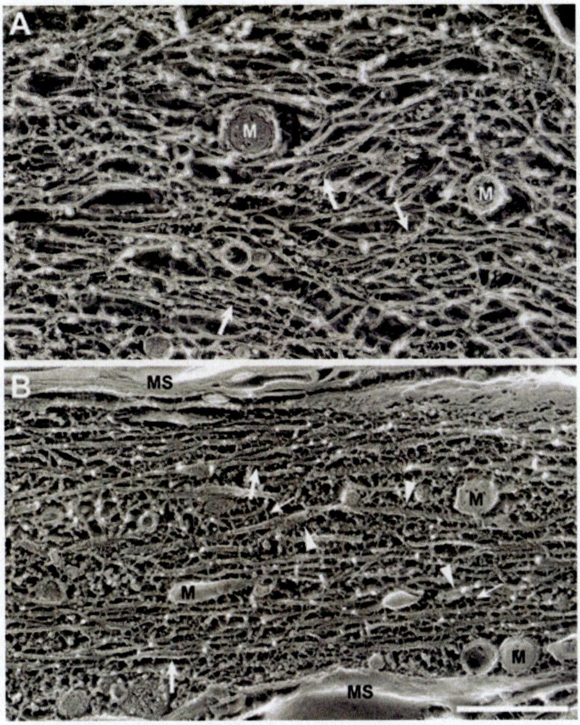

Fig. 10.9 Deep-etch images of NFs in myelinated axons from sciatic nerves in both NF-M and NF-H tail-less mice. **a** In the double tail-less mouse, NFs are devoid of cross-bridges, and core filaments are irregular in alignment and attached frequently to each other (*arrows*). Such image is very similar to that of glial filaments in the astrocyte (see Fig. 10.3). Membrane-bound organelles are frequently seen between NFs, suggesting that their transport is impaired. **b** In the region where the axon is narrower close to the node of Ranvier, NF core filaments are relatively parallel in alignment but almost attached to each other (*arrows*) without cross-bridges However, cross-bridges are still visible between NF core filaments and microtubules (*arrowheads*), indicating clearly that they are formed by MAPs, such as tau and motor proteins. Membrane-bound organelles (M), round or tubular in profile, are also frequently visible here. MS, myelin sheath. Scale bar, 0.5 μm. These micrographs are related to Garcia et al. (2003)

axons. The space between core filaments, which is formed by cross-bridges and thus the NF-specific feature, could be important for also providing the strong resistance of extremely elongated delicate cytoplasmic structure, from the mechanical viewpoint, and essential for both anterograde and retrograde transport of various cell organelles and molecules along the long distance between cell bodies and axon terminals. This may be why the long projection axons of motor and sensory neurons connected to the peripheral nervous system contain more numerous NFs than that of shorter axons in the central nervous system neurons, such as those of cerebellar Purkinje cells and hippocampal CA1 pyramidal cells (Shiozaki et al., 2008; Gotow, 2008). The development of cross-bridges may also be related to augmentation of axon caliber and conduction velocity of impulse because they are not frequent in smaller-caliber, unmyelinated, and autonomic neuronal axons due to less frequent NF density (cf. de Waegh et al., 1992; Cleveland, 1996; our unpublished data). NF-M and also NF-H (i.e., specifically their tail domains), therefore, must be essential for the axonal property in large, elongated, myelinated axons.

10.7 Conclusion

The significance of NFs is still controversial, but they are important especially in longer and thicker myelinated axons. Because NFs are most abundant in projection axons, they play an essential role in a specialized property or a function carried out in these axons. Specifically, the cross-bridge, which is a neuron-specific structure of intermediate filaments, must contribute significantly to such a role by modulating the inter-core-filament space in extremely elongated axons. Although I do not determine whether NF-M tail or NF-H tail is more important for the cross-bridge function, the phosphorylated condition of these tail domains must be related to establishment as well as function of cross-bridges. Especially, the unusually highly phosphorylated condition of NF-H is very attractive for the significance of the cross-bridge in the axon, although this idea is still a puzzle at present. Projections constructed from NF-M and NF-H tails revealed in isolated NFs in vitro are shifted into cross-bridges interconnecting core filaments in the axon, but it is unresolved how the projections, which form a meshwork pattern when interacting with opposed projections extending from different core filaments, convert into cross-bridges. Although NF-H tails form longer cross-bridges than those of NF-M tails in vitro, this seems to be the inverse in the in vivo condition. In any case, because the axonal caliber is reduced and ordered organization of core filaments is destroyed with impairment of axonal transport of membrane-bound organelles in the absence of cross-bridges in NF-M and NF-H double tail-less neurons, NFs with developed cross-bridges may be essential for the regular alignment of core filaments and enlargement of axon caliber, which are related to mechanical rigidity of the delicate axoplasm and axonal transport of various organelles, especially in longer axons and also in large-sized animals including humans, who have much more elongated axons than those of the rodents used for the experiments.

Acknowledgments The data from genetically modified mice described in this chapter were obtained from international collaboration with Drs. Mala V. Rao, Michael L. Garcia, and Don W. Cleveland. I thank Motoko Shiozaki and Naoya Hayakawa for photographic assistance.

References

Chen J, Kanai Y, Cowan NJ, Hirokawa N (1992) Projection domains of MAP2 and tau determine spacings between microtubules in dendrites and axons. Nature 360:674–677

Chen J, Nakata T, Zhang Z, Hirokawa N (2000) The C-terminal tail domain of neurofilament protein-H (NF-H) forms the crossbridges and regulates neurofilament bundle formation. J Cell Sci 113:3861–3869

Cleveland DW (1996) Neuronal growth and death: order and disorder in the axoplasm. Cell 84:663–666 de Waegh SM, Lee VM, Brady ST (1992) Local modulation of neurofilament phosphorylation, axonal caliber, and slow axonal transport by myelinating Schwann cells. Cell 68:451–463

Elder GA, Friedrich VL Jr, Bosco P, Kang C, Gourov A, Tu PH, Lee VM, Lazzarini RA (1998) Absence of the mid-sized neurofilament subunit decreases axonal calibers, levels of light neurofilament (NF-L), and neurofilament content. J Cell Biol 141:727–739

Elder GA, Friedrich VL Jr, Margita A, Lazzarini RA (1999) Age-related atrophy of motor axons in mice deficient in the mid-sized neurofilament subunit. J Cell Biol 146:181–192

Eyer J, Leterrier JF (1988) Influence of the phosphorylation state of neurofilament proteins on the interactions between purified filaments in vivo. Biochem J 252:655–660

Garcia ML, Lobsiger CS, Shah SB, Deerinck TJ, Crum J, Young D, Ward CM, Crawford TO, Gotow T, Uchiyama Y, Ellisman MH, Calcutt NA, Cleveland DW (2003) NF-M is an essential target for the myelin-directed "outside-in" signaling cascade that mediates radial axonal growth. J Cell Biol 163:1011–1020

Garcia ML, Rao MV, Fujimoto J, Garcia VB, Shah SB, Crum J, Gotow T, Uchiyama Y, Ellisman M, Calcutt NA, Cleveland DW (2009) Phosphorylation of highly conserved NF-M KSP repeats is not required for myelin-dependent radial axonal growth. J Neurosci 29: 1277–1284

Gotow T (1995) Dynamic structure and function of neurofilaments. Ann Psychiatr 5:91–111

Gotow T (2000) Neurofilaments in health and disease. Med Electron Microsc 33:173–199

Gotow T (2008) Neurons in the klotho mutant mouse show biochemical and morphological characteristics resembling age-related disorders. Tzu Chi Med J 20:155–160

Gotow T, Hashimoto PH (1988) Deep-etch structure of astrocytes at the superficial glia limitans, with special emphasis on the internal and external organization of their plasma membranes. J Neurocytol 17:399–413

Gotow T, Leterrier JF, Ohsawa Y, Watanabe T, Isahara K, Shibata R, Ikenaka K, Uchiyama Y (1999) Abnormal expression of neurofilament proteins in dysmyelinating axons located in the central nervous system of jimpy mutant mice. Eur J Neurosci 11:3893–3903

Gotow T, Takeda M, Tanaka T, Hashimoto PH (1992) Macromolecular structure of reassembled neurofilaments as revealed by the quick-freeze deep-etch mica method: difference between NF-M and NF-H subunits in their ability to form cross-bridges. Eur J Cell Biol 58:331–345

Gotow T, Tanaka J (1994) Phosphorylation of neurofilament H subunit as related to arrangement of neurofilaments. J Neurosci Res 37:691–713

Gotow T, Tanaka T, Nakamura Y, Takeda M (1994) Dephosphorylation of the largest neurofilament subunit protein influences the structure of crossbridges in reassembled neurofilaments. J Cell Sci 107:1949–1957

Gotow T, Tanaka J, Takeda M (1995) The organization of neurofilaments accumulated in perikaryon following aluminum administration: relationship between structure and phosphorylation of neurofilaments. Neuroscience 64:553–569

Gou JP, Gotow T, Janmey PA, Leterrier JF (1998) Regulation of neurofilament interactions in vitro by natural and synthetic polypeptides sharing Lys-Ser-Pro sequences with the heavy neurofilament subunit NF-H: neurofilament crossbridging by antiparallel sidearm overlapping. Med Biol Eng Comput 36:371–387

Hirokawa N (1982) Cross-linker system between neurofilaments, microtubules, and membranous organelles in frog axons revealed by the quick-freeze, deep-etching method. J Cell Biol 94: 129–142

Hirokawa N, Glicksman MA, Willard MB (1984) Organization of mammalian neurofilament polypeptides within the neuronal cytoskeleton. J Cell Biol 98:1523–1536

Hirokawa N, Noda Y (2008) Intercellular transport and kinesin superfamily proteins, KIFs: structure, function, and dynamics. Physiol Rev 88:1089–1118

Hirokawa N, Shiomura Y, Okabe S (1988) Tau proteins: the molecular structure and mode of binding on microtubules. J Cell Biol 107:1449–1459

Hisanaga S, Hirokawa N (1988) Structure of the peripheral domains of neurofilaments revealed by low angle rotary shadowing. J Mol Biol 202:297–305

Hisanaga S, Hirokawa N (1989) The effects of dephosphorylation on the structure of the projections of neurofilament. J Neurosci 9:959–966

Hisanaga S, Ikai A, Hirokawa N (1990) Molecular architecture of the neurofilament. I. Subunit arrangement of neurofilament L protein in the intermediate-sized filament. J Mol Biol 211: 857–869

Jacomy H, Zhu Q, Couillard-Després S, Beaulieu JM, Julien JP (1999) Disruption of type IV intermediate filament network in mice lacking the neurofilament medium and heavy subunits. J Neurochem 73:972–984

Julien JP (1997) Neurofilaments and motor neuron disease. Trends Cell Biol 7:243–249

Lee MK, Xu Z, Wong PC, Cleveland DW (1993) Neurofilaments are obligate heteropolymers in vivo. J Cell Biol 122:1337–1350

Leterrier JF (2001) Water and the cytoskeleton. Cell Mol Biol 47:901–923

Leterrier JF, Eyer J (1987) Properties of highly viscous gels formed by neurofilaments in vitro. A possible consequence of a specific inter-filament cross-bridging. Biochem J 245:93–101

Leterrier JF, Käs J, Hartwig J, Vegners R, Janmey PA (1996) Mechanical effects of neurofilament cross-bridges. J Biol Chem 271:15687–15694

Lewis SE, Nixon RA (1988) Multiple phosphorylated variants of the high molecular mass subunit of neurofilaments in axons of retinal cell neurons: characterization and evidence for their differential association with stationary and moving neurofilaments. J Cell Biol 107:2689–2701

Matus A (1988) Neurofilament protein phosphorylation – where, when and why. Trends Neurosci 11:291–292

Nakagawa T, Chen J, Zhang Z, Kanai Y, Hirokawa N (1995) Two distinct functions of the carboxyl- terminal tail domain of NF-M upon Neurofilament assembly: cross-bridge formation and longitudinal elongation of filaments. J Cell Biol 129:411–429

Nixon RA, Sihag RK (1991) Neurofilament phosphorylation: a new look at regulation and function. Trends Neurosci 14:501–506

Rao MV, Campbell J, Yuan A, Kumar A, Gotow T, Uchiyama Y, Nixon RA (2003) The neurofilament middle molecular mass subunit carboxyl-terminal tail domains is essential for the radial growth and cytoskeletal architecture of axons but not for regulating neurofilament transport rate. J Cell Biol 163:1021–1031

Rao MV, Garcia ML, Miyazaki Y, Gotow T, Yuan A, Mattina S, Ward CM, Calcutt NA, Uchiyama Y, Nixon RA, Cleveland DW (2002) Gene replacement in mice reveals that the heavily phosphorylated tail of neurofilament heavy subunit does not affect axonal caliber or the transit of cargoes in slow axonal transport. J Cell Biol 158:681–693

Rao MV, Houseweart MK, Williamson TL, Crawford TO, Folmer J, Cleveland DW (1998) Neurofilament-dependent radial growth of motor axons and axonal organization of neurofilaments does not require the neurofilament heavy subunit (NF-H) or its phosphorylation. J Cell Biol 143:171–181

Sánchez I, Hassinger L, Sihag RK, Cleveland DW, Mohan P, Nixon RA (2000) Local control of neurofilament accumulation during radial growth of myelinating axons in vivo: selective role of site-specific phosphorylation. J Cell Biol 151:1013–1024

Shaw G (1986) Neurofilaments: abundant but mysterious neuronal structures. Bioessays 4:161–166

Shaw G (1991) Neurofilament proteins. In: Burgoyne RD (ed) The neuronal cytoskeleton. Wiley-Liss, New York, NY, 185–214

Shea TB, Chan WK-H (2008) Regulation of neurofilament dynamics by phosphorylation. Eur J Neurosci 27:1893–1901

Shiozaki M, Yoshimura K, Shibata M, Koike M, Matsuura N, Uchiyama Y, Gotow T (2008) Morphological and biochemical signs of age-related neurodegenerative changes in klotho mutant mice. Neuroscience 152:924–941

Tsukita S, Usukura J, Tsukita S, Ishikawa H (1982) The cytoskeleton in myelinated axons: a freeze-etch replica study. Neuroscience 7:2135–2147

Willard M, Simon C (1983) Modulations of neurofilament axonal transport during the development of rabbit retinal ganglion cells. Cell 35:551–559

Zhu Q, Couillard-Després S, Julien JP (1997) Delayed maturation of regenerating myelinated axons in mice lacking neurofilaments. Exp Neurol 148:299–316

Zhu Q, Lindenbaum M, Levavasseur F, Jacomy H, Julien JP (1998) Disruption of the NF-H gene increases axonal microtubule content and velocity of neurofilament transport: relief of axonopathy resulting from the toxin β,β'-iminodipropionitrile. J Cell Biol 143:183–193

Chapter 11
Neurofilament Transport

Andrew J. Grierson and Christopher C.J. Miller

Abstract Neurofilaments are the most plentiful intermediate filament proteins found in neuronal cells. They are important cytoskeletal proteins that are made in the soma and transported into axonal and dendritic processes. Neurofilaments are transported by the process of slow axonal transport, which arises from fast transport interrupted by extended pauses. Some of the mechanisms regulating neurofilament assembly and transport have been identified. It is thought that phosphorylation decreases the rate of neurofilament transport by increasing the time neurofilaments spend pausing. There is also increasing evidence for the involvement of neurofilaments in a number of neurodegenerative disorders such as amyotrophic lateral sclerosis, hereditary spastic paraplegia, Charcot–Marie–Tooth disease, and Alzheimer's disease. Disrupted neurofilament transport is thought to be a common mechanism in some or all of these disorders.

Keywords Neurofilaments · Cytoskeleton · Phosphorylation · Axonal transport · Neurodegeneration

Contents

11.1 Introduction

The majority of neuronal proteins are synthesized by ribosomes located in or near the cell body. These proteins are then delivered to their appropriate subcellular

A.J. Grierson (✉)
Department of Neuroscience, University of Sheffield, Sheffield S10 2RX, UK
e-mail: a.j.grierson@sheffield.ac.uk

R.A. Nixon, A. Yuan (eds.), *Cytoskeleton of the Nervous System*,
Advances in Neurobiology 3, DOI 10.1007/978-1-4419-6787-9_11,

location, which presents a unique problem in neurons that have axons that may be more than 1 m long. Transport over such distances is mediated by kinesin and dynein molecular motor proteins, which travel along the microtubule cytoskeleton [reviewed in De Vos et al. (2008); Hirokawa and Takemura (2005)].

Cytoskeletal proteins are a major cargo of the transport machinery, as axons require microtubules, actin, and intermediate filaments to maintain their highly polarized structure and to function normally. Neurofilaments (NFs) are the most abundant intermediate filaments in neuronal cells, and alterations in their pathology are reported in several neurodegenerative diseases. In this chapter, we will present evidence that supports a primary role for NFs in the development of these diseases.

11.2 Neurofilaments

In neurons, the intermediate filaments are composed of NF light chain (NF-L), middle chain (NF-M), and heavy chain (NF-H), peripherin, and α-internexin [reviewed in Lee and Cleveland (1996); Miller et al. (2002)]. NFs and α-internexin are the major intermediate filaments in mature CNS axons, whereas peripherin is expressed mostly in autonomic nerves and sensory neurons (Escurat et al., 1990; Parysek and Goldman, 1988; Troy et al., 1990; Yuan et al., 2006).

Structurally, all intermediate filament proteins have three distinct domains, comprising a central α-helical rod domain, an amino-terminal head domain, and a carboxy-terminal tail domain. Macroscopically, intermediate filaments are 10-nm fibers that are formed by assembly of the rod domains into coiled-coil oligomers. Assembly of the neuronal intermediate filaments is thought to be regulated by the head domains, and the large tail domains of NF-M and NF-H are thought to project from the filaments (Ching and Liem, 1998; Gill et al., 1990; Nakagawa et al., 1995).

Until recently, the precise ways by which intermediate filaments increased their length was not clear. Using a cell fusion assay with differently labeled mCherry- or EGFP-tagged NF subunits and vimentin, Colakoglu and Brown demonstrated that end-to-end annealing is a major mechanism by which intermediate filaments elongate (Colakoglu and Brown, 2009). The same investigators also elucidated one way that intermediate filaments may exchange subunits along their length: a process termed *intercalary subunit exchange* (Colakoglu and Brown, 2009). Direct evidence for both processes was obtained using in vivo imaging techniques.

NFs are phosphoproteins and are phosphorylated on their head and tail domains. The tail domains of NF-M and NF-H have multiple copies of a lys-ser-pro motif, which is a target for phosphorylation. NF-H has about 50 copies of this motif, and most or all of these are phosphorylated by kinases including GSK3β, ERK1/2, JNKs, p38, and cdk5/p35 [reviewed by Sihag et al. (2007)]. The head domains are also phosphorylated, and in the case of NF-L, phosphorylation is proposed to inhibit filament formation (Giasson and Mushynski, 1998; Gonda et al., 1990; Hashimoto et al., 1998; Hisanaga et al., 1994; Nakamura et al., 1990; Yates et al., 2009). In vivo, serines 2, 55, 57, and 66 have been demonstrated to be phosphorylated (Giasson et al., 1996; Hashimoto et al., 2000; Nakamura et al., 2000; Sihag and Nixon, 1991; Trimpin et al., 2004). Protein kinase A (PKA) phosphorylates serines 2 and 55, Rho-associated kinase (ROK) phosphorylates serine 57, Ca2+/calmodulin dependent protein kinase (CAMKII) phosphorylates serines 57 and 66, and protein kinase N (PKN) phosphorylates an undetermined number of site(s) in the head domain (Cleverley et al., 1998; Giasson and Mushynski, 1998; Hashimoto et al., 1998, 2000; Manser et al., 2008; Mukai et al., 1996; Nakamura et al., 2000).

Axonal caliber is known to determine nerve conduction velocity. We know that NFs are particularly abundant in large-diameter axons such as motor neurons, where fast conduction velocities are essential. Studies in both quail and mice have shown that disruption of NFs leads to reduced axonal caliber (Eyer et al., 1998; Ohara et al., 1993; Sakaguchi et al., 1993). The involvement of NF tail domain phosphorylation in determining axon caliber is less clear. Initially, it was proposed that increasing the charge on NF-M and NF-H tail domains by phosphorylation would increase the repulsive forces between the individual filaments (Fuchs and Cleveland, 1998; Lee et al., 1987). However, more recent experiments where tail domain phosphorylation is precluded by deletion, or amino acid substitution of phosphorylated serine residues with alanine residues, suggest that phosphorylation may not play an essential role in determining axon caliber (Garcia et al., 2009; Rao et al., 2002b).

11.3 Axonal Transport of Neurofilaments

The rate of NF transport was first inferred from classical in vivo metabolic labeling studies. Radiolabeled [^{35}S]methionine is injected into mice to label the proteins of the optic or sciatic nerve. After a number of days, the nerve is dissected into ordered segments from the cell body to the nerve terminal, and cytoskeletal fractions are prepared from each segment. These fractions are separated by polyacrylamide gel electrophoresis and the amount of radiolabel in each cytoskeletal component determined by autoradiography. By analyzing how far the radiolabel has traveled during the experiment, one is able to calculate an inferred rate of NF transport, which is typically in the range 0.1–3 mm/day. This is up to two orders of magnitude slower than vesicle transport, which is usually in the region of 1 μm/s (Baas and Brown, 1997; Hirokawa, 1997; Nixon, 1998). This rate of cytoskeletal transport is termed *slow axonal transport*. For a time this led to debate over the precise mechanisms by which there could be a number of different transport rates within cells. For example, were there motors that moved at different speeds? Eventually, this problem was solved by using modern live-imaging methodologies.

To facilitate live imaging, several groups introduced enhanced green fluorescent protein (EGFP) tags onto the N and C termini of plasmid DNA–encoded NF proteins. These were then introduced into primary neurons, which have a limited endogenous NF network, enabling transport of motile EGFP-labeled NFs to be imaged using microscopy focused on the gaps between stationary EGFP-labeled NFs (Roy et al., 2000; Wang et al., 2000). These investigations revealed that slow axonal transport of NFs occurs by fast transport at rates of approximately 1 μm/s that is interrupted by prolonged pauses. Similarly, experiments in extruded squid axoplasm and in vitro demonstrated transport of NFs at fast rates (Prahlad et al., 2000; Shah et al., 2000). Therefore, the basic mechanisms underlying slow axonal transport are now understood.

The ways in which NF transport is regulated is the subject of intense investigation. There is strong evidence to support a role for phosphorylation of NF side-arm domains as an important regulatory mechanism. This was first proposed on the basis of the correlation between the degree of side-arm phosphorylation and overall transport rate (Archer et al., 1994; Nixon et al., 1994). It is possible that phosphorylation of NF side arms regulates attachment or detachment of molecular motors such as kinesin or dynein (Shah et al., 2000; Yabe et al., 2000). In addition, phosphorylation of NFs may lead to bundling and reduced motility (Leterrier et al., 1996; Yabe et al., 2001). Direct evidence for the importance of NF-H side-arm phosphorylation was demonstrated using site-directed mutagenesis. Known in vivo phosphorylation sites within the motif KSPXK were mutated to either alanine residues to preclude phosphorylation or aspartate residues to mimic permanent phosphorylation (Ackerley et al., 2003). These mutations did not disturb filament assembly but had dramatic effects on the transport rate of EGFP-tagged NF-H through axons of transfected cortical neurons. Wild-type NF-H moved at 80 μm/h, whereas the nonphosphorylatable Ala mutant moved at 108 μm/h, and the permanently phosphorylated Asp mutant moved at 52 μm/h (Ackerley et al., 2003). Analysis of individual motile NFs using time-lapse imaging revealed that the velocity and direction of transport of motile NFs was unchanged in Ala and Asp mutants. However, Ala mutant NFs paused significantly less often than Asp mutants (16% compared with 37% of the time spent pausing) (Ackerley et al., 2003). Therefore, NF phosphorylation regulates the time NFs spend pausing. Experiments with photo-activatable forms of GFP–NF-M suggest that there may be two distinct stationary states for NFs, distinguished by average pause durations of 30 s or 60 min (Trivedi et al., 2007). The mechanisms underlying the difference between these two pause states are not yet known. The stationary NF network in axons is thought to determine transport rate to some extent because conditional expression of NF-L in the absence of endogenous axonal NFs led to much faster rates of NF-L transport (10 mm/d) than were observed in the presence of endogenous NFs (1 mm/d) (Millecamps et al., 2007).

We know that NFs and other intermediate filaments bind to motor proteins. Thus, exchange between anterograde and retrograde directed motors may explain the bidirectional transport of NFs. At present, the precise molecular motors for NF transport are unknown, but evidence from a Kif5A conditional knockout mouse supports a role for Kif5A in anterograde transport of NFs (Xia et al., 2003). In these mice, fast axonal transport appeared to be unaffected, however NFs accumulated in the cell bodies of affected neurons, and axonal caliber was reduced, resulting in hindlimb paralysis (Xia et al., 2003). Similarly, inhibition of dynein increases anterograde transport of NFs and ultimately induces the formation of focal accumulations of NFs in neurites. Conversely, inhibition of kinesin slowed anterograde transport of NFs but did not lead to focal accumulation (Motil et al., 2007). The motor myosin

Va has been shown to interact with NFs (Rao et al., 2002a) and to increase their transport rate by a mechanism that involves decreasing the duration of long-term pauses (Alami et al., 2009).

11.4 Neurofilaments and Disease

Evidence for the involvement of NFs in neurodegeneration comes from several sources. First, mutations in NF-L underlie some types of Charcot–Marie–Tooth disease (CMT2A) (De Jonghe et al., 2001; Mersiyanova et al., 2000). Expression of some of these pathogenic forms of NF-L in cultured neurons leads to disruption of axonal transport of NFs and mitochondria (Brownlees et al., 2002; Perez-Olle et al., 2005).

Second, mutations in the highly phosphorylated tail-domain of NF-H are found in a small proportion of amyotrophic lateral sclerosis (ALS) patients, implicating NF-H as a risk factor for developing ALS (Al-Chalabi et al., 1999; Figlewicz et al., 1994; Garcia et al., 2005; Tomkins et al., 1998). Third, altering the expression of NFs, α-internexin, and peripherin in mouse models can lead to the development of neurologic phenotypes, some of which resemble ALS (Beaulieu et al., 1999; Ching et al., 1999; Cote et al., 1993; Lee et al., 1994; Vickers et al., 1994; Xu et al., 1993). Fourth, crossing mouse models of ALS expressing mutant copper/zinc superoxide dismutase (SOD1) transgenes with genetically modified mice with altered NF genes can modify the ALS phenotype (Couillard-Despres et al., 1998; Williamson et al., 1998). Most notably, deletion of the heavily phosphorylated tail-domain of either NF-M or NF-H leads to significant life extension in mice carrying a G37R mutant SOD1 transgene (Lobsiger et al., 2005). Finally, NF pathology is a hallmark of neurodegenerative diseases including ALS, Alzheimer's disease, diabetic neuropathy, Parkinson's disease, and some forms of dementia (Carpenter, 1968; Hirano et al., 1984; Itoh et al., 1992; Rouleau et al., 1996; Schmidt et al., 1987, 1996; Schmidt et al., 1997a, b; Sobue et al., 1990; Trojanowski and Lee, 1998). The proximal accumulation of NFs in these diseases suggests that their axonal transport may be defective. In support of this idea, axonal transport defects have been reported for a number of other cargoes in several of these disorders (De Vos et al., 2007; Lazarov et al., 2007; Pigino et al., 2009; Saha et al., 2004), and in the case of ALS, slowing of NF transport is known to precede the onset of clinical symptoms (Williamson and Cleveland, 1999).

NFs are phosphoproteins, however the tail domains are usually heavily phosphorylated in the axonal compartment and not in the cell body or perikaryon (Carden et al., 1985; Julien and Mushynski, 1982; Lee et al., 1987, 1988; Sternberger and Sternberger, 1983). NF accumulations in neurodegenerative disease may occur in the cell body or perikaryon and are reported to label strongly for phosphorylated NF epitopes (Hirano, 1991; Schmidt et al., 1996, 1997b; Trojanowski et al., 1993). Thus it is likely that altered NF phosphorylation is causally linked to defective NF transport, and this leads to accumulation of phosphorylated NFs in, or close to, the cell body. As discussed earlier, a number of NF kinases have been identified, several of which are activated by increased levels of the neurotransmitter glutamate. Excitotoxic disease mechanisms caused by increased levels of extracellular glutamate have been proposed in ALS. For example, there are changes in glutamate handling (Van Den Bosch et al., 2006); mutant SOD1 has been reported to damage the glutamate transport EAAT2 (Trotti et al., 1999); and EAAT2 levels are reduced in mutant SOD1 mice (Bruijn et al., 1997). Two papers describe how extracellular glutamate reduces the axonal transport of NFs (Ackerley et al., 2000; Hiruma et al., 2003) via a mechanism involving the activation of protein kinases that can phosphorylate NF side arms (Ackerley et al., 2000; Brownlees et al., 2000; Kawasaki et al., 1997; Lee et al., 2000; Schwarzschild et al., 1997). Recently, the prolyl isomerase Pin1 has been demonstrated to modify the effect of glutamate on NF side arms, implicating its potential as a therapeutic target in ALS and Alzheimer's disease (Kesavapany et al., 2007; Rudrabhatla et al., 2008).

In addition to alterations in NFs, it is possible that disruption of specific molecular motor proteins may be involved in the development of NF pathology in diseased states. Mutations in the kinesin motor Kif5A have been reported in hereditary spastic paraplegia (Reid et al., 2002). Because conditional disruption of Kif5A in mice is associated with the accumulation of NFs in the cell body, it is likely that NF transport is disturbed in patients with Kif5A mutations.

Future work will target the mechanisms, such as phosphorylation, that regulate NF transport and how these relate to the pathogenic process in neurodegenerative diseases such as ALS, hereditary spastic paraplegia, and Alzheimer's disease. This work may form the basis for development of rational therapeutic interventions targeting NFs in neurodegenerative disease.

References

Ackerley S, Grierson AJ, Brownlees J, Thornhill P, Anderton BH, Leigh PN, Shaw CE, Miller CC (2000) Glutamate slows axonal transport of neurofilaments in transfected neurons. J Cell Biol 150:165–176

Ackerley S, Thornhill P, Grierson AJ, Brownlees J, Anderton BH, Leigh PN, Shaw CE, Miller CC (2003) Neurofilament heavy chain side arm phosphorylation regulates axonal transport of neurofilaments. J Cell Biol 161:489–495

Al-Chalabi A, Andersen PM, Nilsson P, Chioza B, Andersson JL, Russ C, Shaw CE, Powell JF, Leigh PN (1999) Deletions of the heavy neurofilament subunit tail in amyotrophic lateral sclerosis. Hum Mol Genet 8:157–164

Alami NH, Jung P, Brown A (2009) Myosin Va increases the efficiency of neurofilament transport by decreasing the duration of long-term pauses. J Neurosci 29:6625–6634

Archer DR, Watson DF, Griffin JW (1994) Phosphorylation-dependent immunoreactivity of neurofilaments and the rate of slow axonal transport in the central and peripheral axons of the rat dorsal root ganglion. J Neurochem 62:1119–1125

Baas PW, Brown A (1997) Slow axonal transport: the polymer transport model. Trends Cell Biol 7:380–384

Beaulieu JM, Nguyen MD, Julien JP (1999) Late onset death of motor neurons in mice overexpressing wild-type peripherin. J Cell Biol 147:531–544

Brownlees J, Ackerley S, Grierson AJ, Jacobsen NJ, Shea K, Anderton BH, Leigh PN, Shaw CE, Miller CC (2002) Charcot-Marie-Tooth disease neurofilament mutations disrupt neurofilament assembly and axonal transport. Hum Mol Genet 11:2837–2844

Brownlees J, Yates A, Bajaj NP, Davis D, Anderton BH, Leigh PN, Shaw CE, Miller CC (2000) Phosphorylation of neurofilament heavy chain side-arms by stress activated protein kinase-1b/Jun N-terminal kinase-3. J Cell Sci 113:401–407

Bruijn LI, Becher MW, Lee MK, Anderson KL, Jenkins NA, Copeland NG, Sisodia SS, Rothstein JD, Borchelt DR, Price DL, Cleveland DW (1997) ALS-linked SOD1 mutant G85R mediates damage to astrocytes and promotes rapidly progressive disease with SOD1-containing inclusions. Neuron 18:327–338

Carden MJ, Schlaepfer WW, Lee VM (1985) The structure, biochemical properties, and immunogenicity of neurofilament peripheral regions are determined by phosphorylation state. J Biol Chem 260:9805–9817

Carpenter S (1968) Proximal axonal enlargement in motor neuron disease. Neurology 18:841–851

Ching GY, Chien CL, Flores R, Liem RK (1999) Overexpression of alpha-internexin causes abnormal neurofilamentous accumulations and motor coordination deficits in transgenic mice. J Neurosci 19:2974–2986

Ching GY, Liem RK (1998) Roles of head and tail domains in alpha-internexin's self-assembly and coassembly with the neurofilament triplet proteins. J Cell Sci 111(Pt 3): 321–333

Cleverley KE, Betts JC, Blackstock WP, Gallo JM, Anderton BH (1998) Identification of novel in vitro PKA phosphorylation sites on the low and middle molecular mass neurofilament subunits by mass spectrometry. Biochemistry 37:3917–3930

Colakoglu G, Brown A (2009) Intermediate filaments exchange subunits along their length and elongate by end-to-end annealing. J Cell Biol 185:769–777

Cote F, Collard JF, Julien JP (1993) Progressive neuronopathy in transgenic mice expressing the human neurofilament heavy gene: a mouse model of amyotrophic lateral sclerosis. Cell 73:35–46

Couillard-Despres S, Zhu Q, Wong PC, Price DL, Cleveland DW, Julien JP (1998) Protective effect of neurofilament heavy gene overexpression in motor neuron disease induced by mutant superoxide dismutase. Proc Natl Acad Sci USA 95:9626–9630

De Jonghe P, Mersivanova I, Nelis E, Del Favero J, Martin JJ, Van Broeckhoven C, Evgrafov O, Timmerman V (2001) Further evidence that neurofilament light chain gene mutations can cause Charcot-Marie-Tooth disease type 2E. Ann Neurol 49:245–249

De Vos KJ, Chapman AL, Tennant ME, Manser C, Tudor EL, Lau KF, Brownlees J, Ackerley S, Shaw PJ, McLoughlin DM et al (2007) Familial amyotrophic lateral sclerosis-linked SOD1 mutants perturb fast axonal transport to reduce axonal mitochondria content. Hum Mol Genet 16:2720–2728

De Vos KJ, Grierson AJ, Ackerley S, Miller CC (2008) Role of axonal transport in neurodegenerative diseases. Annu Rev Neurosci 31:151–173

Escurat M, Djabali K, Gumpel M, Gros F, Portier MM (1990) Differential expression of two neuronal intermediate-filament proteins, peripherin and the low-molecular-mass neurofilament protein (NF L), during the development of the rat. J Neurosci 10:764–784

Eyer J, Cleveland DW, Wong PC, Peterson AC (1998) Pathogenesis of two axonopathies does not require axonal neurofilaments. Nature 391:584–587

Figlewicz DA, Krizus A, Martinoli MG, Meininger V, Dib M, Rouleau GA, Julien JP (1994) Variants of the heavy neurofilament subunit are associated with the development of amyotrophic lateral sclerosis. Hum Mol Genet 3:1757–1761

Fuchs E, Cleveland DW (1998) A structural scaffolding of intermediate filaments in health and disease. Science 279:514–519

Garcia ML, Rao MV, Fujimoto J, Garcia VB, Shah SB, Crum J, Gotow T, Uchiyama Y, Ellisman M, Calcutt NA, Cleveland DW (2009) Phosphorylation of highly conserved neurofilament medium KSP repeats is not required for myelin-dependent radial axonal growth. J Neurosci 29: 1277–1284

Garcia ML, Singleton AB, Hernandez D, Ward CM, Evey C, Sapp PA, Hardy J, Brown RH, Cleveland DW (2005) Mutations in neurofilament genes are not a significant primary cause of non-SOD1-mediated amyotrophic lateral sclerosis. Neurobiol Dis 21:102–109

Giasson BI, Cromlish JA, Athlan ES, Mushynski WE (1996) Activation of cyclic AMP-dependent protein kinase in okadaic acid-treated neurons potentiates neurofilament fragmentation and stimulates phosphorylation of Ser2 in the low-molecular-mass neurofilament subunit. J Neurochem 66:1207–1213

Giasson BI, Mushynski WE (1998) Intermediate filament disassembly in cultured dorsal root ganglion neurons is associated with amino-terminal head domain phosphorylation of specific subunits. J Neurochem 70:1869–1875

Gill SR, Wong PC, Monteiro MJ, Cleveland DW (1990) Assembly properties of dominant and recessive mutations in the small mouse neurofilament (NF-L) subunit. J Cell Biol 111: 2005–2019

Gonda Y, Nishizawa K, Ando S, Kitamura S, Minoura Y, Nishi Y, Inagaki M (1990) Involvement of protein kinase C in the regulation of assembly-disassembly of neurofilaments in vitro. Biochem Biophys Res Commun 167:1316–1325

Hashimoto R, Nakamura Y, Goto H, Wada Y, Sakoda S, Kaibuchi K, Inagaki M, Takeda M (1998) Domain- and site-specific phosphorylation of bovine NF-L by Rho-associated kinase. Biochem Biophys Res Commun 245:407–411

Hashimoto R, Nakamura Y, Komai S, Kashiwagi Y, Tamura K, Goto T, Aimoto S, Kaibuchi K, Shiosaka S, Takeda M (2000) Site-specific phosphorylation of neurofilament-L is mediated by calcium/calmodulin-dependent protein kinase II in the apical dendrites during long-term potentiation. J Neurochem 75:373–382

Hirano A (1991) Cytopathology of amyotrophic lateral sclerosis. In: Rowland LP (ed) Amyotrophic lateral sclerosis and other motor neurone diseases. Raven, New York, NY, pp 91–101

Hirano A, Nakano I, Kurland LT, Mulder DW, Holley PW, Saccomanno G (1984) Fine structural study of neurofibrillary changes in a family with amyotrophic lateral sclerosis. J Neuropathol Exp Neurol 43:471–480

Hirokawa N (1997) The mechanisms of fast and slow transport in neurons: identification and characterization of the new kinesin superfamily motors. Curr Opin Neurobiol 7:605–614

Hirokawa N, Takemura R (2005) Molecular motors and mechanisms of directional transport in neurons. Nat Rev Neurosci 6:201–214

Hiruma H, Katakura T, Takahashi S, Ichikawa T, Kawakami T (2003) Glutamate and amyloid beta-protein rapidly inhibit fast axonal transport in cultured rat hippocampal neurons by different mechanisms. J Neurosci 23:8967–8977

Hisanaga S, Matsuoka Y, Nishizawa K, Saito T, Inagaki M, Hirokawa N (1994) Phosphorylation of native and reassembled neurofilaments composed of NF-L, NF-M, and NF-H by the catalytic subunit of cAMP-dependent protein kinase. Mol Biol Cell 5:161–172

Itoh T, Sobue G, Ken E, Mitsuma T, Takahashi A, Trojanowski JQ (1992) Phosphorylated high molecular weight neurofilament protein in the peripheral motor, sensory and sympathetic neuronal perikarya: system-dependent normal variations and changes in amyotrophic lateral sclerosis and multiple system atrophy. Acta Neuropathol (Berl) 83:240–245

Julien JP, Mushynski WE (1982) Multiple phosphorylation sites in mammalian neurofilament polypeptides. J Biol Chem 257:10467–10470

Kawasaki H, Morooka T, Shimohama S, Kimura J, Hirano T, Gotoh Y, Nishida E (1997) Activation and involvement of p38 mitogen-activated protein kinase in glutamate-induced apoptosis in rat cerebellar granule cells. J Biol Chem 272:18518–18521

Kesavapany S, Patel V, Zheng YL, Pareek TK, Bjelogrlic M, Albers W, Amin N, Jaffe H, Gutkind JS, Strong MJ et al (2007) Inhibition of Pin1 reduces glutamate-induced perikaryal accumulation of phosphorylated neurofilament-H in neurons. Mol Biol Cell 18:3645–3655

Lazarov O, Morfini GA, Pigino G, Gadadhar A, Chen X, Robinson J, Ho H, Brady ST, Sisodia SS (2007) Impairments in fast axonal transport and motor neuron deficits in transgenic mice expressing familial Alzheimer's disease-linked mutant presenilin 1. J Neurosci 27:7011–7020

Lee VM, Carden MJ, Schlaepfer WW, Trojanowski JQ (1987) Monoclonal antibodies distinguish several differentially phosphorylated states of the two largest rat neurofilament subunits (NF-H and NF-M) and demonstrate their existence in the normal nervous system of adult rats. J Neurosci 7:3474–3488

Lee MK, Cleveland DW (1996) Neuronal intermediate filaments. Annu Rev Neurosci 19:187–217

Lee MS, Kwon YT, Li M, Peng J, Friedlander RM, Tsai LH (2000) Neurotoxicity induces cleavage of p35 to p25 by calpain. Nature 405:360–364

Lee MK, Marszalek JR, Cleveland DW (1994) A mutant neurofilament subunit causes massive, selective motor neuron death: implications for the pathogenesis of human motor neuron disease. Neuron 13:975–988

Lee VM, Otvos L Jr, Carden MJ, Hollosi M, Dietzschold B, Lazzarini RA (1988) Identification of the major multiphosphorylation site in mammalian neurofilaments. Proc Natl Acad Sci USA 85:1998–2002

Leterrier JF, Kas J, Hartwig J, Vegners R, Janmey PA (1996) Mechanical effects of neurofilament cross-bridges. Modulation by phosphorylation, lipids, and interactions with F-actin. J Biol Chem 271:15687–15694

Lobsiger CS, Garcia ML, Ward CM, Cleveland DW (2005) Altered axonal architecture by removal of the heavily phosphorylated neurofilament tail domains strongly slows superoxide dismutase 1 mutant-mediated ALS. Proc Natl Acad Sci USA 102:10351–10356

Manser C, Stevenson A, Banner S, Davies J, Tudor EL, Ono Y, Leigh PN, McLoughlin DM, Shaw CE, Miller CC (2008) Deregulation of PKN1 activity disrupts neurofilament organisation and axonal transport. FEBS Lett 582:2303–2308

Mersiyanova IV, Perepelov AV, Polyakov AV, Sitnikov VF, Dadali EL, Oparin RB, Petrin AN, Evgrafov OV (2000) A new variant of Charcot-Marie-Tooth disease type 2 is probably the result of a mutation in the neurofilament-light gene. Am J Hum Genet 67:37–46

Millecamps S, Gowing G, Corti O, Mallet J, Julien JP (2007) Conditional NF-L transgene expression in mice for in vivo analysis of turnover and transport rate of neurofilaments. J Neurosci 27:4947–4956

Miller CC, Ackerley S, Brownlees J, Grierson AJ, Jacobsen NJ, Thornhill P (2002) Axonal transport of neurofilaments in normal and disease states. Cell Mol Life Sci 59:323–330

Motil J, Dubey M, Chan WK, Shea TB (2007) Inhibition of dynein but not kinesin induces aberrant focal accumulation of neurofilaments within axonal neurites. Brain Res 1164:125–131

Mukai H, Toshimori M, Shibata H, Kitagawa M, Shimakawa M, Miyahara M, Sunakawa H, Ono Y (1996) PKN associates and phosphorylates the head-rod domain of neurofilament protein. J Biol Chem 271:9816–9822

Nakagawa T, Chen J, Zhang Z, Kanai Y, Hirokawa N (1995) Two distinct functions of the carboxyl-terminal tail domain of NF-M upon neurofilament assembly: cross-bridge formation and longitudinal elongation of filaments. J Cell Biol 129:411–429

Nakamura Y, Hashimoto R, Kashiwagi Y, Aimoto S, Fukusho E, Matsumoto N, Kudo T, Takeda M (2000) Major phosphorylation site (Ser55) of neurofilament L by cyclic AMP-dependent protein kinase in rat primary neuronal culture. J Neurochem 74:949–959

Nakamura Y, Takeda M, Angelides KJ, Tanaka T, Tada K, Nishimura T (1990) Effect of phosphorylation on 68 KDa neurofilament subunit protein assembly by the cyclic AMP dependent protein kinase in vitro. Biochem Biophys Res Commun 169:744–750

Nixon RA (1998) The slow axonal transport of cytoskeletal proteins. Curr Opin Cell Biol 10:87–92

Nixon RA, Paskevich PA, Sihag RK, Thayer CY (1994) Phosphorylation on carboxyl terminus domains of neurofilament proteins in retinal ganglion cell neurons in vivo: influences on regional neurofilament accumulation, interneurofilament spacing, and axon caliber. J Cell Biol 126:1031–1046

Ohara O, Gahara Y, Miyake T, Teraoka H, Kitamura T (1993) Neurofilament deficiency in quail caused by nonsense mutation in neurofilament-L gene. J Cell Biol 121:387–395

Parysek LM, Goldman RD (1988) Distribution of a novel 57 kDa intermediate filament (IF) protein in the nervous system. J Neurosci 8:555–563

Perez-Olle R, Lopez-Toledano MA, Goryunov D, Cabrera-Poch N, Stefanis L, Brown K, Liem RK (2005) Mutations in the neurofilament light gene linked to Charcot-Marie-Tooth disease cause defects in transport. J Neurochem 93:861–874

Pigino G, Morfini G, Atagi Y, Deshpande A, Yu C, Jungbauer L, LaDu M, Busciglio J, Brady S (2009) Disruption of fast axonal transport is a pathogenic mechanism for intraneuronal amyloid beta. Proc Natl Acad Sci USA 106:5907–5912

Prahlad V, Helfand BT, Langford GM, Vale RD, Goldman RD (2000) Fast transport of neurofilament protein along microtubules in squid axoplasm. J Cell Sci 113(Pt 22): 3939–3946

Rao MV, Engle LJ, Mohan PS, Yuan A, Qiu D, Cataldo A, Hassinger L, Jacobsen S, Lee VM, Andreadis A et al (2002a) Myosin Va binding to neurofilaments is essential for correct myosin Va distribution and transport and neurofilament density. J Cell Biol 159:279–290

Rao MV, Garcia ML, Miyazaki Y, Gotow T, Yuan A, Mattina S, Ward CM, Calcutt NA, Uchiyama Y, Nixon RA, Cleveland DW (2002b) Gene replacement in mice reveals that the heavily phosphorylated tail of neurofilament heavy subunit does not affect axonal caliber or the transit of cargoes in slow axonal transport. J Cell Biol 158:681–693

Reid E, Kloos M, Ashley-Koch A, Hughes L, Bevan S, Svenson IK, Graham FL, Gaskell PC, Dearlove A, Pericak-Vance MA et al (2002) A Kinesin Heavy Chain (KIF5A) Mutation in Hereditary Spastic Paraplegia (SPG10). Am J Hum Genet 71:1189–1194

Rouleau GA, Clark AW, Rooke K, Pramatarova A, Krizus A, Suchowersky O, Julien JP, Figlewicz D (1996) SOD1 mutation is associated with accumulation of neurofilaments in amyotrophic lateral sclerosis. Ann Neurol 39:128–131

Roy S, Coffee P, Smith G, Liem RK, Brady ST, Black MM (2000) Neurofilaments are transported rapidly but intermittently in axons: implications for slow axonal transport. J Neurosci 20: 6849–6861

Rudrabhatla P, Zheng YL, Amin ND, Kesavapany S, Albers W, Pant HC (2008) Pin1-dependent prolyl isomerization modulates the stress-induced phosphorylation of high molecular weight neurofilament protein. J Biol Chem 283:26737–26747

Saha AR, Hill J, Utton MA, Asuni AA, Ackerley S, Grierson AJ, Miller CC, Davies AM, Buchman VL, Anderton BH, Hanger DP (2004) Parkinson's disease alpha-synuclein mutations exhibit defective axonal transport in cultured neurons. J Cell Sci 117:1017–1024

Sakaguchi T, Okada M, Kitamura T, Kawasaki K (1993) Reduced diameter and conduction velocity of myelinated fibers in the sciatic nerve of a neurofilament-deficient mutant quail. Neurosci Lett 153:65–68

Schmidt RE, Beaudet LN, Plurad SB, Dorsey DA (1997a) Axonal cytoskeletal pathology in aged and diabetic human sympathetic autonomic ganglia. Brain Res 769:375–383

Schmidt ML, Carden MJ, Lee VM, Trojanowski JQ (1987) Phosphate dependent and independent neurofilament epitopes in the axonal swellings of patients with motor neuron disease and controls. Lab Invest 56:282–294

Schmidt RE, Dorsey D, Parvin CA, Beaudet LN, Plurad SB, Roth KA (1997b) Dystrophic axonal swellings develop as a function of age and diabetes in human dorsal root ganglia. J Neuropathol Exp Neurol 56:1028–1043

Schmidt ML, Martin JA, Lee VM, Trojanowski JQ (1996) Convergence of Lewy bodies and neurofibrillary tangles in amygdala neurons of Alzheimer's disease and Lewy body disorders. Acta Neuropathol 91:475–481

Schwarzschild MA, Cole RL, Hyman SE (1997) Glutamate, but not dopamine, stimulates stress-activated protein kinase and AP-1-mediated transcription in striatal neurons. J Neurosci 17:3455–3466

Shah JV, Flanagan LA, Janmey PA, Leterrier JF (2000) Bidirectional translocation of neurofilaments along microtubules mediated in part by dynein/dynactin. Mol Biol Cell 11:3495–3508

Sihag RK, Inagaki M, Yamaguchi T, Shea TB, Pant HC (2007) Role of phosphorylation on the structural dynamics and function of types III and IV intermediate filaments. Exp Cell Res 313:2098–2109

Sihag RK, Nixon RA (1991) Identification of Ser-55 as a major protein kinase A phosphorylation site on the 70-kDa subunit of neurofilaments. Early turnover during axonal transport. J Biol Chem 266:18861–18867

Sobue G, Hashizume Y, Yasuda T, Mukai E, Kumagai T, Mitsuma T, Trojanowski JQ (1990) Phosphorylated high molecular weight neurofilament protein in lower motor neurons in amyotrophic lateral sclerosis and other neurodegenerative diseases involving ventral horn cells. Acta Neuropathol 79:402–408

Sternberger LA, Sternberger NH (1983) Monoclonal antibodies distinguish phosphorylated and nonphosphorylated forms of neurofilaments in situ. Proc Natl Acad Sci USA 80:6126–6130

Tomkins J, Usher P, Slade JY, Ince PG, Curtis A, Bushby K, Shaw PJ (1998) Novel insertion in the KSP region of the neurofilament heavy gene in amyotrophic lateral sclerosis (ALS). Neuroreport 9:3967–3970

Trimpin S, Mixon AE, Stapels MD, Kim MY, Spencer PS, Deinzer ML (2004) Identification of endogenous phosphorylation sites of bovine medium and low molecular weight neurofilament proteins by tandem mass spectrometry. Biochemistry 43:2091–2105

Trivedi N, Jung P, Brown A (2007) Neurofilaments switch between distinct mobile and stationary states during their transport along axons. J Neurosci 27:507–516

Trojanowski JQ, Lee VM (1998) Aggregation of neurofilament and alpha-synuclein proteins in Lewy bodies: implications for the pathogenesis of Parkinson disease and Lewy body dementia. Arch Neurol 55:151–152

Trojanowski JQ, Schmidt ML, Shin RW, Bramblett GT, Rao D, Lee VM (1993) Altered tau and neurofilament proteins in neuro-degenerative diseases: diagnostic implications for Alzheimer's disease and Lewy body dementias. Brain Pathol 3:45–54

Trotti D, Rolfs A, Danbolt NC, Brown RH Jr, Hediger MA (1999) SOD1 mutants linked to amyotrophic lateral sclerosis selectively inactivate a glial glutamate transporter. Nat Neurosci 2:427–433

Troy CM, Brown K, Greene LA, Shelanski ML (1990) Ontogeny of the neuronal intermediate filament protein, peripherin, in the mouse embryo. Neuroscience 36:217–237

Van Den Bosch L, Van Damme P, Bogaert E, Robberecht W (2006) The role of excitotoxicity in the pathogenesis of amyotrophic lateral sclerosis. Biochim Biophys Acta 1762:1068–1082

Vickers JC, Morrison JH, Friedrich VL Jr, Elder GA, Perl DP, Katz RN, Lazzarini RA (1994) Age-associated and cell-type-specific neurofibrillary pathology in transgenic mice expressing the human midsized neurofilament subunit. J Neurosci 14:5603–5612

Wang L, Ho CL, Sun D, Liem RK, Brown A (2000) Rapid movement of axonal neurofilaments interrupted by prolonged pauses. Nat Cell Biol 2:137–141

Williamson TL, Bruijn LI, Zhu Q, Anderson KL, Anderson SD, Julien JP, Cleveland DW (1998) Absence of neurofilaments reduces the selective vulnerability of motor neurons and slows disease caused by a familial amyotrophic lateral sclerosis-linked superoxide dismutase 1 mutant. Proc Natl Acad Sci USA 95:9631–9636

Williamson TL, Cleveland DW (1999) Slowing of axonal transport is a very early event in the toxicity of ALS-linked SOD1 mutants to motor neurons. Nat Neurosci 2:50–56

Xia CH, Roberts EA, Her LS, Liu X, Williams DS, Cleveland DW, Goldstein LS (2003) Abnormal neurofilament transport caused by targeted disruption of neuronal kinesin heavy chain KIF5A. J Cell Biol 161:55–66

Xu Z, Cork LC, Griffin JW, Cleveland DW (1993) Increased expression of neurofilament subunit NF-L produces morphological alterations that resemble the pathology of human motor neuron disease. Cell 73:23–33

Yabe JT, Chylinski T, Wang FS, Pimenta A, Kattar SD, Linsley MD, Chan WK, Shea TB (2001) Neurofilaments consist of distinct populations that can be distinguished by C-terminal phosphorylation, bundling, and axonal transport rate in growing axonal neurites. J Neurosci 21:2195–2205

Yabe JT, Jung C, Chan WK, Shea TB (2000) Phospho-dependent association of neurofilament proteins with kinesin in situ. Cell Motil Cytoskeleton 45:249–262

Yates DM, Manser C, De Vos KJ, Shaw CE, McLoughlin DM, Miller CC (2009) Neurofilament subunit (NFL) head domain phosphorylation regulates axonal transport of neurofilaments. Eur J Cell Biol 88:193–202

Yuan A, Rao MV, Sasaki T, Chen Y, Kumar A, Veeranna, Liem RK, Eyer J, Peterson AC, Julien JP, Nixon RA (2006) Alpha-internexin is structurally and functionally associated with the neurofilament triplet proteins in the mature CNS. J Neurosci 26:10006–10019

Chapter 12
Knockout Models of Neurofilament Proteins

Rodolphe Perrot and Jean-Pierre Julien

Abstract Neurofilaments (NFs) are the most prominent cytoskeleton components of large myelinated axons from adult central and peripheral nervous systems. In the last 15 years, the gene targeting technique has been widely used to investigate the role of NF proteins in neuronal function. Gene knockout studies have demonstrated that NFs are crucial to expand the caliber of myelinated axons and consequently to increase their conduction velocity. However, the mechanism by which NFs determine the axonal diameter is not yet fully elucidated. NFs also contribute to the dynamic properties of the axonal cytoskeleton during neuronal differentiation, axon outgrowth and regeneration. Perturbations of their metabolism and organization are frequently associated with neurodegenerative disorders, including amyotrophic lateral sclerosis, Charcot-Marie-Tooth disease, Alzheimer's disease and giant axonal neuropathy. Here, we describe how mouse knockout models of NF proteins have been used to study the multiple aspects of NF biology.

Keywords α-Internexin · Axonal caliber · Axonal cytoskeleton · Conduction velocity · Intermediate filaments · Knockout models · Neurofilament · NFH · NFL · NFM · Peripherin · Radial axonal growth

Contents

J.-P. Julien (✉)
Department of Anatomy and Physiology, Centre de Recherche du Centre Hospitalier, Universitaire de Québec, Laval University, Sainte-Foy, QC, Canada G1V 4G2
e-mail: jean-pierre.julien@crchul.ulaval.ca

R.A. Nixon, A. Yuan (eds.), *Cytoskeleton of the Nervous System*, Advances in Neurobiology 3, DOI 10.1007/978-1-4419-6787-9_12,

12.1 Introduction

Three interconnected structures compose the neuronal cytoskeleton: actin microfilaments (MFs), microtubules (MTs) and intermediate filaments (IFs). IF proteins are classified into six types based on similarities in sequence and gene structure. Neurons express differentially several IF proteins depending on their developing stage or their localization in the nervous system: nestin (200 kDa; type IV), NF triplet proteins (called NFL (light, éthique, 68 kDa), NFM (medium, 160 kDa) and NFH (heavy, 205 kDa); type IV), α-internexin (66 kDa; type IV), peripherin (57 kDa; type III) and synemin isoforms (Low synemin (41 kDa), Middle or beta synemin (150 kDa), High or alpha synemin (180 kDa); type IV) (Lariviere and Julien, 2004; Perrot et al., 2008). It was originally believed that NFs were composed only by NFL, NFM and NFH, but recent studies indicated that α-internexin and peripherin are also integral and abundant components of NFs in CNS and PNS, respectively (Beaulieu et al., 1999; Yuan et al., 2006; Yan et al., 2007). These NF proteins share with other members of the IF family a tripartite structure, with non-helical amino and carboxy-terminal regions (called the head and tail domains) flanking a homologous central rod domain of approximately 310 amino acids which is involved in the assembly of 10-nm filaments (Geisler et al., 1983; Fuchs and Weber, 1994). Head and tail domains of the NF subunits are less conserved. Tail domains of NFM and NFH are longer and contain numerous repeats of phosphorylation sites Lys-Ser-Pro (KSP) (Jones and Williams, 1982; Julien and Mushynski, 1983; Geisler et al., 1987; Goldstein et al., 1987; Lee et al., 1988; Pant and Veeranna, 1995). When they are phosphorylated, tail domains of NFM and NFH form lateral projections extending from the filament backbone (Hirokawa et al., 1984; Hisanaga and Hirokawa, 1988). These sidearms participate to the stabilization of the filament network by forming cross-bridges between NFs and other cytoskeletal elements or organelles.

NFs play a central role in growth and maintenance of axonal caliber. This was first suggested by the observation that increased NF numbers and densities are correlated with increased axonal diameters (Friede and Samorajski, 1970). Moreover, the expansion of the axonal caliber coincides with the entry of NFs into axons during axonal development and regeneration (Hoffman et al., 1984). The first evidence for the implication of NFs in axonal caliber expansion in an animal model was provided by Japanese *quiver* quails. These quails displayed a spontaneous mutation in *NFL* gene that generates a truncated protein incapable of forming NFs (Ohara et al.,

1993). In these animals, radial growth of myelinated axons is strongly decreased (Yamasaki et al., 1991), with consequent reduction in axonal conduction velocity (Sakaguchi et al., 1993). The function of NFs in the radial axonal growth was thereafter confirmed by the generation of various mouse models. Here, we will describe mouse knockout models that have been generated to elucidate the functions of NFs. Their analysis revealed an unequivocal role of NFs in determining the diameter of large myelinated axons. We will also discuss other mouse models that express modified NF subunits and which contributed significantly to the study of the role of the NFs.

12.2 Knockout Models of Neurofilament Proteins

12.2.1 NFL Knockout Mice

The de novo assembly of NFs in vivo requires both NFL and either NFM or NFH (Lee et al., 1993). The importance of NFL in NF assembly was confirmed by the targeted disruption of the *Nefl* gene in mice (Zhu et al., 1997). In the absence of NFL, the levels of NFM and NFH subunits are strongly decreased to ~5% of the normal level in sciatic nerve while an increased level of α-tubulin occurs (Table 12.1). As a result, NFL–/– axons have a scarcity of IF structures but display a higher density of MTs, possibly as a compensating effect. In CNS, optic axons from NFL–/– mice contain 52% of the normal level of NFM and smaller proportions of the NFH protein (13%) (Yuan et al., 2003). This suggests that NFM is capable of transport without NFL, probably via its association with α-internexin whose level is unchanged in NFL–/– optic nerve. Indeed, NFM transport is completely abolished by deleting α-internexin (Yuan et al., 2003).

NFL–/– mice exhibit a severe axonal hypotrophy with a ~50% decrease of axonal caliber in L5 ventral root, confirming that NFs are a major determinant of axonal diameter. Note that a ~50% decrease in the NFL mRNA levels of NFL+/– mice provoked a modest reduction of ~15% in the calibers of ventral root axons, indicating that there is no direct correlation between NFL mRNA levels and axon diameter (Zhu et al., 1997). Consistent with axonal atrophy, lower conduction velocities were observed in NFL–/– mice (Kriz et al., 2000b). Despite this, the NFL-null mice exhibit only mild sensorimotor dysfunction and spatial deficits without overt signs of paresis (Dubois et al., 2005b). However, the absence of axonal NFs is not entirely innocuous since altered cytochrome oxidase activities in numerous hindbrain regions and a significant loss of motor axons have been detected in NFL–/– mice (Zhu et al., 1997; Dubois et al., 2005a). The regeneration of myelinated axons is also delayed in the absence of axonal NFs (Zhu et al., 1997). The exact mechanism by which NFs contribute to maturation of regenerating myelinated axons remains to be elucidated. It is conceivable that the NF network could provide a scaffold that contributes to the stabilization of newly growing axons. Alternatively, the IF cytoskeleton could indirectly support axonal regeneration by contributing to MT and/or MF dynamics and functions.

Table 12.1 Targeted disruption of neuronal IF genes

Mice	Axonal caliber	IF content	MT content	Axonal loss	References
NFL–/–	Decreased of ~50% in L5 VR at 2 months old	Scarcity of IFs 5% of normal level of NFM and NFH in SN 52% of normal level of NFM and 13% of normal level of NFH in ON Decreased level of peripherin Normal level of α-internexin	Increased MT density Increased level of α-tubulin in brain and SN	~20% loss of motor axons	Zhu et al. (1997) Levavasseur et al. (1999) Yuan et al. (2003)
NFM–/– (C57BL/6)	Decreased of ~50% in L4 VR at 3 months old	Decreased NF density 68, 35 and 22% of normal level of NFL in SC, SN and ON, respectively 121 and 33% of normal level of NFH in SC and ON, respectively Increased phosphorylation of NFH Decreased level of α-internexin in brain, SC and ON	Increased MT density Normal level of β-tubulin	~10% loss of motor axons	Jacomy et al. (1999)
NFM–/– (129Sv/J)	4 months old: decreased in SC, decreased of 20% in L5 VR, 24% in SN and 20% in ON 2 years old: decreased of ~55% in L5 VR and ~10% in L5 DR	Decreased NF density 13 and 23% of normal level of NFL in cortex and SC, respectively 140% of normal level of heavily phosphorylated NFH in cortex	Increased MT density Normal level of β-tubulin	~10% loss of motor axons	Elder et al. (1998a)

Table 12.1 (continued)

Mice	Axonal caliber	IF content	MT content	Axonal loss	References
NFH–/– (C57BL/6)	Minor change in L5 VR at 2 months old	Normal NF density Normal level of NFL 120–200% of normal level of NFM Increased phosphorylation of NFH	Twofold increased of MT density Fourfold increase of βIII-tubulin	No axonal loss for Zhu et al. (1998) 13% loss of motor axons and 19% loss of sensory axons for Rao et al. (1998)	Rao et al. (1998) Zhu et al. (1998) Jacomy et al. (1999)
NFH–/– (129 Sv/J)	Decreased of ∼20% in L5 VR, SN, SC and ON at 4 months old	Slight decrease of NF density 75–90% of normal level of NFL Normal levels of NFM	Normal MT density Normal level of tubulin	No axonal loss	Elder et al. (1998b)
NFM;NFH–/–	3–4 months old: decreased of 30–50% in L4 and L5 VR 2 years old: decreased of 60% in L5 VR and 41% in L5 DR	Scarcity of IFs 23%, 11–17% and 12% of normal level of NFL in SC, SN and L5 VR, respectively, and barely detectable in ON <10% of normal level of α-internexin in ON Normal level of peripherin	Twofold increased of MT density Normal level of tubulin	24% loss of motor axons	Jacomy et al. (1999) Elder et al. (1999a) Elder et al. (1999b) Yuan et al. (2006)
NFL;NFH–/–	Decreased	Small number of IFs in ON ∼20%, 40% and 85% of normal level of NFM in brain, ON and SC, respectively Normal level of α-internexin	Small increase in MT number	Not determined	Yuan et al. (2003)

Table 12.1 (continued)

Mice	Axonal caliber	IF content	MT content	Axonal loss	References
NFL;M;H+/–	Decreased of ~50% in L5 VR at 7 months old	~60% of normal level of NFL, NF'M and NFH in SC Normal level of α-internexin Normal level of peripherin	Normal level of β-tubulin	No axonal loss	Nguyen et al. (2000)
α-internexin–/–	Normal in L4 VR at 2 months old	Normal NF density Normal levels of NFL, NFM and NFH in SC, ON and SN Normal level of peripherin in SN	Normal MT density Normal level of β-tubulin	No axonal loss	Levavasseur et al. (1999) Yuan et al. (2003)
α-internexin;NFL–/–	Decreased of ~50%	Total absence of IFs in ON Strong decrease of NFM and NFH levels in SC, ON and SN Decreased level of peripherin in SN	Normal level of β-tubulin in SC, ON and SN	~20% loss of motor axons	Levavasseur et al. (1999) Yuan et al. (2003)
peripherin–/–	Normal in L4 VR and DR at 4 months old	Normal levels of NFL and phosphorylated NFH Slight decrease of NFM and hypophosphorylated NFH Increased level of α-internexin in motor axons ~150% of normal level of vimentin	Slight decrease of MT number in some small axons Normal level of β-tubulin	34% loss of small unmyelinated sensory axons	Larivière et al. (2002)

DR dorsal root, *ON* optic nerve, *SC* spinal cord, *SN* sciatic nerve, *VR* ventral root

NFL–/– mice develop abnormal NF accumulations in perikarya and proximal axons from motor neurons (Williamson et al., 1998). The onset of motor neurons loss has been found to begin at about 10 months of age (McLean et al., 2005). These mice were thus used as a neurodegenerative disease model to investigate the temporal relationship between the neuronal aggregates and the activation of glial cells, two common phenomena in neurodegenerative diseases. Three distinct phases of neurodegeneration were identified. First, NF aggregate formation is prominent. In the second phase, microglial proliferation is marked and the numbers of NF aggregate-bearing motor neurons decline. In the last stage, motor neuron numbers decrease while astrocytic proliferation increases (McLean et al., 2005). This temporal pattern of proliferating microglia suggests that they may be responding to signals released by aggregate-bearing motor neurons. The first stage, which is before the onset of motoneuron loss, is characterized by a significant downregulation in the expression of complement receptor type 3 α (CD11b) subunit in microglia (Li et al., 2006). This microglial inhibition might play a particular role in the survival of the abnormal protein aggregate-bearing motoneurons in the early development stage of neurodegeneration in NFL–/– mice.

12.2.2 NFH Knockout Mice

It has long been considered that the extensive phosphorylation of NFs controls radial axonal growth by regulating NF transport (Nixon et al., 1982; Lewis and Nixon, 1988) and/or interfilament spacing (Gotow et al., 1992; Nixon et al., 1994; Sanchez et al., 1996). It has been suggested that phosphorylation of KSP repeats on NFM and NFH sidearms could increase their total negative charge and cause their lateral extension by repulsive attraction, increasing NF spacing and axonal caliber. Many lines of evidence reinforced this assumption, including the presence of a weak repulsive force around the core of the filament detected by using atomic force microscopy (Brown and Hoh, 1997). This force is absent in homopolymers of NFL or trypsinized native filaments which lack the sidearms and attenuate when the filaments are enzymatically dephosphorylated (Kumar and Hoh, 2004). The number of KSP repeats being much more important in NFH than in NFM (51 vs 7 in mice), it was thought that this subunit should contribute more to the radial axonal growth. This idea was reinforced by the observation that phosphorylated NFH sidearms are less pliant and larger structures than dephosphorylated ones (Aranda-Espinoza et al., 2002). However, this hypothesis was invalidated by the generation of three different NFH-null mice (Elder et al., 1998b; Rao et al., 1998; Zhu et al., 1998). Indeed, the absence of NFH has a minor effect on the radial growth of axons, even if subtle differences exist between these models. Rao et al. (1998) and Zhu et al. (1998) reported only a slight reduction in the caliber of myelinated axons from the ventral roots of NFH–/– mice, no modification in NF density, and a twofold increase of MT density, whereas NFH knockout mice described by Elder et al. (1998b) exhibited a more pronounced reduction in the caliber of axons, a slight decrease of NF density and a normal density of MTs. Somehow, there are discrepancies regarding the NFL and NFM levels. Rao et al. (1998) and Zhu et al. (1998) reported normal levels of NFL and an up-regulation of NFM (respectively, 100% and 20%) whereas a decreased level of NFL and no modification in NFM protein level were observed by Elder et al. (1998). This divergence could be explained in part by differences in the mouse genetic backgrounds (C57BL/6 strain for Rao et al. (1998) and Zhu et al. (1998); 129 J strain for Elder et al. (1998)) and by the chronological differences between the data (2 months for Rao et al., 1998 and Zhu et al., 1998; 4 months for Elder et al., 1998), suggesting a later effect of NFH on the radial axonal growth. In any case, the combined results suggest that NFH has a minor effect on the radial expansion of axons. It is also important to note that Rao et al. (1998) and Zhu et al. (1998) described a compensatory increase in MT density and NFM phosphorylation in their NFH–/– mice, complicating the conclusions about the exact role of NFH in the determination of the axonal caliber.

Surprisingly, large myelinated axons in NFH–/– mice showed a significant decrease in conduction velocity despite normal axonal diameter, g ratio and internodal length (Kriz et al., 2000b). Moreover, NFH–/– mice expressed a significant decrease in outward rectification and the refractory period was prolonged, providing strong evidence that NFs are involved in defining not only the structural but also the functional integrity of myelinated axons. It was proposed that NFH may have a specific role in modulating ion channel function (Kriz et al., 2000a, b), but the precise mechanism is still unclear even if it appeared that localization of Na+ and K+ channels is unaffected by the absence of axonal NFs (Perrot et al., 2007). Finally, it cannot be excluded that the absence of NFH, the most extensively phosphorylated and thus negatively charged subunit, also affects the axoplasmic resistance.

12.2.3 NFM Knockout Mice

The generation of two different NFM knockout mice has shown a more preponderant role of this subunit in controlling NF content and axonal caliber (Elder et al., 1998a; Jacomy et al., 1999). Elder et al. (1998a) reported a decrease of the axonal caliber of 20–25% in L5 ventral root, sciatic nerve and optic nerve at 4 months old and a more pronounced atrophy (reduction of 55%) in L5 ventral root at 2 years old, whereas axonal caliber in L5 dorsal root is only reduced by ~10%. In the second model, the diameter of axons is decreased by ~50% in L4 ventral root as soon as 3 months old (Jacomy et al., 1999). In keeping with smaller axon diameter, the conduction velocity was significantly decreased and a prolongation of the refractory period was observed in NFM–/– animals (Kriz et al., 2000b), without phenotypic changes. Both models are characterized by important decreases in NFL levels, resulting in a reduced axonal NF content and thus in axonal atrophy. As a compensatory mechanism, MT density and NFH phosphorylation were increased in the absence of NFM (Elder et al., 1998a; Jacomy et al., 1999).

The simultaneous absence of both NFM and NFH provokes the retention of unassembled NFL subunits in neuronal cell bodies (Jacomy et al., 1999). Consequently, axons from NFM–/–; NFH–/– mice are deprived of NFs, resulting in a reduction of axonal caliber equivalent to the NFL–/– mice. An important decline also occurred in the levels of α-internexin while a twofold increase in the density of MTs is observed (Elder et al., 1999b; Jacomy et al., 1999). As in NFL–/– mice, a significant loss of motor axons (24%) was detected in NFM–/–;NFH–/– mice and these animals developed an age-related atrophy of motor axons accompanied by a hind limb paralysis (Elder et al., 1999a; Jacomy et al., 1999). These results show a requirement of the high molecular weight subunits for the in vivo assembly of NFs. Finally, a 40% decrease of NF content in triple heterozygous mice (NFL±;NFM±;NFH±) was sufficient to induce a 50% decrease of axonal diameter in L5 ventral root, without altering the normal subunit stoichiometry and the levels of other cytoskeletal proteins (Nguyen et al., 2000).

12.2.4 α-Internexin Knockout Mice

α-Internexin is an IF protein that is expressed abundantly in neurons during development of the PNS and CNS (Kaplan et al., 1990; Fliegner et al., 1994), and found mainly in the CNS in adults (Chiu et al., 1989; Kaplan et al., 1990). Some lines of evidence suggested that α-internexin could play a key role in axonal outgrowth (Glasgow et al., 1994; Zhao and Szaro, 1997; Leake et al., 1999; Shea and Beermann, 1999), but the disruption of the *INA* gene in mice did not support this view (Levavasseur et al., 1999). Mice deficient in α-internexin exhibited normal pre- and postnatal development and showed no overt phenotype. The levels of other neuronal IF proteins were unchanged and a normal axonal caliber was achieved in L4 ventral root, indicating that α-internexin is not required for radial axonal growth. In the same way, a normal retinotectal topographic organization was observed in the absence of α-internexin, excluding a role in outgrowth and guidance of axons (Levavasseur et al., 1999). A normal development is also observed in mice deficient in both NFL and α-internexin. No IF structures were observed in their optic nerves, demonstrating that type IV IFs do not play an essential role in axonal outgrowth during development of the nervous system.

Lack of α-internexin did not affect the transport of NFs in optic axons (Yuan et al., 2003), but deletion of α-internexin in the absence of NFH accelerated the transport of NFM and NFL, indicating that α-internexin interacts with NFM or NFL or both during transport under these conditions. This also implied redundancy of function or cooperation between α-internexin and NFH for the axonal transport of NFs (Yuan et al., 2003). It now appears that α-internexin functions in the adult CNS principally as a fourth subunit of NFs, with a stoichiometry of 4:2:2:1 (NFL: α internexin:NFM:NFH), and facilitates the transport of NFs in central axons (Yuan et al., 2006; Yan et al., 2007).

12.2.5 Peripherin Knockout Mice

Peripherin is a type III neuronal IF protein widely expressed in developing neurons, but in the adult it is mainly found in neurons extending to the PNS, as well as in subsets of CNS neurons that have sensory, motor and autonomic functions (Brody et al., 1989; Escurat et al., 1990; Gorham et al., 1990; Barclay et al., 2007). Peripherin can self-assemble or co-assemble with other NF subunits to form IF networks in vitro (Beaulieu et al., 1999). The precise function of peripherin is not clear, but as it is expressed concomitantly with axonal growth and following nerve injury, peripherin was believed to play a role in nerve elongation and in the regeneration process (Escurat et al., 1990; Gorham et al., 1990; Wong and Oblinger, 1990). To investigate the function of peripherin further, mice with a targeted disruption of the peripherin gene were generated (Lariviere et al., 2002). As for α-internexin–/– mice, peripherin knockout mice were viable, reproduced normally and did not exhibit overt phenotypes. No morphological defects in the ventral and dorsal roots, spinal cord and retina were observed. The number and caliber of myelinated motor and sensory axons in the L5 roots remained unchanged in peripherin deficient mice (Lariviere et al., 2002). All these data provided the proof that peripherin, as α-internexin, is dispensable for axogenesis and is not a modulator of axon diameter. However, it should be noted that the ventral root, but not the dorsal root from peripherin-null mice showed increased levels of α-internexin, suggesting that motor neurons from peripherin–/– mice might compensate for the lack of peripherin by increasing their amount of α-internexin (Lariviere et al., 2002). This question could be addressed through the generation of double knockout mice derived by the breeding of peripherin–/– mice with α-internexin–/– mice. It is also interesting to note that L5 dorsal roots from peripherin-null animals revealed a 34% loss of small unmyelinated sensory axons, demonstrating a requirement of peripherin for the proper development of a restricted subset of sensory neurons (Lariviere et al., 2002). Further investigations are necessary to determine its function in peripheral myelinated axons. Recent data showed that peripherin is expressed in the tuberomammillary neurons of the mouse hypothalamus, but no visible differences in the appearance of these neurons or their axons were detected in peripherin–/– mice (Eriksson et al., 2008). Monitoring of locomotor activity, feeding, drinking and energy expenditure in these mice also revealed no difference, so the significance of peripherin in these neurons remains to be elucidated.

12.3 Mice Expressing Modified NF Subunits

12.3.1 NFH-LacZ Mice

NFH-LacZ transgenic mice expressed a fusion protein in which one half of the carboxy-terminal domain of the NFH protein was replaced by β-galactosidase subunit from *Escherichia coli* (Eyer and Peterson, 1994). This transgene was controlled by NFH regulatory sequences, but its expression was obviously low compared to the endogenous NFH protein. The β-galactosidase is composed of four identical subunits which must tetramerize to form a functional protein. So, NFH-β-galactosidase fusion proteins polymerized and caused the perikaryal aggregation of NFs in NFH-LacZ mice (Eyer and Peterson, 1994). Their axons were thus deprived of NFs while MT density was significantly increased. The absence of axonal NFs resulted in a generalized reduction of the axonal caliber in both PNS and CNS (Eyer and

Peterson, 1994; Perrot et al., 2007). In accordance with axonal hypotrophy, the conduction velocity was significantly reduced in NFH-LacZ mice and abnormalities of somesthetic and auditive-evoked potentials were also detected (Zochodne et al., 2004; Perrot et al., 2007). Finally, a significant loss of motor axons and the degeneration of Purkinje cells occurred with age (Eyer and Peterson, 1994; Tu et al., 1997). Interestingly, the replacement of the carboxy-terminal end of NFH by the GFP reporter in transgenic mice did not alter the normal assembly and the axonal transport of the fusion protein together with the endogenous IFs (Letournel et al., 2006). This confirmed that in the NFH-LacZ transgenic model NFs accumulated in neuronal cell bodies because of the presence of β-galactosidase and not because of the removal of the carboxy-terminal end of NFH.

It was recently shown that the synemin L isoform is also sequestered in perikarya of dorsal root ganglion neurons from NFH-LacZ mice, whereas synemin M and H associated with peripherin are present along NF-deficient peripheral axons (Izmiryan et al., 2009). Dorsal root ganglion neurons can be classified into two neuronal subpopulations: large-light neurons and small-dark neurons (Ferri et al., 1990; Goldstein et al., 1996; Fornaro et al., 2008). Peripherin has been found only in the small unmyelinated fibers while NF proteins are expressed in the large myelinated fibers. These data strongly suggest that synemin H/M associated with peripherin define a distinct subpopulation of small neurons within the DRG (Izmiryan et al., 2009). This would also explain why only a subset of unmyelinated sensory axons degenerates in peripherin–/– mice (Lariviere et al., 2002).

12.3.2 NFM$^{tail\Delta}$ and NFH$^{tail\Delta}$ Mice

As mentioned above, the phosphorylated tail domains of NFM and NFH subunits were originally proposed to regulate NF spacing and axon radial growth. In agreement with results obtained in NFH-null mice, no major modifications in NF spacing and axonal diameter were observed in adult mice expressing NFH deprived of its carboxy-terminal domain (NFH$^{tail\Delta}$ mice) (Rao et al., 2002), confirming that phosphorylation of the NFH tail cannot be a primary contributor to radial growth. However, examination of younger NFH$^{tail\Delta}$ animals revealed a delay in the acquisition of a normal diameter for the largest axons, suggesting that the NFH tail may affect the kinetics of growth of large axons. The organization of axoplasm is also subtly altered in NFH$^{tail\Delta}$ mice, with fewer cross-bridges spanning between adjacent NFs and an increase frequency in NFs that longitudinally contact each other. NFM and other linkers would continue to support the three-dimensional array of the axonal cytoskeleton. It should also be mentioned that, in contrast to NFH-null mice, conduction velocity is not altered in NFH$^{tail\Delta}$ mice. The authors proposed that the significant slowing of conduction velocity observed in 3- to 4-month-old NFH–/– mice (Kriz et al., 2000b) arises from a slowed propagation in radial growth distally along the nerves (Rao et al., 2002).

In contrast to mutant NFH$^{tail\Delta}$ mice, NFM$^{tail\Delta}$ mice displayed a strong decrease of the axonal caliber in both motor and sensory nerves (Garcia et al., 2003; Rao et al., 2003). This was accompanied by reduced NF spacing and loss of long cross-bridges without any change in the NF protein content. These results confirmed the preponderant role of NFM in radial axonal growth and showed that NFM tail increased interneurofilament spacing by extending the cross-bridges laterally between filaments. An increased MT density was also reported in $NFM^{tail\Delta}$ axons (Garcia et al., 2003; Rao et al., 2003) but not in $NFH^{tail\Delta}$ mice (Rao et al., 2002), indicating that the NFM tail is essential for maintaining the proper number of MTs in the axon. Finally, it is interesting to note that axonal caliber in double transgenic $NFM/H^{tail\Delta}$ mice from 6 months of age is similar to that of $NFM^{tail\Delta}$ mice, whereas deletion of both tail domains reduced average NF spacing even further and accentuated the disappearance of cross-bridges extending from the surface of NFs and the disorganization of the axoskeleton (Garcia et al., 2003). This suggests a role of the NFH tail in stabilizing NF cross-bridging.

12.3.3 $NFM^{S\rightarrow A}$ Mice

The absence of NFM or the removal of its tail domain resulted in a decreased NF spacing and an impaired radial growth of myelinated axons (Elder et al., 1998a; Jacomy et al., 1999; Garcia et al., 2003; Rao et al., 2003). Since interneurofilament spacing strongly correlates with the extensive phosphorylation of NFM tail domains, it was speculated that the regulation of the NF spacing by the phosphorylation of KSP repeats contained in the NFM tail domain represent the mechanism by which axons modulate their calibers. To test this assumption, mice expressing phosphorylation-incompetent NFM were produced by using gene replacement to substitute all serines of NFM's KSP repeats for alanines (Garcia et al., 2009). Surprisingly, axonal calibers in L5 ventral root from $NFM^{S\rightarrow A}$ mice were indistinguishable from those of wild-type mice. The spacing between NFs is slightly reduced at 2 months of age but not at 6 months old and the axoskeleton is structured similarly in adult $NFM^{S\rightarrow A}$ and wild-type mice. Finally, conduction velocity, locomotor activity and recovery rates after nerve injury were not altered in phospho-incompetent NFM expressing mice. A modest increase in NFH phosphorylation was reported but the authors stated that this cannot be sufficient to serve as a compensatory mechanism (Garcia et al., 2009). All these results indicate that phosphorylation of NFM KSP repeats is not required for a proper organization of the axonal cytoskeleton and for radial axonal growth.

12.4 Mice with Conditional Suppression of NF-L Expression

Recently, mice have been generated with doxycycline control of human NF-L transgene in context of the absence (tTA;hNF-L;NF-L–/–) or presence (tTA;hNF-L; NF-L±) of endogenous mouse NF-L proteins (Millecamps et al., 2007). These mice have been very useful to study the turnover and transport rate of NF proteins in vivo. After treatment with doxycycline, there was rapid loss of human NF-L mRNA within days but the human NF-L proteins remained with a half-life of 3 weeks in the CNS. Interestingly, in the sciatic nerve, the disappearance of NF-L proteins after suppression of transgene expression occurred in synchrony along the nerve, suggesting a proteolytic degradation of NF proteins along the entire axon. The results demonstrated that NF proteins have a very slow turnover in vivo. Another surprising observation was that, after re-induction of transgene expression by cessation of doxycycline treatment, there was no evidence of a leading wave of newly made human NF-L proteins migrating anterogradely into the sciatic nerve axons, as would be predicted from classical pulse chase studies with amino acid precursors. The NF-L proteins reappeared and accumulated in synchrony at similar rates along nerve segments, a phenomenon consistent with a fast transport of NF-L proteins into axons devoid of pre-existing NF structures (Millecamps et al., 2007). So, the transport rate of NF proteins is highly influenced by the presence of the existing NF network and it might differ in neuronal cell types with different NF content. In the absence of a pre-existing NF network, it was estimated that the transport rate of NF proteins was of about 10 mm/day which is 10 fold faster than the transport rate of NF proteins in axons having NF structures.

12.5 Conclusion

Our understanding of the regulation, structure and functions of NFs in neurons has been considerably improved by the characterization of knockout mice for NF proteins. Although NFs appear dispensable for nervous system development, deficiencies in NF proteins are not completely innocuous as significant loss of axons has been noticed in these models (Table 12.1). The gene knockout approach has confirmed a requirement of NFL together with NFM and NFH subunits for the in vivo formation of a NF network and has provided definite proof that NFs are crucial for the radial growth of large myelinated axons and thus to increase the conduction velocity. However, several aspects of NF biology remain elusive and further investigations are required to solve these issues. While it is clearly established that NFs play a central role in the growth and maintenance of the axonal caliber, the exact mechanism employed remains unanswered. Two surprising discoveries were, however, made using mouse models. First, NFM tail domain rather than NFH tail play an essential role in the expansion of the axonal diameter. Second, NFM KSP phosphorylation is not required for radial axonal growth and does not affect the axonal cytoarchitecture. A major challenge will be to elucidate how NFs, and in particular the NFM tail domain, control the radial axonal growth. Future studies are needed to determine the precise role of NF phosphorylation in modulating the axonal caliber and to identify the exact phosphorylation sites and signaling pathways involved.

Acknowledgments J.-P.J. holds a Canada Research Chair in neurodegeneration and R.P. is the recipient of the Tim E. Noël Fellowship (ALS Society of Canada).

References

Aranda-Espinoza H, Carl P, Leterrier JF, Janmey P, Discher DE (2002) Domain unfolding in neurofilament sidearms: effects of phosphorylation and ATP. FEBS Lett 531:397–401

Barclay M, Noakes PG, Ryan AF, Julien JP, Housley GD (2007) Neuronal expression of peripherin, a type III intermediate filament protein, in the mouse hindbrain. Histochem Cell Biol 128: 541–550

Beaulieu JM, Robertson J, Julien JP (1999) Interactions between peripherin and neurofilaments in cultured cells: disruption of peripherin assembly by the NF-M and NF-H subunits. Biochem Cell Biol 77:41–45

Brody BA, Ley CA, Parysek LM (1989) Selective distribution of the 57 kDa neural intermediate filament protein in the rat CNS. J Neurosci 9:2391–2401

Brown HG, Hoh JH (1997) Entropic exclusion by neurofilament sidearms: a mechanism for maintaining interfilament spacing. Biochemistry 36:15035–15040

Chiu FC, Barnes EA, Das K, Haley J, Socolow P, Macaluso FP, Fant J (1989) Characterization of a novel 66 kd subunit of mammalian neurofilaments. Neuron 2:1435–1445

Dubois M, Lalonde R, Julien JP, Strazielle C (2005a) Mice with the deleted neurofilament of low-molecular-weight (Nefl) gene: 1. Effects on regional brain metabolism. J Neurosci Res 80:741–750

Dubois M, Strazielle C, Julien JP, Lalonde R (2005b) Mice with the deleted neurofilament of low molecular weight (Nefl) gene: 2. Effects on motor functions and spatial orientation. J Neurosci Res 80:751–758

Elder GA, Friedrich VL Jr, Bosco P, Kang C, Gourov A, Tu PH, Lee VM, Lazzarini RA (1998a) Absence of the mid-sized neurofilament subunit decreases axonal calibers, levels of light neurofilament (NF-L), and neurofilament content. J Cell Biol 141:727–739

Elder GA, Friedrich VL Jr, Kang C, Bosco P, Gourov A, Tu PH, Zhang B, Lee VM, Lazzarini RA (1998b) Requirement of heavy neurofilament subunit in the development of axons with large calibers. J Cell Biol 143:195–205

Elder GA, Friedrich VL Jr, Margita A, Lazzarini RA (1999a) Age-related atrophy of motor axons in mice deficient in the mid-sized neurofilament subunit. J Cell Biol 146:181–192

Elder GA, Friedrich VL Jr, Pereira D, Tu PH, Zhang B, Lee VM, Lazzarini RA (1999b) Mice with disrupted midsized and heavy neurofilament genes lack axonal neurofilaments but have unaltered numbers of axonal microtubules. J Neurosci Res 57:23–32

Eriksson KS, Zhang S, Lin L, Lariviere RC, Julien JP, Mignot E (2008) The type III neurofilament peripherin is expressed in the tuberomammillary neurons of the mouse. BMC Neurosci 9:26

Escurat M, Djabali K, Gumpel M, Gros F, Portier MM (1990) Differential expression of two neuronal intermediate-filament proteins, peripherin and the low-molecular-mass neurofilament protein (NF-L), during the development of the rat. J Neurosci 10:764–784

Eyer J, Peterson A (1994) Neurofilament-deficient axons and perikaryal aggregates in viable transgenic mice expressing a neurofilament-beta-galactosidase fusion protein. Neuron 12: 389–405

Ferri GL, Sabani A, Abelli L, Polak JM, Dahl D, Portier MM (1990) Neuronal intermediate filaments in rat dorsal root ganglia: differential distribution of peripherin and neurofilament protein immunoreactivity and effect of capsaicin. Brain Res 515:331–335

Fliegner KH, Kaplan MP, Wood TL, Pintar JE, Liem RK (1994) Expression of the gene for the neuronal intermediate filament protein alpha-internexin coincides with the onset of neuronal differentiation in the developing rat nervous system. J Comp Neurol 342:161–173

Fornaro M, Lee JM, Raimondo S, Nicolino S, Geuna S, Giacobini-Robecchi M (2008) Neuronal intermediate filament expression in rat dorsal root ganglia sensory neurons: an in vivo and in vitro study. Neuroscience 153:1153–1163

Friede RL, Samorajski T (1970) Axon caliber related to neurofilaments and microtubules in sciatic nerve fibers of rats and mice. Anat Rec 167:379–387

Fuchs E, Weber K (1994) Intermediate filaments: structure, dynamics, function, and disease. Annu Rev Biochem 63:345–382

Garcia ML, Lobsiger CS, Shah SB, Deerinck TJ, Crum J, Young D, Ward CM, Crawford TO, Gotow T, Uchiyama Y, Ellisman MH, Calcutt NA, Cleveland DW (2003) NF-M is an essential target for the myelin-directed "outside-in" signaling cascade that mediates radial axonal growth. J Cell Biol 163:1011–1020

Garcia ML, Rao MV, Fujimoto J, Garcia VB, Shah SB, Crum J, Gotow T, Uchiyama Y, Ellisman M, Calcutt NA, Cleveland DW (2009) Phosphorylation of highly conserved neurofilament medium KSP repeats is not required for myelin-dependent radial axonal growth. J Neurosci 29: 1277–1284

Geisler N, Kaufmann E, Fischer S, Plessmann U, Weber K (1983) Neurofilament architecture combines structural principles of intermediate filaments with carboxy-terminal extensions increasing in size between triplet proteins. EMBO J 2:1295–1302

Geisler N, Vandekerckhove J, Weber K (1987) Location and sequence characterization of the major phosphorylation sites of the high molecular mass neurofilament proteins M and H. FEBS Lett 221:403–407

Glasgow E, Druger RK, Fuchs C, Lane WS, Schechter N (1994) Molecular cloning of gefiltin (ON1): serial expression of two new neurofilament mRNAs during optic nerve regeneration. EMBO J 13:297–305

Goldstein ME, Grant P, House SB, Henken DB, Gainer H (1996) Developmental regulation of two distinct neuronal phenotypes in rat dorsal root ganglia. Neuroscience 71:243–258

Goldstein ME, Sternberger LA, Sternberger NH (1987) Varying degrees of phosphorylation determine microheterogeneity of the heavy neurofilament polypeptide (Nf-H). J Neuroimmunol 14:135–148

Gorham JD, Baker H, Kegler D, Ziff EB (1990) The expression of the neuronal intermediate filament protein peripherin in the rat embryo. Brain Res Dev Brain Res 57:235–248

Gotow T, Takeda M, Tanaka T, Hashimoto PH (1992) Macromolecular structure of reassembled neurofilaments as revealed by the quick-freeze deep-etch mica method: difference between NF-M and NF-H subunits in their ability to form cross-bridges. Eur J Cell Biol 58:331–345

Hirokawa N, Glicksman MA, Willard MB (1984) Organization of mammalian neurofilament polypeptides within the neuronal cytoskeleton. J Cell Biol 98:1523–1536

Hisanaga S, Hirokawa N (1988) Structure of the peripheral domains of neurofilaments revealed by low angle rotary shadowing. J Mol Biol 202:297–305

Hoffman PN, Griffin JW, Price DL (1984) Control of axonal caliber by neurofilament transport. J Cell Biol 99:705–714

Izmiryan A, Franco CA, Paulin D, Li Z, Xue Z (2009) Synemin isoforms during mouse development: multiplicity of partners in vascular and neuronal systems. Exp Cell Res 315:769–783

Jacomy H, Zhu Q, Couillard-Despres S, Beaulieu JM, Julien JP (1999) Disruption of type IV intermediate filament network in mice lacking the neurofilament medium and heavy subunits. J Neurochem 73:972–984

Jones SM, Williams RC Jr (1982) Phosphate content of mammalian neurofilaments. J Biol Chem 257:9902–9905

Julien JP, Mushynski WE (1983) The distribution of phosphorylation sites among identified proteolytic fragments of mammalian neurofilaments. J Biol Chem 258:4019–4025

Kaplan MP, Chin SS, Fliegner KH, Liem RK (1990) Alpha-internexin, a novel neuronal intermediate filament protein, precedes the low molecular weight neurofilament protein (NF-L) in the developing rat brain. J Neurosci 10:2735–2748

Kriz J, Meier J, Julien JP, Padjen AL (2000a) Altered ionic conductances in axons of transgenic mouse expressing the human neurofilament heavy gene: a mouse model of amyotrophic lateral sclerosis. Exp Neurol 163:414–421

Kriz J, Zhu Q, Julien JP, Padjen AL (2000b) Electrophysiological properties of axons in mice lacking neurofilament subunit genes: disparity between conduction velocity and axon diameter in absence of NF-H. Brain Res 885:32–44

Kumar S, Hoh JH (2004) Modulation of repulsive forces between neurofilaments by sidearm phosphorylation. Biochem Biophys Res Commun 324:489–496

Lariviere RC, Julien JP (2004) Functions of intermediate filaments in neuronal development and disease. J Neurobiol 58:131–148

Lariviere RC, Nguyen MD, Ribeiro-da-Silva A, Julien JP (2002) Reduced number of unmyelinated sensory axons in peripherin null mice. J Neurochem 81:525–532

Leake D, Asch WS, Canger AK, Schechter N (1999) Gefiltin in zebrafish embryos: sequential gene expression of two neurofilament proteins in retinal ganglion cells. Differentiation 65:181–189

Lee VM, Otvos L Jr, Carden MJ, Hollosi M, Dietzschold B, Lazzarini RA (1988) Identification of the major multiphosphorylation site in mammalian neurofilaments. Proc Natl Acad Sci USA 85:1998–2002

Lee MK, Xu Z, Wong PC, Cleveland DW (1993) Neurofilaments are obligate heteropolymers in vivo. J Cell Biol 122:1337–1350

Letournel F, Bocquet A, Perrot R, Dechaume A, Guinut F, Eyer J, Barthelaix A (2006) Neurofilament high molecular weight-green fluorescent protein fusion is normally expressed in neurons and transported in axons: a neuronal marker to investigate the biology of neurofilaments. Neuroscience 137:103–111

Levavasseur F, Zhu Q, Julien JP (1999) No requirement of alpha-internexin for nervous system development and for radial growth of axons. Brain Res Mol Brain Res 69:104–112

Lewis SE, Nixon RA (1988) Multiple phosphorylated variants of the high molecular mass subunit of neurofilaments in axons of retinal cell neurons: characterization and evidence for their differential association with stationary and moving neurofilaments. J Cell Biol 107:2689–2701

Li ZH, Lu J, Tay SS, Wu YJ, Strong MJ, He BP (2006) Mice with targeted disruption of neurofilament light subunit display formation of protein aggregation in motoneurons and downregulation of complement receptor type 3 alpha subunit in microglia in the spinal cord at their earlier age: a possible feature in pre-clinical development of neurodegenerative diseases. Brain Res 1113:200–209

McLean JR, Sanelli TR, Leystra-Lantz C, He BP, Strong MJ (2005) Temporal profiles of neuronal degeneration, glial proliferation, and cell death in hNFL(+/+) and NFL(–/–) mice. Glia 52: 59–69

Millecamps S, Gowing G, Corti O, Mallet J, Julien JP (2007) Conditional NF-L transgene expression in mice for in vivo analysis of turnover and transport rate of neurofilaments. J Neurosci 27:4947–4956

Nguyen MD, Lariviere RC, Julien JP (2000) Reduction of axonal caliber does not alleviate motor neuron disease caused by mutant superoxide dismutase 1. Proc Natl Acad Sci USA 97: 12306–12311

Nixon RA, Brown BA, Marotta CA (1982) Posttranslational modification of a neurofilament protein during axoplasmic transport: implications for regional specialization of CNS axons. J Cell Biol 94:150–158

Nixon RA, Paskevich PA, Sihag RK, Thayer CY (1994) Phosphorylation on carboxyl terminus domains of neurofilament proteins in retinal ganglion cell neurons in vivo: influences on regional neurofilament accumulation, interneurofilament spacing, and axon caliber. J Cell Biol 126:1031–1046

Ohara O, Gahara Y, Miyake T, Teraoka H, Kitamura T (1993) Neurofilament deficiency in quail caused by nonsense mutation in neurofilament-L gene. J Cell Biol 121:387–395

Pant HC, Veeranna (1995) Neurofilament phosphorylation. Biochem Cell Biol 73:575–592

Perrot R, Berges R, Bocquet A, Eyer J (2008) Review of the multiple aspects of neurofilament functions, and their possible contribution to neurodegeneration. Mol Neurobiol 38:27–65

Perrot R, Lonchampt P, Peterson AC, Eyer J (2007) Axonal neurofilaments control multiple fiber properties but do not influence structure or spacing of nodes of Ranvier. J Neurosci 27: 9573–9584

Rao MV, Campbell J, Yuan A, Kumar A, Gotow T, Uchiyama Y, Nixon RA (2003) The neurofilament middle molecular mass subunit carboxyl-terminal tail domains is essential for the radial growth and cytoskeletal architecture of axons but not for regulating neurofilament transport rate. J Cell Biol 163:1021–1031

Rao MV, Garcia ML, Miyazaki Y, Gotow T, Yuan A, Mattina S, Ward CM, Calcutt NA, Uchiyama Y, Nixon RA, Cleveland DW (2002) Gene replacement in mice reveals that the heavily phosphorylated tail of neurofilament heavy subunit does not affect axonal caliber or the transit of cargoes in slow axonal transport. J Cell Biol 158:681–693

Rao MV, Houseweart MK, Williamson TL, Crawford TO, Folmer J, Cleveland DW (1998) Neurofilament-dependent radial growth of motor axons and axonal organization of neurofilaments does not require the neurofilament heavy subunit (NF-H) or its phosphorylation. J Cell Biol 143:171–181

Sakaguchi T, Okada M, Kitamura T, Kawasaki K (1993) Reduced diameter and conduction velocity of myelinated fibers in the sciatic nerve of a neurofilament-deficient mutant quail. Neurosci Lett 153:65–68

Sanchez I, Hassinger L, Paskevich PA, Shine HD, Nixon RA (1996) Oligodendroglia regulate the regional expansion of axon caliber and local accumulation of neurofilaments during development independently of myelin formation. J Neurosci 16:5095–5105

Shea TB, Beermann ML (1999) Neuronal intermediate filament protein alpha-internexin facilitates axonal neurite elongation in neuroblastoma cells. Cell Motil Cytoskeleton 43:322–333

Tu PH, Robinson KA, de Snoo F, Eyer J, Peterson A, Lee VM, Trojanowski JQ (1997) Selective degeneration fo Purkinje cells with Lewy body-like inclusions in aged NFHLACZ transgenic mice. J Neurosci 17:1064–1074

Williamson TL, Bruijn LI, Zhu Q, Anderson KL, Anderson SD, Julien JP, Cleveland DW (1998) Absence of neurofilaments reduces the selective vulnerability of motor neurons and slows disease caused by a familial amyotrophic lateral sclerosis-linked superoxide dismutase 1 mutant. Proc Natl Acad Sci USA 95:9631–9636

Wong J, Oblinger MM (1990) Differential regulation of peripherin and neurofilament gene expression in regenerating rat DRG neurons. J Neurosci Res 27:332–341

Yamasaki H, Itakura C, Mizutani M (1991) Hereditary hypotrophic axonopathy with neurofilament deficiency in a mutant strain of the Japanese quail. Acta Neuropathol 82:427–434

Yan Y, Jensen K, Brown A (2007) The polypeptide composition of moving and stationary neurofilaments in cultured sympathetic neurons. Cell Motil Cytoskeleton 64:299–309

Yuan A, Rao MV, Kumar A, Julien JP, Nixon RA (2003) Neurofilament transport in vivo minimally requires hetero-oligomer formation. J Neurosci 23:9452–9458

Yuan A, Rao MV, Sasaki T, Chen Y, Kumar A, Veeranna, Liem RK, Eyer J, Peterson AC, Julien JP, Nixon RA (2006) Alpha-internexin is structurally and functionally associated with the neurofilament triplet proteins in the mature CNS. J Neurosci 26:10006–10019

Zhao Y, Szaro BG (1997) Xefiltin, a Xenopus laevis neuronal intermediate filament protein, is expressed in actively growing optic axons during development and regeneration. J Neurobiol 33:811–824

Zhu Q, Couillard-Despres S, Julien JP (1997) Delayed maturation of regenerating myelinated axons in mice lacking neurofilaments. Exp Neurol 148:299–316

Zhu Q, Lindenbaum M, Levavasseur F, Jacomy H, Julien JP (1998) Disruption of the NF-H gene increases axonal microtubule content and velocity of neurofilament transport: relief of axonopathy resulting from the toxin beta,beta'-iminodipropionitrile. J Cell Biol 143:183–193

Zochodne DW, Sun HS, Cheng C, Eyer J (2004) Accelerated diabetic neuropathy in axons without neurofilaments. Brain 127:2193–2200

Chapter 13
Neurofilaments and Radial Growth: Deconstruction of a Hypothesis Through the Construction of Gene Targeted Mice

Michael L. Garcia and Devin M. Barry

Abstract Radial axonal growth determines the diameter of myelinated axons, which is one major axonal property that influences the rate of impulse propagation along the axon. Prior to the development of neurofilament gene-targeted mice, it was thought that radial growth was mediated by myelin-dependent phosphorylation of neurofilament heavy and neurofilament medium within the axon. The systematic analysis of gene-targeted mice has furthered our understanding of the role of neurofilaments in radial growth but these analyses have also led to surprising new insights that have challenged the proposed role of neurofilament carboxy terminal phosphorylation in establishing axonal diameter. In this review, the contribution of each neurofilament gene-targeted mouse to our current understanding of the role of neurofilaments and neurofilament phosphorylation in radial axonal growth is discussed.

Keywords Conduction velocity · Gene targeting · Myelination · Neurofilament · Peripheral nerves · Phosphorylation · Radial growth

Contents

M.L. Garcia (✉)
Department of Biological Sciences, Bond Life Sciences Center, University of Missouri-Columbia, Columbia, MO 65211, USA
e-mail: garciaml@missouri.edu

R.A. Nixon, A. Yuan (eds.), *Cytoskeleton of the Nervous System*, Advances in Neurobiology 3, DOI 10.1007/978-1-4419-6787-9_13,

13.1 Introduction – Neurofilaments

Neurofilaments (NFs) are obligate heteropolymers of neurofilament light (NF-L), medium (NF-M), heavy (NF-H) and α-internexin (Yuan et al., 2006b) subunit proteins (Ching and Liem, 1993; Lee et al., 1993), and are type IV members of the intermediate filament (IF) family (Fuchs and Weber, 1994). All three subunits have an amino terminal (NH_2-terminal) head domain, a central rod domain and a carboxy terminal (C-terminal) tail domain (Fig. 13.1). These three domains have different

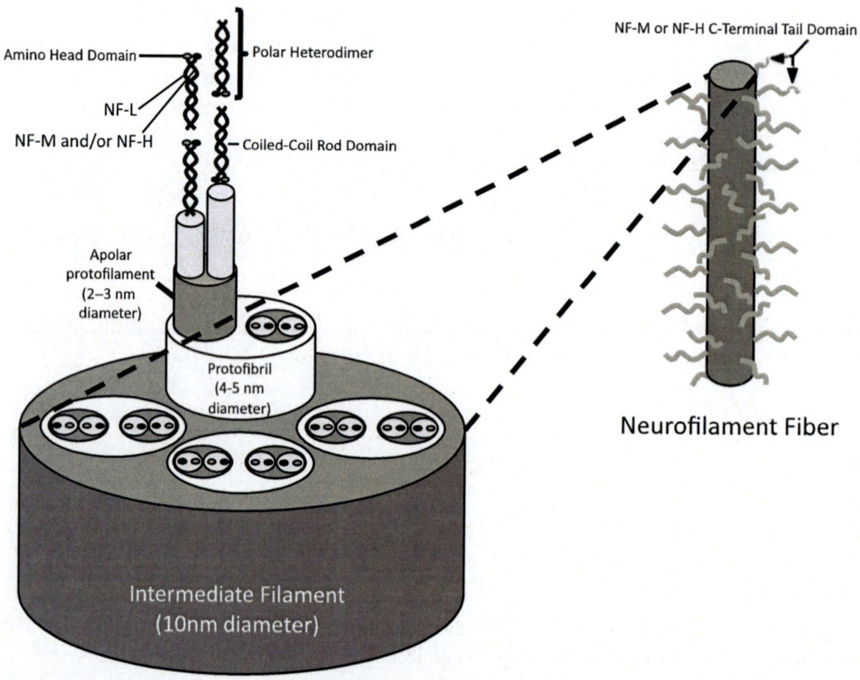

Fig. 13.1 Formation of 10-nm filament from individual neurofilament subunits. **a** Schematic of neurofilament assembly into 10-nm filament that is characteristic of all intermediate filaments. Neurofilament subunits form polar heterodimers that assemble into half-staggered, anti-parallel strands to form protofilaments. Two protofilaments polymerize to form protofibrils. Two to four protofibrils polymerize to form the 10-nm filament that is characteristic of intermediate filaments. **b** Schematic of neurofilament with NF-M and NF-H side arm projections from the core of the filament (adapted from Fuchs and Cleveland, 1998)

effects on assembly into a 10-nm filament that is characteristic of all members of the IF protein family. The NH_2-terminal head domain promotes lateral association of protofilaments into protofibrils and ultimately 10-nm filaments (Heins et al., 1993). The central rod domain contains a highly conserved ~310-amino acid α-helical domain containing a hydrophobic heptad repeat that is essential for assembly into a 10-nm filament (Fuchs and Weber, 1994). The C-terminal tail domain of NF-L controls lateral assembly of protofilaments so that the filament terminates at 10-nm (Heins et al., 1993), whereas the C-terminal tail domains of NF-H and NF-M project from the core of the 10-nm filament (Hirokawa et al., 1984). While neurofilaments consist of all three subunits, NF-L is essential for the assembly of NF-H and NF-M into 10-nm filaments (Lee et al., 1993; Liem and Hutchison, 1982).

Amino acids within the NH_2-terminal head and C-terminal tail domain are post-translationally modified. The NH_2-terminal head domains of all three NF subunits contain multiple serine and threonine residues that are potential sites for post-translational modification. Indeed, O-linked glycosylation sites have been identified within the head domains of all three subunits (Dong et al., 1993, 1996). Moreover, NF-L (Sihag and Nixon, 1991) and NF-M (Sihag and Nixon, 1990) are both serine phosphorylated within the NH_2-terminal head domain. Mimicking constitutive phosphorylation, by mutating serine 55 to aspartate (NF-L^{S55D}) in the NH_2-terminal head domain of NF-L, results in NF-L aggregates in cultured cells (Gibb et al., 1996). Additionally, expression of low levels of NF-L^{S55D} transgene results in central nervous system (CNS) pathology that is associated with the formation of neurofilament inclusions within neuronal cell bodies (Gibb et al., 1998). These data suggest that phosphorylation within the NH_2-terminal head domain of NF-L alters higher order neurofilament organization (Fig. 13.1).

Unlike other IF proteins, neurofilament heteropolymers have side arms that project from the core of the 10 nm filament (Hirokawa et al., 1984). These side arms consist of the C-termini of both NF-M and NF-H proteins (Hisanaga and Hirokawa, 1988). Moreover, both C-termini contain central repeats, the majority of which consist of the trio of amino acids lysine (K), serine (S) and proline (P) in one of three possible combinations: KSP, KXSP and KXXSP (Julien et al., 1988; Levy et al., 1987; Myers et al., 1987). The C-terminus of NF-M also contains a single, highly conserved variant, lysine-serine-aspartate (KSD) (Levy et al., 1987; Myers et al., 1987). (For simplicity, all repeats will be referred to as KSP repeats for the remainder of this review.)

The KSP repeats are posttranslationally modified. The most abundant posttranslational modification of neurofilament C-termini is phosphorylation of the KSP repeat serine residues (Julien and Mushynski, 1982; 1983). However, the C-terminus of NF-H has several O-linked glycosylation sites within the KSP repeat domain (Dong et al., 1996). Recently, O-linked glycosylation sites have also been identified within the KSP repeat domain of NF-M (Deng et al., 2008; Ludemann et al., 2005). Interestingly, O-linked glycosylation is reciprocally regulated with phosphorylation (Deng et al., 2008; Ludemann et al., 2005).

13.2 Introduction – Radial Growth

Motor neurons undergo two distinctive phases of growth each dependent upon different components of the neuronal cytoskeleton. During development, motor neurons extend axons from the ventral horn of the spinal cord to their prospective targets. This phase of growth is dependent upon actin/myosin for growth cone motility and microtubules for timely delivery of proteins from the cell body and to provide direction to the growing axon terminus (Letourneau, 1996). After stable synapse formation, a second phase of growth (Cleveland, 1996), referred to as radial axonal growth, is initiated in which neurofilaments accumulate and axonal diameter increases up to tenfold. Specification of axon diameter is a key component of neuronal function as it is one major axonal property that influences the velocity of electrical signal conduction.

Radial growth initiates with, and is dependent on, formation of compact myelin (de Waegh et al., 1992), which begins on the first postnatal day (PD1), and is complete by PD14 in mice (Mirsky et al., 2002). During radial growth, neurofilaments become the most abundant cytoskeletal proteins in the axon (Lee and Cleveland, 1996). Indeed, neurofilament number has been linearly correlated with the cross sectional area of an axon (Friede and Samorajski, 1970; Hoffman et al., 1985; Nixon et al., 1994; Weiss and Mayr, 1971). However, the generation of transgenic mice suggested that neurofilament composition (Xu et al., 1996), not simply the number of neurofilaments (Monteiro et al., 1990), regulates radial growth.

Mechanistically, it has been proposed that myelination regulates neurofilament phosphorylation, which in turn influences radial growth (Lee and Cleveland, 1996). In myelinated regions of central (CNS) and peripheral nervous system (PNS) axons, NF-H and NF-M are more heavily phosphorylated on their C-terminal tail domains (de Waegh et al., 1992; Hsieh et al., 1994; Nixon et al., 1994; Starr et al., 1996; Yin et al., 1998). Moreover, axonal diameter is increased in myelinated regions of CNS (Nixon et al., 1994) and PNS (de Waegh et al., 1992; Hsieh et al., 1994; Yin et al., 1998) axons relative to unmyelinated regions of the same axon. These data have laid the foundation for the myelin-dependent neurofilament phosphorylation hypothesis of radial growth, which has been the proposed mechanism for over 15 years. Analysis of neurofilament gene-deleted mice has furthered our understanding of the role of neurofilaments in radial growth by identifying the required subunit. However, new gene-targeted mice led to surprising new insights that have challenged the proposed role of neurofilament C-terminal phosphorylation in establishing axonal diameter.

13.3 Preventing Neurofilament Axonal Accumulation: Neurofilaments Are Required for Radial Growth

The first gene deletions of neurofilaments were not the result of gene targeting in embryonic stem cells. In fact, the first model to address unequivocally the role of neurofilaments in myelin-dependent radial axonal growth was not even a mouse.

We will begin our review of neurofilament gene targeting with these non-traditional gene deletion models.

13.3.1 The First Gene Deletions: Intriguing Results from Unlikely Sources

The first gene deletion of neurofilaments occurred in an unlikely source, the Japanese quail *Coturnix coturnix japonica*. Myelinated axons of both the central and peripheral nervous system of the quiver (Quv) quail were hypotrophic (Yamasaki et al., 1991) and devoid of neurofilaments (Yamasaki et al., 1992, 1991). Loss of axonal neurofilament accumulation was due to the introduction of a nonsense mutation in the NF-L gene (Ohara et al., 1993). Failure of radial axonal growth in Quv quail resulted in reduced conduction velocities (Sakaguchi et al., 1993), generalized quivering and mild ataxia (Yamasaki et al., 1991). This established unequivocally that axonal accumulation of neurofilaments is required for radial axonal growth.

The dependence of radial axonal growth on axonal accumulation of neurofilaments was confirmed, serendipitously, by the development of a transgenic line of mice that was originally developed as a means of analyzing NF-H distribution. The C-terminus of NF-H was replaced with the complete enzyme β-galactosidase (NF-H-LacZ) (Eyer and Peterson, 1994). Neurofilaments failed to transport into the majority of axons. Instead they formed large protein aggregates within neuronal cell bodies (Eyer and Peterson, 1994). As in the Quv quail, the axons of this mouse failed to grow radially, resulting in small diameter axons (Eyer and Peterson, 1994). Moreover, NF-H-LacZ mice had a ~24% reduction in the number of ventral root axons in aged mice (Eyer and Peterson, 1994). Unlike the Quv quail, no overt phenotype was observed in these mice. However, when challenged, these mice exhibited abnormalities that were consistent with sensorimotor deficits (Dubois et al., 2002).

13.3.2 Deletion of Neurofilament Genes Determines Which Subunits Are Required for Radial Growth

Gene targeting has been performed for all three neurofilament genes as well as α-internexin, which was recently shown to associate with neurofilament proteins (Yuan et al., 2006b). Moreover, many of these models have been interbred. For simplicity, we will focus on the original deletions of neurofilament genes.

13.3.2.1 Deletion of NF-L Confirms Dependence of Radial Growth on Axonal Accumulation of Neurofilaments

Deletion of NF-L (NF-L$^{-/-}$) in mice recapitulated many of the results observed in both the Quv quail and the NF-H-LacZ transgenic mouse. Specifically, axons of the sciatic nerve were devoid of neurofilaments and failed to grow radially (Zhu et al., 1997). Moreover, conduction velocity in NF-L$^{-/-}$ mice was severely reduced (Kriz et al., 2000). However, there were important differences observed between

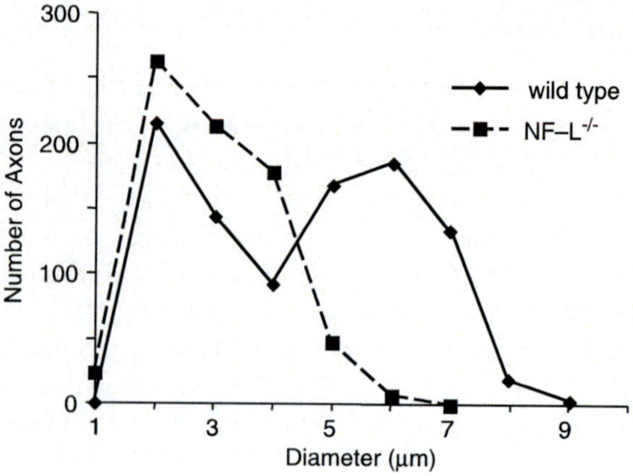

Fig. 13.2 Radial axonal growth is dependent upon neurofilaments. Loss of axonal neurofilaments, through targeted deletion of NF-L, resulted in axons that failed to grow radially. Peak diameter distributions were decreased by ~67% relative to wild type littermates in 8-week-old mice. Moreover, axons of the fifth lumbar ventral root failed to establish a bimodal distribution (reprinted from Zhu et al., 1997)

the NF-L$^{-/-}$ mice and the two previous models. Both Quv quail and NF-H-LacZ mice maintained a bimodal distribution in axons of the sciatic nerve (Sakaguchi et al., 1993) or axons of the fourth lumbar ventral root (Eyer and Peterson, 1994), albeit the peak diameter of the largest axons, in both of these models, was reduced relative to wild type controls (Eyer and Peterson, 1994; Sakaguchi et al., 1993). Deletion of NF-L resulted in the most severe reduction in axonal diameter of any of these three models. Axons of the fifth lumbar ventral root in NF-L$^{-/-}$ mice failed to establish a bimodal distribution with a reduction in peak axonal diameter of ~67% in NF-L$^{-/-}$ mice (Fig. 13.2) as compared to littermate controls (Zhu et al., 1997). The reduction in axonal diameter resulted in a ~70% reduction in neuronal conduction velocity (Kriz et al., 2000). Additionally, NF-L$^{-/-}$ mice displayed an early ~10% reduction in the number of ventral root axons (Zhu et al., 1997). Despite the reduction in conduction velocity and the loss of motor axons, NF-L$^{-/-}$ mice did not display any overt phenotypes. However, as with the NF-H-LacZ mice, when challenged, NF-L$^{-/-}$ mice exhibited abnormalities that were consistent with sensorimotor deficits (Dubois et al., 2005). Finally, NF-L$^{-/-}$ mice had increased α-tubulin content in sciatic nerve axons suggesting a compensatory increase in microtubule content due to loss of neurofilaments (Zhu et al., 1997). Taken together, these three models unequivocally established a role of neurofilaments in determining axonal diameter, thereby influencing the rate of neuronal conduction velocity.

13.3.2.2 Deletion of NF-H and NF-M: Identification of NF-M as the Required Neurofilament Subunit for Radial Growth

Axons of Quv quail, NF-H-LacZ and NF-L$^{-/-}$ mice established that neurofilaments are required for radial growth. However, since axons in all these models are completely devoid of neurofilaments, what remained unclear was which neurofilament protein subunit was required for radial growth. In myelinated axonal segments, axonal diameter is increased relative to unmyelinated segments of the same axon (de Waegh et al., 1992; Hsieh et al., 1994; Yin et al., 1998) and NF-H and NF-M are heavily phosphorylated on their C-terminal tail domains, suggesting a potential role for phosphorylation in radial growth (de Waegh et al., 1992; Hsieh et al., 1994; Nixon et al., 1994). The most heavily phosphorylated subunit is NF-H, which is nearly stoichiometrically phosphorylated on KSP repeat serine residues (Julien and Mushynski, 1982, 1983). If C-terminal phosphorylation of neurofilaments is essential for radial growth, it was reasonable to conclude that the most heavily phosphorylated subunit would contribute the most to radial growth. However, the analysis of NF-M and NF-H gene deleted mice yielded surprising results.

Deletion of NF-M (NF-M$^{-/-}$) resulted in a substantial decrease in radial axonal growth (Elder et al., 1998a). Axons of the fifth lumbar ventral root failed to establish a bimodal distribution (Fig. 13.3) as was observed in NF-L$^{-/-}$ mice (Zhu et al., 1997). Peak axonal diameter was reduced by ~43% in NF-M$^{-/-}$ mice vs littermate controls (Elder et al., 1998a). Overall neurofilament content was reduced in axons of the fifth lumbar ventral root in NF-M$^{-/-}$ relative to control (Elder et al., 1998a). Additionally, axonal diameters were reduced within the ventral spinal cord and optic nerve in NF-M$^{-/-}$ mice (Elder et al., 1998a). As expected, due to decreased axonal diameter, neuronal conduction velocity was ~40% reduced in NF-M$^{-/-}$ mice vs wild type (Kriz et al., 2000). These results suggested that NF-M was required for radial growth.

Deletion of NF-H supported the requirement of NF-M for radial growth. Three independent lines of NF-H gene-targeted (NF-H$^{-/-}$) mice were developed. While the particulars of each experiment differed slightly, the overwhelming message is that loss of NF-H did not result in severe reductions in radial growth as predicted (Elder et al., 1998b; Rao et al., 1998; Zhu et al., 1998). The peak diameter of the largest motor axons of the fifth lumbar ventral root was reduced (Fig. 13.4) by ~10–20%

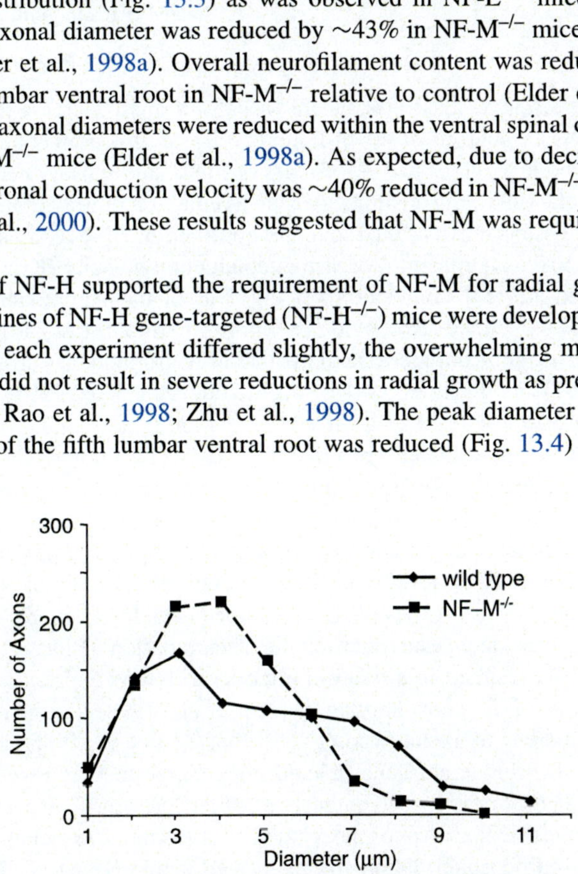

Fig. 13.3 NF-M is essential for radial growth. Targeted deletion of the NF-M subunit resulted in a failure of radial growth. Peak diameter distributions of axons of the fifth lumbar ventral root were ~43% reduced relative to wild type mice in 16-week-old mice. Moreover, similarly to NF-L$^{-/-}$ mice, NF-M$^{-/-}$ mice failed to establish a bimodal distribution. Axon numbers were approximated using the average number of axons (927±99 n=26) reported in (Clement et al., 2003) and the originally reported percentiles (© Elder et al., 1998a)

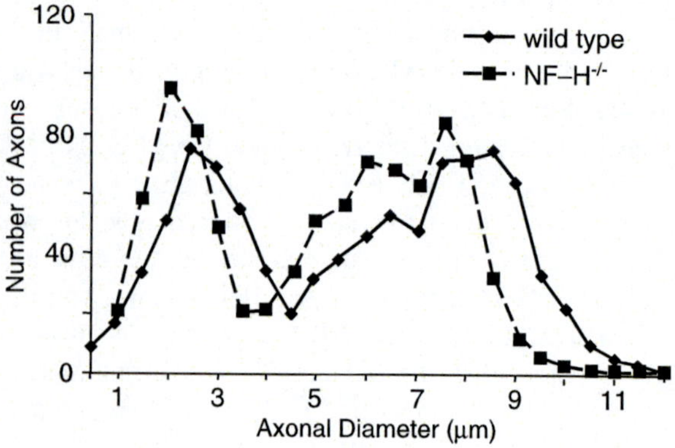

Fig. 13.4 Radial axonal growth is not dependent upon NF-H. Targeted deletion of NF-H resulted in a reduction in radial growth. Peak diameter distributions of axons of the fifth lumbar ventral root were reduced for both large and small caliber axons at 9 weeks (© Rao et al., 1998)

(Rao et al., 1998; Zhu et al., 1998). There was also a small decrease in peak axonal diameter in sensory neurons of the fifth lumbar dorsal root (Rao et al., 1998). Loss of NF-H resulted in a ~13% and ~19% loss of motor and sensory neurons, respectively (Rao et al., 1998). Surprisingly, despite a reduction in peak diameter that was approximately half as much as the reduction observed in NF-M$^{-/-}$ mice (Fig. 13.4), NF-H$^{-/-}$ mice had a significant decrease in conduction velocity (Kriz et al., 2000). The decrease in NF-H$^{-/-}$ conduction velocity was of the same magnitude as that observed in NF-M$^{-/-}$ mice (Kriz et al., 2000). Furthermore, loss of the NF-H subunit resulted in a significant increase in the rate of neurofilament transport in sciatic (Zhu et al., 1998) and optic nerves (Rao et al., 2002; Zhu et al., 1998).

Taken together, these results suggested that the failure of radial growth observed in axons devoid of neurofilaments resulted from loss of NF-M, not NF-H as was predicted, indicating that the NF-M subunit was essential for radial growth. However, interpretation of the NF-M$^{-/-}$ and NF-H$^{-/-}$ phenotypes was complicated by compensatory alterations in the remaining cytoskeletal components. Deletion of NF-M resulted in decreased NF-L levels and decreased phosphorylation of NF-H in spinal cord (Elder et al., 1998a). Axons of NF-M$^{-/-}$ mice have fewer, more widely spaced neurofilaments with increased microtubule accumulation (Elder et al., 1998a). Deletion of NF-H resulted in a twofold increase in NF-M expression (Rao et al., 1998), increased NF-M phosphorylation (Zhu et al., 1998) and increased microtubule accumulations in axons (Rao et al., 1998; Zhu et al., 1998). Despite these complications, neurofilament gene deleted mice supported the myelin-dependent neurofilament phosphorylation hypothesis of radial growth. However, analysis of these mice suggested that if phosphorylation was essential to radial growth, then the critical sites of phosphorylation will be located within NF-M, not NF-H, as was widely expected.

13.3.3 Truncation of NF-H and NF-M Identifies Different Roles for Each Subunit's C-Terminus

Classic gene deletions demonstrated the role for neurofilaments as necessary components of the cytoskeleton for the development of the nervous system. Moreover, gene deleted mice indicated that NF-M, not NF-H, was critical for radial growth. The generation of mice expressing C-terminally truncated neurofilaments not only supported these observations, but also identified the critical domain of NF-M, and suggested a role for the C-terminal tail of NF-H.

13.3.3.1 Expressing C-Terminally Truncated NF-H: No Role for the Highly Phosphorylated Tail Domain in Determining Axonal Diameter

Deletion of NF-H suggested that this subunit does not play a critical role in radial axonal growth. However, loss of a single neurofilament subunit was associated with compensatory increases in the remaining cytoskeletal components. To address directly the role of neurofilament C-terminal tail domains during radial growth, gene-targeted mice were generated that express C-terminally truncated neurofilaments. Additionally, replacing full-length neurofilaments with C-terminally truncated neurofilaments did not result in compensatory increases in the expression of other cytoskeletal proteins.

The first neurofilament gene targeting was performed on NF-H (referred to as NF-H$^{tail\Delta}$ mice). The C-terminal 612 amino acids were replaced with a c-myc tag resulting in loss of all KSP repeats without compensatory increases in remaining cytoskeletal proteins (Rao et al., 2002). Loss of the entire C-terminal tail domain, including all 51 phosphorylation sites, resulted in minor axonal alterations. Radial growth was measured at 2 and 6 months in axons of the fifth lumbar ventral root from both NF-H$^{tail\Delta}$ mice and wild type littermates (Fig. 13.5). At 2 months there was an ~18% decrease in axonal diameter for the largest motor axons relative to wild type (Garcia et al., 2003; Rao et al., 2002). There was also a reduction in the peak diameter of small motor axons at 2 months (Garcia et al., 2003; Rao et al.,

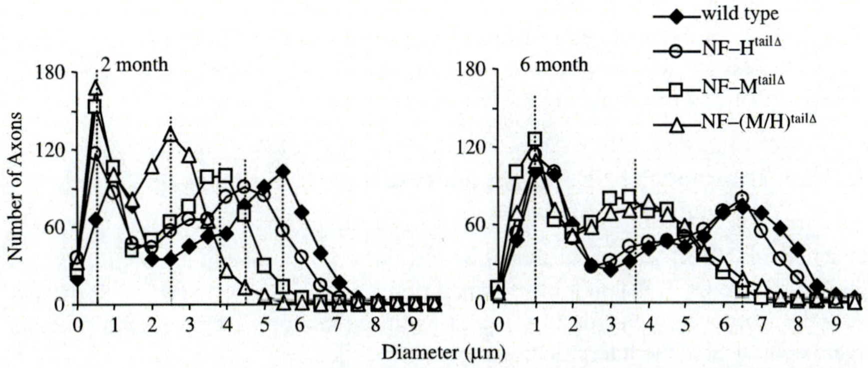

Fig. 13.5 The C-terminal tail domain of NF-M is required for radial growth. Expression of C-terminally truncated neurofilaments indicated that the C-terminal 426 amino acids of NF-M were critical for radial growth. At 6 months, peak diameter distribution was reduced by ~42% in NF-M$^{tail\Delta}$ mice relative to control. However, by 6 months, peak distributions in mice expressing C-terminally truncated NF-H (NF-H$^{tail\Delta}$) were indistinguishable from wild type. At 2 months, NF-(M/H)$^{tail\Delta}$ mice showed the initial delay in radial growth observed in NF-H$^{tail\Delta}$ mice, whereas, by 6 months, NF-(M/H)$^{tail\Delta}$ mice obtained a peak diameter that is indistinguishable from NF-M$^{tail\Delta}$ mice (© Garcia et al., 2003)

2002). However, by 6 months the peak in axonal diameter was indistinguishable from wild type (Fig. 13.5) for both populations of motor axons (Garcia et al., 2003; Rao et al., 2002). At this time point, only the very largest motor axons (greater than 7 μm) were reduced relative to wild type (Garcia et al., 2003; Rao et al., 2002). Neuronal conduction velocity was measured in the sciatic nerve of 5-month-old mice. Unlike loss of the entire NF-H subunit, loss of the NF-H C-terminus did not affect conduction velocity (Rao et al., 2002).

Despite the initial reduction in axonal caliber at 2 months, peak axonal diameters were indistinguishable by 6 months (Fig. 13.5). This suggested that the rate of radial growth in NF-$H^{tail\Delta}$ mice was faster than in wild type littermates, as axons from NF-$H^{tail\Delta}$ mice would have to grow more over the same time span to have overlapping peaks in diameter with their wild type littermates. Taken together, these data suggested that the highly phosphorylated C-terminal tail of NF-H did not influence the overall size of myelinated axons, but did influence the rate of radial growth.

Mechanistically, radial growth and neurofilament transport have been linked (Hoffman et al., 1984; Marszalek et al., 1996). Deletion of the entire NF-H subunit increased the rate of neurofilament transport in both the sciatic (Zhu et al., 1998) and optic nerves (Rao et al., 2002). NF-H dependent slowing of the rate of neurofilament transport has been linked to increased NF-H C-terminal phosphorylation (Ackerley et al., 2003; Jung and Shea, 1999; Jung et al., 2000; Nixon et al., 1994). However, the role of C-terminal NF-H phosphorylation in influencing rates of neurofilament transport has become controversial. Initially, it was reported that truncation of NF-H resulted in increased rates of neurofilament transport in the optic nerve (Rao et al., 2002). However, in a subsequent report, NF-H truncation did not affect the rate of neurofilament transport in the optic nerve (Yuan et al., 2006a). The differences between the two reports were attributed to experimental variation (Yuan et al., 2006a). Alternatively, C-terminal truncation of NF-H may have failed to alter transport rates due to compensatory phosphorylation of NF-M (Rao et al., 2002). Analysis of neurofilament transport in mice expressing both truncated NF-H and NF-M will be necessary to address sufficiently the role of C-terminal phosphorylation in determining the rate of neurofilament transport.

13.3.3.2 Truncation of NF-M Determines that the C-Terminus Is Required for Radial Growth

Deletion of NF-M suggested that this was the critical subunit for radial axonal growth. However, it was not clear which domain of NF-M was critical for radial growth. Moreover, compensatory increases in the remaining cytoskeletal proteins again complicated the interpretation of NF-$M^{-/-}$ mice.

Gene targeting of NF-M was also performed (referred to as NF-$M^{tail\Delta}$ mice). Much like NF-H, the C-terminal 426 amino acid tail domain was replaced with a c-myc tag resulting in loss of all KSP repeats (Garcia et al., 2003; Rao et al., 2003). Truncation of NF-M did not result in increased expression of the remaining cytoskeletal proteins (Garcia et al., 2003; Rao et al., 2003). However, much like truncation of NF-H, preventing NF-M phosphorylation by C-terminal truncation resulted in compensatory phosphorylation of NF-H (Rao et al., 2003) and the microtubule associated protein, tau (Lobsiger et al., 2005).

Radial growth was measured at 2 and 6 months in axons of the fifth lumbar ventral root from both NF-$M^{tail\Delta}$ mice and wild type littermates (Fig. 13.5). The peak axonal diameter was reduced in both small and large motor axons as early as 2 months, with the largest motor axons reduced by ~29% (Garcia et al., 2003). Surprisingly, radial growth did not occur in the largest motor axons from 2 to 6 months in NF-$M^{tail\Delta}$ mice (Fig. 13.5). Therefore, by 6 months, the peak diameter in the largest motor axons was ~42% smaller in NF-$M^{tail\Delta}$ mice vs wild type littermate controls, resulting in a ~30% reduction in sciatic nerve conduction velocity (Garcia et al., 2003). There was no difference in peak diameter in the smallest motor neurons by 6 months (Garcia et al., 2003). Interestingly, this reduction in peak diameter distribution was the same magnitude as that observed in NF-$M^{-/-}$ mice (Elder et al., 1998a), with the exception that NF-$M^{tail\Delta}$ mice established a bimodal distribution (Garcia et al., 2003). Taken together, these data suggest that the NH_2-terminal head and rod domains of NF-M provided sufficient structural support for myelinated axons to distribute into two groups forming the characteristic bimodal distribution of motor axons and the C-terminal tail domain contributed to radial growth after a bimodal distribution was established. Compensatory phosphorylation of cytoskeletal proteins did occur in NF-$M^{tail\Delta}$ mice (Lobsiger et al., 2005; Rao et al., 2003). However, increased C-terminal NF-H phosphorylation was not sufficient to compensate for the loss of the NF-M C-terminus.

Analysis of neurofilament transport in the optic nerve of NF-$M^{tail\Delta}$ mice suggested that phosphorylation of NF-M C-terminal tail domain did not contribute to the rate of neurofilament transport (Rao et al., 2003). Unaltered rates of neurofilament transport may again be due to compensatory phosphorylation of the NF-H C-terminus. Elimination of this possibility will require an analysis of neurofilament transport in mice expressing both truncated NF-H and NF-M.

13.3.3.3 Expressing Truncated NF-H and NF-M: A Compound Phenotype that Supports the Individual Roles of Each C-Terminus

The analysis of mice expressing either C-terminally truncated NF-H or NF-M suggested non-overlapping roles for the NF-H and NF-M C-termini in myelin-dependent radial growth. These roles were confirmed in mice expressing both C-terminally truncated NF-H and NF-M (referred to as NF-$(M/H)^{tail\Delta}$ mice).

Radial growth was measured at 2 and 6 months in axons of the fifth lumbar ventral root from both NF-$(M/H)^{tail\Delta}$ mice and wild type littermates (Fig. 13.5). The peak in axonal diameter distribution was reduced for both large and small axonal populations at 2 months (Garcia et al., 2003). The largest axons were ~55% reduced, which is greater than the reduction observed in NF-$M^{tail\Delta}$ mice at this time point (Garcia et al., 2003). By 6 months, the peak for the smaller motor axons was indistinguishable from wild type (Garcia et al., 2003). Unlike NF-$M^{tail\Delta}$ mice, large axons in NF-$(M/H)^{tail\Delta}$ mice grew from 2 to 6 months (Fig. 13.5). Interestingly, large axons grew to the same size as NF-M$^{tail\Delta}$ mice resulting in a peak diameter distribution that was ~42% smaller than wild type littermates with conduction velocities that were ~30% reduced (Garcia et al., 2003).

Radial growth profiles in NF-(M/H)$^{tail\Delta}$ mice supported the roles for NF-H and NF-M C-terminal tails that were initially suggested in mice expressing single truncations. At 2 months, NF-(M/H)$^{tail\Delta}$ mice had a compound phenotype. Peak diameters in the largest motor axons were reduced in NF-(M/H)$^{tail\Delta}$ mice relative to NF-M$^{tail\Delta}$ mice as the loss of the NF-H C-terminus delayed radial growth (Garcia et al., 2003). The early reduction in peak diameter for the largest motor axons in NF-(M/H)$^{tail\Delta}$ mice was similar to the early reduction observed in NF-H$^{tail\Delta}$ mice (Garcia et al., 2003). However, much like NF-H$^{tail\Delta}$ mice, axons of NF-(M/H)$^{tail\Delta}$ mice grew from 2 to 6 months, obtaining the largest diameter that can be obtained without NF-M C-terminus. Therefore, NF-H C-terminus influenced the rate of radial growth whereas the peak in diameter distribution was determined by NF-M C-terminus.

Phosphorylation of NF-H and NF-M C-termini is associated with increased neurofilament-to-neurofilament spacing or nearest neighbor distance (NND) (de Waegh et al., 1992). Moreover, increased NND is associated with larger axonal diameters (de Waegh et al., 1992). In NF-H$^{tail\Delta}$ mice, there was no difference in axonal diameter and NND due to loss of the NF-H C-terminus supporting a role for NND in determining axonal diameter (Rao et al., 2002). However, analysis of NND in both NF-M$^{tail\Delta}$ and NF-(M/H)$^{tail\Delta}$ mice suggested that NND did not directly correlate with axonal diameter. NND in both mice was analyzed at 2 and 6 months. At 2 months, both mice have reduced axonal diameters (Garcia et al., 2003). Moreover, NF-(M/H)$^{tail\Delta}$ axons were reduced relative to NF-M$^{tail\Delta}$ axons. At 2 months, NND in both mice was reduced relative to control littermates and NND in NF-(M/H)$^{tail\Delta}$ mice was reduced relative to NND in NF-M$^{tail\Delta}$ mice, again supporting a role for NND in establishing axonal diameter (Garcia et al., 2003). However, at 6 months, axonal diameters in NF-M$^{tail\Delta}$ and NF-(M/H)$^{tail\Delta}$ mice were reduced relative to control, but indistinguishable from each other (Garcia et al., 2003). NND, on the other hand, in NF-(M/H)$^{tail\Delta}$ axons, was reduced by the same percentage as was observed at 2 months relative to NF-M$^{tail\Delta}$ mice (Garcia et al., 2003). Neurofilament C-termini did influence neurofilament spacing. However, interfilament spacing was not a primary determinant of axonal diameter.

To this point, analyses of gene-targeted mice identified the critical subunit, NF-M, and the critical domain within this subunit, the C-terminal tail domain. The surprise from these analyses was that the critical subunit was NF-M. However, the requirement of the NF-M C-terminal tail domain for radial growth was consistent with the myelin-dependent neurofilament phosphorylation hypothesis of radial growth. The C-terminus contained all KSP repeats (Levy et al., 1987; Myers et al., 1987), which were phosphorylated in myelinated axonal segments (de Waegh et al., 1992; Hsieh et al., 1994; Starr et al., 1996). Moreover, in cultured dorsal root ganglion neurons, six of the seven KSP repeats were phosphorylated (Xu et al., 1992).

13.3.4 *Full-Length, Phospho-Incompetent NF-M: No Role for KSP Phosphorylation in Radial Growth*

The most recent gene targeted mice provide the most direct test of the myelin-dependent neurofilament phosphorylation hypothesis. Site directed mutagenesis was utilized to mutate all KSP repeat serine residues to alanine residues (referred to as NF-M$^{S\rightarrow A}$ mice). Mutation to alanine residues prevented the phosphorylation of KSP repeats (Julien and Mushynski, 1982, 1983) without deletion of all 426 amino acids of the NF-M C-terminus. Mutation of all seven KSP repeat serine residues resulted in an increased electrophoretic mobility of NF-M (Garcia et al., 2009) that was similar to the increase in mobility observed after prolonged incubation of NF-M with alkaline phosphatase (Julien and Mushynski, 1982). Increased electrophoretic mobility was consistent with the loss of KSP phosphorylation.

Radial growth was measured at 2 and 6 months in axons of the fifth lumbar ventral root from both NF-M$^{S\rightarrow A}$ mice and wild type littermates. The peaks in axonal diameter distribution for both large and small axonal populations were indistinguishable from wild type littermates at both time points examined (Garcia et al., 2009). Interestingly, at 2 months there were more axons in the range of 7–8 μm in NF-M$^{S\rightarrow A}$ mice (Garcia et al., 2009). The increased number of large motor axons resulted in a small, but significant, difference in the overall axonal distributions between NF-M$^{S\rightarrow A}$ and wild type mice at 2 months (Garcia et al., 2009). Additionally, at 2 months the total number of axons was reduced relative to wild type littermates (Garcia et al., 2009). Average interfilament distances of NF-M$^{S\rightarrow A}$ mice and wild type littermates were similar at both time points (Garcia et al., 2009). Neuronal conduction velocities were measured from the sciatic nerve in 5-month-old NF-M$^{S\rightarrow A}$ mice and wild type littermates. As expected, due to similar sized axons, conduction velocities were indistinguishable (Garcia et al., 2009).

Expression of full length, KSP phospho-incompetent NF-M did not affect the expression of other cytoskeletal proteins, including NF-L and NF-H (Garcia et al., 2009). Expression of NF-M$^{S\rightarrow A}$, much like the expression of C-terminally truncated NF-M, resulted in increased phosphorylation of NF-H C-terminus as determined by increased immunoreactivity with phosphorylation specific antibodies (Garcia et al., 2009). In fact, the increased level of NF-H C-terminal phosphorylation in NF-M$^{S\rightarrow A}$ mice was identical to the level observed in NF-M$^{tail\Delta}$ mice (Garcia et al., 2009). Therefore, compensatory phosphorylation of NF-H was insufficient to explain the lack of difference in peak axonal diameter observed in NF-M$^{S\rightarrow A}$ mice relative to wild type littermates.

Taken together, the analysis of NF-H$^{tail\Delta}$, NF-M$^{tail\Delta}$ and NF-M$^{S\rightarrow A}$ mice indicated that the C-terminal tail domain of NF-M was essential for radial growth. However, phosphorylation of C-terminal KSP repeats in both NF-H and NF-M did not play a significant role in radial growth. Yet again, the analysis of a gene targeted mouse yielded unexpected results. However, unlike previous unexpected results, these new results may represent a paradigm shift in understanding neurofilament biology and its role in determining axonal caliber.

These new mice have challenged the long-standing myelin-dependent neurofilament phosphorylation hypothesis of radial growth, which seemed to be supported by previous models. However, it must be stressed that, without the knowledge gained from previous gene-targeted mice, the generation and analysis of the NF-$M^{S\rightarrow A}$ mice would not have been possible. For example, the length and repetitive nature of the KSP repeat region of NF-H C-terminus would have made the generation of NF-$H^{S\rightarrow A}$ mouse extremely difficult. Therefore, the myelin-dependent neurofilament phosphorylation hypothesis of radial growth was put to the test through the systematic generation and analysis of gene-targeted mice.

13.4 Conclusions

Radial axonal growth determines the diameter of myelinated axons, thereby influencing the rate of impulse propagation along the axon. For well over 15 years radial growth was thought to be mediated by myelin-dependent phosphorylation of neurofilament heavy and medium within the axon. The systematic analysis of gene-targeted mice supported a role for neurofilaments in mediating radial growth. These analyses identified the critical subunit, NF-M, and the critical domain within NF-M, the C-terminal 426 amino acids. However, newly developed gene targeted mice indicated that myelin dependent phosphorylation of neurofilament C-termini does not play a significant role in radial axonal growth. This leads to two immediate questions: (1) why phosphorylate neurofilament C-termini and (2) which amino acids, within the C-terminal 426, are required for radial growth? If history is indeed prologue, these important questions will be addressed through careful analysis of newly developed gene-targeted mice.

Acknowledgments We are grateful to Dr. David Schulz, Dr. Christian Lobsiger and Virginia Garcia for comments that improved this manuscript.

References

Ackerley S, Thornhill P, Grierson AJ, Brownlees J, Anderton BH, Leigh PN, Shaw CE, Miller CC (2003) Neurofilament heavy chain side arm phosphorylation regulates axonal transport of neurofilaments. J Cell Biol 161:489–495

Ching GY, Liem RK (1993) Assembly of type IV neuronal intermediate filaments in nonneuronal cells in the absence of preexisting cytoplasmic intermediate filaments. J Cell Biol 122: 1323–1335

Clement AM, Nguyen MD, Roberts EA, Garcia ML, Boillee S, Rule M, McMahon AP, Doucette W, Siwek D, Ferrante RJ, Brown RH Jr, Julien J-P, Goldstein LSB, Cleveland DW (2003) Wild-type nonneuronal cells extend survival of SOD1 mutant motor neurons in ALS mice. Science 302:113–117

Cleveland DW (1996) Neuronal growth and death: order and disorder in the axoplasm. Cell 84:663–666

de Waegh SM, Lee VM, Brady ST (1992) Local modulation of neurofilament phosphorylation, axonal caliber, and slow axonal transport by myelinating Schwann cells. Cell 68: 451–463

Deng Y, Li B, Liu F, Iqbal K, Grundke-Iqbal I, Brandt R, Gong C-X (2008) Regulation between O-GlcNAcylation and phosphorylation of neurofilament-M and their dysregulation in Alzheimer disease. FASEB J 22:138–145

Dong DL, Xu ZS, Chevrier MR, Cotter RJ, Cleveland DW, Hart GW (1993) Glycosylation of mammalian neurofilaments. Localization of multiple O-linked N-acetylglucosamine moieties on neurofilament polypeptides L and M. J Biol Chem 268:16679–16687

Dong DL, Xu ZS, Hart GW, Cleveland DW (1996) Cytoplasmic O-GlcNAc modification of the head domain and the KSP repeat motif of the neurofilament protein neurofilament-H. J Biol Chem 271:20845–20852

Dubois M, Strazielle C, Eyer J, Lalonde R (2002) Sensorimotor functions in transgenic mice expressing the neurofilament/heavy-LacZ fusion protein on two genetic backgrounds. Neuroscience 112:447–454

Dubois M, Strazielle C, Julien JP, Lalonde R (2005) Mice with the deleted neurofilament of low molecular weight (Nefl) gene: 2. Effects on motor functions and spatial orientation. J Neurosci Res 80:751–758

Elder GA, Friedrich VL Jr, Bosco P, Kang C, Gourov A, Tu PH, Lee VM, Lazzarini RA (1998a) Absence of the mid-sized neurofilament subunit decreases axonal calibers, levels of light neurofilament (NF-L), and neurofilament content. J Cell Biol 141:727–739

Elder GA, Friedrich VL Jr, Kang C, Bosco P, Gourov A, Tu PH, Zhang B, Lee VM, Lazzarini RA (1998b) Requirement of heavy neurofilament subunit in the development of axons with large calibers. J Cell Biol 143:195–205

Eyer J, Peterson A (1994) Neurofilament-deficient axons and perikaryal aggregates in viable transgenic mice expressing a neurofilament-beta-galactosidase fusion protein. Neuron 12: 389–405

Friede RL, Samorajski T (1970) Axon caliber related to neurofilaments and microtubules in sciatic nerve fibers of rats and mice. Anat Rec 167:379–387

Fuchs E, Cleveland DW (1998) Science 279:514–519

Fuchs E, Weber K (1994) Intermediate filaments: structure, dynamics, function, and disease. Annu Rev Biochem 63:345–382

Garcia ML, Lobsiger CS, Shah SB, Deerinck TJ, Crum J, Young D, Ward CM, Crawford TO, Gotow T, Uchiyama Y, Ellisman MH, Calcutt NA, Cleveland DW (2003) NF-M is an essential target for the myelin-directed "outside-in" signaling cascade that mediates radial axonal growth. J Cell Biol 163:1011–1020

Garcia ML, Rao MV, Fujimotor J, Garcia VB, Shah SB, Crum J, Gotow T, Uchiyama Y, Ellisman MH, Calcutt NA, Cleveland DW (2009) Phosphorylation of highly conserved NF-M KSP repeats is not required for myelin-dependent radial axonal growth. J Neurosci 29:1277–1284

Gibb BJ, Brion JP, Brownlees J, Anderton BH, Miller CC (1998) Neuropathological abnormalities in transgenic mice harbouring a phosphorylation mutant neurofilament transgene. J Neurochem 70:492–500

Gibb BJ, Robertson J, Miller CC (1996) Assembly properties of neurofilament light chain Ser55 mutants in transfected mammalian cells. J Neurochem 66:1306–1311

Heins S, Wong PC, Muller S, Goldie K, Cleveland DW, Aebi U (1993) The rod domain of NF-L determines neurofilament architecture, whereas the end domains specify filament assembly and network formation. J Cell Biol 123:1517–1533

Hirokawa N, Glicksman MA, Willard MB (1984) Organization of mammalian neurofilament polypeptides within the neuronal cytoskeleton. J Cell Biol 98:1523–1536

Hisanaga S, Hirokawa N (1988) Structure of the peripheral domains of neurofilaments revealed by low angle rotary shadowing. J Mol Biol 202:297–305

Hoffman PN, Griffin JW, Price DL (1984) Control of axonal caliber by neurofilament transport. J Cell Biol 99:705–714

Hoffman PN, Thompson GW, Griffin JW, Price DL (1985) Changes in neurofilament transport coincide temporally with alterations in the caliber of axons in regenerating motor fibers. J Cell Biol 101:1332–1340

Hsieh ST, Kidd GJ, Crawford TO, Xu Z, Lin WM, Trapp BD, Cleveland DW, Griffin JW (1994) Regional modulation of neurofilament organization by myelination in normal axons. J Neurosci 14:6392–6401

Julien JP, Cote F, Beaudet L, Sidky M, Flavell D, Grosveld F, Mushynski W (1988) Sequence and structure of the mouse gene coding for the largest neurofilament subunit. Gene 68:307–314

Julien JP, Mushynski WE (1982) Multiple phosphorylation sites in mammalian neurofilament polypeptides. J Biol Chem 257:10467–10470

Julien JP, Mushynski WE (1983) The distribution of phosphorylation sites among identified proteolytic fragments of mammalian neurofilaments. J Biol Chem 258:4019–4025

Jung C, Shea TB (1999) Regulation of neurofilament axonal transport by phosphorylation in optic axons in situ. Cell Motil Cytoskeleton 42:230–240

Jung C, Yabe JT, Lee S, Shea TB (2000) Hypophosphorylated neurofilament subunits undergo axonal transport more rapidly than more extensively phosphorylated subunits in situ. Cell Motil Cytoskeleton 47:120–129

Kriz J, Zhu Q, Julien JP, Padjen AL (2000) Electrophysiological properties of axons in mice lacking neurofilament subunit genes: disparity between conduction velocity and axon diameter in absence of NF-H. Brain Res 885:32–44

Lee MK, Cleveland DW (1996) Neuronal intermediate filaments. Annu Rev Neurosci 19:187–217

Lee MK, Xu Z, Wong PC, Cleveland DW (1993) Neurofilaments are obligate heteropolymers in vivo. J Cell Biol 122:1337–1350

Letourneau PC (1996) The cytoskeleton in nerve growth cone motility and axonal pathfinding. Perspect Dev Neurobiol 4:111–123

Levy E, Liem RK, D'Eustachio P, Cowan NJ (1987) Structure and evolutionary origin of the gene encoding mouse NF-M, the middle-molecular-mass neurofilament protein. Eur J Biochem 166:71–77

Liem RK, Hutchison SB (1982) Purification of individual components of the neurofilament triplet: filament assembly from the 70 000-dalton subunit. Biochemistry 21:3221–3226

Lobsiger CS, Garcia ML, Ward CM, Cleveland DW (2005) Altered axonal architecture by removal of the heavily phosphorylated neurofilament tail domains strongly slows superoxide dismutase 1 mutant-mediated ALS. Proc Natl Acad Sci USA 102:10351–10356

Ludemann N, Clement A, Hans VH, Leschik J, Behl C, Brandt R (2005) O-Glycosylation of the tail domain of neurofilament protein M in human neurons and in spinal cord tissue of a rat model of amyotrophic lateral sclerosis (ALS). J Biol Chem 280:31648–31658

Marszalek JR, Williamson TL, Lee MK, Xu Z, Hoffman PN, Becher MW, Crawford TO, Cleveland DW (1996) Neurofilament subunit NF-H modulates axonal diameter by selectively slowing neurofilament transport. J Cell Biol 135:711–724

Mirsky R, Jessen KR, Brennan A, Parkinson D, Dong Z, Meier C, Parmantier E, Lawson D (2002) Schwann cells as regulators of nerve development. J Physiol Paris 96:17–24

Monteiro MJ, Hoffman PN, Gearhart JD, Cleveland DW (1990) Expression of NF-L in both neuronal and nonneuronal cells of transgenic mice: increased neurofilament density in axons without affecting caliber. J Cell Biol 111:1543–1557

Myers MW, Lazzarini RA, Lee VM, Schlaepfer WW, Nelson DL (1987) The human mid-size neurofilament subunit: a repeated protein sequence and the relationship of its gene to the intermediate filament gene family. EMBO J 6:1617–1626

Nixon R, Paskevich P, Sihag R, Thayer C (1994) Phosphorylation on carboxyl terminus domains of neurofilament proteins in retinal ganglion cell neurons in vivo: influences on regional neurofilament accumulation, interneurofilament spacing, and axon caliber. J Cell Biol 126:1031–1046

Ohara O, Gahara Y, Miyake T, Teraoka H, Kitamura T (1993) Neurofilament deficiency in quail caused by nonsense mutation in neurofilament-L gene. J Cell Biol 121:387–395

Rao MV, Campbell J, Yuan A, Kumar A, Gotow T, Uchiyama Y, Nixon RA (2003) The neurofilament middle molecular mass subunit carboxyl-terminal tail domains is essential for the radial growth and cytoskeletal architecture of axons but not for regulating neurofilament transport rate. J Cell Biol 163:1021–1031

Rao MV, Garcia ML, Miyazaki Y, Gotow T, Yuan A, Mattina S, Ward CM, Calcutt NA, Uchiyama Y, Nixon RA, Cleveland DW (2002) Gene replacement in mice reveals that the heavily phosphorylated tail of neurofilament heavy subunit does not affect axonal caliber or the transit of cargoes in slow axonal transport. J Cell Biol 158:681–693

Rao MV, Houseweart MK, Williamson TL, Crawford TO, Folmer J, Cleveland DW (1998) Neurofilament-dependent radial growth of motor axons and axonal organization of neurofilaments does not require the neurofilament heavy subunit (NF-H) or its phosphorylation. J Cell Biol 143:171–181

Sakaguchi T, Okada M, Kitamura T, Kawasaki K (1993) Reduced diameter and conduction velocity of myelinated fibers in the sciatic nerve of a neurofilament-deficient mutant quail. Neurosci Lett 153:65–68

Sihag RK, Nixon RA (1990) Phosphorylation of the amino-terminal head domain of the middle molecular mass 145-kDa subunit of neurofilaments. Evidence for regulation by second messenger-dependent protein kinases. J Biol Chem 265:4166–4171

Sihag RK, Nixon RA (1991) Identification of Ser-55 as a major protein kinase A phosphorylation site on the 70-kDa subunit of neurofilaments. Early turnover during axonal transport. J Biol Chem 266:18861–18867

Starr R, Attema B, DeVries GH, Monteiro MJ (1996) Neurofilament phosphorylation is modulated by myelination. J Neurosci Res 44:328–337

Weiss PA, Mayr R (1971) Organelles in neuroplasmic ("axonal") flow: neurofilaments. Proc Natl Acad Sci USA 68:846–850

Xu ZS, Liu WS, Willard MB (1992) Identification of six phosphorylation sites in the COOH-terminal tail region of the rat neurofilament protein M. J Biol Chem 267:4467–4471

Xu Z, Marszalek JR, Lee MK, Wong PC, Folmer J, Crawford TO, Hsieh ST, Griffin JW, Cleveland DW (1996) Subunit composition of neurofilaments specifies axonal diameter. J Cell Biol 133:1061–1069

Yamasaki H, Bennett GS, Itakura C, Mizutani M (1992) Defective expression of neurofilament protein subunits in hereditary hypotrophic axonopathy of quail. Lab Invest 66:734–743

Yamasaki H, Itakura C, Mizutani M (1991) Hereditary hypotrophic axonopathy with neurofilament deficiency in a mutant strain of the Japanese quail. Acta Neuropathol (Berl) 82:427–434

Yin X, Crawford TO, Griffin JW, Tu P, Lee VM, Li C, Roder J, Trapp BD (1998) Myelin-associated glycoprotein is a myelin signal that modulates the caliber of myelinated axons. J Neurosci 18:1953–1962

Yuan A, Nixon RA, Rao MV (2006a) Deleting the phosphorylated tail domain of the neurofilament heavy subunit does not alter neurofilament transport rate in vivo. Neurosci Lett 393:264–268

Yuan A, Rao MV, Sasaki T, Chen Y, Kumar A, Veeranna, Liem RKH, Eyer J, Peterson AC, Julien, J-P, Nixon RA (2006b) {alpha}-Internexin is structurally and functionally associated with the neurofilament triplet proteins in the mature CNS. J Neurosci 26:10006–10019

Zhu Q, Couillard-Després S, Julien JP (1997) Delayed maturation of regenerating myelinated axons in mice lacking neurofilaments. Exp Neurol 148:299–316

Zhu Q, Lindenbaum M, Levavasseur F, Jacomy H, Julien JP (1998) Disruption of the NF-H gene increases axonal microtubule content and velocity of neurofilament transport: relief of axonopathy resulting from the toxin beta,beta'-iminodipropionitrile. J Cell Biol 143:183–193

Chapter 14
Deregulation of Cytoskeletal Protein Phosphorylation and Neurodegeneration

Jyotshnabala Kanungo, Ya-li Zheng, Parvathi Rudrabhatla, Niranjana D. Amin, Bibhutibhushan Mishra, and Harish C. Pant

Abstract Phosphorylation of cytoskeletal proteins is tightly regulated by the activities of multiple protein kinases and phosphatases. Three kinases, Cyclin-dependent kinase 5 (Cdk5), Glycogen synthase kinase 3β (GSK3β) and MAPKs have been implicated in their direct involvement in neuronal cytoskeletal protein phosphorylation. Cdk5 and GSK3β have been identified as prime candidates for pathogenesis. Cdk5 is a proline-directed serine/threonine protein kinase that requires an interaction with its activators, p35 or p39, to be catalytically active. While Cdk5 expression is ubiquitous, p35 and p39 are abundantly expressed in postmitotic neurons which, therefore, exhibit enhanced levels of Cdk5 activity. A pleiotropic kinase, Cdk5 has a multifunctional role in the mammalian central nervous system. Cdk5 was originally identified as a major Tau kinase. It associates with early stages of neurofibrillary tangles (NFTs). NFTs are composed mainly of hyperphosphorylated Tau aggregates, the pathological hallmarks of neurodegenerative tauopathies and Alzheimer's disease (AD). Cdk5, by phosphorylating neuronal cytoskeletal proteins, such as Tau and neurofilaments (NFs), plays a critical role in neurodegeneration. In this review, we focus on the specific roles of Cdk5 phosphorylation of the neuronal cytoskeletal proteins (NFs and Tau) that contribute to neurodegeneration.

Keywords AD · ALS · Neurodegeneration · Neurofilament · Phosphorylation · Proline-directed kinases · Tau

Contents

H.C. Pant (✉)
Laboratory of Neurochemistry, National Institute of Neurological Disorders and Stroke, National Institutes of Health, Bethesda, MD 20892-4130, USA
e-mail: panth@ninds.nih.gov

R.A. Nixon, A. Yuan (eds.), *Cytoskeleton of the Nervous System*, Advances in Neurobiology 3, DOI 10.1007/978-1-4419-6787-9_14,

14.1 Introduction

The cytoskeleton is cellular "scaffolding" or "skeleton" contained within the cytoplasm. The cytoskeleton is present in all cells. Eukaryotic cells contain three main cytoskeletal structures, which are microfilaments, intermediate filaments and microtubules. Microfilaments are composed of two intertwined actin chains, are most concentrated just beneath the cell membrane and are responsible for resisting tension and maintaining cellular shape, forming cytoplasmic protuberances (like pseudopodia and microvilli – although these by different mechanisms) and participation in some cell–cell or cell–matrix junctions. Intermediate filaments (vimentins, keratins, lamins and neurofilaments) are more stable (strongly bound) than actin filaments, and heterogeneous constituents of the cytoskeleton. Like actin filaments, they function in the maintenance of cell-shape by bearing tension. Intermediate filaments organize the internal tridimensional structure of the cell, anchoring organelles and serving as structural components of the nuclear lamina and sarcomeres. They also participate in some cell–cell and cell–matrix junctions. Microtubules are most commonly comprised of 13 protofilaments which, in turn, are polymers of alpha and beta tubulin. They have a very dynamic behavior, binding GTP for polymerization. They are commonly organized by the centrosome and are involved in intracellular transport (associated with dyneins and kinesins, they transport organelles like mitochondria or vesicles) and forming the mitotic spindle.

14.2 Neuronal Cytoskeleton

The eukaryotic neuronal cytoskeleton also comprises three distinct, interacting structural complexes that have very different properties: microtubules (MTs), intermediate filaments (IFs) and microfilaments (MFs). Each has a characteristic composition, structure and organization that may be further specialized in a particular cell type or subcellular domain.

14.2.1 Neuronal Microtubules

Neuronal MTs are structurally similar to those found in other eukaryotic cells. The core structure is a polymer of 50-kDa tubulin subunits. The α- and β-tubulins are the best-known members of a unique protein family, the members of which have significant sequence similarity (Burns and Surridge, 1994). Anchoring and nucleation of MTs appear to require a third class of tubulin, γ-tubulin, which is detectable only as part of the pericentriolar complex (Joshi, 1994). Brain MTs also contain a variety of post-translational modifications.

MTs in vivo invariably include members of a heterogeneous set of proteins known as MAPs (Schoenfeld and Obar, 1994). MAPs interact with MTs rather than with free tubulin and maintain constant stoichiometry with the tubulin in MTs through cycles of assembly and disassembly. Several categories of brain MAPs can be distinguished: the high-molecular-weight MAPs, which include: MAPs 1A, 1B, 2A and 2B (>270 kDa); the Tau proteins; MAPs of intermediate molecular weight, such as MAP3 and MAP4; and the molecular motors kinesin and dynein, which drive the intracellular transport of membrane-bound organelles along MT tracts. Some MAPs are differentially distributed in neurons (Schoenfeld and Obar, 1994). For example, MAP2 is found primarily in cell bodies and dendrites, and Tau is enriched in axons. Additionally, changes in MAP expression and MAP phosphorylation during development suggest that they may play a role in modulating MT function in the developing brain (Schoenfeld and Obar, 1994). For example, MAPs 1A and 1B occur in both axons and somatodendritic domains, but MAP1B is preferentially phosphorylated in axons and especially in developing axons.

Another neuronal MAP is the well-known Tau proteins. As many as six different Tau proteins are derived by alternative splicing from a single *Tau* gene (Goedert et al., 1995; Mandelkow et al., 1996) (Fig. 14.1). The six separate isoforms in adults vary in the number of microtubule-binding domains (three and four) and in the number and size of N-terminal inserts (Himmler et al., 1989). Both the expression and the phosphorylation of the different Tau isoforms are regulated throughout development and in neurodegenerative processes. Fetal Tau is highly phosphorylated (Kanemaru et al., 1992; Watanabe et al., 1993). Characterization of most Tau phosphorylation sites (Fig. 14.2a,b) emerges from using phosphorylation-specific antibodies (Kosik et al., 1988; Goedert et al., 1994). Tau is hyperphosphorylated in Alzheimer's disease (AD) brains (Grundke-Iqbal et al., 1986; Kosik et al., 1986; Wood et al., 1986). However, further analyses of the various patterns of Tau phosphorylation suggested considerable overlap between AD and adult patterns of phosphorylation, although there were quantitative differences in levels (Hasegawa et al., 1993). Surprisingly, most, if not all, of the AD-specific sites on Tau are seen

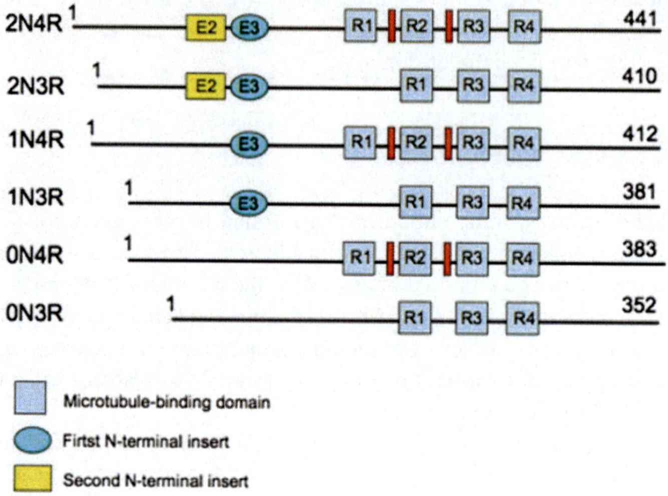

Fig. 14.1 Alternative splicing and domain structure of Tau. There are six isoforms of Tau in the adult human brain that result from a single gene via alternative splicing. Microtubule-binding repeats are located in the C-terminal half of Tau. Two N-terminal inserts and inserts flanking the microtubule-binding domain 2 are characteristics of the larger isoforms. Fetal Tau corresponds to isoform A

in living neurons (Brion et al., 1994; Garver et al., 1994; Matsuo et al., 1994; Seubert et al., 1995). The electrophoretic mobility patterns of insoluble and soluble forms of Tau in the brain extracts of Alzheimer's disease and Amyotropic Lateral Sclerosis (ALS) patients as observed in a 7.5% SDS-PAGE are presented in Fig. 14.2c.

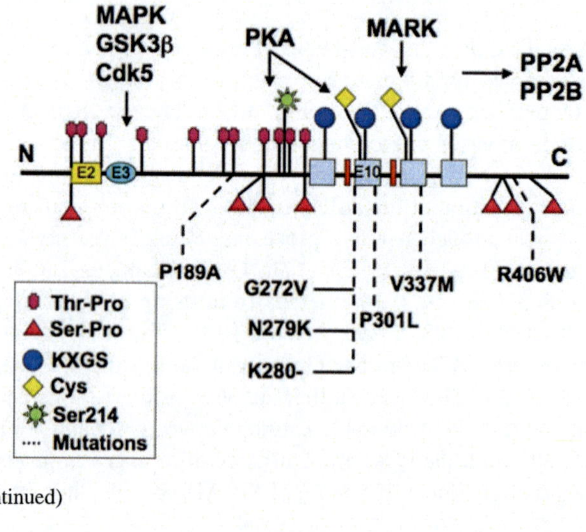

Fig. 14.2 (continued)

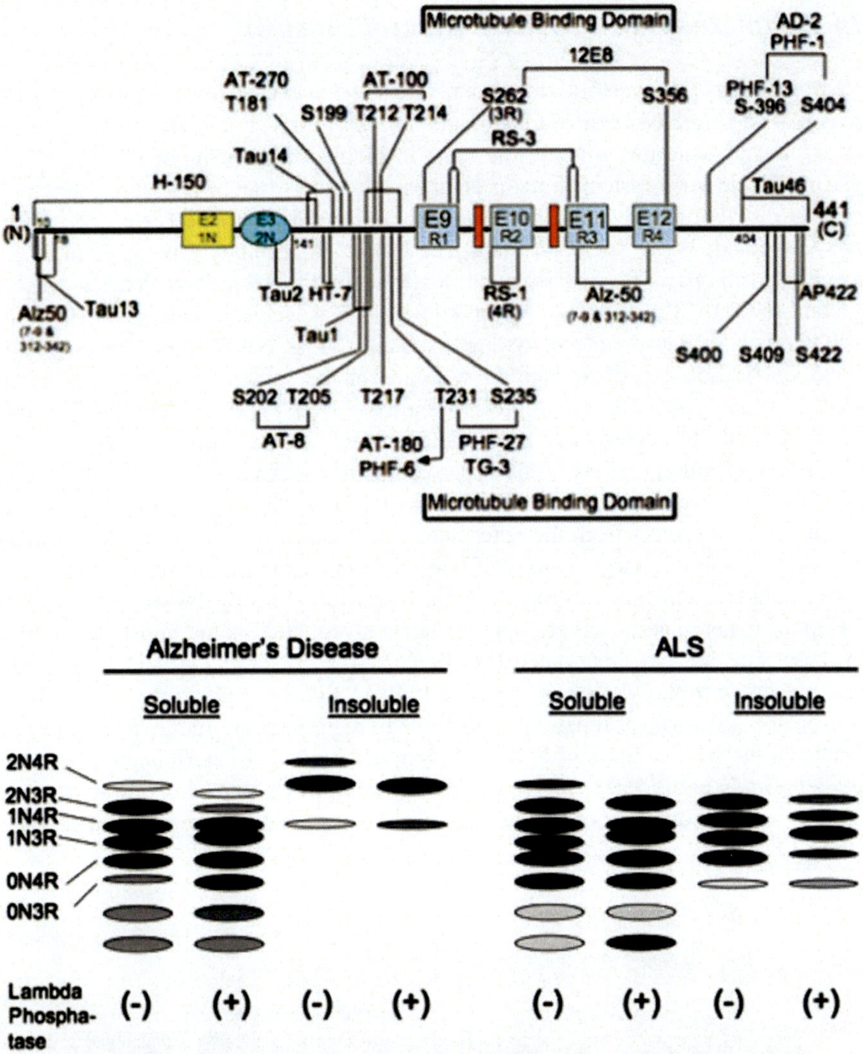

Fig. 14.2 (continued) **a** Schematic diagram of Tau protein shows its domains, phosphorylation sites and reported mutations. The N-terminal inserts (exons 2, 3 – E2 and E3) and the second of the 4 repeats (exon 10 – E10) can be absent due to alternative splicing. The C-terminal half is the 'assembly domain' that binds to the microtubules and stabilizes them. The N-terminal half is the 'projection domain' that protrudes from the microtubule surface. The regions flanking the repeats are rich in Ser-Pro or Thr-Pro motifs that can be phosphorylated by proline-directed kinases, such as MAPK, GSK3β and Cdk5. These sites are epitopes of antibodies diagnostic of Alzheimer Tau. Ser214 and Ser262 (in the KXGS motif of the first repeat) are non-proline motifs phosphorylated in AD and in vitro by PKA or MARK leading to the detachment of Tau from microtubules. Some of the mutations linked to neurodegeneration (tauopathy) are indicated. **b** A variety of antibodies that recognize phospho-specific sites of the Tau protein are shown. **c** Electrophoretic mobility patterns of the soluble and insoluble Tau of brain extracts from Alzheimer's disease and Amyotropic Lateral Sclerosis patients as would be resolved in a 7.5% SDS-PAGE

14.2.2 Intermediate Filaments (Neurofilaments)

Neurofilaments (NFs) are intermediate filaments (IFs) exclusively expressed in neurons that add rigidity, tensile strength and possibly intracellular transport guidance to axons and dendrites. IFs are the most insoluble components of the cell cytoplasm. The nervous system contains an unusually diverse set of IFs with distinctive cellular distributions and developmental expression (Fliegner and Liem, 1991; Lee and Cleveland, 1996; Shea and Flanagan, 2001). The primary type of IF in large myelinated axons is formed from three NF subunits: NF high-molecular-weight subunit (NF-H, 180–200 kDa), NF middle-molecular-weight subunit (NF-M, 130–170 kDa) and NF low-molecular-weight subunit (NF-L, 60–70 kDa) (Fliegner and Liem, 1991; Lee and Cleveland, 1996). Each of the subunit proteins is encoded by a separate gene. The NF triplet proteins are type IV IF proteins, which are expressed only in neurons and have a characteristic domain structure (Fig. 14.3). The amino-terminal regions of all three subunits interact via α-helical coiled coils to form the core of the filament. NF-M and NF-H also have long carboxy-terminal regions, which project from the core filaments as sidearms. NF-H and, to a lesser extent, NF-M have a large number of consensus phosphorylation sites for proline-directed kinases in this carboxy-terminal extension (>50 on NF-H and >10 on NF-M in many species) (Fig. 14.4). In large myelinated axons, most, if not all, of these sites are phosphorylated (Fliegner and Liem, 1991; Lee and Cleveland, 1996). This phosphorylation of NF-H and NF-M side arms alters the charge density on the NF surface, repelling adjacent NFs with similar charge. Such mutual repulsion by the side arms of NFs is thought to be a major determinant of axonal caliber (Brady et al., 1993). There is evidence that critical phosphorylation of sites in head terminal domains of NF-M can impact C-terminal phosphorylation (Zheng et al., 2003).

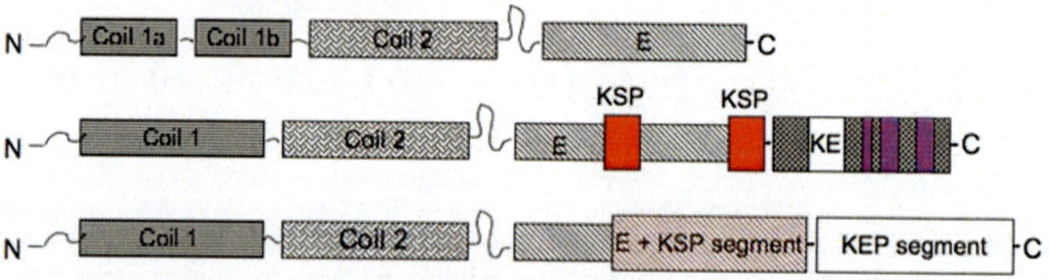

Fig. 14.3 Schema of the neuronal intermediate filaments (NFs). Based on the molecular mass, NFs are classified into three distinct groups. Coil1 and Coil2 regions are conserved α-helical domains that contain heptad repeats of hydrophobic amino acids. These regions are separated by nonhelical 'links'. The NFs contain one or more glutamic acid-rich segments. NF-M and NF-H both contain regions of lysine-serine-proline or KSP repeats, which are highly phosphorylated. The extreme ends of NF-M possess lysine- and glutamic acid-rich regions (KE segments). The extreme ends of NF-H possess lysine-, glutamic acid- and proline-rich segments (KEP segments)

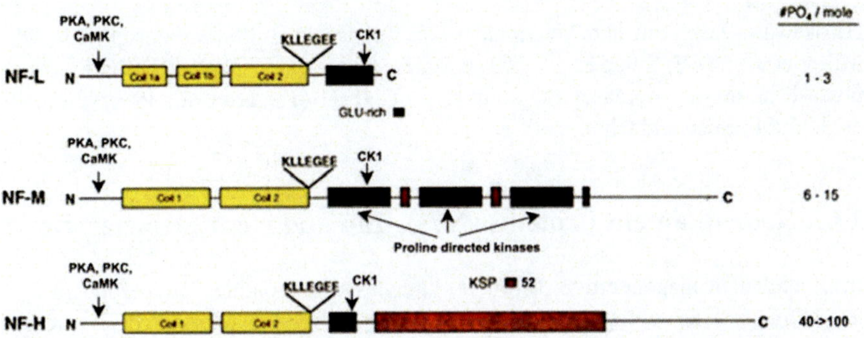

Fig. 14.4 Schema shows phosphorylation of the NF subunit domains by a number of kinases. All three subunits, NF-L, NF-M and NF-H are phosphorylated at their N-terminal domains by PKA, PKC and CamK. NF-L and NF-M are phosphorylated at their KSP repeat regions by ERK1/2 and Cdk5. The number of phosphates incorporated per mole of NF protein is shown on the right. NF-L and NF-M are phosphorylated on 1–3 and 6–15 phosphates per mole, respectively. NF-H has 40–57 phosphates/mole that are potentially phosphorylated by ERK1/2 and Cdk5

14.2.3 Post-Translational Modifications and Neuronal Cytoskeletal Protein Solubility

Two major posttranslational modifications are involved in the formation and functions of neurofilaments, phosphorylation and glycosylation. The mechanisms by which normal soluble cytoskeletal elements, such as Tau and neurofilament proteins, are transformed into insoluble paired helical filaments and aggregates are an important issue (Selkoe et al., 1982; Smith et al., 1996b). Insolubility has been linked to the best known posttranslational change of Tau, abnormal phosphorylation (Goedert et al., 1991; Greenberg et al., 1992) and a number of specific kinases and phosphatases have been implicated (Trojanowski et al., 1993). However, while increased phosphorylation of Tau decreases microtubule stability, a salient feature of the pathology of Alzheimer disease (Perry et al., 1991; Alonso et al., 1994; Iqbal et al., 1994; Alonso et al., 1996; Praprotnik et al., 1996a, b), NFT insolubility is not mediated by phosphorylation (Smith et al., 1996a). Indeed, in vitro phosphorylation of normal Tau or complete dephosphorylation of NFT has no effect on their solubility (Goedert et al., 1991; Greenberg et al., 1992; Gustke et al., 1992; Smith et al., 1996a). Several studies suggest Tau phosphorylation as found in Alzheimer disease may be part of a novel process similar to that seen during mitosis (Pope et al., 1994; Preuss et al., 1995), suggesting that it might be abortively entering the cell cycle (Vincent et al., 1996; McShea et al., 1997). Phosphorylation of NF tail domain sites regulates inter-filament interaction, filament spacing, transport properties, cytoskeleton stability and proteolysis, while phosphorylation of head domain sites appears to be involved in NF assembly and solubility (Hisanaga and

Hirokawa, 1989; Pant and Veeranna, 1995; Tsuda et al., 1997; Elder et al., 1998; Miller et al., 2002; Sihag et al., 2007). Suppression of dephosphorylation by phosphatase inhibitors increases the solubility of NF-H (Tsuda et al., 1997; Fiumelli et al., 2008; Shea and Chan, 2008).

14.3 Neurofilament Proteins (NFs), Tau and Neurodegeneration

In several neurodegenerative disorders, known as 'tauopathies', hyperphosphorylated Tau proteins aggregate. The discovery that highly phosphorylated Tau protein is the major component of the paired helical filaments (PHFs) of AD and other neurodegenerative diseases directed focus onto the post-translational modifications of Tau (Selkoe et al., 1982). Although there are contradictory findings, generally, the degree of dementia correlates with the sites and severity of Tau-based NFTs, where NFTs, a hallmark of AD pathology, comprise Tau-based PHFs and extracellular β-amyloid plaques (Holzer et al., 1994; Feany and Dickson, 1996).

Abnormal accumulation of NFs is detected in many human neurodegenerative disorders, including amyotropic lateral sclerosis (ALS), Alzheimer's disease (AD), and dementia with Lewy bodies, Parkinson's disease (PD), among others. Accumulation of phosphorylated neurofilament proteins (NFPs) may result from many alterations of the behavior of these proteins, such as dysregulation of NFP synthesis, defective axonal transport, aberrant phosphorylation and proteolysis. NFs also affect the dynamics and function of other cytoskeletal proteins, such as microtubules and actin filaments (Zhu et al., 1998; Ahlijanian et al., 2000; Shea and Chan, 2008). NF abnormalities are generally considered as the hallmarks of neurodegenerative disorders.

14.3.1 Alzheimer's Disease (AD)

AD is a CNS neurodegenerative disease, which was first described in 1906 by Dr Alois Alzheimer. The pathological presentation of Alzheimer disease, the leading cause of senile dementia, involves regionalized neuronal death and an accumulation of intraneuronal and extracellular lesions termed neurofibrillary tangles and senile plaques, respectively [reviewed by (Smith and Perry, 1997)]. Several independent hypotheses have been proposed to link the pathological lesions and neuronal cytopathology with, among others, apolipoprotein E genotype (Corder et al., 1993; Roses et al., 1995), hyperphosphorylation of cytoskeletal proteins (Trojanowski et al., 1993) and amyloid-β metabolism (Selkoe, 1997). In AD, neuronal cytoskeleton disruption is a prominent feature along with oxidative damage (Smith and Perry, 1995; Ahlijanian et al., 2000; Nunomura et al., 2001). Many hypotheses have been proposed as determinants of AD, such as amyloid-β cascade (Selkoe, 2000), tauopathy (Lee et al., 2001), inflammation (McGeer and McGeer, 2002; Weiner and Selkoe, 2002) and oxidative stress (Markesbery, 1997; Perry and Smith, 1998; Christen, 2000; Perry et al., 2002; Picklo et al., 2002). NFT, a hallmark of AD, is composed of inappropriately phosphorylated NFs and other cytoskeletal proteins, especially Tau. However, not one of these theories alone is sufficient to explain the diversity of biochemical and pathological abnormalities of Alzheimer disease, which involves a multitude of cellular and biochemical changes. Furthermore, attempts to mimic the disease by a perturbation of one of these elements using cell or animal models, including transgenic animals, do not result in the same spectrum of pathological alterations. The most striking case is that, while amyloid-ß plaques are deposited in some transgenic rodent models over-expressing Aß-protein precursor (Hsiao et al., 1996), there is no neuronal loss – a seminal feature of Alzheimer disease.

In AD, aberrant hyperphosphorylation of proteins, such as Tau and NF is prominent and is considered as the consequence of the perturbation in the balance between the activities of kinases and phosphatases (Gong et al., 2000). MAPK (Trojanowski et al., 1993), GSK-3 (Picklo et al., 2002) and Cdk5 (Lew et al., 1994; Maccioni et al., 2001; Kesavapany et al., 2003) pathways are known to be involved in Tau and NF phosphorylation. One of the earliest changes noted in AD is accumulation of hyperphosphorylated Tau and NF in perikarya (Sternberger et al., 1985; Manetto et al., 1988; Sobue et al., 1990; Cleveland and Rothstein, 2001). This aberrant hyperphosphorylation is thought to be the cause of protein aggregation in NFT in AD neurons, although the role of NFT and senile plaques in AD is still not clear.

14.3.2 Amyotropic Lateral Sclerosis (ALS)

ALS is a motor neuron disease, also called Lou Gehrig's disease. It was first identified in 1869 by the noted neurologist Jean-Martin Charcot. Pathologically, ALS is characterized by the loss of motor neurons in the motor cortex (upper motor neurons), known as Betz cells, brainstem and spinal cord (lower motor neurons), astrogliosis localized to the areas of neurodegeneration, which eventually results in muscle atrophy. The presence of various types of ubiquitinated cytoplasmic inclusions, an extensive array of protein aggregates, including Bunina bodies, and neurofilament-rich 'hyaline conglomerate inclusions' is the basis of a definitive diagnosis of ALS (Ince et al., 2003; Wharton et al., 2003). Neurofilament aggregates intensely colocalize with peripherin, a related intermediate filament protein. Peripherin can co-assemble with the NF subunits in vitro and has been observed to co-aggregate with NF aggregates in ALS (Robertson et al., 2003).

NF aggregate formation is a common phenomenon consequent to aberrant NF hyperphosphorylation. The phosphorylation of cytoskeletal proteins is tightly regulated in the nervous system. Cytoskeletal proteins, especially neurofilaments and microtubules-associated proteins (MAP-2) and Tau, are extensively phosphorylated in the axonal component (Veeranna et al., 1995). Most of this phosphorylation takes place on proline-directed serine and threonine residues of the proteins in a topographically regulated manner. In the cell body of normal neurons, these proteins are little or not at all phosphorylated, whereas extensive phosphorylation is observed in the axonal compartment (Nixon et al., 1994; Sun et al., 1996; Pant et al., 1997; Jaffe et al., 1998).

Proline-directed S/T kinases, such as Cdk5, MAPKs and GSKs, are known to contribute to the pathogenesis of ALS. Regulated topographic phosphorylation induces compartmentally restricted phosphorylation by these kinases (Grant et al., 1999). Extensive phosphorylation of NF-H in human spinal cervical motor neurons of ALS but not in control subjects has been reported (Strong et al., 2001). Mass spectroscopic analysis of NF-H from ALS showed some new phosphorylation sites in ALS when compared with the control subjects. However, studies on hereditary canine spinal muscular dystrophy (HCSMA), an animal model of human motor neuron disease, showed no new S/T phospho-sites were detected (Green et al., 2005). These studies suggest that deregulation of topographic phosphorylation of neuronal cytoskeletal proteins is the major contributor to the development of ALS.

14.3.3 Parkinson's Disease (PD)

PD is a progressive disorder of the CNS. The disease was originally described in 1817 by an English physician, James Parkinson. PD is characterized by degeneration of the pigmented neurons in the substantia nigra of the brain, resulting in decreased dopamine availability. The clinical symptom of the disease involves a decrease in spontaneous movements, gait difficulty, postural instability, rigidity and tremor. The major pathological change of PD is an accumulated protein inclusion called the Lewy body, comprised of a number of proteins, including the essential components of α-synuclein protein, the three NF subunits (Galloway et al., 1992), ubiqutin and proteasome subunits (Trimmer et al., 2004). Electron microscopy and biochemical analyses indicate that the abnormally phosphorylated NFs form a non-membrane-bounded compacted skein in the neuronal soma in the affected neurons.

Mutations in the Parkin gene have been identified in familial PD, in which approximately 20 different mutations have been reported (Abbas et al., 1999; Lucking et al., 2000; Lim et al., 2002). A point mutation resulting in the substitution of serine for glycine at residue 336 in the NF-M gene coding for the rod domain of NF-M in a PD patient has also been reported (Lavedan et al., 2002). Interestingly, in PD, a decrease in NF-L mRNA has also been found (Hill et al., 1993).

14.3.4 Phosphorylation of Tau in Neuropathological Conditions

Neurofibrillary tangles (NFTs) are a common feature of many neurodegenerative diseases, including AD. Human NFTS are composed primarily of Tau in paired helical filaments (PHF), which are hyperphosphorylated at several sites (Grundke-Iqbal et al., 1986; Friedhoff et al., 2000). Tau hyperphosphorylation has been considered to be a pathogenic process that may regulate the initiation and progression of tauopathy. In tauopathy, hyperphosphorylated Tau is less likely to bind to and stabilize microtubules, resulting in the accumulation of insoluble Tau aggregates and the formation of NFTs. The serine/threonine phosphorylation sites of Tau are divided into two main groups: those that can be phosphorylated by proline-directed kinases like Tau protein kinase I (GSK3), Tau protein kinase II (Cdk5), p38 MAPK, JNK and other stress kinases or Cdc2, and those that can be phosphorylated by non-proline-directed kinases like PKA, PKC, CaM kinase II, MARK kinases or CK II (Baudier and Cole, 1987; Correas et al., 1992; Hanger et al., 1992; Scott et al., 1993; Goedert et al., 1997; Imahori and Uchida, 1997; Lucas et al., 2001; Drewes, 2004). Several kinases, including Cdk5, have been implicated as candidates in pathogenic Tau phosphorylation. It has been reported that Cdk5 physically interacts with Tau (Sobue et al., 2000), and phosphorylates several epitopes of Tau that are hyperphosphorylated in AD brains, thus enabling the transformed Tau to form NFTs (Flaherty et al., 2000). Early stage tangles in human AD brains associate with Cdk5 (Augustinack et al., 2002). Studies on transgenic mice that over-express p25/Cdk5 further support the link of NFTs to Cdk5 (Cruz et al., 2003; Noble et al., 2003).

14.3.5 Phosphorylation of NFs in Neuropathological Conditions

NF phosphorylation is topographically regulated within neurons (Grant and Pant, 2002). Phosphorylation of the NF subunits plays a critical role in the regulation of filament translocation, formation and function. Obviously, it is also involved in the pathogenesis of some neurodegenerative diseases.

Almost all of the assembled NFs in the myelinated internodal regions are stoichiometrically phosphorylated in the Lys-Ser-Pro (KSP) repeat domains (Hsieh et al., 1994; Veeranna et al., 1995, 1998). In contrast, the KSP repeats of NFPs in cell bodies, dendrites and nodes of Ranvier are less phosphorylated. The unphosphorylated NFs only account for $\sim$1% of NFs in the neuron as judged by NF density and the relative volumes of the respective compartments (de Waegh et al., 1992; Cole et al., 1994; Hsieh et al., 1994). A hallmark of affected motor neurons in amyotrophic lateral sclerosis (ALS) is the accumulation of filamentous 'spheroids' within proximal axons (Julien and Mushynski, 1998; Rao and Nixon, 2003; Strong et al., 2005; Shea and Beaty, 2007; Sihag et al., 2007). These spheroids are comprised of disorganized NFs that display epitopes in common with those normally segregated to distal axonal NFs. Notably, the pattern of NF phosphorylation in spinal tissue from patients with ALS is identical to that of normal individuals, suggesting that the accumulation of perikaryal/proximal axonal phospho-NFs consists of mislocalized but otherwise normally phosphorylated NFs (Bajaj et al., 1999; Strong et al., 2001). This supports the notion that precocious activation of the kinase(s) resulting in aberrant phosphorylation could perturb association of NFs with their anterograde motor and promote aberrant accumulation by fostering precocious NF–NF associations (Shea and Flanagan, 2001). Several reports support this possibility. The NF kinases Cdk5, GSK3β and p38MAP kinase each phosphorylate NFs to generate epitopes in common with NF spheroids and are associated with ALS (Bajaj et al., 1999; Strong, 1999; Ackerley et al., 2004). Over-expression of MEKK-1, which in turn over-activated the stress-activated/c-jun terminal kinase, inhibited translocation of NFs into growing axonal neurites and fostered accumulation of axonal-specific phospho-NF epitopes within perikarya (DeFuria et al., 2006). Notably, deletion of the C-terminal region of NF-H delays motor neuron pathology in a murine model of ALS (Lobsiger et al., 2005). These findings confirm that the kinase(s) that generate extensive C-terminal NF phosphorylation are present and active within perikarya and proximal axons. This not only suggests that the extent of C-terminal NF phosphorylation within these regions is dependent upon a balance of kinase and phosphatase activities, but also supports the notion that decreased compensatory phosphatase activity could contribute to aberrant NF phosphorylation and mislocalization.

Regarding a phospho-dependent decrease in association of NFs with kinesin (Yabe et al., 2000), a phospho-dependent increase in association of NFs with dynein (Motil et al., 2006) and/or a phospho-dependent increase in association of NFs with each other (Yabe et al., 2001; Shea et al., 2004), all of the phospho-dependent dynamics could contribute to perikaryal accumulation of phospho-NFs, either by inhibiting anterograde NF transport, enhancing a retrograde 'pull' and/or by inducing precocious NF–NF associations within perikarya instead of axons. NF spheroids accumulate kinesin and dynein (Toyoshima, 1998; Toyoshima et al., 1998) and such a motor entrapment may ultimately impair overall axonal transport, including that of non-cytoskeletal elements (Collard et al., 1995). Resultant increased residence time of NFs within perikarya may contribute further to their aberrant phosphorylation (Black and Lee, 1988), detrimental to neuron survival. Even a subtle shift in the balance of NF kinases and phosphatases may, over time, lead to NF phosphorylation-dependent mislocalization that accompanies motor neuron disease. Similarly, comparative studies on axonal transport and models of traffic modeling suggest that a critical imbalance in the association of NFs with their transport motors could generate a long-lasting 'pile up' of NFs (Shea and Beaty, 2007). These studies indicate that restoration of appropriate motor balance could diminish continued delivery of NFs into these and perhaps deplete NF accumulations, as is the case for experimentally induced NF aggregates in cultured cells (Shea et al., 1997). Perikaryal NF accumulations contain NFs that are apparently normally phosphorylated but are mislocated (Bajaj et al., 1999; Strong et al., 2001), suggesting that NF phosphorylation affects both transport and NF–NF associations, major contributing factors in motor neuron pathogenesis.

14.4 Cyclin-Dependent Kinase 5 (Cdk5) and Neuronal Cytoskeletal Protein Phosphorylation

Cdk5 and GSK3β have been identified as prime candidates for neurodegenerative pathogenesis (Sato et al., 2002; Drewes, 2004). Normally, cytoskeletal proteins are phosphorylated on (S/T-P) residues selectively in the axonal compartment of the neuron. Although all kinases, phosphatases, their substrates and their regulators are synthesized in the cell body, no cytoskeletal protein phosphorylation on these proline directed S/T-P residues is detected in the cell body. This compartmentalization of phosphorylation is tightly regulated, and in a number of neuropathological conditions, such as amyotrophic lateral sclerosis (ALS), Alzheimer's disease (AD), Neiman Pick's Type C disease and others, it becomes deregulated. Proline directed S/T-P residues on cytoskeletal proteins are aberrantly hyperphosphorylated within cell bodies, resulting in the accumulation of abnormal cellular aggregates and massive neuronal cell death. The mechanisms of the topographic phosphorylation are not understood.

Studies on kinases responsible for phosphorylation of neuronal cytoskeletal proteins specifically in their S/T-P residues revealed that Cdk5 is one of the principal kinases involved in their phosphorylation (Shetty et al., 1993). Although Cdk5 is ubiquitously expressed in all cells and shares a high degree of homology with other members of the cyclin-dependent kinase family (CDKs), its activity is found specifically in post-mitotic neurons because its activators, p35 and p39, are expressed primarily in neurons (Ko et al., 2001). Cdk5 is a multi-functional S/T protein kinase that is involved in a wide range of neuronal functions from neurite outgrowth and neuronal migration to synaptic activity and cell survival (Cruz and Tsai, 2004).

Studies to identify the consensus sequence in NF-M/H, MAPs/Tau, show that only the first serine with a neutral amino acid followed by a basic residue, VKS (1) PAK, is phosphorylated but not the second, KS(2) PEK, where it is acidic (Sharma et al., 1998). Also, unlike MAPKs (Erk1/2), Cdk5 is more selective for S/T-P K/R motifs of the human NF-M/H proteins.

Cdk5 activity is tightly regulated in the developing nervous system. Cdk5-null (KO) mice are lethal (Ohshima et al., 1996), exhibiting abnormal corticogenesis and other neuronal abnormalities before dying between E16 and P0. Experimental re-expression of Cdk5 in neurons of Cdk5 KO mice in vivo completely restores the wild type phenotype, clearly demonstrating that neuronal and not glial Cdk5 activity is necessary for normal development and survival (Takahashi et al., 2003).

During neuronal insults, increase in intracellular calcium and activation of calpains result in the cleavages of p35–p25, thereby inducing deregulation and hyperactivation of Cdk5. As a result, aberrant hyperphosphorylation of cytoskeletal proteins (e.g. NFs, MAPs, Tau) occurs, forming aggregates of these proteins in the cell body and consequently inducing neuronal death (Fig. 14.5). This process has been associated with a large number of neurodegenerative diseases (Ko et al., 2001). Cdk5 is not only involved in phosphorylating the NFs, MAPs and Tau but also involved directly or indirectly in modulating the other kinase activities that phosphorylate the same proteins as well as other proteins. Cross-talk of Cdk5 with many different signal transduction pathways is involved in nervous system development and neurodegeneration (Kesavapany et al., 2003; Rudrabhatla et al., 2008a). Because of its multifunctional role, as it exerts both positive and negative effects on neuronal function and survival, Cdk5 has been characterized as a "Jekyll and Hyde" kinase (Cruz and Tsai, 2004).

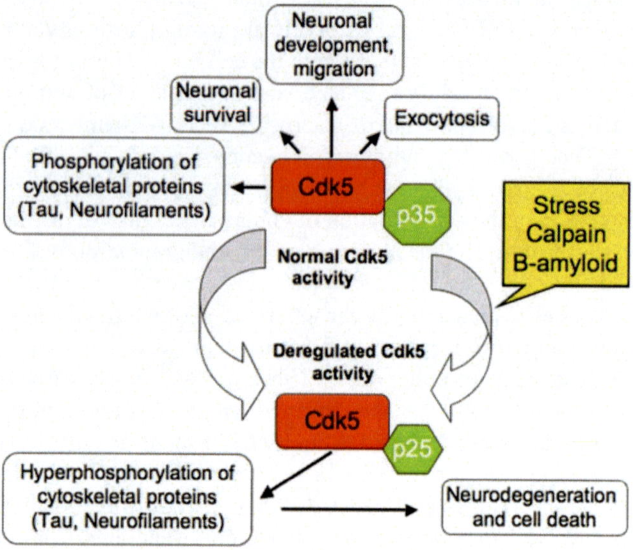

Fig. 14.5 Schematic diagram shows the normal and aberrant Cdk5 activity in the neurons. Cdk5 activity through p35 association assumes a range of function that includes nervous system development, neuronal survival, migration, exocytosis, and neuronal cytoskeletal protein maintenance and stability via phosphorylation. Following various insults, e.g. Ca+2 influx leading to calpain activation, p35 is cleaved and produces p25 protein that lacks the myristoylation site of the N-terminal region that is retained in the cleaved p10 protein. The mislocalized Cdk5/25 in the cytoplasm is hyperactivated, possibly because of its longer half-life than the Cdk5/p35. Cdk5/p25 extensively phosphorylates Tau and NF proteins, which are assumed as the primers neuron death and neurodegeneration

14.4.1 Deregulation of Cdk5 Activity and Other Kinases in NF and Tau Phosphorylation

14.4.1.1 MAPK (Erk1/2)

Earlier, during protein purification steps from rat brain lysates, Cdk5, neurofilament proteins and tubulins were found to coelute with Erk1/2 (Fig. 14.6a) (Veeranna et al., 2000). Furthermore, formation of a multimeric complex of Cdk5, Erk2 and phosphorylated NF-H was shown in microtubule preparations from rat brain lysates (Fig. 14.6b) (Veeranna et al., 2000). Later, it was discovered that Cdk5 inhibits Erk1/2 activity by phosphorylating the upstream kinase MEK1 (Sharma et al., 2002). Immunohistochemical staining of 18-day- old embryonic brain stem sections from Cdk5$^{-/-}$ and wild type mice with SMI31 antibody, which stains phosphorylated S/T-P residues in the neuronal cytoskeletal proteins (e.g. NF-M/H, Tau), shows intense immunostaining of phosphorylated cytoskeletal proteins in the neuronal cell bodies of Cdk5$^{-/-}$ mice compared to the wild type mice (Sharma et al., 2002).

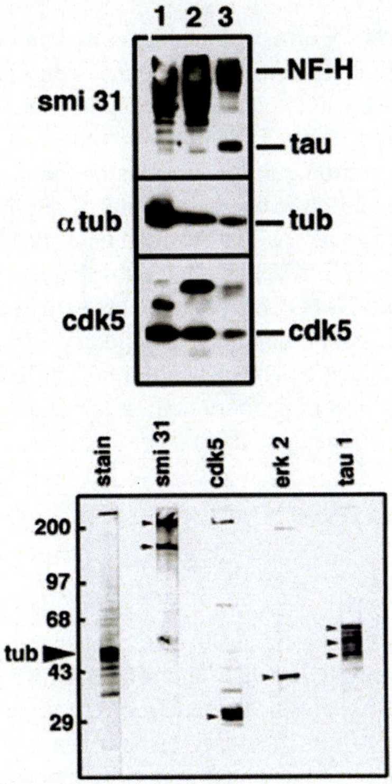

Fig. 14.6 **a** Purified fractions of ERK1/2 from rat brain lysates are enriched in phospho-NF-H/M, tubulin, Tau and Cdk5. Lane 1 is the original (total) lysate. Fractions were collected during the column chromatography. Lane 2 is fraction 25 of the P11 phosphocellulose column that is enriched in MAPK (Erk1/2) activity. Lane 3 is the fraction 35 of a Superdex 200 that is enriched in Erk1/2 activity. Western blotting shows the presence of abundant NF-H/M, tubulin, Tau and Cdk5 in these two fractions. Lane 1 shows levels of NF-H/M, tubulin, Tau and Cdk5 in the crude lysate. **b** Purified fraction from rat brain lysates enriched in tubulin is also enriched in phospho-NF-H/M, Tau, ERK2 and Cdk5. Coomassie Blue staining of the fraction is shown that confirms that the fraction is enriched in tubulin. The same fraction upon Western blotting reveals the co-extraction of NF-H, NF-M, Tau 1, ERK2 and Cdk5 (adapted with modifications from Veeranna et al., 2000)

The expression of the phosphorylated cytoskeletal proteins is restricted to the neuronal processes only. Phosphorylation of cytoskeletal proteins and Cdk5 activity are higher in adult mice brain compared to the embryonic brain. Studies on phosphorylation of middle and high molecular weight neurofilament proteins (NF-M/H) in $p35^{-/-}$ and wild type mice reveals immunoreactivity of NF-M to the SMI31 antibody in the cortex of $p35^{-/-}$ mice is higher than in the wild type-mice (Sharma et al., 2002). The rodent NF-M appears to be a preferred substrate for Erk1/2 phosphorylation (Veeranna et al., 1998). Also in $p35^{-/-}$ mice, there is an inverse relationship between Cdk5 and Erk1/2 activities, while the expression levels of Erk1/2 remained unchanged (Sharma et al., 2002). Such sustained high activity of Erk1/2 in the neuronal cells has been implicated as the cause of cell death

(Cheung and Slack, 2004). Consistent with this assumption, extensive apoptosis in $Cdk^{-/-}$ brains compared to the wild type brains is observed as evidenced by the upregulation of the apoptosis marker, cleaved-caspase-3 and appearance of TUNEL-positive cells (Zheng et al., 2005). Evidence from mouse models, the PC12 cell line and rat primary cortical neuron cultures demonstrate that Cdk5 modulates MAPK (MEK/Erk1/2) pathway and is involved in neuronal survival, in which Cdk5/p35 seemed to act as a "molecular switch" to modulate the duration of Erk1/2 activation, thereby resulting in NF-M/H and Tau phosphorylation (Harada et al., 2001; Sharma et al., 2002). Cdk5-Ras GRF cross talk upstream of the MAPK pathway also regulates neuronal apoptosis through nuclear condensation of neurons (Kesavapany et al., 2004, 2006). Sustained Erk1/2 activation by inhibiting Cdk5 activity using roscovitine induces apoptosis of primary neuron in culture, where Tau and NF are also hyperphosphorylated (Zheng et al., 2007).

14.4.1.2 GSK3β

The other kinase that has been linked to neurodegeneration along with Cdk5 is glycogen synthase kinase 3β (GSK3β). The role of GSK3 in axonal transport has been reported. Extruded axoplasm derived from the giant axon of the squid has been ideal for the visualization and study of both fast and slow axonal transport, making it possible to study the molecular motors involved and their mode of action in transporting vesicles from cell body to terminals. One of these motors is kinesin, which is responsible for fast anterograde transport of vesicles and is also involved in transport of cytoskeletal proteins. It was observed that low concentrations of olomucine, which inhibits Cdk5 activity, also inhibited fast axonal transport and GSK3 phosphorylation of kinesin light chains significantly inhibited anterograde but not retrograde fast transport (Morfini et al., 2002). There is a complex interaction between Cdk5, PP1 phosphatase and GSK3 kinase. Inhibiting Cdk5 with roscovitine or olomucine activates PP1 phosphatase which, in turn, dephosphorylates and activates GSK3β. Active GSK3β then phosphorylates the kinesin light chain, releasing it from its cargo, thereby decreasing vesicle transport. These findings predict a more elaborate cross talk interaction, involving Cdk5, phosphatases and GSK3 kinase, integrated into a network regulating transport of membrane-bound vesicles and targeting them to terminals that may play a role in the neurodegenerative process. Cdk5 and GSK3β both generate disease-associated phospho-epitopes on Tau, and they colocalize with filamentous Tau aggregates in the brains of patients (Imahori and Uchida, 1997; Shelton and Johnson, 2004) and in a transgenic mouse model of tauopathy (Imahori and Uchida, 1997; Ishizawa et al., 2003). Cdk5 and GSK3 also regulate Aβ production in vivo (Phiel et al., 2003; Cruz et al., 2006). There is a high degree of redundancy between the two kinases, and several of the phosphorylation target sites overlap. Therefore the relative impact of each kinase in the context of amyloid precursor protein (APP) processing and Tau phosphorylation is not very clear.

14.4.2 Cdk5 Coactivators and Neurodegeneration

The reported neurotoxic effects of Cdk5 have been linked to p25 production, a proteolytically cleaved product of p35, the major activator of neuronal Cdk5. In primary cortical neuron cultures, p25/Cdk5 complex phosphorylates Tau more efficiently than does the p35/Cdk5 complex (Patrick et al., 1999). In vitro Tau phosphorylation assays have demonstrated that p25 accelerates Cdk5 catalytic activity by ~2.4-fold over p35 (Hashiguchi et al., 2002). Further evidence comes from the preferential increase in Tau phosphorylation in p25 transgenic mice (Ahlijanian et al., 2000; Cruz et al., 2003), while p35 transgenic mice displaying increased Cdk5 catalytic activity do not show increased Tau phosphorylation (Van den Haute et al., 2001). These findings are complemented by the observation that Cdk5-deficient mice show decreased Tau phosphorylation (Takahashi et al., 2003). Surprisingly, however, a new strain of p35-deficient mice displays increased Tau phosphorylation (Hallows et al., 2003). It is possible that in these mice Tau phosphorylation occurred due to the compensatory increases in another Cdk5 activator, p39 level or by Cdk-crosstalk as discussed above. The compensatory increase in p39 expression has been reported in p35-deficient mice (Chae et al., 1997). Moreover, p39-mediated Tau phosphorylation is more efficient than p35-mediated Tau phosphorylation (Iijima et al., 2000). P39-derived p29 is also potent in phosphorylating Tau (Patzke and Tsai, 2002).

Contrary to its reported pathological role, recent studies now implicate p25 as a "normal" player in modulating synaptic function, LTD, learning and memory in specific brain regions in young animals (Fischer et al., 2003, 2005; Angelo et al., 2006). Transgenic mice expressing either low level of p25, or with expression restricted spatiotemporally to specific brain regions, show a Cdk5/p25 positive transient effect on LTD in the hippocampus and water maze learning in these animals. Prolonged expression of p25, in older animals, however, does exert a predictable effect on β-amyloid and Tau phosphorylation and neurodegeneration (Cruz et al., 2003; Cruz and Tsai, 2004). The differential phosphorylation potentials of p25 and p35 are also associated with another Cdk5 substrate, APP, which is involved in neurodegeneration. Cdk5 phosphorylates amyloid precursor protein (APP) in its cytoplasmic domain at Thr668 (Iijima et al., 2000). Increased APP Thr668 phosphorylation is observed in p25 transgenic mice in which p35/Cdk5 activity remains unaltered (Cruz et al., 2003). However, the role of Cdk5/p25 in neuropathological diseases is far more complex than previously assumed.

14.4.3 Cdk5, β-Amyloid and Neuronal Cytoskeletal Proteins

One of the leading hypotheses for the etiology of Alzheimer's disease (AD) is the amyloid hypothesis (Hardy and Selkoe, 2002). In AD, extracellular accumulation of Aβ42 that readily aggregates into amyloid plaques occurs due to an altered ratio of Aβ generation and clearance. One of the downstream events of this elevated Aβ42 levels is the aberrant activation of kinases and inhibition of phosphatases. The resultant imbalance in the kinase/phosphatase activities causes neurofibrillary tangle formation and neuronal death. Myriads of reports now link Cdk5 to Aβ42 toxicity and Tau pathology leading to neurodegeneration. In primary neurons, Aβ42 induces the cleavage of p35–p25 (Alvarez et al., 1999; Lee et al., 2000; Town et al., 2002) (Fig. 14.5). P25 accumulation is found in mutant APP transgenic mice that display elevated AB42 levels (Otth et al., 2002). Inhibition of Cdk5 activity also attenuates Aβ42-induced neuronal death (Alvarez et al., 1999; Lee et al., 2000). These findings indicate that Aβ42 is a potent activator of p25/Cdk5 activity.

14.5 Polypeptide Inhibitors of Deregulated Neuronal Cytoskeletal Protein Phosphorylation

NF- and Tau-directed protein kinases are potential targets for modification when it comes to therapeutic interventions of AD. However, most of the protein kinases that phosphorylate Tau and NFs also phosphorylate other substrates. Inhibitors of protein kinases have not been clinically useful to treat neurodegeneration. However, several inhibitors for the Cdk5/p25, that hyperphosphorylates Tau and NFs aberrantly, have been reported. One such polypeptide CIP (Cdk5/p25 Inhibitory Peptide), a derivative of p35 (Zheng et al., 2002), shows the potential of therapeutic intervention, since it specifically inhibits the Cdk5/p25 deregulated hyperactivity (Fig. 14.7). It

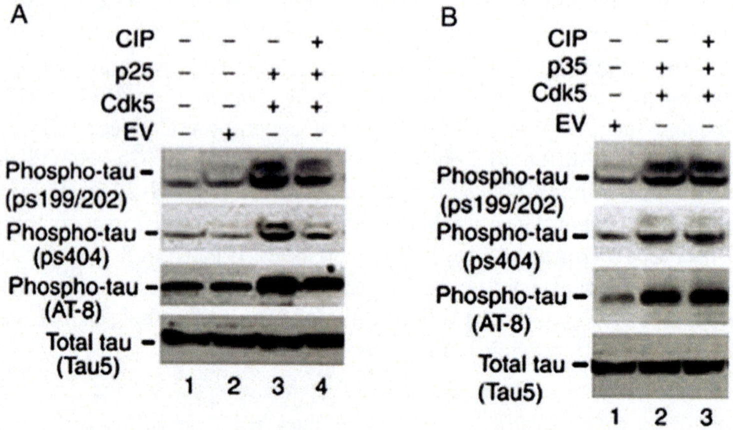

Fig. 14.7 CIP specifically inhibits Tau hyperphosphorylation induced by p25/Cdk5 but not Tau phosphorylation induced by p35/Cdk5. **a** Rat cortical neurons were infected with the following expression constructs in lentivirus: EV (Empty Vector, lane 2), p25 + Cdk5 (lane 3), p25 + Cdk5 + CIP (lane 4). The *top three panels* show phospho-Tau with three different phospho-Tau antibodies (anti-pS199/202, anti-pS404 and AT8), while the *bottom panel* shows total Tau (Tau5). **b** Rat cortical neurons were infected with the following expression constructs: EV (lane 1), p35 + Cdk5 (lane 2), p35 + Cdk5 + CIP (lane 3). The *top three panels* show Western blots for phospho-Tau using the same phospho-Tau antibodies mentioned above and the *bottom panel* shows total Tau (Tau5) (adapted with modifications from Zheng et al., 2005)

is important to note that CIP does not inhibit Cdk5/p35 activity and therefore does not interfere with the normal Cdk5-regulated phosphorylation. Although rare, such inhibitors that do not inhibit normal kinase activity, but specifically inhibit the aberrant, hyperactive kinase activity, are likely to be effective in designing treatment strategies for neurodegeneration. Lately, it has been reported that Pin1, a peptidyl-prolyl *cis/trans*-isomerase substantially increases phosphorylation of NF-H KSP repeats by proline directed kinases, Erk1/2, Cdk5/p35 and JNK3 in a concentration-dependent manner (Kesavapany et al., 2007; Rudrabhatla et al., 2008b). Thus, targeting Pin1 for therapeutic purposes of treating neurodegenerative diseases seems attractive.

14.6 Conclusions

It is surprising that, although Tau and NF phosphorylation has been extensively studied, a clear relationship between their phosphorylation and toxicity leading to neurodegeneration is yet to be established. Since most of the kinases that are involved in Tau and NF hyperphosphorylation in neurodegenerative processes are kinases that are also important for normal physiology, it is difficult to draw a fine line between these two opposing effects of the same kinases. However, there are clues to understand how a normal kinase becomes aberrantly hyperactive (e.g. Cdk5/p35 vs Cdk5/p25), and what the cellular signals are that trigger this transition, be it oxidative stress, awry cellular signals, such as β-amyloid deposits and genetic mutations. Given the complexity of the onset and progression of neurodegenerative processes, understanding kinase-mediated cytoskeletal protein phosphorylation is a step toward elucidating what exactly it is that begins this process and, importantly, whether outlining specific therapies can halt this 'beginning'.

References

Abbas N, Lucking CB, Ricard S, Durr A, Bonifati V, De Michele G, Bouley S, Vaughan JR, Gasser T, Marconi R, Broussolle E, Brefel-Courbon C, Harhangi BS, Oostra BA, Fabrizio E, Bohme GA, Pradier L, Wood NW, Filla A, Meco G, Denefle P, Agid Y, Brice A (1999) A wide variety of mutations in the parkin gene are responsible for autosomal recessive parkinsonism in Europe. French Parkinson's disease genetics study group and the European consortium on genetic susceptibility in Parkinson's disease. Hum Mol Genet 8:567–574

Ackerley S, Grierson AJ, Banner S, Perkinton MS, Brownlees J, Byers HL, Ward M, Thornhill P, Hussain K, Waby JS, Anderton BH, Cooper JD, Dingwall C, Leigh PN, Shaw CE, Miller CC (2004) p38alpha stress-activated protein kinase phosphorylates neurofilaments and is associated with neurofilament pathology in amyotrophic lateral sclerosis. Mol Cell Neurosci 26:354–364

Ahlijanian MK, Barrezueta NX, Williams RD, Jakowski A, Kowsz KP, McCarthy S, Coskran T, Carlo A, Seymour PA, Burkhardt JE, Nelson RB, McNeish JD (2000) Hyperphosphorylated tau and neurofilament and cytoskeletal disruptions in mice overexpressing human p25, an activator of cdk5. Proc Natl Acad Sci USA 97:2910–2915

Alonso AC, Grundke-Iqbal I, Iqbal K (1996) Alzheimer's disease hyperphosphorylated tau sequesters normal tau into tangles of filaments and disassembles microtubules. Nat Med 2:783–787

Alonso AC, Zaidi T, Grundke-Iqbal I, Iqbal K (1994) Role of abnormally phosphorylated tau in the breakdown of microtubules in Alzheimer disease. Proc Natl Acad Sci USA 91:5562–5566

Alvarez A, Toro R, Caceres A, Maccioni RB (1999) Inhibition of tau phosphorylating protein kinase cdk5 prevents beta-amyloid-induced neuronal death. FEBS Lett 459:421–426

Angelo M, Plattner F, Giese KP (2006) Cyclin-dependent kinase 5 in synaptic plasticity, learning and memory. J Neurochem 99:353–370

Augustinack JC, Sanders JL, Tsai LH, Hyman BT (2002) Colocalization and fluorescence resonance energy transfer between cdk5 and AT8 suggests a close association in pre-neurofibrillary tangles and neurofibrillary tangles. J Neuropathol Exp Neurol 61:557–564

Bajaj NP, al-Sarraj ST, Leigh PN, Anderson V, Miller CC (1999) Cyclin dependent kinase-5 (CDK-5) phosphorylates neurofilament heavy (NF-H) chain to generate epitopes for antibodies that label neurofilament accumulations in amyotrophic lateral sclerosis (ALS) and is present in affected motor neurones in ALS. Prog Neuropsychopharmacol Biol Psychiatry 23:833–850

Baudier J, Cole RD (1987) Phosphorylation of tau proteins to a state like that in Alzheimer's brain is catalyzed by a calcium/calmodulin-dependent kinase and modulated by phospholipids. J Biol Chem 262:17577–17583

Black MM, Lee VM (1988) Phosphorylation of neurofilament proteins in intact neurons: demonstration of phosphorylation in cell bodies and axons. J Neurosci 8:3296–3305

Brady ST, Richards BW, Leopold PL (1993) Assay of vesicle motility in squid axoplasm. Methods Cell Biol 39:191–202

Brion JP, Octave JN, Couck AM (1994) Distribution of the phosphorylated microtubule-associated protein tau in developing cortical neurons. Neuroscience 63:895–909

Burns RG, Surridge CD (1994) Functional role of a consensus peptide which is common to alpha-, beta-, and gamma-tubulin, to actin and centractin, to phytochrome A, and to the TCP1 alpha chaperonin protein. FEBS Lett 347:105–111

Chae T, Kwon YT, Bronson R, Dikkes P, Li E, Tsai LH (1997) Mice lacking p35, a neuronal specific activator of Cdk5, display cortical lamination defects, seizures, and adult lethality. Neuron 18:29–42

Cheung EC, Slack RS (2004) Emerging role for ERK as a key regulator of neuronal apoptosis. Sci STKE 2004:PE45

Christen Y (2000) Oxidative stress and Alzheimer disease. Am J Clin Nutr 71:621S–629S

Cleveland DW, Rothstein JD (2001) From Charcot to Lou Gehrig: deciphering selective motor neuron death in ALS. Nat Rev Neurosci 2:806–819

Cole JS, Messing A, Trojanowski JQ, Lee VM (1994) Modulation of axon diameter and neurofilaments by hypomyelinating Schwann cells in transgenic mice. J Neurosci 14:6956–6966

Collard JF, Cote F, Julien JP (1995) Defective axonal transport in a transgenic mouse model of amyotrophic lateral sclerosis. Nature 375:61–64

Corder EH, Saunders AM, Strittmatter WJ, Schmechel DE, Gaskell PC, Small GW, Roses AD, Haines JL, Pericak-Vance MA (1993) Gene dose of apolipoprotein E type 4 allele and the risk of Alzheimer's disease in late onset families. Science 261:921–923

Correas I, Diaz-Nido J, Avila J (1992) Microtubule-associated protein tau is phosphorylated by protein kinase C on its tubulin binding domain. J Biol Chem 267:15721–15728

Cruz JC, Kim D, Moy LY, Dobbin MM, Sun X, Bronson RT, Tsai LH (2006) p25/cyclin-dependent kinase 5 induces production and intraneuronal accumulation of amyloid beta in vivo. J Neurosci 26:10536–10541

Cruz JC, Tsai LH (2004) A Jekyll and Hyde kinase: roles for Cdk5 in brain development and disease. Curr Opin Neurobiol 14:390–394

Cruz JC, Tseng HC, Goldman JA, Shih H, Tsai LH (2003) Aberrant Cdk5 activation by p25 triggers pathological events leading to neurodegeneration and neurofibrillary tangles. Neuron 40: 471–483

de Waegh SM, Lee VM, Brady ST (1992) Local modulation of neurofilament phosphorylation, axonal caliber, and slow axonal transport by myelinating Schwann cells. Cell 68: 451–463

DeFuria J, Chen P, Shea TB (2006) Divergent effects of the MEKK-1/JNK pathway on NB2a/d1 differentiation: some activity is required for outgrowth and stabilization of neurites but overactivation inhibits both phenomena. Brain Res 1123:20–26

Drewes G (2004) MARKing tau for tangles and toxicity. Trends Biochem Sci 29:548–555

Elder GA, Friedrich VL Jr, Kang C, Bosco P, Gourov A, Tu PH, Zhang B, Lee VM, Lazzarini RA (1998) Requirement of heavy neurofilament subunit in the development of axons with large calibers. J Cell Biol 143:195–205

Feany MB, Dickson DW (1996) Neurodegenerative disorders with extensive tau pathology: a comparative study and review. Ann Neurol 40:139–148

Fischer A, Sananbenesi F, Pang PT, Lu B, Tsai LH (2005) Opposing roles of transient and prolonged expression of p25 in synaptic plasticity and hippocampus-dependent memory. Neuron 48:825–838

Fischer A, Sananbenesi F, Spiess J, Radulovic J (2003) Cdk5: a novel role in learning and memory. Neurosignals 12:200–208

Fiumelli H, Riederer IM, Martin JL, Riederer BM (2008) Phosphorylation of neurofilament subunit NF-M is regulated by activation of NMDA receptors and modulates cytoskeleton stability and neuronal shape. Cell Motil Cytoskeleton 65:495–504

Flaherty DB, Soria JP, Tomasiewicz HG, Wood JG (2000) Phosphorylation of human tau protein by microtubule-associated kinases: GSK3beta and cdk5 are key participants. J Neurosci Res 62:463–472

Fliegner KH, Liem RK (1991) Cellular and molecular biology of neuronal intermediate filaments. Int Rev Cytol 131:109–167

Friedhoff P, von Bergen M, Mandelkow EM, Mandelkow E (2000) Structure of tau protein and assembly into paired helical filaments. Biochim Biophys Acta 1502:122–132

Galloway PG, Mulvihill P, Perry G (1992) Filaments of Lewy bodies contain insoluble cytoskeletal elements. Am J Pathol 140:809–822

Garver TD, Harris KA, Lehman RA, Lee VM, Trojanowski JQ, Billingsley ML (1994) Tau phosphorylation in human, primate, and rat brain: evidence that a pool of tau is highly phosphorylated in vivo and is rapidly dephosphorylated in vitro. J Neurochem 63:2279–2287

Goedert M, Hasegawa M, Jakes R, Lawler S, Cuenda A, Cohen P (1997) Phosphorylation of microtubule-associated protein tau by stress-activated protein kinases. FEBS Lett 409:57–62

Goedert M, Jakes R, Crowther RA, Cohen P, Vanmechelen E, Vandermeeren M, Cras P (1994) Epitope mapping of monoclonal antibodies to the paired helical filaments of Alzheimer's disease: identification of phosphorylation sites in tau protein. Biochem J 301(3):871–877

Goedert M, Jakes R, Spillantini MG, Crowther RA, Cohen P, Vanmechelen E, Probst A, Gotz J, Burki K (1995) Tau protein in Alzheimer's disease. Biochem Soc Trans 23:80–85

Goedert M, Sisodia SS, Price DL (1991) Neurofibrillary tangles and beta-amyloid deposits in Alzheimer's disease. Curr Opin Neurobiol 1:441–447

Gong CX, Wegiel J, Lidsky T, Zuck L, Avila J, Wisniewski HM, Grundke-Iqbal I, Iqbal K (2000) Regulation of phosphorylation of neuronal microtubule-associated proteins MAP1b and MAP2 by protein phosphatase-2A and -2B in rat brain. Brain Res 853:299–309

Grant P, Diggins M, Pant HC (1999) Topographic regulation of cytoskeletal protein phosphorylation by multimeric complexes in the squid giant fiber system. J Neurobiol 40:89–102

Grant P, Pant HC (2002) Topographic regulation of kinase activity in Alzheimer's disease brains. J Alzheimers Dis 4:269–281

Green SL, Westendorf JM, Jaffe H, Pant HC, Cork LC, Ostrander EA, Vignaux F, Ferrell JE Jr (2005) Allelic variants of the canine heavy neurofilament (NFH) subunit and extensive phosphorylation in dogs with motor neuron disease. J Comp Pathol 132:33–50

Greenberg SG, Davies P, Schein JD, Binder LI (1992) Hydrofluoric acid-treated tau PHF proteins display the same biochemical properties as normal tau. J Biol Chem 267:564–569

Grundke-Iqbal I, Iqbal K, Tung YC, Quinlan M, Wisniewski HM, Binder LI (1986) Abnormal phosphorylation of the microtubule-associated protein tau (tau) in Alzheimer cytoskeletal pathology. Proc Natl Acad Sci USA 83:4913–4917

Gustke N, Steiner B, Mandelkow EM, Biernat J, Meyer HE, Goedert M, Mandelkow E (1992) The Alzheimer-like phosphorylation of tau protein reduces microtubule binding and involves Ser-Pro and Thr-Pro motifs. FEBS Lett 307:199–205

Hallows JL, Chen K, DePinho RA, Vincent I (2003) Decreased cyclin-dependent kinase 5 (cdk5) activity is accompanied by redistribution of cdk5 and cytoskeletal proteins and increased cytoskeletal protein phosphorylation in p35 null mice. J Neurosci 23:10633–10644

Hanger DP, Hughes K, Woodgett JR, Brion JP, Anderton BH (1992) Glycogen synthase kinase-3 induces Alzheimer's disease-like phosphorylation of tau: generation of paired helical filament epitopes and neuronal localisation of the kinase. Neurosci Lett 147:58–62

Harada T, Morooka T, Ogawa S, Nishida E (2001) ERK induces p35, a neuron-specific activator of Cdk5, through induction of Egr1. Nat Cell Biol 3:453–459

Hardy J, Selkoe DJ (2002) The amyloid hypothesis of Alzheimer's disease: progress and problems on the road to therapeutics. Science 297:353–356

Hasegawa M, Watanabe A, Takio K, Suzuki M, Arai T, Titani K, Ihara Y (1993) Characterization of two distinct monoclonal antibodies to paired helical filaments: further evidence for fetal-type phosphorylation of the tau in paired helical filaments. J Neurochem 60:2068–2077

Hashiguchi M, Saito T, Hisanaga S, Hashiguchi T (2002) Truncation of CDK5 activator p35 induces intensive phosphorylation of Ser202/Thr205 of human tau. J Biol Chem 277: 44525–44530

Hill WD, Arai M, Cohen JA, Trojanowski JQ (1993) Neurofilament mRNA is reduced in Parkinson's disease substantia nigra pars compacta neurons. J Comp Neurol 329:328–336

Himmler A, Drechsel D, Kirschner MW, Martin DW Jr (1989) Tau consists of a set of proteins with repeated C-terminal microtubule-binding domains and variable N-terminal domains. Mol Cell Biol 9:1381–1388

Hisanaga S, Hirokawa N (1989) The effects of dephosphorylation on the structure of the projections of neurofilament. J Neurosci 9:959–966

Holzer M, Holzapfel HP, Zedlick D, Bruckner MK, Arendt T (1994) Abnormally phosphorylated tau protein in Alzheimer's disease: heterogeneity of individual regional distribution and relationship to clinical severity. Neuroscience 63:499–516

Hsiao K, Chapman P, Nilsen S, Eckman C, Harigaya Y, Younkin S, Yang F, Cole G (1996) Correlative memory deficits, Abeta elevation, and amyloid plaques in transgenic mice. Science 274:99–102

Hsieh ST, Crawford TO, Griffin JW (1994) Neurofilament distribution and organization in the myelinated axons of the peripheral nervous system. Brain Res 642:316–326

Iijima K, Ando K, Takeda S, Satoh Y, Seki T, Itohara S, Greengard P, Kirino Y, Nairn AC, Suzuki T (2000) Neuron-specific phosphorylation of Alzheimer's beta-amyloid precursor protein by cyclin-dependent kinase 5. J Neurochem 75:1085–1091

Imahori K, Uchida T (1997) Physiology and pathology of tau protein kinases in relation to Alzheimer's disease. J Biochem 121:179–188

Ince PG, Evans J, Knopp M, Forster G, Hamdalla HH, Wharton SB, Shaw PJ (2003) Corticospinal tract degeneration in the progressive muscular atrophy variant of ALS. Neurology 60: 1252–1258

Iqbal K, Alonso AC, Gong CX, Khatoon S, Singh TJ, Grundke-Iqbal I (1994) Mechanism of neurofibrillary degeneration in Alzheimer's disease. Mol Neurobiol 9:119–123

Ishizawa T, Sahara N, Ishiguro K, Kersh J, McGowan E, Lewis J, Hutton M, Dickson DW, Yen SH (2003) Co-localization of glycogen synthase kinase-3 with neurofibrillary tangles and granulovacuolar degeneration in transgenic mice. Am J Pathol 163:1057–1067

Jaffe H, Veeranna, Shetty KT, Pant HC (1998) Characterization of the phosphorylation sites of human high molecular weight neurofilament protein by electrospray ionization tandem mass spectrometry and database searching. Biochemistry 37:3931–3940

Joshi HC (1994) Microtubule organizing centers and gamma-tubulin. Curr Opin Cell Biol 6:54–62

Julien JP, Mushynski WE (1998) Neurofilaments in health and disease. Prog Nucleic Acid Res Mol Biol 61:1–23

Kanemaru K, Takio K, Miura R, Titani K, Ihara Y (1992) Fetal-type phosphorylation of the tau in paired helical filaments. J Neurochem 58:1667–1675

Kesavapany S, Amin N, Zheng YL, Nijhara R, Jaffe H, Sihag R, Gutkind JS, Takahashi S, Kulkarni A, Grant P, Pant HC (2004) p35/cyclin-dependent kinase 5 phosphorylation of ras guanine nucleotide releasing factor 2 (RasGRF2) mediates Rac-dependent extracellular signal-regulated kinase 1/2 activity, altering RasGRF2 and microtubule-associated protein 1b distribution in neurons. J Neurosci 24:4421–4431

Kesavapany S, Lau KF, Ackerley S, Banner SJ, Shemilt SJ, Cooper JD, Leigh PN, Shaw CE, McLoughlin DM, Miller CC (2003) Identification of a novel, membrane-associated neuronal kinase, cyclin-dependent kinase 5/p35-regulated kinase. J Neurosci 23: 4975–4983

Kesavapany S, Pareek TK, Zheng YL, Amin N, Gutkind JS, Ma W, Kulkarni AB, Grant P, Pant HC (2006) Neuronal nuclear organization is controlled by cyclin-dependent kinase 5 phosphorylation of Ras Guanine nucleotide releasing factor-1. Neurosignals 15:157–173

Kesavapany S, Patel V, Zheng YL, Pareek TK, Bjelogrlic M, Albers W, Amin N, Jaffe H, Gutkind JS, Strong MJ, Grant P, Pant HC (2007) Inhibition of Pin1 reduces glutamate-induced perikaryal accumulation of phosphorylated neurofilament-H in neurons. Mol Biol Cell 18:3645–3655

Ko J, Humbert S, Bronson RT, Takahashi S, Kulkarni AB, Li E, Tsai LH (2001) p35 and p39 are essential for cyclin-dependent kinase 5 function during neurodevelopment. J Neurosci 21:6758–6771

Kosik KS, Joachim CL, Selkoe DJ (1986) Microtubule-associated protein tau (tau) is a major antigenic component of paired helical filaments in Alzheimer disease. Proc Natl Acad Sci USA 83:4044–4048

Kosik KS, Orecchio LD, Binder L, Trojanowski JQ, Lee VM, Lee G (1988) Epitopes that span the tau molecule are shared with paired helical filaments. Neuron 1:817–825

Lavedan C, Buchholtz S, Nussbaum RL, Albin RL, Polymeropoulos MH (2002) A mutation in the human neurofilament M gene in Parkinson's disease that suggests a role for the cytoskeleton in neuronal degeneration. Neurosci Lett 322:57–61

Lee MK, Cleveland DW (1996) Neuronal intermediate filaments. Annu Rev Neurosci 19:187–217

Lee VM, Goedert M, Trojanowski JQ (2001) Neurodegenerative tauopathies. Annu Rev Neurosci 24:1121–1159

Lee MS, Kwon YT, Li M, Peng J, Friedlander RM, Tsai LH (2000) Neurotoxicity induces cleavage of p35 to p25 by calpain. Nature 405:360–364

Lew J, Huang QQ, Qi Z, Winkfein RJ, Aebersold R, Hunt T, Wang JH (1994) A brain-specific activator of cyclin-dependent kinase 5. Nature 371:423–426

Lim KL, Dawson VL, Dawson TM (2002) The genetics of Parkinson's disease. Curr Neurol Neurosci Rep 2:439–446

Lobsiger CS, Garcia ML, Ward CM, Cleveland DW (2005) Altered axonal architecture by removal of the heavily phosphorylated neurofilament tail domains strongly slows superoxide dismutase 1 mutant-mediated ALS. Proc Natl Acad Sci USA 102:10351–10356

Lucas JJ, Hernandez F, Gomez-Ramos P, Moran MA, Hen R, Avila J (2001) Decreased nuclear beta-catenin, tau hyperphosphorylation and neurodegeneration in GSK-3beta conditional transgenic mice. EMBO J 20:27–39

Lucking CB, Durr A, Bonifati V, Vaughan J, De Michele G, Gasser T, Harhangi BS, Meco G, Denefle P, Wood NW, Agid Y, Brice A (2000) Association between early-onset Parkinson's disease and mutations in the parkin gene. N Engl J Med 342:1560–1567

Maccioni RB, Otth C, Concha II, Munoz JP (2001) The protein kinase Cdk5. Structural aspects, roles in neurogenesis and involvement in Alzheimer's pathology. Eur J Biochem 268: 1518–1527

Mandelkow EM, Schweers O, Drewes G, Biernat J, Gustke N, Trinczek B, Mandelkow E (1996) Structure, microtubule interactions, and phosphorylation of tau protein. Ann N Y Acad Sci 777:96–106

Manetto V, Sternberger NH, Perry G, Sternberger LA, Gambetti P (1988) Phosphorylation of neurofilaments is altered in amyotrophic lateral sclerosis. J Neuropathol Exp Neurol 47:642–653

Markesbery WR (1997) Oxidative stress hypothesis in Alzheimer's disease. Free Radic Biol Med 23:134–147

Matsuo ES, Shin RW, Billingsley ML, Van de Voorde A, O'Connor M, Trojanowski JQ, Lee VM (1994) Biopsy-derived adult human brain tau is phosphorylated at many of the same sites as Alzheimer's disease paired helical filament tau. Neuron 13:989–1002

McGeer PL, McGeer EG (2002) Local neuroinflammation and the progression of Alzheimer's disease. J Neurovirol 8:529–538

McShea A, Harris PL, Webster KR, Wahl AF, Smith MA (1997) Abnormal expression of the cell cycle regulators P16 and CDK4 in Alzheimer's disease. Am J Pathol 150:1933–1939

Miller CC, Ackerley S, Brownlees J, Grierson AJ, Jacobsen NJ, Thornhill P (2002) Axonal transport of neurofilaments in normal and disease states. Cell Mol Life Sci 59: 323–330

Morfini G, Szebenyi G, Elluru R, Ratner N, Brady ST (2002) Glycogen synthase kinase 3 phosphorylates kinesin light chains and negatively regulates kinesin-based motility. EMBO J 21:281–293

Motil J, Chan WK, Dubey M, Chaudhury P, Pimenta A, Chylinski TM, Ortiz DT, Shea TB (2006) Dynein mediates retrograde neurofilament transport within axons and anterograde delivery of NFs from perikarya into axons: regulation by multiple phosphorylation events. Cell Motil Cytoskeleton 63:266–286

Nixon RA, Paskevich PA, Sihag RK, Thayer CY (1994) Phosphorylation on carboxyl terminus domains of neurofilament proteins in retinal ganglion cell neurons in vivo: influences on regional neurofilament accumulation, interneurofilament spacing, and axon caliber. J Cell Biol 126:1031–1046

Noble W, Olm V, Takata K, Casey E, Mary O, Meyerson J, Gaynor K, LaFrancois J, Wang L, Kondo T, Davies P, Burns M, Veeranna, Nixon R, Dickson D, Matsuoka Y, Ahlijanian M, Lau LF, Duff K (2003) Cdk5 is a key factor in tau aggregation and tangle formation in vivo. Neuron 38:555–565

Nunomura A, Perry G, Aliev G, Hirai K, Takeda A, Balraj EK, Jones PK, Ghanbari H, Wataya T, Shimohama S, Chiba S, Atwood CS, Petersen RB, Smith MA (2001) Oxidative damage is the earliest event in Alzheimer disease. J Neuropathol Exp Neurol 60: 759–767

Ohshima T, Ward JM, Huh CG, Longenecker G, Veeranna, Pant HC, Brady RO, Martin LJ, Kulkarni AB (1996) Targeted disruption of the cyclin-dependent kinase 5 gene results in abnormal corticogenesis, neuronal pathology and perinatal death. Proc Natl Acad Sci USA 93:11173–11178

Otth C, Concha II, Arendt T, Stieler J, Schliebs R, Gonzalez-Billault C, Maccioni RB (2002) AbetaPP induces cdk5-dependent tau hyperphosphorylation in transgenic mice Tg2576. J Alzheimers Dis 4:417–430

Pant HC, Veeranna (1995) Neurofilament phosphorylation. Biochem Cell Biol 73:575–592

Pant AC, Veeranna, Pant HC, Amin N (1997) Phosphorylation of human high molecular weight neurofilament protein (hNF-H) by neuronal cyclin-dependent kinase 5 (cdk5). Brain Res 765:259–266

Patrick GN, Zukerberg L, Nikolic M, de la Monte S, Dikkes P, Tsai LH (1999) Conversion of p35 to p25 deregulates Cdk5 activity and promotes neurodegeneration. Nature 402: 615–622

Patzke H, Tsai LH (2002) Calpain-mediated cleavage of the cyclin-dependent kinase-5 activator p39 to p29. J Biol Chem 277:8054–8060

Perry G, Kawai M, Tabaton M, Onorato M, Mulvihill P, Richey P, Morandi A, Connolly JA, Gambetti P (1991) Neuropil threads of Alzheimer's disease show a marked alteration of the normal cytoskeleton. J Neurosci 11:1748–1755

Perry G, Nunomura A, Hirai K, Zhu X, Perez M, Avila J, Castellani RJ, Atwood CS, Aliev G, Sayre LM, Takeda A, Smith MA (2002) Is oxidative damage the fundamental pathogenic mechanism of Alzheimer's and other neurodegenerative diseases? Free Radic Biol Med 33: 1475–1479

Perry G, Smith MA (1998) Is oxidative damage central to the pathogenesis of Alzheimer disease? Acta Neurol Belg 98:175–179

Phiel CJ, Wilson CA, Lee VM, Klein PS (2003) GSK-3alpha regulates production of Alzheimer's disease amyloid-beta peptides. Nature 423:435–439

Picklo MJ, Montine TJ, Amarnath V, Neely MD (2002) Carbonyl toxicology and Alzheimer's disease. Toxicol Appl Pharmacol 184:187–197

Pope WB, Lambert MP, Leypold B, Seupaul R, Sletten L, Krafft G, Klein WL (1994) Microtubule-associated protein tau is hyperphosphorylated during mitosis in the human neuroblastoma cell line SH-SY5Y. Exp Neurol 126:185–194

Praprotnik D, Smith MA, Richey PL, Vinters HV, Perry G (1996a) Plasma membrane fragility in dystrophic neurites in senile plaques of Alzheimer's disease: an index of oxidative stress. Acta Neuropathol 91:1–5

Praprotnik D, Smith MA, Richey PL, Vinters HV, Perry G (1996b) Filament heterogeneity within the dystrophic neurites of senile plaques suggests blockage of fast axonal transport in Alzheimer's disease. Acta Neuropathol 91:226–235

Preuss U, Doring F, Illenberger S, Mandelkow EM (1995) Cell cycle-dependent phosphorylation and microtubule binding of tau protein stably transfected into Chinese hamster ovary cells. Mol Biol Cell 6:1397–1410

Rao MV, Nixon RA (2003) Defective neurofilament transport in mouse models of amyotrophic lateral sclerosis: a review. Neurochem Res 28:1041–1047

Robertson J, Doroudchi MM, Nguyen MD, Durham HD, Strong MJ, Shaw G, Julien JP, Mushynski WE (2003) A neurotoxic peripherin splice variant in a mouse model of ALS. J Cell Biol 160:939–949

Roses AD, Saunders AM, Corder EH, Pericak-Vance MA, Han SH, Einstein G, Hulette C, Schmechel DE, Holsti M, Huang D et al. (1995) Influence of the susceptibility genes apolipoprotein E-epsilon 4 and apolipoprotein E-epsilon 2 on the rate of disease expressivity of late-onset Alzheimer's disease. Arzneimittelforschung 45: 413–417

Rudrabhatla P, Kanungo J, Zheng Ya-Li, Amin ND, Kesavapany S, Pant HC (2008a) Cyclin-dependent protein kinase 5 (Cdk5) modulates signal transduction pathways regulating neuronal survival. In: Ip NY, Tsai L-H (eds) Cyclin Dependent Kinase 5 (Cdk5). Springer Science Business Media, New York

Rudrabhatla P, Zheng YL, Amin ND, Kesavapany S, Albers W, Pant HC (2008b) Pin1-dependent prolyl isomerization modulates the stress-induced phosphorylation of high molecular weight neurofilament protein. J Biol Chem 283:26737–26747

Sato H, Nishimoto I, Matsuoka M (2002) ik3-2, a relative to ik3-1/cables, is associated with cdk3, cdk5, and c-abl. Biochim Biophys Acta 1574:157–163

Schoenfeld TA, Obar RA (1994) Diverse distribution and function of fibrous microtubule-associated proteins in the nervous system. Int Rev Cytol 151:67–137

Scott CW, Vulliet PR, Caputo CB (1993) Phosphorylation of tau by proline-directed protein kinase (p34cdc2/p58cyclin A) decreases tau-induced microtubule assembly and antibody SMI33 reactivity. Brain Res 611:237–242

Selkoe DJ (1997) Alzheimer's disease: genotypes, phenotypes, and treatments. Science 275: 630–631

Selkoe DJ (2000) The genetics and molecular pathology of Alzheimer's disease: roles of amyloid and the presenilins. Neurol Clin 18:903–922

Selkoe DJ, Abraham C, Ihara Y (1982) Brain transglutaminase: in vitro crosslinking of human neurofilament proteins into insoluble polymers. Proc Natl Acad Sci USA 79: 6070–6074

Seubert P, Mawal-Dewan M, Barbour R, Jakes R, Goedert M, Johnson GV, Litersky JM, Schenk D, Lieberburg I, Trojanowski JQ et al. (1995) Detection of phosphorylated Ser262 in fetal tau, adult tau, and paired helical filament tau. J Biol Chem 270:18917–18922

Sharma P, Barchi JJ Jr, Huang X, Amin ND, Jaffe H, Pant HC (1998) Site-specific phosphorylation of Lys-Ser-Pro repeat peptides from neurofilament H by cyclin-dependent kinase 5: structural basis for substrate recognition. Biochemistry 37:4759–4766

Sharma P, Veeranna, Sharma M, Amin ND, Sihag RK, Grant P, Ahn N, Kulkarni AB, Pant HC (2002) Phosphorylation of MEK1 by cdk5/p35 down-regulates the mitogen-activated protein kinase pathway. J Biol Chem 277:528–534

Shea TB, Beaty WJ (2007) Traffic jams: dynamic models for neurofilament accumulation in motor neuron disease. Traffic 8:445–447

Shea TB, Chan WK (2008) Regulation of neurofilament dynamics by phosphorylation. Eur J Neurosci 27:1893–1901

Shea TB, Flanagan LA (2001) Kinesin, dynein and neurofilament transport. Trends Neurosci 24:644–648

Shea TB, Wheeler E, Jung C (1997) Aluminum inhibits neurofilament assembly, cytoskeletal incorporation, and axonal transport. Dynamic nature of aluminum-induced perikaryal neurofilament accumulations as revealed by subunit turnover. Mol Chem Neuropathol 32:17–39

Shea TB, Zheng YL, Ortiz D, Pant HC (2004) Cyclin-dependent kinase 5 increases perikaryal neurofilament phosphorylation and inhibits neurofilament axonal transport in response to oxidative stress. J Neurosci Res 76:795–800

Shelton SB, Johnson GV (2004) Cyclin-dependent kinase-5 in neurodegeneration. J Neurochem 88:1313–1326

Shetty KT, Link WT, Pant HC (1993) cdc2-like kinase from rat spinal cord specifically phosphorylates KSPXK motifs in neurofilament proteins: isolation and characterization. Proc Natl Acad Sci USA 90:6844–6848

Sihag RK, Inagaki M, Yamaguchi T, Shea TB, Pant HC (2007) Role of phosphorylation on the structural dynamics and function of types III and IV intermediate filaments. Exp Cell Res 313:2098–2109

Smith MA, Perry G (1995) Free radical damage, iron, and Alzheimer's disease. J Neurol Sci 134(Suppl):92–94

Smith MA, Perry G (1997) The pathogenesis of Alzheimer disease: an alternative to the amyloid hypothesis. J Neuropathol Exp Neurol 56:217

Smith MA, Sayre LM, Monnier VM, Perry G (1996a) Oxidative posttranslational modifications in Alzheimer disease. A possible pathogenic role in the formation of senile plaques and neurofibrillary tangles. Mol Chem Neuropathol 28:41–48

Smith MA, Siedlak SL, Richey PL, Nagaraj RH, Elhammer A, Perry G (1996b) Quantitative solubilization and analysis of insoluble paired helical filaments from Alzheimer disease. Brain Res 717:99–108

Sobue K, Agarwal-Mawal A, Li W, Sun W, Miura Y, Paudel HK (2000) Interaction of neuronal Cdc2-like protein kinase with microtubule-associated protein tau. J Biol Chem 275: 16673–16680

Sobue G, Hashizume Y, Yasuda T, Mukai E, Kumagai T, Mitsuma T, Trojanowski JQ (1990) Phosphorylated high molecular weight neurofilament protein in lower motor neurons in amyotrophic lateral sclerosis and other neurodegenerative diseases involving ventral horn cells. Acta Neuropathol 79:402–408

Sternberger NH, Sternberger LA, Ulrich J (1985) Aberrant neurofilament phosphorylation in Alzheimer disease. Proc Natl Acad Sci USA 82:4274–4276

Strong MJ (1999) Neurofilament metabolism in sporadic amyotrophic lateral sclerosis. J Neurol Sci 169:170–177

Strong MJ, Kesavapany S, Pant HC (2005) The pathobiology of amyotrophic lateral sclerosisa proteinopathy? J Neuropathol Exp Neurol 64:649–664
Strong MJ, Strong WL, Jaffe H, Traggert B, Sopper MM, Pant HC (2001) Phosphorylation state of the native high-molecular-weight neurofilament subunit protein from cervical spinal cord in sporadic amyotrophic lateral sclerosis. J Neurochem 76:1315–1325
Sun D, Leung CL, Liem RK (1996) Phosphorylation of the high molecular weight neurofilament protein (NF-H) by Cdk5 and p35. J Biol Chem 271:14245–14251
Takahashi S, Saito T, Hisanaga S, Pant HC, Kulkarni AB (2003) Tau phosphorylation by cyclin-dependent kinase 5/p39 during brain development reduces its affinity for microtubules. J Biol Chem 278:10506–10515
Town T, Zolton J, Shaffner R, Schnell B, Crescentini R, Wu Y, Zeng J, DelleDonne A, Obregon D, Tan J, Mullan M (2002) p35/Cdk5 pathway mediates soluble amyloid-beta peptide-induced tau phosphorylation in vitro. J Neurosci Res 69:362–372
Toyoshima I (1998) Molecular mechanisms of axonal transport. No To Shinkei 50: 691–702
Toyoshima I, Kato K, Sugawara M, Wada C, Masamune O (1998) Kinesin accumulation in chick spinal axonal swellings with beta,beta'-iminodipropionitrile (IDPN) intoxication. Neurosci Lett 249:103–106
Trimmer PA, Borland MK, Keeney PM, Bennett JP Jr, Parker WD Jr (2004) Parkinson's disease transgenic mitochondrial cybrids generate Lewy inclusion bodies. J Neurochem 88: 800–812
Trojanowski JQ, Schmidt ML, Shin RW, Bramblett GT, Rao D, Lee VM (1993) Altered tau and neurofilament proteins in neuro-degenerative diseases: diagnostic implications for Alzheimer's disease and Lewy body dementias. Brain Pathol 3:45–54
Tsuda M, Tashiro T, Komiya Y (1997) Increased solubility of high-molecular-mass neurofilament subunit by suppression of dephosphorylation: its relation to axonal transport. J Neurochem 68:2558–2565
Van den Haute C, Spittaels K, Van Dorpe J, Lasrado R, Vandezande K, Laenen I, Geerts H, Van Leuven F (2001) Coexpression of human cdk5 and its activator p35 with human protein tau in neurons in brain of triple transgenic mice. Neurobiol Dis 8:32–44
Veeranna, Amin ND, Ahn NG, Jaffe H, Winters CA, Grant P, Pant HC (1998) Mitogen-activated protein kinases (Erk1,2) phosphorylate Lys-Ser-Pro (KSP) repeats in neurofilament proteins NF-H and NF-M. J Neurosci 18:4008–4021
Veeranna, Shetty KT, Link WT, Jaffe H, Wang J, Pant HC (1995) Neuronal cyclin-dependent kinase-5 phosphorylation sites in neurofilament protein (NF-H) are dephosphorylated by protein phosphatase 2A. J Neurochem 64:2681–2690
Veeranna, Shetty KT, Takahashi M, Grant P, Pant HC (2000) Cdk5 and MAPK are associated with complexes of cytoskeletal proteins in rat brain. Brain Res Mol Brain Res 76:229–236
Vincent I, Rosado M, Davies P (1996) Mitotic mechanisms in Alzheimer's disease? J Cell Biol 132:413–425
Watanabe A, Hasegawa M, Suzuki M, Takio K, Morishima-Kawashima M, Titani K, Arai T, Kosik KS, Ihara Y (1993) In vivo phosphorylation sites in fetal and adult rat tau. J Biol Chem 268:25712–25717
Weiner HL, Selkoe DJ (2002) Inflammation and therapeutic vaccination in CNS diseases. Nature 420:879–884
Wharton SB, McDermott CJ, Grierson AJ, Wood JD, Gelsthorpe C, Ince PG, Shaw PJ (2003) The cellular and molecular pathology of the motor system in hereditary spastic paraparesis due to mutation of the spastin gene. J Neuropathol Exp Neurol 62:1166–1177
Wood JG, Mirra SS, Pollock NJ, Binder LI (1986) Neurofibrillary tangles of Alzheimer disease share antigenic determinants with the axonal microtubule-associated protein tau (tau). Proc Natl Acad Sci USA 83:4040–4043
Yabe JT, Jung C, Chan WK, Shea TB (2000) Phospho-dependent association of neurofilament proteins with kinesin in situ. Cell Motil Cytoskeleton 45:249–262

Yabe JT, Wang FS, Chylinski T, Katchmar T, Shea TB (2001) Selective accumulation of the high molecular weight neurofilament subunit within the distal region of growing axonal neurites. Cell Motil Cytoskeleton 50:1–12

Zheng YL, Kesavapany S, Gravell M, Hamilton RS, Schubert M, Amin N, Albers W, Grant P, Pant HC (2005) A Cdk5 inhibitory peptide reduces tau hyperphosphorylation and apoptosis in neurons. EMBO J 24:209–220

Zheng YL, Li BS, Amin ND, Albers W, Pant HC (2002) A peptide derived from cyclin-dependent kinase activator (p35) specifically inhibits Cdk5 activity and phosphorylation of tau protein in transfected cells. Eur J Biochem 269:4427–4434

Zheng YL, Li BS, Kanungo J, Kesavapany S, Amin N, Grant P, Pant HC (2007) Cdk5 Modulation of mitogen-activated protein kinase signaling regulates neuronal survival. Mol Biol Cell 18:404–413

Zheng YL, Li BS, Veeranna, Pant HC (2003) Phosphorylation of the head domain of neurofilament protein (NF-M): a factor regulating topographic phosphorylation of NF-M tail domain KSP sites in neurons. J Biol Chem 278:24026–24032

Zhu Q, Lindenbaum M, Levavasseur F, Jacomy H, Julien JP (1998) Disruption of the NF-H gene increases axonal microtubule content and velocity of neurofilament transport: relief of axonopathy resulting from the toxin beta,beta'-iminodipropionitrile. J Cell Biol 143:183–193

Chapter 15
Neurofilaments in Aged Animals

Shin-ichi Hisanaga, Takahiro Sasaki, and Atsuko Uchida

Abstract Neurofilaments (NFs) are neuronal intermediate filaments assembled mainly from three subunit proteins, NF-L, NF-M and NF-H. NFs are the major cytoskeletal element of axons, particularly large myelinated axons. Accumulation of NFs in the cell body or proximal axons is a hallmark of motor neuron diseases, of which aging is a risk factor. It is not known whether the accumulation of NFs is the primary causative factor of these diseases; however, it is feasible that accumulated NFs are obstacles for axonal trafficking, resulting in a decreased supply of proteins and organelles required for the maintenance of or activity in the distal axons. Axonal transport of NFs slows down with aging and the decreased rate of transport could be a risk factor for disease. Therefore, it is important to determine age-dependent changes in properties of NFs. In this chapter we review the characteristics of NFs in aged animals. The most striking morphological change in NFs is their density within axons. NFs are more than twice as densely packed in the proximal region of aged rat sciatic nerve axons compared with those in young adult rats. A remarkable biochemical change is the reduction of NF-M content in aged NFs. This is partly because of the reduced transcription of NF-M in aged rats. The relationship between NF packing and reduced NF-M is discussed in terms of the age-dependent decrease in axonal transport and neurodegenerative diseases.

Keywords Aging · Amyotrophic lateral sclerosis · Axon · Axonal transport · Neurofilament · Phosphorylation · Protein expression

Contents

S. Hisanaga (✉)
Department of Biological Sciences, Graduate School of Science, Tokyo Metropolitan University, Hachioji, Tokyo 192-0397, Japan
e-mail: hisanaga-shinichi@tmu.ac.jp

R.A. Nixon, A. Yuan (eds.), *Cytoskeleton of the Nervous System*, Advances in Neurobiology 3, DOI 10.1007/978-1-4419-6787-9_15,

15.1 Introduction

Neurofilaments (NFs) are neuronal intermediate filaments composed mainly of three subunit proteins, NF-H, NF-M and NF-L (Shaw, 1991; Pant and Veeranna, 1995; Nixon, 1998; Perrot et al., 2008). Recently, it has been demonstrated that α-internexin is a fourth subunit of NFs in the central nervous system (Yuan et al., 2006b). All these NF subunits have a tripartite molecular structure common to intermediate filament proteins: an amino (N)-terminal head domain, an α-helix-rich central rod domain and a carboxy (C)-terminal tail domain. However, only NF-L can form complete 10-nm filaments in vitro (Liem and Hutchison, 1982; Hisanaga and Hirokawa, 1988), whereas NFs in vivo are obligate heteropolymers requiring NF-L with NF-M or NF-H for filament formation (Ching and Liem, 1993; Lee et al., 1993). Gene disruption studies in mice have revealed different roles for each NF subunit in the formation of the NF network in axons (Hirokawa and Takeda, 1998). Targeted deletion of the *NF-L* gene results in the loss of NFs (Zhu et al., 1997; Lariviere and Julien, 2004), indicating the absolute requirement for NF-L in filament assembly. NF-M also has an important role in maintaining the filamentous organization of the NF network in axons; for example, the number of NFs is significantly reduced in the *NF-M* gene-disrupted mouse (Elder et al., 1998a). In contrast, the *NF-H* null mouse does not exhibit marked changes in the number and organization of axonal NFs compared with wild-type littermates (Elder et al., 1998b; Rao et al., 1998; Zhu et al., 1998).

NFs are the most abundant cytoskeletal element in axons and account for more than 90% of the cytoskeleton in large myelinated axons, such as those of motor neurons. Some motor neuron axons in the human body extend more than 1 m in length. NFs play a role as a structural scaffold of such long axons along their entire length. NFs are also implicated in determining axon caliber (Hoffman et al., 1984), which controls the velocity of electric conduction that travels along the length of the axons. Thus, the distribution of NFs from proximal to distal regions of an axon is very important for maintaining the integrity of axonal structure and function. NF proteins are synthesized in the cell body and conveyed along the axon by slow transport, at an overall rate of about 1 mm/day (Hoffman and Lasek, 1975; Nixon and Logvinenko, 1986; Baas and Brown, 1997; Nixon, 1998). There are two types of NFs in axons, moving NFs and stationary or pausing NFs. The slow overall velocity is because the fast movements of NFs is interrupted by prolonged pausing (Roy et al., 2000; Wang et al., 2000; Uchida and Brown, 2004; Barry et al., 2007), although the relationship between the stationary and pausing states of NFs is unclear (Trivedi et al., 2007; Yuan et al., 2009).

The lack of NFs generated by disruption of the *NF-L* gene does not result in an apparent phenotype in mice (Zhu et al., 1997). However, the Japanese quail that has a natural mutation in the *NF-L* gene displays mild generalized quivering (Yamasaki et al., 1991; Ohara et al., 1993), suggesting that disruption of the NF network becomes defective depending on the animal, probably because of body size or length of axons. In contrast, the absence of NFs or removal of the NF tail domain brings apparent beneficial effects such as delayed onset of amyotrophic lateral sclerosis (ALS)-like pathology in mutant superoxide dismutase 1 (SOD1) transgenic mice (Nguyen et al., 2000; Lobsiger et al., 2005). It is easy to imagine that accumulated NFs become transport barriers in axons. The accumulation of NFs in the proximal axon is often recognized as a hallmark of axonal degeneration, whether the NF accumulation is primary or not (Carpenter, 1968; Schmidt et al., 1987). Aging is a major risk factor for motor neuron diseases such as ALS (Hirano, 1991). The rate of slow axonal transport, particularly the transport of NF proteins, decreases with increasing age (Tashiro and Komiya, 1991a; Uchida et al., 2004). The decreased rate of NF transport would accelerate the accumulation of NFs in axons. It is likely that age-dependent changes in NF transport could be a risk factor for age-related diseases. It is an inevitable fact that mammalian axons naturally contain abundant NFs. However, NFs have not been paid much attention as a risk factor for disease. It is important to determine the age-dependent alterations of NFs and to develop a way to reduce the risks associated with NFs and aging. Here, we review the age-dependent alterations of NFs based mainly on our data.

15.2 Morphological Changes in NFs with Aging

15.2.1 Aged Rat Sciatic Nerve Axons and NFs Within the Axons

The sciatic nerve is composed of axons of ventral horn motor neurons and dendrites of dorsal root ganglion sensory neurons, and has frequently been used for studies on the axonal transport of NFs. Sciatic nerve axons increase in diameter with the thickening of the myelin sheet during development (Hoffman et al., 1988; Uchida et al., 2001). The axons of young rats are round whereas the axons of aged rats show irregular and wavy contours with thicker myelin (Fig. 15.1a,b). Cross-sectional electron micrographs of axons from young and aged rats are shown in Fig. 15.1c,d. It is apparent that NFs are more densely packed in the sciatic nerve of aged rats. The axons of young rats have a density of 200–440 NFs/μm^2 with a mean of about 212 NFs/μm^2 (Uchida et al., 1999). A similar density of 150–300 NFs/μm^2 has been reported for myelinated ventral root axons of cats (Berthold and Rydmark, 1995) and a density of 180 NFs/μm^2 has been reported for mouse ventral root axons (Xu et al., 1996). In contrast, the density of axons in aged rats are distributed broadly from 280 to 920 NFs/μm^2, with two peaks at around 440 and 760 NFs/μm^2 and a

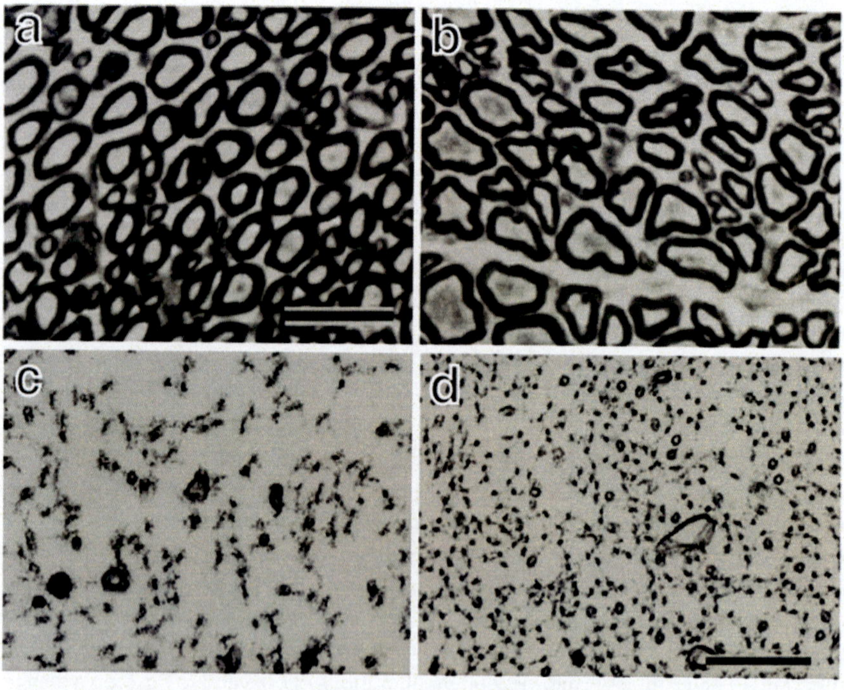

Fig. 15.1 Sciatic nerve axons of young and aged rats. **a,b** Cross-sectional light micrographs of young (5 weeks) and aged (87 weeks) rat sciatic nerve, respectively. *Bar* 20 μm. **c,d**, Thin-section electron micrographs of young (5 weeks) and aged (80 weeks) rat sciatic nerve axons, respectively. *Bar* 250 nm. Reproduced with permission (Uchida et al., 1999)

mean density of about 540 NFs/μm^2. Thus, the density of NFs in axons of the sciatic nerve in aged rats is about twice that of young rats. A high density of NFs in aged rats is also observed in axons of the spinal cord (Uchida et al., 1999). The dense packing of NFs in aged axons appears to be a common phenomenon in the peripheral and central nervous systems.

The increase in NF density accompanies a decrease in the distance between NFs. The inter-NF distance in axons from aged rats is about 28 nm, which is about half that of axons from young rats (55 nm) (Uchida et al., 1999). The inter-NF distance in axons of the mouse sciatic nerve (Wong et al., 1995) and ventral root axons (Xu et al., 1996) is about 50 nm, and about 42 nm in the mouse optic nerve (Nixon et al., 1994). The increased density (and the decreased inter-NF distance) in sciatic nerve and spinal cord axons of rats supports that this change is consistently associated with aging. In contrast, microtubules (MTs), another major cytoskeletal element of axons, do not change in density with aging. MTs are distributed relatively sparsely in rat sciatic nerve axons. The density of MTs shows a broad and rather even distribution of up to 70/μm^2 with no significant difference between young and aged rats (Uchida et al., 1999). Thus, the increase in density of NFs with aging is a unique feature of NFs.

15.2.2 Proximo-Distal Decrease in NF Packing Density in the Sciatic Nerve Axons of Aged Rats

If distal packing is retrogradely propagated to the proximal region of the axon, a high density of NFs should be observed in all regions along axons. However, this is not the case. NF packing has been compared at the proximal region close to the dorsal root ganglion (DRG), and the distal region close to the trifurcating point of the sciatic nerve in young and aged rats. NF density in young rats is about 250/μm^2 in myelinated axons larger than 2 μm^2, and is similar at the proximal and distal level of the sciatic nerve. In contrast, in aged rats NF density is remarkably high in the proximal sciatic nerve (698/μm^2) and reduced in the distal sciatic nerve (209/μm^2) to a level similar to that observed in young rats (Table 15.1). Thus, NF packing

Table 15.1 Axonal area and NF density in the proximal and distal sciatic nerve

	Axonal area (μm^2, ± SD)	Density of NFs (number/μm^2, ± SD)
Yong rat sciatic nerve		
Proximal region	27.4 ± 11.0 (*n*=791)	259 ± 74 (*n*=150)
Distal region	28.5 ± 11.6 (*n*=268)	264 ±75 (*n*=150)
Aged rat sciatic nerve		
Proximal region	20.2 ± 19.4 (*n*=834)	698 ± 166 (*n*=150)
Distal region	7.9 ± 9.8 (*n*=475)	209 ± 63 (*n*=150)

Young rats are 5–10 weeks old and aged rats are 80–138 weeks old. Reproduced with permission (Uchida et al., 1999, 2004)

shows a proximo-distal decrease in sciatic nerve axons of aged rats, suggesting that proximal high-density packing of NFs causes the distal shrinkage of axons. In fact, the average cross-sectional area at the distal sciatic nerve decreases to 7.2 μm^2 in aged rats compared with 28.5 μm^2 in young rats (Table 15.1). The axonal areas of the distal sciatic nerve in young rats are similar or rather larger than those of the proximal sciatic nerve and ventral root, in contrast with aged rats in which axons at the distal sciatic nerve are much smaller than those at the proximal region and the ventral root. Thus, once NF packing begins in the proximal axon, it would create a barrier to the further supply of axonal components, including NFs, to the distal axon, leading to progressive shrinkage of the distal axon.

15.2.3 Aggregates of NFs Isolated from Aged Rat Spinal Cords

The spinal cord is an NF-enriched tissue. NFs prepared from spinal cords of aged rats contain many large aggregates (Uchida et al., 1999). In contrast, only small aggregates or free NFs have been detected in NFs prepared from the spinal cords of young rats. Most of the NFs in the spinal cords of young rats are less than 3.2 μm long with a single peak at around 1 μm. Aged rat NFs contain longer filaments with a peak at 3.6 μm, in addition to a peak at a similar length to young rat NFs, when the length of NFs outside of aggregates were measured (Uchida et al., 1999). Considering that shorter NFs have a tendency to be released from aggregates, NFs inside the aggregates would be longer than those outside of the aggregates; therefore, the length of NFs in the spinal cords of aged rats may be much longer.

15.3 Changes in Biochemical Properties of NFs with Aging

15.3.1 Aggregate Formation of NFs Prepared from the Aged Rat Spinal Cords

NFs are polymers that are sedimentable by ultracentrifugation, and the interaction between NFs that form aggregates can be assessed by sedimentation at lower centrifugation forces. Most NFs from aged rats pellet under centrifugation at 18,000 *g* for 30 min, whereas NFs from young rats do not pellet until the centrifugation force is increased to 40,000 *g* (Uchida et al., 1999). NFs from aged rats form a larger macromolecular complex, suggesting there is a stronger interaction between NFs in aged animals. This is also indicated by dispersion of the pelleted NFs. NFs from young and aged rats pelleted once by centrifugation can be suspended in various solutions containing high NaCl, high KI, or urea, which are often used for the solubilization of proteins, then centrifuged again under the same conditions. All NF proteins of young and aged rats remain in the supernatant when they are disassembled in a solution containing 2 M urea. However, 0.75 M NaCl solubilizes most NFs from young rats and does not solubilize NFs from aged rats. NFs from aged rats can be solubilized in the presence of high concentrations of NaI (Uchida et al.,

1999). When viscosity was compared between young and aged NFs (Gou et al., 1995), plateau levels of viscosity were significantly higher in samples of NFs from 30-month-old rats compared with NFs from 3- to 20-month-old rats. These results indicate that NFs from aged rats interact with each other by stronger forces than those between NFs from young rats.

15.3.2 Subunit Composition of NFs from Aged Rats

NFs form obligate heteropolymers in vivo. Although NF-L alone can assemble into long intermediate-sized filaments in vitro, coexpression of NF-M and/or NF-H is required for proper NF formation in vivo (Chin and Liem, 1993; Lee et al., 1993). NFs prepared from adult bovine spinal cords are composed of triplet proteins in a molar ratio of about 1.5:1:1 for NF-L:NF-M:NF-H (Hisanaga and Hirokawa, 1990). The proportion of NF subunits, however, varies among different types of neurons and depends on developmental stages (Nixon and Shea, 1992; Riederer et al., 1996). Rat cerebellar unipolar brush expresses unusually low levels of NF-M (Harris et al., 1993). NF-L and NF-M are expressed earlier than NF-H during neuronal differentiation (Shaw and Weber, 1982; Julien et al., 1986; Carden et al., 1987). Furthermore, the subunit ratio can also be manipulated experimentally. The amount of NF-H increases relative to NF-L in response to nerve injury (Tsuda et al., 2000). Overexpression of exogenous NF-M causes a decrease in the expression of endogenous NF-H and NF-M (Tu et al., 1995; Wong et al., 1995; Xu et al., 1996). A null mutation in *NF-M* increases the level of NF-H (Elder et al., 1998a, 1999). A particular ratio of NF-M plus NF-H to NF-L may be needed for proper filament formation and/or function. Changes in the subunit ratio might be determined by synthesis of NF proteins, their incorporation into filaments and their degradation; however, the regulatory mechanisms as well as the functional significance of the variation in subunit composition remain largely unknown.

The ratio of NF-L, NF-M and NF-H subunits changes with aging. When the amount of NF-L in the NF fraction of spinal cords is roughly adjusted in Coomassie brilliant blue-stained gels, the reduced amount of NF-M is clearly seen in aged rat NFs (Fig. 15.2). NFs from aged rats contain more NF-H and less NF-M than NFs

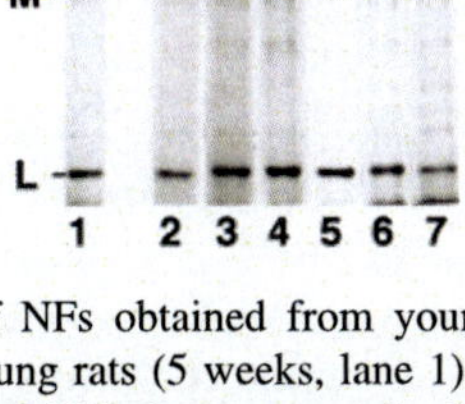

Fig. 15.2 Subunit composition of NFs obtained from young and aged rat spinal cords. The homogenates of spinal cords of young rats (5 weeks, lane 1) and six aged rats (80 weeks, lanes 2–7) were run on 6% SDS-PAGE and NF subunits are stained by Coomassie brilliant blue. NF-H, NF-M and NF-L are indicated by H, M, and L. Reproduced with permission (Uchida et al., 1999)

from young rats. The ratio (0.5) of NF-H to NF-L in aged rat NFs increases a little from the ratio (0.43) of NF-H to NF-L in young rats, whereas the ratio (0.39) of NF-M to NF-L in aged rat NFs decreases by about 60% from that (0.61) of young rat NFs. The molar ratio is calculated to be 0.47; about 2:1 for NF-L:NF-M in NFs from the spinal cords of young rats, and this ratio decreases to 0.31 (that is about 3:1 for NF-L:NF-M) in NFs from the spinal cords of aged rats. The decrease in the NF-M ratio is the most striking biochemical difference between NFs from young and aged rats.

However, it is unlikely that the reduction in NF-M content is directly related to the packing density of NFs because NFs in the distal axons of aged rats are spaced more widely than those in the proximal region, despite the lower NF-M content described below. This is in accordance with several studies on transgenic or knockout mice in which the NF subunit proteins are affected. In transgenic mice overexpressing NF-H, the nearest-neighbor distance of NFs is unaffected (Marszalek et al., 1996) and overexpression of NF-M does not affect NF packing (Wong et al., 1995). Studies on *NF-H* null mutant mice (Elder et al., 1998a, b; Rao et al., 1998; Zhu et al., 1998) also indicate that the change in NF subunit ratio does not affect the packing density of NFs in axons.

15.3.3 Reduced Transcription of NF-M in Aged Rats

The relative reduction of NF-M can be accounted for by a reduction in expression or an enhancement in degradation. However, little is known about the age-dependent regulation of the expression of NF proteins. Several studies report a decrease in NF-L expression with aging (Moskowitz and Oblinger, 1995; Parhad et al., 1995; Kuchel et al., 1996). Transcription of three NF subunit proteins or the relative amounts of mRNA for NF-L, NF-M and NF-H in the spinal cord has been examined by semi-quantitative RT–PCR. NF-L mRNA is reduced by 30% in aged rats when normalized against the expression of GAPDH, and NF-H mRNA is moderately reduced (about 10%) and NF-M mRNA is markedly reduced to 50% of the levels observed in young rats (Uchida et al., 2004). Consistent results have been obtained by incorporation of radiolabeled amino acids into NF proteins in the spinal cords of rats. The amount of radioactivity associated with NF-M in aged rats is about half of that seen in young rats when normalized against NF-L (Uchida et al., 2004). The reduced transcription/translation of NF-M explains, at least in part, the reduced content of NF-M in axons from aged rats.

15.3.4 Phosphorylation of the Tail Domains of NF-M and NF-H

NF-M and NF-H are highly phosphorylated at the C-terminal tail domain in axons (Lee et al., 1988; Grant and Pant, 2000; Sihag et al., 2007). There are about 10 and 50 Lys-Ser-Pro (KSP) phosphorylation sites in NF-M and NF-H, respectively, making them the most phosphorylated proteins in neurons. The inter-NF distance is determined by the tail domains of the NF subunits (Gotow et al., 1994; Leterrier et al., 1996; Brown and Hoh, 1997; Rao et al., 2003) and is controlled by phosphorylation of the C-terminal tail domains of NF-M and NF-H (Hisanaga and Hirokawa, 1988; Gotow et al., 1994; Nixon et al., 1994; Wong et al., 1995; Brown and Hoh, 1997; Uchida et al., 1999; Chen et al., 2000). There are a number of in vivo studies that support this idea. For example, de Waegh et al. (1992) report that demyelination increases the NF density in parallel with a decrease in NF-H phosphorylation. Cole et al. (1994) show there is an association between an increase in NF density and a decrease in NF phosphorylation in hypomyelinated axons of transgenic mice expressing the diphtheria toxin A or SV40 large T antigen.

The large aggregates observed in NFs from aged rats suggest there are increased interactions between NFs. Because the interactions between NFs are mediated by the tail domains of NF-H or NF-M and regulated by their phosphorylation (Hirokawa et al., 1984; Eyer and Leterrier, 1988; Nakagawa et al., 1995), it is possible that the increased packing of NFs in axons from aged rats is caused by dephosphorylation of NF-H. However, the total phosphorylation of NF-H increases in NFs from aged rats (Gou et al., 1995; Uchida et al., 1999). Increased phosphorylation of NF-H has also been reported in brains from aged humans (Blanchard and Ingram, 1989). Dephosphorylated NF-H, which is a minor component, is present to a similar extent in NFs from young and aged rats. The amount of the phosphorylated form of NF-M decreases in aged rats; however, this is because of the decreased expression of NF-M. Dephosphorylated NF-M does not change as much with aging. In addition, phosphorylation with Cdk5, one of the tail domain kinases (Hisanaga et al., 1995), or dephosphorylation with alkaline phosphatase, does not affect the sedimentation properties of NFs from young and aged rats. These results indicate that phosphorylation of the tail domain of NF-H and probably NF-M is not a direct cause for the age-dependent increase in interaction of NFs. Furthermore, replacement of the NF-H tailless mutant or loss of NF-H causes no substantial changes in the nearest-neighbor distance between NFs (Elder et al., 1998b; Rao et al., 1998, 2002; Zhu et al., 1998), indicating that phosphorylation at the tail domain of NF-H is not essential for determination of the spacing between NFs.

15.3.5 Phosphorylation at the N-Terminal Head Domain of NF Proteins

NFs from aged rats contain a population of NFs that are longer than NFs from young rats. The length of NFs can be determined by phosphorylation at the N-terminal head domain with the NF-head domain kinase, such as cAMP-dependent protein kinase (PKA) (Hisanaga et al., 1990; 1994), or by dephosphorylation with NF-head phosphatase, such as protein phosphatase 2A (PP2A) (Saito et al., 1995; Strack et al., 1997). However, there is no difference in the amount of protein and activities of PKA and PP2A between young and aged rat spinal cord homogenates (Uchida et al., 1999). In contrast, the accessibility of the enzyme(s) to the head domain is reduced in NFs from aged rats. When NFs from aged rats are phosphorylated with purified PKA in vitro, they take more time to phosphorylate than NFs from young rats. This may be because of the aggregation of NFs in aged rat axons. The extent of phosphorylation of NF-M is also reduced in aged rats by half or one-third that of young rats (Uchida et al., 1999); however, this is again because of the decreased content of NF-M.

15.3.6 Proximo-Distal Decreases in the Amount of NF Proteins and in the Ratio of NF-M to NF-H in Sciatic Nerve Axons of Aged Rats

The distribution and phosphorylation of NF proteins have been measured in three parts of the sciatic nerve: the proximal sciatic nerve just below the DRG, the mid-sciatic nerve between the DRG and the trifurcation point, and the distal sciatic nerve close to the trifurcation, and compared between young and aged rats (Uchida et al., 2004). All NF subunit proteins decrease in the distal portion of the sciatic nerve; however, the degree of decrease differs between young and aged rats. NF-L moderately decreases proximo-distally in young rats; however, it decreases dramatically in the mid- and the distal-sciatic nerves in aged rats. This marked decrease in NF-L in aged rats agrees with the reduction in the density of NF packing in the distal segments. NF-M is particularly reduced in more distal segments of the sciatic nerve in aged rats. The ratio of NF-M to NF-H is almost constant in the sciatic nerves of young rats; however, the ratio of NF-M to NF-H in aged rats gradually decreases with distance from the spinal cord; in the distal sciatic nerve, it decreases to 45% of the value in the proximal region. Thus, the amount of NF protein, particularly NF-M, decreases significantly in the distal regions of the sciatic nerve in aged rats.

15.4 Axonal Transport of NFs in Aged Animals

15.4.1 Decreased Axonal Transport of NFs in Aged Animals

The observation of NF movement in axons using GFP-tagged NF subunit protein has been established in cultured primary neurons (Roy et al., 2000; Wang et al., 2000). This method has not been applied to axons of aged animals nor in vivo axons. Recently, metabolism of human NF-L has been assessed in the presence or absence of endogenous mouse NF-L by the conditional transgene expression method (Millecamps et al., 2007). Although the movement of NFs themselves has not been observed, the rate of axonal transport has been estimated from the appearance and disappearance of human NF-L in the axon. Millecamps et al. (2007) demonstrate that NF-L is transported much faster than the conventional slow axonal transport rate in the absence of an NF network. Their results clearly indicate that the rate of NF transport is affected by the abundance of NFs in axons. Considering the differences in the abundance of NFs between axons from young and aged rats, it is natural that the NF transport rate is altered with aging.

It is known that the rate of NF transport slows with development (Hoffman et al., 1985; Komiya et al., 1980). The rate is further slowed down in aged animals (Tashiro and Komiya, 1991; 1994). We also confirmed this by the traditional isotope labeling method (Uchida et al., 2004), where the amount of radioactivity associated with each NF protein band was measured and plotted against the distance from the spinal cord to give the transport profiles of each NF subunit. It takes 3 weeks after labeling in young rats and 8 weeks after labeling in aged rats for the peak of radioactivity associated with NF proteins to move 20–25 mm, indicating that the transport rate of NFs decreases considerably in aged rats. Tubulin is another major protein transported at a similar velocity to that of the NFs. In young rats, cytoskeletal tubulin is cotransported with NFs at the rate of about 1.5 mm/day. In aged rats, however, the peak of tubulin with the rate of about 0.8 mm/day slightly precedes that of the NF proteins (<0.5 mm/day). These observations indicate that the transport rate of NFs decreases selectively with aging (Tashiro and Komiya, 1991).

15.4.2 Differential Transport of NF-M and NF-H with Aging

According to the results obtained by the isotope labeling method (Uchida et al., 2004), NF-L and NF-M show similar distribution patterns of radioactivity in young and aged rats, indicating that they are transported coordinately in young or aged axons. In contrast, NF-H is slightly delayed relative to NF-L and NF-M in young rats; however, it is transported slightly faster in aged rats. Thus, the ratio of NF-M to NF-H increases slightly in the leading segments of axons in young rats; however, it gradually decreases towards the leading edge of axons in aged rats. In other words, NF-M is transported slightly faster than NF-H in young rats and more slowly than NF-H in aged rats, suggesting that the transport of NF-M (and NF-L) is more delayed by aging than that of NF-H (Tashiro and Komiya, 1991). The difference in transport rates between NF-H and NF-M/NF-L is even more striking in animals younger than 3 weeks of age (Tsuda et al., 1997, 2000). NF-H transport is regulated in a somewhat different manner from NF-M and NF-L, which are always cotransported. Selective partitioning of NF-H between filamentous and soluble forms may be responsible for this difference.

15.4.3 Decreases in NF-M Content and NF Transport Rate with Aging

The NF subunit ratio is one of the factors influencing axonal transport velocity. NF-M and NF-H appear to work reciprocally in NF axonal transport. It was originally demonstrated that the axonal transport rate is reduced with the increases in NF-H that occur during development (Hoffman et al., 1984; Willard and Simon, 1993). Deletion of the *NF-H* gene results in an increase in the axonal transport rate

(Zhu et al., 1997; Jung et al., 2005). Overexpression of NF-H dramatically slows down NF axonal transport (Collard et al., 1995; Marszalek et al., 1996). A part of this decrease is caused by phosphorylation of the C-terminal tail domain of NF-H (Toyoshima and Komiya, 1995). NFs containing constitutively nonphosphorylated NF-H are transported faster than those containing wild-type NF-H, whereas NFs containing constitutively phosphorylated NF-H are transported slower than those containing wild-type NF-H (Ackerley et al., 2003).

It is now considered that NFs are transported by motor proteins in an ATP-dependent manner (Yabe et al., 1999; Shah et al., 2000; Rao et al., 2002; Xia et al., 2003) at a rate comparable with fast axonal transport, but with long intermittent pauses (Roy et al., 2000; Wang et al., 2000; Wang and Brown, 2001). This also appears to be the case in mature NF-rich axons, in which interaction with the transport machinery determines the velocity of transport. NF-M may be the subunit that interacts with the transport machinery. Oligomers composed of NF-M alone can be transported in axons lacking NFs (Terada et al., 1996), and the overexpression of NF-M accelerates the axonal transport of NFs (Xu and Tung, 2000; Yabe et al., 2001). Targeted deletion of the *NF-M* gene causes a paucity of NFs in axons (Elder et al., 1998a), which can be interpreted as a defect in transport, although its effect on NF assembly cannot be excluded. In contrast, it has also been reported that the average rate of axonal transport of NFs is not altered in axons of mutant mice in which NF-M has been replaced with tailless NF-M (Rao et al., 2003; Yuan et al., 2006). Although it is not known how NF-M interacts with motor proteins, the decrease in NF-M content could be a major factor affecting the reduced rate of NF transport.

15.5 Diseases and Aged NFs

15.5.1 Amyotrophic Lateral Sclerosis (ALS)

Damage to axonal transport may underlie the pathogenic accumulation of organelles and proteins in many neurodegenerative diseases (Roy et al., 2005; De Vos et al., 2008). ALS is a motor neuron disease, primarily characterized by the degeneration of motor neurons in the brain and spinal cord. The most characteristic pathology in ALS is the abnormal accumulation of NFs in the cell body and proximal axons of motor neurons (Carpenter, 1968; Schmidt et al., 1987). Overexpression of NF proteins induces ALS-like phenotypes in transgenic mice (Cote et al., 1993; Xu et al., 1993; Lee et al., 1994; Beaulieu et al., 1999). Mutation studies have not linked NFs to the direct cause of ALS (Vechio et al., 1996; Garcia et al., 2006), although some genetic variations have been found in the tail domain of NF-H in patients with sporadic ALS (Julien, 1997). Mutations in the gene encoding Cu/Zn SOD1 are found in about 20% of familial patients with ALS. Mutant SOD1 affects the axonal transport of NFs as well as the fast transport of vesicles. Expression of mutant SOD1 retards the transport of NFs even before the mice are symptomatic (Zhang et al., 1997; Williamson et al., 1999; De Vos, 2007). In contrast, disruption of the *NF-L* gene delays disease onset and extended survival in SOD mutants

(Williamson et al., 1998). Furthermore, deletion of the NF-M and NF-H tail regions brings beneficial effects against disease caused by mutant SOD1 (Lobsiger et al., 2005). Accumulation of NFs may impair the axonal transport of proteins required for axonal activities in the distal region, resulting in degeneration of the distal axon and eventually of the whole neuron.

ALS is an age-related neuronal disease (Hirano, 1991; Cleaveland, 1999). Age-related cellular factors must influence the pathological consequences of NF perturbation. The rate of slow axonal transport is decreased in aged rats and the decrease is much more remarkable with NF proteins compared with, for example, tubulin, another component involved in slow axonal transport (Komiya, 1980; Tashiro and Komiya, 1991). These age-related changes should associate with NFs. As described above, several properties of NF such as the hyperphosphorylation of NF-H and decreases in NF-M levels change with aging. Experiments with transgenic mice that generate pathological abnormalities resembling ALS by perturbing the NF subunit ratio suggest that decreased levels of NF-M are involved with ALS (Cote et al., 1993; Xu et al., 1993; Eyer and Peterson, 1994; Wong et al., 1995). However, there may also be differences between species and cell type specificities. Mice expressing low levels of human NF-M develop neurofilamentous accumulation in specific subgroups of neurons in an age-dependent manner (Vickers et al., 1994a). In any case, the reduction in the rate of NF transport, the reduction in NF-M content and the proximal packing of NFs are significant features of axons during normal aging and bear some resemblance to the pathological features of axonal degeneration.

The shrinkage of distal axons similar to that observed in patients with ALS has also been reported in patients with toxic neuropathy and diabetes (Medori et al., 1988; Schmidt et al., 1990; Yagihashi et al., 1990; Tamura and Parry, 1994). The age-dependent changes of NFs may also render nerves more susceptible to injury conferred by toxic agents or metabolic conditions such as type II diabetes.

15.5.2 Charcot-Marie-Tooth Disease (CMT)

CMT is the most common inherited peripheral nerve disease (Niemann et al., 2006). CMT is classified into two main types based on nerve conduction velocity: demyelinating CMT1 and axonal CMT2. Mutations of NF-L are associated with dominant axonal (CMT2E) and demyelinating (CMT1F) neuropathies. Mutations are often found in the head domain, as well as the rod domain, of NF-L, which regulates the assembly of filaments by phosphorylation. Mutant NF-L forms aggregates in the cytoplasm (Brownlees et al., 2002; Perez-Olle et al., 2002, 2004; Sasaki et al., 2006) and disrupts axonal transport of NFs and mitochondria (Brownlees et al., 2002; Pérez-Ollé et al., 2005; Zhai et al., 2007). Reduced phosphorylation of NF-L at the head domain is one possibility that enhances aggregate formation (Sasaki et al., 2006), and the effects of tail domain phosphorylation on altered neuronal Schwann-cell signaling has also been proposed to enhance aggregate formation (Niemann et al., 2006). However, it is not yet known how mutations induce the aggregation of NFs.

15.5.3 Neuronal Intermediate Filament Inclusion Disease (NIFID)

Abnormal neuronal aggregates of the four NF subunits, NF-L, NF-M, NF-H and α-internexin, have been identified as the signature lesions of NIFID (Josephs et al., 2003; Cairns et al., 2004). This is a novel neurological disease of early onset (40–50 years of age) with a variable clinical phenotype including frontotemporal dementia and pyramidal and extrapyramidal signs. To date, no pathogenic mutations have been identified in NF proteins (Momeni et al., 2006).

15.5.4 The Early Senescence Mutant Mouse Klotho

Klotho is a mutant mouse strain displaying premature aging symptoms that resemble human aging, including a short lifespan (about 8 weeks), infertility, arteriosclerosis, skin atrophy, osteoporosis and emphysema (Kuro-o et al., 1997). Klotho is a transmembrane protein that functions in the vitamin D/fibroblast growth factor-23 signaling pathway (Nabeshima, 2008). NFs in *klotho* mice have also been studied. NFs are more packed in the sciatic nerve axons of *klotho* mice than wild-type mice, and the transport of axonal cytoskeletal proteins including NF proteins is decreased during the process of premature aging (Uchida et al., 2001). The ratio of NF-M to NF-L is slightly decreased in the spinal cords of *klotho* mice. Furthermore, phosphorylated NFs are increased in the cell bodies of cerebellum Purkinje neurons and anterior horn motor neurons (Anamizu et al., 2005; Shiozaki et al., 2008). These properties of NFs reflect well those of NFs in aged animals. Aging, whether normal or not, may increase phosphorylation and packing of NFs in axons.

15.6 Conclusion

We have described here several changes in NFs that accompany aging, which include NF packing density or aggregate formation with higher phosphorylation of NF-H and reduced NF-M content. These changes may be within acceptable limits for normal neuronal activity. However, densely packed NFs in aged animals may increase the load upon the transport machinery or may prevent the interactions between NFs and transport machinery, resulting in a decreased rate of transport that generates a deleterious feedback loop. NF-M may be the subunit that interacts with the transport machinery (Terada et al., 1996; Wagner et al., 2004). If so, the reduced amount of NF-M in NFs from aged rats would also reduce the interaction with transport machinery, causing a decrease of the transport rate. Phosphorylation of NF-H increases with aging. Phosphorylation increases the interaction between NFs and decreases the interaction with motor proteins. Increased NF-H, compensating for the decreased NF-M, would further enhance the accumulation of NFs in aged animals. When other genetic or environmental factors are imposed (Rouleau et al., 1996; Zhang et al., 1997), these cellular age-dependent alterations may be detrimental to axon maintenance, leading to complete accumulation of NFs, as observed in ALS. Further characterization of age-dependent changes in NFs would provide important clues in understanding the pathological accumulation of NFs found in motor neuron diseases and peripheral neuropathies.

References

Ackerley S, Thornhill P, Grierson AJ, Brownlees J, Anderton BH, Leigh PN, Shaw CE, Miller CC (2003) Neurofilament heavy chain side arm phosphorylation regulates axonal transport of neurofilaments. J Cell Biol 161:489–495

Anamizu Y, Kawaguchi H, Seichi A, Yamaguchi S, Kawakami E, Kanda N, Matsubara S, Kuro-o M, Nabeshima Y, Nakamura K, Oyanagi K (2005) Klotho insufficiency causes decrease of ribosomal RNA gene transcription activity, cytoplasmic RNA and rough ER in the spinal anterior horn cells. Acta Neuropathol 109:457–466

Baas PW, Brown A (1997) Slow axonal transport: the polymer transport model. Trends Cell Biol 7:380–384

Barry DM, Millecamps S, Julien JP, Garcia ML (2007) New movements in neurofilament transport, turnover and disease. Exp Cell Res 313:2110–2120

Beaulieu JM, Nguyen MD, Julien JP (1999) Late onset of motor neurons in mice overexpressing wild-type peripherin. J Cell Biol 147:531–544

Berthold C-H, Rydmark M (1995) Morphology of normal peripheral axons. In: Waxman SG, Kocsis JD, Stys PK (eds) The Axon. Oxford University Press, New York, NY, pp 13–48

Blanchard BJ, Ingram VM (1989) Age-related neurofilament phosphorylation in normal human brains. Neurobiol Aging 10:253–258

Brown HG, Hoh JH (1997) Entropic exclusion by neurofilament sidearms: a mechanism for maintaining interfilament spacing. Biochemistry 36:15035–15040

Brownlees J, Ackerley S, Grierson AJ, Jacobsen NJ, Shea K, Anderton BH, Leigh PN, Shaw CE, Miller CC (2002) Charcot-Marie-Tooth disease neurofilament mutations disrupt neurofilament assembly and axonal transport. Hum Mol Genet 11:2837–2844

Cairns NJ, Grossman M, Arnold SE, Burn DJ, Jaros E, Perry RH, Duyckaerts C, Stankoff B, Pillon B, Skullerud K, Cruz-Sanchez FF, Bigio EH, Mackenzie IR, Gearing M, Juncos JL, Glass JD, Yokoo H, Nakazato Y, Mosaheb S, Thorpe JR, Uryu K, Lee VM, Trojanowski JQ (2004) Clinical and neuropathologic variation in neuronal intermediate filament inclusion disease. Neurology 63:1376–1384

Carden MJ, Trojanowski JQ, Schlaepfer WW, Lee VM-Y (1987) Two-stage expression of neurofilament polypeptides during rat neurogenesis with early establishment of adult phosphorylation patterns. J Neurosci 7:3489–3504

Carpenter D (1968) Proximal axonal enlargement in motor neuron diseases. Neurol 18:841–851

Chen J, Nakata T, Zhang Z, Hirokawa N (2000) The C-terminal tail domain of neurofilament protein-H (NF-H) forms the crossbridges and regulates neurofilament bundle formation. J Cell Sci 113:3861–3869

Chin SS, Liem RK (1989) Expression of rat neurofilament proteins NF-L and NF-M in transfected non-neuronal cells. Eur J Cell Biol 50:475–490

Ching GY, Liem. RK (1993) Assembly of type IV neuronal intermediate filaments in nonneuronal cells in the absence of preexisting cytoplasmic intermediate filaments. J Cell Biol 122: 1323–1335

Cleaveland D (1999) From Charcot to SOD1: mechanisms of selective motor neuron death in ALS. Neuron 24:515–520

Cole JS, Messing A, Trojanowski JQ, Lee VM-Y (1994) Modulation of axon diameter and neurofilaments by hypomyelinating Schwann cells in transgenic mice. J Neurosci 14: 6956–6966

Collard JF, Cote F, Julien J-P (1995) Defective axonal transport in a transgenic mouse model of amyotrophic lateral sclerosis. Nature 375:61–64
Cote F, Collard JF, Julien J-P (1993) Progressive neuronopathy in transgenic mice expressing the human neurofilament heavy gene: a mouse model of amyotrophic lateral sclerosis. Cell 73:35–46
De Vos KJ, Grierson AJ, Ackerley S, Miller CC (2008) Role of axonal transport in neurodegenerative diseases. Annu Rev Neurosci 31:151–173
de Waegh SM, Lee VM-Y, Brady ST (1992) Local modulation of neurofilament phosphorylation, axonal caliber, and slow axonal transport by myelinating Schwann cells. Cell 68:451–463
De Vos KJ, Chapman AL, Tennant ME, Manser C, Tudor EL, Lau KF, Brownlees J, Ackerley S, Shaw PJ, McLoughlin DM, Shaw CE, Leigh PN, Miller CC, Grierson AJ (2007) Familial amyotrophic lateral sclerosis-linked SOD1 mutants perturb fast axonal transport to reduce axonal mitochondria content. Hum Mol Genet 16:2720–2728
Elder GA, Friedrich VL Jr, Bosco P, Kang C, Gourov A, Tu P-H, Lee VM-Y, Lazzarini RA (1998a) Absence of the mid-sized neurofilament subunit decreases axonal calibers, levels of light neurofilament (NF-L) and neurofilament content. J Cell Biol 141:727–739
Elder GA, Friedrich VL Jr, Kang C, Bosco P, Gourov A, Tu P-H, Zhang B, Lee VM-Y, Lazzarini RA (1998b) Requirement of heavy neurofilament subunit in the development of axons with large calibers. J Cell Biol 143:195–205
Elder GA, Friedrich VL Jr, Margita A, Lazzarini RA (1999) Age-related atrophy of motor axons in mice deficient in the mid-sized neurofilament subunit. J Cell Biol 146:181–192
Eyer J, Leterrier JF (1988) Influence of the phosphorylation state of neurofilament proteins on the interactions between purified filaments in vitro. Biochem J 252:655–660
Eyer J, Peterson A (1994) Neurofilament-deficient axons and perikaryal aggregates in viable transgenic mice expressing a neurofilament-beta-galactosidase fusion protein. Neuron 12:389–405
Garcia ML, Singleton AB, Hernandez D, Ward CM, Evey C, Sapp PA, Hardy J, Brown RH Jr, Cleveland DW. (2006) Mutations in neurofilament genes are not a significant primary cause of non-SOD1-mediated amyotrophic lateral sclerosis. Neurobiol Dis 21:102–109
Gotow T, Tanaka T, Nakamura Y, Takeda M (1994) Dephosphorylation of the largest neurofilament subunit protein influences the structure of crossbridges in reassembled neurofilaments. J Cell Sci 107:1949–1957
Gou JP, Eyer J, Leterrier JF (1995) Progressive hyperphosphorylation of neurofilament heavy subunits with aging: possible involvement in the mechanism of neurofilament accumulation. Biochem Biophys Res Commun 215:368–376
Grant P, Pant HC (2000) Neurofilament protein synthesis and phosphorylation. J Neurocytol 29:843–872
Harris J, Moreno S, Shaw G, Mugnaini E (1993) Unusual neurofilament composition in cerebellar unipolar brush neurons. J Neurocytol 22:1039–1059
Hirano A (1991) Cytopathology of amyotrophic lateral sclerosis. Adv Neurol 56:91–101
Hirokawa N, Glicksman MA, Willard MB (1984) Organization of mammalian neurofilament polypeptides within the neuronal cytoskeleton. J Cell Biol 98:1523–1536
Hirokawa N, Takeda S (1998) Gene targeting studies begin to reveal the function of neurofilament proteins. J Cell Biol 143:1–4
Hisanaga S, Gonda Y, Inagaki M, Ikai A, Hirokawa N (1990) Effects of phosphorylation of the neurofilament L protein on filamentous structures. Cell Regul 1:237–248
Hisanaga S, Hirokawa N (1988) Structure of the peripheral domains of neurofilaments revealed by low angle rotary shadowing. J Mol Biol 202:297–305
Hisanaga S, Hirokawa. N (1990) Dephosphorylation-induced interaction of neurofilaments with microtubules. J Biol Chem 265:21852–21858
Hisanaga S, Matsuoka Y, Nishizawa K, Saito T, Inagaki M, Hirokawa N (1994) Phosphorylation of native and reassembled neurofilaments composed of NF- L, NF-M, and NF-H by the catalytic subunit of cAMP-dependent protein kinase. Mol Biol Cell 5:161–172
Hisanaga S, Uchiyama M, Hosoi T, Yamada K, Honma Y, Ishiguro K, Uchida T, Dahl D, Ohsumi K, Kishimoto T (1995) Porcine brain neurofilament-H tail domain kinase: its identification

as cdk5/p26 complex and comparison with cdc2/cyclinB kinase. Cell Motil Cytoskeleton 31: 283–297
Hoffman PN, Cleveland DW, Griffin JW, Landes PW, Cowan NJ, Price DL (1988) Neurofilament gene expression: a major determinant of axonal caliber. Proc Natl Acad Sci USA 84: 3472–3476
Hoffman PN, Griffin JW, Price DL (1984) Control of axonal caliber by neurofilament transport. J Cell Biol 99:705–714
Hoffman PN, Griffin JW, Gold BG, Price DL (1985) Slowing of neurofilament transport and the radial growth of developing nerve fibers. J Neurosci 5:2920–2929
Hoffman PN, Lasek RJ (1975) The slow component of axonal transport. Identification of major structural polypeptides of the axon and their generality among mammalian neurons. J Cell Biol 66:351–366
Josephs KA, Holton JL, Rossor MN, Braendgaard H, Ozawa T, Fox NC, Petersen RC, Pearl GS, Ganguly M, Rosa P, Laursen H, Parisi JE, Waldemar G, Quinn NP, Dickson DW, Revesz T (2003) Neurofilament inclusion body disease: a new proteinopathy? Brain 126:2291–2303
Julien J-P (1997) Neurofilaments and motor neuron disease. Trends Cell Biol 7:243–249
Julien JP, Meyer D, Flavell D, Hurst J, Grosveld F (1986) Cloning and developmental expression of the murine neurofilament gene family. Brain Res 387:243–250
Jung C, Lee S, Ortiz D, Zhu Q, Julien JP, Shea TB (2005) The high and middle molecular weight neurofilament subunits regulate the association of neurofilaments with kinesin: inhibition by phosphorylation of the high molecular weight subunit. Brain Res Mol Brain Res 141:151–155
Komiya Y (1980) Slowing with age of the rate of slow axonal flow in bifurcating axons of rat dorsal root ganglion cell. Brain Res 183:477–480
Kuchel GA, Poon T, Irshad K, Richard C, Julien JP, Cowen T (1996) Decreasedneurofilament gene expression is an index of selective axonal hypotrophy in ageing. Neuroreport 31:1353–1359
Kuro-o M, Matsumura Y, Aizawa H, Kawaguchi H, Suga T, Utsugi T, Ohyama Y, Kurabayashi M, Kaname T, Kume E, Iwasaki H, Iida A, Shiraki-Iida T, Nishikawa S, Nagai R, Nabeshima YI (1997) Mutation of the mouse klotho gene leads to a syndrome resembling ageing. Nature 390:45–51
Lariviere RC, Julien JP (2004) Functions of intermediate filaments in neuronal development and disease. J Neurobiol 58:131–148
Lee MK, Marszalek JR, Cleveland DW (1994) A mutant neurofilament subunit causes massive, selective motor neuron death: implications for the pathogenesis of human motor neuron disease. Neuron 13:975–988
Lee VM-Y, Otvos L Jr, Carden MJ, Hollosi M, Dietzschold B, Lazzarini RA (1988) Identification of the major multiphosphorylation site in mammalian neurofilaments. Proc Natl Acad Sci USA 85:1998–2002
Lee MK, Xu Z, Wong PC, Cleveland DW (1993) Neurofilaments are obligate heteropolymers in vivo. J Cell Biol 122:1337–1350
Leterrier JF, Kas J, Hartwig J, Vegners R, Janmey PA (1996) Mechanical effects of neurofilament cross-bridges. J Biol Chem 271:15687–15694
Liem RK, Hutchison SB (1982) Purification of individual components of the neurofilament triplet: filament assembly from the 70 000-dalton subunit. Biochemistry 21:3221–3226
Lobsiger CS, Garcia ML, Ward CM, Cleveland DW (2005) Altered axonal architecture by removal of the heavily phosphorylated neurofilament tail domains strongly slows superoxide dismutase 1 mutant-mediated ALS. Proc Natl Acad Sci USA 102:10351–10356
Marszalek JR, Williamson TL, Lee MK, Xu Z-S, Hoffman PN, Becher MW, Crawford TO, Cleveland DW (1996) Neurofilament subunit NF-H modulates axonal diameter by selectively slowing neurofilament transport. J Cell Biol 143:171–181
Medori R, Jenich H, Autilio-Gambetti L, Gambetti P (1988) Experimental diabetic neuropathy: similar changes of slow axonal transport and axonal size in different animal models. J Neurosci 8:1814–1821

Millecamps S, Gowing G, Corti O, Mallet J, Julien JP (2007) Conditional NF-L transgene expression in mice for in vivo analysis of turnover and transport rate of neurofilaments. J Neurosci 27:4947–4956

Momeni P, Cairns NJ, Perry RH, Bigio EH, Gearing M, Singleton AB, Hardy J (2006) Mutation analysis of patients with neuronal intermediate filament inclusion disease (NIFID). Neurobiol Aging 27:778.e1–778.e6

Moskowitz PF, Oblinger MM (1995) Transcriptional and post-transcriptional mechanisms regulating neurofilament and tubulin gene expression during normal development of the rat brain. Brain Res Mol Brain Res 30:211–222

Nabeshima Y (2008) The discovery of alpha-Klotho and FGF23 unveiled new insight into calcium and phosphate homeostasis. Cell Mol Life Sci 65:3218–3230

Nakagawa T, Chen J, Zhang Z, Kanai Y, Hirokawa N (1995) Two distinct functions of the carboxyl-terminal tail domain of NF-M upon neurofilament assembly: cross-bridge formation and longitudinal elongation of filaments. J Cell Biol 129:411–429

Nguyen MD, Larivière RC, Julien JP (2000) Reduction of axonal caliber does not alleviate motor neuron disease caused by mutant superoxide dismutase 1. Proc Natl Acad Sci USA 97: 12306–12311

Niemann A, Berger P, Suter U (2006) Pathomechanisms of mutant proteins in Charcot-Marie-Tooth disease. Neuromolecular Med 8:217–242

Nixon RA (1998) Dynamic behavior and organization of cytoskeletal proteins in neurons: reconciling old and new findings. Bioessays 20:798–807

Nixon RA, Logvinenko KB (1986) Multiple fates of newly synthesized neurofilament proteins: evidence for a stationary neurofilament network distributed nonuniformly along axons of retinal ganglion cell neurons. J Cell Biol 102:647–659

Nixon RA, Paskevich PA, Sihag RK, Thayer CY (1994) Phosphorylation on carboxyl terminus domains of neurofilament proteins in retinal ganglion cell neurons in vivo: influences on regional neurofilament accumulation, interneurofilament spacing, and axon caliber. J Cell Biol 126:1031–1046

Nixon RA, Shea TB (1992) Dynamics of neuronal intermediate filaments: a developmental perspective. Cell Motil Cytoskeleton 22:81–91

Ohara O, Gahara Y, Miyake T, Teraoka H, Kitamura T (1993) Neurofilament deficiency in quail caused by nonsense mutation in neurofilament-L gene. J Cell Biol 121:387–395

Pant HC, Veeranna (1995) Neurofilament phoshorylation. Biochem Cell Biol 73:575–594

Parhad IM, Scott JN, Cellars LA, Bains JS, Krekoski CA, Clark AW (1995) Axonal atrophy in aging is associated with a decline in neurofilament gene expression. J Neurosci Res 41:355–366

Perez-Olle R, Jones ST, Liem RK (2004) Phenotypic analysis of neurofilament light gene mutations linked to Charcot-Marie-Tooth disease in cell culture models. Hum Mol Genet 13:2207–2220

Perez-Olle R, Leung CL, Liem RK (2002) Effects of Charcot-Marie-Tooth-linked mutations of the neurofilament light subunit on intermediate filament formation. J Cell Sci 115:4937–4946

Pérez-Ollé R, López-Toledano MA, Goryunov D, Cabrera-Poch N, Stefanis L, Brown K, Liem RK (2005) Mutations in the neurofilament light gene linked to Charcot-Marie-Tooth disease cause defects in transport. J Neurochem 93:861–874

Perrot R, Berges R, Bocquet A, Eyer J (2008) Review of the multiple aspects of neurofilament functions, and their possible contribution to neurodegeneration. Mol Neurobiol 38:27–65

Rao MV, Campbell J, Yuan A, Kumar A, Gotow T, Uchiyama Y, Nixon RA (2003) The neurofilament middle molecular mass subunit carboxyl-terminal tail domains is essential for the radial growth and cytoskeletal architecture of axons but not for regulating neurofilament transport rate. J Cell Biol 163:1021–1031

Rao MV, Engle LJ, Mohan PS, Yuan A, Qiu D, Cataldo A, Hassinger L, Jacobsen S, Lee VM-Y, Andreadis A, Julien JP, Bridgman PC, Nixon RA (2002) Myosin Va binding to neurofilaments is essential for correct myosin Va distribution and transport and neurofilament density. J Cell Biol 159:279–290

Rao MV, Houseweart MK, Williamson TL, Crawford TO, Folmer J, Cleveland DW (1998) Neurofilament-dependent radial growth of motor axons and axonal organization of neurofilaments does not require the neurofilament heavy subunit (NF-H) or its phosphorylation. J Cell Biol 143:171–181

Riederer BM, Porchet R, Marugg RA (1996) Differential expression and modification of neurofilament triplet proteins during cat cerebellar development. J Comp Neurol 364:704–717

Rouleau GA, Clark AW, Rooke K, Pramatarova A, Krizus A, Suchowersky O, Julien JP, Figlewicz D (1996) SOD1 mutation is associated with accumulation of neurofilaments in amyotrophic lateral sclerosis. Ann Neurol 39:128–131

Roy S, Coffee P, Smith G, Liem RK, Brady ST, Black MM (2000) Neurofilament are transported rapidly but intermittently in axon: implications for slow axonal transport. J Neurosci 20: 6849–6861

Roy S, Zhang B, Lee VM, Trojanowski JQ (2005) Axonal transport defects: a common theme in neurodegenerative diseases. Acta Neuropathol 109:5–13

Saito T, Shima H, Osawa Y, Nagao M, Hemmings BA, Kishimoto T, Hisanaga S (1995) Neurofilament-associated protein phosphatase 2A: its possible role in preserving neurofilaments in filamentous states. Biochemistry 34:7376–7384

Sasaki T, Gotow T, Shiozaki M, Sakaue F, Saito T, Julien JP, Uchiyama Y, Hisanaga S (2006) Aggregate formation and phosphorylation of neurofilament-L Pro22 Charcot-Marie-Tooth disease mutants. Hum Mol Genet 15:943–952

Schmidt M, Carden MJ, Lee VM-Y, Trojanowski JQ (1987) Phosphate dependent and independent neurofilament epitopes in the axonal swellings of patients with motor neuron disease and controls. Lab Invest 56:282–294

Schmidt RE, Chae HY, Parvin CA, Roth KA (1990) Neuronaxonal dystrophy in aging human sympathetic ganglia. Am J Pathol 136:1327–1338

Shah JV, Flanagan LA, Janmey PA, Leterrier JF (2000) Bidirectional translocation of neurofilaments along microtubules mediated in part by dynein/dynactin. Mol Biol Cell 11: 3495–3508

Shaw G (1991) Neurofilament proteins: the neuronal cytoskeleton. Wiley-Liss Inc, New York, NY, pp 185–241

Shaw G, Weber K (1982) Differential expression of neurofilament triplet proteins in brain development. Nature 298:277–279

Shiozaki M, Yoshimura K, Shibata M, Koike M, Matsuura N, Uchiyama Y, Gotow T (2008) Morphological and biochemical signs of age-related neurodegenerative changes in klotho mutant mice. Neuroscience 152:924–941

Sihag RK, Inagaki M, Yamaguchi T, Shea TB, Pant HC (2007) Role of phosphorylation on the structural dynamics and function of types III and IV intermediate filaments. Exp Cell Res 313:2098–2109

Strack S, Westphal RS, Colbran RJ, Ebner FF, Wadzinski BE (1997) Protein serine/threonine phosphatase 1 and 2A associate with and dephosphorylate neurofilaments. Brain Res Mol Brain Res 49:15–28

Tamura E, Parry GJ (1994) Severe radicular pathology in rats with longstanding diabetes. J Neurol Sci 127:29–35

Tashiro T, Komiya Y (1991a) Maturation and aging of the axonal cytoskeleton: biochemical analysis of transported tubulin. J Neurosci Res 30:192–200

Tashiro T, Komiya Y (1991b) Changes in organization and axonal transport of cytoskeletal proteins during regeneration. J Neurochem 56:1557–1563

Tashiro T, Komiya Y (1994) Impairment of cytoskeletal protein transport due to aging or beta,beta'-iminodipropionitrile intoxication in the rat sciatic nerve. Gerontology 40:36–45

Terada S, Nakata T, Peterson AC, Hirokawa N (1996) Visualization of slow axonal transport in vivo. Science 273:784–788

Toyoshima I, Komiya Y (1995) Phosphorylation and transport of neurofilament proteins in the rat spinal ganglion. Neurosci Lett 189:69–72

Trivedi N, Jung P, Brown A (2007) Neurofilaments switch between distinct mobile and stationary states during their transport along axons. J Neurosci 27:507–516

Tsuda M, Tashiro T, Komiya Y (1997) Increased solubility of high-molecular-mass neurofilament subunit by suppression of dephosphorylation: its relation to axonal transport. J Neurochem 68:2558–2565

Tsuda M, Tashiro T, Komiya Y (2000) Selective solubilization of high-molecular-mass neurofilament subunit during nerve regeneration. J Neurochem 74:860–868

Tu P-H, Elder G, Lazzarini RA, Nelson D, Troyanowski JQ, Lee VM-Y (1995) Overexpression of the human NFM subunit in transgenic mice modifies the level of endogenous NFL and the phosphorylation state of NFH subunits. J Cell Biol 129:1629–1640

Uchida A, Brown A (2004) Arrival, reversal, and departure of neurofilaments at the tips of growing axons. Mol Biol Cell 15:4215–4225

Uchida A, Komiya Y, Tashiro T, Yorifuji H, Kishimoto T, Nabeshima Y, Hisanaga S (2001) The neurofilament of *Klotho*, the mutant mouse prepaturely displaying symptoms resembling human aging. J Neurosci Res 64:364–370

Uchida A, Tashiro T, Komiya Y, Yorifuji H, Kishimoto T, Hisanaga S (2004) Morphological and biochemical changes of neurofilaments in aged rat sciatic nerve axons. J Neurochem 88: 735–745

Uchida A, Yorifuji H, Lee VM, Kishimoto T, Hisanaga S (1999) Neurofilaments of aged rats: the strengthened interneurofilament interaction and the reduced amount of NF-M. J Neurosci Res 58:337–348

Vechio JD, Bruijn LI, Xu Z, Brown RH Jr, Cleveland DW (1996) Sequence variants in human neurofilament proteins: absence of linkage to familial amyotrophic lateral sclerosis. Ann Neurol 40:603–610

Vickers JC, Morrison JH, Friedrich VL Jr, Elder GA, Perl DP, Katz RN, Lazzarini RA (1994a) Age-associated and cell-type-specific neurofibrillary pathology in transgenic mice expressing the human midsized neurofilament subunit. J Neurosci 14:5603–5612

Vickers JC, Riederer BM, Marugg RA, Buee-Scherrer V, Buee L, Delacourte A, Morrison JH (1994b) Alterations in neurofilament protein immunoreactivity in human hippocampal neurons related to normal aging and Alzheimer's disease. Neurosci 62:1–13

Wang L, Brown A (2001) Rapid intermittent movement of axonal neurofilaments observed by fluorescence photobleacing. Mol Biol Cell 12:3257–3267

Wang L, Ho CL, Sun D, Liem RK, Brown A (2000) Rapid movement of axonal neurofilaments interrupted by prolonged pauses. Nat Cell Biol 2:137–141

Wagner OI, Ascaño J, Tokito M, Leterrier J-F, Janmey PA, Holzbaur ELF (2004) The interaction of neurofilaments with the microtubule motor cytoplasmic dynein. Mol Biol Cell 15: 5092–5100

Williamson TL, Cleveland DW (1999) Slowing of axonal transport is a very early event in the toxicity of ALS-linked SOD1 mutants to motor neurons. Nat Neurosci 2:50–56

Williamson TL, Bruijn LI, Zhu Q, Anderson KL, Anderson SD, Julien JP, Cleveland DW (1998) Absence of neurofilaments reduces the selective vulnerability of motor neurons and slows disease caused by a familial amyotrophic lateral sclerosis-linked superoxide dismutase 1 mutant. Proc Natl Acad Sci U S A 95:9631–9636

Willard M, Simon C (1993) Modulations of neurofilament axonal transport during the development of rabbit retinal ganglion cells. Cell 35:551–559

Wong PC, Marszalek J, Crawford TO, Xu Z, Hsieh ST, Griffin JW, Cleveland DW (1995) Increasing neurofilament subunit NF-M expression reduces axonal NF-H, inhibits radial growth, and results in neurofilamentous accumulation in motor neurons. J Cell Biol 130: 1413–1422

Xia CH, Roberts EA, Her LS, Liu X, Williams DS, Clevelend DW, Goldstein LSB (2003) Abnormal neurofilament transport caused by targeted disruption of neuronal kinesin heavy chain KIF5A. J Cell Biol 161:55–66

Xu Z, Cork LC, Griffin JW, Cleveland DW (1993) Increased expression of neurofilament subunit NF-L produces morphological alterations that resemble the pathology of human motor neuron disease. Cell 73:23–33

Xu Z, Marszalek JR, Lee MK, Wong PC, Folmer J, Crawford TO, Hsieh ST, Griffin JW, Cleveland DW (1996) Subunit composition of neurofilaments specifies axonal diameter. J Cell Biol 133:1061–1069

Xu Z, Tung V (2000) Overexpression of neurofilament subunit M accelerates axonal transport of neurofilaments. Brain Res 866:326–332

Yabe JT, Pimenta A, Shea TB (1999) Kinesin-mediated transport of neurofilament protein oligomers in growing axon. J Cell Sci 112:3799–3814

Yabe JT, Chan WK, Chylinski TM, Lee S, Pimenta AF, Shea TB (2001) The predominant form in which neurofilament subunits undergo axonal transport varies during axonal initiation, elongation, and maturation. Cell Motil Cytoskeleton. 48:61–83

Yagihashi S, Kamijo M, Watanabe K (1990) Reduced myelinated fiber size correlates with loss of axonal neurofilaments in peripheral nerve of chronically streptozotocin diabetic rats. Am J Pathol 136:1365–1373

Yamasaki H, Itakura C, Mizutani M (1991) Hereditary hypotrophic axonopathy with neurofilament deficiency in a mutant strain of the Japanese quail. Acta Neuropathol 82:427–434

Yuan A, Nixon RA, Rao MV (2006a) Deleting the phosphorylated tail domain of the neurofilament heavy subunit does not alter neurofilament transport rate in vivo. Neurosci Lett 393:264–268

Yuan A, Rao MV, Sasaki T, Chen Y, Kumar A, Veeranna, Liem RK, Eyer J, Peterson AC, Julien JP, Nixon RA (2006b) Alpha-internexin is structurally and functionally associated with the neurofilament triplet proteins in the mature CNS. J Neurosci 26:10006–10019

Yuan A, Sasaki T, Rao MV, Kumar A, Kanumuri V, Dunlop DS, Liem RK, Nixon RA (2009) Neurofilaments form a highly stable stationary cytoskeleton after reaching a critical level in axons. J Neurosci 29:11316–11329

Zhai J, Lin H, Julien JP, Schlaepfer WW (2007) Disruption of neurofilament network with aggregation of light neurofilament protein: a common pathway leading to motor neuron degeneration due to Charcot-Marie-Tooth disease-linked mutations in NFL and HSPB1. Hum Mol Genet 16:3103–3116

Zhang B, Tu P, Abtahian F, Trojanowski JQ, Lee VM-Y (1997) Neurofilaments and orthograde transport are reduced in ventral root axons of transgenic mice that express human SOD1 with a G93A mutation. J Cell Biol 139:1307–1315

Zhu Q, Couillard-Despres S, Julien JP (1997) Delayed maturation of regenerating myelinated axons in mice lacking neurofilaments. Exp Neurol 148:299–316

Zhu Q, Lindenbaum M, Levavasseur F, Jacomy H, Julien J-P (1998) Disruption of the NF-H gene increases axonal microtubule content and velocity of neurofilament transport: relief of axonopathy resulting from the toxin β, β'-iminodipropionitrile. J Cell Biol 143:183–193

Chapter 16
Neurofilament Changes in Multiple Sclerosis

Elizabeth Gray and Alastair Wilkins

Abstract Multiple sclerosis is a disease of the central nervous system usually characterised by relapsing episodes of neurological dysfunction, often followed some years later by progressive and irreversible decline. Lesions of multiple sclerosis are characterised by varying degrees of inflammation, myelin and oligodendrocyte loss, astrogliosis and axonal pathology. Axonal loss is seen in progressive phases of the disease and appears to correlate well with clinical disability. Prior to loss of axons, pathological specimens have revealed changes in the immunohistochemical phenotype of axons. Specifically, there may be evidence of dephosphorylation of neurofilaments within axons and transection of axons leading to the formation of axonal spheroids which are rich in dephosphorylated neurofilaments. Evidence of axonal transport defects may also be found in lesions with accumulation of amyloid precursor protein within the axon. Mechanisms of axonal pathology in multiple sclerosis remain unknown, but inflammation in the acute disease phases is likely to cause significant damage. There is also evidence that axonal loss may continue even in the absence of inflammation. It has been postulated that axonal loss in this situation occurs due to loss of trophic support from myelin and oligodendrocytes. Oligodendrocytes and myelin are known to provide trophic support for axons and specifically can influence phosphorylated neurofilament levels. The precise mechanisms are unknown, but recent evidence suggests a combination of contact mediated and soluble factors may increase neurofilament phosphorylation and promote axonal protection. Knowledge of such mechanisms may lead to improved therapies to prevent progressive disease. This chapter will discuss axonal changes in multiple sclerosis, specifically alterations in neurofilament phosphorylation states, and potential mechanisms of axonal protection.

Keywords Axon · Axonal transport · Multiple sclerosis · Myelin · Neurofilament · Phosphorylation

---

A. Wilkins (✉)
Department of Neurology, Institute of Clinical Neurosciences, Clinical Sciences North Bristol, University of Bristol, Bristol BS8 1TH, UK
e-mail: alastair.wilkins@bristol.ac.uk

R.A. Nixon, A. Yuan (eds.), *Cytoskeleton of the Nervous System*,
Advances in Neurobiology 3, DOI 10.1007/978-1-4419-6787-9_16,

Abbreviations

BDNF	Brain-derived neurotrophic factor
CNPase	2′,3′-Cyclic nucleotide 3′-phosphodiesterase
CNS	Central nervous system
CSF	Cerebrospinal fluid
GDNF	Glial cell-line derived neurotrophic factor
IF	Intermediate filament
IGF-1	Insulin-like growth factor type-1
MAG	Myelin associated glycoprotein
MBP	Myelin basic protein
MOG	Myelin-oligodendrocyte glycoprotein
NF	Neurofilament
NGF	Nerve growth factor
NO	Nitric oxide
PLP	Proteolipid protein
ROS	Reactive oxygen species

Contents

16.1 Introduction

Multiple sclerosis is one of the commonest causes of acquired neurological disability in young people in Northern Europe and the United States. It is a disease of the central nervous system (CNS) which shown significant heterogeneity in terms of disease phenotype and course. Once thought the archetypal inflammatory CNS disorder, evidence now suggests the pathophysiology is complex and possibly occurs via multiple mechanisms including neurodegeneration (Compston, 2006). The complexity of the disease is born out by the inadequacy of current treatments, and the progressive phase of the disease, during which patients accrue disability relentlessly, has no effective therapy.

Axonal damage in multiple sclerosis was first described by Charcot in his classical reports of the disease over 100 years ago, but the focus on the immunology of multiple sclerosis and the pathogenesis of inflammatory demyelinating plaques has led to a paucity of research on the subject until several reports in the late 1990s re-emphasised this aspect of pathology. Most notably, Trapp and colleagues (1998) demonstrated axonal transection and neurofilament phosphorylation changes in axons within acute lesions of multiple sclerosis. Subsequent work has re-emphasised the importance of axonal pathology and has revealed that axon loss is a pathological hallmark of progressive multiple sclerosis and is believed to represent a major cause of irreversible disability and progression in the progressive stage of the disease (Cifelli et al., 2002; Dutta and Trapp, 2007; Peterson et al., 2001; Vercellino et al., 2005).

This chapter will discuss the evidence behind axonal pathology in multiple sclerosis, focusing on neurofilament changes, potential mechanisms of the process and possible neuroprotective therapies in the disease.

16.2 The Natural History of Multiple Sclerosis

Multiple sclerosis is an inflammatory CNS disease characterised by axonal degeneration and demyelination (Dutta and Trapp, 2007). Clinically, patients generally experience episodes of acute neurological dysfunction which usually recover, followed at some stage by slow and insidious progression and development of fixed neurological deficits (Compston and Coles, 2002). Most patients experience an initial relapsing and remitting preliminary course of the disease. Relapses are thought to represent recurrent episodes of inflammation and demyelination that are often accompanied by axonal injury (Pittock and Lucchinetti, 2007). During such relapses, neurological deficits are primarily attributable to axonal conduction block and demyelination and cytokine production by immune cells contribute to this process (Smith, 2007). Demyelination is a major cause of axonal dysfunction (Smith et al., 2001), especially if whole internodes of myelin are lost (i.e. segmental demyelination), as commonly occurs in MS (Smith, 2007). Myelin is produced by oligodendrocytes in the central nervous system and allows for rapid saltatory conduction to occur via the formation of node of Ranvier. At such nodes there is grouping of sodium channels allowing for efficient action potential propagation. Furthermore, myelin itself offers high resistance and low capacitance which again improves saltatory conduction. In the absence of myelin, axonal conduction is severely impaired and conduction of electrical impulses between internodes fails by virtue of myelin loss and a redistribution of sodium channels along the length of the axon (Craner et al., 2003, 2004b).

Promoting remyelination at an early stage of MS will restore conduction and, more importantly, may be neuroprotective since chronic demyelination is associated with axonal degeneration (Bjartmar et al., 2003). Therapeutic approaches can involve aiming to reduce inflammation during the acute phase of the disease with drugs such as β-interferon (Jacobs et al., 2000). Additional drugs have now been licensed for the relapse-remitting form of the disease, namely Mitoxantrone and Natalizumab (Tysabri). Mitoxantrone has been shown to slow the accumulation of disability in cases of rapidly worsening multiple sclerosis with frequent relapses, but it has less or no effect on non-relapsing progressive disease (Edan et al., 1997; Hartung et al., 2002). It is probably more efficacious than the interferons, but its use is confined to cases that are characterised by disease which is sufficiently aggressive to justify its toxic effects (Le Page et al., 2008). Natalizumab has been licensed as a monotherapy for severe relapsing-remitting multiple sclerosis (Compston and Coles, 2008).

The remission of symptoms in an acute relapse does not require remyelination to occur. Resolution of inflammation and/or reorganization of sodium channels along the demyelinated axolemma to allow non-saltatory conduction along the demyelinated segment may be sufficient (Felts et al., 1997; Waxman et al., 2004; Franklin and Ffrench-Constant, 2008).

Although there may be extensive remyelination in the early stages of multiple sclerosis (Patani et al., 2007), remyelination ultimately appears to fail in the disease, leading to extensive areas of demyelinated axons within the central nervous system. Reasons for remyelination failure include a primary deficiency in precursor cells, a failure of precursor cell recruitment, a failure of precursor cell differentiation and maturation (Franklin, 2002) or the presence of axonal inhibitory signals (Lubetzki et al., 2005). The best evidence at present suggests that a failure of differentiation and maturation of oligodendrocyte precursor cells may be the most important of these (Chang et al., 2000; Kuhlmann et al., 2008; Reynolds et al., 2002; Wolswijk, 1998). Furthermore, it is possible that chronically demyelinated lesions may contain factors that inhibit precursor differentiation (Back et al., 2005).

Most patients, in due course, will enter a phase of the disease in which remission does not occur and the disease progresses inexorably. This is thought to develop when a threshold of axonal loss is reached and the CNS compensatory resources are exhausted (Bjartmar and Trapp, 2001; Matthews et al., 1998; Trapp et al., 1999; Waxman, 1998). During this progressive phase of the disease disability correlates with radiological markers of axon loss, including magnetic resonance spectroscopic analysis of NAA and brain atrophy (Davie et al., 1997). Reasons as to why remission does not occur at this stage of the disease are the subject of extensive research, but a long-term failure of remyelination and lack of a supportive environment for the axon seem likely.

At the present time, once the progressive phase of the disease has been entered, there are no effective therapies for the disease which would appear to reflect the inability of therapeutic agents to prevent axonal degeneration. A striking failure of current treatments for multiple sclerosis has been their poor efficacy in preventing accumulation of disability in later, progressive stages of the disease (Kappos, 2004). The main reason for this appears to relate to the mode of action of drugs such as β-interferon which target the inflammatory response. Epidemiological data concerning multiple sclerosis has, however, shown that inflammation is unlikely to be the sole cause for clinical deterioration in the progressive phases of the disease.

Specifically the European database for multiple sclerosis (EDMUS) has been highly informative in determining likely mechanisms of disease progression. This has revealed that, once a certain level of disability has been reached in patient cohorts, further disease progression occurs at the same rate in those who do not have superimposed relapses, compared to those who do (Confavreux et al., 2000). This is in contrast to the clinical course early in the disease process, which does seem to be dependent on relapse rate in determining early disease rate. More recent evidence has suggested that there may be dissociation of the inflammatory and degenerative processes even from disease onset, as evidenced by similarities in rates of progression of those patients with primary progressive disease, in which the disease does not commence with a period of relapses and remissions, and those with secondary progressive disease (Confavreux and Vukusic, 2006). Furthermore, evidence from treatment trials concurs with the concept that disease progression may be partially independent of inflammatory activity. CAMPATH-1H (alemtuzimab), a humanised monoclonal antibody, is highly effective in abolishing inflammatory relapses in patients with multiple sclerosis, but has no effect on disease progression in those patients with existing disability (Coles et al., 1999).

Multiple sclerosis has classically been characterised as a demyelinating disease (Lassmann, 1998). However, in recent years attitudes have moved towards a broader view in which the relative contributions of acute and chronic axonal loss, and their dependence on inflammation, also have to be taken into account when reaching a coherent account of the disease pathogenesis. A recent paper by Compston and Coles (2008) proposed the following potential hypotheses that may underlie the pathogenesis of multiple sclerosis. First, inflammation is the exclusive pathogenic event from which all else follows. Second, neurodegeneration occurs first and inflammation is merely a secondary response. Third, inflammation and neurodegeneration both contribute to the clinical course, but are fully independent processes. Finally, inflammation exposes an intrinsic neurodegenerative susceptibility that renders axons vulnerable to cumulative injury. Cumulative loss of axons is likely to be responsible for progression in multiple sclerosis. The extent of injury is dependent on active inflammation, existing neurodegeneration and the dynamic vulnerability of intact axons.

The development of neuroprotectant drugs and also treatments that tackle the neurodegenerative component of disease progression are of pivotal importance in the development of MS therapies; if axons are spared, strategies for the promotion of remyelination and restoration of salutatory conduction can then be commenced, thereby slowing or halting the progression of disability in MS (O'Malley et al., 2009; Waxman, 2006). Possible therapeutic approaches include drugs such as anti-excitotoxic agents (Pitt et al., 2000; Rosin et al., 2004), nitric oxide and iNOS inhibitors (Okuda et al., 1997), anti-oxidants (Gilgun-Sherki et al., 2003), Ca^{2+} channel blockers (Brand-Schieber and Werner, 2004), Na^{+} channel blockers (Waxman, 2002) or Na^{+}/Ca^{2+} exchanger inhibitors and growth/neurotrophic factors (Irvine and Blakemore, 2008; Linker et al., 2002; Webster, 1997).

The blockade of sodium channels may provide an attractive new therapeutic approach in protecting axons from degeneration. Indeed, several studies have reported that phenytoin and flecanide are neuroprotective in experimental autoimmune encephalomyelitis (EAE); preventing degeneration of axons, maintain axonal conduction and improve clinical outcome (Bechtold et al., 2004, 2005, 2006; Lo et al., 2003). However, chronic administration of the drugs would be required in multiple sclerosis because axonal loss appears to occur as a continuous process ("slow burn") in addition to bursts of degeneration associated with the appearance of inflammatory demyelinating lesions (Lassmann and Wekerle, 2006; Smith, 2006; Trapp et al., 1998).

16.3 Axonal Pathology in MS

The classic histological features of acute lesions of multiple sclerosis include perivascular infiltrates of small lymphocytes, loss of myelin and oligodendrocytes, myelin-laden macrophages and axonal damage (Frohman et al., 2006). However there may be extensive variability between lesions. Some authors have recognised four distinct types of acute lesions based on pathological studies (Lassmann et al., 2001; Lucchinetti et al., 2000). The lesions are segregated into these patterns based on plaque geography, extent and pattern of oligodendrocyte pathology, evidence for immunoglobulin deposition and complement activation and pattern of myelin loss (Lassmann et al., 2001; Pittock and Lucchinetti, 2007).

Patterns I and II are both characterised by extensive remyelination and oligodendrocyte injury which suggests that myelin is the target of injury (Pittock and Lucchinetti, 2007). T-cell mediated inflammation with macrophage/microglia activation is present in the lesions. However, pattern II is distinguished by the presence of immunoglobulin and complement activity at sites of active myelin destruction, suggesting complement mediated lysis of antibody-targeted myelin. Patterns III and IV, in contrast, also contain T cells and macrophages but lack immunoglobulin deposition and are characterised by limited remyelination. Pattern III demonstrates oligodendrocyte apoptosis and selective loss of MAG which is thought to represent the inability of the oligodendrocyte to maintain the metabolic demands of the axon. Type IV is characterised by non-apoptotic oligodendrocyte death in the adjacent normal-appearing periplaque white matter (Gendelman et al., 1985; Itoyama et al., 1982; Lassmann et al., 2001; Lucchinetti et al., 2004; Pittock and Lucchinetti, 2007). Thus types III and IV are suggestive of a primary oligodendrocyte injury (Pittock and Lucchinetti, 2007). It was originally thought that each individual with multiple sclerosis had only one pattern of pathological lesions. However, a recent study suggests that acute plaques show pattern 2 pathology in all individuals, and that the other patterns are seen infrequently (Breij et al., 2008). Mechanisms leading to the different lesion patterns may help to explain differences in the extent of demyelination, oligodendrocyte injury, remyelination and axonal damage that are seen across the spectrum of multiple sclerosis (Patrikios et al., 2006) and related disorders, for instance neuromyelitis optica (Lucchinetti et al., 2002). Neuromyelitis optica, which is a distinct variant from multiple sclerosis, is characterised by demyelination and necrosis of white and grey matter of spinal cord, acute axonal injury, antibody deposition, and perivascular complement activation (Lucchinetti et al., 2002) and the presence of a specific circulating antibody in the serum. It typically responds to plasma exchange therapy (Scolding, 2005) which suggests a primary pathogenic role for antibody and complement (Keegan et al., 2005). This pathological and phenotypic subtype of CNS demyelination with a good response to a certain therapy distinguishes it from other inflammatory disease and raises the future possibility of tailoring therapies to immunopathological subtypes of disease.

In chronic active lesions of MS the histological appearances may be somewhat different with hypertrophic astocytes and degenerating axons, as well as perivascular cuffs of infiltrating cells and lipid-laden macrophages. The chronic inactive MS plaque is a sharply circumscribed, hypocellular plaque with no evidence of active myelin breakdown (Pittock and Lucchinetti, 2007). Chronic inactive lesions of MS are characterised by astrogliosis, a paucity of oligodendrocytes and reduced numbers of demyelinated axons. However, variable degrees of inflammation may still be present, particularly in the perivascular region (Lucchinetti et al., 1998). These lesions may show some degree of remyelination and recruitment of cells of the oligodendrocyte lineage, although remyelination is generally poor and incomplete (Dubois-Dalcq et al., 2005). Chronic astrocytic gliosis may well be inhibitory to remyelination (Blakemore et al., 2003). This failure of repair in chronic lesions is thought to be a major reason for disease progression and much active research is targeted at attempts to enhance remyelination (Zhao et al., 2005). Remyelination may help preserve the underlying axons from irreversible damage (Stadelmann and Bruck, 2008).

Axonal pathology was described by Charcot in his initial histopathological descriptions of the disease at the end of the nineteenth century, but it was not until over 100 years later that the subject of axonopathy in MS became the focus of active research. Axon loss is now considered an early and persistent event in the progression of MS pathology (Trapp et al., 1998; 1999). Although evidence of acute axonal injury can be found in association with chronic and early MS lesions, it is also known to occur in the normal appearing and periplaque white matter (Evangelou et al., 2000; Ferguson et al., 1997; Trapp et al., 1998). Acute axon pathology is more frequently seen in association with active demyelination and may already be apparent at the earliest stages of disease evolution (Bitsch et al., 2000; Kuhlmann et al., 2002). In addition, myelin and axonal pathology may occur independently (Bitsch et al., 2000; Pittock and Lucchinetti, 2007).

Early in the disease, acute axon damage and transection are associated with inflammation (Bitsch et al., 2000; Ferguson et al., 1997; Kornek et al., 2000; Kuhlmann et al., 2002; Trapp et al., 1998) but axonal injury and loss are also evident in chronically demyelinated lesions (Dutta and Trapp, 2007) which may occur on a background of residual chronic inflammation (Lassmann and Wekerle, 2006). Acute MS lesions exhibit greater than tenfold more damaged axons than chronic plaques (Kornek et al., 2000; Kuhlmann et al., 2002; Trapp et al., 1998). It remains to be clarified whether this is due to a prior loss of axons in chronic lesions or, alternatively, whether different mechanisms propel axonal degeneration in acute vs chronic lesions (Black et al., 2007). Indeed, histopathological analyses of chronic multiple sclerosis lesions have revealed significant reductions in axon density (DeLuca et al., 2004; Lovas et al., 2000).

Damage to demyelinated axons has been identified in MS by the presence of dephosphorylated neurofilaments (Trapp et al., 1998) and abnormal accumulations of axonally transported proteins such as amyloid precursor protein (APP) (Ferguson et al., 1997). APP is rapidly transported down axons and normal axons do not stain for the protein. The presence of APP staining in axon tracts implies disruption of fast axonal transport and thus dysfunction of the axon. APP accumulation occurs in both acute active lesions and in chronic lesions of MS (Bitsch et al., 2000; Ferguson et al., 1997; Kuhlmann et al., 2002) and appears to correlate with the numbers of inflammatory cells in active lesions (Bitsch et al., 2000). APP accumulation and axon loss may also occur in 'normal appearing white matter' (NAWM) of multiple sclerosis tissue, which has again added to evidence that axonopathy maybe widespread and not exclusively related to inflammation (Bjartmar et al., 2001; Kornek et al., 2000). Indeed, axonopathy in the NAWM not only reflects the Wallerian degeneration of axons but may reflect diffuse lesions not detected by MRI (Pelletier et al., 2001). In the progressive form of the disease a diffuse inflammatory process characterised by perivascular and parenchymal infiltrates has been reported with axonal spheroids and terminal axonal swellings present variably throughout the tissue (Kutzelnigg et al., 2005).

Other axonal changes have been reported, notably changes in ion channel distribution and mitochondrial dysfunction. In normal myelinated axons, sodium channels are clustered at high density at nodes of Ranvier, facilitating effective saltatory conduction. Recent studies have identified changes in the expression pattern of specific sodium channels as an important contributor to remission and progression in MS and revealed that these changes in distribution are associated with areas of axonal injury (Coman et al., 2006; Craner et al., 2004b; Waxman, 2006). In axons positive for APP staining the sodium channel subtype ($Na_V1.6$) which usually clusters at the node of Ranvier is found rather diffusely along the axon. Furthermore, co-localisation of the channel with the sodium/calcium exchanger (NCX) is associated with the axonal injury (Craner et al., 2004a). A more recent study has demonstrated that $Na_V1.6$ is also detectable in demyelinated axons of chronic plaques. APP accumulation occurred to a lesser extent and did not preferentially occur in axons expressing $Na_V1.6$ or NCX. These observations suggest that differing underlying mechanisms may lead to axonal degeneration in acute and chronic plaques (Black et al., 2007).

The diffuse distribution of $Na_V1.6$ along demyelinated axons is proposed to lead to sodium influx mediated by a large persistent current (Brown and Borutaite, 2002; Herzog et al., 2003; O'Malley et al., 2009; Rush et al., 2005; Smith et al., 1998) which would result in the accumulation of intra-axonal Na^+. The consumption of ATP is greatly increased in demyelinating axons, and energy crisis due to ATP depletion impairs the function of ATP-dependent ion channels (e.g. Na/K-ATPase), leading to an increase in intracellular Na^+ concentration (Dutta and Trapp, 2007).

This increased and persistent sodium influx may then lead to activation of reverse Na^+–Ca^{2+} exchange, accumulation of intra-axonal Ca^{2+} and activation of damaging injury cascades (Craner et al., 2004b; O'Malley et al., 2009; Waxman, 2008a, b; Webster, 1997), resulting in lysis and degeneration of the axon (Smith, 2007).

In acute lesions, nitric oxide is present at increased concentrations (Smith and Lassmann, 2002) and is a potent inhibitor of mitochondrial function (Bolanos et al., 1997; Brown and Borutaite, 2002; Duchen, 2004), thereby reducing the production of ATP and causing failure of the Na^+/K^+ ATPase or sodium pump. Thus axons exposed to nitric oxide in acute lesions may be susceptible to destruction by the sodium load resulting from sustained impulse activity (Kapoor et al., 2003; Smith et al., 2001). Indeed, an excessive intra-axonal sodium ion concentration can again result in a lethal influx of calcium ions via reverse action of the sodium/calcium exchanger (Stys, 2004). Thus impulse activity in the presence of NO could initiate a sequence of events that culminates in the digestion of the axon (Bechtold and Smith, 2005).

It was proposed (Kapoor et al., 2003; Smith et al., 2001) that NO injures axons by damaging the mitochondria within them, which leads to a reduction in ATP levels with a subsequent decrease in Na^+/K^+ ATPase activity, thus limiting the ability of axons to extrude Na^+. This hypothesis predicts that Na^+ channel blockers should protect axons from NO-induced injury, and this is indeed the case. It has been revealed that, in a model of inflammatory demyelination, axons exposed to nitric oxide degenerate when electrically activated, a process which is attenuated by sodium channel blockers, such as flecainide (Kapoor et al., 2003; Smith et al., 2001).

Neuronal viability can be affected because inflammation damages mitochondrial DNA and impairs the mitochondrial enzyme complexes activity (Dutta et al., 2006; Lu et al., 2000). Recently it has been suggested that significant neuronal mitochondrial dysfunction occurs in demyelinated axons which leads to reduced ATP production (Dutta et al., 2006) and the possible failure of Na/K-ATPase in chronically demyelinated axons in MS. Mitochondrial dysfunction was also accompanied by a reduction in both γ-aminobutyric acid (GABA)-related gene transcripts and density of inhibitory interneurons (Dutta et al., 2006). These observations are consistent with the concept that a mismatch between energy demand and reduced energy supply of ATP causes degeneration of chronically demyelinated axons (Dutta and Trapp, 2007).

Axonal dysfunction can also be studied by other techniques. Great advances have been made in recent years in imaging the brain and spinal cord. This has led to greater diagnostic clarity but has also provided insight into pathophysiological processes that may be occurring throughout the duration of the disease. Specifically, magnetic resonance spectroscopy (MRS) is a technique which allows quantification of certain tissue metabolites in the brains of live patients. *N*-Acetyl aspartate (NAA) is found predominantly in neurons and the spectroscopic detection of the compound is a marker of axonal density in the nervous system (Bjartmar et al., 2000). Reductions in brain NAA levels have been detected in acute active lesions of MS, chronic active lesions and so-called normal appearing white matter (NAWM) of patients with the disease (Davie et al., 1997; Leary et al., 1999). Furthermore, persistent neurological disability correlates to reductions in NAA levels, an observation which has led to the hypothesis that permanent disability in MS is caused by axonal loss (Bjartmar et al., 2000). A more recent study found that diffusion MRI shows changes consistent with sensitivity to axonal loss, but that relative NAA changes are not necessarily related to this. Furthermore, axonal metabolic function, independent of structural integrity, may be a major determinant of NAA measures in MS (Cader et al., 2007). An 'in vivo' marker of axonal integrity is a vital component of future trials which aim to address neurodegenerative processes in the disease.

Several potential biological markers of axonal injury have been identified in MS and EAE (Gresle et al., 2008). One such marker is the presence of neurofilaments within the cerebrospinal fluid (CSF; Petzold, 2005). It has been postulated that the presence of these molecules within the CSF may be an indicator of axon destruction within the CNS: when axons degenerate neurofilaments are released into the extracellular fluid space and from there enter the CSF. The precise relationship between CSF concentrations of neurofilaments and the degree of axonal degeneration was previously unknown. However, a more recent study has found elevated serum levels of the highly phosphorylated forms of the heavy neurofilament subunit pNF-H in acute, but not chronic, EAE and this elevation is highly correlated with histological verification of axonal loss (Gresle et al., 2008). Although more work is required on changes in neurofilament levels in the CSF as the disease progresses, there is a suggestion that levels increase during progressive phases of multiple sclerosis (Petzold et al., 2005). Indeed, this has been observed in murine MOG-EAE whereby a concurrent dephosphorylation of neurofilaments occurs during disease progression (Herrero-Herranz et al., 2008).

16.4 Neurofilaments and MS

As discussed, changes in neurofilament (NF) phosphorylation states have been detected in axons and taken as evidence of pathological dysfunction. Below we will discuss axonal structure, changes in neurofilament phosphorylation levels in multiple sclerosis and the influence of myelin and oligodendrocytes on neurofilament phosphorylation in relation to putative changes occurring in multiple sclerosis.

The axonal cytoskeleton is composed of scaffolding filaments: actin microfilaments, microtubules and intermediate filaments (IFs). Originally, it was assumed that NF were composed of neuron-specific IFs which consist of only three subunits – a light chain (NF-L), a medium chain (NF-M) and a heavy chain (NF-H) differing in molecular weight by virtue of their hypervariable tail domains (Fuchs and Weber, 1994). Recently it has been suggested that a fourth functional subunit; α-internexin, may be incorporated into the mature fibre as well (Yuan et al., 2006) and is a key determinant for the axonal transport of NF in the CNS (Yuan et al., 2003). Furthermore, another IF protein called peripherin in the peripheral nervous system (PNS) can co-assemble with NFL, NFM and NFH to form NF (Beaulieu et al., 1999; Yuan et al., 2006). NFs also contribute to the dynamic properties of the axonal cytoskeleton during neuronal differentiation, axon outgrowth, regeneration and guidance (Nixon and Shea, 1992).

Phosphorylation is the best documented post-translational modification of NF proteins (Perrot et al., 2008). Phosphorylation of neurofilaments is topographically regulated, with increased phosphorylation in axons and little or no phosphorylation in cell bodies and dendrites (Glicksman et al., 1987; Nixon et al., 1994; Nixon and Shea, 1992; Oblinger et al., 1987; Sternberger and Sternberger, 1983). Shortly after the synthesis of NF proteins, phosphorylation of the head domain occurs mainly in the cell body while the phosphorylation of tail domains coincides with their entry into the axon (Nixon and Lewis, 1986; Nixon et al., 1989, 1987).

The phosphorylation of neurofilaments regulates their axonal transport rate and in doing so provides stability to mature axons (Sihag et al., 2007). The NF-M and NF-H tail domains are the most extensively phosphorylated regions (Perrot et al., 2008). The C-terminal region of NF-M and NF-H contain 10–15 and 40–50 lysine-serine-proline (KSP) repeats, respectively, acting as potential phosphorylation sites (Veeranna et al., 1998) although other Ser-Thr-containing motifs are also phosphorylated (Perrot et al., 2008). The interactions of NFs with each other and other cytoskeletal components is regulated by the phosphorylation of these carboxyl-terminal 'side-arms' which consequently mediates the formation of a cytoskeletal lattice that supports the mature axon (Grant and Pant, 2000; Hirokawa et al., 1984; Julien and Mushynski, 1998; Nixon, 1993, 1998; Nixon and Sihag, 1991). In brief, phosphorylation occurs via a number of kinases and is predominantly regulated within the axonal compartment (Grant and Pant, 2000). Evidence suggests that phosphorylation of KSP motif rich tail domains on NF-M and NF-H might be regulated by the activation of proline-directed kinases (ERK1/2, Cdk5, p38 MAP kinase or SAPK), by signal transduction cascades triggered by growth factors (Li et al., 1999b), Ca^{2+} influx (Li et al., 1999a), extracellular matrix constituents through integrins (Li et al., 2000) and glial factors like myelin-associated glycoprotein (Dashiell et al., 2002; Sihag et al., 2007). NF-H is the most extensively phosphorylated protein in the human brain and the process of phosphorylation and dephosphorylation is highly complex.

During development, NF phosphorylation appears to be a key factor in triggering NF accumulation and thus the formation of axons (Sanchez et al., 2000). Phosphorylation of NFs also leads to slowing of NF transport, promotion of NF alignment and an increase in inter-NF spacing, all of which contribute to axon structure, function and stability (Leterrier et al., 1996; Yabe et al., 2001). Virtually all NF-H KSP sites in mature, healthy axons are phosphorylated in vivo, a process which occurs exclusively in axons (Elhanany et al., 1994).

Neurofilaments are slowly transported down the axon and the velocity of transport is inversely proportional to degree of phosphorylation (Nixon et al., 1994). This has been confirmed by the demonstration that NFs undergo a series of rapid bursts at a fast rate, interspersed with prolonged pauses that average out as slow transport (Roy et al., 2000; Sihag et al., 2007; Wang et al., 2000). The association of NFs with kinesin and dynein (Shah et al., 2000) indicates that they may be the motor proteins utilized during slow axonal transport (Barry et al., 2007). It is thought that when neurofilaments reach the axon terminus they are then degraded through protease dependent mechanisms. Proteases with affinity to NF proteins including calcium-activated proteases are present in axons (Barry et al., 2007). High levels of phosphorylation within neurofilaments are however protective against proteolysis (Goldstein et al., 1987). The extent of NF proteolysis occurring at the synapse and/or locally along the axons in vivo is unknown (Barry et al., 2007), as is whether this process underlies axon degeneration in multiple sclerosis.

Neurofilament dephosphorylation has been used as a marker for axonal pathology in multiple sclerosis and in experimental models of demyelination. The precise relationship between neurofilament dephosphorylation and axonal degeneration is not known. Nor is it known exactly why neurofilaments become dephosphorylated. Antibodies directed against different phosphoisoforms of neurofilament have been used to study the process in multiple sclerosis. Most commonly, antibodies against dephosphorylated neurofilaments (SMI32; Sternberger Monoclonals, USA) have been used and shown to mark axonal abnormalities in the disease (Trapp et al., 1998). In this study, neurofilament dephosphorylation was seen in both active inflammatory lesions and also in chronic active lesions. SMI32 also marked terminal axonal swelling or 'spheroids' which represent transected axons and thus irreversible axonal destruction. Quantification of terminal axonal spheroids identified a strong correlation between axonal transection and the inflammatory activity of the lesion. This provides evidence that axonopathy in the disease may occur early and be related, to some extent, to inflammatory induced demyelination (Trapp et al., 1998) through mediators such as nitric oxide and reactive oxygen species, cytokines and proteases (Aboul-Enein et al., 2006; Garthwaite et al., 2002).

16.5 Mechanisms of Axonal Injury

To date, the mechanisms responsible for axonal loss after demyelination are incompletely understood. However, a broad range of factors including proteolytic enzymes, cytokines, nitric oxide and persistent sodium influx have been suggested to contribute to axonal degeneration after demyelination (for reviews see Bjartmar and Trapp, 2001; Lassmann, 2003; Smith and Lassmann, 2002; Waxman et al., 2004).

A close association exists between axonal damage and the presence of activated microglia/macrophages (Bitsch et al., 2000; Ferguson et al., 1997; Kuhlmann et al., 2002; Trapp et al., 1998). These observations suggest that axons are targeted by macrophages and activated microglial cells. However, whether these cells are directly attacking axons or only removing debris is not certain. It seems unlikely that there is a specific immunologic attack against axons because most axons survive the acute demyelinating process (Dutta and Trapp, 2007). Recently it has been suggested that the axonal injury is due to bystander damage associated with inflammatory mediators such as tumour necrosis factor α (TNF α), other cytokines, proteases, oxidants and nitric oxide playing a key role (Anthony et al., 1997; Benveniste, 1998; Gray et al., 2008; Ruuls et al., 1995; Smith and Lassmann, 2002).

One such oxidant, nitric oxide (NO), serves many physiological functions within the nervous system, including roles in synaptic plasticity, long-term potentiation and neurotransmitter release, but at sites of inflammation high concentrations of NO are thought to mediate cell death (Smith and Lassmann, 2002). Microglial derived NO has been shown to be neurotoxic in vitro (Bal-Price and Brown, 2001; Golde et al., 2002). Mechanisms underlying this neurotoxicity are complex, but NO activates a number of intracellular signalling pathways including MAPkinases and may ultimately lead to neuronal death via inhibition of mitochondrial respiration (Brown and Borutaite, 2002; Ghatan et al., 2000). Recent in vivo work has shown that NO may also mediate activity dependent axon destruction (Smith et al., 2001). Furthermore, pathological studies have demonstrated activity of inducible nitric oxide synthetase (iNOS) within acute (Bagasra et al., 1995) and chronic active multiple sclerosis lesions (Hill et al., 2004).

NO has also been shown to have effects on neurofilament phosphorylation levels within axons in culture (Wilkins and Compston, 2005). In cultured cortical neurons, activation of p38 MAPkinase by NO leads to axonal destruction and reduced levels of phosphorylated neurofilaments (via inhibition of MAPkinase/Erk signalling). Conversely, activation of MAPkinase/Erk and inhibition of p38 MAPkinase signalling by growth factors leads to attenuation of NO mediated axonal destruction.

It is unclear whether axon drop-out occurs in the progressive phase of the disease as a result of previous inflammatory damage to axons, as a result of low grade inflammation occurring causing damage to already vulnerable demyelinated axons, as a result of loss of trophic environment for axons to survive or as part of a completely independent neurodegenerative process. However, it has recently been suggested that progressive axonal degeneration is attributable to adaptive but ultimately damaging changes in sodium channel distribution (Craner et al., 2004b) and oligodendroglial-derived neurotrophic factor deprivation (Gresle et al., 2008; Rasband et al., 2005; Sahenk et al., 1999).

It is possible that axons degenerate in MS without demyelination (Waxman, 2006). Recent evidence has shown that there may be little correlation between plaque load and axon loss in post-mortem specimens of multiple sclerosis tissue, which raises the possibility that demyelination is not necessarily the primary determinant of axonal degeneration in the disease (DeLuca et al., 2006). Indeed, the Na^+–Ca^{2+} exchanger is present at intact nodes where $Na_V1.6$ channels are aggregated in normal white matter (Steffensen et al., 1997). Therefore if the reduction in ATP production described by Dutta et al. (2006) occurs in non-demyelinated axons, the axons might then undergo Ca^{2+} mediated injury (Waxman, 2006).

It has also been suggested that the pathogenesis of axonal loss in normal appearing white matter (NAWM) may occur via two distinct mechanisms. It may result from a Wallerian degenerative process in plaques, where inflammation and demyelination are periodically exacerbated. Alternatively, it may be a neurodegenerative process where a diffuse axonopathy occurs independently of inflammatory demyelination, contributing significantly to loss of axons and producing the inexorable accumulation of disability that is often seen in the progressive phase of the disease

(DeLuca et al., 2006). Certain axonal tracts also appear more susceptible to injury in multiple sclerosis (DeLuca et al., 2004). Taken together, these pathological studies, although only providing a 'snapshot' of a dynamic disease, argue that demyelination may not be the primary determinant of axon loss and axonal degeneration may be driven by an underlying neurodegenerative process.

How to tie up these seemingly disparate pathological mechanisms? One might hypothesise that inflammation, causing widespread tissue damage, contributes to early axon destruction via molecules such as nitric oxide. However, destruction of oligodendrocytes and myelin also occurs at this stage and may 'set up' the later process of chronic axonal drop-out. Repeated or persistent demyelination in lesions which have failed to remyelinate effectively, as well as the lack of trophic support from myelin and oligodendrocytes in chronically demyelinated axons, may contribute to axonal loss in chronic MS.

Late-onset axonal pathology has been observed in mice lacking certain myelin proteins (MAG and PLP) (Sadahiro et al., 2000; Yoshikawa, 2001); therefore, secondary wallerian degeneration may also contribute to diffuse axonal loss (Lassmann, 2003). Myelin mutant studies have shown that myelin is an important determinant of long-term axonal survival (see below). Thus myelin and oligodendrocyte loss may leave axons particularly susceptible to degeneration via this mechanism. Furthermore, such axons may be rendered particularly susceptible to very low levels of inflammation. This may account for the lack of correlation between inflammation and axon loss in later stages of the disease. Thus the microenvironment of the axon may be critical to determine its capacity for long-term survival. In other words, there may be a balance between the deleterious effects of inflammation and other toxic mediators and the trophic, axonoprotective environment of surrounding cells. Once that balance is tipped towards favouring toxic inflammation, axon degeneration may proceed inexorably. The goals of axonoprotective treatments, therefore, may be to tip the balance in the opposite direction by restoring the trophic environment surrounding the axon.

16.6 Oligodendrocyte Influences on Axonal Survival and Neurofilament Phosphorylation

Thus, one interpretation for post-inflammatory axon loss is that the environment in which axons find themselves is no longer sufficient supportive to maintain them. Specifically the severe loss of myelin and oligodendrocytes within chronic lesions may lead to an environment in which the axon lacks necessary survival factors. Indeed there has been increasing evidence for an axonotrophic role of central nervous system myelin and that long-term axonal survival requires the support of ensheathing glial cells, both in the central and in the peripheral nervous system (Griffiths et al., 1998; Lappe-Siefke et al., 2003; Nave and Trapp, 2008). Furthermore, oligodendrocytes have been shown to produce soluble growth factors which influence the survival and phenotype of axons. Although, the molecular mechanisms through which glial cells preserve axons are partially unknown, they are likely to involve a variety of essential axon–glia interactions (Kassmann et al., 2007).

16.6.1 Myelin Has Neurotrophic Properties

Myelin has neurotrophic properties in both the peripheral and central nervous system, although mechanisms differ in these two sites (Yin et al., 2006). Myelin mutants have provided insights into processes by which myelin is important in axon development and maintenance. However, caution should be exercised when translating the study of myelin mutants to consider chronic progressive stages of multiple sclerosis as axons of mutant animals are exposed to abnormal myelin throughout their life.

In the peripheral nervous system, Schwann cells influence peripheral axonal structure, as hypomyelinating sciatic nerves show marked reductions in NF phosphorylation and density, with associated decrease in axonal calibre (Cole et al., 1994). These nerves show no changes in total NF content, suggesting Schwann cell myelin has direct influences on the process of neurofilament phosphorylation. In a similar way, mice deficient in myelin-associated glycoprotein (MAG) have reduced peripheral axonal calibres, reduced NF phosphorylation and decreased inter-NF spacing, again implying a role for myelin components in NF regulation within the peripheral nervous system and evidence that MAG provides trophic support to axons (Yin et al., 1998).

In the central nervous system, myelin also significantly influences axon development. Central myelin is composed of several proteins including proteolipid protein (PLP) and DM20 (an alternate spliced isoform), myelin basic protein (MBP), myelin associated glycoprotein (MAG), myelin-oligodendrocyte glycoprotein (MOG) and 2′,3′-cyclic nucleotide 3′-phosphodiesterase (CNPase). Indeed, several of these proteins contribute to long-term viability of axons (Bjartmar et al., 1999). The absence of central myelin leads to severe disability and pronounced axonal dystrophy, as demonstrated by the *Shiverer* mouse which contains a deletion in the MBP gene and thus produces no MBP protein. This leads to the complete absence of compact CNS myelin and a phenotype of tremor and seizures with decreased life span of the mice. CNS axonal changes occur within these animals, including an increase in slow axonal transport, microtubule density and number, and altered neurofilament assembly and phosphorylation (Brady et al., 1999; Kirkpatrick et al., 2001; Sanchez et al., 2000). The effect appears dose dependent, since mice expressing 25% of wild-type MBP, with thin compact myelin sheaths, show an intermediate phenotype.

Mice containing other mutations in myelin proteins have been produced which show abnormalities in structure of the myelin, but nevertheless some degree of myelination. These mutants may help understand which particular components of myelin are important in the long-term maintenance of axons.

At present, the exact role of PLP or its alternative spliced isoform DM20 in axonal preservation is unknown (Nave and Trapp, 2008). PLP is localised predominantly in compact myelin and is its major structural component. Mice with the *rumpshaker* mutation in Plp1 (Schneider et al., 1992) exhibit late-onset axonal degeneration (Edgar et al., 2004a), suggesting that rumpshaker DM20 may be sufficiently folded to reach the myelin compartment but unable to support fully the integrity of the axon (Nave and Trapp, 2008). These mice are hypomyelinated and are long lived (Nave and Trapp, 2008). Interestingly, mice lacking the Plp gene (X chromosome) are able to synthesise large amounts of myelin, which only show morphological defects in the interperiod line (the area in compact myelin where external leaflets of two adjacent plasma membranes are closely apposed). These mice do however show impairment of axonal transport mechanisms and late onset axonal degeneration with 'ovoid' formation (Griffiths et al., 1998). Such axonal swellings occur in fully myelinated axons and are either organelle rich or filled with non-phosphorylated neurofilaments. Abnormal swellings likely reflect the impairment of fast retrograde transport (Edgar et al., 2004b) leading to a progressive ('Wallerian') degeneration in the presence of normal amounts of CNS myelin (Kassmann and Nave, 2008). However, after transplantation of Plp mutant oligodendrocytes into *shiverer* white-matter tracts, axonal swellings are reduced (Edgar et al., 2004b). Interestingly, it appears that axons may become dependent on oligodendrocyte support sometime after myelination has been completed, as non-myelinated axons within optic nerves from chimeric females (which show patches of normally myelinated and non-myelinated axons) never develop axonal swellings over the expected time course. This implies a role for PLP in maintaining axons that have become oligodendrocyte dependent, although the signal triggering this putative dependency remains unknown. Further studies of chimeric females have revealed a role for PLP in early stages of axon–oligodendrocyte interactions, initiating the ensheathment and myelination of a proportion of small-diameter axons (Yool et al., 2001). The specific role for PLP in the central nervous system has been highlighted by a recent report in which mice were engineered to express the peripheral myelin component P_0 instead of PLP centrally (Yin et al., 2006). Mice lacking PLP (but expressing P_0 centrally) showed a structure and periodicity of myelin which resembled peripheral myelin, yet had severe disability and degeneration of myelinated axons. Those mice expressing equal amounts of PLP and P_0 showed normal phenotype and no axonal degeneration. This implies a specific role of central nervous system myelin (containing PLP) in neuroprotection of central axons, and highlights the differences between the requirements of central and peripheral axons.

Another example of the differences between central and peripheral axonal requirements comes from the study of mice lacking MAG. Axonal abnormalities are not a prominent feature within the central nervous system of these mice, whereas chronic atrophy, changes in calibre and alterations in neurofilament organization of peripheral axons occur (Li et al., 1994; Yin et al., 1998).

A second gene recently associated with oligodendroglial support of axons that is specifically expressed by myelinating glial cells is Cnp1 (Nave and Trapp, 2008). CNPase, a further structural molecule of myelin, has also been shown to have a role in axon–glial interactions (Lappe-Siefke et al., 2003). Mice lacking CNPase have morphologically normal myelin that exhibits a normal periodicity, but show delayed axonal swelling and degeneration, associated with signs of inflammation (microgliosis) again suggesting that specific interactions between myelin components and the axon may determine long-term survival of the axon. Prior to the onset of clinical symptoms many oligodendroglial paranodes become disorganised (Rasband et al., 2005) which implies a crucial role of CNP in the normal communication between axons and oligodendrocytes. The presence of axonal swellings is likely to represent an energy-related problem (Nave and Trapp, 2008).

These observations indicate that chronically demyelinating axons may undergo degeneration due to lack of trophic support from myelin or myelin forming cells (Bjartmar et al., 2003). Furthermore, they suggest that the presence of structurally normal myelin may be required for long-term axonal survival and that its absence will lead to chronic and slow axonal dropout. Whether this translates to the pathophysiology of axonal degeneration in chronic demyelinated lesions is unclear. Axons within myelin mutants are never exposed to structurally normal myelin, and so may develop different trophic requirements to axons ensheathed with structurally normal myelin.

At present the molecular mechanisms that underlie the glial support of axons remain to be clarified. With respect to long axonal tracts, it has recently been hypothesised that mitochondrial function and the generation of ATP at a far distance from the neuronal cell soma could partially rely on the local supply of energy-rich metabolites generated by the axon-associated glial cells (Kassmann and Nave, 2008). This would be likely to require the existence of a complex transport machinery involving Plp1, Cnp 1 and other genes expressed by oligodendrocytes. Indeed, a recent study demonstrated that a primary function of PLP is to serve as a membrane-bound carrier of other proteins into the myelin sheath (Werner et al., 2007). Thus, PLP and perhaps other oligodendrocyte factors may play a key role in the stability of axons (Brinkmann et al., 2008; Franklin and Ffrench-Constant, 2008).

16.6.2 Oligodendrocyte Derived Growth Factors Influence Axonal Survival and Neurofilament Phosphorylation

Influences of myelin on neurofilament phosphorylation have been shown although the exact influence of individual myelin protein on the process are unclear. Furthermore, the mechanism by which neurofilaments become dephosphorylated following demyelination is unknown.

Growth factors have effects on neurofilament phosphorylation levels and there is increasing evidence that growth factors released by oligodendrocytes may influence axonal phenotype in this way. The close proximity of the oligodendrocyte to the axon would imply the possibility of cross talk between the two cell types. Indeed, axon-derived factors are known to influence oligodendrocytes in a way which leads to effective functioning of glial-neuronal sub-units.

Neurons secrete factors which influence oligodendrocyte behaviour and phenotype. Specifically, axons have been shown to exert a proliferative effect on oligodendrocyte precursor cells. This was first shown by transection studies on the rat neonatal optic nerve (David et al., 1984). Experimental transection of axons leads to a decrease in the yield of oligodendrocytes that may be cultured from the nerve. Several factors have subsequently been identified as being important in this process, including PDGF, basic FGF and members of the neuregulin family (Bogler et al., 1990; Canoll et al., 1996). Axons also promote oligodendrocyte survival (Barres and Raff, 1999). This observation is inferred from several studies of post-natal optic nerve transection. For instance, cutting the P8 optic nerve behind the eye leads to a fourfold increase in oligodendrocyte apoptosis, detected 3 days after transection (Barres et al., 1993). However, oligodendrocyte death induced by transection of the optic nerve is abrogated by locally elevated neuregulin concentrations (Fernandez et al., 2000). Neuregulins (Nrg1) are also crucially important in determining myelin sheath thickness and matching it to axon diameter (Michailov et al., 2004). In the CNS, however, the role of neuregulins is less apparent with recent work suggesting that NRG1 is not required as an instructive myelination signal and that other signals must also play a role in the precise relationship between axon and oligodendrocyte (Brinkmann et al., 2008; Franklin and Ffrench-Constant, 2008).

Thus, signalling between neurons and cells of the oligodendrocyte lineage appears to be important in matching oligodendrocyte numbers to axonal surface area and also in ensuring that the correct cellular substrates are in place before the onset of myelination and formation of mature myelinating units.

In a similar way, oligodendrocytes secrete soluble factors which influence axonal survival and phenotype. This phenomenon has been elucidated by a series of in vitro experiments, and their significance in vivo has yet to be fully identified (Du and Dreyfus, 2002). Compared to well documented and extensive studies of astrocyte-derived neurotrophic production, oligodendrocytes have traditionally been considered to be less important providers of such factors in the central nervous system. Neurotrophins compose a family of target-derived growth factors (nerve growth factor (NGF), BDNF, NT3 and NT4/5) well-known for their effect on neuronal survival (Nave and Trapp, 2008). Byravan et al. (1994) initially showed nerve growth factor (NGF) and brain-derived neurotrophic factor (BDNF) mRNA production by an immortalized oligodendrocyte cell line. This observation was prompted by observations of the ability of this line to promote neurite extension in PC12 cells. Furthermore, these workers showed co-localisation of NGF protein in both oligodendrocyte precursor cells and mature oligodendrocytes. A later study showed expression of NGF, BDNF and neurotrophin-3 (NT-3) mRNA in cultures of basal forebrain oligodendrocytes (Dai et al., 2003). Expression of these mRNAs may be altered by factors such as glutamate and potassium, suggesting that neuronal signals may provide feedback to regulate oligodendrocyte-mediated neurotrophism. Shinar and McMorris (Shinar and McMorris, 1995) showed insulin-like growth factor type-1 (IGF-1) mRNA production by cultured oligodendrocytes. An extensive study of glial cell-line derived neurotrophic factor (GDNF) family member mRNA production in cells of oligodendrocyte lineage has shown GDNF mRNA production in oligodendrocyte cell line and within differentiated primary oligodendrocyte cultures (Strelau and Unsicker, 1999).

The functional significance of these findings has been shown by several studies, suggesting that oligodendrocyte derived factors improve the survival of neurons and, more specifically, axons in culture. Meyer-Franke et al. (1995) detected an as yet uncharacterised soluble protein of >30 kDa derived from mature oligodendrocytes that improved the survival of post-natal retinal ganglion cells. Later studies confirmed IGF-1 production by cells of the oligodendrocyte lineage and oligodendrocyte derived IGF-1 improves the survival of stressed neurons under conditions of trophic factor deprivation (Wilkins et al., 2001). Furthermore, oligodendrocyte derived factors also increase levels of phosphorylated neurofilament within cultured neurons and thus aid axonal survival under the same conditions of trophic factor deprivation (Wilkins et al., 2003). Specifically, GDNF is the growth factor that appears to be of importance in this phenomenon and GDNF is only produced by differentiated oligodendrocytes and not precursor cells. GDNF promotes increased levels of neurofilament phosphorylation via MAPkinase/Erk signalling pathways, as inhibition of this pathway attenuates the process and GDNF strongly activates Erk signalling within neurons. Mechanisms by which neurofilaments become phosphorylated are complex, but a role for MAPkinase/Erk has been established (Li et al., 1999b; Pang et al., 1995; Veeranna et al., 1998). Using a similar approach, oligodendrocyte derived growth factors, specifically GDNF, have also been shown to attenuate nitric oxide mediated destruction of axons, again via MAPkinase/Erk signalling (Wilkins and Compston, 2005). Again addition of these factors significantly increases levels of phosphorylated neurofilament within cultured axons. This study emphasises the interaction of several different signalling pathways which are involved in neuroprotective activities. p38 MAPkinase is strongly activated by addition of nitric oxide to neurons, yet inhibition of p38 leads to activation of MAPkinase/Erk pathways. This implies a central role for Erk activation in protection of axons from inflammatory insults.

These studies have shown considerable neurotrophic and axonotrophic effects of oligodendrocytes. Both myelin and oligodendrocyte derived growth factors exert influences on long-term neuronal survival, specifically of the axonal subunit, and influence neurofilament phosphorylation within the axon.

16.6.3 Oligodendrocytes Preserve Axonal Function Through Metabolic Support

At present, the most likely mechanism that underlies the preservation of axons by ensheathing glial cells may be metabolic support (Spencer et al., 1979). Mitochondria have been found to be present in increased numbers in dysmyelinated and demyelinated axons (Andrews et al., 2006; Mutsaers and Carroll, 1998). The majority of ATP is required not only for membrane repolarisation but for the fast axonal transport of organelles (Hollenbeck and Saxton, 2005) such as mitochondria, which travel with frequent stops and restarts (Misgeld et al., 2007). Mitochondria often pause at the node of Ranvier (Fabricius et al., 1993) where they can be found in close proximity to the glial paranodal loops that form a highly specialised axon-glial junction (Nave and Trapp, 2008). Caspr-deficient mice that lack the normal paranodal structure retain even more mitochondria beneath the disrupted paranode, many of which exhibit a swollen morphology (Einheber et al., 2006). This observation suggests that glial contact is partially responsible for the regulation of axonal mitochondrial transport and that myelinating glial cells are required for the normal functioning of mitochondria within axons (Nave and Trapp, 2008). Another physiological function of myelinating oligodendrocytes is hypothesised to be the detoxification of reactive oxygen species (ROS) (Kassmann and Nave, 2008).

A recent study explored the role of peroxisomes in myelinated fibre tracts (Kassmann et al., 2007). Peroxisomes are ubiquitous cellular organelles responsible for the detoxification of reactive oxygen species, plasmalogen synthesis and β-oxidation of fatty acids, in particular that of very long chain fatty acids (VLCFA). *pex-5* deficient mutants lacking an essential peroxisome biogenesis factor (PEX5), selectively in oligodendrocytes, developed normally but within several months showed considerable secondary changes (Kassmann et al., 2007). These mice exhibit axonal swellings followed by progressive subcortical demyelination. Unlike other mouse mutants, an unexpected degree of CNS inflammation occurs, marked not only by microgliosis and astrogliosis, but also by the infiltration of leukocytes. There is no B or T cell infiltration in PLP-deficient and CNP-deficient mice (Griffiths et al., 1998; Lappe-Siefke et al., 2003). Such observations demonstrate that peroxisomal dysfunction in oligodendrocytes may contribute to, if not trigger, inflammation (Kassmann et al., 2007). Indeed, peroxisomes play a major role in the degradation of eicosanoids (Diczfalusy, 1994) that are known to be potent mediators of inflammation (Funk, 2001; Goodarzi et al., 2003). Therefore, it is possible that peroxisomal dysfunction in oligodendrocytes may lead to the local accumulation of these metabolites and the subsequent activation of microglia and to leukocyte infiltration. A challenge will be to discover if the altered lipid metabolism in peroxisome-deficient oligodendrocytes (e.g. reduced β-oxidation of fatty acids) may contribute to a reduced metabolic support of axons and therefore explain the degeneration of long axons. The authors did, however, conclude that peroxisomes provide oligodendrocytes with an essential protective function against axon degeneration and neuroinflammation.

16.7 Conclusions

Axon loss and dysfunction underlies clinical progression in multiple sclerosis. Understanding the mechanisms involved in the process of axonopathy is vital in order to devise therapies which address specific pathophysiological processes. The concept of several mechanisms occurring within the time course of the disease probably belies the failure of current therapies which only address the inflammatory phase of the disease. Non-inflammatory processes are likely to be important in later stages of the disease and may be responsible for clinical disease progression to a large degree. This 'neurodegenerative' phase of the disease is now a major target for research and, in the coming years, treatments will need to address this aspect. Potential treatments include remyelination therapies, delivery of growth factors or neuroprotectant drug therapies. Of these, remyelination has been the subject of extensive research. In theory, remyelination has the potential to address many aspects of disease pathology and offers the hope of recovery from symptoms. In reality, the choice of remyelinating cell, achieving satisfactory remyelination and targeting multiple lesions which are widely disseminated in space, are major obstacles preventing it becoming an imminent therapeutic option. Drugs and growth factor treatments which address axon degeneration and neurofilament dephosphorylation also seem likely to become attractive additions to the clinician's range of options available to treat patients with progressive disease. With the expansion in knowledge concerning axon pathology in multiple sclerosis such treatment options may not be far away.

Acknowledgment EG is supported by a Junior Fellowship from the Multiple Sclerosis Society (UK).

References

Aboul-Enein F, Weiser P, Hoftberger R, Lassmann H, Bradl M (2006) Transient axonal injury in the absence of demyelination: a correlate of clinical disease in acute experimental autoimmune encephalomyelitis. Acta Neuropathol 111:539–547

Andrews H, White K, Thomson C, Edgar J, Bates D, Griffiths I, Turnbull D, Nichols P (2006) Increased axonal mitochondrial activity as an adaptation to myelin deficiency in the Shiverer mouse. J Neurosci Res 83:1533–1539

Anthony DC, Ferguson B, Matyzak MK, Miller KM, Esiri MM, Perry VH (1997) Differential matrix metalloproteinase expression in cases of multiple sclerosis and stroke. Neuropathol Appl Neurobiol 23:406–415

Back SA, Tuohy TM, Chen H, Wallingford N, Craig A, Struve J, Luo NL, Banine F, Liu Y, Chang A, Trapp BD, Bebo BF Jr, Rao MS, Sherman LS (2005) Hyaluronan accumulates in demyelinated lesions and inhibits oligodendrocyte progenitor maturation. Nat Med 11: 966–972

Bagasra O, Michaels FH, Zheng YM, Bobroski LE, Spitsin SV, Fu ZF, Tawadros R, Koprowski H (1995) Activation of the inducible form of nitric oxide synthase in the brains of patients with multiple sclerosis. Proc Natl Acad Sci USA 92:12041–12045

Bal-Price A, Brown GC (2001) Inflammatory neurodegeneration mediated by nitric oxide from activated glia-inhibiting neuronal respiration, causing glutamate release and excitotoxicity. J Neurosci 21:6480–6491

Barres BA, Jacobson MD, Schmid R, Sendtner M, Raff MC (1993) Does oligodendrocyte survival depend on axons? Curr Biol 3:489–497

Barres BA, Raff MC (1999) Axonal control of oligodendrocyte development. J Cell Biol 147:1123–1128

Barry DM, Millecamps S, Julien JP, Garcia ML (2007) New movements in neurofilament transport, turnover and disease. Exp Cell Res 313:2110–2120

Beaulieu JM, Robertson J, Julien JP (1999) Interactions between peripherin and neurofilaments in cultured cells: disruption of peripherin assembly by the NF-M and NF-H subunits. Biochem Cell Biol 77:41–45

Bechtold DA, Kapoor R, Smith KJ (2004) Axonal protection using flecainide in experimental autoimmune encephalomyelitis. Ann Neurol 55:607–616

Bechtold DA, Miller SJ, Dawson AC, Sun Y, Kapoor R, Berry D, Smith KJ (2006) Axonal protection achieved in a model of multiple sclerosis using lamotrigine. J Neurol 253: 1542–1551

Bechtold DA, Smith KJ (2005) Sodium-mediated axonal degeneration in inflammatory demyelinating disease. J Neurol Sci 233:27–35

Bechtold DA, Yue X, Evans RM, Davies M, Gregson NA, Smith KJ (2005) Axonal protection in experimental autoimmune neuritis by the sodium channel blocking agent flecainide. Brain 128:18–28

Benveniste EN (1998) Cytokine actions in the central nervous system. Cytokine Growth Factor Rev 9:259–275

Bitsch A, Schuchardt J, Bunkowski S, Kuhlmann T, Bruck W (2000) Acute axonal injury in multiple sclerosis. Correlation with demyelination and inflammation. Brain 123(6): 1174–1183

Bjartmar C, Kidd G, Mork S, Rudick R, Trapp BD (2000) Neurological disability correlates with spinal cord axonal loss and reduced *N*-acetyl aspartate in chronic multiple sclerosis patients. Ann Neurol 48:893–901

Bjartmar C, Kinkel RP, Kidd G, Rudick RA, Trapp BD (2001) Axonal loss in normal-appearing white matter in a patient with acute MS. Neurology 57:1248–1252

Bjartmar C, Trapp BD (2001) Axonal and neuronal degeneration in multiple sclerosis: mechanisms and functional consequences. Curr Opin Neurol 14:271–278

Bjartmar C, Wujek JR, Trapp BD (2003) Axonal loss in the pathology of MS: consequences for understanding the progressive phase of the disease. J Neurol Sci 206:165–171

Bjartmar C, Yin X, Trapp BD (1999) Axonal pathology in myelin disorders. J Neurocytol 28: 383–395

Black JA, Newcombe J, Trapp BD, Waxman SG (2007) Sodium channel expression within chronic multiple sclerosis plaques. J Neuropathol Exp Neurol 66:828–837

Blakemore WF, Gilson JM, Crang AJ (2003) The presence of astrocytes in areas of demyelination influences remyelination following transplantation of oligodendrocyte progenitors. Exp Neurol 184:955–963

Bogler O, Wren D, Barnett SC, Land H, Noble M (1990) Cooperation between two growth factors promotes extended self-renewal and inhibits differentiation of oligodendrocyte-type-2 astrocyte (O-2A) progenitor cells. Proc Natl Acad Sci USA 87:6368–6372

Bolanos JP, Almeida A, Stewart V, Peuchen S, Land JM, Clark JB, Heales SJ (1997) Nitric oxide-mediated mitochondrial damage in the brain: mechanisms and implications for neurodegenerative diseases. J Neurochem 68:2227–2240

Brady ST, Witt AS, Kirkpatrick LL, de Waegh SM, Readhead C, Tu PH, Lee VM (1999) Formation of compact myelin is required for maturation of the axonal cytoskeleton. J Neurosci 19: 7278–7288

Brand-Schieber E, Werner P (2004) Calcium channel blockers ameliorate disease in a mouse model of multiple sclerosis. Exp Neurol 189:5–9

Breij EC, Brink BP, Veerhuis R, van den Berg C, Vloet R, Yan R, Dijkstra CD, van der Valk P, Bo L (2008) Homogeneity of active demyelinating lesions in established multiple sclerosis. Ann Neurol 63:16–25

Brinkmann BG, Agarwal A, Sereda MW, Garratt AN, Muller T, Wende H, Stassart RM, Nawaz S, Humml C, Velanac V, Radyushkin K, Goebbels S, Fischer TM, Franklin RJ, Lai C, Ehrenreich H, Birchmeier C, Schwab MH, Nave KA (2008) Neuregulin-1/ErbB signaling serves distinct functions in myelination of the peripheral and central nervous system. Neuron 59:581–595

Brown GC, Borutaite V (2002) Nitric oxide inhibition of mitochondrial respiration and its role in cell death. Free Radic Biol Med 33:1440–1450

Byravan S, Foster LM, Phan T, Verity AN, Campagnoni AT (1994) Murine oligodendroglial cells express nerve growth factor. Proc Natl Acad Sci USA 91:8812–8816

Cader S, Johansen-Berg H, Wylezinska M, Palace J, Behrens TE, Smith S, Matthews PM (2007) Discordant white matter *N*-acetylasparate and diffusion MRI measures suggest that chronic metabolic dysfunction contributes to axonal pathology in multiple sclerosis. Neuroimage 36:19–27

Canoll PD, Musacchio JM, Hardy R, Reynolds R, Marchionni MA, Salzer JL (1996) GGF/neuregulin is a neuronal signal that promotes the proliferation and survival and inhibits the differentiation of oligodendrocyte progenitors. Neuron 17:229–243

Chang A, Nishiyama A, Peterson J, Prineas J, Trapp BD (2000) NG2-positive oligodendrocyte progenitor cells in adult human brain and multiple sclerosis lesions. J Neurosci 20:6404–6412
Cifelli A, Arridge M, Jezzard P, Esiri MM, Palace J, Matthews PM (2002) Thalamic neurodegeneration in multiple sclerosis. Ann Neurol 52:650–653
Cole JS, Messing A, Trojanowski JQ, Lee VM (1994) Modulation of axon diameter and neurofilaments by hypomyelinating Schwann cells in transgenic mice. J Neurosci 14:6956–6966
Coles AJ, Wing MG, Molyneux P, Paolillo A, Davie CM, Hale G, Miller D, Waldmann H, Compston A (1999) Monoclonal antibody treatment exposes three mechanisms underlying the clinical course of multiple sclerosis. Ann Neurol 46:296–304
Coman I, Aigrot MS, Seilhean D, Reynolds R, Girault JA, Zalc B, Lubetzki C (2006) Nodal, paranodal and juxtaparanodal axonal proteins during demyelination and remyelination in multiple sclerosis. Brain 129:3186–3195
Compston A (2006) Making progress on the natural history of multiple sclerosis. Brain 129: 561–563
Compston A, Coles A (2002) Multiple sclerosis. Lancet 359:1221–1231
Compston A, Coles A (2008) Multiple sclerosis. Lancet 372:1502–1517
Confavreux C, Vukusic S (2006) Natural history of multiple sclerosis: a unifying concept. Brain 129:606–616
Confavreux C, Vukusic S, Moreau T, Adeleine P (2000) Relapses and progression of disability in multiple sclerosis. N Engl J Med 343:1430–1438
Craner MJ, Hains BC, Lo AC, Black JA, Waxman SG (2004a) Co-localization of sodium channel Nav1.6 and the sodium-calcium exchanger at sites of axonal injury in the spinal cord in EAE. Brain 127:294–303
Craner MJ, Lo AC, Black JA, Waxman SG (2003) Abnormal sodium channel distribution in optic nerve axons in a model of inflammatory demyelination. Brain 126:1552–1561
Craner MJ, Newcombe J, Black JA, Hartle C, Cuzner ML, Waxman SG (2004b) Molecular changes in neurons in multiple sclerosis: altered axonal expression of Nav1.2 and Nav1.6 sodium channels and Na+/Ca2+ exchanger. Proc Natl Acad Sci USA 101:8168–8173
Dai X, Lercher LD, Clinton PM, Du Y, Livingston DL, Vieira C, Yang L, Shen MM, Dreyfus CF (2003) The trophic role of oligodendrocytes in the basal forebrain. J Neurosci 23: 5846–5853
Dashiell SM, Tanner SL, Pant HC, Quarles RH (2002) Myelin-associated glycoprotein modulates expression and phosphorylation of neuronal cytoskeletal elements and their associated kinases. J Neurochem 81:1263–1272
David S, Miller RH, Patel R, Raff MC (1984) Effects of neonatal transection on glial cell development in the rat optic nerve: evidence that the oligodendrocyte-type 2 astrocyte cell lineage depends on axons for its survival. J Neurocytol 13:961–974
Davie CA, Barker GJ, Thompson AJ, Tofts PS, McDonald WI, Miller DH (1997) 1H magnetic resonance spectroscopy of chronic cerebral white matter lesions and normal appearing white matter in multiple sclerosis. J Neurol Neurosurg Psychiatry 63:736–742
DeLuca GC, Ebers GC, Esiri MM (2004) Axonal loss in multiple sclerosis: a pathological survey of the corticospinal and sensory tracts. Brain 127:1009–1018
DeLuca GC, Williams K, Evangelou N, Ebers GC, Esiri MM (2006) The contribution of demyelination to axonal loss in multiple sclerosis. Brain 129:1507–1516
Diczfalusy U (1994) Beta-oxidation of eicosanoids. Prog Lipid Res 33:403–428
Du Y, Dreyfus CF (2002) Oligodendrocytes as providers of growth factors. J Neurosci Res 68: 647–654
Dubois-Dalcq M, Ffrench-Constant C, Franklin RJ (2005) Enhancing central nervous system remyelination in multiple sclerosis. Neuron 48:9–12
Duchen MR (2004) Roles of mitochondria in health and disease. Diabetes 53(Suppl 1):S96–S102
Dutta R, McDonough J, Yin X, Peterson J, Chang A, Torres T, Gudz T, Macklin WB, Lewis DA, Fox RJ, Rudick R, Mirnics K, Trapp BD (2006) Mitochondrial dysfunction as a cause of axonal degeneration in multiple sclerosis patients. Ann Neurol 59:478–489

Dutta R, Trapp BD (2007) Pathogenesis of axonal and neuronal damage in multiple sclerosis. Neurology 68:S22–S31; discussion S43–54

Edan G, Miller D, Clanet M, Confavreux C, Lyon-Caen O, Lubetzki C, Brochet B, Berry I, Rolland Y, Froment JC, Cabanis E, Iba-Zizen MT, Gandon JM, Lai HM, Moseley I, Sabouraud O (1997) Therapeutic effect of mitoxantrone combined with methylprednisolone in multiple sclerosis: a randomised multicentre study of active disease using MRI and clinical criteria. J Neurol Neurosurg Psychiatry 62:112–118

Edgar JM, McLaughlin M, Barrie JA, McCulloch MC, Garbern J, Griffiths IR (2004a) Age-related axonal and myelin changes in the rumpshaker mutation of the Plp gene. Acta Neuropathol 107:331–335

Edgar JM, McLaughlin M, Yool D, Zhang SC, Fowler JH, Montague P, Barrie JA, McCulloch MC, Duncan ID, Garbern J, Nave KA, Griffiths IR (2004b) Oligodendroglial modulation of fast axonal transport in a mouse model of hereditary spastic paraplegia. J Cell Biol 166:121–131

Einheber S, Bhat MA, Salzer JL (2006) Disrupted axo-glial junctions result in accumulation of abnormal mitochondria at nodes of ranvier. Neuron Glia Biol 2:165–174

Elhanany E, Jaffe H, Link WT, Sheeley DM, Gainer H, Pant HC (1994) Identification of endogenously phosphorylated KSP sites in the high-molecular-weight rat neurofilament protein. J Neurochem 63:2324–2335

Evangelou N, Esiri MM, Smith S, Palace J, Matthews PM (2000) Quantitative pathological evidence for axonal loss in normal appearing white matter in multiple sclerosis. Ann Neurol 47:391–395

Fabricius C, Berthold CH, Rydmark M (1993) Axoplasmic organelles at nodes of Ranvier. II. Occurrence and distribution in large myelinated spinal cord axons of the adult cat. J Neurocytol 22:941–954

Felts PA, Baker TA, Smith KJ (1997) Conduction in segmentally demyelinated mammalian central axons. J Neurosci 17:7267–7277

Ferguson B, Matyszak MK, Esiri MM, Perry VH (1997) Axonal damage in acute multiple sclerosis lesions. Brain 120(3):393–399

Fernandez PA, Tang DG, Cheng L, Prochiantz A, Mudge AW, Raff MC (2000) Evidence that axon-derived neuregulin promotes oligodendrocyte survival in the developing rat optic nerve. Neuron 28:81–90

Franklin RJ (2002) Why does remyelination fail in multiple sclerosis? Nat Rev Neurosci 3:705–714

Franklin RJ, Ffrench-Constant C (2008) Remyelination in the CNS: from biology to therapy. Nat Rev Neurosci 9:839–855

Frohman EM, Racke MK, Raine CS (2006) Multiple sclerosis–the plaque and its pathogenesis. N Engl J Med 354:942–955

Fuchs E, Weber K (1994) Intermediate filaments: structure, dynamics, function, and disease. Annu Rev Biochem 63:345–382

Funk CD (2001) Prostaglandins and leukotrienes: advances in eicosanoid biology. Science 294:1871–1875

Garthwaite G, Goodwin DA, Batchelor AM, Leeming K, Garthwaite J (2002) Nitric oxide toxicity in CNS white matter: an in vitro study using rat optic nerve. Neuroscience 109:145–155

Gendelman HE, Pezeshkpour GH, Pressman NJ, Wolinsky JS, Quarles RH, Dobersen MJ, Trapp BD, Kitt CA, Aksamit A, Johnson RT (1985) A quantitation of myelin-associated glycoprotein and myelin basic protein loss in different demyelinating diseases. Ann Neurol 18:324–328

Ghatan S, Larner S, Kinoshita Y, Hetman M, Patel L, Xia Z, Youle RJ, Morrison RS (2000) p38 MAP kinase mediates bax translocation in nitric oxide-induced apoptosis in neurons. J Cell Biol 150:335–347

Gilgun-Sherki Y, Panet H, Holdengreber V, Mosberg-Galili R, Offen D (2003) Axonal damage is reduced following glatiramer acetate treatment in C57/bl mice with chronic-induced experimental autoimmune encephalomyelitis. Neurosci Res 47:201–207

Glicksman MA, Soppet D, Willard MB (1987) Posttranslational modification of neurofilament polypeptides in rabbit retina. J Neurobiol 18:167–196

Golde S, Chandran S, Brown GC, Compston A (2002) Different pathways for iNOS-mediated toxicity in vitro dependent on neuronal maturation and NMDA receptor expression. J Neurochem 82:269–282

Goldstein ME, Sternberger NH, Sternberger LA (1987) Phosphorylation protects neurofilaments against proteolysis. J Neuroimmunol 14:149–160

Goodarzi K, Goodarzi M, Tager AM, Luster AD, von Andrian UH (2003) Leukotriene B4 and BLT1 control cytotoxic effector T cell recruitment to inflamed tissues. Nat Immunol 4:965–973

Grant P, Pant HC (2000) Neurofilament protein synthesis and phosphorylation. J Neurocytol 29:843–872

Gray E, Thomas TL, Betmouni S, Scolding N, Love S (2008) Elevated activity and microglial expression of myeloperoxidase in demyelinated cerebral cortex in multiple sclerosis. Brain Pathol 18:86–95

Gresle MM, Shaw G, Jarrott B, Alexandrou EN, Friedhuber A, Kilpatrick TJ, Butzkueven H (2008) Validation of a novel biomarker for acute axonal injury in experimental autoimmune encephalomyelitis. J Neurosci Res 86:3548–3555

Griffiths I, Klugmann M, Anderson T, Yool D, Thomson C, Schwab MH, Schneider A, Zimmermann F, McCulloch M, Nadon N, Nave KA (1998) Axonal swellings and degeneration in mice lacking the major proteolipid of myelin. Science 280:1610–1613

Hartung HP, Gonsette R, Konig N, Kwiecinski H, Guseo A, Morrissey SP, Krapf H, Zwingers T (2002) Mitoxantrone in progressive multiple sclerosis: a placebo-controlled, double-blind, randomised, multicentre trial. Lancet 360:2018–2025

Herrero-Herranz E, Pardo LA, Gold R, Linker RA (2008) Pattern of axonal injury in murine myelin oligodendrocyte glycoprotein induced experimental autoimmune encephalomyelitis: implications for multiple sclerosis. Neurobiol Dis 30:162–173

Herzog RI, Cummins TR, Ghassemi F, Dib-Hajj SD, Waxman SG (2003) Distinct repriming and closed-state inactivation kinetics of Nav1.6 and Nav1.7 sodium channels in mouse spinal sensory neurons. J Physiol 551:741–750

Hill KE, Zollinger LV, Watt HE, Carlson NG, Rose JW (2004) Inducible nitric oxide synthase in chronic active multiple sclerosis plaques: distribution, cellular expression and association with myelin damage. J Neuroimmunol 151:171–179

Hirokawa N, Glicksman MA, Willard MB (1984) Organization of mammalian neurofilament polypeptides within the neuronal cytoskeleton. J Cell Biol 98:1523–1536

Hollenbeck PJ, Saxton WM (2005) The axonal transport of mitochondria. J Cell Sci 118: 5411–5419

Irvine KA, Blakemore WF (2008) Remyelination protects axons from demyelination-associated axon degeneration. Brain 131:1464–1477

Itoyama Y, Webster HD, Sternberger NH, Richardson EP Jr, Walker DL, Quarles RH, Padgett BL (1982) Distribution of papovavirus, myelin-associated glycoprotein, and myelin basic protein in progressive multifocal leukoencephalopathy lesions. Ann Neurol 11:396–407

Jacobs LD, Beck RW, Simon JH, Kinkel RP, Brownscheidle CM, Murray TJ, Simonian NA, Slasor PJ, Sandrock AW (2000) Intramuscular interferon beta-1a therapy initiated during a first demyelinating event in multiple sclerosis. CHAMPS Study Group. N Engl J Med 343:898–904

Julien JP, Mushynski WE (1998) Neurofilaments in health and disease. Prog Nucleic Acid Res Mol Biol 61:1–23

Kapoor R, Davies M, Blaker PA, Hall SM, Smith KJ (2003) Blockers of sodium and calcium entry protect axons from nitric oxide-mediated degeneration. Ann Neurol 53:174–180

Kappos L (2004) Effect of drugs in secondary disease progression in patients with multiple sclerosis. Mult Scler 10(Suppl 1):S46–S54; discussion S54–45

Kassmann CM, Lappe-Siefke C, Baes M, Brugger B, Mildner A, Werner HB, Natt O, Michaelis T, Prinz M, Frahm J, Nave KA (2007) Axonal loss and neuroinflammation caused by peroxisome-deficient oligodendrocytes. Nat Genet 39:969–976

Kassmann CM, Nave KA (2008) Oligodendroglial impact on axonal function and survival – a hypothesis. Curr Opin Neurol 21:235–241

Keegan M, Konig F, McClelland R, Bruck W, Morales Y, Bitsch A, Panitch H, Lassmann H, Weinshenker B, Rodriguez M, Parisi J, Lucchinetti CF (2005) Relation between humoral pathological changes in multiple sclerosis and response to therapeutic plasma exchange. Lancet 366:579–582

Kirkpatrick LL, Witt AS, Payne HR, Shine HD, Brady ST (2001) Changes in microtubule stability and density in myelin-deficient shiverer mouse CNS axons. J Neurosci 21:2288–2297

Kornek B, Storch MK, Weissert R, Wallstroem E, Stefferl A, Olsson T, Linington C, Schmidbauer M, Lassmann H (2000) Multiple sclerosis and chronic autoimmune encephalomyelitis: a comparative quantitative study of axonal injury in active, inactive, and remyelinated lesions. Am J Pathol 157:267–276

Kuhlmann T, Lingfeld G, Bitsch A, Schuchardt J, Bruck W (2002) Acute axonal damage in multiple sclerosis is most extensive in early disease stages and decreases over time. Brain 125:2202–2212

Kuhlmann T, Miron V, Cuo Q, Wegner C, Antel J, Bruck W (2008) Differentiation block of oligodendroglial progenitor cells as a cause for remyelination failure in chronic multiple sclerosis. Brain 131:1749–1758

Kutzelnigg A, Lucchinetti CF, Stadelmann C, Bruck W, Rauschka H, Bergmann M, Schmidbauer M, Parisi JE, Lassmann H (2005) Cortical demyelination and diffuse white matter injury in multiple sclerosis. Brain 128:2705–2712

Lappe-Siefke C, Goebbels S, Gravel M, Nicksch E, Lee J, Braun PE, Griffiths IR, Nave KA (2003) Disruption of Cnp1 uncouples oligodendroglial functions in axonal support and myelination. Nat Genet 33:366–374

Lassmann H (1998) Pathology of multiple sclerosis. McAlpine's Multiple Sclerosis, 3rd edn. Churchill Livingstone, Edinburgh

Lassmann H (2003) Axonal injury in multiple sclerosis. J Neurol Neurosurg Psychiatry 74: 695–697

Lassmann H, Bruck W, Lucchinetti C (2001) Heterogeneity of multiple sclerosis pathogenesis: implications for diagnosis and therapy. Trends Mol Med 7:115–121

Lassmann H, Wekerle H (2006) The pathology of mulitple sclerosis. Churchill Livingstone, Edinburgh

Le Page E, Leray E, Taurin G, Coustans M, Chaperon J, Morrissey SP, Edan G (2008) Mitoxantrone as induction treatment in aggressive relapsing remitting multiple sclerosis: treatment response factors in a 5 year follow-up observational study of 100 consecutive patients. J Neurol Neurosurg Psychiatry 79:52–56

Leary SM, Davie CA, Parker GJ, Stevenson VL, Wang L, Barker GJ, Miller DH, Thompson AJ (1999) 1H magnetic resonance spectroscopy of normal appearing white matter in primary progressive multiple sclerosis. J Neurol 246:1023–1026

Leterrier JF, Kas J, Hartwig J, Vegners R, Janmey PA (1996) Mechanical effects of neurofilament cross-bridges. Modulation by phosphorylation, lipids, and interactions with F-actin. J Biol Chem 271:15687–15694

Li C, Tropak MB, Gerlai R, Clapoff S, Abramow-Newerly W, Trapp B, Peterson A, Roder J (1994) Myelination in the absence of myelin-associated glycoprotein. Nature 369:747–750

Li BS, Veeranna, Grant P, Pant HC (1999a) Calcium influx and membrane depolarization induce phosphorylation of neurofilament (NF-M) KSP repeats in PC12 cells. Brain Res Mol Brain Res 70:84–91

Li BS, Veeranna, Gu J, Grant P, Pant HC (1999b) Activation of mitogen-activated protein kinases (Erk1 and Erk2) cascade results in phosphorylation of NF-M tail domains in transfected NIH 3T3 cells. Eur J Biochem 262:211–217

Li BS, Zhang L, Gu J, Amin ND, Pant HC (2000) Integrin alpha(1) beta(1)-mediated activation of cyclin-dependent kinase 5 activity is involved in neurite outgrowth and human neurofilament protein H Lys-Ser-Pro tail domain phosphorylation. J Neurosci 20:6055–6062

Linker RA, Maurer M, Gaupp S, Martini R, Holtmann B, Giess R, Rieckmann P, Lassmann H, Toyka KV, Sendtner M, Gold R (2002) CNTF is a major protective factor in demyelinating

CNS disease: a neurotrophic cytokine as modulator in neuroinflammation. Nat Med 8: 620–624

Lo AC, Saab CY, Black JA, Waxman SG (2003) Phenytoin protects spinal cord axons and preserves axonal conduction and neurological function in a model of neuroinflammation in vivo. J Neurophysiol 90:3566–3571

Lovas G, Szilagyi N, Majtenyi K, Palkovits M, Komoly S (2000) Axonal changes in chronic demyelinated cervical spinal cord plaques. Brain 123(2):308–317

Lu F, Selak M, O'Connor J, Croul S, Lorenzana C, Butunoi C, Kalman B (2000) Oxidative damage to mitochondrial DNA and activity of mitochondrial enzymes in chronic active lesions of multiple sclerosis. J Neurol Sci 177:95–103

Lubetzki C, Williams A, Stankoff B (2005) Promoting repair in multiple sclerosis: problems and prospects. Curr Opin Neurol 18:237–244

Lucchinetti CF, Bruck W, Lassmann H (2004) Evidence for pathogenic heterogeneity in multiple sclerosis. Ann Neurol 56:308

Lucchinetti C, Bruck W, Parisi J, Scheithauer B, Rodriguez M, Lassmann H (2000) Heterogeneity of multiple sclerosis lesions: implications for the pathogenesis of demyelination. Ann Neurol 47:707–717

Lucchinetti CF, Brueck W, Rodriguez M, Lassmann H (1998) Multiple sclerosis: lessons from neuropathology. Semin Neurol 18:337–349

Lucchinetti CF, Mandler RN, McGavern D, Bruck W, Gleich G, Ransohoff RM, Trebst C, Weinshenker B, Wingerchuk D, Parisi JE, Lassmann H (2002) A role for humoral mechanisms in the pathogenesis of Devic's neuromyelitis optica. Brain 125:1450–1461

Matthews PM, De Stefano N, Narayanan S, Francis GS, Wolinsky JS, Antel JP, Arnold DL (1998) Putting magnetic resonance spectroscopy studies in context: axonal damage and disability in multiple sclerosis. Semin Neurol 18:327–336

Meyer-Franke A, Kaplan MR, Pfrieger FW, Barres BA (1995) Characterization of the signaling interactions that promote the survival and growth of developing retinal ganglion cells in culture. Neuron 15:805–819

Michailov GV, Sereda MW, Brinkmann BG, Fischer TM, Haug B, Birchmeier C, Role L, Lai C, Schwab MH, Nave KA (2004) Axonal neuregulin-1 regulates myelin sheath thickness. Science 304:700–703

Misgeld T, Kerschensteiner M, Bareyre FM, Burgess RW, Lichtman JW (2007) Imaging axonal transport of mitochondria in vivo. Nat Methods 4:559–561

Mutsaers SE, Carroll WM (1998) Focal accumulation of intra-axonal mitochondria in demyelination of the cat optic nerve. Acta Neuropathol 96:139–143

Nave KA, Trapp BD (2008) Axon-glial signaling and the glial support of axon function. Annu Rev Neurosci 31:535–561

Nixon RA (1993) The regulation of neurofilament protein dynamics by phosphorylation: clues to neurofibrillary pathobiology. Brain Pathol 3:29–38

Nixon RA (1998) The slow axonal transport of cytoskeletal proteins. Curr Opin Cell Biol 10:87–92

Nixon RA, Lewis SE (1986) Differential turnover of phosphate groups on neurofilament subunits in mammalian neurons in vivo. J Biol Chem 261:16298–16301

Nixon RA, Lewis SE, Dahl D, Marotta CA, Drager UC (1989) Early posttranslational modifications of the three neurofilament subunits in mouse retinal ganglion cells: neuronal sites and time course in relation to subunit polymerization and axonal transport. Brain Res Mol Brain Res 5:93–108

Nixon RA, Lewis SE, Marotta CA (1987) Posttranslational modification of neurofilament proteins by phosphate during axoplasmic transport in retinal ganglion cell neurons. J Neurosci 7: 1145–1158

Nixon RA, Paskevich PA, Sihag RK, Thayer CY (1994) Phosphorylation on carboxyl terminus domains of neurofilament proteins in retinal ganglion cell neurons in vivo: influences on regional neurofilament accumulation, interneurofilament spacing, and axon caliber. J Cell Biol 126:1031–1046

Nixon RA, Shea TB (1992) Dynamics of neuronal intermediate filaments: a developmental perspective. Cell Motil Cytoskeleton 22:81–91

Nixon RA, Sihag RK (1991) Neurofilament phosphorylation: a new look at regulation and function. Trends Neurosci 14:501–506

Oblinger MM, Brady ST, McQuarrie IG, Lasek RJ (1987) Cytotypic differences in the protein composition of the axonally transported cytoskeleton in mammalian neurons. J Neurosci 7:453–462

Okuda Y, Sakoda S, Fujimura H, Yanagihara T (1997) Nitric oxide via an inducible isoform of nitric oxide synthase is a possible factor to eliminate inflammatory cells from the central nervous system of mice with experimental allergic encephalomyelitis. J Neuroimmunol 73:107–116

O'Malley HA, Shreiner AB, Chen GH, Huffnagle GB, Isom LL (2009) Loss of Na+ channel beta2 subunits is neuroprotective in a mouse model of multiple sclerosis. Mol Cell Neurosci 40: 143–155

Pang L, Sawada T, Decker SJ, Saltiel AR (1995) Inhibition of MAP kinase kinase blocks the differentiation of PC-12 cells induced by nerve growth factor. J Biol Chem 270: 13585–13588

Patani R, Balaratnam M, Vora A, Reynolds R (2007) Remyelination can be extensive in multiple sclerosis despite a long disease course. Neuropathol Appl Neurobiol 33:277–287

Patrikios P, Stadelmann C, Kutzelnigg A, Rauschka H, Schmidbauer M, Laursen H, Sorensen PS, Bruck W, Lucchinetti C, Lassmann H (2006) Remyelination is extensive in a subset of multiple sclerosis patients. Brain 129:3165–3172

Pelletier J, Suchet L, Witjas T, Habib M, Guttmann CR, Salamon G, Lyon-Caen O, Cherif AA (2001) A longitudinal study of callosal atrophy and interhemispheric dysfunction in relapsing-remitting multiple sclerosis. Arch Neurol 58:105–111

Perrot R, Berges R, Bocquet A, Eyer J (2008) Review of the multiple aspects of neurofilament functions, and their possible contribution to neurodegeneration. Mol Neurobiol 38:27–65

Peterson JW, Bo L, Mork S, Chang A, Trapp BD (2001) Transected neurites, apoptotic neurons, and reduced inflammation in cortical multiple sclerosis lesions. Ann Neurol 50:389–400

Petzold A (2005) Neurofilament phosphoforms: surrogate markers for axonal injury, degeneration and loss. J Neurol Sci 233:183–198

Petzold A, Eikelenboom MJ, Keir G, Grant D, Lazeron RH, Polman CH, Uitdehaag BM, Thompson EJ, Giovannoni G (2005) Axonal damage accumulates in the progressive phase of multiple sclerosis: three year follow up study. J Neurol Neurosurg Psychiatry 76:206–211

Pitt D, Werner P, Raine CS (2000) Glutamate excitotoxicity in a model of multiple sclerosis. Nat Med 6:67–70

Pittock SJ, Lucchinetti CF (2007) The pathology of MS: new insights and potential clinical applications. Neurologist 13:45–56

Rasband MN, Tayler J, Kaga Y, Yang Y, Lappe-Siefke C, Nave KA, Bansal R (2005) CNP is required for maintenance of axon–glia interactions at nodes of Ranvier in the CNS. Glia 50: 86–90

Reynolds R, Dawson M, Papadopoulos D, Polito A, Di Bello IC, Pham-Dinh D, Levine J (2002) The response of NG2-expressing oligodendrocyte progenitors to demyelination in MOG-EAE and MS. J Neurocytol 31:523–536

Rosin C, Bates TE, Skaper SD (2004) Excitatory amino acid induced oligodendrocyte cell death in vitro: receptor-dependent and -independent mechanisms. J Neurochem 90: 1173–1185

Roy S, Coffee P, Smith G, Liem RK, Brady ST, Black MM (2000) Neurofilaments are transported rapidly but intermittently in axons: implications for slow axonal transport. J Neurosci 20: 6849–6861

Rush AM, Dib-Hajj SD, Waxman SG (2005) Electrophysiological properties of two axonal sodium channels, Nav1.2 and Nav1.6, expressed in mouse spinal sensory neurones. J Physiol 564: 803–815

Ruuls SR, Bauer J, Sontrop K, Huitinga I, t Hart BA, Dijkstra CD (1995) Reactive oxygen species are involved in the pathogenesis of experimental allergic encephalomyelitis in Lewis rats. J Neuroimmunol 56:207–217

Sadahiro S, Yoshikawa H, Yagi N, Yamamoto Y, Yanagihara T, Kimura M, Sakoda S (2000) Morphometric analysis of the myelin-associated oligodendrocytic basic protein-deficient mouse reveals a possible role for myelin-associated oligodendrocytic basic protein in regulating axonal diameter. Neuroscience 98:361–367

Sahenk Z, Chen L, Mendell JR (1999) Effects of PMP22 duplication and deletions on the axonal cytoskeleton. Ann Neurol 45:16–24

Sanchez I, Hassinger L, Sihag RK, Cleveland DW, Mohan P, Nixon RA (2000) Local control of neurofilament accumulation during radial growth of myelinating axons in vivo. Selective role of site-specific phosphorylation. J Cell Biol 151:1013–1024

Schneider A, Montague P, Griffiths I, Fanarraga M, Kennedy P, Brophy P, Nave KA (1992) Uncoupling of hypomyelination and glial cell death by a mutation in the proteolipid protein gene. Nature 358:758–761

Scolding N (2005) Devic's disease and autoantibodies. Lancet Neurol 4:136–137

Shah JV, Flanagan LA, Janmey PA, Leterrier JF (2000) Bidirectional translocation of neurofilaments along microtubules mediated in part by dynein/dynactin. Mol Biol Cell 11: 3495–3508

Shinar Y, McMorris FA (1995) Developing oligodendroglia express mRNA for insulin-like growth factor-I, a regulator of oligodendrocyte development. J Neurosci Res 42:516–527

Sihag RK, Inagaki M, Yamaguchi T, Shea TB, Pant HC (2007) Role of phosphorylation on the structural dynamics and function of types III and IV intermediate filaments. Exp Cell Res 313:2098–2109

Smith KJ (2006) Axonal protection in multiple sclerosis–a particular need during remyelination? Brain 129:3147–3149

Smith KJ (2007) Sodium channels and multiple sclerosis: roles in symptom production, damage and therapy. Brain Pathol 17:230–242

Smith KJ, Kapoor R, Hall SM, Davies M (2001) Electrically active axons degenerate when exposed to nitric oxide. Ann Neurol 49:470–476

Smith KJ, Lassmann H (2002) The role of nitric oxide in multiple sclerosis. Lancet Neurol 1: 232–241

Smith MR, Smith RD, Plummer NW, Meisler MH, Goldin AL (1998) Functional analysis of the mouse Scn8a sodium channel. J Neurosci 18:6093–6102

Spencer PS, Sabri MI, Schaumburg HH, Moore CL (1979) Does a defect of energy metabolism in the nerve fiber underlie axonal degeneration in polyneuropathies? Ann Neurol 5: 501–507

Stadelmann C, Bruck W (2008) Interplay between mechanisms of damage and repair in multiple sclerosis. J Neurol 255(Suppl 1):12–18

Steffensen I, Waxman SG, Mills L, Stys PK (1997) Immunolocalization of the Na(+)-Ca2+ exchanger in mammalian myelinated axons. Brain Res 776:1–9

Sternberger LA, Sternberger NH (1983) Monoclonal antibodies distinguish phosphorylated and nonphosphorylated forms of neurofilaments in situ. Proc Natl Acad Sci USA 80: 6126–6130

Strelau J, Unsicker K (1999) GDNF family members and their receptors: expression and functions in two oligodendroglial cell lines representing distinct stages of oligodendroglial development. Glia 26:291–301

Stys PK (2004) White matter injury mechanisms. Curr Mol Med 4:113–130

Trapp BD, Peterson J, Ransohoff RM, Rudick R, Mork S, Bo L (1998) Axonal transection in the lesions of multiple sclerosis. N Engl J Med 338:278–285

Trapp BD, Ransohoff R, Rudick R (1999) Axonal pathology in multiple sclerosis: relationship to neurologic disability. Curr Opin Neurol 12:295–302

Veeranna, Amin ND, Ahn NG, Jaffe H, Winters CA, Grant P, Pant HC (1998) Mitogen-activated protein kinases (Erk1,2) phosphorylate Lys-Ser-Pro (KSP) repeats in neurofilament proteins NF-H and NF-M. J Neurosci 18:4008–4021

Vercellino M, Plano F, Votta B, Mutani R, Giordana MT, Cavalla P (2005) Grey matter pathology in multiple sclerosis. J Neuropathol Exp Neurol 64:1101–1107

Wang L, Ho CL, Sun D, Liem RK, Brown A (2000) Rapid movement of axonal neurofilaments interrupted by prolonged pauses. Nat Cell Biol 2:137–141

Waxman SG (1998) Demyelinating diseases–new pathological insights, new therapeutic targets. N Engl J Med 338:323–325

Waxman SG (2002) Sodium channels as molecular targets in multiple sclerosis. J Rehabil Res Dev 39:233–242

Waxman SG (2006) Axonal conduction and injury in multiple sclerosis: the role of sodium channels. Nat Rev Neurosci 7:932–941

Waxman SG (2008a) Axonal dysfunction in chronic multiple sclerosis: meltdown in the membrane. Ann Neurol 63:411–413

Waxman SG (2008b) Mechanisms of disease: sodium channels and neuroprotection in multiple sclerosis – current status. Nat Clin Pract Neurol 4:159–169

Waxman SG, Craner MJ, Black JA (2004) Na+ channel expression along axons in multiple sclerosis and its models. Trends Pharmacol Sci 25:584–591

Webster HD (1997) Growth factors and myelin regeneration in multiple sclerosis. Mult Scler 3:113–120

Werner HB, Kuhlmann K, Shen S, Uecker M, Schardt A, Dimova K, Orfaniotou F, Dhaunchak A, Brinkmann BG, Mobius W, Guarente L, Casaccia-Bonnefil P, Jahn O, Nave KA (2007) Proteolipid protein is required for transport of sirtuin 2 into CNS myelin. J Neurosci 27: 7717–7730

Wilkins A, Chandran S, Compston A (2001) A role for oligodendrocyte-derived IGF-1 in trophic support of cortical neurons. Glia 36:48–57

Wilkins A, Compston A (2005) Trophic factors attenuate nitric oxide mediated neuronal and axonal injury in vitro: roles and interactions of mitogen-activated protein kinase signalling pathways. J Neurochem 92:1487–1496

Wilkins A, Majed H, Layfield R, Compston A, Chandran S (2003) Oligodendrocytes promote neuronal survival and axonal length by distinct intracellular mechanisms: a novel role for oligodendrocyte-derived glial cell line-derived neurotrophic factor. J Neurosci 23: 4967–4974

Wolswijk G (1998) Chronic stage multiple sclerosis lesions contain a relatively quiescent population of oligodendrocyte precursor cells. J Neurosci 18:601–609

Yabe JT, Chan WK, Chylinski TM, Lee S, Pimenta AF, Shea TB (2001) The predominant form in which neurofilament subunits undergo axonal transport varies during axonal initiation, elongation, and maturation. Cell Motil Cytoskeleton 48:61–83

Yin X, Baek RC, Kirschner DA, Peterson A, Fujii Y, Nave KA, Macklin WB, Trapp BD (2006) Evolution of a neuroprotective function of central nervous system myelin. J Cell Biol 172: 469–478

Yin X, Crawford TO, Griffin JW, Tu P, Lee VM, Li C, Roder J, Trapp BD (1998) Myelin-associated glycoprotein is a myelin signal that modulates the caliber of myelinated axons. J Neurosci 18:1953–1962

Yool DA, Klugmann M, McLaughlin M, Vouyiouklis DA, Dimou L, Barrie JA, McCulloch MC, Nave KA, Griffiths IR (2001) Myelin proteolipid proteins promote the interaction of oligodendrocytes and axons. J Neurosci Res 63:151–164

Yoshikawa H (2001) Myelin-associated oligodendrocytic basic protein modulates the arrangement of radial growth of the axon and the radial component of myelin. Med Electron Microsc 34:160–164

Yuan A, Rao MV, Kumar A, Julien JP, Nixon RA (2003) Neurofilament transport in vivo minimally requires hetero-oligomer formation. J Neurosci 23:9452–9458

Yuan A, Rao MV, Sasaki T, Chen Y, Kumar A, Veeranna, Liem RK, Eyer J, Peterson AC, Julien JP, Nixon RA (2006) Alpha-internexin is structurally and functionally associated with the neurofilament triplet proteins in the mature CNS. J Neurosci 26:10006–10019

Zhao C, Fancy SP, Magy L, Urwin JE, Franklin RJ (2005) Stem cells, progenitors and myelin repair. J Anat 207:251–258

Chapter 17
Intermediate Filament Interactions in Neurons

Kevin G. Young and Rashmi Kothary

Abstract Intermediate filament (IF) proteins constitute a unique part of the cytoskeleton and are present in high abundance in tissues such as skin, muscle, and the central and peripheral nervous systems. In general, their properties suggest that IFs play important roles in maintaining the mechanical stability of tissues undergoing severe mechanical stress. However, the lack of cytoplasmic IFs in the nervous system of some animals and the lack of phenotype in genetic knockouts for various neuronal IFs indicate that IFs may not be absolutely necessary for basic neuronal development and function. Rather, adaptive roles may be a primary importance for IF networks in neurons and other cell types. IFs are integrated with the actin and microtubule networks via structural cross-linking proteins and via signalling mechanisms. Structural proteins linking the IF network also connect it to the plasma membrane and to organelles, including the nucleus. Within the nucleus, IFs play a ubiquitously important role in structuring the nuclear envelope and help to regulate gene expression. Nuclear IF protein function is mediated through interactions with both large structural proteins and small regulatory proteins. This chapter will focus on how IF networks are interconnected to their surrounding environment, and how interactions of IFs in neurons may aid in adaptation of the cell. We will also discuss how IF networks are related to neurological disease etiology, and how IF interacting proteins may play a role in causing disease.

Keywords Bpag1 · Cytoskeleton · Dynein · Dystonin · Intermediate filaments · Kinesin · Lamins · Microtubule-associated proteins · Microtubules · Nesprins · Neurodegenerative disease · Neurofilaments · Nuclear envelope · Nuclear lamina · Plakin · Plectin · Torsina

---

R. Kothary (✉)
Ottawa Hospital Research Institute, Ottawa, ON, Canada K1H 8L6; Department of Cellular and Molecular Medicine, University of Ottawa, Ottawa, ON, Canada; Department of Medicine, University of Ottawa, Ottawa, ON, Canada
e-mail: rkothary@ohri.ca

R.A. Nixon, A. Yuan (eds.), *Cytoskeleton of the Nervous System*,
Advances in Neurobiology 3, DOI 10.1007/978-1-4419-6787-9_17,

Contents

17.1 Introduction

The cell's scaffolding, the cytoskeleton, is composed of three basic networks of protein filaments. These networks have been defined historically by their apparent thicknesses and structural features as thin filaments (6 nm diameter), intermediate filaments (IFs) (~10 nm diameter), and microtubules (15 nm diameter) (Arstila and Hopsu-Havu, 1967; Gall, 1966; Ishikawa et al., 1968; Ledbetter and Porter, 1963; Selby and Bear, 1956). Thin filaments are composed of actin polymers, with myosin cross-bridges commonly associating with these to form contractile actin-myosin filaments. These are the basic components of stress fibers, a part of the cytoskeleton most prominent in fibroblast-like cell types. Actin filaments and myosin filaments (thick filaments) also form the contractile apparatus of muscle. Actin filaments alone provide rigidity to the cell, and actin is involved in numerous cell functions in virtually every region of the cell. Microtubules are the flexible organizers of the cell, and commonly play a key role in trafficking and the positioning of intracellular components. IFs are much more heterogeneous in structure compared to thin filaments and microtubules. A common function assigned to IFs is to increase the mechanical stability of a cell. All three types of cytoskeletal filament network function in association with multiple types of structural and signalling proteins.

Neuronal IFs are the most abundant type of filament in neurons. They are composed of several structurally-related proteins, and generally form as obligate

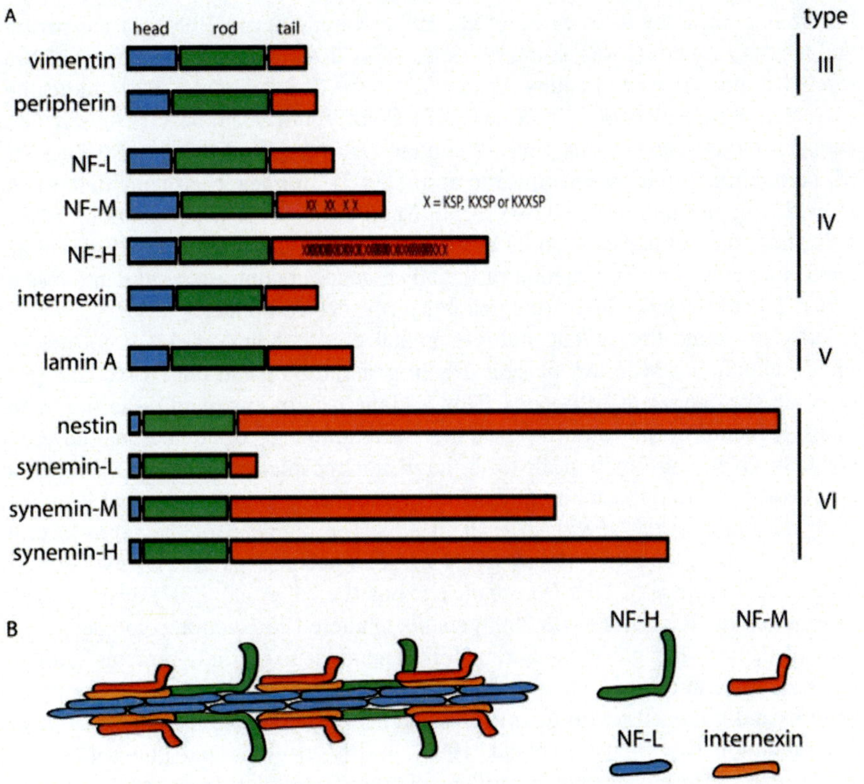

Fig. 17.1 Neuronal intermediate filament proteins. **a** IF proteins consist of head and tail domains separated by a coiled-coil rod domain. The tail domain of some IFs can be quite elongated, and may convey unique functions on these proteins. In NF-M and NF-H, the tail domains protrude from the sides of the neurofilaments, and can affect NF spacing. Though multiple sites can be phosphorylated within these tail domains (Xs), the relationship between this phosphorylation and NF spacing is unclear. **b** Schematic showing that NF-L subunits form the core of the NF structure, with NF-M, NF-H, and α-internexin forming heteropolymers around the NF-L core. The protrusions of the NF-M and NF-H subunits represent their tail domains. Each subunit symbol represents a dimer of the protein. The dimers associate with other dimers in a head-to-tail fashion. Other IF proteins will also polymerize with the NF subunits, as discussed in the text

heteropolymers (Lee et al., 1993). These include the neurofilament (NF) triplet proteins – neurofilament light (NF-L), neurofilament medium (NF-M), and neurofilament heavy (NF-H) – which comprise the bulk of the NF structure (Fig. 17.1a). α-Internexin and peripherin are IF proteins that can also be incorporated into NFs, along with the NF triplet proteins (Lee and Cleveland, 1996; Yuan et al., 2006). In the central nervous system (CNS) of mice, the ratio of NF-L:α-internexin:NF-M:NF-H was estimated to be 4:2:2:1 (Yuan et al., 2006). This ratio is certain to vary, however, as neurons with smaller axon calibre have relatively high α-internexin expression compared to NF proteins, whereas those with larger axons have high

NF protein expression (Fliegner et al., 1994; Kaplan et al., 1990). In the peripheral nervous system (PNS), α-internexin is easily detectable in young animals, with levels being decreased in adults (Chiu et al., 1989; Kaplan et al., 1990), indicating a greater contribution of α-internexin to NFs in younger animals. In this regard, neonatal mouse sympathetic neurons contain NFs composed mainly of NF-L, NF-M, peripherin, and α-internexin (Yan et al., 2007). In mice, transgenic deletion of individual genes coding for NF-H or NF-M still allows for the formation of NFs, indicating that neither is essential to the formation of these filaments (Elder et al., 1998; Rao et al., 1998). Deletion of the NF-L gene, in contrast, results in a loss of NFs (Zhu et al., 1997), indicating that it is a core NF component.

NFs are found throughout multiple animal phyla, though they are not present in all animals. As with cytoplasmic IFs in general, NFs and the NF subunit proteins are not present in arthropods. This, and the lack of any overt disability in NF mouse knockouts, indicates that NFs are not required for basic neuronal development, survival, and function. Instead, the major hypothesis regarding NF function has revolved around the ability of the NFs to affect axon calibre (Eyer and Peterson, 1994; Hoffman et al., 1987; Xu et al., 1996). The diameter of the axon, in turn, can affect the conduction velocity of the nerve impulse propagated by the neurons. Indeed, the removal of NFs from mouse axons (Perrot et al., 2007) or the genetic elimination of NFs (Kriz et al., 2000) results in altered conduction velocities.

Both neurons and glia express vimentin, synemin, and nestin IF proteins at different developmental stages. Vimentin and nestin are most abundant in undifferentiated neurons and glial cells, with their expression being silenced following post-mitotic differentiation (Lee and Cleveland, 1996). Lastly, neurons and glia contain the widely expressed nuclear lamin proteins. These IF proteins serve as core components of the nuclear lamina in virtually all metazoan cells (Melcer et al., 2007). As we will discuss, although vimentin and lamins are widely expressed, unique interacting partners in neurons, and other specialized neuronal components, impart a unique importance to these IFs in neurons.

In the rest of this chapter we will briefly elaborate on the basic characteristics of the individual IF components of neurons. We will then focus on the role of proteins that link the cytoplasmic and nuclear IF networks of neurons to the rest of the cytoskeleton and to other components of the cell. We will assess the contribution of components of the IF cytoskeleton to normal neuronal function, and discuss the consequences of mutation or dysfunction of these components.

17.2 Characteristics of the IF Members in Neurons

17.2.1 The Neurofilament Triplet Proteins, NF-L, NF-M, and NF-H

The NF triplet proteins are class IV IFs first described as the primary components of NFs (Liem et al., 1978). Though atomic level structural analysis with purified NF proteins has not been accomplished to determine precise filament structure, the transgenic deletion of NF-L in mice indicates that it is the core component essential for NF formation. In correspondence with the necessity of the NF-L subunit to form NFs in vivo, NF-L is the only one of the three NF triplet proteins which can self-assemble in vitro to form filaments (Liem and Hutchison, 1982). NF-H and NF-M polymerize only in the presence of NF-L.

NF-M and NF-H are thought to modify the spacing of the filaments. NF-M and NF-H affect filament spacing since these larger NF subunits produce side-branches that protrude out from the filaments (Fig. 17.1b) and are visible by electron microscopy analysis (Hirokawa et al., 1984; Hisanaga and Hirokawa, 1988). These side-branches are made up of the C-terminal tail domains of NF-M and NF-H, which are heavily phosphorylated regions. NF spacing and axonal calibre is correlated with the phosphorylation status of these tail domains (de Waegh et al., 1992). This is a notable feature in myelinated axons within the internodal space. A reduction in phosphorylation of NF-M and NF-H, as with loss of these proteins, was thought to result in a decreased axonal calibre resulting from a decreased spacing of the filaments. However, in more recent work using transgenic mice, removal of the tail domain of NF-H (Rao et al., 2002) or replacement of the phosphorylatable serine residues in the tail of NF-M with non-phosphorylatable alanine residues (Garcia et al., 2009) has no effect on axonal calibre. NF spacing is affected in NF-M null mutants (Garcia et al., 2003, 2009; Rao et al., 2003), with more packed NFs observed in mutant axons. The lack of altered axonal calibre and decreased NF spacing in mice missing the phosphorylatable NF-M tail domain demonstrates that NF spacing can be affected without affecting axonal calibre.

There have, however, been numerous studies correlating a reduction of NFs with reduced axonal calibre (Elder et al., 1998; Jacomy et al., 1999; Kriz et al., 2000; Nukada and Dyck, 1984; Zochodne et al., 2004). This suggests that, while phosphorylation of the NF-M and NF-H tail domain may not regulate axonal calibre, NFs do serve to regulate calibre size. As mentioned above, however, this role does not appear to be absolutely essential for neuronal function in most neurons. Since the loss of NF-H does not cause a decrease in axonal calibre, but does affect conduction velocity, the role of NFs in determining conduction velocity is not strictly limited to altering physically the axon diameter (Kriz et al., 2000). Thus, there is not a strict correlation between NF spacing, axon calibre, and conduction velocity. There is enough evidence to suggest an extensive interplay between these features, however.

17.2.2 α-Internexin

α-Internexin was first described as an IF binding protein expressed in the CNS (Pachter and Liem, 1985) and later shown to be a bona fide IF protein itself. It is a type IV IF that, unlike the NF triplet proteins, can self-assemble (Kaplan et al., 1990). As mentioned above, however, α-internexin may normally assemble with the NF triplet proteins, and is a major component of NFs (Yuan et al., 2006). Though found in neurons throughout the mammalian nervous system, α-internexin expression decreases in the mature PNS in mammals (Chiu et al., 1989; Kaplan et al., 1990). During development, the appearance of α-internexin precedes NF triplet protein expression (Kaplan et al., 1990). α-Internexin is also upregulated following axon damage, possibly contributing to the regenerative process (McGraw et al., 2002). These observations suggest that α-internexin may play a role in the normal development of neurons. Genetic ablation of α-internexin expression in the mouse does not result in any developmental defect, however (Levavasseur et al., 1999).

17.2.3 Peripherin

Peripherin is a relatively small class III IF, similar to vimentin in structure (Fig. 17.1a) (Thompson and Ziff, 1989). First described from neuroblastoma and rat pheochromocytoma (PC12) cells, peripherin is a major component of NFs in the PNS (Portier et al., 1983a, b; Troy et al., 1992). In PC12 cells, peripherin is the major cytoplasmic IF, and its reduced expression leads to defects in neurite growth and maintenance in culture (Helfand et al., 2003). Though it is expressed in some areas of the CNS, this is very limited. In the dorsal root ganglia of the rat PNS, peripherin is only well-expressed in a subset of neurons (Fornaro et al., 2008). It is therefore not ubiquitously important to neuronal structure and function. Correspondingly, peripherin null mice develop normally and are viable (Lariviere et al., 2002). They exhibit only a loss of small, unmyelinated sensory axons, with no corresponding overt neurological phenotype.

17.2.4 Nestin

Nestin is a class VI IF (though it has been formerly grouped with the class IV IFs) that is a well-known marker of undifferentiated neurons (Guerette et al., 2007; Wiese et al., 2004). It also marks progenitor cells destined to become glia, muscle, and various endothelial and epithelial cell types (Shi et al., 2008; Wiese et al., 2004), and is expressed in specific cells of the kidney (Bertelli et al., 2007; Patschan et al., 2007). Nestin is co-expressed with vimentin in neuronal precursors, and co-polymerizes with it (Chou et al., 2003; Steinert et al., 1999). During mitotic cell division, nestin promotes the dissolution of vimentin IFs (Chou et al., 2003). This was suggested to be a mechanism for affecting the distribution of IFs retained within the resulting daughter cells.

17.2.5 Vimentin

Vimentin is a widely-expressed class III IF protein that is found in almost all neuronal precursor cells (Cochard and Paulin, 1984). When these precursor cells cease to divide, vimentin expression is down-regulated. A transient increase in vimentin expression is subsequently observed during axogenesis prior to NFs accumulating in the axon (Shea, 1990). The transition from cells expressing vimentin to those expressing NFs occurs gradually, with an overlap in expression between vimentin and NF proteins (Yabe et al., 2003). During this transition, there is a co-polymerization of vimentin with NFs, with these filaments being replaced by filaments with only the NF proteins of mature neurons. During neuronal regeneration, vimentin expression is up-regulated (Perlson et al., 2005; Shim et al., 2008; Toth et al., 2008). Vimentin associates with signalling proteins which can affect the neuronal regeneration program and affect cytoskeletal structure. Interestingly, vimentin null mice are normal, with only subtle neuronal regeneration defects (Colucci-Guyon et al., 1994; Perlson et al., 2005). It is possible that, while proteins associated with vimentin during neuronal regeneration are important to the regenerative process, vimentin itself is non-essential and its loss is easily compensated for.

17.2.6 Synemin

Synemin is a class VI IF that is produced as three isoforms in mice (synemin-L, -M, and -H) (Fig. 17.1a) and two in humans (synemin α and β). Unlike the NF triplet proteins, which are each produced from separate genes, the different synemin isoforms are produced by alternative splicing from a single gene (Xue et al., 2004). Similar to vimentin and nestin, the expression of synemin occurs in various tissue types. Along with muscle, relatively good expression has been demonstrated in tissues of the nervous system, and several recent papers have detailed the contribution of synemin to the IF network of neurons and glia (Izmiryan et al., 2006, 2009; Mizuno et al., 2007, 2009). Synemin expression is restricted to particular subpopulations of neurons, including serotonergic neurons (Mizuno et al., 2009). Synemin forms obligate heteropolymers with vimentin or NF proteins (Izmiryan et al., 2009). A role for synemin may lie in its ability to link to other cytoskeletal proteins such as dystrophin and plectin (Bhosle et al., 2006; Hijikata et al., 2008), suggesting that synemin may facilitate connections between IFs and other cytoskeletal components.

17.2.7 Lamins

Lamins are a type V IF found throughout the animal kingdom, and are a major constituent of the metazoan nucleus (Mattout et al., 2007; Zimek and Weber, 2008). The lamins likely represent the most ancient type of IF protein found in animals. They form a network in the layer underlying the nuclear envelope. This layer, the nuclear lamina, provides structural support to the nucleus and also provides part of a physical link between chromatin domains within the nucleus and components of the cell outside of the nucleus (Hutchison, 2002; Worman and Gundersen, 2006). In mammals, three lamin genes produce two B-type lamins, and multiple isoforms of A-type lamins (Hutchison, 2002). The lamin B proteins are the most ubiquitously expressed, being found in virtually all nuclei. The lamin A and C proteins, which represent alternatively spliced isoforms produced from the *LMNA* (lamin A/C) gene, are also widely expressed, though this is primarily true for differentiated cell types. Though the nuclear lamina is present in all mammalian cells that contain a nucleus, neuronal-specific abnormalities can arise from mutation of lamin A/C proteins, and from connected proteins that we will discuss later in this chapter.

17.3 Neuronal IF Interacting Proteins

In examining what proteins connect cytoplasmic IFs to other elements of the cell, we can roughly divide the connectors into three groups. These are proteins involved in the transport/assembly of filaments, proteins involved in the disassembly of filaments, and proteins involved in mediating filament function. If a major part of filament function is in regulating the gross structure and mechanical stability of the cell, included in this last group would be proteins that may simply serve as linkers – like nails that hold together the framing of a house. Whether any such proteins are actually limited to a simple linking function is something we will discuss in more detail below.

In the nucleus, a wealth of information has arisen in the past decade regarding the physical linkage that occurs between lamin filaments and chromosomal DNA or nuclear proteins inside the nucleus and the cytoskeleton surrounding the outside of the nucleus. While similar connections likely occur in all nucleated cells, there appears to be more of a requirement for some linking proteins in neurons and muscle. Specific neuromuscular defects can arise when the linkages connecting the nuclear lamina are compromised.

In the next section we will focus on the roles of the proteins responsible for organizing neuronal IFs. We will discuss both the importance of these proteins to mediating IF function and the importance of IFs to the normal functioning of IF-associated proteins.

17.3.1 Connections Between Neurofilaments and Microtubules

While the dephosphorylated tail domain of NF-H can directly interact with microtubules (Miyasaka et al., 1993), connections between NFs and microtubules appear to primarily be mediated indirectly. Proteins that link microtubules to NFs have largely been studied in the context of their role in regulating NF transport. IFs are actively transported in both the anterograde and retrograde direction in neurites (Helfand et al., 2003; Roy et al., 2000; Wang et al., 2000; Yan et al., 2007). In the case of peripherin, the transport occurs with particles (dots) and small protofilaments termed squiggles (Helfand et al., 2003). Dots and squiggles also form for other IFs (Yoon et al., 2001), and may be a general feature in IF formation. NF transport in neurons is representative of slow axonal transport (Hoffman and Lasek, 1975). Interestingly, the apparent slow transport of NFs consists of rapidly transported filament subunits, interrupted by long pauses (Roy et al., 2000; Trivedi et al., 2007; Wang et al., 2000). This is the result of NF subunits switching between "on track" and "off track" states, where the track is microtubules (Trivedi et al., 2007). Correspondingly, coupling to microtubule motor proteins – dyneins for retrograde transport and kinesins for anterograde transport – is important for the movement of NFs (Shah et al., 2000; Wagner et al., 2004; Yabe et al., 2000, 1999). α-Internexin and NF-M are of particular importance to NF transport (Yuan et al., 2003, 2006), suggesting important interactions between motor proteins and these NF subunits. Below, we will focus on the connection of NFs to microtubule motors and microtubule-associated proteins (MAPs), and briefly discuss the potential for actin-myosin based NF transport.

17.3.1.1 Kinesins

Kinesins are a class of motor protein that mediate active transport along microtubules. Cargos transported via kinesins are both membrane bound and non-membrane bound. Kinesin-mediated movement is towards the plus end (growing end) of microtubules, which is generally towards the periphery of a cell. In neurons, this would generally include anterograde transport along most neurites (microtubule plus ends are orientated away from the cell body in axons, though their orientation in dendrites is mixed). The first interaction demonstrated between kinesins and IFs showed that kinesins are required for the association of vimentin filaments with microtubules (Gyoeva and Gelfand, 1991). Vimentin assembly in fibroblasts was subsequently demonstrated to involve the movement of vimentin dots and squiggles along microtubules using a kinesin motor (Prahlad et al., 1998). In the absence of a functional kinesin, vimentin IF assembly cannot occur normally. In the case of NFs, NF-H and NF-M are important for interactions with kinesin (Jung et al., 2005). As mentioned above, NF-M is necessary for NF transport, and may represent the more important interacting subunit with kinesins. Kinesins, then, are generally responsible for the anterograde transport of NFs in neurons.

17.3.1.2 Dyneins

Dyneins mediate the active transport along microtubules in the opposite direction of kinesins, toward the microtubule minus end. In neurons, dyneins would primarily mediate retrograde transport. A dynein/dynactin motor complex can move NFs along microtubules in vitro (Shah et al., 2000; Wagner et al., 2004). This interaction is mediated by a direct interaction between the NF-M subunit of NFs binding to dynein (Wagner et al., 2004). In sympathetic neuron cultures, dynein is required for mediating the retrograde transport of NFs in the axon (He et al., 2005). In intact BHK-21 fibroblast cells, vimentin organization requires movement via dynein/dynactin (Helfand et al., 2002). Vimentin transport in neurons is mediated by dynein during growth of the axon (Shim et al., 2008). Though dynein is likely to generally mediate the retrograde transport of vimentin IFs and NFs, dynein can also be involved in anterograde transport via the movement of microtubules in the anterograde direction. This may account for the observation that dynein can both transport NFs in a retrograde direction and also transport NFs from the cell body into neurites (Motil et al., 2006).

17.3.1.3 MAP2 and MAP6

Along with associations between NFs and microtubule motor proteins, NFs can bind directly to microtubule-associated proteins such as MAP2. MAP2 is a major microtubule-associated protein in neurons, and is enriched in the dendrites. In mice lacking MAP2, the dendrites of hippocampal neurons are shorter and their microtubule density is reduced (Harada et al., 2002). An important function of MAP2 is to aid in the stabilization of microtubules (Kaech et al., 1996; Takemura et al., 1992). MAP2 was demonstrated over two decades ago to bind to the NF-L subunit of NFs (Frappier et al., 1991; Heimann et al., 1985). A cross-linking function for MAP2 has been indicated by electron microscopy analysis (Hirokawa et al., 1988). In dendrites, MAP2 forms cross-bridges between NFs and microtubules. To our knowledge, the relevance of this interaction has not been elaborated any further, however.

In addition to MAP2, a second MAP that may interact with both NFs and microtubules is MAP6 (stable tubule only polypeptide; STOP). MAP6, which occurs as several isoforms in neurons, is important for generating the high stability of microtubules observed in mature neurons (Guillaud et al., 1998). In NF aggregates found in the axons of amyotrophic lateral sclerosis (ALS) patients, MAP6 co-aggregates with NFs (Letournel et al., 2003). Additionally, MAP6 accumulates with NFs in the perikaryon of neurons in a mouse model wherein NFs are restricted from moving into the neurites. This is suggestive of an interaction between NFs and MAP6, though the nature of this interaction has yet to be elaborated on.

17.3.1.4 Actin-Myosin Based NF Transport

Myosin motors have the potential to contribute to NF transport along actin filaments. Depolymerization of actin filaments, or interfering with myosin function in differentiated neuroblastoma cells, affects the transport of NFs (Jung et al., 2004). Specifically, the myosin Va motor interacts with NF-L and peripherin subunits of NFs (Rao et al., 2002). In myosin Va deficient mice, NFs accumulate in axons. While it is possible that the myosin Va interaction with NFs serves primarily to distribute properly the NFs, it is also possible that NFs serve to modulate the myosin based transport of other proteins and organelles in neurons.

17.3.2 Interactions of Neuronal IFs with Scaffolding Proteins

As with the actin and microtubule cytoskeleton networks, IFs can be highly cross-linked and connected to organelles, membranes, and other parts of the cytoskeleton. Cross-linking between IFs, including those found in neurons, has long been suggested based on electron microscopy images (Fig. 17.2) (Hirokawa, 1982; Hirokawa et al., 1988). The connections between IFs and other elements of the cell are mediated in part by large scaffolding proteins. Included among these are the proteins originally described as IF associated proteins, or IFAPs. Most notable among the IFAPs is plectin (Wiche, 1998). Plectin is a member of the plakin family of proteins, which includes relatively large structural proteins that can link multiple elements

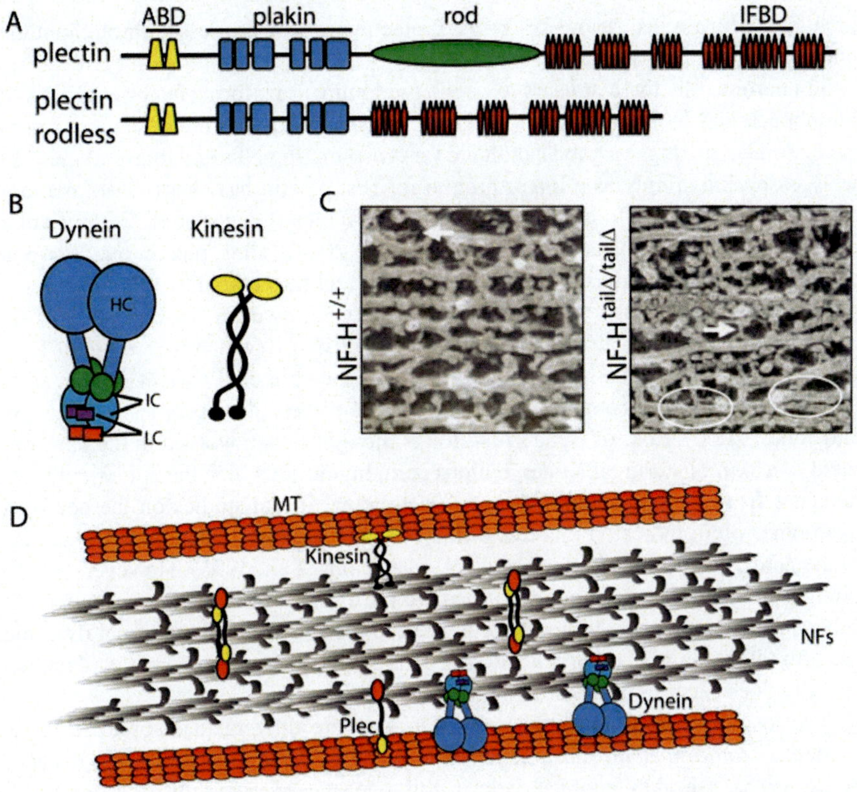

Fig. 17.2 Neuronal intermediate filament-associated proteins in the cytoplasm. **a** Plectin, or plectin-like proteins (such as dystonin), have most commonly been suggested as forming cross-linking structures between individual NFs, and NFs and microtubules. Plectin is a large spectrin-related protein that contains defined actin- and IF-binding domains (ABD and IFBD, respectively), and also has the ability to associate directly with microtubules. **b** Dynein and kinesin protein complexes are involved in the retrograde and anterograde transport of NFs, respectively. Dynein is assembled as a complex of heavy (HC), intermediate (IC), and light chain (LC) components. Kinesin is composed of a dimer of two heavy and light chains. **c** Quick freeze deep etch electron micrographs showing NF networks in axons from wild type (NF-$H^{+/+}$) and NF-H tail domain mutant (NF-$H^{tail\Delta/tail\Delta}$) mice. *Arrows* point to plectin-like linker proteins. Closely apposed NFs found in the NF-$H^{tail\Delta/tail\Delta}$ mouse axons are *circled*. **d** Schematic of NFs separate by NF subunit tail domains, and cross-linked by plectin. Whether plectin actually does cross-link NFs is still speculative. Dynein and kinesin are shown connecting NFs to microtubules. (Electron micrograph images in panel **c** are courtesy of Dr. M.V. Rao, Nathan Kline Institute, Orangeburg, NY)

of the cytoskeleton. The genetic knockout of plectin in mice results in a severe myopathy and skin blistering, with death occurring shortly after birth (Andra et al., 1997). Mutation of the plectin gene in humans results in epidermolysis bullosa simplex with muscular dystrophy (Smith et al., 1996). This disorder is characterized by skin blistering and muscle weakness, with variable severity and age of onset of the symptoms (Shimizu et al., 1999). Plectin is likely to have a role in most tissues, including neurons, as various isoforms of plectin are well-expressed throughout the body.

In neurons, plectin1c appears to be the predominant isoform (Fuchs et al., 2005; Steinboeck and Kristufek, 2005). Plectin1c is associated with microtubules (Andra et al., 2003), making it a good candidate for cross-linking NFs and microtubules. As well as serving simply as a linker protein for IFs, plectin has clearly been demonstrated to serve as scaffolding for signalling molecules (Gregor et al., 2006; Lunter and Wiche, 2002; Osmanagic-Myers and Wiche, 2004). Thus, plectin may serve to anchor signalling complexes along neuronal IFs and microtubules. What exact role plectin plays in neurons remains to be demonstrated, however.

A closely related protein to plectin is dystonin/Bpag1 (which we will refer to as dystonin for the neuronal protein and Bpag1 for the epithelial variant). In the skin, the epithelial Bpag1 protein connects keratin IFs to hemidesmosomes (Hopkinson and Jones, 2000). Loss of Bpag1 function is thought to be involved in the development of a skin blistering disorder, bullous pemphigoid, in which the epidermal layer detaches from the underlying basement membrane. Initial studies on the neuronal dystonin protein indicated that it can link NFs and peripherin-containing filaments to the actin cytoskeleton (Leung et al., 1999a; Yang et al., 1996). However, the IF-binding domain (IFBD) thought to mediate the interaction with neuronal IFs was subsequently demonstrated not to be present in the predominant neuronal dystonin isoform (dystonin-a) (Leung et al., 2001). This called into question whether dystonin actually does serve as an IFAP in neurons.

The loss of dystonin function results in a severe movement disorder in mice, termed *dystonia musculorum* (*dt*) (Brown et al., 1995; Guo et al., 1995; Kothary et al., 1988). One key characteristic in this neurodegenerative disorder, and several major human neurodegenerative disorders, is the presence of NF aggregates in the axon. The observation that dystonin may be an IF organizing/linking protein was intriguing given the role of IF aggregates in neurodegenerative disease. The involvement of NFs in causing the *dt* disorder has, however, been shown to be unlikely (reviewed in Young and Kothary, 2007). Among other observations, elimination of NFs from the axonal compartment, which on its own is not seriously detrimental to animal health and survival, has no effect on the progression of the *dt* disorder (Eyer et al., 1998). Thus, the presence or absence of NFs in the axon is inconsequential to the development of the *dt* disorder.

Another related protein of plectin and dystonin/Bpag1 is Macf1 (acf7) (Bernier et al., 1996; Leung et al., 1999b). Macf1 has been primarily implicated in regulating the connections between actin filaments and microtubules (Karakesisoglou et al., 2000; Kodama et al., 2003; Leung et al., 1999b; Wu et al., 2008). However, both *dystonin* and *Macf1* genes produce isoforms that contain a potential IFBD similar to the IFBD of the epithelial Bpag1 isoform (Gong et al., 2001; Leung et al., 2001; Lin et al., 2005). These isoforms may be expressed in neurons, and therefore dystonin and Macf1 may play some role in connecting neuronal IFs to other parts of the cytoskeleton (as originally suggested for dystonin), though this remains to be demonstrated. As with plectin, regulating the movement of signalling complexes along the cytoskeleton may prove to be an important function of Macf1. Macf1 interacts with members of the Wnt signalling pathway, and can regulate their movement to the plasma membrane (Chen et al., 2006). Thus, Macf1 can serve as a structural linking protein, but it is not limited to a static role in cross-linking cytoskeletal filaments.

It is notable that fruit flies contain an ortholog of dystonin and Macf1, called shot (kakapo; short stop) (Lee et al., 2000; Reuter et al., 2003; Subramanian et al., 2003). Shot plays an important structural role in the epidermis and neuromuscular system, with axonal outgrowth being severely affected in shot mutants. Any interaction between shot and cytoplasmic IFs is not likely to be relevant to its function, however, as fruit flies do not express any cytoplasmic IFs (they contain only nuclear lamin IFs) (Herrmann and Aebi, 2000). While it is possible that a plakin protein or proteins help organize IFs in vertebrate neurons, what form of plakin actually does function in neuronal IF organization, and in what way, is still unclear.

17.3.3 Ndel1, Vimentin, and the Link to DISC1

As mentioned in the previous section, vimentin is expressed in immature neurons and is up-regulated during axonal regeneration. A protein that is up-regulated in neurons along with vimentin, and that associates with vimentin, is Ndel1 (Toth et al., 2008). Ndel1 can associate with mature NFs, but becomes preferentially associated with vimentin following the damage-induced expression of this IF in neurons. The loss of Ndel1 through siRNA gene silencing impairs axonal regeneration. While the loss of vimentin in transgenic mice does not result in an overt pathology, vimentin null neurons do exhibit impaired neuronal regeneration (Perlson et al., 2005).

Interestingly, one of the proteins associated with Ndel1 is DISC1 (disrupted in schizophrenia 1) (Kamiya et al., 2006). DISC1 has gained extensive interest since it has become the first protein whose defective function is definitively linked to a heritable psychiatric disorder (Callicott et al., 2005; Hennah et al., 2003; Millar et al., 2000). As well as initially being linked to schizophrenia, aberrant DISC1 function may play a role in other psychiatric conditions such as schizoaffective disorder, bipolar disorder, major depression, and autism (Hodgkinson et al., 2004; Schosser et al., 2010; Tomppo et al., 2009). A DISC1 "interactome" has been generated, and many of the associated proteins are cytoskeletal interacting proteins, such as dystonin and Macf1 (Camargo et al., 2007). Mutant DISC1 has also been demonstrated to associate with the microtubule organizing center and induce microtubule defects (Morris et al., 2003; Shimizu et al., 2008). An interaction between DISC1 and Ndel1 is important for neurite outgrowth in cultured cells (Kamiya et al., 2006). Whether IF structure is affected by mutant DISC1, or whether IFs serve to modulate DISC1 functions, would be interesting to examine.

17.3.4 Nesprins

Significant progress has been made in recent years on detailing the connections between the nuclear lamina on the inside of the nuclear envelope and the cytoskeleton surrounding the outside of the nucleus. Key among these recent findings was the discovery of the nesprin (syne) proteins. Initially described as proteins involved in tethering subsynaptic nuclei in muscle (Apel et al., 2000), this group of proteins has grown to nesprins-1 to -4, each of which may have several different isoforms. An important common feature of the nesprin proteins is the presence of a C-terminal KASH domain (Starr and Fischer, 2005). This domain contains a transmembrane region that localizes the proteins to membranes surrounding the nucleus and mediates their interactions with SUN proteins (Fig. 17.3) (note: some authors consider the transmembrane region to be separate from the downstream KASH domain).

Studies of nesprin functions have focussed on the nesprins playing connecting roles on both the inside and outside of the nucleus. Inside the nucleus, nesprins-1 and -2 can interact with lamin IF proteins (Libotte et al., 2005; Mislow et al.,

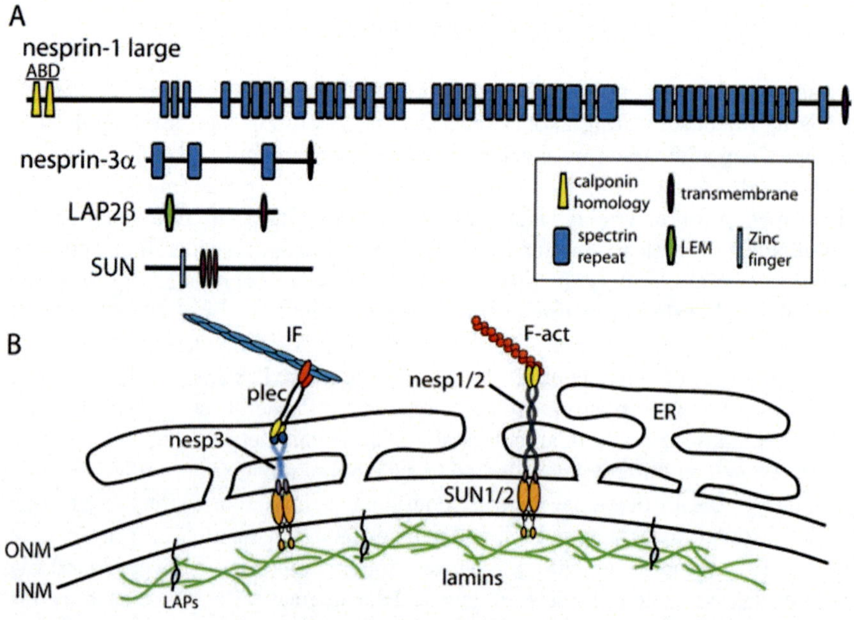

Fig. 17.3 Protein linkers of the nuclear lamina. **a** Nesprin-1 and -2 proteins are large spectrin-related proteins, with a conserved N-terminal actin binding domain (ABD) in the N-terminal region and transmembrane domain at the C-terminal end. Nesprin-3 proteins are much smaller, and the nesprin-3α isoform can link via its N-terminus to plectin. The nesprin transmembrane domains have been implicated in inserting into various membranes, including those of the inner and outer nuclear membranes, and the endoplasmic reticulum. LAP2 proteins are LEM-domain containing proteins that insert in the inner nuclear membrane via their C-terminal transmembrane domain. SUN proteins span the perinuclear space, between the inner and outer nuclear membranes and have their transmembrane domains inserted into the inner nuclear membranes, with their tail domains residing inside of the nucleus. **b** Schematic showing the localizations of the proteins shown in panel **a**. Also shown is the connection of nesprin-3α (nesp3) with plectin and cytoplasmic IFs. Nesprins-1 and -2 (nesp1/2) have been proposed to connect instead to actin filaments. Inside the nucleus, LAPs and SUN proteins connect directly with the nuclear lamins. Various isoforms of nesprins residing inside the nucleus are also likely to connect directly with the lamins (not depicted). ONM = outer nuclear membrane; INM = inner nuclear membrane; ER = endoplasmic reticulum

2002; Wheeler et al., 2007; Zhang et al., 2005). Additionally, they interact with emerin, a small nuclear lamina protein, and may serve as a scaffold for multiple other nuclear lamina proteins. A direct interaction with lamin A/C and emerin has been demonstrated for nesprin-1α (Mislow et al., 2002; Wheeler et al., 2007) and nesprin-2 isoforms (Wheeler et al., 2007; Zhang et al., 2005). Proper nesprin-1 and -2 localizations at the nuclear envelope are dependant on lamin organization, and affect emerin organization (Libotte et al., 2005; Zhang et al., 2007). Nesprin isoforms can vary greatly in size (from under 100 kD to over 800 kD), and it has been suggested that only the shorter isoforms localize within the nucleus (Starr and Fischer, 2005). There have been demonstrations of nesprins primarily within the outer nuclear membrane (and extending into the endoplasmic reticulum) (Ketema et al., 2007; Wilhelmsen et al., 2005) and nesprin-1 and -2 proteins interacting with the SUN proteins most likely reside in the outer nuclear membrane (Crisp et al., 2006; Padmakumar et al., 2005). There have also been clear demonstrations of nesprins within the inner nuclear membrane (the most convincing being the evidence of direct interactions with lamins and emerin), and some data to suggest that they can localize to the intranuclear space. It seems premature to assign particular localizations to the larger isoforms as there is currently a limited means to discriminate between the different isoforms by microscopy, and only the smaller nesprin isoforms have been expressed using recombinant DNA constructs.

Outside the nucleus, the largest isoforms of the nesprins-1 and -2 were initially suggested to link the outer nuclear membrane to actin filaments via an N-terminal actin-binding domain (Padmakumar et al., 2004; Zhen et al., 2002). More recent data question the necessity of the nesprin actin-binding domain for the major functions of nesprin proteins, however (Luke et al., 2008). Following the discovery of nesprin-3, a role in linking the nucleus to IFs was demonstrated (Wilhelmsen et al., 2005). Nesprin-3 (roughly 100 kD; under 1,000 amino acids long) is much shorter than the longest nesprin-1 and -2 isoforms (roughly 600–800 kD; over 6,000 amino acids long), and contains a unique N-terminal domain that binds to plectin. Via this connection to plectin, nesprin-3 mediates a connection between the nucleus and keratin IFs (Wilhelmsen et al., 2005). Plectin can interact with various types of IFs, and a connection between nesprin-3α, plectin, and vimentin filaments has also been demonstrated (Nery et al., 2008). Whether nesprin-3α can mediate a connection with plectin and NFs remains to be demonstrated. NFs tend not to accumulate adjacent to the nucleus to the extent that vimentin or keratin filaments can. However, indicating that tethering NFs to the nucleus may not be a function of nesprin-3 and plectin in neurons.

Nesprin-3, as well as nesprin-1, -2, and -4, are connected to the nuclear lamin proteins via SUN proteins (Crisp et al., 2006; Ketema et al., 2007; Roux et al., 2009). SUN1 and SUN2 are located in the perinuclear space, between the inner and outer nuclear membranes (Fig. 17.3). They bind the C-terminal KASH domains of nesprins within the lumen of the perinuclear space. The SUN proteins span the inner nuclear membrane, and connect via a C-terminal domain directly to lamin proteins in the nuclear lamina. In this way, the SUN proteins can anchor nesprins to the nuclear envelope and prevent mislocalization to the contiguous membranes of the endoplasmic reticulum. Interestingly, the connection between SUN proteins and the nuclear lamins is not required to maintain the localization of the SUN proteins. This is despite the fact that there is a dependency of nesprin-1 and -2 localizations on lamin proteins (Libotte et al., 2005). It may be that much of the nesprin proteins that become mislocalized in the absence of lamin interactions would normally reside within the nuclear lamina, on the inside of the nucleus. Alternatively, the binding of SUN proteins to the lamins may be required for the normal interactions between the nesprins and SUN proteins.

17.3.5 Lamina-Associated Polypeptide (LAP) Proteins

In addition to the nesprins, lamina-associated polypeptide proteins (LAPs) play a role in linking the nuclear lamina to proteins and DNA within the nucleus. Similar to nesprins, multiple LAP isoforms are produced from different genes (known as the *torsin interacting protein 1* and *thymopoietin* genes for LAP1 and LAP2, respectively). LAP1 and LAP2 proteins bind directly to lamins (Dechat et al., 2000; Dorner et al., 2007; Foisner and Gerace, 1993; Furukawa and Kondo, 1998). Unlike the nesprin proteins, LAP proteins are found exclusively within the inner nuclear membrane and intranuclear space (Dechat et al., 2000; Dorner et al., 2007; Furukawa et al., 1998). LAP proteins interact directly with chromatin and connect DNA regulatory proteins to the nuclear lamina, thereby regulating gene expression (Dorner et al., 2007). One important regulatory protein that LAP2α connects to the nuclear lamina is the retinoblastoma protein (Dorner et al., 2006; Markiewicz et al., 2002). In doing so, LAP2α helps regulate cell cycle control (Pekovic et al., 2007). Interactions of LAP2 with barrier-to-autointegration factor mediate chromatin-nuclear envelope interactions (Furukawa, 1999; Segura-Totten et al., 2002). The tethering of chromatin domains to the nuclear envelope has the general effect of repressing transcriptional activity, thus silencing gene expression (Finlan et al., 2008; Reddy et al., 2008; Somech et al., 2005). LAP protein functions are therefore of general importance to gene regulation.

In the next section we will discuss the unique importance of the nuclear lamins and associated proteins within the neuromuscular system, and the so-called laminopathies that result in neuronal death and dysfunction.

17.4 Intermediate Filaments and Interacting Proteins in Neuronal Dysfunction

17.4.1 Intermediate Filament Aggregates as Markers of Neurodegeneration

It has long been recognized that IF aggregates are hallmarks of several major neurological disorders. These include ALS, Charcot-Marie-Tooth (CMT) disease, diabetic neuropathy, Parkinson's disease, and Alzheimer's disease, among others (Table 17.1). Mutations in NF proteins have been proposed as being causative of

Table 17.1 Diseases associated with neurofilament aggregation

Disease	Phenotypic traits	Mutant proteins[a]	Functions
Charcot-Marie-Tooth Types 1 and 2	Neuropathy affecting motor and sensory neurons, with muscle atrophy	NF-L	NF subunit
		hsp-27	Chaperone protein involved in folding NF, tau, and other proteins
Amyotrophic lateral sclerosis	Progressive neuropathy affecting motor neurons	NF-H	NF subunit
		tau	Microtubule stabilization
		Superoxide dismutase 1 (SOD1)	Cu^{2+} and Zn^{2+} binding antioxidant enzyme
Parkinson's disease	Degenerative disease with motor impairment	Parkin	Protein degradation, mitochondrial protection factor
		DJ-1	Antioxidant; transcriptional regulation; protein degradation
		Pink1	Mitochondrial function; protein degradation
Alzheimer's disease	Degenerative disease with cognitive impairment	Amyloid precursor protein (APP)	Synapse formation and plasticity
		Presenilins	Processing of APP
		Tau	Microtubule stabilization
Huntington's disease	Motor and cognitive impairment	Huntingtin	Protein trafficking; transcriptional regulation; mitochondrial regulation
Giant axonal neuropathy	Sensory and motor impairment; death occurs by young adulthood	Gigaxonin	Protein degradation; IF organization
Diabetic neuropathy	Neuropathy affecting primarily sensory neurons; motor neurons also affected	Associated with diabetes	

[a]Note: not all affected proteins are listed for each disease, only representative examples in some cases

ALS and CMT, with mutant NF-L being associated with CMT disease (Jordanova et al., 2003; Mersiyanova et al., 2000; Perez-Olle et al., 2002), and mutant NF-H and peripherin being associated with ALS (reviewed in Robberecht, 2000; Schymick et al., 2007). Expressing human NF proteins or over-expressing mouse NF proteins in transgenic mice can result in the formation of NF aggregates and neuromuscular pathology (Cote et al., 1993; Xu et al., 1993). However, the loss-of-function of NF-L and NF-H proteins do not cause overt pathology in transgenic mice (Zhu et al., 1997, 1998; reviewed in Barry et al., 2007; Perrot et al., 2008).

Abnormalities such as reduced axon calibre and delayed axonal regeneration can be observed in NF protein knockout mouse lines, though these abnormalities are not enough to generate overt neuromuscular dysfunction (reviewed in Barry et al., 2007; Perrot et al., 2008). As is typical of diseases associated with IF mutations, CMT-associated mutations in NF-L may play a role in an autosomal dominant form of the disease (Jordanova et al., 2003; Mersiyanova et al., 2000). Mutations in NF-H have also been suggested to cause the disease in an autosomal dominant manner. Disease-causing NF mutations may then result in a toxic gain-of-function rather than a loss of NF function. However, a more recent analysis of mutant NF associations with ALS has suggested that it is unlikely to be specifically associated with the disease, and NF-L mutations are only rarely associated with CMT (Schymick et al., 2007).

Both in the rare cases of neurological disease associated with NF mutations, and in cases associated with mutations occurring in other genes, NFs can form large aggregates in the axon. Beyond forming aggregates due to mutations in the NF genes, it has been suggested that motor neurons are especially susceptible to NF aggregate formation due to alterations in NF expression levels (Lin and Schlaepfer, 2006). These alterations can lead to the aggregation of NF-L, and provide an environment for the aggregation of neurotoxic agents. The latter phenomenon has been suggested as a mechanism for the selective degeneration of motor neurons in ALS and may explain, for instance, the selective aggregation of mutant SOD1 in the motor neurons of mice (Rakhit et al., 2007). SOD1 is ubiquitously expressed, but the misfolded SOD1 that forms aggregates associated with degeneration is primarily found in motor neurons. This is suggestive of a causative role for NF aggregates in neurological disease, even in cases where the NF genes are not themselves mutated.

Work from the analysis of transgenic mice suggests that NF aggregates are likely a secondary consequence of neurological disease in general rather than a direct cause. The loss of NFs from the axon or loss of NFs due to transgenic deletion of the NF-L gene do not generate an overt phenotype in mice, indicating that loss of NF function is not a mechanism for neurological disease. In addition, when mice lacking axonal NFs were crossed to other mouse lines whose gene mutations lead to either an autosomal recessive or autosomal dominant form of neurodegeneration, the lack of axonal NFs did not affect disease progression (Eyer et al., 1998). Similarly, a reduction of NF levels through genetic manipulation does not affect the progression of disease in SOD1 mutant mice (Nguyen et al., 2000), though a complete loss of NFs does slow disease progression in a different mutant SOD1 mouse line (Williamson et al., 1998). Altering NF content by overexpression of the NF-L subunit does not affect the progression of disease in SOD1 mutant mice (Nguyen et al., 2000). While the presence or absence of NFs may play a modifying role in the disease progression in some neurodegenerative conditions, the mechanisms leading to neuronal degeneration are largely independent of NF expression levels. It is possible that the NF aggregates commonly observed in neurodegenerative conditions are largely tombstones, marking neurons that have already initiated degeneration through non-NF related mechanisms.

17.4.2 Hsp27 and Neurofilament Organization in Neurological Disease

As with NFs, whether defects in NF associated proteins are causative of neurological disorders, and the extent to which they may be so, are not clear. The NF associated protein whose role in NF organization is most directly linked with the generation of a disease phenotype is hsp27. Hsp27 is a small heat shock protein that functions as a chaperone in NF folding. Mutations in hsp27 cause autosomal dominant CMT2 and distal hereditary motor neuropathy (Evgrafov et al., 2004). Mutant hsp27 forms insoluble aggregates, and induces the aggregation of NFs (Evgrafov et al., 2004). In Alzheimer's disease, another neurological disorder that commonly exhibits NF aggregates as well as aggregates of other proteins, hsp27 is up-regulated (Bjorkdahl et al., 2008). That the mutation of a chaperone protein involved in the proper folding of NF subunits can cause neurological disease provides support for the idea that NF aggregates can themselves be causative of disease. However, hsp27 also acts as a chaperone protein for other proteins such as tau (Bjorkdahl et al., 2008; Kosik and Shimura, 2005). While the loss-of-function of NFs is not in itself capable of generating an overt neurological disorder, the loss-of-function of tau may be a mechanism in generating the phenotype of the "tauopathies", a group of neurological disorders that includes Alzheimer's disease (Feinstein and Wilson, 2005). The genetic knockout of tau in mice does not cause disease, though tau-deficient neurons do exhibit neurite extension defects in culture (Dawson et al., 2001). Therefore, mutant hsp27 function may cause disease through its effects on proteins other than the NF subunits.

17.4.3 Lamin A/C and Associated Proteins in Neurological Disease

A second type of IF that has been linked to CMT disease are the nuclear lamins. Over the past decade, mutant forms of lamin A/C protein have been associated with various diseases ranging from progeria to lipodystrophy (reviewed in Stewart et al., 2007; Worman and Bonne, 2007). Collectively these are referred to as the laminopathies. The broad range of phenotypes generated by lamin mutations may be representative of the multiple roles lamins play through interactions with different binding partners (such as the LAPs and nesprins, as mentioned earlier). Given the potential of lamins to affect chromatin tethering at the nuclear envelope and to affect gene expression, it is also possible that the expression of different genes may be affected by the different lamin mutations. A complete loss of lamin A/C in transgenic mice results in a muscular dystrophy phenotype (Sullivan et al., 1999). In addition, these mice exhibit overt signs of peripheral neuropathy, and their peripheral neurons display NF aggregation and a loss of axon myelination similar to that observed in neuropathies such as CMT (De Sandre-Giovannoli et al., 2002). In humans, mutant lamin A/C is associated with CMT2 in an autosomal recessive manner, indicating a loss-of-function phenotype (De Sandre-Giovannoli et al., 2002). These loss-of-function phenotypes are in stark contrast to the lack of disease phenotype observed with the genetic elimination of other neuronal IFs.

The phenotype of the lamin A/C mutant mice indicates the particular susceptibility of cells of the neuromuscular system to nuclear envelope defects. In humans, other diseases associated with nuclear envelope proteins include cerebellar ataxia (Gros-Louis et al., 2007) and early-onset torsion dystonia (Ozelius et al., 1997). Mutation of the *Nesprin-1* (*Syne-1*) gene results in a form of autosomal recessive cerebellar ataxia (Gros-Louis et al., 2007). While nuclear envelope defects may be a feature of this disease, they remain to be shown. It is possible that autosomal recessive cerebellar ataxia affects a neuronal-specific isoform of nesprin, termed CPG2, which is not linked to the nuclear envelope (Cottrell et al., 2004). In contrast, early onset torsion dystonia is more clearly linked to neuronal nuclear envelope abnormalities.

The protein affected in early onset torsion dystonia is a small chaperone protein of the AAA+ family, torsinA (Ozelius et al., 1997). In CNS neurons from torsion dystonia patients, mutant torsinA localizes preferentially within the nuclear envelope (Goodchild and Dauer, 2004). This is in contrast to wild type torsinA, which is found predominantly within the endoplasmic reticulum. Within the nuclear envelope, torsinA interacts with LAP1 and nesprin-3 (Goodchild and Dauer, 2004; Nery et al., 2008). It is possible the mutant torsinA disrupts a mechanical linkage between the nuclear lamina and the cytoskeleton surrounding the nucleus. This would explain the gross abnormalities observed in the nuclear envelope of torsinA mutant mice (Goodchild and Dauer, 2004).

As with mutations in other widely expressed proteins, the tissue-specific defects observed in torsion dystonia patients and torsinA mutant mice raise the question of why certain cell types are preferentially affected by the disease. TorsinA is expressed in tissues throughout the body, with no obvious unique importance in neurons. One possible explanation for the limited effect of mutant torsinA to neurons is that it may affect other proteins that are expressed in a neuronal-specific manner. Among the multiple isoforms of nesprins, some may be better expressed in neurons and have a unique importance in these cells. The one nesprin demonstrated to interact with torsinA thus far, nesprin-3, does not exhibit a tissue-specific distribution (Wilhelmsen et al., 2005). A protein related to the nesprins that is preferentially expressed in neurons and has also been demonstrated to link to nesprin-3 is dystonin-a2 (Young and Kothary, 2008). Dystonin-a2 is a nuclear envelope/endoplasmic reticulum protein in neurons, thus making it an intriguing possibility that the mechanical linkage affected by mutant torsinA at the nuclear envelope may involve dystonin-a2.

17.5 Conclusions

The long-held view that axonal calibre is determined by NF spacing, which is governed by the phosphorylation of the NF-H and NF-M tail domains, now seems doubtful. While there is a clear link between NFs and axon calibre, the exact mechanisms that allow NFs to affect axon calibre are not entirely clear. Beyond a role in the axon, it is interesting to note that much less attention has been focussed on NFs present in dendrites and the perikaryon. In determining the basic functions of neuronal cytoplasmic IFs, it is important to bear in mind that many animals, such as the arthropods, contain no cytoplasmic IFs. Yet these animals are capable of producing neurons with defined neurite sizes. It is, then, not surprising to discover that the genetic elimination of NFs in mice results in only minor abnormalities. While axonal calibre is affected with the elimination of NFs, this has a mostly minor effect on nerve conduction velocity, and nervous function remains by-and-large normal. A reduced regenerative capacity and fragility in some older neurons appear to be the more serious consequences of loss of neuronal IF function. Disease resulting directly from neuronal IF mutations is mostly restricted to dominant gain-of-function toxicity, and is rare.

The importance of IF-interacting proteins in neurons extends beyond their ability to interact with IFs. This is demonstrated by the fact that proteins such as the plakins (plectin, dystonin, Macf1) and IF-associated motor proteins (kinesin, dynein) have arthropod relatives with essential functions in the neuromuscular system, yet arthropods lack cytoplasmic IFs. This negates an essential, conserved role in IF binding for these proteins. In the case of the large structural proteins that link to neuronal IFs, roles in connecting to other elements of the cytoskeleton, organelles, and signalling complexes are possibly more important to their functions than any ability to connect to IFs. In the case of the kinesin and dynein motor proteins, these proteins have well-established roles in transporting many non-IF cargos along microtubules. In this regard, IFs such as NFs may play a mainly modulatory/adaptive role in affecting the functions of structural and motor proteins.

On the other hand, the ability of proteins to connect directly to nuclear lamin IFs seems to be essential throughout the animal kingdom. The loss of lamin A/C function results in neuromuscular defects, among other disorders. Defects caused by mutations in the nesprins and a defective link across the nuclear envelope in early onset torsion dystonia may result in a reduced mechanical integrity of the cell, as well as defects in gene regulation. Unlike other NF networks, most metazoan cell types cannot survive and function normally in the absence of a functional nuclear lamin network. It is no surprise, then, that several neurological disorders have now been described that are associated with nuclear envelope defects. The tissue-specific effect of mutations in nuclear envelope-associated proteins indicates the possibility of tissue-specific proteins being associated with the nuclear envelope and aiding in the generation of disease.

Future studies in the field will hopefully expand on the role of neuronal IFs in regeneration, as this seems to be one of the basic functions most likely to be important to these proteins. The ability of IFs and IF-associated proteins to transport signalling protein complexes is also intriguing, in that it may indicate a necessity of neuronal IFs to mediate adaptive responses in the mature nervous system. The fine-tuning in neuronal connections is dependent on signalling via multiple kinases, phosphatases, and other signalling enzymes. Normal cellular signalling is likely to be compromised in the absence of neuronal IFs and IF-associated proteins. Although the mouse nervous system can function relatively well in the absence of multiple cytoplasmic neuronal IFs, it is hard to imagine that the most abundant filamentous network present in neurons, and the proteins that normally connect to it, would not serve a useful purpose. Finally, the role of NF aggregates in causing human disease is still largely ambiguous. While it is possible that NF aggregates promote disease causing mechanisms, we think it is unlikely that they are a necessary part of the etiology of most neurological diseases. A new focus on mechanisms such as general defects in protein folding and degradation may help to clarify this issue.

Acknowledgements We thank Dr. Venkat Rao for providing us electron microscope images used in Fig. 17.2. R.K. is funded by a grant from the Canadian Institutes of Health Research.

References

Andra K, Kornacker I, Jorgl A, Zorer M, Spazierer D, Fuchs P, Fischer I, Wiche G (2003) Plectin-isoform-specific rescue of hemidesmosomal defects in plectin (-/-) keratinocytes. J Invest Dermatol 120:189–197

Andra K, Lassmann H, Bittner R, Shorny S, Fassler R, Propst F, Wiche G (1997) Targeted inactivation of plectin reveals essential function in maintaining the integrity of skin, muscle, and heart cytoarchitecture. Genes Dev 11:3143–3156

Apel ED, Lewis RM, Grady RM, Sanes JR (2000) Syne-1, a dystrophin- and Klarsicht-related protein associated with synaptic nuclei at the neuromuscular junction. J Biol Chem 275: 31986–31995

Arstila AU, Hopsu-Havu VK (1967) Nuclear and cytoplasmic microfilaments in the pineal chief cells of the rat. Z Zellforsch Mikrosk Anat 80:22–28

Barry DM, Millecamps S, Julien JP, Garcia ML (2007) New movements in neurofilament transport, turnover and disease. Exp Cell Res 313:2110–2120

Bernier G, Mathieu M, De Repentigny Y, Vidal SM, Kothary R (1996) Cloning and characterization of mouse ACF7, a novel member of the dystonin subfamily of actin binding proteins. Genomics 38:19–29

Bertelli E, Regoli M, Fonzi L, Occhini R, Mannucci S, Ermini L, Toti P (2007) Nestin expression in adult and developing human kidney. J Histochem Cytochem 55:411–421

Bhosle RC, Michele DE, Campbell KP, Li Z, Robson RM (2006) Interactions of intermediate filament protein synemin with dystrophin and utrophin. Biochem Biophys Res Commun 346:768–777

Bjorkdahl C, Sjogren MJ, Zhou X, Concha H, Avila J, Winblad B, Pei JJ (2008) Small heat shock proteins Hsp27 or alphaB-crystallin and the protein components of neurofibrillary tangles: tau and neurofilaments. J Neurosci Res 86:1343–1352

Brown A, Bernier G, Mathieu M, Rossant J, Kothary R (1995) The mouse dystonia musculorum gene is a neural isoform of bullous pemphigoid antigen 1. Nat Genet 10:301–306

Callicott JH, Straub RE, Pezawas L, Egan MF, Mattay VS, Hariri AR, Verchinski BA, Meyer-Lindenberg A, Balkissoon R, Kolachana B, Goldberg TE, Weinberger DR (2005) Variation in DISC1 affects hippocampal structure and function and increases risk for schizophrenia. Proc Natl Acad Sci USA 102:8627–8632

Camargo LM, Collura V, Rain JC, Mizuguchi K, Hermjakob H, Kerrien S, Bonnert TP, Whiting PJ, Brandon NJ (2007) Disrupted in Schizophrenia 1 Interactome: evidence for the close connectivity of risk genes and a potential synaptic basis for schizophrenia. Mol Psychiatry 12:74–86

Chen HJ, Lin CM, Lin CS, Perez-Olle R, Leung CL, Liem RK (2006) The role of microtubule actin cross-linking factor 1 (MACF1) in the Wnt signaling pathway. Genes Dev 20:1933–1945

Chiu FC, Barnes EA, Das K, Haley J, Socolow P, Macaluso FP, Fant J (1989) Characterization of a novel 66 kd subunit of mammalian neurofilaments. Neuron 2:1435–1445

Chou YH, Khuon S, Herrmann H, Goldman RD (2003) Nestin promotes the phosphorylation-dependent disassembly of vimentin intermediate filaments during mitosis. Mol Biol Cell 14:1468–1478

Cochard P, Paulin D (1984) Initial expression of neurofilaments and vimentin in the central and peripheral nervous system of the mouse embryo in vivo. J Neurosci 4:2080–2094

Colucci-Guyon E, Portier MM, Dunia I, Paulin D, Pournin S, Babinet C (1994) Mice lacking vimentin develop and reproduce without an obvious phenotype. Cell 79:679–694

Cote F, Collard JF, Julien JP (1993) Progressive neuronopathy in transgenic mice expressing the human neurofilament heavy gene: a mouse model of amyotrophic lateral sclerosis. Cell 73: 35–46

Cottrell JR, Borok E, Horvath TL, Nedivi E (2004) CPG2: a brain- and synapse-specific protein that regulates the endocytosis of glutamate receptors. Neuron 44:677–690

Crisp M, Liu Q, Roux K, Rattner JB, Shanahan C, Burke B, Stahl PD, Hodzic D (2006) Coupling of the nucleus and cytoplasm: role of the LINC complex. J Cell Biol 172:41–53

Dawson HN, Ferreira A, Eyster MV, Ghoshal N, Binder LI, Vitek MP (2001) Inhibition of neuronal maturation in primary hippocampal neurons from tau deficient mice. J Cell Sci 114: 1179–1187

De Sandre-Giovannoli A, Chaouch M, Kozlov S, Vallat JM, Tazir M, Kassouri N, Szepetowski P, Hammadouche T, Vandenberghe A, Stewart CL, Grid D, Levy N (2002) Homozygous defects in LMNA, encoding lamin A/C nuclear-envelope proteins, cause autosomal recessive axonal neuropathy in human (Charcot-Marie-Tooth disorder type 2) and mouse. Am J Hum Genet 70:726–736

de Waegh SM, Lee VM, Brady ST (1992) Local modulation of neurofilament phosphorylation, axonal caliber, and slow axonal transport by myelinating Schwann cells. Cell 68:451–463

Dechat T, Korbei B, Vaughan OA, Vlcek S, Hutchison CJ, Foisner R (2000) Lamina-associated polypeptide 2alpha binds intranuclear A-type lamins. J Cell Sci 113:3473–3484

Dorner D, Gotzmann J, Foisner R (2007) Nucleoplasmic lamins and their interaction partners, LAP2alpha, Rb, and BAF, in transcriptional regulation. Febs J 274:1362–1373

Dorner D, Vlcek S, Foeger N, Gajewski A, Makolm C, Gotzmann J, Hutchison CJ, Foisner R (2006) Lamina-associated polypeptide 2alpha regulates cell cycle progression and differentiation via the retinoblastoma-E2F pathway. J Cell Biol 173:83–93

Elder GA, Friedrich VL Jr, Bosco P, Kang C, Gourov A, Tu PH, Lee VM, Lazzarini RA (1998) Absence of the mid-sized neurofilament subunit decreases axonal calibers, levels of light neurofilament (NF-L), and neurofilament content. J Cell Biol 141:727–739

Evgrafov OV, Mersiyanova I, Irobi J, Van Den Bosch L, Dierick I, Leung CL, Schagina O, Verpoorten N, Van Impe K, Fedotov V, Dadali E, Auer-Grumbach M, Windpassinger C, Wagner K, Mitrovic Z, Hilton-Jones D, Talbot K, Martin JJ, Vasserman N, Tverskaya S, Polyakov A, Liem RK, Gettemans J, Robberecht W, De Jonghe P, Timmerman V (2004) Mutant small heat-shock protein 27 causes axonal Charcot-Marie-Tooth disease and distal hereditary motor neuropathy. Nat Genet 36:602–606

Eyer J, Cleveland DW, Wong PC, Peterson AC (1998) Pathogenesis of two axonopathies does not require axonal neurofilaments. Nature 391:584–587

Eyer J, Peterson A (1994) Neurofilament-deficient axons and perikaryal aggregates in viable transgenic mice expressing a neurofilament-beta-galactosidase fusion protein. Neuron 12:389–405

Feinstein SC, Wilson L (2005) Inability of tau to properly regulate neuronal microtubule dynamics: a loss-of-function mechanism by which tau might mediate neuronal cell death. Biochim Biophys Acta 1739:268–279

Finlan LE, Sproul D, Thomson I, Boyle S, Kerr E, Perry P, Ylstra B, Chubb JR, Bickmore WA (2008) Recruitment to the nuclear periphery can alter expression of genes in human cells. PLoS Genet 4:e1000039

Fliegner KH, Kaplan MP, Wood TL, Pintar JE, Liem RK (1994) Expression of the gene for the neuronal intermediate filament protein alpha-internexin coincides with the onset of neuronal differentiation in the developing rat nervous system. J Comp Neurol 342:161–173

Foisner R, Gerace L (1993) Integral membrane proteins of the nuclear envelope interact with lamins and chromosomes, and binding is modulated by mitotic phosphorylation. Cell 73: 1267–1279

Fornaro M, Lee JM, Raimondo S, Nicolino S, Geuna S, Giacobini-Robecchi M (2008) Neuronal intermediate filament expression in rat dorsal root ganglia sensory neurons: an in vivo and in vitro study. Neuroscience 153:1153–1163

Frappier T, Stetzkowski-Marden F, Pradel LA (1991) Interaction domains of neurofilament light chain and brain spectrin. Biochem J 275:521–527

Fuchs P, Spazierer D, Wiche G (2005) Plectin rodless isoform expression and its detection in mouse brain. Cell Mol Neurobiol 25:1141–1150

Furukawa K (1999) LAP2 binding protein 1 (L2BP1/BAF) is a candidate mediator of LAP2-chromatin interaction. J Cell Sci 112:2485–2492

Furukawa K, Fritze CE, Gerace L (1998) The major nuclear envelope targeting domain of LAP2 coincides with its lamin binding region but is distinct from its chromatin interaction domain. J Biol Chem 273:4213–4219

Furukawa K, Kondo T (1998) Identification of the lamina-associated-polypeptide-2-binding domain of B-type lamin. Eur J Biochem 251:729–733

Gall JG (1966) Microtubule fine structure. J Cell Biol 31:639–643

Garcia ML, Lobsiger CS, Shah SB, Deerinck TJ, Crum J, Young D, Ward CM, Crawford TO, Gotow T, Uchiyama Y, Ellisman MH, Calcutt NA, Cleveland DW (2003) NF-M is an essential target for the myelin-directed "outside-in" signaling cascade that mediates radial axonal growth. J Cell Biol 163:1011–1020

Garcia ML, Rao MV, Fujimoto J, Garcia VB, Shah SB, Crum J, Gotow T, Uchiyama Y, Ellisman M, Calcutt NA, Cleveland DW (2009) Phosphorylation of highly conserved neurofilament medium KSP repeats is not required for myelin-dependent radial axonal growth. J Neurosci 29: 1277–1284

Gong TW, Besirli CG, Lomax MI (2001) MACF1 gene structure: a hybrid of plectin and dystrophin. Mamm Genome 12:852–861

Goodchild RE, Dauer WT (2004) Mislocalization to the nuclear envelope: an effect of the dystonia-causing torsinA mutation. Proc Natl Acad Sci USA 101:847–852

Gregor M, Zeold A, Oehler S, Marobela KA, Fuchs P, Weigel G, Hardie DG, Wiche G (2006) Plectin scaffolds recruit energy-controlling AMP-activated protein kinase (AMPK) in differentiated myofibres. J Cell Sci 119:1864–1875

Gros-Louis F, Dupre N, Dion P, Fox MA, Laurent S, Verreault S, Sanes JR, Bouchard JP, Rouleau GA (2007) Mutations in SYNE1 lead to a newly discovered form of autosomal recessive cerebellar ataxia. Nat Genet 39:80–85

Guerette D, Khan PA, Savard PE, Vincent M (2007) Molecular evolution of type VI intermediate filament proteins. BMC Evol Biol 7:164

Guillaud L, Bosc C, Fourest-Lieuvin A, Denarier E, Pirollet F, Lafanechere L, Job D (1998) STOP proteins are responsible for the high degree of microtubule stabilization observed in neuronal cells. J Cell Biol 142:167–179

Guo L, Degenstein L, Dowling J, Yu QC, Wollmann R, Perman B, Fuchs E (1995) Gene targeting of BPAG1: abnormalities in mechanical strength and cell migration in stratified epithelia and neurologic degeneration. Cell 81:233–243

Gyoeva FK, Gelfand VI (1991) Coalignment of vimentin intermediate filaments with microtubules depends on kinesin. Nature 353:445–448

Harada A, Teng J, Takei Y, Oguchi K, Hirokawa N (2002) MAP2 is required for dendrite elongation, PKA anchoring in dendrites, and proper PKA signal transduction. J Cell Biol 158:541–549

He Y, Francis F, Myers KA, Yu W, Black MM, Baas PW (2005) Role of cytoplasmic dynein in the axonal transport of microtubules and neurofilaments. J Cell Biol 168:697–703

Heimann R, Shelanski ML, Liem RK (1985) Microtubule-associated proteins bind specifically to the 70-kDa neurofilament protein. J Biol Chem 260:12160–12166

Helfand BT, Mendez MG, Pugh J, Delsert C, Goldman RD (2003) A role for intermediate filaments in determining and maintaining the shape of nerve cells. Mol Biol Cell 14:5069–5081

Helfand BT, Mikami A, Vallee RB, Goldman RD (2002) A requirement for cytoplasmic dynein and dynactin in intermediate filament network assembly and organization. J Cell Biol 157:795–806

Hennah W, Varilo T, Kestila M, Paunio T, Arajarvi R, Haukka J, Parker A, Martin R, Levitzky S, Partonen T, Meyer J, Lonnqvist J, Peltonen L, Ekelund J (2003) Haplotype transmission analysis provides evidence of association for DISC1 to schizophrenia and suggests sex-dependent effects. Hum Mol Genet 12:3151–3159

Herrmann H, Aebi U (2000) Intermediate filaments and their associates: multi-talented structural elements specifying cytoarchitecture and cytodynamics. Curr Opin Cell Biol 12:79–90

Hijikata T, Nakamura A, Isokawa K, Imamura M, Yuasa K, Ishikawa R, Kohama K, Takeda S, Yorifuji H (2008) Plectin 1 links intermediate filaments to costameric sarcolemma through beta-synemin, alpha-dystrobrevin and actin. J Cell Sci 121:2062–2074

Hirokawa N (1982) Cross-linker system between neurofilaments, microtubules, and membranous organelles in frog axons revealed by the quick-freeze, deep-etching method. J Cell Biol 94: 129–142

Hirokawa N, Glicksman MA, Willard MB (1984) Organization of mammalian neurofilament polypeptides within the neuronal cytoskeleton. J Cell Biol 98:1523–1536

Hirokawa N, Hisanaga S, Shiomura Y (1988) MAP2 is a component of crossbridges between microtubules and neurofilaments in the neuronal cytoskeleton: quick-freeze, deep-etch immunoelectron microscopy and reconstitution studies. J Neurosci 8:2769–2779

Hisanaga S, Hirokawa N (1988) Structure of the peripheral domains of neurofilaments revealed by low angle rotary shadowing. J Mol Biol 202:297–305

Hodgkinson CA, Goldman D, Jaeger J, Persaud S, Kane JM, Lipsky RH, Malhotra AK (2004) Disrupted in schizophrenia 1 (DISC1): association with schizophrenia, schizoaffective disorder, and bipolar disorder. Am J Hum Genet 75:862–872

Hoffman PN, Cleveland DW, Griffin JW, Landes PW, Cowan NJ, Price DL (1987) Neurofilament gene expression: a major determinant of axonal caliber. Proc Natl Acad Sci USA 84: 3472–3476

Hoffman PN, Lasek RJ (1975) The slow component of axonal transport. Identification of major structural polypeptides of the axon and their generality among mammalian neurons. J Cell Biol 66:351–366

Hopkinson SB, Jones JC (2000) The N terminus of the transmembrane protein BP180 interacts with the N-terminal domain of BP230, thereby mediating keratin cytoskeleton anchorage to the cell surface at the site of the hemidesmosome. Mol Biol Cell 11:277–286

Hutchison CJ (2002) Lamins: building blocks or regulators of gene expression? Nat. Rev Mol Cell Biol 3:848–858

Ishikawa H, Bischoff R, Holtzer H (1968) Mitosis and intermediate-sized filaments in developing skeletal muscle. J Cell Biol 38:538–555

Izmiryan A, Cheraud Y, Khanamiryan L, Leterrier JF, Federici T, Peltekian E, Moura-Neto V, Paulin D, Li Z, Xue ZG (2006) Different expression of synemin isoforms in glia and neurons during nervous system development. Glia 54:204–213

Izmiryan A, Franco CA, Paulin D, Li Z, Xue Z (2009) Synemin isoforms during mouse development: multiplicity of partners in vascular and neuronal systems. Exp Cell Res 315:769–783

Jacomy H, Zhu Q, Couillard-Despres S, Beaulieu JM, Julien JP (1999) Disruption of type IV intermediate filament network in mice lacking the neurofilament medium and heavy subunits. J Neurochem 73:972–984

Jordanova A, De Jonghe P, Boerkoel CF, Takashima H, De Vriendt E, Ceuterick C, Martin JJ, Butler IJ, Mancias P, Papasozomenos S, Terespolsky D, Potocki L, Brown CW, Shy M, Rita DA, Tournev I, Kremensky I, Lupski JR, Timmerman V (2003) Mutations in the neurofilament light chain gene (NEFL) cause early onset severe Charcot-Marie-Tooth disease. Brain 126: 590–597

Jung C, Chylinski TM, Pimenta A, Ortiz D, Shea TB (2004) Neurofilament transport is dependent on actin and myosin. J Neurosci 24:9486–9496

Jung C, Lee S, Ortiz D, Zhu Q, Julien JP, Shea TB (2005) The high and middle molecular weight neurofilament subunits regulate the association of neurofilaments with kinesin: inhibition by phosphorylation of the high molecular weight subunit. Brain Res Mol Brain Res 141:151–155

Kaech S, Ludin B, Matus A (1996) Cytoskeletal plasticity in cells expressing neuronal microtubule-associated proteins. Neuron 17:1189–1199

Kamiya A, Tomoda T, Chang J, Takaki M, Zhan C, Morita M, Cascio MB, Elashvili S, Koizumi H, Takanezawa Y, Dickerson F, Yolken R, Arai H, Sawa A (2006) DISC1-NDEL1/NUDEL protein interaction, an essential component for neurite outgrowth, is modulated by genetic variations of DISC1. Hum Mol Genet 15:3313–3323

Kaplan MP, Chin SS, Fliegner KH, Liem RK (1990) Alpha-internexin, a novel neuronal intermediate filament protein, precedes the low molecular weight neurofilament protein (NF-L) in the developing rat brain. J Neurosci 10:2735–2748

Karakesisoglou I, Yang Y, Fuchs E (2000) An epidermal plakin that integrates actin and microtubule networks at cellular junctions. J Cell Biol 149:195–208

Ketema M, Wilhelmsen K, Kuikman I, Janssen H, Hodzic D, Sonnenberg A (2007) Requirements for the localization of nesprin-3 at the nuclear envelope and its interaction with plectin. J Cell Sci 120:3384–3394

Kodama A, Karakesisoglou I, Wong E, Vaezi A, Fuchs E (2003) ACF7: an essential integrator of microtubule dynamics. Cell 115:343–354

Kosik KS, Shimura H (2005) Phosphorylated tau and the neurodegenerative foldopathies. Biochim Biophys Acta 1739:298–310

Kothary R, Clapoff S, Brown A, Campbell R, Peterson A, Rossant J (1988) A transgene containing lacZ inserted into the dystonia locus is expressed in neural tube. Nature 335:435–437

Kriz J, Zhu Q, Julien JP, Padjen AL (2000) Electrophysiological properties of axons in mice lacking neurofilament subunit genes: disparity between conduction velocity and axon diameter in absence of NF-H. Brain Res 885:32–44

Lariviere RC, Nguyen MD, Ribeiro-da-Silva A, Julien JP (2002) Reduced number of unmyelinated sensory axons in peripherin null mice. J Neurochem 81:525–532

Ledbetter M, Porter K (1963) A "microtubule" in plant cell fine structure. J Cell Biol 19: 239–250

Lee MK, Cleveland DW (1996) Neuronal intermediate filaments. Annu Rev Neurosci 19:187–217

Lee S, Harris KL, Whitington PM, Kolodziej PA (2000) Short stop is allelic to kakapo, and encodes rod-like cytoskeletal-associated proteins required for axon extension. J Neurosci 20:1096–1108

Lee MK, Xu Z, Wong PC, Cleveland DW (1993) Neurofilaments are obligate heteropolymers in vivo. J Cell Biol 122:1337–1350

Letournel F, Bocquet A, Dubas F, Barthelaix A, Eyer J (2003) Stable tubule only polypeptides (STOP) proteins co-aggregate with spheroid neurofilaments in amyotrophic lateral sclerosis. J Neuropathol Exp Neurol 62:1211–1219

Leung CL, Sun D, Liem RK (1999a) The intermediate filament protein peripherin is the specific interaction partner of mouse BPAG1-n (dystonin) in neurons. J Cell Biol 144:435–446

Leung CL, Sun D, Zheng M, Knowles DR, Liem RK (1999b) Microtubule actin cross-linking factor (MACF): a hybrid of dystonin and dystrophin that can interact with the actin and microtubule cytoskeletons. J Cell Biol 147:1275–1286

Leung CL, Zheng M, Prater SM, Liem RK (2001) The BPAG1 locus: alternative splicing produces multiple isoforms with distinct cytoskeletal linker domains, including predominant isoforms in neurons and muscles. J Cell Biol 154:691–697

Levavasseur F, Zhu Q, Julien JP (1999) No requirement of alpha-internexin for nervous system development and for radial growth of axons. Brain Res Mol Brain Res 69:104–112

Libotte T, Zaim H, Abraham S, Padmakumar VC, Schneider M, Lu W, Munck M, Hutchison C, Wehnert M, Fahrenkrog B, Sauder U, Aebi U, Noegel AA, Karakesisoglou I (2005) Lamin A/C-dependent localization of Nesprin-2, a giant scaffolder at the nuclear envelope. Mol Biol Cell 16:3411–3424

Liem RK, Hutchison SB (1982) Purification of individual components of the neurofilament triplet: filament assembly from the 70 000-dalton subunit. Biochemistry 21:3221–3226

Liem RK, Yen SH, Salomon GD, Shelanski ML (1978) Intermediate filaments in nervous tissues. J Cell Biol 79:637–645

Lin CM, Chen HJ, Leung CL, Parry DA, Liem RK (2005) Microtubule actin crosslinking factor 1b: a novel plakin that localizes to the Golgi complex. J Cell Sci 118:3727–3738

Lin H, Schlaepfer WW (2006) Role of neurofilament aggregation in motor neuron disease. Ann Neurol 60:399–406

Luke Y, Zaim H, Karakesisoglou I, Jaeger VM, Sellin L, Lu W, Schneider M, Neumann S, Beijer A, Munck M, Padmakumar VC, Gloy J, Walz G, Noegel AA (2008) Nesprin-2 Giant (NUANCE) maintains nuclear envelope architecture and composition in skin. J Cell Sci 121:1887–1898

Lunter PC, Wiche G (2002) Direct binding of plectin to Fer kinase and negative regulation of its catalytic activity. Biochem Biophys Res Commun 296:904–910

Markiewicz E, Dechat T, Foisner R, Quinlan RA, Hutchison CJ (2002) Lamin A/C binding protein LAP2alpha is required for nuclear anchorage of retinoblastoma protein. Mol Biol Cell 13: 4401–4413

Mattout A, Goldberg M, Tzur Y, Margalit A, Gruenbaum Y (2007) Specific and conserved sequences in D. melanogaster and C. elegans lamins and histone H2A mediate the attachment of lamins to chromosomes. J Cell Sci 120:77–85

McGraw TS, Mickle JP, Shaw G, Streit WJ (2002) Axonally transported peripheral signals regulate alpha-internexin expression in regenerating motoneurons. J Neurosci 22:4955–4963

Melcer S, Gruenbaum Y, Krohne G (2007) Invertebrate lamins. Exp Cell Res 313:2157–2166

Mersiyanova IV, Perepelov AV, Polyakov AV, Sitnikov VF, Dadali EL, Oparin RB, Petrin AN, Evgrafov OV (2000) A new variant of Charcot-Marie-Tooth disease type 2 is probably the result of a mutation in the neurofilament-light gene. Am J Hum Genet 67:37–46

Millar JK, Wilson-Annan JC, Anderson S, Christie S, Taylor MS, Semple CA, Devon RS, Clair DM, Muir WJ, Blackwood DH, Porteous DJ (2000) Disruption of two novel genes by a translocation co-segregating with schizophrenia. Hum Mol Genet 9:1415–1423

Mislow JM, Kim MS, Davis DB, McNally EM (2002) Myne-1, a spectrin repeat transmembrane protein of the myocyte inner nuclear membrane, interacts with lamin A/C. J Cell Sci 115:61–70

Miyasaka H, Okabe S, Ishiguro K, Uchida T, Hirokawa N (1993) Interaction of the tail domain of high molecular weight subunits of neurofilaments with the COOH-terminal region of tubulin and its regulation by tau protein kinase II. J Biol Chem 268:22695–22702

Mizuno Y, Guyon JR, Okamoto K, Kunkel LM (2007) Synemin expression in brain. Muscle Nerve 36:497–504

Mizuno Y, Guyon JR, Okamoto K, Kunkel LM (2009) Expression of synemin in the mouse spinal cord. Muscle Nerve 39:634–641

Morris JA, Kandpal G, Ma L, Austin CP (2003) DISC1 (Disrupted-In-Schizophrenia 1) is a centrosome-associated protein that interacts with MAP1A, MIPT3, ATF4/5 and NUDEL: regulation and loss of interaction with mutation. Hum Mol Genet 12:1591–1608

Motil J, Chan WK, Dubey M, Chaudhury P, Pimenta A, Chylinski TM, Ortiz DT, Shea TB (2006) Dynein mediates retrograde neurofilament transport within axons and anterograde delivery of NFs from perikarya into axons: regulation by multiple phosphorylation events. Cell Motil. Cytoskeleton 63:266–286

Nery FC, Zeng J, Niland BP, Hewett J, Farley J, Irimia D, Li Y, Wiche G, Sonnenberg A, Breakefield XO (2008) TorsinA binds the KASH domain of nesprins and participates in linkage between nuclear envelope and cytoskeleton. J Cell Sci 121:3476–3486

Nguyen MD, Lariviere RC, Julien JP (2000) Reduction of axonal caliber does not alleviate motor neuron disease caused by mutant superoxide dismutase 1. Proc Natl Acad Sci USA 97: 12306–12311

Nukada H, Dyck PJ (1984) Decreased axon caliber and neurofilaments in hereditary motor and sensory neuropathy, type I. Ann Neurol 16:238–241

Osmanagic-Myers S, Wiche G (2004) Plectin-RACK1 (receptor for activated C kinase 1) scaffolding: a novel mechanism to regulate protein kinase C activity. J Biol Chem 279:18701–18710

Ozelius LJ, Hewett JW, Page CE, Bressman SB, Kramer PL, Shalish C, de Leon D, Brin MF, Raymond D, Corey DP, Fahn S, Risch NJ, Buckler AJ, Gusella JF, Breakefield XO (1997) The early-onset torsion dystonia gene (DYT1) encodes an ATP-binding protein. Nat Genet 17:40–48

Pachter JS, Liem RK (1985) Alpha-Internexin, a 66-kD intermediate filament-binding protein from mammalian central nervous tissues. J Cell Biol 101:1316–1322

Padmakumar VC, Abraham S, Braune S, Noegel AA, Tunggal B, Karakesisoglou I, Korenbaum E (2004) Enaptin, a giant actin-binding protein, is an element of the nuclear membrane and the actin cytoskeleton. Exp Cell Res 295:330–339

Padmakumar VC, Libotte T, Lu W, Zaim H, Abraham S, Noegel AA, Gotzmann J, Foisner R, Karakesisoglou I (2005) The inner nuclear membrane protein Sun1 mediates the anchorage of Nesprin-2 to the nuclear envelope. J Cell Sci 118:3419–3430

Patschan D, Michurina T, Shi HK, Dolff S, Brodsky SV, Vasilieva T, Cohen-Gould L, Winaver J, Chander PN, Enikolopov G, Goligorsky MS (2007) Normal distribution and medullary-to-cortical shift of Nestin-expressing cells in acute renal ischemia. Kidney Int 71: 744–754

Pekovic V, Harborth J, Broers JL, Ramaekers FC, van Engelen B, Lammens M, von Zglinicki T, Foisner R, Hutchison C, Markiewicz E (2007) Nucleoplasmic LAP2alpha-lamin A complexes are required to maintain a proliferative state in human fibroblasts. J Cell Biol 176: 163–172

Perez-Olle R, Leung CL, Liem RK (2002) Effects of Charcot-Marie-Tooth-linked mutations of the neurofilament light subunit on intermediate filament formation. J Cell Sci 115:4937–4946

Perlson E, Hanz S, Ben-Yaakov K, Segal-Ruder Y, Seger R, Fainzilber M (2005) Vimentin-dependent spatial translocation of an activated MAP kinase in injured nerve. Neuron 45: 715–726

Perrot R, Berges R, Bocquet A, Eyer J (2008) Review of the multiple aspects of neurofilament functions, and their possible contribution to neurodegeneration. Mol Neurobiol 38:27–65

Perrot R, Lonchampt P, Peterson AC, Eyer J (2007) Axonal neurofilaments control multiple fiber properties but do not influence structure or spacing of nodes of Ranvier. J Neurosci 27: 9573–9584

Portier MM, Brachet P, Croizat B, Gros F (1983a) Regulation of peripherin in mouse neuroblastoma and rat PC 12 pheochromocytoma cell lines. Dev Neurosci 6:215–226

Portier MM, de Nechaud B, Gros F (1983b) Peripherin, a new member of the intermediate filament protein family. Dev Neurosci 6:335–344

Prahlad V, Yoon M, Moir RD, Vale RD, Goldman RD (1998) Rapid movements of vimentin on microtubule tracks: kinesin-dependent assembly of intermediate filament networks. J Cell Biol 143:159–170

Rakhit R, Robertson J, Vande Velde C, Horne P, Ruth DM, Griffin J, Cleveland DW, Cashman NR, Chakrabartty A (2007) An immunological epitope selective for pathological monomer-misfolded SOD1 in ALS. Nat Med 13:754–759

Rao MV, Campbell J, Yuan A, Kumar A, Gotow T, Uchiyama Y, Nixon RA (2003) The neurofilament middle molecular mass subunit carboxyl-terminal tail domains is essential for the radial growth and cytoskeletal architecture of axons but not for regulating neurofilament transport rate. J Cell Biol 163:1021–1031

Rao MV, Garcia ML, Miyazaki Y, Gotow T, Yuan A, Mattina S, Ward CM, Calcutt NA, Uchiyama Y, Nixon RA, Cleveland DW (2002) Gene replacement in mice reveals that the heavily phosphorylated tail of neurofilament heavy subunit does not affect axonal caliber or the transit of cargoes in slow axonal transport. J Cell Biol 158:681–693

Rao MV, Houseweart MK, Williamson TL, Crawford TO, Folmer J, Cleveland DW (1998) Neurofilament-dependent radial growth of motor axons and axonal organization of neurofilaments does not require the neurofilament heavy subunit (NF-H) or its phosphorylation. J Cell Biol 143:171–181

Reddy KL, Zullo JM, Bertolino E, Singh H (2008) Transcriptional repression mediated by repositioning of genes to the nuclear lamina. Nature 452:243–247

Reuter JE, Nardine TM, Penton A, Billuart P, Scott EK, Usui T, Uemura T, Luo L (2003) A mosaic genetic screen for genes necessary for Drosophila mushroom body neuronal morphogenesis. Development 130:1203–1213

Robberecht W (2000) Genetics of amyotrophic lateral sclerosis. J Neurol 247:2–6

Roux KJ, Crisp ML, Liu Q, Kim D, Kozlov S, Stewart CL, Burke B (2009) Nesprin 4 is an outer nuclear membrane protein that can induce kinesin-mediated cell polarization. Proc Natl Acad Sci USA 106:2194–2199

Roy S, Coffee P, Smith G, Liem RK, Brady ST, Black MM (2000) Neurofilaments are transported rapidly but intermittently in axons: implications for slow axonal transport. J Neurosci 20: 6849–6861

Schosser A, Gaysina D, Cohen-Woods S, Chow PC, Martucci L, Craddock N, Farmer A, Korszun A, Gunasinghe C, Gray J, Jones L, Tozzi F, Perry J, Muglia P, Owen MJ, Craig IW, McGuffin P (2010) Association of DISC1 and TSNAX genes and affective disorders in the depression case-control (DeCC) and bipolar affective case-control (BACCS) studies. Mol Psychiatry 15: 844–849

Schymick JC, Talbot K, Traynor BJ (2007) Genetics of sporadic amyotrophic lateral sclerosis. Hum Mol Genet 16:R233–R242

Segura-Totten M, Kowalski AK, Craigie R, Wilson KL (2002) Barrier-to-autointegration factor: major roles in chromatin decondensation and nuclear assembly. J Cell Biol 158:475–485

Selby CC, Bear RS (1956) The structure of actin-rich filaments of muscles according to X-ray diffraction. J Biophys Biochem Cytol 2:71–85

Shah JV, Flanagan LA, Janmey PA, Leterrier JF (2000) Bidirectional translocation of neurofilaments along microtubules mediated in part by dynein/dynactin. Mol Biol Cell 11:3495–3508

Shea TB (1990) Transient increase in vimentin in axonal cytoskeletons during differentiation in NB2a/d1 cells. Brain Res 521:338–342

Shi H, Cui H, Alam G, Gunning WT, Nestor A, Giovannucci D, Zhang M, Ding HF (2008) Nestin expression defines both glial and neuronal progenitors in postnatal sympathetic ganglia. J Comp Neurol 508:867–878

Shim SY, Samuels BA, Wang J, Neumayer G, Belzil C, Ayala R, Shi Y, Shi Y, Tsai LH, Nguyen MD (2008) Ndel1 controls the dynein-mediated transport of vimentin during neurite outgrowth. J Biol Chem 283:12232–12240

Shimizu S, Matsuzaki S, Hattori T, Kumamoto N, Miyoshi K, Katayama T, Tohyama M (2008) DISC1-kendrin interaction is involved in centrosomal microtubule network formation. Biochem Biophys Res Commun 377:1051–1056

Shimizu H, Takizawa Y, Pulkkinen L, Murata S, Kawai M, Hachisuka H, Udono M, Uitto J, Nishikawa T (1999) Epidermolysis bullosa simplex associated with muscular dystrophy: phenotype-genotype correlations and review of the literature. J Am Acad Dermatol 41:950–956

Smith FJ, Eady RA, Leigh IM, McMillan JR, Rugg EL, Kelsell DP, Bryant SP, Spurr NK, Geddes JF, Kirtschig G, Milana G, de Bono AG, Owaribe K, Wiche G, Pulkkinen L, Uitto J, McLean WH, Lane EB (1996) Plectin deficiency results in muscular dystrophy with epidermolysis bullosa. Nat Genet 13:450–457

Somech R, Shaklai S, Geller O, Amariglio N, Simon AJ, Rechavi G, Gal-Yam EN (2005) The nuclear-envelope protein and transcriptional repressor LAP2beta interacts with HDAC3 at the nuclear periphery, and induces histone H4 deacetylation. J Cell Sci 118:4017–4025

Starr DA, Fischer JA (2005) KASH 'n Karry: the KASH domain family of cargo-specific cytoskeletal adaptor proteins. Bioessays 27:1136–1146

Steinboeck F, Kristufek D (2005) Identification of the cytolinker protein plectin in neuronal cells – expression of a rodless isoform in neurons of the rat superior cervical ganglion. Cell Mol Neurobiol 25:1151–1169

Steinert PM, Chou YH, Prahlad V, Parry DA, Marekov LN, Wu KC, Jang SI, Goldman RD (1999) A high molecular weight intermediate filament-associated protein in BHK-21 cells is nestin, a type VI intermediate filament protein. limited co-assembly in vitro to form heteropolymers with type III vimentin and type IV alpha-internexin. J Biol Chem 274:9881–9890

Stewart CL, Roux KJ, Burke B (2007) Blurring the boundary: the nuclear envelope extends its reach. Science 318:1408–1412

Subramanian A, Prokop A, Yamamoto M, Sugimura K, Uemura T, Betschinger J, Knoblich JA, Volk T (2003) Shortstop recruits EB1/APC1 and promotes microtubule assembly at the muscle-tendon junction. Curr Biol 13:1086–1095

Sullivan T, Escalante-Alcalde D, Bhatt H, Anver M, Bhat N, Nagashima K, Stewart CL, Burke B (1999) Loss of A-type lamin expression compromises nuclear envelope integrity leading to muscular dystrophy. J Cell Biol 147:913–920

Takemura R, Okabe S, Umeyama T, Kanai Y, Cowan NJ, Hirokawa N (1992) Increased microtubule stability and alpha tubulin acetylation in cells transfected with microtubule-associated proteins MAP1B, MAP2 or tau. J Cell Sci 103:953–964

Thompson MA, Ziff EB (1989) Structure of the gene encoding peripherin, an NGF-regulated neuronal-specific type III intermediate filament protein. Neuron 2:1043–1053

Tomppo L, Hennah W, Lahermo P, Loukola A, Tuulio-Henriksson A, Suvisaari J, Partonen T, Ekelund J, Lonnqvist J, Peltonen L (2009) Association between genes of disrupted in schizophrenia 1 (DISC1) interactors and schizophrenia supports the role of the DISC1 pathway in the etiology of major mental illnesses. Biol Psychiatry 65:1055–1062

Toth C, Shim SY, Wang J, Jiang Y, Neumayer G, Belzil C, Liu WQ, Martinez J, Zochodne D, Nguyen MD (2008) Ndel1 promotes axon regeneration via intermediate filaments. PLoS ONE 3:e2014

Trivedi N, Jung P, Brown A (2007) Neurofilaments switch between distinct mobile and stationary states during their transport along axons. J Neurosci 27:507–516

Troy CM, Greene LA, Shelanski ML (1992) Neurite outgrowth in peripherin-depleted PC12 cells. J Cell Biol 117:1085–1092

Wagner OI, Ascano J, Tokito M, Leterrier JF, Janmey PA, Holzbaur EL (2004) The interaction of neurofilaments with the microtubule motor cytoplasmic dynein. Mol Biol Cell 15:5092–5100

Wang L, Ho CL, Sun D, Liem RK, Brown A (2000) Rapid movement of axonal neurofilaments interrupted by prolonged pauses. Nat Cell Biol 2:137–141

Wheeler MA, Davies JD, Zhang Q, Emerson LJ, Hunt J, Shanahan CM, Ellis JA (2007) Distinct functional domains in nesprin-1alpha and nesprin-2beta bind directly to emerin and both interactions are disrupted in X-linked Emery-Dreifuss muscular dystrophy. Exp Cell Res 313:2845–2857

Wiche G (1998) Role of plectin in cytoskeleton organization and dynamics. J Cell Sci 111: 2477–2486

Wiese C, Rolletschek A, Kania G, Blyszczuk P, Tarasov KV, Tarasova Y, Wersto RP, Boheler KR, Wobus AM (2004) Nestin expression–a property of multi-lineage progenitor cells? Cell Mol Life Sci 61:2510–2522

Wilhelmsen K, Litjens SH, Kuikman I, Tshimbalanga N, Janssen H, van den Bout I, Raymond K, Sonnenberg A (2005) Nesprin-3, a novel outer nuclear membrane protein, associates with the cytoskeletal linker protein plectin. J Cell Biol 171:799–810

Williamson TL, Bruijn LI, Zhu Q, Anderson KL, Anderson SD, Julien JP, Cleveland DW (1998) Absence of neurofilaments reduces the selective vulnerability of motor neurons and slows disease caused by a familial amyotrophic lateral sclerosis-linked superoxide dismutase 1 mutant. Proc Natl Acad Sci USA 95:9631–9636

Worman HJ, Bonne G (2007) "Laminopathies": a wide spectrum of human diseases. Exp Cell Res 313:2121–2133

Worman HJ, Gundersen GG (2006) Here come the SUNs: a nucleocytoskeletal missing link. Trends Cell Biol 16:67–69

Wu X, Kodama A, Fuchs E (2008) ACF7 regulates cytoskeletal-focal adhesion dynamics and migration and has ATPase activity. Cell 135:137–148

Xu Z, Cork LC, Griffin JW, Cleveland DW (1993) Increased expression of neurofilament subunit NF-L produces morphological alterations that resemble the pathology of human motor neuron disease. Cell 73:23–33

Xu Z, Marszalek JR, Lee MK, Wong PC, Folmer J, Crawford TO, Hsieh ST, Griffin JW, Cleveland DW (1996) Subunit composition of neurofilaments specifies axonal diameter. J Cell Biol 133:1061–1069

Xue ZG, Cheraud Y, Brocheriou V, Izmiryan A, Titeux M, Paulin D, Li Z (2004) The mouse synemin gene encodes three intermediate filament proteins generated by alternative exon usage and different open reading frames. Exp Cell Res 298:431–444

Yabe JT, Chan WK, Wang FS, Pimenta A, Ortiz DD, Shea TB (2003) Regulation of the transition from vimentin to neurofilaments during neuronal differentiation. Cell Motil Cytoskeleton 56:193–205

Yabe JT, Jung C, Chan WK, Shea TB (2000) Phospho-dependent association of neurofilament proteins with kinesin in situ. Cell Motil Cytoskeleton 45:249–262

Yabe JT, Pimenta A, Shea TB (1999) Kinesin-mediated transport of neurofilament protein oligomers in growing axons. J Cell Sci 112(21):3799–3814

Yan Y, Jensen K, Brown A (2007) The polypeptide composition of moving and stationary neurofilaments in cultured sympathetic neurons. Cell Motil Cytoskeleton 64: 299–309

Yang Y, Dowling J, Yu QC, Kouklis P, Cleveland DW, Fuchs E (1996) An essential cytoskeletal linker protein connecting actin microfilaments to intermediate filaments. Cell 86: 655–665

Yoon KH, Yoon M, Moir RD, Khuon S, Flitney FW, Goldman RD (2001) Insights into the dynamic properties of keratin intermediate filaments in living epithelial cells. J Cell Biol 153: 503–516

Young KG, Kothary R (2007) Dystonin/Bpag1–a link to what? Cell Motil Cytoskeleton 64: 897–905

Young KG, Kothary R (2008) Dystonin/Bpag1 is a necessary endoplasmic reticulum/nuclear envelope protein in sensory neurons. Exp Cell Res 314:2750–2761

Yuan A, Rao MV, Kumar A, Julien JP, Nixon RA (2003) Neurofilament transport in vivo minimally requires hetero-oligomer formation. J Neurosci 23:9452–9458

Yuan A, Rao MV, Sasaki T, Chen Y, Kumar A, Veeranna, Liem RK, Eyer J, Peterson AC, Julien JP, Nixon RA (2006) Alpha-internexin is structurally and functionally associated with the neurofilament triplet proteins in the mature CNS. J Neurosci 26:10006–10019

Zhang Q, Bethmann C, Worth NF, Davies JD, Wasner C, Feuer A, Ragnauth CD, Yi Q, Mellad JA, Warren DT, Wheeler MA, Ellis JA, Skepper JN, Vorgerd M, Schlotter-Weigel B, Weissberg PL, Roberts RG, Wehnert M, Shanahan CM (2007) Nesprin-1 and -2 are involved in the pathogenesis of Emery Dreifuss muscular dystrophy and are critical for nuclear envelope integrity. Hum Mol Genet 16:2816–2833

Zhang Q, Ragnauth CD, Skepper JN, Worth NF, Warren DT, Roberts RG, Weissberg PL, Ellis JA, Shanahan CM (2005) Nesprin-2 is a multi-isomeric protein that binds lamin and emerin at the nuclear envelope and forms a subcellular network in skeletal muscle. J Cell Sci 118: 673–687

Zhen YY, Libotte T, Munck M, Noegel AA, Korenbaum E (2002) NUANCE, a giant protein connecting the nucleus and actin cytoskeleton. J Cell Sci 115:3207–3222

Zhu Q, Couillard-Despres S, Julien JP (1997) Delayed maturation of regenerating myelinated axons in mice lacking neurofilaments. Exp Neurol 148:299–316

Zhu Q, Lindenbaum M, Levavasseur F, Jacomy H, Julien JP (1998) Disruption of the NF-H gene increases axonal microtubule content and velocity of neurofilament transport: relief of axonopathy resulting from the toxin beta,beta'-iminodipropionitrile. J Cell Biol 143:183–193

Zimek A, Weber K (2008) In contrast to the nematode and fruit fly all 9 intron positions of the sea anemone lamin gene are conserved in human lamin genes. Eur J Cell Biol 87:305–309

Zochodne DW, Sun HS, Cheng C, Eyer J (2004) Accelerated diabetic neuropathy in axons without neurofilaments. Brain 127:2193–2200

Chapter 18
Tropomyosins in Neuronal Morphogenesis and Development

Nikki Margarita Curthoys, Peter William Gunning, and Thomas Fath

Abstract This chapter discusses different aspects of Tropomyosins (Tms) in neurons. Section 18.1 is a brief outline of the Tm isoforms found in brain, Section 18.2 a detailed description of the expression patterns of these isoforms throughout development in neurons and the brain, Section 18.3 summarises the actin binding proteins (ABPs) which are localised in neurons and which interact with these Tm isoforms, Section 18.4 discusses the growth cone, and how Tm isoforms and ABPs may work together to regulate structure and function in this compartment, Section 18.5 examines the structure of the synapse, and which Tm isoforms and ABPs are implicated in synapse morphology and function, Section 18.6 is a brief overview of the implications of Tms in neurological disorders, and Section 18.7 summarises the reasons why understanding Tm isoform functions in neurons can aid in our understanding of the processes controlling neuronal morphogenesis and development. Tm isoforms each have specific spatial distributions across tissue, cells and subcellular compartments. These distributions are temporally regulated, with expression levels and localisation of Tms changing throughout development. In brain, the repertoire of Tm isoforms expressed changes with maturation, and a number of changes in the expression levels and localisation of specific Tms are associated with the defined cellular processes of neuronal morphogenesis and differentiation. By mapping the expression of Tms in brain and in neurons, and the developmental shifts in expression, a composite picture is formed of how different Tms can augment actin filament function throughout neuronal maturation.

Keywords Actin · Actin binding proteins · Alternative splicing · Brain · Differentiation · Growth cone · Neurological disorders · Neurons · Synapse · Tropomyosin

---

T. Fath (✉)
Department of Anatomy, School of Medical Sciences, University of New South Wales, Sydney, NSW 2052, Australia
e-mail: t.fath@unsw.edu.au

R.A. Nixon, A. Yuan (eds.), *Cytoskeleton of the Nervous System*, Advances in Neurobiology 3, DOI 10.1007/978-1-4419-6787-9_18,

Contents

18.1 Introduction

18.1.1 Developmental Regulation of Tropomyosin Isoforms in Brain

Isoforms from the mammalian αTm, γTm, and δTm genes are expressed in brain. Of these three genes, the γTm gene contributes the largest number of Tropomyosin

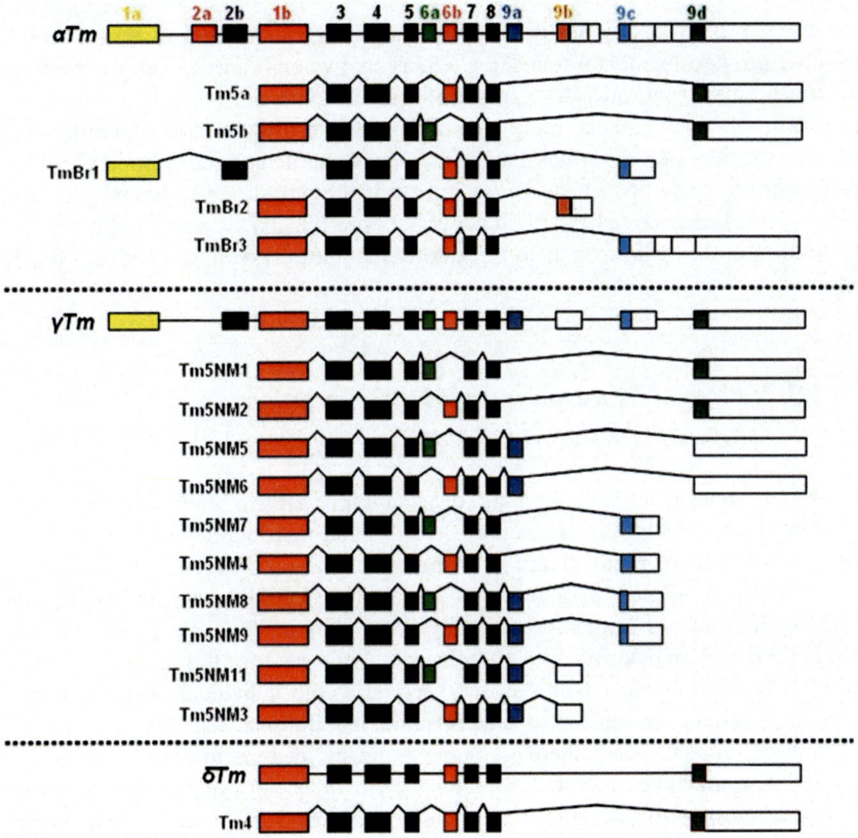

Fig. 18.1 The mammalian Tm isoforms detected in neurons are derived from αTm, γTm and δTm genes. Alternative splicing can give rise to multiple products from a single gene (figure after Gunning et al., 2005). Note that only those isoforms discussed in detail in this review are illustrated in this figure

(Tm) isoforms in brain. The mRNA of at least ten different isoforms arising from the choice of four different exons (C-terminal exons 9a, 9c, 9d, and internal exon 6a or 6b) of the γTm gene have been detected in brain (see Dufour et al., 1998a) (See Fig. 18.1). From the αTm gene, the isoforms TmBr1 and TmBr3 (Schevzov et al., 2005b), TmBr2 (Hannan et al., 1995), and Tm5a and Tm5b (Weinberger et al., 1993) are found in brain, and the mammalian δTm gene product Tm4, is also found throughout rat and mouse brain (Had et al., 1994; Schevzov et al., 2005a). While the βTm gene product Tm1 (Had et al., 1994), and possibly the αTm gene product Tm2 (Nicholson-Flynn et al., 1996; Stamm et al., 1993) are also expressed in brain, their presence and distributions in neurons is not as yet described, and so will not be further discussed in this review.

The following is a discussion on the expression of products from the αTm, γTm, and δTm genes in neurons and brain. The use of different experimental systems with regards to (1) complexity of the model (e.g. *in vitro* vs. *in vivo*), (2) species (e.g.

mouse vs. rat), (3) anatomically distinct regions (e.g. cerebellum vs. whole brain), (4) different developmental times (e.g. embryonal vs. post natal), and (5) variables of measurement (e.g. mRNA vs. protein) render the task of compiling a composite picture of developmental regulation of Tms challenging. A further confounding factor is the presence of various Tms within astrocytes and oligodendrocytes, which can influence some protein measures in neuronal cultures and in brain. For these reasons, we have tried to be explicit about the experimental systems used in each of the studies in this review, so as to be mindful that these systems are not necessarily equivalent.

18.1.2 Neuronal Maturation and Differentiation: What Are We Talking About?

As well as changes in expression and distribution of Tms in brain, development is associated with changes in the neuronal subcellular distributions of Tm isoforms. Here we explore the regional and subcellular shifts in localisation, and the expression changes of different Tms, throughout brain and neuronal development. In fact, defining the nature of "maturational change", and defining the point at which a neuron is no longer immature, is no simple task. While neurons throughout the brain develop at different time points in the lifetime of the individual, there are some general principles we can use to categorise the maturational state. Throughout the literature discussed below, there are some recurring ideas as to what constitutes a mature vs an immature neuron. Generally, the emergence of neurites, and the initial non-polarisation of these neurites are usually considered "immature" stage characteristics, as are the presence of growth cones – the dynamic, distal regions of neurites which sample the environment in search of synaptic targets. Maturation brings the replacement of growth cones with well formed synapses, and subsequent arrangement of axons and dendrites and their synapses with surrounding neurons. Those variables of neurite polarisation, growth cone loss, and synapse formation and persistence with surrounding cells are here considered to be defining characteristics of a mature neuron. The expression of a specific repertoire of Tms at various stages of development, and in various structures of developing (immature) and developed (mature) neurons (e.g. the axon shaft), is subject to complex regulation. For clarity, the different Tm isoforms will initially be considered separately. The exon composition of only those Tm isoforms covered in this review (listed in Section 18.2) are illustrated in Fig. 18.1.

18.2 Neuronally Expressed Tropomyosin Isoforms

The distribution and regulation of Tms is highly isoform-specific. A summary of distributions in immature and mature neurons of the Tm isoforms discussed in this section is given in Fig. 18.2.

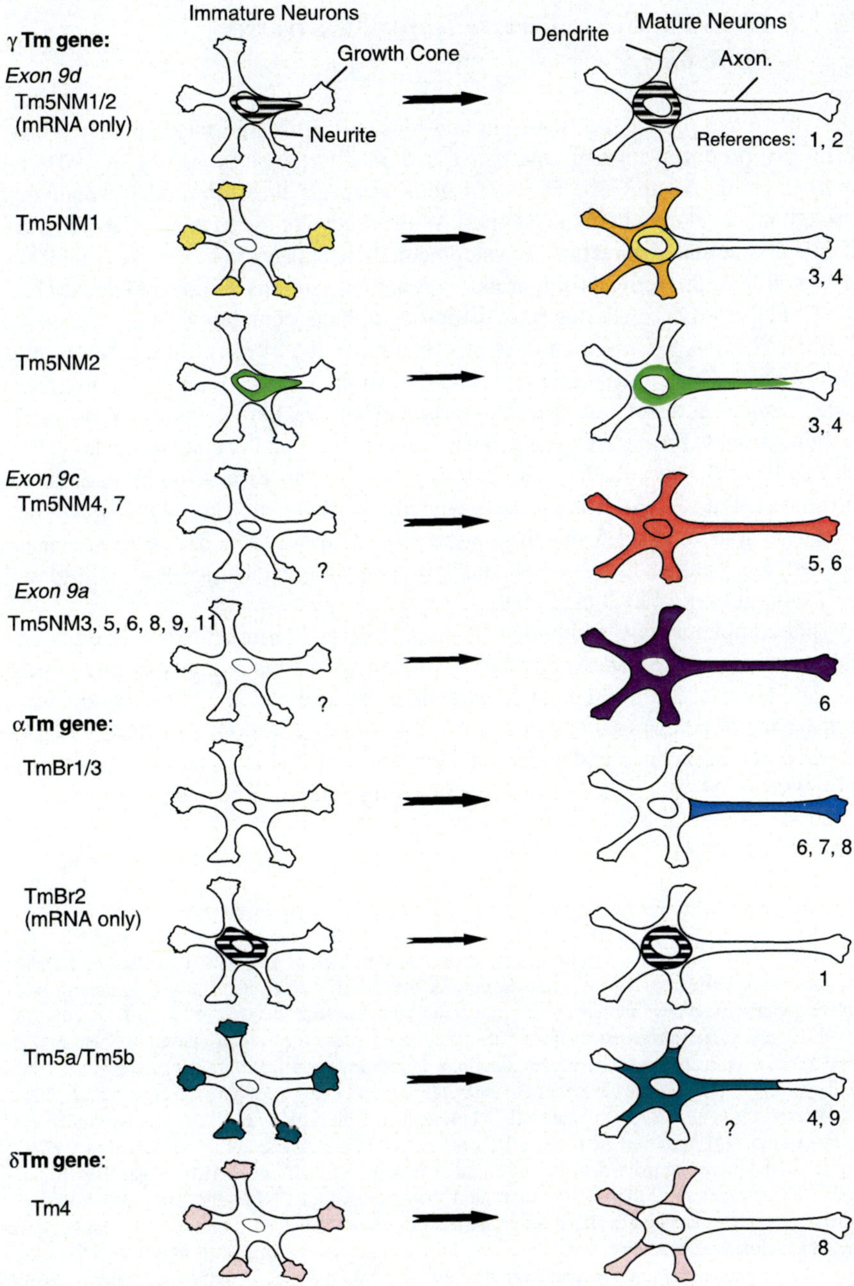

Fig. 18.2 (continued)

18.2.1 The γTm Gene: Diverse Isoforms, Diverse Functions?

The γTm gene produces at least ten non-muscle isoforms in rat brain. These isoforms are produced through alternative use of C-terminal exons (9a, 9c, 9d) and internal exons 6a and 6b. The γTm gene isoforms are differentially regulated throughout development in the rat brain. While no two of the γTm gene isoforms are identically regulated in rat brain development, there appears to be a switch in mRNA in rat whole brain from isoform products containing internal exon 6a (Tm5NM1, 5, 8, 10, 11) in embryonal stages, to isoform products containing internal exon 6b (Tm5NM2, 3, 4, 6, 9) in mature stages (Dufour et al., 1998b). Similarly, proteins from the γTm gene containing C-terminal exon 9a, 9c, or 9d are differently regulated throughout development in mouse brain (Vrhovski et al., 2003). Non-muscle protein products from γTm gene exons 9a and 9c are found at embryonic days (ED) 11.5 and ED15.5 respectively, and both increase in their expression into adulthood (Vrhovski et al., 2003). The products containing γTm gene exon 9d, however, are present in brain at ED11.5, and then increase until birth, at which time they decrease dramatically postnatally to lower adult levels in rat (Weinberger et al., 1996) and mouse brain (Vrhovski et al., 2003).

This complementary expression of the different C-terminal exons (9a, 9c, and 9d) of the γTm gene ensures that the net pool of γTm gene proteins in mouse brain is relatively constant throughout development (Vrhovski et al., 2003). On the basis of these temporal changes in expression, Vrhovski et al. (2003) hypothesise that 9d expression is associated with early brain development and axon outgrowth, whereas 9a and 9c expression are associated with later maturational changes.

Fig. 18.2 (continued) The developmental expression profiles of neuronal Tms. *Colors* indicate protein, *black horizontal stripes* indicate mRNA distribution. Note that these diagrams provide only a general overview, and do not portray some more specific nuances of isoform-specific distributions discussed in the text. Isoforms are listed in the order they are discussed in Section 18.2. "References" indicate the following publications: (1) Hannan et al., 1995, (2) Hannan et al., 1998, (3) Bryce et al., 2003, (4) Schevzov et al., 2005a, (5) Dufour et al., 1998b, (6) Vrhovski et al., 2003, (7) Weinberger et al., 1996, (8) Had et al., 1994, and (9) Schevzov et al., 1997. The two toned coloring of Tm5NM1 in mature neurons reflects the restricted localisation of the endogenous (*yellow*) protein vs the more extensive distribution of the exogenous human Tm5NM1 (*orange*) (Bryce et al., 2003; Schevzov et al., 2005b). Note that the distributions of Tm5NM1 and Tm5NM2 mRNA in immature neurons are identical, as the probe used in these studies could not distinguish between the two isoforms. Not shown is the dynamic shift in Tm5NM1/2 mRNA throughout differentiation where there is initial movement from the cell body into the axonal shaft, and a later recession into the cell body as occurs in the mature neuron (Hannan et al., 1995) (see Section 18.2.1.3 of the text for details). The question marks (?) refer to the fact that the localisation of exon 9c and 9a (γTm gene) products in immature neurons, and Tm5a/5b in mature neurons are yet to be determined

18.2.2 Exon 9d Products from the γTm Gene: Tm5NM1 and Tm5NM2 Isoforms

Early analyses with a particular antibody to exon 9d, the WS5/9d antibody, suggested certain neuronal distributions of Tm5NM1 and Tm5NM2 proteins. We will first discuss these distributions as they were reported, and then the reasons why they need to be re-examined.

18.2.3 Exon 9d Isoforms as Detected by the WS5/9d Antibody

Within embryonal, immature neurons, the exon 9d containing isoforms Tm5NM1 and Tm5NM2 proteins have been located predominantly within the axonal compartment. This has been demonstrated *in vivo* in the rat embryonic (ED13) nervous system: Tm5NM1/2 proteins localised to axons within the spinal cord, tectum, and dorsal root ganglia (DRG) (Weinberger et al., 1996). Even with unipolarity, day ED13 DRG cells localised Tm5NM1/2 throughout the length of the axons, and the protein was absent from the cell body (Hannan et al., 1998). Observations on Tm5NM1/2 distribution in ED14.5 mouse embryonal primary cultured neurons indicated that these isoforms were abundant in the cell body and primary neurite, but invariably absent from growth cones (Schevzov et al., 1997). Throughout maturation, the Tm5NM1/2 isoforms undergo a shift in neuronal subcellular localisation. Within a 2-day period (ED15 to ED17), Tm5NM1/2 were lost from axons in the rat medulla and appeared within neuronal cell bodies (Weinberger et al., 1996). A similar shift in localisation was seen in primary mouse cortical cells, with Tm5NM1/2 protein being initially within the axons and then relocated to the cell body and all neurites (axons and dendrites) post differentiation (Hannan et al., 1995). In adult rat and mouse cerebellum, the Tm5NM1/2 protein was confined to the cell soma and dendrites of Purkinje neurons (Dufour et al., 1998b; Hannan et al., 1998; Vrhovski et al., 2003) but was absent from granule cell bodies (Vrhovski et al., 2003; Weinberger et al., 1996).

These conclusions were all drawn from the use of the WS5/9d antibody. Although the WS5/9d antibody was raised to exon 9d of the γTm gene, it lacks immunoreactivity with the Tm5NM1 isoform (Percival et al., 2004 show none; Schevzov et al., 1997 show only weak reactivity with the Tm5NM1 protein), which would imply that studies using this antibody are reporting Tm5NM2 localisation only. Under this assumption, Tm5NM2 protein would generally be localised in the axons of developing neurons, then shifting to the somatodendritic compartment during development. Unfortunately, in the absence of a positive control for the immunoreactivity and specificity of WS5/9d with Tm5NM2, questions as to its specificity have been raised, and the use of this antibody has been discontinued. Hereafter, this review instead considers the mRNA localisations of these isoforms, and the patterns of protein localisation revealed by verified antibodies raised to the γTm gene 9d exon, as representing the distribution of Tm5NM1 and Tm5NM2. In those cases where probes against the 9d exon, either RNA or antibody, are unable to distinguish between Tm5NM1 and Tm5NM2 isoforms, they will be referred to as Tm5NM1/2.

In primary cultured mouse neurons, Tm5NM1/2 mRNA changes in distribution with the development of the neuron. By day two in culture, mRNA is found in the cell body in a polarised distribution, with enrichment at the base of a single neurite presumed as the precursor to the axon. By day seven in culture, Tm5NM1/2 mRNA is localised within cell bodies and axons, and by day 14 in culture, the mRNA is restricted to the cell body (Hannan et al., 1995). Similarly, Tm5NM1/2 mRNA in neurons of developing embryonic brain is directed towards the axonal pole, even before the outgrowth of neurites (Hannan et al., 1995). And, in mature rat cerebellum, the Tm5NM1/2 mRNA is localised within the cell bodies of developed Purkinje neurons (Hannan et al., 1998).

The localisation of Tm5NM1/2 mRNA to axons in developing neurons indicates a function associated with neuronal polarity during neuronal differentiation. This is further supported by studies of differentiating PC12 cells, which localise Tm5NM1/2 mRNA uniformly throughout the cell body, without polarity in mRNA distribution, consistent with the lack in polarity of these cells (Hannan et al., 1995). Conversely, DRG neurons, which are unipolar and develop neurites into axons only, express Tm5NM1/2 mRNA at the axonal poles (Hannan et al., 1998).

The developmental regulation of Tm5NM1/2 mRNA in polarised cells was further investigated in embryonic rat brain. In cells of the hippocampus and basal ganglia which are proximal to the ventricle and thought to be younger and less developmentally advanced, Tm5NM1/2 mRNA localised to the axonal pole. In cells more developmentally advanced in layers more distal to the ventricle, Tm5NM1/2 mRNA localised to the cell body (Hannan et al., 1995). Despite these differences, using different regions of brain at one developmental time point may be reflective of intrinsic physiological differences between these layers which could affect Tm5NM1/2 distribution, rather than accurately reflecting how Tm5NM1/2 distribution changes with differentiation.

Tm5NM1/2 mRNA localisation is not a universal marker of early neuronal polarity. For example, not all primary mouse cultured neurons localise Tm5NM1/2 mRNA to the axon, not all Tm5NM1/2 mRNA stained neurites are axons, and not all neurons *in vivo* express Tm5NM1 or 2 (Hannan et al., 1995). Evidence as to the differential protein localisation of Tm5NM1 vs. 2 comes from overexpression studies. Human Tm5NM1 (hTm5NM1) differs from its mouse homologue by only one amino acid and is likely to be very similar to the endogenous protein in its physical property and subcellular distribution. In transgenic mice which overexpress exogenous hTm5NM1, the specific distribution of hTm5NM1 can be mapped through different antibodies which together can distinguish between hTm5NM1 and mouse Tm5NM1. In primary cortical neurons cultured from hTm5NM1 transgenic mice (by day 5 in culture), hTm5NM1 localised to the peripheral regions of the growth cone and the filopodia (Bryce et al., 2003; Schevzov et al., 2005a). If we assume the localisation of endogenous Tm5NM1 is the same as this, and compare this to the localisation of total exon 9d protein (detected by the γ9d antibody), then the remaining localisation besides this growth cone/filopodia distribution would be representative of Tm5NM2 distribution only. Under these assumptions, Tm5NM2 protein is found in the axon shaft, and not in the growth cones (Schevzov et al., 2005a). In adult transgenic mice hTm5NM1 was detected in the cell bodies and dendrites of the cortex (Bryce et al., 2003); however, further investigation is required to determine if endogenous Tm5NM1 is in the same regions.

In summary, these studies would indicate that Tm5NM1 is most probably enriched in the growth cones of developing neurons, while Tm5NM2 is within the proximal developing axon shaft and the axonal pole of the cell body. During differentiation, Tm5NM1 relocates to the cell body, so that differentiated cells express only Tm5NM2 in the axon (see Fig. 18.2). In addition, Tm5NM1 in differentiated neurons may also localise to dendrites. Together these studies indicate that Tm5NM2 is associated specifically with the establishment and development of immature axons, whereas Tm5NM1 is associated with neurite outgrowth via involvement in the growth cone and possibly in dendritic function of mature neurons. It would appear that Tm5NM2 mRNA expression is only required in the early stage of axon development; once differentiation is complete, mRNA can shift from the axon to a uniform cell body distribution. The possible functions of Tm5NM1 are discussed in the growth cone and synapse sections of this chapter.

18.2.4 The Exon 9c Products from the γTm Gene: A Contrast to 9d Products

Exon 9c of the γTm gene encodes two isoforms, Tm5NM4 and Tm5NM7, which are specific to brain (Vrhovski et al., 2003). Tm5NM4/7 proteins are expressed in mouse brain starting as early as ED15.5, and levels increase through to adulthood. On a subcellular level these isoforms are widespread throughout cell bodies and dendrites in mature neurons of the mouse hippocampus, cortex, and cerebellum; in the axons of the cortex, and in the adult cerebellum throughout the molecular layer, further indicating an axonal distribution (Vrhovski et al., 2003), and in the Purkinje cell bodies and axons of the molecular layer of adult rat cerebellum (Dufour et al., 1998a). Together this indicates that exon 9c products can localise to dendritic, axonal and cell body compartments in a region dependent manner. It is important to note that the antibody used in the study by Vrhovski et al. (2003) has since been shown to cross react with the αTm gene products, TmBr1 and TmBr3 (Schevzov et al., 2005b). However, the distributions of TmBr1/3 are mainly in the axonal shaft of developed neurons (discussed later in this section), which suggests that the reported distribution of exon 9c products in the cell body and dendrites of mature neurons is a true representation of their location and not due to cross reactivity. The presence of the exon 9c products in axons of the mature cortex and in the molecular layer of the mature cerebellum is less certain due to this cross reactivity, and requires further confirmation.

18.2.5 Exon 9a Products from the γTm Gene

Isoforms containing exon 9a (Tm5NM3, 5, 6, 8, 9, and 11) (Dufour et al., 1998b) are abundant in brain, lung, and spleen, and at lower levels in the kidney, liver, and embryonic primary fibroblasts (Schevzov et al., 2005b). Exon 9a isoforms are found in all brain regions, and whole brain levels increase from ED15.5 to adulthood in mouse (Vrhovski et al., 2003). The subcellular distribution of exon 9a protein products is similar to exon 9c protein products: throughout Purkinje cell bodies and dendrites in the mature mouse cerebellum, and within axons of the molecular layer; within cell bodies, dendrites and axons of neurons in cortex and pyramidal cells in the hippocampus (see Fig. 18.2). Although exon 9c and exon 9a products have similar distributions (see Fig. 18.2), exon 9c containing proteins are found more intensely in the cell body. In general, the distribution of exon 9a and 9c containing isoforms in mouse do not undergo any major shift from neonatal (PND 5) through to adulthood (Vrhovski et al., 2003).

18.2.6 Summary: γTm Gene Products in Brain: Exons 9a, 9c, and 9d

The γTm gene products containing exon 9d change in their predominant subcellular localisations during neuronal differentiation – Tm5NM1 is initially in the growth cone and Tm5NM2 in the axon, and both isoforms shift to the cell body during differentiation. Together these developmental shifts in localisation indicate roles in neuronal differentiation and axonal outgrowth. Conversely, exon 9a and 9c containing products from the γTm gene have more generalized distributions throughout cell bodies, axons, and dendrites, in patterns of expression do not change measurably with development. Within this context it is evident that different exons from the same gene, and different isoforms containing the same exon, can be sorted to very different subcellular domains and, in the cases of Tm5NM1 and 2, that these distributions can change with differentiation. Further evidence of the ability of different Tms to define distinct actin filaments, and their implications for morphological processes associated with differentiation, can be found by examining the complementary expression patterns of other Tms in brain.

18.2.7 αTm Gene Products: TmBr1, TmBr2, TmBr3, Tm5a, and Tm5b

18.2.7.1 Regional Expression in Brain: TmBr1 and 3

The αTm gene produces five isoforms in brain, two of which are brain specific, TmBr1 and TmBr3. TmBr1 is a high molecular weight (HMW) Tm, and TmBr3 is a low molecular weight (LMW) Tm (Lees-Miller et al., 1990; Weinberger et al.,

1993). TmBr2, another LMW isoform produced by the αTm gene, was also hypothesised to be brain specific (Lees-Miller et al., 1990); however transcripts have since been found in fibroblasts (Weinberger et al., 1993). The lack of specific antibodies to TmBr2 makes its protein distribution difficult to ascertain. In brain, TmBr3 is expressed at much greater levels than TmBr1 and TmBr2 (Lees-Miller et al., 1990; Stamm et al., 1993). While TmBr3 protein levels are mainly equivalent throughout different regions of adult rat brain (Stamm et al., 1993), TmBr1 shows some regional variation in expression. Greater abundance of TmBr1 protein was detected in the cortex, hippocampus, striatum, olfactory bulb, and thalamus (areas of brain derived from the prosencephalon) than in the midbrain, cerebellum, spinal cord, and hindbrain, where TmBr1 was almost undetectable (Stamm et al., 1993; Vrhovski et al., 2003).

Unlike Tm5NM1 and Tm5NM2, TmBr1 and TmBr3 isoform expression appears from later developmental stages. TmBr3 protein first appears at ED16 in rat whole brain (Stamm et al., 1993), day ED17 in rat medulla (Weinberger et al., 1996), and birth in mouse whole brain (Vrhovski et al., 2003), whereas TmBr1 protein is not detected until day PND5 (Vrhovski et al., 2003). TmBr1 and TmBr3 protein levels both increase during postnatal rat brain development (Had et al., 1994; Stamm et al., 1993), with transcripts of both showing a dramatic post-natal increase from PND0–PND21 in rat cerebellum, to high adult levels of transcription of these isoforms (Weinberger et al., 1993). Similar to this mRNA profile, minor expression of TmBr3 protein is detected in rat cerebellum at PND0, followed by an increase until PND20, when levels begin to plateau (Had et al., 1994).

Collectively, these investigations into the developmental regulation of TmBr3 expression in rat cerebellum and whole brain shows that, from relatively low levels at birth, protein and mRNA expression increases markedly, and remain relatively high in adulthood. The period of greatest increase in expression of these isoforms (between days PND0 and PND21) coincides with a period of intense neurite outgrowth and synaptogenesis in the rat cerebellum (Altman, 1972). It is possible that TmBr1 and TmBr3 contribute to the establishment of cerebellar neural networks. From PND6, the different layers of the cerebellum are well defined (Had et al., 1994), and so it would seem unlikely that the sole function of TmBr3 in the cerebellum is in the differentiation of cerebellar cells to define these distinct organisational layers. As TmBr3 protein levels in rat whole brain (Stamm et al., 1993), mouse whole brain (Vrhovski et al., 2003), rat cerebellum (Had et al., 1994), and mRNA levels in rat cerebellum (Weinberger et al., 1993) are maintained at relatively high levels from PND20 into adulthood, this suggests that TmBr3 may be responsible for the maintenance of established neural networks in the cerebellum and other brain regions.

The possible roles of TmBr1, TmBr2, and TmBr3 in neurite outgrowth, and the genesis and persistence of the synapse, can be better understood by inspecting the localisation of these isoforms on a subcellular level. Where antibodies or RNA probes are unable to discriminate between the TmBr1 and TmBr3 isoforms, they are collectively termed TmBr1/3.

18.2.7.2 Neuronal Distribution of TmBr1 and TmBr3: Highly Polarised

A primarily, if not entirely, axonal distribution of TmBr1 and TmBr3 is seen in different regions of embryonal and postnatal rat and mouse brain. In the rat medulla, TmBr1/3 protein first appears at day ED17, localised to the axons (Weinberger et al., 1996). In the rat cerebellum, however, levels of TmBr3 protein are almost undetectable until PND10–PND14, when TmBr1/3 appears gradually in the molecular layer and the cerebellar glomeruli of the internal granule layer (Had et al., 1994; Vrhovski et al., 2003). In fully developed (PND35) adult rat cerebellum, TmBr1/3 is present in the pre-synaptic web of cerebellar glomeruli, in the internal granule layer, and clearly in the axons (parallel fibers) of granule cells within the molecular layer (Had et al., 1994; Weinberger et al., 1996) with mildly detectable levels in the dendritic spines of this layer (Had et al., 1994). This predominantly pre-synaptic distribution is echoed in PND14 mouse hippocampus and cortex, where TmBr1/3 has a predominantly axonal distribution (Vrhovski et al., 2003). This illustrates a mainly pre-synaptic distribution of TmBr1/3 that occurs throughout development and persists into adulthood (see Fig. 18.2).

18.2.7.3 TmBr1 – Astrocyte or Neuron Specific?

Specific conclusions about the neuronal specificity of TmBr1 are hampered in part by problems of antibody specificity. To date, the antibodies used recognise both TmBr1 and TmBr3 isoforms, which are only distinguishable on a western blot due to their different molecular weights. Previously it was shown that TmBr1 mRNA was absent from embryonic rat primary neuronal cultures, but present in astrocyte cultures, whereas TmBr3 mRNA was found in these same neuronal cultures but not the astrocytes (Had et al., 1993). On the basis of this, in 1994 Had and others reasoned that their antibody labelling of PND35 rat cerebellum reported TmBr3 localisation only (Had et al., 1994). Using the exon 9c raised antibody, western blotting of both adult rat whole brain lysate and mouse whole brain lysate reveals two bands – one consistent with the molecular weight of TmBr3, and another higher band consistent with TmBr1 molecular weight (Weinberger et al., 1996), although total protein estimates cannot distinguish between different cell types (neurons vs glia) within a brain lysate. There is some controversy surrounded the cell type specificity of TmBr1 either as astrocyte specific (Had et al., 1993) or neuron specific (Weinberger et al., 1993). In fact, as the primary neuronal cultures used in these studies were harvested from animals no later than ED17, and as the TmBr1 protein is not detected before day PND20 (Stamm et al., 1993), it is impossible to use TmBr1 absence from these neuronal cultures as evidence of it being absent from neurons *in vivo* across all developmental time points. Furthermore, the cultures which TmBr1 are absent from are cortical neurons (Had et al., 1993), and may not represent the expression pattern of TmBr1 in other brain regions, such as cerebellum. Because of this, and in the absence of other independent verifications of the cell types in which TmBr1 and 3 are expressed, it is prudent to assume that any localisation revealed by the antibody raised against exon 9c could be reporting the distribution of TmBr1, as well as (or instead of) TmBr3 distribution. Further evidence as to the questionable basis of mRNA levels as a reliable indicator of protein levels can be found in Faivre-Sarrailh et al., 1990. The group found no expression of γTm gene transcripts in rat brain; however, mRNA produced by the γTm gene and protein products from this gene have since then been consistently found in rat brain (see Gunning et al., 1998, 2008 for reviews). In summary, it can not be discounted that TmBr1 is expressed in both astrocytes (Had et al., 1993) and neurons (Weinberger et al., 1993).

18.2.7.4 TmBr3- in the Neurites as They Extend

TmBr3 has been observed in a general distribution throughout the cell body and processes, without reference to neuronal polarity, in rat cortical and midbrain cultures (Stamm et al., 1993). This group confronted problems of antibody specificity in immunohistology by using neurons harvested from day E16 (earlier than the first detected TmBr1 expression at PND20), and confirming TmBr1 was not detected in western blots of these rat cortical and midbrain primary cultures (Stamm et al., 1993). Thus, the immunofluorescence staining of these neurons most probably reflects the subcellular distribution of TmBr3, and not TmBr1. Investigation into a stable cell line, PC12 cells, has shown that differentiation and neurite outgrowth are required precursors to TmBr3 mRNA transcription (Weinberger et al., 1993). In their undifferentiated state, PC12 cells express undetectable levels of TmBr1 and TmBr3. These cells can be induced to differentiate into a quasi-neuronal phenotype lacking neurite polarity – this differentiation induces TmBr3 mRNA in these cells, with transcript level increases mirroring the induction of neurite outgrowth. When cells are forced to reverse morphological differentiation (achieved experimentally by transferring cells to suspension), the levels of TmBr1 and TmBr3 progressively decrease (Weinberger et al., 1993).

In the case of PC12 cells, TmBr1 and TmBr3 expression *in vitro* is correlated with morphological changes in neuronal differentiation. Expression persists once differentiation is complete, but requires the maintenance of those morphological characteristics of differentiation. Induction of TmBr1 and TmBr3 expression in embryonic brain occurs in the developing neuron, at ED17, and persists predominantly in the axons throughout the mature neuron (Weinberger et al., 1996). By at least PND10, TmBr1/3 appears pre-synaptically in rat and mouse cerebellum. In adult mouse and rat cerebellum, TmBr1 and TmBr3 are found predominantly in the axons of Purkinje neurons (Weinberger et al., 1996), in the granule cell axons (parallel fibers) surrounding Purkinje dendrites (Had et al., 1994; Weinberger et al., 1996), and in areas of intense synaptic activity (within the cerebellar glomeruli of the internal granule layer) (Had et al., 1994; Vrhovski et al., 2003).

18.2.7.5 TmBr2: A Uniform Distribution

In the original study by Had et al. (1993), TmBr2 mRNA transcripts were found to be absent from cultured rat neurons. In a subsequent study, TmBr2 mRNA was found in mouse primary cortical neurons, and at day ED13 in dorsal midbrain (Hannan et al., 1995). In contrast to TmBr3 (and possibly TmBr1), TmBr2 mRNA is distributed in a non-polarised, uniform fashion throughout the cell body of embryonic rat hindbrain neurons *in vivo*, and throughout primary mouse cortical neurons *in vitro*, in a distribution that does not change with development (Hannan et al., 1995) (see Fig. 18.2). TmBr2 mRNA expression is present at the initiation of differentiation in PC12 cells, and in contrast to TmBr1 and TmBr3 isoform levels, continues to increase even when cells are forced to lose their morphological characteristics of differentiation, such as neurite outgrowth (Weinberger et al., 1993).

The persistence of TmBr1 and TmBr3 transcription and translation in brain into adulthood (Stamm et al., 1993; Weinberger et al., 1993) suggests a role of these isoforms in the maintenance of neural networks post differentiation. The localisation of TmBr3 pre-synaptically in cerebellum (Had et al., 1994) further implies a function in the plasticity of these networks, perhaps at the level of actin filament remodelling for vesicle release and reuptake at the synapse. The functional significance of TmBr1/3 is further discussed in the synapse section of this chapter.

18.2.7.6 Tm5a and Tm5b: In the Growth Cones

The levels of Tm5a and Tm5b mRNA in developing postnatal rat cerebellum remain mostly constant from high levels at birth, with an increase in expression from PND3 to PND7, and then levels decline slightly to those seen in adult stages (Weinberger et al., 1993). Tm5a and Tm5b proteins are distributed throughout the cell body, axon, and growth cones of primary mouse embryonic cultured neurons (Schevzov et al., 1997). In the growth cones Tm5a and Tm5b are detected in the proximal filopodia, the finger-like projections from the growth cones which project into and sample the environment with rapid turnover of actin (Schevzov et al., 1997). Their presence within the growth cone and proximal filopodia, however, recedes as the growth cone recedes with time in culture (see Fig. 18.2). This localisation shift is accompanied with a reduction in protein levels. Interestingly, within a single growth cone, some but not all filopodia stain for Tm5a and Tm5b (Schevzov et al., 1997). Tm5a is entirely absent from the growth cone of mouse primary cortical neurons at day five in culture (Schevzov et al., 2005a). In PC12 cells, Tm5a and Tm5b mRNAs are the most abundant αTm gene transcripts, and are present when cells are entirely undifferentiated, and increase dramatically in early differentiation. Additionally, when cells are forced to lose the morphological characteristics of differentiation, Tm5a and Tm5b mRNA levels also increase (Weinberger et al., 1993). This suggests that Tm5a and Tm5b may function in the rapid actin remodelling requirements of fast, large scale morphological changes. In addition to its neuronal profile, Tm5a mRNA is also found in glial cultures (Had et al., 1993). The functions of Tm5a are discussed in the growth cone and synapse sections of this chapter.

18.2.8 The δTm Gene Product Tm4

Tm4 is found in a range of tissues, including brain. Tm4 mRNA has been found in cultures of astrocytes, oligodendrocytes, and neurons, with the highest transcription levels corresponding to the immature stages (after 1 day in culture) of neurons and astrocytes, and remaining constant throughout the development of oligodendrocytes (Had et al., 1993). Tm4 mRNA is more concentrated in embryonic than adult whole rat brain (Yamawaki-Kataoka and Helfman, 1987). Tm4 protein is expressed at birth, and expression levels in rat cerebellum increase nearly tenfold from birth to a peak at PND10, which is followed by a rapid decrease to about half this level at PND14, then remaining constant throughout subsequent development into adulthood (Had et al., 1994). Tm4 mRNA shows much the same pattern as protein during rat cerebellum development, but the protein decrease is more marked (a fourfold decrease) (Faivre-Sarrailh et al., 1990). The antibody used for detection of Tm4 protein crossreacts with the αTm gene isoform Tm1. Because of this, immunofluorescence data using this antibody were confirmed by Tm1 peptide inhibition in order to detect specifically Tm4 localisation (Had et al., 1994). In neurons, the subcellular distribution of Tm4 changes with development. Cultured rat cortical neurons express Tm4 protein at 1 day in culture, and it remains in axonal and dendritic growth cones, cell bodies, and neurites until at least 6 days in culture. In these growth cones, Tm4 colocalises with the proximal filopodia, but is not in the distal filopodial tips (Had et al., 1994). In cultured hippocampal rat neurons, both axonal and dendritic growth cones are enriched with Tm4 protein, regardless of the time at which these growth cones emerge in culture (Had et al., 1994).

Within the cerebellum, Tm4 protein was first detected at day ED20, and immunoreactivity increased in intensity until birth, at which time the forming internal granule and molecular layers were immunoreactive, but quite notably, the organising Purkinje cell layer was not. By PND6, Purkinje and granule cell bodies were still not labelled, but the axonal fibers within the internal granular layer (Purkinje cell axons, climbing and mossy fibers) contained Tm4 protein, as did the innermost meningeal layer, the pia mater, and the blood vessels. By PND10 Tm4 was in the Purkinje cell bodies and dendrites, but staining then decreased in intensity in all regions, such that by PND14, only diffuse staining remained in the molecular layer and the cerebellar glomeruli within the internal granule layer, a pattern which persisted into adulthood (PND35). Immuno-electron microscopy (EM) analysis of PND35 rat cerebellum revealed accumulation of Tm4 in the dendrite spines of Purkinje cells in the molecular layer, and especially at the postsynaptic densities of these spines. However, Tm4 was notably absent from the adjacent pre-synaptic parallel fibers (Had et al., 1994). This shift in the localisation of Tm4 from being enriched in the growth cones of axons and dendrites in immature neurons to post-synaptic sites in mature neurons (see Fig. 18.2) represents a potentially substantive shift in function in response to the changing requirements of the neuron throughout differentiation. Tm4 localisation within growth cones indicates a role in the motile events of neurite outgrowth, supported by its rapid increase in mRNA and protein levels between days PND0 and PND10, which coincides with a time of great neurite motility when neurons in the developing cerebellum are seeking and finding targets to build a neural web. The persistence of Tm4 transcription and translation into adulthood, and later localisation to the post synapse, tend to suggest an ongoing involvement in synaptic plasticity (Had et al., 1994). The possible functions of Tm4 are further discussed in the growth cone and synapse sections of this chapter.

18.2.9 Summary: Neuronally Expressed Tms: Isoform-Specific Regulation

Throughout differentiation, the expression of Tms in neurons changes in a highly isoform-specific fashion, as illustrated by comparing the distributions of all Tms discussed here, within immature and mature neurons (see Fig. 18.2). The possible functions of each Tm isoform discussed in this section will be explored in relation to three specific neuronal compartments – the growth cones, neurites, and synapses. To explore why Tms are expressed at different developmental stages, and at different places within the neuron, we will briefly review the functions of those Actin Binding Proteins (ABP) which are associated with these isoforms. By no means do we intend this next section to be an exhaustive review of ABP function, or even of ABP-Tm interactions, but rather an overview of established functional relationships between neuronal Tms and neuronal ABPs.

18.3 How Tropomyosins Differentially Affect Actin Filaments: Actin Binding Proteins

The sheer diversity of Tm isoforms (at least 40) in mammals, and the isoform-specific patterns of localisation throughout different tissues, cells, and subcellular compartments, are indicators of the divergent functions of isoforms *in vivo*. The mechanisms by which actin filaments are regulated by Tms hinge on two general factors: (1) the different abilities of Tm isoforms to interact with other ABPs and (2) the different affinities that Tm isoforms have with actin.

The second factor also means that Tm isoforms can affect the access of other ABPs to the actin filament, and in doing so can differently influence the structural outcome of actin filaments. Tms can augment the interactions between actin and ABPs in a highly isoform-specific manner. We do not aim to provide an exhaustive overview of ABPs, but rather focus only on those which are found within neurons, and have been shown to interact with the Tm isoforms so far covered in this review. The implications of these ABP-Tm interactions will be further discussed in the growth cone and synapse sections of this chapter.

18.3.1 Arp2/3

The actin-related protein 2/3 (Arp2/3) complex is an actin filament nucleator which can initiate lateral branching from already existing filaments, with these new daughter filaments elongating from the barbed ends. Additionally, Arp2/3 may prevent the addition of monomers to the pointed end of filaments (Mullins et al., 1998) and also inhibit the depolymerisation of actin filaments from their pointed ends (Svitkina and Borisy, 1999). Arp2/3 requires activation to nucleate these branches, and WASP/Scar proteins are prominent in activating this nucleation. In turn, the activity of the WASP/Scar proteins is regulated by various Rho-GTPases (see Higgs and Pollard, 2001 for review). Generally, lamellipodia are composed of a diagonal meshwork of actin filaments (Verkhovsky et al., 2003), with barbed ends of these filaments found closest to the leading edge (Svitkina et al., 1997), and Arp2/3 found at the branching points of these filaments (Korobova and Svitkina, 2008; Svitkina and Borisy, 1999). Together these studies suggest a role for Arp2/3 in actin filament assembly which drives protrusion at the leading edge. Aside from roles in the lamellipodial actin network, Arp2/3 may also contribute to filopodial formation, as filaments originating from branching points can then join an actin bundle within filopodia (Korobova and Svitkina, 2008).

Tms can inhibit the Arp2/3 mediated branching of actin filaments. The αTm gene products Tm2 and Tm5a can each reduce the rate of branching by WASP stimulated Arp2/3, although neither Tm isoform can eliminate this branching, even when present in concentrations which saturate the actin filament (Blanchoin et al., 2001). Tm5a is the more effective of these two Tms, and can reduce Arp2/3-mediated branching events by 50% (Blanchoin et al., 2001). Such an interaction has implications for leading edge structures and filopodia in neurons.

18.3.2 Formins

The formins are an extensive family of proteins. Their roles within cells are isoform-specific, and include nucleation of new actin filaments, modulating the rate of elongation at the fast-growing barbed end (by attenuating profilin-actin assembly onto the filament), organising multiple actin filaments into bundles, and even severing actin filaments (see Kovar, 2006 for review). To confuse matters, formins are thought to exist as heterodimers with the potential of having opposing effects on the actin filament – either enhancing polymerisation (elongation) or depolymerisation (dissociation) at the barbed end, depending on their conformational states (Otomo et al., 2005). Some formin isoforms can also inhibit elongation at the barbed end by capping the filament (see Wallar and Alberts, 2003 for further description). At least two neuronally expressed mouse formins, mDia1 and mDia2, are inhibited in this capacity by Tms in an isoform-specific manner (Wawro et al., 2007). While Tm5a can almost completely overcome the inhibition of elongation by the formin construct FRL1-FH1FH2, it has little effect on the inhibition exerted by mDia2. Conversely, Tm5a can stimulate elongation, but this is significantly reduced in the presence of FRL1-FH1FH2 (Wawro et al., 2007). Some formins can also bind to the sides of actin filaments, a process implicated in both filament bundling and severing (see Wallar and Alberts, 2003 for review), and Tm isoforms can also inhibit this side-binding process (Wawro et al., 2007). These particular formins, however, are yet to be located to neurons. Tms can exert their influence on formins without displacing them from the filament (Ujfalusi et al., 2008; Wawro et al., 2007). In fact, the synergistic model of simultaneous Tm-formin association with the actin filament predicts a stable, straight filament, which is protected from severing (Ujfalusi et al., 2008; Wawro et al., 2007).

18.3.3 ADF/Cofilin

Actin-depolymerising factor (ADF)/Cofilin family members exert a range of effects on the actin filament. We will focus on those which are known to be altered by Tms, which include (1) the depolymerisation of actin filaments by enhancing the rate of monomer dissociation from pointed ends, (2) the inhibition of depolymerisation at barbed ends of filaments, and (3) the reduction of filament lengths and the increase in filament numbers by conferring an instability on actin filaments via increasing their chances of severing (Andrianantoandro and Pollard, 2006; Kuhn and Bamburg, 2008). ADF/Cofilin activity is regulated by phosphorylation: ADF/cofilin is active in its dephosphorylated form; ADF/cofilin not bound to actin filaments is typically inactivated by LIM-kinase-mediated phosphorylation (Arber et al., 1998; Yang et al., 1998). These LIM-kinases are in turn regulated in part by Rho-GTPases (see Bernard, 2007 for review). Additionally, the activation of ADF/cofilin can be achieved through its dephosphorylation by Slingshot phosphatases (Niwa et al., 2002).

At its simplest, non-mammalian variants of Tms are antagonists of ADF/cofilin, as Tm binding to the actin filament can stabilise the actin filament by protecting it from the depolymerising effects of ADF/cofilin (e.g. Ono and Ono, 2002). However, the interactions between Tms and ADF/cofilin are much more complex in mammalian systems, as the nature of the relationship is Tm isoform-specific. *In vitro* studies show that overexpressing human Tm5NM1 is sufficient to increase the concentration of unbound, phosphorylated (and therefore inactive) ADF/cofilin, suggesting that hTm5NM1 out-competes ADF/cofilin for binding to the actin filament (Bryce et al., 2003). It must be noted, however, that overexpression of hTm5NM1 induces a concomitant increase in Tm5a concentration, and so it is difficult to specify whether one or both Tms compete with ADF/cofilin. For simplicity, in the following sections we refer to this inhibitory relationship as being between Tm5NM1 and ADF/cofilin. Overexpression of TmBr3 in this same cell line did not induce an increase in phosphorylated ADF/cofilin levels, and in fact TmBr3 co-immunoprecipitates with ADF/cofilin-bound actin filaments, suggesting co-operation between the two, or recruitment of ADF/cofilin by TmBr3 at the actin filament (Bryce et al., 2003). The localisation of ADF/cofilin is also influenced by these isoforms, as shown by the loss of the characteristic localisation of ADF/cofilin in the proximal region of the lamellipodia with hTm5NM1, but not TmBr3, overexpression. This study suggests that the Tm-ADF/cofilin relationship is isoform-specific, with TmBr3 perhaps functioning as a recruiter of ADF/cofilin (or vice versa) to the actin filament in lamellipodia, and Tm5NM1 or Tm5a, or both isoforms (on separate filaments), working to displace ADF/cofilin from actin filaments.

18.3.4 Myosin

Myosins are a superfamily of actin-associated ATPases, originally characterised in striated muscle, but with isoforms now found in all eukaryotic cells. In non-muscle cells, myosin II can act as a molecular motor, translocating actin filaments by mechanical force generated from ATP hydrolysis. To do so, non-muscle myosin requires phosphorylation of its light chain (MLC) domains which is achieved by calcium-calmodulin dependent MLC-kinase (MLCK), Rho-kinase, or AMP-activated protein kinase (AMP-kinase) (see Conti and Adelstein, 2008 for further explanation).

Muscle-specific Tms, of which there are three, are an integral component of the thin filament of the muscle contractile unit, the sarcomere, and the mechanisms by which they regulate the actin-myosin interaction has been well characterised (see Lehman and Craig, 2008 for review). Likewise, there are a number of muscle-specific myosins which are responsible for producing the force required for contraction in skeletal, cardiac, and smooth muscle. In the cytoskeleton, actin and myosin can also work together to produce contractile force. Myosins can also influence the turnover of actin filaments, and myosin can realign, or even translocate actin filaments (Conti and Adelstein, 2008). Myosins are implicated in actin filament stability, such that when myosin II is already bound to actin filaments, phosphorylation of MLC can stimulate ATPase activity to incorporate actin monomers into actin filaments (Sellers, 1981). In these capacities, myosins can influence cell motility, cell polarity, and local processes such as movement of growth cones. The activation of myosin II occurs by phosphorylation of the MLC domains, and can be achieved through multiple pathways, including those using Rho-kinases (e.g. Amano et al., 1996; Chrzanowska-Wodnicka and Burridge, 1996).

Tms affect myosin activity and localisation in an isoform-specific manner. hTm5NM1 overexpression can drive myosin activation by MLC phosphorylation. Overexpression of hTm5NM1 results in myosin enrichment in somatodendritic compartments of cultured mouse neurons, concomitant with an aberrant localisation of myosin IIA to the dendrites (Bryce et al., 2003). Conversely, within growth cones, myosin IIB (and not myosin IIA) colocalises with exogenous hTm5NM1 in filopodia and the leading edge of the lamellipodia, where it is usually absent. This relocation is also true for the αTm gene product, Tm5a, and so the impetus behind such myosin enrichment may be due to either, or both, of these Tm isoforms. In the following sections, where the effect of the specific isoform is indeterminable, we term it Tm5NM1. In contrast, overexpression of TmBr3 does not recruit myosin II in growth cones (Bryce et al., 2003), and thus the process is isoform-specific.

18.3.5 Tropomodulin

Tropomodulin (Tmod) caps the slow-growing, pointed ends of actin filaments, preventing both elongation and depolymerisation at this end of the filament. After being identified as a Tm binding molecule (Fowler, 1987), Tmod was also shown to bind actin filaments (Gregorio and Fowler, 1995; Weber et al., 1994). The polarity of the elongated Tmod allows it to bind to actin at its C-terminal end in a Tm independent manner, while the N-terminus of Tmod contains two Tm binding sites (one for each Tm dimer associated with the actin filament), and one Tm-dependent actin binding site (see Kostyukova, 2008b for review).

Four isoforms of Tmod exist, of which three are expressed in brain: Tmod1, Tmod2, and Tmod3. Tmod1, or E-Tmod (erythrocyte Tropomodulin) is expressed in brain (Watakabe et al., 1996) and has recently been found in horizontal cells of the retina (Yao and Sung, 2009). Tmod1 can bind to LMW products of the δTm, γTm, and αTm genes, but has the highest affinity for αTm gene products, including Tm5a (Kostyukova, 2008a). Through specific binding to Tm5a decorated actin filaments, Tmod1 can specifically stabilise these filament populations.

Neuron-specific Tmod (*N*-tropomodulin, or Tmod2) has also been shown to bind to Tm in a highly isoform-specific manner. TmBr3, Tm5a, and Tm5NM1 isoforms have high affinities for Tmod2 (Watakabe et al., 1996). The binding of these Tm isoforms with Tmod2 enhances the affinity of Tmod2 with the actin filament (see Fischer and Fowler, 2003 for review). Conversely, Tmod2 does not bind with Tm4, Tm2, or TmBr1 isoforms (Watakabe et al., 1996). Tmod2 is found throughout brain, as early as ED11.5, and expression persists through to adulthood. Tmod2 has been implicated in increased hippocampal long term potentiation (LTP) and strain specific learning impairments in mice (Cox et al., 2003); however, the mechanism for this remains to be elucidated. Given the arrangement of actin filaments in the synapse, there is a range of mechanisms by which an actin pointed-end stabiliser such as Tmod could function at this site. Indeed, given the isoform-specific localisation of Tms at the adult synapse (TmBr3 being predominantly pre-synaptic and Tm4 post-synaptic), it is tempting to propose that the differential affinity of Tmods for these Tms produces unique actin filament populations on either side of the synapse. However, this hypothesis precedes data on the localisations of Tmods at the synapse.

The differential affinity of Tm isoforms for Tmods provides a system whereby the local repertoire of Tm isoforms can influence the lengths and stability of actin filaments; for example one could expect a region rich in Tm4 (which has a low affinity for Tmod1 and no detectable affinity for Tmod2) to have actin filaments undergoing pointed-end depolymerisation at greater rates than those in TmBr3-rich regions (as TmBr3 binds to Tmod2, and enhances its affinity for the actin filament).

18.3.6 Drebrin

In mammalian cells, at least two isoforms of drebrin are expressed, drebrin E and drebrin A, characterised by their embryonal and adult expression patterns, respectively. *In vitro* studies have shown that drebrin competes with Tm and α-actinin for the binding to filamentous actin (Ishikawa et al., 1994). These *in vitro* assays were carried out with smooth muscle Tm and a mixture of purified brain Tm. In both cases, actin filament binding of Tm was shown to be inhibited by the presence of drebrin, suggestive of a competitive effect of drebrin that is Tm isoform independent. In fibroblast cultures, drebrin overexpression leads to the displacement of Tm isoforms from stress fibers (Ishikawa et al., 1994), supporting the idea of competitive actin filament binding of drebrin and Tm *in vivo*. Furthermore, it was shown that drebrin bound actin filaments were not protected from the severing activity of gelsolin (Ishikawa et al., 1994) as has been previously demonstrated for Tm-bound filaments (Fattoum et al., 1983; Ishikawa et al., 1989). On the basis of these data, it has been hypothesised that drebrin and Tm generate a mechanism that regulates the stability of actin filaments by differentially allowing different ABPs to exert their activity on the filaments.

18.3.7 Caldesmon

Caldesmon (CaD) can regulate the actin-myosin interaction. CaD isoforms are found within muscle and non-muscle cells. Caldesmon is an elongated molecule, containing binding sites for myosin, Tm, and calmodulin (see Dabrowska et al., 2004 for review). Non-muscle CaD can bind to actin filaments and Tm simultaneously, in the longitudinal groove of actin filaments, much in the same fashion as Tm. Indeed, CaD and Tm have almost identical periodicities (Yamashiro-Matsumura and Matsumura, 1988). In a calcium dependent manner, CaD binds to calmodulin, and through this interaction can regulate actin based motility – in the absence of calcium and calmodulin, CaD can inhibit the actin-myosin interaction. It is hypothesised that CaD regulates actin-myosin in a calcium/calmodulin dependent manner (Kira et al., 1995).

The affinity between Tm and CaD is enhanced by actin (Horiuchi and Chacko, 1988). When bound to actin, CaD inhibits actin-myosin ATPase activity, and this inhibition is synergistically enhanced in the presence of Tm (see Wang, 2008 for review). CaD is found in growth cones and extending filopodia of developing rat cortical neurons, colocalising with myosin II and an LMW Tm, probably Tm4 (Kira et al., 1995).

18.4 Role of Tropomyosins in the Growth Cones

Growth cones are highly motile structures, capable of guiding the corresponding neurite to appropriate targets. The growth cone itself can be subdivided into five general compartments: the filopodia (1) comprised of long, linear bundles of actin projecting distally; the lamellipodia (2) comprised of actin meshworks in "veils" – the growth cone leading edges containing barbed ends of actin filaments spanning between the filopodia; the lamella (3) proximal to the lamellipodia and comprising diagonally opposing filaments in arranged actin meshworks, their barbed end protruding into the lamellipodia; the transition zone (4) where actin bundles are severed, filaments depolymerise from which monomers can be recycled into the meshwork of the lamella and lamellipodia; the transition zone is bordered distally by a series of actin arcs, which encroach onto the actin filament poor central zone (5), at the base of the growth cone. Actin is organised into distinct structures within each of these populations through associations with specific ABPs, which can help assemble the range of higher order superstructures found within the growth cone. Although some previous studies have not been able to resolve the specific location of different ABPs, or Tms, within the growth cone, we have aimed to provide a model of interactions between these proteins at a highly compartmentalised level. We stress that what follows is a series of hypotheses, extrapolated from current knowledge of the actin filament superstructures found within each growth cone compartment, and relationships between specific Tm isoforms and ABPs. Thus far, the Tm isoforms that have been found in growth cones include the αTm gene products Tm5a and Tm5b (we hereafter refer to Tm5a because of a larger understanding of Tm5a relationships with ABPs), the δTm gene product Tm4, and the γTm gene product Tm5NM1. In that case where the effect has not been distinguished between Tm5NM1 and Tm5a (ADF/cofilin inhibition, see Bryce et al., 2003), we refer to the Tm as Tm5NM1.

18.4.1 Architecture of Growth Cones: Initialising Actin Filaments

The initial polymerisation of actin filaments requires an initial nucleation step. Arp2/3 has been localised to growth cones, and contributes to the distinctive cross-hatching diagonal arrangement of actin filaments seen in the lamella and lamellipodia. The only known relationship between Arp2/3 and Tms is an inhibitory

Fig. 18.3 (continued) This figure depicts a newly emerging growth cone (see text in Section 18.4 for details). Tms are associated with actin filaments in the lamellipodia, lamella, and filopodia. In the lamellipodium (**a**) Tm5a associating with actin enhances the barbed end elongation of filaments at the leading edge and in the lamellipodium. Tm5a also inhibits excessive branching by inhibiting Arp2/3 binding to filaments. Tm5a and formin together produce unbranched, stable filaments, which elongate very slowly at their barbed ends. In the lamellum (**b**) Tm4 binds together with Caldesmon, and together they modulate the binding of myosin to form stable filaments. Tm5NM1 bound actin filaments also recruit myosin, and inhibit severing by ADF/cofilin. Myosin acts on these filaments to help pull them rearward in retrograde flow. Tmod2 binding to Tm5NM1 associated actin filaments inhibits depolymerisation from their pointed ends, and results in more stable filaments. Tmod1 binding to Tm5a associated actin filaments slows pointed end depolymerisation, and enhances the stability of the filament. In the filopodium (**c**) Tm5NM1 recruits myosin to the actin filaments in the distal filopodium. The arrangement of these myosin motors produces two directional forces on the filopodium – (1) the rearward pulling of actin filaments, helping to produce retrograde flow and (2) the shearing torque forces which contribute to the breaking of these filaments as they enter the transition zone. Whilst Tm5NM1 may here function indirectly to remodel the filaments of the filopodium, it is also inhibiting ADF/cofilin binding to these filaments, preventing them severing before entering the transition zone. More distally, Tm5a associated actin filaments inhibit excessive Arp2/3 branching of the filopodium one, with Tm5a capable of reducing (but not eliminating) nucleation of branching filaments by Arp2/3. This suggests an interesting role for Tm5a, which is found initially enriched in the growth cone, before retracting back towards the neurite as the neuron develops. We propose the following two roles of Tm5a in initial outgrowth

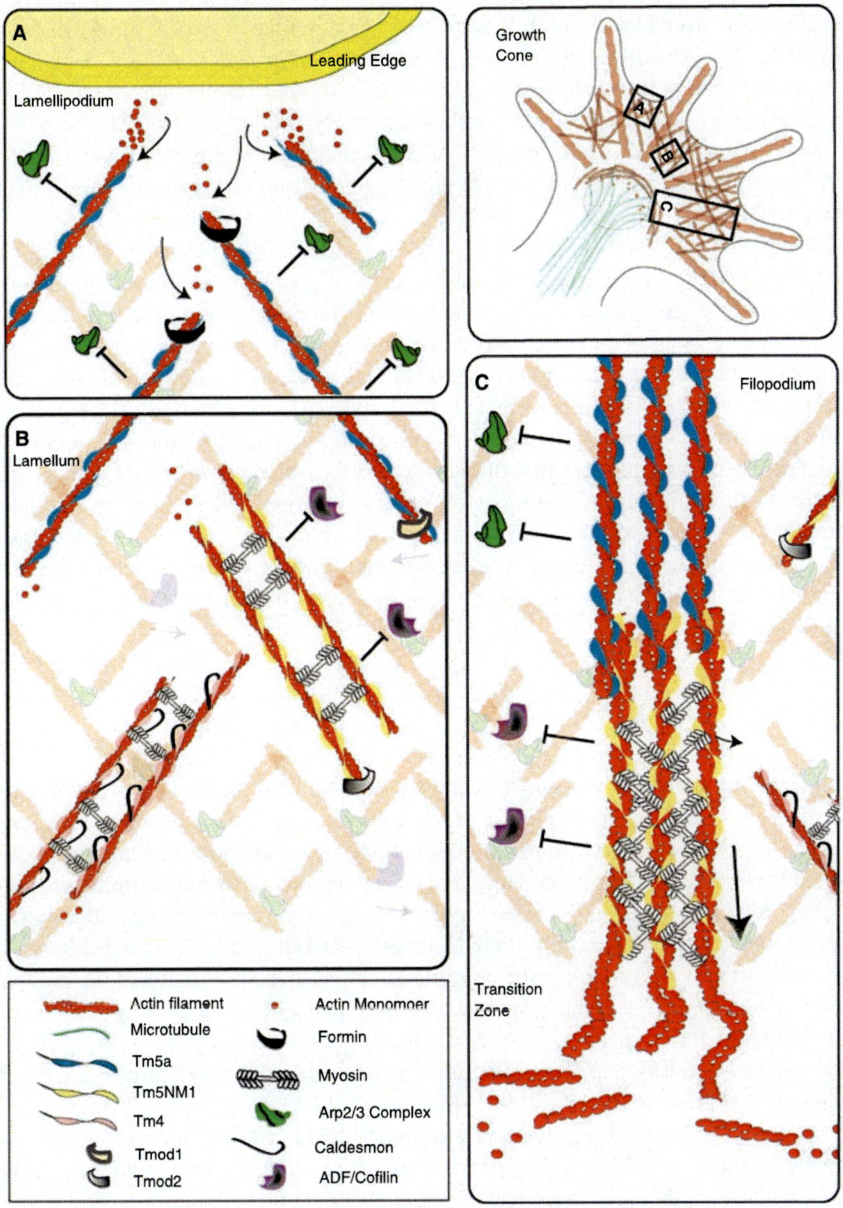

Fig. 18.3 (continued)

of the growth cone: (1) the necessary spike in Arp2/3 activity for the budding of a new population of filaments in the growth cone is held in check by Tm5a acting as a counterweight to this new abundance of Arp2/3 nucleated filaments; and stabilising the number of new filaments being formed and (2) the actin bundles which comprise filopodia, although in cases arising from branched filaments themselves, are prevented from branching further by Tm5a antagonism of Arp2/3 along the bundle length, and into filopodia in early growth cone outgrowth (see Fig. 18.3). This initial stabilisation is superseded when myosin and other stabilising proteins are at levels required to keep the actin bundles linear. By this mechanism, Tm5a being located in some but not all filopodia of the same growth cone (Schevzov et al., 1997) could be explained by a complementary expression of myosin in those Tm5a lacking filopodia.

In each of the above scenarios, Tm5a is a temporary solution to aberrant or excessive filament branching in the early stages of growth cone output, but down-regulated when the growth cone is established. The rapid influx of Tm5a into the establishing growth cone could also reflect the need for a rapid polymerisation of actin to provide a new structural framework, as Tm5a can act alone to enhance the elongation of actin filaments at their barbed ends. As the structure of the growth cone is established, the requirement for such an influx of new filaments is reduced, and Tm5a recedes from the growth cone. Tm5a may also act in co-operation with formin – which could be inhibiting overly extensive barbed end elongations of newly formed Tm5a decorated actin filaments, and also working with Tm5a to increase the stability of those filaments. The preferential binding of Tmod1 could further stabilise Tm5a decorated actin filaments by inhibiting depolymerisation from their pointed ends.

18.4.2 Retrograde Flow—Actin Filaments Move Rearward

Actin superstructures in the growth cone are capable of directional movement and neuron guidance. The ongoing cycle of actin polymerisation and depolymerisation, and the regulation of actin filament contractility by other molecular motors, help provide these forces. Those filaments containing barbed ends which terminate at the leading edge of lamellipodial veils are able to help protrude these veils through a movement of actin. This is commonly referred to as retrograde flow, and describes the rearward movement (i.e., in the opposite direction to the protrusion of the growth cone) of actin filaments in the growth cone, such that actin monomers can associate with the barbed ends of these filaments, immediately proximal to the leading edge. The majority of retrograde flow movement can be achieved through forces generated by myosin ATPases (Lin et al., 1997; Medeiros et al., 2006) and through actin polymerisation (Medeiros et al., 2006). Caldesmon attenuates the function of myosin motors on the actin filament. Myosin is implicated in the cycling of filaments: as actin bundles in filopodia move rearward in retrograde flow, the pointed end (most proximal to the neurite) will buckle as it approaches the central zone, and will ultimately be severed through shearing forces produced in part by myosin (Medeiros et al., 2006). We hypothesise that the role of Tm5NM1 in the growth cone is at least twofold: (1) the Tm5NM1 recruitment of myosin in growth cones allows for the retrograde flow of actin filaments from the meshwork within the lamella, and the retrograde flow and eventual severing of the pointed ends of actin bundles in filopodia, and (2) the simultaneous repulsion of ADF/cofilin from actin filaments confers stability to those Tm5NM1 bound filaments, such that the meshwork and the actin bundles of filopodia remain intact whilst in the lamellipodia and lamella, providing a structural framework of the growth cone. Furthermore, the Tm5NM1 enhancement of Tmod2 affinity for the actin filament prevents or slows the pointed-end depolymerisation of those actin filaments comprising the meshwork in the lamella (see Fig. 18.3). On other actin filaments within the lamella, CaD and myosin II could bind together to Tm4 associated filaments, producing stable filaments which are also pulled rearward in retrograde flow.

18.5 Tropomyosins at the Synapse

Some of the most complex actin filament organisations can be found at the connections that form between mature neurons, the synapses. On both sites of the synapse, the axonal pre-synaptic nerve terminal and the dendritic post-synapse, the actin filament system is crucial for the molecular processes that are involved in synaptic transmission and signal propagation. Previous studies have shown the localisation of Tm at both the pre-synapse and the post-synapse of central nervous system (CNS) synapses (Blitz and Fine, 1974; Had et al., 1994; Mello et al., 2007). Due to isoform-specific segregation of Tms into the two different compartments – with TmBr3 at the pre-synapse and Tm4 at the post-synapse, an isoform-specific function has been suggested (Had et al., 1994). The potential role of Tms on either side of the synapse will be discussed in the following section with brief introductions to the organisation of the actin cytoskeleton in these compartments.

18.5.1 Pre-Synapse

18.5.1.1 Actin at the Pre-Synapse

At the pre-synapse, enrichment of actin filaments has been shown to be close to the endocytic zone (Shupliakov et al., 2002), a region that surrounds the release site (also known as the active zone) for synaptic vesicles (see Cingolani and Goda, 2008 for review). The regulation of actin dynamics has been associated with synaptic activity (Bloom et al., 2003; Shupliakov et al., 2002). The precise role of actin filaments at the pre-synapse during synaptic activity is still unclear. The hypothesised roles of actin filaments in the pre-synapse range from serving as a scaffold structure for recruiting other regulatory proteins (Sankaranarayanan et al., 2003) to providing a physical barrier for the recycling of vesicles (Pilo Boyl et al., 2007; Trifaro et al., 2002). This physical barrier would allow for the tightly controlled release of neurotransmitters during synaptic transmission.

18.5.1.2 The Role of Tms at the Pre-Synapse

As discussed earlier, immuno-EM analysis of rat cerebellum at PND35 detected products of the αTm gene in the pre-synapse. The antibody used in this study is specific for exon 9c of the αTm gene, detecting TmBr1 and TmBr3. To date, TmBr1/3 are the only Tm isoforms that have been localised to the pre-synaptic nerve terminal. Interestingly, the functional characterisation of TmBr3 in stably transfected neuroblastoma cells has revealed isoform-specific properties with regards to regulating actin filament dynamics when compared to γTm gene products (Bryce et al., 2003). The overexpression of TmBr3 allows for the formation of highly dynamic actin filaments by permitting access of specific ABPs, such as ADF/cofilin, to the actin filaments (Bryce et al., 2003). The significance of the enriched expression of TmBr1/3 at the pre-synapse is compelling for its potential to generate an actin filament population that can respond quickly to the need for limiting exocytotic events.

18.5.2 Post-Synapse

18.5.2.1 Actin at the Post-Synapse

The post-synapse is a neuronal compartment with a particularly high density of actin filaments (Matus et al., 1982). The major sites for receiving neuronal excitatory input in the CNS are formed on dendritic spines, small protrusions that occur along the dendrites. The actin filament system has been shown to be critical for the proper development of dendritic structures and signal transduction at the synapse. Data suggest that actin is a driving force in regulating spine morphology, and that it is actively involved in mediating neurotransmitter receptor trafficking such as NMDA and AMPA receptors in excitatory synapses within the dendritic spines (Groc and Choquet, 2006). Neurotransmitter receptors are integrated in the spine head and linked to actin filaments by a large array of scaffolding proteins.

At least three distinct actin filament populations can be defined in the post-synapse both by their ultrastructure and by the analysis of their different dynamic properties. EM analysis has shown these different filament populations in post-synapses throughout different anatomical regions of the CNS (Dillon and Goda, 2005; Landis and Reese, 1983). Expanding from the post-synaptic density (PSD) towards the center of the spine head is a population of straight actin filaments of 4–6 nm in diameter and 20 nm in length. Due to its location this population is considered to be a component of the PSD itself. In the spine body this filament population overlaps with an actin filament population of 8–10 nm in diameter. A third population of filaments of 5–7 nm in diameter can be found close to the spine membrane in the neck and body (Landis and Reese, 1983). Despite this in depth ultrastructural analysis of actin filament organisation in spines of the cerebellum, other data suggest that there may be substantial differences in the organisation of actin filaments between spines of different regions, and between spine populations within one region (Capani et al., 2001) This study reveals a heterogeneity of actin staining in spines of neurons in the hippocampus, cerebellum and striatum with heterogeneity between the regions as well as within a single region.

In parallel to the formation of structurally distinct actin filament populations in the spines, three actin filament pools of different dynamic behavior can be identified in spines of CA1 pyramidal neurons of the rat hippocampus (Honkura et al., 2008). Filamentous actin can be classified in a dynamic, an enlargement, and a stable pool. The dynamic pool of quickly treadmilling actin filaments localises to the distal tip of the spine and is thought to allow for glutamate sensitivity and spine volume regulation. The enlargement pool is considered to be associated with long-term enlargement and LTP of the spines and the stable pool provides the required stability of spines (Honkura et al., 2008).

As in other highly dynamic structures, such as the growth cone, actin filament populations in dendritic spines are regulated by a large cohort of associated proteins (for more detail see Cingolani and Goda, 2008; Sekino et al., 2007). In the following section we will discuss some of the key players of actin filament regulation in the spines with a particular focus on ABPs that have been shown to interact functionally with Tms.

18.5.2.2 The Role of Tms at the Post-Synapse

The regulation of actin filaments in spines during spine morphogenesis, and polymerisation and depolymerisation in the spines of cultured hippocampal neurons, are coordinated throughout development (Hotulainen et al., 2009). In dendritic spines there is a switch from mDia2 to Arp2/3 promoted actin filament nucleation, which is associated with the development of filamentous spines into more mature spines. Such a switch is intriguing in the context of developmentally dependent expression patterns of Tm isoforms as discussed in the previous sections of this chapter. The activity of mDia2 can be affected by the presence of Tms in an isoform-specific manner (Wawro et al., 2007). Tm5a is the only neuronally expressed isoform analyzed in the study that did not reveal any effect on mDia2 function. In contrast, Tm5a has been shown to regulate negatively the actin filament branching activity of Arp2/3 (Blanchoin et al., 2001). Taken together one may speculate that a switch in the local expression profile of Tm isoforms at the sites of spine formation is instrumental in promoting the maturation of spine structures. However, the localisation of Tm5a in the dendrites has yet to be established. One Tm isoform which does localise to the mature dendritic spines is Tm4 (Had et al., 1994). It is therefore important to investigate the potential implications between Tm4 and the actin filament nucleators mDia2 and Arp2/3 to understand the role of Tms in mature spines.

Various studies have identified ADF/cofilin at the post-synapse and shown that ADF/cofilin activity is involved in maintaining spine morphology and function (Carlisle et al., 2008; Hotulainen et al., 2009; Racz and Weinberg, 2006). Again, the choice of Tm isoforms at the post-synapse may be critical to allow for a regulation of ADF/cofilin activity in this structure. The involvement of the actin filament binding protein drebrin in spine plasticity and synaptic activity has been confirmed by a number of recent studies (Ivanov et al., 2009; Takahashi et al., 2006, 2003; Takahashi et al., 2009). Drebrin and Tm have been shown to compete for binding to actin filaments (Ishikawa et al., 1994). Therefore, the presence of both Tm and drebrin at the post-synapse of CNS synapses suggests that these actin filament stabilizing proteins may be working in concert to determine spine morphogenesis and synaptic activity by regulating actin filament dynamics.

Myosin has been localised to the post-synapse of CNS synapses (Ryu et al., 2006; Wang et al., 2008). Myosin IIB is required for regulating both spine morphology and synaptic function in cultures of mature rat hippocampal neurons (Ryu et al., 2006). Since the localisation of myosin IIB can be affected by specific Tm isoforms, as shown for Tm5NM1 (Schevzov et al., 2005a), regulation of myosin IIB localisation and function in spines may be driven by the expression of a specific set of Tms in this compartment. The localisation of Tm5NM1 in mature neurons is still not fully understood. However the detection of γTm gene products in synaptosomes is suggestive for the segregation of products from this gene to synaptic connections (Mello et al., 2007).

Taken together, the observations of compartment specific localisation of Tm isoforms at the CNS synapse and the abundance at this site of proteins regulating actin dynamics, which themselves are affected by Tms in an isoform-specific manner, suggest a regulatory role of Tms in synapse formation and function. For this reason it is fundamental to determine the precise localisation of all Tm isoforms at the synapse of mature neurons and to analyse their interaction with the machinery that locally controls actin filament dynamics in order to understand better how the actin filament system regulates synaptic function.

18.6 Tropomyosins in Neurological Disorders

Involvement of Tms in the pathology of neurodegenerative diseases was first indicated by positive immunoreactivity in Hirano bodies, neuronal cytoplasmic inclusions observed in post mortem Alzheimer's disease (AD) brain tissue samples (Galloway et al., 1987). Shortly after this initial finding, Tm positive immunoreactivity was confirmed by the same laboratory in neurofibrillary tangles (NFT) and dystrophic neurites in AD, progressive supranuclear palsy, Pick's disease and diffuse Lewy body disease (Galloway et al., 1990; Galloway and Perry, 1991). The Tm immunoreactivity in NFTs may be disease specific, since no immunoreactivity has been observed for Tms in NFTs in idiopathic Parkinson's disease samples (Galloway and Perry, 1991). The antibody used in these studies was a polyclonal antibody raised against chicken gizzard and the specific Tms were not identified.

Following these initial studies, altered regulation of Tms has repeatedly been associated with neuronal pathologies. For example, the analysis of gerbil synaptosomes treated with the lipid peroxidation product acrolein – a trigger for oxidative damage and neurotoxicity – showed an upregulation and oxidation of products of the γTm gene (Mello et al., 2007). An increase in products from the γTm gene

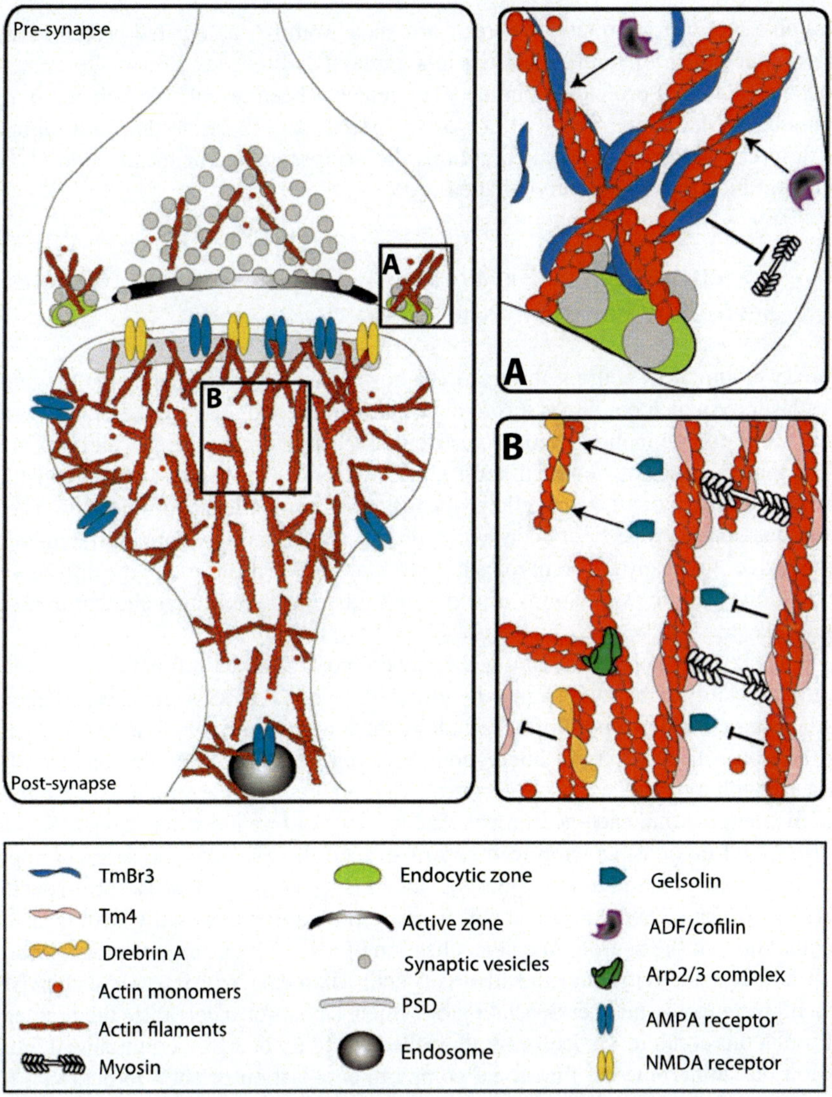

Fig. 18.4 Schematic of a mature excitatory synapse in the CNS. Depicted are individual actin filament populations at the pre-synapse and post-synapse (on the *left*). (see text in Section 18.5 for details). On the *right*, models are shown for Tm decorated actin filaments at the endocytic zone (**a**) and in the spine head (**b**). In the endocytic zone (**a**) decoration of actin filaments with TmBr3 (TmBr1/3 here referred to as TmBr3) allows ADF/cofilin to exert its severing activity on the filament and inhibits the binding of myosin. At the post-synapse, actin filaments are decorated either with Tm4 or drebrin A. In contrast to TmBr3, Tm4 allows myosin to bind to the filament which is critical for processes involved in neurotransmitter receptor trafficking. While Tm4 decorated filaments inhibit gelsolin activity, drebrin A decorated filaments – which block access of Tm4 – allow gelsolin to sever filaments, thereby generating different actin filament populations with distinct dynamic properties

was also found in a proteomic screen of concanavalin-A associated proteins from the hippocampus of human AD tissue samples (Owen et al., 2008). Conversely, downregulation of products from the γTm gene has been revealed in Schizophrenia pathology (Martins-de-Souza et al., 2009). These data illustrate that in a number of different neuropathological conditions the expression and/or localisation of Tm gene products can be severely affected.

18.7 Conclusion – The Big Picture of Tropomyosins in Neurons: Seeing the Forest *and* the Trees

The development of neurons through morphogenesis, differentiation, and the synaptic plasticity which continues throughout ontogeny requires a remarkable diversity of actin filament function. Actin is distributed throughout the neuron, and it is now clear that the pool of neuronal actin is subdivided into populations of actin filaments, each with distinct subcellular localisations, properties, and functions. There is no one single property of actin which provides a diversity of form and dynamics required for the many different roles it performs. Rather, it is through the differential and highly specific associations of actin with other proteins which allow it to be so multifaceted.

Tms play an integral role in this differential control of actin function, by affecting actin stability, by attenuating the effects of other ABPs on the actin filament, and by promoting fidelity of function along the length of a single filament. Because of the isoform-specificity of interactions between Tms and ABPs, the engagement of actin with a particular Tm isoform can be a predictor of the structure and fate of the filament. Within each neuron is a vast array of different compartments, defined by specific functions, and arrangements of structurally distinct actin filament populations. Actin is ubiquitous in neurons, and Tms segregate in an isoform-specific manner to spatially distinct compartments (containing functionally distinct actin populations) of the neuron. The colocalisation of actin with a particular Tm isoform will confer selectivity onto the filament: actin filaments will interact with ABPs in a highly specific manner depending on which Tm isoform decorates the filament. Through this conduit, Tm isoforms allow the selectivity of ABP:actin filament interaction that underpins the functional compartmentalisation of actin in neurons. By visualising the subcellular localisations of different Tms, we can predict the likely ABP–actin interactions.

Clues as to how the actin cytoskeleton is capable of such immense yet specific remodelling can be found in the localisation patterns of different ABPs and Tms, and how these patterns change throughout differentiation. The developmental regulation of Tms and their localisations is so tightly controlled that each neuronal isoform has a distinct spatial pattern of expression within immature and mature neurons. Different Tm isoforms contribute to different aspects of neuronal function: some Tms are important in growth cone morphology (such as Tm5NM1) (Schevzov et al., 2005a), other Tms are not found in growth cones, but are expressed only post differentiation (such as TmBr3) (Vrhovski et al., 2003; Weinberger et al., 1996). Even a single isoform can shift localisation dramatically throughout the development of the neuron. An example is Tm4, which in immature neurons is in the growth cones of both dendrites and axons, but with neuronal maturity shifts to a post synaptic locale (Had et al., 1994). Tms are regulated throughout neuronal development, and are also differently regulated in neurological disorders such as Alzheimer's disease, which underpins the importance of understanding their functions.

We can appreciate the importance of Tms by observing how neurons respond to experimental manipulations of Tm expression profiles. Knock out, knock down, and overexpression studies have helped elucidate the specific relationships different Tms have – whether as competitors, collaborators, or neither – with various ABPs. A single ABP can have different relationships with different Tm isoforms. For example, Tm5NM1 can apparently recruit myosin II to actin filaments, whereas TmBr3 has no such effect (Bryce et al., 2003). Similarly, one Tm isoform can have different relationships with different ABPs – while Tm5NM1 can recruit myosin II to actin filaments, it can also displace active ADF/cofilin from the actin filament (Bryce et al., 2003). The nature of the relationship is defined strictly by the ABP and the Tm isoform.

Tms offer us a way of discriminating one actin population from another. The extrinsic utility of understanding Tm isoform function lies in the information they can tell us about their associated actin filaments – by knowing their properties, we can build a picture of how one protein, actin, can for example be remodelled during the rapid transport of vesicles at the synapse, and simultaneously remain static in bundles within the mature axon. Tms superimpose heterogeneity of function onto the actin pool in the neuron, and allow us to both visualise and understand functionally distinct actin filament populations. It is this level of detail, of local discrimination at the level of actin filament populations and even between filaments themselves, which is so important in understanding neuronal function.

References

Altman J (1972) Postnatal development of the cerebellar cortex in the rat. II. Phases in the maturation of Purkinje cells and of the molecular layer. J Comp Neurol 145:399–463

Amano M, Ito M, Kimura K, Fukata Y, Chihara K, Nakano T, Matsuura Y, Kaibuchi. K (1996) Phosphorylation and activation of myosin by rho-associated kinase (Rho-kinase). J Biol Chem 271:20246–20249

Andrianantoandro E, Pollard. TD (2006) Mechanism of actin filament turnover by severing and nucleation at different concentrations of ADF/cofilin. Mol Cell 24:13–23

Arber S, Barbayannis FA, Hanser H, Schneider C, Stanyon CA, Bernard O, Caroni. P (1998) Regulation of actin dynamics through phosphorylation of cofilin by LIM-kinase. Nature 393:805–809

Bernard O (2007) Lim kinases, regulators of actin dynamics. Int J Biochem Cell Biol 39: 1071–1076

Blanchoin L, Pollard TD, Hitchcock-DeGregori. SE (2001) Inhibition of the Arp2/3 complex-nucleated actin polymerization and branch formation by tropomyosin. Curr Biol 11:1300–1304

Blitz AL, Fine. RE (1974) Muscle-like contractile proteins and tubulin in synaptosomes. Proc Natl Acad Sci USA 71:4472–4476

Bloom O, Evergren E, Tomilin N, Kjaerulff O, Low P, Brodin L, Pieribone VA, Greengard P, Shupliakov. O (2003) Colocalization of synapsin and actin during synaptic vesicle recycling. J Cell Biol 161:737–747

Bryce NS, Schevzov G, Ferguson V, Percival JM, Lin JJ, Matsumura F, Bamburg JR, Jeffrey PL, Hardeman EC, Gunning P, Weinberger. RP (2003) Specification of actin filament function and molecular composition by tropomyosin isoforms. Mol Biol Cell 14:1002–1016

Capani F, Martone ME, Deerinck TJ, Ellisman. MH (2001) Selective localization of high concentrations of F-actin in subpopulations of dendritic spines in rat central nervous system: a three-dimensional electron microscopic study. J Comp Neurol 435:156–170

Carlisle HJ, Manzerra P, Marcora E, Kennedy. MB (2008) SynGAP regulates steady-state and activity-dependent phosphorylation of cofilin. J Neurosci 28:13673–13683

Chrzanowska-Wodnicka M, Burridge. K (1996) Rho-stimulated contractility drives the formation of stress fibers and focal adhesions. J Cell Biol 133:1403–1415

Cingolani LA, Goda. Y (2008) Actin in action: the interplay between the actin cytoskeleton and synaptic efficacy. [Erratum appears in Nat Rev Neurosci. 2008 Jun;9(6):494]. Nat Rev Neurosci 9:344–356

Conti MA, Adelstein. RS (2008) Nonmuscle myosin II moves in new directions. J Cell Sci 121:11–18

Cox PR, Fowler V, Xu B, Sweatt JD, Paylor R, Zoghbi. HY (2003) Mice lacking tropomodulin-2 show enhanced long-term potentiation, hyperactivity, and deficits in learning and memory. Mol Cell Neurosci 23:1–12

Dabrowska R, Kulikova N, Gagola. M (2004) Nonmuscle caldesmon: its distribution and involvement in various cellular processes. review article. Protoplasma 224:1–13

Dillon C, Goda. Y (2005) The actin cytoskeleton: integrating form and function at the synapse. Annu Rev Neurosci 28:25–55

Dufour C, Weinberger RP, Schevzov G, Jeffrey PL, Gunning. P (1998a) Splicing of two internal and four carboxyl-terminal alternative exons in nonmuscle tropomyosin 5 pre-mRNA is independently regulated during development. J Biol Chem 273:18547–18555

Dufour C, Weinberger RP, Gunning. P (1998b) Tropomyosin isoform diversity and neuronal morphogenesis. Immunol Cell Biol 76:424–429

Faivre-Sarrailh C, Had L, Ferraz C, Sri Widada J, Liautard JP, Rabié A (1990) Expression of tropomyosin genes during the development of the rat cerebellum. J Neurochem 55:899–906

Fattoum A, Hartwig JH, Stossel. TP (1983) Isolation and some structural and functional properties of macrophage tropomyosin. Biochemistry 22:1187–1193

Fischer RS, Fowler. VM (2003) Tropomodulins: life at the slow end. Trends Cell Biol 13:593–601

Fowler VM (1987) Identification and purification of a novel Mr 43,000 tropomyosin-binding protein from human erythrocyte membranes. J Biol Chem 262:12792–12800

Galloway PG, Mulvihill P, Siedlak S, Mijares M, Kawai M, Padget H, Kim R, Perry. G (1990) Immunochemical demonstration of tropomyosin in the neurofibrillary pathology of alzheimer's disease. Am J Pathol 137:291–300

Galloway PG, Perry. G (1991) Tropomyosin distinguishes Lewy bodies of Parkinson disease from other neurofibrillary pathology. Brain Res 541:347–349

Galloway PG, Perry G, Gambetti. P (1987) Hirano body filaments contain actin and actin-associated proteins. J Neuropathol Exp Neurol 46:185–199

Gregorio CC, Fowler. VM (1995) Mechanisms of thin filament assembly in embryonic chick cardiac myocytes: tropomodulin requires tropomyosin for assembly. J Cell Biol 129:683–695

Groc L, Choquet. D (2006) AMPA and NMDA glutamate receptor trafficking: multiple roads for reaching and leaving the synapse. Cell Tissue Res 326:423–438

Gunning P, Hardeman E, Jeffrey P, Weinberger. R (1998) Creating intracellular structural domains: spatial segregation of actin and tropomyosin isoforms in neurons. Bioessays 20:892–900

Gunning P, O'Neill G, Hardeman. E (2008) Tropomyosin-based regulation of the actin cytoskeleton in time and space. Physiol Rev 88:1–35

Gunning PW, Schevzov G, Kee AJ, Hardeman. EC (2005) Tropomyosin isoforms: divining rods for actin cytoskeleton function. Trends Cell Biol 15:333–341

Had L, Faivre-Sarrailh C, Legrand C, Mery J, Brugidou J, Rabie. A (1994) Tropomyosin isoforms in rat neurons: the different developmental profiles and distributions of TM-4 and TMBr-3 are consistent with different functions. J Cell Sci 107:2961–2973

Had L, Faivre-Sarrailh C, Legrand C, Rabié. A (1993) The expression of tropomyosin genes in pure cultures of rat neurons, astrocytes and oligodendrocytes is highly cell-type specific and strongly regulated during development. Mol Brain Res 18:77–86

Hannan AJ, Gunning P, Jeffrey PL, Weinberger. RP (1998) Structural compartments within neurons: developmentally regulated organization of microfilament isoform mRNA and protein. Mol Cell Neurosci 11:289–304

Hannan AJ, Schevzov G, Gunning P, Jeffrey PL, Weinberger. RP (1995) Intracellular localization of tropomyosin mRNA and protein is associated with development of neuronal polarity. Mol Cell Neurosci 6:397–412

Higgs HN, Pollard. TD (2001) Regulation of actin filament network formation through ARP2/3 complex: activation by a diverse array of proteins. Annu Rev Biochem 70:649–676

Honkura N, Matsuzaki M, Noguchi J, Ellis-Davies GC, Kasai. H (2008) The subspine organization of actin fibers regulates the structure and plasticity of dendritic spines. Neuron 57:719–729

Horiuchi KY, Chacko. S (1988) Interaction between caldesmon and tropomyosin in the presence and absence of smooth muscle actin. Biochemistry 27:8388–8393

Hotulainen P, Llano O, Smirnov S, Tanhuanpaa K, Faix J, Rivera C, Lappalainen. P (2009) Defining mechanisms of actin polymerization and depolymerization during dendritic spine morphogenesis. J Cell Biol 185:323–339

Ishikawa R, Hayashi K, Shirao T, Xue Y, Takagi T, Sasaki Y, Kohama. K (1994) Drebrin, a development-associated brain protein from rat embryo, causes the dissociation of tropomyosin from actin filaments. J Biol Chem 269:29928–29933

Ishikawa R, Yamashiro S, Matsumura. F (1989) Differential modulation of actin-severing activity of gelsolin by multiple isoforms of cultured rat cell tropomyosin. Potentiation of protective ability of tropomyosins by 83-kDa nonmuscle caldesmon. J Biol Chem 264:7490–7497

Ivanov A, Esclapez M, Pellegrino C, Shirao T, Ferhat. L (2009) Drebrin A regulates dendritic spine plasticity and synaptic function in mature cultured hippocampal neurons. J Cell Sci 122: 524–534

Kira M, Tanaka J, Sobue. K (1995) Caldesmon and low Mr isoform of tropomyosin are localized in neuronal growth cones. J Neurosci Res 40:294–305

Korobova F, Svitkina. T (2008) Arp2/3 complex is important for filopodia formation, growth cone motility, and neuritogenesis in neuronal cells. Mol Biol Cell 19:1561–1574

Kostyukova AS (2008a) Capping complex formation at the slow-growing end of the actin filament. Biochemistry (Mosc) 73:1467–1472

Kostyukova AS (2008b) Tropomodulins and tropomodulin/tropomyosin interactions. Cell Mol Life Sci 65:563–569

Kovar DR (2006) Cell polarity: formin on the move. Curr Biol 16:R535–R538

Kuhn TB, Bamburg. JR (2008) Tropomyosin and ADF/cofilin as collaborators and competitors. Adv Exp Med Biol 644:232–249

Landis DM, Reese. TS (1983) Cytoplasmic organization in cerebellar dendritic spines. J Cell Biol 97:1169 1178

Lees-Miller JP, Yan A, Helfman. DM (1990) Structure and complete nucleotide sequence of the gene encoding rat fibroblast tropomyosin 4. J Mol Biol 213:399–405

Lehman W, Craig. R (2008) Tropomyosin and the steric mechanism of muscle regulation. Adv Exp Med Biol 644:95–109

Lin CH, Espreafico EM, Mooseker MS, Forscher. P (1997) Myosin drives retrograde F-actin flow in neuronal growth cones. Biol Bull 192:183–185

Martins-de-Souza D, Gattaz WF, Schmitt A, Novello JC, Marangoni S, Turck CW, Dias-Neto. E (2009) Proteome analysis of schizophrenia patients wernicke's area reveals an energy metabolism dysregulation. BMC Psychiatry 9:17

Matus A, Ackermann M, Pehling G, Byers HR, Fujiwara. K (1982) High actin concentrations in brain dendritic spines and postsynaptic densities. Proc Natl Acad Sci USA 79:7590–7594

Medeiros NA, Burnette DT, Forscher. P (2006) Myosin II functions in actin-bundle turnover in neuronal growth cones. Nat Cell Biol 8:215–226

Mello CF, Sultana R, Piroddi M, Cai J, Pierce WM, Klein JB, Butterfield. DA (2007) Acrolein induces selective protein carbonylation in synaptosomes. Neuroscience 147: 674–679

Mullins RD, Heuser JA, Pollard. TD (1998) The interaction of Arp2/3 complex with actin: nucleation, high affinity pointed end capping, and formation of branching networks of filaments. Proc Natl Acad Sci USA 95:6181–6186

Nicholson-Flynn K, Hitchcock-DeGregori SE, Levitt. P (1996) Restricted expression of the actin-regulatory protein, tropomyosin, defines distinct boundaries, evaginating neuroepithelium, and choroid plexus forerunners during early CNS development. J Neurosci 16: 6853–6863

Niwa R, Nagata-Ohashi K, Takeichi M, Mizuno K, Uemura. T (2002) Control of actin reorganization by slingshot, a family of phosphatases that dephosphorylate ADF/cofilin. Cell 108:233–246

Ono S, Ono. K (2002) Tropomyosin inhibits ADF/cofilin-dependent actin filament dynamics. J Cell Biol 156:1065–1076

Otomo T, Otomo C, Tomchick DR, Machius M, Rosen. MK (2005) Structural basis of Rho GTPase-mediated activation of the formin mDia1. Mol Cell 18:273–281

Owen JB, Di Domenico F, Sultana R, Perluigi M, Cini C, Pierce WM, Butterfield. DA (2009) Proteomics-determined differences in the concanavalin-A-fractionated proteome of hippocampus and inferior parietal lobule in subjects with Alzheimer's disease and mild cognitive impairment: implications for progression of AD. J Proteome Res 8:471–482

Percival JM, Hughes JA, Brown DL, Schevzov G, Heimann K, Vrhovski B, Bryce N, Stow JL, Gunning. PW (2004) Targeting of a tropomyosin isoform to short microfilaments associated with the Golgi complex. Mol Biol Cell 15:268–280

Pilo Boyl P, Di Nardo A, Mulle C, Sassoe-Pognetto M, Panzanelli P, Mele A, Kneussel M, Costantini V, Perlas E, Massimi M, Vara H, Giustetto M, Witke. W (2007) Profilin2 contributes to synaptic vesicle exocytosis, neuronal excitability, and novelty-seeking behavior. EMBO J 26:2991–3002

Racz B, Weinberg. RJ (2006) Spatial organization of cofilin in dendritic spines. Neuroscience 138:447–456

Ryu J, Liu L, Wong TP, Wu DC, Burette A, Weinberg R, Wang YT, Sheng. M (2006) A critical role for myosin IIb in dendritic spine morphology and synaptic function. Neuron 49:175–182

Sankaranarayanan S, Atluri PP, Ryan. TA (2003) Actin has a molecular scaffolding, not propulsive, role in presynaptic function. Nat Neurosci 6:127–135

Schevzov G, Vrhovski B, Bryce NS, Elmir S, Qiu MR, O'Neill M, Yang G,N, Verrills NM, Kavallaris M, Gunning. PW (2005a) Tissue-specific tropomyosin isoform composition. J Histochem Cytochem 53:557–570

Schevzov G, Gunning P, Jeffrey PL, Temm-Grove C, Helfman DM, Lin JJ, Weinberger. RP (1997) Tropomyosin localization reveals distinct populations of microfilaments in neurites and growth cones. Mol Cell Neurosci 8:439–454

Schevzov G, Bryce NS, Almonte-Baldonado R, Joya J, Lin JJ, Hardeman E, Weinberger R, Gunning. P (2005b) Specific features of neuronal size and shape are regulated by tropomyosin isoforms. Mol Biol Cell 16:3425–3437

Sekino Y, Kojima N, Shirao. T (2007) Role of actin cytoskeleton in dendritic spine morphogenesis. Neurochem Int 51:92–104

Sellers JR (1981) Phosphorylation-dependent regulation of Limulus myosin. J Biol Chem 256:9274–9278

Shupliakov O, Bloom O, Gustafsson JS, Kjaerulff O, Low P, Tomilin N, Pieribone VA, Greengard P, Brodin. L (2002) Impaired recycling of synaptic vesicles after acute perturbation of the presynaptic actin cytoskeleton. Proc Natl Acad Sci USA 99:14476–14481

Stamm S, Casper D, Lees-Miller JP, Helfman. DM (1993) Brain-specific tropomyosins TMBr-1 and TMBr-3 have distinct patterns of expression during development and in adult brain. Proc Natl Acad Sci USA 90:9857–9861

Svitkina TM, Borisy. GG (1999) Arp2/3 complex and actin depolymerizing factor/cofilin in dendritic organization and treadmilling of actin filament array in lamellipodia. J Cell Biol 145:1009–1026
Svitkina TM, Verkhovsky AB, McQuade KM, Borisy. GG (1997) Analysis of the Actin-Myosin II system in fish Epidermal Keratocytes: mechanism of cell body translocation. J Cell Biol 139:397–415
Takahashi H, Mizui T, Shirao. T (2006) Down-regulation of drebrin A expression suppresses synaptic targeting of NMDA receptors in developing hippocampal neurones. J Neurochem 97(Suppl 1):110–115
Takahashi H, Sekino Y, Tanaka S, Mizui T, Kishi S, Shirao. T (2003) Drebrin-dependent actin clustering in dendritic filopodia governs synaptic targeting of postsynaptic density-95 and dendritic spine morphogenesis. J Neurosci 23:6586–6595
Takahashi H, Yamazaki H, Hanamura K, Sekino Y, Shirao. T (2009) Activity of the AMPA receptor regulates drebrin stabilization in dendritic spine morphogenesis. J Cell Sci 122:1211–1219
Trifaro JM, Lejen T, Rose SD, Pene TD, Barkar ND, Seward. EP (2002) Pathways that control cortical F-actin dynamics during secretion. Neurochem Res 27:1371–1385
Ujfalusi Z, Vig A, Hild G, Nyitrai. M (2009) Effect of tropomyosin on formin-bound actin filaments. Biophys J 96:162–168
Verkhovsky AB, Chaga OY, Schaub S, Svitkina TM, Meister J-J, Borisy. GG (2003) Orientational order of the Lamellipodial Actin network as demonstrated in living motile cells. Mol Biol Cell 14:4667–4675
Vrhovski B, Schevzov G, Dingle S, Lessard JL, Gunning P, Weinberger. RP (2003) Tropomyosin isoforms from the gamma gene differing at the C-terminus are spatially and developmentally regulated in the brain. J Neurosci Res 72:373–383
Wallar BJ, Alberts. AS (2003) The formins: active scaffolds that remodel the cytoskeleton. Trends Cell Biol 13:435–446
Wang CL (2008) Caldesmon and the regulation of cytoskeletal functions. Adv Exp Med Biol 644:250–272
Wang Z, Edwards JG, Riley N, Provance DW Jr., Karcher R, Li XD, Davison IG, Ikebe M, Mercer JA, Kauer JA, Ehlers. MD (2008) Myosin Vb mobilizes recycling endosomes and AMPA receptors for postsynaptic plasticity.[see comment]. Cell 135:535–548
Watakabe A, Kobayashi R, Helfman. DM (1996) *N*-Tropomodulin: a novel isoform of tropomodulin identified as the major binding protein to brain tropomyosin. J Cell Sci 109(Pt 9): 2299–2310
Wawro B, Greenfield NJ, Wear MA, Cooper JA, Higgs HN, Hitchcock-DeGregori. SE (2007) Tropomyosin regulates elongation by formin at the fast-growing end of the actin filament. Biochemistry 46:8146–8155
Weber A, Pennise CR, Babcock GG, Fowler. VM (1994) Tropomodulin caps the pointed ends of actin filaments. J Cell Biol 127:1627–1635
Weinberger RP, Henke RC, Tolhurst O, Jeffrey PL, Gunning. P (1993) Induction of neuron-specific tropomyosin mRNAs by nerve growth factor is dependent on morphological differentiation. J Cell Biol 120:205–215
Weinberger R, Schevzov G, Jeffrey P, Gordon K, Hill M, Gunning. P (1996) The molecular composition of neuronal microfilaments is spatially and temporally regulated. J Neurosci 16:238–252
Yamashiro-Matsumura S, Matsumura. F (1988) Characterization of 83-kilodalton nonmuscle caldesmon from cultured rat cells: stimulation of actin binding of nonmuscle tropomyosin and periodic localization along microfilaments like tropomyosin. J Cell Biol 106:1973–1983
Yamawaki-Kataoka Y, Helfman. DM (1987) Isolation and characterization of cDNA clones encoding a low molecular weight nonmuscle tropomyosin isoform. J Biol Chem 262:10791–10800
Yang N, Higuchi O, Ohashi K, Nagata K, Wada A, Kangawa K, Nishida E, Mizuno. K (1998) Cofilin phosphorylation by LIM-kinase 1 and its role in Rac-mediated actin reorganization. Nature 393:809–812
Yao W, Sung. LA (2009) Specific expression of E-Tmod (Tmod1) in horizontal cells: implications in neuronal cell mechanics and glaucomatous retina. Mol Cell Biomech 6:71–82

Chapter 19
The Driving Machinery for Growth Cone Navigation

Takuro Tojima and Hiroyuki Kamiguchi

Abstract The motility of neuronal growth cones plays a crucial role in the formation of neuronal circuit. For growth cone migration, the cytoskeleton and its associated motors generate traction force that is transmitted to the surrounding environment via cell adhesion molecules. The force transmission can be spatiotemporally controlled by a molecular clutch that mediates mechanical coupling of cell adhesion molecules and the actin cytoskeleton. Furthermore, intracellular membrane trafficking may control the supply and removal of membrane components in a spatially defined manner. In this chapter, we will argue that coordinated activity of these molecular events determines the speed and direction of axon growth, with particular emphasis on how Ca^{2+} signals control the driving machinery for growth cone navigation.

Keywords Adhesion · Axon · Calcium · Clutch · Growth cone · Guidance · Membrane

Contents

19.1 Introduction

Neuronal networks are formed by cellular processes, i.e., axons and dendrites, that have emerged and elongated from nascent neurons. The tip of an elongating axon,

H. Kamiguchi (✉)
Laboratory for Neuronal Growth Mechanisms, RIKEN Brain Science Institute, Saitama 351-0198, Japan
e-mail: kamiguchi@brain.riken.jp

R.A. Nixon, A. Yuan (eds.), *Cytoskeleton of the Nervous System*,
Advances in Neurobiology 3, DOI 10.1007/978-1-4419-6787-9_19,

called the growth cone, explores environmental information and migrates along the correct path toward its target. Such extracellular information can be diffusible, substrate-bound, or cell-surface molecules, which serve as attractive/permissive or repulsive/inhibitory cues upon binding to relevant receptors expressed on the growth cone (Huber et al., 2003; Tessier-Lavigne and Goodman, 1996). Activation of the receptors directs axon growth and guidance via generating intracellular signals and regulating cytoskeletal organization. Studying spatiotemporal properties of these events in growth cones is a key step for understanding how the axons elongate in the correct direction and form precise connections with their targets.

Extracellular diffusible cues such as netrin-1, neurotrophins, myelin-associated glycoprotein (MAG), and semaphorins influence the direction of axon elongation through cytosolic Ca^{2+} signals in the growth cone (Henley et al., 2004; Hong et al., 2000; Song and Poo, 1999; Togashi et al., 2008). It has also been shown that a focal and unilateral elevation of cytosolic free Ca^{2+} concentration ($[Ca^{2+}]_c$) in the growth cone is sufficient to induce its turning (Zheng, 2000). Interestingly, extracellular gradients of netrin-1 or MAG, which induce attractive or repulsive growth cone turning respectively, increase $[Ca^{2+}]_c$ that is highest on the side of the growth cone facing the source of the netrin-1/MAG gradients (Henley et al., 2004; Hong et al., 2000). This observation indicates that a $[Ca^{2+}]_c$ gradient across the growth cone can trigger turning to the side with higher $[Ca^{2+}]_c$ (attraction) as well as to the side with lower $[Ca^{2+}]_c$ (repulsion). Then what determines the turning direction with regard to the asymmetric Ca^{2+} signals across the growth cone? It is thought that the amplitude of $[Ca^{2+}]_c$ elevations determines the directional polarity of growth cone turning: high-amplitude Ca^{2+} signals trigger attraction whereas low-amplitude Ca^{2+} signals trigger repulsion (Gomez and Zheng, 2006; Henley and Poo, 2004; Henley et al., 2004). On the other hand, data from our laboratory indicate that the source of Ca^{2+} influx, rather than its amplitude or the baseline Ca^{2+} level, is the critical determinant of the turning direction (Ooashi et al., 2005). Ryanodine receptors are Ca^{2+} channels that mediate secondary Ca^{2+} release from the endoplasmic reticulum into the cytosol in response to primary $[Ca^{2+}]_c$ elevations, a process called Ca^{2+}-induced Ca^{2+} release (CICR). We showed that Ca^{2+} signals that are accompanied by CICR trigger attraction whereas Ca^{2+} signals without CICR trigger repulsion (Ooashi et al., 2005).

Now it is well know that bidirectional axon guidance involves attractive and repulsive Ca^{2+} signals generated asymmetrically across the growth cone. However, it remains less clear what machinery drives bidirectional growth cone turning downstream of the Ca^{2+} signals. In this chapter, we will examine the driving mechanisms for growth cone migration and turning with particular emphasis on cell adhesion and membrane trafficking.

19.2 Driving Mechanisms for Growth Cone Migration

The growth cone makes an adhesive contact with its environment via cell adhesion molecules (CAMs). CAMs expressed by neurons can be divided into three large families: integrins, cadherins, and the immunoglobulin superfamily (IgSF) members. Cadherins and the majority of IgSF CAMs can interact homophilically with the same type of molecules present on adjacent cells. Integrins interact with extracellular matrix molecules such as laminin. The importance of CAMs has been illustrated by the fact that mutations of L1, an IgSF CAM, cause defects in several major axon tracts in humans and mice (Cohen et al., 1998; Dahme et al., 1997; Kamiguchi et al., 1998a). Dynamic interactions of CAMs with cytoskeletal components are critically important for cell migration including axon growth. Spatially localized actin polymerization/depolymerization and actin-myosin interactions generate retrograde flow of filamentous actin (F-actin) that generates driving force for cell migration (Pollard and Borisy, 2003). According to the clutch hypothesis, CAMs on the growth cone transmit this force by linking the F-actin flow with immobile ligands present on adjacent cells or in the extracellular matrix (Mitchison and Kirschner, 1988; Suter and Forscher, 2000). A molecular clutch mediates the engagement between F-actin flow and CAMs in a spatiotemporally regulated manner. The first molecule identified as a component of the clutch module is ankyrin$_B$ that facilitates the initial protrusion of neurites from the soma by coupling L1 with F-actin flow (Nishimura et al., 2003). After the formation of neurites, shootin1 functions as a component of the clutch that promotes growth cone migration in an L1-dependent manner (Shimada et al., 2008). Furthermore, it has been suggested that ezrin-mediated interactions between L1 and F-actin flow transmit traction force for growth cone migration (Sakurai et al., 2008). The clutch mechanism also applies to growth cones that use N-cadherin as an adhesive receptor for their migration (Bard et al., 2008). Traction force generated by F-actin flow is transmitted to N-cadherin adhesions through the clutch module involving catenins. Therefore, the mechanical coupling between CAMs and F-actin flow may be a general mechanism of force transmission during axon growth.

It is thought that, in addition to the CAM-cytoskeletal linkage, a front-vs-rear asymmetry of growth cone adhesion to its environment is required such that the cytoskeletal machinery is able to move the growth cone forward as attachments at its rear are released. To create and maintain such polarized adhesion, CAMs that have been translocated into the growth cone central domain by coupling with F-actin flow should be recycled to the leading front (Kamiguchi and Yoshihara, 2001). One mechanism for CAM recycling is dependent on intracellular membrane trafficking. L1 is endocytosed preferentially at the growth cone central domain via clathrin-mediated pathways (Kamiguchi et al., 1998b). L1-containing endosomes are then transported anterogradely along microtubules (Kamiguchi and Lemmon, 2000), which may be driven by KIF4 (Peretti et al., 2000). Subsequently, L1 is recycled to the growth cone leading front as a result of vesicle fusion with the plasma membrane. It is likely that this exocytic event is mediated by tetanus neurotoxin-insensitive vesicle-associated membrane protein (TI-VAMP) (Alberts et al., 2003). In this way, growth cone migration is driven by spatially regulated endocytic and exocytic events together with cell-surface movement of L1 (Fig. 19.1) (Kamiguchi, 2003). This type of L1 trafficking in growth cones has been confirmed by state-of-the-art molecular imaging techniques (Dequidt et al., 2007). We propose that this type of CAM trafficking is a general driving mechanism for growth cone migration although different recycling pathways and diverse clutch molecules should exist.

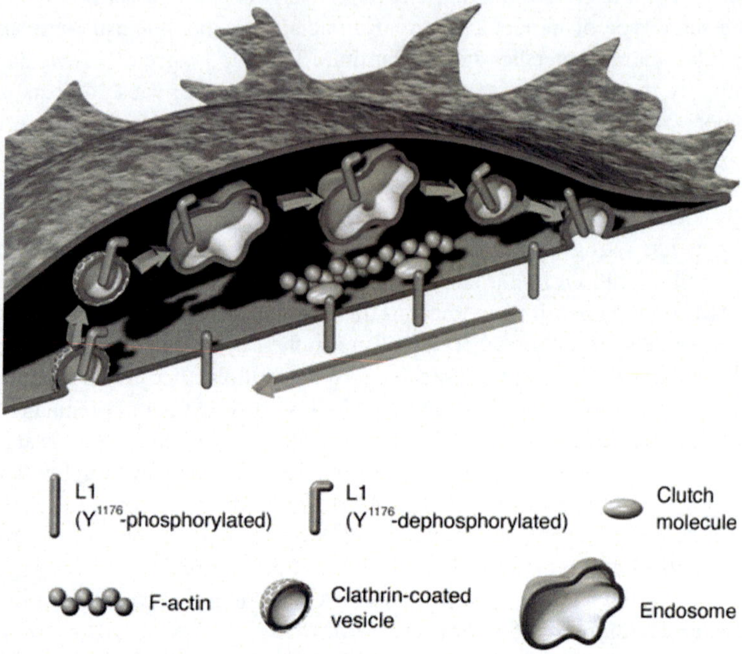

Fig. 19.1 A schematic view of the growth cone sectioned longitudinally, showing a model of L1 trafficking. Dephosphorylation of tyrosine (Y^{1176}) in the L1 cytoplasmic domain triggers clathrin-mediated endocytosis of L1 at the growth cone central domain. Subsequently, the endocytosed L1 is transported into the peripheral domain through sorting and recycling endosomes; this process is dependent on the dynamic ends of microtubules (not shown in this figure). Next, the trafficking L1 is reinserted into the plasma membrane at the leading front. Recycled L1 on the cell surface moves toward the central domain by coupling with retrograde F-actin flow via clutch molecules. Reprinted, with permission of the publisher, from Kamiguchi and Lemmon (2000)

19.3 Driving Mechanisms for Growth Cone Turning

According to the model (Fig. 19.1), the speed of growth cone migration depends on three major factors: (1) cytoskeletal dynamics, (2) clutch engagement, and (3) recycling of CAMs and membrane components. This suggests that asymmetric activity of any of these three factors across the growth cone may drive its turning. Therefore, we have conducted a series of experiments to test for this possibility (Tojima et al., 2007). Attractive Ca^{2+} signals lasting as long as 5 min did not affect microtubule dynamics in the growth cone periphery. Attractive Ca^{2+} signals caused no detectable change in the speed of retrograde F-actin flow or in the number and length of growth cone filopodia. These data indicate that axon-guiding signals do not acutely alter the cytoskeletal dynamics. The clutch engagement was assessed by tracking L1-coupled beads on the growth cone. The bead exhibited two-dimensional Brownian motion when L1 was disconnected from F-actin flow. In contrast, the bead showed retrograde directional movement on the growth cone when the clutch was engaged. Ca^{2+} signals in the growth cone cytosol did not affect the bead behavior (unpublished data) suggesting that the clutch engagement is not an immediate target of axon-guiding signals.

Next we examined the effect of Ca^{2+} signals on membrane recycling by monitoring the movement of FM1-43-labeled endocytic vesicles (Tojima et al., 2007). Within 1 min after the onset of attractive Ca^{2+} signals, the vesicles in the growth cone central domain exhibited a rapid, centrifugal migration into the peripheral domain on the side with elevated Ca^{2+}. In contrast, repulsive Ca^{2+} signals did not cause asymmetric migration of the vesicles. In response to attractive Ca^{2+} signals, exocytosis-competent vesicles that contain vesicle-associated membrane protein-2 (VAMP-2) also exhibited the similar pattern of centrifugal migration. These data indicate that the vesicle migration induced by attractive Ca^{2+} signals precedes any detectable changes in cytoskeletal dynamics and suggest that attractive Ca^{2+} signals affect the transport efficiency of vesicles along pre-existing microtubules rather than the microtubules themselves at this early stage of turning. While it is widely accepted that asymmetric reorganization of the cytoskeleton plays a critical role in growth cone navigation (Dent and Gertler, 2003), such asymmetry provides mechanistic force for turning most likely at later stages. For example, the polarized trafficking of recycling vesicles may recruit the Rac GTPase to restricted areas where actin polymerization is stimulated for directed migration (Palamidessi et al., 2008).

It has been proposed that growth cone turning requires asymmetric plasmalemmal expansion and/or localized recruitment of functional molecules to the plasma membrane (Henley and Poo, 2004). Therefore, the transported vesicles in growth cone periphery should be exocytosed at the site of attractive Ca^{2+} signals. We reported that attractive Ca^{2+} signals induce asymmetric exocytosis of VAMP-2-positive vesicles in the peripheral domain and that the vesicle transport induced by attractive Ca^{2+} signals is responsible for a large asymmetry of exocytosis during attractive turning of the growth cone (Tojima et al., 2007). As expected, growth cone attraction, but not repulsion, was completely prevented by treatment with tetanus neurotoxin that blocks VAMP2-mediated exocytosis. Therefore, attractive growth cone guidance involves asymmetric membrane transport and exocytosis, i.e., the growth cone turns by preferentially supplying membrane components and associated molecules to the side facing the new direction. Also, our results strongly suggest that growth cone attraction and repulsion are driven by distinct mechanisms, rather than using the same molecular machinery with opposing polarities.

Then, what machinery drives growth cone repulsion? Ca^{2+} signals repel growth cones via local activation of intracellular protease calpain that regulates phosphotyrosine signaling at integrin-based adhesion sites (Robles et al., 2003). It has been demonstrated that, in non-neuronal cells, calpain-mediated proteolysis of talin is a critical, rate-limiting step for focal adhesion disassembly (Franco et al., 2004). The linkage between focal adhesion disassembly and growth cone detachment from its environment downstream of Ca^{2+} signals has been suggested experimentally: disappearance of phosphorylated focal adhesion kinase and deadhesion of growth cones from the substrate are induced by Ca^{2+} signals that are expected to repel the growth cones (Conklin et al., 2005). Although the involvement of cytosolic Ca^{2+} signaling is unclear, activation of Roundabout by the repulsive guidance cue Slit inhibits N-cadherin-mediated adhesion (Rhee et al., 2002). Therefore, asymmetric deadhesion may be a general mechanism for repulsion: the growth cone turns to the side with more stable adhesion to the substrate. Given the finding that growth cone collapse is accompanied by endocytic events (Fournier et al., 2000), another mechanism may be that growth cone repulsion is driven by asymmetric removal of the plasma membrane via localized endocytosis triggered by repulsive Ca^{2+} signals. It is possible that the two mechanisms contribute cooperatively or sequentially to repulsive turning, in which detached areas of the plasma membrane are internalized together with cell-surface molecules that are needed for growth cone migration. This would cause asymmetric driving activities across the growth cone leading to its turning.

19.4 Conclusions and Future Prospects

Regulation of the cytoskeleton has been regarded as the primary target of biochemical signals that are generated downstream of receptors for axon guidance cues. Examples of the regulators of cytoskeletal organization during growth cone guidance includes Rho GTPases (Jin et al., 2005; Yuan et al., 2003), Ena/VASP proteins (Drees and Gertler, 2008), LIM kinase and Slingshot phosphatase acting on ADF/cofilin (Wen et al., 2007), microtubule plus-end-tracking proteins (Lee et al., 2004), and local translation machinery for cytoskeleton-associated proteins (Lin and Holt, 2008). In addition, there is increasing evidence that the spatial regulation of membrane-associated functions, i.e., cell adhesion and membrane trafficking, represents an important mechanism for growth cone turning. In this chapter, we have discussed how asymmetric membrane dynamics and adhesion drive growth cone attraction and repulsion. While the involvement of these mechanisms in growth cone steering has been clarified, several important questions remain unanswered. How do growth cones discriminate between attractive and repulsive Ca^{2+} signals and translate them into distinct modes of membrane dynamics? Does asymmetric membrane dynamics trigger growth cone turning by simple alteration of its surface area or by spatial control of molecular compositions? How is altered membrane dynamics linked to cytoskeletal reorganization downstream of Ca^{2+} signals? Answering these questions should be important steps toward understanding of intracellular precise mechanisms of growth cone navigation.

References

Alberts P, Rudge R, Hinners I, Muzerelle A, Martinez-Arca S, Irinopoulou T, Marthiens V, Tooze S, Rathjen F, Gaspar P, Galli T (2003) Cross talk between tetanus neurotoxin-insensitive vesicle-associated membrane protein-mediated transport and L1-mediated adhesion. Mol Biol Cell 14:4207–4220

Bard L, Boscher C, Lambert M, Mege RM, Choquet D, Thoumine O (2008) A molecular clutch between the actin flow and N-cadherin adhesions drives growth cone migration. J Neurosci 28:5879–5890

Cohen NR, Taylor JS, Scott LB, Guillery RW, Soriano P, Furley AJ (1998) Errors in corticospinal axon guidance in mice lacking the neural cell adhesion molecule L1. Curr Biol 8:26–33

Conklin MW, Lin MS, Spitzer NC (2005) Local calcium transients contribute to disappearance of pFAK, focal complex removal and deadhesion of neuronal growth cones and fibroblasts. Dev Biol 287:201–212

Dahme M, Bartsch U, Martini R, Anliker B, Schachner M, Mantei N (1997) Disruption of the mouse L1 gene leads to malformations of the nervous system. Nat Genet 17:346–349

Dent EW, Gertler FB (2003) Cytoskeletal dynamics and transport in growth cone motility and axon guidance. Neuron 40:209–227

Dequidt C, Danglot L, Alberts P, Galli T, Choquet D, Thoumine O (2007) Fast turnover of L1 adhesions in neuronal growth cones involving both surface diffusion and exo/endocytosis of L1 molecules. Mol Biol Cell 18:3131–3143

Drees F, Gertler FB (2008) Ena/VASP: proteins at the tip of the nervous system. Curr Opin Neurobiol 18:53–59

Fournier AE, Nakamura F, Kawamoto S, Goshima Y, Kalb RG, Strittmatter SM (2000) Semaphorin3A enhances endocytosis at sites of receptor-F-actin colocalization during growth cone collapse. J Cell Biol 149:411–422

Franco SJ, Rodgers MA, Perrin BJ, Han J, Bennin DA, Critchley DR, Huttenlocher A (2004) Calpain-mediated proteolysis of talin regulates adhesion dynamics. Nat Cell Biol 6: 977–983

Gomez TM, Zheng JQ (2006) The molecular basis for calcium-dependent axon pathfinding. Nat Rev Neurosci 7:115–125

Henley JR, Huang KH, Wang D, Poo MM (2004) Calcium mediates bidirectional growth cone turning induced by myelin-associated glycoprotein. Neuron 44:909–916

Henley J, Poo MM (2004) Guiding neuronal growth cones using Ca^{2+} signals. Trends Cell Biol 14:320–330

Hong K, Nishiyama M, Henley J, Tessier-Lavigne M, Poo M (2000) Calcium signalling in the guidance of nerve growth by netrin-1. Nature 403:93–98

Huber AB, Kolodkin AL, Ginty DD, Cloutier JF (2003) Signaling at the growth cone: ligand-receptor complexes and the control of axon growth and guidance. Annu Rev Neurosci 26: 509–563

Jin M, Guan CB, Jiang YA, Chen G, Zhao CT, Cui K, Song YQ, Wu CP, Poo MM, Yuan XB (2005) Ca^{2+}-dependent regulation of rho GTPases triggers turning of nerve growth cones. J Neurosci 25:2338–2347

Kamiguchi H (2003) The mechanism of axon growth: what we have learned from the cell adhesion molecule L1. Mol Neurobiol 28:219–228

Kamiguchi H, Hlavin ML, Yamasaki M, Lemmon V (1998a) Adhesion molecules and inherited diseases of the human nervous system. Annu Rev Neurosci 21:97–125

Kamiguchi H, Lemmon V (2000) Recycling of the cell adhesion molecule L1 in axonal growth cones. J Neurosci 20:3676–3686

Kamiguchi H, Long KE, Pendergast M, Schaefer AW, Rapoport I, Kirchhausen T, Lemmon V (1998b) The neural cell adhesion molecule L1 interacts with the AP-2 adaptor and is endocytosed via the clathrin-mediated pathway. J Neurosci 18:5311–5321

Kamiguchi H, Yoshihara F (2001) The role of endocytic L1 trafficking in polarized adhesion and migration of nerve growth cones. J Neurosci 21:9194–9203

Lee H, Engel U, Rusch J, Scherrer S, Sheard K, Van Vactor D (2004) The microtubule plus end tracking protein Orbit/MAST/CLASP acts downstream of the tyrosine kinase Abl in mediating axon guidance. Neuron 42:913–926

Lin AC, Holt CE (2008) Function and regulation of local axonal translation. Curr Opin Neurobiol 18:60–68

Mitchison T, Kirschner M (1988) Cytoskeletal dynamics and nerve growth. Neuron 1:761–772

Nishimura K, Yoshihara F, Tojima T, Ooashi N, Yoon W, Mikoshiba K, Bennett V, Kamiguchi H (2003) L1-dependent neuritogenesis involves ankyrinB that mediates L1-CAM coupling with retrograde actin flow. J Cell Biol 163:1077–1088

Ooashi N, Futatsugi A, Yoshihara F, Mikoshiba K, Kamiguchi H (2005) Cell adhesion molecules regulate Ca^{2+}-mediated steering of growth cones via cyclic AMP and ryanodine receptor type 3. J Cell Biol 170:1159–1167

Palamidessi A, Frittoli E, Garre M, Faretta M, Mione M, Testa I, Diaspro A, Lanzetti L, Scita G, Di Fiore PP (2008) Endocytic trafficking of Rac is required for the spatial restriction of signaling in cell migration. Cell 134:135–147

Peretti D, Peris L, Rosso S, Quiroga S, Caceres A (2000) Evidence for the involvement of KIF4 in the anterograde transport of L1-containing vesicles. J Cell Biol 149:141–152

Pollard TD, Borisy GG (2003) Cellular motility driven by assembly and disassembly of actin filaments. Cell 112:453–465

Rhee J, Mahfooz NS, Arregui C, Lilien J, Balsamo J, VanBerkum MF (2002) Activation of the repulsive receptor Roundabout inhibits N-cadherin-mediated cell adhesion. Nat Cell Biol 4:798–805

Robles E, Huttenlocher A, Gomez TM (2003) Filopodial calcium transients regulate growth cone motility and guidance through local activation of calpain. Neuron 38:597–609

Sakurai T, Gil OD, Whittard JD, Gazdoiu M, Joseph T, Wu J, Waksman A, Benson DL, Salton SR, Felsenfeld DP (2008) Interactions between the L1 cell adhesion molecule and ezrin support traction-force generation and can be regulated by tyrosine phosphorylation. J Neurosci Res 86:2602–2614

Shimada T, Toriyama M, Uemura K, Kamiguchi H, Sugiura T, Watanabe N, Inagaki N (2008) Shootin1 interacts with actin retrograde flow and L1-CAM to promote axon outgrowth. J Cell Biol 181:817–829

Song HJ, Poo MM (1999) Signal transduction underlying growth cone guidance by diffusible factors. Curr Opin Neurobiol 9:355–363

Suter DM, Forscher P (2000) Substrate-cytoskeletal coupling as a mechanism for the regulation of growth cone motility and guidance. J Neurobiol 44:97–113

Tessier-Lavigne M, Goodman CS (1996) The molecular biology of axon guidance. Science 274:1123–1133

Togashi K, von Schimmelmann MJ, Nishiyama M, Lim CS, Yoshida N, Yun B, Molday RS, Goshima Y, Hong K (2008) Cyclic GMP-gated CNG channels function in Sema3A-induced growth cone repulsion. Neuron 58:694–707

Tojima T, Akiyama H, Itofusa R, Li Y, Katayama H, Miyawaki A, Kamiguchi H (2007) Attractive axon guidance involves asymmetric membrane transport and exocytosis in the growth cone. Nat Neurosci 10:58–66

Wen Z, Han L, Bamburg JR, Shim S, Ming GL, Zheng JQ (2007) BMP gradients steer nerve growth cones by a balancing act of LIM kinase and Slingshot phosphatase on ADF/cofilin. J Cell Biol 178:107–119

Yuan XB, Jin M, Xu X, Song YQ, Wu CP, Poo MM, Duan S (2003) Signalling and crosstalk of Rho GTPases in mediating axon guidance. Nat Cell Biol 5:38–45

Zheng JQ (2000) Turning of nerve growth cones induced by localized increases in intracellular calcium ions. Nature 403:89–93

Chapter 20
Glial Fibrillary Acidic Protein: The Intermediate Filament Protein of Astrocytes

Douglas L. Eng and Lawrence F. Eng

Abstract It is now well established that glial fibrillary acidic protein (GFAP) is the principal 8- to 9-nm intermediate filament in mature astrocytes of the central nervous system (CNS). More than 30 years ago, the value of GFAP as a prototype antigen in nervous tissue identification and as a standard marker for fundamental and applied research at an interdisciplinary level was recognized. As a member of the cytoskeletal protein family, GFAP is thought to be important in modulating astrocyte motility and shape by providing structural stability to astrocytic processes. In the CNS of higher vertebrates, following injury, either as a result of trauma, disease, genetic disorders, or chemical insult, astrocytes become reactive and respond in a typical manner, termed astrogliosis. Astrogliosis is characterized by rapid synthesis of GFAP and is demonstrated by increase in protein content or by immunostaining with GFAP antibody. In addition to the major application GFAP antisera for routine use in astrocyte identification in the CNS, the molecular cloning of the mouse gene in 1985 and the subsequent discovery of the gfa-2 promoter have both opened a new and rich realm for GFAP studies. These include antisense, null mice, and numerous promoter studies. GFAP knockout mice studies have shown the important role of GFA cytoskeleton in the astrocyte. The genetic basis for the only disease due to GFAP mutation, Alexander disease, has been demonstrated and confirmed. While the structural function of GFAP has become more acceptable, the use of GFAP antibodies and promoters continue to be valuable in studying CNS injury, disease, and development.

Keywords Alexander disease · Astrocytes · Astrogliosis · CNS intermediate filaments · GFAP · Immunocytochemistry · Knockout mice · Neurodegenerative disorders · Reactive astrogliosis

---

L.F. Eng (✉)
Department of Pathology, Stanford University School of Medicine, Stanford, CA, USA
e-mail: lfeng@stanford.edu

R.A. Nixon, A. Yuan (eds.), *Cytoskeleton of the Nervous System*,
Advances in Neurobiology 3, DOI 10.1007/978-1-4419-6787-9_20,

Contents

20.1 Introduction

With regard to the discovery of GFAP, studies of the glial fibrillary acidic protein plaque protein (Eng et al., 1970), GFA protein or GFAP (Eng et al., 1971b; Uyeda et al., 1972; Bignami et al., 1972), stemmed from our lipid studies in multiple sclerosis (MS) brains. MS is a demyelinating disease of the central nervous system (CNS) which is characterized by demyelination, intense reactive gliosis, and formation of a scar composed of bare axons surrounded by astrocytes filled with glial filaments. To characterize the aqueous-insoluble myelin membrane proteins, we devised an all-glass disk polyacrylamide gel electrophoresis chamber (eight-tube capacity) which would permit the use of corrosive solvents capable of dissolving water-insoluble proteins (Eng et al., 1971a). At that time the now routinely used sodium dodecyl sulfate (SDS) polyacrylamide gel electrophoresis method was already known. However, the use of a mixture of phenol–acetic acid–water or phenol–formic acid–water (PFW) was more commonly employed. The PFW mixture dissolved membrane proteins as well as water-insoluble filamentous proteins.

Initially, GFAP was purified from MS brain tissue fixed in absolute ethanol. A large MS plaque consisting primarily of fibrous astrocytes and demyelinated axons was dissected free of gray and white matter. The proteins denatured by the alcohol fixative were soluble in PFW, but most proteins in brains fixed with aldehyde were not soluble. The plaque proteins were dissolved in PFW and electrophoresced in batches of eight tubes. The proteins in one tube were visualized by an amido black stain. Proteins in the remaining tubes were visualized with 10% perchloric acid, and the individual protein bands were excised from the gel, macerated, and extracted three times with PFW. Pooled extracts of each band were dialyzed against tap water for 3–4 days at room temperature. Gradual dialysis of the PFW extracts removed the PFW and allowed the proteins to aggregate into a filamentous precipitate which could be easily recovered and dried (Fig. 20.1).

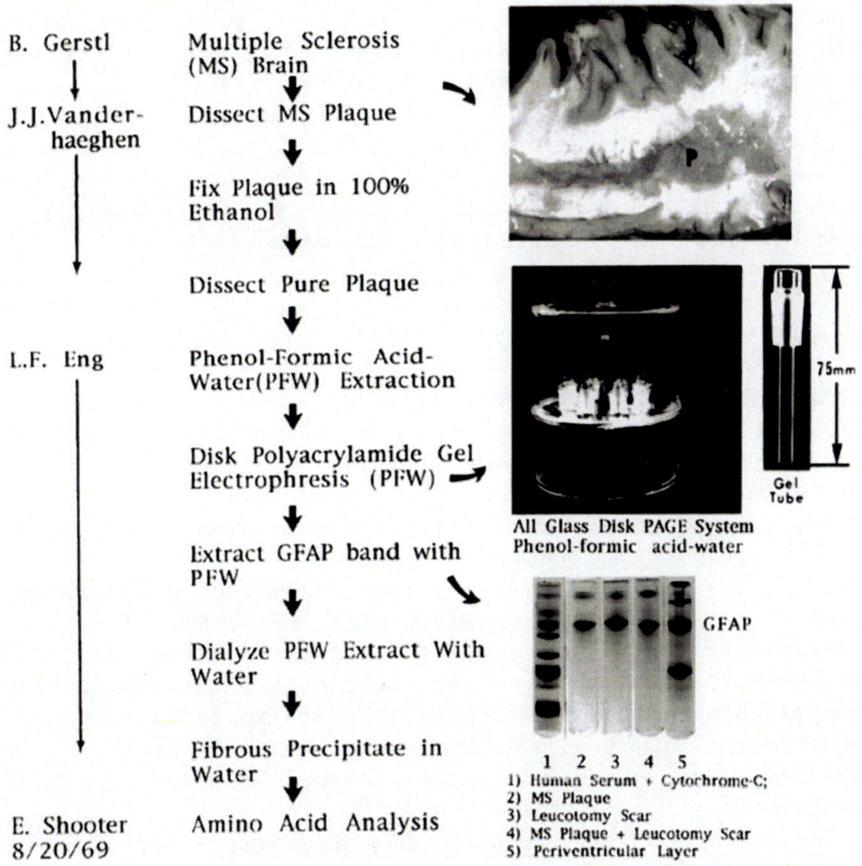

Fig. 20.1 History of GFAP with picture. Initial isolation of the GFAP (cytoskeletal proteins in astroglia). Data presented at the 2nd ISN Meeting in Milan, Italy, 9–3–69, at a round table on "Methods employed in separation, purification, and characterization of neural proteins," Organized by E. R. Einstein. Participants: E. Shooter, L.F. Eng, B.W. Moore, Li-Pen Chao, J. Folch-Pi, Marion Kies, Marjorie Lees, and F. Wolfgram

The amino acid composition of purified GFAP was first determined in the Neurochemistry Laboratory of Dr. Eric Shooter at the Stanford Medical Center and was presented at a round table on "Brain-Specific Proteins" organized by Dr. Elizabeth Roboz-Einstein at the 2nd International Meeting of the International Society for Neurochemistry held in Milan, Italy in September 3, 1969. The participants for this round table on methods employed on separation, purification, and characterization of neural proteins were LiPen Chao, Lawrence Eng, Jordi Folch-Pi, Marion Kies, Marjorie Lees, Eric Shooter, and Fred Wolfgram.

Identification and biochemical properties of GFAP which have impeded its characterization are its insolubility in aqueous solvents, tendency to aggregate or polymerize, susceptibility to neutral proteases, highly specific and antigenic epitopes, and wide distribution of GFAP containing astrocytes. Only a very small pool of aqueous soluble GFAP could be detected at any age in development of the rat brain (Malloch et al., 1987), in cultured astrocytes (Chiu and Goldman, 1984), and in rat spinal cord (Aquino et al., 1988). Early reports of a soluble GFAP fraction utilized human post-mortem tissue and animal tissue where the time between death and homogenization of the tissue was not controlled. The low MW soluble forms in these cases are probably due to a calcium-activated proteinase, which has high substrate specificity for vimentin and desmin (Bigbee et al., 1983; DeArmond et al., 1983; Nelson and Traub, 1983; Schlaepfer and Zimmerman, 1981).

20.2 Function of GFAP

GFAP immunostaining of freeze-substituted paraffin sections of normal mouse cerebellum are illustrated in this figure. Note the fine detail and extensive staining of the astrocytes. Many more astrocytes are present in the brain than one normally observes when the tissue is fixed with aldehyde. Milder tissue-processing methods of human surgical specimens also show intense GFAP immunostaining. However, immunostaining of human autopsy CNS tissue is dependent on the agonal state of the patient at death, the death-autopsy interval, and the time of tissue fixation. Astrocytic gliosis has been demonstrated in specific areas of the CNS by GFAP immunocytochemistry when the well-fixed tissue is treated with protease. While there is a decrease of GFAP epitopes due to aldehyde fixation, the rapid autolysis of GFAP following death disrupts the glial filaments and exposes more epitopes for GFAP staining. Despite these limitations in GFAP immunostaining of CNS tissue, GFAP immunocytochemistry has been and continues to be useful in the study of normal development, injury, and disease in the CNS.

The cytoskeleton is composed of three major filamentous components; microfilaments (mainly actin), microtubules (mainly tubulins), and intermediate filaments (IFs). During the past 40 years the chemistry, biology, and structure of the cytoskeleton and associated proteins have been described in numerous publications (Traub, 1985; Schliwa, 1986; Goldman and Steinart, 1990). Originally, five distinct classes of IF proteins were defined on the basis of the cell type or specific tissue from which the filaments were isolated and characterized: keratin in epithelial cells, vimentin in

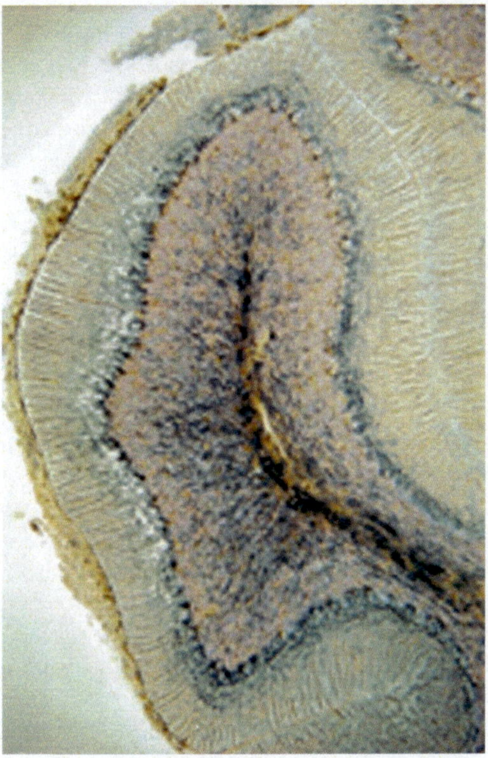

Fig. 20.2 GFAP immunostained picture of mouse cerebellum

cells of mesenchymal origin, desmin in muscle cells, glial fibrillary acidic protein (GFAP) in astrocytes, and neurofilaments in neurons. Molecular biological studies based on amino acid homologies now show that all IF-forming proteins possess a central alpha-helix rod domain of closely conserved length and secondary structure, but with specific differences, which permit classification into distinct sequence types. At present, the IF proteins are divided into five types (Goldman and Steinart, 1990):

Type I: acidic keratins
Type II: neutral-basic keratins
Type III: includes vimentin, desmin, GFAP and/or 57-kD neuronal IF protein
Type IV: classical neurofilaments
Type V: nuclear lamins

Vimentin, desmin, and GFAP are highly homologous throughout their rod and C-terminal domains. GFAP is the major IF protein in mature cerebral astrocytes and we can describe in vitro and in vivo studies relating to GFAP metabolism, phosphorylation, assembly, immunocytochemistry, and molecular biology. The role of GFAP in gliosis and in brain tumor diagnosis has also been included.

The rapid advances in molecular biology and newer techniques such as knockout mice, the use of the gfa-2 promoter to prepare transgenes, antisense RNA methodology, and DNA sequencing have greatly increased our knowledge of GFAP function in CNS development, injury, and disease. The technique of gene knockout (KO) has been used to examine intermediate filaments in mice and has provided the first evidence that intermediate filaments are directly involved in cell resilience and the maintenance of tissue integrity. A putative function for GFAP is as a structural protein in association with the other cytoskeletal proteins–microfilaments (mainly actin) and microtubules (mainly tubulins) (Schliwa, 1986). Studies to determine other possible functions for GFAP have analyzed GFAP null mice (Gomi et al., 1995; McCall et al., 1996; Liedtke et al., 1996). While GFAP null mice exhibit some abnormalities, they survive, reproduce, and live a normal life span (Pekny et al., 1995). There is an intricate relationship between expression of GFAP and of vimentin, another intermediate filament protein, which is widely expressed in embryonic development. Homozygous vimentin knockout (vim-) mice develop and reproduce without an obvious phenotype (Colucci-Guyon et al., 1994) but show a cerebellar defect and impaired motor coordination (Colucci-Guyon et al., 1999). Fibroblasts derived from these mice are also mechanically weak and severely disabled in their capacity to migrate and to contact a 3D collagen network. Wounds in the vim adult animal showed delayed migration of fibroblasts into the wound site (Eckes et al., 2000). The GFAP network in the vim mice has disrupted GFAP and fails to assemble into a filamentous network in astrocytes that normally co-express GFAP and vimentin, i.e., corpus callosum astrocytes and Bergmann glia (Galou et al., 1996). Based on GFAP, vimentin, and double knockout mice studies, Pekny (2001) has provided a concise review on the possible functions of glial filaments. Besides providing structural support (Nawashiro et al., 1998), reactive astrogliosis, and scar formation, there is now evidence that GFAP is involved with long-term depression (Shibuki et al., 1996), long-term potentiation (Tanaka et al., 2002), control of astrocytic glutamine level (Pekny et al., 1999), circadian rhythm (Fernandez-Galaz et al., 1999), and regulation of cell volume and cell motility (Lepekhin et al., 1999; Anderova et al., 2001), and that it promotes normal blood-brain barrier formation (Pekny et al., 1998b). These topics will not be discussed in this chapter; instead an illness which in most cases is caused by a mutation in GFAP, i.e., Alexander disease, will be described later.

20.3 GFAP in Pathology

The pathogenic role of astrocytes is debated hotly in the case of many neurological diseases (see chapters Barger (2004), Brown and Sassoon (2004), Miquel et al. (1982), and Przedborski and Goldman (2004)).

In the central nervous system (CNS) of higher vertebrates, astrocytes become reactive and respond in a typical manner following injury, termed astrogliosis, either as a result of injury or chemical insult. Reactive astrogliosis is characterized by rapid synthesis of GFAP, the principal intermediate filament of mature astrocytes. The

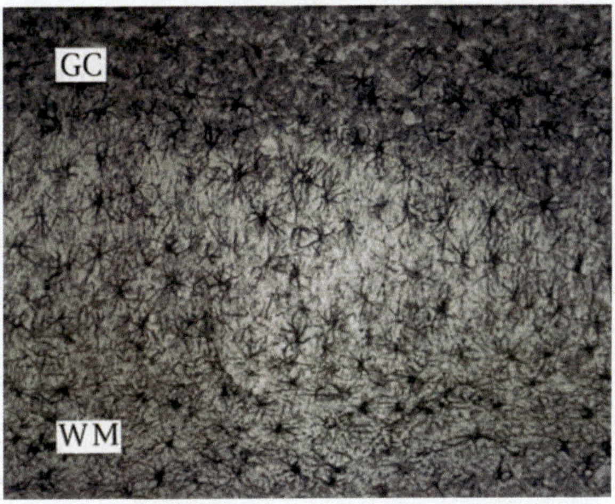

Fig. 20.3 A segment of mouse cerebellum was freeze-substituted in a mixture of chloroform, methanol, and acetone; the tissue was embedded in paraffin; a 5- to 6-μm section was immunostained with a polyclonal antiserum to human GFAP; and the GFAP visualized with the colloidal gold-silver enhancement technique. Note the dark silver-stained astrocytes in the granule cell layer (GC) and white matter (WM)

GFAP astrocytic response serves as a microsensor of the injured microenvironment at any location in the CNS. While GFAP is involved with almost any insult to the CNS, no convincing evidence of a primary astrocyte disease had been demonstrated until the development of the transgenic mouse model expressing a human GFAP transgene. The discovery that these mice formed abundant Rosenthal fibers (RFs) suggested that mutations in the GFAP gene were a cause of Alexander disease. Only in rare cases and in Alexander disease do reactive astrocytes contain RFs. The GFAP mutation is found in many cases of Alexander disease through analysis of patient DNA samples. This non-invasive assay will eliminate morphological autopsy and brain biopsy analyses. As yet, there is only one disease, Alexander disease, which is unequivocally a primary disorder of astrocytes and of glial fibrillary protein (GFAP) (Section 20.10).

GFAP and astrogliosis (Figure 20.1)

Astrocytes account for 25% of the cells and 35% of the mass in the CNS. They have intimate contact with the pia of the brain, neurons, and oligodendrocytes, endothelial cells, pericytes, myelin membrane internodes, synapses, and microglia. Astrocytes in the CNS react to injury by hypertrophy, and in some cases proliferation. A common feature of these cells is enhanced expression of GFAP (Eng et al., 1995). In the CNS of higher vertebrates, astrocytes become reactive and respond in a typical manner, termed astrogliosis following injury either as a result of aging, trauma, disease, neurodegenerative disorders, genetic disorders, or chemical insult. Numerous in vitro and in vivo studies on the molecular profiles of substances, which are upregulated during astrocyte activation, document the complex and varied responses of astrocytes to injury (Eddleston and Mucke, 1993). Reactive astrogliosis is characterized by rapid synthesis of GFAP intermediate filaments, and increased protein content or immunostaining of GFAP has been found in experimental models involving gliosis such as encephalomyelitis (EAE) (Eng et al., 1989), hyperthermia, electrically induced seizures, and toxic lesions. The precise mechanism of this

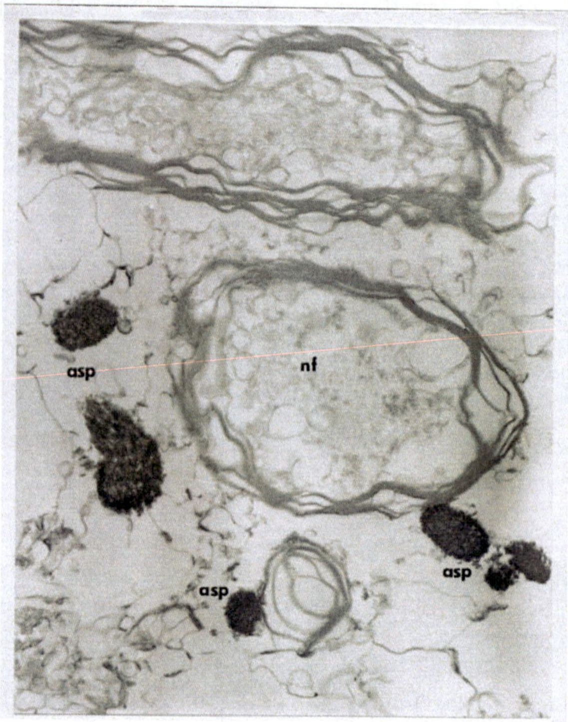

Fig. 20.4 Bovine myelinated axon preparation immunochemically stained for GFA protein showing strong positive staining of astrocytic processes (ASP) adjacent to the axons. Note the lack of staining of any intra-axonal structures including neurofilaments (NF), × 30,000

response is still unknown. Growth factors, hormones, cytokines, and chemokines have been implicated, but no single common factor has been identified (Eng et al., 2000). In the above conditions, the GFAP content of astrocytes at the site of injury or activation increases until the astrocyte cell body and its processes become completely filled. Only in rare cases and in Alexander disease do reactive astrocytes also contain RFs (refer to later part of Section 20.1).

20.4 GFAP in Identification and Diagnosis

With regard to generation of GFAP antiserum, GFAP and the other IFs share some chemical properties and common intramolecular polypeptide domains; however, *GFAP also has some unique, highly immunogenic epitopes. GFAP has proven to be a reliable marker for normal and neoplastic cells of glial lineage*. Rabbit antiserum to human GFAP was first prepared by Uyeda et al. (1972). We have been able to produce specific GFAP antibodies in a rabbit immunized with MS plaque tissue that had been fixed in formalin for several years. High affinity polyclonal antisera and monoclonal antibodies prepared from human and bovine GFAP detect these GFAP-specific epitopes. GFAP immunoreactivity in the mature CNS is restricted to glial filaments within protoplasmic astrocytes in gray matter, fibrous astrocytes in white matter (Fig. 20.4), and radial glial in the cerebellum (Bergmann

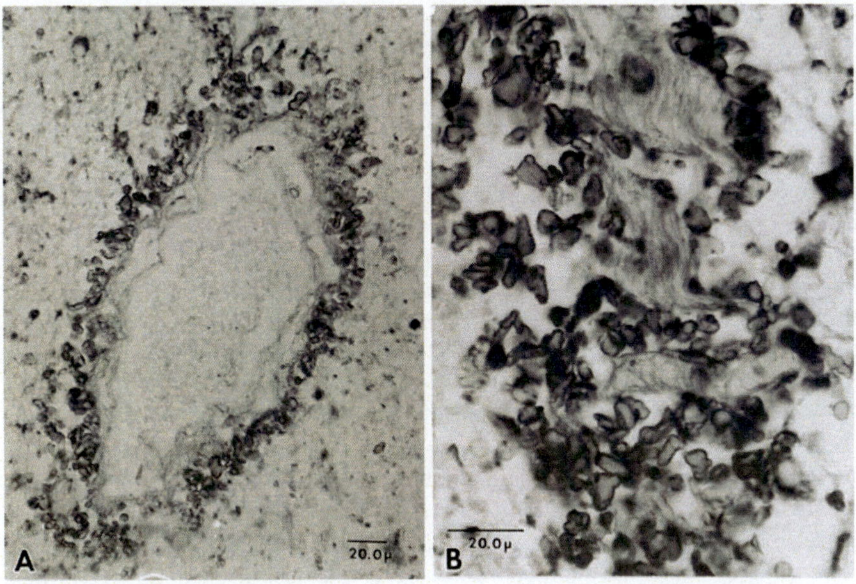

Fig. 20.5 **a** Vessel from Alexander patient. Perivascular astrocytes from Alexander disease stained by anti-GFAP, **b** closeup of astrocytes show Rosenthal fibers. Note the positive GFAP staining surrounding Rosenthal fibers which appear clear

glia, Figs. 20.2 and 20.3) and subependymal astrocytes adjacent to the cerebral ventricles. At the surface of the brain, GFAP immunoreactivity is especially concentrated in astrocytes, which form the outer limiting membrane, the glia limitans. Perivascular astrocytes from Rosenthal's disease patients demonstrate positive GFAP expression (see Fig. 20.5 for perivascular astrocytes around vessel and close up of GFAP positive astrocytes) as visualized using anti-GFAP immunohistology. Note the positive GFAP staining surrounding Rosenthal fibers which appear clear. Mild tissue processing methods (i.e., unfixed, frozen, or freeze-substituted sections) and more sensitive detection procedures (immunogold labeling) have demonstrated GFAP-like immunoreactivity in regenerating Teleost spinal cord, in Schwann cells, glia-like cells in the myenteric plexus, Kupffer cells of the liver, salivary tumors, a pineal astrocytoma, and cells in the pineal gland. GFAP immunoreactivity has also been demonstrated in epiglottic cartilage, pituicytes and pituitary adenomas, immature oligodendrocytes, papillary meningiomas, and metastasizing renal carcinomas. Mouse lens epithelium reacts with both polyclonal and monoclonal anti-GFAP antibodies. Polyclonal and monoclonal antibodies to GFAP have been used extensively for immunochemical and immunocytochemical studies (McLendon and Bigner, 1994; Eng and Lee, 1995).

Regarding applications of GFAP antiserum, immunochemical application of GFAP antiserum has been used to quantitate GFAP in a variety of immunoassays. These include quantitative immunoelectrophoresis, competitive radioimmune assays, 2-site immunoradiometric assays, and enzyme-linked immunoassays (ELISA). Anti-GFAP antisera have been employed to isolate translated GFAP in vitro (Beguin et al., 1980), to immunostain and identify GFAP in transblots, and to identify reassembled GFAP fibers in vitro (Eng and Lee, 1995). Since the initial reports describing the use of GFAP antiserum for tumor diagnosis (Deck et al., 1976, 1978; Duffy et al., 1977; Eng and Rubinstein, 1978), numerous reviews have appeared (McLendon and Bigner, 1994; Eng and Ghirnikar, 1994; Inagaki et al., 1994a, b; Brenner, 1994; Laping et al., 1994; Brenner and Messing, 1996; Eng and Lee, 1995). *Polyclonal and monoclonal antibodies to GFAP are used routinely in medical centers throughout the world to assist in the diagnosis of human neoplasms.* It is always better to use a specific polyclonal antibody than a single monoclonal antibody for tumor diagnosis. A cocktail of three different monclonal antibodies has been used successfully for identifying GFAP containing tumor cells (McLendon and Bigner, 1994).

One must be cautious in interpreting immunocytochemical results when using GFAP antibodies. While positive staining may identify an astrocyte in the CNS, a negative result may be false. It is important to remember that this morphological technique is only indirect evidence for the presence of an antigen. The method of tissue processing is a very important factor. For example, the protoplasmic astrocyte in normal gray matter has a low content of GFAP and does not stain for GFAP when fixed in formaldehyde for an extended period of time. However, pretreatment of the formaldehyde-fixed tissue section with trypsin exposes more GFAP epitopes, which then can be detected by immunostaining. Monoclonal antibodies to GFAP have been generated that are specific and others that bind to both GFAP and vimentin (Eng, 1985).

The intensity of GFAP immunostaining does not always correlate with GFAP content. Two examples are the rapid increase in GFAP staining seen 30 min after a cryogenic lesion of the rat brain (Amaducci et al., 1981) and that seen in the early stages of EAE (Aquino et al., 1988). EAE is a cell-mediated autoimmune disease, which has been the principal experimental model for studying the etiology and pathogenesis of MS (Eng et al., 1989). Astrocyte proliferation and hypertrophy of cell processes appear very early in this model, coincidentally with the first inflammatory foci and is apparent by an increase in immunostaining of the astrocytes with GFAP antibodies without an increase in GFAP content (Smith et al., 1983). Aquino et al. (1988) have shown that this increase in GFAP immunoreactivity of the astrocytes without a corresponding increase in GFAP content is neither due to an increase in GFAP epitopes resulting from limited proteolysis, nor to glial filament dissociation yielding an aqueous soluble fraction, nor to differences in the avidity of a number of different antibody preparations tested. In acute EAE, the increase in GFAP immunostaining of the astrocytes is widespread and not confined to lesion sites (Smith et al., 1983). The reason for astrocyte hypertrophy and increased immunostaining for GFAP without a demonstrable increase in GFAP content is unknown. The onset of (BBB) and leakage of blood-borne substances into the CNS (Cutler et al., 1967; Juhler et al., 1984) may contribute to these phenomena. Swollen astrocytic processes filled with disrupted bundles of glial filaments and glycogen particles have been shown by electron microscopy in edematous brain tissue (Kimelberg et al., 1989, Kimelberg et al., 1982).

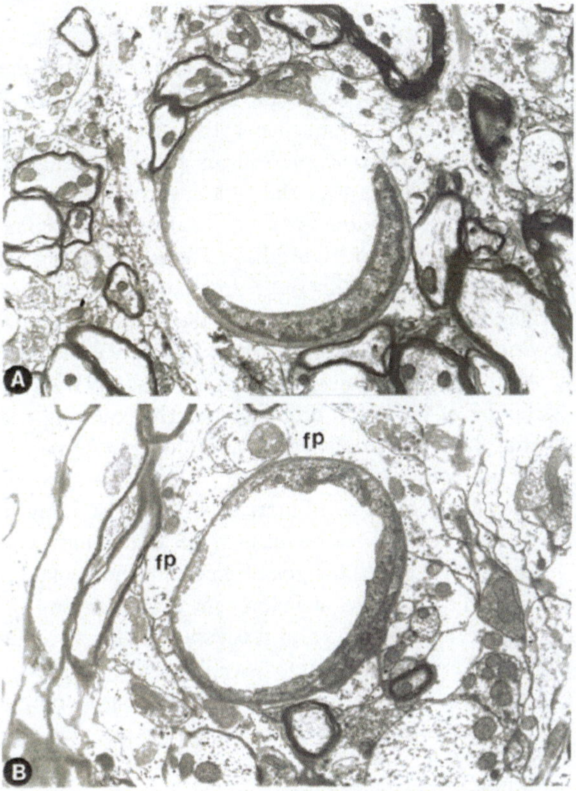

Fig. 20.6 **a** Electron micrograph of normal rat white matter from lumbar spinal cord. **b** Electron micrograph of EAE rat white matter (from lumbar spinal cord) fp shows swelling in astrocytic foot processes

We examined the early EAE lesion by electron microscopy (Fig. 20.6) and obtained results similar to that reported by Kimelberg et al. (1982). The astroglial processes contained many glycogen particles. The glial filaments were arranged in small bundles of loose thin filaments adjacent to the bundles. The glial filaments that normally appear as tight bundles expanded and appeared less dense. The general picture indicated that the first stages of EAE are pathogenically related to an abnormal BBB permeability. The "watery" cytoplasm of astrocytes at this early stage of EAE is most likely expressing a "partial" breakdown of the BBB, resulting in intracellular (astrocytic) edematous fluid (Lee et al., 1982). The pathological changes observed in the spinal cord in EAE seem to indicate that at earlier stages of development the alterations affect mainly the cytoplasm of astrocytes in both soma and processes. Light microscopic observations stained with anti-GFAP reveal proliferation and enlargement of astrocytes, particularly those adjacent to the meninges. Distinctive ultrastructural features of these alterations are a considerable "swelling," the presence of increased amount of glycogen particles and the widespread dispersion of the glial filaments. These changes involve the most slender astrocytic extensions, including the foot processes impinging upon the basement membrane of the capillary walls. Some images show the glial filaments forming electron dense tight bundles reminiscent of Rosenthal fibers. Since some of the capillary walls appear unaffected, it is reasonable to assume that the astroglial changes precede the appearance of perivascular inflammatory cell infiltrates. These were made up of macrophages, mononuclear cells, and lymphocytes which formed perivascular cuffs. Electron microscopic images show the inflammatory cells "dissecting" the vascular walls in between the endothelium and basement membrane. On occasion, mononuclear cells (some of them "activate" microglial cells) were seen in the neuropil enclosing oligodendroglial cells or containing cellular debris. Demyelination was not prominent at this stage. Interlamellar separation and stripping of the myelin sheaths were observed on some of the sections. Axonal alterations were minor and dendritic swelling was occasionally seen. We have suggested that the increase in GFAP immunostaining without an increase in GFAP content is due primarily to the disruption of the blood brain barrier. The resulting edema allows the tight bundles of glial filaments to dissociate and thus expose more antigenic sites (epitopes) to GFAP antibodies (D'Amelio et al., 1990). GFAP intermediate filament dissociation can also explain the rapid increase in GFAP staining following cryogenic lesion of the rat brain without an increase in GFAP content (Amaducci et al., 1981).

The pathological changes observed in the spinal cord in EAE seem to indicate that at earlier stages of development the alterations affect mainly the cytoplasm of astrocytes in both soma and processes. Distinctive ultrastructural features of these alterations are a considerable "swelling," the presence of increased amount of glycogen particles and the widespread dispersion of the glial filaments. These changes involve the most slender astrocytic extensions including the foot processes impinging upon the basement membrane of the capillary walls. Difference in images show the inflammatory cells "dissecting" the vascular walls in between the endothelium and basement membrane. On occasion, mononuclear cells (some of the "activated" microglial cells) were seen in the neuropil enclosing this stage. Interlamellar separation and stripping of the myelin sheaths were observed on some of the sections. Axonal alterations were minor and dendritic swelling was occasionally seen.

20.5 Modification of GFAP in Cultured Astrocytes and GFAP Antisense Studies

Since GFAP accumulation is a prominent feature of astrocytic gliosis, our group focused on inhibiting GFAP synthesis using antisense oligonucleotides with the intention that this might delay scar formation resulting from a CNS injury. The delay in the formation of a gliotic physical barrier might allow the neurons and oligodendrocytes to reestablish a functional environment. In order to test the feasibility of inhibiting GFAP synthesis, antisense GFAP RNA complexed with Lipofectin TM (LF), a cationic liposome, was delivered into cerebral astrocytes in culture (Yu et al., 1991). Results demonstrate that LF facilitated antisense RNA uptake into astrocytes. Dibutyryl cyclic adenosine monophosphate (dbcAMP) is known to induce an increase of GFAP content in cultured astrocytes. The effect of LF/antisense GFAP

RNA on the GFAP content in astrocytes treated with dbcAMP (0.25 mM) was determined. Cultured astrocytes treated with dbcAMP contained almost twice as much GFAP as untreated cultures after 2 days. Similar cultures treated with LF/antisense RNA did not show an increase but a 30–40% decrease in GFAP content 2 days after treatment. A similar decrease in GFAP content was obtained in cultures grown in a chemically defined medium, another condition known to induce an increase in GFAP content in astrocytes. This study demonstrated that antisense GFAP RNA (1,000 nt) can transiently inhibit GFAP synthesis in astrocytes (Yu et al., 1991).

To investigate further the possibility of utilizing antisense oligonucleotides to control the GFAP response in astrocytes after mechanically induced injury, primary astrocytes cultures prepared from newborn rat cerebral cortex were scratched with a plastic pipette tip to produce areas free of astrocytes. This scratched injury model was used to study astrogliotic responses in culture. Scratched injured astrocytes became hyperplastic, hypertrophic, and had an increased GFAP content. These observations demonstrate that scratched injured astrocytes in culture are capable of becoming reactive and exhibit gliotic behavior in culture without the presence of neurons. Electron microscopy showed an increased number of filaments in injured astrocytes. The density of fibrosis is especially high in the cytoplasm of cells along the scratched edge. The changes in the components of these fibrils by double immunostaining for GFAP and vimentin were studied in the injured astrocytes. The control cultures contain cells stained with vimentin and/or GFAP, indicating that these cultures contained astrocytes at different stages of maturation. Within a few hours of scratching, the cytoplasmic processes of cells along the edge became hypertrophic and showed increased immunostaining for these cytoskeletal proteins compared to cells distant from the scratch. The cytoskeletal response was the same when the cells were stained exclusively for GFAP or vimentin. These observations indicate that astrocytes at different stages of maturation respond to injury by sending out hypertrophic processes containing increased GFAP. In astrocytes that stained positively for both cytoskeletal proteins, the vimentin staining was located predominantly close to the cell nucleus. The increase in GFAP content in injured astrocytes could be inhibited by incubating scratched cultures with LF complexed with antisense oligonucleotides (20 nt) targeted to 3′ or 5′ regions of the mouse GFAP gene. The scratch model provided a simple system to examine in more detail the mechanisms involved in triggering glial reactivity and many of the cellular dynamics associated with scar formation. These studies showed that antisense oligonucleotide treatment inhibited GFAP synthesis in scratch injured astrocytes (Yu et al., 1993).

We have also used a recombinant retrovirus expressing antisense GFAP RNA to control the response of mechanically scratch injured astrocytes. A 650-bp fragment from the coding region of mouse GFAP cDNA was cloned in the antisense orientation under the control of long terminal repeat (LTR) promoter of Moloney murine leukemia virus. Increase in GFAP as detected by immunocytochemical staining in injured astrocytes was inhibited by treatment with retrovirus expressing antisense GFAP RNA. Also, astrocytes at the site of injury in these scratched cultures did not show cell body hypertrophy compared to control cultures. These observations demonstrated that the increase in GFAP at the site of injury can be inhibited using retroviral treatment and indicated the potential of retrovirus-mediated gene transfer in modulating scar formation in the CNS in vivo. These studies also shed light on the role of GFAP in maintaining the morphology of astrocytes (Ghirnikar et al., 1994).

20.6 Stable Modification of GFAP in Gliomas

Antisense oligonucleotides to GFAP have been widely used to study neuron-glial interactions (Lafrancois et al., 1997, Chen and Liem, 1994, Yu et al., 1993, Weinstein et al., 1991) through studying the growth, invasion, and adhesion of human astrocytoma cells. Two studies of gliomas cells in which GFAP expression was suppressed have been reported. Using the human astrocytoma cell line, U251, Weinstein et al. demonstrated a requirement for the GFAP in the formation of stable astrocytic processes in response to neurons. Suppression of GFAP expression in U251 human astrocytoma cells was accomplished by stable transfection of cells with plasmid DNA (pSV2i) which constitutively transcribed antisense GFAP RNA, driven by a viral promoter. Cells expressing the antisense oligonucleotides could no longer extend stable processes in the presence of granule cell neurons. These transfected cells were devoid of GFAP and no longer extended their processes in response to coculture with neurons, a stereospecific response of GFAP containing astrocytes. These cells however, retained other astrocytic responses such as proliferation arrest and neuronal support (Weinstein et al., 1991). The antisense GFAP-transfected cells demonstrated marked morphological alterations in the form of flat, epithelioid cells devoid of long, astrocytic glial processes. The antisense GFAP-transfected clones demonstrated a greater degree of cell crowding and piling at confluence, enhanced proliferative potential, formed larger and more numerous colonies when tested for anchorage-independent growth in soft agar. were less adherent to their substratum, and were more readily penetrated Matrigel-coated filters (Rutka et al., 1994). Antisense GFAP-transfected astrocytoma clones also showed a marked increase in vimentin, actin microfilaments, CD44 levels (Rutka et al., 1998), and nestin (Rutka et al., 1999). In vitro studies using neuron-astrocyte co-cultures by Lefrancois et al. (1997) have shown that inhibition in GFAP synthesis leads to a reduction of astroglial hypertrophy and relieves the blockade of neuritic outgrowth that normally is observed after a lesion. These studies also showed that the mechanisms might involve changes in the secretion of extracellular matrix molecules by astrocytes.

In contrast, subsequent studies reintroducing a fully encoding rat brain GFAP cDNA into these antisense GFAP U251 cells resulted in the appearance of rat GFAP as a filamentous network and the ability of these astrocytoma cells to form stable processes when co-cultured with neurons (Chen and Liem, 1994). Another analogous study showed similar results, as a GFAP-negative SF-126 human astrocytoma cell line was stably transfected with a eukaryotic expression vector carrying cDNA for the entire coding sequence of human GFAP. Five stably transfected clones that produced GFAP were found to form elongated processes in cell culture, demonstrated decreased proliferation in a tritiated thymidine uptake assay, and showed marked reduction in the number and growth of colonies in soft agar when compared to controls. The amount of GFAP mRNA expression and immunoreactivity of the stably transfected SF-126 astrocytoma cell clones was found to correlate inversely with cell proliferation and growth in soft agar (Rutka and Smith, 1993). It is evident from these two studies that the presence of GFAP is associated with proliferation arrest, cellular differentiation, and process formation and stabilization.

20.7 Molecular Biology of Glial Fibrillary Acidic Protein

A clone encoding mouse GFAP was initially reported by Lewis et al. (1984). The mouse GFAP gene was first located on chromosome 11 by Bernier et al. (1988) and confirmed by Boyer et al. (1991) and Brownell et al. (1991). The chromosomal location of the human GFAP gene was located on chromosome 17q21 by Bongcam-Rudloff et al. (1991) and also confirmed by Brownell et al. (1991) and Kumanishi et al. (1992). cDNA clones encoding rat GFAP were isolated from rat astrocyte and Schwann cell cultures. Nucleotide sequences from astrocytes and Schwann cells contained identical coding. However, the 5′ untranslated region from the Schwann cell line indicated that the start site for peripheral nervous system GFAP mRNA lies 169 bases upstream from that used in the central nervous system. The data suggest that structural differences between GFAP in these two cell types occur at the nucleic acid and protein level (Feinstein et al., 1992).

The studies of promoter and enhancer elements of GFAP gene provide an understanding of GFAP regulation in glial development and glial response to injury. cDNA clones encoding human GFAP were isolated which contain the complete GFAP coding region (Reeves et al., 1989; Brenner et al., 1990). In vitro transcription analysis showed that the basal level expression of GFAP is controlled by two elements: a TATA box located about 25 base pairs (bp) upstream from the transcription start site, and another element located between +11 and +50 bp downstream from the start site (Nakatani et al., 1990a, b; Nakatani et al., 1990a, b). Miura et al. (1990) found that the *cis* element for astrocyte-specific expression was located within 256 bp of the transcription start-point. They further defined three *trans*-acting factor binding sites: AP-2, NFI, and cAMP-response element motifs, which explain upregulation of GFAP in response to various kinds of injury. Transient transfection studies with a chloramphenicol acetyl transferase reporter gene were used to identify three regions (A, B, and D) responsible for GFAP gene expression (Besnard et al., 1991). The D region is located near the basal promoter, which is similar to the finding of Miura et al. (1990), while A and B are next to each other, about 1,500 bp further upstream. Employing site-directed mutagenesis of the 124-bp B region, the same group from Brenner's laboratory showed that there are multiple active sites in this region. Activation of GFAP transcription in astrocytes involves interaction among factors binding to these sites. The most crucial sequence has been shown to be a consensus AP-1 binding site. This is the first demonstration for a function of a specific transcription site in astrocytes, since polyclonal antibodies to c-Fos and c-Jun recognize proteins binding this site (Masood et al., 1993). Sarkar and Cowan (1991) found two similar positive elements, one is located –1.631

to –1.479 bp and another is located –97 to –80 bp; however, they also found a negative element that is located within the first intron of the GFAP gene. Kaneko and Sueoka (1993) found two negative regulatory regions: GDR1 is in a 2.7-kb region extending from the first intron through the fifth exon, and GDR2 is within 1.7 kb 3′ of the polyadenylation site. GDR1 alone inhibit GFAP expression in nonneuronal tissues, while both GDR1 and GDR2 inhibit the GFAP expression in neuronal cells.

While the major application of GFAP antisera continues to be routine use in astrocyte identification in the CNS, the molecular cloning of the mouse gene by the Cowan laboratory (Lewis and Cowan, 1985) opened a new and rich realm for GFAP studies. The potential for these studies has been highlighted in past reviews (Brenner, 1994; Brenner and Messing, 1996). Genomic clones have been obtained from human, mouse, and rat GFAP genes. Each gene is composed of nine exons distributed over about 10 kb of DNA and yields a mature mRNA of about 3 kb. The coding sequences for the three genes are highly homologous. Strong homology also extends upstream of the RNA start site for about 200 bp, recurs between about –1,300 and 1,700 (RNA startpoint = +1), and is present in some intronic regions. The primary sites for the initiation of RNA and protein synthesis are essentially identical for the three genes, and each contains a TAT-like sequence (CATAAA or AATAA) in the expected 5′-flanking position. In addition to GFAP-alpha, two additional mRNAs that start at different sites have been identified: GFAP-beta and GFAP-gamma. The different tissue distributions of GFAP-alpha, -beta, and -gamma mRNAs suggest that the synthesis of each is subject to unique control. All transcriptional studies to date either have explicitly measured GFAP-alpha or have not distinguished among the possible mRNA isotypes. In the peripheral nervous system, GFAP beta mRNA is thought to be the predominant form of GFAP (Hagiwara et al., 1993) whereas GFAP alpha is the predominant form in the CNS. The GFAP gamma mRNA has been found in both CNS and non-CNS tissues including mouse bone marrow and spleen (Brenner, 1994). Recently, Condorelli et al. (1999) have reported the possible existence in rat brain of a novel GFAP mRNA isoform-GFAP delta that differs in the carboxy-terminal tail domain. GFAP transgenes have been used extensively to study signaling pathways that operate during development disease and injury – all states that increase GFAP gene activity (Brenner, 1994; Brenner and Messing, 1996).

20.8 Uses of the GFAP Promoter

A transgenic vector containing almost the entire GFAP gene plus 5′ and 3′ flanking regions was fused to the *Escherichia coli* lacZ structural gene. Injection of the GFAP-lacZ hybrid gene into the germline of mice yielded six different lines of transgenic mice. The expression of lacZ was astrocyte-specific. The expression of lacZ was astrocyte-specific. Upmodulation of transgene expression showed that induction of GFAP-lacZ expression was detectable within 1 h of a focal mechanical lesion (Mucke et al., 1991). In another study, transgenic mice carrying the lacZ reporter gene linked to a 2.2-kb 5′ flanking sequence derived from the human GFAP gene were produced. This promoter directed expression to astrocytes and was also upregulated following injury to the brain. This approach has been used to target expression of other heterologous genes to astrocytes in vivo, and to study the mechanism for reactive gliosis at the DNA level (Brenner et al., 1994). Cultured astrocytes prepared from GFAP lacZ transgenic pups implanted into brains of nontransgenic mice migrated to specific regions of the host brains and expressed the GFAP lacZ fusion gene for at least a year. Graft-derived astrocytes responded to focal injuries in these regions, but did not invade neural lesions in other areas. Their predictable migration and prolonged injury-responsive expression of GFAP driven transgenes make these astrocytes potent vehicles for delivery of therapeutic agents (Mucke and Rockenstein, 1993).

A major contribution of the GFAP transcription studies has been the identification of DNA sequences that can target transgene expression to astrocytes in mice. Brenner and Messing (1996) have shown that their gfa2 promoter can perform this task, and Mucke's laboratory has done the same for a similar fragment of the mouse GFAP gene, as well as for an entire genomic clone (Mucke and Rockenstein, 1993). Brenner has provided his GFAP promoter to over 170 investigators for various studies. This ability has sparked a range of studies, including in vivo tests of various gene products on astrocyte function, use of astrocytes as factories for bioactive molecules, and creation of disease models. For example, Holland and Varmus (1998) used the gfa2 promoter to express the receptor for avian leukosis virus (ALV) in transgenic mice permitting recombinant ALV to be targeted to astrocytes; and then used this system to study the role of basic fibroblast growth factor in gliomas. Carpenter et al. (1997) used the gfa2 promoter to express human nerve growth factor in mice, and Kordower et al. (1997) prepared progenitor cells from the mice and tested their efficacy in a rodent model of Huntington's disease. McKie et al. (1998) have shown that a gfa2-lacZ transgene delivered by a modified herpes simplex virus is primarily expressed in astrocytes, and suggested using the promoter to inject toxic genes into gliomas. Transgene studies with beta-gal (beta-galactosidase) linked to GFAP promoter have also been used to study neuron–glia interactions and demonstrated that neuron–glia interaction might induce the astrocytic differentiation program (Gomes et al., 1999). GFAP-lacZ transgenic mice have also been used to study astrocyte fate in embryonic neural grafts (Quintana et al., 1998). Sun et al. (1998) and Smith et al. (1998) used the gfa2 promoter to drive expression of apolipoprotein E alleles E3 and E4 in mice to study their involvement with Alzheimer's disease. Delaney et al. have used the gfa2 promoter to examine the role of astrocytes in development by expressing the herpes simplex virus thymidine kinase gene (Delaney et al., 1996). Other experiments with mice expressing HSV-TK from the GFAP promoter have shown that genetic targeting can be used to ablate scar-forming astrocytes. These studies demonstrated roles for astrocytes in regulating leukocyte trafficking, repairing the BBB, protecting neurons, and restricting nerve fiber growth after injury in the adult CNS (Bush et al., 1999). Galbreath et al. have made mouse models for hydrocephalus by expressing TGF beta from the GFAP promoter (Galbreath et al., 1995) and others have obtained behavioral relief in a rat model of Parkinson's disease by lipofection of a gfa2-tyrosine hydroxylase transgene (Segovia et al., 1998; Trejo et al., 1999; Cortez et al., 2000). Studies using GFAP promoter-metallothionein-1 transgene have been used to demonstrate MT protection against acute methylmercury cytotoxicity (Yao et al., 1999). Integrative studies using GFAP-cytokine transgenic mice such as IFN alpha (Carr et al., 1998), IL6 (Carrasco et al., 1998), and IL-1 receptor antagonist (Lundkvist et al., 1999) have provided a more thorough understanding of the actions of cytokines in the CNS (Campbell, 1998). Studies with GFAP-NGF mice have been used to demonstrate a connection between the endogenous ectopic overexpression of NGF and neuropathic pain behavior and sympathetic sprouting in the DRG (Ramer et al., 1998). EGF responsive stem cells, which secrete NGF under the direction of the GFAP promoter, have provided an avenue for delivery of neurotrophins into the CNS (Kordower et al., 1997). Astrocyte specific expression of hamster prion protein has shown that PrP knockout mice are susceptible to hamster scrapie, suggesting that astrocytes could play an important role in scrapie pathogenesis (Raeber et al., 1997).

20.9 GFAP Knockouts

The technique of gene knockout (KO) has been used to examine IFs in mice and has provided the first evidence that intermediate filaments are directly involved in cell resilience and the maintenance of tissue integrity. To investigate the structural role of GFAP in vivo, mice carrying a null mutation in GFAP were generated (Pekny et al., 1995; Gomi et al., 1995; Liedke et al., 1996; McCall et al., 1996).

In vitro studies with GFAP –/– astrocytes showed that they are capable of stellation when cocultured with neurons and exhibit an increased cell saturation density. At the ultrastructural level, the amount of intermediate filaments as revealed by transmission electron microscopy was reduced in GFAP –/– astrocytes compared to that in GFAP +/+ astrocytes. GFAP –/– astrocytes retained the ability to form processes in response to neurons in mixed astrocyte/neuron cultures from the cerebellum. GFAP –/– astrocyte-enriched primary cultures exhibited an increased final cell saturation density. These observations have led to the speculation that the loss of GFAP expression observed focally in a proportion of human malignant gliomas may reflect tumor progression toward a more rapidly growing and malignant phenotype (Pekny et al., 1998). In addition, GFAP –/– astrocytes fail to induce a significant restriction to the passage of potassium and hydrophilic drugs (sucrose, 8-SPT) and fail to induce transendothelial resistance values comparable to control co-cultures. These cells were, however, capable of inducing exclusion of Evans blue by endothelial cells, suggesting that GFAP (and intermediate filaments) may play a role in the induction of BBB properties in non-BBB endothelial cells (Pekny et al., 1998). GFAP –/– mice did not exhibit altered incubation times from untargeted control mice following inoculation with prion protein, suggesting that GFAP did not participate in the pathogenesis of the disease or in the production of prion PrPSc (Tatzelt et al., 1996).

Pekny et al. (1999) have showed that astrocytes in mice deficient in both GFAP and vimentin (GFAP –/– vim –/–) cannot form IFs even when nestin is expressed and are thus devoid of IFs in their reactive state. In studies on the reaction to injury in the CNS in GFAP –/–, vimentin +/+, or GFAP +/+ vim –/– mice, glial scar formation appeared normal after spinal cord or brain lesions in GFAP –/– or vimentin –/– mice, but was impaired in GFAP –/– vim –/– mice that developed less dense scars frequently accompanied by bleeding. These studies suggested that GFAP and vimentin are required for proper glial scar formation in the injured central nervous system and that some degree of functional overlap exists between these IF proteins. Earlier studies on GFAP expression and network formation in vimentin knockout (vim–/–) mice showed that the GFAP network is disrupted in astrocytes that normally coexpress vimentin and GFAP (Galou et al., 1996). Additional studies with the double mutant suggested a role for the cytoskeleton in astrocyte volume regulation and the involvement of intermediate filaments in the process (Ding et al., 1998). The rate of glucose uptake through facilitative hexose transporters was not affected by the depletion of GFAP or vimentin, nor was ascorbate uptake and efflux. However, glutamine levels appeared to correlate inversely with GFAP and the effect of GFAP was dose dependent since the glutamine concentration in GFAP +/– astrocytes falls between those in wild-type and GFAP –/– astrocytes (Pekny et al., 1999).

Homozygous mice created by disrupting the GFAP gene by gene targeting in embryonic stem cells were completely devoid of GFAP but exhibited normal development and showed no obvious anatomical abnormalities in the CNS (Gomi et al., 1995). When these animals were inoculated with scrapie prions, they exhibited neuropathological changes typical of the disease. The degree of accumulation of the prion protein in the various brain regions was similar between the mutant and control mice. These studies suggested that GFAP is not essential for the morphogenesis of the CNS and does not play a role in the pathogenesis of prion disease. Studies examining anatomy and synaptic function of the cerebellum in GFAP deficient mice showed that LTD is clearly deficient in the GFAP mutant cerebellum without any detectable anatomical abnormalities. Furthermore, GFAP mutant mice exhibited a significant impairment of eyeblink conditioning without any detectable deficits in motor coordination tasks. These results suggested that GFAP is required for communications between Bergmann glia and Purkinje cells during LTD induction and maintenance (Shibuki et al., 1996).

In 7/14 mutant animals older than 18 months of age produced by Liedke et al. (1998), hydrocephalus associated with white matter loss was detected. Mutant mice displayed abnormal myelination including the presence of actively myelinating oligodendrocytes in adults, nonmyelinated axons in optic nerve, and reduced myelin thickness in spinal cord. White matter was poorly vascularized and the blood-brain barrier was structurally and functionally impaired. Astrocytic structure and function were abnormal, consisting of shortened astrocytic cell processes, decreased septation of white matter, and increased CNS extracellular space. These studies suggested that GFAP expression is essential for normal white matter architecture and blood-brain barrier integrity, and its absence leads to late-onset CNS dysmyelination. Studies examining the astroglial response in GFAP –/– mice with autoimmune encephalomyelitis (EAE), a model for multiple sclerosis, showed that, clinically, the monophasic disease was more severe in GFAP –/– mice than in wild-type littermates despite increased remyelination in the former. The investigators observed an infiltrative EAE lesion in GFAP –/– mice. GFAP –/– astrocytes had a reduced cytoarchitectural stability as evidenced by less abundant and irregularly spaced hemidesmosomes. The blunt GFAP –/– astrocyte processes possessed intermediate filaments consisting mainly of vimentin, though to a lesser degree than in the wild-type. In contrast, in wild-type littermates, GFAP was most abundant and nestin occurred at lower levels. These studies have suggested that GFAP plays an important role in the control of clinical disease associated with formation of a clearly defined edge to the EAE lesion and that GFAP is operative in the regulation of the intermediate filament components in reactive fibrillary astrogliosis.

GFAP null mice generated by McCall et al. (1996) using homologous recombination in embryonic stem cells showed subtle changes in astrocyte morphology and in addition showed enhanced LTP in hippocampal neurons. These data suggest that GFAP is important for astrocyte-neuronal interactions, and that astrocyte processes pay a vital role in modulating synaptic efficacy in the CNS. Studies on percussive head injury in these GFAP-null mice with a weight drop device showed that when mice were positioned on a foam bed which allowed head movement at impact, all 14 wild-type mice tested survived, but 12 of 15 GFAP-null mice died within a few minutes. When the foam bed was replaced by a firm support, both GFAP-null and wild-type mice survived. These results indicate that mice lacking GFAP are hypersensitive to cervical spinal cord injury caused by sudden acceleration of the head (Nawashiro et al., 1998). Studies on axonal and nonneuronal cell responses to spinal cord injury in these mice have shown that the absence of GFAP in reactive astrocytes does not alter axonal sprouting or regeneration. In addition, in these animals, chondroitin sulfate proteoglycan labeling was generally less intense in the gray matter but the expression of various ECM molecules appeared unaltered in the GFAP –/– mice (Wang et al., 1997).

20.10 Alexander Disease

In 1949 W. Steward Alexander (1949) described the pathologic condition which was given his name by Friede (1964). This illness is characterized clinically by megalencephaly accompanied by progressive spasticity, psychomotor retardation, and dementia. Pathological features include astrocytosis of white matter of the CNS, extensive demyelination, and less pronounced axonal loss. Histologically, it is characterized by the presence of large numbers of homogeneous eosinophilic masses forming elongated tapered rods scattered throughout the cortex and white matter, which are most numerous in the subpial, perivascular, and subependymal regions. Morphologically, the eosinophilic deposits are identical, by both light and electron microscopy, with the so-called Rosenthal fibers (RFs). RFs probably result from degenerative changes, which have taken place in the cytoplasm and cytoplasmic processes of differentiated astrocytes. They are inclusions within astrocytes, which are present in various situations where reactive gliosis has been in progress for a long time, e.g., in pilocytic astrocytomas, in optic nerve gliomas, in astrocytic scars, in multiple sclerosis plaques, in chronic infarcts, in ovarian teratomas, and most prominently in Alexander disease (Herndon et al., 1970; Alexander, 1949; Borrett and Becker, 1985). The RFs appear to be similar among the different disorders. RFs vary in size from round, focal deposits of a few microns to elongated, cigar-shaped bodies, 100 μm or more in length for those that reside in astrocyte processes. At the ultrastructural level, RFs appear as dense, osmophilic masses lying on a meshwork of intermediate filaments (Herndon et al., 1970). The inclusion is composed of two low-molecular-weight heat shock proteins, aB-crystallin and HSP27 (Iwaki et al., 1989; Tomokane et al., 1991). Some of the aB-crystallin is conjugated to ubiquitin (Goldman and Corbin, 1991). Levels of aB-crystallin and HSP27 mRNA are elevated in Alexander disease (Head et al., 1993, 1994). It has been suggested that a variety of 'stresses' might induce the accumulation of RFs in astrocytes (Chin and Goldman, 1996). aB-crystallin is a low-molecular-weight (22-kDa) protein which has a wide distribution of lens and non-lenticular tissues. It is a stress-related protein, which can be induced in culture by heat or hypertonicity, is associated with intermediate filaments, and is thought to stabilize cells in culture (Wisniewski and Goldman, 1998). It is generally soluble in aqueous solutions but exits in an aggregated form in RFs.

We reported that only the periphery of RFs is immunostained with GFAP antibodies (Eng and Bigbee, 1978; Eng and Lee, 2004). Paraffin sections of an infant brain with Alexander disease immunostained with antibody to GFAP by the Sternberger PAP technique is shown in Fig. 20.1. A vessel surrounded by reactive astrocytes filled with RFs is shown in Fig. 20.1a. At higher magnification of an area with abundant RFs (Fig. 20.1b) one can clearly see the unstained fibers surrounded by intense GFAP immunostaining of the glial filaments (Ramsey et al., 1979).

Russo et al. (1976) identified three clinical subgroups of Alexander disease. The first consists of the infantile group, which includes Alexander's original patient. Onset of symptoms is from birth to early childhood, and the course is one of neurological deterioration with psychomotor retardation, seizures, quadriparesis, and megalencephaly. The average duration of the illness is two-and-a-half years. The second is the juvenile group, which has an onset from 7 to 14 years of age. The course is characterized by progressive bulbar symptoms and plasticity. Seizures and cognitive deterioration are less prominent and the average duration of the illness is 8 years. The adult cases comprise the third group. Onset of symptoms may occur between the second and seventh decade. Clinically, these adults, whose pathological picture is similar to that of the infantile and juvenile groups, may follow a course consistent with classical multiple sclerosis or they may be asymptomatic. The pathological hallmark of all groups is the diffuse accumulation of RFs, particularly in the subependymal, subpial, and perivascular regions. Demyelination is extensive in the infantile cases, less severe in the juvenile group, and variable in the adults.

Magnetic resonance (MR) imaging studies in three patients with an autopsy-based diagnosis of Alexander disease were analyzed to define MR criteria for diagnosis. Five MR imaging criteria were defined: (1) extensive cerebral white matter changes with frontal predominance; (2) a periventricular rim with high signal T1-weighted images and low signal on T2-weighted images; (3) abnormality of basal ganglia and thalami; (4) brain stem abnormalities; and (5) contrast enhancement of particular gray and white matter structures. Four of the five criteria had to be met for an MR image-based diagnosis. In a retrospective analysis of 217 children with leukoencephalopathy of unknown origin, 19 were found who fulfilled these criteria. In 4 of the 19 patients, subsequent histologic confirmation was obtained. The clinical symptomatology was the same in the patients with and without histologic confirmation, correlated well with MR abnormalities, and was in close agreement with the known histopathological findings in Alexander disease. The authors conclude that the defined MR imaging criteria are sufficient to diagnose Alexander disease (van der Knaap et al., 2001).

Becker and Teixeira (1988) have stated that pathologic features strongly indicate that Alexander disease represents a non-neoplastic disease of astrocytes. The infant cases may represent a single disease entity. However, questions have been raised as to whether the juvenile and adult cases that have RFs, but different pathologic features, have the same etiology. As recently as 1999, one of the early workers on this subject suggested that, while Alexander disease in infants is a disease entity, the boundaries of the entity remain ill-defined and much of what is called Alexander disease is not that entity (Herndon, 1999). Recent GFAP transgene studies may now help clarify this conclusion (Messing et al., 1998; Brenner et al., 2001).

20.11 Structure of GFAP Intermediate Filaments

GFAP is in the same protein class as keratins and all other intermediate filament proteins. Like those proteins, GFAP monomers consist of a central rod domain flanked by random coiled head and tail regions. The central rod is subdivided into four alpha-helical segments, 1A, 1B, 2A, and 2B, and each segment is separated by short linker segments (see figure). The amino acid sequences of the head and tail regions vary greatly among intermediate filaments, but the central rod has significant conservation, especially at each end (the beginning of the 1A segment and end of the 2B). This alpha helical structure conservation has been attributed to the rod being the primary player in filament formation (reviewed in Fuchs, 1996; Parry and Steinart, 1999). Production of a GFAP intermediate filament involves the parallel and antiparallel association of GFAP monomers, assembly of protofibrils, and combination of protofibrils into the full intermediate filament. Polymerization commences by the parallel association of two intermediate filament monomers to form a dimer. Two dimers then combine in an antiparallel manner to form a tetramer. Considerable evidence indicates that the tetramer is formed by a staggered alignment of dimers, with either the 1A regions of one dimer interacting with the 2A regions of the other, or the 1B and 2B regions interacting (Parry and Steinert, 1999). Two tetramers then intertwine to form a protofibril and, finally, at least four protofibrils interact laterally to form the final 10-nm intermediate filament. Mutations in the alpha helical amino acid structure of the GFAP monomer leads to disfunction and has been the scientific and biochemical basis of explaining one GFAP disease class called Alexander disease (Li et al., 2002). In the past, diagnosis of Alexander disease was based on a combination of clinical, morphological, and MRI imaging examinations. The recent finding that GFAP mutation is found in many cases of Alexander disease offers the possibility of diagnosing most cases of Alexander disease through analysis of patient DNA samples. This non-invasive assay will eliminate morphological autopsy and brain biopsy analyses. The belief that Alexander disease is a primary disease entity of astrocytes (Becker and Teixeira, 1988; Eng et al., 1998) has now been confirmed. Please refer to book chapter and article (Brenner et al., 2008) for in-depth report and description of human genetics work.

20.12 Alternative Spliced Forms of GFAP and Their Function

GFAP is alternatively spliced (Galea et al., 1995; Condorelli et al., 1999; Nielsen et al., 2002; Blechingberg et al., 2007; Hol et al., 2003; Zelenika et al., 1995; see Table 20.1), alpha and beta types involving the 5′ UTR (beta-GFAP), exon 6 (delta-6, delta-135, delta-164; Hol et al., 2003), exon 7 (d-/e-GFAP Condorelli et al., 1999; Nielsen et al., 2002, and k-GFAP Blechingberg et al., 2007). Most intermediate filament genes are directly transcribed, but LAMA (Furukawa et al., 1994; Machiels et al., 1996; McKeon et al., 1986; Lin and Worman, 1993), peripherin (Landon et al., 1989, 2000), and DNM (synemin; Xue et al., 2004) are alternatively spliced during transcription. The relative abundance of alternative GFAP transcripts is usually low (Blechingberg et al., 2007) and can be dependent upon astrocyte anatomy (Roelofs et al., 2005), while the expression of forms (delta-6, delta-135, and delta-164) is induced by specific disorders in neurons rather than astrocytes (Hol et al., 2003). Most of the GFAP products different from alpha and beta transcripts will often be assembly incompetent and the function of these various GFAP splice variants has to be associated with disease and function compared to alpha-GFAP (Pekny and Pekna, 2004), and disease expression is not restricted to astrocytes (Su et al., 2004).

GFAP is expressed in cells besides astrocytes, such as enteric glia, nonmyelinating Schwann cells, liver stellate cells, breast myoepithelial cells, lymphocytes, and respiratory tract chondrocytes. GFAP expression in these other cells is sparse (e.g., Riol et al., 1997), so the primary role is probably not filament formation, but perhaps involved in cell protein expression pathways. In Schwann cells, GFAP protein is required for cell proliferation in a pathway involving the avb8 integrin complex (Triolo et al., 2006).

Beta transcript form of GFAP is expressed normally in Schwann cells (Galea et al., 1995; Riol et al., 1997; Jessen et al., 1984) using a different 5′ UTR from the alpha-GFAP. The beta transcript protein product is the same protein sequence as the alpha-GFAP transcript protein. In the d-/e-GFAP (Condorelli et al., 1999; Nielsen et al., 2002) and k-GFAP (Blechingberg et al., 2007) transcripts the alpha helical rod sections Ia, Ib, IIa, and IIb have conserved the alpha GFAP rod, but these have quite different C-terminal domains (Blechingberg et al., 2007). The d, e, and k transcripts share the same C-terminal antibody epitope as the +1 GFAP product.

Table 20.1 Table of GFAP isoforms. A review of all reported transcripts and their products from the human GFAP gene

Isoforms	Detected by cDNA/RNA	IH	IB	Transcript change	Affected protein feature relative to αGFAP	Human tissue expression pattern	References
α	+	+	+	5′-UTR	Protein product of 432 residues	Astrocytes, enteric glia, nonmyelinating Schwann cells, liver stellate cells, breast myoepithelial cells, lymphocytes and respiratory tract chondrocytes – reviewed in Su et al. (2004)	Reeves et al. (1989)
β	+	+	+	5′-UTR	Product same as α-GFAP	Schwann cells	Galea et al. (1995) and Jessen et al. (1984)
Γ	+	–	–	5′-UTR and exon 1	Coding region predicted to start at residue 275, removing N-terminus and helix 1	Corpus callosum	Zelenika et al. (1995)
Δ	+	+	–	Exon 7$^+$ added	Novel C-terminal sequence, replacing the exons 8 and 9 of αGFAP to give a 431 residue product[a]	Astrocytes of subpial border and subventricular zone	Condorelli et al. (1999)
E	+	+	–	Exon 7$^+$ added,	Same C-terminal sequence as δ-form[b] different polyA to δ		Nielsen et al. (2002) and Roelofs et al. (2005)
K	+	–	–	Both 7 and 7$^+$ spliced together	Novel C-terminal tail sequence to give a 438 residue product[c]	Frontal cortex	Blechingberg et al. (2007)
+1	+	+	–	Molecular misreading in exon 6	Reading frameshifted from residue 307 to generate a unique 113 residue C-terminus and a predicted product 420 residues long[d]	Immunoreactivity in hippocampal neurons from (12/16) Alzheimer disease (4/4) Down syndrome (4/12) epilepsy and (4/20) non-demented control patient brain tissue	Hol et al. (2003)

Table 20.1 (continued)

Isoforms	Detected by cDNA/RNA	IH	IB	Transcript change	Affected protein feature relative to αGFAP	Human tissue expression pattern	References
Δ6	+	–	–	Exon 6 deleted	Exon 6 missing causing a frameshift with some sequences in common with the +1 form forming a predicted product 347 residues long[e]	Transcript found in 1 out of 2 controls and 1 out 2 Alzheimer brain samples	Hol et al. (2003)
Δ164	+	–	–	Exon 6 and new acceptor site	Deletion of 164 bp spanning the end of exon 6 and the start of exon 7 to produce a predicted 366 product with a novel C-terminal sequence[f]	Transcript found in both Down syndrome samples tested, in exon 7 but neither the two controls nor the two Alzheimer samples	Hol et al. (2003)
Δ135	+	–	–	Exon 6	Removal of 135 bp, but results in a predicted "in frame" deletion[g]	Transcript found in both down syndrome samples and one of two Alzheimer brain sample	Hol et al. (2003)

cDNA/RNA, RT-PCR, qPCR or Northern blotting; IH, immunohistochemistry; IB, immunoblotting

[a]Novel C-terminal sequence GKSTKDGENHKVTRYLKSLTIRVIPIQAHQIVNGTPPARG generated from an additional exon (7+), located in intron 6, which incorporates an in frame termination codon and replaces exons 8 and 9 of αGFAP

[b]This transcript utilizes a polyA signal in exon 7+ to the δGFAP, but coding sequence is identical to δGFAP

[c]Uses both exon 7 and 7+ to generate a novel 3′ sequence

[d]Novel C-terminal sequence predicted to correspond to: ADARAGGAARAGGGQLSGGAGAAGGRGAEPQGRDGPPLAGVPGPAQCQAGPGHRDRHLQEAARGRGEPDHHSRADLLQPADSRNQPGHQVCVRRPPQEEHRGEDRGDAGWRGH

[e]C-terminal sequence similar to +1 GFAP predicted to be: DHHSRADLLQPADSRNQPGHQVCVRRPPQEEHRGEDRGDAGWRGH

[f]C-terminal sequence similar to +1 GFAP predicted to be: SRADLLQPADSRNQPGHQVCVRRPPQEEHRGEDRGDAGWRGH

[g]Predicted deleted sequences are: NESLERQMREQEERHVREAASYQEALARLEEEGQSLKDEMARHLQ

Table of GFAP isoforms from page 2083: Experimental Cell Research 313 (2007) 2077–2087

The +1 GFAP product is believed to be caused by molecular misreading of the alpha-GFAP transcript (Hol et al., 2003). The delta135-GFAP alternative transcript retains the reading frame even though exon 6 has been omitted. Due to the low abundance of alternative spliced GFAP products, the identification was done using immunohistochemical techniques, but not confirmed by immunoblotting, except for alpha-GFAP which is abundant.

The expression of alternative transcripts of GFAP occurs during disease (Blechingberg et al., 2007). Alzheimer's disease and Down syndrome show expression of the +1 GFAP product in neurons (Hol et al., 2003). Beta GFAP levels increase in Schwann cells during peripheral nerve injury (Triolo et al., 2006), and astrogliosis due to injury and disease in the CNS greatly increases expression of normal alpha-GFAP (Pekny and Pekna, 2004). Observation in Alzheimer's disease astrogliosis show that there are non-uniform increases in GFAP alternative transcripts, and that makes +1 GFAP a standout marker in both Alzheimer's and Down syndrome (Roelofs et al., 2005). Another example of III intermediate filament gene expression is in motor neuron disease; the splice variant of peripherin, PE-61, is found as a stable product in only motor neurons (Robertson et al., 2003). There is still much to study about GFAP expression and the variants and their interactions with alpha-GFAP. Understanding the interactions of d/e- and k-GFAP with alpha-GFAP function may lead to an understanding of their involvement in Alexander disease as well as the normal nervous system.

20.13 Phosphorylation/Dephosphorylation of Glial Fibrillary Acidic Protein

A number of in vitro assembly studies of GFAP have reported that cations, pH, ionic strength, and ATP influenced assembly (Rueger et al., 1979; Lucas et al., 1980a, b; Yang et al., 1988; Yang and Babitch, 1988; Quinlan et al., 1989; Steward et al., 1989). A recent report employing a fluorescence energy transfer assay method suggests that GFAP filaments appear to be in dynamic equilibrium with subunits in vitro (Nakamura et al., 1991). These authors conclude that factors other than pH, ionic strength, or metal ions modulate the assembly state of GFAP in vivo, one of the factors being phosphorylation. As mentioned previously, all intermediate filament proteins share a common subunit organization consisting of a non-alpha-helical N-terminal lead domain, a central alpha-helical rod domain, and a C-terminal tail domain (Weber and Geisler, 1985). The in vitro assembly process of intermediate filaments seems to consist of several association steps that involve different protein domains. Dimer and tetramer formation depend solely on the alpha-helical rod domain (Kaufmann et al., 1985). Assembly of filaments from tetramers requires the presence of the head domain (Traub and Vorgias, 1983; Kaufmann et al., 1985). The head domain of different intermediate filaments are variable in size and amino acid sequences; however, they are positively charged because of a series of arginine residues, and contain serine residues available for phosphorylation by PKA, PKC, and cdc2 kinase (Inagaki et al.1988). In vitro studies suggest that phosphorylation of vimentin (Inagaki et al., 1987; Evans, 1988), desmin (Geisler and Weber, 1988, Inagaki et al.,1988), and GFAP (Inagaki et al., 1990) inhibit intermediate filament assembly. It has been proposed that phosphorylation could participate in the regulatory processes in assembly and turnover of intermediate filaments (Geisler and Weber, 1988).

Posttranslational modifications to the N- and C-terminal domains of IFs are common, while those within the central alpha-helical domain are rare. The posttranslational modification of greatest functional and structural significance arises from the action of various kinases on serine and threonine embedded within specific recognition sequences. The phosphorylation events that have a regulatory role in the assembly and turnover of most IFs in vivo are confined to the N- and C-terminal domains of the chains (Parry and Steinart, 1999). Phosphatases, which dephosphorylate the chains, have an equally important role in governing assembly and function. In cytokeratins, vimetin, and lamins, the effect of phosphorylation is to disassemble the IFs, whereas dephosphorylation allows spontaneous self-assembly to proceed. These changes typically occur during cell division.

Phosphorylation and dephosphorylation of the IF head regions have a critical role for the maintenance and reconstruction of IFs. Glial fibrillary acidic protein filaments disassemble into a soluble form when the amino acids in the head are phosphorylated. The head region of GFAP is positively charged due to the many arginine residues. The mechanism of GFAP assembly regulation may be related to regulation of the electric charge in the head domain by phosphorylation. The sites of second messenger-dependent protein kinases have been identified in the head region of GFAP. Four serines (Ser-8, Ser-13, Ser-14, and Ser-34) and one threonine (Thr-7) are known to be phosphorylated by different protein kinases; cyclic adenosine monophosphate (AMP)-dependent protein kinase (A kinase), Ca2+ calmodulin-dependent kinase II (CaM kinase II), C kinase, and Rho-kinase. A kinase phosphorylates Thr-7, Ser-8, Ser-13, and Ser-34 residues of the HEAD region. Ca2+ calmodulin-dependent kinase II phosphorylates Ser-8, Ser-17, Ser-34, and Ser-389. cdc2 Kinase phosphorylation of GFAP did not cause disassembly, while CaM kinase II phosphorylation did induce disassembly of the filament (Inagaki et al., 1994). The phosphorylation of vimentin and GFAP by Rho-kinase inhibits filament formation in vitro. The phosphorylation sites of GFAP are Thr-7, Ser-13, and Ser-34, cleavage furrow the same sites that (CF) kinase phosphorylates at the cleavage furrow during cytokinesis. Rho-kinase is a likely candidate for CF kinase phosphorylation of IFs at the cleavage furrow during cytokinesis (Matsuzawa et al., 1998).

A GFAP human wild-type knockin mouse was generated in which the coding region of the HEAD domain of GFAP was replaced with the corresponding human sequences. Employing a series of monoclonal antibodies (mAbs) reactive to human phospho-GFAP, phosphorylated at Thr-7, Ser-8, and/or Ser-13, the distribution of phospho-GFAP was determined by immunocytochemistry in vivo in mice and was shown to increase postnatally in the CNS. Limited populations of GFAP-positive astrocytes were labeled with anti-phospho-GFAP mAbs in most brain areas, while almost all the astrocytes in the optic nerve and spinal cord were labeled. Astrocytes in the ventricular zone and the rostral migratory stream preferentially contained phospho-GFAP. In a cold injury model of the cerebral cortex, phospho-GFAP was detected in reactive astrocytes 2–3 weeks following injury. The authors of this study suggest that the phosphorylation of GFAP plays a role in nondividing astrocytes in vivo (Takemura et al., 2002).

20.13.1 Phosphorylation

Posttranslational phosphorylation may account for the charge heterogeneity of GFAP observed following isoelectric focusing by polyacrylamide gel electrophoresis. In vitro translation of GFAP mRNAs from normal and mutant rodent central nervous system and from a human glioma-derived cell line grown in culture and as a solid tumor yielded single molecular weight polypeptides, which showed ionic charge differences among two to three spots with an isoelectric pH range of 5.7–5.9 (Bigbee and Eng, 1982). Norephinephrine treatment of C6 glioma cells induced the phosphorylation of a number of intracellular proteins, including vimentin and GFAP (Browning and Ruina, 1984). All receptor agonists that have been shown to increase cAMP levels increase phosphorylation of GFAP and vimentin in cultured astrocytes (McCarthy et al., 1985). While cAMP treatment of astrocytes induces a morphological change from the flat to process-bearing cell (Sensenbrenner et al., 1980), phosphorylation of GFAP and vimentin is independent of the morphological changes (Pollenz and McCarthy, 1986).

Phorbol ester-induced changes in astrocyte morphology and GFAP vimentin phosphorylation produce a shift in PKC from the cytosol to the membrane (Harrison and Mobley, 1990). Both the PKC activator, phorbol 12-myristate 13-acetate (PMA), and the cAMP-dependent protein kinase (PKA) activator, 8 bromo-cyclic AMP, phosphorylates several regions of GFAP and vimentin (Harrison and Mobley, 1991). ATP-evoked calcium signal stimulates protein phosphorylation/dephosphorylation in cultured astrocytes. The phosphorylation of a 52-kD protein, which comigrates with GFAP in sodium dodecyl sulfate polyacrylamide gel electrophoresis, supports the premise that calcium-dependent protein kinases and phosphatases are transducing elements for calcium signal brought about by activation of P2 purinergic receptors in astrocytes (Neary et al., 1991). PMA and inhibitors of PKC have been shown to alter GFAP mRNA (Sharma et al., 1991). A cytoskeletal-associated protein activity has been identified that phosphorylates GFAP and vimentin and is distinct from PKC and PKA (Harrison and Mobley, 1992).

Monoclonal antibodies to two synthetic peptides were produced, which react to phosphorylated GFAP. pG1 reacted to the serine in residues 3–13 and pG2 reacted to residues 29–39. The phosphorylation of these two serine residues on intact GFAP induced disassembly of glial filaments in vitro (Inagaki et al., 1990). These two antibodies react specifically with mitotic astroglial cells. The authors suggest that increased phosphorylation during mitosis may directly influence intracellular organization of glial filaments (Nishizawa et al., 1991; Matsuoka et al., 1992).

GFAP was found to be phosphorylated in vivo after intracerebral injection of 32P-orthophosphate, in brain slices and in a cell-free system. In both systems, the GFAP was phosphorylated at the serine and threonine residues. Incubation of brain slices with 32P and the PKC activator PMA, or an activator of cAMP-dependent protein kinase, forskolin, stimulated phosphorylation of GFAP. Phosphorylation of GFAP was also enhanced by calcium/phosphatidyl-serine/diolein and by exogenous cAMP-dependent kinase in a cell-free system. These findings indicate that PKC and cAMP-dependent kinase may play physiological roles in the in situ phosphorylation of GFAP. In cytoskeletal preparations incubated with gamma32P ATP, GFAP was phosphorylated in vitro by two additional protein kinases, Ca2+/calmodulin-dependent and an effector-independent kinase. These data suggest that phosphorylation of GFAP may be regulated by multiple second-message pathways (Noetzel, 1990a). A brain slice preparation was also employed to study GFAP synthesis and phosphorylation. The results indicate that the increase of GFAP during the first month of life cannot be ascribed solely to the rate of GFAP synthesis. The findings are consistent with the hypothesis that during later stages of astrocytic development the accumulation of GFAP may be due to a low rate of protein degradation. The pattern of GFAP phosphorylation in the developing rat brain differed from that observed for the incorporation of [3H]amino acids into GFAP. The peak incorporation of 32P into GFAP occurred on postnatal day 10 at a time when synthesis of the protein had declined by 43%. These results suggest that during GFAP development, phosphorylation of GFAP is mediated by factors different from those directing its synthesis (Noetzel, 1990b).

20.14 Increased GFAP Production of the Human GFAP Gene in a Transgenic Mouse

In order to determine the properties of astrocytes containing increased amounts of GFAP without external stimulation and activation, six lines of transgenic mice were generated which carry added copies of the normal human GFAP (hGFAP) gene and express the human transgene at different levels (Messing et al., 1996, 1998). Mice in lines that expressed high levels of the hGFAP gene (Tg73.2 and 73.8) died young, while mice in lines that expressed lower levels of the transgene (Tg73.2, 73.4, and 73.5) attained adulthood and survived for more than a year. At the light microscopic level, astrocytes in the high-expressing lines were distended by aggregates of globular eosinophilic material. Ultrastructural examination of a transverse section of optic nerve from a 13-day-old high-expressing mouse showed that astrocytes contained abundant cytoplasmic filaments in association with irregular osmophilic deposits resembling RFs.

20.14.1 Astrocytes Cultured from a Low GFAP Overexpressor Mouse, Tg73.2

Astrocytes in primary cultures generally contain larger amounts of GFAP than astrocytes in vivo. Consistent with this, astrocyte cultures prepared from a low overexpressor (Tg73.2) exhibited abnormal cytoplasmic inclusions identical to those seen in vivo in the high overexpressors (Eng et al., 1998). Astrocytes in the Tg73.2 cultures appear oddlyshaped and enlarged, express increased levels of GFAP (both human and mouse, as will be discussed below), and express aB-crystallin protein, HSP27 protein, and vimentin protein. At the light microscopic level, many of the astrocytes in 18-day Tg73.2 cultures exhibited large odd-shaped cells that immunostained with antibody specific hGFAP (SMI-21). Tg73.2, but not wild-type astrocytes, in culture for 18 days, immunostained for aB-crystallin. Vimentin staining was seen in the Tg73.2 cultures, but was also seen in the wild-type cultures. At 14 days in culture, both types of cultures immunostained sparsely for HSP27, but at 18 days the heat shock protein staining became more evident in the Tg73.2 cultures. Staining for hGFAP was already intense by 6 days in culture whereas staining for aB-crystallin continued to increase, suggesting that the overexpression of hGFAP may contribute to the induction of aB-crystallin. Immunostaining of Tg73.2 astrocytes with SMI-21 (hGFAP only) and counterstaining with eosin showed irregular-shaped eosin positive bodies surrounded by SMI-21 immunostain.

Conventional ultrastructural examination of Tg73.2 astrocytes showed numerous osmophilic deposits in a bed of intermediate filaments (Fig. 20.7) identical to that

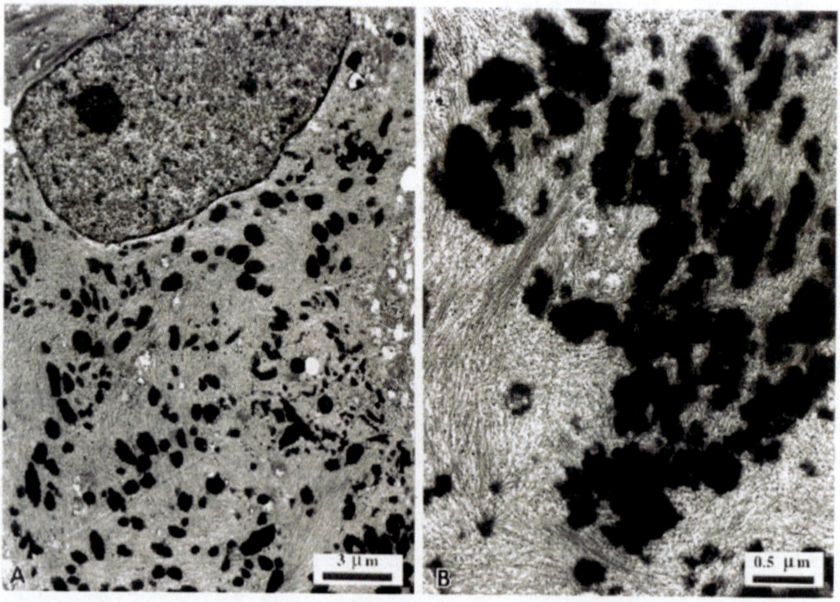

Fig. 20.7 **a,b** Astrocytes in culture for 20 days from a Tg73.2 mouse were analyzed at the ultrastructural level. Note the dense Rosenthal fibers among the glial filaments

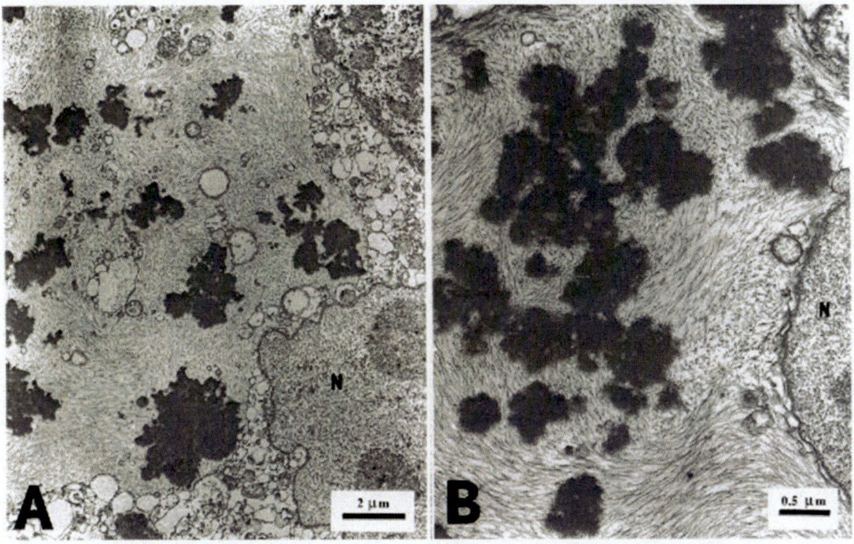

Fig. 20.8 **a,b** Astrocytes from a 17-month-old infant brain with Alexander's disease were examined at the ultrastructural level. Note the dense deposits in the astrocyte cell body which are identical to those seen in the Tg73.2 astrocyte cultures

seen in a case of infantile Alexander disease (Fig. 20.8). Double immunogold staining with SMI-21 (hGFAP only) and 18-nm gold second antibody followed by R-68 (human and mouse GFAP) and 12-nm gold second antibody showed that the GFAP in the wild-type astrocytes bound only to the 12-nm gold particles (Fig. 20.9a), while the GFAP in the Tg73.2 astrocytes bound to both 18- and 12-nm gold particles (Fig. 20.9b). This provided additional evidence demonstrating that both mouse and human GFAP are present in some of the Tg73.2 astrocytes. It appears that Tg73.2 mouse astrocytes in culture do not require additional stress from external sources of contact with other neuroectodermal cells to produce RFs. This suggests that the added hGFAP gene is sufficient to induce RFs and that excess of GFAP in astrocytes and/or the presence of GFAP that differs from species-specific GFAP may be detrimental to normal function. Remember that the GFAP null mouse survives without any GFAP formation, but abnormal accumulation of GFAP leads to Rosenthal fibers.

20.15 Role of GFAP in Alexander Disease

20.15.1 Alexander Disease Is Due to Genetic Mutation of Human GFAP

The overexpressing GFAP mouse suggested (Fig. 20.10a) that a primary alteration in GFAP may be responsible for Alexander disease (Fig. 20.10b) (Messing et al., 1998; Eng et al., 1998). Genomic DNA samples from 11 unrelated patients in whom

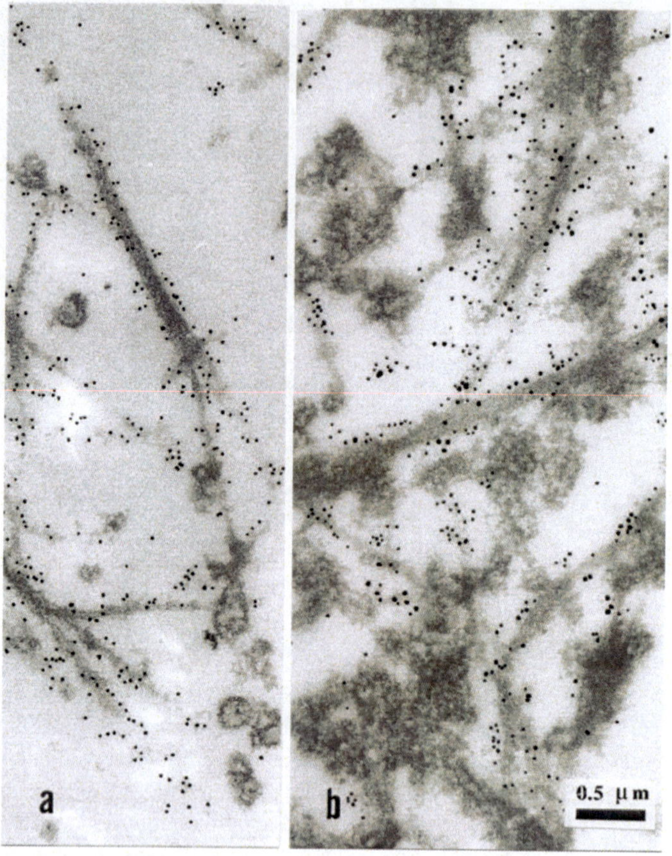

Fig. 20.9 **a**,**b** Ultrastructural analysis of wild-type (**a**) and Tg73.2 (**b**) astrocytes double immunogold staining with SMI-21(anti-human GFAP) with 18-nm gold particle followed by R-68 and 12-nm gold particle anti-sera. The wild-type astrocytes bind only to the 12-nm gold bound antiserum, while the Tg73.2 astrocytes bind to the 12- and 18-nm gold bound antisera

the diagnosis of Alexander disease had been confirmed by autopsy were analyzed by Brenner et al. (2001) For current review of Alexander's disease refer to 2008 book chapter by Brenner et al. Each exon with some adjoining intron segments and 1,717 bp of the 5′ flanking region of GFAP were amplified by PCR and sequenced. Of the 11 DNAs from the Alexander disease patients, 10 contained novel heterozygous mutations of GFAP predicting nonconservative amino acid changes, all involving arginine. Amino acid changes were: arginine 79 to cysteine or to histidine (one case of each); arginine 239 to cysteine (one case); arginine 258 to proline (four cases) or to histidine (one case); and arginine 416 to tryptophan (two cases). None of these mutations was seen in the two non-Alexander disease leukodystrophy control DNAs that were fully sequenced, or in 53 control DNA samples from individuals without neurologic disease that were specifically analyzed for these mutations

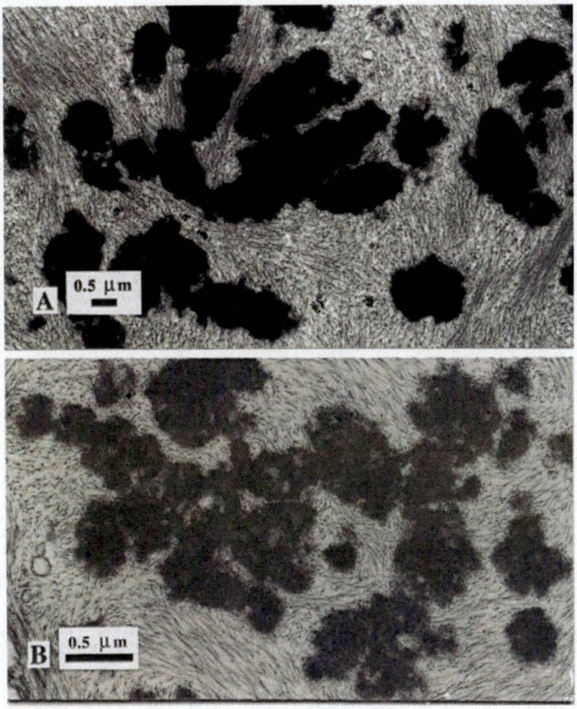

Fig. 20.10 a,b Rosenthal fibers examined at the ultrastructural level in (**a**) Tg73.2 hGFAP expressing astrocyte cultures, and (**b**) astrocytes in 17-month-old human Alexander's disease patient. Note the dense deposits in the astrocyte cell bodies of both the Tg73.2 hGFAP expressing mouse and the mutant human GFAP of the Rosenthal's disease patient

by restriction digestion. In each case, the Alexander disease patient was heterozygous for the mutation, suggesting a dominant mode of action. Since all the parents were phenotypically normal, the authors predicted that these mutations arose de novo. To test this hypothesis, DNA samples from five patients were analyzed by restriction digestion. None of the parental samples contained the mutations found in the affected children. Additional studies showed that of 14 parental DNAs tested, none had the non-conservative mutation found in the affected children. These initial studies showed that most cases of pathologically proven Alexander disease are associated with de novo mutations (occurring in the embryo or in the parental gametes) in the coding region of GFAP. A review (Messing et al., 2001) and several additional studies by other groups have confirmed the original observations.

The DNA of a series of additional infancy-onset patients, who had heterogeneous clinical symptoms but were candidates for Alexander disease on the basis of neuroimaging abnormalities (see above), were analyzed for GFAP mutations by Rodriguez et al. (2001). Missense, heterozygous de novo GFAP mutations were found in exon 1 or 4 for 14 of the 15 patients analyzed, including patients without macrocephaly. Nine patients carried one of the previously established mutations (arginine 79 to histidine (four cases); arginine 239 to cysteine (four cases) or to histidine (one case)). The other five had one of four novel mutations, of which two affected arginine (arginine 88 to cysteine (two cases) or to serine (one case)) and two affected non-arginine residues (leucine 76 to phenylalanine and asparagine 77 to tyrosine (each in one case)). All were located in the central rod domain of GFAP, and there was a correlation between clinical severity and affected amino acid.

Another study on patients suspected of having Alexander disease was conducted by Gorospe et al. (2002) to determine the extent to which clinical and MRI criteria could accurately diagnose affected individuals, using GFAP gene sequencing as the confirmatory assay. Patients showing MRI white matter abnormalities consistent with Alexander disease, unremarkable family history, normal karyotype, and normal metabolic screening, were included in the study. Genomic DNA from patients was screened for mutations in the entire coding region, including the exon-intron boundaries, of the GFAP gene. Twelve of the 13 patients were found to have mutations in GFAP. Seven of the 12 patients presented in infancy with seizure and megalencephaly. Five were juvenile-onset patients with more variable symptoms. Two patients in the latter group were asymptomatic or minimally affected at the time of their initial MRI scan. The mutations were distributed throughout the gene, and all involved sporadic single amino acid heterozygous changes that altered the charge of the mutant protein. Most of the GFAP mutations were sporadic single amino acid heterozygous missense changes: methionine 73 to arginine (one case), arginine 79 to cysteine (one case) or to histidine (two cases), arginine 88 to cysteine (two cases), arginine 239 to cysteine (one case), tyrosine 242 to aspartic acid (one case), glutamic acid 373 to lysine (one case), and arginine 416 to tryptophan (two cases). Moreover, two Japanese juvenile-onset cases have been demonstrated, one in which alanine 244 was mutated to valine, and another in which arginine 239 was changed to cysteine (Aoki et al., 2001; Shiroma et al., 2001).

Presently 124 mutations have been confirmed by GFAP sequencing (Li et al., 2002 and updated at http://www.waisman.wisc.edu/alexander/). The majority of the mutations described involve a change of the positively charged (at pH 7) amino acid arginine to a non-charged residue and there is one change of the non-charged methionine to arginine. In one patient the missense amino acid changed a negative charge on glutamic acid to a positive charge on lysine and in another patient, non-charged tyrosine was altered to negatively charged aspartic acid.

Location of Alexander disease-associated mutations in GFAP in relation to protein domain structure of intermediate filaments. The boxes indicate the four a-helical sub-domains within the central rod domain, separated by non-helical linkers. Each symbol represents the standard single letter code for the mutated amino acid for an individual patient or family (in the case of familial patients or identical twins). Multiple independent occurrences of a mutation are indicated by the number of symbols. Only one symbol is shown for twins or familial patients. A boxed symbol indicates that this mutation was inherited and found in multiple familial members. A circle around a mutation indicates that the patient was asymptomatic at the time of diagnosis. Symbols are color coded for clinical category based on age of onset (infantile, juvenile, adult). Mutations that are considered "disease-causing" are on the right – variants that are presently considered innocuous and not responsible for disease are shown on the left. The diagram includes all patients who have been published as of the date shown, as well as some unpublished patients.

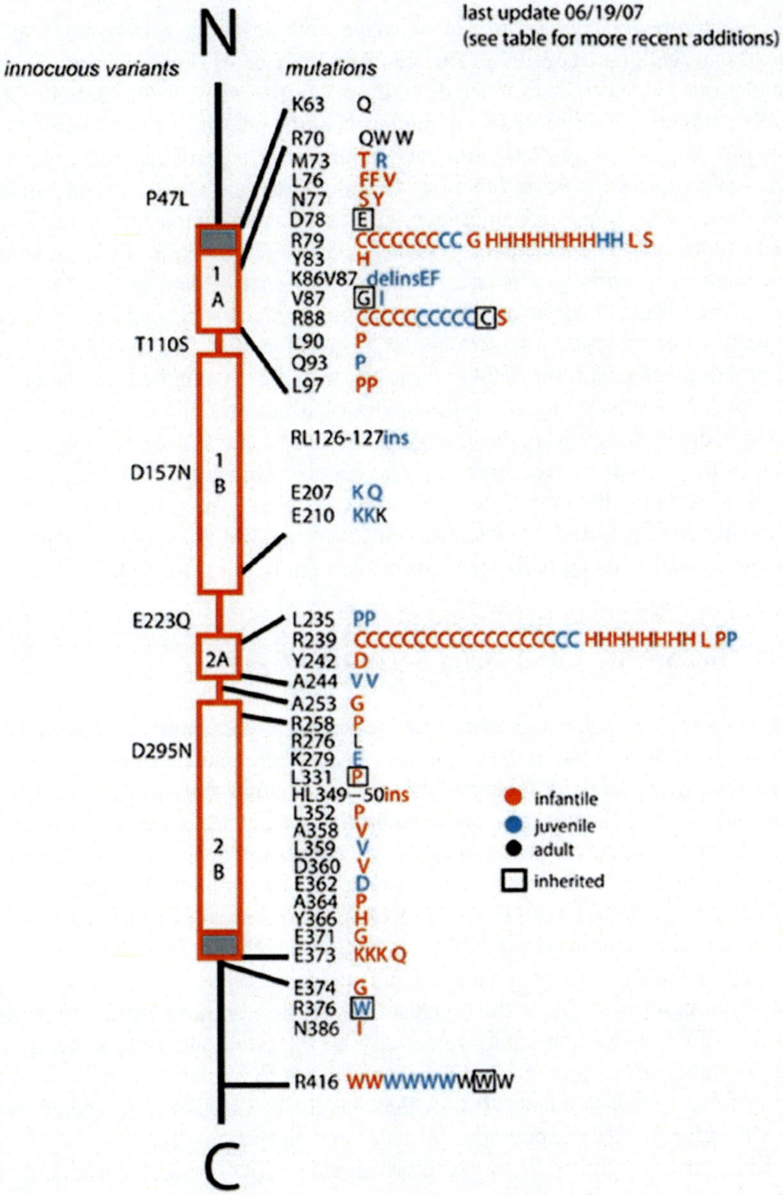

Fig. 20.11 Diagram of GFAP mutations in relation to protein domains and clinical classification

20.16 Consequences of GFAP Mutations

In the review by Messing et al. (2001), several mechanisms have been discussed by which GFAP mutations might lead to Alexander disease. Jaeken (2001) has proposed that Alexander disease is a conformational disease, arising when a constituent protein undergoes a change in size or shape with resultant self-association and intra- or extracellular deposition. The frequent changes in charged amino acids in the mutations linked to Alexander disease are consistent with this notion. GFAP mutations might cause disease by raising levels of the protein and increasing its stability. The mutation might lead to an accumulation of a particular form of GFAP, which compromises astrocyte function. The mutation can lead to accumulation of a toxic form of GFAP, which interferes with its polymerization into normal intermediate filaments. The emerging evidence that GFAP deletion alters functional characteristics of astrocytes is in agreement with such a concept, and a reason for the more severe form in infants may be interference with neuronal-astrocytic and endothelial-astrocytic interactions during development. We hypothesized that the normal mechanism for GFAP turnover may be insufficient to handle the excess GFAP, thus inducing an accumulation of stress proteins (Eng et al., 1998). Alexander disease astrocytes do display properties of physiological stress as evidenced by the elevation of the stress proteins aB-crystallin and HSP27. Brenner and co-workers suggest that Alexander disease likely results from a dominant gain of function that in turn partially blocks filament assembly, which leads to accumulation of an intermediate that participates in toxic interactions (Li et al., 2002).

20.17 Therapeutics and Drug Screening

GFAP mutations have been shown to be the cause of Alexander disease, and it is now possible to work toward developing cures and therapeutics. In AxD, all GFAP mutations are genetically dominant and cause an apparent gain of protein function. As demonstrated, knockout mice show no immediate deleterious effects associated with AxD, and have subtle phenotypes that do not resemble any of AxD. No recessive GFAP mutations have ever been described, and adding normal wild-type GFAP to AxD is unlikely to be useful in AxD. TRH was in one study (Ishigaki et al., 2006) shown to be successful in treating the ataxic gait of cerebellar mutant mice, so they treated a juvenile AxD patient with TRH and claimed success, but the interpretation of this is not confirmed. Since the organism can survive without GFAP expression, the AxD GFAP expression could potentially be inhibited and have a therapeutic effect. We propose to apply antisense RNA methods to inhibit AxD patient GFAP expression and increase the survival of these patients. This is being studied using retroviral antisense RNA expression methods, and experimentally we had used antisense RNA on cell cultures. We propose that stable morpholino analogs to antisense GFAP RNA could be developed as a pharmalogic useful in treating AxD. This is in line with the alternative approach to therapy that currently investigating drug libraries that are approved by the FDA are using to search for the ability to reduce expression of GFAP. Quercetin and some other compounds which have been shown to reduce GFAP expression in cultured astrocytes have been suggested for use as AxD therapeutic drugs (Gorospe et al., 2002). Since the activity of the GFAP promoter is controlled by NFeB, common drugs such as aspirin may be able to suppress the expression of GFAP by acting on its promoter (Bae et al., 2006). But these agents must be able to be effective at low concentrations to be useful and effective for in vivo models and AxD patients. Elevated GFAP function above a toxic threshold causes AxD, so any reduction of GFAP activity below this level may have a therapeutic effect, and even modest reductions of mutant GFAP expression could potentially be beneficial to alleviating or postponing the natural course of the AxD disease.

20.18 Summary and Conclusions

Glial fibrillary acidic protein is the IF protein found in differentiated astrocytes in the CNS. It is now well established that GFAP is the principal 8- to 9-nm intermediate filament in mature astrocytes. Over 40 years ago, GFAP's value as a prototype antigen in nervous tissue identification and a standard marker for fundamental and applied research at an interdisciplinary level was recognized (Raine, 1985). As a member of the cytoskeletal protein family, GFAP is thought to be important in modulating astrocyte motility and shape by providing structural stability to extensions astrocytic processes. *It is evident that GFAP is a very sensitive and specific marker for rapid astrocytic response to injury and disease. Increase in GFAP staining can result either from dissociation of glial filament bundles due to edema or due to increase in GFAP synthesis.* The rapid advances in molecular biology and newer techniques such as knockout mice, the use of the gfa-2 promoter to prepare transgenes, antisense RNA methodology, and DNA sequencing have greatly increased our knowledge of GFAP function in CNS development, injury, and disease. Mutations in the rod or TAIL domain of GFAP have been shown to cause the first known genetic defect in human astrocytes, AxD. The potential functions and GFAP response in astrocytes to stress and injury have been under continuous investigation since GFAP was first identified in 1969. *The astrocyte reacts to any type of insult, whether physical, biochemical, or disease, by enhanced expression GFAP, a process called astrogliosis.* This astrocytic response serves as a microsensor of the injured microenvironment at any location in the CNS. The precise mechanism of this response is still unknown. Growth factors, hormones, cytokines, and chemokines have been implicated, but no one common factor has been identified. *Immunostaining for GFAP is a sensitive method for identifying reactive astrocytes, whether they are due to filament dissembly as a result of kinase activation or to an increase in GFAP synthesis.* Polyclonal and monoclonal antibodies have been used extensively to study astrocytes in CNS development, disease, glial tumors, and experimental injury models (Eng et al., 2000). Based on GFAP, Vimentin, and double knockout mice studies, Pekny and Pekny (2004) has provided a concise review of the possible functions of glial filaments. Besides providing structural support, reactive astrogliosis, and scar formation, there is now evidence that GFAP is involved with LTD, LTP, the circadian rhythm, cell volume, cell motility, and the promotion of normal BBB formation. Specific factors expressed by microglia, blood cells, and injured cells that induce the expression of GFAP require further investigation.

References

Alexander WS (1949) Progressive fibrinoid degeneration of fibrillary astrocytes associated with mental retardatopm in a hydrocephalic infant. Brain 72:373–381

Amaducci L, Forno KI, Eng LF (1981) Glial fibrillary acidic protein in cryogenic lesions of the rat brain. Neurosci Lett 21:27–32

Anderova M, Kubinova S, Mazel T, Chvatal A, Eliasson C, Pekny M, Sykova E (2001) Effect of elevated K+ hypotonic stress, and cortical spreading depression on astrocyte swelling in GFAP-deficient mice. Glia 35:189–203

Aoki Y, Haginoya K, Munakata M, Yokoyama H, Nishio T, Togashi N, Ito T, Suzuki Y, Kure S, Iinuma K, Brenner M, Matsubara Y (2001) A novel mutation in glial fibrillary acidic protein gene in a patient with Alexander disease. Neurosci Lett 312:71–74

Aquino DA, Chiu FC, Brosnan CF, Norton WT (1988) Glial fibrillary acidic protein increases in the spinal cord of Lewis rats with acute experimental autoimmune encephalitis. J Neurochem 51:1085–1096

Bae MK, Kim SR, Lee HJ, Wee HJ, Yoo MA, Ock Oh S, Baek SY, Kim BS, Kim JB, Sik Y, Bae SK (2006) Aspirin-induced blockade of NF-kappa B activity restrains up-regulation of glial fibrillary acidic protein in human astroglial cells. Biochim Biophys Acta 1763: 282–289

Barger SW (2004) Astrocytes and microglia in Alzheimer's disease, in non-neuronal cells of the nervous system: function and dysfunction, Part III: pathological conditions. Elsevier B. V., Amsterdam, Netherlands, pp 883–899

Becker LE, Teixeira F (1988) Alexander's disease. In: Norenberg MD, Hertz L, Schousboe A (eds) The biochemical pathology of astrocytes. Alan R. Liss Inc, New York, NY, pp 179–190

Beguin P, Shooter EM, Eng LF (1980) Cell-free synthesis of glial fibrillary acidic protein. Neurochem Res 5:513–521

Bernier L, Colman DR, D'Eustachio P (1988) Chromosomal locations of genes encoding 2′, 3′ cyclic nucleotide 3′-phosphodiesterase and glial fibrillary acidic protein in the mouse. J Neurosci Res 20:497–504

Besnard F, Brenner M, Nakatani Y, Chao R, Purohit HJ, Freese E (1991) Multiple interacting sites regulate fibrillary acidic protein. J Biol Chem 266:18877–18883

Bigbee JW, Bigner DD, Pegram C, Eng LF (1983) Study of glial fibrillary acidic protein in a human glioma cell line grown in culture and as a solid tumor. J Neurochem 40:460–467

Bigbee JW, Eng LF (1982) Analysis and comparison of in vitro synthesized glial fibrillary acidic protein with rat CNS intermediate filament proteins. J Neurochem 38:130–134

Bignami A, Eng LF, Dahl D, Uyeda CT (1972) Localization of the glial fibrillary acidic protein in astrocytes by immunofluorescence. Brain Res 43:429–435

Blechingberg J, Holm IE, Nielson KB, Jensen TH, Jorgensen AL, Nielsen AL (2007) Identification and characterization of GFAPkappa, a novel glial fibrillary acidic protein isoform. Glia 55: 497–507

Bongcam-Rudloff E, Nister M, Betshotz C, Wang JL, Stenman G, Huebner K, Croce CM, Westermark B (1991) Human glial fibrillary acidic protein: complementary DNA cloning, chromosome localization, and messenger RNA expression in human glioma cell lines of various phenotypes. Cancer Res 51:1553–1560

Borrett G, Becker LE (1985) Alexander's disease: a disease of astrocytes. Brain 108:367–385

Boyer S, Montagutelli X, Gomes D, Simon-Chazottes D, Guenet JL, Dupouey P (1991) Recent evolutionary origin of the expression of the glial fibrillary acidic protein (GFAP) in lens epithelial cells: a molecular and genetic analysis of various mouse species. Mol Brain Res 10:159–166

Brenner M (1994) Structure and transcriptional regulation of the GFAP gene. Brain Pathol 4: 245–257

Brenner M, Goldman JE, Quinlan RA, Messing A (2008) Alexander disease: a genetic disorder of astrocytes. In: Parpura V, Hayden PG(eds) Astrocytes in pathophysiology of the nervous system. Springer Science + Business Media, LLC, doi:10.1007/978-0-387-79492-1

Brenner M, Goldman JE, Quinlan RA, and Messing A (2009) Alexander Disease: A Genetic Disorder of Astrocytes. Chapter 24 Astrocytes in (Patho) Physiology of the Nervous System, V. Parpura and P.G. Haydon (eds.) p 591–648. DOI: 10.1007/978-0-387-79492-1_25, © Springer Science+Business Media, LLC 2009

Brenner M, Johnson AB, Boespflug-Tanguy O, Rodriguez D, Goldman JE, Messing A (2001) Mutations in GFAP, encoding glial fibrillary acidic protein are associated with Alexander disease. Nature Genet 27:117–120

Brenner M, Kisseberth WC, Su Y, Besnard F, Messing A (1994) GFAP promoter directs astrocyte-specific expression in transgenic mice. J Neurosci 14:1030–1037

Brenner M, Lampel K, Nakatani Y, Mill J, Banner C, Mearow K, Dohadwala M, Lipsky R, Freese E (1990) Characterization of human cDNA and genomic clones for glial fibrillary acidic protein. Mol Brain Res 7:277–286

Brenner M, Messing A (1996) GFAP transgenic mice. Methods 10:351–364

Brown DR, Sassoon J (2004) Role of glia in prion disease, in non-neuronal cells of the nervous system: function and dysfunction, Part III: pathological conditions. Elsevier B. V., Amsterdam, Netherlands, pp 1085–1104

Brownell E, Lee AS, Pekar SK, Pravtcheva D, Ruddle FH, Bayney RM (1991) Glial fibrillary acidic protein, an astrocytic-specific marker, maps to human chromosome 17. Genomics 10: 1087–1089

Browning ET, Ruina M (1984) Glial fibrillary acidic protein: norepinephrine stimulated phosphorylation in intact C-6 glioma cells. J Neurochem 42:718–726

Bush TG, Puvanachandra N, Horner CH, Polito A, Ostenfeld T, Svendsen CN, Mucke L, Johnson MH, Sofroniew MV (1999) Leukocyte infiltration, neuronal degeneration and neurite outgrowth after ablation of scar-forming, reactive astrocytes in adult transgenic mice. Neuron 23:297–308

Campbell IL (1998) Transgenic mice and cytokine actions in the brain: bridging the gap between structural and functional neuropathology. Brain Res Brain Res Rev 26:327–336

Carpenter MK, Winkler C, Fricker R, Emerrich DF, Wong SC, Greco C, Chen EY, Chu Y, Kordower. JH, Messing A, Bjorland A, Hammang JP (1997) Generation and transplantation of EGF responsive neural stem cells. Exp Neurol 148:187–204

Carr dJ, Veress LA, Noisakran S, Campbell IL (1998) Astrocyte targeted expression of IFN-alpha 1 protects mice from acute ocular herpes simplex virus type 1 infection. J Immunol 161: 4859–4865

Carrasco J, Nernandez J, Gonzalez B, Campbell IL, Hidalgo J (1998) Localization of metallothionein-I and -III expression in the CNS of transgenic mice with astrocyte-targeted expression of interleukin 6. Exp Neurol 153:184–194

Chen WJ, Liem RL (1994) Reexpression of glial fibrillary acidic protein rescues the ability of astrocytoma cells to form processes in response to neurons. J Cell Biol 127:813–823

Chen WJ, Liem RK (1994) The endless story of the glial fibrillary acidic protein. J Cell Sci. Aug;107 (Pt 8):2299–311

Chin SM, Goldman JE (1996) Glial inculsion in CNS degenerative diseases. J Neuropath Expt Neurol 55:499–508

Chiu F-C, Goldman JE (1984) Synthesis and turnover of cytoskeletal proteins in cultured astrocytes. J Neurochem 42:166–174

Colucci-Guyon E, Gimenez Y, Ribotta M, Maurice T, Babinet C, Privat A (1999) Cerebellar defect and impaired motor coordination in mice lacking vimentin. Glia 25:33–43

Colucci-Guyon E, Portier MM, Dunia I, Paulin D, Pournin S, Babinet C (1994) Mice lacking vimentin develop and reproduce without an obvious phenotype. Cell 79:679–694

Condorelli DF, Nicoletti VG, Barresi V, Conticello SG, Caruso A, Tendi EA, Giuffrida Stella AM (1999) Structural features of the rat GFAP gene and identification of a novel transcript. J Neurosci Res 56:219–228

Cortez N, Trejo F, Vergara P, Segovia J (2000) Primary astrocytes retrovirally transduced with a tyrosine hydroxylase transgene driven by a glial-specific promoter elicit behavioral recovery in experimental Parkinsonism. J Neurosci Res 59:39–46

Cutler RWP, Lorenzo AV, Barlow CF (1967) Brain vascular permeability to I125 gamma globulin and leukocytes in allergic encephalomyelitis. J Neuropathol Exp Neurol 26:558–571

D'Amelio FE, Smith ME, Eng LF (1990) Sequence of tissue responses in the early stages of Experimental Allergic Encephalomyelitis (EAE): immunohistochemical, light microscopic, and ultrastructural observations in the spinal cord. Glial 3:229–240

DeArmond SJ, Fajardo M, Naughton SA, Eng LF (1983) Degradation of glial fibrillary acidic protein by a calcium dependent proteinase: an electroblot study. Brain Res 262:275–282

Deck JHN, Eng LF, Bigbee J (1976) A preliminary study of glioma morphology using the peroxidase-anti-peroxidase method for the GFA protein. J Neuropathol Exp Neurol 35:362

Deck JHN, Eng LF, Bigbee J, Woodcock SM (1978) The role of glial fibrillary acidic protein in the diagnosis of central nervous system tumors. Acta Neuropathol (Berl) 42:183–190

Delaney CL, Brenner M, Messing A (1996) Conditional ablation of cerebellar astrocytes in postnatal transgenic mice. J Neurosci 16:6908–6918

Ding M, Eliasson C, Betsholtz C, Hamberger A, Pekny M (1998) Altered taurine release following hypotonic stress in astrocytes from mice deficient for GFAP and vimentin. Brain Res Molec Brain Res 62:77–81

Duffy PE, Graf L, Rapport MM (1977) Identification of glial fibrillary acidic protein by the immunoperoxidase method in human brain tumors. J Neuropathol Exp Neurol 36:645–652

Eckes B, Colucci,-Guyon E, Smola H, Nodder S, Babinet C, Krieg T, Martin P (2000) Impaired wound healing in embryonic and adult mice lacking vimentin. J Cell Sci 113:2455–2462

Eddleston M, Mucke L (1993) Molecular profile of reactive astrocytes: implications for their role in neurologic diseases. Neuroscience 54:15–36

Eng LF (1985) Glial fibrillary acidic protein: the major protein of glial filament intermediate filaments in differentiated astrocytes. J Neuroimmunol 8:203–214

Eng LF, Bigbee JW (1978) Immunohistochemistry of nervous system-specific antigens. In: Agranoff BW, Aprison MH (eds) Advances in neurochemistry, vol 3. Plenum Press, New York, NY, pp 43–98

Eng LF, Bond P, Gerstl B (1971a) Isolation of myelin proteins from disc acrylamide gels electrophoresed in phenol-formic acid-water. Neurobiol 1:58–63

Eng LF, D'Amelio FE, Smith ME (1989) Dissociation of GFAP intermediate filaments in EAE: observations in the lumbar spinal cord. Glia 2:308–317

Eng LF, Gerstl B, Vanderhaeghen JJ (1970) A study of proteins in old multiple sclerosis plaques. Trans Am Soc Neurochem 1:42

Eng LF, Ghirnikar RS (1994) GFAP and astrogliosis. Brain Pathol 4:229–237

Eng LF, Ghirnikar RS, Lee YL (2000) Glial fibrillary acidic protein: GFAP-thirty-one-years (1969-2000)∗. Neurochem Res 25(9/10):1439–1451

Eng LF, Lee YL (1995) Intermediate filaments in astrocytes. In: Kettenmann H, Ransom BR (eds) Neuroglia. Oxford University Press, New York, NY, pp 650–667

Eng LF, Lee YL (2004) Alexander disease: a primary disease of astrocytes. Mol Cell Biol 31:773–785 Elsevier B. V. (C)2004 ISBN: 0-444-51451-1 refer to page 776 Figure 1 from Eng and Bigbee, 1978

Eng LF, Lee YL, Kwan H, Brenner M, Messing A (1998) Astrocytes cultured from transgenic mice carrying the added human glial fibrillary acidic protein gene contains Rosenthal fibers. J Neurosci Res 53:353–360

Eng LF, Rubinstein LJ (1978) Contribution of immunohistochemistry to diagnostic problems of human cerebral tumors. J Histochem Cytochem 26:513–522

Eng LF, Vanderhaeghen JJ, Bignami A, Gerstl B (1971b) An acidic protein isolated from fibrous astrocytes. Brain Res 28:351–354

Evans RM (1988) Cyclic AMP-dependent protein kinase-induced vimentin filament disassembly involves modification of the N-terminal domain of intermediate filament subunits. FEBS Lett 234:73–78

Feinstein DL, Weinmaster GA, Milner RJ (1992) Isolation of cDNA clones encoding rat glial fibrillary acidic protein: expression in astrocytes and in Schwann cells. J Neurosci Res 32:1–14

Fernandez-Galaz MC, Maartinez Munoz R, Villanua MA, Garcia-Segura LM (1999) Diurnal oscillation in glial fibrillary acidic protein in a perisuprachiasmatic area and its relationship to the luteininzing hormone urge in the female rat. Neuroendocrinology 70:368–376

Friede RL (1964) Alexader's disease. Arch Neurol 11:414–422

Fuchs E (1996) The cytoskeleton and disease: genetic disorders of intermediate filaments. Annu Rev Genet 30:197–231

Furukawa K, Inagaki H, Hotta Y (1994) Identification and cloning of an mRNA coding for a germ cell-specific A-type lamin in mice. Exp Cell Res 212:426–430

Galbreath E, Kim SJ, Park K, Brenner M, Messing A (1995) Overexpression of TGF-beta 1 in the central nervous system of transgenic mice results in hydrocephalus. J Neuropathol Exp Neurol 54:339–349

Galea E, Dupouey P, Feinstein DL (1995) Glial fibrillary acidic protein messenger-RNA isotypes–expression in-vitro and in vivo. J Neurosci Res 41:452–461

Galou M, Colucci-Guyon E, Ensergueix D, Ridet JL, Gimenez Y, Ribotta M, Privat A, Babinet C, Dupouey P (1996) Disrupted glial fibrillary acidic protein network in astrocytes from vimentin knockout mice. J Cell Biol 133:853–863

Geisler N, Weber K (1988) Phosphorylation of desmin in vitro nihibits formation of intermediate filaments; identification of three kinase A sites in the aminoterminal head domain. EMBO J 7:15–20

Ghirnikar RS, Yu AC, Eng LF (1994) Astrogliosis in culture: III. effect of recombinant retrovirus expressing antisense glial fibrillary acidic protein RNA. J Neurosci Res 38: 376–385

Goldman JE, Corbin E (1991) Rosenthal fibers contain ubiquitinated aB-crystallin. Am J Pathol 139:933–938

Goldman RD, Steinart PM (1990) Cellular and molecular biology of intermediate filaments. Plenum, New York, NY

Gomes FC, Garcia,-Abreu J, Galou M, Paulin D, Moura Neto V (1999) Neurons induce GFAP gene promoter of cultured astrocytes from transgenic mice. Glia 26:97–108

Gomi H, Yokoyama T, Fujimoto K, Ikeda T, Katoh A, Itoh T, Itohara S (1995) Properties of astrocytes from mice devoid of the glial fibrillary acidic protein develop normally and are susceptible to scrapie prions. Neuron 14:29–41

Gorospe JR, Naidu S, Johnson AB, Puri V, Raymond GV, Jenkins SD, Pederson RC, Lewis D, Knowles P, Fernandez R, De Vivo D, van de Knaap MS, Messing A, Brenner M, Hoffman EP (2002) Molecular findings in symptomatic and pre-symptomatic Alexander disease patients. Neurology 58:1494–1500

Hagiwara N, Imada S, Sueoka N (1993) Cell type specific segregation of transcriptional expression of glial genes in the rat peripheral neurotumor RT4 cell lines. J Neurosci Res 36: 646–656

Harrison BC, Mobley PL (1990) Phorbol ester-induced change in astrocyte morphology: correlation with protein kinase C activation and protein phosphorylation. J Neurosci Res 25: 71–80

Harrison BC, Mobley PL (1991) Phorbol myristate acetate and 8-bromo-cyclic AMP-induced phosphorylation of glial fibrillary acidic protein and vimentin in astrocytes: comparison of phosphorylation sites. J Neurochem 56:1723 1730

Harrison BC, Mobley PL (1992) Phosphorylation of glial fibrillary acidic protein and vimentin by cytoskeletal-associated intermediate filament protein kinase activity in astrocytes. J Neurochem 58:320–327

Head M, Corbin E, Goldman JE (1993) Overexpression and abnormal modification of the stress proteins aB-crystalline and HSP27 in Alexander disease. Am J Pathol 143:1743–1753

Head M, Corbin E, Goldman JE (1994) Coordination and independent expression of aB-crystallin and HSP27. J Cell Physiol 159:41–50

Herndon RM (1999) Is Alexander's disease a nosologic entity or a common pathologic pattern of diverse etiology? J Child Neurol 14:275–276

Herndon RM, Rubenstein LJ, Freeman JM, Mathieson G (1970) Light and electron microscopic observations on Rosenthal fibers in Alexander disease and in multiple sclerosis. J Neuropathol Exp Neurol 29:524–551

Hol EM, Roelofs RF, Moraal E, Sonnemans JA, Sluijs JA, Proper EA, deGraan PN, Fischer DF, van Leeuwen FW (2003) Neuronal expression of GFAP in patients with Alzheimer pathology and identification of novel GFAP splice forms, Mol. Psychiatry 8:786–796

Holland EC, Varmus HE (1998) Basic fibroblast growth factor induces cell migration and proliferation after glia-specific gene transfer in mice. Proc Natl Acad Sci USA 95: 1218–1223

Inagaki M, Gonda Y, Matsuyama M, Nishizawa K, Nishi Y, Sato C (1988) Intermediate filament reconstitution in vitro. the role of phosphorylation on the assembly-disassembly of desmin. J Biol Chem 263:5970–5978

Inagaki M, Gonda Y, Nishizawa K, Kitamura S, Sato C, Ando S, Tanabe K, Kikuchi K, Tsuiki S, Nishi Y (1990) Phosphorylation sites linked to glial filament disassembly in vitro locate in a non-alpha-helical head domain. J Biol Chem 265:4722–4729

Inagaki M, Nakamura Y, Masatoshi T, Nushimura T, Inagaki N (1994a) Glial fibrillary acidic protein: dynamic property and regulation by phosphorylation. Brain Pathol 4:239–243

Inagaki M, Nakamura Y, Takeda M, Nishimura T, Inagaki N (1994b) Glial fibrillary protein: dynamics property and regulation by phosphorylation. Brain Pathol 4:139–243

Inagaki M, Nishi Y, Nishizawa K, Matsuyama M, Sato. C (1987) Site-specific posphrylation induces dis-assembly of vimentin filaments in vitro. Nature 328:649–652

Ishigaki K, Ito Y, Sawaishi Y, Kodaira K, Funatsuka M, Hattori N, Nakano K, Saito K, Osawa M (2006) TRH therapy in a patient with juvenile Alexander disease. Brain Dev 28:663–667

Iwaki T, Kume-Iwaki A, Liem RKH, Goldman J (1989) aB-crystallin is expressed in non-lenticular tissues and accumulates in Alexander disease. Cell 57:71–78

Jaeken J (2001) Alexander disease and intermediate filaments in astrocytes: a fatal gain of function. Eur J Pediatr Neurol 5:151–153

Jessen KR, Thorpe R, Mirsky R (1984) Molecular identity, distribution and heterogeneity of glial fibrillary acidic protein: an immunoblotting and immunohistochemical study of Schwann cells, satellite cells, enteric glia and astrocytes. J Neurocytol 13:187–200

Juhler M, Barry DL, Offner H, Konat G, Klinken L, Paulson OB (1984) Blood-brain and blood-spinal cord barrier permeability during the course of experimental allergic encephalomyelitis in the rat. Brain Res 302:347–355

Kaneko R, Sueoka N (1993) Tissue-specific versus cell type-specific expression of the glial fibrillary acidic protein. Proc Natl Acad Sci USA 90:4698–4702

Kaufmann E, Weber K, Geisler N (1985) Intermediate filament forming ability of desmin derivatives lacking either the amino-terminal 67 or the carboxy-terminal 27 residues. J Mol Biol 185, 733–742

Kimelberg HK, Bourke RS, Stieg PE, Barron KD, Hirata H, Pelton EW, Nelson LR (1982) Swelling of astroglia after injury to the central nervous system: mechanisms and consequences. In: Grossman RG, Gildenberg PL (eds) Head injury: basic and clinical aspects. Raven Press, New York, NY, pp 31–44

Kimelberg HK, Rose JW, Barron KD, Waniewski RA, Cragoe EJ (1989) Astrocytic swelling in traumatic-hypoxic brain injury. Beneficial effects of an inhibitor of anion exchange transport and glutamate uptake in glial cells. Mol Chem Neuropathol. Aug;11(1): 1–31

Kordower JH, Chen EY, Winkler C, Fricker R, Charles V, Messing A, Mufson EJ, Wong SC, Rosenstein JM, Bjorkland A, Emerich dF, Hammang J, Carpenter MK (1997) Grafts of EGF-responsive neural stem cells derived from GFAP-hNGF transgenic mice: trophic and tropic effects in a rodent model of Huntington's disease. J Comp Neurol 387(1):96–113

Kumanishi T, Usui H, Ichikawa T, Nishiyama A, Katagiri T, Abe S, Yoshida Y, Washiyama K, Kuwano R, Sakimura K, et al. (1992) Human glial fibrillary acidic protein (GFAP): molecular cloning of the complete cDNA sequence and chromosomal localization (chromosome 17) of the GFAP gene. Acta Neuropathol. 83(6):569–78

Lafrancois T, Fages C, Peschanski M, Tardy M (1997) Neuritic outgrowth associated with astroglial phenotypic changes induced by antisense glial fibrillary acidic protein (GFAP) mRNA in injured neuron-astrocyte cocultures. J Neurosci 17:4121–4128
Landon F, Lemonnier M, Benarous R, Huc C, Fiszman M, Gros F, Portier MM (1989) Multiple mRNAs encode peripherin, a neuronal intermediate filament protein. EMBO J 8: 1719–1726
Landon F, Wolff A, de Nechaud B (2000) Mouse peripherin isoforms. Biol Cell 92:397–407
Laping NJ, Teter B, Nichols NR, Rozovsky I, Finch CE (1994) Glial fibrillary acidic protein: regulation by hormones, cytokines, and growth factors. Brain Pathol 4:259–275
Lee JC (1982) Anatomy of the blood-brain barrier under normal and pathological conditions. In: Haymaker W, Adams RD (eds) Histology and histopathology of the nervous system. Charles Thomas, Springfield, Illinois, pp 798–890
Lepekhin EA, Eliasson C, Berthold CH, Berezin V, Bock E, Pekny M (1999) Intermediate filaments regulate astrocyte motility. J Neurochem 79:617–625
Lewis SA, Balcarek JRF, Krek V, Shelanski M, Cowan NJ (1984) Sequence of a cDNA clone encoding mouse glial fibrillary acidic protein: structural conservation of intermediate filaments. Proc Natl, Acad Sci USA 81:2743–2746
Lewis SA, Cowan NJ (1985) Temporal expression of mouse glial fibrillary acidic protein mRNA studied by a rapid in situ hybridization procedure. J Neurochem 45:913–919
Li R, Messing A, Goldman JE, Brenner M (2002) GFAP mutations in Alexander disease. Int J Devl Neurosci 20:259–268
Liedke W, Edelmann W, Chiu FC, Kucherlapati R, Raine CS (1998) Experimental autoimmune encephalomyelitis in mice lacking glial fibrillary acidic protein is characterized by a more severe clinical course and an infiltrative central nervous system lesion. Am J Pathol 152:251–259
Liedtke W, Edelmann W, Bieri PL, Chiu FC, Cowan NJ, Kucherlapati R, Raine CS (1996) GFAP is necessary for the integrity of CNS white matter architecture and long-term maintenance of myelination. Neuron 17:607–615
Lin F, Worman HJ (1993) Structural organization of the human gene encoding nuclear lamin A and nuclear lamin C. J Biol Chem 268:16321–16326
Lucas CV, Bensch KG, Eng LF (1980a) In vitro polymerization of glial fibrillary acidic (GFA) protein extracted from multiple sclerosis (MS) brain. Neurochem Res 5:247–255
Lucas CV, Reaven EP, Bensch KG, Eng LF (1980b) Immunoperoxidase staining of glial fibrillary acidic protein (GFA) protein polymerized in vitro: an ultramicroscopic study. Neurochem Res 5:1199–1209
Lundkvist J, Sundgren-Andersson AK, Tingsborg S, Ostlund P, Engfors C, Alheim K, Bartfai T, Iverfeldt K, Schultzberg M (1999) Acute-phase responses in transgenic mice with CNS overexpression of IL-1 receptor antagonist. Am J Physiol 276:R644–R651
Machiels BM, Zorenc AHG, Endert JM, Kuijpers HJH, V.Eys GJJM, Ramaekers FCS, Proers JLV (1996) An alternative splicing product of the lamin A/C gene lacks exon. J Biol Chem 271:9249–9253
Malloch GDA, Clark JB, Burnet FR (1987) Glial fibrillary acidic protein in the cytoskeletal and soluble protein fractions of the developing rat brain. J Neurochem 48:299–306
Masood K, Besnard F, Su Y, Brenner M (1993) Analysis of a segment of the human glial fibrillary acidic protein gene that directs astrocyte-specific transcription. J Neurochem 61:160–166
Matsuoka Y, Nishizawa K, yano T, Shibata M, Ando S, Takahashi T, Inagaki M (1992) Two different protein kinases act on a different time schedule as glial filament kinases during mitosis. EMBO J 11:2895–2902
Matsuzawa K, Kosako H, Azuma I, Inagaki N, Inagaki M (1998) Possible regulation of intermediate proteins by Rho-binding kinases. In: Herrmann H, Harris R (eds) Subcellular biochemistry, vol 31. Intermediate Filaments. Plenum Press, New York, NY, pp 423–435
McCall MA, Gregg RG, Behringer RR, Brenner M, Delaney CL, Galbreath EJ, Zhang CL, Pearce RA, Chiu SY, Messing A (1996) Targeted deletion in astrocyte intermediate filament (GFAP) alters neuronal physiology. Proc Natl Acad Sci USA 93:6361–6366

McCarthy KD, Prime J, Harmon T, Pollenz RS (1985) Receptor mediated phosphorylation of astroglial intermediate filament proteins in culture. J Neurochem 44:723–730

McKeon FD, Kirschner MW, Caput D (1986) Homologies in both primary and secondary structure between nuclear envelope and intermediate filament proteins. Nature 319:463–468

McKie EA, Graham DI, Brown SM (1998) Selective astrocytic transgene expression in vitro and in vivo from. Gene Therapy 5:440–450

McLendon RE, Bigner DD (1994) Immunohistochemistry of the glial fibrillary acidic protein: basic and applied considerations. Brain Pathol 4:221–228

Messing A, Galbreath EJ, Sijapati KK, Brenner M (1996) Overexpression of GFAP in transgenic mice (abstract). J Neuropathol Exp Neurol 55:620

Messing A, Goldman JE, Johnson AB, Brenner M (2001) Alexander disease: new insights from genetics. J Neuropathol Exp Neurol 60:563–573

Messing A, Head MW, Galles K, Galbreath EJ, Goldman JE, Brenner M (1998) Fatal encephalopathy with astrocyte inclusions in GFAP transgenic mice. Am J Pathol 152:391–398

Miquel J, Foncin JF, Gruner JE, Lee JC (1982) Cerebral edema. In: Haymaker W, Adams RD (eds) Histology and histopathology of the nervous system. Charles Thomas, Bethesda, Maryland, pp 871–919

Miura M, Tamura T, Mikoshhiba K (1990) Cell-specific expression of the mouse glial fibrillary acidic protein Gene: identification of the cis- and trans-acting promoter elements for astrocyte-specific expression. J Neurochem 55:1180–1188

Mucke L, Oldstone MBA, Morris JC, Nerenberg MI (1991) Rapid activation of astrocyte-specific expression of GFAP-lacZ transgene by focal injury. New Biol 3:465–474

Mucke L, Rockenstein EM (1993) Prolonged delivery of transgene products to specific brain regions by migratory astrocyte grafts. Transgene 1:3–9

Nakamura Y, Takeda M, Angelides KJ, Tada D, Hariguchi S, Nishimura T (1991) Assembly, disassembly, and exchange of glial fibrillary acidic protein. Glia 4:101–110

Nakatani Y, Brenner M, Freese E (1990a) An RNA polymerase II promoter containing sequences upstream and downstream from the RNA startpoint that direct initiation of transcription from the same site. Proc Natl, Sci USA 87:4289–4293

Nakatani Y, Horikoshi M, Brenner M, Yamamoto T, Besnard F, Roeder RG, Freese E (1990b) A downstream initiation element required for efficient TATA box bindking and in vitro function of TFIID. Nature 348:86–88

Nawashiro H, Messing A, Azzam N, Brenner M (1998) Mice lacking GFAP are hypersensitive to traumatic cerebrospinal injury. NeuroReport 9:1691–1696

Neary JT, Laskey R, van Breemen C, Blickarska J, Norenberg LO, Norenberg MD (1991) ATP-evoked calcium signal stimulates protein phosphorylation/dephosphorylation in astrocytes. Brain Res 566:89–94

Nelson WJ, Traub P (1983) Proteolysis of vimentin and desmin by the Ca2+ -activated proteinase specific for these intermediate filament proteins. Mol Cell Biol 3:1146–1156

Nielsen AL, Holm iE, Johansen M, Bonven B, Jorgensen P, Jorgensen AL (2002) A new splice variant of glial fibrillary acidic protein, GFAP epsilon, interacts with the presenilin proteins. J Biol Chem 277:29983–29991

Nishizawa K, Yano T, Shibata M, Ando S, Saga S, Takahashi T, Inagaki M (1991) Specific localization of phosphointermediate filament protein in the constricted area of dividing cells. J Biol Chem 266:3074–3079

Noetzel MJ (1990a) Phosphorylation of the glial fibrillary acidic protein. J Neurosci Res 27: 184–192

Noetzel MJ (1990b) Synthesis and phosphorylation of the glial fibrillary acidic protein during brain development: a tissue slice study. Glia 3:450–457

Parry DA, Steinart PM (1999) Intermediate filaments: molecular architecture, assembly, dynamics, and polymorphism. Q Rev Biophys 32:99–187

Pekny M (2001) Astrocytic intermediate filaments: lessons from GFAP and vimentin knock-out mice. Prog Brain Res 132:23–30

Pekny M, Eliasson C, Chien CL, Kindblom LG, Liem R, Hamberger A, Betsholtz C (1998a) GFAP-deficient astrocytes are capable of stellation in vitro when cocultured with neurons and exhibit a reduced amount of intermediate filaments and an increased cell saturation density. Expt Cell Res 239:332–343

Pekny M, Johansson CB, Eliasson C, Stakeberg J, Wallen A, Perlmann T, Lendahl U, Betsholtz C, Berthold CH, Frisen J (1999) Abnormal reaction to central nervous system injury in mice lacking glial fibrillary acidic protein and vimentin. J Cell Biol 145:503–514

Pekny M, Leveen P, Pekna M, Eliasson C, Berthold CH, Westermark B, Betsholtz C (1995) Mice lacking GFAP display astrocytes devoid of intermediate filaments but develop and reproduce normally. EMBO J 14:1590–1598

Pekny M, Pekna M (2004) Astrocyte intermediate filaments in CNS pathologies and regeration. J Pathol 204:428–437

Pekny M, Stanness KA, Eliasson C, Betshotlz C, Janigro D (1998b) Impaired induction of blood-brain barrier properties in aortic endothelial cells by astrocytes from GFAP-deficient mice. Glia 22:390–400

Pollenz RS, McCarthy KD (1986) Analysis of cyclic AMP-dependent changes in intermediate filament protein phosphorylation and cell morphology in cultured astroglia. J Neurochem 47:9–17

Przedborski S, Goldman JE (2004) Pathological role of glial cells in Parkinson's disease, in non-neuronal cells of the nervous system: function and dysfunction, Part III: pathological conditions. Elsevier B. V, Amsterdam, Netherlands, pp 967–982

Quinlan RA, Moir RD, Stewart M (1989) Expression in Escherichia coli of fragments of glial fibrillary acidic protein: characterization, assembly properties and paracrystal formation. J Cell Sci 93:71–83

Quintana JG, Lopez-Colberg I, Cunningham LA (1998) Use of GFAP-lacZ transgenic mice to determine astrocyte fate in grafts of embryonic ventral midbrain. Brain Res Dev Brain Res 105:147–151

Raeber AJ, Race RE, Brandner S, Priola SA, Sailer A, Bessen RA, Mucke L, Manson J, Aguzzi A, Oldstone MB et al. (1997) Astrocyte-specific expression of hamster prion protein (PrP) renders PrP knockout mice susceptible to hamster scrapie. EMBO J 16:6057–6065

Raine CS (1985) Editorial. J Neuroimmunol 8(4–6) p 201

Ramer MS, Kawaja MD, Henderson JT, Roder JC, Bisby MA 1998. Glial overexpression of NGF enhances neuropathic pain and adrenergic sprouting into DRG following chronic sciatic constriction in mice

Ramsey P, Norman M, Eng LF (1979) Chemical study of an Alexander brain (abstract). Trans Am Soc Neurochem 10:125

Reeves SA, Helman LJ, Allison A, Israel. MA (1989) Molecular cloning and primary structure of human glial fibrillary acidic protein. Proc Natl Acad Sci USA 86:5178–5182

Riol H, Tardy M, Rolland B, Levesque G, Murthy MR (1997) Detection of the peripheral nervous system (PNS)-type glial fibrillary acidic protein (GFAP) and its mRNA in human lymphocytes. J Neurosci Res 48:53–62

Robertson J, Doroudchi MM, Nguyen HD, Durham HD, Strong MJ, Shaw G, Julien JP, Mushynski WE (2003) A neurotoxic splice varient in a mouse model of ALS. J Cell Biol 160:939–949

Rodriguez D, Gauthier F, Bertini E, Bugiani M, Brenner M, Nguycn S, Goizct C, Gclot A, Siurtees R, Pedespan JM, Hernandorena X, Troncoso M, Uziel G, Messing A, Ponsot G, Pham-Dinh D, Dautigny A, Boespflug-Tanguy O (2001) Infantile Alexander disease: spectrum of GFAP mutations and genotype-phenotype correlation. Am J Hum Genet 69: 1134–1140

Roelofs RF, Fischer DF, Houtman SH, Sluijs JA, Van Haren W, Van Leeuwen FW, Hol EM (2005) Adult human subventricular, subgranular, and subpial zones contain astrocytes with a specialized intermediate filament cytoskeleton. Glia 52:289–300

Rueger DC, Huston JS, Dahl D, Bignami A (1979) Formation of 100A filaments from purified glial fibrillary acidic protein in vitro. J Mol Biol 135:53–68

Russo LS, Aron A, Anderson PJ (1976) Alexander's disease: a report and reappraisal. Neurology 26:607–614
Rutka JT, Ackerley C, Hubbard SL, Tilup A, Dirks PB, Jung S, ivanchuk S, Kurimoto M, Tsugu A, Becker LE (1998) Chracterization of glial filament cytoskeletal interactions in human astrocytomas: an immuno-ultrastructural analysis. European J Cell Biol 76:279–287
Rutka JT, Hubbard SL, Fukuyama K, Matsuzawa K, Dirks PB, Becker LE (1994) Effects of antisense glial fibrillary acidic protein complementary DNA on the growth, invasion and adhesion of human astrocytoma cells. Cancer Res 54:3267–3272
Rutka JT, Ivanchuk S, Mondal S, Taylor M, Sakai K, Dirks P, Jun P, Jung S, Becker LE, Ackerly C (1999) Co-expression of nestin and vimentin intermediate filaments in invasive human astrocytoma cells. International J of Developmental Neurosci 17:503–515
Rutka JT, Smith SL (1993) Transfection of human astrocytoma cells with glial fibrillary acidic protein complementary DNA: Analysis of expression, proliferation, and tumorigenicity. Cancer Res 53:3624–3631
Sarkar S, Cowan NJ (1991) Intragenic sequences affect the expression of the gene encoding glial fibrillary acidic protein. J Neurochem 57:675–684
Schlaepfer WW, Zimmerman V-JP (1981) Calcium-mediated breakdown of glial filaments and neurofilaments in rat optic nerve and spinal cord. Neurochem Res 6:243–255
Schliwa M (1986) The cytoskeleton: an introductory survey. cell biology monographs, vol 13. Springer, New York, NY
Segovia J, Vergara P, Brenner M (1998) Astrocyte-specific expression of tyrosine hydroxylase after intracerebral gene transfer induces behavioral recovery in experimental Parkinsonism. Gene Therapy 5:1650–1655
Sensenbrenner M, Devilliers G, Bock E, Porte A (1980) Biochemical and ultrastructural studies of cultured rat astroglial cells: effect of brain extract and dibutyryl cyclic AMP on glial fibrillary acidic protein and glial filaments. Differentiation 17:51–61
Sharma N, Norman,-O'Guin K, Shafit-Zagardo B (1991) Phorbol-12-myristate-13-acetate (PMA) and inhibitors of protein kinase C alter glial fibrillary acidic protein (GFAP) mRNA levels. Glia 4:572–579
Shibuki K, Gomi H, Chen L, Bao S, Kim JJ, Wakatsuki H, Fujisaki T, Fujimoto K, Katoh A, Ikeda T, Chen C, Thompson RF, Itohara S (1996) Deficient cerebellar long-term depression, impaired eyeblink conditioning, and normal motor coordination in GFAP mutant mice. Neuron 16:587–599
Shiroma N, Kanazawa N, Izumi M, Sugai K, Fukumizu M, Sasaki M, Hanaoka S, Kaga M, Tsujino S (2001) Diagnosis of Alexander disease in a Japanes patient by molecular genetic analysis. J Hum Genet 46:579–582
Smith JD, Sikes J, Levin AJ (1998) Human apolipoprotein E allele-specific grain expressing transgenic mice. Neurobiol Aging 19:407–413
Smith ME, Somera FP, Eng LF (1983) Immunocytochemical staining for glial fibrillary acidic protein and the metabolism of cytoskeletal proteins in experimental allergic encephalomyelitis. Brain Res 264:241–253
Steward M, Quinlan RA, Moir RD (1989) Molecular interactions in paracrystals of a fragment corresponding to the alpha-helical coiled-coil rod portion of glial fibrillary acidic protein: evidence for an antiparallel packing of molecules and polymorphism related to intermediate filament structure. J Cell Biol 109:225–234
Su M, Hu H, Lee Y, d'Azzo A, Messing A, Brenner M (2004) Expression specificity of GFAP transgenes. Neurochem Res 29:2075–2093
Sun Y, Wu S, Bu G, Onifade MK, Patel SN, LaDu MJ, Fagan AM, Holtzman DM (1998) Glial fibrillary acidic protein-apolipoprotein E (apoE) transgenic. J Neurosci 18:3261–3272
Takemura M, Nishiyama H, Itohara S (2002) Distribution of phosphorylated glial fibrillary acidic protein in the mouse central nervous system. Genes Cells 7:295–307
Tanaka H, Katoh A, Oguro K, Shimazaki K, Gomi H, Itohara S, Masuzawa T, Kawai N (2002) Disturbances of hippocampal long-term potentiation after transient ischemia in GFAP deficient mice. J Neurosci Res 67:11–20

Tatzelt J, Maeda N, Pekny M, Yang SL, Betsholtz C, Eliasson C, Cayetano J, Camerino AP, DeArmond SJ, Prusiner SB (1996) Scrapie in mice deficient in apolipoprotein E or glial fibrillary acidic protein. Neurol 47:449–453

Tomokane N, Iwaki T, Tateishi J, Iwaki A, Goldman JE (1991) Rosenthal fibers share epitopes with aB-crystallin, glial fibrillary acidic protein and ubiquitin, but not with vimentin: immunoelectron microscopy with colloidal gold. Am J Pathol 138:875–885

Traub P (1985) Intermediate filaments: a review. Springer, New York, NY

Traub P, Vorgias CE (1983) Involvement of the N-terminal polypeptide of vimentin in the formation of intermediate filaments. J Cell Sci 63:43–67

Trejo F, Vergara P, Brenner M, Segovia J (1999) Gene therapy in a rodent model of Parkinson's disease using differentiated C6 cells expressing a GFAP-tyrosine hydroxylase transgene. Life Sciences 65:483–491

Triolo D, Dina G, Lorenzetti I, Malaguti M, Morana P, Del Carro U, Comi G, Messing A, Quattrini A,, Previtali SC (2006) Loss of glial fibrillary acidic protein (GFAP) impairs Schwann cell proliferation and delays nerve regeneration after damage. J Cell Sci 119:3981–3993

Uyeda CT, Eng LF, Bignami A (1972) Immunological study of the glial fibrillary acidic protein. Brain Res 37:81–89

Van der Knaap MS, Naidu S, Breiter SN, Blaser S, Strain KH, Springer S, Begeer JC, Van Coster R, Barth PG, Thomas NH, Valk J, Powers JM (2001) Alexander disease: diagnosis with MR imaging. Am J Neuroradiol. March, 22(3): 541–52

Wang X, Messing A, David S (1997) Axonal and non-neuronal cell responses to spinal cord injury in mice lacking glial fibrillary acidic protein. Expt Neurol 148:568–576

Weber K, Geisler N (1985) Intermediate filaments: structural conservation and divergence. Ann N Y Acad Sci 455:126–143

Weinstein DE, Shelanski ML, Liem RK (1991) Suppression by antisense mRNA demonstrates a requirement for the glial fibrillary acidic protein in the formation of stable astrocytic processes in response to neurons. J Cell Biol 112:1205–1213

Wisniewski T, Goldman JE (1998) aB-crystallin is associated with intermediate filaments in astrocytoma cells. Neurochem Res 23:385–392

Xue ZG, Cheraud Y, Brocheriou V, Izmiryan A, Titeux M, Paulin D, Li Z (2004) The mouse synemin gene encodes three intermediate filament proteins generated by alternative exon usage and different open reading frames. Exp Cell Res 298:431–444

Yang ZW, Babitch JA (1988) Factors modulating filament formation by bovine glial fibrillary acidic protein, the intermediate filament component of astroglial cells. Biochemistry 27:7038–7045

Yang ZW, Kong CF, Babitch JA (1988) Characterization and location of divalent cation binding sites in bovine glial fibrillary acidic protein. Biochemistry 27:7045–7050

Yao CP, Allen JW, Conklin DR, Aschner M (1999) Transfection and overexpression of metallothionen-I in neonatal rat primary astrocyte cultures and in astrocytoma cells increases their resistance to methylmercury-induced cytotoxicity. Brain Res 818:414–420

Yu AC, Lee YL, Eng LF (1991) Inhibition of GFAP synthesis by antisense RNA in astrocytes. J Neurosci Res 30:72–79

Yu AC, Lee YL, Eng LF (1993) Astrogliosis in culture. I. The model and the effect of antisense oligonucleotides on glial fibrillary acidic protein synthesis. J Neurosci Res 34:295–303

Zelenika D, Grima B, Brenner M, Pessac B (1995) A novel glial fibrillary acidic protein messenger-rna lacking exon-1, Mol. Brain Res 30:251–258

Chapter 21
Axonal Transport Mechanisms in Cytoskeleton Formation and Regulation

Aidong Yuan and Ralph A. Nixon

Abstract Most axonal and synaptic proteins are synthesized within the cell body and must travel long distances along axons to reach their sites of function. It is widely accepted that membranous organelles are bidirectionally transported by motor proteins, including kinesins and dynein, at fast rates (50–400 mm/day) along axonal microtubules. The mechanisms underlying slow axonal transport of proteins, including components of the cytoskeleton, have been more elusive, although recent genetic and live-cell imaging approaches have yielded general principles about the dynamic behaviors of cytoskeletal elements and formation of the axonal cytoskeleton. Cytoskeletal components may undergo slow transport (0.1–10 mm/day in vivo) as short polymers or oligomeric assemblies of subunits that fully assemble during transport or after they incorporate into the axonal cytoskeleton. Slow transport rates for a particular cargo reflect a pattern of rapid movements along axons interrupted by pauses of varying short durations. In mature myelinated axons, proteins in transport represent a relatively small pool that serve as precursors to a large metabolically stable stationary cytoskeleton composed of neurofilaments networked together by cross bridging proteins along with microtubule and actin filaments. Cytoskeletal components may turn over by local subunit exchange or proteolysis or, alternatively, by detachment of a larger filament fragment that is translocated to a degradative site. Defects in specific aspects of this complex process are the basis for certain neurological disorders.

Keywords Axonal transport · Slow axonal transport · Fast axonal transport · Cytoskeleton · Cytoskeletal protein · Neurofilament · Microtubule · Actin filament

A. Yuan (✉)
Nathan Kline Institute, Center for Dementia Research, New York University School of Medicine, Orangeburg, NY 10962, USA; Department of Psychiatry, New York University School of Medicine, New York, NY 10016, USA
e-mail: yuan@nki.rfmh.org

R.A. Nixon, A. Yuan (eds.), *Cytoskeleton of the Nervous System*,
Advances in Neurobiology 3, DOI 10.1007/978-1-4419-6787-9_21,

Contents

21.1 General Introduction

The movement of materials along the axon from the cell body of the neuron to its axon terminal is referred to as axonal (axoplasmic) transport. In a large neuron, the axon can occupy more than 99% of the cell's volume (Mitsumoto, 1987; Pannese, 1994). Since most of the axonal proteins required to support its structure and function are synthesized in the cell body (Alvarez and Benech, 1983; Eng et al., 1999; Campenot and Eng, 2000; Lee and Hollenbeck, 2003), the bulk synthesis, bidirectional transport, and local assembly of these proteins represent an enormous metabolic energy investment, unique to neurons among all cell types.

Axoplasmic and synaptic constituents move along axons in different states of assembly and at different rates that are generally categorized into three phases of movement: fast or slow anterograde transport and retrograde transport. During the past two decades, the mechanisms underlying fast axonal transport have steadily become elucidated providing insight into the dynamic behaviors of membranous organelles – the main cargoes of this transport phase (Okada et al., 1995). By contrast, slow transport has been more difficult to decipher, although amid considerable controversy over many years, consensus is finally emerging about its underlying mechanisms and what they imply about the structural organization and stability of the axon in health and disease. In this chapter we review the historical development of our understanding of axonal transport and the recent progress made toward solving longstanding puzzles about the assembly forms of slowly transported cytoskeletal elements during transport, the behaviors of the fully assembled axonal cytoskeleton, and the maintenance of the axon by a continuous supply of transported precursors. Various critical reviews have appeared summarizing the state of the field during particular epochs and containing many worthwhile details that cannot be included in this overview (Ochs, 1975a; Grafstein and Forman, 1980; Lasek et al., 1984; Vallee and Bloom, 1991; Hirokawa, 1993; Nixon, 1998a; Brown, 2000; Shah and Cleveland, 2002; Yuan, 2007).

21.2 Historical Development of Axonal Transport Concepts

Augustus Volney Waller, a British neurophysiologist, was the first to suggest the existence of axonal transport between axons and their cell bodies in 1850 (Waller, 1850). Waller described the degenerative changes of frog glossopharyngeal and hypoglossal nerves cut from their cell bodies, showing that axon maintenance and integrity are completely dependent upon the cell body. When an axon is cut, it loses its physiological excitability, resulting in the acute degeneration of the portion distal to the cut, a process known as Wallerian degeneration. Santiago Ramon y Cajal, a Spanish pathologist and physician, observed in sciatic nerve fibers that bulbous enlargements or neuromas form just proximal to sites of experimental ligation, indicating blockage of axonal transport phenomenon (Ramon y Cajar, 1928). Weiss and Hiscoe first demonstrated the existence of axonal transport. In their classic paper "Experiments on the mechanism of nerve growth" (Weiss and Hiscoe, 1948), they showed that axoplasmic materials accumulate just proximal to a point of axonal constriction and, when the nerve constriction is removed, this bolus of materials moves down the axons at a rate of 1–2 mm/day. They suggested that the axoplasm advances as a coherent column that "buckles" when it encounters a reduction in the size of the channel through which it is moving. These experiments led to the view that axon growth, in the sense of production of new axoplasm, occurs solely in the cell body and that material continuously moves slowly from the cell body toward the axon terminals, even in normal mature nerve cells.

In the 1960s, Lubinska and colleagues (Lubinska et al., 1964) identified a much faster rate of axonal transport. After placing two ligatures at varying distances apart along the sciatic nerve, they assayed the content of acetylcholinesterase in segments of the dissected nerve, finding that the enzyme rapidly accumulated over time in the proximal and distal segments adjacent to both ligatures. This important experiment revealed two of the most fundamental properties of fast axonal transport in nerves: namely, that it is bidirectional and that its underlying propulsive mechanism is distributed throughout the whole axon and can proceed independently of the cell body for some period of time.

During the next three decades, researchers used isotopic amino acids microinjected in the vicinity of the nerve cell bodies to radiolabel axonal and synaptic proteins (Ochs and Burger, 1958; Miani, 1960; Droz and Leblond, 1962, 1963; Austin et al., 1966; Grafstein, 1967; Lasek, 1967; Ochs et al., 1967; Bray and Austin, 1968; Lasek, 1968b; Droz and Barondes, 1969; Grafstein et al., 1970; James et al., 1970; Bray et al., 1971; Karlsson and Sjostrand, 1971; Kirkpatrick et al., 1972). These labeled neuronal proteins could be monitored during axonal transport in vivo by analyzing nerves at different intervals after labeling by dissecting the nerves into consecutive segments, subjecting them to one- or two-dimensional gel electrophoresis and autoradiography, and quantifying the distances traversed along the nerves by each labeled protein (e.g., see Fig. 21.1).

Pulse-labeling experiments identified a fast component of anterograde transport (50–400 mm/day) and two general components of slow axonal transport (Lasek, 1968a; Karlsson and Sjostrand, 1971; Willard et al., 1974; Willard and Hulebak,

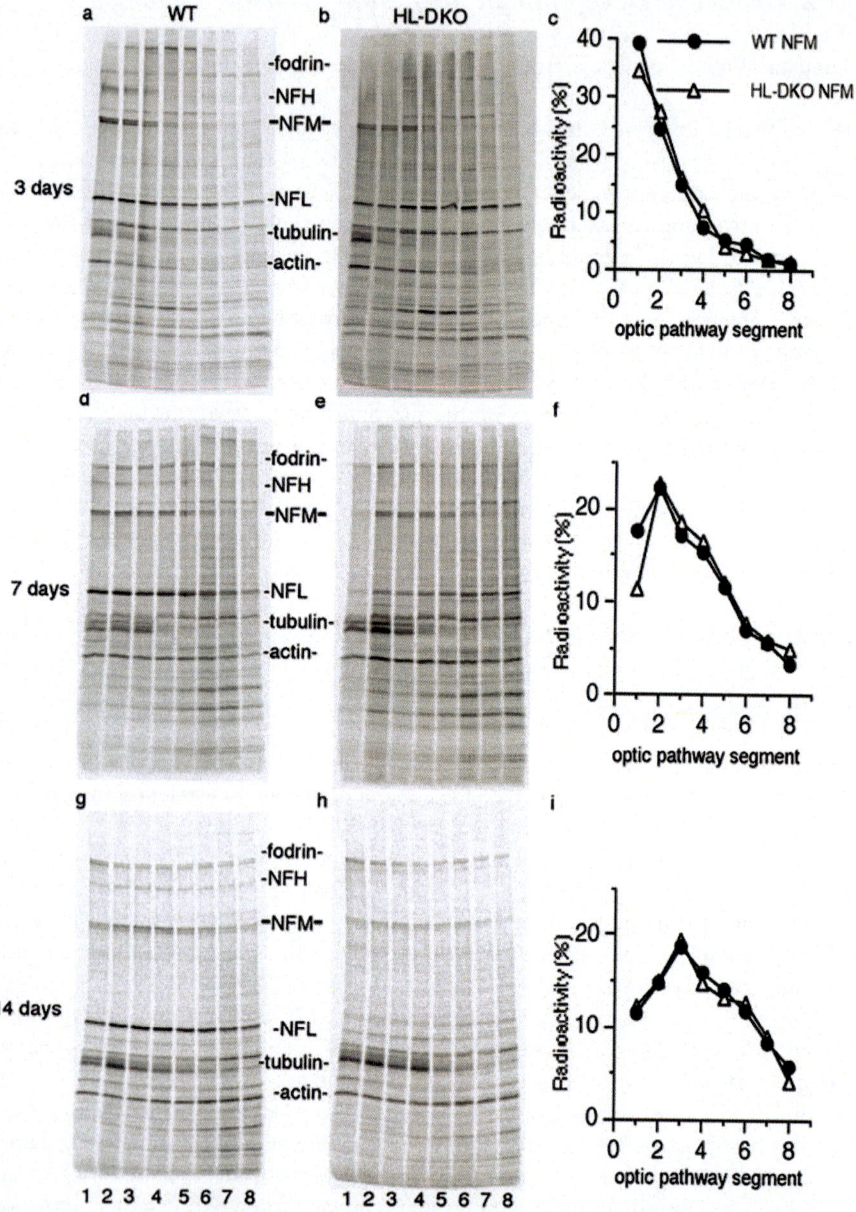

Fig. 21.1 A majority of newly synthesized NFM is translocated at a normal slow transport rate along optic axons in the absence of NFL and NFH and assembled neurofilaments. Similar rates of slow axonal transport are observed in wild-type mice (**a**, **d**, **g**) and HL-DKO mice that lack NFL and NFH (**b**, **e**, **h**) mice receiving intravitreal injection of radiolabeled ^{35}S-methionine. Changes in the distribution of radiolabeled cytoskeletal proteins, including NFM (**c**, **f**, **i**) at consecutive levels along the optic pathway is shown by autoradiography at 3–14 days after injection. Reproduced by permission from Yuan et al. (2003)

1977; Lasek, 1986). Lasek and colleagues classified these slower moving constituents as slow component a (SCa) and slow component b (SCb) (Black and Lasek, 1979; Lasek et al., 1984). SCa, moving at the slower rate (0.1–2 mm/day), has a relatively simple protein composition containing mainly neurofilament triplet proteins and tubulin. SCb, moving two to five times faster (2–10 mm/day), has a much more complex composition that includes more than 200 proteins (Lasek et al., 1984). Further subclassifications of both slow and fast transport groups based on more subtle rate differences and the composition of the materials transported (Willard et al., 1974; Willard, 1977; Willard and Hulebak, 1977; Lorenz and Willard, 1978) highlighted the important points that the range of rates observed for any given transported protein can overlap into several different transport categories and may also vary among different types of nerve fibers and different species (Nixon, 1998a). The detection of proteins moving at the same rate helped to identify subunits of a single organellar structure, such as subunits of the neurofilament; however, the early belief that groups of proteins moving in SCa or SCb represented a preassembled cytoskeleton (e.g., neurofilaments, microtubules, and actin) and an axoplasmic "matrix," respectively, proved not to be valid: for example, neurofilaments, tubulins/microtubules, and their associated proteins may interact in axons but each group is transported as independent structures or oligomeric assemblies at somewhat different rates (Mercken et al., 1995; Yuan et al., 2000, 2006).

21.3 Fast Axonal Transport

Fast axonal transport is the process by which membranous organelles are transported anterogradely or retrogradely along axons (Grafstein and Forman, 1980; Allen et al., 1982; Brady et al., 1982; Hurd and Saxton, 1996; Goodson et al., 1997; Martin et al., 1999). Fast anterograde transport supplies axolemma and synapse with materials synthesized in the cell body, including synaptic vesicles, vesicles containing axolemmal precursors, mRNA and mitochondria, which are also retrogradely transported (Grafstein and Forman, 1980; Sotelo-Silveira et al., 2006). Fast anterograde transport is powered by motor proteins and is vital for the growth and survival of axon and synapse. A diversity of materials move in fast anterograde transport, including enzymes, ion channels, neuropeptides, neurotransmitters, and lipids. A pre-assembled complex of synaptic vesicle proteins, calcium channels, endocytotic machinery, and large dense-core synaptic vesicles has been observed to transport as a unit at a fast rate (Ahmari et al., 2000) while others like vacuolar H^+-ATPase are transported as individual subunit complexes in axons, which assemble in nerve endings (Morel et al., 1998).

Retrograde transport mainly carries vesicular cargoes involved in signaling from postsynaptic terminals on, for example, signaling endosomes (Cui et al., 2007). In addition, obsolete proteins and membranes captured by endocytosis or autophagy are transported retrogradely back to the cell body for processing or degradation (Tsukita and Ishikawa, 1980; Hollenbeck, 1993). Although myosin motors can move these cargoes over short distances (1–5 μm) on actin filaments (Kuznetsov et al., 1992; Langford et al., 1994), microtubules (10–100 μm in length) are the principal tracks for both long-range fast anterograde and retrograde transport.

Retrograde axonal transport comprises two rates. Fast retrograde transport approaches the speed of fast anterograde movements. This mechanism conveys vesicular organelles, lysosomes, mitochondria, autophagic vacuoles, and endosomes and is the route by which toxins, such as tetanus toxin, and herpes and polio viruses, enter the central nervous system via an endocytic/phagocytic process (LaVail and LaVail, 1972; Bisby, 1982; Lasek et al., 1984; DiStefano et al., 1992; Sodeik et al., 1997; Cui et al., 2007). Fast retrograde transport of exogenous substances captured by endocytosis, such as horseradish peroxidase, injected into regions containing nerve terminals has been widely used for tracing neural pathways and is a possible method for gene delivery used in gene therapy for certain CNS neurological diseases (Nauta et al., 1974; Kuo et al., 1995; Mazarakis et al., 2001). Recycled cytoskeletal proteins and other materials like calmodulin (Cena et al., 1984) and serum albumin (Gainer and Fink, 1982) may also undergo slow retrograde transport (Glass and Griffin, 1994). For the rate components of axonal transport, please see Table 21.1.

Table 21.1 Representative constituents within rate components of axonal transport

Rate components	Average rates (mm/day)	Sample proteins identified	References
Fast			
Anterograde	50–400	Snap25	Yuan et al. (2008)
		Vacuolar H^+-ATPase V_0	Morel et al. (1998)
Retrograde	60–200	NGF	Cui et al. (2007)
Slow			
Anterograde			
SCb	2–10	Actin	Mills et al. (1996)
		Cofilin	Mills et al. (1996)
		Profilin	Mills et al. (1996)
		Actin depolymerizing factor	Mills et al. (1996)
		Cyclophilin A	Yuan et al. (1997)
		GAPDH	Yuan et al. (1999)
		Chaperonin CCT	Bourke et al. (2002)
		Hsc73	Bourke et al. (2002)
SCa	0.1–2	NFL, NFM, NFH	Yuan et al. (2003)
		Alpha-internexin	Yuan et al. (2006)
		Peripherin	Chadan et al. (1994)
		Tubulin	Yuan et al. (2006)
Retrograde	0.3–5	Neurofilament proteins, tubulin	Glass and Griffin (1994)
		Albumin	Gainer and Fink (1982)
		Calmodulin	Cena et al. (1984)

21.3.1 Microtubule-Based Transport Mechanisms

The squid giant axon, which was used so successfully to elucidate the biophysical properties of excitable membranes (Hodgkin and Huxley, 1939; Hodgkin, 1976), was also discovered to be an ideal preparation for investigating axonal transport in the early 1980s (Allen et al., 1982). With the development of video-enhanced contrast-differential interference contrast (AVEC-DIC) light microscopy to visualize the extruded axoplasm, organelles could be observed to move rapidly in both the anterograde and retrograde directions (Allen et al., 1982; Brady et al., 1982) on microtubules (Schnapp et al., 1985) in an ATP-dependent manner (Vale et al., 1985b).

Answers to how this organelle translocation is powered and its direction specified came with the characterization of microtubule-based motor proteins (kinesins and cytoplasmic dyneins) and the direct analysis of bidirectional organelle movement on microtubules using AVEC-DIC technology (Brady, 1985; Vale et al., 1985a; Paschal et al., 1987; Schroer et al., 1989; Aizawa et al., 1992). Most known motor proteins move along microtubules in only one direction – either toward the plus end or toward the minus end. This directionality can be analyzed in vitro with polystyrene beads coated with the motor protein and microtubules that have been polymerized on centrosomes. Because the microtubules in such arrays point their plus ends outward, the direction of movement can be readily determined with a light microscope. Beads coated with kinesin move only toward the plus end of the microtubules, while those coated with cytoplasmic dynein move toward the minus ends. The basic findings using this in vitro preparation were later confirmed in studies on intact nerve axons in vivo (Sheetz and Martenson, 1991; Cyr and Brady, 1992; Schroer, 1992; Nangaku et al., 1994; Noda et al., 1995; Okada et al., 1995; Yamazaki et al., 1995; Hirokawa, 1998). Studies in the intact nerves in vivo showed that kinesin binds to anterogradely transported organelles accumulating proximal to a constriction (Dahlstrom et al., 1991; Elluru et al., 1995), while dynein associates with both retrograde cargoes and those moving anterogradely (Hirokawa et al., 1990). Interestingly, only about 15% of dynein in axons is associated with the membranous organelles of anterograde fast transport in pulse-labeling experiments, while nearly 80% of anterogradely moving dynein associates with slow transport (Dillman et al., 1996b, a).

21.3.2 Actin Filament-Based Transport

A few studies have suggested a role for actin filament-based motility in local axonal organelle transport. Some eucaryotic cells have well-developed systems of actomyosin-based organelle transport, such as the giant alga Nitella, which was one of the first systems established for studies of in vitro organelle motility (Sheetz and Spudich, 1983). Microtubule-independent organelle movements seen in extruded axoplasm are also strongly indicative of actomyosin-based motility (Kuznetsov et al., 1992). In excised rat sciatic nerves associated with their dorsal root ganglia axonal organelle transport continues for hours after disrupting microtubules with high concentrations of calcium (Brady et al., 1980) and is partially inhibited by agents that specifically disrupt actin filaments (Goldberg et al., 1980; Goldberg, 1982; Brady et al., 1984). Isolated axoplasmic organelles also have myosin on their surfaces and can move toward the barbed end of actin filaments in vitro (Bearer et al., 1993; Kuznetsov et al., 1994; Langford et al., 1994; Bearer et al., 1996). Short-range organelle transport along actin filaments has been observed in squid giant axons (Tabb et al., 1998; Langford, 2002; Bridgman, 2004), cultured neurons of rat cervical ganglia (Evans and Bridgman, 1995), and chick sympathetic ganglia (Morris and Hollenbeck, 1995).

Neuronal tissue is rich in actin (Fine and Bray, 1971) and myosin (Burridge and Bray, 1975), and several myosin isotypes have been identified in neurons, including myosin I (Li and Chantler, 1992), myosin II (Sun and Chantler, 1991, 1992; Li et al., 1994; Mochida et al., 1994), myosin V (Espreafico et al., 1992; Cheney et al., 1993; Coling et al., 1997; Rao et al., 2002; Lewis et al., 2009), and myosin VI (Hasson and Mooseker, 1994). While it has been suggested that all of these could be involved in neuronal organelle transport (Mooseker, 1993; Mooseker and Cheney, 1995; Prekeris and Terrian, 1997), evidence for long distance organelle movement mediated by actin-myosin systems(s) is still lacking in mammalian nerves. Instead, actin filament-based organelle movement appears to be mainly a mechanism to mediate short-range vesicular translocations, which support synaptic transmission (Correia et al., 2008) or contribute to organizing organelles into functional arrays within the axoplasm (Rao et al. submitted). Because organelle diffusion within cytoplasm is very limited (Hou et al., 1990; Provance et al., 1993), myosin movements on actin filaments may move organelles that dissociate from microtubules back onto these tracks or shuttle them to other local destinations or to the plasma membrane (Morris and Hollenbeck, 1995; Alami et al., 2009).

21.4 The Evolution of Axonal Transport Models

Current thinking on the fundamental properties of axonal constituents and their counterparts undergoing transport has evolved from two different hypotheses advanced by Ochs and colleagues (Ochs, 1975a, b; Stromska and Ochs, 1981; Ochs et al., 1989) and Lasek and colleagues (Black and Lasek, 1980; Lasek et al., 1984; Lasek, 1986).

To account for different rates of axonal transport, Ochs and colleagues advanced a "unitary hypothesis" in which he proposed that transported cargoes have different affinities for a single motor mechanism. The range of transport rates for a given constituent are modulated by rates of attachment and detachment of this cargo from a putative motor and also by steric hindrance from, or interaction with, stationary structures along the axon (Ochs, 1975a; Ochs et al., 1989). Cargoes that are slowly transported dissociate more readily or frequently from the motor mechanism than those that are rapidly transported. This is facilitated, in some cases, by an affinity for stationary axonal structures. In support for this model, Ochs observed that radiolabeled cargoes were recovered mainly in soluble fractions rather than in the Triton-insoluble fraction containing cytoskeletal polymers (Ochs et al., 1967).

Moreover, when nerve fibers were induced to 'bead' by stretching, transported radiolabeled proteins located mainly over the non-constricted (bead) regions of the cytoplasm containing axoplasm, while little or no radioactivity was present over the constricted portions where the axonal cytoskeletal polymers were concentrated (Ochs et al., 1989). These observations suggested that cargoes moved mainly as subunits or small soluble polymers that assembled into preexisting immobile axonal structures, also implying that these transported constituents are a minor proportion of the total pool of these constituents in axons. Various experimental approaches later supported some of the key elements of the unitary hypothesis (for reviews, see Hirokawa et al., 1997; Nixon, 1998a; Galbraith and Gallant, 2000; Shah and Cleveland, 2002; Yuan, 2007; Brown, 2009).

Another set of findings led Lasek and colleagues to advance an alternative transport model (Lasek et al., 1984; Lasek, 1986). Observing that neurofilament and microtubule proteins move within the same slow component of transport and some, such as neurofilament proteins, move cohesively mainly as Triton X-100 insoluble complexes (Black et al., 1986), these investigators proposed that cytoskeletal proteins are preassembled in the cell body and are transported in the form of neurofilaments, microtubules, and actin filaments. Similarities in transport rates of SCa components, and separately of SCb cargoes, were further considered to reflect movement of distinct supermolecular complexes composed of multiple cytoskeletal elements. Later revisions to this model, termed the "structural hypothesis," were made in light of the enormous energy burden that would be required to translocate an assembled cytoskeleton superstructure. A revised model emphasized that individual neurofilaments, microtubules, and actin filaments move by sliding against each other, aided by one or more molecular motors, presumably distinct from those mediating fast transport (Black et al., 1986; Lasek, 1986; Cleveland and Hoffman, 1991).

The two different models provided contrasting perspectives on several critical questions about cytoskeleton dynamic behavior. First, do components of the axon move as subunits and/or oligomeric precursors or as preassembled cytoskeletal organelles? Second, is anterograde transport mainly mediated by a single transport motor or do separate motors exist for slow and fast anterograde transport. Finally, are the components undergoing transport a small "precursor" pool that maintains a large fixed stable axonal cytoskeletal lattice or are the components undergoing slow transport equivalent to the entire cytoskeleton of the axon as viewed ultrastructurally? In the latter case, in other words, is the *entire* cytoskeleton continuously moving, and being turned over, at the rate of slow transport? Significant progress has been made recently toward answering these questions.

21.5 Reconciling Axonal Transport Models

21.5.1 The State of Assembly of Neurofilament Proteins During Transport

The state of assembly of cytoskeletal proteins during axonal transport became an active area of investigation and debate during the 1980s and 1990s, with considerably indirect evidence being amassed to support movement of polymers (Baas and Brown, 1997) or subunit/oligomer assemblies (Terada et al., 1996; Hirokawa et al., 1997), although the possibility that this was not an "either–or" issue was also raised (Nixon, 1998b).

New technical approaches have ultimately clarified this issue for neurofilaments. By transfecting GFP-labeled NF subunits into developing sympathetic neurons in culture, Brown and colleagues were able to visualize directly movements of individual short filaments (1.0–15.8 μm in length) along the relatively NF-poor growing axons (Wang et al., 2000). These studies unequivocally established that some NF do translocate as polymers at least in developing neurons. Recent studies in mice lacking one or more NF subunit genes have shown, however, that transport of NF subunits does not necessarily require formation of neurofilaments. Yuan et al. demonstrated that the number of neurofilaments in optic axons of NFH/NFL double knockout mice is <10% of the number of neurofilaments in wild-type mouse axons and is unlikely to account for the 50% of the normal NFM subunit pool that still moves along optic axons at typical rates in the absence of NFH and NFL proteins (Yuan et al., 2003) (Fig. 21.1). These studies, which involved different combinations of NF subunit deletions, showed that the minimal requirement for transport is hetero-dimer formation between subunits and that the dimer must include either NFM or α-internexin (Fig. 21.1). NFM transport into axons, for example, is completely abolished by deleting α-internexin in the absence of NFL and NFH (Yuan et al., 2003, Yuan et al., unpublished data). These studies also explained why NFL and NFM subunits are detectable in developing rat optic axons before morphologically definable neurofilaments appear (Pachter and Liem, 1984). Movement of fluorescent puncta representing non-filamentous assemblies of GFP-labeled NF subunits have been reported (Yabe et al., 1999; Prahlad et al., 2000; Helfand et al., 2003; Theiss et al., 2005); however, short neurofilaments have also been observed in other types of cultured cells and in the squid giant axon depending on the methods used (Galbraith et al., 1999; Roy et al., 2000; Ackerley et al., 2003; Yan and Brown, 2005). Recent photobleaching analyses of GFP-tagged NFL proteins in cultured cortical neurons reconcile these different observations by demonstrating that both non-filamentous NF subunit assemblies and short NF polymers can be transported in the same axon (Yuan et al., 2009). At proximal axonal levels in these neurons, transport of subunit assemblies predominate while short NF predominate at distal levels of the same axon. These observations established that NF proteins may exist in multiple assembly forms during axonal transport and suggest that transported NF subunits are able to assemble into filaments during axonal transport (Yuan et al., 2009) (Fig. 21.2).

Fig. 21.2 (continued) (**B**). A 5 μm long window of fluorescent NF flanked by two photobleached areas was created. Then, images were recorded every 5 s for 30 min. (**D**) Densitometric analysis of EGFP-NFL levels during the recovery from photobleaching in a 5 μm long window containing fluorescent NF at proximal (*square*), intermediate (*diamond*), and distal regions (*circle*) of a circle (representative of five experiments). Reproduced from Yuan et al. (2009)

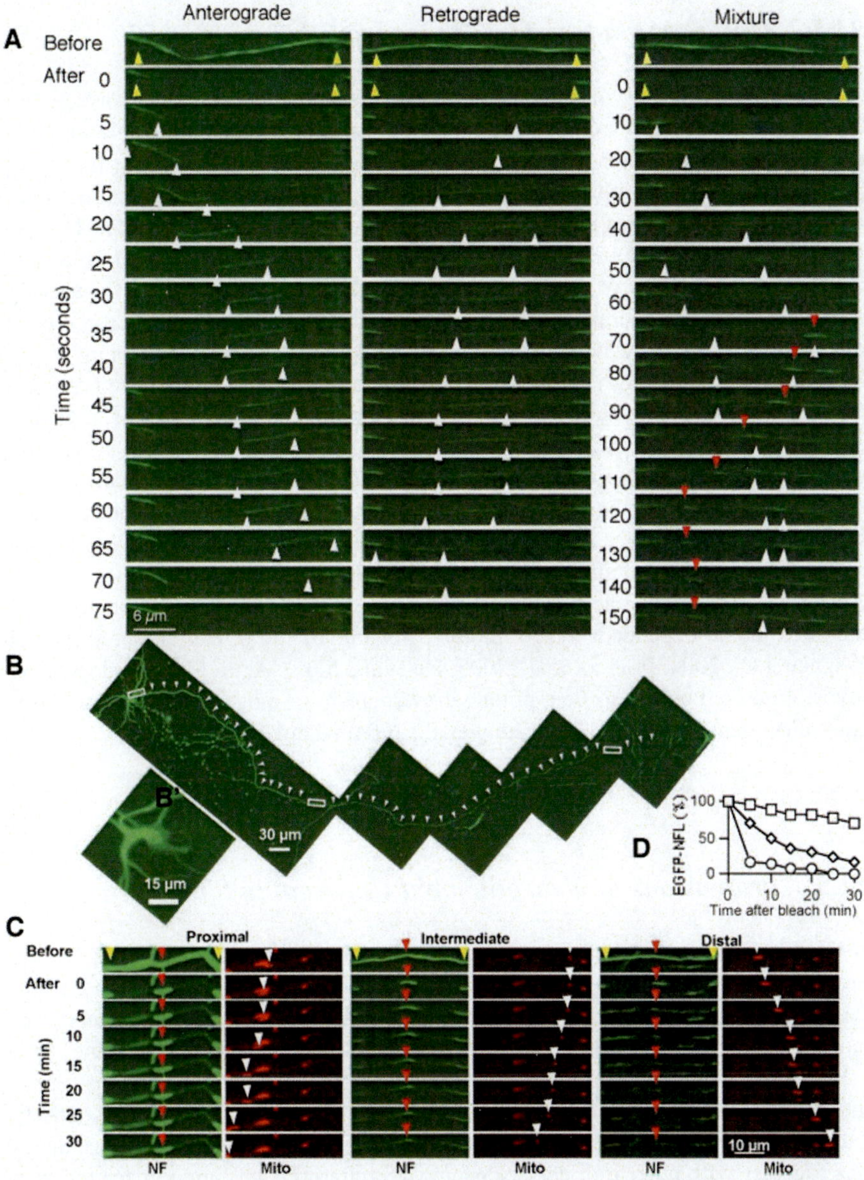

Fig. 21.2 (continued) Different NF dynamics within a single axon of a cultured cortical neuron. Cortical neurons transfected with EGFP-NFL and NFM were examined at 7 DIV, 3 DPT. The distal regions of different axons were examined by time-lapse imaging with 5-s intervals after photobleaching (**A**). (**B**) Distribution of EGFP-NFL in neuron transfected together with NFM. NF are distributed at least 1.5 mm along the length of this axon (*white arrowheads*). The magnified view of the cell body (**B'**) shows that NFs form a normal filamentous network, indicating that the level of NF expression yielded normal patterns of NF organization. (**C**) Differential dynamics of NFs was examined at proximal (*left panels*), intermediate (*center panels*), and distal (*right panels*) regions of the same axon indicated by the white rectangles in

21.5.2 Motors Mediating Slow Transport

A second longstanding question about cytoskeletal protein of slow transport has concerned the motors mediating the rates of slow axonal transport. The identification of a molecular motor with a stroke rate accounting for slow velocity has so far proved elusive. Important insight into this question, however, was gained in studies by Brown and colleagues, who demonstrated that fluorescent-labeled NFs do not move at a continuous slow rate but instead move intermittently at velocities approximating the in vivo rate of fast transport and frequently "pause" for varying short intervals. The average of these movements and pauses describes the typical rate of slow transport, which corresponds to the SCa wave of transport as measured in vivo by radiolabeling techniques. These important observations raised the distinct possibility that the family of fast transport motors, including kinesins, myosins, and dyneins, could potentially support movement of constituents of the slow component of axonal transport. Current available evidence suggests that microtubules and actin filaments are the long- and short-range tracks for neurofilament transport, respectively. Long-range anterograde transport of neurofilaments is probably propelled by the action of microtubule-dependent kinesin motors (Yabe et al., 1999; Xia et al., 2003) whereas the opposite retrograde direction may be powered by the microtubule-dependent dynein motors (Shah et al., 2000; Helfand et al., 2003; Wagner et al., 2004; Francis et al., 2005; He et al., 2005). Actin filament-dependent myosin motors could also be a potential mechanism to move neurofilaments over short distances (Rao et al., 2002; Jung et al., 2004; Alami et al., 2009).

21.5.3 Proportions of Neurofilament Undergoing Slow Transport

A third longstanding debate has concerned the proportions of axonal cytoskeletal proteins that are undergoing slow transport in axons (Nixon and Logvinenko, 1986; Lasek et al., 1992). The earliest in vivo pulse-labeling studies of cytoskeletal protein transport by Lasek and colleagues demonstrated the progressive advance of a "slow transport wave" along rat optic axons, implying, though not demonstrating, that, if longer survival times had been studied, the wave would have exited entirely from the axons at a rate predicted by the velocity of slow transport. These analyses, together with other indirect evidence, were interpreted as support for the idea that the labeled neurofilaments undergoing transport are the entire axonal cytoskeleton. The cytoskeleton was, therefore, in continuous slow motion and continually degraded at axonal terminals replaced by new cytoskeletal polymers at the proximal end as the cytoskeleton advanced and was degraded at nerve terminals. A different model, however, was advanced by Nixon and colleagues who carried out both shorter and exceptionally long-term (6 months) pulse-labeling studies in the mouse optic system. The relatively short length of this pathway in mice facilitated a test of whether newly synthesized neurofilament proteins clear completely from axons by axonal transport. These quantitative studies demonstrated that a significant proportion of the labeled NF never cleared from the axons over the long term of these experiments, but instead persisted in a non-uniform distribution along the length of the axon and turned over locally and very slowly (>75 day half-life). Because of their slow turnover relative to neurofilaments that are transported out of the axons, it could be determined that these stationary neurofilaments constituted a much larger proportion of the steady-state level of neurofilaments in axons than those that reach the axonal terminal in the slow transport wave. These studies supported the idea that neurofilaments (or oligomeric assemblies) undergoing slow transport in myelinated axons are a small precursor pool that maintains a preexisting large fixed NF lattice. This lattice exists within a complex stationary cytoskeletal network composed of the various cytoskeletal elements (NF, microtubules, actin filaments) seen in ultrastructural image of the cytoskeleton (Fig. 21.3). Later studies added support to this general model, including an important recent in vivo study by Julien and colleagues involving the behavior of existing neurofilaments in axons after NFL expression was acutely shut off in a genetic mouse model engineered to contain a conditionally regulated NFL gene (Millecamps et al., 2007). In these studies, the half-lives of preexisting NFL in axons were >2.5 months, similar to the most recent estimates obtained in pulse labeling studies in vivo (Yuan et al., 2009). The exceptionally slow rate of loss of neurofilaments established that NFs are present predominantly in axons as an immobile cytoskeletal network.

Although the foregoing studies strongly favored a stationary cytoskeleton in axons, Brown and colleagues in studies of fluorescently-labeled NF subunits in axons of cultured sympathetic neurons described a single population of NF in axons characterized by slow wave rates of transport, achieved by rapid movements interrupted by frequent pauses of varying duration (Brown et al., 2005). To reconcile the seemingly disparate pictures of NF dynamics from in vivo and in vitro studies,

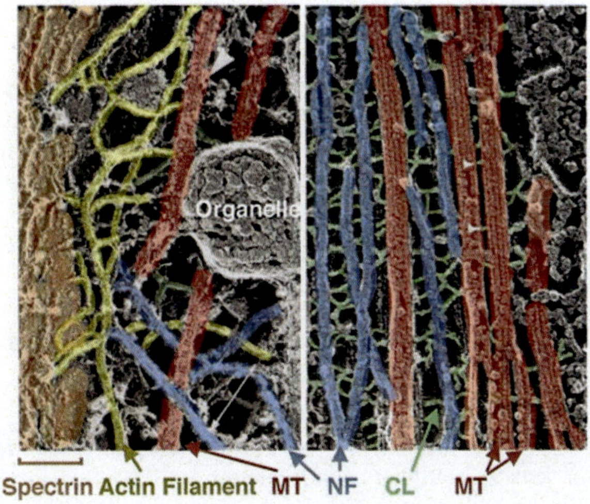

Fig. 21.3 Organization of the neuronal cytoskeleton network. Quick-frozen and rotary shadowed images of the neuronal cytoskeleton (adapted from Hirokawa, 1982 by copyright permission) reveal interconnected actin filaments, neurofilaments, and microtubules joined together by cross-linker proteins in the cell's cortical region (*left*) and in the deeper axoplasm (*right*). NF: neurofilament; MT: microtubule; CL: cross-linker

Yuan et al. recently analyzed EGPP-tagged NFL transport in primary cortical neurons (Yuan et al., 2009), which, in comparison to sympathetic neurons, achieve a more advanced state of axonal maturity. These studies showed unequivocally that primary cortical neurons form a distinct and substantial immobile NF network, coexisting with a smaller pool of moving NF. The findings accounted for the single moving population in sympathetic neurons that express low amounts of NF by showing that formation of a stable stationary NF network requires a critical level of NF, which is not achieved in relatively NF-poor developing neurons. In addition, the primary cortical neurons, unlike cultured sympathetic neurons, express all four CNS NF subunits, which, in turn, display the higher state of neurofilament protein phosphorylation seen at later stages of neuronal maturation. The findings in primary cortical neurons accord with in vivo studies (Nixon and Logvinenko, 1986; Yuan et al., 2009) that show during several months after labeling no net proximal to distal redistribution of the radiolabeled NFs left behind in the optic nerve after the slow transport wave had passed. Both in vivo and in vitro, the stationary NF network was observed to be highly stable metabolically. These findings in primary neurons were consistent with estimates that more than 90% of the NF in mature optic axons in vivo are stationary based on the amount of labeled NF retained in axons and their turnover rate.

The stability and stationary nature of the NF cytoskeleton is compatible with ultrastructural evidence in mature axons that NFs form extensive crosslinks with each other through their long C-terminal tail domains and through additional cross-bridging proteins that integrate the filaments into a network with other stationary cytoskeletal elements in axons (Nixon, 1998a; Perkins et al., 2008), such as microtubules (Lim et al., 1990; Okabe and Hirokawa, 1993) and the membrane skeleton (Frappier et al., 1991; Svitkina et al., 1996; Yang et al., 1996, 1999) (Fig. 21.3). NF proteins have binding sites for microtubule-associated proteins, such as tau (Miyata et al., 1986) and MAP 1 and 2 (Leterrier et al., 1982; Miyata et al., 1986; Frappier et al., 1991; Yang et al., 1999; Ma et al., 2000), which are believed to mediate reversible interactions between NFs and microtubules (Leterrier and Eyer, 1987). Moreover, spectrins bind NF proteins (Leterrier and Eyer, 1987; Frappier et al., 1991), actin (Frappier et al., 1992), and other cytoskeletal compartments of a stationary axolemma. In some neurons, BPAG (Yang et al., 1996) or plectin (Svitkina et al., 1996) link together NFs, microtubules, and actin filaments possibly tethering the cortical cytoskeleton to the stationary axolemma and membrane skeleton (Hirokawa, 1982).

21.5.4 Physiological Implications of a Stationary Neurofilament Network

A model envisioning relatively small transport components as providing precursors for a large stationary cytoskeleton as illustrated in Fig. 21.4 differs fundamentally from the single NF population transport model of Lasek and, more recently,

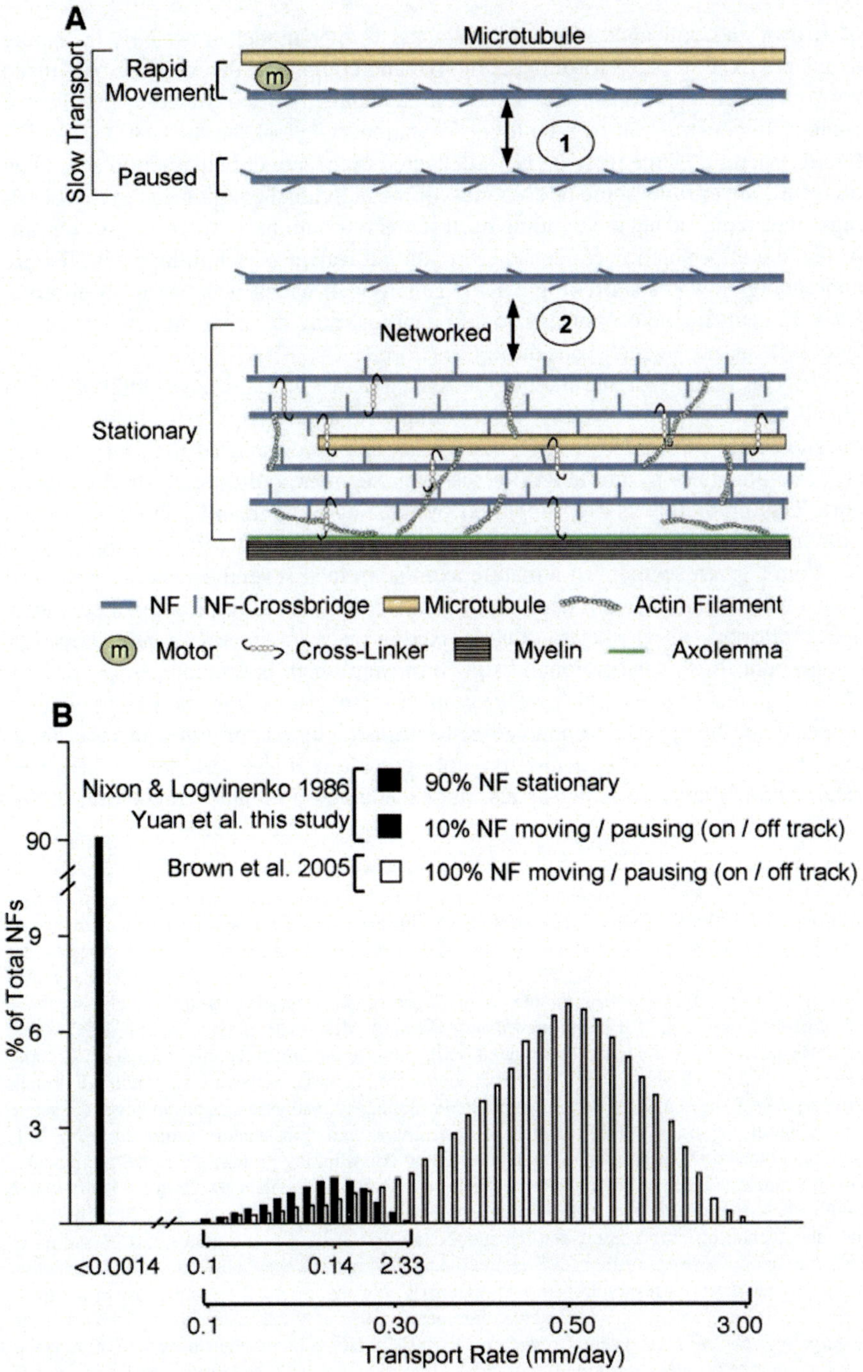

Fig. 21.4 (continued)

of Brown and colleagues. According to the former model, most NFs in mature axons are fixed in place for months in vivo and compose a distinctive cross-linked network in axons with other cytoskeletal elements. This can be compared by the analogy to construction of a building. A stationary cytoskeleton is built from different structural elements (e.g., NFs) delivered to the site of construction (i.e., slow transport). Over time, some or even most of the individual components of this building can be replaced but the building itself remains a stationary structure. By contrast, in the single population model, there is no equivalent of a building: all NFs are undergoing slow transport (Fig. 21.4). This model, which was based on observations on neurofilament-poor developing neurons, may describe the features of the cytoskeleton in growing axons during development.

The two models of the axonal cytoskeleton in Fig. 21.4 have different functional/mechanistic implications. For example, a model in which all NFs are in transport implies that NF in a segment of an axon are replaced (i.e., turned-over) at a rate governed by the length of the axon segment and the rate of slow transport. This prediction is not supported by existing experimental data (Nixon and Logvinenko, 1986; Millecamps et al., 2007; Yuan et al., 2009). The turnover of NF from a given segment of a mature axon is, in fact, several orders of magnitude slower than that predicted if neurofilaments in axons were a single population undergoing transport. A cytoskeleton that is fixed in place over many months, however, can account for this exceptionally slow turnover, which is a fundamental property of NF in axons. A stationary cytoskeleton can also account for many of the known properties of the cytoskeleton including its ultrastructural appearance in axons as an extensively crosslinked structure, the considerable regional variation in NF abundance and architecture along axons, the exceptionally complex regulation of NF

Fig. 21.4 (continued) Distinctions between models of NF behavior in axons. (**A**) Based on data in Yuan et al. (2009)and other studies referenced in the text, the formation and maintenance of the NF cytoskeleton in axons consists of two major processes. 1. Slow transport reflecting the net rate of NF translocation achieved by rapid movements, presumably involving a molecular motor moving on microtubules (*on track*), and "pauses" (*off track*) of varying duration, including those of longer duration (e.g., 1 h in cultured neurons recently described in Trivedi et al., 2007). This process corresponds to the transport model described by Brown and colleagues (Wang et al., 2000; Brown et al., 2005) 2. A second process that occurs seen in axons achieving sufficient NF protein expression (e.g., most mature axons) but not in developing sympathetic neurons involves the active integration of "off-track" NF into a distinct stationary axonal cytoskeleton composed of NF, MT, actin, and other skeletal proteins often connected by cross-linking proteins (e.g., BPAG, plectin). The diagram in (**A**) and a schematic representation of experimental data (**B**) from our study (*black* columns) and that of Brown et al. (*open columns*) (Brown et al., 2005; Trivedi et al., 2007) illustrate the fundamental differences between our model and that of Brown and colleagues. Based on in vivo data in optic axons, >90% of NF are stationary, defined as a rate below the limit of detection (<0.0014 mm/day), corresponding to the stationary network in (**A**). This stationary structure is maintained by a small pool of filaments/oligomers (<10% of total NF in optic axons) that are undergoing slow transport. In the Brown model, 100% of NFs move in the slow phase of transport at a broad range of rates averaging 0.5 mm/day (Trivedi et al., 2007) (*white* columns). This rate is governed by the balance of rapid (*on track*) movements and pauses, and most of the NF in slow transport may be pausing or off-tract at any given time. Reproduced from Yuan et al. (2009)

phosphorylation needed to coordinate cytoskeletal assembly, and the ability of NF to act as a stable scaffold for vesicles and membrane bound receptors (Rao et al., unpublished observations).

21.5.5 The State of Assembly of Tubulin and Actin During Transport

The kinetics of transport for other cytoskeletal proteins is increasingly believed to obey principles similar to those governing NF. In the case of tubulins, rapid movement of short microtubules (range 1.1–5.0 μm, average 2.7 μm) has been observed in growing axons of cultured sympathetic neurons using a modified fluorescence photobleaching combined with difference imaging (also called image subtraction) (Wang and Brown, 2002). However, subsequent studies from three independent groups using more sensitive methods—microtubule plus-end tracking and live cell imaging of GFP-tubulin movement in axons – all failed to demonstrate transport of axonal microtubules (Ma et al., 2004; Kimura et al., 2005; Kim and Chang, 2006). By contrast, only tubulin-containing structures termed tubulin parcels were observed to move along axon (Ma et al., 2004). These findings support the view that axonal microtubules are stationary. Thus, while transport of short microtubules may occur in some growing axons, tubulin is most likely transported primarily in the form of dimers or oligomers. In vivo analyses also suggest that actin is transported primarily in unpolymerized form together with actin monomer-binding proteins profilin, cofilin, and actin-depolymerizing factor (Mills et al., 1996). In cultured rat hippocampal neurons, movement of GFP labeled actin puncta was recently observed in non-growing axons (Colicos et al., 2001; Sankaranarayanan et al., 2003), but in growing axons, anterograde saltatory movement along axons of growth-cone-like structures containing actin filaments was seen, indicating that actin filaments may be translocated under certain conditions (Ruthel and Banker, 1998, 1999; Flynn et al., 2009). Collectively, the findings on kinetics of cytoskeletal protein transport underscore the importance of developmental state of neurons as a determinant of assembly form and other properties of slow axonal transport.

21.6 Conclusions and Future Perspectives

We have reviewed here a century of progress leading to our currently substantial, though still incomplete, understanding of cytoskeletal protein kinetics in axons. Although mechanisms of fast axonal transport were the first to be clarified, considerable progress has recently been made toward elucidating the molecular properties of slow transport. Evidence now indicates that aspects of both the unitary hypothesis and structural hypotheses, of transport advanced earlier, are correct. Based on current evidence, the cytoskeleton is viewed as a large stationary, metabolically stable structure that is assembled from transported elements. These elements can either be short polymers or oligomers. The proportions of these assembly forms may possibly vary depending on cell type and developmental state (Nixon and Shea, 1992), which is why, previously, a consensus was difficult to build among investigators studying different systems. Regardless of the assembly form that cytoskeletal proteins take during transport (dimer, oligomer, or short filament), they subsequently undergo additional steps of integration within a stationary cytoskeleton, which involves regulatory events that still remain to be fully explored. The complexity of the transport process and the subsequent integration steps provides a useful context for considering how disruption of these events could generate the characteristic patterns of degradation in various neurological disorders. Dysregulation of axonal transport, such as that arising when cytoskeletal proteins are hyper-phosphorylated, could lead to interruption or collapse of such a transport process and potentially contribute to neuropsychiatric disorders.

Acknowledgments We gratefully acknowledge the assistance of Nicole Piorkowski in preparing the manuscript for publication. Work cited from this laboratory has been supported by Grant AG05604 from the National Institutes on Aging.

References

Ackerley S, Thornhill P, Grierson AJ, Brownlees J, Anderton BH, Leigh PN, Shaw CE, Miller CC (2003) Neurofilament heavy chain side arm phosphorylation regulates axonal transport of neurofilaments. J Cell Biol 161:489–495

Ahmari SE, Buchanan J, Smith SJ (2000) Assembly of presynaptic active zones from cytoplasmic transport packets. Nat Neurosci 3:445–451

Aizawa H, Sekine Y, Takemura R, Zhang Z, Nangaku M, Hirokawa N (1992) Kinesin family in murine central nervous system. J Cell Biol 119:1287–1296

Alami NH, Jung P, Brown A (2009) Myosin Va increases the efficiency of neurofilament transport by decreasing the duration of long-term pauses. J Neurosci 29:6625–6634

Allen RD, Metuzals J, Tasaki I, Brady ST, Gilbert SP (1982) Fast axonal transport in squid giant axon. Science 218:1127–1129

Alvarez J, Benech CR (1983) Axoplasmic incorporation of amino acids in a myelinated fiber exceeds that of its soma: a radioautographic study. Exp Neurol 82:25–42

Austin L, Bray JJ, Young RJ (1966) Transport of proteins and ribonucleic acid along nerve axons. J Neurochem 13:1267–1269

Baas PW, Brown A (1997) Slow axonal transport: the polymer transport model. Trends Cell Biol 7:380–384

Bearer EL, DeGiorgis JA, Bodner RA, Kao AW, Reese TS (1993) Evidence for myosin motors on organelles in squid axoplasm. Proc Natl Acad Sci USA 90:11252–11256

Bearer EL, DeGiorgis JA, Medeiros NA, Reese TS (1996) Actin-based motility of isolated axoplasmic organelles. Cell Motil Cytoskeleton 33:106–114

Bisby MA (1982) Functions of retrograde axonal transport. Fed Proc 41:2307–2311

Black MM, Keyser P, Sobel E (1986) Interval between the synthesis and assembly of cytoskeletal proteins in cultured neurons. J Neurosci 6:1004–1012

Black MM, Lasek RJ (1979) Axonal transport of actin: slow component b is the principal source of actin for the axon. Brain Res 171:401–413

Black MM, Lasek RJ (1980) Slow components of axonal transport: two cytoskeletal networks. J Cell Biol 86:616–623

Bourke GJ, El Alami W, Wilson SJ, Yuan A, Roobol A, Carden MJ (2002) Slow axonal transport of the cytosolic chaperonin CCT with Hsc73 and actin in motor neurons. J Neurosci Res 68:29–35

Brady ST (1985) A novel brain ATPase with properties expected for the fast axonal transport motor. Nature 317:73–75

Brady ST, Crothers SD, Nosal C, McClure WO (1980) Fast axonal transport in the presence of high Ca^{2+}: evidence that microtubules are not required. Proc Natl Acad Sci USA 77:5909–5913

Brady ST, Lasek RJ, Allen RD (1982) Fast axonal transport in extruded axoplasm from squid giant axon. Science 218:1129–1131

Brady ST, Lasek RJ, Allen RD, Yin HL, Stossel TP (1984) Gelsolin inhibition of fast axonal transport indicates a requirement for actin microfilaments. Nature 310:56–58

Bray JJ, Austin L (1968) Flow of protein and ribonucleic acid in peripheral nerve. J Neurochem 15:731–740

Bray JJ, Kon CM, Breckenridge BM (1971) Reversed polarity of rapid axonal transport in chicken motoneurons. Brain Res 33:560–564

Bridgman PC (2004) Myosin-dependent transport in neurons. J Neurobiol 58:164–174

Brown A (2000) Slow axonal transport: stop and go traffic in the axon. Nat Rev Mol Cell Biol 1:153–156

Brown A (2009) Slow axonal transport. In: Squire L (ed) Encyclopedia of neuroscience. Elsevier, New York, NY, pp 1–9

Brown A, Wang L, Jung P (2005) Stochastic simulation of neurofilament transport in axons: the "stop-and-go" hypothesis. Mol Biol Cell 16:4243–4255

Burridge K, Bray D (1975) Purification and structural analysis of myosins from brain and other non-muscle tissues. J Mol Biol 99:1–14

Campenot RB, Eng H (2000) Protein synthesis in axons and its possible functions. J Neurocytol 29:793–798

Cena V, Garcia AG, Gonzalez-Garcia C, Kirpekar SM (1984) Orthograde and retrograde axonal transport of calmodulin in a cat noradrenergic neurone. Br J Pharmacol 82:143–149

Chadan S, Le Gall JY, Di Giamberardino L, Filliatreau G (1994) Axonal transport of type III intermediate filament protein peripherin in intact and regenerating motor axons of the rat sciatic nerve. J Neurosci Res 39:127–139

Cheney RE, O'Shea MK, Heuser JE, Coelho MV, Wolenski JS, Espreafico EM, Forscher P, Larson RE, Mooseker MS (1993) Brain myosin-V is a two-headed unconventional myosin with motor activity. Cell 75:13–23

Cleveland DW, Hoffman PN (1991) Slow axonal transport models come full circle: evidence that microtubule sliding mediates axon elongation and tubulin transport. Cell 67:453–456

Colicos MA, Collins BE, Sailor MJ, Goda Y (2001) Remodeling of synaptic actin induced by photoconductive stimulation. Cell 107:605–616

Coling DE, Espreafico EM, Kachar B (1997) Cellular distribution of myosin-V in the guinea pig cochlea. J Neurocytol 26:113–120

Correia SS, Bassani S, Brown TC, Lise MF, Backos DS, El-Husseini A, Passafaro M, Esteban JA (2008) Motor protein-dependent transport of AMPA receptors into spines during long-term potentiation. Nat Neurosci 11:457–466

Cui B, Wu C, Chen L, Ramirez A, Bearer EL, Li WP, Mobley WC, Chu S (2007) One at a time, live tracking of NGF axonal transport using quantum dots. Proc Natl Acad Sci USA 104: 13666–13671

Cyr JL, Brady ST (1992) Molecular motors in axonal transport. cellular and molecular biology of kinesin. Mol Neurobiol 6:137–155

Dahlstrom AB, Pfister KK, Brady ST (1991) The axonal transport motor 'kinesin' is bound to anterogradely transported organelles: quantitative cytofluorimetric studies of fast axonal transport in the rat. Acta Physiol Scand 141:469–476

Dillman JF 3rd, Dabney LP, Karki S, Paschal BM, Holzbaur EL, Pfister KK (1996b) Functional analysis of dynactin and cytoplasmic dynein in slow axonal transport. J Neurosci 16:6742–6752

Dillman JF 3rd, Dabney LP, Pfister KK (1996a) Cytoplasmic dynein is associated with slow axonal transport. Proc Natl Acad Sci USA 93:141–144

DiStefano PS, Friedman B, Radziejewski C, Alexander C, Boland P, Schick CM, Lindsay RM, Wiegand SJ (1992) The neurotrophins BDNF, NT-3, and NGF display distinct patterns of retrograde axonal transport in peripheral and central neurons. Neuron 8:983–993

Droz B, Barondes SM (1969) Nerve endings: rapid appearance of labeled protein shown by electron microscope radioautography. Science 165:1131–1133

Droz B, Leblond CP (1962) Migration of proteins along the axons of the sciatic nerve. Science 137:1047–1048

Droz B, Leblond CP (1963) Axonal migration of proteins in the central nervous system and peripheral nerves as shown by radioautography. J Comp Neurol 121:325–346

Elluru RG, Bloom GS, Brady ST (1995) Fast axonal transport of kinesin in the rat visual system: functionality of kinesin heavy chain isoforms. Mol Biol Cell 6:21–40

Eng H, Lund K, Campenot RB (1999) Synthesis of beta-tubulin, actin, and other proteins in axons of sympathetic neurons in compartmented cultures. J Neurosci 19:1–9

Espreafico EM, Cheney RE, Matteoli M, Nascimento AA, De Camilli PV, Larson RE, Mooseker MS (1992) Primary structure and cellular localization of chicken brain myosin-V (p190), an unconventional myosin with calmodulin light chains. J Cell Biol 119:1541–1557

Evans LL, Bridgman PC (1995) Particles move along actin filament bundles in nerve growth cones. Proc Natl Acad Sci USA 92:10954–10958

Fine RE, Bray D (1971) Actin in growing nerve cells. Nat New Biol 234:115–118

Flynn KC, Pak CW, Shaw AE, Bradke F, Bamburg JR (2009) Growth cone-like waves transport actin and promote axonogenesis and neurite branching. Dev Neurobiol 69:761–779

Francis F, Roy S, Brady ST, Black MM (2005) Transport of neurofilaments in growing axons requires microtubules but not actin filaments. J Neurosci Res 79:442–450

Frappier T, Derancourt J, Pradel LA (1992) Actin and neurofilament binding domain of brain spectrin beta subunit. Eur J Biochem 205:85–91

Frappier T, Stetzkowski-Marden F, Pradel LA (1991) Interaction domains of neurofilament light chain and brain spectrin. Biochem J 275(Pt 2):521–527

Gainer H, Fink DJ (1982) Evidence for slow retrograde transport of serum albumin in rat sciatic nerve. Brain Res 233:404–408

Galbraith JA, Gallant PE (2000) Axonal transport of tubulin and actin. J Neurocytol 29:889–911

Galbraith JA, Reese TS, Schlief ML, Gallant PE (1999) Slow transport of unpolymerized tubulin and polymerized neurofilament in the squid giant axon. Proc Natl Acad Sci USA 96: 11589–11594

Glass JD, Griffin JW (1994) Retrograde transport of radiolabeled cytoskeletal proteins in transected nerves. J Neurosci 14:3915–3921

Goldberg DJ (1982) Microinjection into an identified axon to study the mechanism of fast axonal transport. Proc Natl Acad Sci USA 79:4818–4822

Goldberg DJ, Harris DA, Lubit BW, Schwartz JH (1980) Analysis of the mechanism of fast axonal transport by intracellular injection of potentially inhibitory macromolecules: evidence for a possible role of actin filaments. Proc Natl Acad Sci USA 77:7448–7452

Goodson HV, Valetti C, Kreis TE (1997) Motors and membrane traffic. Curr Opin Cell Biol 9:18–28

Grafstein B (1967) Transport of protein by goldfish optic nerve fibers. Science 157:196–198

Grafstein B, Forman DS (1980) Intracellular transport in neurons. Physiol Rev 60:1167–1283

Grafstein B, McEwen BS, Shelanski ML (1970) Axonal transport of neurotubule protein. Nature 227:289–290

Hasson T, Mooseker MS (1994) Porcine myosin-VI: characterization of a new mammalian unconventional myosin. J Cell Biol 127:425–440

He Y, Francis F, Myers KA, Yu W, Black MM, Baas PW (2005) Role of cytoplasmic dynein in the axonal transport of microtubules and neurofilaments. J Cell Biol 168:697–703

Helfand BT, Loomis P, Yoon M, Goldman RD (2003) Rapid transport of neural intermediate filament protein. J Cell Sci 116:2345–2359

Hirokawa N (1982) Cross-linker system between neurofilaments, microtubules, and membranous organelles in frog axons revealed by the quick-freeze, deep-etching method. J Cell Biol 94: 129–142

Hirokawa N (1993) Axonal transport and the cytoskeleton. Curr Opin Neurobiol 3:724–731

Hirokawa N (1998) Kinesin and dynein superfamily proteins and the mechanism of organelle transport. Science 279:519–526

Hirokawa N, Sato-Yoshitake R, Yoshida T, Kawashima T (1990) Brain dynein (MAP1C) localizes on both anterogradely and retrogradely transported membranous organelles in vivo. J Cell Biol 111:1027–1037

Hirokawa N, Terada S, Funakoshi T, Tekeda S (1997) Slow axonal transport: the subunit transport model. Trends Cell Biol 7:384–388

Hodgkin AL (1976) Chance and design in electrophysiology: an informal account of certain experiments on nerve carried out between 1934 and 1952. J Physiol 263:1–21

Hodgkin AL, Huxley AF (1939) Action potentials recorded from inside a nerve fibres. Nature (Lond) 144:710–711

Hollenbeck PJ (1993) Products of endocytosis and autophagy are retrieved from axons by regulated retrograde organelle transport. J Cell Biol 121:305–315

Hou L, Lanni F, Luby-Phelps K (1990) Tracer diffusion in F-actin and Ficoll mixtures. toward a model for cytoplasm. Biophys J 58:31–43

Hurd DD, Saxton WM (1996) Kinesin mutations cause motor neuron disease phenotypes by disrupting fast axonal transport in Drosophila. Genetics 144:1075–1085

James KA, Bray JJ, Morgan IG, Austin L (1970) The effect of colchicine on the transport of axonal protein in the chicken. Biochem J 117:767–771

Jung C, Chylinski TM, Pimenta A, Ortiz D, Shea TB (2004) Neurofilament transport is dependent on actin and myosin. J Neurosci 24:9486–9496

Karlsson JO, Sjostrand J (1971) Synthesis, migration and turnover of protein in retinal ganglion cells. J Neurochem 18:749–767

Kim T, Chang S (2006) Quantitative evaluation of the mode of microtubule transport in Xenopus neurons. Mol Cells 21:76–81

Kimura T, Watanabe H, Iwamatsu A, Kaibuchi K (2005) Tubulin and CRMP-2 complex is transported via Kinesin-1. J Neurochem 93:1371–1382

Kirkpatrick JB, Bray JJ, Palmer SM (1972) Visualization of axoplasmic flow in vitro by Nomarski microscopy. comparison to rapid flow of radioactive proteins. Brain Res 43:1–10

Kuo H, Ingram DK, Crystal RG, Mastrangeli A (1995) Retrograde transfer of replication deficient recombinant adenovirus vector in the central nervous system for tracing studies. Brain Res 705:31–38

Kuznetsov SA, Langford GM, Weiss DG (1992) Actin-dependent organelle movement in squid axoplasm. Nature 356:722–725

Kuznetsov SA, Rivera DT, Severin FF, Weiss DG, Langford GM (1994) Movement of axoplasmic organelles on actin filaments from skeletal muscle. Cell Motil Cytoskeleton 28: 231–242

Langford GM (2002) Myosin-V, a versatile motor for short-range vesicle transport. Traffic 3: 859–865

Langford GM, Kuznetsov SA, Johnson D, Cohen DL, Weiss DG (1994) Movement of axoplasmic organelles on actin filaments assembled on acrosomal processes: evidence for a barbed-end-directed organelle motor. J Cell Sci 107(8):2291–2298

Lasek RJ (1967) Bidirectional transport of radioactively labelled axoplasmic components. Nature 216:1212–1214

Lasek R (1968a) Axoplasmic transport in cat dorsal root ganglion cells: as studied with [3-H]-L-leucine. Brain Res 7:360–377

Lasek RJ (1968b) Axoplasmic transport of labeled proteins in rat ventral motoneurons. Exp Neurol 21:41–51

Lasek RJ (1986) Polymer sliding in axons. J Cell Sci Suppl 5:161–179

Lasek RJ, Garner JA, Brady ST (1984) Axonal transport of the cytoplasmic matrix. J Cell Biol 99:212s–221s

Lasek RJ, Paggi P, Katz MJ (1992) Slow axonal transport mechanisms move neurofilaments relentlessly in mouse optic axons. J Cell Biol 117:607–616

LaVail JH, LaVail MM (1972) Retrograde axonal transport in the central nervous system. Science 176:1416–1417
Lee SK, Hollenbeck PJ (2003) Organization and translation of mRNA in sympathetic axons. J Cell Sci 116:4467–4478
Leterrier JF, Eyer J (1987) Properties of highly viscous gels formed by neurofilaments in vitro. A possible consequence of a specific inter-filament cross-bridging. Biochem J 245:93–101
Leterrier JF, Liem RK, Shelanski ML (1982) Interactions between neurofilaments and microtubule-associated proteins: a possible mechanism for intraorganellar bridging. J Cell Biol 95:982–986
Lewis TL Jr, Mao T, Svoboda K, Arnold DB (2009) Myosin-dependent targeting of transmembrane proteins to neuronal dendrites. Nat Neurosci 12:568–576
Li D, Chantler PD (1992) Evidence for a new member of the myosin I family from mammalian brain. J Neurochem 59:1344–1351
Li D, Miller M, Chantler PD (1994) Association of a cellular myosin II with anionic phospholipids and the neuronal plasma membrane. Proc Natl Acad Sci USA 91:853–857
Lim SS, Edson KJ, Letourneau PC, Borisy GG (1990) A test of microtubule translocation during neurite elongation. J Cell Biol 111:123–130
Lorenz T, Willard M (1978) Subcellular fractionation of intra-axonally transport polypeptides in the rabbit visual system. Proc Natl Acad Sci USA 75:505–509
Lubinska L, Niemierko S, Oderfeld Nowak B, Szwarc L (1964) Behaviour of acetylcholinesterase in isolated nerve segments. J Neurochem 11:493–503
Ma D, Himes BT, Shea TB, Fischer I (2000) Axonal transport of microtubule-associated protein 1B (MAP1B) in the sciatic nerve of adult rat: distinct transport rates of different isoforms. J Neurosci 20:2112–2120
Ma Y, Shakiryanova D, Vardya I, Popov SV (2004) Quantitative analysis of microtubule transport in growing nerve processes. Curr Biol 14:725–730
Martin M, Iyadurai SJ, Gassman A, Gindhart JG Jr., Hays TS, Saxton WM (1999) Cytoplasmic dynein, the dynactin complex, and kinesin are interdependent and essential for fast axonal transport. Mol Biol Cell 10:3717–3728
Mazarakis ND, Azzouz M, Rohll JB, Ellard FM, Wilkes FJ, Olsen AL, Carter EE, Barber RD, Baban DF, Kingsman SM, Kingsman AJ, O'Malley K, Mitrophanous KA (2001) Rabies virus glycoprotein pseudotyping of lentiviral vectors enables retrograde axonal transport and access to the nervous system after peripheral delivery. Hum Mol Genet 10: 2109–2121
Mercken M, Fischer I, Kosik KS, Nixon RA (1995) Three distinct axonal transport rates for tau, tubulin, and other microtubule-associated proteins: evidence for dynamic interactions of tau with microtubules in vivo. J Neurosci 15:8259–8267
Miani N (1960) Proximo-distal movement along the axon of protein synthesized in the perikaryon of regenerating neurons. Nature 185:541
Millecamps S, Gowing G, Corti O, Mallet J, Julien JP (2007) Conditional NF-L transgene expression in mice for in vivo analysis of turnover and transport rate of neurofilaments. J Neurosci 27:4947–4956
Mills RG, Minamide LS, Yuan A, Bamburg JR, Bray JJ (1996) Slow axonal transport of soluble actin with actin depolymerizing factor, cofilin, and profilin suggests actin moves in an unassembled form. J Neurochem 67:1225–1234
Mitsumoto HaW B (1987) Structure and development of nerves. In: Vinken P, Bruyn GW, Klawans HL (eds) Handbook of clinnical neurology. Elsevier Health Sciences, Philadelphia, PA, pp 1–22
Miyata Y, Hoshi M, Nishida E, Minami Y, Sakai H (1986) Binding of microtubule-associated protein 2 and tau to the intermediate filament reassembled from neurofilament 70-kDa subunit protein. Its regulation by calmodulin. J Biol Chem 261:13026–13030
Mochida S, Kobayashi H, Matsuda Y, Yuda Y, Muramoto K, Nonomura Y (1994) Myosin II is involved in transmitter release at synapses formed between rat sympathetic neurons in culture. Neuron 13:1131–1142
Mooseker M (1993) Myosin superfamily: a multitude of myosins. Curr Biol 3:245–248
Mooseker MS, Cheney RE (1995) Unconventional myosins. Annu Rev Cell Dev Biol 11:633–675

Morel N, Gerard V, Shiff G (1998) Vacuolar H+-ATPase domains are transported separately in axons and assemble in Torpedo nerve endings. J Neurochem 71:1702–1708

Morris RL, Hollenbeck PJ (1995) Axonal transport of mitochondria along microtubules and F-actin in living vertebrate neurons. J Cell Biol 131:1315–1326

Nangaku M, Sato-Yoshitake R, Okada Y, Noda Y, Takemura R, Yamazaki H, Hirokawa N (1994) KIF1B, a novel microtubule plus end-directed monomeric motor protein for transport of mitochondria. Cell 79:1209–1220

Nauta HJ, Pritz MB, Lasek RJ (1974) Afferents to the rat caudoputamen studied with horseradish peroxidase. an evaluation of a retrograde neuroanatomical research method. Brain Res 67: 219–238

Nixon RA (1998a) The slow axonal transport of cytoskeletal proteins. Curr Opin Cell Biol 10: 87–92

Nixon RA (1998b) The slow axonal transport debate. Trends Cell Biol 8:100

Nixon RA, Logvinenko KB (1986) Multiple fates of newly synthesized neurofilament proteins: evidence for a stationary neurofilament network distributed nonuniformly along axons of retinal ganglion cell neurons. J Cell Biol 102:647–659

Nixon RA, Shea TB (1992) Dynamics of neuronal intermediate filaments: a developmental perspective. Cell Motil Cytoskeleton 22:81–91

Noda Y, Sato-Yoshitake R, Kondo S, Nangaku M, Hirokawa N (1995) KIF2 is a new microtubule-based anterograde motor that transports membranous organelles distinct from those carried by kinesin heavy chain or KIF3A/B. J Cell Biol 129:157–167

Ochs S (1975a) A unitary concept of axoplasmic transport based on the transport filament hypothesis. In: Bradley WG, Gardner-Medwin D, Walton JN (eds) Third international congress on muscle disease. Eccerpta Medica, Amsterdam, pp 128–133

Ochs S (1975b) Retention and redistribution of proteins in mammalian nerve fibres by axoplasmic transport. J Physiol 253:459–475

Ochs S, Burger E (1958) Movement of substance proximo-distally in nerve axons as studied with spinal cord injections of radioactive phosphorus. Am J Physiol 194:499–506

Ochs S, Jersild RA Jr., Li JM (1989) Slow transport of freely movable cytoskeletal components shown by beading partition of nerve fibers in the cat. Neuroscience 33:421–430

Ochs S, Johnson J, Ng MH (1967) Protein incorporation and axoplasmic flow in motoneuron fibres following intra-cord injection of labelled leucine. J Neurochem 14:317–331

Okabe S, Hirokawa N (1993) Do photobleached fluorescent microtubules move? Re-evaluation of fluorescence laser photobleaching both in vitro and in growing Xenopus axon. J Cell Biol 120:1177–1186

Okada Y, Yamazaki H, Sekine-Aizawa Y, Hirokawa N (1995) The neuron-specific kinesin superfamily protein KIF1A is a unique monomeric motor for anterograde axonal transport of synaptic vesicle precursors. Cell 81:769–780

Pachter JS, Liem RK (1984) The differential appearance of neurofilament triplet polypeptides in the developing rat optic nerve. Dev Biol 103:200–210

Pannese E (1994) Neurocytology. Thieme Publishers, New York, NY

Paschal BM, Shpetner HS, Vallee RB (1987) MAP 1C is a microtubule-activated ATPase which translocates microtubules in vitro and has dynein-like properties. J Cell Biol 105:1273–1282

Perkins GA, Sosinsky GE, Ghassemzadeh S, Perez A, Jones Y, Ellisman MH (2008) Electron tomographic analysis of cytoskeletal cross-bridges in the paranodal region of the node of Ranvier in peripheral nerves. J Struct Biol 161:469–480

Prahlad V, Helfand BT, Langford GM, Vale RD, Goldman RD (2000) Fast transport of neurofilament protein along microtubules in squid axoplasm. J Cell Sci 113(22):3939–3946

Prekeris R, Terrian DM (1997) Brain myosin V is a synaptic vesicle-associated motor protein: evidence for a Ca^{2+}-dependent interaction with the synaptobrevin-synaptophysin complex. J Cell Biol 137:1589–1601

Provance DW Jr, McDowall A, Marko M, Luby-Phelps K (1993) Cytoarchitecture of size-excluding compartments in living cells. J Cell Sci 106(2):565–577

Ramon y Cajar S (1928) Degeneration and regeneration of the nervous sytem. Oxford University Press, London

Rao MV, Engle LJ, Mohan PS, Yuan A, Qiu D, Cataldo A, Hassinger L, Jacobsen S, Lee VM, Andreadis A, Julien JP, Bridgman PC, Nixon RA (2002) Myosin Va binding to neurofilaments is essential for correct myosin Va distribution and transport and neurofilament density. J Cell Biol 159:279–290

Roy S, Coffee P, Smith G, Liem RK, Brady ST, Black MM (2000) Neurofilaments are transported rapidly but intermittently in axons: implications for slow axonal transport. J Neurosci 20: 6849–6861

Ruthel G, Banker G (1998) Actin-dependent anterograde movement of growth-cone-like structures along growing hippocampal axons: a novel form of axonal transport? Cell Motil Cytoskeleton 40:160–173

Ruthel G, Banker G (1999) Role of moving growth cone-like "wave" structures in the outgrowth of cultured hippocampal axons and dendrites. J Neurobiol 39:97–106

Sankaranarayanan S, Atluri PP, Ryan TA (2003) Actin has a molecular scaffolding, not propulsive, role in presynaptic function. Nat Neurosci 6:127–135

Schnapp BJ, Vale RD, Sheetz MP, Reese TS (1985) Single microtubules from squid axoplasm support bidirectional movement of organelles. Cell 40:455–462

Schroer TA (1992) Motors for fast axonal transport. Curr Opin Neurobiol 2:618–621

Schroer TA, Steuer ER, Sheetz MP (1989) Cytoplasmic dynein is a minus end-directed motor for membranous organelles. Cell 56:937–946

Shah JV, Cleveland DW (2002) Slow axonal transport: fast motors in the slow lane. Curr Opin Cell Biol 14:58–62

Shah JV, Flanagan LA, Janmey PA, Leterrier JF (2000) Bidirectional translocation of neurofilaments along microtubules mediated in part by dynein/dynactin. Mol Biol Cell 11:3495–3508

Sheetz MP, Martenson CH (1991) Axonal transport: beyond kinesin and cytoplasmic dynein. Curr Opin Neurobiol 1:393–398

Sheetz MP, Spudich JA (1983) Movement of myosin-coated fluorescent beads on actin cables in vitro. Nature 303:31–35

Sodeik B, Ebersold MW, Helenius A (1997) Microtubule-mediated transport of incoming herpes simplex virus 1 capsids to the nucleus. J Cell Biol 136:1007–1021

Sotelo-Silveira JR, Calliari A, Kun A, Koenig E, Sotelo JR (2006) RNA trafficking in axons. Traffic 7:508–515

Stromska DP, Ochs S (1981) Patterns of slow transport in sensory nerves. J Neurobiol 12:441–453

Sun WD, Chantler PD (1991) A unique cellular myosin II exhibiting differential expression in the cerebral cortex. Biochem Biophys Res Commun 175:244–249

Sun W, Chantler PD (1992) Cloning of the cDNA encoding a neuronal myosin heavy chain from mammalian brain and its differential expression within the central nervous system. J Mol Biol 224:1185–1193

Svitkina TM, Verkhovsky AB, Borisy GG (1996) Plectin sidearms mediate interaction of intermediate filaments with microtubules and other components of the cytoskeleton. J Cell Biol 135:991–1007

Tabb JS, Molyneaux BJ, Cohen DL, Kuznetsov SA, Langford GM (1998) Transport of ER vesicles on actin filaments in neurons by myosin V. J Cell Sci 111(21):3221–3234

Terada S, Nakata T, Peterson AC, Hirokawa N (1996) Visualization of slow axonal transport in vivo. Science 273:784–788

Theiss C, Napirei M, Meller K (2005) Impairment of anterograde and retrograde neurofilament transport after anti-kinesin and anti-dynein antibody microinjection in chicken dorsal root ganglia. Eur J Cell Biol 84:29–43

Trivedi N, Jung P, Brown A (2007) Neurofilaments switch between distinct mobile and stationary states during their transport along axons. J Neurosci 27:507–516

Tsukita S, Ishikawa H (1980) The movement of membranous organelles in axons. Electron microscopic identification of anterogradely and retrogradely transported organelles. J Cell Biol 84:513–530

Vale RD, Reese TS, Sheetz MP (1985a) Identification of a novel force-generating protein, kinesin, involved in microtubule-based motility. Cell 42:39–50

Vale RD, Schnapp BJ, Reese TS, Sheetz MP (1985b) Movement of organelles along filaments dissociated from the axoplasm of the squid giant axon. Cell 40:449–454

Vallee RB, Bloom GS (1991) Mechanisms of fast and slow axonal transport. Annu Rev Neurosci 14:59–92

Wagner OI, Ascano J, Tokito M, Leterrier JF, Janmey PA, Holzbaur EL (2004) The interaction of neurofilaments with the microtubule motor cytoplasmic dynein. Mol Biol Cell 15: 5092–5100

Waller AV (1850) Experiments on the section of the glossopharyngeal and hypoglossal nerves of the frog, and observations of the alterations produced thereby in the structure of their primitive fibres. Philos Trans R Soc London 140:423–429

Wang L, Brown A (2002) Rapid movement of microtubules in axons. Curr Biol 12:1496–1501

Wang L, Ho CL, Sun D, Liem RK, Brown A (2000) Rapid movement of axonal neurofilaments interrupted by prolonged pauses. Nat Cell Biol 2:137–141

Weiss P, Hiscoe HB (1948) Experiments on the mechanism of nerve growth. J Exp Zool 107: 315–395

Willard M (1977) The identification of two intra-axonally transported polypeptides resembling myosin in some respects in the rabbit visual system. J Cell Biol 75:1–11

Willard M, Cowan WM, Vagelos PR (1974) The polypeptide composition of intra-axonally transported proteins: evidence for four transport velocities. Proc Natl Acad Sci USA 71:2183–2187

Willard MB, Hulebak KL (1977) The intra-axonal transport of polypeptide H: evidence for a fifth (very slow) group of transported proteins in the retinal ganglion cells of the rabbit. Brain Res 136:289–306

Xia CH, Roberts EA, Her LS, Liu X, Williams DS, Cleveland DW, Goldstein LS (2003) Abnormal neurofilament transport caused by targeted disruption of neuronal kinesin heavy chain KIF5A. J Cell Biol 161:55–66

Yabe JT, Pimenta A, Shea TB (1999) Kinesin-mediated transport of neurofilament protein oligomers in growing axons. J Cell Sci 112(21):3799–3814

Yamazaki H, Nakata T, Okada Y, Hirokawa N (1995) KIF3A/B: a heterodimeric kinesin superfamily protein that works as a microtubule plus end-directed motor for membrane organelle transport. J Cell Biol 130:1387–1399

Yan Y, Brown A (2005) Neurofilament polymer transport in axons. J Neurosci 25:7014–7021

Yang Y, Bauer C, Strasser G, Wollman R, Julien JP, Fuchs E (1999) Integrators of the cytoskeleton that stabilize microtubules. Cell 98:229–238

Yang Y, Dowling J, Yu QC, Kouklis P, Cleveland DW, Fuchs E (1996) An essential cytoskeletal linker protein connecting actin microfilaments to intermediate filaments. Cell 86:655–665

Yuan A (2007) Neurofilament protein partnership, export, transport, phosphorylation and neurodegeneration. In: Arlen RK (ed) New research on neurofilament proteins. Nova Science Publishers, Inc, New York, NY, pp 53–79

Yuan A, Kumar A, Peterhoff C, Duff K, Nixon RA (2008) Axonal transport rates in vivo are unaffected by tau deletion or overexpression in mice. J Neurosci 28:1682–1687

Yuan A, Mills RG, Bamburg JR, Bray JJ (1997) Axonal transport and distribution of cyclophilin A in chicken neurones. Brain Res 771:203–212

Yuan A, Mills RG, Bamburg JR, Bray JJ (1999) Cotransport of glyceraldehyde-3-phosphate dehydrogenase and actin in axons of chicken motoneurons. Cell Mol Neurobiol 19: 733–744

Yuan A, Mills RG, Chia CP, Bray JJ (2000) Tubulin and neurofilament proteins are transported differently in axons of chicken motoneurons. Cell Mol Neurobiol 20:623–632

Yuan A, Rao MV, Kumar A, Julien JP, Nixon RA (2003) Neurofilament transport in vivo minimally requires hetero-oligomer formation. J Neurosci 23:9452–9458

Yuan A, Rao MV, Sasaki T, Chen Y, Kumar A, Veeranna, Liem RK, Eyer J, Peterson AC, Julien JP, Nixon RA (2006) Alpha-internexin is structurally and functionally associated with the neurofilament triplet proteins in the mature CNS. J Neurosci 26:10006–10019

Yuan A, Sasaki T, Rao MV, Kumar A, Kanumuri V, Dunlop DS, Liem RK, Nixon RA (2009) Neurofilaments form a highly stable stationary cytoskeleton after reaching a critical level in axons. J Neurosci 29:11316–11329

Chapter 22
Axonal Transport and Motor Neuron Disease

Anna-Lena Ström, Lawrence J. Hayward, Edward J. Kasarskis, and Haining Zhu

Abstract Transport of material between the cell body and neuronal processes is crucial to neuronal function and survival. Growing evidence shows that deficits in axonal transport contribute to the pathogenesis of multiple neurodegenerative diseases including amyotrophic lateral sclerosis (ALS). Here we review recent data indicating that defects in anterograde and retrograde axonal transport are involved in ALS etiology. We discuss how ALS-linked mutations of copper-zinc superoxide dismutase (SOD1) could affect the molecular transport motor proteins and how this might lead to decreased function and viability of motor neurons in ALS.

Keywords Amyotrophic lateral sclerosis · Axonal transport · Dynein · Kinesin · Motor neuron · SOD1

Contents

H. Zhu (✉)
Department of Molecular and Cellular Biochemistry, College of Medicine, University of Kentucky, Lexington, KY 40536, USA
e-mail: haining@uky.edu

R.A. Nixon, A. Yuan (eds.), *Cytoskeleton of the Nervous System*,
Advances in Neurobiology 3, DOI 10.1007/978-1-4419-6787-9_22,

22.1 Introduction

Amyotrophic lateral sclerosis (ALS) is a progressive and fatal neurodegenerative disease characterized by loss of motor neurons in the spinal cord, brainstem, and motor cortex, leading to paralysis and death. Although most forms of ALS are sporadic, approximately 10% of ALS cases are familial, of which 20% are caused by autosomal, dominantly inherited mutations in the *Cu, Zn-superoxide dismutase 1 (SOD1)* gene (Deng et al., 1993; Rosen et al., 1993). More recently, a growing fraction of familial ALS cases have been attributed to mutations in other genes, including the *vesicle associated membrane protein-associated protein B* (VAPB) gene (Nishimura et al., 2004, 2005), the *TAR DNA-binding protein 43* (TPD43) gene (Kabashi et al., 2008; Sreedharan et al., 2008; Van Deerlin et al., 2008), the *Senataxin* (SETX) gene (Chen et al., 2004), and the *fused in sarcoma/translated in liposarcoma* (FUS/TLS) gene (Kwiatkowski et al., 2009; Vance et al., 2009).

Several studies have shown that neurons are highly sensitive to defects in axonal transport. Mutations in the anterograde transport motor protein kinesin (KIF5A) can cause spastic paraplegia (Reid et al., 2002) and mutations in the retrograde motor complex dynein–dynactin cause motor neuron degeneration in humans and mice (Hafezparast et al., 2003; Puls et al., 2005). Decreased kinesin-mediated (anterograde) and dynein-mediated (retrograde) axonal transport have been observed both in ALS patients and in transgenic animal models (Breuer and Atkinson, 1988; Breuer et al., 1987; Collard et al., 1995; Ligon et al., 2005; Sasaki and Iwata, 1996; Williamson and Cleveland, 1999). The detailed mechanisms by which axonal transport is affected in ALS have yet to be established. However, in the last few years, ALS-related mutant SOD1 proteins have been shown to interact with both the anterograde motor protein kinesin-2 complex via KAP3 and the retrograde motor protein complex dynein–dynactin (Zhang et al., 2007; Tateno et al., 2009).

In this review we will discuss recent data linking defects in axonal transport as a factor contributing to motor neuron degeneration in ALS. We will focus on how mutations in the SOD1 protein could directly or indirectly impair axonal transport and thereby influence neuronal survival.

22.2 Axonal Transport System in Neurons

Neurons have extensive dendritic arbors and axonal processes that can extend far from the cell body. The ability of the neuron to maintain this highly specialized morphology depends on continuous transport of proteins and organelles to and from the cell body. The two major components of the transport machinery are the microtubules, which act as tracks on which the second component, the molecular motors, can transport cargos. Microtubules are polarized with a fast growing end (the plus end) and a slower growing end (the minus end). In axons, the orientation of microtubules is almost uniform, with the plus end localized toward the synapse and the minus end toward the cell body. Members of the kinesin motor family transport cargos in the anterograde direction (toward plus end of microtubules) whereas cytoplasmic dynein is the major motor protein driving retrograde transport (toward the minus end of microtubules) in cells.

To date, more than 45 kinesin motor proteins have been identified, and they are subdivided into 14 kinesin families (see reviews Miki et al., 2001, 2005). In mice, 38 of the 45 kinesins are expressed in neurons. Two members of the kinesin-1 family, KIF5A and KIF5C, are neuron specific and play essential roles in axonal transport (Hirokawa and Noda, 2008). The KIF5 family motor proteins contain a cargo-binding domain and also associate with a kinesin light chain (KLC), which can mediate cargo binding. Phosphorylation of KLC by kinases including GSK-3β can regulate KLCs′ binding affinity for cargos (Hirokawa and Noda, 2008). The KIF5 motor proteins have been suggested to transport neurofilaments, SNARE proteins essential for synaptic vesicle docking, mRNAs and mitochondria (Hirokawa and Noda, 2008). A member of the kinesin-2 family, KIF-3, has also been shown to play a role in anterograde transport of mitochondria in neurons (Hirokawa and Noda, 2008). Other known KIF3 cargos include choline acetyltrasferase, acetylcholinesterase, and voltage-dependent potassium (Kv) channels. KIF3 is known to be closely associated with the KAP3 adaptor that can mediate cargo binding (Hirokawa and Noda, 2008).

Cytoplasmic dynein is the major motor driving retrograde transport in cells. Dynein is a large complex (approximately 2 million Da) and consists of two dynein heavy chains (DHC), two dynein intermediate chains (DIC), four dynein light intermediate chains (DLIC), and various dynein light chains (DLC). The dimeric DHC forms the core of the dynein complex. Each DHC subunit folds to form a globular head containing the motor domain and a flexible stalk that is involved in dimerization of the two heavy chains as well as interaction with other dynein subunits (reviewed in Pfister et al., 2006). In vivo, dynein requires the co-complex dynactin for most functions. Dynactin functions both to increase the motor efficiency of dynein and to serve as an adaptor between dynein and various cargos (reviewed in Schroer, 2004). The dynactin assembly comprises multiple subunits that form a distinct structure: a filament base and a projecting sidearm linked to the base by a shoulder domain. The base of dynactin is formed by a small filament of the actin-related protein Arp1 capped by other subunits such as CapZ, Arp11, p25, and p27. The projecting sidearm of dynactin is formed by two dimeric p150^{Glued} subunits and is involved in the interaction with dynein via the DIC, but it also interacts with microtubules. A tetramer of the dynamitin-p50 subunit forms the shoulder of dynactin that links the sidearm to the filament base. Over-expression of p50 competitively dissociates the p150^{Glued} sidearm from the Arp1 filament base (Melkonian et al., 2007).

22.3 Coordination of Axonal Transport

Axonal transport is usually divided into fast and slow axonal transport on the basis of speed of net cargo movement. Vesicles and mitochondria move by fast axonal transport (FAT) at a speed of ~1 μm/s, whereas cytoskeletal components move by slow axonal transport at ~1 mm/day (De Vos et al., 2008). Fast and slow transport are both carried out by kinesins and dynein. The slower transport of some cargos can be attributed to prolonged pauses between movements. Moreover, when axonal transport is observed in living neurons, cargos do not necessary move unidirectionally, but rather migrate back and forth within the axon (Welte, 2004). In some cases the overall direction of motility is regulated, but in other cases the movement appears to be stochastic, with multiple apparently random changes in direction. Several models have been proposed to explain the bi-directionality and interdependency of transport by kinesin and dynein motors (reviewed in Gross et al., 2002; Holzbaur, 2004). In the "tug-of-war" model, kinesins and dynein are suggested to bind to the cargo simultaneously, and the direction of movement is determined by the dominant motor at any given time. In another model, only kinesin or dynein is active at a certain time point. The slow transport of neurofilaments has been suggested to occur in the "tug-of-war" fashion, while cargos such as peroxisomes and vesicles have been suggested to be transported in a coordinated fashion (Gross et al., 2002; Holzbaur, 2004).

The coordination of motor activities could be regulated by interaction between different motors, adapter proteins linking different motors, or other regulatory pathways. Recent studies have supplied evidence for all of these mechanisms. For instance, a direct interaction between the KIF5-KLC complex and dynein intermediate chain has been shown (Ligon et al., 2004). Furthermore, dynactin has been observed to interact not only with dynein, but also with several members of the KIF3 kinesin family via the adapter KAP3 (Blangy et al., 1997; Deacon et al., 2003). In addition, microtubule associated proteins (MAPs) such as Tau can also have significant, and potentially differential, influence on the motility of kinesin and dynein motors (Dixit et al., 2008). Because of the complex interplay between anterograde and retrograde transport, disruption of axonal transport in one direction can also affect movement in the opposite direction (Brady et al., 1990; Martin et al., 1999; Waterman-Storer et al., 1997).

22.4 Impaired Axonal Transport in ALS

Studies of both ALS patients and transgenic animals have revealed decreased axonal transport in both anterograde and retrograde directions (Breuer and Atkinson, 1988; Breuer et al., 1987; Collard et al., 1995; Ligon et al., 2005; Sasaki and Iwata, 1996; Williamson and Cleveland, 1999). The early appearance of this inhibition supports the hypothesis that impaired axonal transport is a primary contributor to the disease process rather than an associated but secondary phenomenon. It is noteworthy that some ALS-causing SOD1 mutant proteins can differentially affect the axonal transport of distinct cargos. Anterograde transport of cytoskeletal elements, including neurofilaments, is slowed (Williamson and Cleveland, 1999; Warita et al., 1999), while fast transport of vesicles is inhibited in both the anterograde and the retrograde direction (De Vos et al., 2007), and inhibition of mitochondrial movement is anterograde-specific (De Vos et al., 2007).

22.4.1 Defects in Mitochondrial Transport

Impaired axonal transport of mitochondria in neurons could lead to energy deprivation in the extremities of the neuron and an inability to maintain normal synaptic structure and neurotransmission function. Axonal transport is especially important to sustain the normal dynamics and distribution of mitochondria. Because mitochondria cannot be synthesized de novo in cells, they depend upon transport, fission, and fusion to maintain proper morphology and function (see Frazier et al., 2006 for review). Furthermore, transport is required to ensure that mitochondria accumulate at high density in regions with high metabolic demands, such as the axon hillock, the nodes of Ranvier, and the synapse. Loss of dynein-mediated retrograde transport could not only affect the distribution of mitochondria but also affect mitochondrial homeostasis, since dynein is important for mitochondrial fission (Frazier et al., 2006). Moreover, damaged mitochondria with reduced membrane potential are degraded by autophagy, an intracellular turnover process in which dynein has been suggested to play a role (Twig et al., 2008). Abnormal and degenerating mitochondria have been observed from early stages in both sporadic and familial ALS patients and animal models (see review in Baron et al., 2007).

22.4.2 Defects in Transport of Neurotrophic Factors

Neurotrophic factors including NGF, BDNF, and NT3 are secreted by target tissues (e.g., muscles) and then bind to tyrosine kinase receptors (Trks) on the surface of neurons. The Trk receptor-neurotrophin complex is then internalized and initiates signaling cascades that regulate neuronal cell growth, survival and repair pathways (Bronfman et al., 2007; Campenot and MacInnis, 2004). In order for the neurotrophins to act effectively and promote long-term survival, transport of the neurotrophin-receptor complex to the cell body by the dynein complex may be required (Delcroix et al., 2003; Heerssen et al., 2004; Yano et al., 2001; Ye et al., 2003). Some ALS studies in mouse models have demonstrated a benefit upon administration of exogenous neurotrophic factors supporting this hypothesis (Azari et al., 2003; Kaspar et al., 2003; Wang et al., 2002; Azzouz et al., 2004; Zheng et al., 2004; Dobrowolny et al., 2005; Storkebaum et al., 2005; Pun et al., 2006; Li et al., 2007). However, other studies have shown little or no effect (Feeney et al., 2003; Raoul et al., 2005; Li et al., 2007) and to date no viable therapy based on neurotrophic factors has been established for patients.

22.4.3 Defects in Transport and ER/Golgi Fragmentation

The structural and functional integrity of the endoplasmic reticulum (ER)-Golgi network is dependent on microtubule-mediated vesicular membrane trafficking. Blockage of dynein–dynactin function by p50 over-expression has been shown to affect endosome trafficking and to cause Golgi fragmentation (Burkhardt et al., 1997; Valetti et al., 1999). Fragmentation of the Golgi has been observed in both sporadic and SOD1-mediated familial ALS (Gonatas et al., 2006). In ALS mouse models, Golgi fragmentation can also be observed before onset of paralysis (Gonatas et al., 2006).

22.4.4 *Defects in Transport and Autophagy*

Macroautophagy (or autophagy) is an intracellular process in which proteins, organelles, or protein aggregates are degraded. During autophagy the substrate is engulfed by membranous structures to form autophagosomes, which then fuse with lysosomes containing hydrolytic enzymes (Mizushima, 2007). Dynein has been linked to autophagic clearance of aggregated proteins in several studies. First, dynein-mediated transport has been shown to target misfolded proteins from the cell periphery and transport them to the perinuclear area, where they form inclusions called aggresomes; it is suggested that these structures may become membrane-bound and associate with lysosomes (Johnston et al., 2002; Taylor et al., 2003). Dynein has also been suggested to play an important role in lysosomal transport and to mediate the fusion of the autophagosome and the lysosome (Burkhardt et al., 1997; Ravikumar et al., 2005). Mutant SOD1 is believed to be degraded by both the ubiquitin proteasome system and autophagy (Kabuta et al., 2006). Supporting the importance of autophagic clearance of aggregates in preventing neuronal toxicity, a recent study showed that blocking of dynein caused decreased autophagic clearance of mutant huntingtin protein, resulting in increased inclusion/aggresome burden and toxicity in Huntington's disease model mice (Ravikumar et al., 2005). The role of autophagy in ALS etiology remains to be further elucidated.

22.5 Mutant SOD1 and Axonal Transport Defects

Several different mechanisms regarding the potential effect of mutant SOD1 on axonal transport have been proposed, and it is likely that more than one may contribute to the disease process. These mechanisms include (1) damage of mitochondria causing reduced energy supply to the motor proteins, (2) altered phosphorylation of motor protein complexes affecting their activity, cargo or microtubule binding capabilities, (3) direct interaction of mutant SOD1 with motor proteins causing alterations of motor complex integrity and or masking of cargo binding sites, (4) physical blockade of motors by mutant SOD1 aggregates, or (5) disruption of microtubule dynamics affecting formation and or stability (Fig. 22.1).

Many studies have demonstrated that mutant SOD1 can selectively associate with and damage mitochondria (Boillee et al., 2006). This damage is believed to impair the electron transfer chain and ATP synthesis (Mattiazzi et al., 2002). Injury to mitochondria has been shown to decrease their anterograde transport, and insufficient energy production in the axon could then lead to even further reduction in transport in both directions (Miller and Sheetz, 2004; De Vos et al., 2007).

Recently, mutant SOD1 has been shown to interact with both dynein–dynactin and kinesin-2 complexes. We showed that mutant SOD1 interacted more stably with dynein compared to WT SOD1 in both ALS cell culture and animal models (Zhang et al., 2007). It is still unclear which subunit(s) of dynein and/or dynactin mutant SOD1 interacts with. If mutant SOD1 is a bona fide cargo of dynein, the interaction could be mediated by dynein light chain (DLC) or dynactin, which are

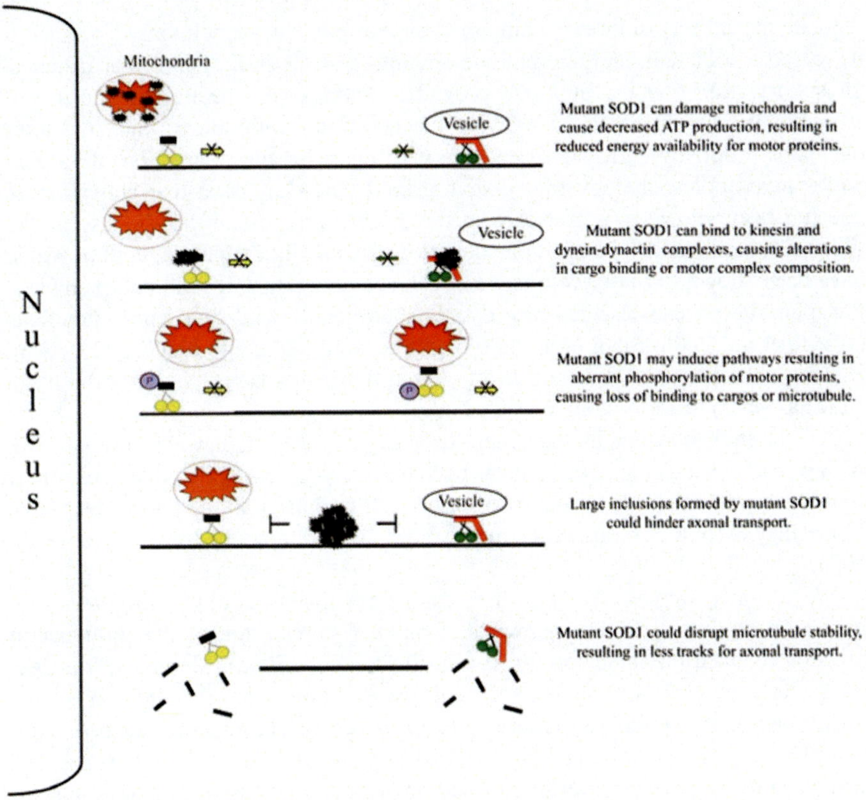

Fig. 22.1 Possible mechanisms by which mutant SOD1 could interfere with axonal transport

known to mediate the binding of other cargos. However, it is also possible that this aberrant interaction involves novel mechanisms different from linkage to DLC or dynactin. The interaction of mutant SOD1 with kinesin-2 was shown to be mediated by the KAP3 adapter protein (Tateno et al., 2009). The interaction of mutant SOD1 with both dynein–dynactin and KAP3–kinesin-2 was shown to start at a presymptomatic stage of ALS. Moreover, in ALS transgenic mice, dynein–dynactin and KAP3 was shown to co-localize with mutant SOD1 aggregates in motor neuron axons (Ligon et al., 2005; Zhang et al., 2007; Tateno et al., 2009). The interactions of mutant SOD1 with dynein and kinesin complexes could cause alterations of the motor complex integrity and or change cargo or microtubule binding capabilities.

Alternatively, mutant SOD1 conceivably can interfere with the function of motor proteins by altering the phosphorylation status of the motors. Experimental evidence supporting this hypothesis in ALS is scarce. However, misregulation of fast axonal transport induced by imbalance in specific kinase phosphorylation events have been implicated in other neurodegenerative diseases such as Alzheimer's and polyglutamine disease (Pigino et al., 2003; Morfini et al., 2006). Phosphorylation can regulate the activity of kinesin complexes on multiple levels, including motor activity, cargo association, and microtubule binding (De Vos et al., 2008). For instance, phosphorylation of KLC by a p38-dependent pathway has been shown to inhibit mitochondria-bound kinesin-1 activity. Furthermore, inhibition of kinesin-1 after phosphorylation by JNK has been shown to decrease the interaction of kinesin with microtubules, thereby impeding transport. In ALS, persistent activation of p38 has been shown to correlate with the disease progression in a mouse model (Tortarolo et al., 2003). In addition, excitotoxicity and inflammation, both of which have been shown to play a role in ALS, can activate both JNK and p38 pathways. Phosphorylation can also regulate dynein intermediate and light chain functions (Whyte et al., 2008; Song et al., 2007) and possibly the cargo binding of dynein. However, the possible involvement of altered dynein-dynactin phosphorylation in ALS has yet to be thoroughly investigated.

Axonal inclusions, including spheroids as large as 20 μm, have been reported in ALS patients (Kato et al., 2003). It has been reported that dynein-mediated transport could transport misfolded proteins from the cell periphery to the perinuclear area, where they form inclusions called aggresomes (Johnston et al., 2002; Taylor et al., 2003).

We recently showed that the interaction with dynein could contribute to the formation of mutant SOD1 aggregates, and that suppression of dynein-dynactin function reduced the inclusion formation (Ström et al., 2008). Moreover, it is possible that mutant SOD1 aggregates could hinder transport of cargos on axonal microtubules. A similar mechanism was proposed in Huntington's disease after huntingtin aggregates were observed with diameters exceeding those of axons; consistent with this, axonal swelling around aggregates and loss of axonal transport were reported in the *Drosophila* models of Huntington's disease (Gunawardena et al., 2003; Lee et al., 2004).

A significantly altered microtubule turnover in G93A ALS mice has recently been reported. Treating G93A mice with noscapine, a drug that increases the time microtubules spend in the attenuated state (neither growing or shortening), had a positive effect on disease onset and progression (Fanara et al., 2007). These data suggest that decreased axonal transport in ALS could be due to a decrease in microtubules suitable as tracks for molecular motors.

22.6 Modulation of ALS Phenotype by Dynein and Dynactin Mutations

Several murine models with altered dynein-dynactin function have been shown to develop motor neuron symptoms. Impaired dynein–dynactin function after overexpression of the p50 dynactin subunit in mice has been shown to cause motor neuron disease (LaMonte et al., 2002). A mutation in the p150^{Glued} dynactin subunit (G59S) was also recently linked to a slowly progressive form of motor neuron disease in a North American family (Puls et al., 2003, 2005). Heterozygous knock-in mice, as well as transgenic mice carrying this mutation, also developed motor neuron degeneration (Lai et al., 2007; Laird et al., 2008). Furthermore, transgenic mice with neuron-specific expression of Bicaudal D2 N-terminus (BICD2-N) have also been shown to have chronic impairment of dynein–dynactin function and to develop impaired retrograde trafficking (Teuling et al., 2008). Finally, two independent *N*-ethyl-*N*-nitrosourea (ENU)-induced missense mutations in the stalk domain of the dynein heavy chain (DHC) gene, *Loa* and *Cra1*, caused decreased retrograde transport and produced late-onset motor neuron degeneration in heterozygous mice (Hafezparast et al., 2003).

Surprisingly, crossing heterozygous *Loa* and *Cra1* mutant mice with G93A SOD1 mutant mice ameliorated the transport defect, delayed disease onset, and slowed disease progression in G93A mice (Kieran et al., 2005; Teuchert et al., 2006). While the precise mechanism(s) leading to this effect remain(s) unclear, several explanations have been suggested. One hypothesis states that the dynein mutations alter the intracellular transport and thereby change the subcellular localization of SOD1 or the interaction of SOD1 with other proteins or organelles (El-Kadi et al., 2007; Kieran et al., 2005). For example, it is possible that decreased interaction of mutant SOD1 with mitochondria could improve cell survival by reducing apoptosis or other downstream consequences (Liu et al., 2004; Menzies et al., 2002; Pasinelli et al., 2004; Wong et al., 1995).

In addition to motor neuron phenotype, the *Loa* and *Cra1* mutations were recently shown to cause early postnatal loss of sensory axons including those responsible for proprioception (Chen et al., 2007). Subsequently, the *Loa* mutation was suggested to extend survival of G93A mice by decreasing the number of proprioceptive sensory axons and their glutamatergic input on motor neurons (Ilieva et al., 2008). However, crossing G93A mice with mice carrying another DHC mutation *Swl*, which developed proprioceptive sensory neuropathy, but no motor neuron phenotype, did not improve ALS disease onset or progression (Chen et al., 2007).

A third hypothesis for how decreased retrograde transport by *Loa* or *Cra1* mutations could improve G93A ALS is by counterbalancing an inhibition of anterograde transport caused by G93A mutant SOD1, thereby restoring the balance between anterograde and retrograde transport (El-Kadi et al., 2007; Kieran et al., 2005), This hypothesis is supported by the observation that crossing G93A mice with mice having decreased dynein-dynacting function due to BICD2-N over-expression also improved the lifespan of the G93A mice (Teuling et al., 2008). However, inhibition of retrograde transport by other dynein–dynactin approaches did not influence survival or disease progression in G93A mice. Lai et al. reported that crossing mice with the G59S mutation in the dynactin subunit p150^{Glued} which show signs of reduced motor neuron axonal transport did not ameliorate the G93A phenotype (Lai et al., 2007).

The studies crossing different dynein and dynactin mutant mice with G93A mice are summarized in Table 22.1. Two possible mechanisms could be envisioned to explain the various effects on motor neuron viability and G93A SOD1 mice phenotype by the different dynein and dynactin mutants. First, the various dynein and dynactin mutations could inhibit dynein transport of all cargos, but to different degrees. Compared to the *Swl* mutation, *Loa* and *Cra1* mutation could cause a

Table 22.1 The effect of dynein and dynactin mutations on axonal transport and protein aggregation in the G93A ALS mouse model

Mice	Retrograde transport in motor neurons	SOD1 aggregates in motor neurons	Development of motor neuron disease	Phenotypic outcome of crossing	References
G93A	Reduced	Yes	Yes	N.A.	Kieran et al. (2005); Gurney et al. (1994)
Loa +/+	Reduced	Yes, WT SOD1 positive inclusions seen	Yes, severe phenotype, mice die at P0–P1	N.A.	Hafezparast et al. (2003); Kieran et al. (2005)
Loa –/+	No change	N.D.	Yes, but normal lifespan	N.A.	Hafezparast et al. (2003); Kieran et al. (2005)
Loa –/+ × G93A	No change	N.D.	Yes	Delayed onset and extended survival of G93A	Kieran et al. (2005)
Cra1 +/+	Reduced	No	Yes, severe phenotype, mice die at P0–P1	N.A.	Hafezparast et al. (2003)
Cra1–/+	N.D.	N.D.	Yes, but normal lifespan	N.A.	Hafezparast et al. (2003)
Cra1 –/+ × G93A	N.D.	N.D.	Yes	Delayed onset and extended survival of G93A	Teuchert et al. (2006)
Swl +/+	N.D.	N.D.	Embryonic lethal	N.A.	Chen et al. (2007)
Swl –/+	N.D.	N.D.	No	N.A.	Chen et al. (2007)
Swl –/+ × G93A	N.D.	N.D	Yes	No effect on G93A phenotype	Chen et al. (2007)
G59S p150^{Glued} +/+	N.D.	N.D.	Embryonic lethal	N.A.	Lai et al. (2007)
G59S p150^{Glued}–/+	Reduced (accumulation of cargos, but transport rate not measured)	N.D.	Yes	N.A.	Lai et al. (2007)
G59S p150^{Glued} × G93A	N.D.	N.D.	Yes	No effect on G93A phenotype	Lai et al. (2007)
BICD2-N	Reduced	N.D.	No	N.A.	Teuling et al. (2008)
BICD2-N × G93A	N.D.	Yes, also positive for p150^{Glued} and DHC		Delayed onset and extended survival of G93A	Teuling et al. (2008)

N.D. = Not Determined; N.A. = Not Applicable.

higher level of dynein inhibition, thereby causing both sensory and motor neuron degeneration. It is possible that only reduction of dynein transport within a certain range could have beneficial effect on G93A SOD1 mice. To address this hypothesis, the rate of axonal transport in motor and sensory neurons in the *Swl* and G59S $p150^{Glued}$ mice needs to be determined. Alternatively, different dynein and dynactin mutations could affect binding and transport of specific dynein–dynactin cargos. Cargos can bind to dynein in multiple ways, via the DIC interacting complex dynactin or via DIC, DLIC, or DLC (reviewed in Chevalier-Larsen and Holzbaur, 2006; Karcher et al., 2002). It is plausible that the effect of *Loa*, *Cra1*, and BICD2-N on G93A SOD1 mice could be due to reduced transport of certain cargos, which are not affected by the *Swl* and G59S-$p150^{Glued}$ mutations. In support of this hypothesis, Teuling et al. showed that BICD2-N can reduce the retrograde transport of injury signals in G93A motor neurons, suggesting that attenuation of injury response might underlie the improvement of BICD2-N on G93A mice (Teuling et al., 2008).

22.7 Conclusions and Future Studies

It is clear that axonal transport is essential to motor neurons and that both kinesin-mediated anterograde and dynein-mediated retrograde axonal transport are perturbed in ALS. The underlying mechanism(s) remain(s) to be defined more specifically and various possibilities have been discussed here. Some mutations in dynein or dynactin, all of which alter dynein-dynactin function and retrograde transport, have been shown to modulate the ALS phenotype caused by SOD1 mutations. Crossing *Loa*, *Cra1*, or BICD2-N mice with G93A SOD1 mice surprisingly lead to amelioration of the ALS disease process. In contrast, crossing *Swl* or G59S-$p150^{Glued}$ mice with G93A mice did not alter the ALS phenotype. Future studies in the area are needed to provide a better understanding of how transport of specific cargos is regulated. It also remains to be determined how the balance of either retrograde or anterograde transport modulates the development of a given phenotype caused by deleterious mutations.

Acknowledgments This study was in part supported by NIH grants R01-NS049126 and R21-AG032567 (to H.Z.), R01-NS045087 (to E. J. K), R01-NS44170 (to L. J. H.), and a Muscular Dystrophy Association grant (to L. J. H.).

References

Azari MF, Lopes EC, Stubna C, Turner BJ, Zang D, Nicola NA, Kurek JB, Cheema SS (2003) Behavioural and anatomical effects of systemically administered leukemia inhibitory factor in the SOD1(G93A G1H) mouse model of familial amyotrophic lateral sclerosis. Brain Res 982:92–97

Azzouz M, Ralph GS, Storkebaum E, Walmsley LE, Mitrophanous KA, Kingsman SM, Carmeliet P, Mazarakis ND (2004) VEGF delivery with retrogradely transported lentivector prolongs survival in a mouse ALS model. Nature 429:413–417

Baron M, Kudin AP, Kunz WS (2007) Mitochondrial dysfunction in neurodegenerative disorders. Biochem Soc Trans 35:1228–1231

Blangy A, Arnaud L, Nigg EA (1997) Phosphorylation by p34cdc2 protein kinase regulates binding of the kinesin-related motor HsEg5 to the dynactin subunit p150. J Biol Chem 272:19418–19424

Boillee S, Vande Velde C, Cleveland DW (2006) ALS: a disease of motor neurons and their nonneuronal neighbors. Neuron 52:39–59

Brady ST, Pfister KK, Bloom GS (1990) A monoclonal antibody against kinesin inhibits both anterograde and retrograde fast axonal transport in squid axoplasm. Proc Natl Acad Sci USA 87:1061–1065

Breuer AC, Atkinson MB (1988) Fast axonal transport alterations in amyotrophic lateral sclerosis (ALS) and in parathyroid hormone (PTH)-treated axons. Cell Motil Cytoskeleton 10:321–330

Breuer AC, Lynn MP, Atkinson MB, Chou SM, Wilbourn AJ, Marks KE, Culver JE, Fleegler EJ (1987) Fast axonal transport in amyotrophic lateral sclerosis: an intra-axonal organelle traffic analysis. Neurology 37:738–748

Bronfman FC, Escudero CA, Weis J, Kruttgen A (2007) Endosomal transport of neurotrophins: roles in signaling and neurodegenerative diseases. Dev Neurobiol 67:1183–1203

Burkhardt JK, Echeverri CJ, Nilsson T, Vallee RB (1997) Overexpression of the dynamitin (p50) subunit of the dynactin complex disrupts dynein-dependent maintenance of membrane organelle distribution. J Cell Biol 139:469–484

Campenot RB, MacInnis BL (2004) Retrograde transport of neurotrophins: fact and function. J Neurobiol 58:217–229

Chen YZ, Bennett CL, Huynh HM et al. (2004) DNA/RNA helicase gene mutations in a form of juvenile amyotrophic lateral sclerosis (ALS4). Am J Hum Genet 74:1128–1135

Chen XJ, Levedakou EN, Millen KJ, Wollmann RL, Soliven B, Popko B (2007) Proprioceptive sensory neuropathy in mice with a mutation in the cytoplasmic Dynein heavy chain 1 gene. J Neurosci 27:14515–14524

Chevalier-Larsen E, Holzbaur EL (2006) Axonal transport and neurodegenerative disease. Biochim Biophys Acta 1762:1094–1108

Collard JF, Cote F, Julien JP (1995) Defective axonal transport in a transgenic mouse model of amyotrophic lateral sclerosis. Nature 375:61–64

De Vos KJ, Chapman AL, Tennant ME et al. (2007) Familial amyotrophic lateral sclerosis-linked SOD1 mutants perturb fast axonal transport to reduce axonal mitochondria content. Hum Mol Genet 16:2720–2728

De Vos KJ, Grierson AJ, Ackerley S, Miller CC (2008) Role of axonal transport in neurodegenerative diseases. Annu Rev Neurosci 31:151–173

Deacon SW, Serpinskaya AS, Vaughan PS, Lopez Fanarraga M, Vernos I, Vaughan KT, Gelfand VI (2003) Dynactin is required for bidirectional organelle transport. J Cell Biol 160:297–301

Delcroix JD, Valletta JS, Wu C, Hunt SJ, Kowal AS, Mobley WC (2003) NGF signaling in sensory neurons: evidence that early endosomes carry NGF retrograde signals. Neuron 39:69–84

Deng HX, Hentati A, Tainer JA et al. (1993) Amyotrophic lateral sclerosis and structural defects in Cu,Zn superoxide dismutase. Science 261:1047–1051

Dixit R, Ross JL, Goldman YE, Holzbaur EL (2008) Differential regulation of dynein and kinesin motor proteins by tau. Science 319:1086–1089

Dobrowolny G, Giacinti C, Pelosi L, Nicoletti C, Winn N, Barberi L, Molinaro M, Rosenthal N, Musaro A (2005) Muscle expression of a local Igf-1 isoform protects motor neurons in an ALS mouse model. J Cell Biol 168:193–199

El-Kadi AM, Soura V, Hafezparast M (2007) Defective axonal transport in motor neuron disease. J Neurosci Res 85:2557–2566

Fanara P, Banerjee J, Hueck RV, Harper MR, Awada M, Turner H, Husted KH, Brandt R, Hellerstein MK (2007) Stabilization of hyperdynamic microtubules is neuroprotective in amyotrophic lateral sclerosis. J Biol Chem 282:23465–23472

Feeney SJ, Austin L, Bennett TM, Kurek JB, Jean-Francois MJ, Muldoon C, Byrne E (2003) The effect of leukaemia inhibitory factor on SOD1 G93A murine amyotrophic lateral sclerosis. Cytokine 23:108–118

Frazier AE, Kiu C, Stojanovski D, Hoogenraad NJ, Ryan MT (2006) Mitochondrial morphology and distribution in mammalian cells. Biol Chem 387:1551–1558
Gonatas NK, Stieber A, Gonatas JO (2006) Fragmentation of the Golgi apparatus in neurodegenerative diseases and cell death. J Neurol Sci 246:21–30
Gross SP, Welte MA, Block SM, Wieschaus EF (2002) Coordination of opposite-polarity microtubule motors. J Cell Biol 156:715–724
Gunawardena S, Her LS, Brusch RG, Laymon RA, Niesman IR, Gordesky-Gold B, Sintasath L, Bonini NM, Goldstein LS (2003) Disruption of axonal transport by loss of huntingtin or expression of pathogenic polyQ proteins in Drosophila. Neuron 40:25–40
Gurney ME, Pu H, Chiu AY et al. (1994) Motor neuron degeneration in mice that express a human Cu,Zn superoxide dismutase mutation. Science 264:1772–1775
Hafezparast M, Klocke R, Ruhrberg C et al. (2003) Mutations in dynein link motor neuron degeneration to defects in retrograde transport. Science 300:808–812
Heerssen HM, Pazyra MF, Segal RA (2004) Dynein motors transport activated Trks to promote survival of target-dependent neurons. Nat Neurosci 7:596–604
Hirokawa N, Noda Y (2008) Intracellular transport and kinesin superfamily proteins, KIFs: structure, function, and dynamics. Physiol Rev 88:1089–1118
Holzbaur EL (2004) Motor neurons rely on motor proteins. Trends Cell Biol 14:233–240
Ilieva HS, Yamanaka K, Malkmus S, Kakinohana O, Yaksh T, Marsala M, Cleveland DW (2008) Mutant dynein (Loa) triggers proprioceptive axon loss that extends survival only in the SOD1 ALS model with highest motor neuron death. Proc Natl Acad Sci USA 105:12599–12604
Johnston JA, Illing ME, Kopito RR (2002) Cytoplasmic dynein/dynactin mediates the assembly of aggresomes. Cell Motil Cytoskeleton 53:26–38
Kabashi E, Valdmanis PN, Dion P et al. (2008) TARDBP mutations in individuals with sporadic and familial amyotrophic lateral sclerosis. Nat Genet 40:572–574
Kabuta T, Suzuki Y, Wada K (2006) Degradation of amyotrophic lateral sclerosis-linked mutant Cu,Zn-superoxide dismutase proteins by macroautophagy and the proteasome. J Biol Chem 281:30524–30533
Karcher RL, Deacon SW, Gelfand VI (2002) Motor-cargo interactions: the key to transport specificity. Trends Cell Biol 12:21–27
Kaspar BK, Llado J, Sherkat N, Rothstein JD, Gage FH (2003) Retrograde viral delivery of IGF-1 prolongs survival in a mouse ALS model. Science 301:839–842
Kato S, Shaw P, Wood-Allum C, Leigh PN, Shaw C (2003) Amyotrophic lateral sclerosis. In: Neurodegeneration:the molecular pathology of dementia and movement disorders. ISN Neuropath Press, Basel, pp 350–368
Kieran D, Hafezparast M, Bohnert S, Dick JR, Martin J, Schiavo G, Fisher EM, Greensmith L (2005) A mutation in dynein rescues axonal transport defects and extends the life span of ALS mice. J Cell Biol 169:561–567
Kwiatkowski TJ Jr, Bosco DA, Leclerc AL et al. (2009) Mutations in the FUS/TLS gene on chromosome 16 cause familial amyotrophic lateral sclerosis. Science 323:1205–1208
Lai C, Lin X, Chandran J, Shim H, Yang WJ, Cai H (2007) The G59S mutation in p150(glued) causes dysfunction of dynactin in mice. J Neurosci 27:13982–13990
Laird FM, Farah MH, Ackerley S et al. (2008) Motor neuron disease occurring in a mutant dynactin mouse model is characterized by defects in vesicular trafficking. J Neurosci 28:1997–2005
LaMonte BH, Wallace KE, Holloway BA, Shelly SS, Ascano J, Tokito M, Van Winkle T, Howland DS, Holzbaur EL (2002) Disruption of dynein/dynactin inhibits axonal transport in motor neurons causing late-onset progressive degeneration. Neuron 34:715–727
Lee WC, Yoshihara M, Littleton JT (2004) Cytoplasmic aggregates trap polyglutamine-containing proteins and block axonal transport in a Drosophila model of Huntington's disease. Proc Natl Acad Sci USA 101:3224–3229
Li W, Brakefield D, Pan Y, Hunter D, Myckatyn TM, Parsadanian A (2007) Muscle-derived but not centrally derived transgene GDNF is neuroprotective in G93A-SOD1 mouse model of ALS. Exp Neurol 203:457–471

Ligon LA, LaMonte BH, Wallace KE, Weber N, Kalb RG, Holzbaur EL (2005) Mutant superoxide dismutase disrupts cytoplasmic dynein in motor neurons. Neuroreport 16:533–536

Ligon LA, Tokito M, Finklestein JM, Grossman FE, Holzbaur EL (2004) A direct interaction between cytoplasmic dynein and kinesin I may coordinate motor activity. J Biol Chem 279:19201–19208

Liu J, Lillo C, Jonsson PA et al. (2004) Toxicity of familial ALS-linked SOD1 mutants from selective recruitment to spinal mitochondria. Neuron 43:5–17

Martin M, Iyadurai SJ, Gassman A, Gindhart JG Jr, Hays TS, Saxton WM (1999) Cytoplasmic dynein, the dynactin complex, and kinesin are interdependent and essential for fast axonal transport. Mol Biol Cell 10:3717–3728

Mattiazzi M, D'Aurelio M, Gajewski CD, Martushova K, Kiaei M, Beal MF, Manfredi G (2002) Mutated human SOD1 causes dysfunction of oxidative phosphorylation in mitochondria of transgenic mice. J Biol Chem 277:29626–29633

Melkonian KA, Maier KC, Godfrey JE, Rodgers M, Schroer TA (2007) Mechanism of dynamitin-mediated disruption of dynactin. J Biol Chem 282:19355–19364

Menzies FM, Cookson MR, Taylor RW, Turnbull DM, Chrzanowska-Lightowlers ZMA, Dong L, Figlewicz DA, Shaw PJ (2002) Mitochondrial dysfunction in a cell culture model of familial amyotrophic lateral sclerosis. Brain 125:1522–1533

Miki H, Okada Y, Hirokawa N (2005) Analysis of the kinesin superfamily: insights into structure and function. Trends Cell Biol 15:467–476

Miki H, Setou M, Kaneshiro K, Hirokawa N (2001) All kinesin superfamily protein, KIF, genes in mouse and human. Proc Natl Acad Sci USA 98:7004–7011

Miller KE, Sheetz MP (2004) Axonal mitochondrial transport and potential are correlated. J Cell Sci 117:2791–2804

Mizushima N (2007) Autophagy: process and function. Genes Dev 21:2861–2873

Morfini G, Pigino G, Szebenyi G, You Y, Pollema S, Brady ST (2006) JNK mediates pathogenic effects of polyglutamine-expanded androgen receptor on fast axonal transport. Nat Neurosci 9:907–916

Nishimura AL, Al-Chalabi A, Zatz M (2005) A common founder for amyotrophic lateral sclerosis type 8 (ALS8) in the Brazilian population. Hum Genet 118:499–500

Nishimura AL, Mitne-Neto M, Silva HC et al. (2004) A mutation in the vesicle-trafficking protein VAPB causes late-onset spinal muscular atrophy and amyotrophic lateral sclerosis. Am J Hum Genet 75:822–831

Pasinelli P, Belford ME, Lennon N, Bacskai BJ, Hyman BT, Trotti D, Brown RH Jr (2004) Amyotrophic lateral sclerosis-associated SOD1 mutant proteins bind and aggregate with Bcl-2 in spinal cord mitochondria. Neuron 43:19–30

Pfister KK, Shah PR, Hummerich H, Russ A, Cotton J, Annuar AA, King SM, Fisher EM (2006) Genetic analysis of the cytoplasmic dynein subunit families. PLoS Genet 2:e1

Pigino G, Morfini G, Pelsman A, Mattson MP, Brady ST, Busciglio J (2003) Alzheimer's presenilin 1 mutations impair kinesin-based axonal transport. J Neurosci 23:4499–4508

Puls I, Jonnakuty C, LaMonte BH et al. (2003) Mutant dynactin in motor neuron disease. Nat Genet 33:455–456

Puls I, Oh SJ, Sumner CJ et al. (2005) Distal spinal and bulbar muscular atrophy caused by dynactin mutation. Ann Neurol 57:687–694

Pun S, Santos AF, Saxena S, Xu L, Caroni P (2006) Selective vulnerability and pruning of phasic motoneuron axons in motoneuron disease alleviated by CNTF. Nat Neurosci 9: 408–419

Raoul C, Abbas-Terki T, Bensadoun JC, Guillot S, Haase G, Szulc J, Henderson CE, Aebischer P (2005) Lentiviral-mediated silencing of SOD1 through RNA interference retards disease onset and progression in a mouse model of ALS. Nat Med 11:423–428

Ravikumar B, Acevedo-Arozena A, Imarisio S, Berger Z, Vacher C, O'Kane CJ, Brown SD, Rubinsztein DC (2005) Dynein mutations impair autophagic clearance of aggregate-prone proteins. Nat Genet 37:771–776

Reid E, Kloos M, Ashley-Koch A et al. (2002) A kinesin heavy chain (KIF5A) mutation in hereditary spastic paraplegia (SPG10). Am J Hum Genet 71:1189–1194

Rosen DR, Siddique T, Patterson D et al. (1993) Mutations in Cu/Zn superoxide dismutase gene are associated with familial amyotrophic lateral sclerosis. Nature 362:59–62

Sasaki S, Iwata M (1996) Impairment of fast axonal transport in the proximal axons of anterior horn neurons in amyotrophic lateral sclerosis. Neurology 47:535–540

Schroer TA (2004) Dynactin. Annu Rev Cell Dev Biol 20:759–779

Song Y, Benison G, Nyarko A, Hays TS, Barbar E (2007) Potential role for phosphorylation in differential regulation of the assembly of dynein light chains. J Biol Chem 282:17272–17279

Sreedharan J, Blair IP, Tripathi VB et al. (2008) TDP-43 mutations in familial and sporadic amyotrophic lateral sclerosis. Science 319:1668–1672

Storkebaum E, Lambrechts D, Dewerchin M et al. (2005) Treatment of motoneuron degeneration by intracerebroventricular delivery of VEGF in a rat model of ALS. Nat Neurosci 8:85–92

Ström AL, Shi P, Zhang F, Gal J, Kilty R, Hayward LJ, Zhu H (2008) Interaction of amyotrophic lateral sclerosis (ALS)-related mutant copper-zinc superoxide dismutase with the dynein-dynactin complex contributes to inclusion formation. J Biol Chem 283:22795–22805

Tateno M, Kato S, Sakurai T, Nukina N, Takahashi R, Araki T (2009) Mutant SOD1 impairs axonal transport of choline acetyltransferase and acetylcholine release by sequestering KAP3. Hum Mol Genet 18:942–955

Taylor JP, Tanaka F, Robitschek J, Sandoval CM, Taye A, Markovic-Plese S, Fischbeck KH (2003) Aggresomes protect cells by enhancing the degradation of toxic polyglutamine-containing protein. Hum Mol Genet 12:749–757

Teuchert M, Fischer D, Schwalenstoecker B, Habisch HJ, Bockers TM, Ludolph AC (2006) A dynein mutation attenuates motor neuron degeneration in SOD1G93A mice. Exp Neurol 198:271–274

Teuling E, van Dis V, Wulf PS, Haasdijk ED, Akhmanova A, Hoogenraad CC, Jaarsma D (2008) A novel mouse model with impaired dynein/dynactin function develops amyotrophic lateral sclerosis (ALS)-like features in motor neurons and improves lifespan in SOD1-ALS mice. Hum Mol Genet 17:2849–2862

Tortarolo M, Veglianese P, Calvaresi N, Botturi A, Rossi C, Giorgini A, Migheli A, Bendotti C (2003) Persistent activation of p38 mitogen-activated protein kinase in a mouse model of familial amyotrophic lateral sclerosis correlates with disease progression. Mol Cell Neurosci 23:180–192

Twig G, Elorza A, Molina AJ et al. (2008) Fission and selective fusion govern mitochondrial segregation and elimination by autophagy. Embo J 27:433–446

Valetti C, Wetzel DM, Schrader M, Hasbani MJ, Gill SR, Kreis TE, Schroer TA (1999) Role of dynactin in endocytic traffic: effects of dynamitin overexpression and colocalization with CLIP-170. Mol Biol Cell 10:4107–4120

Van Deerlin VM, Leverenz JB, Bekris LM et al. (2008) TARDBP mutations in amyotrophic lateral sclerosis with TDP-43 neuropathology: a genetic and histopathological analysis. Lancet Neurol 7:409–416

Vance C, Rogelj B, Hortobagyi T et al. (2009) Mutations in FUS, an RNA processing protein, cause familial amyotrophic lateral sclerosis type 6. Science 323:1208–1211

Wang LJ, Lu YY, Muramatsu S et al. (2002) Neuroprotective effects of glial cell line-derived neurotrophic factor mediated by an adeno-associated virus vector in a transgenic animal model of amyotrophic lateral sclerosis. J Neurosci 22:6920–6928

Warita H, Itoyama Y, Abe K (1999) Selective impairment of fast anterograde axonal transport in the peripheral nerves of asymptomatic transgenic mice with a G93A mutant SOD1 gene. Brain Res 819:120–131

Waterman-Storer CM, Karki SB, Kuznetsov SA, Tabb JS, Weiss DG, Langford GM, Holzbaur EL (1997) The interaction between cytoplasmic dynein and dynactin is required for fast axonal transport. Proc Natl Acad Sci USA 94:12180–12185

Welte MA (2004) Bidirectional transport along microtubules. Curr Biol 14:R525–R537

Whyte J, Bader JR, Tauhata SB et al. (2008) Phosphorylation regulates targeting of cytoplasmic dynein to kinetochores during mitosis. J Cell Biol 183:819–834

Williamson TL, Cleveland DW (1999) Slowing of axonal transport is a very early event in the toxicity of ALS-linked SOD1 mutants to motor neurons. Nat Neurosci 2:50–56

Wong PC, Pardo CA, Borchelt DR, Lee MK, Copeland NG, Jenkins NA, Sisodia SS, Cleveland DW, Price DL (1995) An adverse property of a familial ALS-linked SOD1 mutation causes motor neuron disease characterized by vacuolar degeneration of mitochondria. Neuron 14:1105–1116

Yano H, Lee FS, Kong H, Chuang J, Arevalo J, Perez P, Sung C, Chao MV (2001) Association of Trk neurotrophin receptors with components of the cytoplasmic dynein motor. J Neurosci 21:RC125

Ye H, Kuruvilla R, Zweifel LS, Ginty DD (2003) Evidence in support of signaling endosome-based retrograde survival of sympathetic neurons. Neuron 39:57–68

Zhang F, Strom AL, Fukada K, Lee S, Hayward LJ, Zhu H (2007) Interaction between familial ALS-linked SOD1 mutants and the dynein complex: implications of retrograde axonal transport in ALS. J Biol Chem 282:16691–16699

Zheng C, Nennesmo I, Fadeel B, Henter JI (2004) Vascular endothelial growth factor prolongs survival in a transgenic mouse model of ALS. Ann Neurol 56:564–567

Chapter 23
In Vivo Imaging of Axonal Transport in Aging and Alzheimer's Disease

Donna J. Cross and Satoshi Minoshima

Abstract In neurons, the process of axonal transport helps maintain health and homeostasis for normal neuronal functions. Protein products, synthesized in the cell body and organelles such as mitochondria, are transported via molecular "motors" to the synapse for incorporation into membranes or cytoskeletal architecture, release into the synaptic cleft, or use in synaptic functioning. Retrograde transport occurs when substances, such as trophic factors, are taken up from the synaptic cleft via endocytosis and are transported back to the cell body to modulate further protein synthesis. Although retrograde transport is an important neuronal process, this chapter will focus on imaging of anterograde axonal transport and the effects of aging and neurodegeneration. Because axonal transport mechanisms become even more critical to maintain neuronal function in neuronal populations with long axonal projections, perturbations of axonal transport may contribute to the selective vulnerability of cortical projection neurons in neurodegenerative processes such as Alzheimer's disease (AD). With the increasing longevity of the world's population, aging and age-related diseases are becoming a major healthcare issue. Research in the areas of brain aging is critical, not only to maintaining life, but also to maintaining quality of life and sustaining a fully functional, independent, elderly population. Aging is also known to be a major risk factor for AD. According to pathological observations, cortical projection neurons are particularly vulnerable to AD pathophysiology (Hof et al., 1990). In such neurons, the process of axonal transport is particularly crucial to sustaining neuronal homeostasis and functions. Certain systems can be challenging to investigate non-invasively due to the relative inaccessibility of some brain regions and the technical challenges inherent in studying dynamic, sub-cellular processes. Therefore, the methodology to study axonal transport in living brains has been limited. This chapter will present on-going efforts to investigate this critical process using recently developed in vivo imaging techniques.

Keywords Aging · Alzheimer's disease · Axonal transport · Manganese · MRI

D.J. Cross (✉)
Department of Radiology, University of Washington, Seattle, WA 98195, USA
e-mail: dcross@u.washington.edu

R.A. Nixon, A. Yuan (eds.), *Cytoskeleton of the Nervous System*,
Advances in Neurobiology 3, DOI 10.1007/978-1-4419-6787-9_23,

Abbreviations

AD	Alzheimer's disease
APP	Amyloid precursor protein
ATP	Adenosine triphosphate
Aβ	Amyloid-beta
FDG-PET	(F-18) Flurodeoxyglucose and positron emission tomography
GSK-3	Glycogen synthase kinase 3
MEMRI	Manganese-enhanced magnetic resonance imaging
VOI	Volume of interest

Contents

23.1 Overview of Axonal Transport

The movement of materials from neuronal cell bodies to synapse was originally demonstrated by Weiss and Hiscoe (1948). A substantial amount of research into the mechanisms and proteins involved with axoplasmic transport has been conducted over the last few decades. Although many of the proteins, particularly in the area of transport regulation, have yet to be understood fully, the overall mechanistic picture of axonal transport is beginning to emerge.

Many of the components involved in axonal transport processes have been identified. Like a molecular superhighway, cytoskeletal elements provide the framework for the transport system, the most important of which are microtubules (Kreutzberg, 1969; Morfini et al., 2002; Stokin and Goldstein, 2006b). Microtubules are polarized proteins that are oriented with the plus ends towards the presynaptic terminal of the axon. The molecular motor proteins responsible for movement have been identified as the kinesin superfamily, primarily kinesin-1 proteins (anterograde movement) and a multiprotein complex called dynein (retrograde movement) (Brady, 2000; Stokin and Goldstein, 2006a). Previously it was thought that so-called "fast transport" of vesicles and synaptic proteins was a separate process from "slow transport" of certain cargos, such as cytoskeletal components. Current indications are that rate differences between slow and fast transport might be a product of the amount of time a particular type of cargo spends associated with the motor proteins. Transport may not be a continuous smooth process, but may comprise many small steps and pauses (Ochs, 1975; Shea, 2000). If the cargo spends more time associated with the motor and less time pausing, then the overall transport rate from cell body to synapse is faster. The interaction of multiple motor molecules with cargo, such as vesicles, may promote the subsequent movement along microtubules (Tomishige et al., 2002; Rashid et al., 2005), with a greater number of motors on a single cargo resulting in less pausing rather than faster movement velocity. Although the basic mechanism for vesicular, protein, or organelle transport may be similar, the large variation in reported rates has been attributed to different cargos having more pausing or, in other words, spending less time associated with the motor proteins.

In vivo studies using peripheral nerve models have indicated fast axonal transport rates of normal subjects in a range from 100 to 400 mm/day (Ochs, 1972; Stromska and Ochs, 1982; Brunetti et al., 1987; Frolkis et al., 1997; Verdu et al., 2000; Uchida et al., 2004). Ochs et al. pioneered a method in which radiolabeled amino acids were injected into the cell bodies of the cat sciatic nerve. In this method, sciatic nerves were removed at time points of 2–10 h after radiolabeling and cut into 5-mm sections. Radioactivity was counted for each segment and plotted against distance of the segment from the dorsal root ganglia. A rate was calculated by dividing the distance the radioactivity profile curve traveled over the time interval. Ochs also showed that different cell types exhibit different transport rates and that transport itself is ATP dependent (Ochs, 1972).

Research indicates that cortical axonal transport rates may vary from those measured in peripheral nerves or even in the spinal cord. Indeed, Vahlsing et al. hypothesized that transport rates may even vary along the length of very long axons in their study of transport in the corticospinal tract (Vahlsing et al., 1981). That study reported rates of 460 mm/day over the early time points and 240 mm/day for the later time points over the more distal section of the tract. Jacob and O'Donoghue calculated a rate of 303 mm/day in the corticospinal tract (Jacob and O'Donoghue, 1995). In contrast, cortical studies such as one by Fibiger et al., that used radio-labeled amino acid injected into nigro-striatal and nigro-thalamic tracts, reported rates that ranged from 25 to 50 mm/day (Fibiger et al., 1972; Levin, 1977). Levin injected noradrenergic neurons with tritiated glycoproteins and measured a fast axonal transport rate in 2 waves of 96 and 48 mm/day respectively (Levin, 1977). The large disparity in the rates measured from these various studies may be due to differences in methodology or technique. However, the hypothesis that absolute rates of transport may depend on the type of neuron, the microenvironment, the length of the tract, and which section along the length is under study, as well as the type of cargo being transported, must be considered when interpreting investigations of axonal transport. This issue highlights a major challenge for researchers of this critical neuronal process and makes it difficult to compare the rates from one system to another across various investigations. Although axonal transport has been studied extensively with in vitro cell cultures and in vivo peripheral neuronal processes, in vivo studies of axonal transport in the cortex remain limited.

23.2 Aging Effect on Transport Rates

Previous studies using traditional methods to quantify transport rates have indicated a decrease with age under normal conditions (Ochs, 1973; Stromska and Ochs, 1982; Brunetti et al., 1987; McQuarrie et al., 1989; Frolkis et al., 1997; Verdu et al., 2000; Uchida et al., 2004). Several factors have been suggested that may contribute to aged-related decrease in axonal transport. Aging neurons exhibit a decline in metabolic function, which directly impacts transport, as the process is ATP dependent (Frolkis et al., 1997; Verdu et al., 2000). With a higher potential impact on transport mechanisms, many investigators have shown an age-related decrease in the production of the cytoskeletal framework, such as microtubules and associated proteins (e.g., neurofilaments and tau) (Ochs, 1973; McQuarrie et al., 1989; Parhad et al., 1995; Cash et al., 2003; Uchida et al., 2004). Other investigations have noted an increase in axonal swellings in aged neurons (Masuoka et al., 1979; McNeill et al., 1984). Fiala et al. examined dystrophic neurites near primitive (senile) plaques in aged monkey brains and found diverticula of looping connected microtubules that eventually appear to accumulate mitochondria and other organelles (Fiala et al., 2007). Recently there have also been indications that several age-associated neurodegenerative diseases may show impairment of axonal transport at a very early pathophysiological phase of the disease process (Cash et al., 2003; Pigino et al., 2003; Stokin and Goldstein, 2006a, b).

23.3 Axonal Transport and AD

Alzheimer's disease is characterized by the early and selective involvement of projection neurons in the brain. In AD it is the cortico-cortical glutamatergic projection neurons in the brain that are probably the earliest affected in the disease process (Hof et al., 1990; Stokin and Goldstein, 2006a). Impaired axonal transport may play a synergic role in pathophysiology of Alzheimer's disease by altering the delivery of various materials from the cell body to axon and synapse including cytoskeletal components, vesicular cargoes, and mitochondria for energy production, which in turn may contribute to the cascade of neurodegeneration. Several groups have suggested a link between axonal transport misregulation and the development of AD pathophysiology (Praprotnik et al., 1996; Kasa et al., 2000; Tesseur et al., 2000; Kamal et al., 2001; Lewis et al., 2001; Dai et al., 2002; Morfini et al., 2002; Pigino et al., 2003; Zhang et al., 2004; Stokin et al., 2005).

In 2007, in the area of potential AD treatments, NIH announced a clinical trial under the Alzheimer's Disease Cooperative Study (ADCS) of lithium, a drug best known for treatment of bi-polar disease (NIH, 2007). Lithium is an inhibitor of a protein known as glycogen synthase kinase-3 (GSK-3), which has been suggested as having a fundamental role in the cascade of AD pathophysiology (Takashima, 2006). There are two important isoforms, GSK-3α and GSK-3ß that have been shown to be involved in both APP processing and tau phosphorylation. Lithium treatment has been shown to block Aß production through an effect on APP cleavage by γ-secretase (Phiel et al., 2003). GSK-3ß also is known to phosphorylate tau and lithium treatment has been shown to reduce tauopathy and degeneration (Noble et al., 2005). It has been shown that Aß activates GSK-3ß through phosphatidylinositol-3 Akt signaling and this promotes hyperphosphorylation of tau (Takashima, 2006). Several recent studies, in particular one from Rockenstein and Masliah et al., indicate improved pathological outcome in mice transgenic for AD pathology after lithium-derived inhibition of GSK-3 (Phiel et al., 2003; Noble et al., 2005; Engel et al., 2006; Caccamo et al., 2007; Rockenstein et al., 2007). One study, published by Mudher et al., describes reversal of axonal transport deficits in *Drosophila* following treatment by lithium (Mudher et al., 2004).

With mounting evidence to indicate the impairment of axonal transport in the progression of AD, the critical scientific question is do these deficits precede or concur with the appearance of AD pathology? Interesting research in the area of early plaque and tangle formation is beginning to shed some light on this important question. Pathology investigations of AD have indentified abnormal axons with bulging focal swellings as well as dystrophic neurites. Axonal defects can be found associated with amyloid in plaques as well as with neurofibrillary tangles (Stokin and Goldstein, 2006a). Other axonal swellings that are not associated with plaques or tangles appear in regions known to be affected by AD. Adalbert et al. examined dystrophic axons adjacent to plaques in brains of transgenic mice and found that the axons remained continuous and cell bodies viable despite the defects (Adalbert et al., 2009). However, some reduction in associated presynaptic synaptophysin was seen. The investigators hypothesized that plaque-associated axonal swellings impaired but did not disrupt transport and did not have an immediate effect of initiating cell death or degeneration. Therefore, it is possible that pathology evidence of axonal swelling and dystrophic neurites seen in AD may or may not represent actual transport disruption in vivo. In order to answer definitively the question posed above, a method to examine transport concurrently with the development of pathophysiological lesions is required.

23.4 Manganese-Enhanced MRI of In Vivo Axonal Transport

23.4.1 Development and Early Applications

In the last decade, several groups have used radioisotopes of manganese (Mn^{2+}) to study the brain (Tjalve and Henriksson, 1999; Brenneman et al., 2000). In particular, Mn^{2+} can be taken up by neurons and, because it can traverse synapses, can be used to trace tracts through the brain (Tjalve et al., 1996; Gianutsos et al., 1997). Although Mn^{2+} is essential for the function of several endogenous proteins, including mitochondrial products, it is known to have toxic effects as it accumulates in the brain (Gavin et al., 1990; Sloot and Gramsbergen, 1994). These original studies of radiolabeled ^{54} Mn applied to the olfactory system showed transient accumulation of the metal in regions that were synaptically connected to primary olfactory neurons.

Manganese ions act as a calcium analog and enter neurons through voltage-gated Ca^{2+} channels (Narita et al., 1990; Takeda et al., 1998; Pautler and Koretsky, 2002). Manganese uptake into neuronal cells via calcium channels has been investigated (Cross et al., 2007). Uptake was modulated and then blocked in the olfactory bulb (as assessed by dynamic changes in MR enhancement) by increasing doses of the calcium channel blocker, verapamil. In vitro cell culture combined with mass spectrometry was used to confirm that manganese is an intracellular contrast agent for MRI and not merely inhabiting the interstitial space.

Studies using cell fractionation techniques have determined that the largest portion of intracellular Mn^{2+} was in the microsomal fraction, which is consistent with the hypothesis that it is packaged into vesicles for transport (Pautler et al., 1998). Several groups showed interruption of Mn^{2+} transport after application of the microtubule disrupter, colchicine (Sloot and Gramsbergen, 1994; Pautler and Koretsky, 2002). These results indicated that Mn^{2+} is transported down the axon in a microtubule-dependent manner. Once released from presynaptic neurons, Mn^{2+} crosses the synapse, enters post-synaptic neurons, and thus is distributed through interconnecting regions by selectively anterograde transport. In contrast, studies using intravenously injected Mn^{2+} have indicated that Mn^{2+} crosses the blood-brain barrier in very small amounts by a transport-mediated process, resulting in a generally diffuse uptake throughout the brain (Gallez et al., 1998; Aschner et al., 1999).

Based on the early investigations using radioactive manganese, several groups proposed to develop in vivo imaging of neuroanatomical connectivity by exploiting the paramagnetic properties of Mn^{2+}, which has a shortening effect on the MRI relaxation constant T1. The resultant MR image shows increased signal intensity in regions containing the manganese. These studies showed enhancement of the mouse olfactory and visual systems (Pautler et al., 1998), the rat optic pathway (Watanabe et al., 2001), functional plasticity in the songbird brain (Van der Linden et al., 2002), as well as monkey striatal connections (Saleem et al., 2002) and rat somatosensory cortical projections (Leergaard et al., 2003).

Some of these initial studies using manganese-enhanced MRI (MEMRI) reported rough axonal transport rate estimates. In the first study to pioneer this technique, Pautler et al. used intravitreal injection of $MnCl_2$ in a mouse model and reported enhanced contrast in the optic nerve within 2 h (2 mm/h) (Pautler et al., 1998). Another group, Watanabe et al., performed an intravitreal injection in the rat and saw a "weak but unequivocal enhancement of the superior colliculus" at 8 h post administration to yield an approximate rate of 2.8 mm/h (Watanabe et al., 2001). Saleem et al. calculated rates ranging from 0.64 to 1.42 mm/h after intracortical injections in the caudate or putamen to globus pallidus and substantia nigra destinations in monkeys (Saleem et al., 2002). In a study investigating the songbird vocal system, Van der Linden et al. estimated transport of manganese injected into the high vocal center and traveling to the nucleus robustus archistriatalis as ranging from 2 to 6 mm/h (Van der Linden et al., 2002). Leergaard et al. employed a similar method to calculate intracortical manganese transport in the rat brain at 2.1–2.6 mm/h and at 4.6–6.1 mm/h in different descending corticofugal pathways (Leergaard et al., 2003). While all of these studies reported transport rates that are within the expected range for fast axonal transport, they employed only a rough estimation using arbitrarily defined changes in MR signal intensity at a predetermined time point for the calculation. The analytical methods used to estimate axonal transport rates in those studies were not accurate enough to quantify rate changes between groups or longitudinally.

23.4.2 MEMRI Reveals Age-Associated Decline in Axonal Transport

We have investigated age-associated axonal transport deficits including the evidence of longitudinal decline in living rat brains using manganese-enhanced MRI (Cross et al., 2008). Customized statistical algorithms were developed to quantify transport rate of Mn^{2+} through the olfactory tract. Subjects underwent serial T1-weighted MR imaging after administration of $MnCl_2$ to the nasal cavity. Statistical modeling of group-wise transport through the olfactory tract indicated delay in uptake and bulk transport through olfactory system with aging (Fig. 23.1).

The aged rat group did not show statistically significant MR enhancement of the posterior olfactory tract until the 36 h post-administration scan. In contrast, the young group showed significant posterior tract enhancement by 11–12 h after administration; when rescanned longitudinally as mid-age rats, significant posterior tract enhancement was not indicated until 24 h. In order to quantify these findings into a transport rate, tracer kinetic analysis of signal changes in the olfactory tract was applied to individual animals using the bulb intensity as an input function to the tract. Kinetic analysis also allowed quantification of transport rate decrease in aged rat brains. In summary, VOIs were carefully selected to measure the average change in signal intensity in the bulb (the origin of the tract input) as well as two locations in the tract: one more anterior and the other at the most posterior portion of the tract. Tracer kinetic analysis that was derived from the theory of longitudinal dispersion and travel time of contaminants in rivers and streams was performed on the time-intensity data (Ho et al., 2002). Using individual estimates of time to peak flow in both anterior and posterior regions of the lateral olfactory tract, rate of neuronal transport of Mn^{2+} was estimated. Both longitudinally scanned mid-aged rats and the separate aged group had significantly decreased rates of Mn^{2+} transport as compared to young rats. Rate of transport in the longitudinally scanned, mid-aged group was decreased by 58% and that in the separate very aged group was decreased by 71% of the rate in young rats. The decrease in neuronal transport rate correlated with increased age (Fig. 23.2).

23.4.3 Axonal Transport in Brains Expressing Alzheimer Pathology

Initial in vivo studies using MEMRI in AD transgenic mice indicate that axonal transport may be affected prior to Aß deposition. Smith and Pautler et al. showed that uptake and transport time of manganese in the olfactory nerve was decreased

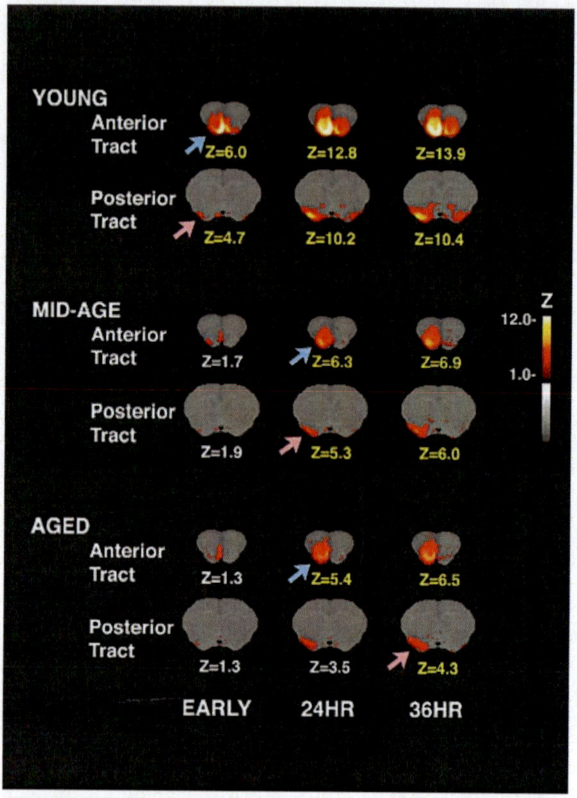

Fig. 23.1 Bulk transport in the olfactory tract is delayed with aging. Group-wise statistical maps of Mn-induced changes of intensity in the lateral olfactory tract indicated significant enhancement (threshold $Z = 4.0$, $Z>4.0$ in *yellow* text, $Z<4.0$ in *white* text). Young subjects reached threshold in both anterior (*blue arrow*) and posterior (*purple arrow*) tract by the early time of 11–12 h. Mid-age group statistical maps indicate sub-threshold enhancement of both anterior and posterior tract until 24 h. In contrast, aged group peak z-value exceeded the threshold by 24 h in the anterior volume of interest (VOI), but was still sub-threshold in the posterior VOI. By 36 h post-administration, aged group posterior VOI reached statistical threshold. Statistical maps are shown in coronal orientation superimposed onto a template MR image. *Anterior tract* = +4.6 mm, *posterior tract* = +0.2 mm from bregma landmark. (Reprinted with permission from Cross et al., 2008)

in 7–8 month Tg2576 transgenic mice compared to controls (Smith et al., 2007). Further pathological investigation indicated that these deficits were concurrent with an increase in insoluble Aß but prior to the appearance of plaques in the olfactory bulb. One limitation of that study was that axonal transport rates could not be quantified directly as the relative differences seen included the uptake via calcium channels in the nasal epithelium. The possibility that calcium influx into the olfactory nerve alters with age or disease must be considered.

In a small preliminary study, axonal transport rates in ^{SWE}APP/PS1 transgenic mice were investigated (Minoshima and Cross, 2008). Using statistical mapping of

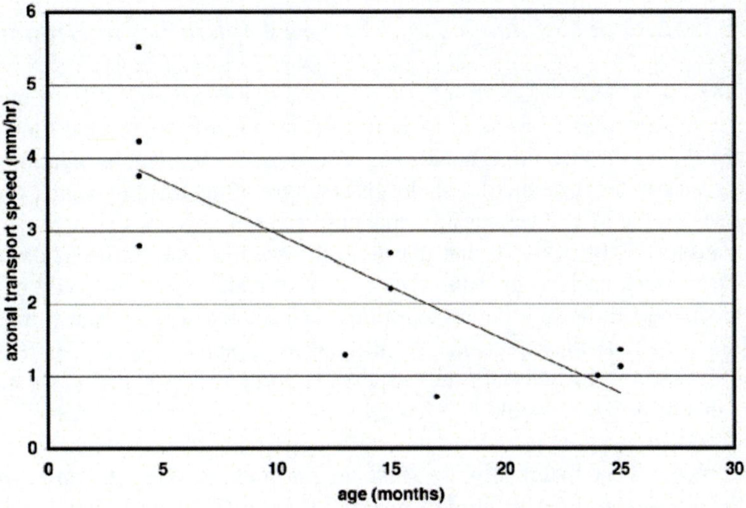

Fig. 23.2 Aging is correlated to decreased rate of axonal transport. Following time to peak flow estimation by curve-fitting, axonal transport rate between the two tract VOI was calculated by dividing the scaled tract distance by the time difference (dist(P – A))/(timepeakflowP–timepeakflowA). Trendline indicates axonal transport rates decreased with increasing age (Pearson's correlation coefficient = –0.82). (Reprinted with permission from Cross et al., 2008)

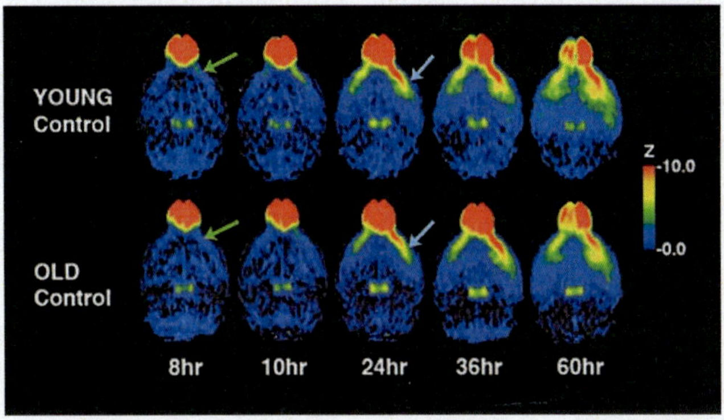

Fig. 23.3 Transverse projection images of age-related transport decline. Maximum pixel intensity in the superior inferior direction is projected in the transverse view to demonstrate rostral-caudal progression of Mn^{2+} transport. After statistical comparison of pre and post administration ($n = 5$ for 8–60 h), from *left* to *right*, images show rostral to caudal progression of Mn^{2+} through olfactory system. Early time points (8–10 h, *green arrows*) indicate differences between young and old controls, while later time points (24–60 h, *blue arrows*) show a more similar Mn^{2+} enhancement, indicating delayed but not reduced transport

serial images, the results indicated transport delays that were age-related (Fig. 23.3). Preliminary tracer kinetic analysis of young transgenics at 8–10 weeks of age showed transport rates in the lateral olfactory tract that were decreased by 46% from young controls.

23.4.4 Technical Challenges and Future Application to Humans

Regarding the technical aspects of using manganese-enhanced MRI to quantify axonal transport rates in vivo, it is important to appreciate the real-time, non-invasive value of the technique, but also to understand the limitations regarding how precisely we can localize and define the exact nature of the imaging signal changes that we see on the MRI. One question that has been raised is does manganese itself affect transport?" To evaluate that question, it would be interesting to develop a single-cell model, such as the squid giant axon, in which transport could be evaluated independently with a fluorescent probe and a microscope. Then it might be possible to examine the effect, if any, of manganese uptake and transport within that single neuron. However, for the purposes of further investigations, we have to assume that all subjects would be affected similarly by manganese. Therefore, the direct effects of manganese on transport are not differentially modulated by specific disease processes and that relative decreases in transport rates are valid observations. Also, the dose of manganese used in imaging experiments is very small, approximately 5–10 μL of 1 M $MnCl_2$, and is assumed to have minimal physiological effects. Another question has been posed regarding the biological "reason" as to why manganese is packaged into vesicles for transport down the axon. For further characterization of manganese destinations within a neuron, membrane compartmental models could be constructed and used to verify/analyze manganese sub-cellular behavior. While these are very important questions that need to be answered, the ultimate promise and power of MEMRI in the investigation of axonal transport lies in the potential application to human imaging. One technical disadvantage of this approach is that manganese has to be administered locally into the brain, which limits the application of MEMRI to a system, such as the olfactory system, that is accessible without invasive procedures for tracer administration.

The potential application of MEMRI to human studies of neurodegenerative disease as well as olfactory dysfunction is under development. Such studies will require rigorous investigation into potential toxicity of Mn^{2+}. However, the exciting scientific advantages include the potential to estimate axonal transport rates directly in living human brains non-invasively and longitudinally, which has not been shown to date. This would provide a powerful tool to estimate directly this critical process in an array of human neurodegenerative diseases as well as in normal subjects. Longitudinal estimation would allow the comparison of early, preclinical disease, such as mild cognitive impairment (MCI) in AD, to the rate of axonal transport once the disease progresses to a clinical diagnosis. These longitudinal imaging results could be correlated to human imaging data, such as studies of glucose metabolism (F-18) flurodeoxyglucose and positron emission tomography (FDG-PET) (Minoshima et al., 1997), diffusion tensor imaging of white matter tracts (DTI) (Firbank et al., 2007), and amyloid deposition (PET, Pittsburg compound B) (Klunk et al., 2004). In addition, the olfactory system has been shown to be particularly vulnerable both to age-related changes and disease and to environmentally originated pathologies (Enwere et al., 2004; Albers et al., 2006; Attems and Jellinger, 2006; Barrios et al., 2007; Luzzi et al., 2007).

23.5 Conclusion

The maintenance of health and homeostasis is critical to continuing neuronal viability. The processes involved in axonal transport and regulation are crucial to such maintenance functions. In humans, disruption of transport may underlie, in part, the loss of brain function that occurs through aging and disease. Previous investigators have employed various methods to study the general mechanisms of transport dynamics and the proteins involved. However, the very nature of axonal transport, in that the rates may depend on the cargo undergoing transport as well as the localized microenvironment in which it occurs, makes this particular process difficult to investigate. In order to uncover the subtle alterations and impairments that may be present in very early disease pathophysiology, a non-invasive imaging method of estimating transport is under development. Investigations using manganese-enhanced MRI indicated that axonal transport rates in the brain decrease with aging and in Alzheimer's disease prior to the appearance of pathology. Further development and application of this technology to humans will provide exciting scientific advantages, including the potential to estimate axonal transport rates directly in living human brains non-invasively and longitudinally, which has not been done to date. This would provide a powerful tool to estimate directly this critical process in an array of human neurodegenerative diseases.

References

Adalbert R, Nogradi A, Babetto E, Janeckova L, Walker SA, Kerschensteiner M, Misgeld T, Coleman MP (2009) Severely dystrophic axons at amyloid plaques remain continuous and connected to viable cell bodies. Brain 132(Pt 2):402–416

Albers MW, Tabert MH, Devanand DP (2006) Olfactory dysfunction as a predictor of neurodegenerative disease. Curr Neurol Neurosci Rep 6(5):379–386

Aschner M, Vrana KE, Zheng W (1999) Manganese uptake and distribution in the central nervous system (CNS). Neurotoxicology 20(2–3):173–180

Attems J, Jellinger KA (2006) Olfactory tau pathology in Alzheimer disease and mild cognitive impairment. Clin Neuropathol 25(6):265–271

Barrios FA, Gonzalez L, Favila R, Alonso ME, Salgado PM, Diaz R, Fernandez-Ruiz J (2007) Olfaction and neurodegeneration in HD. Neuroreport 18(1):73–76

Brady ST (2000) Neurofilaments run sprints not marathons. Nat Cell Biol 2(3):E43–E45

Brenneman KA, Wong BA, Buccellato MA, Costa ER, Gross EA, Dorman DC (2000) Direct olfactory transport of inhaled manganese ((54)MnCl(2)) to the rat. Toxicol Appl Pharmacol 169(3):238–248

Brunetti M, Miscena A, Salviati A, Gaiti A (1987) Effect of aging on the rate of axonal transport of choline-phosphoglycerides. Neurochem Res 12(1):61–65

Caccamo A, Oddo S, Tran LX, LaFerla FM (2007) Lithium reduces tau phosphorylation but not A beta or working memory deficits in a transgenic model with both plaques and tangles. Am J Pathol 170(5):1669–1675

Cash AD, Aliev G, Siedlak SL, Nunomura A, Fujioka H, Zhu X, Raina AK, Vinters HV, Tabaton M, Johnson AB, Paula-Barbosa M, Avila J, Jones PK, Castellani RJ, Smith MA, Perry G (2003) Microtubule reduction in Alzheimer's disease and aging is independent of tau filament formation. Am J Pathol 162(5):1623–1627

Cross DJ, Flexman JA, Anzai Y, Maravilla KR, Minoshima S (2008) Age-related decrease in axonal transport measured by MR imaging in vivo. Neuroimage 39(3):915–926

Cross DJ, Flexman JA, Anzai Y, Sasaki T, Treuting PM, Maravilla KR, Minoshima S (2007) In vivo manganese MR imaging of calcium influx in spontaneous rat pituitary adenoma. AJNR Am J Neuroradiol 28(10):1865–1871

Dai J, Buijs RM, Kamphorst W, Swaab DF (2002) Impaired axonal transport of cortical neurons in Alzheimer's disease is associated with neuropathological changes. Brain Res 948(1–2): 138–144

Engel T, Goni-Oliver P, Lucas JJ, Avila J, Hernandez F (2006) Chronic lithium administration to FTDP-17 tau and GSK-3beta overexpressing mice prevents tau hyperphosphorylation and neurofibrillary tangle formation, but pre-formed neurofibrillary tangles do not revert. J Neurochem 99(6):1445–1455. Epub 2006 Oct 24

Enwere E, Shingo T, Gregg C, Fujikawa H, Ohta S, Weiss S (2004) Aging results in reduced epidermal growth factor receptor signaling, diminished olfactory neurogenesis, and deficits in fine olfactory discrimination. J Neurosci 24(38):8354–8365

Fiala J, Feinberg M, Peters A, Barbas H (2007) Mitochondrial degeneration in dystrophic neurites of senile plaques may lead to extracellular deposition of fine filaments. Brain Struct Funct 212(2):195–207

Fibiger HC, Pudritz RE, McGeer PL, McGeer EG (1972) Axonal transport in nigro-striatal and nigro-thalamic neurons: effects of medial forebrain bundle lesions and 6-hydroxydopamine. J Neurochem 19(7):1697–1708

Firbank MJ, Blamire AM, Krishnan MS, Teodorczuk A, English P, Gholkar A, Harrison RM, O'Brien JT (2007) Diffusion tensor imaging in dementia with Lewy bodies and Alzheimer's disease. Psychiatry Res 155:135–145

Frolkis VV, Tanin SA, Gorban YN (1997) Age-related changes in axonal transport. Exp Gerontol 32(4–5):441–450

Gallez B, Baudelet C, Geurts M (1998) Regional distribution of manganese found in the brain after injection of a single dose of manganese-based contrast agents. Magn Reson Imaging 16(10):1211–1215

Gavin CE, Gunter KK, Gunter TE (1990) Manganese and calcium efflux kinetics in brain mitochondria. relevance to manganese toxicity. Biochem J 266(2):329–334

Gianutsos G, Morrow GR, Morris JB (1997) Accumulation of manganese in rat brain following intranasal administration. Fundam Appl Toxicol 37(2):102–105

Ho DT, Schlosser P, Caplow T (2002) Determination of longitudinal dispersion coefficient and net advection in the tidal hudson river with a large-scale, high resolution SF_6 tracer release experiment. Environ Sci Technol 36(15):3234–3241

Hof PR, Cox K, Morrison JH (1990) Quantitative analysis of a vulnerable subset of pyramidal neurons in Alzheimer's disease: I. Superior frontal and inferior temporal cortex. J Comp Neurol 301(1):44–54

Jacob JM, O'Donoghue DL (1995) Direct measurement of fast axonal transport rates in corticospinal axons of the adult rat. Neurosci Lett 197(1):17–20

Kamal A, Almenar-Queralt A, LeBlanc JF, Roberts EA, Goldstein LS (2001) Kinesin-mediated axonal transport of a membrane compartment containing beta-secretase and presenilin-1 requires APP. Nature 414(6864):643–648

Kasa P, Papp H, Kovacs I, Forgon M, Penke B, Yamaguchi H (2000) Human amyloid-beta1-42 applied in vivo inhibits the fast axonal transport of proteins in the sciatic nerve of rat. Neurosci Lett 278(1–2):117–119

Klunk WE, Engler H, Nordberg A, Wang Y, Blomqvist G, Holt DP, Bergstrom M, Savitcheva I, Huang GF, Estrada S, Ausen B, Debnath ML, Barletta J, Price JC, Sandell J, Lopresti BJ, Wall A, Koivisto P, Antoni G, Mathis CA, Langstrom B (2004) Imaging brain amyloid in Alzheimer's disease with pittsburgh compound-B. Ann Neurol 55(3):306–319

Kreutzberg GW (1969) Neuronal dynamics and axonal flow. IV. Blockage of intra-axonal enzyme transport by colchicine. Proc Natl Acad Sci USA 62(3):722–728

Leergaard TB, Bjaalie JG, Devor A, Wald LL, Dale AM (2003) In vivo tracing of major rat brain pathways using manganese-enhanced magnetic resonance imaging and three-dimensional digital atlasing. Neuroimage 20(3):1591–1600

Levin BE (1977) Axonal transport of [3H]fucosyl glycoproteins in noradrenergic neurons in the rat brain. Brain Res 130(3):421–432

Lewis J, Dickson DW, Lin WL, Chisholm L, Corral A, Jones G, Yen SH, Sahara N, Skipper L, Yager D, Eckman C, Hardy J, Hutton M, McGowan E (2001) Enhanced neurofibrillary degeneration in transgenic mice expressing mutant tau and APP. Science 293(5534):1487–1491

Luzzi S, Snowden JS, Neary D, Coccia M, Provinciali L, Lambon Ralph MA (2007) Distinct patterns of olfactory impairment in Alzheimer's disease, semantic dementia, frontotemporal dementia, and corticobasal degeneration. Neuropsychologia 45(8):1823–1831

Masuoka DT, Jonsson G, Finch CE (1979) Aging and unusual catecholamine-containing structures in the mouse brain. Brain Res 169(2):335–341

McNeill TH, Koek LL, Haycock JW (1984) The nigrostriatal system and aging. Peptides 5(Suppl 1):263–268

McQuarrie IG, Brady ST, Lasek RJ (1989) Retardation in the slow axonal transport of cytoskeletal elements during maturation and aging. Neurobiol Aging 10(4):359–365

Minoshima S, Cross D (2008) In vivo imaging of axonal transport using MRI: aging and Alzheimer's disease. Eur J Nucl Med Mol Imaging, 35(Suppl 1):S89–92

Minoshima S, Giordani B, Berent S, Frey KA, Foster NL, Kuhl DE (1997) Metabolic reduction in the posterior cingulate cortex in very early Alzheimer's disease. Ann Neurol 42(1):85–94

Morfini G, Pigino G, Beffert U, Busciglio J, Brady ST (2002) Fast axonal transport misregulation and Alzheimer's disease. Neuromolecular Med 2(2):89–99

Mudher A, Shepherd D, Newman TA, Mildren P, Jukes JP, Squire A, Mears A, Drummond JA, Berg S, MacKay D, Asuni AA, Bhat R, Lovestone S (2004) GSK-3beta inhibition reverses axonal transport defects and behavioural phenotypes in Drosophila. Mol Psychiatry 9(5):522–530

Narita K, Kawasaki F, Kita H (1990) Mn and Mg influxes through Ca channels of motor nerve terminals are prevented by verapamil in frogs. Brain Res 510(2):289–295

NIH (2007) news briefs: new Alzheimer's clinical trials to be undertaken by NIA nationawide consortium. Am J Alzheimers Dis Other Demen 21:460–462. doi:10.1177/1533317506297573

Noble W, Planel E, Zehr C, Olm V, Meyerson J, Suleman F, Gaynor K, Wang L, LaFrancois J, Feinstein B, Burns M, Krishnamurthy P, Wen Y, Bhat R, Lewis J, Dickson D, Duff K (2005) Inhibition of glycogen synthase kinase-3 by lithium correlates with reduced tauopathy and degeneration in vivo. Proc Natl Acad Sci USA 102(19):6990–6995. Epub 2005 May 2

Ochs S (1972) Fast transport of materials in mammalian nerve fibers. Science 176(32):252–260

Ochs S (1973) Effect of maturation and aging on the rate of fast axoplasmic transport in mammalian nerve. Prog Brain Res 40(0):349–362

Ochs S (1975) A unitary concept of axoplasmic transport based on the transport filament hypothesis. In: Bradley W et al. (eds) Recent advances in myology, vol 360. Excerpta Medica, Amsterdam, pp 189–194

Parhad IM, Scott JN, Cellars LA, Bains JS, Krekoski CA, Clark AW (1995) Axonal atrophy in aging is associated with a decline in neurofilament gene expression. J Neurosci Res 41(3): 355–366

Pautler RG, Koretsky AP (2002) Tracing odor-induced activation in the olfactory bulbs of mice using manganese-enhanced magnetic resonance imaging. Neuroimage 16(2):441–448

Pautler RG, Silva AC, Koretsky AP (1998) In vivo neuronal tract tracing using manganese-enhanced magnetic resonance imaging. Magn Reson Med 40(5):740–748

Phiel CJ, Wilson CA, Lee VM, Klein PS (2003) GSK-3alpha regulates production of Alzheimer's disease amyloid-beta peptides. Nature 423(6938):435–439

Pigino G, Morfini G, Pelsman A, Mattson MP, Brady ST, Busciglio J (2003) Alzheimer's presenilin 1 mutations impair kinesin-based axonal transport. J Neurosci 23(11):4499–4508

Praprotnik D, Smith MA, Richey PL, Vinters HV, Perry G (1996) Filament heterogeneity within the dystrophic neurites of senile plaques suggests blockage of fast axonal transport in Alzheimer's disease. Acta Neuropathol (Berl) 91(3):226–235

Rashid DJ, Bononi J, Tripet BP, Hodges RS, Pierce DW (2005) Monomeric and dimeric states exhibited by the kinesin-related motor protein KIF1A. J Pept Res 65(6):538–549

Rockenstein E, Torrance M, Adame A, Mante M, Bar-on P, Rose JB, Crews L, Masliah E (2007) Neuroprotective effects of regulators of the glycogen synthase kinase-3beta signaling pathway in a transgenic model of Alzheimer's disease are associated with reduced amyloid precursor protein phosphorylation. J Neurosci 27(8):1981–1991

Saleem KS, Pauls JM, Augath M, Trinath T, Prause BA, Hashikawa T, Logothetis NK (2002) Magnetic resonance imaging of neuronal connections in the macaque monkey. Neuron 34(5):685–700

Shea TB (2000) Microtubule motors, phosphorylation and axonal transport of neurofilaments. J Neurocytol 29(11/12):873–887

Sloot WN, Gramsbergen JB (1994) Axonal transport of manganese and its relevance to selective neurotoxicity in the rat basal ganglia. Brain Res 657(1/2):124–132

Smith KD, Kallhoff V, Zheng H, Pautler RG (2007) In vivo axonal transport rates decrease in a mouse model of Alzheimer's disease. Neuroimage 35(4):1401–1408. Epub 2007 Feb 12

Stokin GB, Goldstein LS (2006a) Axonal transport and Alzheimer's disease. Annu Rev Biochem 75:607–627

Stokin GB, Goldstein LS (2006b) Linking molecular motors to Alzheimer's disease. J Physiol Paris 99(2/3):193–200

Stokin GB, Lillo C, Falzone TL, Brusch RG, Rockenstein E, Mount SL, Raman R, Davies P, Masliah E, Williams DS, Goldstein LS (2005) Axonopathy and transport deficits early in the pathogenesis of Alzheimer's disease. Science 307(5713):1282–1288

Stromska DP, Ochs S (1982) Axoplasmic transport in aged rats. Exp Neurol 77(1):215–224

Takashima A (2006) GSK-3 is essential in the pathogenesis of Alzheimer's disease. J Alzheimers Dis 9(3 Suppl):309–317

Takeda A, Kodama Y, Ishiwatari S, Okada S (1998) Manganese transport in the neural circuit of rat CNS. Brain Res Bull 45(2):149–152

Tesseur I, Van DJ, Bruynseels K, Bronfman F, Sciot R, Van LA, Van LF (2000) Prominent axonopathy and disruption of axonal transport in transgenic mice expressing human apolipoprotein E4 in neurons of brain and spinal cord. Am J Pathol 157(5):1495–1510

Tjalve H, Henriksson J (1999) Uptake of metals in the brain via olfactory pathways. Neurotoxicology 20(2–3):181–195

Tjalve H, Henriksson J, Tallkvist J, Larsson BS, Lindquist NG (1996) Uptake of manganese and cadmium from the nasal mucosa into the central nervous system via olfactory pathways in rats. Pharmacol Toxicol 79(6):347–356

Tomishige M, Klopfenstein DR, Vale RD (2002) Conversion of Unc104/KIF1A kinesin into a processive motor after dimerization. Science 297(5590):2263–2267

Uchida A, Tashiro T, Komiya Y, Yorifuji H, Kishimoto T, Hisanaga S (2004) Morphological and biochemical changes of neurofilaments in aged rat sciatic nerve axons. J Neurochem 88(3): 735–745

Vahlsing HL, Hirschl RB, Feringa ER (1981) Axoplasmic flow of tritiated proline in the corticospinal tract of the rat. Cell Tissue Res 214(2):279–287

Van der Linden A, Verhoye M, Van Meir V, Tindemans I, Eens M, Absil P, Balthazart J (2002) In vivo manganese-enhanced magnetic resonance imaging reveals connections and functional properties of the songbird vocal control system. Neuroscience 112(2):467–474

Verdu E, Ceballos D, Vilches JJ, Navarro X (2000) Influence of aging on peripheral nerve function and regeneration. J Peripher Nerv Syst 5(4):191–208

Watanabe T, Michaelis T, Frahm J (2001) Mapping of retinal projections in the living rat using high-resolution 3D gradient-echo MRI with Mn^{2+}-induced contrast. Magn Reson Med 46(3):424–429

Weiss P, Hiscoe HB (1948) Experiments on the mechanism of nerve growth. J Exp Zool 107(3):315–395

Zhang B, Higuchi M, Yoshiyama Y, Ishihara T, Forman MS, Martinez D, Joyce S, Trojanowski JQ, Lee VM (2004) Retarded axonal transport of R406W mutant tau in transgenic mice with a neurodegenerative tauopathy. J Neurosci 24(19):4657–4667

Chapter 24
Regulation of Cytoskeletal Composition in Neurons: Transcriptional and Post-transcriptional Control in Development, Regeneration, and Disease

Ben G. Szaro and Michael J. Strong

Abstract The neuronal cytoskeleton consists of microfilaments, microtubules, and neurofilaments, which are composed of actins, tubulins, and neurofilament proteins, respectively. Each of these polymers plays a distinctive role that is subserved from embryo to adult by modulating the relative mix of the different subtypes and isoforms of their constituent monomeric subunits. Expression levels of each of these cytoskeletal protein subunits are under tight spatial and temporal control, especially during neuronal process outgrowth. In the adult, aberrations in the normal patterns of expression of cytoskeletal subunits accompany the failure of axon outgrowth during regeneration and neurodegenerative disease. In some cases, these abnormal expression patterns directly contribute to the pathological state, whereas in others, they reflect defects in regulatory modules that couple expression of cytoskeletal subunits to that of other essential intracellular elements. This control is governed by a complex interplay of transcriptional and post-transcriptional gene regulatory mechanisms. This chapter focuses on current knowledge of how the composition of each neuronal cytoskeletal polymer changes during the life cycle of the neuron, as well as the *cis*-acting elements and *trans*-acting factors that operate at the transcriptional and post-transcriptional levels.

Keywords Amyotrophic lateral sclerosis · Axon regeneration · Microfilament · Microtubule · Neurofilament · Neuronal development · Post-transcriptional control · Transcriptional control

Contents

B.G. Szaro (✉)
Department of Biological Sciences, Neuroscience Research Center, University at Albany, Albany, NY 12222, USA
e-mail: bgs86@albany.edu

R.A. Nixon, A. Yuan (eds.), *Cytoskeleton of the Nervous System*,
Advances in Neurobiology 3, DOI 10.1007/978-1-4419-6787-9_24,

24.1 Introduction – Why Cytoskeletal Composition Matters

Ever since Ramon y Cajal, neuroanatomists have been struck by the diversity of neuronal morphologies, which are defined by the size, shape, branching, and orientation of dendrites and axons. Morphometric parameters such as axon caliber and dendritic branching influence variations in neuronal physiology among neurons, defining such properties as the conduction velocity of the action potential, and synaptic connectivity. As neuronal morphologies emerge in development, they first serve to define neuronal interconnectivity. Later in development and throughout life, neuronal morphologies continue to evolve in response to epigenetic cues and in accordance with each neuron's intrinsic phenotype to provide the structural underpinnings of neuronal homeostasis and plasticity. The molecular actions of these cues converge upon the cytoskeleton. Studies of the cytoskeleton have largely concentrated on the dynamics of cytoskeletal polymer assembly and disassembly and on the movements of the assembled polymers themselves. Today, the molecular composition of the cytoskeleton is increasingly seen as playing an important ancillary role in defining the cytoskeleton's effects on cellular morphology and physiology.

The neuronal cytoskeleton, like that of all cells, is composed of three structural polymers: microfilaments, microtubules, and intermediate filaments, which in neurons are called neurofilaments (NFs). Especially during axonal and dendritic outgrowth, the assembly of these polymers, fueled by the synthesis of their constituent monomeric subunits, must be precisely timed and spatially controlled to facilitate each polymer's particular role. During neuronal process outgrowth, the microfilaments are the first to appear. They are directly responsible for the filopodial and lamellar movements that initiate process outgrowth and guide growth cones (Smith, 1994). Next to emerge are the microtubules, which play an essential role in process extension and serve as conveyor tracks for the transport of intracellular cargoes (Mitchison and Kirschner, 1988; Tanaka and Kirschner, 1991; Tanaka et al., 1995; Gallo and Letourneau, 1999). The NFs emerge last. During process outgrowth, they serve to consolidate and promote process elongation over long distances (Jiang et al., 1996; Zhu et al., 1997; Shea and Beermann, 1999; Walker et al., 2001). After synapses are established, and on through adulthood, NFs serve to establish and maintain axon caliber, especially in large diameter, myelinated PNS axons (Sakaguchi et al., 1993; Eyer and Peterson, 1994). NFs also play a critical role in localizing neurotransmitter receptors at synapses. This tethering function affects such physiological, synaptic properties as desensitization to dopaminergic stimulation (Kim et al., 2002). Finally, a number of neurological disorders, including Alzheimer's disease and amyotrophic lateral sclerosis (ALS), have significant cytoskeletal pathologies, further emphasizing the importance of appropriate cytoskeletal protein expression.

Each neuronal cytoskeletal polymer is a heteropolymeric mix of monomers that come in multiple subtypes and isoforms. By modulating which of these are used in assembly, the polymers adopt different dynamic and structural properties, many of whose functions are still being elucidated. Beginning in development and continuing throughout life, cytoskeletal subunit composition evolves continually, especially with axon development. In the adult, some of the most dramatic changes in cytoskeletal subunit composition occur in response to injury and in neurodegenerative disease. Successfully regenerating axons recapitulate many of the progressive changes in cytoskeletal composition that occurred in development. Aberrant changes that disrupt this normal progression frequently accompany the failure of axons to regenerate. Also, aberrant cytoskeletal subunit compositions are frequently hallmarks of neuropathology.

Understanding how neuronal cytoskeletal subunit protein synthesis is controlled has important implications not only for the emergence of neuronal morphologies and the establishment of neuronal connectivity during development but also for the maintenance and restoration of function in injury and disease. This chapter focuses on the interrelationship between cytoskeletal composition and neuronal shape and function and on how this synthesis is governed in healthy and diseased neurons. Although much of the early work in this area concentrated on transcriptional control, increasingly there is evidence that post-transcriptional regulation of gene expression plays an equally important role.

24.2 Microfilaments and Actins

Actins are the most ancient, abundant, and highly conserved of the three superfamilies of cytoskeletal subunit proteins. They arose from a bacterial ancestor and are expressed by all eukaryotic cells, making up as much as 1–5% of the total protein in non-muscle cells (Jones et al., 2001; van den Ent et al., 2001; Doolittle and York, 2002). The human genome contains six actin genes. Four of these genes encode isotypes of α-actin and are expressed in muscle. The other two genes encode the β- and γ-subtypes and are found in all non-muscle cells, including neurons. These proteins differ in sequence at only a few amino acids, placing them among the most highly conserved of all eukaryotic proteins.

Actin subtypes can form both homo- and heteropolymers. During development, subtypes exhibit varying preferences for co-polymerizing with one another and for differentially localizing within cells (Hayakawa et al., 1996; Rommelaere et al., 2004). Because β- and γ-actins are found in all cells and at all stages of development, little attention was paid at first to what role variations in their gene expression might play in neuronal development. Then β-actin expression in particular was found to decline in adult brain, suggesting that this subtype plays a special role during development (Weinberger et al., 1996). The strongest candidate for this role was to support process outgrowth. As discussed below, the localization of β-actin and its synthesis are now known to be crucial for growth cone motility and guidance.

24.2.1 Local β-Actin Protein Synthesis in Growing Neuronal Processes

The discovery that β-actin mRNA, along with that of other selective proteins, is locally translated in growing axons has helped to stimulate a growing interest in post-transcriptional control of gene expression in neurons. The conclusion that proteins are synthesized within the axon took a long time to be accepted, chiefly because early studies had difficulty identifying ribosomes and other components of the translational machinery that are located there (Koenig and Giuditta, 1999). Direct evidence for actin as one of these proteins was first obtained through SDS PAGE analysis of proteins labeled by local application of [^{35}S]-methionine to goldfish Mauthner cell axons and to rat spinal roots in vitro (Koenig, 1991). β-Actin was then found enriched in neuritic filopodia and growth cones, whereas γ-actin was found to be more uniformly distributed throughout the cell. This difference is reflected in the localization of their respective mRNAs as well (Bassell et al., 1998). γ-Actin mRNA is largely restricted to perikarya; its protein moves into the axon by axonal transport. Neuronal perikarya also contain the bulk of β-actin mRNA, but roughly 2% of it is transported into the neurites, where some finds its way into the growth cone.

Neuritic β-actin mRNA, along with various components of the translational machinery (e.g., EF1α and 60S ribosomal protein), is found in distinct mRNA-ribonucleoprotein (mRNP) granules whose distributions correlate spatially with microtubules. This co-localization with the translational machinery argued heavily for β-actin protein being locally synthesized within the neurite. Also, the co-localization of the mRNA with microtubules indicated that it arrives by transport rather than by leakage from the cell body and subsequent diffusion. Within the growth cone, the mRNP granules containing β-actin mRNA further concentrate within the central domain, occasionally extending into the growth cone periphery and filopodia, primarily at the leading edge. Moreover, transport is stimulated by cAMP and neurotrophins. These observations further indicate that the physiologic functions of neuritic β-actin mRNA are regulated rather than constitutive (Bassell et al., 1998; Hannan et al., 1998).

Similarity with other motile cell-types implies conservation of mechanism, which suggested early on that the localization and translation of β-actin mRNA plays a direct role in growth cone motility. β-Actin mRNA, for example, is also found at the leading edges of migrating cardiac myoblasts and fibroblasts (Sundell and Singer, 1991; Hill and Gunning, 1993). We now know that β-actin local protein synthesis is indeed crucial for growth cone motility and guidance (Leung et al., 2006; Yao et al., 2006). Local protein synthesis of it and other proteins mediates the tropic activities of a number of axon guidance cues applied locally to growth cones (Campbell and Holt, 2001; Wu et al., 2005). Application of attractive cues stimulates translation of β-actin mRNA, whereas application of repulsive cues stimulates the translation of proteins, such as cofilin, which depolymerize and fragment F-actin (reviewed in Lin and Holt, 2008). One concept for the intracellular molecular mechanism driving the attractive activity is that newly synthesized actin raises the local monomer concentration sufficiently to nucleate F-actin polymerization (Lin and Holt, 2008). Localized synthesis may also supply the continued retrograde movement of F-actin (which takes place in growth cones and is essential for its motility) with sufficient monomer to fuel the continued assembly that is needed at the distal, plus-ends of the polymers (Lin et al., 1997; Bassell et al., 1998).

24.2.2 cis-*Acting Elements and* trans-*Acting Factors Governing Local Synthesis of β-Actin*

The many parallels seen between neuronal and non-neuronal cells ultimately led to the identification of shared *cis*-acting elements in the β-actin 3′-UTR and the *trans*-acting factors that bind them. In chick fibroblasts, an RNA binding protein (RBP) called zipcode binding protein (ZBP1) was discovered to bind specifically to an ACACCC sequence at the single-stranded, exposed tip of a hairpin loop within the β-actin 3′-UTR (Kislauskis et al., 1994; Ross et al., 1997). The association between ZBP-1 and β-actin mRNA begins in the nucleus, soon after transcription (Oleynikov and Singer, 2003). Once exported from the nucleus, the mRNP complex interacts with myosin motors in the fibroblast, translocating the β-actin mRNA to where it is needed at the leading edge of the moving cell. Other non-neuronal RNAs are also transported by ZBP1-mediated movement. For example, the *Xenopus* ZBP1 ortholog, Vera, is essential for the translocation of Vegl mRNA in oocytes. This transport is necessary to specify the ventral, vegetal side of the embryo as endoderm (Deshler et al., 1998). The Vegl 3′-UTR has a zipcode sequence closely resembling that of β-actin mRNA. Thus, the mechanisms responsible for translocating β-actin mRNA are both evolutionarily conserved and utilized by a variety of cell types and different messages. ZBP1, and another ribonucleoprotein (RNP), ZBP2, mediate translocation of β-actin mRNA to growth cones, where its localized translation in response to extracellularly applied directional cues such as neurotrophins and netrin-1 helps to steer growth cones (Gu et al., 2002; Leung et al., 2006).

One early-recognized difference between β-actin mRNA localization in fibroblasts vs neurons was that in fibroblasts translocation is microfilament-dependent whereas in neurons it is microtubule-dependent (Bassell et al., 1998). This conundrum was resolved by the discovery that ZBP1 can interact with both types of motors, enabling the transport and localization of mRNA cargoes to be linked with either actin- or microtubule-based transport, depending on the local circumstances (Condeelis and Singer, 2005). The translocation of mRNAs is directly intertwined with translational control, although precisely how this is achieved is not yet fully understood. In general, mRNA cargoes are translationally silenced until they reach their destinations (Chang et al., 2006). This localized translation is probably cap-dependent, since the eukaryotic translation initiation factors eIF-4E and the binding proteins are, like actin synthesis, asymmetrically activated by local application of attractive cues like netrin-1 or BDNF.

β-Actin mRNA is just one of a much larger set of transcripts found in growth cones. Microarray analysis of laser-captured growth cones of cultured dorsal root ganglionic (DRG) neurons has identified more than 200 such transcripts (Willis et al., 2007). These additional transcripts include mRNAs encoding factors that regulate actin polymerization and de-polymerization (van Kesteren et al., 2006; Piper et al., 2006), as well as subunits of the other cytoskeletal polymers (see Sections 24.3.2 and 24.4.1.2). These transcripts can be grouped according to their differential responses to externally applied tropic molecules (e.g., NGF, BDNF, NT3, MAG, and Sema3A), arguing that distinct pathways for activating transcripts may underlie the differential responses to external cues (Willis et al., 2007). Shared binding to different sets of RNPs, possibly modulated by microRNAs (miRNAs), has been envisioned as a means to coordinate expression of functionally inter-related genes that are otherwise under separate transcriptional control (Keene and Tenenbaum, 2002). This model predicts that the grouping of transcripts according to their differential response to tropic cues should reflect differential associations with various sets of RNPs, an idea that remains to be tested empirically. Post-transcriptional mechanisms for orchestrating gene expression are now being investigated more widely for other sets of functionally inter-related proteins in the nervous system, including, as discussed later in this chapter, the NFs.

24.3 Microtubules and Tubulins

Like microfilaments, microtubules are also ubiquitous among eukaryotic cells, where they play critical roles in cellular process outgrowth, motility, and intracellular transport. Microtubules are made from a 1:1 stoichiometric mixture of α- and β-tubulin subunits. Tubulin isoforms are slightly more diverse than actins, with six α- and seven β-tubulin genes in the human genome. α-Tubulin isotypes 1, 2, and 4 and β-tubulin isotypes I, II, III, and IV are expressed in the nervous system (Ludueña, 1998). Of these, α1 and βIII are exclusively neuronal.

In general, neuronal microtubules are heteropolymeric mixtures of the neuron-specific tubulins combined with the more generally expressed ones. For example, β-tubulin in bovine brain is approximately 3% βI, 58% βII, 25% βIII, and 13% βIV (Banerjee et al., 1988). Incorporating differing mixes of tubulin isotypes into microtubules confers distinct properties on the polymers. For example, microtubules containing βIII-tubulin are more stable than those containing βII-tubulin (Schwarz et al., 1998), and altering the mix of the different α- and β-isotypes influences the assembly dynamics of microtubule plus-ends as well as the association of microtubules with different motor proteins (Carvalho et al., 2003; Howard and Hyman, 2003). This section will deal with the importance of the neuronal tubulins for process outgrowth and current knowledge about the control of their expression.

24.3.1 Tubulins in Neuronal Differentiation and Process Outgrowth

Because βIII tubulin is exclusively neuronal, it has been used extensively in studies of how the neuronal phenotype becomes established during development (McKerracher et al., 1993). This isotype is recognized by the Tuj1 monoclonal antibody, which is used widely as a marker for differentiated neurons. Both mRNA and protein are detectable in neuroepithelia as soon as neurons begin to differentiate (Geisert Jr. and Frankfurter, 1989; Caccamo et al., 1989; Moody et al., 1989; Easter Jr et al., 1993). Occasionally they are found in specific populations of dividing neuroblasts, where they are considered to represent a sign of neuronal commitment (Menezes and Luskin, 1994).

In the frog (*Xenopus laevis*), which appears to lack the βIII isotype, it is the βII isotype that is neuron specific. In this species, this isotype is referred to as N-tubulin (Dworkin-Rastl et al., 1986; Richter et al., 1988; Good et al., 1989; Moody et al., 1996). In all vertebrates, βII tubulin is generally the most abundant β-tubulin isotype in the nervous system. In frog it is exclusively neuronal, but in mammal it is found in both neurons and glia (Burgoyne et al., 1988). In mammal, βII-tubulin also appears in a few other organs and cell-types, such as lung and fibroblasts (Lewis et al., 1985; Lopata and Cleveland, 1987). The onset of the expression of frog βII, N-tubulin mRNA, but not the protein, occurs in animal cap cells, which are precursors of both neural and epidermal cells. During mid-neural plate stages, N-tubulin mRNA expression becomes progressively more restricted to neuronal precursors, and soon after these cells withdraw from the mitotic cycle, N-tubulin mRNA expression becomes exclusively neuronal from then on (Oschwald et al., 1991). This gradually more restrictive expression of the N-tubulin mRNA has become one of the most used markers for neuronal commitment in studies of the cascade of transcription factor activities culminating in neuronal differentiation (Oschwald et al., 1991).

Expression of the N-tubulin protein itself is delayed until neurite outgrowth, which occurs after neural tube closure and coincides with pioneer axon outgrowth,

CNS tract, and PNS nerve formation (Moody et al., 1996). The delay between mRNA and protein expression, which is not seen for βIII tubulin, suggests that expression of the βII isotype is under stronger post-transcriptional control than is βIII.

Expression of βII-tubulin in mammals, although not exclusively neuronal, is nonetheless also associated with axonal outgrowth. For example, in PC12 cells and mammalian DRG neurons, βII tubulin protein expression increases with the onset of neurite outgrowth, after which it ultimately becomes the predominant β-tubulin isoform. βIII-Tubulin also increases in expression in these cells during neurite outgrowth, but it is under-utilized compared to other isotypes for making polymerized microtubules within the axon (Joshi and Cleveland, 1989). This under-utilization may reflect its tendency to stabilize microtubules, which may need to be limited and under tight control during assembly. PNS axotomy re-induces both βII- and βIII-tubulin mRNA expressions during regeneration (Hoffman and Cleveland, 1988), further suggesting that both these isotypes play important roles in process outgrowth, although their precise individual functional roles are still under investigation.

Increased synthesis of tubulins was among the first changes in gene expression during axon regeneration to be documented (Heacock and Agranoff, 1976). Subsequent studies demonstrated that the principal α-tubulin isotype involved in these axotomy-induced increases was α1, both in rat PNS, where it is referred to as Tα1, and in fish CNS (Miller et al., 1989; Hieber et al., 1998). During both fish and rodent brain development, the onset of α1-tubulin expression, like that of βIII-tubulin, coincides with neuronal differentiation and early process outgrowth, and its expression remains high throughout brain development, after which it declines in the adult (Miller et al., 1987; Hieber et al., 1998). For these reasons, *cis*-elements and *trans*-factors that govern expression of α1, βII, and βIII tubulins have become major foci for studying both the establishment of the neuronal phenotype and the activation of a successful axonal growth program after axotomy.

24.3.2 cis-*Acting Elements and* trans-*Acting Factors Governing Neuronal Tubulin Expression*

During neuronal process outgrowth, post-transcriptional control of tubulin expression undoubtedly plays a role, although relatively little is known about it at this time. In addition to the aforementioned time delay between N-tubulin mRNA and protein expressions, there is also local synthesis of tubulins during axon outgrowth, and the presence of α1-tubulin mRNA in growth cones (Koenig, 1991; Moody et al., 1996; Willis et al., 2007). Despite these intriguing observations, most work on the control of tubulin expression in neurons has, to date, concentrated on transcriptional control. This is undoubtedly due to the impetus received from the success of a line of transgenic mice (Tα1:*lacZ*) in which the *lacZ* reporter gene is driven by 1.1 kb of upstream regulatory sequence (URS) of the rat Tα1 α-tubulin gene (Gloster et al., 1994). Expression of this reporter gene faithfully reconstitutes many of the features of α1-tubulin expression seen during neural development and axon regeneration, including the early onset of neuron-specific expression, the up-regulation after axotomy, and the later suppression of its expression by target-contact at the conclusion of axon regeneration (Wu et al., 1997; Gloster et al., 1999). The activity of the α1-tubulin promoter is also evolutionarily conserved, since 1.7 kb of URS and the first intron from the goldfish α1-tubulin gene used to drive a green fluorescent protein (eGFP) reporter in zebrafish exhibits analogous behavior to that seen with the Tα1-promoter in transgenic mice (Hieber et al., 1998; Goldman et al., 2001).

Comparison of the sequences from the rodent and fish α1-promoters led to the identification of several shared sequence motifs. The first was a set of three motifs having binding and transcriptional activities for the CCAAT enhancer binding protein (C/EBPβ) transcription factor. These motifs contain 6–11 nucleotides each, 2 of which closely resemble the consensus sequence for C/EBPβ binding (Menard et al., 2002). C/EBPβ is essential for the differentiation of cortical progenitor cells into neurons, which is accompanied by increased expression of endogenous α1-tubulin. Co-transfection analyses of cell lines demonstrated that the two most distal C/EBPβ sites (located between –176 and –110 in the mammalian Tα1 promoter), but not the third site, are essential for C/EBPβ-activation of a reporter gene. Gel shift analyses demonstrated that all three sites nonetheless bind C/EBPβ. Further mutational analyses confirmed that they are all needed for the robust, C/EBP-mediated transcriptional activation of the Tα1 promoter. When the Tα1:*lacZ* reporter gene is introduced into a C/EBPβ-null background, these transgenic mice up-regulate neither endogenous α1-tubulin nor the reporter gene after facial nerve axotomy, confirming that C/EBPβ is also essential for the injury-induced expression (Nadeau et al., 2005). Thus, C/EBPβ regulates increased expression of α1-tubulin in both neuronal differentiation and in axon regeneration.

C/EBPβ is one of six members of a family of transcription factors containing a similar C-terminal basic region-leucine zipper (bZIP) DNA-binding domain (Nerlov, 2008). These transcription factors play important roles in regulating cell metabolism, as well as in mediating differentiation and regenerative responses within a variety of neural and non-neural tissues alike, including liver (Wang et al., 2008). Thus, this pathway is likely to be important for regulating the growth, differentiation, and injury responses of many cell types. Because C/EBPs are subject to a number of post-translational modifications, including sumoylation, acetylation, and phosphorylation by MEK-family kinases and others, they have become targets of intense study for how the effects of numerous trophic and differentiation factors are mediated (Nerlov, 2008). Thus, C/EBPβ likely helps to couple the injury-induced expression of α1-tubulin with that of other genes whose products are needed for axon regeneration.

The expense and time needed for the production of multiple transgenic mouse lines has limited the pace of research on promoter elements involved in tubulin isotype expression. The reduced cost of producing transgenic lower vertebrates, such as zebrafish, provides a cost-effective alternative for analyzing transcriptional control elements in a physiologically relevant context (Goldman et al., 2001). Transgenic zebrafish have now been used to analyze the activities of several additional control elements conserved between the fish and mammalian α1-promoters. One of these is a repetitive homeodomain consensus sequence flanked by a basic helix-loop-helix binding E-box sequence, upstream of the C/EBPβ elements (Hieber et al., 1998). In transgenic zebrafish, this bHLH-domain has proved necessary for re-inducing increased α1-tubulin transcription in retinal ganglion cells after axotomy, which are capable of regenerating fully functional retinotectal projections in this animal (Senut et al., 2004). This bHLH E-box element was, however, not essential for either α1-tubulin transcription in retinal ganglion cells during development or in proliferating Müller radial glia after retinal damage.

In zebrafish, damaging the retina stimulates replacement of lost cells by proliferating Müller cells, which in fish express α1-tubulin. A separate E-box element located even further upstream than the two aforementioned elements proved to be essential for inducing this increased α1-tubulin expression in Müller cells following injury (Fausett et al., 2008). This element mediates the activity of the transcription factor, achaete-scute complex-like 1a (ascl1a). Knockdown of ascl1a expression by electroporated antisense morpholino oligonucleotides blocks not only induction of α1-tubulin expression, but also pax6 expression and Müller glia proliferation. Hence, as was the case with C/EBPβ, a factor that regulates α1-tubulin expression also regulates other genes involved in the same cellular behavior. Utilizing the *α1-tubulin:GFP* transgenic zebrafish in a mutagenesis screen has since identified 32 additional recessive genetic loci that alter reporter gene expression and affect neuronal differentiation and axonal branching (Gulati-Leekha and Goldman, 2006). Further analyses of these mutant lines will undoubtedly reveal additional insights into not only the control of tubulin gene expression but also how it relates to the complex, upstream neurogenic and injury-related pathways that impinge upon it.

Analysis of the promoters of neuronally expressed β-tubulins has progressed more slowly than that of the α1 promoter, but the appropriate tools for doing so are becoming available. Transgenic mice expressing YFP under control of the βIII promoter (Liu et al., 2007) and transgenic *Xenopus* expressing either eGFP (Marsh-Armstrong et al., 1999) or alkaline phosphatase (Huang et al., 2007) reporter genes under control of the N-tubulin (frog βII) promoter now exist. In both cases, expression of these reporter genes is indistinguishable from that of the endogenous β-tubulin isotypes. Thus, the URS of these genes must possess elements sufficient for temporally correct, neuron-specific expression. These lines will undoubtedly facilitate further analyses of the neuronal β-tubulin genes and determine whether they share similar control elements with the neuronally expressed α1-tubulin gene.

24.4 Neurofilaments and Their Intermediate Filament Subunit Proteins

NFs constitute the most genetically diverse of the three superfamilies of cytoskeletal genes. Mammalian NFs comprise a set of neuron-specific genes, most of which belong to the Type IV class of intermediate filaments (Steinert and Roop, 1988). The

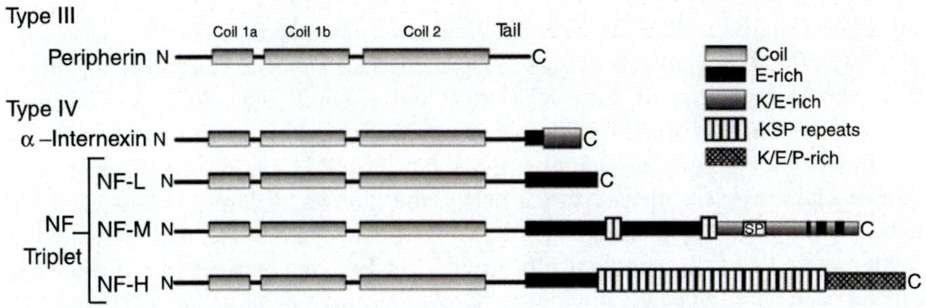

Fig. 24.1 Schematic illustration of neurofilaments

Type IV NF proteins are α-internexin and the three NF triplet genes, which are designated, based on their relative sizes, as the light-, medium-, and heavy-NF subunits (NF-L, NF-M, and NF-H, respectively; Fig. 24.1). An additional NF subunit, called peripherin, is a Type III intermediate filament, a class that includes intermediate filament proteins expressed in other, non-neuronal cell types (e.g., vimentin, desmin, and GFAP). In mammal, peripherin is expressed primarily in neurons having peripherally projecting axons and in some phylogenetically ancient CNS neurons, such as reticular neurons. Another Type III intermediate filament protein, vimentin, is also sometimes found in neurons, especially early in development and during injury-induced axon regeneration (Bignami et al., 1982; Cochard and Paulin, 1984; Boyne et al., 1996; Willis et al., 2007). Yet another intermediate filament, nestin, is classified as a Type VI intermediate filament gene (Lendahl et al., 1990). It is not typically found in differentiated neurons, but instead is found in neuroepithelia and neural stem cells, as well as in embryonic myoblasts (Zimmerman et al., 1994).

Vertebrate NFs are of relatively recent origin, having evolved separately from invertebrate NFs. Both the placement of introns and exons and the sequences of invertebrate NF genes more closely resemble those of the non-neuronal invertebrate intermediate filament genes than they do any of the vertebrate NFs, whose sequences, in turn, more closely resemble other vertebrate non-neuronal intermediate filaments (Szaro et al., 1991; Way et al., 1992; Dodemont et al., 1994; Adjaye et al., 1995; Erber et al., 1998). NF-M appears to be the most ancient of the three vertebrate NF triplet genes. It alone of these genes has been identified in virtually all vertebrates from lamprey and torpedo fish to human (Shaw, 1992). Lamprey expresses only this single NF triplet gene, plus several, as yet unclassified lighter subunits (Pleasure et al., 1989; Jacobs et al., 1995; Jin et al., 2005). These lighter subunits lack a number of properties characteristic of vertebrate NF-Ls, and therefore possibly represent peripherin or α-internexin-like genes.

Homologs of the different mammalian NF genes are more generally found among gnathostomes. In addition to NF-M, teleosts (i.e., goldfish and zebrafish) have orthologs of α-internexin (gefiltin) and NF-L, as well as peripherin (plasticin) and vimentin (Quitschke and Schechter, 1986; Glasgow et al., 1992, 1994a, b; c). The frog (i.e., *Xenopus laevis*) has NF-M and NF-L, two distinctly different α-internexin-like subunits [xefiltin and *Xenopus* neuronal intermediate filament (XNIF)], as well as orthologs of vimentin, peripherin (XIF3), and nestin (tannabin) orthologs (Sharpe, 1988; Sharpe et al., 1989; Herrmann et al., 1989; Charnas et al., 1992; Hemmati-Brivanlou et al., 1992; Zhao and Szaro, 1997a; Gervasi et al., 2000). Comparable homologs of these NFs are found in chick and reptile, too (Phillips et al., 1983; Lasek et al., 1985; Zopf et al., 1987; Rodger et al., 2001).

Among the various NF subunits, the status of NF-H among the different vertebrate classes is less clear. Among fishes, the number of larger (>100 kDa) NF subunits varies greatly (Lasek et al., 1985). Frog (*Xenopus*) has an intermediate filament protein that is immunologically similar to NF-H, but because its sequence has not yet been determined, its phylogenetic relationship to mammalian and avian NF-H is still uncertain (Szaro and Gainer, 1988a). This is because both urodeles and reptiles generally lack an NF-H ortholog, although most birds (except parakeets) and all mammals have one (Phillips et al., 1983; Lasek et al., 1985).

24.4.1 Changes in Neurofilament Protein Subunit Expressions During Development

The developmental expression patterns of these various orthologs of mammalian NF genes are remarkably well conserved phylogenetically. This conservation of expression suggests that the differential expression of the individual NF subunits is an important component of the well orchestrated program of neuronal differentiation in vertebrates. The temporal succession of NF subunits during development begins with undifferentiated neural precursor cells (e.g., embryonic neuroectodermal cells, radial glia, and stem cells) expressing vimentin and nestin. This pattern is seen in zebrafish, *Xenopus*, chick, and mammal (Schmid et al., 1979; Tapscott et al., 1981; Cochard and Paulin, 1984; Szaro and Gainer, 1988b; Lendahl et al., 1990; Hemmati-Brivanlou et al., 1992; Yuan et al., 1997; Napier et al., 1999; Mahler and Driever, 2007). Expression of these two genes diminishes as neuroblasts differentiate into neurons and glia (Bignami et al., 1982). The next subunit to appear is peripherin. Its expression accompanies early neurite outgrowth, remaining high throughout axonogenesis, then diminishing to varying degrees among separate neuronal populations as it becomes replaced by NF-L (Escurat et al., 1990; Gorham et al., 1990; Troy et al., 1990a; Canger et al., 1998; Goldstone and Sharpe, 1998; Gervasi et al., 2000). This exchange of NF-L for peripherin is thought to reflect different structural properties of the two subunits – with NF-L more conducive for a stable cytoskeleton and expansion of axon caliber in adults and peripherin for growing neuronal processes (Oblinger et al., 1989). Peripherin's expression in frog embryos differs somewhat from that in mammal. In frog, peripherin is expressed at the time of axon initiation equally in embryonic CNS and PNS, whereas in mammal its expression is largely restricted to the PNS, although it does appear in phylogenetically old CNS neurons, which are homologous to the early larval frog CNS neurons. After metamorphosis, peripherin's neuronal expression in frogs becomes more restricted to PNS, and thus closely resembles that of mammal.

During the period of most rapid axon elongation, CNS neurons in zebrafish, *Xenopus*, chick, and mammal next begin to express NF-M and α-internexin-like subunits (Carden et al., 1987; Bennett et al., 1988; Szaro et al., 1989; Kaplan et al., 1990; Charnas et al., 1992; Fliegner et al., 1994; Benson et al., 1996; Zhao and Szaro, 1997b; Asch et al., 1998; Gervasi et al., 2000). Although in these animals, α-internexin subunits continue to be expressed into adulthood, NF-L eventually emerges to become the dominant light subunit in the CNS, especially in Golgi Type I, projection neurons (Shaw and Weber, 1982; Pachter and Liem, 1984; Szaro and Gainer, 1988a; Szaro et al., 1989; Muma et al., 1991; Charnas et al., 1992; Glasgow et al., 1994a; Zhao and Szaro, 1997a; Rodger et al., 2001; Yuan et al., 2006). This increased expression of NF-L depends directly upon the establishment of synaptic contacts and is thus influenced by external cues encountered by axons (Schwartz et al., 1990; Undamatla and Szaro, 2001).

Myelination, which follows synaptogenesis, triggers still further increases in NF triplet expressions. In those organisms that express it, NF-H is always the last subunit to emerge during development (Shaw and Weber, 1982; Carden et al., 1987; Szaro et al., 1989). During postnatal mammalian brain development, the onset of NF-H expression is generally accompanied by parallel increases in expressions of the other two NF triplets (Scott et al., 1985; Schlaepfer and Bruce, 1990; Kost et al., 1992). This increased NF triplet content sets the stage for increased phosphorylation of the tail domains of NF-H and NF-M, which directly correlates with the final expansion of axon caliber (de Waegh and Brady, 1990, 1991; de Waegh et al., 1992).

This orchestrated progression of NF subunit expression is directly coupled with changes in neuronal morphology. The correlation between axon caliber, which sets action potential conduction velocity, and NF content has been known for decades (Friede and Samorajski, 1970; Hoffman et al., 1984, 1987; Cleveland et al., 1991). The first empirical evidence that this correlation is causal came from two animals lacking NFs altogether: a Japanese quail with an NF-L null mutation and a transgenic mouse lacking NFs altogether because of a transport defect arising from the fusion of *lacZ* sequences to the NF-H tail domain (Yamasaki et al., 1991; 1992; Sakaguchi et al., 1993; Eyer and Peterson, 1994). Myelinated PNS axons of both animals are significantly reduced in caliber. The phosphorylation of NF-M and NF-H tail domains also directly influences axon caliber (de Waegh et al., 1992). This relationship has led to much of the subsequent work being focused on which of these two larger subunits has the greater influence on caliber. Although altering NF-H expression in transgenic mice does affect caliber, the effects are mild compared to those arising from altering NF-M expression (Marszalek et al., 1996; Elder et al., 1998a, b; Rao et al., 2002; Garcia et al., 2003). The currently prevailing view is that axon caliber is controlled by modulating expression and phosphorylation levels of NF-M, and to a lesser degree, NF-H. Because NF-H and NF-M require the addition of light NF subunits to polymerize (Lee et al., 1993), the expressions of the latter must also be coupled in some way to the two heavier subunits.

Altering the relative expressions of the three NF triplet genes also influences dendritic arborization, as seen in spinal motor neurons (Kong et al., 1998). Thus, heterogeneous expression levels of NF subunits are likely to play fundamental roles in the morphological variation that is seen among populations of neurons.

Understanding the genetic control of NF subunit expression can therefore provide insights into the molecular mechanisms that regulate both neuronal differentiation and morphological variation.

24.4.1.1 Transcriptional Control of Neurofilament Genes During Development

Studies to date indicate that the onset of NF mRNA expression and its specificity to neurons largely depends upon transcriptional control. Differences in the level of NF protein expression among neuronal populations generally reflect differences in mRNA expression. Although mRNA levels can be affected by stability and turnover, even in the absence of changes in transcription, in situ hybridization done with probes to NF-M introns has demonstrated that neurons in the adult expressing high levels of NF protein express similarly high levels of primary transcript (hnRNA) (Ananthakrishnan et al., 2008). Thus, phenotypic differences in NF subunit expression among different populations of neurons are also likely to be dominated by strong transcriptional components.

Early studies of how the elaborately orchestrated succession of NF subunits is controlled in development focused first on upstream regulatory elements. The first NF gene to be analyzed for such was peripherin, primarily because it is the first and most abundant subunit expressed in PC12 cells exposed to nerve growth factor (NGF) (Leonard et al., 1988). A deletion-series analysis of reporter genes that were transfected into these cells identified two positive upstream regulatory elements that are required for full induction of peripherin by NGF: a proximal constitutive element lying within the first 111 basepairs (bp) upstream of the transcription start site, and an additional distal positive element (–2660 to –2308 bp). A negative regulatory element is found between these two, centered at –173 bp. This negative element plays an important role in repressing peripherin expression in undifferentiated PC12 cells. The dissociation of a repressor protein from this negative element and a positive signal mediated via the distal element are believed to be jointly responsible for de-repression of this gene upon application of NGF. The distal positive element is constitutively active and binds a complex of proteins that includes an Ets family member (Chang and Thompson, 1996). The peripherin URS also has additional elements resembling those found in other NGF-responsive genes, reiterating a common theme that control of cytoskeletal genes is coupled with that of other genes involved in similar cell behaviors through shared *cis*-elements and *trans*-factors (Thompson and Ziff, 1989).

The success of these studies in PC12 cells led to analogous ones for the Type IV NF genes in other cell lines. Because the expression of these genes is typically limited to neurons, the strategy was to compare upstream sequences among them for shared elements, which would be candidates for elements governing neuronal expression. In vitro binding assays of the URS of NF-L, NF-H, avian and mammalian NF-M, and α-internexin were also done to identify and characterize further additional candidate elements and associated transcription factors (Zopf et al., 1990; Ching and Liem, 1991; Shneidman et al., 1992; Elder et al., 1992a, c, d;

Yazdanbakhsh et al., 1994; Pospelov et al., 1994; Schoenherr et al., 1996). From these analyses, a variety of elements were identified. These included consensus sequences for transcription factor binding, direct and indirect repeats, and palindromic sequences, which appear to be involved in DNA–protein interactions, as well as several neuron-specific DNase I hypersensitive sites. Transcription factors that were implicated from these studies include Sp1 (NF-L, -M, -H, and α-internexin), Oct-1 (α-internexin), a CAAT-box-like factor (NF-M), REST1 (NF-M), Krox-24 (NF-L), AP-1 (NF-L and α-internexin), AP-2 and AP-4 (α-internexin), and Brn-3a, b, and c (α-internexin).

Another strategy was to compare the non-coding domains of an individual NF subunit gene across vertebrates. The strong degree of phylogenetic conservation of NF-M led to a comparison of fish, frog, avian, and mammalian NF-Ms (Zopf et al., 1990; Shneidman et al., 1992; Elder et al., 1992d; Glasgow et al., 1994c; Roosa et al., 2000). Although the sequences proved too disparate to align directly, shared motifs, including some that matched consensus sequences for transcription factor binding, were nonetheless present. These included a consensus sequence for the PEA3 transcription factor, which is known to stimulate expression of the neuron-specific gene, synapsin II (Petersohn et al., 1995). In chicken NF-M, the PEA3-elements lie within a long string of GAAAG direct repeats, which occurs a striking 37 times (Zopf et al., 1990). The same motif is repeated six times in human, four times in mouse, five times in *Xenopus*, and twice in fish (Roosa et al., 2000). Another transcription factor consensus element, a PTF1 site, was also present in all NF-Ms. It seems unlikely that bona fide PTF1 itself would target NF-M, since PTF1 is generally considered to be pancreas-specific (Krapp et al., 1996). Nonetheless, this site may represent a binding site for related basic helix-loop-helix proteins, which are clearly involved in neuronal differentiation and in the expression of α1-tubulin (Section 24.3.2). A palindromic (PAL) sequence motif that was identified by footprint analysis of human NF-M is also present in the frog at a comparable position within the NF-M URS (Elder et al., 1992d; Roosa et al., 2000). PAL sites also occur in the URS of NF-H and α-internexin (Ching and Liem, 1991; Elder et al., 1992a).

Attempts to study the functional activity of these various sites using transfected cell lines have proved less informative than had the studies of peripherin in PC12 cells. One strategy was to infer the activity of promoter elements by comparing expression of reporter genes transfected into neuronal vs non-neuronal cell lines. Although this method succeeded in identifying URS regions that modulate overall levels of expression, it was unable to demonstrate that any of these are responsible for neuron-specific expression, either because the transgenes lack important control elements from the endogenous gene or because neuron-specific expression requires alterations in chromatin structure not reproduced in transfected cells (Elder et al., 1992a, b). Transcription factors that functionally modulated NF expression in these assays included Brn-3a, b, and c (for α-internexin) and Krox-24 and AP-1 (for NF-L) (Pospelov et al., 1994; Budhram-Mahadeo et al., 1995). Overall, the general conclusion from these studies was that Type IV NF genes possess a strong, ubiquitously active proximal promoter, which is then further modulated by a distal positive element and an intervening negative regulatory element lying within the first few kilobases upstream. This conclusion is similar to that reached for peripherin in PC12 cells.

Evidence that this conclusion is likely to reflect properties of these NF upstream sequences in vivo came from a deletion analysis of a 1.5-kb NF-M URS:*lacZ* reporter plasmid injected into *Xenopus* embryos (Roosa et al., 2000). Because injecting plasmid DNA into *Xenopus* embryos leads to transient and mosaic expression, this method shares many similarities with the transfection of cell lines (Vize et al., 1991). This approach found domains within the URS that modulated reporter gene expression in the same way and that were located at equivalent positions as those found for chicken and rodent NF-M in the cell line studies (Zopf et al., 1990; Shneidman et al., 1992). Thus, the activity of the NF-M URS, like that of α1-tubulin, appears to be evolutionarily conserved.

A method for producing transgenic *Xenopus* that yields stable, non-mosaic expression in embryos and is more analogous to transgenic mice was then employed to determine unambiguously whether the 1.5-kb NF-M URS of *Xenopus* could produce neuron-specific expression (Amaya and Kroll, 1999; Roosa et al., 2000). The conclusion from this study was that the NF-M 1.5-kb URS, while capable of producing neuronal expression, is highly susceptible to overriding position effects originating from elements surrounding the site of integration within the chromosome, essentially acting as an enhancer trap. Thus, elements that repress non-neuronal expression are likely to lie either further upstream or downstream of the URS. Transgenic mouse studies indicate that these elements are mostly downstream of the URS, presumably in introns. A transgene extending from 2.5 kb of mammalian NF-M URS through 0.5 kb of the 3′-UTR is neuron-specific; however, each line of mice expressed the construct in different neuronal populations (Elder et al., 1994). Thus, although the elements restricting expression to neurons appear to be intragenic, additional flanking elements must be needed to direct expression to specific populations of neurons.

Clearly, the proper study of NF transcriptional elements requires the use of transgenic organisms. The conclusions reached from these studies are so far rather limited, but one consistent theme to emerge is that neuron-specific expression of all the NF genes requires intragenic elements. Including all the introns and exons of NF-L that are downstream of the 300 bp of URS into a reporter gene yields neuron-specific expression in transgenic mice (Beaudet et al., 1992, 1993). Similarly, intronic sequences must be combined with the promoters of peripherin and nestin to achieve precise, neuron- and neuroepithelia-specific expressions, respectively (Belecky-Adams et al., 1993; Zimmerman et al., 1994).

Finding the precise, intragenic elements responsible for tissue-specific expression has moved faster for peripherin and nestin than for the NF triplet genes. With peripherin, although 5.8 kb of URS is sufficient to drive temporally correct reporter gene expression in the nervous system, targeting expression to the correct populations of neurons requires additional sequences from intron 1 (Belecky-Adams et al., 1993; Uveges et al., 2002). This region binds Sp1-related proteins. Interestingly, these intron 1 sequences are unnecessary for up-regulating expression in peripheral neurons after axotomy (they were also unnecessary for NGF-induced increases in PC12 cells). Thus, tissue-specific elements are different from those needed to respond to injury.

Analogous requirements for intragenic sequences have emerged for nestin. Human and rat nestin genes contain highly conserved sequences in the 3′ portion of intron 2. *LacZ*-expressing transgenic mice utilizing the promoter and 714 bp from this intron, either from human or from rat, exhibit spatially and temporally correct expression in neural tube (Lothian and Lendahl, 1997). The human sequences also yield expression in neural crest whereas those of rat do not, preserving a species difference that is seen for the endogenous nestin genes. Further analyses demonstrated that combining a 374-bp region of human intron 2 with 120 bp of URS in the same construct is sufficient for achieving expression throughout the developing CNS. This region from intron 2 includes putative binding sites for nuclear hormone receptors, which in vitro bind thyroid hormone receptors, retinoic acid receptors (RXR and RAR), and an orphan receptor, COUP-TF. Another enhancer, further upstream, controls midbrain expression at E10.5. Transgenic mice utilizing 636 bp of rat intron 2 precisely reproduce endogenous nestin expression in the embryonic CNS and in cultured, embryonal neural stem cells (Kawaguchi et al., 2001). The human intron 2 enhancer also yields expression in cultured stem cells. This rat enhancer is also sufficient for expression in periventricular progenitor cells of the intact CNS, whereas the equivalent region of human intron 2 is not, but it yields expression in astrocytes (Johansson et al., 2002). The rat enhancer is also sufficient for elevated expression in injured spinal cord but not in injured brain, which requires the full nestin gene, including 5 kb of URS. These experiments reveal that a complex, combinatorial system of controls generates the full array of nestin expression. Although a similarly detailed analysis has yet to be carried out for the other NF subunits, what has been learned so far points to a similarly complex, combinatorial control mechanism. Future studies are needed to characterize these elements and their associated transcription factors more fully. When finished, they should reveal much about the combinatorial control of cell phenotype among different populations of neurons and how it emerges during development.

24.4.1.2 Post-transcriptional Control of Neurofilaments in Development

NF genes possess 3′-UTRs that are exceptionally well conserved in evolution, much more so than are the introns and the URS of these genes (Duret et al., 1993; Roosa et al., 2000; Thyagarajan and Szaro, 2004). Because the 3′-UTR is often the principal target of mRNPs and miRNAs that regulate mRNAs at the post-transcriptional level, this conservation suggests that post-transcriptional control of NF subunit expression must play an important role. We begin by addressing this role in development.

Although NGF clearly initiates peripherin expression in PC12 cells through activating transcriptional control elements, it acts on the other NF triplet genes chiefly by influencing RNA stability and translation (Lindenbaum et al., 1998). As is the case in embryos, expression of NF triplet proteins in PC12 cells exposed to NGF increases after the onset of peripherin expression. After onset, NF-H protein expression continues to increase in the absence of any further increase in mRNA levels. NGF-treatment of PC12 cells also stabilizes NF-M mRNA. Post-transcriptional mechanisms also modulate levels of NF triplet expression in primary neuronal culture. For example, immediately after axotomy of peripheral nerve in the intact animal, NF triplet mRNA and protein expression diminishes. This decrease is also seen when DRG cells are first placed into culture and has been thought to involve de-stabilization of the NF mRNAs as this process is inhibited by both transcriptional and translational inhibitors, which are believed to act by blocking expressions of mRNPs (Schwartz et al., 1992). In the intact animal, the increased levels of NF triplet mRNA that occur post-natally in PNS neurons after synaptogenesis also accompany increased stability (or more accurately, the decreased de-stabilization) of their RNAs (Schlaepfer and Bruce, 1990).

Another observation indicating that NF triplet RNAs are under strong post-transcriptional control is that gene dosage in transgenic mice has less effect on NF-L than on NF-H protein levels (Beaudet et al., 1993). Labeling studies with [^{35}S]-methionine have confirmed that this disparity arises from translational repression of NF-L rather than from faster NF-L protein turnover. Even more direct evidence for post-transcriptional control of NF triplet expression in brain development comes from studies of postnatal cerebral cortex. Here, NF triplet mRNA levels increase with age while transcription of these genes, as demonstrated by nuclear run-off experiments, is decreasing (Moskowitz and Oblinger, 1995).

NF subunit mRNAs are also subject to differential intracellular localization. This happens in both vertebrates and invertebrates. In squid, three NF proteins (NF60, NF70, and NF220) arise from alternative splicing of a single transcript, each having a different 3′-UTR (Szaro et al., 1991; Way et al., 1992). By in situ hybridization, the NF70 transcript is found in the axon hillock, whereas the NF220 transcript is more evenly distributed throughout the perikaryon (Way et al., 1992). In vertebrates, NF-M mRNA has been found in goldfish Mauthner cell axons, and all three NF triplet mRNAs are present in axons of mammalian sciatic nerve (Weiner et al., 1996; Sotelo-Silveira et al., 2000). In all these instances, the localized NF mRNAs are used for protein synthesis within the axon (Koenig, 1991; Crispino et al., 1993; Koenig and Giuditta, 1999). Microarray analysis of laser-captured growth cones has also found both peripherin and vimentin but not NF triplet or α-internexin mRNAs localized there (Willis et al., 2007). As the peripherin mRNA is translocated along microtubules, it is translationally silent, indicating that, as for other RNAs, the localization and transport of peripherin mRNA are intertwined with its translational control (Chang et al., 2006).

As a general rule, post-transcriptional control mechanisms appear to modulate levels of expression, complementing the role of transcriptional control mechanisms and providing neurons with an additional degree of fine control. This fine control also allows neurons to direct NF protein subunit synthesis differentially to locations within the neuron. Later in this chapter, we will see how such control is especially important in the neuronal response to injury and in neurodegenerative disease.

24.4.2 Neurofilament Protein Expression During Axon Regeneration

Changes in NF subunit expression after traumatic injury are dramatic and correlate highly with the success or failure of axon regeneration. Peripheral nerve axotomy induces increases in peripherin and α-internexin expressions, which remain elevated throughout regeneration (Oblinger et al., 1989; Muma et al., 1990; Wong and Oblinger, 1990; Troy et al., 1990b; Chadan et al., 1994; McGraw et al., 2002). Axotomy of peripheral nerve also induces a decline in the expressions of NF triplet mRNAs (Goldstein et al., 1988; Muma et al., 1990; Wong and Oblinger, 1990; Troy et al., 1990b). Both sets of changes are influenced by external cues encountered by regenerating axons. For example, increased α-internexin expression is prolonged by preventing axons from reaching their targets and NF triplet expressions rise prematurely when centrally projecting, and regenerating dorsal root axons contact the CNS glial limitans (Liuzzi and Tedeschi, 1992; McGraw et al., 2002).

Within the mammalian CNS, which has less capacity for repair than does the PNS, these axotomy-induced changes either fail to occur or do so with inappropriate timing (Mikucki and Oblinger, 1991; Bates and Meyer, 1993). In contrast, axotomy of lower vertebrate CNS axons that successfully regenerate elicits changes in peripherin, α-internexin, and NF-triplet expressions analogous to those seen in axotomized peripheral neurons (Glasgow et al., 1992; Fuchs et al., 1994; Jacobs et al., 1997; Asch et al., 1998; Gervasi et al., 2003). Such changes have been studied in detail in the optic nerve of frog and fish, and in lamprey spinal cord. Although there are many similarities between these changes in regenerating CNS axons and those in the PNS, there is one important distinction. In lower vertebrate CNS, after an early initial decline, NF-M expression subsequently rises during the peak period of axonal regrowth rather than at its end, as it does in PNS (Zhao and Szaro, 1994; Jacobs et al., 1997; Gervasi et al., 2003). Most likely, this difference reflects the higher caliber of PNS axons, which is accompanied by a much higher NF content, as well as the special role played by NF-M in determining axon caliber (Parhad et al., 1987; Garcia et al., 2003). In both frog and lamprey CNS, this increase in NF-M expression is contingent upon regeneration progressing successfully, for it fails when regenerating frog optic axons follow inappropriate pathways and when lamprey spinal cord axons fail to regenerate (Zhao and Szaro, 1995; Jacobs et al., 1997).

Further evidence that the axotomy-induced, successive changes in NF subunit expression are coupled with the progress of axon regeneration comes from studies of α-internexin expression. This expression declines at the end of lower vertebrate optic nerve regeneration, as well as in mammalian PNS axon regeneration. The decline in both situations is target-dependent (Hall and Schechter, 1991; Niloff et al., 1998). The expression of α-internexin also increases during optic nerve regeneration in lizard, but in this animal it never returns to normal (Rodger et al., 2001). Lizard optic axons regenerate into the tectum, but make synapses at inappropriate target-loci, leaving the animal functionally blind (Beazley et al., 1997). Thus, the failure of α-internexin to return to normal levels of expression in lizard can be attributed to these axons remaining in a constant state of growth. These studies collectively form a powerful argument supporting the idea that the control of NF subunit expression is directly coupled with the axonal growth program. Because NF subunit composition continually shifts throughout regeneration, understanding the molecular regulatory mechanisms underlying the control of these shifts should provide insights into the axonal growth program at multiple phases of successful regeneration.

24.4.3 Changes in Neurofilament Subunit Expression Accompanying Disease

Since most organisms are unlikely to survive the kinds of trauma necessitating axonal regeneration, it is difficult to envision how such a complex, orchestrated set of changes in gene expression has persisted throughout vertebrate evolution. One possibility is that these injury-induced changes represent an amplification of modulations in NF subunit expression that are ongoing throughout life. Retaining such control could have substantial survival benefits for recovery from relatively mild injuries or from exposure to neurotoxins and disease. A gradual decline in the ability to activate this response or correctly regulate it is seen in aging animals. Decreased NF-L RNA expression occurs selectively in populations of neurons whose axons undergo hypotrophy during aging (Tang et al., 1996). Also, NF-M mRNA levels, as opposed to NF-L or -H, are selectively diminished in aging motor neurons (Hisanaga et al., 2004). Defects in the control of NF expression could therefore contribute to the increased susceptibility to neurodegenerative changes and the decreased ability to recover from brain trauma that occurs with aging.

A correlation between abnormalities in NFs and declining neuronal function is especially evident in motor neuron disease, perhaps because the motor neurons express such high levels of NFs. Inclusion bodies containing NF proteins are an early pathogenic hallmark of degenerating motor neurons in both sporadic and familial ALS cases (Manetto et al., 1988; Leigh et al., 1989; Murayama et al., 1992; Strong, 1999). Mutations within the coding domains of peripherin and NF-H genes that lead to NF aggregation or defects in NF assembly have been associated with amyotrophic lateral sclerosis (ALS) and other neurodegenerative disorders for some time (Omary et al., 2004; Xiao et al., 2006).

In ALS, such mutations seem to represent risk factors rather than direct causes of disease. Defects in NF subunit expression are more generally found in human disease than are mutations within NF gene coding domains. Levels of both NF-L and peripherin mRNA are elevated in homogenates of human ALS spinal cord (Ge et al., 2003; Strong et al., 2004). Within degenerating spinal motor neurons, however, the steady state levels of NF-L mRNA are selectively suppressed, while those of NF-M and NF-H are spared (Bergeron et al., 1994; Wong et al., 2000; Menzies et al., 2002). Also, peripherin RNA exhibits splicing defects that are found in human ALS (Robertson et al., 2003). Some of these defects have been shown to contribute directly to neurodegeneration in a murine model of ALS. The traditionally known splice form of peripherin is translated into a 58-kDa protein, but several additional isoforms have recently been characterized (Landon et al., 1989, 2000; McLean et al., 2008). One such isoform results from the retention of intron 4, leading to the production of a 61-kDa protein. This form is not normally expressed in motor neurons, but it is expressed in motor neurons of mutant SOD1 transgenic mice, a disease model for familial ALS. Cultured neurons transfected with this isoform develop neurotoxic aggregates containing both it and native peripherin (Robertson et al., 2003). Another 28-kDa isoform, resulting from the retention of introns 3 and 4, is normally expressed at low levels in human motor neurons, but in ALS patients it is expressed at high levels and co-aggregates with the other NF subunits.

Empirical evidence that further links levels of NF subunit expression with the continued health of neurons has come from work with transgenic mice (Larivíere and Julien, 2004). Mice that over-express peripherin four-to sevenfold, due to an increase in transgene copy number, develop a late-onset, progressive neurodegenerative disorder that is characterized by NF inclusion bodies in motor neurons. These inclusion bodies resemble those seen in ALS. Neurodegeneration occurs even faster when peripherin is over-expressed in an NF-L null background, implicating imbalances in NF subunit stoichiometry as an important factor contributing to NF aggregation (Beaulieu et al., 1999). Over-expression of human NF-H in an NF-L null background also leads to aggregation of NFs. Although these aggregates are less neurotoxic than are peripherin-aggregates, they still induce motor dysfunction (Beaulieu et al., 2000). Increasing NF-L expression in the NF-H over-expressing mice reduces the swelling in perikarya and rescues the motor neuropathy, again, presumably by restoring NF subunit stoichiometry (Meier et al., 1999).

The importance of controlling the expression of each subunit separately is highlighted by the different severities of motor dysfunction and neurodegeneration that occur in the different transgenic mice. Perikaryal inclusions of NFs sometimes even improve survival. Over-expressing NF-H, which by itself leads to only mild motor dysfunction, actually increases survival of mutant SOD1 mice (Couillard-Després et al., 1998). Such studies suggest that altering NF subunit expression can, depending on the circumstances, both contribute to and protect against neurodegeneration in disease.

24.4.4 Regulating Changes in NF Protein Expression in Regeneration and Disease

Regardless of the physiological role played by changes in NF expression during injury and disease, the gene control mechanisms that regulate them provide another window into the complexities of the axonal growth program and neurodegenerative disease mechanisms at the molecular level. The first step in such a research program is to separate the relative contributions of transcriptional vs post-transcriptional mechanisms. One way to do this is to compare changes in the level of primary transcript, or heterogeneous nuclear RNA (hnRNA), to those in the mature message

(Fremeau Jr et al., 1986; Perrone-Bizzozero et al., 1991; Yue et al., 2006). This can be done with probes to introns vs the exons. By this approach, NF-M and α-internexin hnRNAs are seen to rise dramatically during frog optic nerve regeneration in both the operated and the unoperated eye, even while levels of NF mRNA undergo their initial decline in only the operated eye (Ananthakrishnan et al., 2008; Ananthakrishnan and Szaro, 2009). Throughout regeneration, hnRNA levels remain elevated in both eyes, never fully reflecting either the increases in mRNA and protein that are seen in the operated eye or the maintained levels seen in the unoperated eye. Thus, increases in protein and mRNA specific for axonal regrowth must derive largely from post-transcriptional control mechanisms acting in conjunction with transcriptional increases, which appear to arise from the disruption of the visual circuitry caused by the injury. The efficiency of translation of these NF mRNAs, as determined by polysome profiling, are also higher in the operated eye during the peak period of axon regrowth. These data indicate that the regenerative changes that occur in levels of NF-M and α-internexin mRNA and protein after optic nerve crush result from post-transcriptional control operating at several points in the life of the RNAs.

The situation for the peripherin subunit in frog optic nerve regeneration is slightly different. It is not normally expressed in mature retinal ganglion cells, but its expression is induced in retinal ganglion cells of the operated eye after axotomy. The differences in its mRNA levels, particularly during later periods of regeneration, more strongly parallel the differences seen in hnRNA levels between the two eyes. Thus, peripherin appears to be under more direct transcriptional control during regeneration than are the Type IV subunits. This difference may reflect the fact that its expression rises early during regeneration from levels that are virtually undetectable prior to injury, whereas changes in the Type IV subunits occur later and represent increases in expression that is already ongoing prior to injury.

Alterations in the post-transcriptional processing of NF mRNAs also appear to play a role in neurodegeneration. In both sporadic and familial ALS, steady state levels of NF-L mRNA are selectively suppressed in degenerating motor neurons (Bergeron et al., 1994; Strong, 1999; Wong et al., 2000; Menzies et al., 2002; Ge et al., 2003). In contrast, when NF-L mRNA levels are assayed in whole tissue homogenates, NF-L mRNA levels are significantly higher in ALS spinal cord than in controls (Strong et al., 2004). These differences reflect differences in the assay technique, but also highlight that the mechanisms of regulating NF expression in motor neurons are likely to be different from those found in non-motor neuronal cells. Because heat denaturation or proteinase K digestion of homogenates from normal individuals increased NF-L mRNA stability to approach that of the ALS patients, *trans*-acting factors that normally destabilize the mRNA appear to be lost in the disease (Ge et al., 2003). The stability of NF-L mRNA is regulated by increasing number of RNA binding proteins, the expression of which is altered in ALS. These include the TAR DNA binding protein of 43-kDa (TDP-43) (Strong et al., 2007), 14-3-3 proteins (Ge et al., 2007), either wild-type or mutant copper zinc superoxide dismutase (Ge et al., 2005) and the recently described Rho guanine nucleotide exchange factor RGNEF (Volkening et al., 2009a). Of note, each of these proteins has been associated with the formation of intra-neuronal aggregates in ALS, further suggesting that alterations in the metabolism of RNA binding proteins may be either directly or indirectly associated with the formation of NF aggregates. Adding to this complexity of regulation, it is also increasingly clear that these RNA stability determinants do not act independently of each other (Volkening et al., 2009b). Moreover, in the NFL –/– transgenic mouse model of ALS, the altered stoichiometry of NF expression is associated with not only the formation of intra-neuronal NF aggregates but also microglial activation and astrocytic proliferation, which are hallmarks of ALS (Lariviére and Julien, 2004; McLean et al., 2005). In vitro, deficiencies in NF-L protein place motor neurons at risk for oxidative injury (Strong et al., 2003), whereas the presence of NF aggregates in vitro leads to an enhanced NMDA mediated excitotoxicity (Sanelli et al., 2004, 2007; Sanelli and Strong, 2007). Both mechanisms are thought to be critical to the pathogenesis of ALS.

In a transgenic mouse disease model, over-expression of the NF-L 3′-UTR by itself also leads to neurodegeneration (Nie et al., 2002). With age, these mice develop impaired motor function and hypotrophied motor axons. Because endogenous levels of NF-L protein are normal in these mice, the defect must arise from over-expression of the NF-L 3′-UTR itself. The degenerative activity of the NF-L 3′-UTR requires a destabilizing element, found at the 5′-end of the UTR, to stimulate aggregates (Lin et al., 2005). One possibility is that the NF-L 3′-UTR, perhaps in co-operation with trans-acting factors that bind it, facilitates aggregate nucleation. Another possibility is that over-expressing the 3′-UTR may titrate out trans-acting factors that regulate not only NF-L RNA but other RNAs as well, leading to misexpression of critical proteins and contributing to aggregate formation. Either way, these observations suggest that certain feedback loops involved in regulating NF-L expression through its RNA go awry, leading to neuropathology. Also, the involvement of post-transcriptional control of NFs in neurodegenerative disease and in successful axon regeneration suggests that the two may be interrelated.

24.4.4.1 mRNPs and the Post-transcriptional Control of NF Expression

The above observations have led to searches for *cis*-acting elements and trans-acting factors involved in NF post-transcriptional control. Current ideas concerning how RNAs are regulated propose that they exist not in isolation within the cell, but as components of complexes with a continually evolving set of RNPs. These RNPs govern the fate of a RNA as it moves from the nucleus, into mRNP granules, and on to the translational machinery (Moore, 2005). Genome wide screens of the targets of RNA binding proteins have led to the idea that these RNPs may coordinate the post-transcriptionally regulated expression of multiple genes in much the same way as operons regulate the transcription of multiple functionally related genes in prokaryotes (Keene and Tenenbaum, 2002). According to this post-transcriptional operon model, the set of RNPs that associate with a given mRNA changes with the physiological state of the cell and collectively determines its fate. The same set of proteins may then bind multiple RNA targets to control jointly their expression. By placing these mRNAs under similar post-transcriptional control, cells could coordinately regulate expressions of subsets of functionally inter-related genes, even if their transcriptions are controlled independently. Because NFs are heteropolymers, such coordinated post-transcriptional control may be one way to regulate filament formation both spatially and temporally (Chang et al., 2006). The precise composition of these trans-acting modules and how they change under different circumstances, whether more than one such module exists for each NF RNA, and whether the composition of these modules is the same for the different NF RNAs, are important questions to which partial answers already exist.

In the case of NF-M, for example, several mRNPs that associate with its 3′-UTR have now been identified. These include canonical RNA binding proteins such as HuB and hnRNPs K, E1, and E2, all of which bind the NF-M 3′-UTR (Antic et al., 1999; Thyagarajan and Szaro, 2004). HuB is a member of the embryonic lethal, abnormal vision (*elav*) family of RNA binding proteins. hnRNP K and hnRNPs E1 and E2 (also known as αCP1 and αCP2) are members of the K-homology (KH) domain family of RNA binding proteins. Each has three KH-domains, which interact directly with RNAs. These proteins regulate nucleocytoplasmic export, stability, and translational activity of multiple RNAs (Bomsztyk et al., 1997; Ostareck-Lederer et al., 1998; Makeyev and Liebhaber, 2002; Bomsztyk et al., 2004; Mikula et al., 2006a).

HuB and hnRNP K are directly involved in regulating NF-M translation. Over-expressing HuB in human embryonic teratocarcinoma cells (hNT2) stimulates both neuritic outgrowth and increased translation of NF-M. Suppressing hnRNP K expression during neural development in *Xenopus* reduces the efficiency of NF-M RNA export from the nucleus and blocks its translation (Liu et al., 2008b). These effects are specific to a subset of RNA targets, because loss of hnRNP K has no effect on peripherin, GAPDH, EF1α, or N-tubulin protein expression. Loss of hnRNP K also inhibits axonal outgrowth in intact embryos and completely blocks the initiation of neurites in culture. In cultured neurons, microtubule and peripherin-containing intermediate filament organizations are also disrupted. Because such defects are not seen when NF-M expression by itself is suppressed (Lin and Szaro, 1995; Elder et al., 1998a), hnRNP K must target additional RNAs needed for axon development. In mammalian brain, NF-L, NF-H, p21, and GAP-43 RNAs are all known targets of hnRNP K and therefore possible candidates (Irwin et al., 1997; Yano et al., 2005; Thyagarajan and Szaro, 2008).

Furthermore, hnRNP K's influence on NF triplet RNAs appears to be regulated rather than constitutive. Its association with all three NF triplet mRNAs during rat postnatal cortical development varies over time (Thyagarajan and Szaro, 2008). There are several possibilities for regulating hnRNP K. For example, hnRNP K is targeted by many kinases and phosphatases (Bomsztyk et al., 2004). Its sequence contains as many as 73 potential phosphorylation sites (S/T or Y residues), including ones for casein kinases (CK1 and CK2), CAMKII, GSK3, Cdk1, PKC, Src-like kinase, and p70S6. By mass spec analyses, many, but not all, of the sites are utilized in vivo, favoring a model in which hnRNP K exists in a limited number of phosphorylation states (Mikula et al., 2006b). These phosphorylation states respond to extracellular factors, including insulin and IL-1, which influence the interactions between hnRNP K and mRNAs and with other proteins (Ostrowski et al., 2001). In addition, hnRNP K mRNA exists in four different splice forms and is subject to RNA editing in tumor cells (Dejgaard et al., 1994; Klimek-Tomczak et al., 2006). Such diversity has given rise to a model in which hnRNP K functions as a highly regulated docking platform for recruiting multiple RNPs that dictate the fate of target RNAs (Bomsztyk et al., 2004).

Although hnRNP K binds multiple targets, its effects on these targets may be regulated separately. For example, its association with NF-M RNA varies independently of its associations with NF-L and -H mRNAs in post natal cortical development (Thyagarajan and Szaro, 2008). This observation suggests that hnRNP K may participate with different co-factors for each NF subunit's RNA, forming distinct regulatory modules that nonetheless share some individual components. The idea that different collective sets of shared and unique *trans*-acting factors regulate each NF subunit separately is further indicated by the distinct behaviors of NF-M, -L, and -H RNAs on gel shift assays and by their differing stabilities in ALS patients (Cañete-Soler et al., 1998; Cañete-Soler and Schlaepfer, 2000; Ge et al., 2003). Because these studies also demonstrate that disrupting the factors involved in post-transcriptional control of NF subunits have wider effects on neuron development and health, the NF mRNAs are likely to be co-targets of post-transcriptional regulatory modules targeting other, functionally interrelated neuronal mRNAs. How these ideas resolve into a single, unitary mechanism remains to be seen.

The idea that NF mRNAs are parts of post-transcriptional regulatory modules affecting a range of neuronal functions has important implications for disease, too. The observation that the stability of NF-L mRNA differs between ALS and control spinal cord, with a selective suppression in degenerating motor neurons, has led to searches for candidate stability determinants of the NF-L mRNA. These include not only a number of well known RNA binding proteins [e.g., TAR DNA-binding protein (TDP-43) (Strong et al., 2007; Buratti and Baralle, 2008)], but also additional ones, not previously thought of as binding RNAs. The latter include mutant (but not native) SOD1 (mSOD1), 14-3-3 proteins, p190RhoGEF, and aldolases A and C (Cañete-Soler et al., 2001, 2005; Ge et al., 2002, 2005, 2007). Both TDP-43 and 14-3-3 are scaffolding proteins, like hnRNP K, and thus are ideally suited for roles in mRNP complex formation and RNA targeting within the cell. Although p190RhoGEF is only expressed in mouse tissue, the human homolog (termed RGNEF) has recently been described and demonstrates differential interaction with human NF-L mRNA in ALS as opposed to control spinal cord homogenates (Volkening et al., 2009b).

TDP-43, mSOD1, and p190RhoGEF also participate directly in NF aggregate formation (Johnston et al., 2000; Lin et al., 2005; Neumann et al., 2006; Arai et al., 2006), and p190RhoGEF interacts with 14-3-3 (Zhai et al., 2001). One possibility is that these proteins help recruit NF-L to the aggregates through their associations with its mRNA, where the translated protein then participates in aggregate formation (Lin et al., 2005). Proteins like p190RhoGEF and aldolases are better known for their roles as important metabolic enzymes and thus could potentially couple changes in NF post-transcriptional control to a variety of cell signaling and metabolic pathways. Another alternative is that NF stoichiometry, as well as the splicing of peripherin RNA, could become perturbed through the effects of signaling pathways activated in disease that act on the RNPs that influence them.

24.4.4.2 Role of RNA Granules and miRNAs in Regulating mRNA Processing

Ultimately, alterations in the steady state levels of mRNA are mediated by either enhanced rates of decay of the mRNA species, or stabilization in a process that is governed by the composition of the RNA granule in which the mRNA finds itself incorporated. In the mature neuron, mRNA can be maintained in a translationally quiescent state as either transport granules during transport to the site of nascent protein synthesis or as stress granules in response to neuronal injury. mRNA can also be targeted for degradation within processing bodies (P-bodies), RNA granules that exist in a state of dynamic flux with stress granules and which can be modified in an activity-dependent manner with dynamic interchange between the two (Kedersha et al., 2005; Barbee et al., 2006; Anderson and Kedersha, 2008) (Fig. 24.2). To a significant extent, the determination of whether an mRNA will be maintained in a translationally quiescent state, or targeted for degradation, is governed by microRNAs (miRNAs).

Increasingly, miRNAs are recognized to be key determinants of RNA expression through either leading to translational repression or enhanced rates of degradation. These highly conserved non-coding RNAs are abundant in the central nervous system (Lowery et al., 2008) where they play a critical role in the regulation of neuronal development (Gao, 2008; Sreedharan et al., 2008), including dendritic synaptic spine development (Schratt et al., 2006). MiRNAs are also increasingly recognized to participate in neuronal degeneration (Nelson and Keller, 2007; Nelson et al., 2008).

Although the mechanisms regulating the expression of miRNAs and the rules ultimately governing their interactions with a host of potential mRNAs remain to be clarified, the genesis of miRNAs is clearer. They are transcribed by RNA polymerase II from either noncoding DNA or, less commonly, from the introns of protein

Fig. 24.2 (continued) In the next step, proteins that lack mRNA binding properties are recruited, which then integrate the SG with specific signaling pathways (e.g., TIA-1 with binding proteins SRC3, FAST, PMR1; G3BP with binding protein plakophilin 3) and provide the SG with the ability to integrate aspects of cellular metabolism with the translational response to stress. At this stage, RNA is triaged either into a translationally-quiescent compartment, which protects the RNA from degradation until translation is reinitiated, or into a pathway in which the RNA is destabilized and degraded in P-bodies. The interchange between SGs and P-bodies is dynamic with reversible exchange of mRNA, with the current view being that the determination of whether an mRNA is stabilized or degraded being in part dependant on the nature of the miRNA interaction with its respective MRE

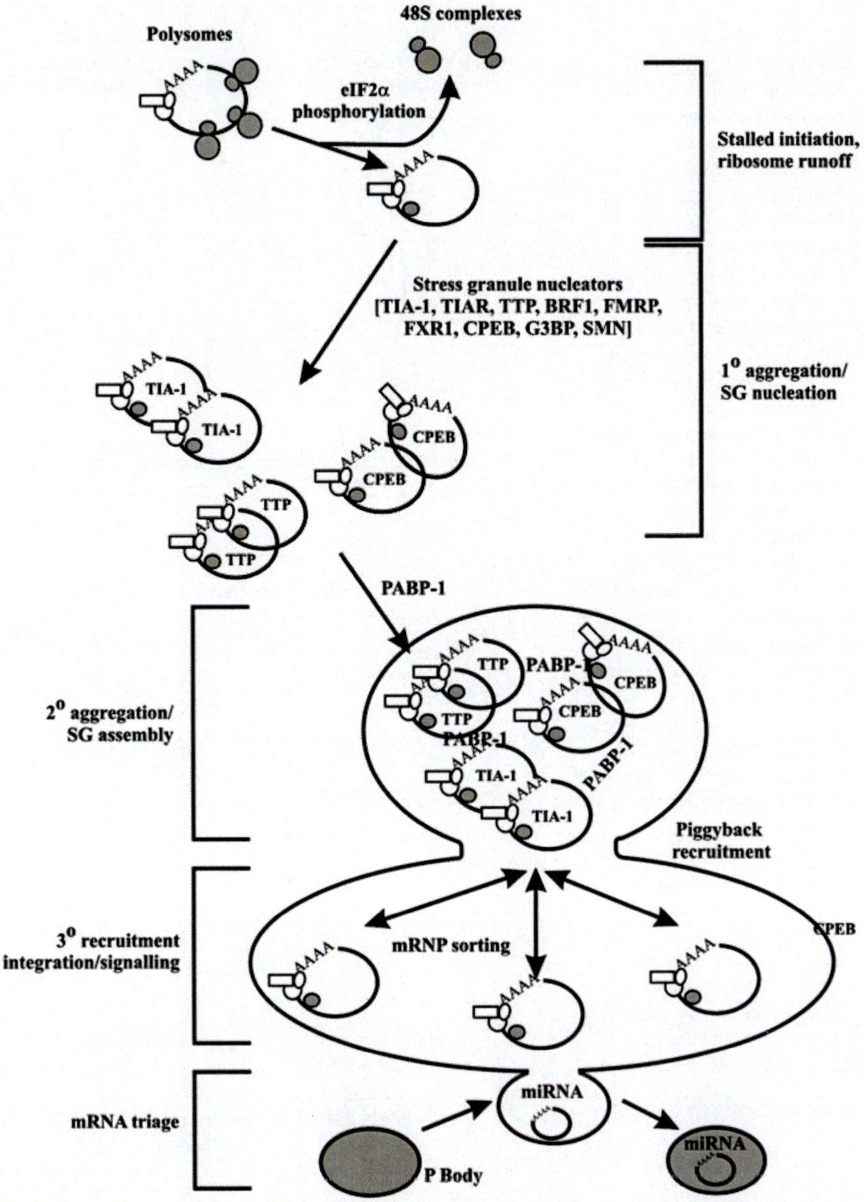

Fig. 24.2 (continued) Schematic illustration of the five stages of stress granule (SG) formation (modified from Anderson and Kedersha, 2008; Strong, 2010). In response to cellular injury or stress, phosphorylation of eIF2α results in abortive initiation complexes with stalled initiation and the conversion of polysomes into 48S ribosomes. Primary aggregation and SG nucleation is dependent on the presence of free 48S complexes and can be initiated by multiple proteins, which then become part of the SG that they nucleate. Secondary aggregation through protein–protein interactions (e.g., PABP-1 mediated) results in the progressive fusion of SG to form larger aggregates (visible at the μm size).

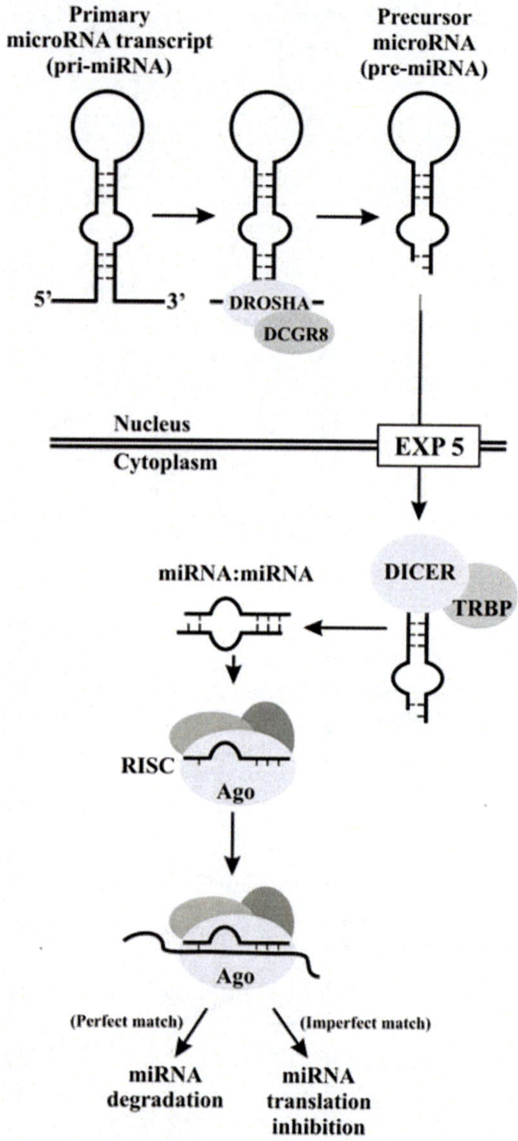

Fig. 24.3 MicroRNA (miRNA) processing. The processing of miRNA includes both nuclear and cytoplasmic components. The primary miRNA transcript is predominantly transcribed by RNA polymerase II from intronic DNA, may be kilobases in length, and forms stem–loop structures that are cleaved by a microprocessor complex, the composition of which includes Drosha, TDP-43 and FUS/TLS. Following nuclear export, the resultant pre-miRNAs are cleaved by Dicer/TRBP to yield transient miRNA duplexes which are then associated with argonaute proteins (Ago), the catalytic component of the RNA-induced silencing complex (RISC) giving rise to the functionally mature miRNA. With Ago, the miRNA becomes associated with RNPs. If the miRNA has complete complementarity to its mRNA recognition element, the mRNA is directed to P-bodies for degradation. However, if the complementarity is incomplete, then translation inhibition is induced and the mRNA preferentially targeted to stress granules (from Strong, 2010)

coding genes (Pillai, 2005) (Fig. 24.3). These pri-miRNAs, which can be kilobases long, form imperfect stem-loop structures (also known as hairpins) that are cleaved within the nucleus by the RNase III enzyme Drosha and its binding partner DGCR8 into 70–85-nucleotide (nt) precursor miRNAs (pre-miRNAs) which are then exported from the nucleus by Exportin-5 (Liu et al., 2008a). Pre-miRNA is cleaved by Dicer and TRBP into a transient miRNA duplex containing a mature miRNA sequence of 22 nt and the complementary 21- to 25-nt miRNA. The functionally mature and active miRNA is formed by the association of this miRNA duplex with argonaute (Ago) proteins, which form the catalytic component of the RNA-induced silencing complex (RISC) (Jabri, 2005; Rossi, 2005). Interestingly, only 6–7 nt at the 5′ end of the miRNA (the "seed region") is required for interactions with the miRNA recognition element (MRE) – the cis-acting binding element with its target (the "seed region"). If the miRNA has complete complementarity to its target mRNA MRE, the mRNA is preferentially directed to processing bodies (P-bodies), and the mRNA is targeted for degradation (deadenylation followed by decapping and 5′–3′ exonuclease digestion) (Liu et al., 2008a). However, the majority of miRNAs base-pair with imperfect complementarity, leading to translational silencing, possibly by the inhibition of translational initiation (Nelson and Keller, 2007). It is also likely that any one miRNA does not act in isolation in governing RNA expression in that multiple MREs (for the same, or different, miRNAs) within the same 3′-UTR can function co-operatively to enhance repression. There is, in addition, a certain degree of lack of fidelity to this process in that a single miRNA can interact with many different mRNA species, whereas some miRNAs can lead to increased translational activity. To date, little is known with regards to the regulation of miRNA interactions and the mechanisms by which specific MREs are targeted for interaction, while others may not be even within the same mRNA. Nonetheless, it is now becoming evident that "miRNA profiles" of individual disease states can be generated, such as those already being defined for Alzheimer's disease (Boissonneault et al., 2009), prion-based neurodegeneration (Saba et al., 2008), and several malignancies (Lowery et al., 2008; Sun and Tsao, 2008).

24.5 Integration of Concepts and Future Directions

We have seen how the control of neuronal cytoskeletal gene expression, at both the transcriptional and post-transcriptional levels, is fundamental to neural development and to the neuronal response to trauma and disease. Neurons use this control to modulate cytoskeletal structure and function, both spatially and temporally, by determining the precise mix of monomers available for making each polymer. Much of this regulation is directly linked to the outgrowth and maturation of axons. At present, our understanding of how this control is achieved is emerging piecemeal. Although much progress has been made toward understanding how cytoskeletal composition relates to specific facets of neuronal development, more progress remains to be made, especially in the area of how aberrant cytoskeletal protein synthesis and its control relates to disease. So far, only a few of the cis-elements and trans-acting factors involved have been identified. Even less is understood about their partnering co-factors and the specific mechanisms they use to act on their RNA targets. Even as these details continue to come to light, some common themes are now emerging.

First, the temporal onset and cell type specificity of the expression of each cytoskeletal gene appears to be largely under direct transcriptional control. This control is exercised through sets of factors that act combinatorially to determine overall levels of expression at a particular phase in the life of a neuron and in particular neuronal subpopulations. In some cases, such as that of frog N-tubulin, transcription of the gene reflects the prospects of becoming a neuron, whereas expression of the protein itself is delayed until after neuronal differentiation. Thus, post-transcriptional control mechanisms act further to sculpt and fine-tune the final temporal and spatial expression patterns of cytoskeletal proteins. Post-transcriptional control becomes especially important for the temporal and spatial control of cytoskeletal protein expression within different subcellular domains of the neuron itself, frequently in response to extracellular cues and in association with axonal outgrowth.

Second, none of the mRNPs and transcription factors identified to date are neuron-specific, nor do they regulate cytoskeletal targets exclusively. For example, the same ZBPs that target β-actin mRNA to the leading edge of growth cones do so in fibroblasts, and yet in both cell-types this control is coupled to cell motility. These same ZBPs can also participate in the localization of other mRNAs, functioning as adapters to link them to motor proteins. C/EBPβ, a transcription factor that regulates α1-tubulin expression during axon regeneration, regulates additional genes as well. In both neurons and non-neurons alike, C/EBPβ targets share a common characteristic of being involved in injury-induced, regenerative responses. TDP-43, hnRNP K, and HuB target not only NF mRNAs but also a wide range of other RNAs in multiple cell types. Yet, in neuronal development, loss of hnRNP K specifically interferes with axon outgrowth. Thus, neurons make use of the very same transcription factors and mRNPs that other cells use, not only to regulate cytoskeletal genes but also other genes involved in similar cellular behaviors and functions, marshaling them for very specific cellular functions.

How neurons utilize these common pathways to interface the expressions of cytoskeletal proteins with these other gene products, and do so in ways that subserve the unique needs of neurons, is an important remaining question. Undoubtedly, neurons employ additional co-factors, some of which may be neuron-specific. Trans-acting factors may also confer specificity on their actions by forming unique combinations with each other ones. These additional trans-acting factors will undoubtedly include both RNPs and miRNAs. The ultimate aim of these studies is to learn how neurons coordinate the actions of multiple proteins to control their development, regulate their physiology, and maintain homeostasis, as well as how this control goes awry in disease. In this chapter we have seen examples of how the control of cytoskeletal subunit expression is integral to these processes. Its study has and should continue to provide great insights into them.

Acknowledgments Work from the authors' laboratories is supported by grants from the National Science Foundation (BGS), the ALS Association (MJS), the Muscular Dystrophy Association – Tucson (MJS), the Canadian Institutes of Health Research (MJS), and the ALS Society of Canada (MJS). The authors thank John Schmidt, Kurt Gibbs, and Yuanyuan Liu for editorial comments.

References

Adjaye J, Plessmann U, Weber K, Dodemont H (1995) Characterisation of neurofilament protein NF70 mRNA from the gastropod *Helix aspersa* reveals that neuronal and non-neuronal intermediate filament proteins of cerebral ganglia arise from separate lamin-related genes. J Cell Sci 108:3581–3590

Amaya E, Kroll KL (1999) A method for generating transgenic frog embryos. Methods Mol Biol 97:393–414

Ananthakrishnan L, Gervasi C, Szaro BG (2008) Dynamic regulation of middle neurofilament (NF-M) RNA pools during optic nerve regeneration. Neuroscience 153:144–153

Ananthakrishnan L, Szaro BG (2009) Transcriptional and translational dynamics of light neurofilament subunit RNAs during *Xenopus laevis* optic nerve regeneration. Brain Res 1250:27–40

Anderson P, Kedersha N (2008) Stress granules: the Tao of RNA triage. Trends Biochem Sci 33:141–150

Antic D, Lu N, Keene JD (1999) ELAV tumor antigen, Hel-N1, increases translation of neurofilament M mRNA and induces formation of neurites in teratocarcinoma cells. Genes Dev 13:449–461

Arai T, Hasegawa M, Akiyama H, Ikeda K, Nonaka T et al. (2006) TDP-43 is a component of ubiquitin-positive tau-negative inclusions in frontotemporal lobar degeneration and amyotrophic lateral sclerosis. Biochem Biophys Res Commun 351:602–611

Asch WS, Leake D, Canger AK, Passini MA, Argenton F, Schechter N (1998) Cloning of zebrafish neurofilament cDNAs for plasticin and gefiltin: increased mRNA expression in ganglion cells after optic nerve injury. J Neurochem 71:20–32

Banerjee A, Roach MC, Wall KA, Lopata MA, Cleveland DW, Luduena RF (1988) A monoclonal antibody against the type II isotype of β-tubulin. J Biol Chem 265:1794–1799

Barbee SA, Estes PS, Cziko AM, Hillebrand J, Luedeman RA et al. (2006) Staufen- and FMRP-containing neuronal RNPs are structurally and functionally related to somatic P bodies. Neuron 52:997–1009

Bassell GJ, Zhang H, Byrd AL, Femino AM, Singer RH et al. (1998) Sorting of β-actin mRNA and protein to neurites and growth cones in culture. J Neurosci 18:251–265

Bates CA, Meyer RL (1993) The heavy neurofilament protein is expressed in regenerating adult but not embryonic mammalian optic fibers in vitro. Exp Neurol 119:249–257

Beaudet L, Charron G, Houle D, Tretjakoff I, Peterson A, Julien J-P (1992) Intragenic regulatory elements contribute to transcriptional control of the neurofilament light gene. Gene 116: 205–214

Beaudet L, Cote F, Houle D, Julien J-P (1993) Different post transcriptional controls for the human neurofilament light and heavy genes in transgenic mice. Mol Brain Res 18:23–31

Beaulieu J-M, Jacomy H, Julien J-P (2000) Formation of intermediate filament protein aggregates with disparate effects in two transgenic mouse models lacking the neurofilament light subunit. J Neurosci 20:5321–5328

Beaulieu J-M, Nguyen MD, Julien J-P (1999) Late onset death of motor neurons in mice overexpressing wild-type peripherin. J Cell Biol 147:531–544

Beazley LD, Sheard PW, Tennant M, Starac D, Dunlop SA (1997) Optic nerve regenerates but does not restore topographic projections in the lizard *Ctenophorus ornatus*. J Comp Neurol 377:105–120

Belecky-Adams T, Wight DC, Kopchick JJ, Parysek LM (1993) Intragenic sequences are required for cell type-specific and injury-induced expression of the rat peripherin gene. J Neurosci 13:5056–5065

Bennett GS, Hollander BA, Laskowska D (1988) Expression and phosphorylation of the mid-sized neurofilament protein NF-M during chick spinal cord neurogenesis. J Neurosci Res 21:376–390

Benson DL, Mandell JW, Shaw G, Banker G (1996) Compartmentation of alpha-internexin and neurofilament triplet proteins in cultured hippocampal neurons. J Neurocytol 25:181–196

Bergeron C, Beric-Maskarel K, Muntasser S, Weyer L, Somerville MJ, Percy ME (1994) Neurofilament light and polyadenylated mRNA levels are decreased in amyotrophic lateral sclerosis motor neurons. J Neuropathol Exp Neurol 53:221–230

Bignami A, Raju T, Dahl D (1982) Localization of vimentin, the nonspecific intermediate filament protein, in embryonal glia and in early differentiating neurons. Dev Biol 91: 286–295

Boissonneault V, Plante I, Rivest S, Provost P (2009) MicroRNA-298 and microRNA-328 regulate expression of mouse β-amyloid precursor protein converting enzyme 1. J Biol Chem 284: 1971–1981

Bomsztyk K, Denisenko O, Ostrowski J (2004) hnRNP K: one protein multiple processes. Bioessays 26:629–638

Bomsztyk K, Van Seuningen I, Suzuki H, Denisenko O, Ostrowski J (1997) Diverse molecular interactions of the hnRNP K protein. FEBS Lett 403:113–115

Boyne LJ, Fischer I, Shea TB (1996) Role of vimentin in early stages of neuritogenesis in cultured hippocampal neurons. Int J Dev Neurosci 14:739–748

Budhram-Mahadeo V, Morris PJ, Lakin ND, Theil T, Ching GY et al. (1995) Activation of the alpha-internexin promoter by the Brn-3a transcription factor is dependent on the N-terminal region of the protein. J Biol Chem 270:2853–2858

Buratti E, Baralle RE (2008) Multiple roles of TDP-43 in gene expression, splicing regulation, and human disease. Front Biosci 13:867–878

Burgoyne RD, Cambray-Deakin MA, Lewis SA, Sarkar S, Cowan NJ (1988) Differential distribution of β-tubulin isotypes in cerebellum. EMBO J 7:2311–2319

Caccamo D, Katsetos CD, Herman MM, Frankfurter A, Collins VP, Rubinstein LJ (1989) Immunohistochemistry of a spontaneous murine ovarian teratoma with neuroepithelial differentiation. Neuron-associated β-tubulin as a marker for primitive neuroepithelium. Lab Invest 60:390–398

Campbell DS, Holt CE (2001) Chemotropic responses of retinal growth cones mediated by rapid local protein synthesis and degradation. Neuron 32:1013–1026

Cañete-Soler R, Reddy KS, Tolan DR, Zhai J (2005) Aldolases A and C are ribonucleolytic components of a neuronal complex that regulates the stability of the light neurofilament mRNA. J Neurosci 25:4353–4364

Cañete-Soler R, Schlaepfer WW (2000) Similar poly(C)-sensitive RNA-binding complexes regulate the stability of the heavy and light neurofilament mRNAs. Brain Res 867: 265–279

Cañete-Soler R, Schwartz ML, Hua Y, Schlaepfer WW (1998) Characterization of ribonucleoprotein complexes and their binding sites on the neurofilament light subunit mRNA. J Biol Chem 273:12655–12661

Cañete-Soler R, Wu J, Zhai J, Shamin M, Schlaepfer WW (2001) p190RhoGEF binds to a destabilizing element in the 3′UTR of NF-L mRNA and alters the stability of the transcript. J Biol Chem 276:32046–32050

Canger AK, Passini MA, Asch WS, Leake D, Zafonte BT et al. (1998) Restricted expression of the neuronal intermediate filament protein plasticin during zebrafish development. J Comp Neurol 399:561–572

Carden MJ, Trojanowski JQ, Schlaepfer WW, Lee VMY (1987) Two-stage expression of neurofilament polypeptides during rat neurogenesis with early establishment of adult phosphorylation patterns. J Neurosci 7:3489–3504

Carvalho P, Tirnauer JS, Pellman D (2003) Surfing on microtubule ends. Trends Cell Biol 13: 229–237

Chadan S, Moya KL, Portier M-M, Filliatreau G (1994) Identification of a peripherin dimer: changes during axonal development and regeneration of the rat sciatic nerve. J Neurochem 62:1894–1905

Chang L, Shav-Tal Y, Trcek T, Singer RH, Goldman RD (2006) Assembling an intermediate filament network by dynamic cotranslation. J Cell Biol 172:747–758

Chang L, Thompson MA (1996) Activity of the distal positive element of the peripherin gene is dependent on proteins binding to an Ets-like recognition site and a novel inverted repeat site. J Biol Chem 271:6467–6475

Charnas LR, Szaro BG, Gainer H (1992) Identification and developmental expression of a novel low molecular weight neuronal intermediate filament protein in *Xenopus laevis*. J Neurosci 12:3010–3024

Ching GY, Liem RKH (1991) Structure of the gene for the neuronal intermediate filament protein alpha internexin and functional analysis of its promoter. J Biol Chem 266:19459–19468

Cleveland DW, Monteiro MJ, Wong PC, Gill SR, Gearhart JD, Hoffman PN (1991) Involvement of neurofilaments in the radial growth of axons. J Cell Sci 15:85–95

Cochard P, Paulin D (1984) Initial expression of neurofilaments and vimentin in the central and peripheral nervous system of the mouse embryo in vivo. J Neurosci 4:2080–2094

Condeelis J, Singer RH (2005) How and why does β-actin mRNA target? Biol Cell 97:97–110

Couillard-Després S, Zhu Q, Wong PC, Price DL, Cleveland DW, Julien J-P (1998) Protective effect of neurofilament heavy gene overexpression in motor neuron disease induced by mutant superoxide dismutase. Proc Natl Acad Sci USA 95:9626–9630

Crispino M, Capano CP, Kaplan BB, Giuditta A (1993) Neurofilament proteins are synthesized in nerve endings from squid brain. J Neurochem 61:1144–1146

de Waegh S, Brady ST (1990) Altered slow axonal transport and regeneration in a myelin-deficient mutant mouse: the trembler as an in vivo model for Schwann cell-axon interactions. J Neurosci 10:1855–1865

de Waegh SM, Brady ST (1991) Local control of axonal properties by Schwann cells: neurofilaments and axonal transport in homologous and heterologous nerve grafts. J Neurosci Res 30:201–212

de Waegh S, Lee VMY, Brady ST (1992) Local modulation of neurofilament phosphorylation, axonal caliber, and slow axonal transport by myelinating Schwann cells. Cell 68:451–463

Dejgaard K, Leffers H, Rasmussen HH, Madsen P, Kruse TA et al. (1994) Identification, molecular cloning, expression and chromosome mapping of a family of transformation upregulated hnRNP-K proteins derived by alternative splicing. J Mol Biol 236:33–48

Deshler JO, Highett MI, Abramson T, Schnapp BJ (1998) A highly conserved RNA-binding protein for cytoplasmic mRNA localization in vertebrates. Curr Biol 8:489–496

Dodemont H, Riemer D, Ledger N, Weber K (1994) Eight genes and alternative RNA processing pathways generate an unexpectedly large diversity of cytoplasmic intermediate filament proteins in the nematode Caenorhabditis elegans. EMBO J 13:2625–2638

Doolittle RF, York AL (2002) Bacterial actins? An evolutionary perspective. Bioessays 24:293–296

Duret L, Dorkeld F, Gautier C (1993) Strong conservation of non-coding sequences during vertebrate evolution: potential involvement in post-transcriptional regulation of gene expression. Nucleic Acids Res 21:2315–2322

Dworkin-Rastl E, Kelley DB, Dworkin MB (1986) Localization of specific mRNA sequences in *Xenopus laevis* embryos by in situ hybridization. J Embryol Exp Morphol 91:153–168

Easter SS Jr, Ross LS, Frankfurter A (1993) Initial tract formation in the mouse brain. J Neurosci 13:285–299

Elder GA, Friedrich VL Jr, Bosco P, Kang C, Giourov A et al. (1998a) Absence of the mid-sized neurofilament subunit decreases axonal calibers, levels of light neurofilament (NF-L) and neurofilament content. J Cell Biol 141:727–739

Elder GA, Friedrich VL Jr, Kang C, Bosco P, Gourov A et al. (1998b) Requirement of heavy neurofilament subunit in the development of axons with large calibers. J Cell Biol 143:195–205

Elder GA, Friedrich VL Jr, Zuozong L, Li X, Lazzarini RA (1994) Enhancer trapping by a human mid-sized neurofilament transgene reveals unexpected patterns of neuronal enhancer activity. Mol Brain Res 26:177–188

Elder G, Liang Z, Lee N, Fiedrich VL Jr, Lazzarini RA (1992a) Novel DNA binding proteins participate in the regulation of human neurofilament H gene expression. Mol Brain Res 15: 85–98

Elder GA, Liang Z, Li C, Lazzarini RA (1992c) Targeting of Sp1 to a non-Sp1 site in the human neurofilament (H) promoter via an intermediary DNA-binding protein. Nucleic Acids Res 20:6281–6285

Elder G, Liang Z, Snyder SE, Lazzarini RA (1992b) Multiple nuclear factors interact with the promoter of the human neurofilament M gene. Mol Brain Res 15:99–107

Elder GA, Liang Z, Snyder SE, Lazzarini RA (1992d) Multiple nuclear factors interact with the promoter of the human neurofilament M gene. Mol Brain Res 15:99–107

Erber A, Riemer D, Bovenschulte M, Weber K (1998) Molecular phylogeny of metazoan intermediate filament proteins. J Mol Evol 47:751–762

Escurat M, Djabali K, Gumpel M, Gros F, Portier M-M (1990) Differential expression of two neuronal intermediate-filament proteins, peripherin and the low-molecular-mass neurofilament protein (NF-L), during the development of the rat. J Neurosci 10:764–784

Eyer J, Peterson A (1994) Neurofilament-deficient axons and perikaryal aggregates in viable transgenic mice expressing a neurofilament-beta-galacotsidase fusion protein. Neuron 12:389–405

Fausett BV, Gumerson JD, Goldman D (2008) The proneural basic helix-loop-helix gene *Ascl1a* is required for retina regeneration. J Neurosci 28:1109–1117

Fliegner KH, Kaplan MP, Wood TL, Pintar JE, Liem RKH (1994) Expression of the gene for the neuronal intermediate filament protein alpha-internexin coincides with the onset of neuronal differentiation in the developing rat nervous system. J Comp Neurol 342:161–173

Fremeau RT Jr, Lundblad JR, Pritchett DB, Wilcox JN, Roberts JL (1986) Regulation of pro-opiomelanocortin gene transcription in individual cell nuclei. Science 234:1265–1269

Friede RL, Samorajski T (1970) Axon caliber related to neurofilaments and microtubules in sciatic nerves of rats and mice. Anat Rec 167:379–388

Fuchs C, Glasgow E, Hitchcock PF, Schechter N (1994) Plasticin, a newly identified neurofilament protein, is preferentially expressed in young retinal ganglion cells of adult goldfish. J Comp Neurol 350:452–462

Gallo G, Letourneau PC (1999) Different contributions of microtubule dynamics and transport to the growth of axons and collateral sprouts. J Neurosci 19:3860–3873

Gao F-B (2008) Posttranscriptional control of neuronal development by microRNA networks. Trends Neurosci 31:20–26

Garcia ML, Lobsiger CS, Shah SB, Deerinck TJ, Crum J et al. (2003) NF-M is an essential target for the myelin-directed "outside-in" signaling cascade that mediates radial axonal growth. J Cell Biol 163:1011–1020

Ge WW, Leystra-Lantz C, Wen W, Strong MJ (2003) Selective loss of trans-acting instability determinants of neurofilament mRNA in amyotrophic lateral sclerosis spinal cord. J Biol Chem 278:26558–26563

Ge WW, Volkening K, Leystra-Lantz C, Jaffe H, Strong MJ (2007) 14-3-3 protein binds to the low molecular weight neurofilament (NFL) mRNA 3′UTR. Mol Cell Neurosci 34:80–87

Ge WW, Wen W, Strong WL, Leystra-Lantz C, Strong MJ (2005) Mutant copper/zinc superoxide dismutase binds to and destabilizes human low molecular weight neurofilament mRNA. J Biol Chem 280:118–124

Ge W, Wu J, Zhai J, Nie Z, Lin H et al. (2002) Binding of p190RhoGEF to a destabilizing element on the light neurofilament mRNA is competed by BC1 mRNA. J Biol Chem 277:42701–42705

Geisert EE Jr, Frankfurter A (1989) The neuronal response to injury as visualized by immunostaining of class III β-tubulin in the rat. Neurosci Lett 102:137–141

Gervasi C, Stewart C-B, Szaro BG (2000) *Xenopus laevis* peripherin (XIF3) is expressed in radial glia and proliferating neural epithelial cells as well as in neurons. J Comp Neurol 423:512–531

Gervasi C, Thyagarajan A, Szaro BG (2003) Increased expression of multiple neurofilament mRNAs during regeneration of vertebrate central nervous system axons. J Comp Neurol 461:262–275

Glasgow E, Druger RK, Fuchs C, Lane WS, Schechter N (1994a) Molecular cloning of gefiltin (ON1): serial expression of two new neurofilament mRNAs during optic nerve regeneration. EMBO J 13:297–305

Glasgow E, Druger RK, Fuchs C, Levine EM, Giordano S, Schechter N (1994b) Cloning of multiple forms of goldfish vimentin: differential expression in CNS. J Neurochem 63:470–481
Glasgow E, Druger RK, Levine EM, Fuchs C, Schechter N (1992) Plasticin, a novel type III neurofilament protein from goldfish retina: increased expression during optic nerve regeneration. Neuron 9:373–381
Glasgow E, Hall CM, Schechter N (1994c) Organization, sequence, and expression of a gene encoding goldfish neurofilament medium protein. J Neurochem 63:52–61
Gloster A, El-Bizri H, Bamji SX, Rogers D, Miller FD (1999) Early induction of Tα1 α-tubulin transcription in neurons of the developing nervous system. J Comp Neurol 405:45–60
Gloster A, Wu W, Speelman A, Weiss S, Causing C et al. (1994) The T alpha 1 alpha-tubulin promoter specifies gene expression as a function of neuronal growth and regeneration in transgenic mice. J Neurosci 14:7319–7330
Goldman D, Hankin M, Li Z, Dai X, Ding J (2001) Transgenic zebrafish for studying nervous system development and regeneration. Transgenic Res 10:21–33
Goldstein ME, Weiss SR, Lazzarini RA, Shneidman PS, Lees JF, Schlaepfer WW (1988) mRNA levels of all three neurofilament proteins decline following nerve transection. Brain Res 427:287–291
Goldstone K, Sharpe CR (1998) The expression of *Xif3* in undifferentiated anterior neuroectoderm, but not in primary neurons, is induced by the neuralizing agent noggin. Int J Dev Biol 42: 757–762
Good PJ, Richter K, Dawid IB (1989) The sequence of a nervous system-specific, class II β-tubulin gene from *Xenopus laevis*. Nucleic Acids Res 17:8000
Gorham JD, Baker H, Kegler D, Ziff EB (1990) The expression of the neuronal intermediate filament protein peripherin in the rat embryo. Dev Brain Res 57:235–248
Gu W, Pan F, Zhang H, Bassell GJ, Singer RH (2002) A predominantly nuclear protein affecting cytoplasmic localization of β-actin mRNA in fibroblasts and neurons. J Cell Biol 156:41–51
Gulati-Leekha A, Goldman D (2006) A reporter-assisted mutagenesis screen using α1-tubulin-GFP transgenic zebrafish uncovers missteps during neuronal development and axonogenesis. Dev Biol 296:29–47
Hall CM, Schechter N (1991) Expression of neuronal intermediate filament proteins ON_1 and ON_2 during goldfish optic nerve regeneration: effect of tectal ablation. Neuroscience 41:695–701
Hannan AJ, Gunning P, Jeffrey PL, Weinberger RP (1998) Structural compartments within neurons: developmentally regulated organization of microfilament isoform mRNA and protein. Mol Cell Neurosci 11:289–304
Hayakawa K, Ono S, Nagaoka R, Saitoh O, Obinata T (1996) Differential assembly of cytoskeletal and sarcomeric actins in developing skeletal muscle cells in vitro. Zoolog Sci 13:509–517
Heacock AM, Agranoff BW (1976) Enhanced labeling of a retinal protein during regeneration of optic nerve in goldfish. Proc Natl Acad Sci USA 73:828–832
Hemmati-Brivanlou A, Mann RW, Harland RM (1992) A protein expressed in the growth cones of embryonic vertebrate neurons defines a new class of intermediate filament protein. Neuron 9:417–428
Herrmann H, Fouquet B, Franke WW (1989) Expression of intermediate filament proteins during development of Xenopus laevis. I. cDNA clones encoding different forms of vimentin. Development 195:279–298
Hieber V, Dai X, Foreman M, Goldman D (1998) Induction of α1-tubulin gene expression during development and regeneration of the fish central nervous system. J Neurobiol 37:429–440
Hill MA, Gunning P (1993) Beta and gamma actin mRNAs are differentially located within myoblasts. J Cell Biol 122:825–832
Hisanaga S-I, Uchida A, Tashiro T, Komiya Y, Yorifuji H et al. (2004) Morphological and biochemical changes of neurofilaments in aged rat sciatic nerve axons. J Neurochem 88:735–745
Hoffman P, Cleveland DW (1988) Neurofilament and tubulin expression recapitulates the developmental program during axonal regeneration: induction of a specific β-tubulin isotype. Proc Natl Acad Sci USA 85:4530–4533

Hoffman PN, Cleveland DW, Griffin JW, Landes PW, Cowan NJ, Price DL (1987) Neurofilament gene expression: a major determinant of axonal caliber. Proc Natl Acad Sci USA 84:3472–3476

Hoffman PN, Griffin JW, Price DL (1984) Control of axonal caliber by neurofilament transport. J Cell Biol 99:705–714

Howard J, Hyman AA (2003) Dynamics and mechanics of the microtubule plus end. Nature 422:753–758

Huang JK, Dorey K, Ishibashi S, Amaya E (2007) BDNF promotes target innervation of *Xenopus* mandibular trigeminal axons in vivo. BMC Dev Biol 7:59

Irwin N, Baekelandt V, Goritchenko L, Benowitz LI (1997) Identification of two proteins that bind to a pyrimidine-rich sequence in the 3′-untranslated region of GAP-43 mRNA. Nucleic Acids Res 25:1281–1288

Jabri E (2005) P-bodies take a RISC. Nat Struct Mol Biol 12:564

Jacobs AJ, Kamholz J, Selzer ME (1995) The single lamprey neurofilament subunit (NF-180) lacks multiphosphorylation repeats and is expressed selectively in projection neurons. Mol Brain Res 29:43–52

Jacobs AJ, Swain GP, Snediker JA, Pijak DS, Gladstone LJ, Selzer ME (1997) Recovery of neurofilament expression selectively in regenerating reticulospinal neurons. J Neurosci 17:5206–5220

Jiang XM, Zhao JX, Ohnishi A, Itakura C, Mizutani M et al. (1996) Regeneration of myelinated fiber after crush injury is retarded in sciatic nerves of mutant Japanese quails deficient in neurofilaments. Acta Neuropathol (Berl) 92:467–472

Jin L-Q, Zhang G, Selzer ME (2005) Lamprey neurofilaments contain a previously unreported 50-kDa protein. J Comp Neurol 483:403–414

Johansson CB, Lothian C, Molin M, Okano H, Lendahl U (2002) Nestin enhancer requirements for expression in normal and injured adult CNS. J Neurosci Res 69:784–794

Johnston JA, Dalton MJ, Gurney ME, Kopito RR (2000) Formation of high molecular weight complexes of mutant, Cu,Zn-superoxide dismutase in a mouse model for familial amyotrophic lateral sclerosis. Proc Natl Acad Sci USA 97:12571–12576

Jones LJF, Carballido-Lopez R, Errington J (2001) Control of cell shape in bacteria: helical, actin-like filaments in *Bacillus subtilis*. Cell 104:913–922

Joshi HC, Cleveland DW (1989) Differential utilization of the available β-tubulin isotypes in differentiating neurites. J Cell Biol 109:663–673

Kaplan MP, Chin SSM, Fliegner KH, Liem RKH (1990) Alpha-internexin, a novel neuronal intermediate filament protein, precedes the low molecular weight neurofilament protein (NF-L) in the developing rat brain. J Neurosci 10:2735–2748

Kawaguchi A, Miyata T, Sawamoto K, Takashita N, Murayama A et al. (2001) Nestin-EGFP transgenic mice: visualization of the self-renewal and multipotency of CNS stem cells. Mol Cell Neurosci 17:259–273

Kedersha N, Stoecklin G, Ayodele M, Yacono P, Lykke-Andersen J et al. (2005) Stress granules and processing bodies are dynamically linked sites of mRNP remodeling. J Cell Biol 169:871–884

Keene JD, Tenenbaum SA (2002) Eukaryotic mRNPs may represent posttranscriptional operons. Mol Cell 9:1161–1167

Kim O-K, Ariano MA, Lazzarini RA, Levine MS, Sibley DR (2002) Neurofilament-M interacts with the D1 dopamine receptor to regulate cell surface expression and desensitization. J Neurosci 22:5920–5930

Kislauskis EH, Zhu X, Singer RH (1994) Sequences responsible for intracellular localization of β-actin messenger RNA also affect cell phenotype. J Cell Biol 127:441–451

Klimek-Tomczak K, Mikula M, Dzwonek A, Paziewska A, Karczmarski J et al. (2006) Editing of hnRNP K protein mRNA in colorectal adenocarcinoma and surrounding mucosa. Br J Cancer 94:586–592

Koenig E (1991) Evaluation of local synthesis of axonal proteins in the goldfish Mauthner cell axon and axons of dorsal and ventral roots of the rat in vitro. Mol Cell Neurosci 2: 384–394

Koenig E, Giuditta A (1999) Protein-synthesizing machinery in the axon compartment. Neurosci 89:5–15

Kong JM, Tung VWY, Aghajanian J, Xu ZS (1998) Antagonistic roles of neurofilament subunits NF-H and NF-M against NF-L in shaping dendritic arborization in spinal motor neurons. J Cell Biol 140:1167–1176

Kost SA, Chacko K, Oblinger MM (1992) Developmental patterns of intermediate filament gene expression in the normal hamster brain. Brain Res 595:270–280

Krapp A, Knofler M, Frutiger S, Hughes GJ, Hagenbuchle O, Wellauer PK (1996) The p48 DNA-binding subunit of transcription factor PTF1 is a new exocrine pancreas-specific basic helix-loop-helix protein. EMBO J 15:4317–4329

Landon F, Lemonnier M, Benarous R, Huc C, Fiszman M et al. (1989) Multiple mRNAs encode peripherin, a neuronal intermediate filament protein. EMBO J 8:1719–1726

Landon F, Wolff A, de Nechaud B (2000) Mouse peripherin isoforms. Biol Cell 92:397–407

Lariviére RC, Julien J-P (2004) Functions of intermediate filaments in neuronal development and disease. J Neurobiol 58:131–148

Lasek RJ, Phillips L, Katz MJ, Autilo-Gambetti L (1985) Function and evolution of neurofilament proteins. Ann NY Acad Sci 455:462–478

Lee MK, Xu Z, Wong PC, Cleveland DW (1993) Neurofilaments are obligate heteropolymers in vivo. J Cell Biol 122:1337–1350

Leigh PN, Dodson A, Swash M, Brion JP, Anderton BH (1989) Cytoskeletal abnormalities in motor neuron disease. An immunocytochemical study. Brain 112:521–535

Lendahl U, Zimmerman LB, McKay RDG (1990) CNS stem cells express a new class of intermediate filament protein. Cell 60:585–595

Leonard DGB, Gorham JD, Cole P, Greene LA, Ziff EB (1988) A nerve growth factor-regulated messenger RNA encodes a new intermediate filament protein. J Cell Biol 106: 181–193

Leung K-M, van Horck FPG, Lin AC, Allison R, Standart N, Holt CE (2006) Asymmetrical β-actin mRNA translation in growth cones mediates attractive turning to netrin-1. Nat Neurosci 9:1247–1256

Lewis SA, Lee MG, Cowan NJ (1985) Five mouse tubulin isotypes and their regulated expression during development. J Cell Biol 101:852–861

Lin CH, Espreafico EM, Mooseker MS, Forscher P (1997) Myosin drives retrograde F-actin flow in neuronal growth cones. Biol Bull 192:183–185

Lin AC, Holt CE (2008) Function and regulation of local axonal translation. Curr Opin Neurobiol 18:60–68

Lin W, Szaro BG (1995) Neurofilaments help maintain normal morphologies and support elongation of neurites in *Xenopus laevis* cultured embryonic spinal cord neurons. J Neurosci 15:8331–8344

Lin H, Zhai J, Schlaepfer WW (2005) RNA-binding protein is involved in aggregation of light neurofilament protein and is implicated in the pathogenesis of motor neuron degeneration. Hum Mol Genet 14:3463–3659

Lindenbaum MH, Carbonetto S, Grosveld F, Flavell D, Mushynski WE (1998) Transcriptional and post-transcriptional effects of nerve growth factor on expression of the three neurofilament subunits in PC-12 cells. J Biol Chem 263:5662–5667

Liu X, Fortin K, Mourleatos Z (2008a) MicroRNAs: biogenesis and molecular functions. Brain Pathol 18:113–121

Liu L, Geisert EE, Frankfurter A, Spano AJ, Jiang CX et al. (2007) A transgenic mouse class-III β-tubulin reporter using yellow fluorescent protein. Genesis 45:560–569

Liu Y, Gervasi C, Szaro BG (2008b) A crucial role for hnRNP K in axon development in *Xenopus laevis*. Development 135:3125–3135

Liuzzi FJ, Tedeschi B (1992) Axo-glial interactions at the dorsal root transitional zone regulate neurofilament protein synthesis in axotomized sensory neurons. J Neurosci 12: 4783–4792

Lopata MA, Cleveland DW (1987) In vivo microtubules are copolymers of available beta-tubulin isotypes: localization of each of six vertebrate beta-tubulin isotypes using polyclonal antibodies elicited by synthetic peptide antigens. J Cell Biol 105:11707–11720

Lothian C, Lendahl U (1997) An evolutionarily conserved region in the second intron of the human nestin gene directs gene expression to CNS progenitor cells and to early neural crest cells. Eur J Neurosci 9:452–462

Lowery AJ, Miller N, McNeill RE, Kerin MJ (2008) MicroRNAs as prognostic indicators and therapeutic targets: potential effect on breast cancer management. Clin Cancer Res 14:360–365

Ludueña RF (1998) Multiple forms of tubulin: different gene products and covalent modifications. Int Rev Cytol 178:207–275

Mahler J, Driever W (2007) Expression of the zebrafish intermediate neurofilament nestin in the developing nervous system and in neural proliferation zones at postembryonic stages. BMC Dev Biol 7:89

Makeyev AV, Liebhaber SA (2002) The poly(C)-binding proteins: a multiplicity of functions and a search for mechanisms. RNA 8:265–278

Manetto V, Sternberger NH, Perry G, Sternberger LA, Gambetti P (1988) Phosphorylation of neurofilaments is altered in amyotrophic lateral sclerosis. J Neuropathol Exp Neurol 47:642–653

Marsh-Armstrong N, Huang H, Berry DL, Brown DS (1999) Germ-line transmission of transgenes in *Xenopus laevis*. Proc Natl Acad Sci USA 96:14389–14393

Marszalek JR, Williamson TL, Lee MK, Xu ZS, Hoffman PN et al. (1996) Neurofilament subunit NF-H modulates axonal diameter by selectively slowing neurofilament transport. J Cell Biol 135:711–724

McGraw TS, Mickle JP, Shaw G, Streit WJ (2002) Axonally transported peripheral signals regulate alpha-internexin expression in regenerating motoneurons. J Neurosci 22:4955–4963

McKerracher L, Essagian C, Aguayo AJ (1993) Marked increase in beta-tubulin mRNA expression during regeneration of axotomized retinal ganglion cells in adult mammals. J Neurosci 13:5294–5300

McLean JR, Leystra-Lantz C, He BP, Strong MJ (2005) Temporal profiles of neuronal degeneration, glial proliferation, and cell death in hNFL+/+ and NFL–/– mice. Glia 52:59–69

McLean J, Xiao S, Miyazaki K, Robertson J (2008) A novel peripherin isoform generated by alternative translation is required for normal filament network formation. J Neurochem 104:1663–1673

Meier J, Couillard-Després S, Jacomy H, Gravel C, Julien J-P (1999) Extra neurofilament NF-L subunits rescue motor neuron disease caused by overexpression of the human NF-H gene in mice. J Neuropathol Exp Neurol 58:1099–1110

Menard C, Hein P, Paquin A, Savelson A, Yang XM et al. (2002) An essential role for a MEK-C/EBP pathway during growth factor-regulated cortical neurogenesis. Neuron 36: 597–610

Menezes JRL, Luskin MB (1994) Expression of neuron-specific tubulin defines a novel population in the proliferative layers of the developing telencephalon. J Neurosci 14:5399–5416

Menzies FM, Grierson AJ, Cookson MR, Heath PR, Tomkins J et al. (2002) Selective loss of neurofilament expresion in Cu/Zn superoxide dismutase (SOD1) linked amyotrophic lateral sclerosis. J Neurochem 82:1118–1128

Mikucki SA, Oblinger MM (1991) Corticospinal neurons exhibit a novel pattern of cytoskeletal gene expression after injury. J Neurosci Res 30:213–225

Mikula M, Dzwonek A, Karczmarksi J, Rubel T, Dadlez M et al. (2006a) Landscape of the hnRNP K protein-protein interactome. Proteomics 6:2395–2406

Mikula M, Karczmarski J, Dzwonek A, Rubel T, Hennig W et al. (2006b) Casein kinases phosphorylate multiple residues spanning the entire hnRNP K length. Biochim Biophys Acta 1764:299–306

Miller FD, Naus CCG, Durand M, Bloom FE, Milner RJ (1987) Isotypes of α-tubulin are differentially regulated during neuronal maturation. J Cell Biol 105:3065–3073

Miller FD, Tetzlaff W, Bisby MA, Fawcett JW, Milner RJ (1989) Rapid induction of the major embryonic α-tubulin, Tα1, during nerve regeneration in adult rats. J Neurosci 9: 1452–1463

Mitchison T, Kirschner M (1988) Cytoskeletal dynamics and nerve growth. Neuron 1:761–772

Moody SA, Miller V, Spanos A, Frankfurter A (1996) Developmental expression of a neuron-specific beta-tubulin in frog (*Xenopus laevis*): a marker for growing axons during the embryonic period. J Comp Neurol 364:219–230

Moody SA, Quigg MS, Frankfurter A (1989) Development of peripheral trigeminal system in the chick revealed by an isotype-specific anti-β-tubulin monoclonal antibody. J Comp Neurol 279:567–580

Moore MJ (2005) From birth to death: the complex lives of eukaryotic mRNAs. Science 309: 1514–1518

Moskowitz PF, Oblinger MM (1995) Transcriptional and post-transcriptional mechanisms regulating neurofilament and tubulin gene expression during normal development of the rat brain. Mol Brain Res 30:211–222

Muma NA, Hoffman PN, Slunt HH, Applegate MD, Lieberburg I, Price DL (1990) Alterations in levels of mRNAs coding for neurofilament protein subunits during regeneration. Exp Neurol 107:230–235

Muma NA, Slunt HH, Hoffman PN (1991) Postnatal increases in neurofilament gene expression correlate with the radial growth of axons. J Neurocytol 20:844–854

Murayama S, Bouldin TW, Suzuki K (1992) Immunocytochemical and ultrastructural studies of upper motor neurons in amyotrophic lateral sclerosis. Acta Neuropathol 83: 518–524

Nadeau S, Hein P, Fernandes KJL, Peterson AC, Miller FD (2005) A transcriptional role for C/EBP β in the neuronal response to axonal injury. Mol Cell Neurosci 29:525–535

Napier A, Yuan A, Cole GJ (1999) Characterization of the chicken transitin gene reveals a strong relationship to the nestin intermediate filament class. J Mol Neurosci 12:11–22

Nelson PT, Keller JN (2007) RNA in brain disease: no longer just "the messenger in the middle". Neuropathol Exp Neurol 66:461–468

Nelson PT, Wang W-X, Rajeev BW (2008) MicroRNAs (miRNAs) in neurodegenerative diseases. Brain Pathol 18:130–138

Nerlov C (2008) C/EBPs: recipients of extracellular signals through proteome modulation. Curr Opin Cell Biol 20:180–185

Neumann M, Sampathu DM, Kwong LK, Truax AC, Micsenyi MC et al. (2006) Ubiquitinated TDP-43 in frontotemporal lobar degeneration and amyotrophic lateral sclerosis. Science 314:130–133

Nie Z, Wu J, Shai J, Lin H, Ge W et al. (2002) Untranslated element in neurofilament mRNA has neuropathic effect on motor neurons of transgenic mice. J Neurosci 22: 7662–7770

Niloff MS, Dunn RJ, Levine RL (1998) The levels of retinal mRNA for gefiltin, a neuronal intermediate filament protein, are regulated by the tectum during optic nerve regeneration in the goldfish. Mol Brain Res 61:78–89

Oblinger MM, Wong J, Parysek LM (1989) Axotomy-induced changes in the expression of a type III neuronal intermediate filament. J Neurosci 9:3766–3775

Oleynikov Y, Singer RH (2003) Real-time visualization of ZBP1 association with β-actin mRNA during transcription and localization. Curr Biol 13:199–207

Omary MB, Coulombe PA, McLean WH (2004) Intermediate filament proteins and their associated diseases. N Engl J Med 351:2087–2100

Oschwald R, Richter K, Grunz H (1991) Localization of a nervous system-specific class II β-tubulin gene in *Xenopus laevis* embryos by whole-mount in situ hybridization. Int J Dev Biol 35:399–405

Ostareck-Lederer A, Ostareck DH, Hentze MW (1998) Cytoplasmic regulatory functions of the KH domain proteins hnRNPs K and E1/E2. Trends Biochem Sci 23:409–411

Ostrowski J, Kawata Y, Schullery DS, Denisenko ON, Higaki Y et al. (2001) Insulin alters heterogenous nuclear ribonucleoprotein K protein binding to DNA and RNA. Proc Natl Acad Sci USA 98:9044–9049

Pachter JS, Liem RKH (1984) The differential appearance of neurofilament triplet polypeptides in the developing rat optic nerve. Dev Biol 103:200–210

Parhad IM, Clark AW, Griffin JW (1987) Effect of changes in neurofilament content on caliber of small axons: the beta-beta' iminodipropionitrile model. J Neurosci 7:2256–2263

Perrone-Bizzozero NI, Neve RL, Irwin N, Lewis S, Fischer I, Benowitz LI (1991) Post-transcriptional regulation of GAP-43 mRNA levels during neuronal differentiation and nerve regeneration. Mol Cell Neurosci 2:402–409

Petersohn D, Schoch S, Brinkman DR, Thiel G (1995) The human synapsin II gene promoter. Possible role for the transcription factor zif268/egr-1, polyoma enhancer activator 3, and AP2. J Biol Chem 270:24361–24369

Phillips L, Autilo-Gambetti L, Lasek RJ (1983) Bodian's silver method reveals molecular variation in the evolution of neurofilament proteins. Brain Res 278:219–223

Pillai RS (2005) MicroRNA function: multiple mechanisms for a tiny RNA? RNA 11:1753–1761

Piper M, Anderson R, Dwivedy A, Weihl C, van Horck F et al. (2006) Signaling mechanisms underlying Slit2-induced collapse of Xenopus retinal growth cones. Neuron 49: 215–228

Pleasure SJ, Selzer ME, Lee VMY (1989) Lamprey neurofilaments combine in one subunit the features of each mammalian NF triplet protein but are highly phosphorylated only in large axons. J Neurosci 9:698–709

Pospelov VA, Pospelova TV, Julien J-P (1994) AP-1 and Krox-24 transcription factors activate the neurofilament light gene promoter in P19 embryonal carcinoma cells. Cell Growth Differ 5:187–196

Quitschke W, Schechter N (1986) Homology and diversity between intermediate filament proteins of neuronal and nonneuronal origin in the goldfish optic nerve. J Neurochem 46: 545–555

Rao MV, Garcia ML, Miyazaki Y, Gotow T, Yuan A et al. (2002) Gene replacement in mice reveals that the heavily phosphorylated tail of neurofilament heavy subunit does not affect axonal caliber or the transit of cargoes in slow axonal transport. J Cell Biol 158:681–693

Richter K, Grunz H, Dawid IB (1988) Gene expression in the embryonic nervous system of *Xenopus laevis*. Proc Natl Acad Sci USA 85:8086–8090

Robertson J, Doroudchi MM, Nguyen MD, Durham HD, Strong MJ et al. (2003) A neurotoxic peripherin splice variant in a mouse model of ALS. J Cell Biol 160:939–949

Rodger J, Bartlett CA, Harman AM, Thomas C, Beazley LD, Dunlop SA (2001) Evidence that regenerating optic axons maintain long-term growth in the lizard *Ctenophorus ornatus*: growth-associated protein-43 and gefiltin expression. Neurosci 102:647–654

Rommelaere H, Waterschoot D, Neirynck K, Vanderkerckhove J, Ampe C (2004) A method for rapidly screening functionality of actin mutants and tagged actins. Biol Proced Online 6: 235–249

Roosa JR, Gervasi C, Szaro BG (2000) Structure, biological activity of the upstream regulatory sequence, and conserved domains of a middle molecular mass neurofilament gene of *Xenopus laevis*. Mol Brain Res 82:35–51

Ross AF, Oleynikov Y, Kislauskis EH, Taneja KL, Singer RH (1997) Characterization of a β-actin mRNA zipcode-binding protein. Mol Cell Biol 17:2158–2165

Rossi JJ (2005) RNAi and the P-body connection. Nat Cell Biol 7:643–644

Saba R, Goodman CD, Huzarewich RLCH, Robertson C, Booth SA (2008) miRNA signature of prion induced neurodegeneration. PLoS One 3:e3652

Sakaguchi T, Okada M, Kitamura T, Kawasaki K (1993) Reduced diameter and conduction velocity of myelinated fibers in the sciatic nerve of a neurofilament-deficient mutant quail. Neurosci Lett 153:65–68

Sanelli T, Ge W, Leystra-Lantz C, Strong MJ (2007) Calcium mediated excitotoxicity in neurofilament aggregate bearing neurons in vitro is NMDA receptor dependent. J Neurol Sci 256:39–51

Sanelli T, Sopper MM, Strong MJ (2004) Sequestration of nNOS in neurofilamentous aggregate bearing neurons leads to enhanced glutamate-mediated calcium influx. Brain Res 1004: 8–17

Sanelli T, Strong MJ (2007) Loss of nitric oxide-mediated down-regulation of NMDA receptors in neurofilament aggregate-bearing motor neurons in vitro: implications for motor neuron disease. Free Radic Biol Med 42:143–151

Schlaepfer WW, Bruce J (1990) Simultaneous up-regulation of neurofilament proteins during the postnatal development of the rat nervous system. J Neurosci Res 25:39–49

Schmid E, Tapscott S, Bennett GS, Croop J, Fellini SA et al. (1979) Differential location of different types of intermediate-sized filaments in various tissues of the chicken embryo. Differentiation 15:27–40

Schoenherr CJ, Paquette AJ, Anderson DJ (1996) Identification of potential target genes for the neuron-restrictive silencer factor. Proc Natl Acad Sci USA 93:9881–9886

Schratt GM, Tuebing F, Nigh EA, Kane CG, Sabatini ME et al. (2006) A brain-specific microRNA regualtes dendritic spine development. Nature 439:283–289

Schwartz ML, Shneidman PS, Bruce J, Schlaepfer WW (1990) Axonal dependency of the postnatal upregulation in neurofilament expression. J Neurosci Res 27:193–201

Schwartz ML, Shneidman PS, Bruce J, Schlaepfer WW (1992) Actinomycin prevents the destabilization of neurofilament mRNA in primary sensory neurons. J Biol Chem 267: 24596–24600

Schwarz PM, Liggins JR, Ludueña RF (1998) Beta-tubulin isotypes purified from bovine brain have different relative stabilities. Biochemistry 37:4687–4692

Scott D, Smith KE, O'Brien BJ, Angelides KJ (1985) Characterization of mammalian neurofilament triplet proteins: subunit stoichiometry and morphology of native and reconstituted filaments. J Biol Chem 260:10736–10747

Senut M-C, Gulati-Leekha A, Goldman D (2004) An element in the α1-tubulin promoter is necessary for retinal expression during optic nerve regeneration but not after eye injury in the adult zebrafish. J Neurosci 24:7663–7673

Sharpe CR (1988) Developmental expression of a neurofilament-M and two vimentin- like genes in *Xenopus laevis*. Development 103:269–277

Sharpe CR, Pluck A, Gurdon JB (1989) XIF3, a *Xenopus* peripherin gene, requires an inductive signal for enhanced expression in anterior neural tissue. Development 107:701–714

Shaw G (1992) A neurofilament-specific sequence motif. Trends Biochem Sci 17:3405

Shaw G, Weber K (1982) Differential expression of neurofilament triplet proteins in brain development. Nature 298:277–279

Shea TB, Beermann ML (1999) Neuronal intermediate filament protein alpha-internexin facilitates axonal neurite elongation in neuroblastoma cells. Cell Motil Cytoskeleton 43:322–333

Shneidman PS, Bruce J, Schwartz ML, Schlaepfer WW (1992) Negative regulatory regions are present upstream in the three mouse neurofilament genes. Mol Brain Res 13:127–138

Smith CL (1994) The initiation of neurite outgrowth by sympathetic neurons grown in vitro does not depend on assembly of microtubules. J Cell Biol 127:1407–1418

Sotelo-Silveira JR, Calliari A, Kun A, Benech JC, Sanguinetti C et al. (2000) Neurofilament mRNAs are present and translated in the normal and severed sciatic nerve. J Neurosci Res 62:65–74

Sreedharan J, Blair IP, Tripathi VB, Hu X, Vance C et al. (2008) TDP-43 mutations in familial and amyotrophic lateral sclerosis. Science 319:1668–1672

Steinert PM, Roop DR (1988) Molecular and cellular biology of intermediate filaments. Annu Rev Biochem 57:593–625

Strong MJ (1999) Neurofilament metabolism in sporadic amyotrophic lateral sclerosis. J Neurol Sci 169:170–177

Strong MJ (2010) The evidence for altered RNA metabolism in amyotrophic lateral sclerosis (ALS). J Neurol Sci 288:1–12

Strong MJ, Leystra-Lantz C, Ge WW (2004) Intermediate filament steady-state mRNA levels in amyotrophic lateral sclerosis. Biochem Biophys Res Commun 316:317–322

Strong MJ, Sopper MM, He BP (2003) In vitro reactive nitrating species toxicity in dissociated spinal motor neurons from NFL (–/–) and hNFL (+/+) transgenic mice. Amyotroph Lateral Scler Other Motor Neuron Disord 4:81–89

Strong MJ, Volkening K, Hammond R, Yang W, Strong W et al. (2007) TDP43 is a human low molecular weight neurofilament (hNFL) mRNA-binding protein. Mol Cell Neurosci 35: 320–327

Sun BK, Tsao H (2008) Small RNAs in development and disease. J Am Acad Dermatol 59: 725–737

Sundell CL, Singer RH (1991) Requirements of microfilaments in sorting of actin messenger RNA. Science 253:1275–1277

Szaro BG, Gainer H (1988a) Identities, antigenic determinants, and topographic distributions of neurofilament proteins in the nervous systems of adult frogs and tadpoles of *Xenopus laevis*. J Comp Neurol 273:344–358

Szaro BG, Gainer H (1988b) Immunocytochemical identification of non-neuronal intermediate filament proteins in the developing *Xenopus laevis* nervous system. Dev Brain Res 43:207–224

Szaro BG, Lee VMY, Gainer H (1989) Spatial and temporal expression of phosphorylated and non-phosphorylated forms of neurofilament proteins in the developing nervous system of *Xenopus laevis*. Dev Brain Res 48:87–103

Szaro BG, Pant HC, Way J, Battey J (1991) Squid low molecular weight neurofilament proteins are a novel class of neurofilament protein. J Biol Chem 266:15035–15041

Tanaka E, Ho T, Kirschner MW (1995) The role of microtubule dynamics in growth cone motility and axonal growth. J Cell Biol 128:139–155

Tanaka EM, Kirschner MW (1991) Microtubule behavior in the growth cones of living neurons during axon elongation. J Cell Biol 115:345–363

Tang DM, Lee KY, Qi Z, Matsuura I, Wang JH (1996) Neuronal Cdc2-like kinase: from cell cycle to neuronal function. Biochem Cell Biol 74:419–429

Tapscott SJ, Bennett GS, Toyama Y, Kleinbart F, Holtzer H (1981) Intermediate filament proteins in the developing chick spinal cord. Dev Biol 86:40–54

Thompson MA, Ziff EB (1989) Structure of the gene encoding peripherin, an NGF-regulated neuronal-specific type III intermediate filament protein. Neuron 2:1043–1053

Thyagarajan A, Szaro BG (2004) Phylogenetically conserved binding of specific KH domain proteins to the 3′ untranslated region of the vertebrate middle neurofilament mRNA. J Biol Chem 279:49680–49688

Thyagarajan A, Szaro BG (2008) Dynamic endogenous association of neurofilament mRNAs with K-homology domain ribonucleoproteins in developing cerebral cortex. Brain Res 1189:33–42

Troy CM, Brown K, Greene LA, Shelanski ML (1990a) Ontogeny of the neuronal intermediate filament protein, peripherin, in the mouse embryo. Neuroscience 36:217–237

Troy CM, Muma NA, Greene LA, Price DL, Shelanski ML (1990b) Regulation of peripherin and neurofilament expression in regenerating rat motor neurons. Brain Res 529:232–238

Undamatla J, Szaro BG (2001) Differential expression and localization of neuronal intermediate filament proteins within newly developing neurites in dissociated cultures of *Xenopus laevis* embryonic spinal cord. Cell Motil Cytoskeleton 49:16–32

Uveges TE, Shan Y, Kramer BE, Wight DC, Parysek LM (2002) Intron 1 is required for cell type-specific, but not injury responsive, peripherin gene expression. J Neurosci 22:7959–7967

van den Ent F, Amos LA, Lowe J (2001) Prokaryotic origin of the actin cytoskeleton. Nature 416:39–44

van Kesteren RE, Carter C, Dissel HM, van Minnen J, Gouwenberg Y et al. (2006) Local synthesis of actin-binding protein β-thymosin regulates neurite outgrowth. J Neurosci 26: 152–157

Vize PD, Hemmati-Brivanlou A, Harland RM, Melton DA (1991) Assays for gene function in developing *Xenopus* embryos. In: Kay BK, Peng HB (ed) Xenopus laevis: practical uses in cell and molecular biology. Academic Press, San Diego, CA, pp 368–388

Volkening K, Leystra Lantz C, Yang WC, Jaffe H, Strong MJ (2009b) Tar DNA binding protein of 43 kDa (TDP-43), 14-3-3 proteins and copper/zinc superoxide dismutase (SOD1) interact to modulate NFL mRNA stability. Implications for altered RNA processing in amyotrophic lateral sclerosis (ALS). Brain Res 1305:168–182

Volkening K, Leystra-Lantz C, Strong MJ (2009a) Human low molecular weight neurofilament (NFL) mRNA interacts with a predicted p190RhoGEF homologue (RGNEF). Amyotroph lateral Scler 1:1–7 [Epub ahead of print]

Walker KL, Yoo H-K, Undamatla J, Szaro BG (2001) Loss of neurofilaments alters axonal growth dynamics. J Neurosci 21:9655–9666

Wang H, Peiris H, Mowery A, Le Lay J, Gao Y, Greenbaum LE (2008) CCAAT/Enhancer binding protein-β is a transcriptional regulator of peroxisome-proliferator-activated receptor-γ coactivator-1α in the regenerating liver. Mol Endocrinol 22:1596–1605

Way J, Hellmich MR, Jaffe H, Szaro BG, Pant HC et al. (1992) A high-molecular-weight squid neurofilament protein contains a lamin-like rod domain and a tail domain of Lys-Ser-Pro repeats. Proc Natl Acad Sci USA 89:6963–6967

Weinberger R, Schevzov G, Jeffrey P, Gordon K, Hill M, Gunning P (1996) The molecular composition of neuronal microfilaments is spatially and temporally regulated. J Neurosci 16:238–252

Weiner OD, Zorn AM, Krieg PA, Bittner GD (1996) Medium weight neurofilament mRNA in goldfish Mauthner axoplasm. Neurosci Lett 213:83–86

Willis DE, van Niekerk EA, Sasaki Y, Mesngon M, Merianda TT et al. (2007) Extracellular stimuli specifically regulate localized levels of individual neuronal mRNAs. J Cell Biol 178: 965–980

Wong N, He BP, Strong MJ (2000) Characterization of neuronal intermediate filament protein expression in cervical spinal motor neurons in sporadic amyotrophic lateral sclerosis (ALS). J Neuropathol Exp Neurol 59:972–982

Wong J, Oblinger MM (1990) Differential regulation of peripherin and neurofilament gene expression in regenerating rat DRG neurons. J Neurosci Res 27:332–341

Wu W, Gloster A, Miller FD (1997) Transcriptional repression of the growth-associated T alpha1 tubulin gene by target contact. J Neurosci Res 48:477–487

Wu KY, Hengst U, Cox LJ, Macosko EZ, Jeromin A et al. (2005) Local translation of RhoA regulates growth cone collapse. Nature 436:1020–1024

Xiao S, McLean J, Robertson J (2006) Neuronal intermediate filaments and ALS: a new look at an old question. Biochim Biophys Acta 1762:1001–1012

Yamasaki H, Bennett GS, Itakura C, Mizutani M (1992) Defective expression of neurofilament protein subunits in hereditary hypotrophic axonopathy of quail. Lab Invest 66:734–743

Yamasaki H, Itakura C, Mizutani M (1991) Hereditary hypotrophic axonopathy with neurofilament deficiency in a mutant strain of the Japanese quail. Acta Neuropathol 82:427–434

Yano M, Okano HJ, Okano H (2005) Involvement of Hu and heterogeneous nuclear ribonucleoprotein K in neuronal differentiation through p21 mRNA post-transcriptional regulation. J Biol Chem 280:12690–12699

Yao J, Sasaki Y, Wen Z, Bassell GJ, Zheng JQ (2006) An essential role for beta-actin mRNA localization and translation in Ca^{2+}-dependent growth cone guidance. Nat Neurosci 9: 1265–1273

Yazdanbakhsh K, Fraser P, Kioussis D, Vidal M, Grosveld F, Lindenbaum M (1994) Functional analysis of the human neurofilament light chain gene promoter. Nucl Acids Res 21: 455–461

Yuan Y, Lee JA, Napier A, Cole GJ (1997) Molecular cloning of a new intermediate filament protein expressed by radial glia and demonstration of alternative splicing in a novel heptad repeat region located in the carboxy-terminal tail domain. Mol Cell Neurosci 10:71–86

Yuan A, Rao MV, Sasaki T, Chen Y, Kumar A et al. (2006) Alpha-internexin is structurally and functionally associated with the neurofilament triplet proteins in the mature CNS. J Neurosci 26:10006–10019

Yue C, Mutsuga N, Scorkalakes EM, Gainer H (2006) Studies of oxytocin and vasopressin gene expression in the rat hypothalamus using exon- and intron-specific probes. Am J Physiol Regul Integr Comp Physiol 290:R1233–R1241

Zhai J, Lin H, Shamin M, Schlaepfer WW, Cañete-Soler R (2001) Identification of a novel interaction of 14-3-3 with p190RhoGef. J Biol Chem 276:41318–41324

Zhao Y, Szaro BG (1994) The return of phosphorylated and nonphosphorylated epitopes of neurofilament proteins to the regenerating optic nerve of *Xenopus laevis*. J Comp Neurol 343:158–172

Zhao Y, Szaro BG (1995) The optic tract and tectal ablation influence the composition of neurofilaments in regenerating optic axons of *Xenopus laevis*. J Neurosci 15:4629–4640

Zhao Y, Szaro BG (1997a) Xefiltin, a new low molecular weight neuronal intermediate filament protein of *Xenopus laevis*, shares sequence features with goldfish gefiltin and mammalian alpha-internexin and differs in expression from XNIF and NF-L. J Comp Neurol 377:351–364

Zhao Y, Szaro BG (1997b) Xefiltin, a *Xenopus laevis* neuronal intermediate filament protein, is expressed in actively growing optic axons during development and regeneration. J Neurobiol 33:811–824

Zhu Q, Couillard-Despres S, Julien J-P (1997) Delayed maturation of regenerating myelinated axons in mice lacking neurofilaments. Exp Neurol 148:299–316

Zimmerman L, Parr B, Lendahl U, Cunningham M, McKay R et al. (1994) Independent regulatory elements in the nestin gene direct transgene expression to neural stem cells or muscle precursors. Neuron 12:11–24

Zopf D, Dineva B, Betz H, Gundelfinger ED (1990) Isolation of the chicken middle molecular weight neurofilament (NF-M) gene and characterization of its promoter. Nucleic Acids Res 18:521–529

Zopf D, Hermans-Borgmeyer I, Gundelfinger ED, Betz H (1987) Identification of gene products expressed in the developing chick visual system: characterization of a middle-molecular weight neurofilament cDNA. Genes and Dev 1:699–708

Chapter 25
Crosstalks Between Myelinating Cells and the Axonal Cytoskeleton

Rodolphe Perrot and Joël Eyer

Abstract A constant and dynamic communication between axons and myelinating cells is necessary for the correct development, function, and maintenance of myelinated fibers. Recently, several studies highlighted the pivotal role of the axonal cytoskeleton in this reciprocal communication. In particular, myelinating cells control the radial axonal growth by regulating the expression, transport, and organization of the axonal cytoskeleton. Conversely, this latter modulates dimensions of the myelin sheath by controlling the axonal caliber. Here, we will review the main investigations contributing to a better understanding of how the axoskeleton and myelinating cells influence each other to optimize conduction properties of myelinated fibers.

Keywords Axo-glial interactions · Axonal caliber · Axonal cytoskeleton · Internodal length · Myelin · Myelin thickness · Myelinating cells · Neuregulin-1 · Neurofilament · Oligodendrocytes · Radial axonal growth · Schwann cells

Contents

J. Eyer (✉)
Laboratoire Neurobiologie et Transgenèse, UPRES-EA3143, INSERM, Bâtiment Monteclair, CHU, 49033 Angers, France
e-mail: joel.eyer@univ-angers.fr

R.A. Nixon, A. Yuan (eds.), *Cytoskeleton of the Nervous System*, Advances in Neurobiology 3, DOI 10.1007/978-1-4419-6787-9_25,

25.1 Introduction

In the vertebrate nervous system, the myelination of axons by Schwann cells in PNS and by oligodendrocytes in CNS ensures the fast nerve-impulse conduction by allowing saltatory conduction of the action potential (Stampfli, 1954). The ultrastructure of myelinated fibers is finely regulated in order to maximize the conduction velocity. This includes an increase of the axonal caliber and the adjustment of the myelin sheath dimensions (length and thickness) to the axonal size. Thus, internodal lengths are approximately 100-fold the diameter of the ensheathed axons (Rushton, 1951; Fried et al., 1982) while the optimal g ratio (axon diameter/fiber diameter) for optimizing internodal spread of current has been evaluated to range between 0.60 and 0.80, with species and nerve-specific variations (Goldman and Albus, 1968; Williams and Chalupa, 1983; Fraher and O'Sullivan, 2000).

Originally, interactions between myelinating cells and axons were considered as unidirectional, axons transmitting information to the surrounding glial cells, which respond appropriately. This view emerged from the observation that dimensions of the myelin sheath are proportionally adjusted to changes in axon caliber (Friede and Miyagishi, 1972). On the other hand, the axonal diameter was considered to be dependent only on neuronal factors, such as the neuron's developmental program and factors that affect the density and distribution of axonal neurofilaments (NFs), the most abundant cytoskeletal component of large myelinated axons. However, many studies showed, thereafter, that myelinating cells could also modulate morphological characteristics of axons. It was demonstrated that myelination determines the axonal caliber of dorsal root ganglion neurons in culture (Windebank et al., 1985), whereas demyelination causes a significant decrease of axonal diameter in the demyelinated regions (Aguayo et al., 1977; Pollard and McLeod, 1980; Perkins et al., 1981), further indicating that axonal caliber is not a predetermined neuronal parameter. It was then shown that modifications of axonal caliber during development, regeneration, and in certain pathological conditions are accompanied by changes in expression, transport, and organization of the axonal cytoskeleton (Lasek et al., 1983; Griffin et al., 1984; Hoffman et al., 1985; Monaco et al., 1985; Komiya et al., 1986; Perrot et al., 2008). It appeared that myelinating cells directly influence the properties of the axoskeleton, in particular the NF network (de Waegh and Brady, 1990; de Waegh et al., 1992; Kirkpatrick et al., 2001). Conversely, by controlling the axonal caliber, NFs indirectly determine dimensions of the myelin sheath. Together, these data revealed that a tight reciprocal interaction between the axonal cytoskeleton and myelinating cells is necessary to shape adequately the geometry of myelinated fibers in order to optimize their conduction velocity. Here, we will review current knowledge about these relationships between axoskeleton and myelinating cells.

25.2 Myelinating Cells Influence the Organization, Composition, Expression, and Transport of Axonal Cytoskeletal Components

25.2.1 Myelinating Cells Control the Radial Axonal Growth by Modulating the NF Network

Myelin is produced by two distinct cell types. In the PNS each Schwann cell surrounds only one axonal segment, while in the CNS one oligodendrocyte can myelinate up to 50 axons. Peripheral and central myelin also differ in composition, structure, and embryonic origin. Both myelinating cells differently influence the axonal cytoskeleton and more particularly the NF network (Table 25.1).

NFs are the major intermediate filaments present in adult neurons and they are composed of three NF subunits (called NFL (light, 68 kDa), NFM (medium, 160 kDa), and NFH (heavy, 205 kDa)) but also of α-internexin in the CNS (Yuan et al., 2006, 2007), and peripherin in the PNS (Beaulieu et al., 1999; Yan et al., 2007). As demonstrated by several animal models lacking NFs or expressing a disorganized NF network, these cytoskeletal components play a crucial role in the radial axonal growth (for review see Perrot et al., 2008). During development and axonal regeneration, the increased axonal caliber is closely related to the composition of the axonal cytoskeleton and to the level of myelination (Hoffman et al., 1985; Sanchez et al., 1996). A close link between myelination, radial axonal growth, and the NF network has been proposed following the analysis of the relationship between the reduced axonal caliber at nodes of Ranvier and the changes in the structure of the axonal cytoskeleton. Indeed, the axonal constrictions at the nodes of Ranvier in mouse sciatic nerve correspond quantitatively to a decreased number of nodal NFs and an increased number of MTs, resulting in an enriched MT network in nodal profiles (Reles and Friede, 1991).

The analysis of the dysmyelinated mouse mutant Trembler reinforced the view that myelin locally control the properties of the axonal cytoskeleton (Table 25.1). Trembler mice are characterized by a missense mutation in the Peripheral Myelin Protein 22 (PMP22), which encodes a membrane protein of PNS myelin. Their Schwann cells begin to envelop the axon but myelin undergoes rapid degeneration and is phagocytosed. Interestingly, the mean axonal diameter is significantly reduced in PNS from Trembler mice (Aguayo et al., 1977). Moreover, the graft of a segment of Trembler sciatic nerve into a normal mouse resulted in a local caliber decrease of normal axons regrowing through the hypomyelinating environment of the Trembler graft. In contrast, when a segment of normal sciatic nerve was grafted into a Trembler mouse, atrophied Trembler axons regrowing through the normal graft become myelinated and their diameter increased in the grafted region (Aguayo et al., 1977). These results illustrate the tight relationship between a correct myelination and the radial axonal growth. Since axonal calibers depend mainly on their NF network, its properties were compared in control and Trembler sciatic nerves. Morphometric analysis revealed a twofold increase in the density of NFs in axons from Trembler mice without modification in the number of NFs (de Waegh

Table 25.1 Effects of myelination defects on NF and MT network

PNS				
Mutant	Myelination defect	Effect on NFs	Effect on MTs	References
Trembler mice (missense mutation in the PMP22 gene)	Dysmyelination	Twofold increase in NF density	Normal MT density	de Waegh and Brady (1990)
		Normal number of NFs Strong decrease of NF phosphorylation	Decreased MT stability Normal levels of tubulin	de Waegh and Brady (1991) de Waegh et al. (1992)
		Normal levels of NFL, NFM, and NFH Decreased velocity of axonal transport of NF proteins	Increased velocity of axonal transport of tubulin	Kirkpatrick and Brady (1994)
CMT1A patients or CMT1A xenograft in nude mice (PMP22 duplication)	Decreased g ratio and some demyelination in patients	1.9-Fold increase in NF density	Depletion in the MT number	Watson et al. (1994)
	The onset of myelination is delayed in CMT1A grafts	Decreased NF phosphorylation		Sahenk (1999)
CMTX xenograft in nude mice (mutations in the Cx32 gene)	Decreased myelin thickness in the CMTX-graft distal segments	1.75-Fold increase in NF density	Decreased MT density	Sahenk and Chen (1998)
		Decreased NF phosphorylation		

Table 25.1 (continued)

PNS				
Mutant	Myelination defect	Effect on NFs	Effect on MTs	References
DT transgenic mice (diphtheria toxin A chain under the control of the P0 gene promoter)	Strong hypomyelination	2.8-Fold increase in NF density Strong decrease of NF phosphorylation	Not determined	Cole et al. (1994)
SV40 transgenic mice (simian virus 40 large T antigen under the control of the P0 gene promoter)	Modest hypomyelination	1.7-Fold increase in NF density Small decrease of NF phosphorylation	Not determined	Cole et al. (1994)
MAG–/– mice	Normal myelination but the compact myelin is never closely opposed to the axolemma	Increased NF density Decreased NF phosphorylation Decreased NFL and NFH levels	Not determined	Yin et al. (1998)

Table 25.1 (continued)

CNS				
Mutant	**Myelination defect**	**Effect on NFs**	**Effect on MTs**	**References**
Shiverer mice (deletion in the MBP gene)	Lack of compact myelin	Increased NF density	Increased MT density	Brady et al. (1999)
		Decreased NF phosphorylation	Decreased MT stability	Kirkpatrick et al. (2001)
		Decreased NF stability	Increased levels of α- and β-tubulin	
		Increased NFL level and decreased NFM and NFH levels	Increased velocity of axonal transport of tubulin	
		Increased velocity of axonal transport of NF proteins		
MBP/MBP transgenic mice (25% of Wt MBP levels)	Decreased myelin thickness	Small increase of NF density	Increased MT density	Brady et al. (1999)
		Decreased NF phosphorylation	Decreased MT stability	Kirkpatrick et al. (2001)
		Normal NF stability	Increased levels of α- and β-tubulin	
		Decreased NFH level and slight increased NFM level	Normal velocity of axonal transport of tubulin	
		Normal velocity of axonal transport of NF proteins		

et al., 1992). This was correlated with a strong decrease of NF phosphorylation. Moreover, in control sciatic nerves with grafted Trembler nerve segments, these changes are confined to hypomyelinated axonal segments (de Waegh et al., 1992). Concordant results were obtained when sural nerves from Charcot-Marie-Tooth X (CMTX) patients were grafted into nude mice (Sahenk and Chen, 1998). CMTX results from mutations in the connexin-32 (Cx32) gene, encoding a gap junction protein localized in regions of uncompacted myelin, such as paranodal loops and Schmidt-Lantermann incisures. Sahenk and Chen (1998) found in CMTX regions an increased number of NFs and a reduced level of their phosphorylation. It was thus postulated that the extensive phosphorylation of NFs increases their spacing by repulsive interactions between the negative charges of the phosphate groups. The correlation between the absence of myelin and the hypophosphorylation of NFs was thereafter extended to regions normally deprived of myelin, such as the nodes of Ranvier (Mata et al., 1992), the stem processes of primary sensory neurons in the dorsal root ganglion (Hsieh et al., 1994), the initial segment of the optic nerve (Sanchez et al., 1996), and also inherited neuropathies. For example, sural nerves from Charcot-Marie-Tooth disease type 1A (CMT1A) patients exhibited marked hypophosphorylation of NFs and decreased axonal caliber (Watson et al., 1994; Sahenk, 1999).

Using a different experimental approach, it was also demonstrated that changes in NF organizations are proportional to the extent of hypomyelination (Cole et al., 1994). For instance, the expression of the diphtheria toxin A chain under the control of the myelin protein zero gene (P0) promoter caused a strong hypomyelination of the mouse sciatic nerve, which consequently induced a dramatic decrease of NF phosphorylation, a marked increase of NF density, and an important reduction of the axonal caliber. On the other hand, the modest hypomyelination induced by the expression of the simian virus 40 large T antigen under the control of the *P0* gene promoter provokes only a small decrease of NF phosphorylation and axonal caliber, and a modest increase in NF density (Cole et al., 1994). All these results support the view that myelinating Schwann cells exert a significant influence on axon caliber by modulating NF phosphorylation and their packing density in the PNS.

Several pieces of evidence also demonstrated that oligodendrocytes exert similar effects in the CNS. Shiverer mutant mice were particularly used to determine the possible consequences of dysmyelination on the NF network in the CNS (Sanchez et al., 1996; Brady et al., 1999). Due to the deletion in the Myelin Basic Protein (MBP) gene (Roach et al., 1985), which codes for a major structural protein involved in the compaction of the CNS myelin (Campagnoni and Macklin, 1988), these mice lack compact CNS myelin (Readhead and Hood, 1990). Although MBP is also expressed in PNS myelin, other proteins, including the P0 PNS-specific protein, compensate for the absence of MBP. Therefore myelination appears normal in the PNS (Rosenbluth, 1980). While shiverer oligodendrocytes do not form compact myelin, they ensheath axons with a cytoplasm containing process. This allowed one to assess in the optic nerve the influence of oligodendrocytes in the absence of myelin (Sanchez et al., 1996). Such axons achieved the same caliber, with a similar NF number and inter-NF spacing as myelinated axons, suggesting that a prolonged oligodendroglial contact is sufficient to trigger full-caliber expansion without myelin formation. However, contradictory results were thereafter obtained (Brady et al., 1999). Indeed, it was reported that axonal caliber and phosphorylation of NFH are significantly reduced in the shiverer optic nerve while NF packing density is increased, suggesting that formation of compact myelin is required for maturation of NF network in CNS. This discrepancy could be explained by the fact that Sanchez et al. (1996) focused on retinal ganglion cell axons as they emerge from the lamina cribosa, a transitional myelination region not representative of mature myelinated fibers. Brady et al. (1999) also used MBP/MBP transgenic mice in which the wild-type MBP gene was reintroduced in a shiverer background (Popko et al., 1987; Readhead et al., 1987). These mice express only 25% of wild-type MBP levels and their central axons are surrounded by correspondingly thinner myelin sheaths. This modest myelination is sufficient to restore an NF density intermediate between shiverer and wild-type nerves, but not to restore normal phosphorylation of NFs (Brady et al., 1999). This incomplete recovery of normal phenotype suggests that myelin modulate the axonal cytoskeleton by several regulatory pathways.

25.2.2 Possible Signalling Pathways that Controls the Radial Axonal Growth

It is well documented that myelinating cells regulate the radial axonal growth by modulating the phosphorylation state of NFs. It was first thought that the extensive phosphorylation of the carboxy-terminal domains of NFM and more particularly of NFH (the most phosphorylated subunit) could increase their total negative charges and cause their lateral extension by repulsive interactions, increasing NF spacing and axonal caliber. However, production of NFH-null mice (Elder et al., 1998b; Rao et al., 1998; Zhu et al., 1998), as well as mice expressing NFH deprived of its carboxy-terminal domain (Rao et al., 2002), demonstrated no major modification of NF spacing and axonal caliber. In contrast, disruption of the *NFM* gene (Elder et al., 1998a), or deletion of its carboxy-terminal domain (Garcia et al., 2003; Rao et al., 2003), reduced the interfilament spacing and axonal caliber, showing a preponderant role of NFM in determining axonal diameter.

The discovery that NF phosphorylation is reduced in regions deprived of myelin suggested the existence of a local glial signal regulating the activity of kinases/phosphatases in the axon. The activation of kinases and/or the inhibition of phosphatases in myelinated regions would support the phosphorylation of NFs, leading to the expansion of the axonal caliber. In contrast, phosphatase activities would predominate in region deprived of myelin, decreasing the axonal diameter by reducing phosphorylation and spacing of NFs. Molecules responsible for transducing the myelin signal to the axon have long been elusive. The Myelin-Associated Glycoprotein (MAG) was suspected of playing a critical role in the transduction of this signal because it is enriched in periaxonal membrane (Paivalainen and Heape, 2007), binds to neuronal membranes (Poltorak et al., 1987; Sadoul et al., 1990), and starts to be expressed when the axonal caliber expands. This was supported by the generation of MAG-null mice (Li et al., 1994; Montag et al., 1994). These mice develop normal myelin sheaths during the first months of life, but then a reduced caliber of myelinated axons was reported in their sciatic nerve concomitant with a reduced NF phosphorylation and a decreased NF spacing (Yin et al., 1998; Pan et al., 2005). Similarly, sural nerves from patients with anti-MAG paraproteinaemic neuropathies display reduced inter-NF spacing (Lunn et al., 2002). As MAG increases NFM and NFH phosphorylation in vitro (Dashiell et al., 2002), all these data strongly supported the hypothesis that MAG represents a key player in the control of NF spacing through their sidearm phosphorylation. Furthermore, axoskeleton abnormalities observed in fibers with impaired myelination could be due to MAG dysfunction or mislocalization. In agreement with this, MAG immunoreactivity is decreased or not detectable in some fibers from Trembler-J mice (Heath et al., 1991), and P0-deficient mice displayed a strong reduction of MAG at the periaxonal membrane (Carenini et al., 1999).

The exact axonal ligand of MAG and the downstream effectors involved in the phosphorylation of NFs remain uncertain. Nevertheless, possible axonal MAG receptors were first identified as the Nogo receptor (NgR) (Domeniconi et al., 2002; Liu et al., 2002) and/or the neuronal gangliosides GT1b and GD1a (Yang et al., 1996; Vyas et al., 2002; Pan et al., 2005), in association with the low affinity nerve growth factor receptor p75NTR (Wang et al., 2002; Wong et al., 2002; Yamashita et al., 2002; Quarles, 2009). β1-Integrin and paired immunoglobulin-like receptor B (PirB) were also recently identified as functional receptors for MAG (Atwal et al., 2008; Goh et al., 2008). If it is known that these receptors function in the inhibition of neurite outgrowth, it remains to be established whether they are also involved in the radial axonal growth. MAP1B was also proposed as a probable axonal receptor for MAG. MAP1B is expressed as a plasma membrane glycoprotein in neurons and is a neuronal binding partner for MAG (Franzen et al., 2001). Although the authors hypothesized that MAG-MAP1B interaction could provide a structural link between the myelinating cell and the axoskeleton, more investigations are necessary to clarify their reciprocal functions.

ERK1/2 and cdk5 are strong candidates for mediating the phosphorylation of NFs in response to MAG-dependent signalling. These kinases are known to phosphorylate NFM and NFH (Pant et al., 1997; Veeranna et al., 1998) and their activities are increased in PC12 neurons treated with a soluble MAG Fc-chimera or co-cultured with MAG-expressing COS cells (Dashiell et al., 2002). The precise phosphorylation sites involved in the radial axonal growth and how their phosphorylation provokes the expansion of the axonal caliber are still unclear (Garcia et al., 2009). Finally, MAG could also play an indirect role by maintaining a close proximity between the myelinating cell and the axon, thereby facilitating others signalling pathways.

25.2.3 Myelinating Cells Influence the MT Network

NFs are not the only components of the axonal cytoskeleton to be subject to the influence of myelinating cells. The density of axonal MTs is increased and their stability is reduced in shiverer CNS axons (Kirkpatrick et al., 2001). The partial myelination in MBP/MBP transgenic mice failed to restore MT density and stability, indicating that a thin myelin sheath is insufficient for normal regulation of the MT network. Since the density of NFs is partially restored in MBP/MBP mice but not the density of MT, these results argue that mechanisms controlling the axonal density of NFs and MTs are fundamentally different in CNS.

The influence of myelinating Schwann cells in PNS seems to differ from that of oligodendrocytes in CNS. Indeed, no modification of MT spacing or number had been seen in dysmyelinated Trembler nerves. However, this could be due to the fact that Trembler peripheral axons are myelinated before myelin degenerates, whereas shiverer central axons are never ensheathed by compact myelin. Thus, a brief myelination would be sufficient to produce a normal MT network. Moreover, the graft of sural nerves from CMTX patients into nude mice revealed modification in MT density in abnormally myelinated axons (Sahenk and Chen, 1998). A decreased MT density is observed in CMTX grafts, in opposition to the normal MT density in Trembler nerves. This discrepancy could be explained by a species difference or by different degrees of demyelination in these nerves. The reduced MT density in CMTX grafts is also in opposition with the increased MT density in CNS from shiverer mice, suggesting the opposite effect of myelinating Schwann cells and oligodendrocytes on MT density.

The stability of MTs is also influenced by myelinating cells. A significant decrease in the proportion of cold insoluble tubulin in Trembler mice was reported (Kirkpatrick and Brady, 1994), reflecting a significant reduction of MT stability in this mutant (Brady et al., 1984). A similar effect is observed in shiverer mice and the formation of a thin myelin sheath in MBP/MBP transgenic mice is not sufficient to restore a normal MT stability (Kirkpatrick et al., 2001). Generation of stable MTs usually correlates with reduced plasticity and maturation of the axoskeleton, suggesting that the MT network remains immature in abnormally myelinated axons. Note that NF stability evaluated by cold-calcium fractionation is also decreased in optic nerves from shiverer mice, but not in MBP/MBP mice (Kirkpatrick et al., 2001), indicating that a partial myelination is sufficient to stabilize NFs, whereas a higher level of myelination is required to achieve a normal MT stabilization. As the stability of NFs could depend on their composition, the proportions of NFM and NFH were shown to be reduced in CNS from shiverer mice while MBP/MBP mice displayed a reduced amount of NFH but a normal level of NFM. This compositional difference could explain the disparity in NF stability, and thus it would appear that adequate levels of NFM are necessary for increasing the stability of axonal NFs.

25.2.4 Myelinating Cells Influence the Expression and the Axonal Transport of Cytoskeletal Components

Myelinating cells change not only the organization of the axonal cytoskeleton but also its composition. The relative amounts of NFM and NFH are significantly reduced in hypomyelinated shiverer optic nerves, while NFL levels are slightly but significantly increased (Brady et al., 1999). NFH mRNA is also decreased but NFM mRNA level is not significantly different from Wt, suggesting that either translation of NFM mRNA or the stability of NFM protein is sensitive to the myelination. It also indicates that each NF subunit is differently regulated by the level of myelin, suggesting different regulation pathways.

Concerning MTs, a significant increase in the amount of α- and β-tubulin was reported in brain from shiverer mice (Kirkpatrick et al., 2001). Analysis of tubulin mRNA also revealed substantial increases in this mutant, indicating that the absence of myelin directly affects the transcription in neuronal cell bodies. This reveals the existence of signals produced during myelination and retrogradely transported to the neuronal perikarya to regulate the expression of cytoskeletal components. The expression of tubulin and NF proteins was also examined in MBP/MBP transgenic mice (Brady et al., 1999; Kirkpatrick et al., 2001). Although the level of NFH expression is reduced as in shiverer mice, NFH mRNA level is not modified. Moreover, the amount of NFM protein is slightly increased in MBP/MBP mice, without modification of the mRNA level. Finally, α- and β-tubulin mRNA and protein levels are increased in MBP/MBP mouse brain as in shiverer mice. These results indicate that 25% of normal compact myelin levels allows the restoration of some parameters but is not sufficient to induce a full response from neurons in CNS. They also imply that some signals are produced from the myelinated segments, and then retrogradely transported to the neuronal nucleus in order to modulate the expression of the cytoskeletal components.

The axonal transport of cytoskeletal elements is also perturbed in CNS from shiverer mice (Brady et al., 1999; Kirkpatrick et al., 2001). An increased velocity of axonal transport for tubulin and NF proteins was measured in shiverer optic nerve. Interestingly, normal velocities were observed in MBP/MBP mice, indicating that the presence of a thinner than normal compact myelin sheath is sufficient to restore normal transport rates. These results also suggest that myelination reduces slow axonal transport velocities during neuronal development. A possible mechanism could be the myelin-induced phosphorylation of axonal components, as shown for the phosphorylation of NFs which decreases their rate of transport (Archer et al., 1994; Jung et al., 2000; Ackerley et al., 2003; Yates et al., 2009), probably by controlling their association with molecular motors (Yabe et al., 2000; Jung et al., 2005), or increasing their association with the stationary pool of NFs. However, more investigations are required to identify the exact molecular pathways by which myelination influences the axonal transport machinery.

Taken together, these data show that composition of the axonal cytoskeleton in central axons from shiverer mice displays multiple characteristics of immature axoskeleton, including a lower content of NFH subunit (Shaw and Weber, 1982; Willard and Simon, 1983; Pachter and Liem, 1984; Carden et al., 1987), increased tubulin levels (Lewis et al., 1985), and higher velocities for the slow axonal transport (Hoffman et al., 1983). The failure of shiverer mice to produce a mature CNS axoskeleton and the fact that maturation of the axonal cytoskeleton in CNS from normal animals occurs during the period of intense myelination suggest that formation of compact myelin generate signal(s) essential for normal differentiation of neurons. The intermediate phenotype observed in MBP/MBP mice (normal slow axonal transport velocity and NFM expression but decreased expression and phosphorylation of NFH) reinforced this view.

Interestingly, the composition of the axonal cytoskeleton is differently modified by hypomyelination in PNS. No change for NF content was observed in dysmyelinated Trembler mice (de Waegh et al., 1992) as well as in other mutants characterized by a hypomyelinated PNS (Cole et al., 1994). Similarly, levels of tubulin were not affected in Trembler nerves (de Waegh and Brady, 1991; Kirkpatrick and Brady, 1994). These results are in sharp contrast to those obtained in CNS from shiverer mice, and reflect fundamental differences in axo-glial interactions between CNS and PNS. In agreement with this, it is known that neurons react differently to PNS vs CNS glial environments (Vidal-Sanz et al., 1987; Schwab and Caroni, 1988). Another possibility could be that Trembler peripheral axons are transiently myelinated and then lose their myelin sheath, whereas compact myelin never surrounds shiverer central axons. The myelination level attained before demyelination in Trembler mice would be sufficient to induce normal expression of tubulin and NF proteins. Surprisingly, MAG-deficient mice displayed a significant decreased of NFL and NFH levels in their PNS despite the formation of a normal compact myelin (Yin et al., 1998). However, although MAG-deficient axons are myelinated, the compact myelin is never closely opposed to the axolemma. This could prevent or perturb the transmission of an adequate myelin-signal involved in the regulation of NF gene expression, while the transitory myelination observed in Trembler mice would be sufficient to transduce this signal. Altogether, these results indicate that NF gene expression can also be modulated by myelination in PNS.

Transport of NF proteins and tubulin is altered differentially in Trembler peripheral nerves. As in shiverer mice, tubulin is transported faster in Trembler mice compared to normal, but the rate for NF transport is significantly reduced (de Waegh and Brady, 1990). Similar results are obtained in normal axons surrounded by Trembler Schwann cells (de Waegh and Brady, 1990; de Waegh et al., 1992). Finally, the axoskeleton from Trembler mice also displayed typical signs of immaturity, like a decreased stability of MTs. The expression pattern of microtubule associated proteins, including tau, MAP1A, and MAP1B, also presents similarities with that of immature axons (Kirkpatrick and Brady, 1994).

25.3 Role of the NF Network in the Ultrastructure of Myelin Sheaths

While myelinating cells regulate the radial axonal growth by modulating multiple properties of the axoskeleton, it is also known that several characteristics of the myelin sheath, including the myelin thickness and the internodal length, are adjusted to the size of the axon in order to optimize the conduction velocity. Recent studies highlighted how these parameters are regulated and reported fundamental differences between CNS and PNS (Table 25.2).

Table 25.2 Myelin ultrastructure in NF mutant mice

	Axonal caliber		Myelin ultrastructure		
Mice	PNS	CNS	PNS	CNS	References
NFH-LacZ	Decrease of ~50% in L4VR and SN	Decrease of ~50% in SC	Decreased g ratio	Normal g ratio	Eyer and Peterson (1994)
			Normal internodal length	Normal intermodal length	Perrot et al. (2007)
NFL–/–	Decrease of ~50% in L5VR	Decrease in SC	Decreased g ratio	Decreased g ratio	Zhu et al. (1997) Wu et al. (2008)
NFM–/–	Decrease of ~20% in L5 VR	Decrease in SC Decreased of 20% in ON	Decreased g ratio	Normal g ratio	Elder et al. (1998a) Elder et al. (2001)
NFH–/–	Decrease of ~20% in L5 VR and SN	Decrease of ~20% in SC and in ON	Decreased g ratio	Normal g ratio	Elder et al. (1998b) Elder et al. (2001)
NFM;NFH–/–	Decrease of ~30% in L5 VR	Decrease of ~50% in SC	Decreased g ratio	Normal g ratio	Elder et al. (2001)

25.3.1 NF and Control of the Myelin Thickness

The myelin thickness is highly regulated to achieve g ratio values of between 0.6 and 0.8 (Goldman and Albus, 1968; Williams and Chalupa, 1983; Fraher and O'Sullivan, 2000). One of the first pieces of evidence for implication of the NF network in the determination of the myelin thickness was provided by the analysis of NFH-LacZ mice (Eyer and Peterson, 1994). In their PNS, the 50% reduction of the axonal caliber, due to the deficiency of axonal NFs, is not accompanied by a proportional decrease in the myelin thickness, resulting in a decreased g ratio typical of hypermyelinated axons (Perrot et al., 2007). One explanation could be that Schwann cells synthesize a myelin sheath with a thickness that would be appropriate for the size that the axon should have been. In contrast, while axonal caliber is also reduced in CNS from NFH-LacZ mice, these axons are invested with proportionally thinner myelin, resulting in a normal g ratio (Perrot et al., 2007). Thus, in contrast to Schwann cells, oligodendrocytes adjust myelin thickness according to the actual size of the axon. A similar disparity was reported in NFM and NFM/H null mutant mice (Elder et al., 2001). Thus, it appears that both myelinating cells read differently the same axonal signal, or that the axonal signal is different between CNS and PNS. However, contradictory results were recently obtained from NFL–/– mice. A significant decrease of the g ratio, due to a decreased axonal caliber cumulated to an increased myelin thickness, was reported in the CNS from these mice (Wu et al., 2008). This result suggests that oligodendrocytes do not adjust their production of myelin in response to the decreased axonal diameter. While the authors concluded that signalling communication between the myelinating cells and neurons may not be determined by NF contents in axons, further investigations are necessary to elucidate fully the contribution of NFs in the determination of the myelin thickness.

One fundamental question is to determine how myelinating cells perceive the diameter of the axon they engulfed. Several studies have demonstrated the implication of Neuregulin 1 (NRG1)/ErbB signalling in the control of the myelin thickness. The neuronal growth factor NRG1 comprises a family or more than 15 transmembrane and secreted proteins. Neuregulin1 types I and III are the major isoforms expressed in the nervous system. Their receptors on myelinating cells are a heterodimer ErbB2/ErbB3 in PNS and also ErbB2/ErbB4 in CNS. The first evidence showing the implication of NRG1/ErbB signalling in the control of myelin thickness came from the analysis of mice with conditional disruption of ErbB2 in Schwann cells (Garratt et al., 2000). These mutants exhibit a widespread peripheral neuropathy characterized by abnormally thin myelin sheaths, indicating that an NRG1 signal is required to form adequate myelin thickness. This was confirmed by the manipulation of the NRG1 expression. First, the decreased expression of NRG1 in mice heterozygous for the *Nrg1* null allele caused a reduction of the myelin thickness in PNS (Michailov et al., 2004). The extent of hypomyelination is similar in *Nrg1/erbB2/erbB3* triple heterozygous mice and no reduced myelination was observed in *erbB2* or *erbB3* heterozygous mice, indicating that axonal NRG1 expression is the rate-limiting factor for myelination in PNS, whereas

Schwann cells possess saturating amounts of ErbB receptors (Michailov et al., 2004). Heterozygous mice with reduced *Nrg1 type III* gene dosage also exhibit abnormally thin myelin in their PNS and the neuronal overexpression of this isoform caused an important hypermyelination. In contrast, no modification of the myelin thickness occurred when NRG1 type I is overexpressed, showing that only the type III isoform is able to elicit the proper response (Michailov et al., 2004). All these results suggest that the expression of NRG1 type III serves as a molecular tool by which Schwann cells evaluate the size of the axons. This model implies that the amount of NRG1 type III expressed on the axolemma is proportional to the diameter of the axon, and thus to the NF content, but this assumption needs to be confirmed. The amount of NRG1 type III accumulated in peripheral nerves from adult NFH-LacZ mice is reduced proportionally to the decrease of the axonal caliber, indicating that the expression of this isoform is regulated in function of the axonal size (Perrot et al., 2007). However, it is intriguing to note that peripheral axons deprived of NFs are hypermyelinated despite the reduced expression of NRG1 type III. Because NRG1 type III activity is dispensable for the maintenance of myelin sheaths in the adult (Atanasoski et al., 2006), sufficient signalling to establish normal myelin levels might be achieved during early formation of myelin in NFH-LacZ mice. Moreover, it is well admitted that myelinating cells perceive axonal diameter at the initial stages of myelination, because a minimum initial diameter is necessary for axons to be myelinated (Peters and Muir, 1959; Friede and Samorajski, 1968; Matthews, 1968). In the future, it will be crucial to understand how neurons regulate NRG1 expression as a function of their own axonal caliber. One interesting finding is that the intracellular domain of NRG1 is proteolytically cleaved and returns to the nucleus to modulate the expression of different genes (Bao et al., 2003). Such a mechanism would constitute a back-propagation of axonal size information.

In agreement with the disparity between central and peripheral myelination observed in mice deficient in for axonal NFs, the signalling pathways involved in the control of the myelin thickness differ between these regions. Indeed, detailed ultrastructural analysis demonstrated normal g ratios in the optic nerve and spinal cord of adult NRG1 type III $\pm$ mice (Taveggia et al., 2008). However, a significant reduction in the number of myelin lamellae was observed in their corpus callosum. These findings suggest regional differences for the effect of NRG1 type III on myelination. Such diversities could be due to the differential expression of NRG1 isoforms within the CNS (Kerber et al., 2003) and/or by intrinsic differences in the responsiveness of oligodendrocytes to NRG1. However, Brinkmann et al. (2008) recently showed that the complete absence of neuronal NRG1 or oligodendroglial ErbB3/ErbB4 does not perturb oligodendrocytic development and myelination in all CNS regions. While these conflicting results remain unexplained, it should be noted that the overexpression of NRG1 type I or type III isoforms induced a significant increase of the myelin thickness in neocortex and in spinal cord (Brinkmann et al., 2008). To test whether NRG1 would have a similar stimulating effect on CNS remyelination, a focal demyelination into the ventrolateral region of the spinal cord was induced. Surprisingly, while axons in NRG1 overexpressing mice were clearly hypermyelinated on the contralateral side, the extent of axonal remyelination in the lesion was not different between NRG1 transgenic mice and Wt mice, indicating that oligodendrocytes lose their responsiveness to NRG1 with age or during remyelination (Brinkmann et al., 2008). Interestingly, in contrast to PNS, the amount of NRG1 type III accumulated in spinal cord from NFH-LacZ mice is not reduced in response to the axonal atrophy, and these mice displayed a normal expression of NRG1 type I in their CNS (Perrot et al., 2007). The myelin thickness being reduced proportionally to the decrease of the axonal caliber in CNS from NFH-LacZ mice, these results indicated that normal levels of NRG1 are not sufficient for oligodendrocytes to produce a normal myelin thickness.

All these studies suggest different effects of NFs on the control of myelin thickness. In PNS, NFs indirectly determine the myelin thickness by increasing the amount of NRG1 type III exposed on the axolemma via the increase of the axonal caliber. In CNS, myelin thickness is not precisely correlated with the level of NRG1. Brinkmann et al. (2008) suggested that NRG1 type III/ErbB signalling is an ancestral signal necessary and sufficient for myelination by Schwann cells and that evolution has made vertebrate oligodendrocytes independent from NRG1. One explanation could be that the electrical activity of axons might have evolved as a myelination signal in the CNS. Indeed, several studies showed that electrical activity-induced release of promyelinating factors is a positive signal necessary for myelination onset in CNS (Gyllensten and Malmfors, 1963; Tauber et al., 1980; Barres and Raff, 1993; Demerens et al., 1996; Stevens et al., 2002). The axonal diameter determined by NFs being a crucial determinant of the conduction properties, this could explain the correlation existing between the axonal caliber and the myelin thickness in normal and NF-deficient axons.

25.3.2 NF and Internodal Length

Like myelin thickness, internodal length positively correlates with axonal caliber and thus with axonal NF content. The length of myelin sheath is approximately 100-fold the diameter of the ensheathed axons (Fried et al., 1982). If the physiological relationships between fiber diameter, myelin thickness, and conduction velocity are well established, our understanding of how internodal length influences conduction velocity has long been theoretical (Rushton, 1951; Brill et al., 1977). However, experimental clues were recently obtained from periaxin-null mice (Court et al., 2004). The absence of Cajal band in their Schwann cells decreased their capacity to elongate, resulting in an important reduction of the internodal length without modification of the axonal caliber and the myelin thickness. In agreement with the theoretical models, nerve conduction velocities are strongly reduced in their peripheral nerves, confirming that internodal length is a key determinant of the conduction velocity. It remained to identify the cellular mechanisms that regulate the length of the myelinated segments and whether similar signalling pathways govern both length and thickness of the myelin sheath. A major issue was to establish whether regular spacing of nodes results from regularly spaced glial contacts or is instead intrinsically specified by the axonal cytoskeleton. Two distinct mechanisms control the internodal length: the initial positioning of myelinating cells along the axon and their elongation. This elongation depends principally on the nerve growth, explaining that internodes are longer in larger animals, but is independent from axonal properties (Court et al., 2004). In contrast, the positive correlation between axonal caliber and internodal length suggests that axonal signals determine the initial location of nodes of Ranvier. To determine whether the absolute size of axons plays an important role in this relationship, the internodal length was measured in CNS and PNS from NFH-LacZ mice (Perrot et al., 2007). Despite the 50% decrease of the axonal caliber due to the withdrawal of NFs from the axonal compartment, the mean internodal length is unchanged in both CNS and PNS. Consequently, internodal lengths are ~200 times the diameter of NF-deficient axons when compared to normal axons. Moreover, the clustering of nodal components and the nodal ultrastructure are not affected in the absence of axonal NFs, indicating that NFs are not essential for the formation of nodes of Ranvier (Perrot et al., 2007). Although the mechanism used by fibers to position nodes remains elusive, these data demonstrate that the lack of axonal NFs and alterations in the radial dimensions of axons and myelin sheaths are not disruptive. It should also be noted that, although internodal length is not decreased in response to the reduced axonal caliber in NFH-LacZ mice, a positive correlation between these parameters persists (Perrot et al., 2007), suggesting that myelinating cells always perceive the size of the axon they engulf and regulate their length in consequence. This probably occurs at the initial stage of myelination, independently of the radial axonal growth. In agreement with this, a minimum initial diameter is required for axons to be myelinated (Peters and Muir, 1959; Friede and Samorajski, 1968; Matthews, 1968), indicating that myelinating glia detect the axonal size early during the development. Today, substantial evidence suggests that the position of nodes in the PNS is determined by Schwann cells and is not intrinsically specified (for review see Hedstrom and Rasband, 2006).

In the CNS, an alternative scenario has been suggested from experiments on retinal ganglion cells in culture. The addition of soluble factors released by oligodendrocytes in these cultures induced formation of some regularly spaced channel clusters along neurites (Kaplan et al., 1997), suggesting that nodal and internodal domains are initially specified by the axon, with glial contact occurring subsequently. However, in vivo, nodes of optic nerve axons are not regularly spaced along the length of the axon but become progressively more widely spaced further away from the retina (Baker and Stryker, 1990; Hildebrand et al., 1993), suggesting that axons do not control the location of nodes of Ranvier. Further investigations are thus necessary to determine the specific mechanisms governing the positioning of nodes of Ranvier in the CNS. To this end, the identification of soluble factors released by oligodendrocytes will be crucial.

Several pieces of evidence showed that length and thickness of the myelin sheath are independently regulated. If the amount of NRG1 type III expressed on the axolemma controls the myelin thickness in the PNS, the spacing between nodes is unchanged when NRG1 type III is reduced (Michailov et al., 2004). Similarly, the internodal length was increased by 17% by the forced elongation of rat sciatic nerve, without modification of the myelin thickness (Hara et al., 2003). Finally, the specific silencing of p120 catenin in myelinating Schwann cells in vivo results in a marked reduction in the thickness of the myelin sheath without a change in internodal length (Perrin-Tricaud et al., 2007). Conversely, periaxin-null mice displayed shorter internodes but normal myelin thickness in their PNS (Court et al., 2004). So, different mechanisms inform Schwann cells about the axonal diameter to regulate independently the properties of myelin sheath. This does not require axonal NFs since myelin thickness and internodal length are not altered in sciatic nerve from NFH-LacZ mice (Perrot et al., 2007), supporting the view that these parameters are defined before the expansion of the axonal caliber. Despite few data available in CNS due to the difficulty to measure internodal distance in this region, length and thickness of myelin produced by oligodendrocytes also appeared to be independently regulated. Indeed, only myelin thickness is reduced proportionally to the decreased axonal caliber in CNS from NFH-LacZ mice (Perrot et al., 2007).

25.4 Conclusions

An elaborated communication between axons and their myelinating cells is crucial to increase conduction velocity by promoting the radial growth of axons and their correct insulation. It is now recognized that the axonal cytoskeleton is a key element involved in this bidirectional communication. While it is widely accepted that oligodendrocytes and Schwann cells control the expansion of the axonal caliber by regulating the expression, the transport and the organization of the axoskeleton, future investigations are necessary to identify the molecular mechanisms involved in these processes. At least two distinct signalling pathways originating from myelinating cells exist. One is a local signal regulating locally the organization of the axoskeleton, in particular by mediating the phosphorylation of cytoskeletal proteins. MAG seems to be one of the key components of this signalling pathway. The other could be a retrogradely transported signal able to modulate the expression of cytoskeletal components. The identification of these regulatory pathways represents a challenging and promising research area for the understanding of the axo-glial communication.

Dimensions of the myelin sheath are adjusted to the axonal size, and consequently to the composition of their axonal cytoskeleton. However, the absence of axonal NFs differently affects the myelination process in CNS and PNS, reflecting crucial differences in axo-glial interactions between these regions. In particular, it seems that Schwann cells regulate their dimensions in function of the initial axonal size, thus independently of the reorganization of the NF network occurring during the radial axonal growth. Consistent with this, a minimum initial diameter is required for axons to be myelinated by Schwann cells and the dimensions of the myelin sheaths are unchanged in PNS when the radial axonal growth is decreased. This suggests that the immature axoskeleton, enriched for MTs, could play a more important role in the determination of the myelin sheath dimensions than the mature one, enriched for NFs. In contrast, oligodendrocytes adjust their myelin thickness, but not their internodal length, when the axonal growth is restricted. We can reasonably speculate that the ability of oligodendrocytes to elaborate and maintain several myelin sheaths (usually 5–15) requires a more elaborate and strict local control. Each oligodendroglial process could "measure" locally the diameter of the ensheathed axon and inform the oligodendrocyte in order to regulate the production of myelin. A major challenge lies in identifying the precise molecular basis for this process.

References

Ackerley S, Thornhill P, Grierson AJ, Brownlees J, Anderton BH, Leigh PN, Shaw CE, Miller CC (2003) Neurofilament heavy chain side arm phosphorylation regulates axonal transport of neurofilaments. J Cell Biol 161:489–495

Aguayo AJ, Attiwell M, Trecarten J, Perkins S, Bray GM (1977) Abnormal myelination in transplanted trembler mouse Schwann cells. Nature 265:73–75

Archer DR, Watson DF, Griffin JW (1994) Phosphorylation-dependent immunoreactivity of neurofilaments and the rate of slow axonal transport in the central and peripheral axons of the rat dorsal root ganglion. J Neurochem 62:1119–1125

Atanasoski S, Scherer SS, Sirkowski E, Leone D, Garratt AN, Birchmeier C, Suter U (2006) ErbB2 signaling in Schwann cells is mostly dispensable for maintenance of myelinated peripheral nerves and proliferation of adult Schwann cells after injury. J Neurosci 26:2124–2131

Atwal JK, Pinkston-Gosse J, Syken J, Stawicki S, Wu Y, Shatz C, Tessier-Lavigne M (2008) PirB is a functional receptor for myelin inhibitors of axonal regeneration. Science 322:967–970

Baker GE, Stryker MP (1990) Retinofugal fibres change conduction velocity and diameter between the optic nerve and tract in ferrets. Nature 344:342–345

Bao J, Wolpowitz D, Role LW, Talmage DA (2003) Back signaling by the Nrg-1 intracellular domain. J Cell Biol 161:1133–1141

Barres BA, Raff MC (1993) Proliferation of oligodendrocyte precursor cells depends on electrical activity in axons. Nature 361:258–260

Beaulieu JM, Robertson J, Julien JP (1999) Interactions between peripherin and neurofilaments in cultured cells: disruption of peripherin assembly by the NF-M and NF-H subunits. Biochem Cell Biol 77:41–45

Brady ST, Tytell M, Lasek RJ (1984) Axonal tubulin and axonal microtubules: biochemical evidence for cold stability. J Cell Biol 99:1716–1724

Brady ST, Witt AS, Kirkpatrick LL, de Waegh SM, Readhead C, Tu PH, Lee VM (1999) Formation of compact myelin is required for maturation of the axonal cytoskeleton. J Neurosci 19: 7278–7288

Brill MH, Waxman SG, Moore JW, Joyner RW (1977) Conduction velocity and spike configuration in myelinated fibres: computed dependence on internode distance. J Neurol Neurosurg Psychiatry 40:769–774

Brinkmann BG, Agarwal A, Sereda MW, Garratt AN, Muller T, Wende H, Stassart RM, Nawaz S, Humml C, Velanac V, Radyushkin K, Goebbels S, Fischer TM, Franklin RJ, Lai C, Ehrenreich H, Birchmeier C, Schwab MH, Nave KA (2008) Neuregulin-1/ErbB signaling serves distinct functions in myelination of the peripheral and central nervous system. Neuron 59:581–595

Campagnoni AT, Macklin WB (1988) Cellular and molecular aspects of myelin protein gene expression. Mol Neurobiol 2:41–89

Carden MJ, Trojanowski JQ, Schlaepfer WW, Lee VM (1987) Two-stage expression of neurofilament polypeptides during rat neurogenesis with early establishment of adult phosphorylation patterns. J Neurosci 7:3489–3504

Carenini S, Montag D, Schachner M, Martini R (1999) Subtle roles of neural cell adhesion molecule and myelin-associated glycoprotein during Schwann cell spiralling in P0-deficient mice. Glia 27:203–212

Cole JS, Messing A, Trojanowski JQ, Lee VM (1994) Modulation of axon diameter and neurofilaments by hypomyelinating Schwann cells in transgenic mice. J Neurosci 14:6956–6966

Court FA, Sherman DL, Pratt T, Garry EM, Ribchester RR, Cottrell DF, Fleetwood-Walker SM, Brophy PJ (2004) Restricted growth of Schwann cells lacking Cajal bands slows conduction in myelinated nerves. Nature 431:191–195

Dashiell SM, Tanner SL, Pant HC, Quarles RH (2002) Myelin-associated glycoprotein modulates expression and phosphorylation of neuronal cytoskeletal elements and their associated kinases. J Neurochem 81:1263–1272

de Waegh S, Brady ST (1990) Altered slow axonal transport and regeneration in a myelin-deficient mutant mouse: the trembler as an in vivo model for Schwann cell-axon interactions. J Neurosci 10:1855–1865

de Waegh SM, Brady ST (1991) Local control of axonal properties by Schwann cells: neurofilaments and axonal transport in homologous and heterologous nerve grafts. J Neurosci Res 30:201–212

de Waegh SM, Lee VM, Brady ST (1992) Local modulation of neurofilament phosphorylation, axonal caliber, and slow axonal transport by myelinating Schwann cells. Cell 68: 451–463

Demerens C, Stankoff B, Logak M, Anglade P, Allinquant B, Couraud F, Zalc B, Lubetzki C (1996) Induction of myelination in the central nervous system by electrical activity. Proc Natl Acad Sci USA 93:9887–9892

Domeniconi M, Cao Z, Spencer T, Sivasankaran R, Wang K, Nikulina E, Kimura N, Cai H, Deng K, Gao Y, He Z, Filbin M (2002) Myelin-associated glycoprotein interacts with the Nogo66 receptor to inhibit neurite outgrowth. Neuron 35:283–290

Elder GA, Friedrich VL Jr, Bosco P, Kang C, Gourov A, Tu PH, Lee VM, Lazzarini RA (1998a) Absence of the mid-sized neurofilament subunit decreases axonal calibers, levels of light neurofilament (NF-L), and neurofilament content. J Cell Biol 141:727–739

Elder GA, Friedrich VL Jr, Kang C, Bosco P, Gourov A, Tu PH, Zhang B, Lee VM, Lazzarini RA (1998b) Requirement of heavy neurofilament subunit in the development of axons with large calibers. J Cell Biol 143:195–205

Elder GA, Friedrich VL Jr, Lazzarini RA (2001) Schwann cells and oligodendrocytes read distinct signals in establishing myelin sheath thickness. J Neurosci Res 65:493–499

Eyer J, Peterson A (1994) Neurofilament-deficient axons and perikaryal aggregates in viable transgenic mice expressing a neurofilament-beta-galactosidase fusion protein. Neuron 12:389–405

Fraher JP, O'Sullivan AW (2000) Interspecies variation in axon-myelin relationships. Cells Tissues Organs 167:206–213

Franzen R, Tanner SL, Dashiell SM, Rottkamp CA, Hammer JA, Quarles RH (2001) Microtubule-associated protein 1B: a neuronal binding partner for myelin-associated glycoprotein. J Cell Biol 155:893–898

Fried K, Hildebrand C, Erdelyi G (1982) Myelin sheath thickness and internodal length of nerve fibres in the developing feline inferior alveolar nerve. J Neurol Sci 54:47–57

Friede RL, Miyagishi T (1972) Adjustment of the myelin sheath to changes in axon caliber. Anat Rec 172:1–14

Friede RL, Samorajski T (1968) Myelin formation in the sciatic nerve of the rat. A quantitative electron microscopic, histochemical and radioautographic study. J Neuropathol Exp Neurol 27:546–570

Garcia ML, Lobsiger CS, Shah SB, Deerinck TJ, Crum J, Young D, Ward CM, Crawford TO, Gotow T, Uchiyama Y, Ellisman MH, Calcutt NA, Cleveland DW (2003) NF-M is an essential target for the myelin-directed "outside-in" signaling cascade that mediates radial axonal growth. J Cell Biol 163:1011–1020

Garcia ML, Rao MV, Fujimoto J, Garcia VB, Shah SB, Crum J, Gotow T, Uchiyama Y, Ellisman M, Calcutt NA, Cleveland DW (2009) Phosphorylation of highly conserved neurofilament medium KSP repeats is not required for myelin-dependent radial axonal growth. J Neurosci 29: 1277–1284

Garratt AN, Voiculescu O, Topilko P, Charnay P, Birchmeier C (2000) A dual role of erbB2 in myelination and in expansion of the Schwann cell precursor pool. J Cell Biol 148:1035–1046

Goh EL, Young JK, Kuwako K, Tessier-Lavigne M, He Z, Griffin JW, Ming GL (2008) beta1-integrin mediates myelin-associated glycoprotein signaling in neuronal growth cones. Mol Brain 1:10

Goldman L, Albus JS (1968) Computation of impulse conduction in myelinated fibers; theoretical basis of the velocity-diameter relation. Biophys J 8:596–607

Griffin JW, Anthony DC, Fahnestock KE, Hoffman PN, Graham DG (1984) 3,4-Dimethyl-2,5-hexanedione impairs the axonal transport of neurofilament proteins. J Neurosci 4:1516–1526

Gyllensten L, Malmfors T (1963) Myelinization of the optic nerve and its dependence on visual function–a quantitative investigation in mice. J Embryol Exp Morphol 11:255–266

Hara Y, Shiga T, Abe I, Tsujino A, Ichimura H, Okado N, Ochiai N (2003) P0 mRNA expression increases during gradual nerve elongation in adult rats. Exp Neurol 184:428–435

Heath JW, Inuzuka T, Quarles RH, Trapp BD (1991) Distribution of P0 protein and the myelin-associated glycoprotein in peripheral nerves from trembler mice. J Neurocytol 20:439–449

Hedstrom KL, Rasband MN (2006) Intrinsic and extrinsic determinants of ion channel localization in neurons. J Neurochem 98:1345–1352

Hildebrand C, Remahl S, Persson H, Bjartmar C (1993) Myelinated nerve fibres in the CNS. Prog Neurobiol 40:319–384

Hoffman PN, Lasek RJ, Griffin JW, Price DL (1983) Slowing of the axonal transport of neurofilament proteins during development. J Neurosci 3:1694–1700

Hoffman PN, Thompson GW, Griffin JW, Price DL (1985) Changes in neurofilament transport coincide temporally with alterations in the caliber of axons in regenerating motor fibers. J Cell Biol 101:1332–1340

Hsieh ST, Kidd GJ, Crawford TO, Xu Z, Lin WM, Trapp BD, Cleveland DW, Griffin JW (1994) Regional modulation of neurofilament organization by myelination in normal axons. J Neurosci 14:6392–6401

Jung C, Lee S, Ortiz D, Zhu Q, Julien JP, Shea TB (2005) The high and middle molecular weight neurofilament subunits regulate the association of neurofilaments with kinesin: inhibition by phosphorylation of the high molecular weight subunit. Brain Res Mol Brain Res 141:151–155

Jung C, Yabe JT, Shea TB (2000) C-terminal phosphorylation of the high molecular weight neurofilament subunit correlates with decreased neurofilament axonal transport velocity. Brain Res 856:12–19

Kaplan MR, Meyer-Franke A, Lambert S, Bennett V, Duncan ID, Levinson SR, Barres BA (1997) Induction of sodium channel clustering by oligodendrocytes. Nature 386:724–728

Kerber G, Streif R, Schwaiger FW, Kreutzberg GW, Hager G (2003) Neuregulin-1 isoforms are differentially expressed in the intact and regenerating adult rat nervous system. J Mol Neurosci 21:149–165

Kirkpatrick LL, Brady ST (1994) Modulation of the axonal microtubule cytoskeleton by myelinating Schwann cells. J Neurosci 14:7440–7450

Kirkpatrick LL, Witt AS, Payne HR, Shine HD, Brady ST (2001) Changes in microtubule stability and density in myelin-deficient shiverer mouse CNS axons. J Neurosci 21:2288–2297

Komiya Y, Cooper NA, Kidman AD (1986) The long-term effects of a single injection of beta,beta'-iminodipropionitrile on slow axonal transport in the rat. J Biochem 100:1241–1246

Lasek RJ, Oblinger MM, Drake PF (1983) Molecular biology of neuronal geometry: expression of neurofilament genes influences axonal diameter. Cold Spring Harb Symp Quant Biol 48(2):731–744

Lewis SA, Lee MG, Cowan NJ (1985) Five mouse tubulin isotypes and their regulated expression during development. J Cell Biol 101:852–861

Li C, Tropak MB, Gerlai R, Clapoff S, Abramow-Newerly W, Trapp B, Peterson A, Roder J (1994) Myelination in the absence of myelin-associated glycoprotein. Nature 369:747–750

Liu BP, Fournier A, GrandPre T, Strittmatter SM (2002) Myelin-associated glycoprotein as a functional ligand for the Nogo-66 receptor. Science 297:1190–1193

Lunn MP, Crawford TO, Hughes RA, Griffin JW, Sheikh KA (2002) Anti-myelin-associated glycoprotein antibodies alter neurofilament spacing. Brain 125:904–911

Mata M, Kupina N, Fink DJ (1992) Phosphorylation-dependent neurofilament epitopes are reduced at the node of Ranvier. J Neurocytol 21:199–210

Matthews MA (1968) An electron microscopic study of the relationship between axon diameter and the initiation of myelin production in the peripheral nervous system. Anat Rec 161: 337–351

Michailov GV, Sereda MW, Brinkmann BG, Fischer TM, Haug B, Birchmeier C, Role L, Lai C, Schwab MH, Nave KA (2004) Axonal neuregulin-1 regulates myelin sheath thickness. Science 304:700–703

Monaco S, Autilio-Gambetti L, Zabel D, Gambetti P (1985) Giant axonal neuropathy: acceleration of neurofilament transport in optic axons. Proc Natl Acad Sci USA 82:920–924

Montag D, Giese KP, Bartsch U, Martini R, Lang Y, Bluthmann H, Karthigasan J, Kirschner DA, Wintergerst ES, Nave KA et al. (1994) Mice deficient for the myelin-associated glycoprotein show subtle abnormalities in myelin. Neuron 13:229–246

Pachter JS, Liem RK (1984) The differential appearance of neurofilament triplet polypeptides in the developing rat optic nerve. Dev Biol 103:200–210

Paivalainen S, Heape AM (2007) Myelin-associated glycoprotein and galactosylcerebroside expression in Schwann cells during myelination. Mol Cell Neurosci 35:436–446

Pan B, Fromholt SE, Hess EJ, Crawford TO, Griffin JW, Sheikh KA, Schnaar RL (2005) Myelin-associated glycoprotein and complementary axonal ligands, gangliosides, mediate axon stability in the CNS and PNS: neuropathology and behavioral deficits in single- and double-null mice. Exp Neurol 195:208–217

Pant AC, Veeranna, Pant HC, Amin N (1997) Phosphorylation of human high molecular weight neurofilament protein (hNF-H) by neuronal cyclin-dependent kinase 5 (cdk5). Brain Res 765:259–266

Perkins CS, Aguayo AJ, Bray GM (1981) Behavior of Schwann cells from trembler mouse unmyelinated fibers transplanted into myelinated nerves. Exp Neurol 71:515–526

Perrin-Tricaud C, Rutishauser U, Tricaud N (2007) P120 catenin is required for thickening of Schwann cell myelin. Mol Cell Neurosci 35:120–129

Perrot R, Berges R, Bocquet A, Eyer J (2008) Review of the multiple aspects of neurofilament functions, and their possible contribution to neurodegeneration. Mol Neurobiol 38:27–65

Perrot R, Lonchampt P, Peterson AC, Eyer J (2007) Axonal neurofilaments control multiple fiber properties but do not influence structure or spacing of nodes of Ranvier. J Neurosci 27: 9573–9584

Peters A, Muir AR (1959) The relationship between axons and Schwann cells during development of peripheral nerves in the rat. Q J Exp Physiol Cogn Med Sci 44:117–130

Pollard JD, McLeod JG (1980) Nerve grafts in the Trembler mouse. An electrophysiological and histological study. J Neurol Sci 46:373–383

Poltorak M, Sadoul R, Keilhauer G, Landa C, Fahrig T, Schachner M (1987) Myelin-associated glycoprotein, a member of the L2/HNK-1 family of neural cell adhesion molecules, is involved in neuron-oligodendrocyte and oligodendrocyte-oligodendrocyte interaction. J Cell Biol 105:1893–1899

Popko B, Puckett C, Lai E, Shine HD, Readhead C, Takahashi N, Hunt SW 3rd, Sidman RL, Hood L (1987) Myelin deficient mice: expression of myelin basic protein and generation of mice with varying levels of myelin. Cell 48:713–721

Quarles RH (2009) A hypothesis about the relationship of myelin-associated glycoprotein's function in myelinated axons to its capacity to inhibit neurite outgrowth. Neurochem Res 34:79–86

Rao MV, Campbell J, Yuan A, Kumar A, Gotow T, Uchiyama Y, Nixon RA (2003) The neurofilament middle molecular mass subunit carboxyl-terminal tail domains is essential for the radial growth and cytoskeletal architecture of axons but not for regulating neurofilament transport rate. J Cell Biol 163:1021–1031

Rao MV, Garcia ML, Miyazaki Y, Gotow T, Yuan A, Mattina S, Ward CM, Calcutt NA, Uchiyama Y, Nixon RA, Cleveland DW (2002) Gene replacement in mice reveals that the heavily phosphorylated tail of neurofilament heavy subunit does not affect axonal caliber or the transit of cargoes in slow axonal transport. J Cell Biol 158:681–693

Rao MV, Houseweart MK, Williamson TL, Crawford TO, Folmer J, Cleveland DW (1998) Neurofilament-dependent radial growth of motor axons and axonal organization of neurofilaments does not require the neurofilament heavy subunit (NF-H) or its phosphorylation. J Cell Biol 143:171–181

Readhead C, Hood L (1990) The dysmyelinating mouse mutations shiverer (shi) and myelin deficient (shimld). Behav Genet 20:213–234

Readhead C, Popko B, Takahashi N, Shine HD, Saavedra RA, Sidman RL, Hood L (1987) Expression of a myelin basic protein gene in transgenic shiverer mice: correction of the dysmyelinating phenotype. Cell 48:703–712

Reles A, Friede RL (1991) Axonal cytoskeleton at the nodes of Ranvier. J Neurocytol 20:450–458

Roach A, Takahashi N, Pravtcheva D, Ruddle F, Hood L (1985) Chromosomal mapping of mouse myelin basic protein gene and structure and transcription of the partially deleted gene in shiverer mutant mice. Cell 42:149–155

Rosenbluth J (1980) Peripheral myelin in the mouse mutant Shiverer. J Comp Neurol 193:729–739

Rushton WA (1951) A theory of the effects of fibre size in medullated nerve. J Physiol 115:101–122

Sadoul R, Fahrig T, Bartsch U, Schachner M (1990) Binding properties of liposomes containing the myelin-associated glycoprotein MAG to neural cell cultures. J Neurosci Res 25:1–13

Sahenk Z (1999) Abnormal Schwann cell-axon interactions in CMT neuropathies. The effects of mutant Schwann cells on the axonal cytoskeleton and regeneration-associated myelination. Ann NY Acad Sci 883:415–426

Sahenk Z, Chen L (1998) Abnormalities in the axonal cytoskeleton induced by a connexin32 mutation in nerve xenografts. J Neurosci Res 51:174–184

Sanchez I, Hassinger L, Paskevich PA, Shine HD, Nixon RA (1996) Oligodendroglia regulate the regional expansion of axon caliber and local accumulation of neurofilaments during development independently of myelin formation. J Neurosci 16:5095–5105

Schwab ME, Caroni P (1988) Oligodendrocytes and CNS myelin are nonpermissive substrates for neurite growth and fibroblast spreading in vitro. J Neurosci 8:2381–2393

Shaw G, Weber K (1982) Differential expression of neurofilament triplet proteins in brain development. Nature 298:277–279

Stampfli R (1954) Saltatory conduction in nerve. Physiol Rev 34:101–112

Stevens B, Porta S, Haak LL, Gallo V, Fields RD (2002) Adenosine: a neuron-glial transmitter promoting myelination in the CNS in response to action potentials. Neuron 36:855–868

Tauber H, Waehneldt TV, Neuhoff V (1980) Myelination in rabbit optic nerves is accelerated by artificial eye opening. Neurosci Lett 16:235–238

Taveggia C, Thaker P, Petrylak A, Caporaso GL, Toews A, Falls DL, Einheber S, Salzer JL (2008) Type III neuregulin-1 promotes oligodendrocyte myelination. Glia 56:284–293

Veeranna, Amin ND, Ahn NG, Jaffe H, Winters CA, Grant P, Pant HC (1998) Mitogen-activated protein kinases (Erk1,2) phosphorylate Lys-Ser-Pro (KSP) repeats in neurofilament proteins NF-H and NF-M. J Neurosci 18:4008–4021

Vidal-Sanz M, Bray GM, Villegas-Perez MP, Thanos S, Aguayo AJ (1987) Axonal regeneration and synapse formation in the superior colliculus by retinal ganglion cells in the adult rat. J Neurosci 7:2894–2909

Vyas AA, Patel HV, Fromholt SE, Heffer-Lauc M, Vyas KA, Dang J, Schachner M, Schnaar RL (2002) Gangliosides are functional nerve cell ligands for myelin-associated glycoprotein (MAG), an inhibitor of nerve regeneration. Proc Natl Acad Sci USA 99:8412–8417

Wang KC, Kim JA, Sivasankaran R, Segal R, He Z (2002) P75 interacts with the Nogo receptor as a co-receptor for Nogo, MAG and OMgp. Nature 420:74–78

Watson DF, Nachtman FN, Kuncl RW, Griffin JW (1994) Altered neurofilament phosphorylation and beta tubulin isotypes in Charcot-Marie-Tooth disease type 1. Neurology 44:2383–2387

Willard M, Simon C (1983) Modulations of neurofilament axonal transport during the development of rabbit retinal ganglion cells. Cell 35:551–559

Williams RW, Chalupa LM (1983) An analysis of axon caliber within the optic nerve of the cat: evidence of size groupings and regional organization. J Neurosci 3:1554–1564

Windebank AJ, Wood P, Bunge RP, Dyck PJ (1985) Myelination determines the caliber of dorsal root ganglion neurons in culture. J Neurosci 5:1563–1569

Wong ST, Henley JR, Kanning KC, Huang KH, Bothwell M, Poo MM (2002) A p75(NTR) and Nogo receptor complex mediates repulsive signaling by myelin-associated glycoprotein. Nat Neurosci 5:1302–1308

Wu YJ, Tang YF, Xiao ZC, Bao ZM, He BP (2008) NG2 cells response to axonal alteration in the spinal cord white matter in mice with genetic disruption of neurofilament light subunit expression. Mol Neurodegener 3:18

Yabe JT, Jung C, Chan WK, Shea TB (2000) Phospho-dependent association of neurofilament proteins with kinesin in situ. Cell Motil Cytoskeleton 45:249–262

Yamashita T, Higuchi H, Tohyama M (2002) The p75 receptor transduces the signal from myelin-associated glycoprotein to Rho. J Cell Biol 157:565–570

Yan Y, Jensen K, Brown A (2007) The polypeptide composition of moving and stationary neurofilaments in cultured sympathetic neurons. Cell Motil Cytoskeleton 64:299–309

Yang LJ, Zeller CB, Shaper NL, Kiso M, Hasegawa A, Shapiro RE, Schnaar RL (1996) Gangliosides are neuronal ligands for myelin-associated glycoprotein. Proc Natl Acad Sci USA 93:814–818

Yates DM, Manser C, De Vos KJ, Shaw CE, McLoughlin DM, Miller CC (2009) Neurofilament subunit (NFL) head domain phosphorylation regulates axonal transport of neurofilaments. Eur J Cell Biol 88:193–202

Yin X, Crawford TO, Griffin JW, Tu P, Lee VM, Li C, Roder J, Trapp BD (1998) Myelin-associated glycoprotein is a myelin signal that modulates the caliber of myelinated axons. J Neurosci 18:1953–1962

Yuan A, Rao MV, Sasaki T, Chen Y, Kumar A, Veeranna, Liem RK, Eyer J, Peterson AC, Julien JP, Nixon RA (2006) Alpha-internexin is structurally and functionally associated with the neurofilament triplet proteins in the mature CNS. J Neurosci 26:10006–10019

Zhu Q, Couillard-Despres S, Julien JP (1997) Delayed maturation of regenerating myelinated axons in mice lacking neurofilaments. Exp Neurol. 148:299–316

Zhu Q, Lindenbaum M, Levavasseur F, Jacomy H, Julien JP (1998) Disruption of the NF-H gene increases axonal microtubule content and velocity of neurofilament transport: relief of axonopathy resulting from the toxin beta,beta'-iminodipropionitrile. J Cell Biol 143:183–193

Chapter 26
Topographic Regulation of Neuronal Intermediate Filament Proteins by Phosphorylation: In Health and Disease

Parvathi Rudrabhatla and Harish C. Pant

Abstract Neurofilaments (NFs) belong to type 1 V family of intermediate filaments and are the most abundant proteins of the nervous system. Mammalian NF triplet proteins constitute three subunits, low molecular weight (NF-L), medium molecular weight (NF-M), and high molecular weight (NF-H) subunit proteins. The NF-M/H carboxy terminal tail domain has multiple KSP repeats ($40 \geq 100$) depending upon species, the phosphorylation of which regulates axonal caliber and axonal transport. The NF tail domain contains phosphorylation sites for proline directed kinases such as mitogen activated protein kinases (MAPKs), cyclin dependent protein kinase 5 (Cdk5), c-Jun amino terminal kinase (JNKs), and glycogen synthase kinase-3 (GSK-3). The signaling cascades involved in NF phosphorylation like MAPKs and Cdk5 can also be affected by myelin associated glycoprotein (MAG). MAG activates the Erk1/2 and Cdk5 activities due to the glial/axon interaction and increases the proline directed Ser/Thr phosphorylation of C-terminal domain of NF-M/H. The NF phosphorylation is topographically regulated. In normal neurons, NFs are phosphorylated only in the axonal compartment. However, in degenerative neurons such as Alzheimer's disease (AD), spinal cord motor neuron inclusions of amyotrophic lateral sclerosis (ALS), Lewy bodies of Parkinson's disease (PD), and Pick's disease neurofilament proteins are aberrantly phosphorylated in the cell bodies. Aberrant phosphorylation of NFs in neurodegeneration is either due to the deregulation of proline directed kinases such as MAPKs and Cdk5 or the downregulation of protein phosphatases such as protein phosphatase 2A (PP2A), calcineurin (PP2B), or both. Recently, studies from our laboratory have shown that peptidyl prolyl isomerase 1 (Pin1) stabilizes the NF phosphorylation in normal and stressed neurons. Pin1 selectively binds to the phosphorylated Ser/Thr-Pro residues and converts the *cis* isomers to the more stable *trans* isomers. The multiple SP repeats of NF M/H are stabilized by Pin1 in a phosphorylation specific manner. Pin1 modulates the excitotoxic and oxidative stress induced perikaryal phosphorylation of NF-M/H. Here, we have

H.C. Pant (✉)
Laboratory of Neurochemistry, National Institutes of Neurological Disorders and Stroke, National Institutes of Health, Bethesda, MD 20892, USA
e-mail: panth@ninds.nih.gov

R.A. Nixon, A. Yuan (eds.), *Cytoskeleton of the Nervous System*, Advances in Neurobiology 3, DOI 10.1007/978-1-4419-6787-9_26,

discussed the factors regulating the topographic phosphorylation of NFs in health and disease.

Keywords ALS · Alzheimer's disease · Axonal transport · Neurofilament · Proline-directed kinases

Contents

26.1 Introduction

Cell cytoskeleton is mainly composed of microtubules, microfilaments and the intermediate filaments. Intermediate filaments (IFs) are 10-nm-thick cytoskeletal structures formed by members of a family of related proteins, which vary in size

(40–280 kDa) and their primary structure (Parry and Steinert, 1992; Fuchs and Weber, 1994; Herrmann et al., 2003; Herrmann and Aebi, 2004). IFs are classified into five classes based on the structure and amino acid sequence (Fuchs and Weber, 1994; Herrmann et al., 2003; Herrmann and Aebi, 2004; Coulombe et al., 2001; Omary et al., 2006; Steinert and Parry, 1985; Blumenberg, 1989; Herrmann and Aebi 2000). Types I and II IFs comprise of acidic and basic keratins that are typically expressed in epithelial cells. Type III IF proteins include Vimentin (mesenchymal cells), desmin (muscle cells), glial fibrillary acidic protein (GFAP) (glia and astrocytes), and peripherin (neurons of the peripheral nervous system). Type IV IF proteins are expressed in neurons and include the neurofilament triplet proteins, NF-L, NF-M, and NF-H and *a*-internexin. In contrast to the cytoplasmic location of the four classes of IF proteins described above, type V IF proteins, also known as lamins, are expressed exclusively in the nucleus. Proteins belonging to the IF superfamily but not included in the five types, such as nestin and synemins, are in some classifications referred to as type VI IFs (Coulombe et al., 2001).

The IF proteins exhibit a common tripartite domain structure, with non-helical amino (head)- and carboxyl-terminal (tail) domains flanked by a central α-helical core region of 310-amino-acid long in cytoplasmic IF, and 352-amino-acid long in the type V nuclear lamins (Herrmann et al., 2003; Omary et al., 2006; Steinert and Parry, 1985; Blumenberg, 1989; Herrmann and Aebi, 2000). The size and amino acid of the amino-terminal head and carboxyl-terminal tail domains vary extensively between individual IF sequences. In this chapter, we mainly review the progress made in the past in topographic regulation of NFs in normal and diseased neurons.

26.2 NF Domain Structure

Neurofilaments are the most abundant proteins of the myelinated neurons and are specifically expressed in neurons. NF triplet proteins are used to identify the post-mitotic cells. They are expressed either independently or together with other IF proteins such as nestin, vimentin, alpha-internexin, and peripherin. Mammalian NF triplet proteins (NFTPs) comprise of three subunits of low, NF-L (60–70 kDa), medium, NF-M (145–160 kDa), and high NF-H (200–220 kDa) molecular weights. All the NF subunits share a tripartite structure typical of IF proteins consisting of highly conserved central alpha helical coil-coiled rod domain flanked by a amino terminal globular head domain and a hyper variable carboxy-terminal tail domain which differ in length among subunits. The central rod domain is involved in coiled-coil formation of filamentous structure while the globular head domain is involved in NF assembly (Fliegner and Liem, 1991; Shaw, 1991; Lee and Cleveland, 1996; Heins and Aebi, 1994). NFs are obligate heteropolymers in vivo, with assembly depending on a core NF-L subunit interacting with NF-M or NF-H subunits (Ching and Liem, 1993; Lee et al.,1993). Human NF-L can partially self-assemble into filaments, whereas rodent NF-L cannot (Ching and Liem, 1993; Lee et al., 1993; Carter et al., 1998; Perez-olle et al., 2002). Together with microtubules, microtubule associated proteins (MAP1A and B), Tau, actin, and other associated proteins, NFs make up the dynamic axonal cytoskeleton. As the development continues, the expression of NFTPs gradually increases. In the adult, NFTPs are abundantly expressed in both central nervous system (CNS) and peripheral nervous system (PNS). NFTPs are present in large levels in axons. The head domain of NFs contains multiple phosphorylation sites for second-messenger related phosphorylation sites which might regulate head domain function in assembly of NFs (Gonda et al., 1990; Inagaki et al., 1990; Kitamura et al., 1989). The carboxy-terminal domain of NF-M and NF-H consists of a large number of KSP repeats (30–100) depending upon species, the phosphorylation of which determines axonal caliber (Lee et al., 1988). The domain structure of NF-L, NF-M, and NF-H is shown in Fig. 26.1. The phosphorylation of NF-M may be more important than NF-H in determining the axonal caliber. The KSP repeats of NF-M/H are phosphorylated by proline directed kinases such as MAPKs, Cdk5, GSK3, JNKs, and dephosphorylated by protein phosphatases such as PP2A. For the most part, NF proteins follow a sequential order of expression in

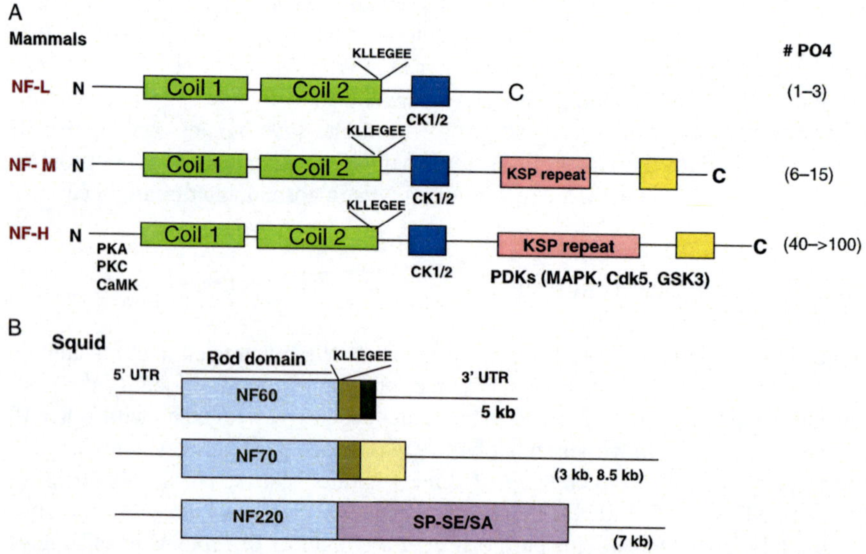

Fig. 26.1 Domain structure of NF. **a** Mammalian NF protein subunits NF-L, NF-M, and NF-H. Neurofilament subunit domains and sites that are phosphorylated by known kinases. Schematic representation of NF-L, NF-M, and NF-H proteins and sites that are phosphorylated by different kinases. NF-subunits are phosphorylated at their amino termini by Protein Kinase A (PKA), Protein Kinase C (PKC), and Calcium–Calmodulin kinase II (Ca/CAMII). All three subunits are phosphorylated in the glutamic acid-rich region (*blue*) by Casein Kinase I (CKI/II). NF-M and NF-H are phosphorylated in their KSP repeat regions by ERK1/2, GSK3, and Cdk5. The number of phosphates incorporated per mole of neurofilament protein are indicated on the *right*. NF-L and NF-M have 1–3 and 6–15 phosphates per mole, respectively, and NF-H has between (40≥100) phosphates incorporated. **b** Schematic diagram of squid cDNAs encoding NF-60, NF70, and NF-220 proteins. Coding regions are represented by *rectangles*; untranslated (UT) regions are represented by *straight lines*. Identical amino acid residues in the rod region are shown in *blue*. An *arrow* marks the beginning of divergent tail domains (*green*)

vertebrate development, with NF-L and NF-M appearing initially as neurites differentiates. Later, as the nervous system matures, NF-H is expressed in brain and spinal cord neurons as the cytoskeleton is stabilized (Willard and Simon, 1983; Carden et al., 1987). In squid, NFs make up about 13% of total axoplasm protein (Brown and Lasek, 1990) and consist of three subunit proteins, a large, highly phosphorylated NF-220, and two smaller subunits, NF60 and NF70. These have been cloned and shown to arise by alternative splicing from a single gene, in contrast to the three independent genes that code for mammalian NF subunits (Way et al., 1992). The NF-220 is phosphorylated only in the axon, whereas a non-phosphorylated NF-180, together with NF60/70, is detected in the perikarya of the stellate ganglia and axons (Tytell et al., 1988). Phosphorylation occurs predominantly on the multiple K/RSP, SAR/K, and SEK/R repeat sites (>47) on the C-terminal tail of the NF-220 subunit (Jaffe et al., 2001).

26.3 Developmental Expression of NF-L/M/H

Sequential forms of NFs are expressed by developing and maturing neurons throughout the nervous system (Carden et al., 1987). NF proteins are first detected on the embryonic day 12, by E 13 the coexpression of NF-L and NF-M is widespread, and levels of NF-H are extremely low at E15. NF-H expression increases very slowly and is well below those of NF-M and NF-L for several weeks beyond birth. An "immature" form of NFs, composed of NF-M and NF-L, appears to function in establishing the neuronal phenotype and in initiating and maintaining neurite outgrowth. NF-H phosphorylation confers a "mature" state to the NF. This delayed expression of NF-H is a slow and gradual process that coincides in time with the stabilization of neuronal circuitries and may be important in modulating axonal events, such as the slowing of axonal transport and the growth of axonal caliber.

26.4 Phosphorylation of NFs

NF protein phosphorylation plays a key role in the dynamic remodeling of cytoskeletal architecture during axonal growth, guidance, and synaptogenesis (Carden et al., 1987; Nixon and Shea, 1992; Grant and Pant, 2000). The NF phosphorylation is topographically regulated and in normal neurons NFP synthesis occurs in the cell bodies and NF phosphorylation occurs once they are transported into axons. The N-terminal head domains are transiently phosphorylated by second messenger-dependent protein kinases, the cyclic AMP-dependent protein kinase (protein kinase A, PKA), calcium/calmodulin protein kinase (CaMPK), and protein kinase C (PKC) (Sihag and Nixon, 1989, 1990; Dosemeci and Pant, 1992; Hisanaga et al., 1994) in the cell bodies. The C-terminal tails of NF-M and NF-H, enriched in numerous Lys-Ser-Pro (KSP) repeats, on the other hand, are preferred substrates for the second messenger-independent kinases associated with neurofilaments, such as CKI (Floyd et al., 1991) and GSK3 (Guidato et al., 1996), and the proline-directed kinases such as Erk1/2 and Cdk5 (Nixon and Sihag, 1991; Veeranna et al., 1998; Li et al., 1999b). N-terminal head domain phosphorylation by PKA at Ser^{55} in NFL or Ser^{46} in NF-M inhibits assembly into a heteropolymer in vitro and in vivo (Sihag and Nixon, 1991; Nakamura et al., 2000). It has been suggested that this transient head domain phosphorylation in cell bodies inhibits NF polymerization and filament formation (Ching and Liem, 1999). Extensive phosphorylation of KSP repeats in the tail domain of NF-M and NF-H occurs primarily in axons (Lee et al., 1988; Sternberger and Sternberger, 1983). This results in sidearm formation, increased inter-neurofilament spacing, radial growth of axons, and increased conduction velocity (de Waegh et al., 1992; Nakagawa et al., 1995; Yin et al., 1998). NF-M/H tail domains contain KSPXK, KSPXXK, and KSPXXXK repeats.

26.5 Signaling Cascades Involved in NF Phosphorylation

26.5.1 MAP Kinase Pathway

Neurofilament function depends on the state of phosphorylation of the numerous serine/threonine residues in these proteins. Most phosphorylation occurs in the Lys-Ser-Pro (KSP) repeats in the C-terminal tail domains of NF-H and NF-M (Veeranna et al., 1998). A comparative kinetic study of Erk1/2 and Cdk5 phosphorylation of KSPXK and KSPXXXK peptides revealed that Cdk5 phosphorylated mostly the KSPXK peptide; however, Erk1/2 could phosphorylate both. The preferred substrate for Erk1/2 was KSPXXXK peptide. Of the repeats of rodent NF-H, 80% are of the KSPXXXK type. The MEK inhibitor PD98059 inhibited the phosphorylation of NF-H, NF-M, and microtubule-associated protein (MAP) in primary rat hippocampal cells that caused a reduction in neurite outgrowth, suggesting that Erk1 and Erk2 may play an important role in neurite growth and branching. Epidermal growth factor (EGF), which induces the MAP kinase cascade in NIH 3T3 cells, activated endogenous Erk1 and Erk2 that phosphorylate NF-M tail domain phosphorylation in the transfected cells. Furthermore, in transfected NIH3T3 cells, constitutively active mitogen-activated Erk activating kinase (MEK1), but not the dominant negative mutant, induced phosphorylation of NF-M (Li et al., 1999a). Activation of endogenous Erk1 and Erk2 by membrane depolarization and calcium influx through L-type calcium channels resulted in phosphorylation of the NF-M tail domain in PC12 cells (Li et al., 1999b). This phosphorylation was inhibited in the presence of nifedipine, an L-type calcium channel inhibitor, and PD98059, a specific MEK1 inhibitor, a mechanism linking calcium influx through voltage-gated calcium channels with the MAP kinase pathway and NF-M tail domain phosphorylation and neurite outgrowth.

26.5.2 Signaling Pathway Involving Cdk5/p35

Cdk5 has been shown to play an important role in neuronal development and neurogenesis. Cdk5, unlike other CDKs, is active only in neuronal cells where its neuron-specific activators p35 and p39 are present. Cdk5 specifically phosphorylates the serine residues within the KSPXK sites. In human neuroblastoma SHSY5Y cells, inhibition of Cdk5 levels by antisense treatment resulted in a decrease in phosphorylation of NF-H that correlated with a decline in neurite outgrowth (Li et al., 2000). Upon differentiation of SHSY5Y cells with retinoic acid, we found that the phosphorylation of NF-H and NF-M is increased. Cdk5 modulates neurofilament metabolism in axon outgrowth, cytoskeletal stabilization, and radial growth. Intracellular signaling by the alpha(1)beta(1) integrin-induced activation of Cdk5 is involved in neurite outgrowth and human neurofilament protein H (hNF-H) Lys-Ser-Pro (KSP) tail domain phosphorylation in differentiated human SH-SY5Y cells (Li et al., 2000). The integrin alpha(1) and beta(1) monoclonal antibodies and butyrolactone-I (BL-1), a specific Cdk5 inhibitor, inhibited these effects. Cdk5 activity and hNF-H KSP tail domain phosphorylation were increased in Cdk5/p35 and hNF-H tail domain co-transfected HEK293 cells grown on laminin. This increased hNF-H tail domain phosphorylation was triggered by Cdk5 activation. Cdk5 may play an important role in promoting neurite outgrowth and hNF-H tail KSP domain phosphorylation through the integrin alpha(1)beta(1) signaling pathway.

26.5.3 Signaling Pathway Involving MAG

Phosphorylation of NF proteins occurs in the axonal compartment due to the glial axonal interaction in close proximity to myelin sheaths (Dashiell et al., 2002). The more recent report of similarly decreased axonal caliber and reduced neurofilament phosphorylation in myeline associated glycoprotein (MAG)-null mice supported this hypothesis that MAG-mediated signaling could affect the axonal cytoskeleton phosphorylation. Decreased phosphorylation of neurofilaments in mice lacking myelin-associated glycoprotein (MAG) was shown to be associated with decreased activities of extracellular-signal regulated kinases (ERK1/2) and cyclin-dependent kinase-5 (Cdk5). Total amounts of NF-M, microtubule-associated protein 1B (MAP1B), MAP2, and Tau were up-regulated significantly in DRG neurons in the presence of MAG (Dashiell et al., 2002). There was also increased expression of phosphorylated NF-H, NF-M, and MAP1B. Additionally, in similar in vitro paradigms, total and phosphorylated NF-M were increased significantly in PC12 neurons co-cultured with MAG-expressing COS cells or treated with a soluble MAG Fc-chimera. The increased expression of phosphorylated cytoskeletal proteins in the presence of MAG in vitro was associated with increased activities of Erk 1/2 and Cdk5. These findings have let us to hypothesize that the interaction of MAG with axonal receptor or the glia-axonal interaction induces signal transduction pathways that regulates expression of cytoskeletal proteins and their phosphorylation by proline-directed Ser/Thr kinases (Fig. 26.2). On the basis of these studies, glial/axonal interaction inducing signal transduction and activation of proline-directed S/T-P phosphorylation of neuronal cytoskeleton proteins is proposed in Fig. 26.2.

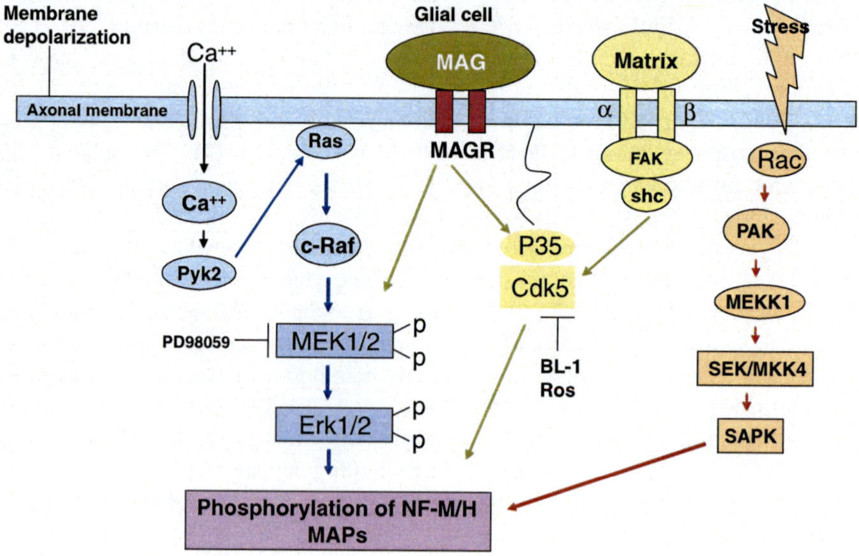

Fig. 26.2 Schematic diagram summarizing signal transduction pathways of cytoskeletal phosphorylation (NFs and MAP). Growth factor (GF) binds to the growth factor receptor to initiate the Ras-mediated signaling through Grb2/SOS that culminates in the activation of MEK1/2 and ERK1/2 leading to the phosphorylation of the cytoskeletal proteins such as Tau, MAPs, and NF. Cdk5, on the other hand, can be activated through signaling originating at the integrin receptor, leading to the phosphorylation of cytoskeletal proteins. Cdk5 is also involved in 'crosstalk' between the MAP kinase pathway through its phosphorylation of MEK1/2 and its involvement with the MAG pathway where an absence of MAG leads to a decrease in ERK1/2 and Cdk5 activities. During stress, NFs and MAPs are phosphorylated by stress activated protein kinase pathway (SAPK) pathway

26.5.4 Signaling Pathway Involving Casein Kinase 1/II (CK1/II)

Second messenger independent kinases such as CKI and CKII are also firmly bound to NF preparations from mammalian and squid nervous tissues. CKI/CKII are constitutively active kinases associated with NF preparations from squid axoplasm and bovine and chicken spinal cord (Dosemeci et al., 1990; Hollander et al., 1996). It is also an endogenous kinase that effectively phosphorylates native NF subunits in vitro. In vitro, CKI actively phosphorylates Ser/Thr residues in the glutamic acid rich region of the C-terminal tail domain shared by all three subunits (Sihag and Nixon, 1989, 1990; Dosemeci and Pant, 1992; Hisanaga et al., 1994; Dosemeci et al., 1990; Hollander et al., 1996). A VEEIIEET sequence in this region is conserved in all three NF proteins from lamprey to humans, and at least 5 serine residues, C-terminal to this consensus sequence, are phosphorylated by CKI in chicken NF-M (Shaw et al., 1997). It is possible that phosphorylation/dephosphorylation reactions at these sites may contribute to the dynamic exchanges between oligomers and polymers within the axon, particularly during early labile periods of axon growth. In mature neurons, after tail domains of NF-M and NF-H are phosphorylated to stabilize the cytoskeleton, CKI may become involved in transient exchanges of phosphorylated and non-phosphorylated NF proteins between soluble and NF polymers during slow axon transport. Together with other kinases, CKI may be integrated into a multisite phosphorylation cascade where phosphorylation at one site by CKI, for example, may be essential for phosphorylation at other sites by other kinases such as proline directed kinases and GSK3β (Kennelly and Krebs, 1991). For example, Tau contains many phosphate acceptor sites and is phosphorylated by a wide variety of kinases, among which are CKI, PKA, GSK3, and Cdk5. Prior phosphorylation of Tau by CKI, CDK5, or PKA potentiates phosphorylation by GSK3 (Sengupta et al., 1997) as if conformational changes in Tau increase the accessibility of other kinase motifs. It is a common observation that native NF preparations are readily phosphorylated by CKI and PKA in vitro but are not phosphorylated by proline directed kinases (Pant et al., unpublished). Moreover, if NFs are completely dephosphorylated in vitro by alkaline phosphatase, they become relatively poor substrates for both CKI and PKA, as if a proper sequence of site-specific phosphorylation is required for complete phosphorylation to occur. CKII, a microtubule associated kinase that phosphorylates β-tubulin (Serrano et al., 1987, 1989; Crute and Van Buskirk, 1992) is also associated with NF preparations from squid (Link et al., 1992). It is less active than CKI with NF substrates, particularly NF-H, but a recent study has shown that, in vitro, CKII phosphorylates NF-L at Ser-473, in the tail domain (Nakamura et al., 1999). Moreover, an antibody specific for the NF-L phosphorylated epitope exhibited robust expression in perikarya of the rat cortex, suggesting that CKII phosphorylation at this site occurs in vivo. CAMK II, another kinase that associates with brain microtubule and cytoskeletal preparations containing NF proteins (Goldenring et al., 1984), phosphorylates tubulins and NFs. The kinase is enriched in squid NF preparations (Dosemeci et al., 1990) and in synaptic regions (PSD) where it phosphorylates NF proteins among other substrates in the complex of PSD proteins (Yoshimura et al., 2000).

26.5.5 NF Phosphorylation by Stress Activated Protein Kinases (SAPK)

Sustained activation of p38 kinase, a stress activated protein kinase by NGF promotes neurite outgrowth which can be blocked by a specific inhibitor of p38. It is noteworthy that PC 12 cells with constitutively active MEK also stimulates p38 activation as well as Erk, which suggests that multiple signal transduction cascades may be simultaneously activated. Another important kinase in NF phosphorylation is SAP kinase. SAP kinase is a major stress kinase in the nervous system, expressed in CNS and PNS, and its specific pathway is modulated by its upstream regulator SEK1 and is expressed in diverse neuronal tissues in the developing nervous system (Martin et al., 1996). Under stress conditions (osmotic, UV), PC 12 and DRG cells exhibited a marked stimulation of SAP kinase activity accompanied by NF-H phosphorylation in perikarya, a condition resembling the perikaryal accumulations of phosphorylated NF-H in neurodegenerative disorders such as ALS (Giasson and Mushynski, 1996). Apparently, under these stress conditions, Erk1/2 was not activated. The preferred motif on NF-H for SAPK phosphorylation is KSPXE rather than KSPXK. Mushinski and co-workers found that cotransfection into NIH 3T3 cells of constitutively active MEK3, the principal activator in the SAP kinase cascade, together with the mouse NF-H gene, resulted in activation of SAP kinase and phosphorylation of the expressed NF-H (Giasson and Mushynski, 1996). Among the JNK isoforms, JNK3 (SAPK1β) is the major stress activated protein kinase widely expressed in the nervous system (Giasson and Mushynski, 1996), specifically in neuronal cell bodies (Martin et al.,1996). The NF-H side-arms are phosphorylated by JNK3 in vitro and in transfected cells. Glutamate stress induces activation of SAPK1beta in primary cortical neurons and increases the phosphorylation of NF-H in cell bodies (Brownlees et al., 2000).

26.5.6 NF Phosphorylation by GSK3

Another kinase implicated in axonal NF phosphorylation is glycogen synthase kinase (GSK3α and GSK3β) which phosphorylates some of the KSP sites in bovine NF-M. A comparison of NF-H phosphorylation in COS cells cotransfected with NF-H, Cdk5, or GSK3 further illustrated the in vivo phosphorylation by GSK3 and found differences in site specificity between the two kinases (Dosemeci and Pant, 1992). Cdk5 promoted a significant mobility shift after phosphorylation; GSK3 produced only a partial retardation, suggesting that fewer sites were phosphorylated. Phosphorylation of cytoskeletal protein like Tau by GSK3 is potentiated by synergistic prior phosphorylation by non-proline directed kinases such as PKA, CKI, PKC, and by Cdk5 (Sengupta et al., 1997). Prior phosphorylation of Ser/Thr residues adjacent to KSP repeats by one kinase may induce conformational changes that facilitate GSK3 phosphorylation, an example of hierarchical, sequential phosphorylation of multi-acceptor sites in molecules like Tau or NF proteins (Roach, 1991). Figure 26.2 summarizes the signaling pathways associated with NF.

26.6 Proposed Mechanisms of NF Tail Domain KSP Phosphorylation

26.6.1 NF Head Domain Phosphorylation by Second Messenger Kinases, Protein Kinase C/Protein Kinase A (PKC/PKA) Inhibits the Tail Domain Phosphorylation by PDKs

NF proteins are known to be phosphorylated on their head and tail domains and the dynamics of their phosphorylation/dephosphorylation plays a major role in regulating the structural organization and function of NFs. Neurofilaments are very dynamic structures; as discussed above, they contain phosphorylation sites for a large number of protein kinases, including protein kinase A (PKA), protein kinase C (PKC), cyclin-dependent kinase 5 (Cdk5), extra cellular signal regulated kinase (ERK), glycogen synthase kinase-3 (GSK-3), and stress-activated protein kinase gamma (SAPKγ). Most of the neurofilament phosphorylation sites, located in tail regions of NF-M and NF-H, consist of the repeat sequence motif, Lys-Ser-Pro (KSP). We proposed that phosphorylation of some sites on the KSP repeat motifs by proline-directed kinases might be influenced by the presence or absence of phosphate groups on the head domain of the NF subunit. Our group demonstrated that, in transfected HEK293 cells, the head domain phosphorylation of endogenous NF-M by forskolin-activated protein kinase A inhibited NF-M tail domain phosphorylation, suggesting that assembly and KSP phosphorylation of NF-M in axons depend on prior dephosphorylation of head domain sites (Zheng et al., 2003). These findings support the idea that phosphorylation of head domain sites of NFs not only regulates the NF assembly/disassembly, they might also play a role in determining the phosphorylation state of specific carboxyl-terminal tail domain phosphorylation sites.

26.6.2 Involvement of Protein Phosphatases in Topographic Phosphorylation of NF

Protein kinases that phosphorylate NF proteins have been characterized, but much less is known about the protein phosphatases (PPs) in the nervous system (Sim, 1991), especially those acting on NFs. Earlier studies demonstrated the presence of a PP2A activity in high-salt extracts of NF preparations from bovine and rat spinal cord that could dephosphorylate NF proteins (Guru et al., 1991; Shetty and Guru, 1992). The reported turnover of phosphate groups on NFs during axonal transport (Nixon et al., 1987) and the decrease in number of phosphorylated epitopes of NF proteins at the nodes of Ranvier relative to the internodal myelinated regions (de Waegh et al., 1992; Mata et al., 1992) demonstrate the physiological significance of PP(s) in regulating the state of NF phosphorylation.

Phosphatases do play an important role in the labile stages of axonal outgrowth and in the more stable phases of mature axon function during impulse conduction. In mature neurons, NFs are abundantly phosphorylated selectively in the axonal compartment and phosphatases may be regulating the topographic phosphorylation of cytoskeletal proteins, particularly during more labile stages of axon outgrowth.

An indication of phosphatase regulation of NF assembly in vivo is seen in the effect of okadaic acid, a specific phosphatase inhibitor on the cytoskeleton of cultured DRG cells (Sacher et al., 1992). NF proteins isolated from OA treated DRG neurons migrated slower on SDS-PAGE compared to non-treated, suggesting higher phosphorylation of NF subunits upon inhibiting the phosphatase (PP1 and PP2A) (Sacher et al., 1992). The NFs shifted into the Triton X soluble fraction and exhibited a decreased electrophoretic mobility suggesting increased phosphorylation. Inhibition of phosphatase activity enhanced the phosphorylation of NF-L and NF-M head domains and fragmentation of neurofilaments, a condition reversed when okadaic acid was removed. Subsequent studies identified phosphatase 2A as the active phosphatase (Sacher et al., 1994). Mushynski and co-workers have reported an increase in phosphorylation of the amino-terminal domain of NF-L and NF fragmentation in OA treated DRG neurons. The punctuate appearance of axonal NF could be due to the fragmentation of hyperphosphorylation of these molecules as suggested (Sacher et al., 1992, 1994). The phosphatase and kinase are associated with NF-preparation isolated from nervous tissue (de Freitas et al., 1995; Strack et al., 1997). More than 75% of the endogenous phosphatase activity in these fractions was attributed to PP2A. It is speculated that its function is to preserve the filament structure of NFs by its regulation of head domain phosphorylation (Saito et al., 1995).

PP2A was also shown to dephosphorylate the KSP sites in the tail domain of NF-M/H that are phosphorylated by Cdk5 and MAPKs, implicating PP2A in the regulation of sidearm formation and MT interactions (Veeranna et al., 1995). Cdk5 and PP2A may regulate phosphorylation of the KSPXK repeats in NF-M and NF-H, possibly during axonal transport as sidearms extend. At the terminals, turnover of NFs are facilitated by prior dephosphorylation of these sites and degradation by calpains, a calcium-calmodulin dependent phosphatase PP2B before digestion by proteases (Pant, 1988).

The catalytic subunits of PP1 and PP2A as well as regulatory subunits of PP2A have been detected in the NF fraction isolated from bovine spinal cord, whereas PP2B and PP2C are found exclusively in the low molecular weight soluble fraction. PP2A accounts for 60% of NF dephosphorylation and PP1 contributes 10–20% of the same process (Veeranna et al., 1995). In cultured N2a and SHSY5Y cells, hyperphosphorylation and accumulation of NF in cell bodies have been induced by treatment with OA (Shea et al., 1993).

26.7 Function of NFs

26.7.1 NFs Regulate Axonal Caliber

The conduction of a nerve impulse along the axon is directly proportional to the caliber of axon (Hoffman et al., 1987; Sakaguchi et al., 1993; Nixon et al., 1994). One of the main functions of the NFs is to regulate the axonal caliber. NF formation was affected in mutant quiver Quail, *Coturnix japonica,* where NF-L gene was disrupted, suggesting that NFs are involved in caliber determination. Axonal diameter was significantly reduced in mutant quiver quail as was the conduction velocity of nerve impulses.

26.7.2 Axonal Transport of Neurofilaments

NF phosphorylation is known to modulate NF axonal transport (Hoffman et al., 1983; Komiya et al., 1987; Lewis and Nixon, 1988; de Waegh et al., 1992; Nixon,

1993; Shea et al., 1997; Watson et al., 1993; Collard et al., 1995; Tu et al., 1995; Marszalek et al., 1996; Zhang et al., 1997; Zhu et al., 1998; Jung and Shea, 1999; Jung et al., 2000). Axonal transport is divided into two major categories based on the rate of the transport. The fast component which includes membranous components and slow axonal transport includes cytoskeletal components such as NFs, microtubules, and their associated proteins. The developmental appearance of NF-H is accompanied by slowing of neurofilament axonal transport (Hoffman et al., 1983; Jung et al., 2000). Hypophosphorylated neurofilaments are transported more quickly than extensively phosphorylated ones (Lewis and Nixon, 1988; Jung et al., 2000). NF-H overexpression and increased phosphorylation retard transport of neurofilaments, and elimination of NF-H hastens it (Marszalek et al., 1996; Zhu et al., 1998; Jung and Shea, 1999). In situ phosphatase inhibition increases NF-H C-terminal phosphorylation in retina and proximal axons and decreased axonal transport in the proximal axons; however, the overall axonal transport is not decreased (Jung and Shea, 1999).

26.8 Squid as a Model System to Study Topographic Phosphorylation of NFs

Squid neurofilaments in the giant axon, like those in mammalian axons, contribute to the large axon caliber responsible for rapid conduction velocity, a property essential to the jet propulsive function of mantle muscle. Neurofilaments, together with microtubules and associated proteins, are organized into a hexagonal array that makes up a three-dimensional cytoskeletal network in the giant axon (Metuzals and Izzard, 1969; Way et al., 1992). NFs make up about 13% of total axoplasm protein (Brown and Lasek, 1990) and consist of three subunit proteins – a large, highly phosphorylated NF-220 and two smaller subunits, NF60 and NF70. These have been cloned and shown to arise by alternative splicing from a single gene, in contrast to the three independent genes that code for mammalian NF subunits (Way et al., 1992). Biochemically and immunocytochemically, the NF-220 is phosphorylated only in the axon, whereas a non-phosphorylated NF-180, together with NF60/70, is detected in the perikarya of the stellate ganglia (Tytell et al., 1990; Cohen et al., 1987). Phosphorylation occurs predominantly on the multiple K/RSP, SAR/K, and SEK/R repeat sites (>50) on the C-terminal tail of the NF-220 subunit (Jaffee et al., 2001). This topographic pattern of NF phosphorylation resembles that seen in mammalian neurons, with the axon as the principal site of NF phosphorylation. A comparison of kinase assays of axoplasm and GFL lysates illustrates the differences in compartment activities. Except for NF-220, cytoskeletal protein substrates are indeed present in both compartments (Grant et al., 1999). Although the protein expression profiles of axoplasm and GFL are similar, the endogenous kinase activity of axoplasm is significantly higher than in the GFL, with NF220 and other proteins highly phosphorylated. The GFL lysate exhibits fewer phosphorylated proteins, with most phosphorylation seen in low-molecular-weight bands. Although cell body tubulin is moderately phosphorylated, the NF220 is not, in spite of the fact that the overall phosphorylation activity is similar in both compartments. Although this suggests cytoskeletal protein phosphorylation activity is low within cell bodies, equivalent phosphorylation of exogenous substrates such as histone and casein is found. Kinases are indeed present and equally active in the GFL lysate.

Data from our laboratory and others on mammalian neurofilament phosphorylation, provides the model of NF biogenesis and processing in the squid giant fiber system (Fig. 26.3) (Cohen et al., 1987). It is assumed that all "players" in these processes – kinases, phosphatases, regulators, and cytoskeletal substrates – are synthesized in cell bodies of the GFL. The polymerization of NF monomers and oligomers is transiently affected by phosphorylation of NF60 and NF 180 head domain sites by PKA, PKC, and CaMPK. A cell-body-specific multimeric complex of "incomplete" filaments and dense bodies is assembled with little or no evidence of sidearm formation. The complex consists of substrates; kinases, including proline-directed kinases (Erk1/2, cdc2-like kinase, SAP kinase), and phosphatases. Because of the three-dimensional configuration of the complex, the numerous phosphate acceptor sites in the C-terminal tail domain of NF 180 are inaccessible to these kinases. Accordingly, they are not phosphorylated. Furthermore, the higher levels of phosphatase may also contribute to an inactive complex by down-regulating the activity of MAP kinases within the complex. Under these conditions, the non-phosphorylated NFs readily associate with microtubules within the axon initial segment for transport within the axon. In this region, Ser/Thr phosphatases are

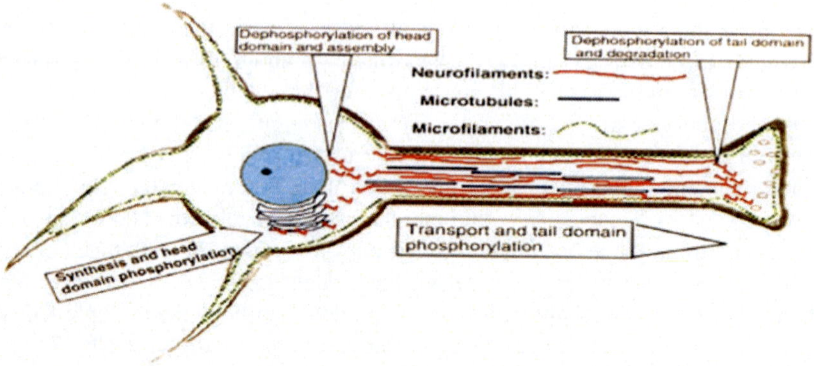

Fig. 26.3 Model of normal topographic regulation of neurofilament (NF) phosphorylation. (1) Synthesis of NF monomers in cell body, coupled to transient head domain phosphorylation by PKA, PKC, and/or CAMKII. (2) Axon hillock region in which Ser/Thr phosphatases dephosphorylate NF head domains and promote NF assembly and association with motors to form an activated axon transport complex. (3) Within the axon, proline-directed kinases (Erk1/2, SAPK, Cdk5) along with CK1, CKII, and GSK3 progressively phosphorylate newly accessible tail domain phosphate acceptor sites, including the numerous KSP repeats, resulting in sidearm extension and formation of a stable cytoskeletal lattice. As the complex is slowly transported, small NF oligomers are progressively added to the growing NF-core complex, where they are phosphorylated. (4) Finally, at the axon tip near the synapse, Ser/Thr phosphatases are activated dephosphorylating tail domain sites, followed by degradation of the complex by calpains

subsequently activated and NF head domain sites are dephosphorylated, thereby promoting more stable polymerization, changes in confirmation of the multimeric complex, and exposure of tail domains to proline-directed kinases. As polymeric NFs are slowly transported within the axon, proline-directed kinases sequentially phosphorylate tail domain KSP repeat sites; this promotes further extension of sidearms and crossbridge formation between NFs and microtubules, which assemble into a stable cytoskeletal core. Inasmuch as the giant axons grow continuously throughout the 1-year life of the squid, axon growth is dynamic, with a cross-bridged cytoskeletal core moving slowly, if at all, surrounded by more soluble tubulins and smaller moving NF polymers that become integrated into the expanding cytoskeleton (Galbraith et al., 1999). Sustaining this dynamic process of assembly and transport is a more robust phosphorylation complex of active kinases and phosphatases to ensure continuous growth in length and diameter of the underlying cytoskeletal network.

26.9 Aberrant Phosphorylation of NF and Neurodegeneration

A number of human neurodegenerative disorders such as Alzheimer's, ALS, Neiman Pick's disease with Lewy bodies, dementia with Lewy bodies, Parkinsons and others are characterized by a common neuronal pathology, an accumulation of relatively insoluble filamentous aggregates within perikarya (Julien and Mushynski, 1983, 1998). The relationship between NFs and neurodegeneration has been extensively studied in transgenic mouse. Transgenic mice that overexpress NFTPs exhibit extensive NF protein accumulations in motor neurons. With overexpression of mutated mouse NF-L or WT NF-M/H, the motor neuron pathologies are accompanied by skeletal muscle atrophy or motor dysfunction, reminiscent of those found in human ALS. It was shown that overexpression of mutated NF-L leads to motor neuron death but not wild-type NF-L. Interestingly, over expression of human WT NF-M/L, but not mouse NF-M/L, causes motor neuron death. Defects in the compartmentalization of cytoskeletal protein phosphorylation in neurons may lead to neuronal cell death and functional pathology.

26.9.1 ALS

Amyotrophic Lateral Sclerosis (ALS), a fatal disorder marked by loss of motor activity, is characterized by NF-containing filamentous aggregates primarily in perikarya and proximal axons of spinal motor neurons (Julien and Mushynski, 1983). Moreover, the perikaryal NFs are extensively phosphorylated, particularly in tail domains of NF-M and NF-H, suggesting that topographic regulation of phosphorylation has been seriously disturbed. It is assumed that neurons exhibiting this pathology are destined to die, which accounts for neuron loss, nerve degeneration, and muscle dystrophy characteristic of the disease. Changing the stoichiometry of NF protein subunits within neurons has produced a number of transgenic mouse models of ALS. Diverse, often contradictory, neuronal and behavioral phenotypes were created. Although they share some common features, the factors responsible for neuronal cell loss, axon degeneration, and dystrophy are still not understood. For example, a two- to fourfold overexpression of NF-H (mouse, human or a β-gal-tagged) usually produces a massive accumulation of neurofilaments within spinal motor neuron perikarya, but only the human NF-H (hNF-H) transgenic produces overt motor dysfunction (Cote et al., 1993; Eyer and Peterson, 1994; Marszalek et al., 1996; Beaulieu et al., 2000). Likewise, some transgenic mice overexpressing NF-L also display a spinal neuron pathology resembling ALS, with hyperphosphorylated NFs in perikarya (Xu et al., 1993; Houseweart and Cleveland, 1999). Surprisingly, neuron loss was seen only in the case of the mNF-L transgenics though both showed signs of motor dysfunction (Houseweart and Cleveland, 1999). In contrast, transgenics expressing low levels of human NF-L (hNF-L) show neither perikaryal accumulations nor motor dysfunction (Julien et al., 1987). In the latter case, the level of endogenous mouse NF-L expression may have compensated for the presence of hNF-L. Overexpression of mutant or normal mouse NF-M, or hNF-M, however, may lead to some perikaryal accumulations in motor neurons (Wong et al., 1995), but in most instances no massive perikaryal accumulations, neuronal loss, nor motor dysfunction were observed (Xu et al., 1996). In some instances only selected neurons in the brain were affected while sensory neurons in CNS and PNS were relatively unaffected. NF-M seems to play a dominant role in regulation of NFL protein stoichiometry and the phosphorylation state of NF-H. This is based on observations of NF-M KO mice in which the NF-L protein decreases while NF-H protein and NF transport velocity increase (Elder et al., 1998; Jacomy et al., 1999). In addition, NF-L increases and NF-H and phospho-NF-H decrease as if to compensate. Presumably as long as the correct ratio of NF-L to NF-H is maintained, normal filaments assemble and are transported. Stoichiometric imbalances among NF-subunits are primarily responsible for the abnormal spinal motor neurons since inclusion of extra NF-L into mice overexpressing NF-H restores a normal phenotype (Meier et al., 1999). Presumably, neurons with high levels of axonal NFs are most sensitive to these imbalances in the ratio of subunit proteins. This was most apparent in mice with overexpressed peripherin; no massive accumulations of filaments in spinal motor neurons were seen but significant proximal axonal accumulations accompanied by motor dysfunction occurred in some animals only after 6–28 months (Beaulieu et al., 2000). Neuronal cell death does not invariably result from aggregates of phosphorylated NFs. Not all aggregates within perikarya or in proximal axons are identical; some are spherical while others are filamentous and the nature of associated proteins is unknown. Nor is the state of NF phosphorylation within perikarya a predictor of neuronal cell death. In fact, in a recent comparison of the sites specific pattern of phosphorylation of NF-H extracted from human ALS and normal spinal cords, the levels of phosphorylation and the sites phosphorylated were similar (Xu et al., 1993). The filamentous accumulations of NF are not hyperphosphorylated as it has been assumed. Rather, the pattern of phosphorylation resembles that seen in axonal NFs. Moreover, in some instances, abnormal axonal accumulations are more closely correlated with motor neuron loss and dysfunction than massive perikaryal accumulations, at least with respect to overexpression of peripherin (Beaulieu et al., 2000). What then is responsible for neuronal cell death, axon degeneration, and muscle dystrophy, which account for the clinical features of ALS? Significant changes in the ratio of NF subunits within the cell body may affect NF assembly, NF transport, and/or NF phosphorylation, or possibly all three. What seems to be common to most of these mouse ALS models is that NF transport is seriously obstructed, perhaps as a result of premature assembly of abnormal polymer within the cell body, which prevents it from associating with the MT transport machinery (Marszalek et al., 1996). Phosphorylation of NF-M and NF-H tail domains in such cell body filaments would further block transport because of sidearm formation, which might also mechanically interfere with the transport machinery. The state of NF assembly within the cell body may control slow axonal transport. Consistent with this observation is the behavior of superoxide dismutase 1 (SOD1) mutations in transgenic mice. SOD1 autosomal dominant mutations are a primary cause of ALS in a small number of human families, inducing a toxic mediated motor neuron death. Motor neurons display the typical perikaryal accumulations of phosphorylated neurofilaments. Mice transgenic with these mutations exhibit an early onset of reduced tubulin transport in large axons months before the appearance of motor neuron pathology and degeneration (Strong et al., 2001).

26.9.2 Alzheimer's Disease (AD)

AD is a CNS neurodegenerative disease. Cytoskeleton disruption is a prominent feature and secondary event followed by oxidative damage in AD (Smith et al., 1995; Weiwad et al., 2004; Nunomura et al., 2001). Neufibrillary tangles (NFT) are the hallmark of AD. The abnormally modified Tau, NFPs is the main pathological hallmark of AD. In AD, aberrant hyperphosphorylation of Tau and NF is likely due to perturbation of imbalance of kinase and phosphatase activities like MAP kinase (Trojanowski et al., 1993), GSK-3 (Mandelkow et al., 1992), and Cdk5 (Lew et al., 1994). In AD hippocampus, PP2A mRNA levels are shown to be downregulated (Vogelsberg-Ragaglia et al., 2001).

26.9.3 Parkinson's Disease (PD)

PD is a progressive disorder of CNS. PD is caused by degeneration of the neurons in the substantia niagra of the brain, with reduced dopamine availability. The major pathological alteration in PD is the accumulation of protein inclusions called Lewy bodies composed of alpha-synuclein, three NF subunit proteins (Galloway et al., 1992), ubiquitin, and proteosome subunits (Trimmer et al., 2004). Electron microscopy and biochemical evidence indicate that the abnormally phosphorylated NFs form a non-membrane bound compact skein in the neuronal cell bodies in the affected neurons. A point mutation has been reported in the NF-M gene encoding the rod domain 2B of NF-M in an individual with PD (Lavedan et al., 2002). The base pair change 336 Ser for Gly, possible disrupts assembly. The patient with this mutation developed PD at the very young age of 16. It is possible that this mutation is responsible for the young onset of PD. NF-L mRNA is decreased in PD, correlating with severity of the disease (Hill et al., 1993).

26.9.4 Diabetic Neuropathy

Diabetes is a disease wherein the body doesn't produce insulin. Diabetes is associated with a symmetrical distal axonal neuropathy. The neuropathy is mainly in dorsal root ganglion (DRG) neurons. Rat DRG neurons treated with streptozocin-induced diabetes have a hyperphosphorylation of NFs in lumbar DRG (Fernyhough et al., 1999). This is probably activation of c-JNK which is an NF kinase. Pathologically, neurites from sympathetic ganglia are swollen with disorganized aggregates of NFs and peripherin (Schmidt et al., 1997). The diabetic neuropathy is associated with impairment of axonal transport, a reduction in axonal caliber, and a reduced capacity for nerve regeneration.

26.10 Mutations in Neuronal IFs and Neurodegeneration

Evidence of sequence anomalies in the NF genes were promising, with the identification of deletion/insertion mutations within the multiphosphorylation domain of NF-H occurring in 1% of ALS cases. This corresponds with the finding of abnormal topographic localization of hyperphosphorylated NFs to perikarya of diseased motor neurons. Mutations in NF-L gene are causative of Charcot-Marie-Tooth (CMT) disease, an inherited peripheral neuropathy, and are related to defects in NF assembly and axonal properties (Jordanova et al., 2003). The listing of NF-L mutants are found at http://www.interfil.org. Although there are few mutations within the C-terminal domain, most are within N-terminal head and alpha-helical rod domain, consistent with defects in NF assembly properties. Transgenic mice overexpressing an NF-L mutant harboring a synthesized leucine to Proline mutant (L394P) within the helix termination sequence of the rod domain causes a selective degeneration of motor neurons characterized by abnormal accumulations of NFs. These findings demonstrate that a primary defect in NFs could cause motor neuron disease.

Disturbances in the assembly of the NF triplet impacts on neural function (Gill et al., 1990). N-terminal deletions of NF-L gene yields assembly-incompetent NF-L polypeptides (Gill et al., 1990). Cleavage of NF-L N-terminus inhibits assembly of NF-L into 10-nm filaments (Schmalbruch et al., 1991). A single missense mutation in the rod domain (Leu394Pro) of NF-L (Lee et al., 1994) leads to profound accumulation of NF within motor neurons, resulting in motor neuron death and atrophy of muscle fibers. In the latter transgenic mice, mutation disrupts the binding of a ribonucleoprotein complex to the 3′ UTR of the NF-L mRNA and thereby alters mRNA stability (Cañete-Soler et al., 1999). Mutant but not WT SOD1 is a transacting-binding protein capable of destabilizing NF-L mRNA (Ge et al., 2005), thereby providing a potential linkage between the expression of a familial ALS-associated mutation and the genesis of alteration in NF stoichiometry. Deletions of the NF-H tail were also found in amyotrophic lateral sclerosis (Al-Chalabi et al., 1999). Peripherin is an NIF protein found associated with pathological aggregates in motor neurons of patients with ALS and of transgenic mice overexpressing mutant SOD-1, and induces a selective degeneration of motor neurons when overexpressed in transgenic mice. Three alternate splice variants of peripherin exist, Per 58, Per 56, and Per 61 (Robertson et al., 2001). Of the three isoforms, Per 61 proved to be distinctly neurotoxic, being assembly incompetent and inducing degeneration of motor neurons in culture. Accumulating evidence suggests that expression of neurotoxic splice variants of peripherin may contribute to the neurodegenerative mechanism of ALS. The Per 61 is specifically detected in motor neurons and axonal spheroids in spinal cord of ALS.

26.10.1 Charcot-Marie-Tooth Disease (CMT)

CMT is the most common inherited neurological disorder, affecting both sensory and motor neurons. There are several types of CMT. Types 1 and 3 are due to demyelination, and Type 2 is an axonal disease. Several families are identified in which heterozygosity for mutations of NF-L gene on chromosome 8 are associated with CMT2 (Mersiyanova et al., 2000; De Jonghe et al., 2001; Georgiou et al., 2002; Jordanova et al., 2003). Leu333Pro in conserved rod domain 2B is found in CMT. The NF-L CMT2 mutations disrupt NF assembly and axonal transport. A mouse model Leu394Pro resulted in selective motor neuron death due to disruption of NF assembly (Lee et al., 1994). NF-L missence mutation (Cys64Thr) caused this disease in Slovenian CMT2 family. A similar Pro22 Thr was detected in Japanese patients with CMT disease (Jordanova et al., 2003; Yoshihara et al., 2002). Some of the CMT types are caused by defects in connexin 32 genes that result in aberrant myelination and altered NF phosphorylation. In such cases, aberrant NF phosphorylation occurs downstream of primary pathological process but is still upstream of the end point.

26.11 Stabilization of NF Phosphorylation by Pin1

Within the last decade, a novel level of modulation of protein phosphorylation has emerged, namely the factors that regulate the structural conformation and stability of proteins, particularly those phosphorylated at Ser/Thr-Pro (S/T-P) sites by proline-directed kinases. Most proline-directed kinases and phosphatases are highly selective for *trans* S/T-P bonds (Wulf et al., 2005). Peptidyl prolyl isomerase 1 (Pin1) is a *cis trans* isomerase that converts the *cis* peptidyl prolyl bonds to the *trans* bonds. Pin1 is a small protein of 18 kDa, comprised of two domains, N-terminal

WW domain and the C-terminal isomerase domain. Pin1 specifically targets phosphorylated S/T-P sites and, by virtue of the proline residue, can "toggle" an inactive *cis* isomer to the more stable *trans* form, often with altered function (Lu et al., 1996). Pin1 is known to regulate important targets with phospho-S/T-P residues such as Tau, RNA polymerase, and transcription factors such as c-Jun and β-catenin. The multiple S/T-P repeats in the NF-M/H have prompted us to understand the role of Pin1 in NF structure. In neurons, however, Pin1 is distributed in both nucleus and cytoplasm and increases during neuronal differentiation. Neurofilaments contain many S/T-P phosphate acceptor sites that are targeted by several different kinases and phosphatases. In contrast to Tau, however, NFs, particularly NF-H, are enriched with numerous KSP repeat motifs in the tail domain (43–100 depending on species), sites for proline-directed kinase phosphorylation. We showed that Pin1, distributed throughout the neuron, is directly involved in modulating NF phosphorylation to affect its distribution in the cell body and axon. Pin1 increased the phosphorylation of NF by PDKs such as MAPKs, Cdk5, and JNK3 in a dose dependent manner in vitro. Pin1 is associated with phospho-NF in brain. By deregulating the neuronal system via induction of neurodegeneration by glutamate excitotoxicity or oxidative stress such as hydrogen peroxide and heat stresses that evoke NF pathologies resembling those seen in AD and ALS, we have been able to identify a role for Pin1 in the regulation of NF phosphorylation. Both excitotoxic stress and oxidative stress increase the perikaryal phosphorylation of NF, and Pin1 co-localizes with phosphorylated NF in cell bodies and leads to cell death. Inhibition of Pin1 inhibits the perikaryal phosphorylation of NF neuronal apoptosis (Kesavapany et al., 2007; Rudrabhatla et al., 2008). Based on our findings, we proposed a model for topographic phosphorylation of NF by Pin1. A model for the NF hyperphosphorylation is proposed (Fig. 26.4). The phosphorylation "unwinds" the outer repeat units and

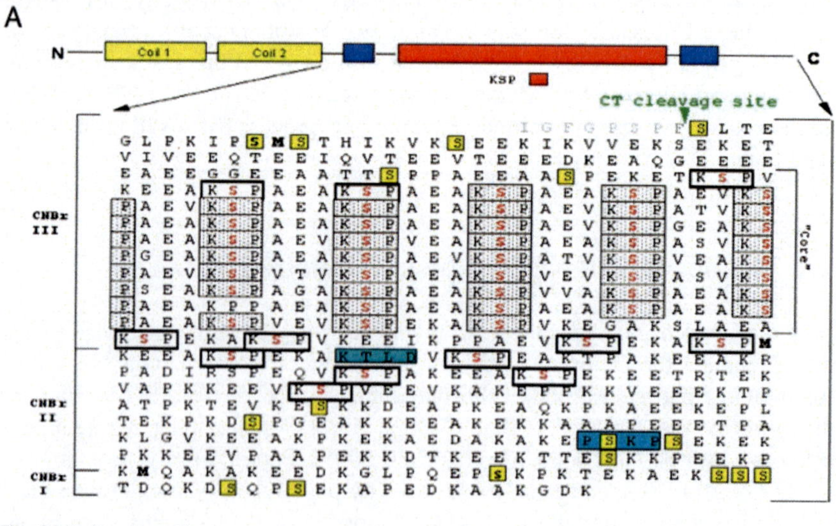

Fig. 26.4 (continued)

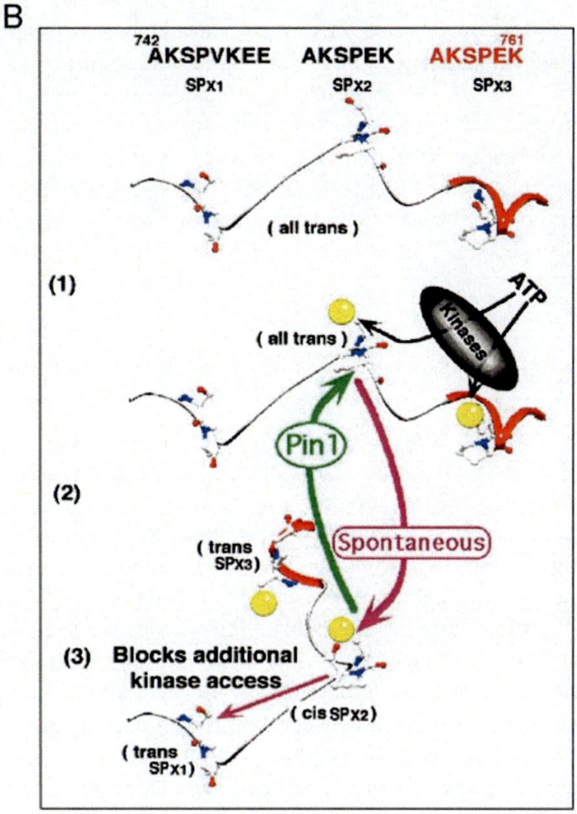

Fig. 26.4 (continued) Hypothetical mechanism for Pin1 regulation of NF-H stable phosphorylation. **a** The structure of Rat NF-H showing the KSP repeats. Rat NF-H has 52 SP repeats and almost all of them are phosphorylated in vivo. Some of the phosphorylated SP motifs identified by mass spectrometry are *boxed*. **b** Model for the stabilization of NF-H phosphorylation by *Pin1*. The three adjacent KSP repeat units used to illustrate this model are the human NF-H sequence 742–761 in **b**. We arbitrarily diagram the kinase phosphorylation of NF tail domain repeats as normally starting at the most C-terminal repeat unit and proceeding toward their N-terminus. The more N-terminal repeat units are assumed to be sterically shielded from kinase access by burial within the tail domain until the adjacent C-terminal units are phosphorylated. (1) Tail domain phosphorylation occurs at kinase-accessible C-terminal repeats (*yellow spheres*) in the *trans* form; phosphorylation induces spontaneous transition to *cis* form the outer repeat units and permits kinase access to additional repeat units. (2) Phosphorylation of initial KSP motif will initiate a *trans* to *cis* isomerization of the last phosphorylated S-P bond. This causes a local conformation that does not expose further tail domain phosphorylation sites. However, if active *Pin1* is available it will rapidly return the *cis* p-S/T-P, to *trans* and allow additional phosphorylation to proceed normally in axons. (3) If this occurs in perikarya, premature extension of sidearms may prevent neurofilament subunit transport out of perikarya and cause aggregation of p-NF-H subunits

permits kinase access to additional repeat units. In some cases, two or more phosphorylations will initiate a *trans* to *cis* isomerization of the last phosphorylated S-P bond. This causes a local conformation that does not expose further tail domain phosphorylation sites. However, if active Pin1 is available it will rapidly return the *cis* p-S/T-P, to *trans* and allow additional phosphorylation to proceed normally in axons. However, if this occurs in perikarya, premature extension of sidearms may prevent neurofilament subunit transport out of perikarya and cause aggregation of p-NF-H subunits. Thus, isomerization of Lys-Ser-Pro repeat residues that are abundant in NF-H tail domains by Pin1 can regulate NF-H phosphorylation, which suggests that Pin1 inhibition may be an attractive therapeutic target to reduce pathological accumulations of p-NF-H.

26.12 Conclusion

Neurofilaments, although similar to genes of IFs, are, however, unique in many respects, specific to nervous system. Developmental regulation of NFs is necessary for normal development of the nervous system and survival. They are not essential for early development of the organism, but their expression is topographically and developmentally regulated. Defects in the NF developmental regulation is lethal for the nerve cell. NF phosphorylation is topographically regulated. In normal neurons NFs are phosphorylated in the axons. However, in degenerative neurons in diseases such as AD, PD, and ALS, NF proteins are phosphorylated in the cell bodies. The stable and dynamic properties of NF are regulated by phosphorylation, dephosphorylation equilibria between kinase and phosphatase activities. The topographic phosphorylation of NF regulates axonal caliber and axonal transport. Increase in NF phosphorylation in the cell body results in the blockage of axonal transport. Multiple signal transduction pathways are involved in topographic phosphorylation of NFs. The multiple KSP repeats of the tail domain of the NF are phosphorylated by signaling cascades involving MAP kinases, Cdk5 and GSK3. The head domain phosphorylation by second messenger kinases like PKA inhibits the tail domain phosphorylation. The glial axonal interactions activate ERK1/2 and Cdk5, thus increasing NF phosphorylation. The mutations in NF genes are associated with AD and CMT diseases. Pin1 modulates the topographic phosphorylation in normal and stressed neurons. Inhibition of Pin1 inhibits the glutamate and oxidative stress induced perikaryal phosphorylation of NF proteins, suggesting that Pin1 could be an attractive therapeutic target to reduce NF protein inclusions in AD, PD, and ALS. Therefore, the factors regulating these processes are important to understand neurophysiology and pathology. A large number of review articles have been published on different aspects of neuronal IFs. In this review, we have included most of the work carried out by author's laboratory. This review in no way claims to be complete and comprehensive on the subject. However, we have tried to provide a general view of several aspects of the neuronal IFs. We may have omitted to refer to some of the investigators' work; this was in no way intentional but was due only to space limitations.

Acknowledgment This work is supported by NINDS, NIH intramural funding.

References

Al-Chalabi A, Andersen PM, Nilsson P, Chioza B, Andersson JL, Russ C, Shaw CE, Powell JF, Leigh PN (1999) Deletions of the heavy neurofilament subunit tail in amyotrophic lateral sclerosis. Hum Mol Genet 8:157–164

Beaulieu JM, Jacomy H, Julien JP (2000) Formation of intermediate filament protein aggregates with disparate effects in two transgenic mouse models lacking the neurofilament light subunit. J Neurosci 20:5321–5328

Blumenberg M (1989) Evolution of homologous domains of cytoplasmic intermediate filament proteins and lamins. Mol Biol Evol 6:53–65

Brown A, Lasek RJ (1990) The cytoskeleton of the squid giant axon. In: Gilbert DS Adelman WJJ Arnold JM (ed) Squid as experimental animals. Plenum Press, New York, NY, pp 235–302

Brownlees J, Yates A, Bajaj NP, Davis D, Anderton BH, Leigh PN, Shaw CE, Miller CC (2000) Phosphorylation of neurofilament heavy chain side-arms by stress activated protein kinase-1b/Jun N-terminal kinase-3. J Cell Sci 113:401–407

Cañete-Soler R, Silberg DG, Gershon MD, Schlaepfer WW (1999) Mutation in neurofilament transgene implicates RNA processing in the pathogenesis of neurodegenerative disease. J Neurosci 19:1272–1283

Carden MJ, Trojanowski JQ, Schlaepfer WW, Lee VM (1987) Two-stage expression of neurofilament polypeptides during rat neurogenesis with early establishment of adult phosphorylation patterns. J Neurosci 7:3489–3504

Carter J, Gragerov A, Konvicka K et al. (1998) Neurofilament (NF) assembly; divergent characteristics of human and rodent NF-L subunits. J Biol Chem 273:5101–5108

Ching GY, Liem RK (1993) Assembly of type IV neuronal intermediate filaments in nonneuronal cells in the absence of preexisting cytoplasmic intermediate filaments. J Cell Biol 122: 1323–1335

Ching GY, Liem RK (1999) Analysis of the roles of the head domains of type IV rat neuronal intermediate filament proteins in filament assembly using domain-swapped chimeric proteins. J Cell Sci 112:2233–2240

Cohen RS, Pant HC, House S, Gainer H (1987) Biochemical and immunocytochemical characterization and distribution of phosphorylated and nonphosphorylated subunits of neurofilaments in squid giant axon and stellate ganglion. J Neurosci 7:2056–2074

Collard JF, Côté F, Julien JP (1995) Defective axonal transport in a transgenic mouse model of amyotrophic lateral sclerosis. Nature 375:61–64

Cote F, Collard JF, Julien JP (1993) Progressive neuronopathy in transgenic mice expressing the human neurofilament heavy gene: a mouse model of amyotrophic lateral sclerosis. Cell 73:35–46

Coulombe PA, Ma L, Yamada S, Wawersik M (2001) Intermediate filaments at a glance. J Cell Sci 114:4345–4347

Crute BE, Van Buskirk RG (1992) A casein kinase-like kinase phosphorylates beta-tubulin and maybe a microtubule-associated protein. J Neurochem 59:2017–2023

Dashiell SM, Tanner SL, Pant HC, Quarles RH (2002) Myelin-associated glycoprotein modulates expression and phosphorylation of neuronal cytoskeletal elements and their associated kinases. J Neurochem 81:1263–1272

de Freitas MS, de Mattos AG, Camargo MM, Wannmacher C, Pessoa-Pureur R (1995) Cytoskeletal-associated protein kinase and phosphatase activities from cerebral cortex of young rats. Neurochem Res 20:951–956

De Jonghe P, Mersivanova I, Nelis E, Del Favero J, Martin JJ, Van Broeckhoven C et al. (2001) Further evidence that neurofilament light chain gene mutations can cause Charcot-Marie-Tooth disease type 2E. Ann Neurol 49:245–249

de Waegh SM, Lee VM, Brady ST (1992) Local modulation of neurofilament phosphorylation, axonal caliber, and slow axonal transport by myelinating Schwann cells. Cell 68: 451–463

Dosemeci A, Floyd CC, Pant HC (1990) Characterization of neurofilament-associated protein kinase activities from bovine spinal cord. Cell Mol Neurobiol 10:369–382

Dosemeci A, Pant HC (1992) Association of cyclic-AMP-dependent protein kinase with neurofilaments. Biochem J 282:477–481

Elder GA, Friedrich VLJR, Bosco P, Kang C, Gourov A, Tu PH, Lee VM, Lazzarini RA (1998) Absence of the mid-sized neurofilament subunit decreases axonal calibers, levels of light neurofilament (NF-L), and neurofilament content. J Cell Biol 141:727–739

Eyer J, Peterson A (1994) Neurofilament-deficient axons and perikaryal aggregates in viable transgenic mice expressing a neurofilament-beta-galactosidase fusion protein. Neuron 12:389–405

Fernyhough P, Gallagher A, Averill SA, Priestley JV, Hounsom L, Patel J et al. (1999) Aberrant neurofilamentphosphorylation in sensory neurons of rats with diabetic neuropathy. Diabetes 48:881–889

Fliegner KH, Liem RK (1991) Cellular and molecular biology of neuronal intermediate filaments. Int Rev Cytol 131:109–167

Floyd CC, Grant P, Gallant PE, Pant HC (1991) Principal neurofilament-associated protein kinase in squid axoplasm is related to casein kinase. J Biol Chem 266:4987–4994

Fuchs E, Weber K (1994) Intermediate filaments: structure, dynamics, function, and disease. Annu Rev Biochem 63:345–382

Galbraith JA, Reese TS, Schlief ML, Gallant PE (1999) Slow transport of unpolymerized tubulin and polymerized neurofilament in the squid giant axon. Proc Natl Acad Sci USA 96: 11589–11594

Galloway PG, Mulvihill P, Perry G (1992) Filaments of Lewy bodies contain insoluble cytoskeletal elements. Am J Pathol 140:809–822

Ge WW, Wen W, Strong WL, Leystra-Lantz C, Strong MJ (2005) Mutant copper/zinc superoxide dismutase binds to and destabilizes human low molecular weight neurofilament mRNA. J Biol Chem 280:118–124

Georgiou DM, Zidar J, Korosec M, Middleton LT, Kyriakides T, Christodoulou K (2002) A novel NF-L mutation Pro22Ser is associated with CMT2 in a large Slovenian family. Neurogenetics 4:93–96

Giasson BI, Mushynski WE (1996) Aberrant stress-induced phosphorylation of perikaryal neurofilaments. J Biol Chem 271:30404–30409

Gill SR, Wong PC, Monteiro MJ, Cleveland DW (1990) Assembly properties of dominant and recessive mutations in the small mouse neurofilament (NF-L) subunit. J Cell Biol 111: 2005–2019

Goldenring JR, Casanova JE, Delorenzo RJ (1984) Tubulin-associated calmodulin-dependent kinase: evidence for an endogenous complex of tubulin with a calcium-calmodulin-dependent kinase. J Neurochem 43:1669–1679

Gonda Y, Nishizawa K, Ando S, Kitamura S, Minoura Y, Nishi Y, Inagaki M (1990) Involvement of protein kinase C in the regulation of assembly-disassembly of neurofilaments in vitro. Biochem Biophys Res Commun 167:1316–1325

Grant P, Diggins M, Pant HC (1999) Topographic regulation of cytoskeletal protein phosphorylation by multimeric complexes in the squid giant fiber system. J Neurobiol 40: 89–102

Grant P, Pant HC (2000) Neurofilament protein synthesis and phosphorylation. J Neurocytol 29:843–872

Guidato S, Bajaj NP, Miller CC (1996) Cellular phosphorylation of neurofilament heavy-chain by cyclin-dependent kinase-5 masks the epitope for monoclonal antibody. Neurosci Lett 217: 157–160

Guru SC, Shetty KT, Shankar KS (1991) Effect of chronic ethanol ingestion on phosphate content of NF proteins and NF associated protein phosphatase in rat spinal cord. Neurochem Res 16:1193–1197

Heins S, Aebi U (1994) Making heads and tails of intermediate filaments assembly, dynamics and networks. Curr Opin Cell Biol 6:25–33

Herrmann H, Aebi U (2000) Intermediate filaments and their associates: multi-talented structural elements specifying cytoarchitecture and cytodynamics. Curr Opin Cell Biol 12:79–90

Herrmann H, Aebi U (2004) Intermediate filaments: molecular structure, assembly mechanism, and integration into functionally distinct intracellular scaffolds. Annu Rev Biochem 73: 749–789

Herrmann H, Hesse M, Reichenzeller M, Aebi U (2003) Functional complexity of intermediate filament cytoskeletons: from structure to assembly to gene ablation. Int Rev Cytol 223:83–175

Hill WD, Arai M, Cohen JA, Trojanowski JQ (1993) Neurofilament mRNA is reduced in Parkinson's disease substantia nigra pars compacta neurons. J Comp Neurol 329:328–336

Hisanaga S, Matsuoka Y, Nishizawa K, Saito T, Inagaki M, Hirokawa N (1994) Phosphorylation of native and reassembled neurofilaments composed of NF-L, NF-M, and NF-H by the catalytic subunit of cAMP-dependent protein kinase. Mol Biol Cell 5:705–709

Hoffman DW, Cleveland JW, Griffin PW, Landes NJ, Cowan PDL (1987) Neurofilament gene expression: a major determinant of axonal caliber. Proc Natl Acad Sci USA 84: 3472–3476

Hoffman PN, Lasek RJ, Griffin JW, Price DL (1983) Slowing of the axonal transport of neurofilament protein during development. J Neurosci 3:1694–1700

Hollander BA, Bennett GS, Shaw G (1996) Localization of sites in the tail domain of the middle molecular mass neurofilament subunit phosphorylated by a neurofilament-associated kinase and by casein kinase. J Neurochem 66:412–420

Houseweart MK, Cleveland DW (1999) Cytoskeletal linkers: new MAPs for old destinations. Curr Biol 9:R864–R866

Inagaki M, Gonda Y, Nishizawa K, Kitamura S, Sato C, Ando S, Tanabe K, Kikuchi K, Tsuiki S, Nishi Y (1990) Phosphorylation sites linked to glial filament disassembly in vitro locate in a non-alpha-helical head domain. J Biol Chem 265:4722–4729

Jacomy H, Zhu Q, Couillard-Despres S, Beaulieu JM, Julien JP (1999) Disruption of type IV intermediate filament network in mice lacking the neurofilament medium and heavy subunits. J Neurochem 73:972–984

Jaffe H, Sharma P, Grant P, Pant HC (2001) Characterization of the phosphorylation sites of the squid (*holigs pealei*) high-molecular-weight neurofilament protein from giant axon axoplasm. J Neurochem 76:1022–1031

Jordanova A, De Jonghe P, Boerkoel CF, Takashima H, De Vriendt E, Ceuterick C et al. (2003) Mutations in the neurofilament light chain gene (NEFL) cause early onset severe Charcot-Marie-tooth disease. Brain 126:590–597

Julien JP, Mushynski WE (1983) The distribution of phosphorylation sites among identified proteolytic fragments of mammalian neurofilaments. J Biol Chem 258:4019–4025

Julien JP, Mushynski WE (1998) Neurofilaments in health and disease. Prog Nucleic Acid Res Mol Biol 61:1–23

Julien JP, Tretjakoff I, Beaudet L, Peterson A (1987) Expression and assembly of a human neurofilament protein in transgenic mice provide a novel neuronal marking system. Genes Dev 1:1085–1095

Jung C, Shea TB (1999) Regulation of neurofilament axonal transport by phosphorylation in optic axons in situ. Cell Motil Cytoskeleton 43:230–240

Jung C, Yabe JT, Shea TB (2000) Hypophosphorylated neurofilament subunits undergo axonal transport more rapidly than more extensively phosphorylated subunits in situ. Cell Motil Cytoskeleton 47:120–129

Kennelly PJ, Krebs EG (1991) Consensus sequences as substrate specificity determinants for protein kinases and protein phosphatases. J Biol Chem 266:15555–15558

Kesavapany S, Patel V, Zheng YL, Pareek TK, Bjelogrlic M, Albers W, Amin N, Jaffe H, Gutkind JS, Strong MJ, Grant P, Pant HC (2007) Inhibition of pin1 reduces glutamate-induced perikaryal accumulation of phosphorylated neurofilament-H in neurons. Mol Biol Cell 18:3645–3655

Kitamura S, Ando S, Shibata M, Tanabe K, Sato C, Inagaki M (1989) Protein kinase C phosphorylation of desmin at four serine residues within the non-alpha-helical head domain. J Biol Chem 264:5674–5678

Komiya Y, Cooper NA, Kidman AD (1987) The recovery of slow axonal transport after a single intraperitoneal injection of beta, beta-iminodipropionitrile in the rat. J Biochem (Tokyo) 102:869–873

Lavedan C, Buchholtz S, Nussbaum RL, Albin RL, Polymeropoulos MH (2002) A mutation in the human neurofilament M gene in Parkinson's disease that suggests a role for the cytoskeleton in neuronal degeneration. Neurosci Lett 322:57–61

Lee MK, Cleveland DW (1996) Neuronal intermediate filaments. Ann Rev Neuroscience 19: 187–217

Lee MK, Marszalek JR, Cleveland DW (1994) A mutant neurofilament subunit causes massive, selective motor neuron death: implications for the pathogenesis of human motor neuron disease. Neuron 13:975–988

Lee MK, Xu Z, Wong PC et al. (1993) Neurofilaments are obligate heteropolymers in vivo. J Cell Biol 122:1337–1350

Lee VM, Otvos L Jr, Carden MJ, Hollosi M, Dietzschold B, Lazzarini RA (1988) Identification of the major multiphosphorylation site in mammalian neurofilaments. Proc Natl Acad Sci USA 85:1998–2002

Lew J, Huang QQ, Qi Z, Winkfein RJ, Aebersold R, Hunt T et al. (1994) A brain-specific activator of cyclindependent kinase 5. Nature 371:423–426

Lewis SE, Nixon RA (1988) Multiple phosphorylated variants of the high molecular mass subunit of neurofilaments in axons of retinal cell neurons: characterization and evidence for their differential association with stationary and moving neurofilaments. J Cell Biol 107:2689–2701

Li BS, Veeranna, Grant P, Pant HC (1999a) Calcium influx and membrane depolarization induce phosphorylation of neurofilament (NF-M) KSP repeats in PC12 cells. Brain Res Mol Brain Res 70:84–91

Li BS, Veeranna, Gu J, Grant P, Pant HC (1999b) Activation of mitogen-activated protein kinases (Erk1 and Erk2) cascade results in phosphorylation of NF-M tail domains in transfected NIH 3T3 cells. Eur J Biochem 262:211–217

Li BS, Zhang L, Gu J, Amin ND, Pant HC (2000) Integrin alpha(1) beta(1)-mediated activation of cyclin-dependent kinase 5 activity is involved in neurite outgrowth and human neurofilament protein H Lys-Ser-Pro tail domain phosphorylation. J Neurosci 20:6055–6062

Link WT, Grant P, Hidaka H, Pant HC (1992) Casein kinase I and II from squid brain exhibit selective neurofilament phosphorylation. Mol Cell Neurosci 3:548–558

Lu KP, Hanes SD, Hunter T (1996) A human peptidyl-prolyl isomerase essential for regulation of mitosis. Nature 380:544–547

Mandelkow EM, Drewes G, Biernat J, Gustke N, Van Lint J, Vandenheede JR et al. (1992) Glycogen synthase kinase-3 and the Alzheimer-like state of microtubule-associated protein tau. FEBS Lett 314:315–321

Marszalek JR, Williamson TL, Lee MK, Xu Z, Hoffman PN, Becher MW, Crawford TO, Cleveland DW (1996) Neurofilament subunit NF-H modulates axonal diameter by selectively slowing neurofilament transport. J Cell Biol 135:711–724

Martin JH, Mohit AA, Miller CA (1996) Developmental expression in the mouse nervous system of the p493F12 SAP kinase. Brain Res Mol Brain Res 35:47–57

Mata M, Kupina N, Fink DJ (1992) Phosphorylation dependent neurofilament epitopes are reduced at the node of Ranvier. J Neurocytol 21:199–210

Meier J, Couillard-Despres S, Jacomy H, Gravel C, Julien JP (1999) Extra neurofilament NF-L subunits rescue motor neuron disease caused by overexpression of the human NF-H gene in mice. J Neuropathol Exp Neurol 58:1099–1110

Mersiyanova IV, Perepelov AV, Polyakov AV, Sitnikov VF, Dadali EL, Oparin RB et al. (2000) A new variant of Charcot-Marie-Tooth disease type 2 is probably the result of a mutation in the neurofilament-light gene. Am J Hum Genet 67:37–46

Metuzals J, Izzard CS (1969) Spatial patterns of threadlike elements in the axoplasm of the giant nerve fiber of the squid (*Loligo pealii* L.) as disclosed by differential interference microscopy and by electron microscopy. J Cell Biol 43:456–479

Nakagawa T, Chen J, Zhang Z, Kanai Y, Hirokawa N (1995) Two distinct functions of the carboxyl-terminal tail domain of NF-M upon neurofilament assembly: cross-bridge formation and longitudinal elongation of filaments. J Cell Biol 129:411–429

Nakamura Y, Hashimoto R, Kashiwagi Y, Aimoto S, Fukusho E, Matsumoto N, Kudo T, Takeda M (2000) Major phosphorylation site (Ser^{55}) of neurofilament L by cyclic AMP-dependent protein kinase in rat primary neuronal culture. J Neurochem 74:949–959

Nakamura Y, Hashimoto R, Kashiwagi Y, Wada Y, Sakoda S, Miyamae Y, Kudo T, Takeda M (1999) Casein kinase II is responsible for phosphorylation of NF-L at Ser-473. FEBS Lett 455:83–86

Nixon RA (1993) The regulation of neurofilament protein dynamics by phosphorylation: clues to neurofibrillary pathology. Brain Pathol 3:29–38

Nixon RA, Lewis SE, Marotta CA (1987) Posttranslational modification of neurofilament proteins by phosphate during axoplasmic transport in retinal ganglion cell neurons. J Neurosci 7: 1145–1158

Nixon RA, Paskevich PA, Sihag RK, Thayer CY (1994) Phosphorylation on carboxyl terminus domains of neurofilament proteins in retinal ganglion cell neurons in vivo: influences on regional neurofilament accumulation, interneurofilament spacing, and axon caliber. J Cell Biol 126:1031–1046

Nixon RA, Shea TB (1992) Respective roles of neurofilaments, microtubules, MAP1B, and tau in neurite outgrowth and stabilization. Cell Motil Cytoskeleton 22:81–91

Nixon RA, Sihag RK (1991) Neurofilament phosphorylation: a new look at regulation and function. Trends Neurosci 14:501–506

Nunomura A, Perry G, Aliev G, Hirai K, Takeda A, Balraj EK et al. (2001) Oxidative damage is the earliest eventin Alzheimer disease. J Neuropathol Exp Neurol 60:759–767

Omary MB, Ku NO, Tao GZ, Toivola DM, Liao J (2006) Heads and tails of intermediate filament phosphorylation: multiple sites and functional insights. Trends Biochem Sci 31: 383–394

Pant HC (1988) Dephosphorylation of neurofilament proteins enhances their susceptibility to degradation by calpain. Biochem J 256:665–668

Parry DA, Steinert PM (1992) Intermediate filament structure. Curr Opin Cell Biol 4:94–98

Perez-olle R, Leung CL, Liem RK (2002) Effect of Charcot-Marie-Tooth-linked mutations of the neurofilament light subunit on intermediate filament formation. J Cell Sci 115:4937–4946

Roach PJ (1991) Multisite and hierarchal protein phosphorylation. J Biol Chem 266: 14139–14142

Robertson J, Beaulieu JM, Doroudchi MM, Durham HD, Julien J-P, Mushynski WE (2001) Apoptotic death of neurons exhibiting peripherin aggregates is mediated by the proinflammatory cytokine tumor necrosis factor-a. J Cell Biol 55:217–226

Rudrabhatla P, Zheng YL, Amin ND, Kesavapany S, Albers W, Pant HC (2008) Pin1-dependent prolyl isomerization modulates the stress-induced phosphorylation of high molecular weight neurofilament protein. J Biol Chem 283:26737–26747

Sacher MG, Athlan ES, Mushynski WE (1992) Okadaic acid induces the rapid and reversible disruption of the neurofilament network in rat dorsal root ganglion neurons. Biochem Biophys Res Commun 186:524–530

Sacher MG, Athlan ES, Mushynski WE (1994) Increased phosphorylation of the amino-terminal domain of the low molecular weight neurofilament subunit in okadaic acid-treated neurons. J Biol Chem 269:18480–18484

Saito T, Shima H, Osawa Y, Nagao M, Hemmings BA, Kishimoto T, Hisanaga S (1995) Neurofilament-associated protein phosphatase 2A: its possible role in preserving neurofilaments in filamentous states. Biochemistry 34:7376–7384

Sakaguchi T, Okada M, Kitamura T (1993) Reduced diameter and conduction velocity of myelinated fibers in the sciatic nerve of a neurofilament-deficient mutant quail. Neurosci Lett 153:65–68

Schmalbruch H, Jensen H-J S, Bjaerg M, Kamieniecka Z, Kurland L (1991) A new mouse mutant with progressive motor neuronopathy. J Neuropathol Exp Neurol 50:192–204

Schmidt RE, Beaudet LN, Plurad SB, Dorsey DA (1997) Axonal cytoskeletal pathology in aged and diabetic human sympathetic autonomic ganglia. Brain Res 769:375–383

Sengupta A, Wu Q, Grundke-iqbal I, Iqbal K, Singh TJ (1997) Potentiation of GSK-3-catalyzed Alzheimer-like phosphorylation of human tau by cdk5. Mol Cell Biochem 167:99–105

Serrano L, Diaz-nido J, Wandosell F, Avila J (1987) Tubulin phosphorylation by casein kinaseII is similar to that found in vivo. J Cell Biol 105:1731–1739

Serrano L, Hernandez MA, Diaz-nido J, Avila J (1989) Association of casein kinase II with microtubules. Exp Cell Res 181:263–272

Shaw G (1991) Neurofilament proteins. In: Burgoyne RD (ed) The neuronal cytoskeleton. Wiley-Liss, New York, NY, 185–214

Shaw G, Miller R, Wang DS, Tang D, Hollander BA, Bennett GS (1997) Characterization of additional casein kinase I sites in the C terminal "tail" region of chicken and rat neurofilament-M. J Neurochem 69:1729–1737

Shea TB, Paskevich PA, Beermann ML (1993) The protein phosphatase inhibitor okadaic acid increases axonal neurofilaments and neurite caliber, and decreases axonal microtubules in NB2a/d1 cells. Neurosci Res 35:507–521

Shea TB, Wheeler E, Jung C (1997) Aluminum inhibits neurofilament assembly, cytoskeletal incorporation and axonal transport: dynamic nature of aluminum-induced perikaryal neurofilament accumulations as revealed by subunit turnover. Mol Chem Neuropathol 32:17–39

Shetty KT, Veeranna, Guru SC (1992) Effect of chronic ethanol ingestion on phosphate content of neurofilament proteins and neurofilament associated protein phosphatase in rat spinal cord. Neurosci Lett 137:83–86

Sihag RK, Nixon RA (1989) In vivo phosphorylation of distinct domains of the 70-kilodalton neurofilament subunit involves different protein kinases. J Biol Chem 264: 457–464

Sihag RK, Nixon RA (1990) Phosphorylation of the amino-terminal head domain of the middle molecular mass 145-kDa subunit of neurofilaments. Evidence for regulation by second messenger-dependent protein kinases. J Biol Chem 265:4166–4171

Sihag RK, Nixon RA (1991) Identification of Ser-55 as a major protein kinase A phosphorylation site on the 70-kDa subunit of neurofilaments. Early turnover during axonal transport. J Biol Chem 266:18861–18867

Sim ATR (1991) The regulation and function of protein phosphatases in the brain. Mol Neurobiol 5:229–246

Smith MA, Rudnicka-Nawrot M, Richey PL, Praprotnik D, Mulvihill P, Miller CA et al. (1995) Carbonyl-related posttranslational modification of neurofilament protein in the neurofibrillary pathology of Alzheimer's disease. J Neurochem 64:2660–2666

Steinert PM, Parry DA (1985) Intermediate filaments: conformity and diversity of expression and structure. Annu Rev Cell Biol 1:41–65

Sternberger LA, Sternberger NH (1983) Monoclonal antibodies distinguish phosphorylated and nonphosphorylated forms of neurofilaments in situ. Proc Natl Acad Sci USA 80: 6126–6130

Strack S, Westphal RS, Colbran RJ, Ebner FF, Wadzinski BE (1997) Protein serine/threonine phosphatase 1 and 2A associate with and dephosphorylate neurofilaments. Brain Res Mol Brain Res 49:15–28

Strong MJ, Strong WL, Jaffe H, Tragert B, Sopper MM, Pant HC (2001) Phosphorylation state of native high molecular weight human neurofilament sub-unit protein (NF-H) from cervical spinal cord in sporadic amyotrophic lateral sclerosis. J Neurochem 76: 1315–1325

Trimmer PA, Borland MK, Keeney PM, Bennett JP Jr, Parker WD Jr (2004) Parkinson's disease transgenic mitochondrial cybrids generate Lewy inclusion bodies. J Neurochem 88:800–812

Trojanowski JQ, Mawal-Dewan M, Schmidt ML, Martin J, Lee VM (1993) Localization of the mitogen activated protein kinase ERK2 in Alzheimer's disease neurofibrillary tangles and senile plaque neurites. Brain Res 618:333–337

Tu PHG, Elder RA, Lazzarini D, Nelson JQ, Trojanowski LVMY (1995) Overexpression of the human NFM subunit in transgenic mice modifies the level of endogenous NFL and the phosphorylation state of NFH subunits. J Cell Biol 129:1629–1640

Tytell M, Zackroff RV, Hill WD (1988) Axonal neurofilaments differ in composition and morphology from those in the soma of the squid stellate ganglion. Cell Motil Cytoskeleton 9:349–360

Tytell MHC, Pant H, Gainer H, Hill WD (1990) Characterization of the distinctive neurofilament subunits of the soma and axon initial segments in the squid stellate ganglion. J Neurosci Res 25:153–161

Veeranna, Amin ND, Ahn NG, Jaffe H, Winters CA, Grant P, Pant HC (1998) Mitogen-activated protein kinases (Erk1,2) phosphorylate Lys-Ser-Pro (KSP) repeats in neurofilament proteins NF-H and NF-M. J Neurosci 18:4008–4021

Veeranna, Shetty KT, Link WT, Jaffe H, Wang J, Pant HC (1995) Neuronal cyclin-dependent kinase-5 phosphorylation sites in neurofilament protein (NF-H) are dephosphorylated by protein phosphatase 2A. J Neurochem 64:2681–2690

Vogelsberg-Ragaglia T, Schuck JQ, Trojanowski LVM (2001) PP2A mRNA expression is quantitatively decreased in Alzheimer's disease hippocampus. Exp Neurol 168:402–412

Watson DF, Glass JD, Griffin JW (1993) Redistribution of cytoskeletal proteins in mammalian axons disconnected from their cell bodies. J Neurosci 13:4354–4360

Way J, Hellmich MR, Jaffe H, Szaro B, Pant HC, Gainer H, Battey J (1992) A high-molecular-weight squid neurofilament protein contains a lamin-like rod domain and a tail domain with Lys-Ser-Pro repeats. Proc Natl Acad Sci USA 89:6963–6967

Weiwad M, Werner A, Rucknagel P, Schierhorn A, Kullertz G, Fischer G (2004) Catalysis of proline-directed protein phosphorylation by peptidyl-prolyl *cis/trans* isomerases. J Mol Biol 339:635–646

Willard M, Simon C (1983) Modulations of neurofilament axonal transport during the development of rabbit retinal ganglion cells. Cell 35:551–559

Wong PC, Marszalek J, Crawford TO, Xu Z, Hsieh ST, Griffin JW, Cleveland DW (1995) Increasing neurofilament subunit NF-M expression reduces axonal NF-H, inhibits radial growth, and results in neurofilamentous accumulation in motor neurons. J Cell Biol 130: 1413–1422

Wulf G, Finn G, Suizu F, Lu KP (2005) Phosphorylation-specific prolyl isomerization: is there an underlying theme? Nat Cell Biol 7:435–441

Xu Z, Cork C, Griffin JW, Cleveland DW (1993) Increased expression of neurofilament subunit NF-L produces morphological alterations that resemble the pathology of human motor neuron disease. Cell 73:23–33

Xu Z, Marszalek JR, Lee MK, Wong PC, Folmer J, Crawford TO, Hsieh ST, Griffin JW, Cleveland DW (1996) Subunit composition of neurofilaments specifies axonal diameter. J Cell Biol 133:1061–1069

Yin X, Crawford TO, Griffin JW, Tu P, Lee VM, Li C, Roder J, Trapp BD (1998) Myelin-associated glycoprotein is a myelin signal that modulates the caliber of myelinated axons. J Neurosci 18:1953–1962

Yoshihara T, Yamamoto M, Hattori N, Misu K, Mori K, Koike H et al. (2002) Identification of novel sequence variants in the neurofilament-light gene in a Japanese population: analysis of Charcot-Marie-Tooth disease patients and normal individuals. J Peripher Nerv Syst 7:221–224

Yoshimura Y, Aoi C, Yamauchi T (2000) Investigation of protein substrates of Ca(2+)/calmodulindependent protein kinase II translocated to the postsynaptic density. Brain Res Mol Brain Res 81:118–128

Zhang B, Tu P, Abtahian F, Trojanowski JQ, Lee VMY (1997) Neurofilaments and orthograde transport are reduced in ventral root axons of transgenic mice that express human SOD1 with a G93A mutation. J Cell Biol 139:1307–1315

Zheng YL, Li BS, Veeranna, Pant HC (2003) Phosphorylation of the head domain of neurofilament protein (NF-M): a factor regulating topographic phosphorylation of NF-M tail domain KSP sites in neurons. J Biol Chem 278:24026–24032

Zhu Q, Lindenbaum M, Levavasseur F, Jacomy H, Julien JP (1998) Disruption of the NF-H gene increases axonal microtubule content and velocity of neurofilament transport: relief of axonopathy resulting from the toxin β′β′-iminodipropionitrile. J Cell Biol 43:183–193

Chapter 27
Cytoskeleton, Axonal Transport, and the Mechanisms of Axonal Neuropathy

Hsinlin T. Cheng, Brian Callaghan, Jacqueline R. Dauch, and Eva L. Feldman

Abstract Axonal neuropathy, or axonopathy, is a major category of neuropathy in the central and peripheral nervous systems. Axonopathy is characterized by axonal degeneration and dysfunctional axonal transport. Peripheral axonopathies are more common than central axonopathies due to their lack of protection from the blood–brain barrier and resultant vulnerability to metabolic challenges. Although the pathogenic mechanisms of peripheral axonal neuropathy are still unclear, the dying-back pattern of the axonal damage suggests axons, rather than neuronal cell bodies, are the primary targets of the disease. Recent studies have revealed that defects of the cytoskeleton and axonal transport are associated with several types of peripheral neuropathy and some central neurological diseases. Direct evidence from genetic studies demonstrates that mutations in major components of the cytoskeleton and axonal transport result in axonal defects in several types of Charcot-Marie-Tooth disease, amyotrophic lateral sclerosis, Alzheimer disease, and other types of genetic neurological disorders. In addition, post-translational modifications of cytoskeleton proteins also result in axonal defects in metabolic diseases like diabetic neuropathy. In this condition, phosphorylation and excess glycation of the axonal cytoskeletal components induce abnormal axonal functions. Advanced glycation end products (AGEs) and their receptors are most likely responsible for the axonal dysfunction. Taken together, understanding the defects in the axonal cytoskeleton and transport mechanisms provides important information for developing new treatments to prevent cytoskeletal damage in axonal neuropathy.

Keywords Axonal transport · Cytoskeleton · Peripheral neuropathy

Contents

H.T. Cheng (✉)
Department of Neurology, University of Michigan, Ann Arbor, MI 48109-2200, USA
e-mail: chengt@umich.edu

R.A. Nixon, A. Yuan (eds.), *Cytoskeleton of the Nervous System*,
Advances in Neurobiology 3, DOI 10.1007/978-1-4419-6787-9_27,

27.1 Introduction

Axonal neuropathy consists of conditions that primarily affect axons of the central and/or peripheral nervous system. In comparison to axons of the central nervous system, which are protected by the blood brain barrier, peripheral nervous system axons are more prone to toxic and metabolic injuries, and other systemic conditions. Peripheral nerves extend from motor, sensory, and autonomic neuronal cell bodies located in the central nervous system or peripheral ganglia. Each peripheral nerve has an axon which can be classified by its diameter. The large myelinated axons from motor neurons and large sensory nerves that mediate position and vibration sense. In comparison, nociceptive and autonomic nerve fibers are thinly myelinated or unmyelinated and are usually the first to be affected in peripheral neuropathy. In general, peripheral neuropathies can be divided into axonal and demyelinating neuropathies, based on the underlying pathogenic mechanisms (Table 27.1). In addition, neuropathies can also be classified as affecting small or large fibers (Thomas et al., 1975; Thomas and Ochoa, 1993).

The symptoms of peripheral neuropathy usually start with pain, sensory loss, and/or autonomic dysfunctions. The sensory symptoms often derive from the involvement of small, thinly myelinated or nonmyelinated fibers that transmit sensations of pain and temperature. Most peripheral neuropathies are length-dependent and the symptoms, including numbness and pain, usually present in the distribution of the longest nerves (those innervating the feet). Over time, peripheral neuropathy extends toward the truck. Once the symptoms pass the level of the knees, patients frequently develop symptoms in the fingers, which then extend upward toward the elbows. This process of gradually developing symmetric sensory dysfunction in distal parts of the limbs is referred to as a stocking and glove distribution. As the neuropathy progress, sensory axons with larger calibers and even motor axons, are affected in a similar fashion, albeit at late stages of the disease (Thomas and Ochoa, 1993).

Table 27.1 Causes of peripheral neuropathy

Predominantly axonal disorders
Diabetic neuropathy
Alcoholic neuropathy
Medication-related neuropathy (e.g., metronidazole, colchicine, nitrofurantoin, isoniazid)
Systemic disease-related neuropathy (e.g., chronic renal failure, inflammatory bowel disease, connective tissue disease)
Thyroid neuropathy
Heavy metal toxic neuropathy (lead, arsenic, cadmium)
Porphyric neuropathy
Paraneoplastic neuropathy
Syphilitic, lyme neuropathy
Sarcoid neuropathy
Human immunodeficiency virus-related neuropathy
Hereditary neuropathies (Charcot-Marie-Tooth, type 2; familial amyloid; mitochondrial)
Critical illness neuropathy
Predominantly demyelinating disorders
Idiopathic chronic inflammatory demyelinating polyradiculoneuropathy (CIDP)
CIDP associated with monoclonal proteins
Antimyelin-associated glycoprotein neuropathy (a form of CIDP)
Antisulfatide antibody-associated neuropathy (a form of CIDP)
Human immunodeficiency virus-associated CIDP
Guillain-Barré syndrome
Hereditary (Charcot-Marie-Tooth, types 1 and 3)

Abbreviations: CIDP, chronic inflammatory demyelinating polyneuropathy
From Frontera: Essentials of physical medicine and rehabilitation, 2nd edn

Allodynia, hyperalgesia, and sensory loss are commonly associated with small fiber abnormalities at the beginning of peripheral neuropathy. Allodynia is defined as normal, nonpainful stimuli that inappropriately induce a painful response. Hyperalgesia means that normally painful stimuli elicit a greater nociceptive response than usual. In addition to sensory abnormalities, autonomic symptoms are frequent in neuropathies associated with diabetes or amyloidosis and include urinary retention or incontinence, impotence, abnormalities of sweating, abnormal bowel functions, and orthostatic hypotension. Large fiber sensory dysfunction develops at a later stage and is usually associated with loss of proprioception. Weakness in peripheral nerve disease is often distal and more severe in the legs than the arms (Thomas and Ochoa, 1993).

The etiology of peripheral neuropathy is complex and includes a wide range of toxic, metabolic, neoplastic, paraneoplastic, autoimmune, and genetic mechanisms that can affect the integrity of peripheral axons, the myelin sheath, and/or the neuronal cell body (Table 27.1). The length-dependent feature of peripheral neuropathy suggests the major site of the disease process is the nerve fibers instead of the cell body. Most of the functions of axons require normal axonal transport, a functional cytoskeleton, and an intact myelin sheath. Peripheral neuropathy, almost without exception, affects at least one of these three critical components of axonal function. It has also been suggested that the reason shorter axons can tolerate neuropathic insults better than longer axons is that they have less cytoskeleton to damage and have more capacity to withstand axonal transport defects due to their shorter distance from the cell body (Duncan and Goldstein, 2006). Recently, numerous publications have reported that mutations of cytoskeletal or axonal transport proteins can cause features of axonal neuropathy. These new findings suggest that a better understanding of the involvement of cytoskeletal integrity and axonal transport in the pathogenesis of axonal neuropathy could provide new approaches to control potentially or even reverse the disease process.

27.2 Cytoskeleton and Axonal Transport

27.2.1 The Roles of the Cytoskeleton in Normal Axonal Function

The axonal cytoskeleton is the key structure maintaining the physical integrity of axons. Moreover, it determines the intracellular dynamics of organelles and is essential for neuronal function. It consists of three major filamentous structures: microtubules, neurofilaments, and actin microfilaments (Fig. 27.1).

27.2.1.1 Microtubules

Microtubules are the main elements responsible for the polarity of the axon. Microtubules are constructed from 13 protofilaments oligomerized in a tubular array with a diameter of 25–28 nm. The protofilaments are formed from alternate stacking of α- and β-tubulin proteins which have a molecular weight of about 50 kd. The order of this stacking determines the polarity of each microtubule. The α-tubulin side of the microtubule defines the minus end which is located closer to the cell body whereas the β-tubulin side defines the plus end which it is located closer to the synapse. This orientation not only gives the microtubule polarity but also the axon (Conde and Caceres, 2009). The polarity of the microtubule directs motors of axonal transport to undergo anterograde (toward the plus end) or retrograde (toward the minus end) transport.

Tubulins are not only structural proteins, but also enzymes with GTPase activity. The binding of GTP to a tubulin monomer promotes the recruitment of another tubulin monomer to the end of the polymer, and thus elongates the microtubule. When the GTP-bound monomers hydrolyze GTP, the now GDP-bound tubulins lose their ability to maintain the stability of the microtubule which leads to depolymerization. The stability of microtubules also depends on microtubule-associated proteins (MAPs). MAPs bind to microtubules and promote the oriented polymerization and assembly of microtubules.

27.2.1.2 Neurofilaments

Neurofilaments are the most abundant cytoskeletal component of the axons. Neurofilaments belong to the family of intermediate filaments, which also includes glial fibrillary acidic protein, desmin, vimentin, and keratin. Neurofilaments are

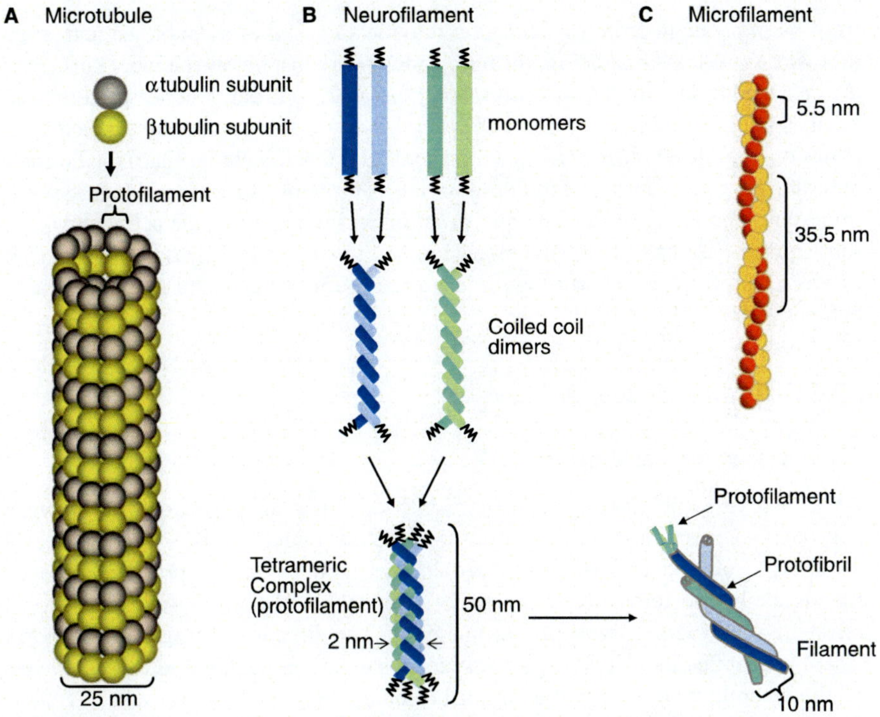

Fig. 27.1 The microstructures of the axonal cytoskeleton. **a** Microtubules are the largest-diameter fibers (25 nm) with helical structures consisting of 13 protofilaments. Each protofilament is formed by linearly arranged pairs of α- and β-tubulin subunits. **b** Neurofilaments are twists of protofibrils. Each protofibril consists of bundles of protofilaments which are tetrametric complexes from coiled-coil dimers of monomers. **c** Microfilaments are composed of two linear actin polymers that are formed by globular actin monomers. Adapted from Principles of Neural Science, 4th Edition

stable structures which do not undergo continuous polymerization and depolymerization like microtubules. There are three proteins, NF-H (~200 kd), NF-M (~150 kd) and NF-L (~70 kd), that comprise the vast majority of the neurofilament (Sharp et al., 1982). NF-L and either NF-M or NF-H combine to make a side to side coiled coil parallel dimer through the interaction of their conserved rod domains. These dimers form anti-parallel tetramers with each other in a half staggered manner. These tetramers then combine to form protofilaments. Two protofilaments form a protofibril. Three twisted protofibrils then become a neurofilament with size of 10 nm. Neurofilaments provide most of the structural support for maintaining axonal caliber and integrity (Perrot et al., 2008).

27.2.1.3 Microfilaments

Microfilaments, 3–5 nm in diameter, are the thinnest fiber of the cytoskeleton. Microfilaments are polar polymers of actin monomers. Most axonal actins are made from β- and γ-actin proteins. The actin polymers consist of a double helical structure with significant branching, forming a meshwork just below the inner surface of the cell membrane. In this sub-cell membrane actin scaffold, bundles of actin filaments aggregate and intermingle with special transmembrane structures called focal adhesion complexes, which anchor the axons to the extracellular matrix. The focal adhesion complex consists of multiple structural proteins and kinases which serve as important regulators for the dynamic structure and functions of the actin cytoskeleton, and are essential for axonal extension and regeneration. In addition, the actin network also provides short-range transportation for vesicles and mitochondria to cell periphery.

27.2.2 Axonal Transport

27.2.2.1 Fast Axonal Transport

Axons provide the linear connections between individual neurons or between neurons and their muscular, sensory, or secretory targets. Carrier proteins travel on specific cytoskeletal structures within the axon to deliver proteins, or even membrane-bound intracellular organelles, from the cell body towards the axonal terminals (anterograde transport), or in the opposite direction (retrograde transport). These organelles include synaptic vesicle precursors, mitochondria, and elements of the smooth endoplasmic reticulum. In general, axonal transport can be classified by the rate of transport. Fast axonal transport (greater than 400 mm/day) is essential for both anterograde and retrograde transport and functions via distinct carriers (Fig. 27.2). The motor molecules for anterograde transport are kinesins and kinesin-related proteins called KIFs (Hirokawa and Noda, 2008). The kinesins and KIFs consist of two heavy chains and two light chains. Each heavy chain consists of a globular motor domain, a coiled-coil stalk domain, and a C-terminal tail domain (Goldstein and Yang, 2000; Almenar-Queralt and Goldstein, 2001). The motor domains bind to microtubules, the coiled domain regulates motor activity, and the tail domains interact with cargo proteins (Lamb et al., 1995). Each KIF protein is location- and cargo-specific in neurons (Hirokawa and Noda, 2008).

Retrograde fast axonal transport primarily transports endosomes generated from endocytosis at the nerve terminals. Some of these endosomes contain ligand-bound growth-factor receptors, and thus deliver the associated signals to the cell body. Retrograde fast transport is about one-third to one-half as fast as anterograde fast axonal transport. The motor for the retrograde transport is dynein. Dynein consists of two globular heads, the motor domains, which are connected to a basal structure by two stalks. The globular heads bind to the microtubules and move the molecule toward the negative end of the microtubule. The dynactin complex, comprised of nine proteins, binds to dynein and is required for dynein-mediated retrograde transport of vesicles and organelles along the microtubules. It provides a link between specific cargos, the microtubule, and dynein during vesicle transport (LaMonte et al., 2002). Dynein-mediated retrograde transport is essential for the delivery of neurotrophic factors from the axon terminal to the cell body. For example, nerve

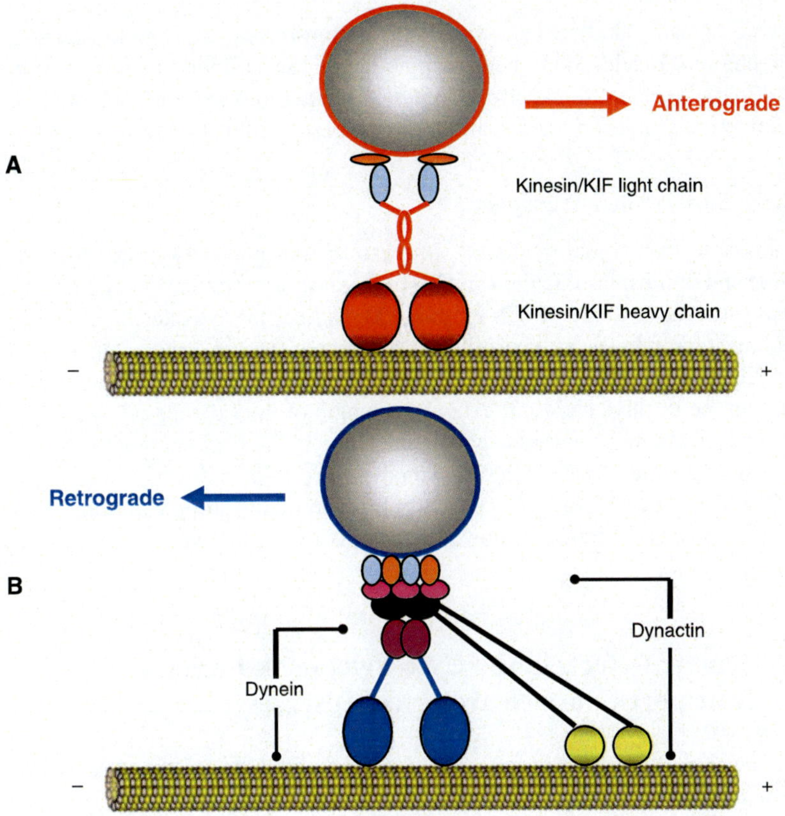

Fig. 27.2 Anterograde and retrograde axonal transport. **a** The motor molecules for anterograde transport are kinesins and kinesin-related proteins called KIFs. The kinesins and KIFs consist of two *heavy* chains and two *light* chains. Each *heavy* chain consists of a globular motor domain, a coiled-coil stalk domain, and a C-terminal tail domain. The motor domains bind to microtubules, the coiled domains regulate motor activity, and the tail domains interact with cargo proteins. Each motor protein is location- and cargo-specific in neurons. **b** Retrograde fast axonal transport primarily transports endosomes generated from endocytosis at the nerve terminals. Some of these endosomes contain ligand-bound growth-factor receptors, and thus deliver the associated signals to the cell body. The motor for the retrograde transport is dynein. Dynein consists of two globular heads, the motor domains, which are connected to a basal structure by two stalks. The globular heads bind to the microtubules and move the molecule toward the negative end of the microtubule. The dynactin complex, comprised of nine proteins, binds to dynein and is required for dynein-mediated retrograde transport of vesicles and organelles along the microtubules. It provides a link between specific cargos, the microtubule, and dynein during vesicle transport

growth factor (NGF) binds to the cell membrane high affinity Trk A receptor in the axonal terminals (Chao, 2003). The NGF-Trk A complex is then internalized via endocytosis, and the endosomes are transported via retrograde axonal transport to the cell body. This retrograde transport through the dynein-dependent mechanism is an essential process to transmit the NGF-activated signal to the cell body

(Reynolds et al., 2000), where the signal is transduced to promote survival and differentiation. In addition to neurotrophic factors, several viral pathogens also utilize retrograde axonal transport for infecting the nervous system, including herpes simplex, varicella-zoster, rabies, and polio viruses (Griffin and Watson, 1988).

27.2.2.2 Slow Axonal Transport

In contrast to fast axonal transport, slow axonal transport only moves towards the positive end of the microtubules (anterograde transport) (Nixon, 1992). Slow axonal transport moves at a rate of 0.2–2.5 mm/day and carries cytoskeletal components, including microtubule and neurofilament elements (Nixon, 1998). Once reaching the positive ends of the cytoskeleton, the newly-arrived cytoskeletal subunits associate with the positive ends and extend the length of the existing cytoskeleton. It is now clear that slow axonal transport is mediated by the same motors (kinesin and cytoplasmic dynein) involved in fast axonal transport. Recently, it has been reported that the reduced rate in slow axonal transport is caused by multiple prolonged pauses during the process of transport down the axon (Roy et al., 2000; Wang et al., 2000; Xia et al., 2003).

27.3 Genetic Defects in the Cytoskeleton and Axonal Transport Cause Neurological Diseases

The known neurological diseases associated with mutations of cytoskeletal proteins or axonal transport motor proteins are listed in Table 27.2.

Table 27.2 Mutations of cytoskeletal and axonal transport components in neurological diseases

Disease	Gene	References
ALS	KIF1Bβ	Pantelidou et al. (2007)
	KIF3Aβ	Pantelidou et al. (2007)
	Dynactin, cytoplasmic dynein heavy chain	Munch et al. (2004)
CMT 2A	KIF1B	Saito et al. (1997), Zhao et al. (2001)
	MFN2	Baloh et al. (2007)
CMT2E and CMT1F	Neurofilament light chain	Fabrizi et al. (2007), Shin et al. (2008), Bhagavati et al. (2009)
CMT2F/CMT2L, dHMN II	Heat shock protein 27	Houlden et al. (2008)
GAN	Gigaxonin	Yang et al. (2007)
HSP	Atlastin	Reviewed in Lau et al. (2009)
	Spastin	Reviewed in Lau et al. (2009)
	KIF5A	Lo Giudice et al. (2006)

Abbreviations: Dhmn Ii, Autosomal Dominant Distal Hereditary Motor Neuropathy Type Ii; Gan, Giant Axonal Neuropathy; Hsp, Hereditory Spastic Paraparesis

27.3.1 Amyotrophic Lateral Sclerosis

The molecular mechanisms underlying the selective neurodegeneration of motor neurons in amyotrophic lateral sclerosis (ALS) are unclear. ALS can be divided into familiar (fALS) or sporadic (sALS) subtypes. Several recent reports have demonstrated impaired axonal transport, caused by defective expression of several molecular motors, as key elements for the disease onset and progression of sALS, but not in fALS.

There are at least 15 KIFs expressed in the human motor cortex. Among these KIFs, KIF1Bβ and KIF3Aβ are two major isoforms enriched in the motor cortex (Pantelidou et al., 2007). The gene expression of both KIF1Bβ and KIF3Aβ is reduced in the brains of sALS patients. This is accompanied by a significant reduction in KIF3Aβ protein levels. In comparison, KIF1Bβ and KIF3Aβ mRNAs are not altered throughout disease progression in the motor cortex of SOD1 G93A mice, a model of fALS (Conforti et al., 2003; Pantelidou et al., 2007). This suggests these two motor proteins as possible candidates in sALS, but not fALS pathology (Pantelidou et al., 2007).

Lamonte and colleagues reported that mice overexpressing dynamitin demonstrate a late-onset progressive motor neuron degenerative disease. Dynamitin overexpression inhibits retrograde axonal transport by disassembling dynactin, a required activator of cytoplasmic dynein (LaMonte et al., 2002). Munch and colleagues detected heterozygous missense mutations of p150 subunit of dynactin (DCTN1) and the cytoplasmic dynein heavy chain (DNCHC1) genes in both sALS and fALS patients. Their findings suggest that this defect in the machinery of retrograde axonal transport is involved in patients with ALS (Munch et al., 2004). Defective retrograde axonal transport from mutations of DCTN1 and DNCHC1 may lead to decreased neurotrophic support from the end organ-derived neurotrophic factors to related motor neurons. Deprivation of this neurotrophism has been proposed as a cause of cell death in the pathogenesis of ALS (Ekestern, 2004).

In addition, mutations of superoxide dismutase 1 (SOD1) could directly or indirectly affect axonal transport. Several potential mechanisms have been proposed: (1) interfering with cargo-binding by either disruption of dynactin (or another adaptor) or by competitive binding by mutant SOD1; (2) physical blockade of retrograde transport by mutant SOD1 aggregates; (3) disruption of microtubule formation and stability; (4) disruption of dynein motor activity by mutant SOD1; (5) disruption of dynein or dynactin complex integrity; (6) disruption of dynein-dynactin microtubule binding (reviewed in (Strom et al., 2008)).

27.3.2 Charcot-Marie-Tooth Disease

Charcot-Marie-Tooth disease (CMT), the most common inherited peripheral neuropathy in humans with a prevalence of 1/2,500, is clinically characterized by weakness and atrophy of distal muscles, depressed or absent deep tendon reflexes, and mild sensory loss (Shy, 2004). It has been classified into types I (myelinopathy) and II (axonopathy) by using motor nerve conduction velocity (MNCV) as a marker of myelin degeneration (Harding and Thomas, 1980). Most CMTs are caused by mutations in myelin proteins (Shy, 2004). CMT2A, an autosomal dominant subtype of type II CMT, was mapped to human chromosome 1p35-36 (Ben Othmane et al., 1993). Zhao and colleagues analyzed the KIF1B locus in CMT2A patients, and discovered a Q98L missense mutation in the ATP binding consensus of motor domain. The Q98L mutant KIF1Bβ protein decreases ATPase activity and its ability to mediate anterograde mitochondrial transport, suggesting that a haplo-insufficiency of this motor protein is responsible for CMT2A neuropathy (Nangaku et al., 1994; Zhao et al., 2001). In addition, Saito and colleagues also reported a Japanese family with CMT2A carrying a mutation of KIF1B (Saito et al., 1997).

Mitofusin 2 (MFN2) is an outer mitochondrial membrane protein, and was previously known as mitochondrial assembly regulatory factor (MARF) or hyperplasia suppressor gene (HSG) (Bach et al., 2003; Chen et al., 2004). MFN2 has many roles in regulating normal cellular functions, such as oxidative metabolism, cell cycle, cell death, and mitochondrial axonal transport. It has been reported that mutations in MFN2 cause the autosomal dominant neurodegenerative disease CMT2A [30]. The mutated MFN2 protein interferes with the transport of mitochondria distally into axons, where they are required for maintenance of Na^+-K^+ ATPase activity and normal synapse functions (Baloh et al., 2007). It has been proposed that the mutant MFN2 interferes with binding between the mitochondria and microtubules (Baloh, 2008).

A heterozygous, missense mutation of the neurofilament light chain polypeptide (*NEFL*) gene has been found to associate with a mixed axonal and demyelinating neuropathy, with widespread demyelination involving both proximal and distal nerve segments. Mutations at this site in the *NEFL* gene have been linked to CMT2E and CMT1F (Previtali et al., 2003; Fabrizi et al., 2007; Shin et al., 2008; Bhagavati et al., 2009).

Mutations in the myotubularin-related protein 2 and 13 (*MTMR2* and *MTMR13*) genes lead to CMT4B. MTMR2 acts as a phosphoinositide D3-phosphatase which dephosphotizes phosphatidylinositol (PtdIns) 3-phosphate and PtdIns 3,5-bisphosphate. MTMR13/SBF2 forms a tetrameric complex with MTMR2 to enhance its enzymatic activity. Mutations of these two proteins cause NEFL aggregation and reduced ability to undergo phosphorylation and form dimers, implicating impaired neurofilament formation in myotubularin-related pathology (Goryunov et al., 2008).

27.3.3 Alzheimer's Disease

Alzheimer's disease (AD) is characterized pathologically by neurofibrillary tangles and neuritic plaques. The neurofibrillary tangles consist of tau and amyloid β protein. The amyloid β protein is produced from the proteolytic cleavage of amyloid precursor protein (APP). Recently, growing evidence has indicated that AD is associated with defects in axonal transport related to interference of microtubule structure. A key feature of AD is the accumulation of phosphorylated tau in the neuronal cells and axons. Tau is a microtubule-binding protein that promotes microtubule stability. The excessive phosphorylation prevents the binding of tau to microtubule and thus damages microtubule integrity (Lee et al., 1994; Higuchi et al., 2002).

In addition, the impaired binding of APP to kinesin-I is a potential mechanism for reduced axonal transport in AD. There is evidence to suggest that APP knockout could interfere with axonal transport (Gunawardena and Goldstein, 2004). The link of APP to axonal transport is related to its ability to bind kinesin-I (Kamal et al., 2000). Mutations of the C-terminal domain of APP impairs its affinity for kinesin-I and reduces axonal transport (Gunawardena and Goldstein, 2001). Furthermore, evidence also suggests that soluble amyloid β, the end product of APP, disrupts both neurotransmitter and neurotrophin signaling by affecting axonal transport (Schindowski et al., 2008).

Mutations in presenilin cause a familial form of AD. Presenilin interacts with glycogen synthase kinase 3 β (GSK3 β), which phosphorylates kinesin light chains. In a mouse line that carries a presenilin mutation, the mutant presenilin increases the activity of GSK3 β and increases the phosphorylation of kinesin light chains. As a result, the kinesin detaches from its cargo and disrupts axonal transport (Pigino et al., 2003).

27.3.4 Hereditary Spastic Paraplegia

Hereditary spastic paraplegia (HSP) or Strümpell-Lorrain syndrome is a heterogeneous group of inherited disorders associated with mutations of several proteins linked to the cytoskeleton and axonal transport (Fink, 2006). A common feature of these disorders is the slowly progressive degeneration of the corticospinal tract and associated weakness and spasticity. HSP is subdivided into pure spastic paraplegia or complicated HSP, depending on the presence of other neurological deficits in addition to spastic paraparesis. Inheritance may be autosomal dominant, autosomal recessive, or, rarely, X-linked. Genetic studies have revealed as many as 31 different chromosomal HSP loci. Mutations in the gene encoding the microtubule-severing protein spastin are the most common cause of HSP (40% of all HSP cases). Spastin is not only a microtubule-associated protein, but is also recruited to endosomes as part of the secretory pathway (Connell et al., 2009). Kasher et al. reported that mutant spastin perturbs anterograde transport of both mitochondria and amyloid precursor protein-containing membrane-bound organelles, and reduces both anterograde and retrograde transport (Kasher et al., 2009).

About 10% of HSP cases are due to mutations in *SPG3A*, which encodes atlastin (Lau et al., 2009). Atlastin is a GTPase that functions prominently in both ER and Golgi morphogenesis, but does not appear to be required for anterograde ER-to-Golgi trafficking or other axonal transport functions (Rismanchi et al., 2008).

In addition to spastin and atlastin, mutations in the *KIF5A* gene, a member of the kinesin superfamily, have been found to cause early-onset autosomal dominant hereditary spastic paraparesis (ADHSP). It is recommended that the *KIF5A* gene should be routinely analyzed in patients with hereditary spastic paraplegia who are negative for spastin and atlastin mutations (Lo Giudice et al., 2006).

27.3.5 Giant Axonal Neuropathy

Giant axonal neuropathy (GAN) is a rare genetic disease associated with axonopathy of the peripheral and central nervous system. The characteristic pathological features of GAN are giant axons with accumulation of a disorganized cytoskeletal network. The known mutations that cause GAN are found in the gene *GAN,* which encodes the protein gigaxonin. Gigaxonin appears to play an important role in cytoskeletal functions and dynamics by directing ubiquitin-mediated degradations of cytoskeletal proteins, including the microtubule-associated protein MAP8 (Ding et al., 2006). Mutations of *GAN* cause axonopathy by inducing defects in the degradation of cytoskeletal proteins and causing abnormal accumulation of toxic products (Yang et al., 2007).

27.4 Diabetic Neuropathy

Distal sensory symmetric polyneuropathy (DPN) is the most common of the various peripheral nerve disorders associated with diabetes. This syndrome affects the extremities in a "stocking and glove" distribution and is usually more severe in the legs. The symptoms are variable, but usually begin with sensory symptoms, including pain, numbness, and/or autonomic dysfunctions. Similarly, the nerve fibers succumb to the toxic effects of diabetes in a size-dependent order, with small caliber Aδ and C fibers affected first (Fig. 27.3). Motor fibers with thick myelin sheaths are affected later, and thus, accordingly, motor symptoms usually occur late in the course of diabetic polyneuropathy. However, diabetes-induced plexopathy, mononeuropathy, and radiculopathy do not have this size- and length-dependency.

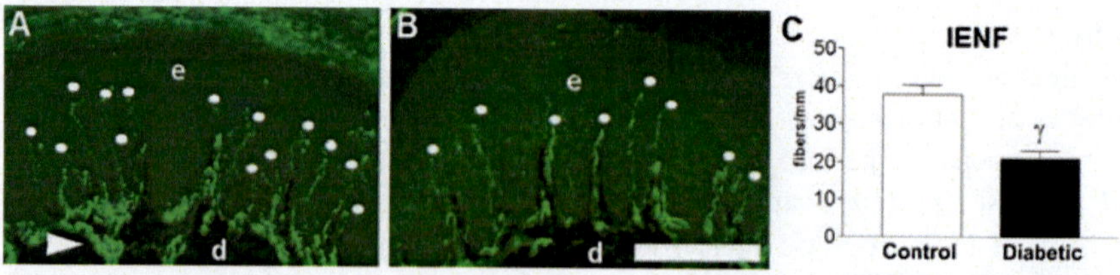

Fig. 27.3 Intraepidermal nerve fiber density (IENF) measures functional innervation of the skin. **a** PGP9.5 immunofluorescence within a control non-diabetic ($n = 5$) BKS-db$^+$ foot pad demonstrating a normal pattern of innervation (*dots*). **b** PGP9.5 immunofluorescence within a diabetic ($n = 5$) BKS-db/db foot pad demonstrating a decrease in the number of fibers (*dots*). **c** Quantitation of IENF is presented as the number of fibers/linear millimeter of epidermis. d = dermis, e = epidermis, *dots* = intraepidermal nerve fiber, *arrowhead* = dermal fiber bundles. $^{\gamma}p < 0.01$. *Scale bar* = 100 μm. Open *bars* represent the control non-diabetic measurements and the *black bars* represent the diabetic measurements. (From Sullivan et al., 2007)

In these conditions, both motor and sensory functions are affected simultaneously and appear to have distinct pathomechanisms. Diabetic polyneuropathy has been proposed to be caused by several metabolic mechanisms; in contrast, plexopathy, mononeuropathy, and radiculopathy are considered to be vasculitic in nature and thus do not have the length-dependent features of polyneuropathy (Dyck and Windebank, 2002; Edwards et al., 2008).

This dying back pattern (distal–proximal direction) is characteristic of a failure in fast axonal transport, but not of injury to the neuronal cell body. Axonal transport is known to be affected in experimental models of diabetes. Retrograde fast axonal transport is reduced in diabetic animals (Sidenius and Jakobsen, 1981). In addition, the slow component of anterograde axonal transport for actin, tubulin, and the two lightest subunits of the neurofilament triplet are also impaired in diabetic mice with a leptin receptor mutation (Vitadello et al., 1985). Similarly, in streptozotocin (STZ)-treated rats (a model of type I diabetes), slow axonal transport of neurofilament and microtubule components is also reduced and results in decreased axonal caliber (Medori et al., 1988). Interestingly, the impairment of axonal transport in diabetic nerves is more significant in sensory than in motor fibers (Macioce et al., 1989), which is consistent with the sensory predominant neuropathy observed in patients with diabetes.

Potential mechanisms of hyperglycemia that affect axonal transport could include: (1) the induction of metabolic abnormalities in the neuronal cell body that affect the synthesis of elements necessary for axonal transport; (2) increased glycation or phosphorylation of axonal transport elements, which leads to deterioration in the efficiency of axonal transport (McLean, 1997); and/or (3) deleterious glycation in Schwann cells, affecting the myelin sheath stability and inducing demyelinating neuropathy (Toth et al., 2008). This chapter focuses on the first and second mechanisms, which primarily affect the axons.

27.4.1 Reduced Levels of Cytoskeletal Proteins in Diabetes

The expression of cytoskeletal proteins is significantly affected by diabetes. In diabetic nerves, impairment of slow anterograde axonal transport causes smaller axonal calibers, which is correlated with reduced local levels of neurofilament (Medori et al., 1985; Yagihashi et al., 1990). The gene expression of the neurofilament proteins is also reduced in diabetes with the expression of NF-L and NF-H, but not NF-M (Mohiuddin et al., 1995). In addition, the reduced ability for axonal regeneration in diabetes could result from impaired expression of Tα1α-tubulin and GAP-43, which are growth-associated proteins (Mohiuddin et al., 1995).

27.4.2 Advanced Glycation in Diabetic Neuropathy

As a result of the chronic hyperglycemia of diabetes, tissue glucose levels are persistently increased. This in turn results in the generation of advanced glycation endproducts (AGEs) in the peripheral nervous tissue. Increased tissue and cellular glucose levels stimulate glycolytic and polyol pathways in the peripheral nerve (Sugimoto et al., 2008). Enhanced activation of the glycolytic pathway produces the intermediate metabolite glyceraldehyde-3-phosphate (GA3P), which is further metabolized to methylglyoxal (MG), and then N^{ε}-(carboxyethyl)lysine (CML). GA3P and MG are precursors of AGEs. In parallel, activation of the polyol pathway increases intracellular fructose via aldose reductase-mediated sorbitol synthesis. Fructose is further converted to fructose-3-phosphate (F3P), fructose-1-phosphate, and fructose-6-phosphate. 3-Deoxyglucosone (3-DG) is an end product of fructose and F3P metabolism and also serves as a precursor of AGEs. AGEs are known to accumulate in the peripheral nerves of STZ-induced diabetic animals and patients with diabetes (Sugimoto et al., 1997; Wada et al., 2001). In STZ-induced diabetic rats, increased CML immunoreactivity was detected in the perineurium, Schwann cells, axons, and vascular walls, while only the perineurium was slightly positive for CML in normal peripheral nerve (Wada et al., 2001). In the human diabetic peripheral nerve, CML was detected in vascular endothelial cells, pericytes, the basement membrane, axons, and Schwann cells (Sugimoto et al., 1997). The distribution of CML in the peripheral nerve appeared to be consistent with the components of structural proteins, which are demonstrated to be glycated by biochemical analysis (King, 2001). The AGE pentosidine is formed by glucose auto-oxidation (Sell et al., 1991). Pentosidine is relatively easy to quantify and is frequently used as a measure of total AGE products in experimental studies of the effects of glycation on the PNS (King, 2001). Pentosidine levels are upregulated in nerves of both diabetic humans and STZ rats (Ryle and Donaghy, 1995; Ryle et al., 1997). The degree of AGE accumulation also reflects the severity of diabetic neuropathy as well as other diabetic complications (Meerwaldt et al., 2005). The axonal levels of AGEs correlate with the severity of axonal pathology, including perineural thickening and axonal loss in type 2 diabetes (Misur et al., 2004).

As AGE precursors and AGEs accumulate in the peripheral nerve, several cytoskeletal proteins, the myelin sheath, and extracellular matrix are affected by glycation (McLean, 1997). In diabetic animals, residues of cytoskeleton proteins are glycated three times more than in control animals (Vlassara et al., 1981). This post-translational non-enzymatic glycation can be reversed by insulin treatment, suggesting it is from chronic hyperglycemia (Pekiner et al., 1993). In addition, glycation of tubulin and neurofilament in the axons have been demonstrated in STZ-treated rats (McLean et al., 1992; Ryle et al., 1997). Myelin proteins in Schwann cells and extracellular matrix components including laminin and collagens are also affected by non-enzymatic glycation (Toth et al., 2007, 2008). In the central nervous system, glycation of microtubule-associated tau proteins is also detected and might contribute to the increased risk of Alzheimer's disease in diabetic patients (Toth et al., 2007). Recent experimental work suggests that the peripheral nervous system cytoskeleton is more vulnerable to non-enzymatic glycation than the central nervous system cytoskeleton (Ryle et al., 1997). This difference could also contribute to preferential damage of sensory dorsal root ganglion neurons compared with motor neurons. Additional support for the importance of AGEs in the development of PNS damage is provided by experiments on STZ diabetic rats using AGE inhibitors (Thornalley, 2002). Sensory and motor nerve conduction velocities are reduced in this animal model of diabetes and are improved by treatment with aminoguanidine, an AGE inhibitor (Cameron et al., 1992).

In addition to the accumulation of AGEs in tissues, receptors for AGEs also contribute to the development of diabetic complications (Ahmed, 2005). Several receptors are known to bind to AGEs: the receptor for AGE (RAGE), galectin-3 (Vlassara et al., 1995), scavenger receptor class A (SR-A), CD36, SR-BI, and LOX-1 (the latter three belonging to scavenger receptor class B) (Toth et al., 2007). RAGE has been localized to the peripheral nerve in experimental diabetic animals [76]. In the rat peripheral nerve, RAGE was expressed in endothelial cells both of perineural and endoneural vessels. Schwann cells were also found to be positive for RAGE (Wada and Yagihashi, 2005). RAGE expression was of greater intensity within axons and Schwann cells of both the sciatic nerve and, especially, the sural nerve of diabetic mice (Wada and Yagihashi, 2005; Toth et al., 2008). In addition, expression of RAGE was also noted within endothelial cells of the vasa nervorum (Toth et al., 2008). In RAGE knockout mice, the absence of RAGE attenuated both structural and electrophysiological changes within the peripheral nerves and the DRG after 5 months of STZ-induced diabetes. (Toth et al., 2008).

There are multiple mechanisms proposed for how AGE and RAGE cause dysfunction of the peripheral axons:

1. AGE administration has been shown to induce apoptosis in neuronal cells and Schwann cells (Takeuchi et al., 2000; Sekido et al., 2004; Vincent et al., 2007). Studies using cultured DRG neurons have demonstrated that ligand-activated RAGE induces increased NAD(P)H oxidase activity and elevated levels of reactive oxidative species (Fig. 27.4). As a result, the levels of oxidative stress are significant enough to induce apoptosis (Fig. 27.5) (Vincent et al., 2007). The RAGE-dependent oxidative stress could result in upregulation of nuclear factor (NF)-kappaB and several NF-kappaB-dependent proinflammatory genes (Bierhaus et al., 2004; Toth et al., 2008). These studies suggest that the fiber loss from neuronal and myelin damage in human diabetic peripheral nerve may in part be accounted for by the accumulation of AGE.
2. Modification of neurofilament and tubulin with AGEs may possibly interfere with axonal transport (Williams et al., 1982). These axonal transport defects then contribute to the development of atrophy and degeneration of nerve fibers.
3. Modification of myelin proteins with AGEs may serve as a basis for demyelination of nerve fibers (Vlassara et al., 1981). AGE-modified peripheral nerve myelin is susceptible to phagocytosis by macrophages and contributes to segmental demyelination (Vlassara and Palace, 2002). This finding provides another potential mechanism for the slow nerve conduction velocity, in addition to the myelin defects, observed in diabetic patients.
4. Glycation of the extracellular matrix protein laminin leads to impaired regenerative activity in diabetic neuropathy (Sugimoto et al., 2008).

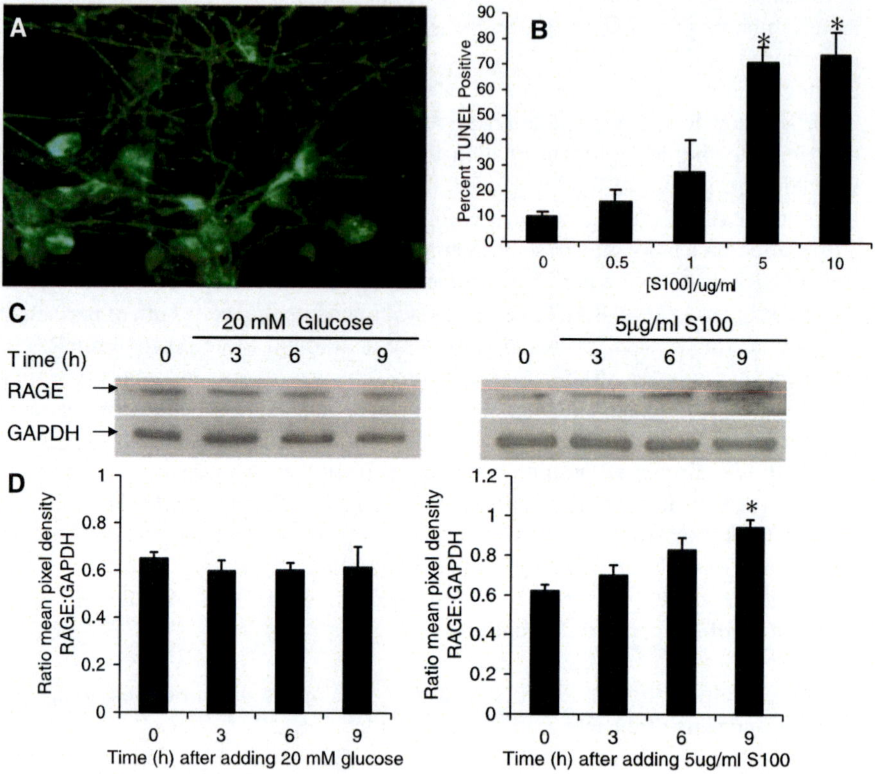

Fig. 27.4 DRG neurons express RAGE. **a** Cultured DRG neurons were fixed and stained using a RAGE antibody, with a fluorescein isothiocyanate-conjugated secondary antibody. **b** Cultured DRG neurons were exposed to increasing concentrations of S100 for 24 h and then TUNEL stained. Concentrations of 5 μM S100 and higher significantly increased DRG neuron injury. **c,d** Cultured DRG neurons were treated with 20 mM glucose or 5 μg/mL S100 for 0, 3, 6, or 9 h and processed for Western blotting. **c** Blots were probed using antibodies against RAGE or glyceraldehyde-3-phosphate dehydrogenase (GAPDH) as a loading control. The figure contains a representative blot from three replicates performed on separate occasions and producing the same pattern of RAGE expression. **d** The mean pixel density for each band in the three replicate blots was determined using Scion Image software. The graphs illustrate the mean and SE for all the blots. RAGE bands were normalized against GAPDH in each blot. RAGE expression was significantly increased over control after 9 h S100 treatment.*, $p < 0.05$. (From Vincent et al., 2007)

27.4.3 Phosphorylation of Cytoskeletal Proteins in Diabetic Neuropathy

In addition to non-enzymatic glycation, post-translational phosphorylation of cytoskeleton proteins is commonly associated with axonal neuropathy (Pestronk et al., 1990; Arias et al., 1993; McLean et al., 1995). Phosphorylation of microtubule and neurofilament proteins affects axonal growth, caliber, and regeneration (McLean et al., 1995). In STZ-induced type 1 diabetes, phosphorylation

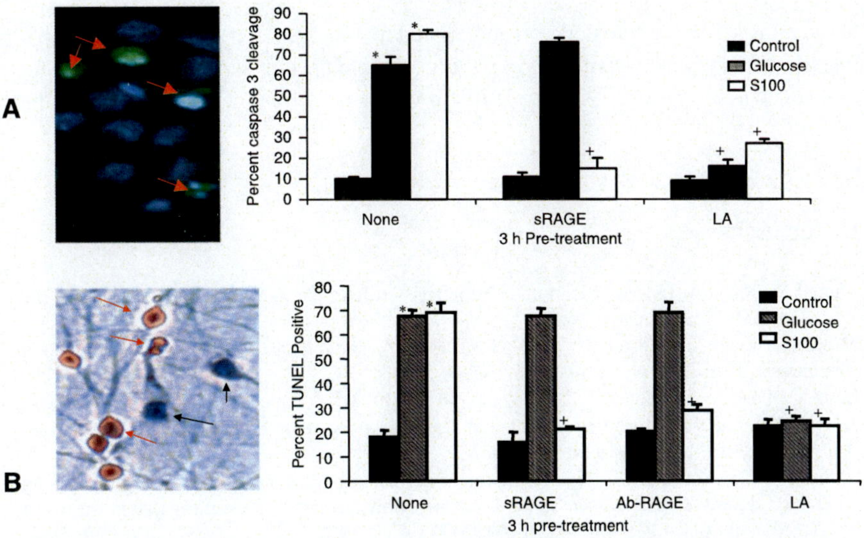

Fig. 27.5 Activation of RAGE leads to DRG neuron injury via oxidative stress. The development of programmed cell death was assessed in DRG cultures exposed to S100 (5 μg/mL) or hyperglycemia (20 mM added glucose) ± pretreatment with sRAGE (500 μg/mL), α-lipoic acid (AO; 100 μM), Ab-RAGE (1 μg/mL), or controls nonimmune IgG (1 μg/mL) or BSA (20 μg/mL). **a** Caspase-3 activation was determined using a fluorescent substrate (CaspaTag) after 6 h exposure to S100. **b** DNA fragmentation was determined using the TUNEL assay 24 h after exposure to S100.*, $p < 0.001$. S100-induced death was prevented by pretreatment with sRAGE, Ab-RAGE, or AO. Only AO prevented hyperglycemia-induced injury. + TUNEL labeling was significantly decreased, compared with glucose or S100 alone. (From Vincent et al., 2007)

of neurofilament proteins is increased after 6–8 weeks of diabetes (Pekiner and McLean, 1991; Fernyhough et al., 1999; Fernyhough and Schmidt, 2002). The hyperphosphorylation of NF-H could affect its interactions with other elements of the cytoskeleton and disrupt the normal spacing between neurofilaments (Carden et al., 1985). This phosphorylation is associated with increased c-jun N-terminal protein kinase activity in DRG and peripheral nerves of the diabetic animals (Fernyhough et al., 1999). Taken together, the post-translational phosphorylation of neuronal cytoskeletal proteins may contribute to the altered axonal transport and subsequent nerve dysfunction in experimental diabetes.

27.5 Conclusions

The cytoskeletal scaffolding not only provides structural support for the axons but is also essential to axonal transport. Neuronal dysfunction occurs when the integrity of the cytoskeleton and axonal transport are compromised in various conditions, including genetic and metabolic disorders such as diabetes. Although various causes of peripheral neuropathy have distinct pathologies, they often share common features of defective cytoskeleton and axonal transport systems. Further understanding of the mechanisms underlying the cytoskeletal and axonal transport defects in axonal neuropathies will hopefully provide new treatments to improve these currently irreversible conditions.

References

Ahmed N (2005) Advanced glycation endproducts – role in pathology of diabetic complications. Diabetes Res Clin Pract 67:3–21

Almenar-Queralt A, Goldstein LS (2001) Linkers, packages and pathways: new concepts in axonal transport. Curr Opin Neurobiol 11:550–557

Arias C, Sharma N, Davies P, Shafit-Zagardo B (1993) Okadaic acid induces early changes in microtubule-associated protein 2 and tau phosphorylation prior to neurodegeneration in cultured cortical neurons. J Neurochem 61:673–682

Bach D, Pich S, Soriano FX, Vega N, Baumgartner B et al. (2003) Mitofusin-2 determines mitochondrial network architecture and mitochondrial metabolism. A novel regulatory mechanism altered in obesity. J Biol Chem 278:17190–17197

Baloh RH (2008) Mitochondrial dynamics and peripheral neuropathy. Neuroscientist 14:12–18

Baloh RH, Schmidt RE, Pestronk A, Milbrandt J (2007) Altered axonal mitochondrial transport in the pathogenesis of Charcot-Marie-Tooth disease from mitofusin 2 mutations. J Neurosci 27:422–430

Ben Othmane K, Middleton LT, Loprest LJ, Wilkinson KM, Lennon F et al. (1993) Localization of a gene (CMT2A) for autosomal dominant Charcot-Marie-Tooth disease type 2 to chromosome 1p and evidence of genetic heterogeneity. Genomics 17:370–375

Bhagavati S, Maccabee PJ, Xu W (2009) The neurofilament light chain gene (NEFL) mutation Pro22Ser can be associated with mixed axonal and demyelinating neuropathy. J Clin Neurosci 16:830–831

Bierhaus A, Haslbeck KM, Humpert PM, Liliensiek B, Dehmer T et al. (2004) Loss of pain perception in diabetes is dependent on a receptor of the immunoglobulin superfamily. J Clin Invest 114:1741–1751

Cameron NE, Cotter MA, Dines K, Love A (1992) Effects of aminoguanidine on peripheral nerve function and polyol pathway metabolites in streptozotocin-diabetic rats. Diabetologia 35: 946–950

Carden MJ, Schlaepfer WW, Lee VM (1985) The structure, biochemical properties, and immunogenicity of neurofilament peripheral regions are determined by phosphorylation state. J Biol Chem 260:9805–9817

Chao MV (2003) Retrograde transport redux. Neuron 39:1–2

Chen KH, Guo X, Ma D, Guo Y, Li Q et al. (2004) Dysregulation of HSG triggers vascular proliferative disorders. Nat Cell Biol 6:872–883

Conde C, Caceres A (2009) Microtubule assembly, organization and dynamics in axons and dendrites. Nat Rev 10:319–332

Conforti L, Dell'Agnello C, Calvaresi N, Tortarolo M, Giorgini A et al. (2003) Kif1Bbeta isoform is enriched in motor neurons but does not change in a mouse model of amyotrophic lateral sclerosis. J Neurosci Res 71:732–739

Connell JW, Lindon C, Luzio JP, Reid E (2009) Spastin couples microtubule severing to membrane traffic in completion of cytokinesis and secretion. Traffic (Copenhagen, Denmark) 10:42–56

Ding J, Allen E, Wang W, Valle A, Wu C et al. (2006) Gene targeting of GAN in mouse causes a toxic accumulation of microtubule-associated protein 8 and impaired retrograde axonal transport. Hum Mol Genet 15:1451–1463

Duncan JE, Goldstein LS (2006) The genetics of axonal transport and axonal transport disorders. PLoS Genet 2:e124

Dyck PJ, Windebank AJ (2002) Diabetic and nondiabetic lumbosacral radiculoplexus neuropathies: new insights into pathophysiology and treatment. Muscle Nerve 25:477–491

Edwards JL, Vincent AM, Cheng HT, Feldman EL (2008) Diabetic neuropathy: mechanisms to management. Pharmacol Ther 120:1–34

Ekestern E (2004) Neurotrophic factors and amyotrophic lateral sclerosis. Neurodegener Dis 1: 88–100

Fabrizi GM, Cavallaro T, Angiari C, Cabrini I, Taioli F et al. (2007) Charcot-Marie-Tooth disease type 2E, a disorder of the cytoskeleton. Brain 130:394–403

Fernyhough P, Gallagher A, Averill SA, Priestley JV, Hounsom L et al. (1999) Aberrant neurofilament phosphorylation in sensory neurons of rats with diabetic neuropathy. Diabetes 48:881–889

Fernyhough P, Schmidt RE (2002) Neurofilaments in diabetic neuropathy. Int Rev Neurobiol 50:115–144

Fink JK (2006) Hereditary spastic paraplegia. Curr Neurol Neurosci Rep 6:65–76

Goldstein LS, Yang Z (2000) Microtubule-based transport systems in neurons: the roles of kinesins and dyneins. Annu Rev Neurosci 23:39–71

Goryunov D, Nightingale A, Bornfleth L, Leung C, Liem RK (2008) Multiple disease-linked myotubularin mutations cause NFL assembly defects in cultured cells and disrupt myotubularin dimerization. J Neurochem 104:1536–1552

Griffin JW, Watson DF (1988) Axonal transport in neurological disease. Ann Neurol 23:3–13

Gunawardena S, Goldstein LS (2001) Disruption of axonal transport and neuronal viability by amyloid precursor protein mutations in Drosophila. Neuron 32:389–401

Gunawardena S, Goldstein LS (2004) Cargo-carrying motor vehicles on the neuronal highway: transport pathways and neurodegenerative disease. J Neurobiol 58:258–271

Harding AE, Thomas PK (1980) Genetic aspects of hereditary motor and sensory neuropathy (types I and II). J Med Genet 17:329–336

Higuchi M, Lee VM, Trojanowski JQ (2002) Tau and axonopathy in neurodegenerative disorders. Neuromolecular Med 2:131–150

Hirokawa N, Noda Y (2008) Intracellular transport and kinesin superfamily proteins, KIFs: structure, function, and dynamics. Physiol Rev 88:1089–1118

Houlden H, Laura M, Warrant-De Vrieze F, Blake J et al. (2008) Mutations in the HSP27 (HSPB1) gene cause dominant, recessive, and sporadic distal HMN I CMT type 2. Neurology 71:1660–8

Kamal A, Stokin GB, Yang Z, Xia CH, Goldstein LS (2000) Axonal transport of amyloid precursor protein is mediated by direct binding to the kinesin light chain subunit of kinesin-I. Neuron 28:449–459

Kasher PR, De Vos KJ, Wharton SB, Manser C, Bennett EJ et al. (2009) Direct evidence for axonal transport defects in a novel mouse model of mutant spastin-induced hereditary spastic paraplegia (HSP) and human HSP patients. J Neurochem 110:34–44

King RH (2001) The role of glycation in the pathogenesis of diabetic polyneuropathy. Mol Pathol 54:400–408

LaMonte BH, Wallace KE, Holloway BA, Shelly SS, Ascano J et al. (2002) Disruption of dynein/dynactin inhibits axonal transport in motor neurons causing late-onset progressive degeneration. Neuron 34:715–727

Lamb JR, Tugendreich S, Hieter P (1995) Tetratrico peptide repeat interactions: to TPR or not to TPR? Trends Biochem Sci 20:257–259

Lau KK, Ching CK, Mak CM, Chan YW (2009) Hereditary spastic paraplegias. Hong Kong Med J 15:217–220

Lee VM, Daughenbaugh R, Trojanowski JQ (1994) Microtubule stabilizing drugs for the treatment of Alzheimer's disease. Neurobiol Aging 15(Suppl 2):S87–S89

Lo Giudice M, Neri M, Falco M, Sturnio M, Calzolari E et al. (2006) A missense mutation in the coiled-coil domain of the KIF5A gene and late-onset hereditary spastic paraplegia. Arch Neurol 63:284–287

Macioce P, Filliatreau G, Figliomeni B, Hassig R, Thiery J et al. (1989) Slow axonal transport impairment of cytoskeletal proteins in streptozocin-induced diabetic neuropathy. J Neurochem 53:1261–1267

McLean WG (1997) The role of axonal cytoskeleton in diabetic neuropathy. Neurochem Res 22:951–956

McLean WG, Pekiner C, Cullum NA, Casson IF (1992) Posttranslational modifications of nerve cytoskeletal proteins in experimental diabetes. Mol Neurobiol 6:225–237

McLean WG, Roberts RE, Mullins FH (1995) Post-translational modifications of microtubule- and growth-associated proteins in nerve regeneration and neuropathy. Biochem Soc Trans 23:76–80

Medori R, Autilio-Gambetti L, Jenich H, Gambetti P (1988) Changes in axon size and slow axonal transport are related in experimental diabetic neuropathy. Neurology 38:597–601

Medori R, Autilio-Gambetti L, Monaco S, Gambetti P (1985) Experimental diabetic neuropathy: impairment of slow transport with changes in axon cross-sectional area. Proc Natl Acad Sci USA 82:7716–7720

Meerwaldt R, Links TP, Graaff R, Hoogenberg K, Lefrandt JD et al. (2005) Increased accumulation of skin advanced glycation end-products precedes and correlates with clinical manifestation of diabetic neuropathy. Diabetologia 48:1637–1644

Misur I, Zarkovic K, Barada A, Batelja L, Milicevic Z et al. (2004) Advanced glycation endproducts in peripheral nerve in type 2 diabetes with neuropathy. Acta Diabetol 41: 158–166

Mohiuddin L, Fernyhough P, Tomlinson DR (1995) Reduced levels of mRNA encoding endoskeletal and growth-associated proteins in sensory ganglia in experimental diabetes. Diabetes 44:25–30

Munch C, Sedlmeier R, Meyer T, Homberg V, Sperfeld AD et al. (2004) Point mutations of the p150 subunit of dynactin (DCTN1) gene in ALS. Neurology 63:724–726

Nangaku M, Sato-Yoshitake R, Okada Y, Noda Y, Takemura R et al. (1994) KIF1B, a novel microtubule plus end-directed monomeric motor protein for transport of mitochondria. Cell 79:1209–1220

Nixon RA (1992) Slow axonal transport. Curr Opin Cell Biol 4:8–14

Nixon RA (1998) The slow axonal transport of cytoskeletal proteins. Curr Opin Cell biol 10:87–92

Pantelidou M, Zographos SE, Lederer CW, Kyriakides T, Pfaffl MW et al. (2007) Differential expression of molecular motors in the motor cortex of sporadic ALS. Neurobiol Dis 26: 577–589

Pekiner C, Cullum NA, Hughes JN, Hargreaves AJ, Mahon J et al. (1993) Glycation of brain actin in experimental diabetes. J Neurochem 61:436–442

Pekiner C, McLean WG (1991) Neurofilament protein phosphorylation in spinal cord of experimentally diabetic rats. J Neurochem 56:1362–1367

Perrot R, Berges R, Bocquet A, Eyer J (2008) Review of the multiple aspects of neurofilament functions, and their possible contribution to neurodegeneration. Mol Neurobiol 38:27–65

Pestronk A, Watson DF, Yuan CM (1990) Neurofilament phosphorylation in peripheral nerve: changes with axonal length and growth state. J Neurochem 54:977–982

Pigino G, Morfini G, Pelsman A, Mattson MP, Brady ST et al. (2003) Alzheimer's presenilin 1 mutations impair kinesin-based axonal transport. J Neurosci 23:4499–4508

Previtali SC, Zerega B, Sherman DL, Brophy PJ, Dina G et al. (2003) Myotubularin-related 2 protein phosphatase and neurofilament light chain protein, both mutated in CMT neuropathies, interact in peripheral nerve. Hum Mol Genet 12:1713–1723

Reynolds AJ, Bartlett SE, Hendry IA (2000) Molecular mechanisms regulating the retrograde axonal transport of neurotrophins. Brain Res Brain Res Rev 33:169–178

Rismanchi N, Soderblom C, Stadler J, Zhu PP, Blackstone C (2008) Atlastin GTPases are required for Golgi apparatus and ER morphogenesis. Hum Mol Genet 17:1591–1604

Roy S, Coffee P, Smith G, Liem RK, Brady ST et al. (2000) Neurofilaments are transported rapidly but intermittently in axons: implications for slow axonal transport. J Neurosci 20: 6849–6861

Ryle C, Donaghy M (1995) Non-enzymatic glycation of peripheral nerve proteins in human diabetics. J Neurol Sci 129:62–68
Ryle C, Leow CK, Donaghy M (1997) Nonenzymatic glycation of peripheral and central nervous system proteins in experimental diabetes mellitus. Muscle Nerve 20:577–584
Saito M, Hayashi Y, Suzuki T, Tanaka.H, Hozumi.I(1997) Linkage mapping of the gene for Charcot-Marie-Tooth disease type 2 to chromosome 1p (CMT2A) and the clinical features of CMT2A. Neurology 49:1630–1635
Schindowski K, Belarbi K, Buee L (2008) Neurotrophic factors in Alzheimer's disease: role of axonal transport. Genes Brain Behav 7(Suppl 1):43–56
Sekido H, Suzuki T, Jomori T, Takeuchi M, Yabe-Nishimura C et al. (2004) Reduced cell replication and induction of apoptosis by advanced glycation end products in rat Schwann cells. Biochem Biophys Res Commun 320:241–248
Sell DR, Nagaraj RH, Grandhee SK, Odetti P, Lapolla A et al. (1991) Pentosidine: a molecular marker for the cumulative damage to proteins in diabetes, aging, and uremia. Diabetes Metab Rev 7:239–251
Sharp GA, Shaw G, Weber K (1982) Immunoelectronmicroscopical localization of the three neurofilament triplet proteins along neurofilaments of cultured dorsal root ganglion neurones. Exp Cell Res 137:403–413
Shin JS, Chung KW, Cho SY, Yun J, Hwang SJ et al. (2008) NEFL Pro22Arg mutation in Charcot-Marie-Tooth disease type 1. J Hum Genet 53:936–940
Shy ME (2004) Charcot-Marie-Tooth disease: an update. Curr Opin Neurol 17:579–585
Sidenius P, Jakobsen J (1981) Retrograde axonal transport. A possible role in the development of neuropathy. Diabetologia 20:110–112
Strom AL, Gal J, Shi P, Kasarskis EJ, Hayward LJ et al. (2008) Retrograde axonal transport and motor neuron disease. J Neurochem 106:495–505
Sugimoto K, Nishizawa Y, Horiuchi S, Yagihashi S (1997) Localization in human diabetic peripheral nerve of N(epsilon)-carboxymethyllysine-protein adducts, an advanced glycation endproduct. Diabetologia 40:1380–1387
Sugimoto K, Yasujima M, Yagihashi S (2008) Role of advanced glycation end products in diabetic neuropathy. Curr Pharm Des 14:953–961
Sullivan KA, Hayes JM, Wiggin TD, Backus C, Oh S et al. (2007) Mouse models of diabetic neuropathy. Neurobiol Dis 28:276–85
Takeuchi M, Bucala R, Suzuki T, Ohkubo T, Yamazaki M et al. (2000) Neurotoxicity of advanced glycation end-products for cultured cortical neurons. J Neuropathol Exp Neurol 59: 1094–1105
Thomas PK, Eliasson S, Dyck PJ, Thomas PK, Lambert EM (1975) Peripheral neuropathy. Saunders, Philadelphia, PA
Thomas PK, Ochoa J (1993) Clinical features and differential diagnosis. In: Dyck PJ, Thomas PK Griffin JW, Low PA, Poduslo JF(ed) Peripheral neuropathy, vol 2. Saunders, Philadelphia, PA, pp 749–774
Thornalley PJ (2002) Glycation in diabetic neuropathy: characteristics, consequences, causes, and therapeutic options. Int Rev Neurobiol 50:37–57
Toth C, Martinez J, Zochodne DW (2007) RAGE, diabetes, and the nervous system. Curr Mol Med 7:766–776
Toth C, Rong LL, Yang C, Martinez J, Song F et al. (2008) Receptor for advanced glycation end products (RAGEs) and experimental diabetic neuropathy. Diabetes 57:1002–1017
Vincent AM, Perrone L, Sullivan KA, Backus C, Sastry AM et al. (2007) Receptor for advanced glycation end products activation injures primary sensory neurons via oxidative stress. Endocrinology 148:548–558
Vitadello M, Filliatreau G, Dupont JL, Hassig R, Gorio A et al. (1985) Altered axonal transport of cytoskeletal proteins in the mutant diabetic mouse. J Neurochem 45:860–868
Vlassara H, Brownlee M, Cerami A (1981) Nonenzymatic glycosylation of peripheral nerve protein in diabetes mellitus. Proc Natl Acad Sci USA 78:5190–5192

Vlassara H, Li YM, Imani F, Wojciechowicz D, Yang Z et al. (1995) Identification of galectin-3 as a high-affinity binding protein for advanced glycation end products (AGE): a new member of the AGE-receptor complex. Mol Med 1:634–646

Vlassara H, Palace MR (2002) Diabetes and advanced glycation endproducts. J Intern Med 251: 87–101

Wada R, Nishizawa Y, Yagihashi N, Takeuchi M, Ishikawa Y et al. (2001) Effects of OPB-9195, anti-glycation agent, on experimental diabetic neuropathy. Eur J Clin Invest 31:513–520

Wada R, Yagihashi S (2005) Role of advanced glycation end products and their receptors in development of diabetic neuropathy. Ann NY Acad Sci 1043:598–604

Wang L, Ho CL, Sun D, Liem RK, Brown A (2000) Rapid movement of axonal neurofilaments interrupted by prolonged pauses. Nat Cell Biol 2:137–141

Williams SK, Howarth NL, Devenny JJ, Bitensky MW (1982) Structural and functional consequences of increased tubulin glycosylation in diabetes mellitus. Proc Natl Acad Sci USA 79:6546–6550

Xia CH, Roberts EA, Her LS, Liu X, Williams DS et al. (2003) Abnormal neurofilament transport caused by targeted disruption of neuronal kinesin heavy chain KIF5A. J Cell Biol 161:55–66

Yagihashi S, Kamijo M, Watanabe K (1990) Reduced myelinated fiber size correlates with loss of axonal neurofilaments in peripheral nerve of chronically streptozotocin diabetic rats. Am J Pathol 136:1365–1373

Yang Y, Allen E, Ding J, Wang W (2007) Giant axonal neuropathy. Cell Mol Life Sci 64:601–609

Zhao C, Takita J, Tanaka Y, Setou M, Nakagawa T et al. (2001) Charcot-Marie-Tooth disease type 2A caused by mutation in a microtubule motor KIF1Bbeta. Cell 105:587–597

Chapter 28
Effects of Insulin on Tau and Neurofilament

R. Schechter and K.E. Miller

Abstract We have demonstrated the synthesis of insulin by neurons within the central nervous system in vivo and in vitro. We showed that neuronal insulin induces neuronal differentiation and axonal growth. In insulin type 1 and type 2 diabetes mellitus, one major complication is nervous system neuropathy. Lack of insulin has been postulated to be the cause of the pathological alterations seen in diabetic neuropathy and diabetes mellitus has also been associated with increased risk for Alzheimer's disease and other types of dementia. We studied the role of insulin in the regulation of neurofilament and tau phosphorylation in vivo and in vitro. Neuronal insulin was shown to regulate neurofilament and tau phosphorylation via mitogen activated protein kinase and c-Jun N-terminal kinase. Lack of neuronal insulin promoted hyperphosphorylation of tau and neurofilament in vivo by increasing c-Jun N-terminal kinase activity and decreasing mitogen activated protein kinase. In vitro, the lack of insulin action or inhibition of mitogen activated protein kinase promoted alterations in the nonphosphorylated and phosphorylated neurofilament within the axons. Thus, insulin needs a constant presence to have a role on neurofilament and tau metabolism. Pancreatic insulin secretion is induced by high blood glucose concentration in blood, has a short half life, and the crossing of the blood brain barrier is regulated by receptors. This would make pancreatic insulin a difficult source to regulate the neuronal cytoskeleton. In conclusion, de novo synthesis of insulin by neurons is the way in which the brain has a proper continual source of insulin. These studies are also in favor of insulin's action in the brain and possible applications in patients with diabetes mellitus and Alzheimer's disease.

Keywords Alzheimer's disease · Brain development · Diabetes mellitus · GSK-3β · Immunohistochemistry · Insulin · Insulin receptor · JNK · MAPK · Neurofilament · Neurofilament phosphorylation · PI-3 K · Tau · Western blots

R. Schechter (✉)
Department of Anatomy and Cell Biology, Oklahoma State University CHS, Tulsa, OK 74107, USA
e-mail: ruben.schechter@oustate.edu

R.A. Nixon, A. Yuan (eds.), *Cytoskeleton of the Nervous System*, Advances in Neurobiology 3, DOI 10.1007/978-1-4419-6787-9_28,

Contents

28.1 Introduction

Insulin is related to cell growth (Clarke et al., 1985; Aizenman and de Vellis, 1987), differentiation (Saneto and deVellis, 1984; Puro and Agardh, 1984), synaptic modulation (Boyd et al., 1985; Catalan et al., 1992), and glucose metabolism (Clarke et al., 1984) in the nervous system. We showed in vivo and in vitro that insulin is produced and secreted within the nervous system by neurons (Schechter et al., 1994, 1996). We have demonstrated that neuronal synthesized insulin [I(n)] promoted axonal length and neurofilament distribution to the axons via mitogen activated protein kinase (MAPK) phosphorylation in fetal neuron cell cultures (Schechter et al., 1999, 1999). In insulin type 1 (insulin dependent) and type 2 (non-insulin dependent) diabetes mellitus, one major complication is nervous system neuropathy (Biessels et al., 1999, 2002). Lack of insulin has been postulated to be the cause of the pathological alterations seen in diabetic neuropathy (Yagihashi et al., 1990; Sugimoto et al., 2000; Singhal et al., 1997; Grant and Pant, 2000).

The central nervous system neuropathy is characterized by decreased conduction velocity with changes in the EEG, cerebral atrophy (global subcortical and cortical atrophy), decrease in cognition, and increased risk for stroke (Biessels et al., 2002; Sharma et al., 2003; Lovestone and Reynolds, 1997). Furthermore, diabetes mellitus has also been associated with increased risk for Alzheimer's disease and other types of dementia (Arvanitakis et al., 2004). Insulin regulates the concentration of neurofilament mRNA in neuroblastoma cells (Wang et al., 1992) and, in animals with diabetes mellitus, neurofilament mRNA and neurofilament content have been shown to be decreased (Yagihashi et al., 1990; McLean, 1997).

Insulin is related to the regulation of tau phosphorylation in cell cultures and the lack of insulin inhibition of glycogen synthase kinase 3 β (GSK-3β) at serine 9 causes hyperphosphorylation of tau. This promotes microtubule depolymerization and possible induction of neurodegenerative diseases such as Alzheimer's (Hong and Lee, 1997). Animal studies may provide important clues into the cause of diabetes mellitus brain complications. Insulin is related to the regulation of tau and neurofilament in vitro and in vivo (Schechter et al., 1998, 2005). Schubert et al. described the hyperphosphorylation of tau at threonine 231 in the brain of neuronal insulin receptor knockout mouse by the inability of the neurons to activate Akt and inhibition of GSK-3β at serine 9 (Schubert et al., 2004). We, in the insulin knockout mice [I (–/–)] that lack insulin and develop hyperglycemia (Schechter et al., 2005), demonstrated the hyperphosphorylation of c-Jun N-terminal kinase (JNK), hyperphosphorylation of tau at threonine 231 and neurofilament, and severe cellular damage (Schechter et al., 2005). In contrast, GSK-3β showed phosphorylation of the inhibitory site, serine 9 and MAPK was decreased in activity. The studies that will be described below aim to demonstrate the role of I(n) in the maintenance of neuronal cytoskeleton, especially tau and neurofilament.

28.2 The Cytoskeleton Role Within the Nervous System

28.2.1 Role of the Neurofilament

Neurofilaments are involved in axonal length growth by stabilizing the cytoskeleton and allowing the continuation of axonal length growth (Lin and Szaro, 1995). Phosphorylated neurofilament contributes to the distribution of neurofilament to the axon and dendrites and to the neurite's stabilization (Schechter et al., 1998; Schechter and Abboud, 2001). Neurofilament distribution and content play a role as a determining factor in normal axonal length and caliber growth (Schechter et al., 1999). and also determine the caliber of the axon (Hoffman et al., 1984; Pant and Veeranna, 1995). The caliber of the axon plays an important role in determining nerve conduction velocity (Hoffman et al., 1987; Pestronk et al., 1990). The importance of neurofilament in determining axonal caliber has been documented in mice lacking different neurofilament subunits and with neurofilament deficiency in quail (Elder et al., 1998; Ohara et al., 1993). The phosphorylation of neurofilament is important for the transportation of the neurofilament throughout the axon and for inducing sidearm formation for neurofilament anchoring, axon extension, and axon caliber (Li et al., 2001). The phosphorylation of the tail domain of NF-M allows for the stabilization of the filament by promoting lateral extension of the sidearm that increases neurofilament spacing, axon caliber, and conduction velocity of the nerve (Li et al., 2001). The phosphorylation of neurofilament is regulated by different kinases. Our interest was to study the mitogen activated protein kinases. Mitogen activated protein kinases are a family of kinases that include mitogen activated protein kinases 1 and 2 (MAPK 1 and 2) and JNK (Fernyhough et al., 1999). These kinases are responsible for inducing different pathways that participate in the normal neuronal function, such as phosphorylation of neurofilament at different sites (Fernyhough et al., 1999; Reynolds et al., 2000). MAPK and JNK are regulated by growth factors such as insulin (Schechter et al., 1998; Desbois-Mouthon et al., 1998; Morino et al., 2001). JNK phosphorylation can be induced by different stress factors, and also has a role in normal microtubule assembly and proliferation (Chang et al., 2003). Neurofilaments may also be involved in axonal length growth by stabilizing the cytoskeleton and allowing the continuation of axonal length growth (Lin and Szaro, 1995). Phosphorylated neurofilament contributes to the distribution of neurofilaments to the axon and dendrites and to the neurite's stabilization (Pant and Veeranna, 1995). Thus, this is the importance in studying I(n) and exogenous insulin's role in neurofilament phosphorylation.

28.2.2 Role of Tau

Tau is part of the microtubule associated protein family. It is found mainly within neurons in the brain of different species including rat, mouse, and human (Buee et al., 2000).

Tau's most important function is to participate in the microtubule assembly (Buee et al., 2000). Microtubules are part of the neuronal cytoskeleton involved in axon caliber and cellular transport. Tau plays a role in neuron growth and differentiation during brain development (Avila et al., 2004). It is present as six different isoforms resulting from mRNA splitting and the isoforms range from 352 to 441 amino acids (Buee et al., 2000). Tau is divided into regions of projection and binding domain (Buee et al., 2000) (Mandelkow et al., 1995). In these domains are specific serine and threonine sites that are phosphorylated during non-pathological or pathological conditions (Buee et al., 2000; Avila et al., 2004). During brain development, several amino acids sites are hyperphosphorylated compared to adult tau and this increase in phosphorylation is lost in adulthood (Schechter et al., 2005). The binding domain has three or four regions that interact within the microtubules to induce filament assembly (Buee et al., 2000). The projection domain interacts with other microtubules to increase the inter-microtubule space and contribute to increased axonal caliber (Buee et al., 2000). Tau action is regulated by different kinases. MAPK, GSK-3β, and JNK are kinases that regulate tau phosphorylation during non-pathological conditions (Fernyhough et al., 1999; Reynolds et al., 2000; Chang et al., 2003; Schechter et al., 2005). The balance of tau phosphorylation status is important for facilitating transport from the neuronal body to the axons and binding of organelles (Mandelkow et al., 1995; Takuma et al., 2003). Deregulation of the phosphorylation of the different sites in tau causes improper microtubule assembly that induces the development of taupathies.

28.3 Insulin Effects Within the Neurofilament and Tau in the Central Nervous System

Possible functions of insulin within the central nervous system (CNS) include cell growth (Clarke et al., 1985; Aizenman and de Vellis, 1987), cell differentiation (Saneto and deVellis, 1984; Puro and Agardh, 1984; Schechter and Abboud, 2001), glucose metabolism (Clarke et al., 1984), synaptic modulation (Boyd et al., 1985; Catalan et al., 1992), trophic factor (Huck, 1983; Aizenman and de Vellis, 1987;

Schechter et al., 1994), feeding behavior regulation (Schwartz et al., 1992), and promoting axonal growth (Schechter et al., 1999). Insulin and insulin mRNA has been found in vivo within the CNS (Havrankova et al., 1978; Duve and Thorpe, 1979; De Pablo et al., 1982; Dorn et al., 1982; Young, 1986; Schechter et al., 1992; Devaskar et al., 1993b, 1994; Schechter et al., 1996). In addition, the insulin receptor has been shown within the fetal and adult nervous system and demonstrated to be functional (Kenner and Heidenreich, 1991; Devaskar et al., 1993a; Schechter et al., 1994). We have shown the presence of insulin in the rabbit brain, which occurs in higher concentrations during the fetal and early postnatal stages (Schechter et al., 1992). We demonstrated the "de novo" synthesis of insulin within the fetal nervous system in vivo (Schechter et al., 1996) and in vitro (Schechter et al., 1988, 1990; Schechter et al., 1994, 1998; Schechter et al., 1999) by showing the presence of preproinsulin I and II mRNA and insulin immunoreaction in the rough endoplasmic reticulum and Golgi apparatus within the fetal rat brain, spinal cord and dorsal root ganglia (Schechter et al., 1996). Secretion of insulin by the neurons was also revealed using Western blot and radioimmunoassay within neuron cell cultures incubated in an insulin-free defined medium (Schechter et al., 1994, 1998; Schechter et al., 1999).

Devaskar et al. showed the presence of the insulin mRNA within the adult rat brain (Devaskar et al., 1993b) and we showed the presence of I(n) immunoreaction in the rough endoplasmic reticulum, Golgi apparatus of neurons and axons within the newborn (Schechter et al., 1996) and adult rat brain (Beju and Schechter, unpublished data), confirming the synthesis of I(n) by the brain. Furthermore, the sequence of the preproinsulin I and II mRNA showed 100% homology with pancreatic insulin (Schechter et al., 1996). Wang et al. showed that in neuroblastoma cells exogenous insulin in physiological concentrations is able to increase neurofilament mRNA (Wang et al., 1992).

Pancreatic insulin has been demonstrated to cross the blood-brain barrier (Baskin et al., 1983; Pardridge et al., 1985; Wallum et al., 1987). Therefore, insulin is present in the CNS as an endogenous and exogenous protein, which may have different actions depending on its origin. It appears that exogenous insulin plays a role in appetite regulation (Schwartz et al., 1992; Schwartz and Porte, 2005), whereas I(n) is involved in the regulation of neurofilament and tau (Schechter et al., 2005), cell growth (Clarke et al., 1985; Aizenman and de Vellis, 1987), cell differentiation (Saneto and deVellis, 1984; Puro and Agardh, 1984; Schechter and Abboud, 2001), glucose metabolism (Clarke et al., 1984), and synaptic modulation (Boyd et al., 1985; Catalan et al., 1992). In addition, insulin is related to the regulation of tau phosphorylation in cell cultures at serine 202 and at threonine 205, 231, and 180 (Hong and Lee, 1997). Insulin regulated the activity of numerous kinases, e.g., ERK 1, 2, JNK 1,2, 3, and GSK-3β (White and Kahn, 1994; Desbois-Mouthon et al., 1998; Schechter et al., 1998; Morino et al., 2001; Schechter et al., 1998, 2005). These kinases are known to regulate the phosphorylation of neurofilament and tau. Insulin is known to phosphorylate GSK-3β at serine 9, thereby inhibiting its function. Insulin via phosphatidylinositol 3 kinase (PI-3 K) inhibits the action of GSK-3β, dephosphorylates tau in neurons derived from a human teratocarcinoma, and stabilizes microtubules (Hong and Lee, 1997). We demonstrated that insulin regulates the phosphorylation of tau in the mouse during brain development (Schechter et al., 2005).

In summary, insulin plays a role in the balance of transport, phosphorylation, and structure of microtubules and neurofilaments. Tau and neurofilament play a role in neuron growth and differentiation during brain development and stability of the neuron in adulthood.

28.4 Lack of Insulin Effects Within the Central Nervous System

In insulin type 1 (insulin dependent) and type 2 (non-insulin dependent) diabetes mellitus, one of the major complications is nervous system neuropathy (Biessels et al., 1999, 2002). Lack of insulin, hyperglycemia, hypoglycemia, and temporal combinations of all three have been postulated to be the cause of the pathological alterations seen in diabetic neuropathy (Yagihashi et al., 1990; Singhal et al., 1997; Grant and Pant, 2000; Sugimoto et al., 2000). The central nervous system neuropathy is characterized by decreased conduction velocity with changes in the EEG, cerebral atrophy (global subcortical and cortical atrophy), decrease in cognition, and increased risk for stroke (Biessels et al., 2002; Sharma et al., 2003). Furthermore, diabetes mellitus has also been associated with increased risk for Alzheimer's disease and other types of dementia (Arvanitakis et al., 2004).

In peripheral diabetic neuropathy, alterations occur in neurofilament, such as hyperphosphorylation, leading to concomitant decrease in nerve diameter and axonal shrinkage (Mohseni et al., 2000; Sugimoto et al., 2000; Liu et al., 2004), decrease in protein transport, and excessive accumulation of neurofilament within the axons (Arvanitakis et al., 2004). Axonal damage may be induced by decrease of neurofilament flow to the axon (Yagihashi et al., 1990) or/and aberrant phosphorylation of neurofilament (Fernyhough et al., 1999), and insulin is capable of improving this pathophysiologic alteration (Singhal et al., 1997; Biessels et al., 1999). The mechanism that insulin uses to promote this improvement is unknown. In addition, insulin is related to the regulation of tau phosphorylation in cell cultures. The lack of insulin inhibition of GSK-3β at serine 9 causes hyperphosphorylation of tau promoting microtubule depolymerization and possible induction of neurodegenerative diseases such as Alzheimer's (Hong and Lee, 1997). Animal studies may provide important clues into the cause of diabetes mellitus brain complications. Schubert et al. described the hyperphosphorylation of tau at threonine 231 in the brain of the neuronal insulin receptor knockout mouse by the inability of neurons to activate Akt and inhibit GSK-3β at serine 9 (Schubert et al., 2004). We, in the insulin knockout mouse [I(–/–)] that lacks insulin and develops hyperglycemia, demonstrated the hyperphosphorylation of JNK, tau at threonine 231 and neurofilament, and severe cellular damage (Schechter et al., 2005). In contrast, GSK-3β showed phosphorylation of the inhibitory site, serine 9, and MAPK was decreased in activity (Schechter et al., 2005) (Fig. 28.1).

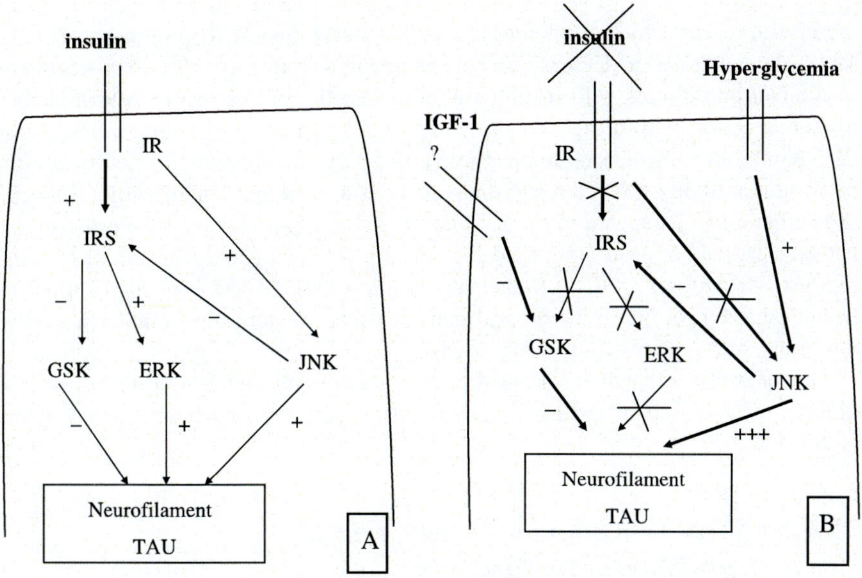

Fig. 28.1 Possible role of neuronal insulin [I(n)] within the brain. **A**) I(n) binding of the neuronal insulin receptor promotes the phosphorylation of insulin receptor substrate 1 (IRS) that activates the MAPK or PI-3 K pathway. The activation of the PI-3 K pathway inhibits GSK-3β (GSK) at serine 9 site. This event inhibits the phosphorylation of tau and neurofilament. MAPK activation phosphorylates tau and neurofilament. JNK may also be phosphorylated by insulin and regulate the phosphorylation of IRS. **B**) The lack of insulin and/or hyperglycemia causes hyperphosphorylation of JNK that in time hyperphosphorylates tau and neurofilament. MAPK is hypophosphorylated, losing control of the balance of the cytoskeleton. GSK, in spite of the lack of insulin, remains inhibited

28.4.1 Effects of Insulin on Non-phosphorylated Neurofilament In Vitro

Neurofilament distribution and content determines the caliber of the axon (Hoffman et al., 1984; Pant and Veeranna, 1995) and axonal caliber is important in establishing nerve conduction velocity (Hoffman et al., 1987)(Pestronk et al., 1990). Neurofilament also induces the longitudinal growth of the axons (Lin and Szaro, 1995; Schechter et al., 1999). Lack of insulin promotes pathological changes in the central and peripheral nervous system by affecting neurofilaments (Fernyhough et al., 1999; Schechter et al., 2005). We studied the effect of inhibiting I(n) in neuron cell cultures derived from 19 days gestational age rat brain incubated in an insulin free medium (IFM) (Schechter et al., 1998). The effects of different treatments were studied using a mouse monoclonal antibody to non-phosphorylated neurofilament (SMI-311) (Schechter et al., 1998, 1999). These studies demonstrated that inhibition of I(n) by anti-insulin antibody decreased the transport of neurofilament to the axon (Schechter et al., 1998), and consequently significantly decreased axonal length (Schechter et al., 1999). Furthermore, inhibition of MAPK by PD98059 or the insulin receptor by isoproterenol (via phosphorylation of Atk) (Schechter et al., 1998) demonstrated a decrease in neurofilament within the axon. Post-treatment of the neuron cultures with insulin was followed by an increase of neurofilament immunoreaction within the axon (Fig. 28.2). The treatment of the neurons with PI-3 K inhibitor resulted in no changes (Fig. 28.2). In addition, I(n) neutralization by insulin antibody showed a significant decrease in axonal length within the neuron cell cultures incubated in the IFM compared to non-treated cells. Furthermore, I(n) was capable of promoting axonal growth beyond exogenous insulin or insulin-like growth factor 1 (IGF-1) (Schechter et al., 1999). MAPK was also shown to be activated by insulin in a temporal manner using Western blots emphasizing the inhibitory studies (Schechter et al., 1998).

Our data show that I(n) within the fetal neuron cell cultures promotes neurofilament distribution to the axon and neurite outgrowth (Schechter et al., 1998, 1999).

28.4.2 Effects of Insulin on Phosphorylated Neurofilament In Vitro

Neurofilaments are involved in growth of axonal length by stabilizing the cytoskeleton and allowing the continuation of axonal lengthening (Lin and Szaro, 1995). Phosphorylated neurofilament contributes to the distribution of neurofilaments to the axon and dendrites and to the neurite's stabilization (Pestronk et al., 1990; Wang et al., 1992). Neurofilament in the phosphorylated state forms sidearms (Li et al., 2001) allowing the neurofilaments to link and determine axonal caliber by increasing the distance between the filaments (Li et al., 2001). The effect of the lack of insulin in inducing neurofilament phosphorylation was studied using inhibitory drugs to MAPK and insulin receptor and binding I(n) with an anti-insulin antibody in neuron cell cultures derived from 19 days gestational age rat brain incubated in

Fig. 28.2 (continued) These microphotographs represent the immunostain with mouse monoclonal anti-pan non-phosphorylated neurofilament antibody (SMI-311) in fetal neuron cell cultures from 19 day gestational age rat brains incubated in IFM (I(n) is produced and secreted by the neurons in this medium). **a** No treatment. **b** High magnification of cells with no treatment. **c** Neurons treated with anti-insulin antibody. **d** Neuron treated with wortmannin. **e** Neurons treated with PD98059, a MAPK inhibitor. **f** Neurons treated with PD98059 followed by 5 ng/mL of insulin for 45 min. **g** Neurons treated with isoproterenol, an insulin receptor inhibitor. **h** Neurons treated with isoproterenol followed by 5 ng/mL of insulin for 45 min. Note the positive diffuse immunoreaction within the axons in the non-treated neurons. The treatment with anti-insulin antibody, PD98059 and isoproterenol inhibited the non-phosphorylated neurofilament immunoreaction within the axons. Addition of insulin promoted non-phosphorylated neurofilament immunoreaction along the axons. The effect of anti-insulin antibody treatment was irreversible. In addition, neurons treated with anti-insulin antibody become swollen. The treatment with PI-3 K inhibitor, wortmannin, did not inhibit the non-phosphorylation neurofilament immunoreaction. Original magnification **a, c–g, h** × 200, **b** × 400

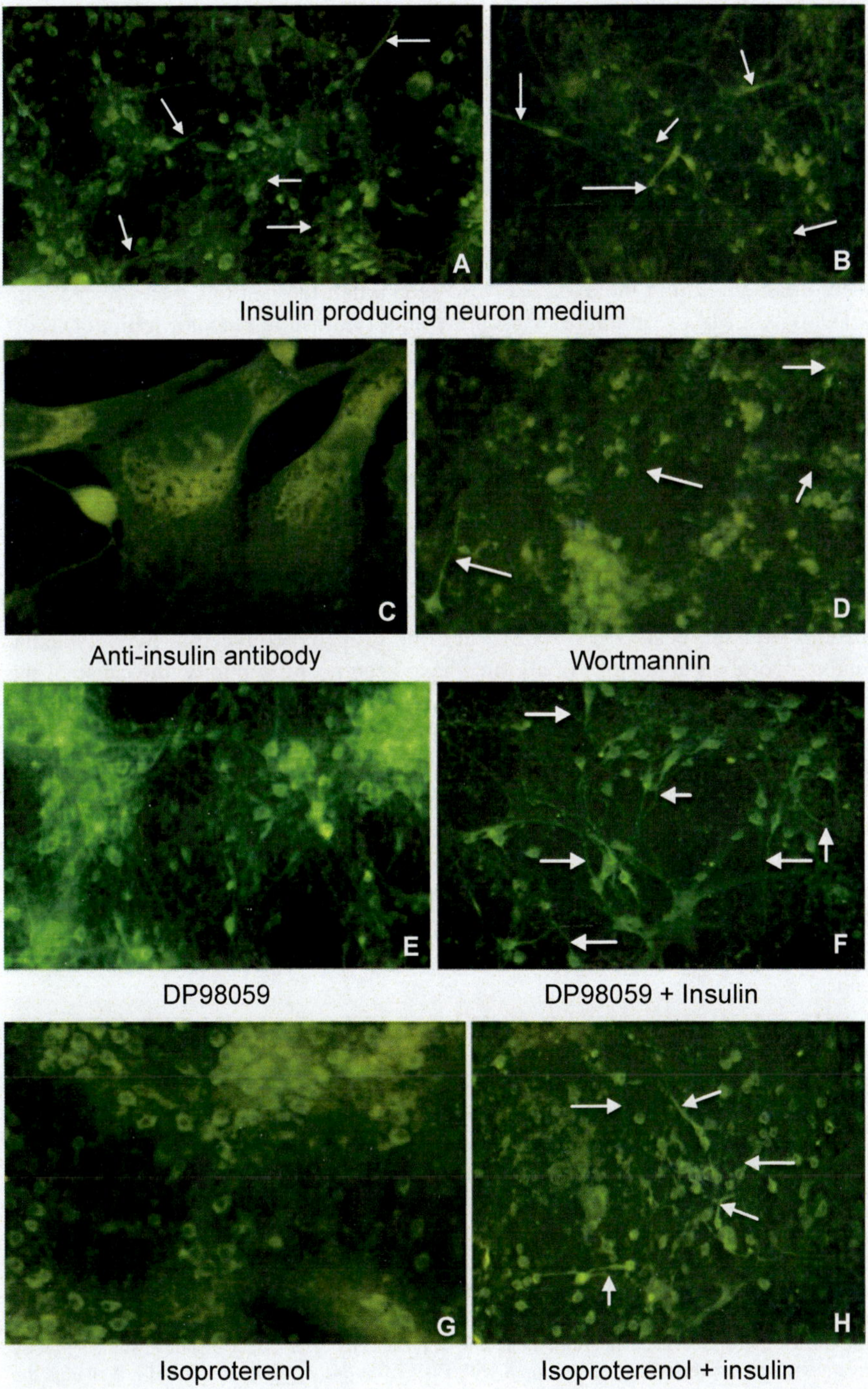

Fig. 28.2 (continued)

an insulin free medium (IFM) (Schechter et al., 1998). The effects of the different treatments were studied using a mouse monoclonal antibody to phosphorylated neurofilament (SMI-312) (Schechter et al., 1998, 1999). These studies demonstrated that phosphorylated neurofilament in the IFM (in which neurons produce and secrete insulin) is present within the neuron body and is seen diffusely within the axon (unpublished data). When the neurons were treated with an anti-insulin antibody that neutralizes I(n), the reaction for phosphorylated neurofilament in the axon became punctuated. This punctuated immunoreaction of the phosphorylated neurofilament within the axon was also seen when the neurons were treated with PD98059, a MAPK inhibitor, or isoproterenol (an insulin receptor inhibitor). Post treatment of the neuron cultures with insulin was followed by the return of a diffuse immunoreaction in the axon (Fig. 28.3) except in the insulin antibody treated neurons.

28.4.3 *Effects of Insulin on Tau and Neurofilament In Vivo*

As mentioned above, in insulin type 1 (insulin dependent) and type 2 (non-insulin dependent) diabetes mellitus, central and peripheral neuropathy are major complications (Biessels et al., 1999, 2002). Lack of insulin, hyperglycemia, hypoglycemia, and temporal combinations of all three have been postulated to be the cause of the pathological alterations seen in diabetes neuropathy (Yagihashi et al., 1990; Singhal et al., 1997; Grant and Pant, 2000; Sugimoto et al., 2000). The central nervous system neuropathy is characterized by decreased conduction velocity with changes in the EEG, cerebral atrophy (global subcortical and cortical atrophy), decrease in cognition, and increased risk for stroke (Lovestone and Reynolds, 1997; Biessels et al., 2002; Sharma et al., 2003). Furthermore, diabetes mellitus has also been associated with increased risk for Alzheimer's disease and other types of dementia (Arvanitakis et al., 2004). In addition, Sharma et al. have described cerebral atrophy in young

Fig. 28.3 (continued) These microphotographs represent the immunostain with mouse monoclonal anti-phosphorylated neurofilament antibody (SMI-312) in fetal neuron cell cultures from 19 day gestational age rat brains incubated in IFM. (I(n) is produced and secreted by the neurons in this medium). **a** No treatment. **b** High magnification of cells with no treatment. **c** Neurons treated with anti-insulin antibody. **d** Neuron treated with wortmannin. **e** Neurons treated with PD98059, a MPAK inhibitor. **f** Neurons treated with PD98059 followed by 5 ng/mL of insulin for 45 min. **g** Neurons treated with isoproterenol, an insulin receptor inhibitor. **h** Neurons treated with isoproterenol followed by 5 ng/mL of insulin for 45 min. Note the positive immunoreaction within the axons in the non treated neuron. The treatment with anti-insulin antibody, PD98059 and isoproterenol induce a loss of the diffuse phosphorylated neurofilament immunoreaction to a punctuated reaction within the axons. Addition of insulin to the treated neurons promoted a restoration of the diffuse phosphorylated neurofilament immunoreactions. The effect of anti-insulin antibody treatment was irreversible. The treatment with PI-3 K inhibitor, wortmannin, did not change the phosphorylated neurofilament immunoreaction. Original magnification **a**, **c–g**, **h** × 200, **b**: × 400

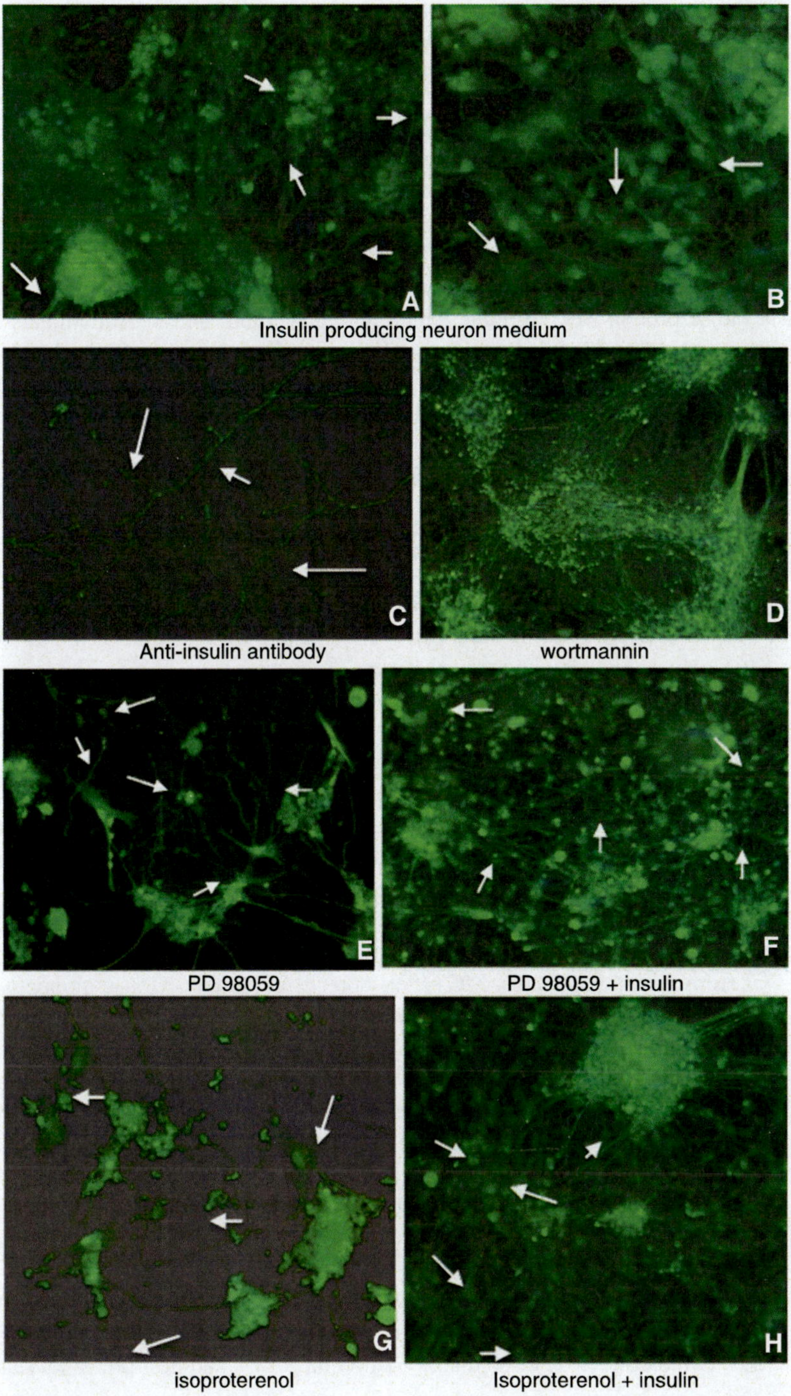

Fig. 28.3 (continued)

patients with type 1 diabetes mellitus who are otherwise healthy (Biessels et al., 2002; Sharma et al., 2003).

Animal studies may provide important clues into the cause of cerebral atrophy. For example, Schubert et al. described the hyperphosphorylation of tau at threonine 231 in the brain of neuronal insulin receptor knockout mouse (Schubert et al., 2004). This effect of insulin resistance was due to the inability of the neurons to activate Akt and inhibition of GSK-3β at serine 9 (Schubert et al., 2004). To approach the question of the effects of the lack of insulin within the brain, we used an insulin knockout mouse that develops diabetes mellitus after the first feeding (Duvillie et al., 1997). Tau, a protein that facilitates microtubule binding, and neurofilaments are present in a phosphorylated status during brain development to facilitate transport from the neuronal body to the axons and dendrites (Hornung and Riederer, 1999; Takuma et al., 2003). Neurofilament and tau are involved in neuronal differentiation and axonal growth during brain development (Schechter et al., 1998, 1999; Grant and Pant, 2000; Schechter and Abboud, 2001). In the newborn insulin knockout mouse hyperglycemia was diagnostic by measuring the glucose in urine (Duvillie et al., 1997; Schechter et al., 2005). The brains from the insulin knockout mice were obtained at 36 h after birth and processed for Western blot and electron microscopy; wild type mice were used for controls (Schechter et al., 2005). We studied neurofilament and tau phosphorylation status and the phosphorylation level of the kinases known to phosphorylate tau and neurofilament (Schechter et al., 2005). Tau is phosphorylated at different sites during the developmental stages studied (Lovestone and Reynolds, 1997; Takuma et al., 2003).

Western blot analysis was performed employing specific antibodies to phosphorylated sites of tau, neurofilament, MAPK, JNK, and GSK-3β, as well as antibodies to non-phosphorylated tau and neurofilament. Antibodies against phosphorylated tau were: mouse monoclonal AT-180 (threonine 231), 12E8 (serine 262, 356), a gift from Elan Pharm (Dublin, Ireland), AT-8 (serine 202, threonine 205), AT-270 (threonine 181) from Innogenetics (Ghent, Belgium), a gift from Elan Pharm (Dublin, Ireland), and rabbit anti-Tau-1 that recognizes non-phosphorylated tau at epitopes 189–207 from Sigma (St. Louis, MO) (Schechter et al., 2005). Mouse monoclonal anti-medium molecular weight neurofilament RMO-281 that recognizes phosphorylated neurofilament was from Zymed (San Francisco, CA). Neuronal insulin was recognized using a guinea pig anti-porcine insulin antibody from Linco (St. Louis, MO) (Schechter et al., 2005). The kinases were studied using rabbit anti-JNK to threonine 183 and tyrosine 185 (New England Biolabs, Beverly, MA), rabbit anti-GSK-3β serine 9 (New England Biolabs, Beverly, MA), mouse monoclonal anti-phosphotyrosine GSK α (tyrosine 279) and β (tyrosine 216) (UBI, Lake Placid, NY), and rabbit anti-active MAPK that recognizes ERK 1 and 2 (Promega, Madison, WI) (Schechter et al., 2005).

We studied tau phosphorylation in the insulin knockout brains at epitopes sites that are known to be phosphorylated in neurodegenerative diseases: threonine 231, serine 262, 356, serine 202, threonine 205, threonine 181, and non-phosphorylated tau at epitopes 189–207, sites that correspond to the big tau amino acids sequence (Lovestone and Reynolds, 1997). The Western blots of the brain homogenates showed that tau was only significantly hyperphosphorylated at threonine 231, whereas the rest of the sites were not hyperphosphorylated when compared to wild type mice (Schechter et al., 2005). Total tau showed no significant increase as recognized by Tau 1 antibody (Schechter et al., 2005). These data are important because hyperphosphorylated threonine at 231 site can become a marker for brain insults like in diabetes mellitus, ischemia, Alzheimer's disease, or other neurodegenerative diseases (Schechter et al., 2005). Buerger et al. demonstrated hyperphosphorylation of threonine 231 in tau in spinal fluid of Alzheimer's patients, suggesting that this site is an early marker for Alzheimer's disease (Buerger et al., 2002). Schubert et al. showed that tau was hyperphosphorylated at serine 202 in the insulin receptor substrate-2 knockout mouse (Schubert et al., 2003) and at threonine 231 in brain/neuron-specific insulin receptor knockout (NIRKO) (Schubert et al., 2004). Furthermore, the authors showed, in both studies, that GSK-3β was phosphorylated at serine 9, the inhibitory site of the kinase. The studies by Schubert et al. related to insulin resistance showed differences within the capacity of insulin to stimulate the insulin receptor. Our data fits most closely to the NIRKO animals, where tau is hyperphosphorylated in threonine 231 and GSK-3β is phosphorylated at serine 9, but with the difference in that GSK-3β at serine 9 site is hyperphosphorylated (Schechter et al., 2005). By further investigating the three kinases involved in tau phosphorylation, we showed that JNK in the two isoforms present in the brain (46,000 and 54,000 molecular weight) were significantly hyperphosphorylated; the 46,000 molecular weight been highly phosphorylated compared to the 54,000 molecular weight (Schechter et al., 2005). MAPK 1 phosphorylation in the insulin knockout mouse showed significant decrease when compared to the wild type, whereas MAPK 2 showed no changes (Schechter et al., 2005).

We also studied the status of medium molecular weight neurofilament (NF-M) (160,000 molecular weight). Studies in our laboratory demonstrated the absence of high molecular weight neurofilament, corresponding with the age of the animals. The NF-M was found to be hyperphosphorylated using a phosphorylated dependent antibody (ROM 281) that recognizes the tail domain of the NF-M (Schechter et al., 2005). This domain of the neurofilament facilitates the formation of sidearms that induce the space between the neurofilaments (Li et al., 2001). The data of the hyperphosphorylation of tau and NF-M in Western blots caused us to study possible ultrastructural damage provoked by this event.

Ultrastructural investigation revealed swollen organelles: mitochondria, endoplasmic reticulum and Golgi apparatus, and a disorganized nucleus (Schechter et al., 2005). Microtubules and neurofilaments showed increased spacing and decreased number (Schechter et al., 2005). The electron microscopy studies are in agreement with studies of cells in stressful environments such as the lack of insulin and hyperglycemia.

28.5 Conclusion and Perspective

Our data reveal that neuronal insulin's role within the brain is to promote neuronal differentiation by balancing the phosphorylation of neurofilament and tau. This role promotes axonal growth and cell differentiation. Lack of insulin disrupts this balance, inducing severe cellular alterations that are acute in the newborn mouse. There is, however, the possibility that these events may occur slowly in humans with diabetes mellitus, especially in young patients where the central nervous system is still in development. We believe that these data are in favor of the role of neuron producing insulin. Pancreatic insulin is secreted only by high blood glucose concentration, has a short half life, and the crossing of the blood brain barrier is regulated by receptors. This would make pancreatic insulin a difficult source to maintain the balance of the cytoskeleton. In conclusion, de novo synthesis of insulin by neurons is the way in which the brain has a proper continual source of insulin.

These studies also are in favor of continued research of insulin's action in the brain and possible applications in patient with diabetes mellitus and Alzheimer's disease.

References

Aizenman Y, de Vellis J (1987) Brain neurons develop in a serum and glial free environment: effects of transferrin, insulin, insulin-like growth factor-I and thyroid hormone on neuronal survival, growth and differentiation. Brain Res 406:32–42

Arvanitakis Z, Wilson RS, Bienias JL, Evans DA, Bennett DA (2004) Diabetes mellitus and risk of Alzheimer disease and decline in cognitive function. Arch Neurol 61:661–666

Avila J, Lucas JJ, Perez M, Hernandez F (2004) Role of tau protein in both physiological and pathological conditions. Physiol Rev 84:361–384

Baskin DG, Woods SC, West DB, van Houten M, Posner BI, Dorsa DM, Porte D Jr (1983) Immunocytochemical detection of insulin in rat hypothalamus and its possible uptake from cerebrospinal fluid. Endocrinology 113:1818–1825

Biessels GJ, Cristino NA, Rutten GJ, Hamers FP, Erkelens DW, Gispen WH (1999) Neurophysiological changes in the central and peripheral nervous system of streptozotocin-diabetic rats. course of development and effects of insulin treatment. Brain 122(4): 757–768

Biessels GJ, van der Heide LP, Kamal A, Bleys RL, Gispen WH (2002) Ageing and diabetes: implications for brain function. Eur J Pharmacol 441:1–14

Boyd FT Jr, Clarke DW, Muther TF, Raizada MK (1985) Insulin receptors and insulin modulation of norepinephrine uptake in neuronal cultures from rat brain. J Biol Chem 260:15880–15884

Buee L, Bussiere T, Buee-Scherrer V, Delacourte A, Hof PR (2000) Tau protein isoforms, phosphorylation and role in neurodegenerative disorders. Brain Res Brain Res Rev 33:95–130

Buerger K, Teipel SJ, Zinkowski R, Blennow K, Arai H, Engel R, Hofmann-Kiefer K, McCulloch C, Ptok U, Heun R, Andreasen N, DeBernardis J, Kerkman D, Moeller H, Davies P, Hampel H (2002) CSF tau protein phosphorylated at threonine 231 correlates with cognitive decline in MCI subjects. Neurology 59:627–629

Catalan RE, Martinez AM, Aragones MD, Fernandez I, Miguel BG (1992) Inhibitory effect of insulin and cytoplasmic factor(s) on brain (Na(+) + K+) ATPase. Neurosci Res 13:139–145

Chang L, Jones Y, Ellisman MH, Goldstein LS, Karin M (2003) JNK1 is required for maintenance of neuronal microtubules and controls phosphorylation of microtubule-associated proteins. Dev Cell 4:521–533

Clarke DW, Boyd FT Jr, Kappy MS, Raizada MK (1984) Insulin binds to specific receptors and stimulates 2-deoxy-D-glucose uptake in cultured glial cells from rat brain. J Biol Chem 259:11672–11675

Clarke DW, Boyd FT Jr, Kappy MS, Raizada MK (1985) Insulin stimulates macromolecular synthesis in cultured glial cells from rat brain. Am J Physiol 249:C484–C489

De Pablo F, Roth J, Hernandez E, Pruss RM (1982) Insulin is present in chicken eggs and early chick embryos. Endocrinology 111:1909–1916

Desbois-Mouthon C, Blivet-Van Eggelpoel MJ, Auclair M, Cherqui G, Capeau J, Caron M (1998) Insulin differentially regulates SAPKs/JNKs and ERKs in CHO cells overexpressing human insulin receptors. Biochem Biophys Res Commun 243:765–770

Devaskar SU, Giddings SJ, Rajakumar PA, Carnaghi LR, Menon RK, Zahm DS (1994) Insulin gene expression and insulin synthesis in mammalian neuronal cells. J Biol Chem 269: 8445–8454

Devaskar SU, Sadiq HF, Holtzclaw L, George M (1993a) The developmental pattern of rabbit brain insulin and insulin-like growth factor receptor expression. Brain Res 605:101–109

Devaskar SU, Singh BS, Carnaghi LR, Rajakumar PA, Giddings SJ (1993b) Insulin II gene expression in rat central nervous system. Regul Pept 48:55–63

Dorn A, Bernstein HG, Rinne A, Hahn HJ, Ziegler M (1982) Insulin-like immunoreactivity in the human brain – a preliminary report. Histochemistry 74:293–300

Duve H, Thorpe A (1979) Immunofluorescent localization of insulin-like material in the median neurosecretory cells of the blowfly, Calliphora vomitoria (Diptera). Cell Tissue Res 200: 187–191

Duvillie B, Cordonnier N, Deltour L, Dandoy-Dron F, Itier JM, Monthioux E, Jami J, Joshi RL, Bucchini D (1997) Phenotypic alterations in insulin-deficient mutant mice. Proc Natl Acad Sci USA 94:5137–5140

Elder GA, Friedrich VL Jr, Bosco P, Kang C, Gourov A, Tu PH, Lee VM, Lazzarini RA (1998) Absence of the mid-sized neurofilament subunit decreases axonal calibers, levels of light neurofilament (NF-L), and neurofilament content. J Cell Biol 141:727–739

Fernyhough P, Gallagher A, Averill SA, Priestley JV, Hounsom L, Patel J, Tomlinson DR (1999) Aberrant neurofilament phosphorylation in sensory neurons of rats with diabetic neuropathy. Diabetes 48:881–889

Grant P, Pant HC (2000) Neurofilament protein synthesis and phosphorylation. J Neurocytol 29:843–872

Havrankova J, Schmechel D, Roth J, Brownstein M (1978) Identification of insulin in rat brain. Proc Natl Acad Sci USA 75:5737–5741

Hoffman PN, Cleveland DW, Griffin JW, Landes PW, Cowan NJ, Price DL (1987) Neurofilament gene expression: a major determinant of axonal caliber. Proc Natl Acad Sci USA 84: 3472–3476

Hoffman PN, Griffin JW, Price DL (1984) Control of axonal caliber by neurofilament transport. J Cell Biol 99:705–714

Hong M, Lee VM (1997) Insulin and insulin-like growth factor-1 regulate tau phosphorylation in cultured human neurons. J Biol Chem 272:19547–19553

Hornung JP, Riederer BM (1999) Medium-sized neurofilament protein related to maturation of a subset of cortical neurons. J Comp Neurol 414:348–360

Huck S (1983) Serum-free medium for cultures of the postnatal mouse cerebellum: only insulin is essential. Brain Res Bull 10:667–674

Kenner KA, Heidenreich KA (1991) Insulin and insulin-like growth factors stimulate in vivo receptor autophosphorylation and tyrosine phosphorylation of a 70 K substrate in cultured fetal chick neurons. Endocrinology 129:301–311

Li BS, Daniels MP, Pant HC (2001) Integrins stimulate phosphorylation of neurofilament NF-M subunit KSP repeats through activation of extracellular regulated-kinases (Erk1/Erk2) in cultured motoneurons and transfected NIH 3T3 cells. J Neurochem 76:703–710

Lin W, Szaro BG (1995) Neurofilaments help maintain normal morphologies and support elongation of neurites in Xenopus laevis cultured embryonic spinal cord neurons. J Neurosci 15:8331–8344

Liu Q, Xie F, Siedlak SL, Nunomura A, Honda K, Moreira PI, Zhua X, Smith MA, Perry G (2004) Neurofilament proteins in neurodegenerative diseases. Cell Mol Life Sci 61:3057–3075

Lovestone S, Reynolds CH (1997) The phosphorylation of tau: a critical stage in neurodevelopment and neurodegenerative processes. Neuroscience 78:309–324

Mandelkow EM, Biernat J, Drewes G, Gustke N, Trinczek B, Mandelkow E (1995) Tau domains, phosphorylation, and interactions with microtubules. Neurobiol Aging 16:355–362; discussion 362–353

McLean WG (1997) The role of axonal cytoskeleton in diabetic neuropathy. Neurochem Res 22:951–956

Mohseni S, Lillesaar C, Theodorsson E, Hildebrand C (2000) Hypoglycaemic neuropathy: occurrence of axon terminals in plantar skin and plantar muscle of diabetic BB/Wor rats treated with insulin implants. Acta Neuropathol 99:257–262

Morino K, Maegawa H, Fujita T, Takahara N, Egawa K, Kashiwagi A, Kikkawa R (2001) Insulin-induced c-Jun N-terminal kinase activation is negatively regulated by protein kinase C delta. Endocrinology 142:2669–2676

Ohara O, Gahara Y, Miyake T, Teraoka H, Kitamura T (1993) Neurofilament deficiency in quail caused by nonsense mutation in neurofilament-L gene. J Cell Biol 121:387–395

Pant HC, Veeranna (1995) Neurofilament phosphorylation. Biochem Cell Biol 73:575–592

Pardridge WM, Eisenberg J, Yang J (1985) Human blood-brain barrier insulin receptor. J Neurochem 44:1771–1778

Pestronk A, Watson DF, Yuan CM (1990) Neurofilament phosphorylation in peripheral nerve: changes with axonal length and growth state. J Neurochem 54:977–982

Puro DG, Agardh E (1984) Insulin-mediated regulation of neuronal maturation. Science 225: 1170–1172

Reynolds CH, Betts JC, Blackstock WP, Nebreda AR, Anderton BH (2000) Phosphorylation sites on tau identified by nanoelectrospray mass spectrometry: differences in vitro between the mitogen-activated protein kinases ERK2, c-Jun N-terminal kinase and P38, and glycogen synthase kinase-3beta. J Neurochem 74:1587–1595

Saneto RP, deVellis J (1984) Hormonal regulation of the proliferation and differentiation of astrocytes and oligodendrocytes in primary cultures. In: Schubert D (ed) Developmental biology of cultured nerve, muscle and glia, vol 4. John Wiley & Sons, New York, NY, pp 125–127

Schechter R, Abboud M (2001) Neuronal synthesized insulin roles on neural differentiation within fetal rat neuron cell cultures. Brain Res Dev Brain Res 127:41–49

Schechter R, Abboud M, Johnson G (1999) Brain endogenous insulin effects on neurite growth within fetal rat neuron cell cultures. Brain Res Dev Brain Res 116:159–167

Schechter R, Beju D, Gaffney T, Schaefer F, Whetsell L (1996) Preproinsulin I and II mRNAs and insulin electron microscopic immunoreaction are present within the rat fetal nervous system. Brain Res 736:16–27

Schechter R, Beju D, Miller KE (2005) The effect of insulin deficiency on tau and neurofilament in the insulin knockout mouse. Biochem Biophys Res Commun 334:979–986

Schechter R, Holtzclaw L, Sadiq F, Kahn A, Devaskar S (1988) Insulin synthesis by isolated rabbit neurons. Endocrinology 123:505–513

Schechter R, Sadiq HF, Devaskar SU (1990) Insulin and insulin mRNA are detected in neuronal cell cultures maintained in an insulin-free/serum-free medium. J Histochem Cytochem 38: 829–836

Schechter R, Whitmire J, Holtzclaw L, George M, Harlow R, Devaskar SU (1992) Developmental regulation of insulin in the mammalian central nervous system. Brain Res 582:27–37

Schechter R, Whitmire J, Wheet GS, Beju D, Jackson KW, Harlow R, Gavin JR 3rd (1994) Immunohistochemical and in situ hybridization study of an insulin-like substance in fetal neuron cell cultures. Brain Res 636:9–27

Schechter R, Yanovitch T, Abboud M, Johnson G 3rd, Gaskins J (1998) Effects of brain endogenous insulin on neurofilament and MAPK in fetal rat neuron cell cultures. Brain Res 808:270–278

Schubert M, Brazil DP, Burks DJ, Kushner JA, Ye J, Flint CL, Farhang-Fallah J, Dikkes P, Warot XM, Rio C, Corfas G, White MF (2003) Insulin receptor substrate-2 deficiency impairs brain growth and promotes tau phosphorylation. J Neurosci 23:7084–7092

Schubert M, Gautam D, Surjo D, Ueki K, Baudler S, Schubert D, Kondo T, Alber J, Galldiks N, Kustermann E, Arndt S, Jacobs AH, Krone W, Kahn CR, Bruning JC (2004) Role for neuronal insulin resistance in neurodegenerative diseases. Proc Natl Acad Sci USA 101:3100–3105

Schwartz MW, Figlewicz DP, Baskin DG, Woods SC, Porte D Jr (1992) Insulin in the brain: a hormonal regulator of energy balance. Endocr Rev 13:387–414

Schwartz MW, Porte D Jr (2005) Diabetes, obesity, and the brain. Science 307:375–379

Sharma J, Bakshi R, Lee D, Hachinski V, Chan. RKT (2003) Cerebral atrophy in young, otherwise healthy patients with type 1 diabetes mellitus. Neurology (suppl) 35:3–29

Singhal A, Cheng C, Sun H, Zochodne DW (1997) Near nerve local insulin prevents conduction slowing in experimental diabetes. Brain Res 763:209–214

Sugimoto K, Murakawa Y, Zhang W, Xu G, Sima AA (2000) Insulin receptor in rat peripheral nerve: its localization and alternatively spliced isoforms. Diabetes Metab Res Rev 16:354–363

Takuma H, Arawaka S, Mori H (2003) Isoforms changes of tau protein during development in various species. Brain Res Dev Brain Res 142:121–127

Wallum BJ, Taborsky GJ Jr, Porte D Jr, Figlewicz DP, Jacobson L, Beard JC, Ward WK, Dorsa D (1987) Cerebrospinal fluid insulin levels increase during intravenous insulin infusions in man. J Clin Endocrinol Metab 64:190–194

Wang C, Li Y, Wible B, Angelides KJ, Ishii DN (1992) Effects of insulin and insulin-like growth factors on neurofilament mRNA and tubulin mRNA content in human neuroblastoma SH-SY5Y cells. Brain Res Mol Brain Res 13:289–300

White MF, Kahn CR (1994) The insulin signaling system. J Biol Chem 269:1–4

Yagihashi S, Kamijo M, Watanabe K (1990) Reduced myelinated fiber size correlates with loss of axonal neurofilaments in peripheral nerve of chronically streptozotocin diabetic rats. Am J Pathol 136:1365–1373

Young WS 3rd (1986) Periventricular hypothalamic cells in the rat brain contain insulin mRNA. Neuropeptides 8:93–97

Chapter 29
Ethanol Effects on the Cytoskeleton of Nerve Tissue Cells

Sergio G. Evrard and Alicia Brusco

Abstract Ethanol (EtOH) is the most ancient drug and one of those most used and abused by human beings. Although its effects are better known and are best studied on the cytoskeleton of hepatocytes (due to the EtOH role in the etiology of liver cirrhosis), its effects on nerve tissue cells' cytoskeleton are beginning to be elucidated. We review in this chapter the known mechanisms by which EtOH affects microtubules, intermediate filaments, and microfilaments. EtOH disrupts the cytoskeleton in many different and still incompletely known ways (spontaneous chemical reactions with derivatives of EtOH metabolism, changes in the expression level of cytoskeletal proteins, functional modification of proteins with regulatory actions on cytoskeleton assembly or disassembly, etc.). EtOH's deleterious effects, by acting on neuronal and glial cytoskeleton during embryonic and/or fetal development, are a critical factor in the induction of alcohol-related neurodevelopmental disorders (ARND) and fetal alcohol syndrome (FAS). Its cytoskeletal actions during adulthood are also critical in the induction of different neuropsychiatric disorders (such as alcoholic dementia). However, existing evidence points to the fact that EtOH disruption of cytoskeleton in neurons is shared with the effects of other noxious stimuli (hypoxia-asphyxia, traumatic brain injuries, electrical discharges, neurotoxicants, other drugs of abuse, neurodegenerative disorders, etc.). We propose here that a presently forgotten and disregarded (but classically described) form of chronic neuronal disease (*Zellschrumpfung*) is an ancient morphological evidence of neuronal cytoskeletal involvement in the action of those many noxious stimuli. Taking into account all the existing present and historical evidence, we may conclude that the damage to the cytoskeleton is a common final pathway for neuronal damage.

Keywords Alcohol · Alcoholism · Alcohol-related neurodevelopmental disorders · Cytoskeleton · Dark neuron · Ethanol · Fetal alcohol spectrum disorders · Fetal alcohol syndrome · Glial fibrillary acidic protein · Intermediate filaments · MAP-2 ·

A. Brusco (✉)
Instituto de Biología Celular y Neurociencias "Prof. Eduardo De Robertis", Facultad de Medicina, Universidad de Buenos Aires, Paraguay 2155 3er piso, (C1121ABG) Buenos Aires, Argentina
e-mail: hbrusco@fmed.uba.ar

R.A. Nixon, A. Yuan (eds.), *Cytoskeleton of the Nervous System*, Advances in Neurobiology 3, DOI 10.1007/978-1-4419-6787-9_29,

Microfilaments · Microtubules · Microtubule associated proteins · Moniliform state of neurons · Neurofilaments · Vimentin · Zellschrumpfung

Contents

29.1 Introduction

Without a proper skeleton, a mammal would be nothing but an amorphous jelly mass unable to perform any type of goal-directed activities in its terrestrial environment. Each cell constituting the mammalian organism depends on a proper cytoskeleton in order to acquire the optimal morphology, which will allow it to perform its specific activities.

One of the distinctive features of the nervous system is that all cell types, both neurons and glia, display highly specific and complex morphologies upon which their functions critically depend. Consider, for example, the very long axonal process involving the spinal cord α-motoneurons, or the highly developed dendritic tree of the Purkinjě cells in the cerebellar cortex (which can receive up to 500,000 synaptic contacts!), or the numerous cellular processes of the oligodendrocytes in the optic nerve, each able to ensheath up to 50 different axons in order to provide them with internodal myelin segments, or the profuse arborization of the protoplasmatic astrocytes in the cellularly crowed cortical gray matter, isolating neurons in such a fashion that they do not maintain intimate contact except for at the synapses. Consider also that the axonal speed of conduction of the action potential directly depends on axon diameter and the latter directly depends, in turn, on the number and composition of its neurofilaments. The neuronal and glial cytoskeleton is the cytosolic component that provides the cellular scaffolding upon which depends the morphology of all those cells. Thus, the cytoskeleton is important in every mammalian cell, but in the nervous tissue it is deadly important.

Of all the psychotropic drugs human beings use or abuse, ethanol (EtOH) is the most ancient and, nowadays, the second most used (only surpassed by caffeine). The effects of EtOH on different organs and behaviors have been known for a long time, from both lay and scientific viewpoints (see for example, Huss, 1852; Korsakoff, 1890; Magnan, 1871; Marchiafava and Bignami, 1903; Wernicke, 1881). Generally speaking, there is a lot of available information about the effects of EtOH, mainly on liver metabolism and on different aspects of the central nervous system (CNS) function. The effects of EtOH on the cytoskeleton, in particular, are beginning to be known in hepatocytes. However, surprisingly, and contrary to expectations, there is not a great bulk of available knowledge regarding EtOH and the cytoskeleton in the CNS. There has been almost no scientific works conducted in order to evaluate specifically the cytoskeleton in the CNS but only secondary work in relation to other aspects of CNS physiology and/or pathology.

Discussion on the effects of EtOH on different cytoskeletal components will be organized here relative to its main components, i.e., microtubules (MT, called neurotubules in neurons), and its main associated proteins (MAPs), intermediate filaments (IF), and microfilaments (MF). The separate consideration of these elements is only for didactic reasons but we should always take into account that the three cytoskeletal elements not only interact with each other but also are inseparable – they really form a unity. We will not review here the structure of each component of the normal cytoskeleton since this is outside the scope of this chapter, but we refer interested readers to the many and very good existing books on molecular cell biology (see, for example, Karp, 2007 or Lodish et al., 2007) and some reviews on specific issues (see below).

A common and generally known mechanism for cytoskeletal alteration is that observed under conditions of cellular oxidative stress. At least four mechanisms are involved in the oxidative stress-induced cytoskeletal damage: ATP depletion, altered phosphorylation of cytoskeletal proteins, Ca^{2+}-dependent dissociative and degradative reactions, and thiol oxidation (Bellomo and Mirabelli, 1992). On the other hand, EtOH is a long known source of cellular oxidative stress (for recent reviews, see Albano, 2006; Das and Vasudevan, 2007). Although not much explored in nerve tissue specifically, this physiopathogenic connection (EtOH–oxidative stress–cytoskeletal damage) does exist and we must keep it in mind during the rest of the present discussion.

Besides the specific molecular effects of EtOH on the cytoskeleton, we should bear in mind here, as a general principle, that these effects will probably vary with the age of the animals subjected to EtOH exposure (whether it be prenatal, neonatal, infantile, adolescent, adult, or old-age), with the sex of the animal, with the levels of blood ethanol concentrations (BECs) that could be eventually reached (high, medium, or low levels), with the specific region of the brain that was analyzed, and with the pattern of intoxication and the moment of evaluation (acute or chronic, during intoxication, or after withdrawal). This is, perhaps, one of the reasons why one can sometimes find different and even contradictory or completely opposed findings in the literature.

29.2 Ethanol Effects on Microtubules and Associated Proteins

At present, the intimate molecular mechanisms involved in EtOH action on MTs are beginning to be elucidated, mainly in hepatocytes or in hepatocyte-derived cell lines. Probably the first review of cytoskeletal pathology induced by EtOH is one from French et al. (1980) but it was exclusively centered on hepatocyte cytoskeleton. Furthermore, by the time it was written, many of the cytoskeletal-associated proteins that we now know had not yet been discovered. However, the review by French et al. (1980) is very useful from many aspects, mainly on the ultrastructure of Mallory bodies.

In comparison with that of the liver, almost nothing is known regarding EtOH effects on neurons and/or glia MTs. However, some indications exist of what may occur in the EtOH-exposed CNS. These indications come from experimental studies carried out mainly on liver and hepatocytes, and to a lesser extent from work on intestinal epithelium and nerve tissue.

The microtubule-associated proteins (MAPs) are a group of several types of proteins with MT-binding capacity such as MT-based motor proteins (kinesins and dyneins), MT-plus-end-binding protein, centrosome-associated proteins, enzymatically active MAPs, and structural MAPs. The group of structural MAPs includes proteins such as severing proteins (e.g., katanin-p60, spastin), catastrophe factors (e.g., stathmin), and MT assembly-promoting proteins generically referred to as MT-associated proteins (MAPs, e.g., those of the MAP-2/Tau family and doublecortin, among others). These latter proteins seem to modulate polymerization, stability, and arrangement of cytoplasmatic MTs and play a pivotal role in dendritic branching by directly linking MTs to the actin cytoskeleton (for useful reviews on MAPs see Dehmelt and Halpain, 2005; Garrett and Kapoor, 2003; Sánchez et al., 2000). The group of MAPs is a very crowded one. The most studied proteins of those collectively called MAPs are of the MAP-2/Tau family. Unfortunately, regarding EtOH effects on them, there has not been much research made to date.

MAP-2 is the major MAP in brain. It is almost exclusively expressed in dendrites and neuronal soma. Its expression coincides with neurite outgrowth and dendritic branching; it is critical for dendritogenesis and plays a pivotal role in dendrite branching by directly linking MTs to the actin MFs (Georges et al., 2008). Consequently, this protein is used as a dendritic marker and as an index of dendrite level of development and health in order to assess dendritic density and thus the extension of synaptic circuitry in the belief that, the higher the expression of MAP-2, the higher the dendritic density and the higher the complexity of possible cognitive and behavioral performances (Di Stefano et al., 2006; Mukaetova-Ladinska et al., 2004; Whitaker-Azmitia et al., 1997). MAP-2 function is tightly and finely regulated by its phosphorylation level; complex kinase and phosphatase enzymatic pathways are involved in this process (see Sánchez et al., 2000 for a comprehensive review).

29.2.1 Microtubules (MTs)

After chronic EtOH consumption, a delay in the secretory activity of hepatocytes has been associated with impairment of MT-based vesicular transport. In the liver of adult male Sprague-Dawley rats chronically exposed (8–17 weeks) to a liquid diet containing 36% of EtOH-derived calories (EDC), Yoon et al. (1998) have found that most of the hepatocytes' tubulin mRNA and tubulin protein levels were unchanged after treatment. MTs exhibited normal rates of assembly and showed normal length and structure. However, a smaller but significant portion of the tubulin was "EtOH-sensitive" and assembly incompetent. EtOH is enzymatically oxidized into acetaldehyde. This molecule is highly reactive and spontaneously forms both stable and unstable (mainly Schiff bases) adducts with the free ϵ-amino group of α-tubulin lysyl residues (Tuma et al., 1991). Yoon et al. (1998) found that acetaldehyde, in substoichiometric concentrations, significantly inhibits tubulin polymerization into functional MTs by binding the soluble monomers. Free tubulin binds $\sim$20% more acetaldehyde than MT-polymerized tubulin and it has been established that an adduct on only 1 of 20 tubulin dimers is sufficient to inhibit MT polymerization (Tuma et al., 1991; Yoon et al., 1998).

In a hepatoma hybrid cell line (WIF-B cells) it was found in vitro that, when the cells were cultured in a medium containing a concentration of 50 mM EtOH ($\sim$230 mg/dL), they showed a threefold increase in acetylation of α-tubulin and impaired MT polymerization but that, once the MTs were formed, hyperstabilization of them was noted (Kannarkat et al., 2006). Another study carried out with the same type of cells (WIF-B) Joseph et al. (2008) showed that specific pathways of protein trafficking dependent on MTs (i.e., the internalization from the basolateral membrane), but not other types of protein trafficking, are altered after a 72-h exposure to 50 mM EtOH.

In 2004, in the context of the increasingly studied fetal alcohol syndrome (FAS; a human condition seen in children born to mothers who drank heavily during pregnancy), Azorín et al. (2004) claimed that, as far as they knew, there was no previous information regarding the impairment of liver cytoskeleton by prenatal ethanol exposure (PEE). They assessed newborn Wistar rats whose mothers received an EtOH-containing liquid diet before and during pregnancy. Hepatocytes's actin levels in the PEE pups were unchanged; its cytosolic distribution was constant as well. In contrast, there was an increase in the levels of cytokeratin, an epithelial type of IF present in hepatocytes. As Yoon et al. (1998), Azorín et al. (2004) have found, tubulin levels were increased by 33% but its polymerization was reduced by 50% and its cytosolic distribution was altered. Besides the effects of PEE on the cytoskeleton, Azorín et al. (2004) have found microheterogeneity in several glycoproteins due to an altered glycosylation process and intracellular traffic of proteins.

Acetaldehyde is one of the intermediate metabolites in EtOH oxidation, carried out in the liver for the most part. Acetaldehyde has been attributed a great toxic capacity and many authors make it responsible for most of the toxic effects primarily ascribed to EtOH. In tubulin, lysine residues that possess highly spontaneous reactivity towards acetaldehyde are called "highly reactive lysine" (HRL) and are located on the α-subunit, while those lysine residues with a normal reactivity are called "bulk" lysine (Tuma et al., 1987). McKinnon et al. (1987) showed that concentrations of acetaldehyde above 0.5 mM have an inhibitory effect on the in vitro polymerization of calf brain microtubular proteins; long-term incubation of tubulin with acetaldehyde leads to a significant loss of polymerization ability that cannot be reversed with acetaldehyde removal. In 1989, Smith et al. (1989), working in vitro with purified tubulin from bovine brains, found that MT formation is very sensitive to even small mole fractions of acetaldehyde-modified tubulin (0.08 mol of acetaldehyde per mol of tubulin). In 2000, Rintala et al. (2000), by means of immunocytochemistry, demonstrated in the brain of both alcohol-preferring and alcohol non-preferring rats the presence of acetaldehyde–protein adducts after life-long EtOH consumption. These adducts were found in layers IV and V of the frontal cortex, as well as in the subjacent white matter, in the molecular layer of the cerebellum, and in the hepatocytes of zone 3 of Rappaport's acinus. Unfortunately, in this study, the authors did not discriminate between the types of proteins showing the acetaldehyde-adduct formation.

Besides the acetaldehyde-mediated acetylation of α-tubulin, EtOH can alter MTs in exposed tissues by means of still other biochemical pathway. In an in vitro model of the gastrointestinal epithelium it was suggested by Banan et al. (2000) that EtOH-induced damaged on the cytoskeleton of intestinal cells is mediated by iNOS activation (the inducible form of the nitric oxide synthase), followed by nitric oxide (NO) overproduction and consequently the formation of the reactive peroxynitrite anion ($ONOO^-$). In turn, $ONOO^-$ induces MTs damage by means of tubulin nitrosilation and oxidative injury. In the CNS it was found that this mechanism (protein nitrosilation in tyrosine residues by NO-derived $ONOO^-$) can similarly alter other proteins such as, for example, tryptophan hydroxylase (TPH), the first and rate-limiting enzyme in the serotonin biosynthetic pathway (Kuhn and Arthur, 1997; Kuhn and Geddes, 1999).

In contrast to the already described EtOH tendency to reduce MTs polymerization, there is a report by Reiter-Funk and Dhrman (2005) showing that, when PC12 cells were exposed in vitro for 96 h to 100 mM EtOH (~460 mg/dL), MTs content increased while free tubulin content decreased. Thus, EtOH appears to be enhancing polymerization of tubulin into MTs. For this effect to be seen in PC12 cells, the protein kinase C (PKC)-Δ isoform seems to be necessary, given that this kinase activation phosphorylates tubulin and MTs containing phosphorylated-tubulin are more stable than those with dephosphorylated-tubulin. PC12 cells are a type of rat chromaffin cell line that differentiates into neuronal-like cells and extends neurites in the presence of nerve growth factor (NGF). Neurites contain a core of MTs that are formed from polymerization of free-tubulin. However, neurite outgrowth does not correlate with enhanced MT polymerization in these cells.

In 2005, Tomás et al. (2005), from a renowned Spanish group that has been working on astrocytes biology for many years, exposed in vitro E21 (21st embryonic or gestational day) rat astrocytes to different EtOH concentrations for 7 days in order to explore their secretory pathway. Along with other alterations in the secretory pathway from the rough endoplasmic reticulum (RER) to the Golgi complex (GC), the authors found a dose-dependent MTs disorganization. At EtOH concentrations of 30 and 50 mM (~138 and ~230 mg/dL) MTs organization did not differ from that of control astrocytes, but at 100 mM (~460 mg/dL) MTs appeared disrupted and disorganized. The same group showed similar results in a previous article (Tomás et al., 2003) dealing mainly with astrocytes glucose uptake under EtOH exposure (see below).

As we have seen, spontaneous acetylation of α-tubulin and MTs can occur in EtOH-exposed cells as part of protein–acetaldehyde adducts formation. Acetylation of α-tubulin and/or MTs may alter their normal dynamics. However, normal acetylation is one of the well-known post-translational modifications of α-tubulin. This post-translational modification is a unique type since it occurs on lysine 40 of the aminoacidic chain (Hammond et al., 2008). Lysine 40 is postulated to reside on the luminal face of MTs. It is still unclear how the enzymes that carry out acetylation/deacetylation would have access to this site. Similarly, it is not known how this inner modification could influence the MT-based functions occurring on the outer, cytoplasmatic side of MTs. Whatever may be the case, it is known that in many cells there exist two pools of MTs: long-lived MTs are correlated with differentiation and stable cell shape while transient-lived MTs are best correlated with proliferation, migration, and cell remodeling. Long-lived MTs become increasingly acetylated with time although it is still unclear whether this post-translational modification affects stability or is simply a consequence of such acetylation (Southwood et al., 2007). It is also known that acetylated α-tubulin subunits are normally absent from dynamic cellular structures such as neuronal growth cones (Hubbert et al., 2002; Robson and Burgoyne, 1989). It was not until recently that the enzymes with tubulin-acetylation/deacetylation activity were discovered. Tubulin-deacetylation activity was first ascribed to histone deacetylase 6 (HDAC6) (Hubbert et al., 2002) and, later, also to sirtuin type 2 (SIRT2), a human ortholog of the *Saccharomyces cerevisiae* silent information regulator 2 protein (Sir2) (North et al., 2003). Tubulin-acetylation activity was very recently demonstrated to be carried out by elongator, a histone-acetylation enzyme (Creppe et al., 2009).

Histone deacetylases are a group of enzymes with lysine-deacetylase activity first described as involved in the regulation of cell cycle, chromatin condensation and aging processes (Southwood et al., 2007). Histone deacetylase 6 (HDAC6) is a cytoplasmatic enzyme pertaining to the class II of mammalian HDAC, known to regulate many important biological processes such as cell migration, synapse formation in immune cells, viral infections, and degradation of misfolded proteins. By deacetylating α-tubulin, HDAC6 is required in the regulation of cell morphology and cell migration (Valenzuela-Fernández et al., 2008). Sirtuin type 2 (SIRT2) is a class III type of HDAC that requires NAD^+ for its function (as well as alcohol and aldehyde dehydrogenase, the EtOH metabolizing enzymes) (North et al., 2003). In some cell types, HDAC6 and SIRT2 colocalize with tubulin (Hubbert et al., 2002; Shepard et al., 2008), and are expressed in CNS cells (Southwood et al., 2007).

Iwata et al. (2005) have shown in vitro in a Neuro2a huntingtin-inducible cell line that intact MT cytoskeleton and HDAC6 are necessary to the autophagic degradation of aggregated proteins (such as huntingtin) that have escaped surveillance by the normally acting ubiquitin proteasome system (this latter situation is commonly seen in neurodegenerative diseases such as Huntington disease). The authors suggested that HDAC6-dependent retrograde transport of these aggregated proteins on MTs is used by cells to increase the efficiency and selectivity of autophagic degradation (Iwata et al., 2005).

Experimental overexpression of either HDAC6 or SIRT2 has led to a loss of acetylated MTs and a decreased MT stability; in contrast, when these deacetylases are inactivated or their expression is knocked down, MTs were found to be hyper-acetylated and, consequently, more stable (Hubbert et al., 2002; North et al., 2003; Zhang et al., 2003).

Southwood et al. (2007) have found in mice that HDAC6 is expressed in neurons (particularly in Purkinjě cells), that SIRT2 is expressed in the cytoplasm of differentiated oligodendrocytes, and Schwann cells when they start axons' myelination. SIRT2 may also be abundantly expressed in pre-myelinating cells. In the mature CNS and peripheral nervous system (PNS), SIRT2 is localized in the cytoplasmatic compartment of mature myelin sheaths, including myelin paranodes. SIRT2 is developmentally regulated with a peak expression around P16 and decreasing to adult levels by about P30.

Except for two recent pieces of work (Joseph et al., 2008; Shepard et al., 2008), there is still no study assessing the impact of EtOH exposure on these MT deacetylases. In in vitro experiments carried out on WIF-B cells, Joseph et al. (2008) showed that trichostatin A (an HDAC6 inhibitor) induced MT acetylation and stability to the same extent (~threefold increase) as did 50 mM EtOH (see above). On the other hand, Shepard et al. (2008) showed that a chronic exposure to 50 mM EtOH (~230 mg/dL) did not alter HDAC6 subcellular distribution in WIF-B cells, but led to a decrease in its protein levels. Moreover, HDAC6 binding to MTs was significantly impaired in EtOH-exposed cells. The authors attributed this binding decrease to EtOH-induced modifications in tubulin structure that prevented associations, and to the increased MT acetylation and stability in this hepatoma cell type (at least in part due to the decreased HDAC6 protein levels and the decreased MT binding).

EtOH exposure is known to affect prenatal and postnatal extension of neurites and the subsequent establishment of neuronal circuits supporting cognitive and behavioral performances. Since it has been proved that EtOH can modify at least one of the enzymes involved in MT acetylation/deacetylation in WIF-B cells (Shepard et al., 2008), and keeping in mind that acetylated α-tubulin subunits are normally absent from dynamic cellular structures such as neuronal growth cones (Hubbert et al., 2002; Robson and Burgoyne, 1989), it would be interesting to carry out proper experiments in order to assess whether the same would occur in nerve tissue cells, in vitro and in vivo.

Elongator is the other protein involved in MT acetylation/deacetylation with an acetylation activity (Creppe et al., 2009). Elongator is a highly conserved protein complex of six subunits (Elp1-Elp6) in which its Elp3 catalytic subunit has been found to acetylate histone H3 and play a role in exocytosis and tRNA cytoplasmatic modification. Elongator has been found to be defective in the human disease familial dysautonomia (FD), an autosomal recessive neurological disease characterized by poor development and progressive degeneration of sensory and autonomic neurons (Slaugenhaupt and Gusella, 2002). The full Elp1-Elp3 elongator complex is associated with α-tubulin in the cytoplasm. Evidence exists that, by means of its Elp3 subunit, elongator may directly acetylate α-tubulin on lysine 40 and counteract its HDAC6-mediated deacetylation (Creppe et al., 2009). This post-translational modification of α-tubulin is required, for example, for the anchoring of motor proteins (kinesins and dyneins) and thus critical for MT-based molecular transport in the neuronal soma, dendrites, and axons. Elongator is expressed in the cytoplasm of cortical projection neurons where it regulates developmental radial migration and branching (Creppe et al., 2009). Consequently, its deficiency may have an important role in neuronal cell shape remodeling during migration and terminal branching in corticogenetic processes. Unfortunately, nothing is known yet about a possibly deleterious prenatal (or postnatal) action of EtOH on elongator.

During mouse brain development, elongator expression appears primarily at E12.5 throughout the cortical wall and later on (E14.5 and E18.5) becomes restricted to the cortical plate and the two germinal zones (ventricular and subventricular zones). Cajal-Retzius cells (located in the marginal zone, pivotal cellular elements in the migration of postmitotic neuroblasts) show the strongest Elp1 expression level among cortical cells. Elp1 subunit seems to regulate the migration speed of bipolar and multipolar cortical neuroblasts since its functional experimental silencing slows down (but does not block) neuroblasts migration speed. However, these retarded neuroblasts eventually reach their terminal destination in the cortex. Interestingly, it has been shown previously that PEE induces a decrease in neuroblast generation and a notorious delay in neuroblasts migration during rat corticogenesis (Miller, 1986). Given that elongator is a seemingly important factor determining neuroblast migration speed, it would be very interesting to determine experimentally whether PEE is able to modify its cytoplasmatic levels during this critical period of corticogenesis.

On the other hand, Creppe et al. (2009) has shown that Elp1 subunit silencing does not interfere with neuroblast fate specification *but does affect cell shape* in such a way that many Elp1-deficient neurons entering into the intermediate zone during migration "show highly twisted leading processes" and a defective branching in terminally differentiated projection neurons (see below). As Creppe et al. (2009) in their excellent article have said, "moreover, the ones that reach the cortical plate fail to acquire apical dendrite tree characteristic of pyramidal neurons but instead adopt leading processes harboring several bulges" (see legend of Fig. S6 in Creppe et al., 2009).

Elongator seems to be involved in neuronal branching since Elp1 silencing during development also impairs the dendrite total length growth and degree of branching that is noticeable even in P17 fully differentiated projections neurons. A physical interaction between Elp1 and Elp3 seems to be required for neuronal branching of postmigratory projection neurons said Creppe et al. (2009), and the lack of activity of Elp3 could be responsible for the defective branching observed by the authors in their Elp1 silencing experiments. The data presented by Creppe et al. (2009), as the authors stated, suggests that Elp3 catalytic activity underlies the regulation of neuroblasts migration speed and the branching of differentiating cortical projection neurons. Neurons defective in elongator show reduced levels of acetylated α-tubulin (Creppe, et al., 2009). The acetylation of α-tubulin is known to increase the binding of motor proteins that regulate bidirectional molecular transport along axons and dendrites (Reed et al., 2006) and thus may underlie the transport of cytoskeleton elements and/or proteins that are required during neuroblast migration or that regulate the growth and maturation of axons or dendrites in cortical neurons.

We will see below that PEE alters neuronal and astrocytic morphology and, given the results presented by Creppe et al. (2009), it is probable that elongator disruptions could be involved in its genesis by disrupting, among other elements, MT cytoskeleton.

In summary, EtOH does induce chemical alterations on MTs by altering both its structural and functional properties. EtOH may modify acetylation levels of α-tubulin by inducing the spontaneous formation of α-tubulin–acetaldehyde adducts on HRL residues. The formation of α-tubulin–acetaldehyde adducts is a known process occurring in the liver where acetaldehyde synthesis (and concentration) is high. In the brain, the main metabolizing enzymes for EtOH are catalase and cytochrome P450 2E1 (CYP450 2E1) and they account for up to 80% of brain EtOH metabolism. Alcohol dehydrogenase is responsible for less than 20% of its metabolism, in contrast to what happens in the liver. This is the reason why acetaldehyde local formation in the brain is possibly low; but it certainly exists. On the other hand, peripherally formed acetaldehyde does not seem to reach brain tissue in appreciable levels. Although the blood–brain barrier permeability to acetaldehyde is significant, it does not really cross the barrier because of the presence of aldehyde dehydrogenase in endothelial cells. Thus, as yet we really do not know whether α-tubulin–acetaldehyde adducts formation on HRL would be relevant inside the CNS where knowledge about acetaldehyde formation and concentration levels is still controversial (Dietrich, 2004; Hipólito et al., 2007). However, given the existence of evidence for the presence of acetaldehyde–protein adduct formation inside the brain after a long EtOH consumption (Rintala et al., 2000), it is highly probable that there also exist in vivo formations of these types of adducts on brain MTs.

We have still seen that another mechanism for physiologically modifying the acetylation levels on lysine 40 of tubulin act through enzymes with α-tubulin–acetylation/deacetylation capacity (such as HADC6, SIRT2, and elongator). The alteration of this latter mechanism has been proven in vitro in WIF-B cells whose HADC6 levels were decreased by an EtOH exposure. Similar reports regarding SIRT2 or elongator are lacking. More important, we are also lacking studies on the effects of EtOH on these three enzymes in the CNS.

Whatever the case may be, the disruption of MT cytoskeleton may affect in turn several critical cell processes such as cell morphology, proliferation, migration, differentiation, structural remodeling, neurite extension and synaptogenetic capacities, axonal and dendritic transport of organelles and molecules from and toward the perikaryon, autophagic degradation of aggregated protein, etc. All these

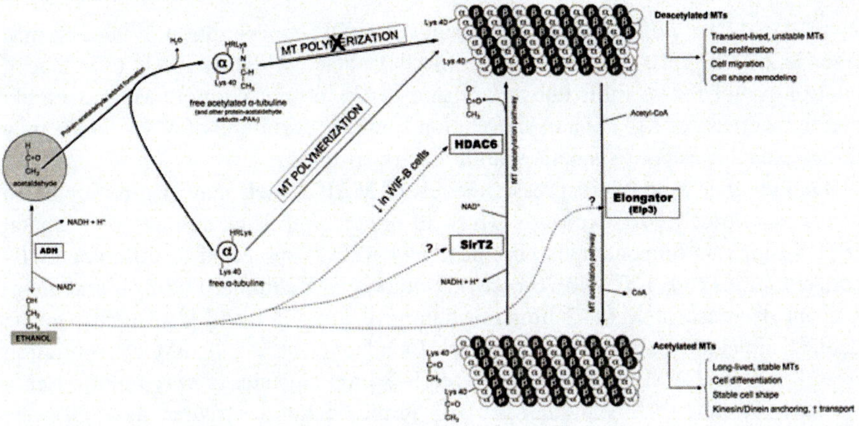

Fig. 29.1 Summary representation of EtOH actions on MTs. Note that not all the pathways have been shown to occur in nerve tissue. The α-tubulin–acetaldehyde adducts formation and the MT polymerization inhibition has been proven to occur in the liver, but not yet in the CNS where, however, protein–acetaldehyde adducts had indeed been shown (see text). The physiological acetylating/deacetylating enzymes with proven activities within the CNS are elongator (especially its Elp3 catalytic subunit), HDAC6 (histone deacetylase type 6), and SIRT2 (sirtuin type 2, a human ortholog of the *Saccharomyces cerevisiae* silent information regulator 2 protein, Sir2). The only enzyme in which a deleterious EtOH action was proven is HDAC6 but in non-nervous cells (it was shown in WIF-B cells, a hepatoma hybrid cell type)

processes have been reported repeatedly to be impaired after EtOH exposure, both during development, childhood, adolescence, and adulthood in human subjects and in animals.

For a graphic summary about known effects of EtOH and acetaldehyde on MTs, see Fig. 29.1.

29.2.2 Microtubule-Associated Proteins (MAPs)

Among all the existing MAPs we will only consider here MAP-2, tau protein, doublecortin, and motor MAPs because these proteins are the only ones for which information is available regarding nerve tissue.

29.2.2.1 MAP-2

EtOH may modify MAP-2 phosphorylation levels as was shown in 1993 by Tan et al. (1993), who found in the brain of rats subjected to PEE that MAP-2 in vitro phosphorylation was decreased in the frontal cortex but not in the hippocampus. Ahluwalia et al. (2000), in an in vitro study with whole brain homogenates obtained from male Sprague-Dawley rats, found a biphasic dose-response effect on the phosphorylation of MAP-2 molecules. EtOH concentrations of 6–48 mM (∼27.64–221.1 mg/dL) increased MAP-2 phosphorylation showing a peak at

24 mM (~110 mg/dL). Further increments on EtOH concentration of 96–768 mM (~442–3537 mg/dL) did not increase MAP-2 phosphorylation levels over that of control except for 96 mM. Phosphorylated MAPs do not promote MTs assembly and the increase of MAPs phosphorylation causes a disruption of MTs, which may affect axonal transport, synaptic viability and plasticity.

On the other hand, the expression levels of MAP-2 itself may also be subjected to changes after EtOH exposure. Thus, in organotypic slice cultures of neonatal (P5) Wistar rats hippocampus chronically (4 weeks) subjected to different EtOH concentrations (50, 100, and 200 mM), Noraberg and Zimmer (1998) found a significant decrease of MAP-2 immunostaining in the CA1 and CA3 areas; in the dentate molecular layer a decrease was found only at a 200 mM concentration (~921 mg/dL). A slight increase of MAP-2 immunostaining was found after 4 weeks at 50 mM (~230 mg/dL) yet only in the dentate molecular layer. Of note, these authors observed regional differences as would be seen later by others.

In 1998, Putzke et al. (1998) subjected a group of adult rats (18 months of age) of an AA (alcohol-preferring) strain to chronic EtOH exposure for 16 months in a free choice paradigm. The authors found a regionally-differentiated brain response in the levels of MAP-2 mRNA. In almost all brain regions MAP-2 mRNA was decreased, most prominently in the dorso-lateral putamen and the nucleus accumbens, the substantia nigra pars compacta, and the globus pallidus. Other components of the extrapyramidal system were less affected. In hypothalamic regions, such as the paraventricular and septohypothalamic nuclei, there was also a significant reduction of MAP-2 mRNA. In another experiment with a group of heterogeneous Wistar rats chronically exposed to EtOH for 8 months in a free choice paradigm, Putzke et al. (1998) found a significant and strong reduction of MAP-2 mRNA in the nucleus accumbens/striatum complex while the hippocampus and cerebellum showed smaller reductions and the frontoparietal cortex was not affected. However, 72 h after EtOH withdrawal, MAP-2 mRNA levels in the nucleus accumbens/striatum and the hippocampus returned to normal while the frontoparietal cortex and the cerebellum showed an increase of MAP-2 mRNA levels, surpassing that of controls (i.e., a rebound phenomenon after EtOH withdrawal).

In 2002, our laboratory reported (Tagliaferro et al., 2002) a significant 41.6% reduction of MAP-2 immunostaining in the CA1 hippocampal area (Hipp) of adult Wistar rats after 6 weeks of a chronic 6.6% (v/v) EtOH exposure in drinking water. Dendritic processes not only showed a decrease in MAP-2 but also a qualitative disruption of dendrites' morphology under MAP-2 immunostaining (fragmented and abnormal structure).

Another piece of work from our laboratory (Evrard et al., 2006) assessed the morphological alterations induced by a low chronic EtOH exposure on Wistar adolescent male rats. Animals were exposed to 6.6% (v/v) EtOH in drinking water for 6 weeks, starting in their late adolescence (P45–P50) and reaching low BECs (16–25 mg/dL = 3.47–5.42 mM). One group of rats was studied (by immunocytochemical and morphometric methods) immediately after intoxication in order to assess the EtOH-induced damage. Another group of rats was allowed to withdraw from EtOH and recover by drinking water for 10 further weeks in order to assess if their youth conferred upon them an appreciable (partial or total) recovering capacity. Apart from other markers, we studied MAP-2 positive dendrites and the relative area of brain tissue they covered in three prosencephalic areas: CA1 hippocampal area (Hipp), corpus striatum (Strt), and frontal cortex (FCx). After the intoxication period, MAP-2 expression was consistently decreased in all three areas, albeit at different levels (51% of control value in the Hipp, 31% in the Strt, and 41% in the FCx). As expected, after the abstinence period we found a recovered MAP-2 expression in the Hipp (92% of controls) and also in the Strt with even a slight but not significant rebound phenomenon (123% of controls). We should remember here that a similar rebound phenomenon was observed by Putzke et al. (1998) in MAP-2 mRNA, not in the Strt but in the frontoparietal cortex and cerebellum. However, unexpectedly for us and unlike the results of Putzke et al. (1998), we found in the FCx (a philogenetically newer area) only a tendency to recover (58% of controls) that did not amount to a full recovery as was seen in the other areas. In that work we speculated on the possibility that the concomitant alterations we observed in the serotoninergic system and in S-100B protein expression levels were partially involved in the altered levels of MAP-2 (a prior decrease and later increase), as was previously seen by other authors in different experimental conditions (Azmitia et al., 1995; Whitaker-Azmitia et al., 1997). Of note, S-100B is a neurite extension factor secreted by astrocytes under 5-HT stimulation of their 5-HT_{1A} surface receptors. S-100B displays important actions on neuronal and astroglial cytoskeletons acting during both development and adulthood as a promoting- and maintenance-factor of the mature and differentiated form of the CNS. S-100B is an intracytoplasmatic Ca^{2+} sensor whose actions are achieved, after Ca^{2+} binding and dimerization, acting through direct cross-linking of different cytoskeletal proteins (MTs and Nfs) by means of special hydrophobic regions that the S-100B dimers possess at their sides (Donato, 2003). In this way, S-100B acts as, for example, a neurite extension-promoting factor; it also induces neuronal and astroglial cytoskeletal stability. It promotes the acquisition of a mature phenotype in several types of neurons and has prosurvival, antiapoptotic effects at low, physiological concentrations (in the nM range). However, it is pro-apoptotic at higher concentrations (in the mM range), it is a synaptic plasticity promoter, and it induces the expression of astroglial GFAP, etc. (Azmitia, 2001; Donato, 2003; Santamaria-Kisiel et al., 2006). However, in our study, since the expression levels of S-100B and MAP-2 were not strictly parallel in the three areas after the abstinence period, it became obvious to us that other factors might be involved in the production of the observed effects. Beyond the quantitative, region-dependent, post-intoxication decreased and post-abstinence recovered MAP-2 expression levels, we also observed qualitative, morphological alterations in the parallel-ordered dendritic trees of the Hipp and FCx: a waving, corkscrew-like shape that persisted after abstinence was present in both areas (see below for further discussion).

Consistent with the previous studies we have already mentioned here, but working in vitro with E14 neural and glial cells exposed to 0, 25, 50, and 100 mM EtOH (0, ~115, ~230, ~460 mg/dL), Tateno et al. (2005) found a dose-dependent increase in the number of $GFAP^{+}$ cells and a parallel decrease in MAP-2^{+} cells while, at a concentration of 100 mM (~460 mg/dL), they found a time-dependent increase in GFAP expression over 24 h of exposure and a parallel decreased in MAP-2 expression measured in Western blot assays.

29.2.2.2 Tau Protein

Tau (τ) protein, the other major component of the MAP/Tau family of MAPs, is expressed in axons where it promotes MT assembly and stabilization. Tau is such an important MAP for human neuropathology that it has its own group of diseases – tauopathies (Delacourte, 2005; Hasegawa, 2006; Hernández and Ávila, 2007). As this protein is so important in the field of neuropsychiatric diseases, it is surprising that there are almost no studies carried out in the field of experimental alcoholism. In this regard, to our knowledge there is only one piece of work searching for the effects of PEE on τ-protein, the above-mentioned work by Tan et al. (1993) that mainly investigated the alterations on MAP-2 phosphorylation. These authors did not find any change in the in vitro phosphorylation of τ-protein, nor in the FCx or in the Hipp. In contrast, on the human neuropathological ground, Cullen and Halliday (1995a, 1995b), by means of τ-protein immunocytochemistry and a modified Bielschovsky silver stain, found τ-positive granular inclusions and neurofibrillary tangles within the magnocellular cholinergic neurons of the nucleus basalis of Meynert in human chronic alcoholic subjects also affected by Wernicke's encephalopathy. The authors suggested in those studies that the accumulation of phosphorylated τ-protein and the cell death they have observed could be related with a thiamine deficiency-dependent mechanism. On the other hand, Morikawa et al. (1999), on clinical grounds, found no differences in the cerebrospinal fluid (CSF) content of τ-protein when comparing demented and non-demented alcoholics against normal control subjects (such levels were much lower and significantly different than that of patients with Alzheimer's disease), suggesting that alcoholic dementia is not accompanied by τ-protein CSF elevations and that CSF examination could help in differentiating both types of dementia (Kapaki et al., 2005; Morikawa et al., 1999). It was further suggested that τ-protein CSF levels, used as a trait marker, could help not only in that differentiation but also in the diagnosis of acute Wernicke's encephalopathy (Matsushita et al., 2008).

29.2.2.3 Doublecortin

Doublecortin is another MAP which associates preferentially with 13 protofilament MTs by recognizing a site between protofilaments and thus stabilizes MTs. Doublecortin interacts with adapter molecular complexes specifically at the ends of neuritic and leading processes and is involved in MT-based vesicle trafficking. This way, it plays a role in the growth of neuronal processes, downstream of directional or guidance signals. Doublecortin is widely expressed in migrating neurons (both radially and non-radially migrating ones) during prenatal and postnatal development in the CNS and PNS and during adult neurogenesis as well. Mutations in this protein have been shown in a large proportion of type I lyssencephaly and in subcortical laminar heterotopia (or double cortex syndrome), two human neuronal migration disorders causing epilepsy and severe mental retardation, among other neuropsychiatric manifestations (Brown et al., 2003; Friocourt et al., 2003; Gleeson et al., 1999; Moores et al., 2004). Doublecortin expression was certainly altered after EtOH exposure in the following four works by the group of Nixon and Crews that we will next discuss, although its study was not the main object of interest (it was evaluated only as a marker of newborn, migrating, and differentiating neurons).

In a chronic 4-day binge model, Nixon and Crews (2004) exposed adult male Sprague-Dawley rats to 25% (w/v) EtOH in a nutritionally complete liquid diet, intragastrically. They showed that doublecortin expression peaked at 14 days after abstinence began, while neural progenitor cells (assessed by means of BrdU staining) peaked 7 days after abstinence began (a fourfold burst in cell proliferation). These observations suggest that, after withdrawal from EtOH and with protracted abstinence, a compensatory neurogenesis is possible, but its functional implications remain speculative.

Male Sprague-Dawley rats were EtOH-exposed by means of a liquid diet for 1, 2, or 4 weeks in a study by He et al. (2005). The authors studied Hipp neurogenesis at the dentate gyrus and found that immunoreactive doublecortin in EtOH-exposed animals showed a decrease tendency at 1 week of treatment, but at 2 and 4 weeks, they showed a significant and timely dependent reduction. Moreover, by means of doublecortin immunoreactivity, He et al. (2005) were able to determine that 4 weeks of EtOH treatment decreased the number of dendritic nodes by 60%, of dendritic endings by 49%, and the total length of apical dendrites by 50%, albeit not affecting the cell area in the soma. Thus, the authors suggested that EtOH treatment suppresses the dendritic growth of the newborn neurons in the Hipp by affecting doublecortin. These altered newborn neurons will probably not be fully competent in order to take part in normal neuronal circuits since its dendritic trees are altered from the very beginning.

In the context of adult neurogenesis and adolescent alcoholism, Crews et al. (2006) exposed male Sprague-Dawley rats, intragastrically, with a single dose of 25% (v/v) EtOH (1.0, 2.5, or 5.0 g/kg) which rendered peak BECs of 33, 72, and 131 mg/dL (~7.16, 15.6, and 28.4 mM), respectively. After 3 days of exposure, doublecortin immunoreactivity was reduced by 38 $\pm$ 4% in the Hipp of rats exposed to a 5.0 g/kg dose. As expectedly, neurogenesis was also significantly reduced both in the subventricular zone and in the dentate gyrus in a dose-dependent fashion.

Stevenson et al. (2009) have recently published work in which they allowed adult C57BL/6 J mice to self-administer 10% (v/v) EtOH in drinking water for 28 days. Then mice were withdrawn from EtOH for 1 or 14 days. When tested on the forced swimming test, mice showed an increased depression-like behavior after 14 days of abstinence (but not after 1 day). This behavior was associated with a reduction of neural stem cells proliferation and of immature migrating neurons in the dentate gyrus of the Hipp. These latter cells were assessed by means of doublecortin immunoreactivity (which was accordingly decreased).

To our knowledge, there is, at present, only one piece of work dealing with doublecortin during PEE but, again, it was studied only secondarily as a marker of neuroblast migration. Mooney et al. (2004) exposed in vitro organotypic cultures taken from E17 fetal rat brain cortices to 0, 200, 400, or 800 mg/dL EtOH (~0, 43.4, 86.8, or 173.7 mM) for 2, 6, 8, 16, 24, or 32 h. Suprapial heterotopias (marginal glioneuronal heterotopias) or "warts" appeared as a function of time and EtOH concentration. Along with other findings, the authors showed that doublecortin was present in the ectopic neurons from both the suprapial and subventricular warts, suggesting that neuronal migration (and, consequently, neuronal cytoskeleton) was implicated in the formation of this cortical microdysplasia.

29.2.2.4 Motor Microtubule-Associated Proteins

Dyneins and kinesins are families of motor MAPs involved in many aspects of intracellular transportation (Hirokawa et al., 1998; Karp, 2007; Lodish et al., 2007). As with other cytoskeletal proteins, detailed knowledge is lacking regarding EtOH effects on them. However, some indications exist. Azorín et al. (2004), in their mentioned paper (see above), have found that the levels of MT motors dynein and kinesin were unchanged in vivo by PEE on newborn Wistar rat hepatocytes. By contrast, Tomás et al. (2005) (see above) have found that the levels of dynein and kinesin were significantly reduced in in vitro EtOH-exposed astrocytes taken from E21 normal rat fetuses. Transport of plasma membrane proteins from the GC to the cell surface is completely dependent on MT motor activity in which dynein and kinesin are involved. There is evidence that EtOH affects both anterograde and retrograde vesicle transport in astrocytes; MTs and MAPs sustain several aspects of this transport (as it is the axonal transport). In this regard, it was shown almost 20 years earlier that acute and chronic ethanol exposure decreased axonal transport in rat nerves (McLane, 1987).

In summary, about MAPs and EtOH it could be said that all the available information points towards a noticeable effect of EtOH exposure in modifying both their phosphorylation and expression levels. Some reports (still scant) already exist on that issue. However, in most of the existing works showing altered expression levels in MAPs we cannot strictly state whether those levels are altered due to an absolute alteration of the protein level within each cell or whether the alterations have to be attributed to a relative altered number of cells. In any case, further studies are granted in order to grasp the innermost mechanism/s by which EtOH disrupts this group of proteins (whether it be due to transcriptional, translational, or regulatory disruptions to direct molecular damage as it is the case for MTs or by means of indirect mechanisms mediated by other molecules). As a desirable arrival point, of course, these studies should be able to design suitable molecular or pharmacological tools aiming to prevent or restore EtOH-induced damage.

29.3 Ethanol Effects on Intermediate Filaments (IF)

One of the first indications that IF could be involved in EtOH-induced pathology were the works by Tinberg's group who detected the presence of keratin-like moieties in Mallory bodies obtained from human alcoholic livers (Tinberg and Mednick, 1980, 1981; Wessely et al., 1981). Mallory bodies (as well as other intracellular protein aggregates such as neurofibrillary tangles in Alzheimer's disease or Lewy bodies in Parkinson's disease) are inclusion bodies typically composed of misfolded and ubiquitinylated structural proteins together with different quantities of chaperones and the ubiquitin-binding protein sequestosome 1/p62 (Strnad et al., 2008).

In dealing with IF in nerve tissue cells, we must differentiate between those of neurons and those of astrocytes. In neurons, neurofilament proteins (Nfs), a class IV IF, are the most prominently expressed and in astrocytes, the characteristic IF is glial fibrillary acidic protein (GFAP), a class III IF and main determinant of astrocytes cell shape. While there is only one type of GFAP, Nfs are differentiated mainly by their molecular weights into three types: light Nf of 68 kDa (Nf-L or Nf-68), medium Nf of 160 kDa (Nf-M or Nf-160), and heavy Nf of 200 kDa (Nf-H or Nf-200). In forming a complete and mature IF, Nf proteins assembly in a stoichiometric relation of 7:3:2 (for Nf-68, Nf-160, and Nf-200, respectively). This proportion may vary in developing neurons or diseased ones. Thus, it is generally accepted that Nf-68 predominates during development and in regenerative phenomena (Hoffman et al., 1987; Hoffman and Cleveland, 1988) while Nf-200 is characteristic for mature neurons (Nixon and Sihag, 1991). Nfs, among other functions, are a major determinant of axonal diameter and, in turn, axonal diameter is one of the main determinants of the speed of conduction of the action potential in myelinated axons (Hoffman et al., 1987).

29.3.1 Neurofilaments (Nfs)

In the context of experimental FAS studies, Poltorak et al. (1990) evaluated the non-phosphorylated (nPNfs) and phosphorylated Nfs (PNfs) in the developing cerebellum of C57/B1/6 J mice subjected to PEE. Early in postnatal life, nPNfs were reduced in primary and secondary dendrites of Purkinjě cells; this reduction disappeared by P60. The authors suggested that a developmental delay existed in the maturation of nPNfs. A similar delay has been observed in other experimental FAS studies (see above in relation with neuroblast migration and below in relation with GFAP).

In 1997, Saunders et al. (1997) worked in vitro with primary cultured hippocampal neurons taken from the brain of normal E18 Long-Evans rat fetuses. They found that after exposure to 21.8, 54.5, and 109 mM EtOH (~100, ~251, and ~501 mg/dL, respectively), Nf-68 and Nf-200 were reduced up to 47% while Nf-160 showed a reduction up to 32%, both in a dose-response fashion. A 72-h exposure of these cells to 218 mM EtOH (~1002 mg/dL), surprisingly, did not render any obvious alterations in neurite extension or explant morphology, and there were no visual signs of cell death, although in the clinical ground human death is almost granted with BECs acutely higher than 500 mg/dL (~109 mM).

In 2002, our laboratory carried out a study on the Hipp of adult male Wistar rats exposed to low EtOH for 6 weeks (Tagliaferro et al., 2002). We observed that Nf-200 expression (the "mature" form of Nf proteins) was decreased by 50% in the *stratum radiatum* of the CA1 area, a region where Nf-200 fibers correspond to the apical dendrites of pyramidal neurons located at the *stratum pyramidale*.

In the context of a Wistar rat model of the alcohol-related neurodevelopmental disorder (ARND), a slighter disease form of FAS (Bertrand et al., 2004; Hoyme et al., 2005), our laboratory found in 2003 (Evrard et al., 2003) that Nf-200 expression was altered in the prosencephalon of P21 male offspring PEE to low EtOH levels through the placenta. Rat mothers were given 6.6% EtOH in drinking water; BECs were in the range of 16–25 mg/dL = 3.47–5.42 mM. In contrast to what was seen by Poltorak et al. (1990) in the cerebellum of mice, we found a significant parallel 1.3-fold increased expression of Nf-200 both in the Hipp and Strt. In the FCx we did not observe such an increase but Nf-200 fibers were qualitatively altered in the dendrites of the pyramidal neurons traversing the *stratum radiatum* in the Hipp, in the axons located in the striosomes of the Strt, and in the apical dendrites of the pyramidal cortical neurons: an abnormal waving, corkscrew-like morphological pattern was clearly observed.

In the same work from our laboratory (Evrard et al., 2006), mentioned above, in which we evaluated Wistar adolescent male rats exposed to EtOH which later spontaneously recovered by means of drinking water, we found that Nf-200 expression was significantly decreased after EtOH exposure in all the three prosencephalic areas we assessed (by 82% in the Hipp, by 50% in the Strt, and by 64% in the FCx). After the recovery period, we observed a great 3.67-fold increase in the Nf-200 expression of the Hipp that, however, did not reach control values (it was still reduced by 34%). In contrast, in the Strt a full recovery was observed with a slight but non-significant tendency to surpass control values (by 19%); the recovery increase was of a 2.38-fold magnitude. Moreover, in the FCx a significant and surpassing recovery capacity was observed since Nf-200 expression reached a value that was 1.33-fold that of controls, and, at the same time, a 3.65-fold increase from the previous EtOH-exposed value. This way, the recovery capacity was not equal in all the studied areas and even a rebound phenomenon could be observed in the FCx. However, as was the case with MAP-2, we could observe persisting qualitative alterations (waving, corkscrew-like Nf-200 fibers) in the three areas (albeit not always and not in all animals) both after EtOH exposure and abstinence. Thus, despite the observed quantitative recovery in Nf-200 expression, morphology did not recover.

In one recent study from our laboratory (Aronne et al., 2008) we studied the effects of high EtOH doses administered intraperitoneally (3.5 g/kg/day) to pregnant Wistar rats from E10 to E18 (this time, it was a rat model of FAS). High BECs were obtained (ranging from 119 to >300 mg/dL = 25 to > 65 mM). Along with definitive FAS signs (decreased litter size, lower body and brain weights, thinner dorsal brain wall, less developed cortical plate, defects in caudal neural tube closure –dysraphia–), we observed, at E18, a significant decreased expression of Nf-68 (the "immature" form of Nf proteins) both in the dorsal telencephalic wall (where cortex is developing; a 20% reduction was observed) and in the mesencephalon (the region were serotoninergic neurons reside; a 32% reduction was observed).

In summary, Nfs are effectively altered after EtOH exposure in different experimental paradigms, albeit in diverse ways: they may be increased or decreased, in a transient or long-lasting fashion; they may sometimes recover from the damage, and sometimes not. However, in PEE studies, it seems to be, as a general rule, that Nfs appear decreased. Moreover, it is common to observe that a morphological disruption may fall on them.

29.3.2 Glial Fibrillary Acidic Protein (GFAP)

In relation to EtOH exposure, GFAP was more extensively studied than Nfs. There are studies reporting that, after EtOH exposure, in different experimental conditions, GFAP expression, and/or its mRNA, may be decreased, increased, or not altered.

In the context of experimental FAS, in vitro studies were done on cortical astrocytes obtained for E21 fetuses of female Wistar rats subjected to EtOH exposure during pregnancy. Astrocytes were maintained in culture for 28 days in the absence or presence of 25 mM EtOH (~115 mg/dL). The authors (Renau-Piqueras et al., 1989) observed that astrocytes from EtOH-exposed fetuses failed to developed cellular processes or to acquire a filamentous IF distribution pattern, and showed less GFAP content than control astrocytes. These abnormal characteristics were also observed when astrocytes were in vitro newly exposed to EtOH. For this reason, the authors concluded that the initial damage had to be inflicted on precursor cells during intrauterine gestation. The same Spanish group reported some years later, in two papers (Vallés et al., 1996, 1997), that PEE delayed the appearance of GFAP and its mRNA and significantly decreased GFAP expression both in vivo and in vitro. GFAP gene transcription rate was significantly decreased and the stability of GFAP mRNA was slightly reduced. GFAP DNA was hypermethylated and methylation-mediated repression could be a mechanism involved in EtOH-induced reduction of GFAP expression, the authors stated. In culture, some morphological alterations were observed and a delay of ~2 days was also apparent in the appearance of GFAP immunofluorescence. The authors suggested, from their results, that abnormalities in astrogliogenesis might underlie the neuronal migration disorders observed in PEE. Of note, as we mentioned below, Miller (1986) also observed a 1-day delay in the neuroblast generation period and a 2-day extension of it.

In an in vivo study, again in the context of FAS research, Tajjudin et al. (2003) found that the offspring of Sprague-Dawley rats, aged P5 and P19, showed significantly less GFAP-immunoreactive astrocytes in the dorsal and median raphe nucleus region (both nuclei are the source of the prosencephalic serotonin innervation). On the other hand, ipsapirone (a 5-HT_{1A} receptor agonist, such as buspirone, able to induce the astroglial secretion of S-100B; see below) prevented EtOH effects on astrocytic density in the dorsal raphe and blunted EtOH effects in the median raphe when evaluated at P5 and P19. Although interesting and promising, these results told us nothing about the expression levels of GFAP on those astrocytes because, in this study, GFAP immunoreactivity was only used as an astrocytic marker.

Aside from FAS experimental studies, Rintala et al. (2001) carried out studies on aged rats. They studied the cerebellum of EtOH-preferring Alko, Alcohol (AA) rats, aged 3 and 24 months. Both females and males were given 12% (v/v) EtOH in drinking water for 21 months. Different effects were observed in both sexes. Decreased expression levels of GFAP in different areas of female cerebellum correlated well with increased EtOH daily consumption (the higher the intake, the higher the GFAP expression level decrease). On the other hand, in male cerebellum different cerebellar regions exhibited decreased and increased levels, mostly depending on aging. Fibers of Bergmann glia appeared tortuous and unevenly spaced in old control animals but much more in EtOH-exposed ones.

In a postmortem study on alcoholic human subjects without Wernicke encephalopathy or Korsakoff psychosis, with postmortem intervals until fixation ranging from 6 to 36 h, Miguel-Hidalgo et al. (2002) found that, relative to controls, alcoholics showed an astrocytic density reduced by 11–14% in layers V and VI of the dorsolateral prefrontal cortex. If the alcoholic subjects were further depressed, then they showed lower density values. The authors did not find any differences in GFAP immunoreactivity when comparing both groups, although the lowest values were found in the alcoholic subjects.

Another group of studies found mainly increased GFAP expression levels. In 1993, Fletcher and Shain (1993) found that, when postnatal rats between ages P5 and P7 were briefly and moderately EtOH-exposed, GFAP and its mRNA increased both in vivo and in vitro, suggesting that even brief EtOH exposures may alter GFAP gene expression in astrocytes. The same year, Goodlett et al. (1993) reported that a cortical transient astrogliosis may be induced by low to high EtOH exposure during the brain growth spurt or synaptogenic period in rat pups aged P4–P9. When studied on P10, cortical GFAP levels were increased as a function of BECs up to 325% of the level observed in controls and an astroglial reaction (hypertrophy and hyperplasia) was evident. Nevertheless, by P15 the astroglial effects were no longer evident, suggesting that astroglial reaction was transient after a binge-like exposure.

Satriotomo et al., (2000) subjected a group of adult mice from the BALB/C strain for 4–5 days to a 6.6% (v/v) EtOH liquid diet exposure. They observed a different level increase of GFAP-ir astrocytes (reactive astrocytes) in different hippocampal regions (CA3 > CA2 > CA1 > dentate gyrus).

In the already mentioned study by Tagliaferro et al. (2002) carried out on adult male Wistar rats exposed to EtOH for 6 weeks, we found a 92% increase in GFAP expression in the CA1 Hipp. In a Wistar rat model of ARND, we (Ramos et al., 2002) exposed a group of females for 6 weeks prior to mating, during pregnancy and postnatally in the lactation period, to 6.6% (v/v) EtOH in drinking water. This way, offspring have had their brain development completely under EtOH exposure. At age P21, male offspring, prenatally exposed through the placenta and postnatally through maternal milk, showed a near twofold increase in GFAP immunoreactivity, i.e., astrocytes' cell area (reactive astrocytes), in the CA1 hippocampal region along with a 40% increase in the cell area of Strt astrocytes. Concomitantly with the effects of EtOH on astrocytes, in that same study we found that there was also an EtOH-induced increase in the relative area of brain tissue covered by innervating 5-HT fibers (studied through the expression of the immunoreactive 5-HT transporter

–5-HTT–), that increase being higher in the Hipp (55% more) than in the Strt (33% more). Intracytoplasmatic S-100B protein was also increased in Hipp astrocytes (a 57% increase) but in the Strt it showed a non-significant 15% increase in the Strt.

In a later study from our laboratory (Evrard et al., 2003), we found grossly similar results in a slightly different model of ARND. This time we did not exposed rat pups through their mother's milk; instead, rat mothers were withdraw from EtOH at the moment of delivery. The exposure level being low (BECs ranging from 16 to 25 mg/dL = 3.47–5.42 mM), none of them exhibited any abstinence sign. In P21 male offspring, GFAP expression in the Hipp showed a 2.5-fold increase (astrocytes were reactive: hyperplastic, with thicker and more tortuous cellular processes), a slight but not significant increase in the Strt, and a 1.7-fold increase in FCx. Certain concomitant changes in the serotoninergic system (increased TPH and 5-HT expression in the dorsal and median raphe nuclei), along with increased astrocytic S-100B expression in the same prosencephalic areas, led us to speculate with the possibility that, at least partly, 5-HT-S-100B relationship (a proven neuron-glia relationship) were intimately involved in the observed effects on GFAP and Nfs (see above). However, we highlighted the fact that, given that the EtOH-effects were not strictly parallel in all regions, other factors might be involved in the production of the cytoskeletal alterations we observed.

It has previously been mentioned in relation to MAP-2 (see above) that, in 2005, working in vitro with E14 neural and glial cells exposed to 0, 25, 50, and 100 mM EtOH (0, ~115, ~230, ~460 mg/dL), Tateno et al. (2005) found a dose-dependent increase in the number of $GFAP^+$ cells while, at a concentration of 100 mM (~460 mg/dL), they found a time-dependent increase in GFAP expression over 24 h of exposure.

In adolescent rats chronically exposed to EtOH (Evrard et al., 2006; see above), we found after the intoxication period that GFAP expression was increased in Hipp, Strt, and FCx by 62–73%. After the 10-week abstinence period, male rats showed a still increased expression in GFAP (11–26%) in Hipp, Strt, and FCx, but there was a noticeable but slow tendency to a reduction in the astrocytic cell area. This could be indicating that chronic EtOH effects on GFAP from adolescent brains are long lasting.

Not only may the expression levels of GFAP be subjected to changes under EtOH exposure (both prenatal and postnatal) but its timed expression could be altered as well. Miller and Robertson (1993) found that EtOH accelerates the normal transformation (and disappearance) of radial glia cells into astrocytes when PEE rat offspring were studied by means of GFAP-immunostaining from P0 to P45. These authors found that, between P5 and P12, an increased number of astrocytes (hyperplasia) were also detected in the cortex. Moreover, the authors stated that the EtOH-induced premature degradation of the network of radial glial fibers might underlie the migration of late-generated neurons to ectopic sides. In partial accordance with the work of Miller and Robertson (1993) is one recent study from our laboratory mentioned above (Aronne et al., 2008) in which we observed in Wistar rat fetuses PEE, at E18, a significant decreased expression of vimentin both in the cortex (a ~62% decrease) and in the mesencephalon (a ~48% decrease). Vimentin is a class III IF used as a marker of immature glial cells in the developing CNS, mainly of radial glia cells. It was concomitantly observed to be a decreased expression of S-100B and of *Pax6*, a transcription factor with multiple functions, involved in eye and cortical development, which plays an essential role in the differentiation of cortical radial glia cells during development (Götz, 1998; Simpson and Price, 2002), and probably involved in the generation of cortical dyplasias under PEE.

In summary, as was the case with Nfs, GFAP expression may be increased, decreased, or not altered after EtOH exposure. However, a predominant increase seems to be mostly observed in accordance with the normal answer observed in astrocytes under the action of many noxious stimuli: *astroglial reactivity*, their common response to nerve tissue injuries.

29.4 Ethanol Effects on Microfilaments (MFs)

Microfilaments are not only composed by F-actin but also for many acting-binding proteins (ABPs). The ABPs regulate the filamentous network in which polymeric actin assembles (for a review, see dos Remedios et al., 2003). Foreseeably, EtOH may adversely act on both elements. Unfortunately, almost nothing is known regarding ABPs.

29.4.1 F-Actin Microfilaments

To our knowledge, actin MFs were first observed to be altered after EtOH exposure in skeletal muscle biopsies from three human neonates affected by fetal alcohol myopathy. Ultrastructural alterations were found in the sarcomeric I band (Adickes and Schuman, 1983); in the heart muscle of PEE Sprague-Dawley rat pups, it was found that actin was disarrayed in thin myofilaments (Adickes and Mollner, 1986). In an in vitro study carried out on myocytes from chick embryos heart, actin content was found to be decreased (Ni et al., 1992). Immortal rat gastric mucosal cell monolayers (RGM-1 cells) were cultured by Bidel et al. (2006) for 15 min in the presence of increasing EtOH concentrations (17.13, 34.25, 85.64, 171.3, and 256.92 mM, 78.9, 157.8, 394.5, 789.0, and 1183.5 mg/dL). These authors observed increasing degradation and irregularity in the actin filament organization in a dose-dependent fashion. At higher concentrations, cells were detached from each other and MF damage was severe. A similar picture was evident when zonula adherens and migration rate were studied under the same conditions. Since allopurinol (a xantine oxidase inhibitor which inhibits the production of intracellular reactive oxygen species) partially prevented EtOH effects (except at high doses), Bidel et al. (2006) suggested that EtOH action on MF cytoskeleton could be due to oxidative stress (see above, and Bellomo and Mirabelli, 1992).

In 1986, Hassler and Moran (1986a, b) were probably the first to study the actin cytoskeleton in nerve tissue cells after EtOH exposure when exposed in vitro neural crest cells from spotted salamander (*Ambystoma maculatum*) embryos both acutely (2 h 20 min) and chronically (6 days) to 0.05, 0.10, 0.15, or 0.20% EtOH (39.45, 78.9, 118.3, or 157.8 mg/dL, 8.56, 17.12, 25.68, or 34.25 mM). They found that cells migrated in vitro as did control cells, but they did not complete their morphological differentiation. Cells were contracted; their shape and cell-to-cell contacts were altered and exhibited a less developed dendritic tree. Long treatment induced the loss of substratum adherence. Given that altered, disrupted actin (and tubulin) fibers were found in those EtOH-exposed cells, the authors concluded that EtOH could exert teratogenic effects by means of the disruption of cytoskeletal components. Many years later, Rovasio and Battiato (2002) made grossly similar in vivo and in vitro experiments with neural crest cells of chick embryos. They observed that, after in vitro exposure to 150 mM EtOH, the actin cytoskeleton was disorganized, with condensation of the cell cortex, and that there was a subpopulation of blebbing apoptotic-like cells (see Fig. 29.8 in the paper by Rovasio and Battiato, 2002). In fact, membrane blebbing is one of the morphological signs of MF damage seen in apoptosis (Charras, 2008) and in cells suffering oxidative stress (Bellomo and Mirabelli, 1992). Accompanying dynamic alterations were observed in the migration capacity of these cells: minor velocity of migration, significant abnormal motility behavior, and disturbed cell capacity to maintain the directionality of locomotion (linearity). All these findings clearly point to a disrupted actin cytoskeleton as one of the physiopathogenic mechanisms that may underlie the generation of physical malformations in FAS-affected children (Rovasio and Battiato, 2002).

Continuing with their research on EtOH-exposed astrocytes, Tomás et al. (2003) exposed astrocytes from E21 rat fetuses in vitro to 30, 50, or 100 mM EtOH (~138.2, 230.3, or 460.6 mg/dL) for 7 days. They found that these cells "showed fewer actin fibers, which, in addition, were rearranged in a circular structure beneath plasma membrane, leaving the central portion of cells devoid of actin." This morphological disruption of the actin cytoskeleton was dose-dependent and could be prevented by adding lysophosphatidic acid (LPA) to the culture medium. This phospholipid mediates a number of biological responses (such as actin reorganization and survival, amongst many others), via G protein-coupled serpentine receptors. One of the responses that LPA may induce is the activation of RhoA, one of the many members of the Rho GTPase family of proteins (Jaffe and Hall, 2005; Koh, 2006), which regulates the formation of actin stress fibers. Rho-GTPases (RhoA, Cdc42, and Rac1) are also very important molecules in regulating dendrites morphology (Georges et al., 2008). Since LPA prevented EtOH actions on actin cytoskeleton, the authors suggested that EtOH alters MFs arrangement by interfering with the RhoA signaling pathway. EtOH also inhibits some forms of PKC (α, β1, and γ), and PKC normally leads to the stimulation of actin polymerization (via phosphatidylinositol kinases and the formation of PIP_2).

Laas and Hagel (1994) obtained evidence for actin involvement in EtOH-induced damage in humans' nerve tissue in 1994. They found in autopsy samples from the Hipp of 123 chronic alcoholics that Hirano bodies had increased in the *stratum lacunosum* when compared to the 197 controls they evaluated. Moreover, Hirano bodies were the most frequent neuropathological finding in the alcoholics, and when the number of Hirano bodies was over a certain value, it was highly indicative of chronic alcoholism. Laas and Hagel (1994) proposed that increased numbers of Hirano bodies in the *stratum lacunosum* indicate alterations of the apical dendrites of the pyramidal neurons of the CA1 area that are probably due to direct neurotoxic effects of EtOH. Hirano bodies are bright, eosinophilic, intracytoplasmatic inclusions with a characteristic crystalloid structure, the main components of which are F-actin and several ABPs (Galloway et al., 1987) but which also contain many other cytoskeletal proteins (such as α-actinin, cofilin, MAP-1 and MAP-2, Nf-160, and tropomyosin among others) (Hirano, 1994). It was described in 1965 in relation to the histopathology of amyotrophic lateral sclerosis. Since then, it was found to occur preferentially in the neuronal processes of the pyramidal cells of the CA1 Hipp area. These bodies were later observed in the context of other neurodegenerative diseases (e.g., Alzheimer's disease, amyotrophic lateral sclerosis, and Parkinsonism-dementia complex of Guam, Pick disease) and even in "normal" elderly individuals. It is accepted that they are largely originated from alterations of the MFs cytoskeleton (Hirano, 1994).

In summary, EtOH actions on MFs cytoskeleton may imply the disruption of many very important cellular activities dependent on F-actin. For example: intracellular vesicle trafficking, receptor-mediated endocytosis, cell shape maintenance and reshaping, cell motility and migration, neurite extension, dendrite branching, dendritic spine sprouting and stabilization, membrane anchoring of neurotransmitter receptor complexes (e.g., the NMDA receptor complex), and many others. As we have seen above, it is understandable that the deleterious effects of EtOH are more strongly displayed during development in the context of PEE.

29.4.2 Acting-Binding Proteins (ABPs)

In searching for up-regulated genes in the human prefrontal cortex of alcoholic human subjects, a new protein was isolated, cloned, and characterized in 2001: the human neuronal protein 22 (hNP22) (Fan et al., 2001). hNP22 was later found to be increased in all human cortical layers of the prefrontal cortex and in CA3 and CA4 hippocampal regions but not in other cortical areas (such as motor cortex) nor in other hippocampal areas (Depaz et al., 2003). In the rat, this protein (rNP22) has been found to be restricted to the cytoplasm and processes of neurons and to colocalize with proteins of the MF and MT matrices (such as F-actin, α-tubulin, τ-protein, and MAP-2). After exposing Wistar rats to EtOH vapors (a method able to induce very high BECs), acute withdrawal induced rNP22 and its mRNA to increase in the cortex, CA2 area, and dentate gyrus of the Hipp. In contrast, it decreased in the Strt. rNP22 colocalization with F-actin, α-tubulin, or MAP-2 was not markedly altered after chronic EtOH exposure although colocalization at the periphery of the neuronal soma with F-actin was observed only after chronic EtOH exposure and withdrawal. During withdrawal, rNP22 colocalization with MAP-2 reduced whereas its association with F-actin and α-tubulin was maintained (Depaz et al., 2005). The authors of this study suggested that (similar to what was observed in many other studies) the effect of chronic EtOH exposure and subsequent withdrawal on rNP22 expression is region selective and that its disordered levels could alter the neuroplastic cytoskeletal changes associated with the development of EtOH dependence and physical withdrawal (Depaz et al., 2005). Further characterization of this protein in the rat showed that rNP22 expression is evident as early as E15 in the cortical plate (future cortex). However, rNP22 levels are significantly lower during the embryonic than during the postnatal period. Its expression levels slightly increase at P0, then increase significantly at P4 and P16, and remain elevated until P24 at least. Given the temporal expression pattern of rNP22, and since it was observed in growth cones, it was speculated that this protein could be involved in neuronal branching and synaptogenesis (Depaz and Wilce, 2006). Later it was found that the hNP22 (also known as NP12 or transgelin 3) binds to F-actin in vitro and in vivo. hNP22 directly binds to α-tubulin and colocalizes with actin and is possibly involved in mediating (and regulating) an association between MTs and MFs in order to control neuronal processes formation and morphology. hNP22 requires phosphorylation at Ser-180 by PKC in order to induce cytoskeletal rearrangements. Thus, hNP22 was found to be part of a signaling complex that associates with cytoskeletal elements to regulate neuronal morphology (de las Heras et al., 2007).

As we have seen, much more research is needed in order to unravel better the possible contribution of the numerous existing ABPs to the distorted cytoskeleton after EtOH exposure.

In Table 29.1 are summarized the relevant works on EtOH effects on the different components of the cytoskeleton of neurons and glia.

29.5 *Zellschrumpfung*, or a Shared Morphological (Cytoskeletal) Type of Chronic Neuronal Damage

In several previous pieces of work from our laboratory we have found a morphological irregularity in both dendrites and axons appearing mainly as a sinuous, waving, corkscrew-like shape. Sometimes this dysmorphology was also accompanied by immunostaining irregularities in the form of discontinuities. This alteration was found when neuronal processes were evidenced by both MAP-2 and Nf-200 immunostaining. Incidentally, it was also observed with other cytoplasmatic immunomarkers (e.g., tryptophan hydroxylase or 5-HT). We were able to observe this corkscrew-like shape in male Wistar rats subjected to low chronic EtOH exposure at different ages: in PEE postnatal offspring (Evrard et al., 2003), in adolescents (Evrard et al., 2006), and in adults (Tagliaferro et al., 2002) (Fig. 29.2). It was observed immediately after intoxication, after a long abstinence, and even when a quantitative recovery was evident in the expression level of such immunomarkers, suggesting that a long-lasting damage had taken place in the cytoskeleton.

Table 29.1 Effects of EtOH on cytoskeletal proteins[a]

Model of EtOH exposure	In vivo/in vitro model	Level of exposure/BEC	Effect on cytoskeletal protein expression	References[b]
		TUBULIN		
Chronic (8–17 weeks), adult	In vivo, male Sprague-Dawley rats, hepatocytes	High, 36% EDC	Unchanged protein and mRNA; normal assembly, length and structure. A smaller proportion of tubulin: EtOH sensitive and assembly incompetent	Yoon et al. (1998)
Acute, 30 min	In vitro model of gastrointestinal epithelium	Low (1%) to high (2.5 and 15 vol%) EtOH	Tubulin nitrosilation and oxidative injury of cytoskeleton induced by iNOS activation, NO overproduction and $ONOO^-$ formation	Banan et al. (2000)
PEE	In vivo, newborn Wistar rats	High, EtOH in liquid diet	Tubulin levels increased 33% in hepatocytes, polymerization reduced 50%, altered cytosolic distribution	Azorín et al. (2004)
Chronic, 72 h	In vitro, WIF-B cells (hepatoma hybrid cell line)	High, 50 mM (~230 mg/dL)	Threefold increase in α-tubulin acetylation, impaired MT polymerization, hyperstabilization of polymerized MTs	Kannarkat et al. (2006)

Table 29.1 (continued)

Model of EtOH exposure	In vivo/in vitro model	Level of exposure/BEC	Effect on cytoskeletal protein expression	References[b]
Chronic, 72 h	In vitro, WIF-B cells (hepatoma hybrid cell line)	High, 50 mM (~230 mg/dL)	Defects in MT-dependent membrane trafficking; threefold increase in α-tubulin acetylation and hyperstabilization of polymerized MTs with an HDAC6 inhibitor (trichostatin A)	Joseph et al. (2008)
Chronic, 72 h	In vitro, WIF-B cells (hepatoma hybrid cell line)	High, 50 mM (~230 mg/dL)	Decreased levels of HDAC6 with unaltered subcellular distribution	Shepard et al. (2008)
Chronic, 19.5 h	In vitro, calf brain MT proteins	EtOH up to 100 mM/acetaldehyde 0.5 mM	Loss of polymerization ability with acetaldehyde only	**McKinnon et al. (1987)**
Acute exposure	In vitro, tubulin from bovine brain	0.02–0.2 mol acetaldehyde per mol of tubulin	Complete inhibition of tubulin polymerization with 0.08 mol acetaldehyde/mol tubulin	**Smith et al. (1989)**
Chronic, lifelong exposure	In vivo, AA preferring and non-preferring rats	Low, EtOH 10–12 vol% in drinking water	Acetaldehyde–proteins (tubulin) adduct formation in layers IV–V of frontal cortex, molecular layer of cerebellum and zone 3 of Rappaport's hepatic acinus	**Rintala et al. (2000)**
Chronic, 96 h	In vitro, PC12 cells	High, 100 mM (~460 mg/dL)	Increased MTs content, decreased free tubulin	**Reiter-Funk and Dhrman (2005)**
Chronic, 7 days	In vitro, E21 Wistar rat astrocytes	Medium to high (30–100 mM ≈ 138–460 mg/dL)	Dose-dependent MTs disorganization	**Tomás et al. (2005)**

Table 29.1 (continued)

Model of EtOH exposure	In vivo/in vitro model	Level of exposure/BEC	Effect on cytoskeletal protein expression	References[b]
		MAP-2		
Chronic, 4 weeks	In vitro, organotypic slice culture of P5 Wistar rats Hipp	Medium to high (50–200 mM ≈ 230–921 mg/dL)	Decreased expression in CA1 and CA3 areas; decreased expression in dentate molecular layer only at 200 mM and slight increase at 50 mM	**Noraberg and Zimmer (1998)**
Chronic, 16 months (AA), 8 months (Wistar)	In vivo, adult AA and Wistar rats	Low (10 vol% EtOH in drinking water)	AA: mRNA decreased in almost all brain regions, most prominently in putamen, nucleus accumbens, substantia nigra and globus pallidus. Wistar: mRNA decreased in nucleus accumbens, striatum complex, Hipp and cerebellum (less affected), FCx not affected	**Putzke et al. (1998)**
PEE	In vitro, rats, isolated phosphoproteins from FCx and Hipp	High	Decreased phosphorylation in FCx but not in Hipp	**Tan et al. (1993)**
Acute	In vitro, male Sprague-Dawley rats brain	Low (6 mM) to high (768 mM) (≈27.6–3,537 mg/dL)	Biphasic dose-response effect on MAP-2 phosphorylation (increasing until 24 mM and decreasing from 96 mM on)	**Ahluwalia et al. (2000)**

Table 29.1 (continued)

Model of EtOH exposure	In vivo/in vitro model	Level of exposure/BEC	Effect on cytoskeletal protein expression	References[b]
Chronic, 6 weeks	In vivo, adult male Wistar rats	Low (6.6 vol% in drinking water; BECs: 16–25 mg/dL ≈ 3.47–5.42 mM)	41.6% reduction of MAP-2 expression in CA1 Hipp; altered morphology (corkscrew-like apical dendrites)	**Tagliaferro et al. (2002)**
Chronic, 24 h	In vitro, neural and glial cells from E14 Wistar rat fetuses	Medium to High (25–100 mM ≈ 155–460 mg/dL)	Decrease number of MAP-2$^+$ cells (dose-dependently)	**Tateno et al. (2005)**
Chronic, 6 weeks intoxication + 10 weeks abstinence	In vivo, adolescent male Wistar rats	Low (6.6 vol% in drinking water; BECs: 16–25 mg/dL ≈ 3.47–5.42 mM)	After intoxication: decreased expression in CA1 Hipp, Strt and FCx. After abstinence: recovered expression in CA1 Hipp and Strt but not in FCx; altered morphology (corkscrew-like apical dendrites)	**Evrard et al. (2006)**
		τ-PROTEIN		
PEE	In vitro, rats, isolated phosphoproteins from FCx and Hipp	High	Phosphorylation not affected	**Tan et al. (1993)**
Chronic	In vivo, post-mortem human alcoholic subjects with Wernicke's encephalopathy	?	τ-Positive granular inclusions and neurofibrillary tangles in magnocellular cholinergic neurons of the nucleus basalis of Meynert	**Cullen and Halliday, (1995a, b)**
Chronic	In vivo, CSF from demented and non-demented human alcoholics	?	No elevation of τ-protein levels in CSF	**Morikawa et al. (1999)**

Table 29.1 (continued)

Model of EtOH exposure	In vivo/in vitro model	Level of exposure/BEC	Effect on cytoskeletal protein expression	References[b]
Chronic	In vivo, CSF from human subjects with Wernicke's encephalopathy, withdrawal delirium, korsakoff psychosis and non-alcoholic Alzheimer's disease patients	?	Elevation of τ-protein levels in CSF from patients with acute Wernicke's encephalopathy and non-alcoholics with Alzheimer's disease	**Matsushita et al. (2008)**
		DOUBLECORTIN		
Acute/Chronic, 2–32 h	In vitro, organotypic slice cultures of fetal (E17) cortex	Medium to high (200–800 mg/dL ≈ 43.4–173.7 mM)	Marginal (suprapial) glioneuronal heterotopias or "warts" appeared as function of time and EtOH concentration. doublecortin present in warts	**Mooney et al. (2004)**
Chronic, 4-day binge	In vivo, adult male Sprague-Dawley rats, Hipp neurogenesis	High, 25% w/v EtOH in liquid diet (by intragastric administration)	Doublecortin expression peaked at 14 days after abstinence began	**Nixon and Crews (2004)**
Chronic, 1, 2, or 4 weeks	In vivo, adult male Sprague-Dawley rats, Hipp neurogenesis	High, 7% w/v EtOH in liquid diet	Decrease tendency at 1 week, significant and timely dependent decrease at 2 and 4 weeks of treatment; decreased dendritic nodes and endings, and decreased dendritic length of apical dendrites	**He et al. (2005)**

Table 29.1 (continued)

Model of EtOH exposure	In vivo/in vitro model	Level of exposure/BEC	Effect on cytoskeletal protein expression	References[b]
Acute	In vivo, adolescent male Sprague-Dawley rats, Hipp adult neurogenesis	Low to Medium (intragastric single dose: 1.0, 2.5, and 5.0 g/kg; BECs: 33–131 mg/dL ≈ 7.16–28.4 mM)	Reduced expression in the Hipp of rats exposed to a 5.0 g/kg dose	**Crews et al. (2006)**
Chronic, 28 days	In vivo, adult C57BL/6 J mice	Low to medium (10 vol% EtOH in drinking water)	Decreased expression in dentate gyrus of Hipp	**Stevenson et al. (2009)**
MOTOR MAPS (KINESINS AND DYNEINS)				
PEE	In vivo, newborn Wistar rats	High, EtOH in liquid diet	Unchanged levels of kinesin and dynein in hepatocytes	**Azorín et al. (2004)**
Chronic, 7 days	In vitro, E21 Wistar rat astrocytes	Medium to high (30–100 mM ≈ 138–460 mg/dL)	Reduced levels of kinesin and dynein in astrocytes	**Tomás et al. (2005)**
NEUROFILAMENTS (NF-68, NF-160, AND NF-200)				
PEE	In vivo, C57BL/6 J mice	High	Reduced non-phosphorylated Nfs in primary and secondary dendrites of cerebellar Purkinjě neurons; reduction disappeared by P60	**Poltorak et al. (1990)**
Chronic, 72 h	In vitro, primary cultured Hipp neurons from E18 Long-Evans rat fetuses	Medium to high (21.8–109 mM ≈ 100–501 mg/dL)	Reduced expression of Nf-68, Nf-160, and Nf-200	**Saunders et al. (1997)**
Chronic, 6 weeks	In vivo, adult male Wistar rats	Low (6.6 vol% in drinking water; BECs: 16–25 mg/dL ≈ 3.47–5.42 mM)	Reduced expression of Nf-200 in *stratum radiatum* of CA1 Hipp (apical dendrites of pyramidal neurons)	**Tagliaferro et al. (2002)**

Table 29.1 (continued)

Model of EtOH exposure	In vivo/in vitro model	Level of exposure/BEC	Effect on cytoskeletal protein expression	References[b]
PEE (ARND model)	In vivo, P21 Wistar male offspring	Low (6.6 vol% in drinking water; BECs: 16–25 mg/dL ≈ 3.47–5.42 mM)	Increased Nf-200 expression in Hipp and Strt; no change in FCx. However, altered morphology (corkscrew-like apical dendrites in FCx and Hipp, and axons in Strt)	**Evrard et al. (2003)**
Chronic, 6 weeks intoxication + 10 weeks abstinence	In vivo, adolescent male Wistar rats	Low (6.6 vol% in drinking water; BECs: 16–25 mg/dL ≈ 3.47–5.42 mM)	After intoxication: decreased expression of Nf-200 in CA1 Hipp, Strt and FCx. After abstinence: recovered expression (but still reduced) in CA1 Hipp; full recovery and slightly increased in Strt; recovered and surpassed levels in FCx; altered morphology (corkscrew-like apical dendrites in Hipp and FCx, and axons in Strt)	**Evrard et al. (2006)**
PEE (FAS model)	In vivo, E18 Wistar fetuses	High (EtOH dose: 3.5 g/kg i.p. from E10-E18; BECs: 119–>300 mg/dL ≈ 25–>65 mM)	Decreased expression of Nf-68 in telencephalon (dorsal wall, cortical plate) and in mesencephalon	**Aronne et al. (2008)**
		GFAP		
PEE, Chronic, 28 days	In vitro, E21 Wistar rats cortical astrocytes	Medium (25 mM ≈ 115 mg/dL)	Reduced expression	**Renau-Piqueras et al. (1989)**

Table 29.1 (continued)

Model of EtOH exposure	In vivo/in vitro model	Level of exposure/BEC	Effect on cytoskeletal protein expression	References[b]
Acute, FAS	In vivo, P5–P7 pup rats	Medium	Increased expression of protein and mRNA	**Fletcher and Shain (1993)**
Acute, FAS	In vivo, P4–P9 pup rats	Low to High (artificial rearing, EtOH doses: 4.5–6.6 g/kg/day; BECs: 50–300 mg/dL ≈ 10.8–65.1 mM)	Increased immunoreactivity; transient cortical astrogliosis	**Goodlett et al. (1993)**
PEE	In vivo, E1–P4 Wistar fetuses and neonates	Low to medium (5% w/v EtOH in liquid diet; BECs:10–34 mM ≈ 46–156 mg/dl)	Delayed appearance of GFAP expression (and of its mRNA)	**Vallés et al. (1996)**
PEE	In vivo, E21 Wistar rats cortical astrocytes	Medium (5% w/v EtOH in liquid diet; BECs: 60–150 mg/dL ≈ 13–32.5 mM)	GFAP gene transcription significantly reduced, stability of GFAP mRNA slightly reduced, GFAP DNA hypermethylated	**Vallés et al. (1997)**
Acute, 4–5 days	In vivo, adult BALB/C mice	Medium (6.6 vol% EtOH liquid diet)	Increase of GFAP-immunoreactive astrocytes in Hipp (CA3 > CA2 > CA1 > dentate gyrus)	**Satriotomo et al. (2000)**

Table 29.1 (continued)

Model of EtOH exposure	In vivo/in vitro model	Level of exposure/BEC	Effect on cytoskeletal protein expression	References[b]
Chronic, 21 months	In vivo, AA preferring and non-preferring rats	Low, EtOH 2 vol% in drinking water	Decreased expression in different areas of female cerebellum (intake level-related): decreased and increased expression in different areas of male cerebellum (age-related); altered morphology (corkscrew-like fibers in Bergmann glia).	**Rintala et al.** (2001)
Chronic	In vivo, chronic human alcoholics (postmortem delay to fixation: 6–36 h)	?	Reduced astrocytes density in layer V–VI of dorsolateral prefrontal cortex with lower densities if the subjects were also depressed. No differences in GFAP immunoreactivity	**Miguel-Hidalgo et al.** (2002)
Chronic, 6 weeks	In vivo, adult male Wistar rats	Low (6.6 vol% in drinking water; BECs: 16–25 mg/dL ≈ 3.47–5.42 mM)	Increased expression in CA1 Hipp; astrogliosis	**Tagliaferro et al.** (2002)
PEE (ARND model)	In vivo, P21 Wistar male offspring	Low (6.6 vol% in drinking water; BECs: 16–25 mg/dL ≈ 3.47–5.42 mM)	Increased expression in CA1 Hipp and Strt	**Ramos et al.** (2002)
PEE (ARND model)	In vivo, P21 Wistar male offspring	Low (6.6 vol% in drinking water; BECs: 16–25 mg/dL ≈ 3.47–5.42 mM)	Increased GFAP expression in Hipp and FCx, but not in Strt	**Evrard et al.** (2003)

Table 29.1 (continued)

Model of EtOH exposure	In vivo/in vitro model	Level of exposure/BEC	Effect on cytoskeletal protein expression	References[b]
PEE	In vivo,P5–P19 Sprague-Dawley offspring	Medium (6.6 vol% EtOH in liquid diet; BECs: 80–123 mg/dL ≈ 17.4–26.7 mM)	Decreased number of GFAP immunoreactive astrocytes in dorsal and median raphe nucleus mesencephalic region; partial or total reversion by ipsapirone (a 5-HT_{1A} agonist)	**Tajjudin et al. (2003)**
Chronic, 24 h	In vitro, neural and glial cells from E14 Wistar rat fetuses	Medium to High (25–100 mM ≈ 155–460 mg/dL)	Increased expression of GFAP (time-dependently) and increased number of $GFAP^+$ cells (dose-dependently)	**Tateno et al. (2005)**
Chronic, 6 weeks intoxication + 10 weeks abstinence	In vivo, adolescent male Wistar rats	Low (6.6 vol% in drinking water; BECs: 16–25 mg/dL ≈ 3.47–5.42 mM)	After intoxication: increased expression in CA1 Hipp, Strt and FCx. After abstinence: tendency to reduction but still increased expression in CA1 Hipp, Strt and FCx	**Evrard et al. (2006)**
		VIMENTIN		
PEE (FAS model)	In vivo, E18 Wistar fetuses	High (EtOH dose: 3.5 g/kg i.p. from E10-E18; BECs: 119–>300 mg/dL ≈ 25–>65 mM)	Decreased expression in telencephalon (dorsal wall, cortical plate) and in mesencephalon	**Aronne et al. (2008)**
		F-ACTIN		
Chronic, FAS	In vivo, human neonates with fetal alcohol myopathy	?	Ultrastructural alterations in sarcomeric band I of skeletal muscles	Adickes and Schumann (1983)

Table 29.1 (continued)

Model of EtOH exposure	In vivo/in vitro model	Level of exposure/BEC	Effect on cytoskeletal protein expression	References[b]
Chronic	In vivo, Sprague-Dawley rat pups	High	Actin disarray in thin myofilaments of heart muscle	Adickes and Mollner (1986)
Acute (2 h 20 min) and Chronic (6 days)	In vitro, neural crest cells of *Ambystoma maculatum*	Low to Medium (0.05–02% 39.45–157.8 mg/dL ≈ 8.56–34.25 mM)	Disrupted actin fibers; incomplete morphological differentiation, cells were contracted, altered shape and cell-to-cell contacts, loss of substratum adherence	**Hassler and Moran (1986b)**
Acute, binge	In vitro, myocytes from chick embryos heart	High	Decreased expression of actin	Ni et al. (1992)
Chronic	In vivo, autopsy samples of human alcoholics' Hipp	?	Increased Hirano bodies in the *stratum lacunosum*	**Laas and Hagel (1994)**
PEE	In vivo/In vitro, neural crest cells of chick embryos	High (150 mM ≈ 691 mg/dL)	Disorganized actin cytoskeleton, condensation of cell cortex, subpopulation of blebbing apoptotic-like cells; minor velocity of migration, abnormal motility behavior, disturbed linearity	**Rovasio and Battiato, 2002**
Chronic, 7 days	In vitro, E21 Wistar rat cortical astrocytes	Medium to high (30–100 mM ≈ 138–460 mg/dL)	Dose-dependent decrease of actin fibers, rearranged in a circular structure beneath plasma membrane; effect prevented by lysophosphatidic acid	**Tomás et al. (2003)**

Table 29.1 (continued)

Model of EtOH exposure	In vivo/in vitro model	Level of exposure/BEC	Effect on cytoskeletal protein expression	References[b]
PEE	In vivo, newborn Wistar rats	High, EtOH in liquid diet	No changes in hepatocytes	Azorín et al. (2004)
Acute, 15 min	In vitro, RGM-1 cells (rat gastric mucosal cells)	Medium to high (17.1–256.9 mM ≈ 78.9–1,183 mg/dL)	Increased degradation and irregularity of actin filament organization (dose-dependently)	Bidel et al. (2006)
ACTIN-BINDING PROTEINS (ABPS)				
Chronic	In vivo, alcoholic human subjects (postmortem delay to fixation: 5–75 h)	?	Up-regulated gene expression of hNP22 in superior FCx, but not in primary motor cortex or cerebellum	**Fan et al. (2001)**
Chronic	In vivo, alcoholic human subjects	?	Increased expression of hNP22 in all cortical layers of prefrontal cortex, CA3 and CA4 Hipp	**Depaz et al. (2003)**
Acute withdrawal	In vivo, Wistar rats	High (exposure to EtOH vapor)	Increased rNP22 and its mRNA in cerebral cortex, CA2 and dentate gyrus of Hipp; decreased rNP22 and its mRNA in Strt	**Depaz et al. (2005)**

[a]Listed in chronological order

[b]References in bold type referred to nerve tissue cells (both neurons and astrocytes or oligodendrocytes)

Abbreviations: ARND, alcohol-related neurodevelopmental disorders; BEC, blood ethanol concentrations; CA1 Hipp, CA1 area of the hippocampus; CSF, cerebrospinal fluid; EDC, ethanol-derived calories; EtOH, ethanol; FAS, fetal alcohol syndrome; FCx, frontal cortex; GFAP, glial fibrillary acidic protein; PEE, prenatal ethanol exposure; Strt, corpus striatum

In a recent work from our laboratory (Evrard, 2008) we have found a striking morphological alteration in the cerebral cortex of PEE P21 male offspring. In some (but not all) animals of different (but not all) litters we observed in Nissl-stained preparations a structural alteration in pyramidal neurons located in the lateral orbital region of the FCx, according to plate 170 of a rat brain atlas (Paxinos and Watson, 2007) (Fig. 29.3). Cytoarchitectonically, this region corresponds to Brodmann's area 10, according to the classic characterization of the rat cerebral cortex made by Wendell Krieg (1946a, b). Despite a very careful search, we were unable to observe this finding in other cortical regions or in other prosencephalic deep nuclei. These abnormal pyramidal neurons were located spanning from layer II (external granular) to layer V (internal pyramidal), completely sparing layer I (molecular) and most of layer VI (fusiform). All those affected neurons showed a columnar (rather than a laminar) organization. Depending on the tissue slice that was examined, the cortical region of abnormal neurons was comprised of 15–20 columns at its maximum width to 3–5 columns at its minimum. This region showed the shape of a narrow and elongated trapezoid of which the major base was oriented to the pia mater and the minor to the subjacent white matter (Fig. 29.3). This two-dimensional arrangement, apparent in serial vibratome slices, suggested that, tridemensionally, the affected cortex must have had the form of a truncated cone. Except for some isolated neurons, which seemed to be disoriented (see the neuron marked with a double black arrow in Fig. 29.3c4), the majority of them conserved their normal apical-basal orientation within their pertaining column. At higher amplifications, the morphological feature that stood out was the clear cytoplasmatic hyperchromasia in which the normal separation between the Nissl lumps had been lost. This hyperchromasia, however, varied along a spectrum from a nearly normal staining to a highly basophilic dark appearance. In the darkest neurons, the soma were so retracted that it represented only a thin envelop to the equally retracted nucleus. In these cases, both the soma and the nucleus showed clear indentations and undulations, one accompanying the other. Due to the cytoplasmatic retraction, many of these neurons acquired a "spiny"

Fig. 29.2 (continued) Immunohistochemical pictures of EtOH-affected nerve tissue cells in male Wistar rats at different ages. **a** Striatal patches in the Strt of P21 offspring subjected to PEE. **b**,**c** Two cortical neurons from the FCx of the same animals. **d** Apical dendrites of pyramidal neurons in the CA1 Hipp, traversing the *stratum radiatum*, from P60 offspring subjected to PEE. **e** Astrocytes in the same animals and region as in **d**; *inset* shows a corkscrew-like shaped astrocyte. **f** Pyramidal neurons from the FCx of adult males chronically (6 weeks) exposed to EtOH during adulthood. **g** Pyramidal neurons from the FCx of adult males, immediately after a chronic (6 weeks) EtOH exposure during their adolescence; note the highly waving shape of apical dendrites (ad) and the relative sparing of the secondary dendrites (sd). **h** A pyramidal neuron from the FCx of adult males (as in **g**) who were allowed to recover for 10 weeks drinking water (studied after the recovery or abstinence period); note that the corkscrew-like apical dendrite (ad) did not recover its normal *shape* albeit in this region was observed a quantitative recovery in the Nf-200 expression; note also that the secondary (sd) and basal dendrites (bd) are spared. **i** The same region and animals as in **h**, showing corkscrew-like apical dendrites from neighbor pyramidal neurons surrounding a stellate interneuron displaying a smooth and regular *shape*. Immunostains: Nf-200 (in **a–c** and **f–i**); MAP-2 (in **d**); GFAP (in **e**). Primary amplification: 400× (**a–i**) (Modified from Evrard et al., 2003 (**a–c**); Evrard, 2008 (**d–f**); Evrard et al., 2006 (**g–i**); used with permission from Elsevier (**a–c** and **g–i**))

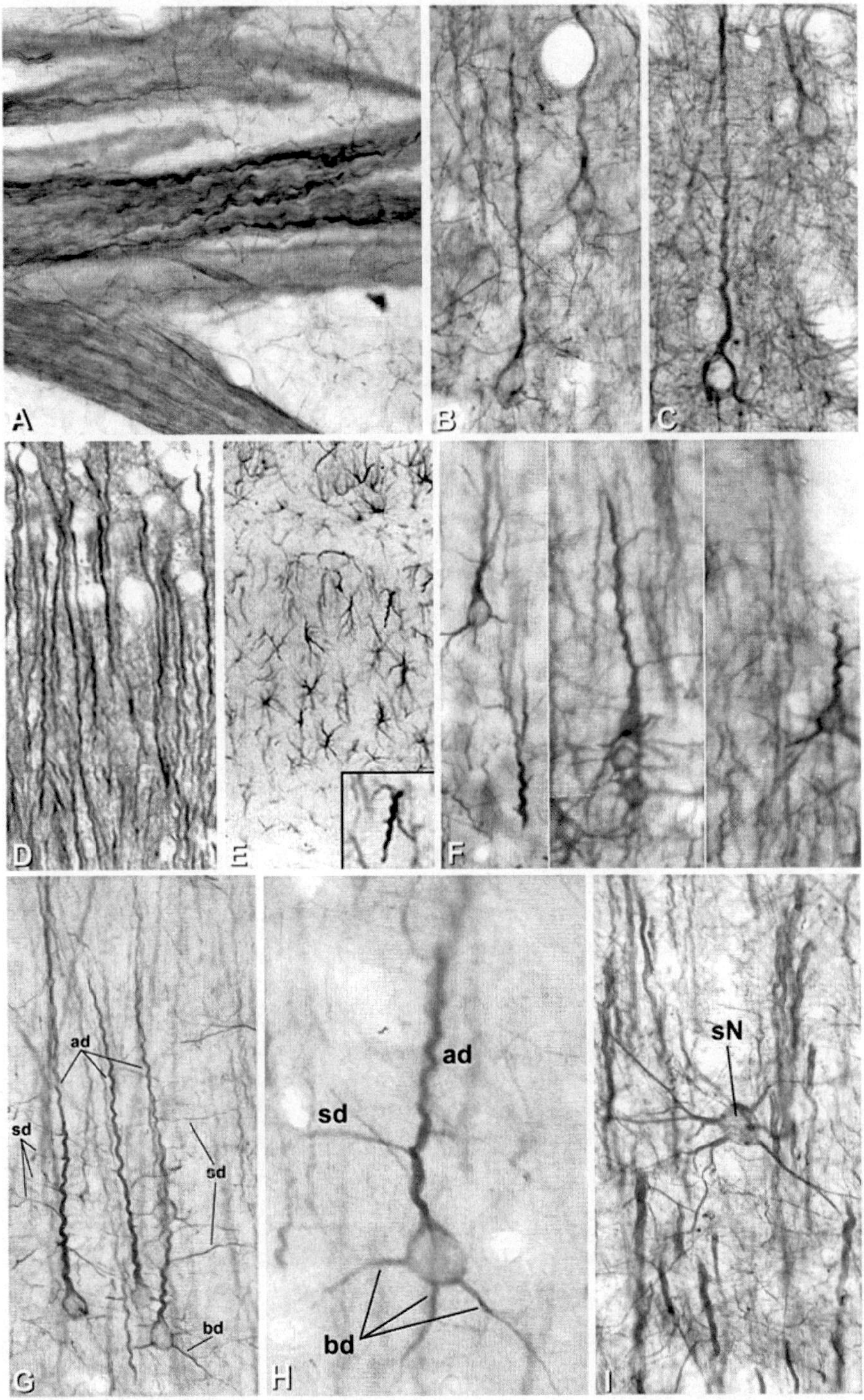

Fig. 29.2 (continued)

aspect (Fig. 29.3c$_{1,2,4}$), each spine corresponding to the cytoplasmatic emergence of axon and dendrites. Globally considered, the cytoplasm tended to show a thinned, rough, and elderly aspect, reminiscent of the shape of raisins. Usually present were the apical dendrites with a sinuous, waving shape. They displayed a corkscrew-like appearance (see neurons marked with a single black arrow in Fig. 29.3c$_{1–4}$,d). The initial segment of the axon was sometimes visible, with a corkscrew-like shape also (see the white arrows in Fig. 29.3c$_{1–3}$). When a waving dendrite (or an axon) changed its direction, they usually showed an obtuse angle, rarely an acute one. The nuclei showed varying degrees of altered morphology, from those almost normal to those with a gradually transverse narrowing and superficial indentations (see Fig. 29.3c$_{1–3}$,e–i). Their chromatin was always lax, with an opaque glass-like look; a nucleolus was always constantly present, even in the more abnormal neurons (see Fig. 29.3h,i).

In our recent work (Evrard, 2008), we have encountered another remarkable finding for the brain of P21 rats: the presence of isolated bipolar cells (marked with an asterisk in Fig. 29.3c$_5$) with a slightly hyperchromatic cytoplasm, an indented nucleus with lax chromatin, and an evident nucleolus, perpendicularly oriented relative to the pia mater. The soma sends two perfectly aligned apical and basal processes (each marked with white arrows in Fig. 29.3c$_5$). Furthermore, the apical process showed a slightly waving shape. These morphological features reminded us of a migrating postmitotic neuroblast still in the process of migration along the apical fiber of a radial glia cell. However, by P21, neuroblast migration has normally finished in the rat cortex (Parnavelas, 1999; Sauvageot and Stiles, 2002). In this context, we should now remember the recent work by Creppe et al., (2009) (see above) who showed that, by altering the Elp3 catalytic subunit of elongator (the tubulin acetylating enzyme), migrating neuroblasts slowed down (but did not stop) their migrating speed. The same authors have also shown that the shape of cells with a silencing in the elongator Elp1 subunit "show highly twisted leading processes." These cells also showed a defective branching in terminally differentiated projection neurons (see above) which correspond to the pyramidal neurons that we have observed in a columnar array in other cortical layers. Morphologically, our results with the Nissl stain (Evrard, 2008) are in full agreement with those of Creppe et al. (2009). What they called "leading processes harboring several bulges" could be nothing but the corkscrew-like shape that we have revealed with this simple stain and with different cytoskeletal immunomarkers (see Fig. 29.2) in different models of EtOH exposure (compare Figure #4D,K and S6P,R from the work of Creppe et al., 2009 with Fig. 29.3, and especially 29.3c$_5$, in the present work).

Fig. 29.3 (continued) Nissl stain and cytoarchitectonic aspect of the FCx – lateral orbital area – from P21 offspring subjected to PEE. **a** The plate 170 from the Paxinos and Watson (2007) rat brain atlas corresponding to the cortical region in which the morphological alterations were observed. **b** The frontal pole and, highlighted, the trapezoid area depicted in **c**. **c** Highlighted the areas shown in **c**$_{1–5}$ and **f**. See the text for further descriptions. *Scale bars*: 675 μm in **b**; 250 μm in **c**; 50 μm in **c**$_1$ (applies to **c**$_{1–5}$, and **d–i**) (Modified from Evrard, 2008)

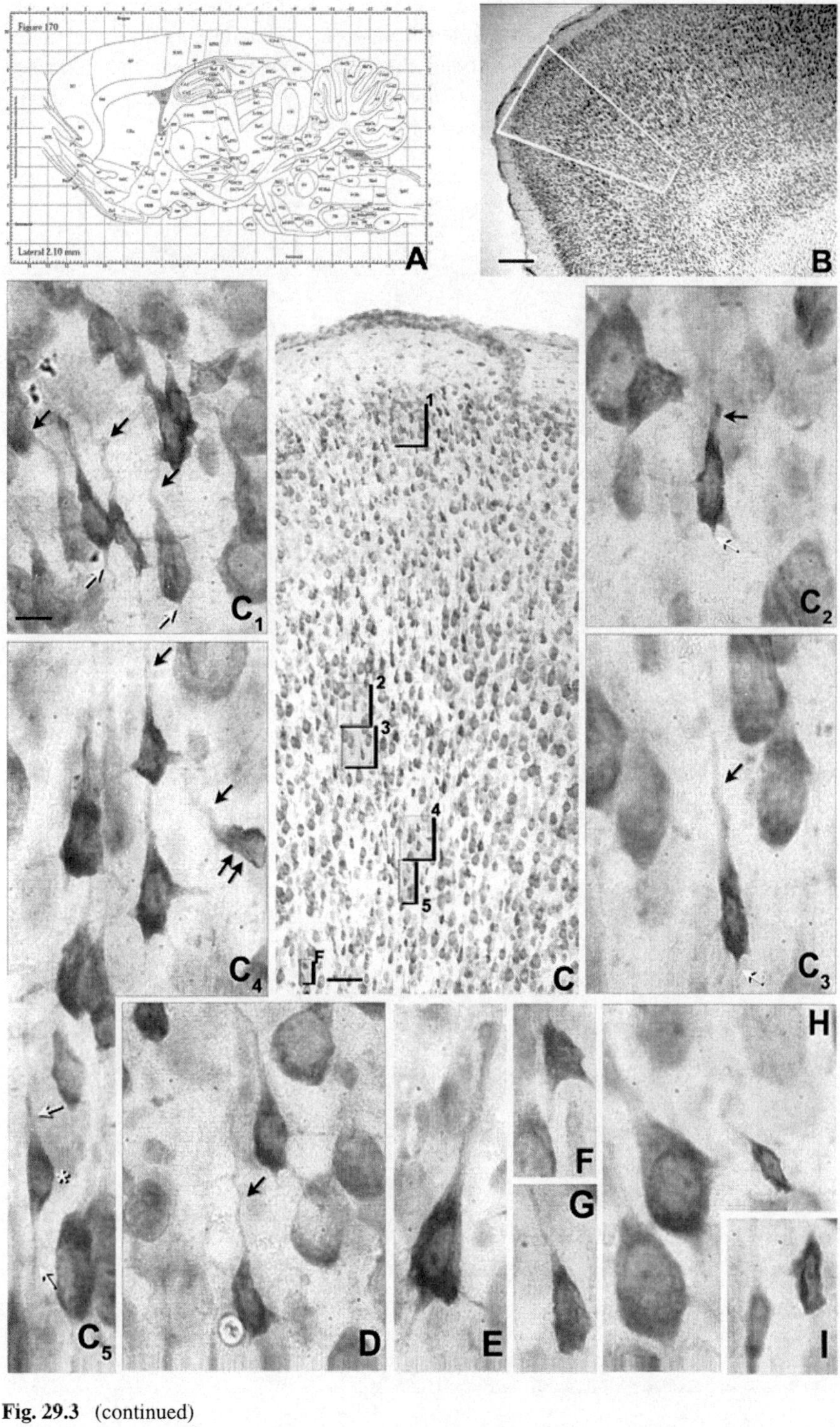

Fig. 29.3 (continued)

When we studied the EtOH-affected brains of P21 male rat offspring under the electron microscope, we also found the corkscrew-like axons and dendrites (Fig. 29.4) and they both showed a disorganized cytoskeleton (Evrard, 2008). Despite the normal pleomorphic morphology of mitochondria (Cragg, 1976), in

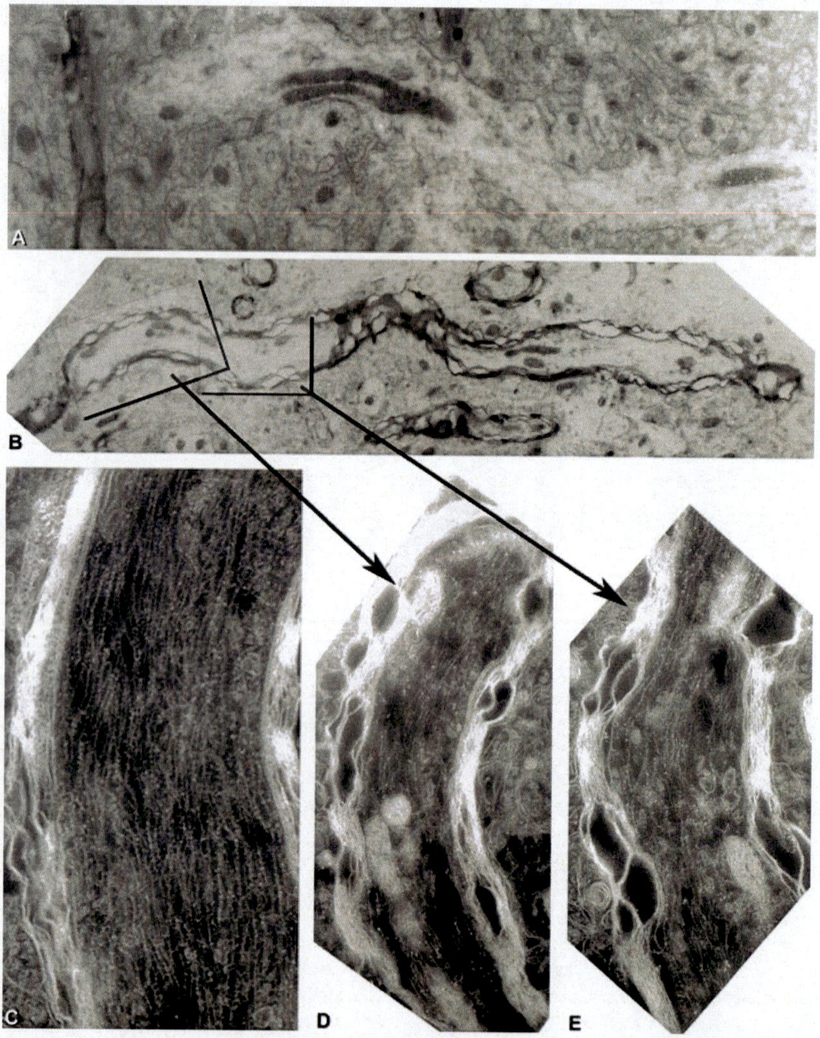

Fig. 29.4 An abnormal apical dendrite (**a**) from a pyramidal neuron in the FCx and an abnormal axon (**a**) from the Strt (both from P21 PEE male offspring). Both processes show a clear corkscrew-like appearance. Note in **a**, inside the dendrite, one mitochondria with an unusual Y-bifurcated shape. A normal axon from the Strt of control animals (**c**) and two amplified portions (**d,e**) from the axon depicted in **b**, showing a disorganized cytoskeleton. Images in **c–e** were digitally inverted to their negative form for a better visualization of the cytoskeleton. Primary amplification: **a**: 7000×; **b**: 5000×; **c–e**: 31500×. (Modified from Evrard, 2008)

many of the dendrites we have observed unusually frequent mitochondria with abnormal, sometimes aberrant, shapes (see, for example, Fig. 29.4a). It has been known for many years that EtOH and oxidative stress are able to alter mitochondria morphology and function (Bakeeva et al., 2001; Cahill et al., 2002; de la Monte and Wands, 2001; Flax and Tisdale, 1964; Koch and Gamboni, 1977; Koch et al., 1977; Lane and Lieber, 1966).

As may be observed in Fig. 29.3, *not all the columns* in the affected region showed altered pyramidal neurons. Similarly, within an affected column *not all the pyramidal neurons* were affected. Thus, besides a more or less affected pyramidal neuron there were many completely normal, non-affected neurons in very close proximity. This fact suggests that the putative noxious factor (EtOH, for instance) did not act unselectively during cortical development. Indeed, it did not act indiscriminately even within the affected region. Instead, we should consider a more selectively induced damage. First, we should ask how this strictly columnar array of the affected neurons in a restricted region of the FCx could be explained? The answer to this question, we think, comes through a necessary developmental consideration. As Kornack and Rakic (1995) and Rakic (1995) pointed out, the deployment of clonally related neuroblasts during corticogenesis in the primate neocortex may take place by means of two mechanisms: a radial deployment or a horizontal one (Fig. 29.5). According to Rakic, in the *radial deployment*, "sibling" neuroblasts generated sequentially from an asymmetrically dividing progenitor cell located in the ventricular zone migrate later in a tandem sequential fashion. Once their migration is stopped, they finally locate within the same column, keeping the normal known inside-out pattern of cortical layers development, according to the radial unit hypothesis (Rakic, 1988). Similar radially oriented clones have also been observed in the rodent cerebral cortex (Luskin et al., 1988; Walsh and Cepko, 1988). In the *horizontal deployment*, "cousin" neuroblasts migrate synchronously in a wave fashion after been generated almost simultaneously by related symmetrically dividing progenitor cells. In this latter mode of deployment, cells locate within the same cortical layer. Because of these two modes of neuroblast deployment in the cortical plate, and considering the strict columnar arrangement of the altered neurons that we have observed, it is very reasonable to conclude that the deleterious action of EtOH must have taken place not in the mature affected neurons themselves, but in a small number of progenitor cells within the ventricular zone. Moreover, given that the altered neurons were located in layers II–V and the apical part of layer VI, we could conclude that the alteration of the progenitor cells must have taken place early during corticogenesis. Those altered progenitor cells, in turn, generated neuroblasts that were in some way altered from their very beginning. After having migrated, those neuroblasts located within the cortical column to which they belong, extended cytoplasmatic processes (axon and dendrites) and probably made synaptic contact with other neurons, thus entering into the formation of new (probably altered) neural circuits. Second, why were not all the neurons affected within a cortical column? Again, in accordance with the radial unit hypothesis (Rakic, 1988), the neurons located within a single column have a polyclonal origin. Thus only those neurons descending from EtOH-affected progenitors would show the altered morphology we

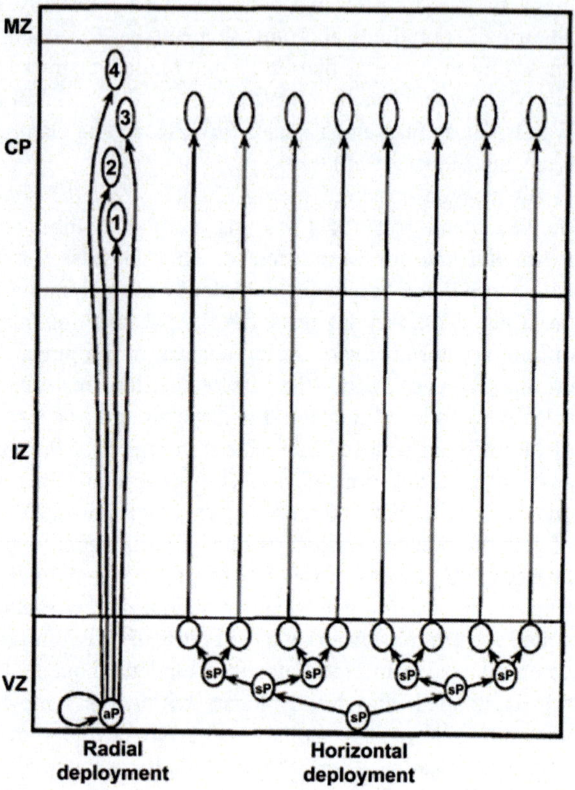

Fig. 29.5 Modes of deployment of clonally related neuroblasts in the developing primate cortex. See description in the text. *VZ*: ventricular zone; *IZ*: intermediate zone; *CP*: cortical plate; *MZ*: marginal zone; aP: asymmetrically dividing progenitor; sP: symmetrically dividing progenitor. (Modified from Kornack and Rakic, 1995; used with permission from Elsevier)

have observed and those neurons that were clearly not affected, but located even in the closest vicinity to the affected ones within the same cortical column, must have descended from non-affected progenitor cells.

The morphological alteration we have found, however, is long and strongly considered merely as of artifactual origin by some authors (Cammermeyer, 1961; Ebels, 1975; Jortner, 2005, 2006). The morphological picture presented by these aberrant neurons is called "*dark neurons*" by those authors because of the hyperchromatic and retracted appearance of the affected cells. The dark neurons would be induced by a defective or delayed brain fixation, by an unskillful manipulation before or after brain fixation (postfixation compressive trauma, for example), by a modified pH or osmolarity of the solutions in which the tissues are processed, etc. For example, when commenting on other authors' work, Jortner (2006), in his figure reproduced at 3, showed dark neurons located in layers V and III of the cerebral cortex of rats chronically intoxicated with a pesticide; Jortner ascribed the origin of these dark neurons to the postmortem manipulation of "unfixed tissue or with perfused, but not fully fixed material" (Jortner, 2006). Jan Cammermeyer wrote many articles on the issue of dark neurons but his most important and cited is one of 1961 (Cammermeyer, 1961) in which he tried to establish unquestionably the artifactual origin of those neurons. As he stated: "Unquestionably, the "dark" neurons have been ignored by many of the authors studying materials fixed by perfusion; they have been regarded as normal cell structure by some, pathological by others, and their causation by failures of the procedure has scarcely been discussed, except by Cotte." Cammermeyer additionally and carefully reviewed the neurohistological and neuropathological history citing Ramón y Cajal, Nissl, von Economo, or Spielmeyer, among many others. Cammermeryer strikingly disbelieved all of them, considering their explanations as erroneous. This author also compiled the most extensive list of dark neuron synonymy ever made (see the note in his page 262) dating its synonyms back to 1886. Incidentally, and inexplicably to us, in his Figure #1c, reproduced here as Fig. 20.6, he confounded a clear pyramidal retracted dark neuron in the cerebral cortex of a guinea pig with an acutely "swollen" oligodendrocyte. In addition, as in our recent work, Cammermeyer even showed pyramidal dark neurons of varying size with a clear columnar array in the cerebral cortex of a cat; some of those neurons also presented a disrupted orientation within their column (Fig. 29.6). Ebel Ebels (Ebels, 1975), a neuropathologist from the University of Gröningen in The Netherlands, easily found dark Purkinjě neurons in the cerebellum of young rats (P16 and older) killed by decapitation. Noteworthy, he could *never* observe them in younger postnatal animals (P0–P12). From the very title, and from the text of his article, it is evident too that Ebels suspected that there was something else in addition to a simple histological artifact in the production mechanism of dark neurons. He noted the dependence on the maturational state of the neurons as a necessary condition to allow the appearance of this morphology. Ebels agreed with Cammermeyer in that "dark neurons are an avoidable artifact" but he made the important statement "that the possibility of the occurrence of

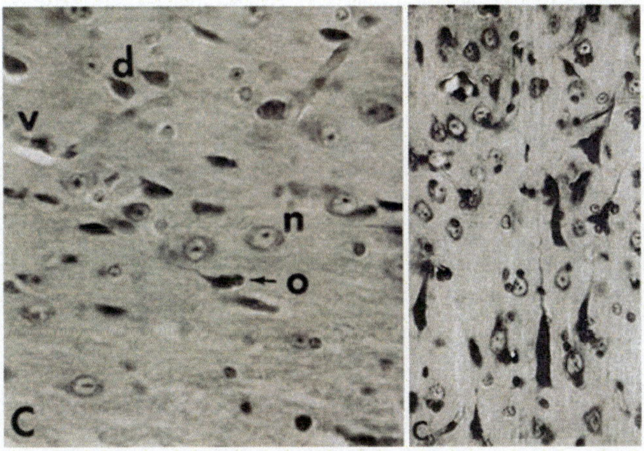

Fig. 29.6 Figures 1**c** (*left*) and 2**c** (*right*) modified from the article by Cammermeyer (1961) showing "dark" (d) and normal (n) neurons and what the author said is an acute "swollen" oligodendrocyte (o); perivascular space (v). See further description in the text. Used with permission from Springer Verlag

this type of artifact depends on the maturational state of the neurons concerned." Ebels ended his article with an ironic statement: "But the fact that some cells (in our material: mature Purkinje cells) can give rise to this type of artifact whereas others (immature Purkinjě cells) cannot proves that calling dark neurons an artifact is not the final verdict. If an artifact, then a significant artifact!" As we now known, neurofilaments, a neuronal cytoskeletal component of outstanding importance, assemble from different Nf proteins in a developmental-ordered way being, as we have seen above, Nf-68 predominant during development and regenerative phenomena (Hoffman et al., 1987; Hoffman and Cleveland, 1988), while Nf-200 is characteristic of mature neurons (Nixon and Sihag, 1991).

There exists abundant direct and indirect recent and historical scientific evidence suggesting the central role of the cytoskeleton in establishing the altered neuronal morphology that we found in our experiments using low EtOH doses. We should stress here that, in our opinion, what certainly suggests (and points to) a cytoskeletal impairment is the constant presence of corkscrew-like, waving dendrites and/or axons. This type of neuronal morphological lesion has already been found in Argentina by Tagliaferro et al. (2006) with MAP-2 immunostaining in the apical dendrites of Hipp neurons in rat chronically subjected to a cannabinoid receptor agonist (WIN-55,212–2); by Ramos et al. (2000) with Nf-200 immunostaining in the axons running in the striosomes of the Strt, after the chronic exposure of Wistar rats to a drug (*p*-chlorophenylalanine) that induce a 5-HT depletion by irreversibly inhibiting the enzyme TPH; by Loidl et al. (1997) with the histochemical NADPH-diaphorase reactivity of nNOS containing neurons in the brain of adult rats subjected to perinatal asphyxia; by Benítez et al. (1996) with the simple Nissl or Tolivia stains in layers III and V of a pachygyric region in the anterior orbitofrontal cortex of a 30-year-old thief and assassin with a severe antisocial, psychopathic personality (the case of M.G., known as "La Iguana"); by Orlando (1952) (Fig. 29.7) in sclerotic and atrophic pyramidal neurons in layers III, V, and VI of the cerebral cortex of subjects affected by the Marchiafava-Bignami disease (a systematic degeneration of the commissural fibers observed in chronic alcoholism); by Christfried Jakob, the initiator and highest exponent of the Argentine-German neurobiological school (Triarhou and del Cerro, 2006a, b) in the pyramidal neurons surrounding the so-called "lacunar disintegration foci" found in the FCx of schizophrenic subjects (Fig. 29.7) (Jakob and Pedace, 1938); and finally by Braulio Moyano, a neuropsychiatrist and best disciple of Jakob, who found these neurons in layer III of the cerebral cortex of demented patients (Fig. 29.7) (Moyano, 1933). In the international literature, neurons with this type of neuronopathy were reported in the following articles: in experimental models of cerebral hypoxia-ischemia with posterior reperfusion (Iizuka et al., 1989; Kövesdi et al., 2007; Nedergaard et al., 2003; Onizuka et al., 1996); models of traumatic brain injury (Gallyas et al., 2004, 2006; Ooigawa et al., 2006); in a transgenic rat model of familial amyotrophic lateral sclerosis (Rafalowska et al., 2006); in a transgenic R6/2 mouse model of Huntington diseases (Petersén et al., 2005); in a model of Alzheimer's disease carried out with transgenic mice overexpressing human presenilin-1 (Chui et al., 1999); in a model of faecal peritonitis in pigs (Papadopoulos et al., 1999); in a model of chronic neuropathic pain induced by peripheral nerve constriction (Mayer et al., 1999); in a model of

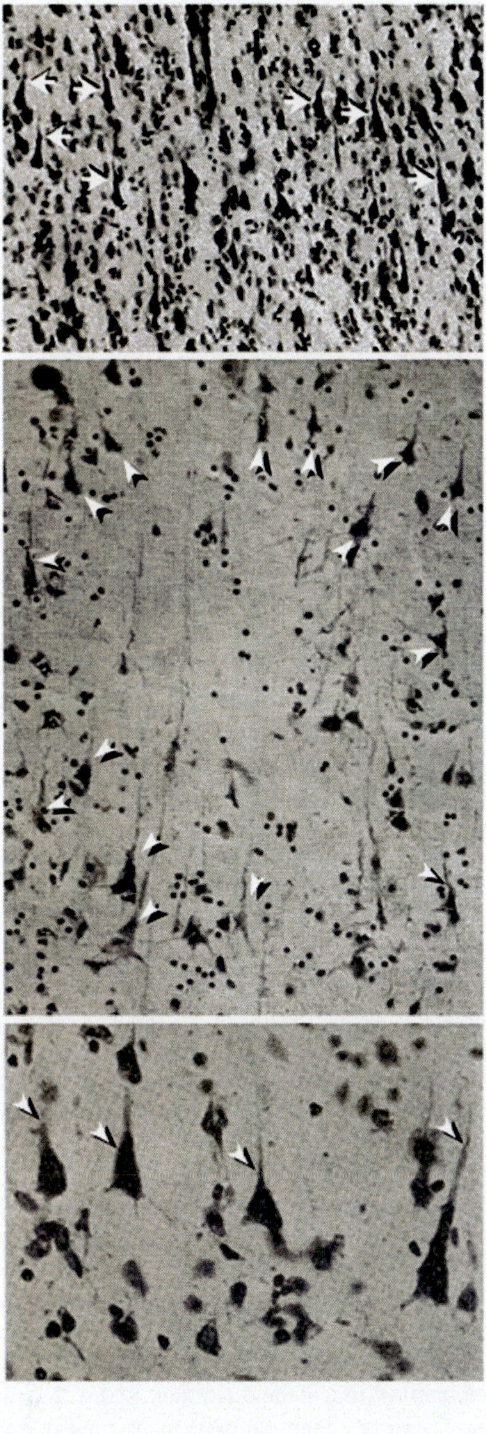

Fig. 29.7 Different pictures from sclerotic and atrophic neurons (*Zellschrumpfung*). *Above*, detail of the figure 20 from the doctoral thesis of Orlando (Orlando, 1952). *Middle*, a "lacunar disintegration focus" or "lacunar desert" observed in the FCx of schizophrenic patients as observed by Jakob (Jakob and Pedace, 1938); note the neuronal desert in which pyramidal neurons are completely absent and the atrophic peripheral neurons with corkscrew-like apical dendrites (*arrowheads*). *Below*, neuronal atrophy in Pick disease such as Moyano has shown in his doctoral thesis (Moyano, 1933)

mild neuronal injury induced by forced swimming or severe injury induced by local injection of ibotenic acid in the Hipp (Ishida et al., 2004); in a model of neuroexcitotoxicity by the iontoforetic injection of kainic acid in the rat amygdala (Hajnal et al., 1997); in models of injury by different types of electrical discharges (Islam et al., 1994; Kellermayer et al., 2006; Zsombok et al., 2005); in a report studying the brain of otherwise normal old men in which the accumulation of τ-protein was studied, more prominently in layers II–IV in the frontal and enthorrinal cortices (Yang et al., 2005); and finally in normal old rats (20 months of age), this morphological feature being found by means of Nf-68, Nf-160, and Nf-200 immunostaining in pyramidal neurons of layers III and V of the FCx in the context of a study of cerebral infarct in rats (Schroeder et al., 2003). In a renowned book by Eugen Bleuler (Bleuler, 1967), this Swiss psychiatrist showed, under the heading of "Alzheimer fibrillary degeneration," two Bielschowski-stained neurons with processes adopting a "twisted appearance or wired-like." Of course, Santiago Ramón y Cajal also showed similar morphological alterations in his renowned book about degeneration and regeneration of the nervous system (Ramón Cajal, 1913). The most remote antecedent we were able to find in the literature is one that Ramón y Cajal himself cited. It corresponds to a work of the Belgian neurohistologist and educator Jean Demoor (1867–1941) who, in a work of 1898 (Demoor, 1898), wrote about the so-called "moniliform neuronal state" (Fig. 29.8) (*état moniliforme*), named after the similar appearance of fungi from the *Candida* genre (formerly called *Monilia*). After having realized experiments with amoebas, leucocytes, heliozoans, *Actinospaerium*, and other inferior organisms, Demoor obtained the same morphological alterations in dogs, frogs, and mice subjected to intoxication with cocaine, alcohol, chloral hydrate, chloroform, ether, or gaslight, subjected to electrical discharges or by cold exposure and in animals in their normal period of hibernation. He also observed this *état moniliforme* in the brain of imbeciles, men affected by "consecutive dementia," in epileptics and in general parethics. Demoor said in his work that this rare morphology was also called "pearly state" of Renaut (*état perlé*) or "granular disintegration" of Verworn (*körnige Zerfall*). Besides the explanation of this state that was current by the time Demoor wrote his article (it was related to processes of neuronal plasticity) (Berlucchi and Buchtel, 2009; Demoor, 1896), for Demoor this state started in the piriform appendices of Stefanowska (our dendritic spines) and propagated from them to the rest of the dendrite. More than 100 years ago he stated that "in the neuron, the moniliform state is a reaction of contraction" and envisioned that "the in-depth study of the physiology of spines or piriform appendices will probably bring the solution to the problem" (Demoor, 1898, page 231). We now know that in dendritic spines there exists a very complex cytoskeleton composed of F-actin and MAP-2, among other cytoskeletal proteins (Di Stefano et al., 2006; Georges et al., 2008). In addition, we have seen above what EtOH does to these proteins.

With regard to the histopathological appearance of dark neurons, Jortner stated that "the changes referenced above, which purport to represent neuronal injury and death, *do not match any of the classically described histopathological forms of neuronopathy*" (our italics) (Jortner, 2006). In spite of this categorical statement, we have seen that there do exist many present and historical antecedents if one goes back even to the nineteenth century in the (mostly European) literature. In our

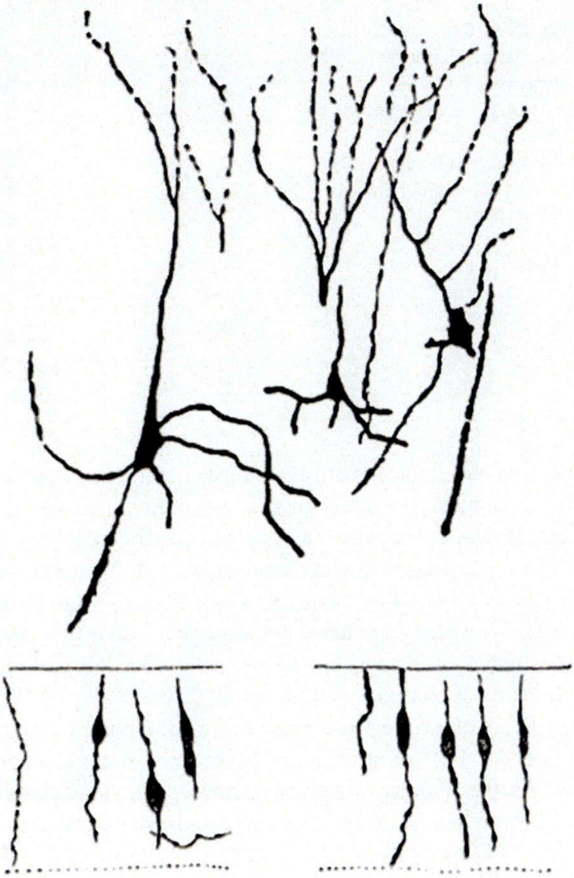

Fig. 29.8 Moniliform state of neurons (*état moniliforme*). *Above*, neurons in the cerebral cortex of "morphinized" dogs (modified from Demoor, 1896) and *below*, olfactory neurons from frogs subjected to the action of cocaine (modified from Demoor, 1898)

opinion, the best description was that made by Walther Spielmeyer (1879–1935) in his classical and influential treatise in textbook style (and, at present, almost neglected) on the histopathology of the human nervous system (Spielmeyer, 1922). Under the heading *Zellschrumpfungen* (cell shrinkage), from page 63–67, in the section "The particular disease forms of neurons" he comprehensively described this morphological neuronal change.[1] Spielmeyer stressed the fact that it is possible to

[1] The name "disease forms" used by Spielmeyer is a clear reminiscence of the work of Alfred Erich Hoche (1865–1943) for whom he was assistant in Freiburg im Breisgau (Hoche, 1912). Spielmeyer, neuropathologist and neuropsychiatrist, later worked in Münich at the Deutsche Forschungsanstalt für Psychiatrie-DFA (German Institute for Psychiatric Research and former name of the present Max-Plank-Institut für Psychiatrie) that was founded and directed until his death by Emil Kräpelin. There, he was Alois Alzheimer's successor, and worked with Korbinian Brodmann, Franz Nissl, Felix Plaut, and Kurt Schneider, among others. Spielmeyer's department at the DFA was known at the time as the "international Mecca of neuropathology" (Weber, 2000).

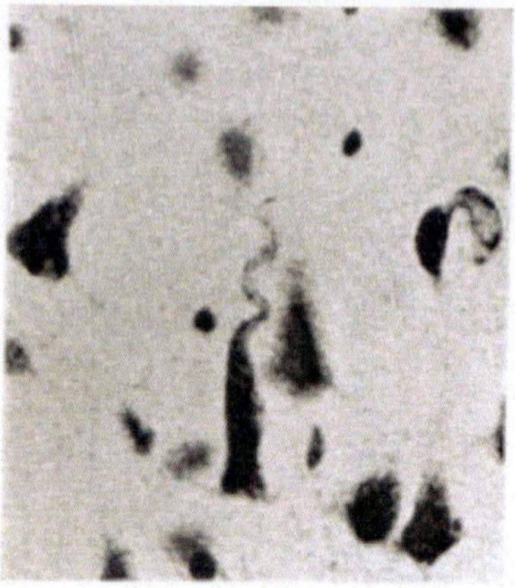

Fig. 29.9 Corresponds to Fig. 30 of Spielmeyer's *Histopathologie des Nervensystem* (Spielmeyer, 1922). Legend: Fig. 30: Sclerotic ganglion cell with a very sinuous apical process (Nissl stain)

observe this neuronopathy (as Jortner called it) in certain acute circumstances; however, the most common causes are chronic processes observed in the cerebral cortex and, in particular, types of neurons, mainly located in layers II and III. This pathological picture was mainly indicative, to Spielmeyer, of a "chronic cell disease" (as it was first called by Franz Nissl) of a slow progression type. The outcome of this *Zellschrumpfung* is "neuronal sclerosis" (*Zellsklerose*), a long-lasting morphology in which neurons may persist for years since it is common to observe them in almost all forms of chronic organic mental diseases (Fig. 29.9). Max Bielschowsky (1869–1940), another great German neuropathologist, wrote (in English) a chapter on the histopathology of nerve cells in a renowned book edited by Wilder Penfield (1891–1976), the famous Canadian neurosurgeon. This chapter contained the description of "shrinkage of nerve cells" (that closely followed that of Spielmeyer's) in pages 173–174 (Bielschowsky, 1932).

Nowadays, for the study of this type of degenerating neurons, there are useful histological techniques; for example, two silver impregnation techniques, such as the argyrophil III method (Gallyas et al., 1990), the amino-cupro-silver or de Olmos technique (de Olmos et al., 1994), and vanadium acid fuchsin (Victorov et al., 2000) and fluoro-jade C (Schmued et al., 2005). However, as we could see, the simple microscopic observation of morphology, as with any histological method, is the most useful tool in order to diagnose this neuronopathy.

We want to call attention here to the fact that none of the authors devoted to the theory of dark neurons as histological artifacts could explain why such notable neuronal damage would be induced with a laminar distribution (mainly in layers II–III and V, as we have seen above) or with a columnar array (as we have observed in Evrard, 2008). In fact, they never intended to do it. They neither explained why a particular neuron could be strongly affected while another, very close in the same layer or column, could not. Similarly, they never advanced a theory to explain why only a few affected neurons were observed in selective, often deep cortical layers, and why not in all neurons in all layers located under the postulated point of pressure, if such a pressure was applied to the surface of the cerebral cortex by an unskilled postmortem manipulation of the brain. Apart from Ebels (1975) who pointed to the developmental factor, no author was able to explain why the dark neurons are not observed under a certain age in animals or why they are more frequently seen in adults, both animals and humans.

Finally, we would like to advance here an alternative possible explanation for all these morphological changes that we and others have observed under EtOH exposure. We have seen above that there are many indications pointing to cytoskeletal involvement in the production of this type of neuronal damage. Ferenc Gallyas and his colleagues at the University of Pécs in Hungary made important indications in that regard. After many years of research on the dark neurons, Gallyas advanced and developed his explanatory theory mainly in two works (Gallyas et al., 1992, 2004). He called it the *gel-to-gel phase transition theory of whole-cell ultrastructural compaction* (Gallyas et al., 2004). In a more recent work (Kellermayer et al., 2006), Gallyas summarized his theory in two paragraphs to which the interested reader is referred. This theory has the great advantage of explaining how a particular neuron may be affected while their immediate neighbors may not. Once the phase transition had taken place in one particular point of the cytoplasmatic gel, it spreads out to all over the soma and cellular processes in an all-or-nothing fashion, Gallyas hypothesized (Gallyas et al., 2004). Based on Spielmeyer's description and on our own observations, but unlike Gallyas, we think that the process might not be strictly of an all-or-nothing nature but one of a graded nature, since an entire spectrum of morphologically altered neurons could be observed. Gallyas once thought that Nfs network could be the main cytoskeletal component responsible for this morphology based on the assumed storage of torsional energy (Gallyas et al., 1992). However, he has disregarded this idea in favor of his proposed cytoplasmatic gel structure (Gallyas et al., 2004). We have previously seen that EtOH may alter Nfs structure, and not only EtOH but many other physical or chemical factors as well. These factors may all provide the required "activation energy" that Gallyas predicts as necessary to activate the irreversible "domino principle" that would initiate the morphological change (Gallyas et al., 2004). However, we propose here that another (possibly related) situation is possible, based on the structure and internal disposition of the Nfs network. Theoretically, all those factors might induce oxidative stress on the cytoskeleton (Albano, 2006; Bellomo and Mirabelli, 1992; Das and Vasudevan, 2007; Haorah et al., 2008), which in turn might trigger an autocatalytic conformational transition or misfolding of Nfs structure such that it would subsequently induce an aggregation prone state and thus a metastable morphological change (Gallyas et al., 2004; Morris et al., 2009). This change could be of a similar nature to that observed in the unstable hemoglobin of patients with sickle cells disease under oxidative stress (Morris et al., 2009; Williamson, 1993) or with the aggregation of prion proteins (see for example, Pinheiro, 2006; Soto and Estrada, 2008). In such cases, a structurally abnormal protein possessing sticky ends, once in contact with other similar protein/s, finally leads to the global morphological change of the whole cell.

Moreover, the EtOH-disordered structure of Nfs could explain why there were no recoveries of the morphological changes in adult rats exposed during adolescence after abstinence, despite the increased recovered expression of Nfs (Evrard et al.,

2006). Thus, the newly aggregated Nf protein would be incorporated to already structurally disordered Nfs. On the other hand, taking into account the developmental differential expression and stoichiometric varying proportion of the different Nf proteins in developing neurons (Hoffman et al., 1987; Hoffman and Cleveland, 1988; Nixon and Sihag, 1991), Nfs involvement in the action of those noxious factors could also explain why certain neurons (with a columnar or laminar array) appear as *Zellschrumpfung* only after a certain postnatal age (Ebels, 1975; Kornack and Rakic, 1995; Evrard, 2008). If a developing neuron possesses primarily altered and still immature Nfs, the addition of subsequent monomers of Nfs protein would possibly be capable of increasing the damage. Regarding the different intracellular regional compartments (soma, apical, and basal or primary and secondary dendrites, axon), Nfs involvement in EtOH damage could additionally explain why, in a given particularly affected neuron, there are different levels of corkscrew-like impairments, such as those observed in apical, basal, or secondary dendrites in our work (Evrard et al., 2006). For instance, initial damage in an apical dendrite could spread out all over it; however, it could stop initially at the emergence of secondary dendrites or at the soma. Increasing damage to the Nfs could imply an increased level of morphological damage, and the level of damage could be characteristic of each neuron itself, independently of how near or far they are located from each other.

From our results, it does not seem likely that we will observe a recovery of cytoskeletal damage once it is EtOH-damaged, as other authors have observed under different conditions. The key question is, however, whether that damage will allow the neuron to function normally, subnormally, abnormally, or not at all. If a damaged neuron functions abnormally, its consequences will depend, of course, on the type and number of neurons, on its brain location, on the developmental period during which it was damaged, and on the age of the animal which carries the altered neuron.

Many questions are still to be resolved and many experiments still have to be done in order to answer those questions.

29.6 Conclusions

At the moment, all the existing data that have been jointly presented here as pieces of a still incomplete four-dimensional puzzle (data from our and other authors' laboratories) seem to point towards the following statement: *relative to the cytoskeleton, EtOH, as a poisoning drug, acts during prenatal development, childhood, adolescence, or adulthood, as a disturbing element which can modify its fine composition and morphology (in a more subtle or gross fashion), and concomitantly impairs (to a higher or lesser degree) the exquisite normal functionality of both neurons and glia.*

Zellschrumpfung, or the dark neuron, is certainly not a histological artifact possibly observed after EtOH exposure, but nothing more than the expression of cytoskeletal damage that can be induced by many different causes of a physical or chemical nature. The molecular mechanisms by which EtOH ultimately affects the neuronal and glial cytoskeleton are, in all probability, not exclusive to this drug. In contrast, as the morphological evidence (both from pathological and experimental, current and historical, sources) seems to suggest, EtOH and many other types of injuries acting on the CNS share those mechanisms.

In any case, the cytoskeletal disruption would be the final common pathway, the generally shared mechanism by which many chronic noxious stimuli may alter cellular morphology and functionality in the nervous tissue.

> En realidad, la disposición de una neurona adulta representa el término de una serie de movimientos, de impulsos interiores y exteriores, que obraron durante la época embrionaria y juvenil (. . .) La razón de la forma está, pues, por entero en la función actual ó pasada.
>
> Santiago Ramón Cajal (1852–1934)[2]

> Äußerten wir oben, daß die Geschichte des Menschen den Menschen darstelle, so laßt sich hier auch wohl behaupten, daß die Geschichte der Wissenschaft die Wissenschaft selbst sei
>
> Johann Wolfgang von Goethe (1749–1832)[3]

References

Adickes ED, Mollner TJ (1986) Ethanol-induced cytoskeletal dysgenesis with dietary protein manipulation. Alcohol Alcohol 21(4):347–355

Adickes ED, Schuman RM (1983) Fetal alcohol myopathy. Pediatr Pathol 1(4):369–384

Ahluwalia B, Ahmad S, Adeyiga O, Wesley B, Rajguru S (2000) Low levels of ethanol stimulate and high levels decrease phosphorylation in microtubule-associated proteins in rat brain: an in vitro study. Alcohol Alcohol 35(5):452–457

Albano E (2006) Alcohol, oxidative stress and free radical damage. Proc Nutr Soc 65(3):278–290

Aronne MP, Evrard SG, Mirochnic S, Brusco A (2008) Prenatal ethanol exposure reduces the expression of the transcriptional factor Pax6 in the developing rat brain. Ann NY Acad Sci 1139:478–498

Azmitia EC (2001) Modern views on an ancient chemical: serotonin effects on cell proliferation, maturation, and apoptosis. Brain Res Bull 56(5):413–424

Azmitia EC, Rubinstein VJ, Strafaci JA, Ríos JC, Whitaker-Azmitia PM (1995) 5-HT1A agonist and dexamethasone reversal of para-chloroamphetamine induced loss of MAP-2 and synaptophysin immunoreactivity in adult rat brain. Brain Res 677:181–192

Azorín I, Portolés M, Marín P, Lázaro-Diéguez F, Megías L, Egea G, Renal-Piqueras J (2004) Prenatal ethanol exposure alters the cytoskeleton and induces glycoprotein microheterogeneity in rat newborn hepatocytes. Alcohol Alcohol 39(3):203–212

Bakeeva LE, Skulachev VP, Sudarikova YV, Tsyplenkova VG (2001) Mitochondria enters the nucleus (one further problem in chronic alcoholism). Biochemistry (Moscow) 66(12): 1335–1341

Banan A, Fields JZ, Decker H, Zhang Y, Keshavarzian A (2000) Nitric oxide and its metabolites mediate ethanol-induced microtubule disruption and intestinal barrier dysfunction. J Pharmacol Exp Ther 294(3):997–1008

[2] "In fact, the disposition of an adult neuron represents the end of a series of movements, of internal and external impulses, which acted during the embryonic and juvenile epochs (. . .). The reason for the form is, then, entirely in the present or past function". Ramón Cajal (1899; p. viii).

[3] "If we said above that the history of the person explains the person, then it can probably also be maintained here that the history of science is science itself".

Bellomo G, Mirabelli F (1992) Oxidative stress and cytoskeletal alterations. Ann NY Acad Sci 663:97–109

Benítez I, Montero LO, Affani JM (1996) Alteraciones de la corteza orbitaria anterior en un sujeto con grave comportamiento antisocial. Alcmeon 5(2):117–120 (This article may be accessed at http://www.alcmeon.com.ar/5/18/a18_01.htm). [Alterations of the anterior orbital cortex in a subject with a severe antisocial behavior]

Berlucchi G, Buchtel HA (2009) Neuronal plasticity: historical roots and evolution of meaning. Exp Brain Res 192:307–319

Bertrand J, Floyd RL, Weber MK, O'Connor M, Riley EP, Johnson KA, Cohen DE, National Task Force on FAS/FAE (2004) Fetal alcohol syndrome: guidelines for referral and diagnosis. Centers for Disease Control and Prevention, Atlanta, GA

Bidel S, Mustonen H, Khalighi-Sikaroudi G, Lehtonen E, Puolakkainen P, Kiviluoto T, Kivilaakso E (2006) Effect of ulcerogenic agents ethanol, acetylsalicylic acid and taurocholate on actin cytoskeleton and cell motility in cultures rat gastric mucosal cells. World J Gastroenterol 11(26):4032–4039

Bielschowsky M (1932) Histopathology of nerve cells. In: Penfield W (ed) Cytology and cellular pathology of the nervous system. Paul B. Hoeber (publisher), New York, NY, pp 145–188

Bleuler E (1967) Tratado de Psiquiatría. 2ª edición. (Spanish translation of from the 10th German edition of the "Lehrbuch der Psychiatrie", revised by Manfred Bleuler). Espasa-Calpe, Madrid, p 764

Brown JP, Couillard-Després S, Cooper-Kuhn CM, Winkler J, Aigner L, Kuhn HG (2003) Transient expression of doublecortin during adult neurogenesis. J Comp Neurol 467(1):1–10

Cahill A, Cunningham CC, Adachi M, Ishii H, Bailey SM, Fromenty B, Davies A (2002) Effects of alcohol on oxidative stress on liver pathology: the role of the mitochondrion. Alcohol Clin Exp Res 26(6):907–915

Cammermeyer J (1961) The importance of avoiding "dark" neurons in experimental neuropathology. Acta Neuropathol 1:245–270

Charras GT (2008) A short history of blebbing. J Microsc 231(3):466–478

Chui D-H, Tanahashi H, Ozawa K, Ikeda S, Checler F, Ueda O, Suzuki H, Araki W, Inoue H, Shirotani K, Takahashi K, Gallyas F, Tabira T (1999) Transgenic mice with Alzheimer presenilin 1 mutations show accelerated neurodegeneration without amyloid plaque formation. Nat Med 5(5):560–564

Cragg BG (1976) Ultrastructural features of human cerebral cortex. J Anat 121(2):331–362

Creppe C, Malinouskaya L, Volvert M-L, Gillard M, Close P, Malaise O, Laguesse S, Cornez I, Rahmouni S, Ormenese S, Belachew S, Malgrange B, Chapelle J-P, Siebenlist U, Moonen G, Chariot A, Nguyen L (2009) Elongator controls the migration and differentiation of cortical neurons through acetylation of α-tubulin. Cell 136(3):551–564

Crews FT, Mdzinarishvili A, Kim D, He J, Nixon K (2006) Neurogenesis in adolescent brain is potently inhibited by ethanol. Neuroscience 137(2):437–445

Cullen KM, Halliday GM (1995a) Mechanisms of cell death in cholinergic basal forebrain neurons in chronic alcoholics. Metab Brain Dis 10(1):81–91

Cullen KM, Halliday GM (1995b) Neurofibrillary tangles in chronic alcoholics. Neuropathol Appl Neurobiol 21(4):312–318

Das SK, Vasudevan DM (2007) Alcohol-induced oxidative stress. Life Sci 81:177–187

Dehmelt L, Halpain S (2005) The MAP2/Tau family of microtubule-associated proteins. Genome Biol 6(1):204

Delacourte A (2005) Tauopathies: recent insights into old diseases. Folia Neuropathol 43(4): 244–257

de la Monte SM, Wands JR (2001) Mitochondrial DNA damage and impaired mitochondrial function contribute to apoptosis of insulin-stimulated ethanol exposed neuronal cells. Alcohol Clin Exp Res 25(6):898–906

De las Heras R, Depaz I, Jaquet V, Kroon P, Wilce PA (2007) Neuronal protein 22 colocalises with both the microtubule and microfilament cytoskeleton in neurite-like processes. Brain Res 1128:12–20

Demoor J (1896) La plasticité morphologique des neurones cérébraux. Arch Biol Bruxelles 14:723–752

Demoor J (1898) Le mécanisme et la significatión de l'etat moniliforme des neurones. Ann Soc Roy Sci Med et Nat Bruxelles 7:205–250

de Olmos J, Beltramino CA, de Olmos de Lorenzo S (1994) Use of an amino-cupro-silver technique for the detection of early and semiacute neuronal degeneration caused by neurotoxicants, hypoxia, and physical trauma. Neurotoxicol Teratol 16(6):545–561

Depaz IM, Ito M, Matsumoto I, Niwa S, Kroon P, Wilce PA (2003) Expression of hNP22 is altered in the frontal cortex and hippocampus of the alcoholic human brain. Alcohol Clin Exp Res 27(9):1481–1488

Depaz IM, de las Heras R, Kroon PA, Wilce PA (2005) Changes in neuronal protein 22 expression and cytoskeletal association in the alcohol-dependent and withdrawn rat brain. J Neurosci Res 81(2):253–260

Depaz IM, Wilce PA (2006) The novel cytoskeleton-associated protein Neuronal protein 22: elevated expression in the developing rat brain. Brain Res 1081:59–64

Dietrich RA (2004) Acetaldehyde: déjà vu du jour. J Stud Alcohol 65(5):557–572

Di Stefano G, Casoli T, Fattoretti P, Balietti M, Grossi Y, Giogetti B, Bertoni-Freddari C (2006) Level and distribution of microtubule-associated protein-2 (MAP2) as an index of dendritic structural dynamics. Rejuvenation Res 9(1):94–98

Donato R (2003) Intracellular and extracellular roles of S100 proteins. Microsc Res Tech 60: 540–551

dos Remedios CG Chhabra D Kekic M Dedova IV Tsubakihara M Berry DA Nosworthy NJ (2003) Acting binding proteins: regulation of cytoskeletal microfilaments. Physiol Rev 83: 433–473

Ebels EJ (1975) Dark neurons. A significant artifact: the influence of the maturational state of neurons on the occurrence of the phenomenon. Acta Neuropathol (Berlin) 33:271–273

Evrard SG 2008. Alteraciones del desarrollo cerebral en el alcoholismo materno-fetal: rol del sistema serotoninérgico y de la astroglía. Tesis de doctorado. Facultad de Medicina, Universidad de Buenos Aires. Pp 367. [Alterations of brain development in maternal-fetal alcoholism: role of the serotoninergic system and astroglia. Doctoral dissertation]

Evrard SG, Duhalde Vega M, Ramos AJ, Tagliaferro P, Brusco A (2003) Altered neuron-glia interactions in a low, chronic prenatal ethanol exposure. Dev Brain Res 147(1–2): 119–133

Evrard SG, Duhade Vega M, Tagliaferro P, Mirochnic S, Caltana LR, Brusco A (2006) A low chronic ethanol exposure induces morphological changes in the adolescent rat brain that are not fully recovered even after a long abstinence: an immunohistochemical study. Exp Neurol 200(2):438–459

Fan L, Jaquet V, Dodd PR, Chen W, Wilce PA (2001) Molecular cloning and characterization of hNP22: a gene up-regulated in human alcoholic brain. J Neurochem 76:1275–1281

Flax MH, Tisdale WA (1964) An electron microscopic study of alcoholic hyalin. Am J Pathol 44(3):441–453

Fletcher TL, Shain W (1993) Ethanol-induced changes in astrocyte gene expression during rat central nervous system development. Alcohol Clin Exp Res 17(5):993–1001

French SW, Katsuma Y, Ray MB, Swierenga SHH (1987) Cytoskeletal pathology induced by ethanol. Ann NY Acad Sci 492:262–276

Friocourt G, Koulakoff A, Chafey P, Boucher D, Fauchereau F, Chelly J, Francis F (2003) Doublecortin functions at the extremities of growing neuronal processes. Cerebral Cortex 13:620–626

Galloway PG, Perry G, Gambetti P (1987) Hirano body filaments contain actin and actin-associated proteins. J Neuropathol Exp Neurol 46(2):185–199

Gallyas F, Güldner FH, Zoltay G, Wolff JR (1990) Golgi-like demonstration of "dark" neurons with an argyrophil III method for experimental neuropathology. Acta Neuropathol 79(6):620–628

Gallyas F, Zoltay G, Dames W (1992) Formation of "dark" (argyrophilic) neurons of various origins proceeds with a common mechanism of biophysical nature (a novel hypothesis). Acta Neuropathol 83:504–509

Gallyas F, Farkas O, Mázló M (2004) Gel-to-gel phase transition may occur in mammalian cells: mechanism of formation of "dark" (compacted) neurons. Biol Cell 96:313–324

Gallyas F, Pál J, Farkas O, Dóczi T (2006) The fate of axons subjected to traumatic ultrastructural (neurofilament) compaction: an electron-microscopic study. Acta Neuropathol 111: 229–237

Garrett S, Kapoor TM (2003) Microtubule assembly: catastrophe factors to the rescue. Curr Biol 13:R810–R812

Georges PC, Hadzimichalis NM, Sweet ES, Firestein BL (2008) The yin-yang of dendrite morphology: unity of actin and microtubules. Mol Neurobiol 38(3):270–284

Gleeson JG, Lin PT, Flanagan LA, Walsh CA (1999) Doublecortin is a microtubule-associated protein and is expressed widely by migrating neurons. Neuron 23:257–271

Goodlett CR, Leo JT, O'Callaghan JP, Mahoney JC, West JR (1993) Transient cortical astrogliosis induced by alcohol exposure during the neonatal brain growth spurt in rats. Dev Brain Res 72(1):85–97

Götz M (1998) How are neurons specified: master or positional control? Trends Neurosci 21: 135–136

Hajnal A, Lénárd L, Czurkó A, Sándor P, Karádi Z (1997) Distribution and time course of appearance of "dark" neurons and EEG activity after amygdaloid kainate lesion. Brain Res Bull 43(2):235–243

Hammond JW, Cai D, Verhey KJ (2008) Tubulin modifications and their cellular functions. Curr Opin Cell Biol 20:71–76

Haorah J, Ramírez SH, Floreani N, Gorantla S, Morsey B, Persidsky Y (2008) Mechanism of alcohol-induced oxidative stress and neuronal injury. Free Radic Biol Med 45(11): 1542–1550

Hasegawa M (2006) Biochemistry and molecular biology of tauopathies. Neuropathology 26(5):484–490

Hassler JA, Moran DJ (1986a) The effects of ethanol on embryonic actin: a possible role in teratogenesis. Experientia 42(5):575–577

Hassler JA, Moran DJ (1986b) Effects of ethanol on the cytoskeleton of migrating and differentiating neural crest cells: possible role in teratogenesis. J Craniofac Genet Dev Suppl 2:129–136

He J, Nixon K, Shetty AK, Crews FT (2005) Chronic alcohol exposure reduces hippocampal neurogenesis and dendritic growth of newborn neurons. Eur J Neurosci 21(10):2711–2720

Hernández F, Ávila J (2007) Tauopathies. Cell Mol Life Sci 64(17):2219–2233

Hipólito L, Sánchez MJ, Polache A, Granero L (2007) Brain metabolism of ethanol and alcoholism: an update. Curr Drug Metab 8(7):716–727

Hirano A (1994) Hirano bodies and related neuronal inclusions. Neuropathol Appl Neurobiol 20(1):3–11

Hirokawa N, Noda Y, Okada Y (1998) Kinesin and dynein superfamily proteins in organelle transport and cell division. Curr Opin Cell Biol 10(1):60–73

Hoche AE (1912) Die Bedeutung der Symptomenkomplexe in der Psychiatrie. Ztsch f d ges Neurol u Psychiat 12:540–551 [The significance of symptom complexes in psychiatrie]. (There exists an English translation: Hoche A. E. 1991. Die Bedeutung der Symptomenkomplexe in der Psychiatrie. Hist Psychiatry 2: 334–343. There also exist a Spanish translation: El significado de los complejos sintomáticos en psiquiatría. En: Hoche A., Kräpelin E., Bumke O. 1999. Los síntomas de la locura. Editorial Triacastela, Madrid. 179 pp.)

Hoffman PN, Cleveland DW, Griffin JW, Landes PW, Cowan NJ, Price DL (1987) Neurofilament gene expression: a major determinant of axonal caliber. Proc Natl Acad Sci USA 84: 3472–3476

Hoffman PN, Cleveland DW (1988) Neurofilament and tubulin expression recapitulates the developmental program during axonal regeneration: induction of a specific beta-tubulin isotype. Proc Natl Acad Sci USA 85:4530–4533

Hoyme HE, May PA, Kalberg WO, Kodituwakku P, Gossage JP, Trujillo PM, Buckley DG, Miller JH, Aragón AS, Khaole N, Viljoen DL, Jones KL, Robinson LK (2005) A practical clinical approach to diagnosis of fetal alcohol spectrum disorders: clarification of the 1996 institute of medicine criteria. Pediatrics 115:39–47

Hubbert C, Guardiola A, Shao R, Kawaguchi Y, Ita A, Nixon A, Yoshida M, Wang XF, Yao TP (2002) HDAC6 is a microtubule-associated deacetylase. Nature 417(6887):455–458

Huss M 1852. Chronische alkoholskrankheit oder alcoholismus chronicus. Ein beitrag zur Kenntniss der vergiftungs-krankheiten, nach eigener und andener Erfahrung. Stockholm und Leipzig, C. E. Fritze Verlag. 574 pp. [Chronic alcohol illness or alcoholismus chronicus. A contribution to the study of dyscrasias based on my personal experience and the experience of others]

Iizuka H, Sakatani K, Young W (1989) Selective cortical neuronal damage after middle cerebral artery occlusion in rats. Stroke 20:1516–1523

Ishida K, Shimizu H, Hida H, Urakawa S, Ida K, Nishino H (2004) Argyrophilic dark neurons represent various states of neuronal damage in brain insults: some come to die and others to survive. Neuroscience 125:633–644

Islam N, Moriwaki A, Hattori Y, Hori Y (1994) Appearance of dark neurons following anodal polarization in the rat brain. Acta Med Okayama 48(3):123–130

Iwata A, Riley BE, Johnston JA, Kopito RR (2005) HDAC6 and microtubules are required for autophagic degradation of aggregated huntingtin. J Biol Chem 280(48):40282–40292

Jaffe AB, Hall A (2005) Rho GTPases: biochemistry and biology. Annu Rev Cell Dev Biol 21: 247–269

Jakob C, Pedace EA (1938) Estudio anátomopatológico de la esquizofrenia. Rev Asoc Méd Arg 52:326–334 [Anatomo-pathological study of schizophrenia]

Jortner BS (2005) Neuropathological assessment in acute neurotoxic states. The "dark" neuron. J Med Chem Radiol Defense 3:1–5

Jortner BS (2006) The return of the dark neuron. A histological artifact complicating contemporary neurotoxicologic evaluation. Neurotoxicology 27(4):628–634

Joseph RA, Shepard BD, Kannarkat GT, Rutledge TM, Tuma DJ, Tuma PL (2008) Microtubule acetylation and stability may explain alcohol-induced alterations in hepatic protein trafficking. Hepatology 47:1745–1753

Kannarkat GT, Tuma DJ, Tuma PL (2006) Microtubules are more stable and more highly acetylated in ethanol-treated hepatic cells. J Hepatol 44:963–970

Kapaki E, Liappas I, Paraskevas GP, Theotoka I, Rabavilas A (2005) The diagnostic value of tau protein, beta-amyloid (1–42) and their ratio for the discrimination of alcohol-related cognitive disorders from Alzheimer's disease in the early stages. Int J Geriatr Psychaitry 20(8):722-729

Karp G (2007) Cell and molecular biology. Concepts and experiments, 5th edn. Wiley, New York, NY

Kellermayer R, Zsombok A, Auer T, Gallyas F (2006) Electrically induced gel-to-gel phase-transition in neurons. Cell Biol Int 30:175–182

Koch OR, Gamboni M (1977) Alteraciones ultraestructurales de las mitocondrias del hígado en alcoholistas sin enfermedad hepática. Análsis estereológico. Medicina (Buenos Aires) 37:351–357 [Ultrastructural alterations of the mitochondria in the liver of alcoholics without liver disease. Stereological analysis]

Koch OR, Roatta de Conti LL, Pacheco Bolaños L, Bugari G, Stoppani AOM (1977) Alteraciones morfo-funcionales de las mitocondrias hepáticas en ratas tratadas crónicamente con alcohol. Regresión con la abstinencia. Medicina (Buenos Aires) 37:21–32 [Morpho-functional disorders of the liver mitochondria in rats chronically treated with alcohol]

Koh CG (2006) Rho GTPases and their regulators in neuronal functions and development. Neurosignals 15(5):228–237

Kornack DR, Rakic P (1995) Radial and horizontal deployment of clonally related cells in the primate neocortex: relationship to distinct mitotic lineages. Neuron 15:311–321

Korsakoff SS (1890) Über eine besondere Form psychiser Störung kombiniert mit multipler Neuritis. Arch für Psych und Nerv 21:669–704 [On a special form of psychic disorder in combination with multiple neuritis]

Kövesdi E, Pál J, Gallyas F (2007) The fate of "dark" neurons produced by transient focal cerebral ischemia in a non-necrotic and non-excitotoxic environment: neurobiological aspects. Brain Res 1147:272–283

Krieg WJS (1946a) Connections of the cerebral cortex. I. The albino rat. A. Topography of the cortical areas. J Comp Neurol 84(3):221–275

Krieg WJS (1946b) Connections of the cerebral cortex. I. The albino rat. B. Structure of the cortical areas. J Comp Neurol 84(3):277–323

Kuhn DM, Arthur JrR (1997) Molecular mechanism of the inactivation of tryptophan hydroxylase by nitric oxide: attack on critical sulphydryls that spare the enzyme iron center. J Neurosci 17(19):7245–7251

Kuhn DM, Geddes TJ (1999) Peroxynitrite inactivates tryptophan hydroxylase via sulfhydryl oxidation. J Biol Chem 274(42):29726–29732

Laas R, Hagel C (1994) Hirano bodies and chronic alcoholism. Neuropathol Appl Neurobiol 20(1):12–21

Lane BP, Lieber CS (1966) Ultrastructural alterations in human hepatocytes following ingestion of ethanol with adequate diets. Am J Pathol 49(4):593–603

Lodish H, Berk A, Kaiser CA, Krieger M, Scout MP, Bretscher A, Ploegh H, Matsudaira P (2007) Molecular Cell Biology. 6th ed. W.H. Freeman, New York

Loidl CF, Capani F, López-Costa JJ, Selvín-Testa A, López EM, Goldstein J, Pecci Saavedra J (1997) Short-term changes in NADPH-diaphorase reactivity in rat brain following perinatal asphyxia. Neuroprotective effects of cold treatment. Mol Chem Neuropathol 31(3):301–316

Luskin MB, Pearlman AL, Sanes JR (1988) Cell lineage in the cerebral cortex of the mouse studied in vivo and in vitro with a recombinant retrovirus. Neuron 1(8):35–47

Magnan V (1871) Étude expérimentale et clinique sur l'alcoolisme. Renou et Maulde, Paris [Experimental and clinical study on alcoholism]

Marchiafava E, Bignami A (1903) Sopra una alterazione del corpo calloso osservata in soggeti alcoolisti. Rivista de Patologia Nervosa e Mentale 8:544–549 [On an alteration of the corpus callosum observed in an alcoholic subject]

Matsushita S, Miyakawa T, Maesato H, Matsui T, Yokoyama A, Arai H, Higuchi S, Kashima H (2008) Elevated cerebrospinal fluid tau protein levels in Wernicke's encephalopathy. Alcohol Clin Exp Res 32(6):1091–1095

Mayer DJ, Mao J, Holt J, Price DD (1999) Cellular mechanism of neuropathic pain, morphine tolerance, and their interactions. Proc Natl Acad Sci USA 96:7731–7736

McKinnon G, Davidson M, De Jersey J, Shanley B, Ward L (1987) Effects of acetaldehyde on polymerization of microtubule proteins. Brain Res 416(1):90–99

McLane JA (1987) Decreased axonal transport in rat nerve following acute and chronic ethanol exposure. Alcohol 4(5):385–389

Miguel-Hidalgo JJ, Wei J, Andrew M, Overholser JC, Jurjus G, Stockmeier CA, Rajkowska G (2002) Glia pathology in the prefrontal cortex in alcohol dependence with and without depressive symptoms. Biol Psychiatry 52:1121–1133

Miller MW (1986) Effects of alcohol on the generation and migration of cerebral cortical neurons. Science 233(4770):1308–1311

Miller MW, Robertson S (1993) Prenatal exposure to ethanol alters the postnatal development and transformation of radial glia to astrocytes in the cortex. J Comp Neurol 337(2):253–266

Mooney SM, Siegenthaler JA, Miller MW (2004) Ethanol induces heterotopias in organotypic cultures of rat cerebral cortex. Cereb Cortex 14(10):1071–1080

Moores CA, Perderiset M, Francis F, Chelly J, Houdusse A, Milligan RA (2004) Mechanism of microtubule stabilization by doublecortin. Mol Cell 14(6):833–839

Morikawa Y, Arai H, Matsushita S, Kato M, Higuchi S, Miura M, Kawakami H, Higuchi M, Okamura N, Tashiro M, Matsui T, Sasaki H (1999) Cerebrospinal fluid tau protein levels in demented and nondemented alcoholics. Alcohol Clin Exp Res 23(4):575–577

Morris AM, Watzky MA, Finke RG (2009) Protein aggregation kinetics, mechanism, and curve-fitting: a review of the literature. Biochim Biophys Acta 1794:375–397

Moyano BA 1933. Demencia senil y demencias preseniles. Tesis de Doctorado. Facultad de Medicina, Universidad de Buenos Aires. Acta Nº 4686. Buenos Aires. 84 pp. [Senile dementia and presenile dementias]

Mukaetova-Ladinska EB, Jaros AH, Perry R, Perry E (2004) Depletion of MAP2 expression and laminar cytoarchitectonic changes in dorsolateral prefrontal cortex in adult autistic individuals. Neuropathol Appl Neurobiol 30(6):615–623

Nedergaard M, Ransom B, Goldman SA (2003) New roles for astrocytes: redefining the functional architecture of the brain. Trends Neurosci 26(10):523–530

Ni Y, Feng-Chen KC, Hsu L (1992) A tissue culture model for studying ethanol toxicity on embryonic heart cells. Cell Biol Toxicol 8(1):1–11

Nixon K, Crews FT (2004) Temporally specific burst in cell proliferation increases hippocampal neurogenesis in protracted abstinence from alcohol. J Neurosci 24(43):9714–9722

Nixon RA, Sihag RK (1991) Neurofilament phosphorylation: a new look at regulation and function. Trends Neurosci 14:501–506

Noraberg J, Zimmer J (1998) Ethanol induces MAP2 changes in organotypic hippocampal slice cultures. Neuroreport 9:3177–3182

North BJ, Marshall BI, Borra MT, Denu JM, Verdin E (2003) The Sir2 ortholog, SirT2, is an NAD^+-dependent tubulin deacetylase. Mol Cell 11(2):437–444

Onizuka K, Fukuda A, Kunimatsu M, Kumazaki M, Sasaki M, Takaku A, Nishino H (1996) Early cytopathic features in rat ischemia model and reconstruction by neural graft. Exp Neurol 137:324–332

Ooigawa H, Nawashiro H, Fukui S, Otani N, Osumi A, Toyooka T, Shima K (2006) The fate of Nissl-stained dark neurons following traumatic brain injury in rats: difference between neocortex and hippocampus regarding survival rate. Acta Neuropathol 112:471–481

Orlando JC 1952. Enfermedad de Marchiafava-Bignami. Sobre la degeneración sistemática de las comisuras cerebrales en el alcoholismo crónico. Tesis de Doctorado. Facultad de Medicina de la Universidad de Buenos Aires. Buenos Aires, 61 pp. [Marchiafava-Bignami disease. On the systematic degeneration of brain commisures in chronic alcoholism. Later published in Neuropsiquiatría (1952); 3 (3–4): 97]

Papadopoulos MC, Lamb FJ, Moss RF, Davies DC, Tighe D, Bennett ED (1999) Faecal peritonitis causes oedema and neuronal injury in pig cerebral cortex. Clinical Sci 96:461–466

Parnavelas JG (1999) Glial cell lineages in the rat cerebral cortex. Exp Neurol 156(2):418–429

Paxinos G, Watson C (2007) The rat brain in stereotaxic coordinates, 6th edn. Academic Press, Amsterdam, 462 pp

Petersén Å, Gil J, Maat-Schiemann MLC, Björkvist M, Tanila H, Araújo I,, Smith R, Popovic N,, Wierup N,, Norlén P, Li J-Y, Roos RAC, Sundler F, Mulder H, Brundin P (2005) Orexin loss in Huntington's disease. Hum Mol Gen 14(1):39–47

Pinheiro TJ (2006) The role of rafts in the fibrillization and aggregation of prions. Chem Phys Lipids 141(1/2):66–71

Poltorak M, Freed WJ, Casanova MF (1990) Prenatal exposure to ethanol causes a delay in the developmental expression of neurofilament epitopes in cerebellum. Pharmacol Biochem Behav 35(3):693–698

Putzke J, De Beun R, Schreiber R, De Vry J, Tölle TR, Zieglgänsberger W, Spanagel R (1998) Long-term alcohol self –administration and alcohol withdrawal differentially modulate microtubule-associated protein 2 (MAP2) gene expression in the rat brain. Mol Brain Res 62:196–205

Rafalowska J, Fidziańska A, Dziewulska D, Gadamski R, Ogonowska W, Grieb P (2006) Progression of morphological changes within CNS in a transgenic rat model of familial amyotrophic lateral sclerosis. Folia Neuropathol 44(3):162–174

Rakic P (1988) Specification of cerebral cortical areas. Science 241(4862):170–176

Rakic P (1995) Radial versus tangential migration of neuronal clones in the developing cerebral cortex. Proc Natl Acad Sci USA 92:11323–11327

Ramón Cajal S 1899–1904. Textura del sistema nervioso del hombre y los vertebrados. Imprenta y Librería de Nicolás Moya, Madrid. t. I: 566 pp.; t. II: 1209 pp

Ramón Cajal S 1913. Sobre la degeneración y regeneración del sistema nervioso. Imprenta de Hijos de Nicolás Moya, Madrid. 1913. 2 tomos. [On degeneration and regeneration of the nervous system]

Ramos AJ, Tagliaferro P, López EM, Pecci Saavedra J, Brusco A (2000) Neuroglial interactions in a model of para-chlorophenylalanine-induced serotonin depletion. Brain Res 883:1–14

Ramos AJ, Evrard SG, Tagliaferro P, Tricárico MV, Brusco A (2002) Effects of chronic maternal ethanol exposure on hippocampal and striatal morphology in offspring. Ann NY Acad Sci 965:343–353

Reed NA, Cai D, Blasius TL, Jih GT, Meyhofer E, Gaertig J, Verhey KJ (2006) Microtubule acetylation promotes kinesin-1 binding and transport. Curr Biol 16(21):2166–2172

Reiter-Funk CK, Dhrman DP (2005) Chronic ethanol exposure increases microtubule content in PC12 cells. BMC Neuroscience 6:16

Renau-Piqueras J, Zaragoza R, De Paz P, Baguena-Cervellera R, Megías L, Guerri C (1989) Effects of prolonged ethanol exposure on the glial fibrillary acidic protein-containing intermediate filaments of astrocytes in primary culture: a quantitative immunofluorescence and immunogold electron microscopy study. J Histochem Cytochem 37(2):229–240

Rintala J, Jaatinen P, Parkkila S, Sarviharju M, Kiianmaa K, Hervonen A, Niemela O (2000) Evidence of acetaldehyde-protein adduct formation in rat brain after lifelong consumption of ethanol. Alcohol Alcohol 35(5):458–463

Rintala J, Jaatinen P, Kiianmaa K, Riikonen J, Kemppainen O, Sarviharju M, Hervonen A (2001) Dose-dependent decrease in glial fibrillary acidic protein immunoreactivity in rat cerebellum after lifelong ethanol consumption. Alcohol 23:1–8

Robson SJ, Burgoyne RD (1989) Differential localization of tyrosinated, detyrosinated, and acetylated α-tubulin in neurites and growth cones of dorsal root ganglion neurons. Cell Motil Cytoskeleton 12(4):273–282

Rovasio RA, Battiato NL (2002) Ethanol induces morphological and dynamic changes on in vivo and in vitro neural crest cells. Alcohol Clin Exp Res 26(8):1286–1298

Sánchez C, Díaz-Nido J, Ávila J (2000) Phosphorylation of microtubule-associated protein 2 (MAP2) and its relevance for the regulation of the neuronal cytoskeleton function. Prog Neurobiol 61:133–168

Santamaria-Kisiel L, Rintala-Dempsey AC, Shaw GS (2006) Calcium-dependent and-independent interactions of the S100 protein family. Biochem J 396:201–214

Satriotomo I, Miki T, Itoh M, Ameno K, Ijiri I, Takeuchi Y (2000) Short-term ethanol exposure alters calbindin D28k and glial fibrillary acidic protein immunoreactivity in hippocampus of mice. Brain Res 879:55–64

Saunders DE, DiCerbo JA, Williams JR, Hannigan JH (1997) Alcohol reduces neurofilament protein levels in primary cultured hippocampal neurons. Alcohol 14(5):519–526

Sauvageot CM, Stiles CD (2002) Molecular mechanisms controlling cortical gliogenesis. Curr Opin Neurobiol 12:244–249

Schmued LC, Stowers CC, Scallet AC, Xu L (2005) Fluoro-Jade C results in ultra high resolution and contrast labeling of degenerating neurons. Brain Res 1035(1):24–31

Schroeder E, Vogelgesang S, Popa-Wagner A, Kessler C (2003) Neurofilament expression in the rat brain after cerebral infarction: effect of age. Neurobiol Aging 24:135–145

Shepard BD, Joseph RA, Kannarkat GT, Rutledge TM, Tuma DJ, Tuma PM (2008) Alcohol-induced alterations in hepatic microtubule dynamics can be explain by impaired histone deacetylase 6 function. Hepatology 48:1671–1679

Simpson TI, Price DJ (2002) *Pax6*: a pleiotropic player in development. Bioessays 4:1041–1051

Slaugenhaupt SA, Gusella JF (2002) Familial dysautonomia. Curr Opin Genet Dev 12(3): 307–311

Smith SL, Jennett RB, Sorrell MF, Tuma DJ (1989) Acetaldehyde substoichiometrically inhibits bovine neurotubulin polymerization. J Clin Invest 84:337–341

Soto C, Estrada LD (2008) Protein misfolding and neurodegeneration. Arch Neurol 65(2):184–189

Southwood CM, Peppi M, Dryden S, Tainsky MA, Gow A (2007) Microtubule deacetylases, SirT2 and HDAC6, in the nervous system. Neurochem Res 37:187–195

Spielmeyer W (1922) Histopathologie des nervensystems. Verlag von Julius Springer, Berlin, p 493 [Histopathology of the nervous system]

Stevenson JR, Schroeder JP, Nixon K, Besheer J, Crews FT, Hodge CW (2009) Abstinence following alcohol drinking produces depression-like behavior and reduced hippocampal neurogenesis in mice. Neuropsychopharmacology 34(5):1209–1222

Strnad P, Zatloukal K, Stumptner C, Kulaksiz H, Denk H (2008) Mallory-Denk bodies: lesson from keratin-containing hepatic inclusion bodies. Biochim Biophys Acta 1782:764–774

Tagliaferro P, Duhalde Vega M, Evrard SG, Ramos AJ, Brusco A (2002) Alcohol exposure during adulthood induces neuronal and astroglial alterations in the hippocampal CA-1 area. Ann NY Acad Sci 965:334–342

Tagliaferro P, Ramos AJ, Onaivi ES, Evrard SG, Lujilde J, Brusco A (2006) Neuronal cytoskeleton and synaptic densities are altered after a chronic treatment with the cannabinoid receptor agonist WIN 55,212-2. Brain Res 1085(1):163–176

Tajjudin NF, Orrico LA, Eriksen JL, Druse MJ (2003) Effects of ethanol and ipsapirone on the development of midline raphe glial cells and astrocytes. Alcohol 29:157–164

Tan SE, Abel EL, Berman RF (1993) Brain MAP-2 phosphorylation is decreased following prenatal alcohol exposure in rats. Alcohol 10(5):391–396

Tateno M, Ukai W, Yamamoto M, Hashimoto E, Ikeda H, Saito T (2005) The effect of ethanol on cell fate determination of neural stem cells. Alcohol Clin Exp Res 29(12):225S–229S

Tinberg HM, Mednick DL (1980) Intermediate filaments: detection of lectin-like activity in purified alcoholic hyaline. Biochem Biophys Res Commun 94(3):940–947

Tinberg HM, Mednick DL (1981) Immunochemical comparison of prekeratin and alcoholic hyaline intermediate filaments. Biochem Biophys Res Commun 102(3):867–876

Tomás M, Lázaro-Diéguez F, Durán JM, Marín P, Renal-Piqueras J, Egea G (2003) Protective effects of lysophosphatidic acid (LPA) on chronic ethanol-induced injuries to the cytoskeleton and on glucose uptake in rat astrocytes. J Neurochem 87:220–229

Tomás M, Marín P, Megías L, Egea G, Renal-Piqueras J (2005) Ethanol perturbs the secretory pathway in astrocytes. Neurobiol Dis 20:773–784

Triarhou LC, del Cerro M (2006a) Semicentennial tribute to the ingenious neurobiologist Christfried Jakob (1866-1956). 1. Works from Germany and the first Argentina period, 1891-1913. Eur Neurol 56:176–188

Triarhou LC, del Cerro M (2006a) Semicentennial tribute to the ingenious neurobiologist Christfried Jakob (1866–1956). 2. Publications from the second Argentina period, 1913–1949. Eur Neurol 56:189–198

Tuma DJ, Jennett RD, Sorrell M (1987) The interaction of acetaldehyde with tubuline. Ann NY Acad Sci 282:227–286

Tuma DJ, Smith SL, Sorrell MF (1991) Acetaldehyde and microtubules. Ann NY Acad Sci 625:786–792

Valenzuela-Fernández A, Cabrero JR, Serrador JM, Sánchez-Madrid F (2008) HDAC6: a key regulator of cytoskeleton, cell migration and cell-cell interactions. Trends Cell Biol 18(6): 291–297

Vallés S, Sancho-Tello M, Miñana R, Climent E, Reanu-Piqueras J (1996) Glial fibrillary acidic protein expression in rat brain and in radial glia culture is delayed by prenatal ethanol exposure. J Neurochem 67(6):2425–2433

Vallés S, Pitarch J, Renau-Piqueras J, Guerri C (1997) Ethanol exposure affects glial fibrillary acidic protein gene expression and transcription during rat brain development. J Neurochem 69(6):2484–2493

Victorov IV, Prass K, Dirnagl U (2000) Improved selective, simple, and contrast staining of acidophilic neurons with vanadium acid fuchsin. Brain Res Protoc 5(2):135–139

Walsh C, Cepko CL (1988) Clonally related cortical cells show several migration patterns. Science 241(4871):1342–1345

Weber MM (2000) Psychiatric research and science policy in Germany. The history of the Deutsche Forschungsanstalt für Psychiatrie (German Institute for Psychiatric Research) in Munich from 1917 to 1945. Hist Psychiatry 11:235–258

Wernicke C, 1881. Lehrbuch der Gehirnkrankheiten für Ärzte und Studierende. Theodor Fischer, Kassel und Berlin, vol. 2, pp. 229–242. [Handbook of brain diseases for physicians and students]

Wessely Z, Shapiro SH, Klavins JV, Tinberg HM (1981) Identificaton of mallory bodies with rhodamine B fluorescence and other stains for keratin. Stain Technol 56(3):169–176

Whitaker-Azmitia PM, Wingate M, Borella A, Gerlai R, Roder J, Azmitia EC (1997) Transgenic mice overexpressing the neurotrophic factor S-100$ show neuronal cytoskeletal and behavioral signs of altered aging processes: implications for Alzheimer's disease and Down's syndrome. Brain Res 776:51–60

Williamson D (1993) The unstable hemoglobins. Blood Rev 7(3):146–163

Yang W, Ang LC, Strong MJ (2005) Tau protein aggregation in the frontal and enthorrinal cortices as a function of aging. Dev Brain Res 156:127–138

Yoon Y, Török N, Krueger E, Oswald B, McNiven MA (1998) Ethanol-induced alterations of the microtubule cytoskeleton in hepatocytes. Am J Physiol (Gastrointest Liver Physiol 37) 274:G757–G766

Zhang Y, Li N, Caron C, Matthias G, Hess D, Kochbin S, Matthias P (2003) HDAC6 interacts with and deacetylates tubulin and microtubules in vivo. EMBO J 22(5):1168–1179

Zsombok A, Tóth Z, Gallyas F (2005) Basophilia, acidophilia and argyrophilia of "dark" (compacted) neurons during their formation, recovery or death in an otherwise undamaged environment. J Neurosci Meth 142:145–152

Index

R.A. Nixon, A. Yuan (eds.), *Cytoskeleton of the Nervous System*, Advances in Neurobiology 3, DOI 10.1007/978-1-4419-6787-9,